International Encyclopedia of
Composites

Editorial Board

International Encyclopedia of Composites

Stuart M. Lee
Editor

VCH
New York

Stuart M. Lee
3718 Cass Way
Palo Alto, California 94306

Library of Congress Cataloging-in-Publication Data

International encyclopedia of composites.

Includes bibliographical references.
Contents: v. 1. Acetal resins and composites--cyanate ester.
1. Composite materials--Dictionaries. I. Lee, Stuart M.
TA418.9.C6E53 1989 620.1'18'03 89-24893
ISBN 0-89573-290-4 (set)

British Library Cataloguing in Publication Data

International encyclopedia of composites.
Vol. 1, Acetal resins and composites
1. Composites materials
I. Lee, Stuart M.
620.1'18

ISBN 3-527-26852-9

Printed in the United States of America

ISBN 0-89573-731-0 (volume 1) VCH Publishers
ISBN 0-89573-290-4 (set) VCH Publishers
ISBN 3-527-27947-4 (volume 1) VCH Verlagsgesellschaft
ISBN 3-527-26852-9 (set) VCH Verlagsgesellschaft

Printing History:
10 9 8 7 6 5 4 3 2 1

Published jointly by:

VCH Publishers, Inc.
220 East 23rd Street
Suite 909
New York, NY 10010

VCH Verlagsgesellschaft mbH
P.O. Box 10 11 61
D-6940 Weinheim
Federal Republic of Germany

VCH Publishers (UK) Ltd.
8 Wellington Court
Cambridge CB1 1HW
United Kingdom

Foreword

The field of composite materials, as a subject of both academic and technological interest, has grown remarkably within the last few decades. It can be considered truly multi- and interdisciplinary. An optimum development of structural or functional components fabricated from composite materials requires knowledge, ideas, and close collaboration of people working in areas such as materials science (including, polymer physics and chemistry, metallurgy, and ceramics engineering), processing and manufacturing technology, fluid and solid mechanics, and engineering construction and design. VCH Publishers' new *International Encyclopedia of Composites* is a remarkable multivolume handbook addressing many of these important topics.

The encyclopedia is one of the largest, most up-to-date reference sources to consult when seeking in-depth engineering information on any topic in composite technology. More than 300 distinguished international researchers, theoreticians, end-users, and technologists have contributed to make it the most authoritative and accurate work of its kind. For recruiting this outstanding list of authors, special thanks and congratulations have to be extended to the editor, Stuart M. Lee, and his colleagues of the editorial board who are recognized as international experts in composite science and technology. Their contribution guaranteed high-quality articles of great relevance to the composite community.

The individual subjects covered by this book series include almost all essential areas necessary to understand this group of materials in detail, and how to use them for different applications from both a practical and an economical point of view. Topics discussed can be subdivided roughly into four groups:

- Group one deals with such special types of composites as engineering materials and their components. Included are, for example, polymeric matrices, such as liquid crystal polymers, interpenetrating networks, high-temperature thermoplastics, and newly developed thermosetting resins. Reinforcement materials under consideration include carbon and other ceramic fibers, tungsten and SiC fibers, particulate fillers, and corresponding coupling agents. The different composites discussed range from natural composites (including wood), elastomeric and other polymer matrix composites to carbon/carbon-, ceramic and metal matrix composites, and aramid/aluminum laminates.
- The second group of topics focuses on the behavior of composite materials under different types of loading conditions and on relevant characterization methods. Creep, fatigue, impact, fractography, environmental stress cracking, and wear performance of composites are typical entries. In addition, issues relating to lamination theory, heat transfer, flamability, nondestructive evaluation, and quality control are presented.
- Another set of articles concentrates on composites with special property profiles: electronic, conductive, piezoelectric, microwave, or low-observability (Stealth aircraft) composites. This third group also covers materials developed or designated for particular applications: circuit board laminates, orthopedic devices, air- and spacecraft structures, sports and recreation uses, automotive and marine composites, and tires. In some chapters, future growth opportunities are discussed (market evaluation and costs).
- Finally, group four combines design aspects of composite materials (CAD, CAM, CAE; design methodology) with their manufacturing and fabrication processes. Key words for the latter subjects are: prepreg processing science, autoclave technology, resin transfer molding, braiding, filament winding, pulforming, tape placement, robotics, thermoforming, joining, machining, squeeze casting, and powder metallurgy.

As a consequence of this exhaustive review, the encyclopedia is expected to be of considerable value to materials scientists as well as structural designers and manufacturing engineers as a helpful handbook giving insight into the technical possibilities of tailoring composite materials to particular applications. In addition, it will also be of interest to senior undergraduate and postgraduate students who want to enhance their knowledge in this increasingly complex, multi/interdisciplinary subject.

Klaus Friedrich
Professor for Polymers and Composites
Technical University Hamburg-Harburg
Hamburg, Federal Republic of Germany

Preface

Just as people have progressed from the Stone Age to the Composites Age, so have composites evolved from the primitive chopped straw and brick used in biblical times to today's sophisticated ceramic matrix composites and metal matrix composites. We are rapidly approaching almost a century of modern composite usage, which began with the introduction of particulate or fibrous reinforcement materials into thermoset phenolics in the early 1900s. During recent times we have seen an extraordinary explosion in composites usage, research, publishing activity, and interest. Witness the unusual and exotic application of composites use in Stealth aircraft and the recent arrival of superconductive composites. Composites are one of the fastest growing industries and continue to demonstrate a marked impact on the materials world. A 20 billion dollar composites industry is projected by the year 2000.

"Composite" can have many meanings and arbitrary classifications. There appears to be no completely acceptable universal definition for a composite material. One group believes that a composite can consist only of a continuous matrix surrounding a reinforcement, which precludes the inclusion of materials such as core composites, laminates, felts, etc. Another school would like composites to include everything except homogeneous and single-phase materials. Some even classify composites into microcomposites—covering reinforced and toughened thermoplastics, sheet-molding compounds, and metallic alloys—and macrocomposites, which include helicopter blades, reinforced concrete, and even galvanized steel. Composites have also been classified as "in situ" or "artificial." The in situ composites are defined as materials derived from directionally transformed poly-phase materials containing the fibrous reinforcing phase usually in chemical equilibrium with the matrix. In contrast, an "artificial composite" is defined as one in which the fiber and matrix composites are not in chemical equilibrium with the solid fibers and are blended mechanically with the matrix.

Other approaches to classification of composites include attempts to differentiate between composite materials and composite structures such as laminates, and the consideration not to classify bone or wood as composites because they do not contain structures of "arbitrary variation."

Interestingly, materials such as graphite fibers and boron fibers are themselves considered composites. Graphite fibers are composites because only part of the carbon has been converted to graphite in tiny crystalline platelets, specifically oriented to the fiber axis; boron fibers consist of a thin layer of boron coating a tungsten or carbon substrate. Other unconventional composites are safety-laminated glass, which contain a polyvinylbutyral interlayer that acts as a safety net when impacted; the thermoformed, six-layered, high-oxygen plastic laminate barrier used in beverage bottles; and solid golf balls, which contain only polybutadiene rubber reinforced with silica pigment.

In the Encyclopedia we have strived to avoid adhering to any pre-chosen guidelines for classification or selection of the articles on composites. The composite materials included herein are in agreement with the following definition of a composite:

Composite*: A multiphase material from a combination of materials, differing in composition or form, which remain bonded together, but retain their identities and properties. The components do not dissolve or completely merge. They maintain an interface between each other and act in concert to provide improved, specific or synergistic characteristics not obtainable by any of the original components acting singly. Included are (1) the fibrous composites composed of fibers, usually in a matrix; (2) the laminar composites or layers of materials; (3) the particulate composites containing particles or flakes, usually in a matrix; and (4) the hybrid composites, which can be combinations of any of the above.

Besides the traditional polymer matrix composites, today's composites include metal matrix composites and ceramic matrix composites. As a result modern composite technologists are not bounded by classical training and disciplines and become interdisciplinary and knowledgeable in organic and polymer chemistry, inorganic chemistry, ceramics, metallurgy, engineering, manufacturing and automation, and textiles. It is these scientists, engineers, chemists, physicists, fabricators, and analysts, as well as students, that this multiple-volume Encyclopedia will serve, and to whom it will be a valuable source. The typical user of this reference book may be an expert in one or more aspects of science and engineering. However, in consulting the Encyclopedia, information may be found in unfamiliar areas. The reader may then choose to proceed to the primary references cited in the bibliography for further details.

The articles in this encyclopedia have been written by

authors selected from international universities, industries, research institutions, and government laboratories, and represent worldwide expertise in their disciplines. Their submitted articles have been subjected to peer review and editorial processing to ensure clarity, conciseness, and stylistic consistency.

The articles in this encyclopedia are arranged in alphabetical order. Cross-references at the end of an article refer the reader to related articles in the encyclopedia. Most of the articles include a list of references or a bibliography that provide additional sources for pursuing a subject in greater depth. Units of measurement are given in SI (Système Internationale) units and a conversion chart is provided on pages xv–xvi.

No work of this magnitude is possible without a committed team effort. I would acknowledge the assistance of the many outstanding contributors who generously devoted their valuable time in creating this work: the referees, those "unsung heroes" who diligently reviewed each article and rendered their constructive and critical comments; the Editorial Board for their dedication and assistance; Barbara Chernow for her outstanding editing; Liz Geller for her valuable help; and my wife Miriam for her unlimited patience and understanding.

STUART M. LEE
Palo Alto, California

*Stuart M. Lee, *Dictionary of Composite Materials*, Technomic Publishing Co., Lancaster, PA, 1989, p. 32.

Contents

Main Entries

Contributors

Donald F. Adams, University of Wyoming
Composite Micromechanics

Ronald E. Allred, Michigan Molecular Institute
Aramid Fiber Composites

Paul Becher, Oakridge National Laboratory
Ceramic Matrix Composites, SiC Whisker-Reinforced

Clifford R. Boldt, Hoechst Celanese Corporation
Acetal Resins and Composites

Selby M. Brannon, Plasticolors Inc.
Colorants for Composites

J. P. Brazel, General Electric Company
Composite Materials, Ablative

Raju S. Davé, Michigan Molecular Institute
Autoclave Process Modeling

T. A. DeVilbiss, Dow Chemical Company
Carbon Fibers: Property Testing and Analysis

John G. Dillard, Virginia Polytechnic Institute and State University
Carbon Fibers: Property Testing and Analysis

P. Dreyfuss, Michigan Molecular Institute
Amino Resins

M. V. Gandhi, Michigan State University
Articulating Mechanical and Robotic Systems Fabricated from Advanced Composite Materials

Richard B. Graver, Hi-Tek Polymers, Inc.
Cyanate Ester Resins

Hemant K. Gupta, Ferro Corporation
Chemical Analysis and Characterization of Polymer Composite Materials

John Haggerty, Massachusetts Institute of Technology
Ceramic Matrix Composites, Reaction Bonded

Joseph D. Halford, E.I. Dupont de Nemours and Company
CAD/CAM/CAE

Dale E. Hall, National Institute of Standards and Technology
Conductive Composites

Frances I. Hurwitz, NASA Lewis Research Center
Ceramic Matrix Composites; Ceramic Fiber Reinforced

P. W. Juneau, Jr., General Electric Company
Composite Materials, Ablative

Melvin R. Kantz, Ferro Corporation
Chemical Analysis and Characterization of Polymer Composite Materials

Fred W. Kear, Siemens Transmission Systems, Inc.
Circuit Board Laminates

L. Brian Keller, Programmed Composites Incorporated
Autoclave Processing

David E. Kranbuehl, College of William and Mary
Cure Monitoring

W. J. Lackey, Georgia Institute of Technology
Ceramic Matrix Composites CVI: Chemical Vapor Infiltration

Ronald N. Lee, Naval Surface Warfare Center
Carbon Fiber Surface Properties

David Lewis III, Naval Research Laboratory
Ceramic Matrix Composites

James S. Margolis, Margolis Marketing and Research Company
Automotive Composites

Andrew C. Marshall, Marshall Consulting
Core Composite and Sandwich Structures

Derrick B. McKie, Hoechst Celanese Corporation
Acetal Resins and Composites

P. E. McMahon, Hoechst Celanese Corporation
Carbon Fiber, Evolving Technology

Roger J. Morgan, Lawrence Livermore National Laboratory
Aging of Polymer Matrix Fibrous Composites and *Aramid Fiber Composites*

Larry N. Mueller, Alcoa Laboratories
Arall® Laminates (Aramid Aluminum Laminates)

Thomas Ngai, Lockheed Missiles & Space Company
Carbon-Carbon Composites

Igor Palley, Allied-Signal Corporation
Characterization

Charles W. Paul, National Starch and Chemical
Characterization

Edwin P. Plueddermann, Dow Corning
Coupling Agents

Peter Popper, DuPont Experimental Station
Braiding

John P. Riggs, Hoechst Celanese Corporation
Carbon Fibers

Peter Schwartz, Cornell University
Creep-Rupture

Surendra P. Shah, Northwestern University
Concrete Composites, Fiber Reinforced

T. P. Sivam, Boeing Military Airplanes
Bending Analysis of Composite Plates and Built-Up Sections

E. R. Stover, General Electric Company
Composite Materials, Ablative

R. Talreja, Technical University of Denmark
Continuum Damage Mechanics

B. S. Thompson, Michigan State University
Articulating Mechanical and Robotic Systems Fabricated from Advanced Composite Materials

Judy Morgon Thuen, International Plastics Selector
Curing Agents

Terry Tiegs, Oak Ridge National Laboratory
Ceramic Matrix Composites, SiC Whisker-Reinforced

James W. Warren, Composites Innovations Corporation
Ceramic Matrix Composites, Carbon Fiber Reinforced

Robert N. Wieland, DuPont Engineering Department
CAD/CAM/CAE

James P. Wightman, Virginia Polytechnic Institute and State University
Carbon Fibers: Property Testing and Analysis

Ernest G. Wolff, Oregon State University
Creep

Edwina Ying, Hoechst Celanese Corporation
Acetal Resins and Composites

Conversion Table from SI Units

Property	From	To	Multiply By
Dimension			
	cubic meter (m^3)	cubic inch (in^3)	6.102×10^4
	cubic meter (m^3)	cubic foot (ft^3)	35.315
	cubic meter (m^3)	cubic yard (yd^3)	1.308
	cubic meter per second (m^3/s)	cubic foot per minute (cfm)	2.119×10^3
	meter (m)	inch (in)	39.37
	meter (m)	foot (ft)	3.281
	meter (m)	micron (μm)	1.0×10^6
	meter (m)	mil	3.937×10^4
	meter (m)	mile (statute)	6.214×10^{-4}
	meter (m)	yard (yd)	1.094
	micron (μm)	microinch	39.37
	micron (μm)	angstrom (Å)	1.0×10^4
	newton (N)	kip	2.248×10^{-4}
	square meter (m^2)	square inch (in^2)	1.55×10^3
	square meter (m^2)	square foot (ft^2)	10.76
	square meter (m^2)	square mile (mi^2)	3.861×10^{-7}
	square meter (m^2)	square yard (yd^2)	1.196
Electrical			
	joule (J)	electron volt	6.241×10^{18}
	joule (J)	erg	1.0×10^7
	megajoule (MJ)	kilowatt hour (kW·h)	0.2778
	newton (N)	dyne	1.0×10^5
	newton per meter (N/m)	dyne per centimeter (dyne/cm)	1.0×10^3
Dielectric Strength	millivolts per meter (mV/m)	volts per mil ($V/10^{-3}$ in)	2.539×10^{-2}
Force			
	joule (J)	foot-pound force (lbf)	0.7376
	kilogram per square meter (kg/m^2)	ounce mass per square yard	29.412
	meter per second (m/s)	mile per hour (mph)	2.237
	meter per square sec (m/s^2)	gal	1.0×10^2
	newton (N)	ounce-force	3.597
	weber (Wb)	maxwell	1.0×10^8
Mass/Volume			
	cubic meter (m^3)	liter (for fluids only)	1.0×10^3
	cubic meter (m^3)	ounce (U.S. fluid)	3.381×10^4
	cubic meter (m^3)	quart (U.S. liquid)	1.057×10^3
	kilogram (kg)	ounce (avoirdupois)	35.27
	kilogram (kg)	ounce (troy)	32.15
	kilogram (kg)	pound (avoirdupois)	2.205
	kilogram (kg)	pound (troy)	2.679
	kilogram (kg)	ton (long, 2240 pounds)	9.842×10^{-4}
	kilogram (kg)	ton (metric)	1.0×10^{-3}
	kilogram (kg)	ton (short, 2000 pounds)	1.102×10^{-3}
	newton (N)	poundal	7.233
	newton (N)	pound-force (lbf)	0.2248
Core Density	kilogram per cubic meter (kg/m^3)	pounds per cubic foot (lb/ft^3)	6.243×10^{-2}
Density	milligram per cubic centimeter (mg/cm^3)	pound per cubic foot (lb/ft^3)	6.243×10^{-2}
Specific Volume	cubic meter per milligram (m^3/mg)	cubic inches per pound (in^3/lb)	2.768×10^{10}
Mechanical			
	kilogram per meter (kg/m)	denier	9.0×10^6
	kilogram per meter (kg/m)	tex	1.0×10^6
	newton per tex (N/tex)	gram force per denier	11.33
	tex	denier	9
Fracture Energy (G_{IC})	joules per square meter (J/m^2)	inch pounds per square inch (in lbs/in^2)	5.710×10^{-3}
Izod Impact	kilojoules per meter (kJ/m)	foot pounds per inch (ft lb/in)	18.73
Modulus	gigapascal (GPa)	pound force per square inch (psi)	1.450×10^5
Strength	megapascal (MPa)	pound force per square inch (psi)	1.450×10^2

Conversion Table from SI Units

Property	From	To	Multiply By
Pressure			
	newton (N)	kilogram-force (kgf)	0.1020
	pascal (Pa)	atmosphere (normal)	9.869×10^{-6}
	pascal (Pa)	inch of mercury (32°F)	2.953×10^{-4}
	pascal (Pa)	inch of mercury (60°F)	4.019×10^{-3}
	pascal (Pa)	inch of water (39°F)	4.015×10^{-3}
	pascal (Pa)	millibar	1.0×10^{-2}
	pascal (Pa)	pound force per square inch (psi)	1.450×10^{-4}
	pascal (Pa)	torr (mm Hg, 0°C)	7.501×10^{-3}
Radiation			
	coulomb per kilogram (c/kg)	roentgen	3.876×10^{3}
	joule per kilogram (J/kg)	rad	1.0×10^{2}
Thermal			
	joule (J)	Btu (International Table)	9.478×10^{-4}
	joule (J)	Btu (mean)	9.471×10^{-4}
	joule (J)	Btu (thermochemical)	9.485×10^{-4}
	joule (J)	calorie (International Table)	0.2388
	joule (J)	calorie (mean)	0.2387
	joule (J)	calorie (thermochemical)	0.2390
	joule per square meter (J/m^2)	Btu per square foot (Btu/ft^2)	8.805×10^{-5}
	kilojoule (kJ)	kilocalorie (kcal)	0.2388
Linear Coefficient of Thermal Expansion	millimeter per millimeter K	inch per inch °F	0.5556
Specific Heat	kilojoule per kilogram K (kJ/kg·K)	Btu per pound °F	0.2388
Temperature	kelvin (K)	centigrade (°C)	subtract 273.
	centigrade (°C)	Fahrenheit (°F)	multiply by 1, then add 32
Thermal Conductance	watt per meter Kelvin (w/m·K)	Btu·foot per hour·square foot·°F	6.933
Viscosity			
	cubic meter per second (m^2/s)	centistoke	1.0×10^{6}
	pascal second (Pa·s)	centipoise	1.0×10^{3}
	pascal second (Pa·s)	poise (absolute viscosity)	10.0

International Encyclopedia of
Composites

A

Ablative Materials

See Composite Materials, Ablative

Abrasion

See Polymer Composites with High-Wear Resistance; Wear

Absorption

See Environmental Resistance; Environmental Stress Cracking; Polymer Composites, Corrosion

Acetal Resins and Composites

The evolutionary development of numerous composites among engineering resins is fueled by improvements in the technology to produce these cost-effective products. Improvements not only in the production machinery available but also in the materials, resins, and modifiers have paved the way for continued rapid growth in this area. The challenges of demanding requirements of current and emerging applications are continually met by innovative material suppliers and custom compounders.

Among the engineering resins, acetals have a unique property profile of high mechanical strength, stiffness, and toughness, as well as chemical resistance and electrical properties over a broad temperature range for long periods of time [1]. These properties are attributable to the highly crystalline morphology of these polymers in the solid state. It is this high crystallinity that often makes acetal difficult to modify [2]. Polyacetals are available as homopolymers and copolymers. Each is used in most applications, which include the appliance, plumbing, transportation, industrial, materials handling, electrical, and electronics as well as the consumer product market sectors.

Specialty acetal composites with a base matrix of acetal and an appropriate reinforcement can offer a balance of properties with unique polyacetal characteristics and modifications that extend the limits of performance.

This introduction to acetal composites consists of a discussion of acetal polymerization, commercial acetal base resins, and resin additives used to create composites, as well as a survey on how composites are produced and their properties. Also included are sections on converting the composite into finished articles, regulatory issues, and selecting a composite for an application.

Acetal Polymerization [1,3]

Acetal copolymer and homopolymer both begin with formaldehyde as the raw material feedstock. In the homopolymer case, for the polymerization to be successfully controlled, formaldehyde, the polymerization monomer, must be rigorously purified to remove water, formic acid, metals, and methanol (these and other impurities can act as chain transfer agents or reaction-terminating agents). The purified formaldehyde is then polymerized to form the acetal homopolymer backbone.

$$CH_2O \xrightarrow{\rightarrow} [CH_2O]_n \quad (1)$$

The polymer end groups are stabilized by reacting them with acetic anhydride to form acetate end groups [4].

$$HO—CH_2—O[CH_2—O]_nCH_2—OH + (CH_3CO_2)O \rightarrow CH_3COO[CH_2—O]_nC—OOCCH_3 \quad (2)$$

Without the acetate end groups, the polyoxymethylene backbone would depolymerize back to monomeric formaldehyde. This depolymerization is quite rapid at elevated temperatures.

In the case of the acetal copolymer, moderately pure aqueous formaldehyde is reacted in the presence of strong mineral acids to form trioxane, the cyclic trimer of formaldehyde.

$$CH_2O \overset{H^+}{\rightleftarrows} \overline{CH_2—O—CH_2—O—CH_2—O} \quad (3)$$

After rigorous purification to ensure a successful and controllable polymerization reaction, the trioxane is reacted with a cyclic ether (the most common being ethylene oxide, 1,3-dioxolane, or butanediol formal) in the presence of a cationic catalyst (such as boron trifluoride etherates) to form the acetal copolymer. Inter- and intramolecular transacetalization reactions randomly distribute the oxyethylene bond throughout the polymer backbone.

$$\overline{CH_2—O—CH_2—O—CH_2—O} + \overline{CH_2—CH_2—O} \xrightarrow{\text{catalyst}}$$

$$R—CH_2—O\Big[[CH_2—O]_nCH_2—CH_2—O[CH_2—O]_m\Big]_xCH_2—R$$

$$R = O—O—CH,\ O—CH_2—OH \text{ (unstable groups), or } O—CH_3 \text{ (stable group)} \quad (4)$$

Catalyst residues and unreacted monomers are neutralized and removed. The copolymer is stabilized by either end-capping or hydrolyzing the unstable polyoxymethylene ends back to the first oxyethylene group that is encountered. This group thus becomes the stable polymer chain end cap.

$$P—CH_2—O\Big[[CH_2—O]_nCH_2—CH_2—O[CH_2—O]_m\Big]_xCH_2—P \xrightarrow[\Delta H]{OH^-}$$

$$OH—CH_2—CH_2—O\Big[[CH_2—O]_nCH_2—CH_2—O[CH_2—O]_m\Big]_{x-2}CH_2—O—CH_2—CH_2—OH \quad (5)$$

$$P = O—O—CH \text{ or } O—CH_2—OH$$

Precisely controlled levels of chain termination agents are added during polymerization to produce final polymers having various molecular weights. Stabilizers and antioxidants are added to both the homopolymer and the copolymer to create the basic polymer grades. Fillers, modifiers, pigments, UV stabilizers, processing enhancers, and reinforcements can be added to tailor the polymers to suit many different applications.

Market Volume and Trends

There are extensive market surveys on polyacetal resins [3,5]. Since the introduction of polyacetal in the 1960s, production and consumption have been increasing steadily, both domestically and for export. In 1960 production was in the tens of millions of pounds; in 1970 the production exceeded thousands of metric tons; and in 1985 total production was approximately seventy thousand metric tons. Most of the polyacetal resin produced is consumed domestically. The acetal production capacity in the United States is expected to grow in the next few years, with DuPont increasing the capacity of their plant at Parkersburg, West Virginia, and BASF completing a production facility in Mobile, Alabama.

A small portion of polyacetal resin production goes into composites. The majority of these acetal composites are reinforced with glass. This glass–acetal composite occupies less than 10% of the total polyacetal resin capacity; and less than 5% of the glass fiber reinforced thermoplastics (e.g., polyolefins, polyamide) market. The demand for glass fiber reinforced acetal is higher in Western Europe than in the United States. The estimated growth rate in Western Europe is 4% [6].

Although the volume of acetal composites is relatively small, the potential of its applications is quite large. With the new development in fibers and fillers, the improvement in blend technology, and the optimization in coupling agent and surface treatments, acetal composites can increase market share and create new business opportunities.

Additives

Today there exist a variety of acetal composite products. The formulations of these products are carefully designed with respect to many considerations and often are in the proprietary domain. The efficient selection of an additive to create a successful composite should be a systematic process. Of course "efficiency" implies bridging the gap from concept to commercial reality in the shortest lead time possible. Also a successful composite satisfies the requirements of the processor and the ultimate user in a cost-effective manner. The following issues should be considered when selecting and screening an acetal composite additive:

- Safety and toxicity
- Price
- Variety of commercial grades available and their recommended uses
- Additive dimensions and size distribution
- Additive chemistry
- Surface modification
- Processibility with respect to composite production
- Additive physical properties and desired composite physical properties

All these issues are considered with the appropriate personnel, including the additive supplier, the composite customer, and executives of the manufacturing operation that would produce the composite.

Safety and toxicity are of chief concern with respect to material handling and processing during composite manufacture, all subsequent operations, and ultimate use. As such, when additives are selected for use with acetal, their toxicity and chemical activity are considered with respect to interaction with acetal and the environment in which the composite is expected to function.

The price of an additive should be considered, since ultimate composite pricing must be competitive to gain market share. A variety of similar additives (several suppliers) can allow the compound manufacturer flexibility and degrees of freedom if a supplier changes or modifies an additive.

Additive dimensions (or aspect ratio), size distribution, and surface area can have a significant effect on composite processibility and physical properties. The chemistry of the additive can cause interaction with acetals. If the additive is a strong enough acid or base ($pH < 3$, $pH > 10$), polymer degradation can occur.

Often additives are available with surface coating or treatment to enhance interfacial wetting and adhesion with the polymer, which is usually desirable for mechanical property improvement. Typically compound extrusion is used to introduce an additive into acetal. The intensity of the mixing can be controlled through equipment selection and screw design. This is important to obtain optimum dispersion and distribution of an additive throughout the acetal matrix for desired composite physical properties.

A generalized description of additive particle geometry and effects on composite properties is provided by Ferrigno [2]. Table 1 describes particle characteristics of various additives and classifies them. These additives can be naturally occurring ground classified rocks or ores such as talc or fabricated materials, including glass fibers and microspheres, calcium carbonate, carbon fibers, and aramid fibers [2]. These materials are widely used to create specialty acetal composites for reasons that are given below.

Talc, or hydrated magnesium silicate, is mined in California and Montana; it occurs in fibrous, platy, and granular forms. Platy talcs are used to improve polymer stiffness and reduce creep at ambient and elevated temperatures.

Borosilicate glass fibers are used in acetal to improve stiffness and tensile strength while maintaining notched Izod impact strength. The fibers are introduced into the polymer as chopped strand (typically 3–12 mm long) or a continuous roving. Individual filaments that constitute the strand range from 6 to 20 μm in diameter. Proprietary chemical coating or sizing is applied to glass fiber to maintain strand integrity during handling and to provide improved interfacial adhesion between the acetal matrix and the rigid glass filaments.

Solid glass microspheres used in acetal composites can range in diameter from about 5 to 700 μm. Their spheroidal geometry provides unique property profile enhancement in acetal of increased stiffness and heat distortion temperature, while sharply reducing warpage in thin, large, molded products. This is achieved through uniform shrinkage at lower levels when compared with unmodified acetal. The microspheres are available with chemical coatings to improve adhesion with the acetal matrix and thus improve physical properties in comparison to composites based on unmodified glass spheres.

Particulate calcium carbonate in a rectangular solid form is available in a wide range of sizes depending on the grinding and classification operations used to prepare these additives. Generally these materials, when surface-treated, increase product stiffness without compromising impact strength of the acetal matrix.

Carbon fibers are commonly prepared from pitch or polyacrylonitrile (PAN). Carbon fibers have three to four times the tensile modulus of glass fiber, about one-eighth the thermal expansion, at least twice the thermal conductivity, and only three-fourths of the density. PAN carbon fibers are available in multifilament bundles and, like glass fibers, are surface-treated to promote bonding between the fiber and the polymer matrix.

Aromatic polyamide fibers sold under the DuPont trademark Kevlar offer unique advantages over glass fiber. With tensile strength and stiffness essentially equivalent to glass fiber, this organic fiber has significantly lower density; thus lighter weight components can be produced. Kevlar fibers are more resistant than glass fiber to hydrofluoric acid and sodium hydroxide but less resistant than glass to hydrochloric acid. Both fiber types are resistant to common industrial solvents, fuels, and lubricants at ambient temperature. Composites reinforced with polyamide fiber generally offer higher abrasion and wear resistance than glass fiber reinforced composites.

TABLE 1
Particle Characteristics

Idealized Shape Class:

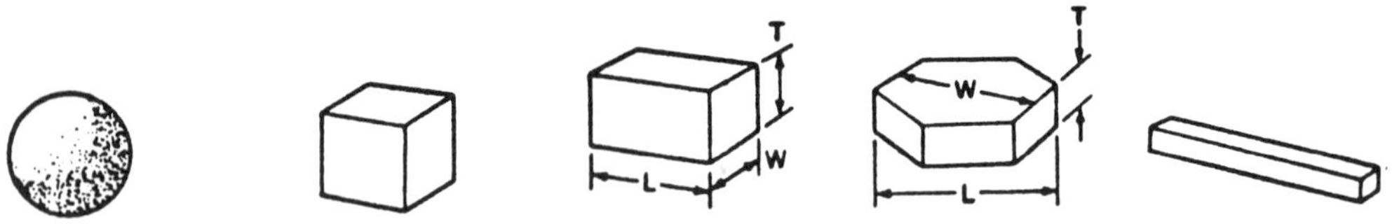

Particle Class:	**Sphere**	**Cube**	**Block**	**Flake**	**Fiber**
Descriptor[a]:	**Spheroidal[b]**	**Cubic[c]** **Prismatic** **Rhombohedral**	**Tabular** **Prismatic** **Pinacoid** **Irregular**	**Platy** **Flaky[d]**	**Acicular** **Elongated** **Fibrous**
Shape ratios					
Length (L)	1	~ 1	1.4–4	1	1
Width (W)	1	~ 1	1	< 1	< 0.1
Thickness (T)	1	~ 1	1–< 1	0.25–0.01	< 0.1
Sedimentation diameter[e]	1	esd	esd	esd[f]	esd[f]
Surface area equivalence[g]	1	1.24[h]	1.26–1.5[i]	1.5–9.9[j]	1.87 for 0.1 2.3 for 0.2[k]
Examples	Glass Spheres Microspheres	Calcite[l] Feldspar	Calcite Feldspar Silica Barite Nephelite	Kaolin Mica Talc Graphite Hydrous alumina	Wollastonite Tremolite Wood flour

[a] Preferred to particle class since this is based on relative surface area. First descriptor is preferred.
[b] In the sense that a spheroid approaches a true sphere.
[c] Generally distorted cubes; more nearly prismatic.
[d] Generally having the nature of hexagonal platelets, as illustrated.
[e] According to Stokes' law, esd = equivalent spherical diameter or the diameter of a sphere having the same volume as that of the particle.
[f] Must be modified for dissymmetry of greater than 4-1, maximum to minimum particle dimensions.
[g] Equivalent to a spherical diameter of 1; an approximation of the area when the particle has a volume equivalent to an esd of 1.
[h] About the same for cubic and prismatic shapes.
[i] For lengths of 1.4–4, respectively.
[j] Based on hexagonal platelets as follows:

Length/Thickness	Area Factor
4/1	1.47
6/1	1.78
8/1	2.09
10/1	2.34
100/1	9.88

[k] For a square cross section.
[l] There are more than 300 crystal shapes for calcite alone, but generally it is an irregular, low surface area particle.
From Ref. 2.

Molybdenum disulfide (MoS_2) is used primarily to reduce the coefficient of friction of composites. Also, electrical insulatability and stiffness are generally increased.

Composite Production Technology

The high crystallinity of acetals that provides them with many of their advantageous properties also makes compounding them with additives very difficult. The dense structure imparted by the crystals tends to exclude foreign material from the polymer matrix. For optimum additive improvement of the base resin property, it is desirable to obtain bonding or compatibilization with the acetal, not its exclusion. Also, the polymer–modifier blend must be well mixed to uniformly disperse the modifier throughout the polymer. This is to ensure that uniform properties are obtained throughout the finished product and to minimize the usage of what is many times an expensive modifier. If good compounding design is not practiced, this mixing energy can thermally and mechanically degrade the base polymer, and an unusable product will result.

The modifier–acetal compounding design should reflect the following goals.

- To uniformly disperse the modifier throughout the polymer blend.
- To completely wet all particles of modifier with polymer.

- To minimize temperature increases that will result in thermal degradation.
- To control the intensity and duration of the mechanical work, to prevent the mechanical degradation of the polymer.
- To maintain positive pumping, hence a predictable residence time and thermal–shear history in the final product, and to produce uniform product pellets.

Acetal–modifier compounding can be done in four broad classes of compounding processes: single screw, twin screw, reciprocating single screw, and pultrusion. The latter is limited to use with continuous-strand modifier feeds (i.e., glass rovings). The other three processes find utilization in virtually all types of modifier–acetal compounding.

Feed Systems

Consistent feed introduction is critical if uniform, predictable final properties are to be maintained. Large swings in feed/additive ratios cannot be blended out by the compounder. Specific energy input will vary inversely with feed rate, and polymer degradation is probable if feed rate falls. In single-screw compounders where mixing is a function of die backpressure and the generated melt pool, the modifier may not be completely wetted out with acetal, and inferior properties will again result.

Acetal–additive feeds can be introduced as a dry preblend, as individual feedstreams, or as a combination of the two. Combinations are used when several additives in small individual concentrations are preblended to give one feed that is added separately to the acetal and modifier components. Individual modifier and acetal feeds allows downstream modifier feed addition. High shear acetal melting can be done without reducing the size of the modifier particle (important with aspect-ratio-dependent modifiers like glass or carbon fibers), and compounder wear can be reduced by feeding the modifier into the melt.

Feed addition can be controlled by flood feeding or starve feeding through volumetric or mass feeders. Flood feed is the simplest, but feed rate depends on compounder speed. Compounding control cannot be done independent of feed rate. Volumetric and mass feeders allow feed rate control to be independent of compounder control. However, volumetric feeders are susceptible to changes in feed density. Mass feeders provide the best feed rate control. Master–slave systems can adjust component rates based on input component ratios and an overall feed rate. Component inputs are maintained as closely as possible to the desired final compositions, modifier use is optimized, and product property variations are minimized.

In addition to feeders, feed enhancements are sometimes needed for less dense modifiers. Without them, modifiers like glass or carbon fibers can alternately bridge and flood. Screw crammers push the material into the extruder, and feed rates are maintained. Steep-sided feed hoppers with smooth walls also help to maintain consistent feed flow. Crammers are not without drawbacks, however. Particle size reduction can be a problem, and care must be practiced when using crammers with aspect-ratio-dependent modifiers.

Modifiers can be added as continuous strands or discrete particles. Strands like fiberglass rovings are fed into the compounder with the acetal. The strands are pulled into the compounder and chopped by the compounder screw. Such a system is labor intensive. Moreover, feed rate is a function of screw speed, strand length is difficult to control, and compounder wear is high. Modifier in the form of discrete particles offers the advantages of independent feed control, better particle size distribution, and reduced machine wear; in addition, it requires less handling [7].

Compounding

The compounding process can be divided into five unit operations, which all must be successfully completed to achieve a marketable product [8].

- Melt the acetal polymer.
- Feed the modifier into the polymer melt.
- Uniformly disperse the modifier throughout the polymer melt and wet each modifier particle with polymer.
- Allow for degassing of entrained moisture and air.
- Provide a uniform pressure at the die for pelletizing.

Melting the acetal feed requires intense shear work input to the polymer. However, excessive shear intensity or duration will degrade the acetal. The melt section should be designed to only just melt the acetal. This is one reason that downstream modifier feed is preferred to upstream feed with the polymer feed. Particularly when several modifiers are run on the same screw configuration, the various modifiers all create different melt blend viscosities. A melt section designed to melt a low viscosity blend will overwork and degrade an acetal blend that has a higher viscosity. Downstream modifier feed allows the designer to design the melt section much more accurately, and a uniform acetal melt with little or no excessive heat history will result.

The second step of compounding is to introduce the modifier to the melt. With upstream feed, this is done as the acetal melts. In downstream modifier feeding, the screw has a low pressure section that allows the modifier to fall into the melt. This is the preferred method for aspect ratio modifiers, since high pressure feeding can break up the modifier particles.

Dispersion and wetting of the modifier particles is done with a lower intensity version of the acetal melt section. It is very easy in this section to mechanically and/or thermally degrade the melt. Because the viscosity of the melt has been increased by the additive, higher temperatures and more input shear will result unless the screw design is milder here than in the melting zone. The objectives in this zone are to separate the modifier feed as completely as possible into the individual particles, to disperse them throughout the melt, and to complete any bonding reactions that may take place between the modifier and acetal. Positive conveying through this zone is also important to ensure a uniform heat history on the polymer and to reduce modifier degradation, both via backmixing.

Any compounder feed will entrain air and moisture, which must be vented from the melt. Modifier compounding requirements for venting are even more important. The modifier also entrains gases that must be removed from the melt. This can be done by an atmospheric vent or a vacuum vent system. The latter is more effective because the pressure drop driving forces for devolatilization are greater, but vacuum systems can also be more prone to melt backups into the vents. Vent stuffers can be used to guard against this occurrence.

The compounder then becomes a melt pump to move the product melt through the die for stranding and pelletizing. Die pressures should not be excessive. High die pressures require more pumping and can lead to modifier particle attrition, large melt pools, and longer residence times in the machine (and potential melt pool backups into the vents), as well as temperature excursions from excessive shear through the die. Die design must take into account the higher viscosity of the modifier–acetal melt. Transitions must be more gradual, to allow the melt to completely fill all the die holes. When transitions are sharp, only the center holes of the die will be filled, and there will be weeping at the peripheral holes.

Strand cooling and pelletizing operations are very similar to unfilled acetal systems. Quenching can be done in water baths or air. Pelletizing can be done by die face systems or strand choppers.

Materials of Construction

Production of acetal composites creates higher wear and abrasion on compounder elements, dies, and choppers than do unfilled acetal product compounding operations. Screw and barrel materials and construction methods must address this reality to ensure a cost-effective operation.

Screws can be coated or tipped with high carbide content, nickel-based metals or through-hardened; in some machine designs, the screw elements can be replaced in segments (this is an advantage in both screw design and maintenance). Barrels can be constructed using similar techniques or by using removable liner sections made of the resistant alloy. In the cases of some materials, increased hardness also means increased brittleness. Feed systems should be designed to guard against foreign material that can chip or break the compounder internals, and product systems should protect the finished product from damage if such debris cannot be prevented.

As mentioned earlier, feed configuration can reduce wear. When modifier and rerun feed into the downstream melt, where the particles are coated with polymer melt, wear rates are greatly reduced. Upstream rerun feed essentially introduces modifier back into the feed zone and defeats some of the gains of downstream virgin modifier feed. Roving feed creates high wear rates because the rovings must be chopped by the screw.

Wear in inevitable. To reduce downtime and maintenance costs, repair and/or replacement of the compounder internals should be considered in the compounder design and selection. For instance, replaceable screw elements can be designed to wear faster than solid barrel sections that are difficult to remove and costly to refurbish.

Similar design considerations are made at the pelletizer. Through-hardened or tipped blades are used and replacement/repair is included in the design. Once again, hardened materials tend to be more brittle, and care is necessary in the handling and use of such treated blades.

Typical Equipment

Single-screw compounders, probably the simplest modified acetal compounding machines, were widely used as continuous operations replaced older batch operations. However, as compounding requirements became more rigorous, particularly as blends, alloys, and modifiers were introduced, limitations of the single-screw approach were revealed. Recent innovations in screw design have introduced some improvements in residence time control, unmelted polymer conveying, controlled shear intensity, and reduced overshearing and excessive temperatures [8–11].

However, twin-screw and reciprocating-single-screw compounders are widely recognized as being superior in mixing, dispersing, and controlling excessive temperatures in acetal–modifier compounding. The intermeshing of the twin-screw flights and the flights and barrel pins of the reciprocating-single-screw machines induces positive forward conveying, increases additive mixing and devolatilization, reduces backmixing and excessive residence time in the machine, and reduces dead spots in the machine, which can char acetal and modifier if allowed to accumulate. Intensive mixing, with minimum temperature increase and polymer degradation, is the result [8,9,12–14]. These machines can be purchased with modular barrel sections and with screws that are assembled from several basic elements. Compounder modifications to customize a screw and barrel to different types of modifier and mixing requirements is much simplified.

Pultrusion

The pultrusion process involves pulling a continuous fiber roving through a melt pool of the base polymer to impregnate the roving with the polymer. Pultrusion has long been used for reinforcing thermosets, but the high viscosity of thermoplastics had prevented intimate wetting of the fibers by the polymer. Technology patented by ICI uses plasticizers or lower viscosity variants of thermoplastics than normally could be utilized to achieve fiber–polymer wetting and improved product properties [15–17]. A later press release announcing a long strand acetal composite using ICI technology implies the use of similar technology [18].

Another series of patents details the pultrusion process [19,20]. A special forming die is mounted on the discharge end of a typical extruder. The die has ports allowing fiber rovings to be pulled through it. A polymer melt pool under pressure is maintained in the die by the extruder, and the rovings are pulled through the pool and to a shaping die. The polymer-coated roving is pulled through a cooler and cutter to cut the product to the desired length. By modifying the shaping die and the number of fiber rovings pulled through the melt pool, shaped products and chopped strands containing various levels of fiber can be manufactured. The shaped products can then be used to reinforce modified or unmodified plastic shapes by inserting them in standard injection molds and molding a second polymer around the insert. Bonding compatibilizers can be added to either plastic to aid the plastic–plastic bonding in the mold. The patents do not specifically mention the use of modified-viscosity thermoplastic feeds to aid roving wetting. Reference 20 specifically names acetal as a good second polymer.

Analytical Techniques

The goal of modified acetals is to enhance one, or a set of, the original acetal properties. Directly measuring such a property or properties may at first seem to be the most reasonable method for measuring the effectiveness of the modifier and the compounding process. During product development, the property is usually directly measured and correlated to modifier content, which may be also measured or calculated from the feed ratios. However in some cases, the property cannot be easily and quickly measured for real-time quality assurance at the plant (i.e., absence of warp or shrinkage). Also, an analytical technique is sometimes desired to check that specified feed ratios have been maintained.

An analytical technique to be used for quality assurance should meet the following criteria:

- It should be a quick method, to allow timely feedback to the processing unit.
- It should be reproducible across several technicians, so that it is applicable for multishift use.
- It should use standard equipment, either dedicated for that technique or shared among other analytical routines.
- It should be easy to calibrate the method against some standard sample or measurement.

Glass fiber and glass bead modified acetals can be ashed to remove the polymer and leave behind the incorporated glass, which is measured gravimetrically. Typical ashing conditions for acetals are 725°C and 20 minutes for fiber and somewhat longer for beads. Ashing in an atmospheric oven will remove all nonglass components, so additives and pigments are not a problem.

Carbon fibers are isolated by heating the sample at 600°C in an inert atmosphere. At these conditions, acetal will depolymerize and gas off as formaldehyde. The fibers can then be analyzed gravimetrically. In this case, though, additives like carbon black will also remain and confound the results.

Glass and carbon modified acetal properties also depend on the retention of the fiber aspect ratio and orientation within the polymer. Evaluation of these characteristics involves optical and electron microscopy. A novel approach to sample preparation and the use of automatic image analysis has been demonstrated with glass modified acetal copolymer to provide a rapid, repeatable, and accurate measure of fiber length distribution [21]. In situ modifier distribution and orientation can be analyzed by a number of microscopic techniques. Reflected light microscopy, scanning electron microscopy, X-ray imaging, and transmitted light microscopy can each be used, depending on the information desired [22]. These techniques are more complex and time-consuming than those mentioned previously, and real-time process control with such methods is not practical. However, these techniques are very useful in evaluating new and modified processes for the effects of the process on fiber distribution and orientation. Optimization programs to increase modified acetal properties with new feedstocks and revised screw designs and extruder conditions can be guided by the results of such analytical techniques.

Acetal Composite Properties

Composites form a class of material derived from the combination of two or more constituents. Unlike a homogeneous material, a composite can be tailored to a given application by varying the constituents, resulting in properties that are roughly the product of the components' properties. Other factors that affect composite performance include processing parameters and interfacial properties of the component materials in the composite [23–27]. With all these variables, the calculation of the composite properties becomes an approximation at best. Without performing the actual test measurements, the composite properties are generally approximated [28] by the rule of mixtures:

$$P\text{composite} = P\text{constituent} \times V\text{constituent}$$

where P represents property and V stands for volume percent.

Acetal composites are combinations of acetal polymer (homopolymer or copolymer) and fillers of various types. Each combination has unique properties constituted by both the matrix and the reinforcement. The acetal polymer matrix gives the composite its chemical resistance and processibility; and the fillers compensate for the shortcomings of the matrix for a given application.

Since different applications have different requirements, the properties of the acetal composite family are quite diversified. In general, these properties can be classified into five categories: physical, mechanical, chemi-

TABLE 2
Typical Physical Properties of Acetal Composites

Property	ASTM Test Method	Materials: Glass [29] Fiber	Glass [29] Bead	Carbon [36]	Mineral: Molybdenum [31]	Mineral: Coupled [32]	Teflon Fiber [33]
Specific gravity, g/cm³	D 792	1.58	1.58	1.46	1.41	1.48	1.54
Water absorption, 24 hour immersion, %	D 570	0.27	0.20	0.50	0.20		0.20
Filler content, %		25	25	20			

cal, thermal, and electrical. The sections that follow summarize the common properties under these categories for most of the commercially available acetal composites.

Physical Properties

Most of the properties (physical, mechanical, chemical, thermal, and electrical) are a function of test procedure and test condition. Therefore, in material comparison, the evaluation work is performed under a certain guideline. The most commonly known methods are those of the international, U.S., West German, and British standards organizations: ISO, and ASTM, DIN, and IEC, respectively. Table 2 summarizes some of the physical properties of acetal composites based on glass, carbon, inorganic mineral, and organic fiber (PTFE). Since there is more than one formulation for each composite set, the actual property range is a spectrum of data. Table 2 presents typical values only. In this physical property category, the rule of mixtures governs the specific gravity but not the water absorption parameter.

Mechanical Properties

Table 3 summarizes the typical mechanical properties of the composites listed in Table 2. These data are collected at room temperature using ASTM test methods. However, when the test temperature changes, the properties measured will be different. This temperature-dependent nature of properties is derived from the matrix (acetal) portion of the composite. Table 4 illustrates this effect on tensile properties of 25 wt % glass bead filled acetal copolymer (GB25) [29]. However, in certain tests, temperature has a minimal effect.

Table 5 shows the effect of testing temperature on the impact properties of GB25.

Other long-term mechanical properties include heat aging, wear, creep, fatigue, and relaxation. The general creep and fatigue properties of glass fiber reinforced acetal copolymer are shown in Figures 1 and 2, respectively [35]. Because of the glass fiber reinforcement, creep and fatigue properties of the composite are superior to those of the unfilled polymer. On the other hand, the coeffi-

TABLE 3
Typical Mechanical Properties of Acetal Composites at Room Temperature [29–33]

Property	ASTM Test Method	Materials: Glass Fiber	Glass Bead	Carbon Fiber	Mineral: Molybdenum	Mineral: Coupled	Teflon Fiber
Filler level, wt %		25	25				
Tensile strength at yield, MPa	D 638		46.2		64.1[a]	53.8	
Strength at break, MPa		110.3	41.4	75.9			52.4
Elongation at break, %		2	10	5	20	60	22
Modulus, MPa		8274	3964		2999	2931	2895
Flexural stress at 5% defor-	D 790					89.6	72.4
mation/yield, MPa						13.0	10.5
At break, MPa		158.6	84.1	113.8			
Modulus, MPa		7584	3240	6206	2689	2965	2340
Rockwell hardness	D 785	M84	M75		M85		M78
Izod impact, notched, J/m	D 256	64	37	43	60	61	64
Tensile impact, kJ/m²	D 1822	105	42				105
Shear strength, MPa	D 732	58.6	48.3			49.6	55.2

[a] Test rate at 50 mm/min as opposed to standard 5 mm/min.

TABLE 4
Effect of Testing Temperature on Tensile Properties of 25 wt % Glass Bead Filled Acetal Copolymer (GB25)

Temperature	Tensile Strength at Yield[a]	Strength at Break
°C	MPa	MPa
−40	73.4	67.6
23	46.2	41.4
82	26.2	29.3

[a] ASTM D 638.
From Ref. 29.

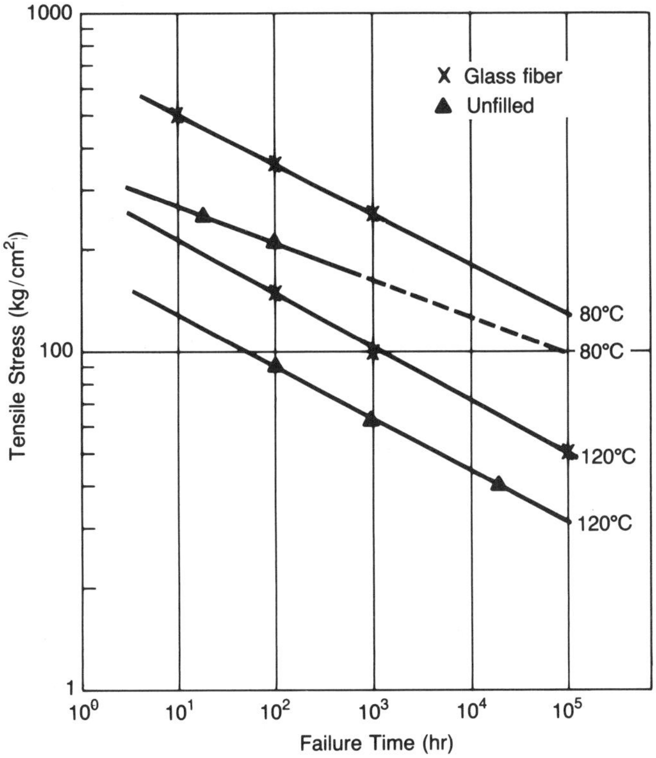

FIGURE 1 Creep properties of acetal copolymer [13].

cients of friction increase with the incorporation of glass fiber, and an increase in wear of the mating surface occurs. Unlike glass fiber, carbon fiber filled acetal composite has good fatigue endurance and low coefficients of friction; and acetal filled with PTFE fiber (or powder) has good wear performance. The wear properties of composites filled with glass fiber, carbon fiber, and PTFE fiber are summarized in Table 6 [29,36]. These studies are performed with a thrust washer test apparatus, which applies a constant force on the test specimen when it is moving against a mating surface at a certain speed. The wear factor can be calculated as follows:

$$\text{wear factor} = \frac{\text{volume wear}}{(\text{force} \times \text{speed} \times \text{testing time})}$$

Volume wear is the change in the volume of the specimen (dimension) due to wear. It can be estimated by the change in weight and the material's density.

Besides the nature and the dimension of the fillers, molding conditions play an important role in determining the wear performance of the composites. In general, high mold temperature gives a resin-rich surface, which improves the wear properties.

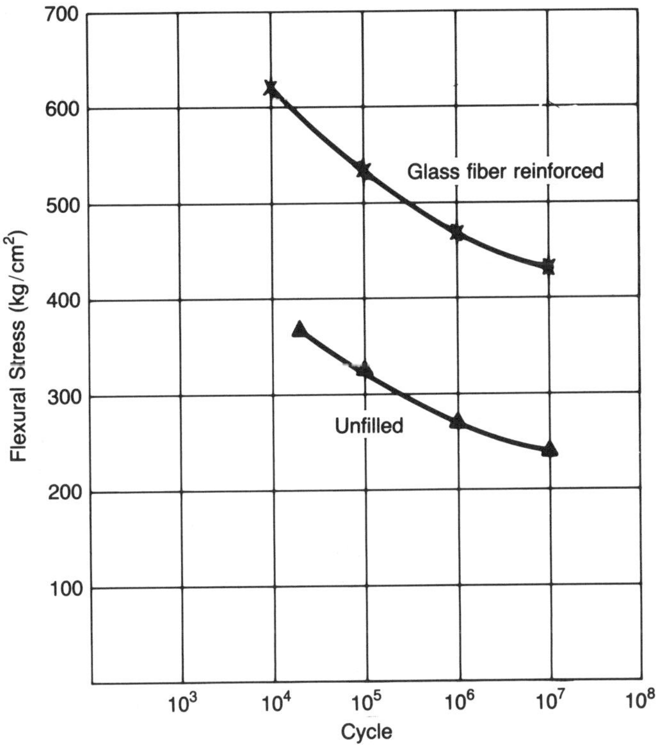

FIGURE 2 Flexural fatigue of acetal copolymer—unfilled and glass fiber reinforced [13].

TABLE 5
Effect of Testing Temperature on Impact Properties of 25 wt % Glass Bead Filled Acetal Copolymer (GB25)

Temperature	Izod Impact Strength
°C	J/m
−40	38
23	38
82	38

From Ref. 29.

TABLE 6
Wear and Other Mechanical Properties of Acetal Composites Based on Glass Fiber, Carbon Fiber, and Teflon Fiber [33,36]

Property	ASTM Test Method	Reinforcement[a]		
		Glass	Carbon	Teflon
Fiber content, wt %		30	20	
Tensile strength, MPa	D 638	89.6	62.1	52.4
Flexural modulus, MPa	D 790	8964	9309	2340
Notched Izod impact, J/M	D 256	42.7	53.3	64.0
Wear factor, 10^{-10} cm^3 min/cm/kg/h		289	47	
Coefficient of friction, 253 mm/s				
Static, 0.28 MPa		0.25	0.11	0.08[b]
Dynamic 0.28 MPa		0.34	0.14	0.14[c]

[a] Mating surface is steel.
[b] Measured at 2 MPa.
[c] Measured at 2 MPa, 50 mm/s.

Chemical Properties

Chemical properties are a function of the chemical type and concentration, and the exposure time and stress level imposed on the composite. In general, acetal polymer exhibits excellent resistance over a wide range of chemical exposures (e.g., solvents, lubricants, automotive fluids, many weak acids and weak alkalies, and water) [31,33–35]. However, properties are significantly affected by oxidizing agents (e.g., hypochlorite, zinc chloride solution). Homopolymer properties are also affected by strong bases.

Since most of the modifiers are encapsulated by the polymer, the chemical properties of the composites are similar to that of the polymer. However, there is still a percentage of exposed additive, and the additive–matrix interface may be subject to chemical reaction.

Thermal Properties

As acetal crystallizes or melts, the physical dimensions change. By incorporating a second phase, usually a more rigid (higher modulus) filler, the designer causes the temperature dependency of the composite to deviate from the thermal profile of the polymer. Because of the rigidity of the filler, in most cases, the shrinkage experienced during cooling of the composite (as the part starts to crystallize) is significantly reduced: up to 90% improvement is observed in glass fiber filled acetal, and up to 35% improvement in glass bead filled acetal. However, being a heterogeneous system, acetal composites exhibit a certain anisotropy, which is caused by the orientation of fillers, especially those having an aspect ratio greater than 1. For example, the coefficient of linear thermal expansion of an injection-molded, carbon fiber filled acetal shows different values in the directions perpendicular and parallel to the carbon fiber. This difference in coefficients constitutes the uneven shrinkage of the molded parts. The anisotropical behavior can be reduced by using a filler with aspect ratio that approaches 1, such as glass bead. However, some degree of uneven mold shrinkage of the injection-molded parts, due to the polymer flow characteristics, will remain.

Besides the physical dimension, the rigidity of the filler improves the high temperature performance of composites. For example, the heat deflection temperature at 1.8 MPa increased from 110°C to 163°C after glass fiber was incorporated into acetal polymer. Some of these thermal properties of acetal composites are listed in Table 7. (These data are collected using the ASTM procedure, which minimizes the effect of anisotropy.)

Electrical Properties

Pure acetal polymer is a good insulator. It has good dielectric strength, a low dielectric constant, and a good dissipation factor. By adding fillers to the acetal polymer, and depending on the nature of the additives, the electrical properties may be significantly altered. For example, because of the conductive nature of carbon, the incorporation of carbon in fiber or powder form will bring the electrical properties of acetal more toward metal. Table 8 summarizes some of the electrical properties.

Processing the Composites

Precision acetal composite components are produced by several fabrication processes. Glass fiber reinforced acetal is converted by profile extrusion into sheet, rods, and tubes, which are subsequently machined into finished parts [37]. However, acetal composites are predominantly processed by screw injection molding.

Procedures and conditions are similar for unfilled acetal and acetal composites [38]. Generally, the composites have higher viscosity and therefore are more resistant to flow than the unfilled counterparts. Therefore, higher injection pressures are often required, but a higher melt temperature can facilitate flow. Conventional and hot runner molds can be used with acetal composites because these materials have good process-

TABLE 7
Typical Thermal Properties of Acetal Composites [29–33]

Property	ASTM Test Method	Materials: Glass Fiber	Glass Bead	Carbon Fiber	Molybdenum	Mineral Filled	Teflon Fiber
Filler level, wt %		25	25				
Melting point, °C		166	166		164–168		175
Heat deflection temperature at	D 648						
1.8 MPa, °C		163	118	161	110	93	118
0.5 MPa, °C		166	158		160	154	168
Coefficient of linear thermal expansion x 10^{-5} cm/cm/°C	D 696	5.0	8.4		19.7	19.9	12.2
Vicat softening, °C	D 1525	162	162			160	

ing stability. However, significant polymer degradation will occur if they are severely overheated. General processing melt temperatures (checked by hand-held pyrometer) for acetal copolymer and homopolymer are 182–210°C and 204–227°C, respectively. Acetal resins and PVC should never be mixed or used sequentially in a molding machine or other supplementary equipment without thorough cleaning. These two materials are highly chemically interactive, causing rapid material destruction. Also, acetal composites should never be heated above 249°C. Recommended injection molding factors are shown in Table 9 [39]. The injection molding screw used with acetal should have a length-to-diameter ratio between 16 : 1 and 24 : 1, with a compression ratio in the range of 3 : 1–4 : 1. When composites are processed, the screw should be coated or hard faced with tungsten carbide or Colmonoy 56.

Acetal resins do not normally need to be dried before molding, but if they have been stored in an open bag under humid conditions, drying may be desirable to obtain optimum properties and finish in a molded part. Drying in a dehumidifying oven with good air circulation through the bed of pellets for 3 hours at 82°C is recommended. In the molding process, backpressure is often used to develop a more homogeneous melt. This can be desirable for unfilled acetal composites reinforced with particulates such as glass beads and talc. However, it is not recommended when acicular reinforcement (glass fibers, carbon fibers, etc.) is involved, since backpressure can cause significant reduction in aspect ratio (fiber length/diameter), hence a reduction in tensile and stiffness properties of the molded object. Also, use of regrind blended with virgin material should be limited (see Table 9) for the same reason.

The cycle time for acetal composites can be estimated [40] by identifying the thickest section in the molded part (in millimeters) and multiplying by 13 seconds. Thus if a molded part has its thickest wall at 2.5 mm, cycle time can be estimated to be 32.5 seconds. Optimization of part design and nucleation by some fillers may result in significant reduction in cycle time compared to unfilled acetal.

Applications

The selection of an acetal composite for a particular application is based on the special features of the composite. The sections that follow classify some of the current applications of acetal composites by industries.

TABLE 8
Typical Electrical Properties of Acetal Composites [29,30–33]

Property	ASTM Test Method	Materials: Glass Fiber	Carbon Fiber	Mineral, Molybdenum	Teflon Fiber
Filler level, wt %		25			
Dielectric constant	D 150	3.9			3.1
Dissipation factor at 106 Hz	D 150	0.0062		0.0024	0.009
Surface resistivity, Ω	D 257	3.8×10^{15}	7.0×10^{3}		
Volume resistivity, Ω · cm	D 257	1.2×10^{14}	1.0	10^{15}	3.0×10^{16}
Arc resistivity, s	D 495	136			183
Dielectric strength (2.29 mm sheet), V/0.0254 mm	D 149	600			400

TABLE 9
Injection-Moldable Acetal Resin-Recommended Processing Factors

Grade	Melt Temperature Range [°C]	Melt Viscosity (Pa-S)[a]	Injection Speed	Mold Temperature [°C]	Maximum Regrind (%)	Gating Diameter [cm][b]	Shrinkage Range [cm/cm]
Homopolymer							
Unmodified base resin (includes glass-filled PTFE filled, all viscosities)	205–225	240–1300	Slow to moderate	80 to 105	30	0.16 to 0.190	0.051
Impact modified	195–215	330–1300	Slow to moderate	10 to 70	35	0.160 to 0.190	0.051
Copolymers							
Unmodified base resin							
Low viscosity (45 melt index)	175–200	180	Medium to fast	150 to 200	25	0.203 by 0.38	0.051 to 0.059
Medium viscosity (9.0 melt index)	180–200	250	Medium			0.254 by 0.38	0.051 to 0.059
High viscosity (2.5 melt index)	185–205	400	Medium			0.318 by 0.508	0.046 to 0.059
Glass-fiber reinforced							
12.5%	185–205	325	Medium			0.254 by 0.38	0.015 to 0.051
25%	195–210	480	Medium			0.318 by 0.508	0.010 to 0.046
Mineral filled							
Low filler, high flow	180–200	200	Slow			0.254 by 0.38	0.040 to 0.051
High filler, low flow	185–205	200	Slow			0.318 by 0.508	0.035 to 0.046
Impact modified							
Medium	180–200		Medium			0.254 by 0.38	0.046 to 0.051
High	185–205		Medium	65 to 95	25	0.318 by 0.508	0.043 to 0.051

[a] Melt viscosity measured at 1000 s^{-1} at the medium temperature.
[b] Data based on 0.318 cm wall thickness.
From *Plastics Engineering*, September 1985.

Automotive

Most of the acetal composites for the automotive industry are based on glass-filled materials, both fiber and beads. The selection of these composites is based on their strength and low warp properties, respectively, as well as their rigidity and inertness to most automotive fluids. Some applications [34,40–42] are heater control base plates, adjustment pedals, sunroof cranks, wiper pivots, fuel pumps, window brackets, and speedometer frames.

The acetal used in another automotive application [43], suspension linkage bushings, is filled with PTFE fiber. Instead of the rigidity offered by glass, this composite provides low wear and friction.

Consumer Goods

The diversity of the properties of acetal composites accounts for a wide range of applications in various segments of the consumer products industry. The applications, the materials selected, and the material attributes are listed in Table 10.

Plumbing

The major reason for the usage of acetal composites in plumbing applications is the performance of acetal in a water-contact environment. Such applications [34] include water meter pressure plates, which also require some rigidity, and hot water drain valves.

Polyacetal Suppliers

The three major manufacturers of acetal composites in the United States are Hoechst Celanese Corporation, Engineering Plastics Division (formerly Celanese Engineering Resins, Inc.), DuPont, and BASF. Outside the United

TABLE 10
Acetal Composites Used in Consumer Industries [34,40,44]

Applications	Additive	Composite Attributes
Adjustable computer workstations	Mineral filled	Flatness, strength, lubricity
Printing plates, lettering discs, broadcast cartridge hubs, electric toothbrush motor housings	Glass bead filled	Low warpage, rigidity
Spin drive gear	Glass fiber reinforced	Strength, rigidity
Conveyor systems, gear/ratchets	PTFE filled	Lubricity
Timing gears	PTFE/glass fiber filled	Rigidity, strength, lubricity

States, there are Asahi, Mitsubishi, and Polyplastics of Japan, and Hoechst Ticona of West Germany. These composites suppliers are also producers of acetal polymer (homopolymer by DuPont and Asahi; copolymer by Hoechst Celanese, Hoechst Ticona, BASF, Polyplastics, and Asahi). Besides these acetal producers, acetal composites are formulated by various specialty compounders (LNP, Fiberfil, Thermofil, etc.). Table 11 lists some of the acetal composites currently available, along with the general formulations, product abbreviations, and product features.

Edwina Ying, Derrick B. McKie, and Clifford R. Boldt

TABLE 11
Trademarks and Features of Acetal Composites [34–36,40,45–47]

	Acetal Polymer			
Filler	Name	Abbreviation	Feature	Manufacturer
Glass fiber	Ultraform copolymer	N2200G5	High strength and rigidity	BASF
Molybdenum disulfide	Ultraform copolymer	11005MO	Friction bearings	BASF
Glass fiber	Duracon copolymer	GC25 GR25	High strength and rigidity	Polyplastics
Glass bead	Duracon copolymer	GB25	Low warpage, improved rigidity	Polyplastics
Milled glass fiber	Duracon copolymer	GM20	Low warpage, improved rigidity	Polyplastics
Glass fiber	Duracon copolymer	GS25	Electrical contact contamination prevention	Polyplastics
Molybdenum disulfide	Duracon copolymer	MSO2	Improved wear resistance	Polyplastics
Carbon fiber	Duracon copolymer	CE20	Antistatic, electrical conductivity, excellent wear in water	Polyplastics
Carbon fiber	Duracon copolymer	CR20	Excellent wear, improved strength and rigidity	Polyplastics
Mineral	Celcon copolymer	MC series	Low shrinkage, improved stiffness	Hoechst–Celanese
Glass fiber	Celcon copolymer	GC25A	High strength and rigidity	Hoechst–Celanese
Glass bead	Celcon copolymer	GB25	Low warpage, improved stiffness	Hoechst–Celanese
Carbon fiber	Celcon copolymer	EF25	Improved conductivity	Hoechst–Celanese
Glass fiber	Hostaform copolymer	C9021 (GV1)	High strength and rigidity	Hoechst AG
Glass bead	Hostaform copolymer	C9021 (GV3)	Improved stiffness, low warpage	Hoechst AG
Mineral	Hostaform copolymer	C9021K	Low wear for unlubricated slides	Hoechst AG
Molybdenum	Hostaform copolymer	C9021M	Low wear for high surface load	Hoechst AG
Milled glass fiber	Delrin homopolymer	570	High stiffness	Du Pont
Teflon fiber	Delrin homopolymer	100AF, 500AF	Low friction, low wear	Du Pont
Aramid fiber	Delrin homopolymer		Improved wear and tensile strength	Du Pont
Glass fiber	Tenac homopolymer	GA510 GA520	High stiffness	Asahi
Glass bead	Tenac homopolymer	GT525	Low warpage, improved rigidity	Asahi
Milled glass–Teflon		KFL-4016M	Rigidity with improved wear	LNP
Glass fiber–Teflon		KFL-4036	Strength, rigidity, improved wear	LNP

References

1. S. M. Sinker, "Acetal," in *Modern Plastics Encyclopedia*, Vol. 64, no. 10A, McGraw-Hill, New York, 1987, p. 11.
2. T. H. Ferrigno, "Principles of Filler Selection and Use," in H. S. Katz and J. V. Milewski, Eds., *Handbook of Fillers and Reinforcements for Plastics*, Van Nostrand Reinhold, New York, 1978, pp. 11–58.
3. Polyacetals, CEH Product Review, *Chemical Economics Handbook*, SRI International, Chicago, August 1986.
4. S. DalNogare and J. O. Punderson, U.S. Patent 2,998,409 (Aug. 29, 1961), to E. I. du Pont de Nemours & Co., Inc.
5. "Performance Plastics I: Thermoplastics—A Technical, Economic and Marketing Report," Margolis Marketing Research Report, 1986.
6. "Reinforced Thermoplastics—U.S.A.," SRI International, Chicago, July 1982.
7. "Reinforced Thermoplastics—U.S.A. Specialty Chemicals—Strategies for Success," SRI International, Chicago, July 1982.
8. D. F. Mielcarek, *Chem. Eng. Prog.*, June 1987, pp. 59–67.
9. S. Jakopin, *Adv. Polym. Technol.*, *3*(4), 365 (1983).
10. R. Barr, *Plast. Eng.*, January 1981, pp. 37–41.
11. *Plast. World*, December 1985, pp. 61–63.
12. K. Eise, J. Curry, and J. Farrell, *Proceedings of the Society of Plastics Engineers RETEC*, 1984.
13. D. B. Todd and D. K. Baumann, *Chem. Eng. Prog.*, January 1977, pp. 65–68.
14. K. Stade, *Polym. Eng. Sci.*, *17*(1), 50 (1977).
15. F. N. Cogswell et al., U.S. Patent 4,541,884 (Sept. 17, 1985), to Imperial Chemical Industries, PIC, England.
16. F. N. Cogswell et al., U.S. Patent 4,549,920 (Oct. 29, 1985), to Imperial Chemical Industries, PIC, England.
17. F. N. Cogswell et al., U.S. Patent 4,559,262 (Dec. 17, 1985), to Imperial Chemical Industries, PIC, England.
18. "LNP Looks to Replace Metals with Plastic Compounds," *Plast. Busi. News*, Oct. 14, 1987.
19. R. C. Hawley, U.S. Patent 4,312,917 (Jan. 26, 1982), to Polymer Composites, Inc.
20. R. C. Hawley, U.S. Patent 4,439,387 (March 27, 1984), to Polymer Composites, Inc.
21. L. C. Sawyer, *Polym. Sci. Eng.*, *19*(5), 377 (1979).
22. L. C. Sawyer and D. T. Grubb, *Polymer Microscopy*, Chapman and Hall, London, 1987.
23. *Interface in Composites*, ASTM STP 452
24. A. T. DiBeredetto and L. Nicolais, "Interfaces in Composites," in G. Piatti, Ed., *Advances in Composite Materials*, Applied Science Publishers, Barking, England, 1978, Ch. 8.
25. L. J. Broutman and B. D. Agarwal, *Polym. Eng. Sci.*, *14*(8), 581 (1974).
26. L. J. Broutman and B. D. Agarwal, *Proceedings of the 28th Annual Technical Conference*, SPI, 1973, 5B.
27. P. Yeung and L. J. Broutman, *Proceedings of the 32nd Annual Technical Conference*, SPI, 1977, 9B.
28. W. B. Hillig, "Composite Materials," in *Encyclopedia of Composite Materials and Composites*, Wiley-Interscience, New York, 1983.
29. "Glass Filled Products," Celcon Acetal Copolymer, Celanese Engineering Resins, 1987.
30. "Preliminary Data Sheet for Conductive Grades," Celcon Acetal Copolymer, Celanese Engineering Resins, November 1988.
31. "Product Line Properties, Applications" Ultraform Polyoxymethylene POM, BASF, 1986.
32. "Celcon Mineral Coupled MC90 and MC90-HM," Celcon Data Sheet, Celanese Engineering Resins, 1983.
33. *Design Handbook for Du Pont Engineering Plastics*, Module III, Delrin Acetal Resins, DuPont, E-62619.
34. "Bulletin C1A: Properties," Celcon Acetal Copolymer, Celanese Engineering Resins, 1985.
35. "Technical Series of Duracon, Design Data," Duracon Acetal Copolymer, Polyplastics Company, Ltd., 1986.
36. "LNP—Internally Lubricated Reinforced Thermoplastics," LNP Bulletin 254–782, LNP, 1982.
37. Celcon Acetal Copolymer, Bulletin C3B, *Extrusion*, Hoechst–Celanese Corporation, Engineering Plastics Division, Chatham, NJ, 1983.
38. Celcon Acetal Copolymer, Bulletin C3A, *Injection Molding*, Hoechst–Celanese Corporation, Engineering Plastics Division, Chatham, NJ, 1986.
39. V. Wigotsky, "Engineering Resins: Producing Precision Parts," in G. D. Shook, Ed., *Reinforced Plastics for Commercial Composites*, American Society for Metals, Metals Park, OH, 1986, pp. 31–49.
40. "Hostaform, Acetal Copolymer Resins, Product Summary, ASTM Values," Hoechst Engineering Plastics, Hoechst, 1986.
41. "LNP: Internally Lubricated Reinforced Thermoplastics," LNP Bulletin 254779.
42. "Asahi Polyacetal—Tenac," Asahi Polyacetal, Asahi Chemical Industry Company, Ltd.
43. "General Guide to Products and Properties," Delrin Acetal Resin, du Pont Engineering Plastics.
44. C. Gottfried, "Reinforced and Filled Acetal Copolymer for Extended Performance in Various End Uses," presented at SPI/SPE Plastics—West, 1987.
45. "Product Line Properties, Applications," Ultraform Polyoxymethylene POM, BASF, 1986.
46. L. Nazarenko, Ed., "Aramid Fiber Gives 'Super' Wear Resistance," *Mod. Plast.*, June 1987.
47. "Asahi Polyacetal—Tenac," Asahi Polyacetal, Asahi Chemical Industry Company, Ltd.

Adhesive Joining

See Joining, Adhesives; Joining, High-Temperature Adhesives

Aerospace Materials Specifications

See Standards and Specifications

Aging of Polymer Matrix Fibrous Composites

Chemical Aging

Hydrolyses

The hydrolytically induced chain scission of typical thermoset or thermoplastic polymeric composite matrices is not a significant aging phenomenon. The rare exceptions involve impurity-catalyzed hydrolyses—for example, glass filler alkali surface impurity catalyses of polycarbonate matrices.

In the case of fibers, aromatic polyamides, such as Kevlar 49, are susceptible to hydrolytic degradation, which is accelerated by impurities associated with the fiber fabrication processes.

The amide linkage in the poly(*p*-phenylene terephthalamide) (PPTA) macromolecule is susceptible to hydrolytically induced chain scission as illustrated in Eq. (1).

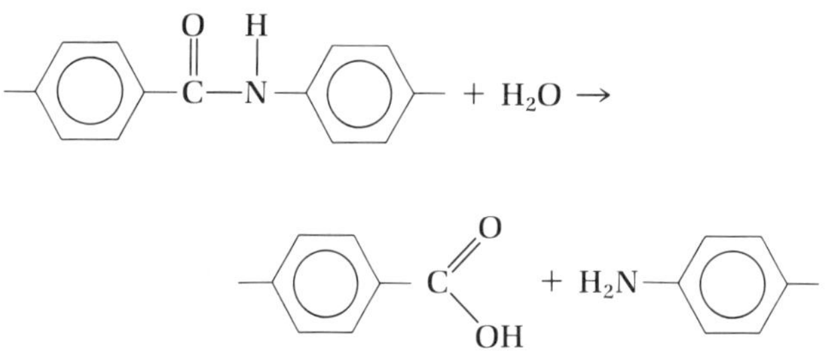
(1)

From the law of mass action, the rate of formation of new macromolecular ends is expressed as

$$\frac{d[NH_2]}{dt} = k[\text{amide}][H_2O] \tag{2}$$

where k is the rate constant. The hydrolysis reaction rate is a constant for fiber strength deterioration because (a) only 4% of the amide linkages have to hydrolyze to produce zero strength [1–3] and (b) the H_2O consumed in the reaction will readily be replaced in normal service environment conditions. Smith [4] has reported that the equilibrium moisture content of Kevlar 49 yarns ($[H_2O]$ in Eq. 2) is directly proportional to the relative humidity (RH) at 23°C. Hence, from Eq. (2), the rate of hydrolysis will also be directly proportional to RH. The rate of hydrolysis can be accelerated by any H_2SO_4 and NaOH impurities present in the fiber. The H_2SO_4 will act as a true catalyst and will remain chemically unmodified after each hydrolytic reaction, whereas NaOH will accelerate the hydrolytic reaction and then react to form a COO^- Na^+ salt macromolecular end group. From exposing Kevlar 49 yarn to H_2SO_4 and NaOH solutions and monitoring yarn strength degradation, the rates of hydrolysis have been found to be $\sim 10^5$ and $\sim 10^3$ times faster in the presence of H_2SO_4 and NaOH, respectively [5].

The rate of the hydrolytic degradation has been shown [1] to be related to temperature T by the relationship

$$\ln[d\sigma/dt] = -E/RT + C \tag{3}$$

where E is the activation energy for hydrolytic chain scission of the PPTA amide linkage and C and R are constants. Based on the assumption that applied stress lowers the activation energy for hydrolysis by $B\sigma_{\%}$, where B is a constant and $\sigma_{\%}$ is the applied stress expressed in terms of a percentage of the fiber strength at a specific instant in time t, then the magnitude of fiber strength degradation Δ_t after a time t in a specific environment is related to t, RH, E, $\sigma_{\%}$, and T by

$$\Delta_t = (\text{RH}/100)\, Dt \exp - \left(\frac{E - B\sigma_{\%}}{RT}\right) \tag{4}$$

where D is a constant [1].

Hydrolytic degradation of bare Kevlar 49 yarns has been reported (Table 1) for exposure conditions of 100% RH in the 65–200°C range as a function of time. Little spool-to-spool variation was found in the hydrolytic degradation rate at 100°C/100% RH exposure, which indicates that potential H_2SO_4 or NaOH impurities that could result from fiber fabrication and would catalyze the hydrolytic degradation are essentially absent or immobilized in the "as received" fiber.

TABLE 1
Rates of Hydrolytically Induced Strength Degradation of Kevlar 49 Fiber as a Function of Temperature at 100% RH

Exposure Conditions at 100% RH		Strength Degradation
Temperature (°C)	**Time (days)**	**(%/year)**
200	2	12,426
175	4	5,585
175	4	5,466
150	14	1,563
125	32	199
125	40	264
100	162	64
Each Test Performed on a Different Spool		
100	281	77
100	281	77
100	281	76
100	281	78
100	281	86
100	281	83
65	440	3[a]
65	440	1.5[a]
65	440	5[a]
65	440	4[a]

[a] Statistically meaningful data.

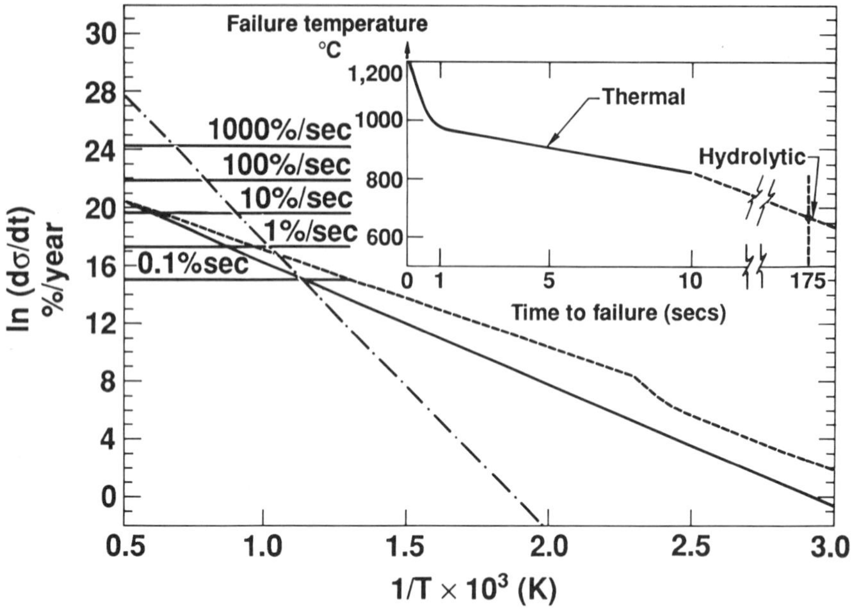

FIGURE 1 Log of rate of fiber strength degradation $d\sigma/dt$ versus $1/T$ for Kevlar 49 fibers: dashed curve, hydrolytic; solid curve, oxidative; dot-dash curve, thermal. (Inset plot is failure temperature for 100% strength loss, chain scission model, versus time to failure.)

Figure 1 plots $\ln(d\sigma/dt)$ versus $1/T$ for the hydrolytic aging data, together with other literature data on hydrolytic and oxidative degradation [4,6–10]. Kevlar 49 fibers degrade ~200 times faster in a 100% RH environment compared to a 0% RH environment at the same temperature. The hydrolytic aging data exhibit a discontinuity from linearity at 100°C in the $\ln(d\sigma/dt)$ versus $1/T$ plot because of the onset of physically induced fiber damage as a result of high steam pressures, which is indicated by higher moisture sorption in such fibers [1]. Table 2 gives some estimated rates of fiber degradation in a variety of service environment conditions, as determined from extrapolation of the $\ln(d\sigma/dt)$ versus $1/T$ plot to lower temperatures together with utilization of Eq. (4). The data indicate that in typical service environment conditions, hydrolytically induced fiber strength degradation is not a serious concern, with rates of degradation of 5% or less a year at 65°C, 100% RH in the nonstressed state.

Oxidation

There are experimental difficulties in studying the oxidation of polymer matrix fibrous composites (PMFCs) because the reactions occur between a solid and a gas. The effects of O_2 concentration and its associated gaseous diffusion through the composite make it difficult to analyze oxidation on a quantitative basis.

Epoxy matrices are susceptible to oxidation in the 150–200°C range. Tetraglycidyl-4,4′-diaminodiphenyl methane (TGDDM) epoxies cured with diaminodiphenyl sulfone (DDS), the most common composite matrices used in high performance fibrous composites, oxidize above 200°C. Indeed, for these 177°C cured systems there is a narrow window between the temperatures associated with cure conditions and the onset of degradation. Figure 2 presents data on the cure and degradation reactions of a TGDDM-DDS epoxy in the 177–300°C temperature range, obtained by Fourier transform infrared difference spectrometry [11]. The disappearance of the epoxide and primary amine groups (%) and the appearance of the secondary amine, hydroxyl, and ether groups in the form of normalized infrared band intensities $I_{NH{:}SO_2}$, R_{OH}, and R_{ether}, respectively, are plotted as a function of DDS concentration and cure conditions. In the 177–300°C temperature range, the ether band intensity

TABLE 2
Estimated Kevlar 49 Strength Degradation as a Result of Hydrolyses

Exposure Temperature (°C)	Imposed Stress Level (%)	Strength Degradation (%/year)
65	0	3.0
	20	4.6
	75	16.3
38	0	0.5
	20	0.8
	75	2.8
23	0	0.2
	20	0.3
	75	1.1

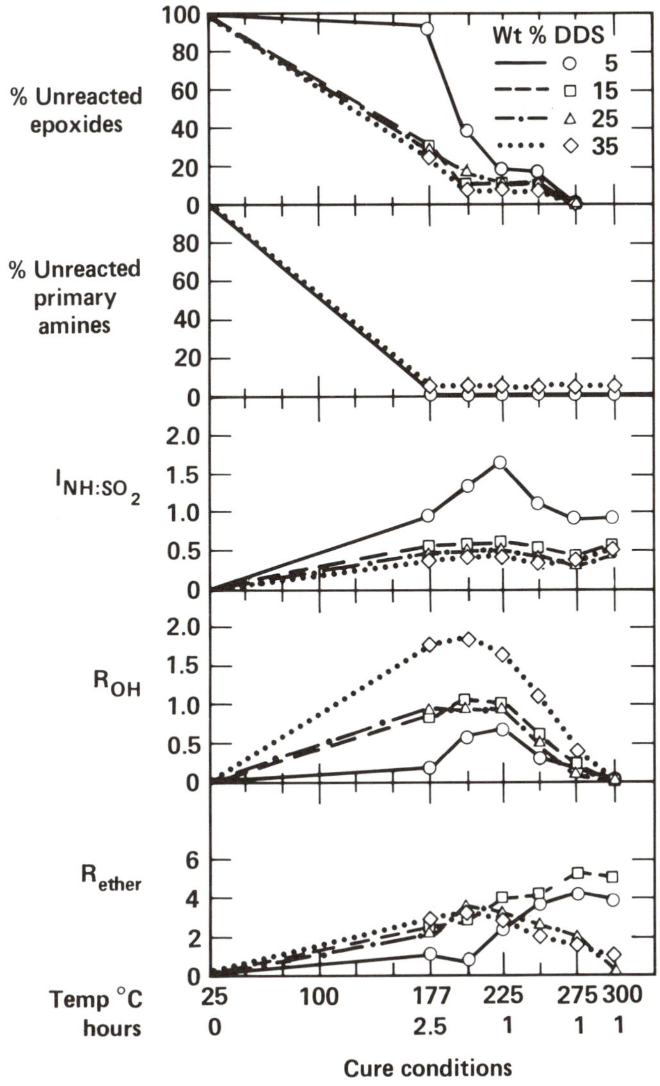

FIGURE 2 **Percent of unreacted epoxide and PA groups, $I_{NH:SO_2}$, R_{ether}, and R_{OH} values versus cure conditions in the 177–300°C range and DDS concentration for TGDDM-DDS (0 wt % $BF_3:NH_2C_2H_5$) epoxies.**

(R_{ether}) increases with increasing cure temperature as a result of epoxy network oxidation and/or dehydration, both of which result in the formation of ether cross-links. The hydroxyl infrared band intensity, plotted in the form of R_{OH}, decreases with increasing temperature above 200°C, which indicates the TGDDM-DDS epoxies do dehydrate in this temperature range. A number of studies on the degradation of epoxies indicate that dehydration will be the principal degradation mechanism in the 225–300°C temperature range over time periods of hours [12–15]. These network degradation reactions, over periods of hours, cause loss in the mechanical strength of TGDDM-DDS epoxies at 250°C and higher [16]. In the absence of O_2, systematic differential scanning calorimetry studies, at a heating rate of 10°C/min, indicate that the TGDDM-DDS network undergoes thermal decomposition, principally by scission of the

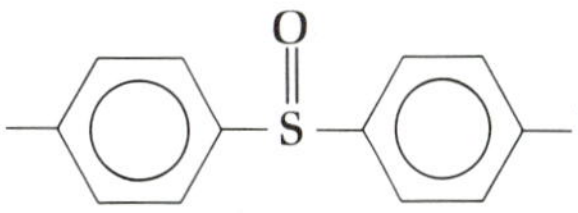

bonds in the DDS portion of the network [17].

Higher temperature composite matrices, of which the most common are polyimide based, oxidatively degrade in air, with associated strength loss in the 250–300°C, 10^3–10^4 hour temperature–time exposure regime [18]. Partial fluorination of polyimide matrices enhances oxidation stability [19]. Figure 3 illustrates the effect of 24 hour anneal temperature in air on the oxidatively induced deterioration in the tensile strength of poly(4,4′-oxydiphenylene) pyromellitimide above 400°C. (The effects of quenching rapidly from above T_g and annealing below T_g, illustrated in Figure 3, are discussed below under PHYSICAL AGING.)

In the case of fibers used in composites, glass fibers are not susceptible to oxidation, whereas carbon and Kevlar 49 lose their mechanical strength due to oxidation after 1 hour exposure to air in the ~550° and ~250°C temperature regions, respectively.

The oxidative stability of carbon fibers based on polyacrylonitrile (PAN) is deteriorated by the presence of residual Na impurities that originate from the PAN precursor fabrication procedures [20]. However, spectrochemical analysis of the latest PAN-based carbon fibers indicates Na impurities are now below 200 ppm; see Table 3 [21]. The residual nitrogen content in PAN-based carbon fibers also enhances oxidation as a result of the lower oxidative stability of nitrogen containing six-membered rings relative to the graphite structure [21].

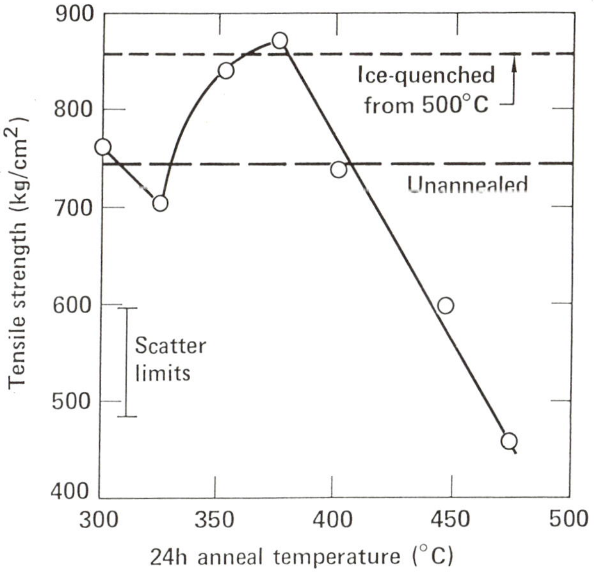

FIGURE 3 **Tensile strength of poly(4,4′-oxydiphenylene) pyromellitimide as a function of 24 hour anneal temperature in air.**

TABLE 3
Impurities in Carbon Fibers from Spectrochemical Analysis

Element	Impurities (ppm) in Fibers[a,b] Celion 6000	Thornel 300	Hercules ASX12	Hercules IM6	Thornel Special
Ca	200	30	100	80	80
Hg	i	i	i	i	100
Na	i	i	200	100	i
P	100	i	i	i	i
Si	100	200	300	200	300
Mg	60	20	60	20	10
Cu	30	20	30	30	60
Fe	ii	10	20	20	10
Al	ii	ii	20	20	20
Ni	ii	10	ii	ii	ii
Sr	3	iii	iii	iii	iii
Be	iv	iv	iv	iv	iv
Zn	v	v	v	v	v
Ti	ii	ii	ii	ii	ii

[a] Sizing has been removed.
[b] Ranges of detectability:
i not detectable ≤ 100 ppm
ii not detectable ≤ 10 ppm
iii not detectable ≤ 3 ppm
iv not detectable ≤ 1 ppm
v not detectable ≤ 30 ppm

Kevlar 49 fibers exhibit a 100 wt % loss at 700°C in an oxygen environment for a heating rate of 10°C/min [22]. The rate of loss strength $\ln(d\sigma/dt)$ as a function of $1/T$ for oxidative degradation of Kevlar fibers is illustrated in Figure 1. The rate of oxidative degradation of Kevlar 49 fibers will vary with the concentrations of small quantities of H_2SO_4 and Na impurities within the fibers that catalyze oxidative degradation reactions [23]. Impurity-catalyzed oxidative degradation of Kevlar 49 is used to identify yarns with "unacceptable" impurity levels. In this procedure bare Kevlar 49 yarns are placed in a calibrated air-convection oven for 3 hours at 240°C and their residual strengths are determined at 23°C as twisted bare yarns [24]. A loss in strength of ~10% is considered to be indicative of the presence of an unacceptable concentration of impurities [25].

In the case of the composite, impurities at the fiber surface, such as Na or K, can accelerate oxidative degradation at the fiber–matrix interphase region [20,26]. The sizings on carbon fibers degrade in a nitrogen atmosphere in the 300–400°C temperature range, at a heating range of 10°C/min, with an associated 0.5–1.0% fiber weight loss [21]. In the presence of oxygen the sizing would degrade at lower temperatures.

Radiation

Carbon fiber–TGDDM-DDS epoxy composites have been proposed for use in space structures in geosynchronous orbit for a period of 30 years. The radiation level in geosynchronous orbit is considerably higher than at the earth's surface [27]. Fornes et al. [28,29] have simulated the effect of radiation in a 30 year lifetime in geosynchronous orbit, by use of 0.5 MeV electrons and 1.17 and 1.33 MeV gamma rays, on the mechanical integrity of carbon fiber–TGDDM-DDS epoxy, polyimide, and polysulfone composites. These composites exhibited no deleterious mechanical integrity losses after exposure to 0.5 MeV electron radiation in vacuum for up to 5,000 Mrad exposure levels. The carbon fiber–TGDDM-DDS epoxy composites, also, showed little change in mechanical properties when exposed to several hundred megarads of gamma radiation. These observations indicate that carbon fiber–TGDDM-DDS epoxy composites would have a considerable lifetime (>30 years) in space without their mechanical integrity being significantly affected by high energy radiation.

In earth environments, ultraviolet radiation does not deteriorate the mechanical integrity of PFMCs, except for Kevlar 49 fibers. Aromatic polyamides, such as Kevlar 49, are susceptible to loss of strength upon exposure to UV radiation [30–34]. Kevlar 49 absorbs sunlight in the near-UV and in part of the visible region in the 300–450 nm range [31]. Only small amounts of this light are present in artificial light sources such as incandescent and fluorescent bulbs, or in sunlight filtered by window glass [31]. Brown and Sandreczki [33] have used electron paramagnetic resonance (EPR) techniques to study the free radicals that are produced in Kevlar 49 fibers after UV irradiation in vacuum (photodegradative radicals) and in air (photooxidative radicals). These UV-induced radical formation mechanisms involve either permanent chain scission or radical recombination to form structural irregularities in the PPTA chains. Brown and Sandreczki [33] also observed that the rate of radical production decreased with irradiation time, which suggests that photolytic products absorb in UV wavelengths, hence act as a light-protective shield. Photooxidative degradation causes a more rapid loss of fiber strength than in the absence of oxygen. In composites, UV-induced fiber aging is not considered serious because fiber strength deterioration only occurs in the surface layers.

Thermal Decomposition

Thermally induced bond dissociation, in the absence of direct chemical reaction, is not generally a significant aging mechanism for PFMCs except in extreme service environment conditions. For example, in the absence of radical recombination, rapid heating of a composite by a laser beam in the 0.1–10 second time range will cause complete loss of strength via thermal-induced bond dissociation, for organic macromolecular structures in the 825–1,200°C temperature range (see Fig. 1, above) [35]. At lower temperatures and for longer time periods, chemical degradation reactions, such as oxidation or hydrolyses, predominate.

However, for composites that are under stress in a service environment, thermal exposure can cause fiber fracture prior to any thermal-induced chemical or physical fiber or matrix aging, principally because of the decrease in fiber strength with increasing temperature. From studies on the failure temperature of stressed, twisted bare Kevlar 49 yarns as a function of percent stress level of the 23°C fiber strength it is observed (Fig. 4) that Kevlar 49 yarns stressed at 50% of ultimate 23°C strength fail at 275°C [35]. The failure temperature for carbon fibers stressed 50% of their ultimate 23°C strength is near 500°C. This decrease in fiber strength with increasing temperature can initiate catastrophic composite failure for stressed composites, such as filament-wound composite rocket motor casings, when the composites are exposed to the high temperatures generated, for example, by laser exposure or aerodynamic heating effects. [The composite failure is rapid because upon fracture of the exterior composite fibers, the remaining load-bearing fibers carry a higher load and therefore are (a) susceptible to failure at lower temperatures but (b) exposed to an increasing temperature field.]

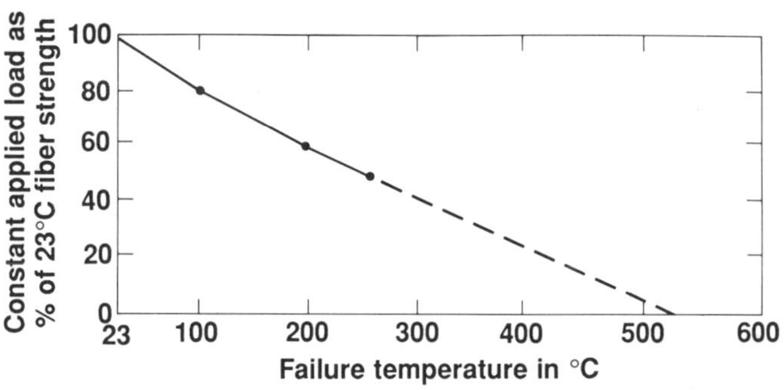

FIGURE 4 Failure temperature of stressed, bare, twisted Kevlar 49 yarn as a function of percent stress level of the 23°C fiber strength.

At lower temperatures, in addition to dehydration (see Oxidation), above 125°C thermal decomposition of ring structures present in TGDDM-DDS epoxies can occur [11,36,37], as indicated in Eq. (5).

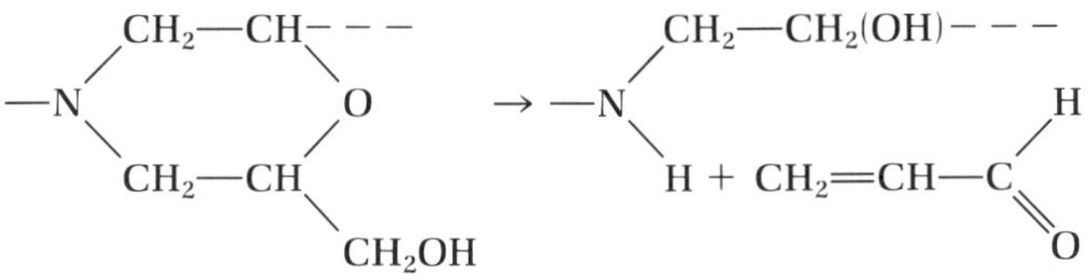

(5)

In Figure 2 the $I_{NH:SO_2}$ intensity increases in the 177–225°C region, despite the completion of the primary amine–epoxide cure reaction that would result in additional secondary amine groups. This $I_{NH:SO_2}$ intensity increase is more pronounced for systems exhibiting lower DDS concentration. From vaporization gas chromatography–mass spectroscopy studies, Grayson and Wolf [37] report that propenal is the principal degradation product of TGDDM-DDS epoxies in the 125–215°C temperature region. These observations are consistent with the decomposition mechanism illustrated in Eq. (5) [11]. Because of the complexity of the TGDDM-DDS networks [11], there is no evidence that this thermal-induced structural network modification causes any deterioration in the mechanical integrity of the TGDDM-DDS epoxy.

Physical Aging

Free Volume

The mechanical properties of thermoset and thermoplastic polymeric composite matrices exhibit a free-volume and mechanical property dependence as a function of thermal history [38–47]. Changes in free volume, or local order, in the glassy state can occur as a result of the extension to temperatures below T_g of packing changes associated with the liquid state. The liquid–volume temperature plot extrapolated to below T_g in Figure 5 represents the lower free-volume, equilibrium state of the glass. The time necessary to achieve the equilibrium state at a given temperature below T_g depends on the glassy state mobility. Below a specific temperature, the glassy state mobility is too small to allow any changes in free volume. A decrease in free volume that occurs in the glassy state results in inhibition of the flow processes that occur during deformation and a more brittle mechanical response. In the case of tough, thermoplastic polymeric matrices, service environment exposure in the T_g −50°C region can cause nearly an order of magnitude decrease in impact properties [38–44]. This embrittlement is a result of cessation of large volume shear yielding and reversion to a crazing failure mode with a corresponding decrease in molecular flow and energy to failure [48]. In the case of thermoset cross-linked composite matrices, such as epoxies, the effect of thermal

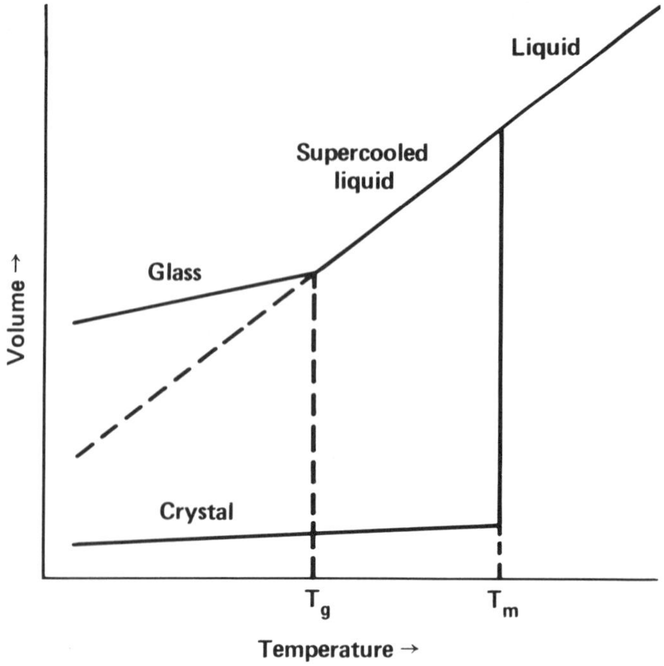

FIGURE 5 Volume–temperature plot for a polymeric composite matrix.

exposure in the T_g −50°C range does not decrease the matrix toughness to the same extent as in thermoplastic matrices [46]. (The effects of thermal history on the mechanical properties of a polyimide are shown in Fig. 3, above.)

Relaxation

One of the more serious aging concerns in future applications of polymer matrix composites is the relaxation of any "frozen in" matrix orientations that originate during composite fabrication. High polymer matrix toughness necessarily requires high macromolecular mobility, which also, however, leads tc the relaxation of fabrication strains and, under load, to stress-induced creep rupture. In the composite processing arena, little effort has been devoted to correlating composite fabrication conditions with the generation of "frozen in" polymer orientations. This concern is particularly valid for thermoplastics that exhibit higher viscosities and longer melt relaxation times than thermosets during the critical stages of composite fabrication. Relaxation of fabrication strains in thermoplastic matrices can cause severe cavitation and microcracking [49].

Low molecular weight organic species that are soluble in the polymer matrix will plasticize the polymer and lower its T_g, which will accelerate relaxation of fabrication strains.

The lowering of the T_g by a diluent is a result of a decrease in the free volume (total volume minus molar volume) of the system caused by the addition of the lower free-volume diluent. The free volume at T_g for a wide variety of glass-forming materials has been shown to be a constant and to have a critical free-volume fraction of 0.025 of the total volume [50–52]. Kelley and Bueche [53] have derived an expression relating the T_g of a polymer–diluent system to that of the T_g's of the two components. The Kelley–Bueche equation assumes that the free volume contributed by the diluent is additive to that of the polymer and that the free volumes of the mixture and components at their T_g's are a universal constant. The Kelley–Bueche equation for the T_g of the polymer–diluent system is

$$T_g = \frac{\alpha_p V_p T_{gp} + \alpha_d (1 - V_p) T_{gd}}{\alpha_p V_p + \alpha_d (1 - V_p)} \tag{6}$$

where

T_{gp} = T_g of the polymer
T_{gd} = T_g of the diluent
α_p = coefficient of expansion of polymer
α_d = coefficient of expansion of diluent
V_p = volume fraction of the polymer

Inhomogeneous distribution of the diluent in the polymer matrix can cause crack initiation. For example, finger-grease can cause surface microcracking of polycarbonate [49]. The regions of the polycarbonate surface that have come into contact with finger-grease are plasticized, and fabrication stresses in such regions relax near T_g at a faster rate than those in the unplasticized surroundings. The plasticized, relaxed regions separate from their surroundings, producing microcracks, which serve as sites for craze initiation and growth, and possible embrittlement of the polymer.

Effect of Diluents on Matrix Deformation and Failure Modes

Absorbed low molecular weight liquids generally cause failure in polymers by inducing crazing or cracking at stresses much lower than those observed in their absence. Numerous studies on the interaction of low molecular weight diluent molecules on the crazing process have illustrated that this phenomenon is complex and is not completely understood on a molecular level [45].

The modification of craze initiation and propagation in polymer matrices by the presence of diluents can occur as a result of plasticization [54] and/or surface energy reduction [55]. The presence of the diluent at the craze tip can reduce the surface energy of the polymer, facilitating the creation of a new surface [56]. For an absorbed diluent layer at least a few molecules thick, the interfacial surface energy participating in craze cavitation can be reduced from the higher polymer–air surface energy value. The plasticization of the polymer at the craze tip by the diluent lowers the shear yield stress, which is also active in the craze cavitation process. The stress concentration at the craze tip increases the fractional free volume of the polymer, which both enhances absorption of the diluent and lowers the T_g of the polymer [56,57].

Andrews and coworkers [58,59] have shown that solvent craze work formation W is governed by the cavitation properties of a solvated zone of polymer at the craze tip:

$$W = 4.84(h\gamma/p)f^{2/3} + 0.66Y\psi hf \quad (7)$$

where h is the craze thickness, γ the interfacial energy of the voids, p the mean distance between the void centers, f the void fraction, Y the shear yield stress, and ψ a constant with a value of 4.4. These threshold conditions for craze cavitation depend on the work necessary to produce plastic yielding and the work necessary to create void interfaces. The yield stress is strongly temperature dependent relative to the surface energy term. Andrews and coworkers [58,59] showed that W decreases with temperature and that, above a critical temperature T_c, W assumes a minimum constant value W_0. The temperature T_c has been associated with the T_g of the plasticized swollen polymer at the craze tip. The yield stress term in Eq. (7) reduces to near zero at T_g because $Y \simeq 0$ at or above T_g. When $Y \simeq 0$, W is dependent only on the temperature-independent surface energy term.

Kambour and coworkers [60] have suggested that the plasticization effect is the more important factor in solvent crazing. They reported that the solvent-crazing resistance of undiluted polystyrene is the same as that of the preplasticized polystyrene in air. This finding implies that the presence of a liquid–polymer interface is not critical to crazing effectiveness of a given diluent. These workers, therefore, concluded that liquids do not appear to reduce crazing resistance by flowing into and wetting the surface of holes as they form.

There have been a number of efforts to correlate the modification of the craze–crack behavior of the polymer in the presence of a diluent with the solubility parameters of the polymer and diluent. Bernier and Kambour [61] studied the effect of a variety of liquids on the failure of poly(2,6-dimethyl-1,4-phenylene) oxide. They concluded that a liquid acts as a solvent when its solubility parameter is close to that of the polymer. When the difference between the solubility parameters of the liquid and polymer is small, the liquid promotes cracking; when this difference is large, crazing is enhanced. Vincent and Raha [62], however, found that a more effective representation of the solvent-cracking versus crazing behavior is obtained by considering the capacity of each liquid-to-hydrogen bond in addition to its solubility parameter.

Kambour and coworkers [61,63] have also shown a close correlation between the critical strain for craze or crack formation e_c and the solubility parameter of the diluent. Subsequently, Andrews and Bevan [58] found that the work of solvent craze formation (W) above the T_g of the plasticized polymer at the craze tip is a smooth function of the difference in the solubility parameters of the polymer (δ_p) and the diluent (δ_d), that is, $\delta_p - \delta_d$. These workers also found that W exhibits a minimum at $\delta_p \simeq \delta_d$. Mai [64] has reported that $W - \delta_d$ relations are not affected by hydrogen bonding. Despite these reported correlations between the solubility parameters of polymer and diluent and the work of craze cavitation formation, the modification of craze initiation and propagation in a specific polymer by a given diluent cannot be predicted a priori. The polymer–diluent interactions on a molecular level and the effect of these interactions on plasticization and swelling behavior at the craze tip are not understood. For example, we do not fully understand the role of factors such as the size and shape of the diluent molecule [65], the rotational isomeric configurations of the polymer, and the flexibility of the polymer chains in polymer–diluent interactions.

Recrystallization

Semicrystalline, thermoplastic composite matrices are susceptible to recrystallization in service environment conditions near the T_g of the amorphous phase and/or in the presence of organic chemicals. For example, polycarbonate crystallization is enhanced by the presence of plasticizers that allow greater macromolecular chain mobility necessary for crystallization [66–74]. Solvent-induced crystallization of thermoplastic composite matrices at bulk surfaces and in thin films has been directly observed after exposure to organics for short time periods (i.e., 10 seconds) [74,75]. The shrinkage associated with crystallization causes cavitation and the growth of surface crazes, which could deteriorate the mechanical integrity of the composite [75].

Absorbed Moisture

Absorbed moisture in PMFCs has been one of the principal sources of physical aging of these composites. The moisture-induced aging mechanisms can involve both the polymer matrix and the fiber–matrix interface and, also, the fiber, in the case of Kevlar 49.

In the case of a polymeric composite matrix that does absorb significant quantities of moisture, such as the TGDDM-DDS epoxy system (>5 wt % moisture), a number of physical mechanisms can deteriorate the matrix and, hence, the composite performance.

The absorption of moisture by epoxies lowers their T_g's and correspondingly causes them to soften at lower temperatures and to undergo deterioration in their mechanical response. For an epoxy–water system, the Kelley–Bueche equation (Eq. 6) must be used in a simplified form. The coefficient of expansion of amorphous water between its T_g and melting point T_M at 0°C is unknown. Therefore, assuming $\alpha_p \simeq \alpha_d$, the Kelley–Bueche equation simplifies to

$$T_g = V_p T_{gp} + (1 - V_p)T_{gd} \quad (8)$$

Experimentally, the T_g of water has been reported in the 128–142 K range [76–81]. A good empirical rule [82] for most substances is

$$T_g/T_M \simeq \tfrac{1}{2} \text{ to } \tfrac{2}{3} \quad (9)$$

which places the T_g of water in the 137–182 K range. Figure 6 compares experimentally determined T_g's of a TGDDM-DDS epoxy–moisture system [83], containing equilibrium amounts of absorbed moisture, with the theoretically calculated values of T_g determined from Eq. (8). The upper limits of the theoretical plot are calculated assuming a T_g of water of 182 K, whereas the lower limits

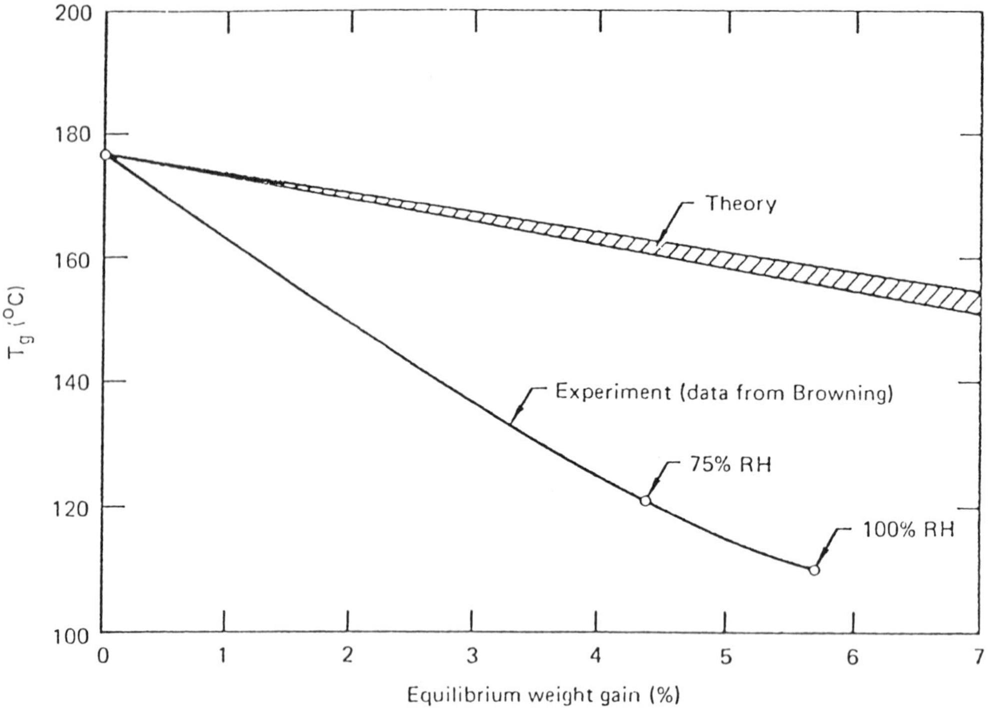

FIGURE 6 **Theoretical and experimental plots of T_g versus equilibrium moisture weight gain for a TGDDM-DDS epoxy–moisture system.**

are determined using a T_g of 135 K. The experimental data points are well below the theoretical predictions. This discrepancy is caused by the strong hydrogen-bonding capability of water.

The plasticization of epoxies by absorbed moisture has generally been reported to be reversible, since desorption of moisture from the polymer regenerates the original physical and mechanical properties [45].

The aging of epoxy composite matrices in many service environments depends on the degree of deterioration of the high temperature mechanical properties caused by the plasticizing effect of absorbed moisture and on whether this phenomenon is adequately considered during the composite design stage. The PFMC's previous exposure to relative humidity, time, and temperature and the diffusion characteristics of moisture in these composites determine the moisture profile and resultant mechanical response of the material. Hence, the diffusion characteristics of moisture in the PMFC are critical factors for predicting mechanical response and aging in a specific service environment.

Many diffusion processes can be adequately described by Fick's laws of diffusion [84]. A number of workers have successfully applied Fick's second law of diffusion to the absorption and desorption of moisture in epoxies and epoxy composites and have predicted the moisture profiles as a function of time for specific environmental conditions [83,85–91]. Fick's second law for one-dimensional diffusion is given by

$$\frac{dc}{dt} = D\frac{d^2c}{dx^2} \tag{10}$$

where c is concentration, D is the diffusion coefficient, and x is the distance the moisture has advanced from a given boundary. Equation (10) has been solved for the case of a flat sheet absorbing vapor at both faces under constant environmental conditions. The solution for the fraction of the equilibrium amount of moisture absorbed in time t, $M_{(t)}$, is given by

$$M_{(t)} = \frac{4}{L}\left(\frac{Dt}{\pi}\right)^{1/2} \tag{11}$$

where L is the plate thickness [92]. From linear plots of $M_{(t)}$ versus $(t)^{1/2}$, the Fickian behavior in composites, corrected for volume effects, was identical with that of the neat polymer matrix [85,86]. The rates of moisture absorption increased with temperature, and the equilibrium amount of moisture was directly proportional to the relative humidity.

At high moisture contents ($M_{(t)} \geq 0.6$), Eq. (11) does not hold, and a number of methods have been used to determine the moisture profiles in epoxy matrices and their composites. Estimates within 15% of the moisture content and its distribution have been obtained from a series solution of Eq. (10) by Shen and Springer [89] and from a hyperbolic tangent method developed by McKague [93]. Bohlmann and Derby [91] have noted that a numerical methods technique using a computer program [94] is necessary to account for the moisture profile for transient conditions with different relative humidities at each surface.

Absorbed moisture causes epoxy matrices to swell [83]. The swelling stresses generated by the absorbed moisture can significantly affect the durability of these composite matrices in many environments. Swelling stresses caused by moisture gradients, together with other stresses inherent in the material, such as fabrication stresses, can be large enough to cause localized fracture of the polymer matrix [83]. Although the moisture-induced swelling of epoxies generally results in only a 1–2% thickness increase, these dimensional changes can result in high stresses in a composite where the fibers constrain the swelling. For durability predictions of epoxy matrices, the local magnitude of the swelling stresses and strains and the threshold values that produce permanent damage must be known. Halpin [95] has calculated the dilatational strains in epoxies assuming volume additivity and a homogeneous moisture distribution. However, the experimentally observed swelling strains were greater than predicted by theory by a factor of 3 at moisture contents of 2% or less.

Absorbed moisture lowers the tensile strength and moduli of TGDDM-DDS epoxies and enhances cavitation [96].

The rapid heating of a moist epoxy matrix that can occur at the surface of an epoxy composite of a fighter aircraft as a result of a supersonic maneuver [96] can cause enhanced moisture absorption and surface microcracking. The moisture-induced swelling stresses together with the enhanced mobility of the water molecules within the epoxy–moisture system during a thermal spike produce free-volume increases that involve rotational–isomeric configurational changes within the epoxy network. Such changes are fixed in the epoxy during the rapid cooling after the thermal spike. This additional free volume allows water molecules access to previously inaccessible active sites within the epoxy [96]. To a lesser extent, the rupture of cross-links, crazing and/or cracking, and the loss of unreacted material can also contribute to enhanced moisture absorption after thermal spike exposure.

Thermal spike exposure can cause surface crazing and/or cracking of epoxies if the moisture-induced swelling stresses together with the stresses that result from temperature gradients and relaxation of fabrication stresses exceed the craze initiation stress at the maximum thermal spike temperature [96].

For fibers that absorb moisture, such as Kevlar 49 fibers, composite moisture-induced physical aging can occur. Kevlar 49 fibers absorb 4 wt % moisture at 23°C, 100% relative humidity, which is generally more than that of the polymer matrix. Furthermore, these fibers readily fibrillate in steam or microbial environments [97,98]. Kevlar 49 fibers consist of 60 nm diameter fibrils with Na_2SO_4 impurities and microvoids concentrating at the interfibrillar boundaries, where moisture will preferentially accumulate [23]. Cracks can readily propagate along the fiber axis at the interfibrillar boundaries. When Kevlar 49 polymer matrix composites are exposed to thermal cycle conditions that involve cycling from hot, humid conditions to below freezing as in, for example, the latest commercial aircraft service environments, ice formation within the fiber is possible. Ice formation within the fiber could readily force interfibrillar microcracking at the fiber surface that then initiates fiber–matrix interfacial failure. Recently, Kevlar 49–epoxy composites have been observed to exhibit microcracking parallel to the fiber direction upon exposure to thermal–humidity cycling from −54 to +38°C (3–5 cycles/day) associated with commercial aircraft operating up to 40,000 feet [99].

For fibers that absorb more moisture than the polymer matrix, such as Kevlar 49, at the high temperature portion of a thermal cycle, moisture migrates from the fiber to the matrix. However, if cool-down is sufficiently rapid, at lower temperatures the polymer matrix will possess moisture levels greater than the equilibrium values at these temperatures. The effect of this thermal-cycle-induced nonequilibrium moisture concentration on composite fiber–matrix interfacial integrity has not yet been investigated.

Macromolecular Chain Scission or Slippage

Exposure of polymeric matrices or fibers to thermal and/or stress exposure can cause macromolecular chain scission and/or slippage that creates defects that deteriorate the mechanical integrity of the composite.

In the absence of stress, as discussed in the THERMAL DECOMPOSITION section, thermally induced bond dissociation is not generally a significant aging mechanism for PMFCs except in extreme service environment conditions. However, under load, stress-induced macromolecular chain scission or slippage can decrease particularly fiber strength at higher service environment temperatures.

In the case of cross-linked, thermoset composite matrices, the crack resistance of the network can decrease during plastic deformation as a result of network chain scission. The amount of plastic flow and the toughness exhibited by a cross-linked polymer glass is limited by the network extensibility. The segments with the lowest extensibilities at any given time in the failure process will carry a significant portion of the load and will undergo scission, resulting in a process of progressive scission of the least extensible segments. For example, when epoxies were strained up to 30%, then the load removed, and this strain completely annealed out, the ductility of such epoxies decreased by 50% and their T_g's were lowered 8°C (Fig. 7) compared with reference specimens that were exposed to the same thermal history but were not initially strained [100]. Such deterioration in the mechanical response and T_g are consistent with network deterioration via chain scission. In more direct evidence, permanent chemical changes were observed in epoxy films that were deformed directly in the Fourier transform infrared spectrometer after 10% deformation and subsequent removal of the load [100]. Also, Brown and Sandreczki [101] have detected free radicals in epoxies, indicative of network chain scission, by electron paramagnetic resonance studies after ballmilling. Cross-linked polymer networks with segments of equal extensibilities would be the most desirable networks for optimizing the stress-induced aging of tough composite matrices.

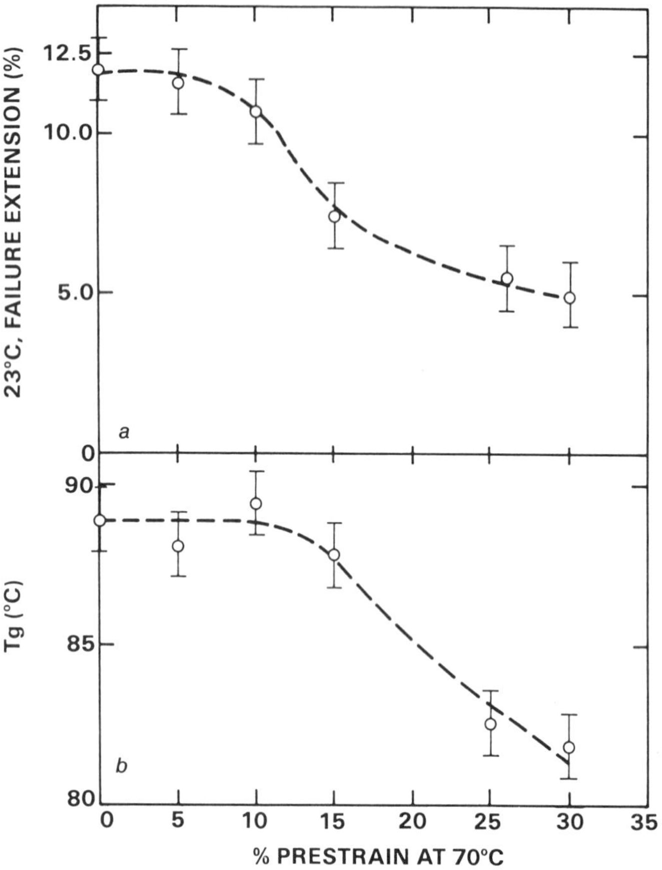

FIGURE 7 (*a*) Tensile failure extension and (*b*) T_g versus prestrain for subsequent annealed epoxy.

Thermal Expansion Mismatch

In the fabricated composite, a significant aging mechanism results from residual stresses caused by expansional mismatch between the fibers and matrix [102]. These stresses, which result in interply cracking, delamination, and loss of stiffness, become larger as the service temperature is further removed from the original composite fabrication temperature. These stresses will become more significant as (a) higher temperature fabrication conditions are used for more thermally stable matrices, (b) composites are exposed to lower temperatures, as in space applications, and (c) rapid heating of composites occurs, such as upon high energy laser exposure.

The interfacial shear strength between Kevlar 49 fibers and epoxy matrices has been reported to be in the 15–30 MPa range [103–105]. However, for oxidative surface treated carbon fibers, the interfacial shear strength with epoxies is considerably higher—in the 80–100 MPa range. The most promising approach to further increasing the resistance of the composite to fiber–matrix interfacial failure and composite ply delamination involves generating residual macroscopic compressive stresses in the composite by innovative composite processing procedures [106].

Roger J. Morgan

[*See also* Environmental Resistance; Environmental Stress Cracking; Space Environmental Effects.]

References

1. R. J. Morgan, C. O. Pruneda, N. Butler, F. M. Kong, L. Caley, and R. L. Moore, *Proceedings of the 29th Natl. SAMPE Symposium*, 1984, pp. 891–902.
2. R. S. Jones, E. W. Choe, H. Noether, I. L. Hay, P. K. Sullivan, M. J. Hannon, M. Jaffe, J. Soehngen, and M. Tan, "Structure–Property Relationships of Aramids," Celanese Research Company (unpublished).
3. M. Arpin and C. Strazielle, *Polymer, 18,* 591 (1977).
4. W. S. Smith, "Environmental Effects of Aramid Composites," E. I. du Pont, Technical Paper, 1980.
5. R. J. Morgan and F. M. Kong, Michigan Molecular Institute (unpublished).
6. L. Penn and R. J. Sherry, Lawrence Livermore Laboratory Report UCRL-79499, 1977.
7. Kevlar 49 Technical Data Report K-2, E. I. du Pont, 1978.
8. W. D. Humphrey, *Proceedings of the 29th Mechanical Failure Composites Meeting*, 1979, pp. 177–197.
9. J. W. Mead, K. E. Mead, I. Auerbach, and R. H. Ericksen, *ACS Prepr. Org. Coat. Plastics Chem., 44,* 336 (1981).
10. H. Golopol, J. Clarkson, R. Moore, and N. Hetherington, Lawrence Livermore Laboratory Report UCRL-86034, 1981.
11. R. J. Morgan and E. T. Mones, *J. Appl. Polym. Sci., 33,* 999 (1987).
12. L. H. Lee, *J. Polym. Sci., A3,* 859 (1965).
13. D. P. Bishop and D. A. Smith, *J. Appl. Polym. Sci., 14,* 205 (1970).
14. E. C. Leisegang, A. M. Stephen, and J. C. Paterson-Jones, *J. Appl. Polym. Sci., 14,* 1961 (1970).
15. J. C. Paterson-Jones, *J. Appl. Polym. Sci., 19,* 1539 (1975).
16. R. J. Morgan, J. E. O'Neal, and D. B. Miller, *J. Mater. Sci., 14,* 109 (1979).
17. R. J. Morgan, C. M. Walkup, and T. H. Hoheisel, *J. Appl. Polym. Sci., 30,* 289 (1985).
18. T. T. Serafini, Ed., *High Temperature Polymer Matrix Composites*, Noyes, Park Ridge, NJ, 1987.
19. R. J. Jones, G. E. Chang, S. H. Powell, and H. E. Green, in T. T. Serafini, Ed., *High Temperature Polymer Matrix Composites*, Noyes, Park Ridge, NJ, 1987, pp. 267–281.
20. W. B. Alston, in T. T. Serafini, Ed., *High Temperature Polymer Matrix Composites*, Noyes, Park Ridge, NJ, 1987, pp. 186–196.
21. J. A. Hodson, L. J. Gregory, and R. J. Morgan, *Proceedings of the 17th Natl. SAMPE Technical Conference*, 1985, pp. 261–273.
22. T. Kunugi, H. Watanabe, and M. Hashimoto, *J. Appl. Polym. Sci., 24,* 1039 (1979).
23. R. J. Morgan and C. O. Pruneda, *Polymer, 28,* 340 (1987).

24. L. E. Caley, P. K. Ambalal, J. F. Carley, T. S. Ford, R. L. Moore, D. J. Noecker, M. E. Ross, R. J. Sherry, and J. M. Yoshiyama, "Quality Assurance Tests of Five Y-12 Kevlar 49 Spools Used to Fabricate Strands and Reliability Specimens for Stress–Rupture Tests," Lawrence Livermore Laboratory Report UCID-20078, 1984.
25. M. Wardle, Textile Fibers Division, E. I. du Pont (private communication).
26. D. A. Scola, in T. T. Serafini, Ed., *High Temperature Polymer Matrix Composites*, Noyes, Park Ridge, NJ, 1987, pp. 214–222.
27. K. M. V. Apparao, *Composition of Cosmic Radiation*, Gordon & Breach, New York, 1975, pp. 6–13.
28. R. E. Fornes, J. D. Memory, and N. Naranong, *J. Appl. Polym. Sci., 26*, 2061 (1981).
29. K. R. Schaffer, R. E. Fornes, R. D. Gilbert, and J. D. Memory, *Polymer, 25*, 54 (1984).
30. W. L. Harper and C. E. McAlister, "The Effect of Radiation on Kevlar 49," U.S. Department of Energy Report Y-DA-6527, 1975.
31. Du Pont Technical Report C-29, "The Effect of Ultraviolet Light on Products Based on Fibers of Kevlar 29 and Kevlar 49 Aramid," 1977.
32. D. J. Carlsson, L. H. Gan, and D. M. Wiles, *J. Polym. Sci. (Polym. Chem. Ed.), 16*, 2353 (1978).
33. I. M. Brown and T. C. Sandreczki, "Characterization of Kevlar 49 Fibers by Electron Paramagnetic Resonance," McDonnell Douglas Research Laboratories, Lawrence Livermore National Laboratory Contract P.O. 3115901, Final Report, 1982.
34. J. Cook, PERME, Waltham Abbey, England (unpublished data).
35. R. J. Morgan, F. M. Kong, and J. K. Lepper, *J. Compos. Mater. 22*, 1026 (1988).
36. C. David, in C. H. Bamford and C. F. H. Typer, Eds., *Comprehensive Chemical Kinetics*, Vol. 14, *Degradation of Polymers*, Elsevier, New York, 1975, p. 1.
37. M. A. Grayson and C. J. Wolf, *J. Polym. Sci. (Polym. Chem. Ed.), 22*, 1897 (1984).
38. G. Peilstocker, *Kunstoffe, 51*, 509 (1961).
39. G. Peilstocker, *Br. Plast. 35*, 365 (1962).
40. J. H. Golden, B. L. Hammant, and E. A. Hazell, *J. Appl. Polym. Sci., 11*, 1571 (1967).
41. D. G. LeGrand, *J. Appl. Polym. Sci., 13*, 2129 (1969).
42. D. G. LeGrand, *J. Appl. Polym. Sci., 16*, 1367 (1972).
43. R. E. Robertson and C. W. Joynson, *J. Appl. Polym. Sci., 16*, 733 (1972).
44. G. Allen, D. C. W. Morley, and T. Williams, *J. Mater. Sci., 8*, 1449 (1973).
45. R. J. Morgan and J. E. O'Neal, *Polym. Plast. Technol., 10*(1), 49 (1978).
46. R. J. Morgan, *J. Appl. Polym. Sci., 23*, 2711 (1979).
47. E. S. Kong, G. L. Wilkes, J. E. McGrath, A. K. Banthia, Y. Mohojer, and M. R. Tant, *Polym. Eng. Sci., 21*, 943 (1981).
48. R. J. Morgan and J. E. O'Neal, *J. Polym. Sci. (Polym. Phys. Ed.) 14*, 1053 (1976).
49. R. J. Morgan and J. E. O'Neal, *Polymer, 20*, 375 (1979).
50. M. L. Williams, R. F. Landel, and J. D. Ferry, *J. Am. Chem. Soc., 77*, 3701 (1955).
51. F. Bueche, *J. Chem. Phys., 24*, 418 (1956).
52. M. H. Cohen and D. Turnbull, *J. Chem. Phys., 31*, 1164 (1959).
53. F. N. Kelley and F. Bueche, *J. Polym. Sci., 50*, 549 (1961).
54. B. Maxwell and L. F. Rahm, *Ind. Eng. Chem., 41*, 1988 (1949).
55. R. J. Bergen, Jr., *SPE J., 24*, 77 (1968).
56. A. Peterlin and H. G. Olf, *J. Polym. Sci., Polym. Symp., 50*, 243 (1975).
57. A. N. Gent, *J. Mater. Sci., 5*, 925 (1970).
58. E. H. Andrews and L. Bevan, *Polymer, 13*, 337 (1972).
59. E. H. Andrews, G. M. Levey, and J. Willis, *J. Mater. Sci., 8*, 1000 (1973).
60. R. P. Kambour, C. L. Gruner, and E. E. Romagosa, *J. Polym. Sci., (Polym. Phys. Ed.), 11*, 1879 (1973).
61. G. A. Bernier and R. P. Kambour, *Macromolecules, 1*, 393 (1968).
62. P. I. Vincent and S. Raha, *Polymer, 13*, 283 (1972).
63. R. P. Kambour, E. E. Romagosa, and C. L. Cuner, *Macromolecules, 5*, 335 (1972).
64. Y. W. Mai and A. G. Atkins, *J. Mater. Sci., 11*, 677 (1976).
65. R. J. Morgan and L. E. Nielsen, *J. Polym. Sci. (Polym. Phys. Ed.), 10*, 1575 (1972).
66. W. R. Moore and R. P. Sheldon, *Polymer, 2*, 315 (1961).
67. R. P. Kambour, F. R. Karasz, and J. M. Duane, *J. Polym. Sci., A-2, 4*, 327 (1966).
68. J. P. Mercier, G. Groeninckx, and M. Lesne, *J. Polym. Sci., C, 16*, 2059 (1967).
69. A. Siegmann and P. H. Geil, *J. Macromol. Sci.-Phys., B4*(2), 239, 273 (1970).
70. G. W. Miller, S. A. D. Visser, and A. S. Morecroft, *Polym. Eng. Sci., 11*, 73 (1971).
71. M. G. Wyzgoski and G. S. Y. Yeh, *J. Macromol. Sci.-Phys., B10*(3), 441, 647 (1974).
72. W. V. Titow, M. Braden, B. R. Currell, and R. J. Loneragan, *J. Appl. Polym. Sci., 18*, 867 (1974).
73. F. Gallez, R. Legras, and J. P. Mercier, *J. Polym. Sci. (Polym. Phys. Ed.), 14*, 1367 (1976).
74. G. L. Wilkes and J. Parlapiano, *ACS Prepr. Polym., 17*(2), 937 (1976).
75. R. J. Morgan and E. T. Mones, *ACS Prepr. Org. Coat. Plast. Chem., 41*, 463 (1979).
76. L. G. Dowell and A. P. Rinfret, *Nature, 188*, 1144 (1960).
77. M. Sugisaki, H. Suga, and S. Seki, *Bull. Chem. Soc. Japan, 41*, 2591 (1968).
78. A. Korosi and B. M. Fabus, *Anal. Chem., 40*, 157 (1968).
79. N. H. Fletcher, *The Chemical Physics of Ice*, Cambridge University Press, Cambridge, 1970, Ch. 3.
80. R. Frank, *Water*, Vol. 1, Plenum, New York, 1972.
81. Y. Y. Tan and G. Challa, *Polymer, 17*, 739 (1976).
82. L. E. Nielsen, *Mechanical Properties of Polymers*, Reinhold, New York, 1962, p. 27.
83. C. E. Browning, "The Mechanism of Elevated Temperature Property Losses in High Performance Structural Epoxy

Resin Matrix Materials After Exposures to High Humidity Environments," Ph.D. thesis, University of Dayton, 1976.

84. J. Crank, *The Mathematics of Diffusion*, Oxford University Press (Clarendon), London/New York, 1957.
85. J. Hertz, "Investigation into the High-Temperature Strength Degradation of Fiber-Reinforced Resin Composites During Ambient Aging," General Dynamics, Convair, Final Report GDCA-DBG73-005, Contract NAS, 8-27435, 1973.
86. J. F. Carpenter, "Moisture Sensitivity of Epoxy Composites and Structural Adhesives," McDonnell Aircraft Report MDC A2640, 1973.
87. R. M. Verette, "Temperature/Humidity Effects on the Strength of Graphite/Epoxy Laminates," AIAA Paper 75-1011, 1975.
88. E. L. McKague, Jr., J. E. Halkias, and J. D. Reynolds, *J. Compos. Mater.*, *9*, 2 (1975).
89. C. H. Shen and G. S. Springer, *J. Compos. Mater.*, *10*, 2 (1976).
90. Air Force Conference on the Effects of Relative Humidity and Temperature on Composite Structures, University of Delaware, March 1976, AFOSR-TR-77-0030, 1977.
91. R. E. Bohlmann and E. A. Derby, *Proceedings of the AIAA*, 1977, Paper 77-399.
92. C. J. Van Amerongen, *Rubber Chem. Technol.*, *37*, 871 (1964).
93. E. L. McKague, "Life Assurance of Composite Structures," Technical Report AFML-TR-75-51, 1975.
94. A. J. King, "Multicomponent Diffusion (1-D)," in Computer Program User's Manual 61014, Hercules Corp., 1976.
95. J. C. Halpin, AFML/AFOSR, *Mechanics of Composites Review*, Dayton, 1976.
96. R. J. Morgan, J. E. O'Neal, and D. L. Fanter, *J. Mater. Sci.*, *15*, 751 (1980).
97. R. J. Morgan and R. E. Allred, "Aramid Fiber Reinforcements," in S. M. Lee, Ed., *Reference Book for Composites Technology*, Vol. 1, Technomic, Lancaster, PA, 1989, Chap. 8.
98. T. Watanabe, *Sen-I Gakkaishi*, *43*, 192 (1987).
99. H. Y. Loken, D. A. Nollen, M. W. Wardle, and G. E. Zahr, "Water Ingression Resistant Thin Faced Honeycomb Cored Composite Systems with Facesheets Reinforced with Kevlar Aramid Fiber and Kevlar with Carbon Fibers," E. I. du Pont de Nemours Report, 1987.
100. R. J. Morgan and C. M. Walkup, *J. Appl. Polym. Sci.*, *34*, 37 (1987).
101. I. M. Brown and T. C. Sandreczki, McDonnell Douglas Research Laboratories, St. Louis (personal communication).
102. H. T. Hahn, *J. Astronaut. Sci.*, *32*, 253 (1984).
103. N. A. Mumford, P. C. Hopkins, and B. A. Lloyd, *Proceedings of the 18th Joint AIAA-SAE-ASME Properties Conference*, 1982.
104. L. T. Drzal, *Proceedings of the 15th Natl. SAMPE Technical Conference*, 1983, pp. 190–210.
105. L. S. Penn, F. A. Bystry, and H. J. Marchionni, *ACS Prepr. Polym.* *24*, 189 (1983).
106. R. R. Johnson, R. M. Bluck, A. M. C. Holmes, and M. H. Kural, *Proceedings of the 31st SAMPE Symposium*, 1986, pp. 90–102.

Aircraft Composites

See Spacecraft Applications; Spacecraft, Structural Development; Structural Aircraft, Noncommercial and Homebuilt

Alumina Fiber

See Ceramic Matrix Composites, Ceramic Fiber Reinforced; Ceramic Fiber Tiles

Aluminum Matrix Composites

See Metal Matrix Composites, Aluminum

Amino Resins

Amino resins are a broad class of materials made by the controlled reaction of an aldehyde with a compound containing an amino group. The topic has been surveyed in many books and reviews (e.g., Refs. 1–12). The terms "aminoplast" and "aminoplastic" commonly are applied to thermosetting molding compounds based on amino resins. These compounds have been in use for at least 50 years, and applications such as molding powders, laminating compounds, adhesives, textile treatments, and surface coatings are now firmly established. Amino resins combine easily with a variety of fillers and are used widely in composites. Compared with other segments of the plastics industry, amino resins are mature resins and their market share is fairly stable. The two most important resins are prepared from formaldehyde and either urea or melamine. Combined annual usage is about 600 kt. Over the past 10 years, the annual growth rate of urea resins has been about 3%, while the amount of melamine resin used has been practically unchanged [13,14]. Urea–formaldehyde resins (U/F) account for more than 80% of amino resins and melamine–formaldehyde resins (M/F) for most of the rest. This brief article is devoted to U/F and M/F and their applications in composites.

The reaction between urea and formaldehyde has been known since 1884 [6], but commercial interest did not begin until 1918, when patents were granted to scientists in several countries. An early compound was cast into beautiful clear transparent sheets, and the use of U/F as an organic glass was proposed; but the new material had poor water resistance and was soon displaced by thermoplastic materials such as polystyrene and polymethyl methacrylate. A molding compound reinforced with cellulose fiber was developed in England in 1925. Patent rights were acquired by American Cyanamid Company, and by 1930 a similar filled resin was marketed in the United States [7]. The use of U/F in composites had begun. M/F resins were introduced about 10 years later. The resins are similar in appearance to U/F but their properties, especially water resistance, are superior.

Monomers

Formaldehyde

Formaldehyde (HCHO) is an aliphatic aldehyde prepared by the controlled catalytic oxidation of methanol (CH_3OH). At room temperature, formaldehyde is a gas; at $-19°C$ it is a boiling liquid. The normal procedure for its preparation is to pass a mixture of methanolic vapor and air over a heated catalyst, usually copper.

$$CH_3OH \rightleftharpoons HCHO + H_2$$

Vapors of formaldehyde and of water are then mixed and condensed to yield formalin. Several grades are marketed, but for resinification purposes it is normal to use solutions containing between 35 and 40% by weight of formaldehyde [6]. Impurities in the formalin include methanol and formic acid (HCOOH) [6]. The methanol (usually 5–10%) stabilizes the solution and allows storage at room temperature without precipitation of solid formaldehyde polymers. Some formalin stabilized with urea is sold. The latter is particularly useful for the preparation of U/F resins, because the user need only adjust the ratio of urea to formaldehyde by adding more urea to produce a urea–formaldehyde solution ready for use [7].

Urea

Urea (NH_2CONH_2) is a water-soluble, white crystalline solid melting at 132.6°C. Synthetic methods for urea have been known since 1828 when Wöhler first obtained it accidentally, while preparing ammonium cyanate. Urea is now made by reaction under pressure of carbon dioxide and ammonia in silver-lined or steel autoclaves [6]. Ammonium carbamate is obtained as a by-product and must be recycled.

$$CO_2 + 2NH_3 \rightarrow NH_2CONH_2 + H_2O \rightarrow NH_2COONH_4$$

Because urea is hygroscopic, it is frequently pelletized or prilled (made into little beads) to avoid caking and to make the material easier to handle. The raw materials are based on the element nitrogen, not on fossil fuels, so urea should continue to be available and cheap. Moreover, since urea is an important agricultural chemical and only about 10% of total production is used for amino resins, its future availability as a continued source of low cost raw materials seems assured.

Melamine

Melamine (2,3,6-triamino-1,3,5-triazine) was first prepared in 1843 by Liebig. It is a white crystalline solid, only sparingly soluble in water, and it melts just above 350°C with vaporization. It has been produced on a commercial scale by heating dicyandiamide under pressure with ammonia and by reacting lime with coke, but in recent years the trend has been toward manufacturing melamine from urea.

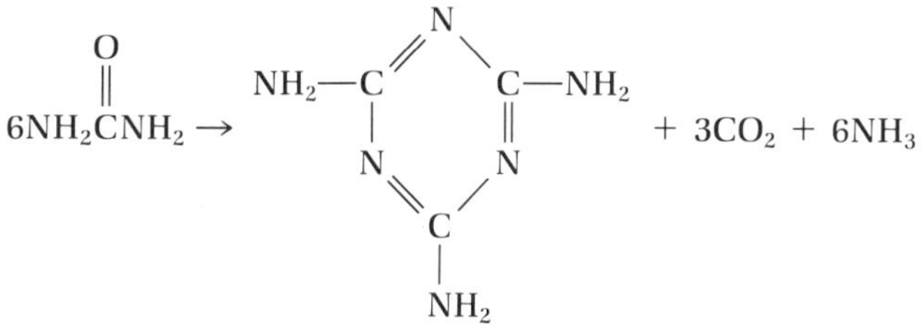

Essentially all the melamine produced is used for making amino resins and plastics. Melamine was neglected as a commercial material until Henkel in 1935 patented its resinous condensations with formaldehyde. Since melamine is produced from urea, it is more expensive than urea and uses of its resins are restricted to applications calling for superior performance.

Resin Formation

The first step in the formation of resins and plastics from formaldehyde and amino compounds is the addition of formaldehyde to the amino group to form a hydroxymethyl group.

$$R—NH_2 + HCHO \rightarrow R—NH—CH_2OH$$

The second step is the condensation of the groups formed in the first step with additional amine and the liberation of water to form a dimer, a polymer chain, or a network. In the case of urea, the type of structure

$$RNH—CH_2OH + H_2NR \rightarrow RNH—CH_2—NHR + H_2O$$

formed might be the one shown in structure **1.**

Another reaction that can occur is combination of two methylol groups to produce a dimethylene ether linkage, $RNHCH_2—O—CH_2NHR$. The $—CH_2—O—CH_2—$ linkage is less stable than the $—CH_2—$ linkage and may rearrange to form a $—CH_2—$ bridge and liberate a molecule of formaldehyde.

Melamine can combine with 1–6 moles of formalde-

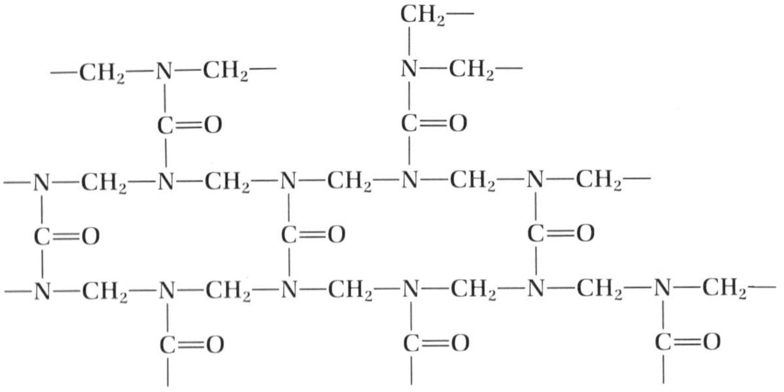

hyde and so can form very stable networks. Polymerization and curing reactions produce an M/F that has been represented by structure **2** [4,8]. In contrast to U/F resins, which have almost exclusively diaminomethylene linkages, M/F resins have both diaminomethylene and dimethylene ether linkages and there may be some unreacted methylol groups. This is based on the observation that in curing M/F up to 100°C, no substantial amounts of formaldehyde are split off, and only small quantities are detected on curing up to 150°C. Curing U/F under the same conditions liberates copious quantities of formaldehyde [8].

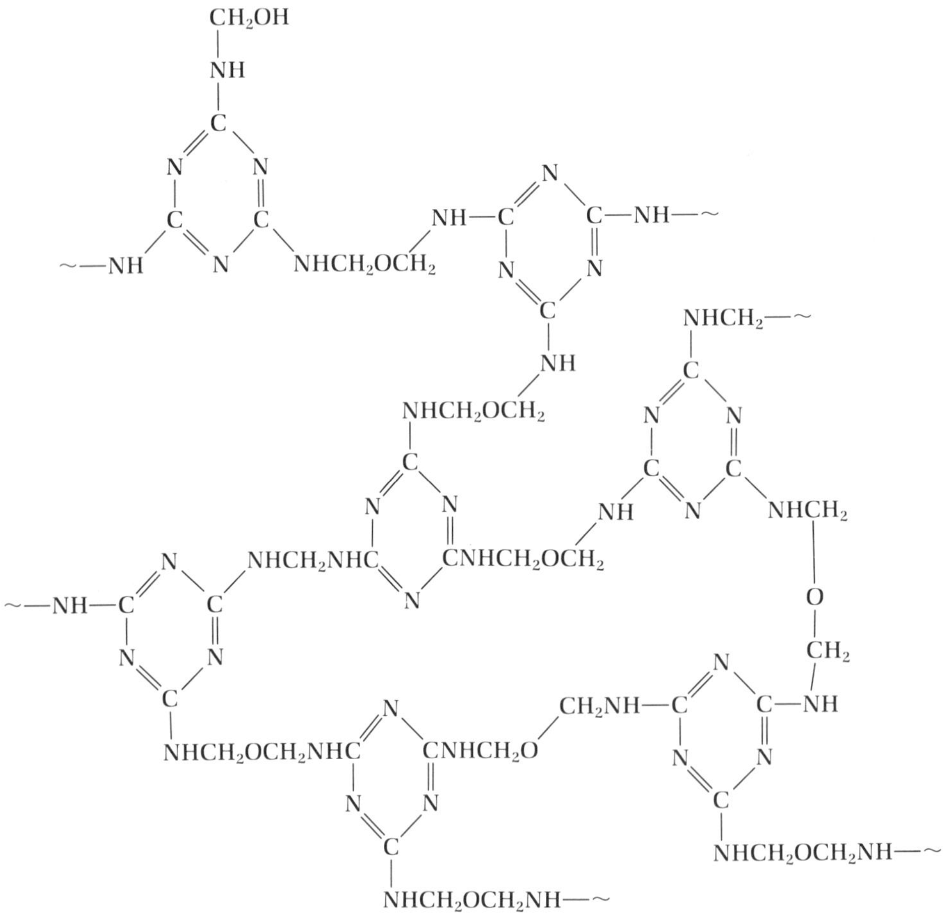

Successful manufacture and use of amino resins depends on precise control of the two main steps involved in the reaction of formaldehyde with amino compounds to form resins and plastics. The first reaction is catalyzed by either acids or bases, but the methylol compounds are stable only under neutral or alkaline conditions. The second step is catalyzed only by acids. It is common practice for the resin supplier to carry out the first step under alkaline conditions and for the resin user to complete the reaction later by addition of an acidic substance and frequently also heat.

Care must be taken to avoid a troublesome side reaction encountered in the first step, when formaldehyde may be converted to formic acid. Especially when the reaction must be heated under alkaline conditions, circumstances are favorable for a reaction in which two molecules of formaldehyde interact to yield one molecule of methanol and one of formic acid (Cannizzaro reaction). If sufficient formic acid forms, the solution may become acidic enough to catalyze the condensation reaction and cause abnormally high viscosity or premature gelling of the resin solution. The factors that control the type of product obtained include mole ratio of reactants, catalyst (pH of reaction mixture), reaction time, and temperature.

Properties

Some typical properties of U/F and M/F resins and filled molding compounds are given in Table 1 [6,7]. These representative data are from the late 1970s. Similar data appear almost annually in the current issue of *Modern Plastics Encyclopedia* (in the 1988 edition [1], e.g., prop-

TABLE 1
Typical Properties of Amino Resins and Amino Resin Molding Compounds [6,7]

	Resins		Compounds				
Properties	U/F	M/F	U/F α-Cellulose	M/F α-Cellulose	M/F Macerated Fabric	M/F Asbestos	M/F Glass Fiber
Physical							
Specific gravity	1.48–1.52	1.47–1.52	1.47–1.52	1.47–1.52	1.5	1.7–2.0	1.8–2.0
Water absorption (24 h, 3.2 mm thick), %			0.48	0.1–0.6	0.3–0.6	0.08–0.14	0.09–0.21
Mechanical							
Tensile strength, MPa	34.5–48.3	48.3–70.0	38–48	48–90	55–69	38–45	35–70
Elongation, %	0.6	0.7	0.5–1.0	0.6–0.9	0.6–0.8	0.3–0.45	
Tensile modulus, GPa			9–9.7	9.3	9.7–11	13.5	16.5
Hardness, Rockwell M			110–120	120	120	110	115
Flexural strength, MPa	70–110	70–110	70–124	83–104	83–104	52–69	90–165
Flexural modulus, GPa			9.7–10.3	7.6	9.7	12.4	16.5
Impact strength, J/m of notch	14	16	14–18	13–19	32–53	16–21	32–1000
Thermal							
Thermal conductivity, $\times 10^{-4}$ J · cm/s · cm^2 · °C^a	29–41	29–41	42.3	29.3–42.3	44.3	54.4–71	48.1
Coefficient of thermal expansion, $\times 10^{-5}$ cm/(cm · °C)			2.2–3.6	2.0–5.7	2.5–2.8	2.0–4.5	1.5–1.7
Deflection temperature at 1.8 MPa, °C	132–138	193	130	182	154	129	204
Flammability class, UL			VO^a	VO^a			VO
Continuous no-load service temperature, °C			77^b	99^b	121	149	149–204
Electrical							
Dielectric strength, V/0.00254 cm	300–400	300–400					
Short time, 3.2 mm thick			330–370	270–300	250–350	410–430	170–300
Step by step			220–250	240–270	200–300	280–320	170–240
Dielectric constant at 22.8°C							
At 60 Hz			7.7–7.9	8.4–9.4	7.6–12.6	6.4–10.2	9.76–11.1
At 10^3 Hz	6–7	7–8		7.8–9.2	7.1–7.8	9.0	
Dissipation factor at 22.8°C							
At 60 Hz			0.034–0.043	0.030–0.083	0.07–0.34	0.07–0.17	0.14–0.23
At 10^3 Hz				0.015–0.036	0.03–0.05	0.07	
Volume resistivity at 22.8°C, 50% RH, Ω · cm	10^{12}–10^{13}	10^{12}–10^{14}	$0.5–5.0 \times 10^{11}$	$0.8–2.0 \times 10^{12}$	$1.0–3.0 \times 10^{11}$	1.2×10^{12}	$0.9–2.0 \times 10^{11}$
Arc resistance, s	—	—	80–100	125–136	122–128	20–180	180–186

^{a} Applies to specimens thicker than 1.6 mm.
^{b} Based on no color change.

erty data could be found on pages 519, 551, 559). The resins from the two amines are similar in appearance, water-white syrups or white powders, but those from melamine are superior in water resistance. They accept colorants well. Fillers improve molding and physical characteristics and lower the cost of the resins. Both have good mechanical strength and rigidity and can be readily compression molded. Injection molding requires the use of special technology developed for use with thermosets. Cured moldings are tasteless, odorless, and nontoxic, and those from melamine will withstand staining from coffee, tea, and fruit juices. Electrical properties are excellent. M/F is the hardest commercial plastic material and it is scratch resistant [4,5]. Both U/F and M/F are self-extinguishing materials; with some fillers the composites become less fire resistant, but they are usually still nonburning [5]. Resistance to many chemicals is good, but both resins are attacked by acids and alkalies.

Processing

How an amino resin is processed depends on the application [6]. Since U/F and M/F molding compounds are thermosetting resins, traditionally they have been processed by compression and transfer molding [5]. Transfer molding is more successful with M/F than with U/F resins, which cure very rapidly. Allowance must also be made for the bulk factor of the molding powder. Special reciprocating-screw injection machines are also used. The great advantage of injection molding is that it reduces costs by eliminating some of the hand labor.

The automatic injection molding process is similar to that used for thermoplastics except for the thermal cycle. Instead of being plasticized in a hot cylinder and then injected into a much cooler mold cavity, the thermosets are plasticized in a warm cylinder and then injected into a hot mold cavity, where the curing reaction sets the resin [4]. The process is best applied to small parts. Larger items like melamine plastic dinnerware are still molded by standard compression molding techniques. Some properties of amino molding compounds are given in Table 2 [4].

U/F adhesives are processed by a variety of techniques depending on the application. Plywood is made by spreading an adhesive/hardener mixture on successive sheets of base wood veneers with the grain crossed each time, followed by hot-press bonding at 95–110°C and 0.2–1.4 MPa [6]. In certain construction applications it is not always convenient to apply heat to the structure. Cold-setting U/F adhesives have been developed for these situations, and pressure is applied by jigs and clamps. Chipboard is made by mixing 5–10% of U/F with wood chips and pressing at 1.4–6.9 MPa and 140–160°C. Further details about these and processing techniques for other applications can be found in Refs. 1–12 and in original papers not cited in this brief review.

Economic Aspects

In 1984 about 4% of U.S. plastics and resin production consisted of amino resins [13,14]. In 1961 the production figure was 6.5% [8]. Thus although production of amino resins has grown steadily along with that of all plastics, the market share of amino resins has decreased slightly during the past 20 years or so. (An excellent summary of the economic history of amino resins from 1944 to 1961 can be found in Ref. 8.) In 1984 M/F for molding cost about \$1.80/kg, while U/F for the same purpose sold for about \$1.50/kg. Liquid urea adhesive resins were about \$1.10/kg [4]. The prices quoted are for carload or truckload quantities.

There are many manufacturers of amino resins (Ref. 7 lists about 50). Many of these manufacturers specialize in only one or two particular types or resins for particular applications. Among the large producers of amino resins are American Cyanamid Company, Borden Chemical Company, Georgia-Pacific Corporation, Gulf Oil Company, Monsanto Company, and Reichhold Company [1,4].

Analysis

Usually the urea or melamine is supplied in a sufficient degree of purity to be used without analysis. Formalde-

TABLE 2
Typical Processing Properties of Filled Amino Compounds [4]

	Urea	Melamine			
Property	α-Cellulose	α-Cellulose	Macerated Fabric	Glass Fiber	Phenolic Resin Modifier
Molding qualities	Excellent	Excellent	Good	Good	Good
Compression molding temperature, °C	135–175	135–190	135–165	135–170	150–175
Compression molding pressure, MPa	14–65	10–55	28–55	14–55	14–34
Specific gravity, density	1.47–1.52	1.47–1.52	1.5	1.8–2.0	1.5–1.7
Mold shrinkage, linear, cm/cm	0.006–0.014	0.005–0.015	0.003–0.004	0.001–0.004	0.004–0.010
Machining qualities	Fair	Fair	Good	Good	Fair

hyde may be checked for pH and impurities by standard analytical techniques. Fillers can be identified by microscopy after the resin in a molded article has been dissolved away by heating in a slightly acid solution of formaldehyde under pressure. A burning test can be used as qualitative means to distinguish an amino resin from other resins. U/F materials smell somewhat like burning hair or horn and have an odor of formaldehyde. Burning melamine resins also smell like burning horn or hair and also have the odor of ammonia. Urea or melamine content can be determined by an elemental analyzer or by degrading the resin and making a derivative of the amine [4].

Health and Safety Factors [4]

Urea and melamine monomers are nontoxic. Urea is an important agricultural product used as feed supplement for animals and as fertilizer. No ill effects have been observed when rats were fed large quantities of melamine. Molded amino resins are considered nontoxic also. Formaldehyde is toxic. It combines easily with the proteins of the human body, and its vapors can cause irritation to the eyes, nose, and throat. Formaldehyde has been found to cause cancer in laboratory animals. Care to avoid exposure to formaldehyde must be taken during synthesis of amino resins. Urea–formaldehyde foam used in insulation of some older houses and mobile homes can release formaldehyde and is now considered to be unsafe for this application. Combustion of cured amino resins can release toxic gases; fortunately, flammability is low and especially M/F is quite heat resistant.

Uses

U/F and M/F resins are used in a variety of applications. Table 3 lists many of these uses, as well as the amount of amino resin used in the United States in 1986 and 1987 for that application [15]. Each application requires a particular resin formulated especially for that application. A typical molding resin contains resin, filler, pigment, catalyst, stabilizer, plasticizer, and lubricants. The largest single use of α-cellulose-filled M/F is in the manufacture of tableware, but it is also used in buttons, ashtrays, lavatory bowls, utensil handles, shaver housings, appliance components, control buttons, and knobs. α-Cellulose-filled U/F resins are used in wiring devices (circuit breakers, wall plates, receptacles), closures, electric blanket control housings, buttons, toilet seats, knobs, and toothpaste tube inserts. Wood-flour-filled M/F resins are used for industrial electrical parts and military specifications. Glass- and mineral-filled M/F resins find exclusive use in military applications [1]. A large use of U/F is in adhesives for furniture, plywood, and chipboard. Water resistance is poor, so in the latter applications, the products are suitable for interior use only. M/F resins are often used to make waterproof exterior plywood and marine plywood.

TABLE 3
Consumption of Urea and Melamine Resins [15]

Use	Amount (kt) 1986	Amount (kt) 1987
Bonding and adhesive resins for		
Fibrous and granulated wood	458.1	499.0
Laminating	20.4	23.1
Plywood	29.0	29.9
Molding compounds	28.6	30.8
Paper treatment and coating resins	45.4	39.5
Protective coatings	46.7	50.3
Textile treatment and coating resins	9.5	10.9
Export	9.1	10.4
Other	15.0	15.9
Total	661.8	709.8
Urea molding powder for		
Closures	3.2	3.6
Electrical devices	15.0	15.9
Other	0.9	1.4
Total	19.4	20.9
Melamine molding powder for		
Buttons	0.9	0.9
Dinnerware	6.8	6.8
Sanitary ware	0.9	0.9
Other	0.9	1.4
Total	9.5	10.0

Other uses of amino resins are in bonding of glass or mineral fiber mats, in binders for foundry cores and coated abrasive papers, and in orthopedic casts and bandages. As textile finishes, U/F and M/F are employed to make drip-dry fabrics and fabrics with permanent glazing, rotproofing, and shrinkage control. Laminating resins are used for kitchen counters and table tops. The use of etherified or partially etherified (alkylated) M/F coating resins is a major market in the automobile industry, but these resins are not used in composite applications [1,16].

P. Dreyfuss

References

1. *Modern Plastics Encyclopedia*, McGraw-Hill, New York, 1988, pp. 110, 112, 519, 559, 733, 743.
2. T. Richardson, *Composites*, Industrial Press, New York, 1987, pp. 68–71.
3. A. L. Wooten, in S. H. Goodman, Ed., *Handbook of Thermoset Plastics*, Noyes, Park Ridge, NJ, 1986, pp. 45–58.
4. I. H. Updegraff, *Encyclopedia of Polymer Science and Engineering*, 2nd ed., Vol. 1, 1985, pp. 752–789.
5. T. L. R. Richardson, *Industrial Plastics: Theory and Application*, South-Western Publishing, Cincinnati, OH, 1983.
6. D. C. Miles and J. H. Briston, *Polymer Technology*, Chemical Publishing, New York, 1979, pp. 73–90.

7. I. H. Updegraff, S. T. Moore, W. F. Herbes, and P. B. Roth, *Encyclopedia of Chemical Technology*, 3rd ed., Vol. 2, Wiley, New York, 1978, p. 440.
8. G. Widmer, in *Encyclopedia Polymer Science and Engineering*, Vol. 2, Wiley, New York, 1965, p. 1.
9. C. P. Vale and W. G. K. Taylor, *Aminoplastics*, Iliffe Books (for the Plastics and Rubber Institute), London, 1964.
10. T. J. Suen, in C. E. Schildknecht, Ed., *Polymer Processes*, Wiley-Interscience, New York, 1956, Ch. VIII.
11. A. Brookes and C. P. Vale, "Resins from Urea, Melamine and Related Materials," Plastics Institute Monograph, 1954.
12. B. Meyer, *Urea Formaldehyde Resins*, Addison-Wesley Publishing, Reading, MA, 1979.
13. *Chem. Eng. News, 61*(18), 13 (1983).
14. *Chem. Eng. News, 61*(24), 33 (1983).
15. *Mod. Plast.*, January 1988, p. 101.
16. R. J. Schupp, American Cyanamid Company, private communication.

Analysis

See Bending Analysis of Composite Plates and Built-up Sections; Carbon Fibers, Property Testing and Analysis; Characterization; Chemical Analysis and Characterization of Polymer Composite Materials; Finite Element Analysis; Fractography; Nondestructive Evaluation; Stress Analysis; Structural Analysis; Surface Analysis; Thermal Analysis

Antistatic Agents

See Static Electrical Agents

Applications

See Automotive Composites; Electrical and Electronic Composites; Large Space Composite Mirrors; Low-Observability Composites; Powder Metal Composites; Propellants, Composite Rocket; Propulsion Structures; Sound Control, Tires; Spacecraft Applications; Sports and Recreational Use; Structural Aircraft, Noncommercial and Homebuilt; Structures, Marine; Superconductive Composites, High-Pressure Fabrication; System Avionics Applications; Woven Composite Structures; Industrial and Commercial; Propulsion Structures; Spacecraft Applications; Sports and Recreational Use; Structural Aircraft, Noncommercial and Homebuilt

ARALL® Laminates (Aramid Aluminum Laminates)

ARALL® laminates are a new family of hybrid composite materials developed in the Netherlands and being commercialized by Alcoa with the assistance of 3M [1–10]. Originally an acronym for aramid-reinforced aluminum laminates, as derived by Delft University, ARALL laminates are now a registered trademark of Alcoa.

ARALL laminates are made of high strength aramid fibers embedded in a structural epoxy adhesive sandwiched between multiple layers of thin gage anodized and primed aluminum alloy sheets and processed to have controlled residual stresses. A schematic of a typical 3/2 layup (three layers of aluminum and two layers of prepreg) is shown in Figure 1. Uniting high strength aluminum and strong aramid fibers in this adhesive-bonded arrangement creates a new material with a unique set of properties. ARALL laminates have the high strength and excellent fatigue resistance of advanced composites while retaining the traditional durability, supportability, and fabricability advantages of metals.

A first-generation set of standard ARALL laminates are being commercialized. These are simply called ARALL-1, -2, -3 and -4 laminates. Their main attribute is the ability to retard and self-arrest fatigue cracking by transferring loads from the metal to the stronger unbroken fibers in the crack wake. They also possess 15–20% lower density and up to 60% higher directional strength than aerospace aluminum alloys. ARALL laminates have fabricability characteristics similar to those of aluminum sheets. They can be cut, drilled, formed, joined, and inspected by conventional aluminum practices. The outer metal layers protect the fiber–resin systems against damage by moisture, thermal attacks, lightning strikes, and impacts [12–15].

Because of these features, some candidate aircraft applications include fatigue-critical lower wing and fuselage structure, as well as skin areas prone to vibration and lightning strikes. Design trade studies and actual built-up structures have shown cost-effective weight savings of 15–40% over monolithic aluminum for these applications [16–20].

The next generation of aircraft, both commercial and military, will be made of many different structural materials, including the many composite materials covered in this encyclopedia. Traditional metals will continue to have a significant role, and ARALL laminates, a combination of the best from both these materials worlds, will also have a significant role. One study [6] estimates that 5–10% of aircraft structures in the late 1990s will be made of ARALL laminates.

Product Forms and Descriptions

The idea of putting fiber reinforcement in the adhesive bond lines between aluminum alloy has been researched at many laboratories over the past 30 years.

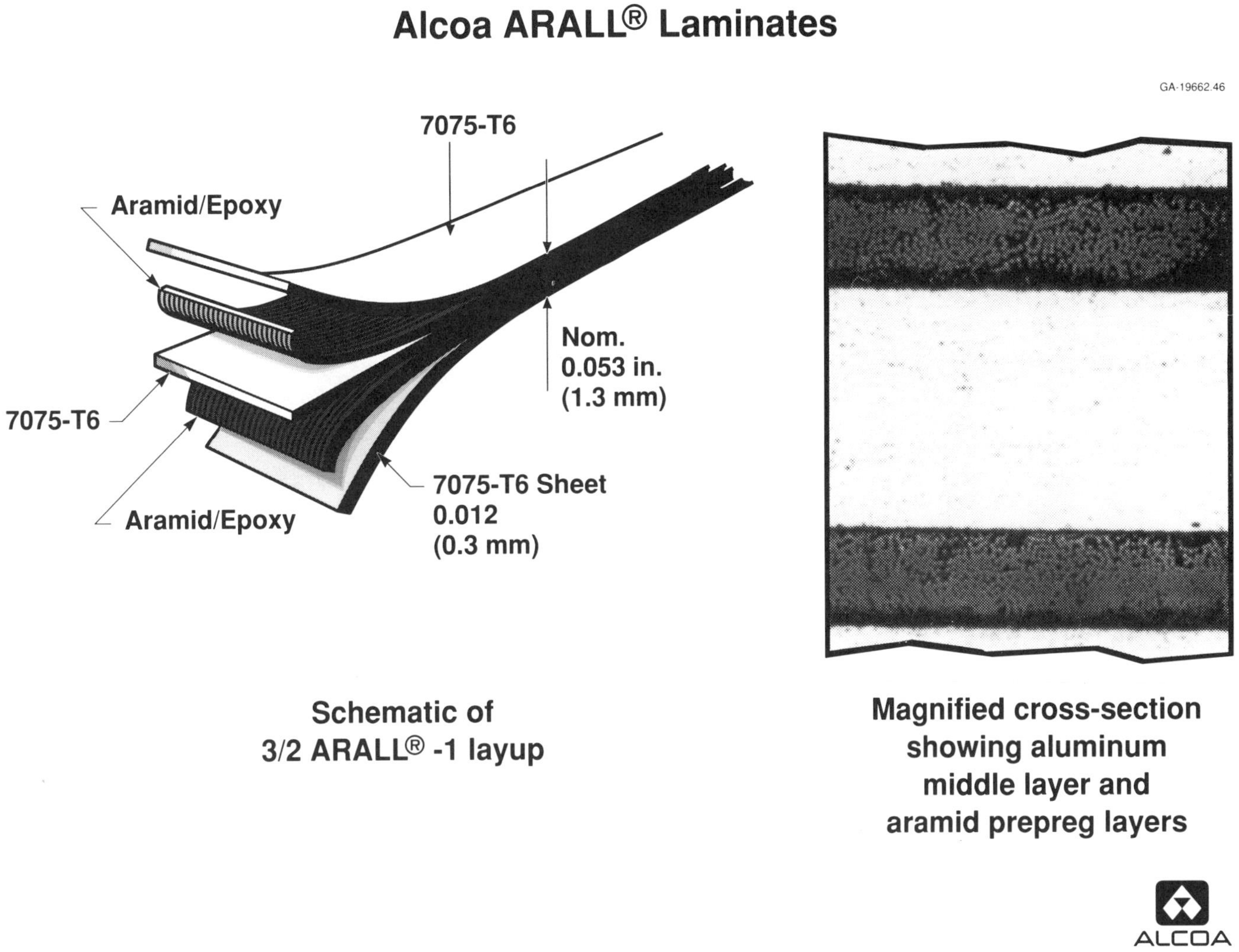

FIGURE 1 Typical 3/2 layup of ARALL-1 laminate. (Schematic courtesy of Alcoa ARALL Laminates.)

As indicated previously, Delft University optimized this concept in the early 1980s and coined the name ARALL laminates [1–6]. Many different combinations of ARALL laminates can be made by varying fiber systems, resin systems, aluminum sheet alloys, surface preparation techniques, stacking sequences, fiber orientation, etc. The number of plies and the type of postcure mechanical treatment may also be varied.

To produce a standard product for which industry specifications with guaranteed minimum properties can be written, it was necessary to select the most promising sets of variables and commercialize these as a first generation of materials; these ARALL-1, -2, -3, and -4 laminates are described in Table 1. All have high static strength and can be considered fatigue insensitive for normal aerospace applications.

ARALL-1 and -2 laminates have been the focus of most efforts to date. ARALL-1 is 7075-T6 based, has 0.4% nominal postcure stretch, and is intended for structures that will bear high fatigue loading. ARALL-2 is suited to applications requiring higher damage tolerance and formability. ARALL-3 is an ARALL-1 upgrade that uses controlled-toughness alloy 7475 in an exfoliation-resistant T76 temper condition in lieu of 7075-T6. ARALL-4 is 2024-T8 based and employs a special prepreg system to achieve higher temperature capability.

All four standard ARALL products employ a thermoset adhesive system impregnated with unidirectional aramid fibers in a fiber-to-resin weight ratio of 50 : 50. The fibers are oriented parallel to the aluminum sheet rolling direction. ARALL-1, -2, and -3 use a 120°C cure resin system with a recommended service temperature of 90°C. The ARALL-4 prepreg cures at 175°C and is intended for use up to 160°C. Nominal thickness of the individual aluminum sheet layers is 0.3 mm, and the prepreg nominal thickness is 0.2 mm. Standard laminate configurations vary from a 2/1 (two layers of aluminum surrounding one layer of prepreg), which is 0.83 mm thick, to a 5/4 (five layers of aluminum with four prepreg layers interspaced), which is 2.4 mm thick. Thicker laminates have also been made. Because of the low density of the aramid prepreg layers, the net density of ARALL laminate is (15–20%) lower than that of monolithic aluminum.

TABLE 1
ARALL Laminate Standard Product Forms

Description	Characteristics	Tentative Dates for Aerospace Material Specifications
ARALL®-1 laminate alloy 7075-T6[a,b] 120°C cure prepreg,[c] 0.4% permanent stretch	Superior fatigue resistance High strength	Fall 1988
ARALL-2 laminate alloy 2024-T3[a,b] 120°C cure prepreg,[c] with or without 0.4% stretch	Excellent fatigue resistance Increased formability Damage tolerant	Fall 1988
ARALL-3 laminate alloy 7475-T76[a,b] 120°C cure prepreg,[c] 0.4% permanent stretch	Superior fatigue resistance Controlled toughness Good exfoliation High strength	Spring 1989
ARALL-4 laminate alloy 2024-T8[a,b] 175°C cure prepreg,[c] with or without 0.4% stretch	Excellent fatigue resistance Increased elevated temperature properties	Fall 1989

[a] Also produced in a one- or two-sided clad condition for added corrosion protection.
[b] Bonding surfaces anodized and primed.
[c] With 50% by volume unidirectional aramid fibers.

Processing

ARALL laminates are produced under very rigid quality control standards. The aluminum sheets are anodized and primed according to existing aerospace industry specifications. Both chromic acid anodizing and phosphoric acid anodizing treatments are used. The outer aluminum surfaces can be left bare for later processing after an ARALL part has been formed. The prepreg used in ARALL-1, -2, and -3 laminates is made from AF-163-2 120°C cure resin. The prepreg used in ARALL-4 laminates is made from AF-191 175°C cure resin. Both these resins have industry specifications. The prepreg is 50% by volume unidirectional aramid fibers. The ARALL panels are laid up and cured under the appropriate times and temperatures either in a mechanical press or an autoclave. After curing, the aluminum layers are in a slight tensile residual stress state and the prepreg layers are in a compensating compressive residual stress state [6]. The ARALL-1 and -3 laminates are given a postcure 0.4% nominal stretch to reverse these residual stress signs. The ARALL-2 and -4 laminates are not postcured stretched as a standard practice, although they can be supplied that way.

Properties

Table 2 compares ARALL laminate room temperature typical mechanical properties with those of monolithic aluminum sheet alloys and two common carbon fiber composite layouts [11]. As is the case with most fibrous composites, the properties of ARALL laminates are anisotropic. Many of these properties can be estimated by the rule of mixtures and the appropriate aluminum and directional prepreg properties. Because of space limitations, the ARALL properties are given only for the 3/2 configuration. Alternate layup properties are available from Refs 12–15. Elevated temperature properties can be found in Ref 21.

ARALL laminates were developed primarily to obtain a material with good fatigue resistance. Their primary attribute is the ability to slow and self-arrest fatigue crack growth, particularly when tested in the stretched condition. Extensive fatigue testing on ARALL laminates -1 and -2 has been done under both constant load amplitude and realistic flight simulation loading conditions [1–11]. These tests include smooth and notched coupons, cracked specimens, lugs, and mechanically fastened and bonded joints in combination with a variety of environ-

TABLE 2
Typical Mechanical Properties of ARALL Laminate, Aluminum Sheet, and Carbon Fiber Composite Room Temperature

Property	Test Direction	3/2 ARALL Laminate, 1.3 mm Thick				Aluminum Alloy Sheet, 1.6 mm Thick				Carbon Fiber Composite[a]	
		ARALL-1	ARALL-2	ARALL-3[b]	ARALL-4[b]	7075-T6	2024-T3	7475-T76	2024-T8	Unidirectional[c]	Typical Structure[d]
Tensile ultimate	L	800	717	828	731	572	455	545	497	1241	655
strength, MPa	LT	386	317	373	338	572	448	552	497	55	276
0.2% Off, tensile yield	L	641	359	587	373	510	359	476	442		
strength, MPa	LT	331	228	317	317	496	324	476	442		
Tensile elastic modulus,	L	68	64	68	61	71	72	69	72	145	76
GPa	LT	48	49	51	49	71	72	69	72	12	35
Tensile elongation, %	L	[e]	1.4[e]	1.0[e]	1.4[e]	11	18	12	8		
	LT	[e]	12.0[e]		3.9[e]	11	18	12	8		
Tensile total strain to	L	1.9	2.5	2.2	2.6	12	19	13	9	0.9	0.5
failure, %	LT	7.9	12.7		4.6	12	19	13	9	0.5	0.5
Poisson ratio (tension)	L	0.33	0.32			0.33	0.33	0.33	0.33	0.31	0.48
	LT	0.25	0.26			0.33	0.33	0.33	0.33	0.017	0.23
0.2% Off. compressive	L	372	262			503	304	442	435	1241	655
yield strength, MPa	LT	393	234			524	345	469	442	207	290
Compressive elastic	L	70	67			72	74	72	74	145	76
modulus, GPa	LT	52	52			72	74	72	74	12	35
0.2% Off. shear yield	L-LT		117[g]							83[h]	255[h]
strength, MPa	LT-L		114[g]								
Shear modulus, msi	L-LT	17[g]	17[g]			27	28	26	28	4.5	21
(Gpa)	LT-L		16[g]			27	28	26	28		
Bearing ultimate	L	655	531			890	759	828	738		
strength	LT	703	545			890	759	828	738		
at e/D = MPa	L	738	565			1145	945	1055	938		
at e/D = MPa	LT	724	545			1145	945	1055	938		
Bearing yield strength	L	586	386			759	538	690	607		
at e/D = MPa	LT	607	386			759	538	690	607		
at e/D = MPa	L	703	455			883	649	780	690		
	LT	669	441			883	649	780	690		

[a] DoD/NASA *Advanced Composites Design Guide*, Vol. 1A, generic F180 graphite/epoxy composite, 60% fiber content, 1.55 g/m^3 (e.g., AS/3501-6, T300/5208).
[b] Single laboratory lot.
[c] 100% fibers in L direction.
[d] Fiber orientations: 42% 0 degree, 50% ± 45°, 8% 90°.
[e] Plastic strain determined from test record by back-extrapolating elastic slope from point of fracture.
[f] Compressive ultimate strength.
[g] Iosipescu shear test method; L-LT means fibers parallel to long specimen axis, LT-L means fibers normal to long specimen axis.
[h] In-plane shear ultimate.
Table reprinted from Ref. 10.

mental exposures. Under these conditions, ARALL laminates always showed significantly longer life than monolithic aluminum when the principal fatigue loading was tension dominated and in the direction of reinforcement. Naturally initiated cracks were arrested after a small amount of crack travel, or grew very slowly, depending on the loading conditions. Crack arrest was observed, even when ARALL specimens were subjected to higher stress than the aluminum specimens. Preliminary design studies for aircraft lower wing and fuselage components have shown that, in general, ARALL laminate fatigue properties are so good that governing design criteria shift away from fatigue to other factors, such as compression and static notch strength [6,16–18].

The environmental test results of Table 3 show that exposure to high humidity, thermal cycle testing, salt fog, and seacoast atmosphere has little effect on the tensile and compressive strengths of ARALL laminates -1 and -2. Where strength loss is shown, surface corrosion of the outer aluminum faces is in large part responsible. Cladding the outer aluminum faces offers a possible solution to this problem.

Though the aramid prepreg layer is susceptible to moisture absorption, the ARALL laminate weight gain is negligible compared with that of pure composites because of the aluminum barrier properties [22,23]. In bond line exposure tests, the fiber–adhesive interface, not the metal–adhesive interface, has been shown to be the weakest link. Since moisture can gain access only at free edges, it is reasonable to expect that sealants could be relied on for bond line protection in actual structure. Additionally, real structure will have a much smaller free edge–volume fracture than is present in the small-coupon simulations.

Applications

ARALL laminates were developed to be fatigue-resistant materials, and thus are ideally suited for such aerospace structures as lower wings and fuselages [6]. In addition, the impact resistance of ARALL laminates [11] makes it a candidate in areas where that property is critical. It also has been shown that lightning strike resistance [20] for ARALL laminates is better than for monolithic aluminum, and the areas around fuel tanks in commuter aircraft are sized according to this.

TABLE 3
Effect of Environmental Presoak on Tensile and Compressive Properties of ARALL Laminate[a] and Aluminum Sheet[b]

	Percent Property Retention After Exposure[c]															
	Tensile Ultimate Strength[d]								Compressive Yield Strength[e]							
	ARALL-1		ARALL-2		7075-T6		2024-T3		ARALL-1		ARALL-2		7075-T5		2024-T3	
Conditions	L	LT	L	LT	L	LT	L	LT	L	LT	L	LT	L	LT	L	LT
100% RH at 60°C																
1 month	697	683	704	683	690	690	697	690	738	738	697	683	690	704	697	690
3 months	697	683	690	649	697	697	697	690	711	718	676	676	683	697	704	697
6 months	683	697	683	621	704	704	711	669	718	690	683	697	711	704	704	683
Hot-Wet/Cold/Hot-Dry, Cycled Daily[f]																
1 month	697	683	704	628	690	690	690	676	718	690	669	676	697	697	690	683
3 months	676	669	683	559	683	690	662	642	725	697	669	676	718	725	697	676
6 months	697	669	676	624	676	683	624	621	725	690	669	697	718	718	731	704
5% Salt Fog at 35°C																
500 hours	697	669	683	607	683	683	676	635	690	676	676	676	697	704	690	690
1000 hours	683	614	676	573	676	669	635	587	718	725	683	683	683	690	690	690
Point Judith (RI) seacoast																
1 year	683	676	656	580	690	683	662	621	676	669	662	662	690	690	683	676

[a] 3/2 layup, 1.3 mm thick.
[b] 1.3 mm sheet gage.
[c] Average of triplicate tests.
[d] 21.6 × 1.9 cm tensile blank, edges unprotected.
[e] 6.6 × 1.6 cm compression blank; edges unprotected.
[f] Daily cycle of hours at 100% RH/5 hours at −23°C/hours at 82°C.

Summary

ARALL laminates were developed to fill a materials need in the aerospace industry, namely, the requirement for a material that couples the higher specific strength and fatigue performance capabilities of advanced composites with the classical durability, fabricability, and supportability characteristics of aluminum. In addition, the material's enhanced vibrational damping and lightning protection abilities provide new options for innovative aircraft usage. Since the early 1980s, ARALL laminates have shown themselves to be a most exciting new material for aerospace structure. The unique ARALL laminate properties match up well with existing structural integrity requirements and the more demanding performance needs of next-generation aircraft. Through use of good design practices, goals of 30–40% weight savings and no-repair structure have been demonstrated to be within reach.

Larry N. Mueller

References

1. R. Marissen, and L. B. Vogelesang, *Proceedings of the Intercontinental SAMPE Meeting*, 1981.
2. L. B. Vogelesang, R. Marissen, and J. Schijve, *Proceedings of the 11th ICAF Symposium*, 1981.
3. J. W. Gunnink, L. B. Vogelesang, and J. Schijve, *Proceedings of the ICAS-82-2.6.1, 13th Congress of the Int. Council of the Aeronautical Sciences*, 1982, p. 990.
4. L. B. Vogelesang, and J. W. Gunnink, "ARALL®, A Material for the Next Generation of Aircraft; A State of the Art," Delft University Report LR-400, August 1983.
5. J. W. Gunnink, M. L. C. E. Verbruggen, and L. B. Vogelesang, *Vertica*, *10*(2), 241 (1986) [*Proceedings of the 10th European Rotocraft Forum*, 1984].
6. L. B. Vogelesang, and J. W. Gunnink, *Mater Design*, 7(6) (November/December 1986) [*Proceedings of the International Federation of Airworthiness*, Amsterdam, 1985].
7. L. N. Mueller, J. L. Prohaska, and J. W. Davis, *Proceedings of the AIAA Aerospace Engineering Conference and Show*, 1985.
8. R. J. Bucci, L. N. Mueller, R. W. Schultz, and J. L. Prohaska, *Proceedings of the 32nd Int. SAMPE Symposium*, 1987, pp. 902–916.
9. R. J. Bucci, and L. N. Mueller, *Proceedings of the ARALL® Laminates Technical Conference*, 1987.
10. R. J. Bucci, L. N. Mueller, L. B. Vogelesang, and J. W. Gunnink, *Proceedings of the 33rd Int. SAMPE Symposium*, 1988, pp. 1237–1248.
11. M. A. Gregory, "ARALL® 1 Laminates Preliminary Technical Information," Alcoa Technical Center, October 1987.

12. M. A. Gregory, "ARALL® 2 Laminates Preliminary Technical Information," Alcoa Technical Center, October 1987.

13. M. A. Gregory, "ARALL® 3 Laminates Preliminary Technical Information," Alcoa Technical Center, October 1987.

14. M. A. Gregory, "ARALL® 4 Laminates Preliminary Technical Information," Alcoa Technical Center, October 1987.

15. L. H. van Veggel, A. A. Jongebreur, and J. W. Gunnink, *Proceedings of the 14th ICAF Symposium,* Ottawa, 1987.

16. J. W. Gunnink, and P. A. v. d. Schee, *Proceedings of the Fourth Int. Conference on Composite Structures,* 1987.

17. L. van Veggel, *Proceedings of the ARALL® Laminates Technical Conference,* 1987.

18. J. W. Gunnink, *Proceedings of the ICCM 6/ECCM 2,* 1987.

19. M. Ioannou, L. J. Kok, T. M. Fielding, and N. J. McNeill, *Proceedings of the 14th ICAF Symposium,* 1987.

20. H. F. Wu, *Proceedings of the 33rd SAMPE Int. Symposium,* 1988, pp. 1249–1259.

21. M. L. C. E. Verbruggen, "Aramid Reinforced Aluminum Laminates: ARALL Adhesion Problems and Environmental Effects," Reports LR 504 and LR 505, Ph.D. thesis, Faculty of Aerospace Engineering, Delft University, the Netherlands, 1987.

22. S. Lee, and T. R. Kipp, "ARALL® Laminates Environmental Durability Test Results," Alcoa Laboratories Progress Report, Oct. 25, 1987.

Aramid Fiber Composites

Aramid fibers are aromatic polyamide fibers in which at least 85% of the amide linkages are attached directly to two aromatic rings [1].

The path to the development of aramid fibers can be traced to the research of P. W. Morgan [2] at du Pont in the 1950s on low temperature processes for the preparation of condensation polymers. This research involved solution polymerization in nonaqueous media using a new range of amide and amide salt media that were especially applicable to the preparation of aromatic polyamides. In the early 1960s du Pont introduced a commercial flame-resistant aromatic polyamide fiber, Nomex, which is a poly(*m*-phenylene isophthalamide) [3–7]. Later Teijin produced a similar commercial fiber under the trade name Conex [8]. In 1965 Kwolek [9–11], working with Morgan, discovered that synthetic aromatic polyamides with para-oriented aromatic groups, such as polybenzamide (PBA), formed liquid crystalline solutions and that relatively strong, high modulus fibers could be spun from these solutions. This work led to the production of high performance aramid fibers, which were made available to selected companies and government agencies in 1969. The first series of these para-phenylene high modulus aramid fibers were based on PBA [5,7], but later versions introduced commercially in 1972 under the trade name Kevlar are generally considered to be comprised of poly(*p*-phenylene terephthalamide) (PPTA) [7,12–19]. An aramid fiber similar to Kevlar is manufactured by Akzo N.V. under the trade name Twaron.

In the late 1960s and early 1970s, Preston and coworkers [20–22] at Monsanto developed a high modulus aromatic polyamide–hydrazide, X-500, which is not strictly an aramid; this fiber was never fully commercialized. Also, Jaffe and coworkers [23] of Celanese conducted considerable research on aramid fibers but, again, none were commercialized. In more recent developments, Teijin announced the planned commercialization of a high performance aramid fiber, HM-50, in 1987. This fiber is reported to be a copolyterephthalamide [24]. Also, since the early 1980s a polyamidobenzimidazole fiber has been produced by the USSR [25].

Table 1 presents the chemical structures and trade names of aromatic polyamides that were developed for commercial fiber production.

At present, and into the foreseeable future, the principal aramid fiber for composite reinforcement for high performance composites will be Kevlar. The use of Kevlar aramid fibers as a reinforcement in composites has grown dramatically over the past 5 years, with aircraft and aerospace representing 50% of the applications.

Kevlar aramid fiber is available in four forms: Kevlar, Kevlar 29, Kevlar 49, and Kevlar 149. Kevlar is manufactured specifically for reinforcement of elastomers for tires, hoses, and belts. Kevlar 29 is used in cables, ropes, coated fabrics, tapes, webbings, and ballistics. Kevlar 49, which has a strength similar to Kevlar 29 but 50% higher modulus, is principally used as a reinforcement for polymeric organic matrix composites used in aircraft, aerospace, marine, sporting goods, and electrical applications. Recently, Kevlar 149 has become commercially available, offering a 40% higher modulus and 70% lower moisture absorption than Kevlar 49.

As summarized in Table 2, the combined attributes of low density, high strength, and stiffness of Kevlar fibers has led to numerous current aerospace composite uses, including exterior structures of aircraft and helicopters, interior aircraft structures, and missile and space applications.

The fiber failure longitudinal splitting mode has led to the unique ballistic resistance of Kevlar composites.

Also, an increasingly important use of Kevlar fibers is in low thermal expansion composite circuit boards, which take advantage of the negative thermal coefficient of expansion of the fibers to produce boards with coefficients of expansion similar to those of the alumina leadless chip carriers. This design minimizes stresses on the surface mount solder joints during thermal cycling.

Fiber Fabrication

Polymerization and Fiber Fabrication

An understanding of the chemical and physical processes involved in the PPTA polymerization and fiber fabrication and the role these processes play in the final fiber structure is necessary for meaningful structure–property correlations. The polymerization and fiber fabrication of PPTA have been described in the patent literature [10,26,27]. It is generally accepted in the scientific literature that Kevlar 29 and 49 fibers are basically PPTA [7,12–19].

PPTA is prepared by reaction (1), a condensation polymerization of the exact stoichiometric ratio of *p*-phenylenediamine with terephthaloyl chloride in *N*-methylpyrrolidone [10,27,28],

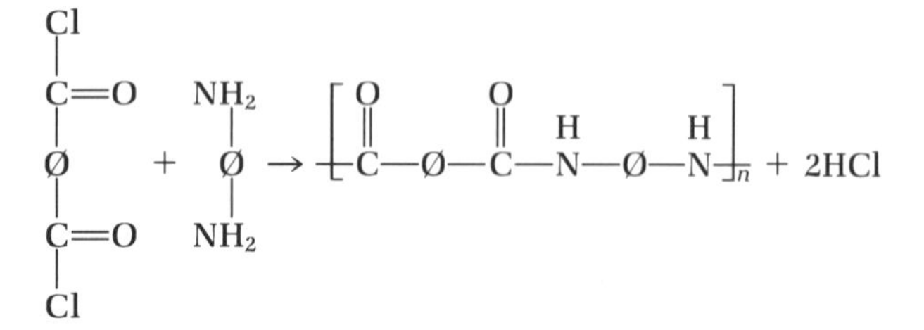

Terephthaloylchloride — *p*-Phenylenediamine — Poly(*p*-phenylene terephthalamide) (PPTA) (1)

and the resulting polymer is washed with dilute NaOH to neutralize the HCl formed during polymerization. At this stage the PPTA macromolecules are rodlike, with a reported molecular weight M_W of 40,000 [29], containing nearly equal amounts of —NH_2 and ionic COO^- Na^+ end groups.

The PPTA fibers are extruded at about 80°C in the form of liquid–crystalline PPTA-H_2SO_4 dopes (~ 20 wt % PPTA) from a spinneret at 0.1–6 m/s. The spinneret face is separated from a 1°C water coagulation bath by an air gap of about 5 mm. The resulting yarn is washed with water, neutralized with NaOH, and subsequently washed again with water to remove the Na_2SO_4. The yarn is then dried under tension on rollers in the 65–130°C temperature range, at which stage it is equivalent to Kevlar 29-type yarns. Further drawing at ~ 550°C for 1–6 seconds produces yarns of higher modulus without significantly enhancing the fiber strength. Such higher modulus yarns can be associated with the higher modulus Kevlar 49-type yarns.

Figure 1 illustrates the PPTA fiber fabrication processes and the corresponding chemical and physical structures that result from these processes. We now consider in more detail the processes involved in each stage of the fiber fabrication process.

PPTA-H_2SO_4 Dopes

PPTA can be dissolved in concentrated sulfuric acid [10,27,28]. However 100% H_2SO_4 is preferred because water in the presence of H_2SO_4 can induce polymer chain scission by hydrolysis [27,30,31].

PPTA-H_2SO_4 dopes, upon flow, can exhibit liquid–crystalline properties. Liquid crystal states are fluid states of intermediate order between the crystalline solid and the isotropic or random liquid states. PPTA-H_2SO_4 solutions are isotropic below 9 wt % PPTA and anisotropic liquid-crystalline dopes above 9.5 wt % PPTA [10,29,32]. The PPTA-H_2SO_4 (20 wt % PPTA) dopes that are used to spin PPTA fibers are solids at room temperature and melt in the 65–70°C range.

Fabrication Processes

Before fiber fabrication, the PPTA-H_2SO_4 (20 wt % PPTA) dope is heated above its crystalline melting point to 80°C. Magat [33] reports that at 100°C the viscosity of PPTA-H_2SO_4 dopes exhibits a minimum as a function of PPTA concentration at about 20 wt %. (This minimum results from two phenomena affecting the viscosity of liquid–crystalline dopes as a function of concentration. The viscosity of liquid–crystalline dopes decreases with increasing concentration as the anisotropic phase forms at the expense of the isotropic phase. However, at higher concentrations, solid polymer formation occurs in the dope, causing a subsequent upturn in the viscosity [34].) Hence, one reason PPTA-H_2SO_4 (20 wt % PPTA) dopes are chosen for fiber fabrication at 80°C is their inherent low viscosity.

When PPTA macromolecules are exposed to 100% H_2SO_4 during the polymer dope dissolution process and the fiber spinning operation, sulfonation of the PPTA phenyl groups can occur, producing impurities in the form of benzene sulfonic acid side groups (Fig. 1). The sulfonating power of the H_2SO_4 is a maximum during the polymer dope dissolution process. The addition of water in the later stages of fiber fabrication will dramatically decrease the potential for sulfonation of the PPTA macromolecules [35,36].

The dopes are extruded from the spinneret orifice at speeds in the range of 0.1–6 m/s [27,28]. Extensional stresses can be imposed on the liquid dope as it enters the spinneret die. Such stresses will depend on the die geometry. Despite the low viscosity of the dope, shear stresses will be imposed on the portion of the dope that is adjacent to the wall in the spinneret capillary, which will produce a fiber whose exterior is more highly oriented. In the air gap after the dope exits the spinneret orifice and before its solidification in the 1°C coagulation bath, it will be exposed to extensional forces in the liquid dope state.

The air gap between the face of the spinneret and the coagulation bath in the dry-jet/wet-spin process is critical to the production of fibers with good mechanical and structural integrity for a number of reasons.

1. This gap eliminates the possibility of H_2SO_4-H_2O attack on the steel spinneret.

2. It allows fibers to be spun from high dope concentrations because the dope temperature can be higher than the coagulation bath temperature, hence above the dope solidification temperature. Higher dope concentra-

TABLE 1
Aromatic Polyamides Developed for Commercial Fiber Production

Chemical Name	Chemical Structure	Trade Name
Poly(*m*-phenylene isophthalamide) (PmPI)	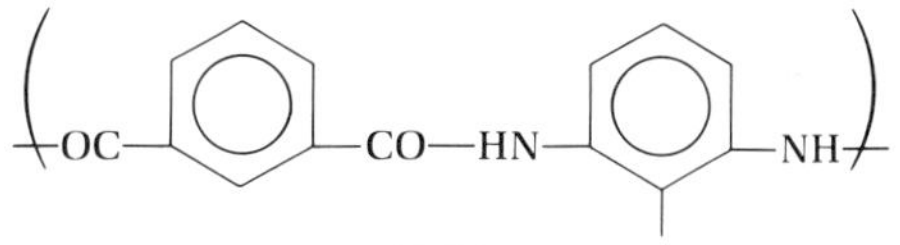	Nomex (du Pont); Conex (Teijin)
Polybenzamide (PBA)		PRD 49-1[a] (du Pont)
Poly(*p*-phenylene terephthalamide) (PPTA)	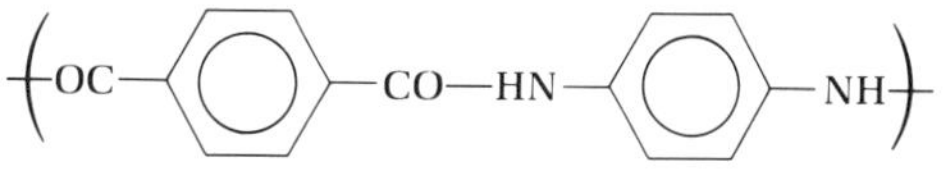	Kevlar (du Pont); Twaron (Akzo N.V.)
Polyterephthaloyl-*p*-aminobenzhydrazide (PABH-T)	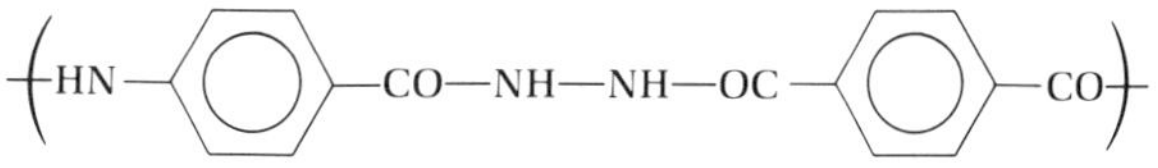	X-500[a] (Monsanto)
Copolyterephthalamide of *p*-phenylenediamine and 3,4′-diaminodiphenyl ether (CPTA)	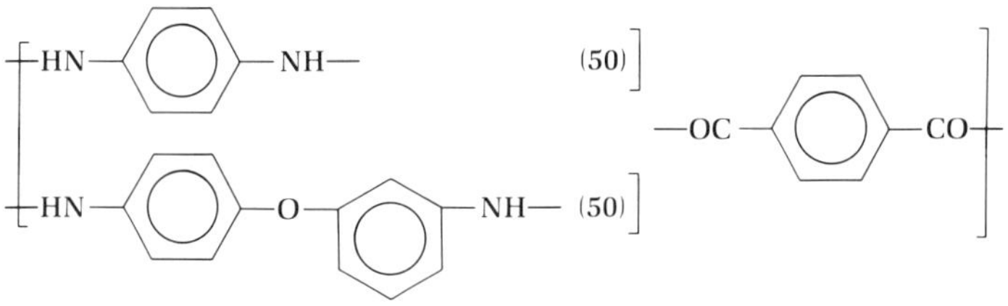	HM-50, Technora[a] (Teijin)
Polyamidobenzimidazole (PABI)	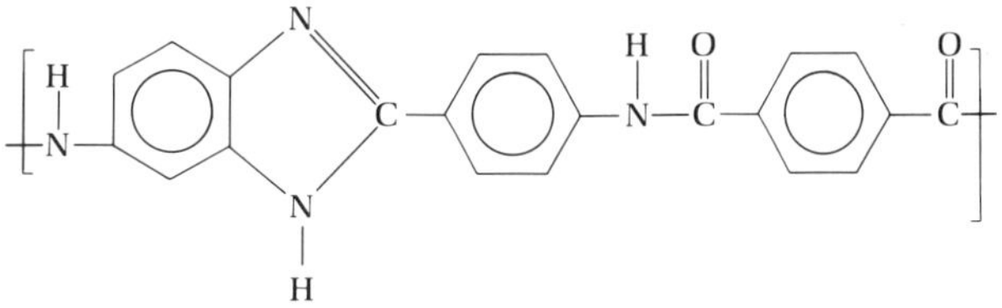	FVM (USSR)

[a] No longer commercially produced.

TABLE 2
Aerospace Composite Uses for Kevlar

Aircraft: Exterior

Wing-to-body fairings
Landing gear doors
Leading and trailing edges of wings and control panels
Access panels
Radomes
Engine nacelles, cowlings, and pylons
Helicopter blades
Propellers

Aircraft: Interior

Window reveals
Overhead and side panels
Cargo liner panels
Floors
Partitions, lavatories, galleys, and bulkheads
Pressure bottles for escape slides
Air ducting
Passenger seat pedestals

Missiles and Space

Filament-wound rocket engine cases
Pressure bottles
Launch tube reinforcement
Air ducting

From Ref. 26.

tions produce fibers with fewer flaws because there is less acid to remove from the solidified dope.

3. Most important, the air gap also allows the imposition of additional extensional stresses on the fibrous dope before solidification. The ratio of the speed of the fiber as it exits the coagulation bath to that when it exits the spinneret capillary (spin stretch factor) [27] determines the extensional stresses imposed on the fibrous dope. The ability to apply these stresses just before dope solidification is critical in producing high orientation in the fiber.

4. The air gap also minimizes die-swell disorientation effects before solidification.

The optimum length of about 5 mm for the air gap is determined by two conflicting requirements:

1. The coagulation bath must be sufficiently removed from the spinneret face to minimize contact between the coagulant and the spinneret orifices resulting from turbulence in the bath. Any contact would cause blockage of the spinneret orifices.

2. However, the longer the air gap, the more likely it is that surface tension effects will transform the cylindrical dope into spherical droplets.

On entering the coagulation bath, the rate of cooling of the fibrous dope determines the final fiber morphol-

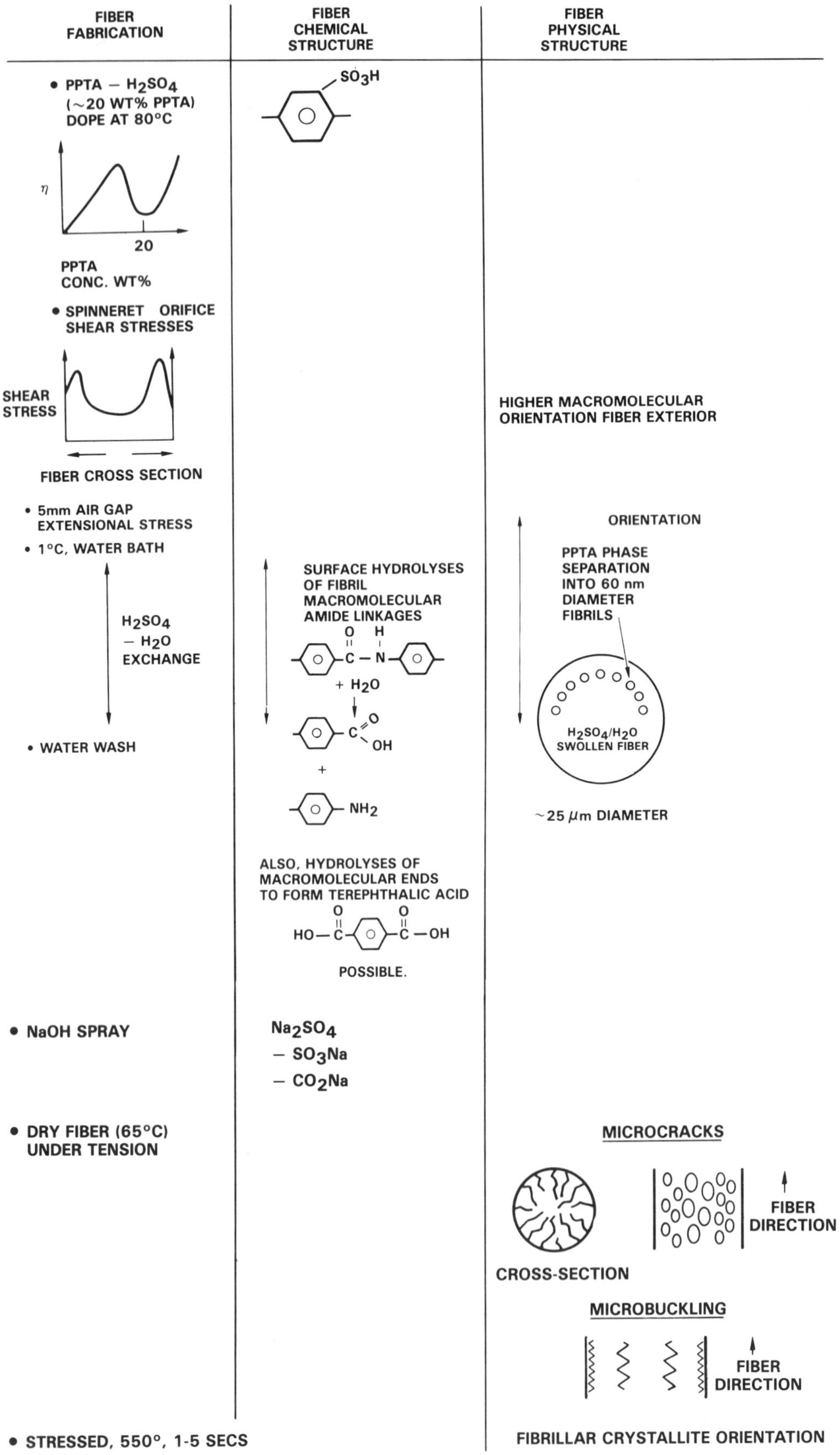

FIGURE 1 PPTA fiber fabrication processes and the corresponding chemical and physical structures that result from these processes.

ogy. It is during this cool-down period that the dope has the opportunity to crystallize under stress. The maximum rate of crystallization of PPTA-H_2SO_4 (20 wt % PPTA) should occur near its melting point at 68°C.

The rate of cooling of the fibrous dope in the coagulation bath will be nonuniform, with the exterior of the fiber cooling at a significantly faster rate than the interior, which could result in a less crystalline fiber skin. The dope will, most likely, crystallize before the diffusion of H_2O into the dope. Upon crystallization, the PPTA macromolecules crystallize as a separate phase and exclude H_2SO_4. A PPTA-H_2SO_4 dope that contains 20 wt % PPTA is at a polymer concentration just below the concentration at which solid polymer precipitation occurs at 80°C, and the effect of lowering the temperature under stress should favor PPTA phase separation. This phase separation should facilitate subsequent removal of the H_2SO_4 from the fiber.

From structural studies of hydrolytically degraded Kevlar 49, there is evidence that the PPTA macromolecules phase separate into fibrils of 60 nm diameter. (Upon this separation, the PPTA fibrillar crystallite boundaries are spaced 120 nm apart in a continuous H_2SO_4 media.) These fibrils are clearly evident in the fracture topographies of Kevlar 49 fibers that have been swollen in superheated steam and hydrolytically degraded [31], as illustrated in Figure 2. The fibril diameter is observed to be constant in fibers that originated from a variety of Kevlar 49 spools. Careful control of the PPTA-H_2SO_4 dope concentration and initial temperature in the spinneret, the air gap length, and the water bath temperature would produce PPTA phase-separated fibrils of similar diameters.

After crystallization of the PPTA-H_2SO_4 dope, subsequent fiber fabrication also plays a critical role in the final fiber structure. A large portion of the H_2SO_4 is washed out from the swollen fiber by the water in the coagulation bath and then by water and/or dilute NaOH sprays [27,28]. During the period when there is an acidic water mixture in the swollen fiber, acid-catalyzed hydrolyses of the PPTA macromolecular amide linkages can occur at the PPTA fibril surfaces (Fig. 1).

A dilute 1% NaOH spray is used to neutralize the H_2SO_4 that resides between the PPTA fibrils, resulting in the formation of Na_2SO_4. The copious quantities of Na_2SO_4 formed upon neutralization are washed from the fiber by water. The PPTA COOH macromolecular end groups and SO_3H side groups will also be neutralized by NaOH to form the corresponding Na salts if the NaOH can penetrate the polymer fibrils and the free volume in the local molecular environment of the acid groups is sufficient to sterically allow the neutralization reaction to occur. In this regard, the spacing in PPTA crystals based on the reported unit cell lattice constants [7] suggests that H_2O and NaOH molecules could penetrate the PPTA fibrillar crystals but sterically prohibits penetration or mobility of Na_2SO_4 or H_2SO_4 within these crystals.

After the neutralization–washing process, the fiber still contains significant quantities of water, which is removed from the fiber by slowly drying the yarn under slight stress in the 65–130°C range. To produce a higher modulus PPTA fiber, such as Kevlar 49, the still partially moist fiber is oriented at ~ 550°C for 1–6 seconds.

The heat treatment at ~ 550°C under stress to produce higher modulus fibers brings about more perfect molecular and crystallite alignment.

The distribution of any residual Na_2SO_4 impurities throughout the fiber and the fiber microvoid characteristics will intimately depend on the dynamics of fiber drying. The diameter of the H_2SO_4 swollen fiber is reduced by half upon final fiber drying and "solidification." This shrinkage causes skin–core radial and hoop stresses and longitudinal kinking of the fibrillar crystallites under compression forces.

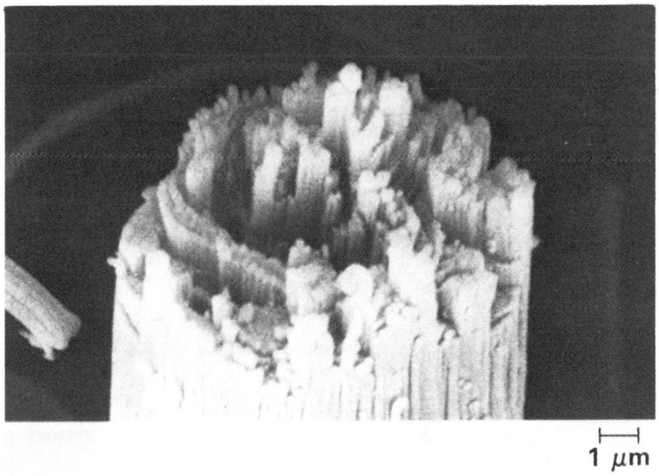

FIGURE 2 Scanning electron micrograph of hydrolytically degraded Kevlar 49 fiber exhibiting a 60 nm diameter fibrillar morphology.

Chemical Structure

Analyses

The chemical structure of poly(*p*-phenylene terephthalamide) (PPTA) was illustrated in Table 1. Carbon, nitrogen, and hydrogen values for Kevlar 49 samples dried at 100°C before introduction into the analyzer and samples dried at 100°C inside the analyzer are compared to the theoretical values for pure PPTA in Table 3 [37]. The maximum C value of 69.39 wt % obtained when specimens

TABLE 3
Carbon, Nitrogen, and Hydrogen Analyses (wt %) of Kevlar 49 Fibers

Element	Theoretical	Dried Outside Analyzer	Dried Inside Analyzer
C	70.59	68.71	69.39
N	11.77	11.51	11.66
H	4.20	4.53	4.29

TABLE 4
Ash, Na, SO_4^{2-}, and S in Kevlar 49 Fibers

Impurity	Average (ppm)	Range (ppm)
Ash	10,950	6,200–15,700
Na	3,750	2,000– 5,100
SO_4^{2-}	5,050	800– 8,650
S	4,050	3,600– 4,600

were dried inside the analyzer indicates that Kevlar 49 fibers contain 17,000 ppm impurities.

From chemical analyses, the average values of the ash, Na, SO_4^{2-}, and S, in parts per million by weight (ppm) and the range of these values observed in 25 Kevlar 49 spools [37–39] are documented in Table 4.

Other minor impurities of Al, Fe, Si, Cl, Ca, and K have been detected in Kevlar 49 fibers in concentrations in the 10–100 ppm range. If present as their oxides, the total concentration of these minor impurities is ~ 350 ppm.

The Na and SO_4^{2-} impurities, in the form of Na_2SO_4, can be partially leached out of Kevlar 49 fibers upon immersion in water [37]. At 23°C, after 6 hours immersion in water ~ 15 wt % of the fiber impurities can be leached out, with little additional loss of impurities occurring after 15 days of further water immersion. However, more than 50% of the impurities can be leached out upon exposure to boiling water for 30 days.

Impurity Model

A model of the characteristics of the impurities in Kevlar 49 fibers has been proposed [37]. This model assumes that (a) complete PPTA polymer phase separation occurs during the fiber spinning operation and all H_2SO_4 is completely excluded from the fibrillar crystallites, (b) all measured SO_4^{2-} is in the form of Na_2SO_4, (c) S in excess of that associated with SO_4^{2-} is in the form of —SO_3H PPTA side groups, and (d) Na in excess of that associated with Na_2SO_4 is in the form of —COO^-Na^+ PPTA end groups or —$SO_3^-Na^+$ side groups. Based on these assumptions, together with the average values of the impurity analyses data in Table 4, the nature of the chemical impurities in Kevlar 49 fibers is documented in Table 5.

Fifty percent of the impurities in Kevlar 49 fibers are in the form of Na_2SO_4, which is situated between the fibrillar, 60 nm diameter crystallites. If this Na_2SO_4 were uniformly distributed on the crystalline surfaces, a Na_2SO_4 molecule would be spaced every 45 Å along each surface PPTA macromolecule (Fig. 3). The Na_2SO_4 molecules are more likely to exist in the form of small clusters that get trapped upon shrinkage during the fiber drying stage at molecular "log-jams" in the preferential diffusion paths that develop within the three-dimensional aggregating fibrillar crystallite array. Such Na_2SO_4 clusters will be a source of microvoid formation, particularly in the later stages of drying when trapped aqueous Na_2SO_4 concentrations will impose osmotic pressures upon their surroundings, resulting in spherical defects.

TABLE 5
Chemical Impurities in Kevlar 49 Fibers

Chemical	Concentration (ppm)
Na_2SO_4	7,450
—$SO_3^-Na^+$	4,550
—SO_3H	2,400
—Na^+(COO^-Na^+ end groups)	350
Metallic oxides	350
Total	15,100

The chemical impurity analyses in Table 5 indicate there are three —SO_3H groups per PPTA macromolecule, which together with the single —COOH macromolecular end group result in four acid groups per PPTA macromolecule (Fig. 3). The Na present in the fiber in excess of that associated with Na_2SO_4 is only sufficient for 60% of the acid groups to be in the form of their Na salts —$SO_3^-Na^+$ and —COO^-Na^+. (It is assumed that there is equal probability of the —COOH and —SO_3H acid groups being neutralized by NaOH.) There are 1.6 acid groups per PPTA macromolecule that are not neutralized.

The characteristics of the impurities in Kevlar 49 fibers may be more complex than is implied by the impurity model. It is possible that H_2SO_4 is (a) not completely excluded from the PPTA crystals upon phase separation, particularly that in the form of $NH_3^+HSO_4^-$ macromolecular end groups, and (b) not neutralized or completely washed from the fiber. Any mobile residual H_2SO_4, specifically that residing in the interfibrillar regions, will accelerate hydrolytic and oxidative degradation of PPTA. Penn and Larsen [39] have detected traces of H_2SO_4 in Kevlar 49 fibers by mass spectrometry. Acid-catalyzed oxidative degradation of Kevlar 49 yarns is used to iden-

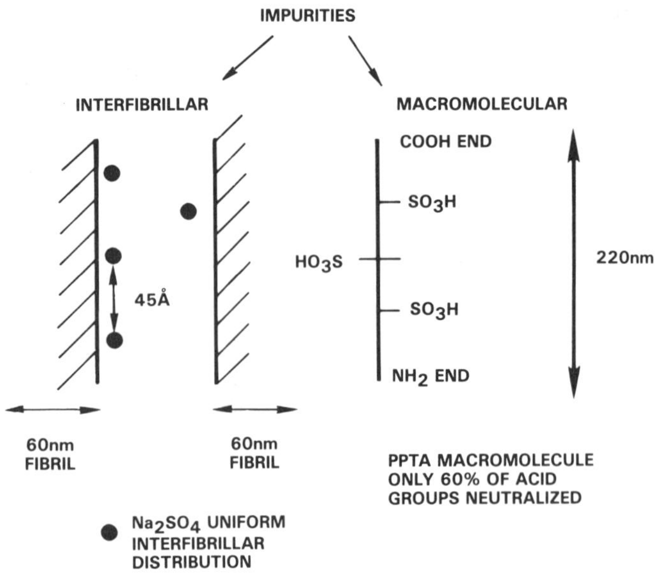

FIGURE 3 Impurities in Kevlar 49 fibers.

tify yarns with "unacceptable" acid levels. In this procedure bare Kevlar 49 yarns are placed in a calibrated air-convection oven for 3 hours at 240°C and their residual strengths are determined at 23°C as twisted bare yarns [40]. A loss in strength exceeding 10% is considered to indicate the presence of unacceptable impurity concentrations [41].

Surface Chemistry

The surface chemical composition and structure of polyaramid filaments has received little attention compared with their graphite and glass counterparts. Two studies of the surface chemical composition of Kevlar filaments have been conducted with X-ray photoelectron spectroscopy (XPS) [39,42]. Those studies show that the surface chemistry of commercial PPTA filaments differs substantially from that of the bulk as a result of finish, fiber surface modifications, and/or manufacturing conditions. The surface consists of an oxidized hydrocarbon with approximately twice the oxygen content and half the nitrogen content of the bulk. Table 6 gives chemical compositions, normalized to 100% C, O, and N, of polyaramid filaments [39,42]. Table 6 also gives the composition of a PPTA model compound, 1,4-bis(*N,N'*-benzamido)benzene. Such models often are used in XPS analyses to establish reliability of composition and line shapes for elements in known bonding environments. The aramid model reflects the typical accuracy seen in composition analysis with XPS (i.e., within a few percent of absolute values).

As seen in Table 6, Kevlar 49 exhibits a higher surface oxygen content than Kevlar 29. This oxide overlayer is unchanged by refluxing in organic solvents, but heating at 350°C does remove a portion of it.

The chemical structure of the Kevlar 49 surface oxide has been examined by XPS line shape analysis, which indicates that the oxide primarily consists of hydroxyl and ether linkages, with smaller amounts of carbonyl and carboxyl groups [42,43]. This oxide layer, whose thickness is on the order of 20 Å, plays a major role in the wetting and interfacial bonding of PPTA filaments with matrix resins.

TABLE 6
XPS Surface Composition of Aramids

Aramid	Composition[a] C (%)	O (%)	N (%)
PPTA (theoretical)	78	11	11
Kevlar 29 yarn	70	18	12
Kevlar 49 yarn	68	25	7
Kevlar 49 fabric	73	22	5
1,4-Bis(*N,N'*-benzamido)benzene			
Theoretical	83	8	8
Measured	80	10	10

[a] Normalized to 100% C, O, and N.

Physical Structure

Macroscopic Features

Kevlar 49 filaments are circular in cross section with a smooth surface that exhibits little texture except for isolated scuffed fibrillar regions caused during fiber fabrication and handling by, for example, rough areas on guide rolls. Kevlar 49 has a density of 1.45 g/cc [1,38,44] compared to a theoretical density of 1.50 g/cc [7] based on crystal unit cell dimensions. The fiber diameter of Kevlar 49 is nominally 12 μm [40,45], but Dobb and Robson [45] report fluctuations of 20% in fiber diameter along the fiber length. Such variations can cause a high degree of scatter in surface- or cross-section-dependent measurements such as wetting or tensile tests.

Morphology

Since the introduction of Kevlar fibers more than 15 years ago, there has been a considerable effort to elucidate their physical structure. The fibers consist of extended-chain, rodlike macromolecules [7] aligned in the fiber direction. Hydrogen-bonded sheets of these macromolecules (Fig. 4) exist in the lateral fiber direction [46]. Both X-ray and electron diffraction studies indicate that Kevlar fibers consist of crystallites with average dimensions of approximately 7 nm lateral to the fiber direction [15,16,46–50]. From birefringence and electron diffraction studies, Ballou [49] reported that the lateral order of the crystalline relative to the phenyl group plane can vary from radial to tangential to completely random packing in any given fiber. Dobb et al. [16] suggested from electron diffraction and electron microscope dark-field image techniques that Kevlar 49 fibers consist of a system of H-bonded sheets, regularly pleated along their axes and arranged radially (Fig. 5). The angle between adjacent components of the pleat is 170°. These workers state that this supermolecular structure leads to a periodic banding along the fiber axis with periodicities of approximately 250 and 500 nm.

Since the early physical structure investigations, evidence has accumulated that Kevlar 49 fibers have a different physical structure in the fiber skin relative to the

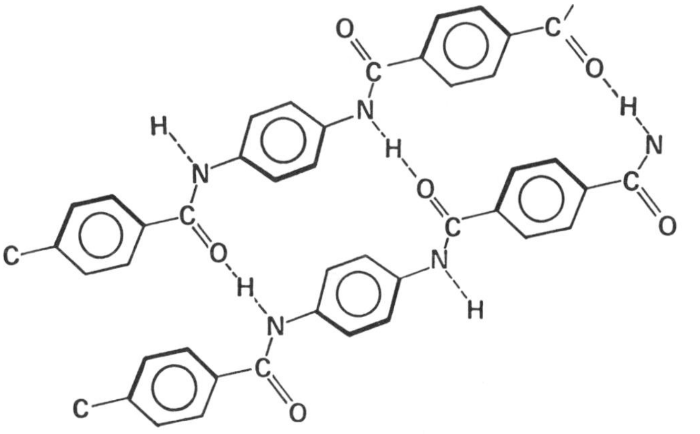

FIGURE 4 Hydrogen bonding in Kevlar 49.

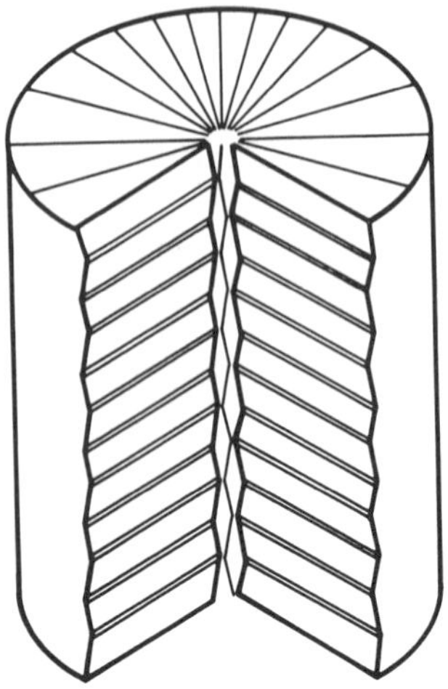

FIGURE 5 Pleated hydrogen-bond sheets in Kevlar 49. (From Ref. 16.)

core. Fracture topography studies of Kevlar 49 epoxy composite strands and both bare yarns and single filaments of Kevlar 49 that failed in tension revealed that the fibers contain a skin that can tear off the fiber in the longitudinal fiber direction to form a continuous ribbon, which often coils into a helix (Fig. 6) [14,51]. Fracture topographies of Kevlar 49 fibers that were etched with HCl and subsequently fractured, also, indicate a difference in physical structure between the fiber core and skin [51]. The scanning electron micrographs in Figure 7 illustrate that the fiber core fails in a number of parallel planes perpendicular to the fiber axis, whereas the skin exhibits a smaller sized, micalike topography as indicated by the arrow in Figure 7*a*. Often the skin retains its structural integrity in the longitudinal fiber direction, suggesting the more continuous nature of the fiber skin relative to the core in this direction. In the skin, the planes of the micalike topographies have areas of ~ 0.1 μm^2. The orientations of these planes are generally perpendicular to the longitudinal fiber axis and parallel to neighboring planes. The thickness of the fiber exterior that exhibits the micalike fracture topography varies considerably from one failure surface to another and is in the 0.1–0.6 μm range [51]. X-Ray diffraction and electron microscopy studies of AgS-impregnated fibers have

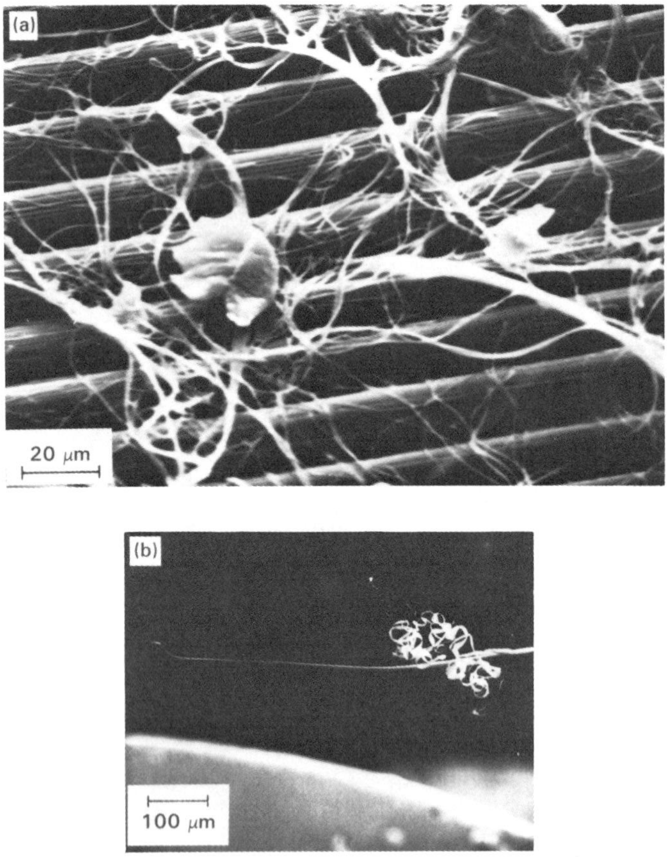

FIGURE 6 Scanning electron micrographs of torn skin of Kevlar 49 fiber in the form of a helical ribbon.

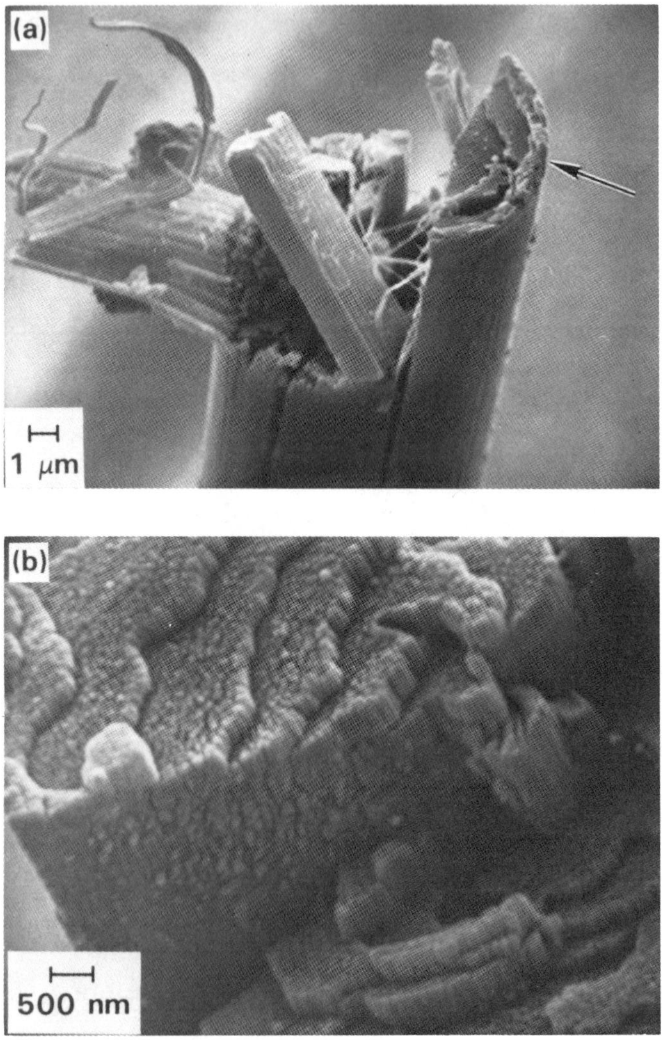

FIGURE 7 Scanning electron micrographs of the fracture topographies of HCl-etched Kevlar 49 fibers; arrow indicates skin topography.

revealed variabilities in the "skin" dimensions both in the fiber cross section and along the fiber axis. Stress/optical microscopy studies of the bare Kevlar 49 fibers also show that the fiber skin and core deform by different modes (see the section of this article on deformation and failure modes), which further supports the suggestion that there is a different physical structure between the skin and core [32,52,53]. Panar et al. [54] observed a skin–core structure of Kevlar fibers from polarized optical microscopy, and scanning electron microscopy of heavily etched fibers.

Factors that occur during fiber fabrication that may contribute to a skin–core morphology include (a) the higher orientations imposed on the fiber exterior as a result of the shear stresses imposed on that portion of the PPTA-H_2SO_4 dope adjacent to the spinneret orifice walls, (b) shrinkage-induced skin–core radial and hoop stresses and longitudinal kinking of the fibrillar crystallites that occur during fiber drying, being more severe in the fiber exterior, and (c) the faster rate of cooling of the fiber exterior, which decreases the three-dimensional crystalline packing efficiency, but not the molecular orientation.

In the fiber core, the crystallites are less aligned than in the skin. Periodicities in the fiber core in the 200–1000 nm range have been observed along the fiber axis by electron diffraction [55], electron and optical microscopy [14–16,30,51,56,57], and chemical degradation studies [55,56,58]. Such periodicities have been associated with the H-bonded pleat structure [14–16,30,49,56,57] or with planes of high concentrations of molecular defects such as chain ends or conformational defects [51,59]. Shorter periodicities (~ 20–40 nm) also have been reported from small-angle X-ray scattering, electron microscopy, and chemical degradation studies [55,56,58]. The latter periodicities have been associated with defect layers whose characteristics have been described only vaguely.

The fracture topographies in Figure 7*b* of the fiber core of an HCl-etched Kevlar 49 fiber clearly show areas of ~ 1 μm^2 oriented perpendicular to the longitudinal fiber axis and interconnected by perpendicular steps of ~ 200 nm or multiples thereof. These observations indicate that the fiber core contains periodic transverse weak planes that occur roughly every 200 nm along the fiber axis. This spacing (~ 200 nm) is equivalent to the reported average molecular length of a PPTA macromolecule, which led to the suggestion [51] that these planes may be inherently weak because of a clustering of the chain ends within their vicinity. The absence of such transverse weak planes in the fiber skin indicates that it contains a more random chain-end distribution, as a result of exposure to higher shear stresses.

Slutsker [59] has found that the periodic "defect" zones in Kevlar 49 fibers are not detectable by small-angle X-ray scattering, which indicates the absence of a significant concentration of microvoids in these regions. However, Slutsker [59] did show that these zones exhibit a lower modulus than that of Kevlar 49 crystallites and that this modulus increases with increasing strain rate and lower test temperature. These observations further suggest that the periodic zones consist of kinked, poorly packed PPTA macromolecules, which would allow preferential diffusion of small molecules such as water through such zones [51]. When interfibrillar interactions are destroyed upon swelling in superheated steam, well-defined transverse fracture planes are not observed [37]. Also, the kinks within these transverse planes can be reoriented at high stress levels.

The transverse defect planes most likely originate from the ionic clustering of $NH_3^+HSO_4^-$ PPTA macromolecular chain ends at PPTA-H_2SO_4 liquid crystal domain surfaces. These charged domain surfaces would energetically align in planes in the fiber cross section [51], which would be subsequent weak planes for microbuckling upon relaxation of the macromolecular array after exit from the spinneret orifice.

Microvoid Characteristics

Small-angle X-ray scattering studies [60] indicate that Kevlar 29 fibers contain 0.8% of 10 nm diameter spherical microvoids. This concentration of microvoids is equivalent to 10 microvoids per 250 nm length of each 60 nm diameter PPTA fibrillar crystal. The presence of spherical microvoids in a highly anisotropic structure indicates that their formation is governed by osmotic pressure effects rather than by structural effects [37]. Upon orientation of Kevlar 29 fibers to form Kevlar 49 fibers, the anisotropic structural changes cause the microvoids to change shape to 10 × 5 × 5 nm ellipsoids aligned along the fiber axis. Treating Kevlar 49 fibers with superheated steam [31] causes the ellipsoidal microvoids to revert to their original spherical shape. Absorbed moisture within a Kevlar 49 fiber will preferentially accumulate at hygroscopic Na_2SO_4 concentrations. Superheated steam exposure clearly reveals the 60 nm diameter crystallites in fractured fibers (Fig. 2), which suggests that H_2O preferentially accumulates in the interfibrillar regions in which the Na_2SO_4 clusters and their associated microvoids reside and forces the crystallites apart. It is assumed these small microvoids preferentially reside in the fiber exterior because they can be filled with low molecular weight compounds, such as bisphenol —A diglycidyl ether epoxide monomer under pressure [60].

Kevlar 49 fibers, whose density indicates that they contain 3.4% voids, do contain larger needle-shaped microvoids, up to 10 μm longer than those observed by small-angle X-ray scattering studies [45,61]. These larger voids are a result of fiber handling, sample preparation techniques, and shrinkage stresses, and we do not associate such voids with Na_2SO_4 impurities, although such voids could have propagated initially from Na_2SO_4-induced smaller spherical microvoids.

Deformation and Failure Modes

Deformation

Stress/optical microscopy studies of Kevlar 49 bare yarns and single filaments revealed the fiber skin deformed in tension in the form of periodic helical deformation bands oriented at ≤ 45° to the fiber axis [51,52,62] (Fig. 8).

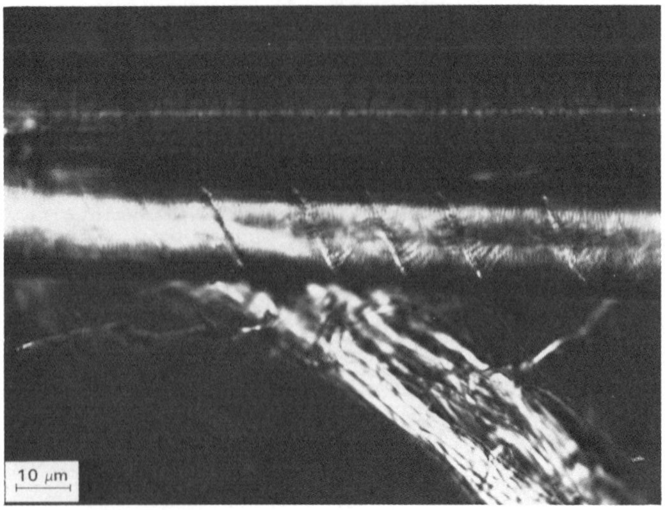

FIGURE 8 **Optical micrograph of stressed Kevlar 49 single filament exhibiting helical deformation bands in the fiber skin and periodic transverse defect planes in the fiber core.**

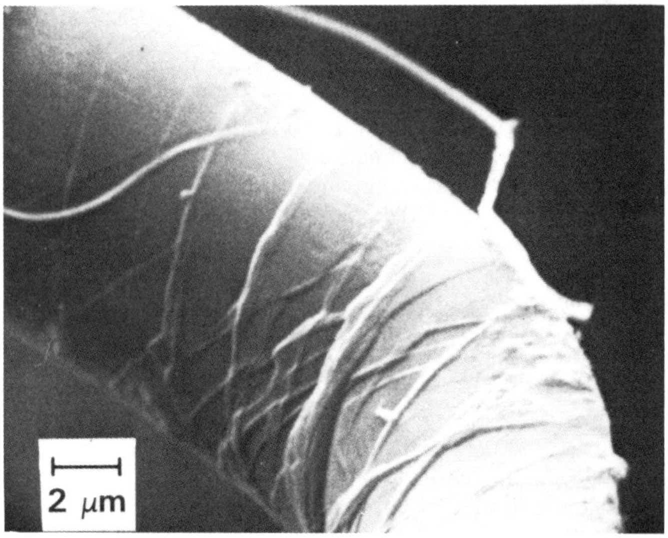

FIGURE 9 **Scanning electron micrograph of compression-induced kink bands in Kevlar 49 fiber.**

The crystallites in the fiber core elastically deform and eventually reorient, causing a permanent modulus increase [63]. This core reorientation process most likely occurs in the periodically spaced, lower modulus transverse fiber "defect" planes. Stress/optical microscopy studies revealed that the spacing of the transverse defect zones in the fibers were in the 600–800 nm range, which suggests that these zones are more deformable than the reported shorter spaced transverse planes. The deformation bands observed in the fiber skin must involve primarily crystallite and macromolecular pullout rather than the reorientation processes that occur in the fiber core, because of the higher skin orientation. The angle at which the helical skin deformation modes are oriented relative to the fiber axis will be determined principally by the spacing of the core deformation zones and the relative deformability of each zone. Both the skin and core tensile deformation modes will result in microvoid formation.

When Kevlar 49 fibers are placed in compression along their longitudinal axis, they readily buckle and form kink bands [64–69] (Fig. 9). Both the weak lateral H-bonding of Kevlar 49 fibers and the presence of residual impurities in the interfibrillar regions contribute to the buckling susceptibility of the fiber. Deteresa et al. [69] have reported an apparent yielding of Kevlar 49 fibers in compression and have shown that previously compressed fibers after deformations up to 3% compressive strain exhibited only a 10% loss in tensile strength. These observations suggest that under compression, cooperative rotational isomeric deformations occur in the PPTA macromolecules, with associated minimal chain scission. Such cooperative rotational isomeric changes would be most likely to occur in the more poorly packed, kinked, transverse core planes within the fiber and/or in the helical skin deformation zones.

Failure Modes

In the fracture topographies of Kevlar 49 composites and also bare yarns, the fibers fail in tension by axially splitting 20 to 50 times their diameter D (20–50D or about 200–500 μm) along their lengths [14,51,53,62,66,70] (Fig. 10). This splitting exhibits no correlation with strain rate within the experimental scatter [70].

Stress/electron paramagnetic resonance studies of Kevlar 49 fibers by Brown et al. [71] revealed no evidence of PPTA macromolecular chain scission until catastrophic failure (Fig. 11). Furthermore, stress/small-angle X-ray scattering experiments [60] revealed the generation of new 10 nm spherical microvoids upon application of

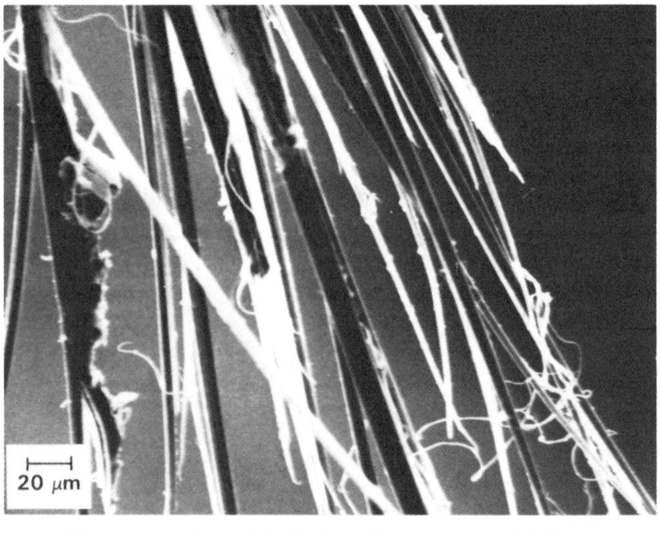

FIGURE 10 **Optical micrograph of fractured split Kevlar 49 fibers that failed in tension.**

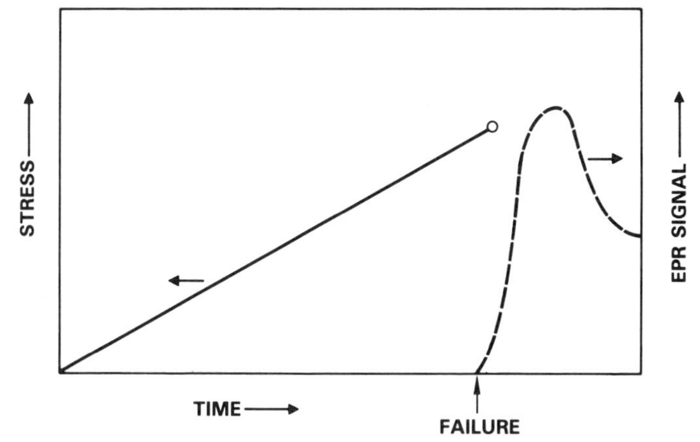

FIGURE 11 **Electron paramagnetic resonance signal as a function of tensile stress for Kevlar 49 fibers.**

stress. The growth in concentration of these microvoids is linear, with increasing stress level up to ~ 70% of ultimate, and at higher stress levels increases at a slower rate (Fig. 12). The apparent rate of microvoid growth decreases at stress levels exceeding 70% because the 10 nm microvoids propagate and coalesce along the interfibrillar regions faster than new ones form. (As these 10 nm microvoids increase in dimensions, they move outside the detectability range of the small-angle X-ray scattering technique.) These observations suggest that, in tension, fiber failure involves shear-induced microvoid growth along the fiber axis, followed by catastrophic, fast macroscopic crack propagation through such microvoids, resulting in axially split failed fibers.

Fractured Kevlar 49 single filaments exhibit considerably less fiber splitting with split lengths of less than *5D*, because of the absence of external shear stresses imposed on these filaments. Single-filament failure modes have been described in terms of skin–core transverse and longitudinal crack propagation paths [52,53]. Single-filament failure modes can involve essentially transverse crack propagation through the fiber skin and core or a more complex combination of skin–core transverse and longitudinal crack propagation. Failure also can be initiated by transverse crack propagation through a portion of the fiber skin, though not generally round the whole circumference, followed by longitudinal crack propagation along the fiber axis, resulting in the skin peeling longitudinally along the fiber axis and separating from the body of the fiber. The presence of the skin deformation bands causes the separated skin to coil up. This skin deformation and separation process has been monitored directly by stress/optical microscopy [52,53]. The three experimentally observed failure paths are summarized in Figure 13. There was no detectable correlation between the strength of a Kevlar 49 filament [72] and its fracture topography features.

Chemically induced PPTA macromolecular chain scission will create additional molecular defects in the fiber that will lower the tensile strength. These additional defects decrease the transverse crack resistance of the fiber, hence modify the failure crack propagation path. For degraded fibers in a composite, catastrophic failure occurs before significant shear-induced microvoid growth along the fiber axis, resulting in progressively shorter split lengths in the fractured fibers as the degree of degradation increases. Studies of hydrolytically degraded

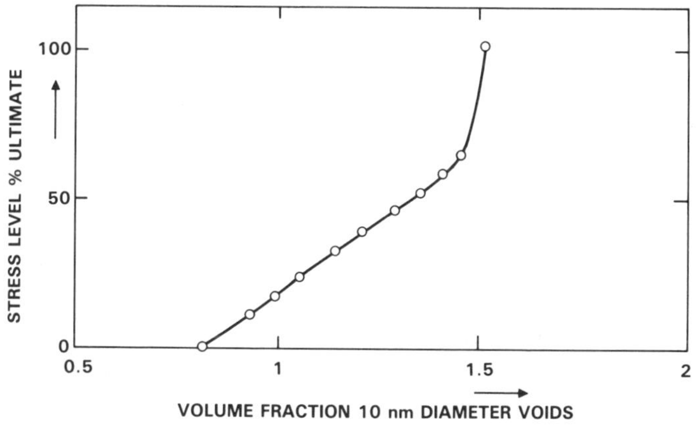

FIGURE 12 **Growth in volume fraction of 10 nm microvoids, determined by small-angle X-ray scattering, in Kevlar 49 fibers as a function of tensile stress level.**

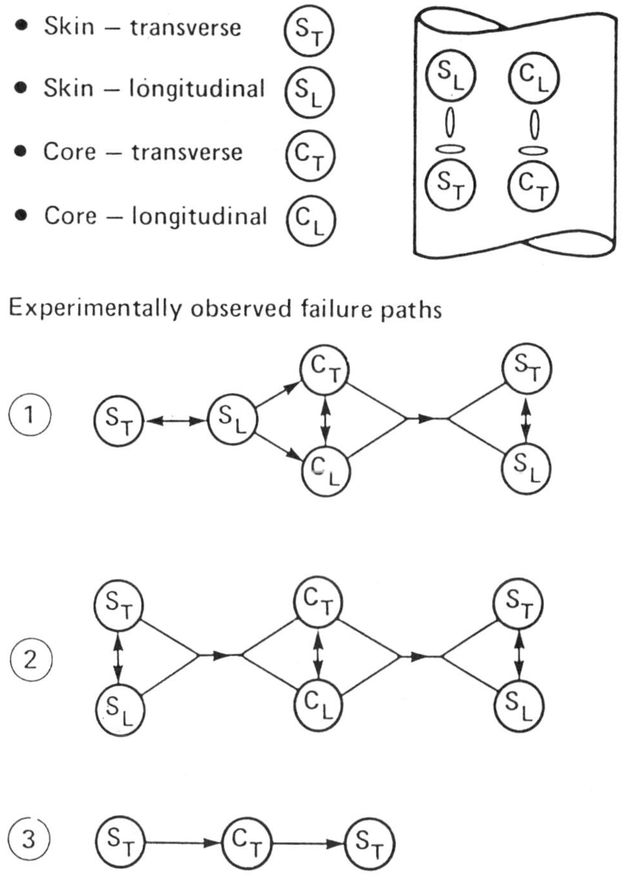

FIGURE 13 **Schematic of Kevlar single-filament failure paths.**

FIGURE 14 **Scanning electron micrograph of the fracture topography of a Kevlar 49–epoxy composite strand with hydrolytically degraded fibers.**

Kevlar 49 fibers [31] indicate that for decreases in tensile strengths exceeding 30%, transverse fiber failures with zero split lengths will occur. Figure 14 illustrates the fracture topography of Kevlar 49–epoxy strands with degraded fibers. From topographical features that consist of transverse fiber failures, it can be determined whether chemically induced fiber degradation is responsible for any premature composite failures in a specific service environment.

Compression of Kevlar 49 composites generally does not result in complete fiber failure; rather, a sequence of buckling and fibrillation takes place. However, after severe buckling, the fibers readily fail in tension with short split lengths as a result of compression-induced PPTA macromolecular chain scission [66].

Mechanical Properties

The mechanical properties of Kevlar 49 yarns at 23°C are documented in Table 7. The tensile property data in this table are based on the most recent, thorough mechanical property study on Kevlar 49 yarns reported to date [40], in which 20 twisted bare yarns and epoxy-impregnated strands were determined for five different Kevlar 49 spools. The average coefficient of variation (CV%) for each spool for the mechanical properties in this study was in the 2–6% range. The stress–strain plots were essentially linear.

Wagner et al. [73] have studied the statistical variability in the strength of single Kevlar 49 filaments. They report that both the failure load and the tenacity of filaments for a given gage length and yarn cross section can be fitted to a two-parameter Weibull distribution. Within a yarn cross section, they observed that filament density (mass/unit length) and diameter varied significantly, which contributed to the variability of filament failure load.

Kevlar 49 yarn exhibits a loss in mechanical properties with increasing temperature; for example, at 200°C there is a 20% loss in tensile modulus and a 25–40% loss in tensile strength relative to the corresponding 23°C mechanical properties [1,74,75]. At low temperatures, Kevlar 49 yarn exhibits essentially no loss in strength, even at −196°C [26].

Electrical Properties

The electrical properties of Kevlar 49 fabric–epoxy composites are presented in Table 8.

Environmental Susceptibility

Chemical Resistance

Except for a few strong acids and alkalies, the chemical resistance of Kevlar 49 fibers is good [1]. For the details of acid and/or alkaline susceptibility of Kevlar 49 fibers, see the section of this article on hydrolyses.

Moisture Sorption

Unlike other common composite reinforcements, Kevlar 49 filaments are hygroscopic and can sorb up to 3.5–4.5 wt % moisture at ambient temperatures in 100% relative humidity [1,39,74,76]. Moisture sorption from humid air occurs rapidly. Equilibrium moisture content is attained in 16–36 hours. Smith [74] and Moss [77] have reported that the equilibrium moisture content in Kevlar 49 yarns is directly proportional to RH at 23°C. For a specific RH, Moss [77] reported spool-to-spool variations in moisture sorption of up to 30%. The moisture sorption isotherms for Kevlar 49 yarn are nearly linear, which is unusual for most hydrophilic polymers [78]. Most polymers exhibit nonlinear isotherms because moisture-induced swelling opens additional binding sites via changes in free volume. No indications of swelling effects were observed in moisture sorption data for Kevlar 49 even after long-term exposures ($>$ 4000 hours) at ambient conditions. (Above 100°C, however, superheated steam exposure does cause permanent fiber structural damage, resulting in fibers that can subsequently sorb more than 10 wt % moisture [31].)

Augl [76] has shown that the moisture diffusion coefficient of Kevlar filaments is concentration dependent and much lower than typical polymers ($\sim 5 \times 10^{-12}$ cm^2/s at 28°C). Thus, in resin matrix composites, the matrix controls moisture diffusion rates, and the Kevlar filaments only contribute to the overall moisture uptake. The moisture diffusion coefficient shows a maximum in concentration dependence near a value that corresponds to 60% relative humidity. Small-angle X-ray scattering studies of dry and wet Kevlar filaments indicate that moisture is absorbed into the system of microvoids within the filament [61]. Moisture most likely diffuses rapidly through a network of microvoids in the fiber interfibrillar regions where hygroscopic Na_2SO_4 impurities reside, although penetration by water molecules into the "molecular" holes in the PPTA crystals is sterically possible [37].

TABLE 7
Mechanical Properties of Kevlar 49 Yarns at 23°C

Mechanical Property	Value	Average CV% for Each Spool	Ref.
Tensile strength			
Twisted bare yarn	3024 MPa	3.0	40
Epoxy-impregnated strands	3414 MPa	5.5	40
Tensile elongation			
Twisted bare yarn	2.48%	3.1	40
Epoxy-impregnated strands	2.62%	5.7	40
Tensile modulus			
Twisted bare yarn	121.9 GPa	1.8	40
Epoxy-impregnated strands	130.3 GPa	2.0	40
Bending modulus	106 GPa		1
Calculated axial compression modulus	76 GPa		1
Theoretical PPTA crystallite tensile modulus	186 GPa		60
Transverse filament compressive modulus	0.76 GPa		

Absorbed moisture has little effect on the tensile properties of Kevlar filaments [39,44] or their composites [79]. There are few data on the effects of moisture on the shear, compression, or radial tensile properties of polyaramid filaments. These properties are difficult to test but are expected to be sensitive to internal defects and moisture content. Indications of such effects are seen in composite behavior. Flexural strength, which is determined by compressive buckling in Kevlar–epoxy, has been shown to be a strong function of moisture content [80]. Moisture may aid the buckling of polymer chains by interrupting hydrogen bonds. Transverse tensile properties of Kevlar–epoxy composites are also a strong function of moisture content [81]. At high moisture contents, the filaments appear to become more radially compliant and tend to crack internally under stress.

Hydrolyses

The amide linkage in the PPTA macromolecule is susceptible to hydrolytically induced chain scission as illustrated in Eq. (2).

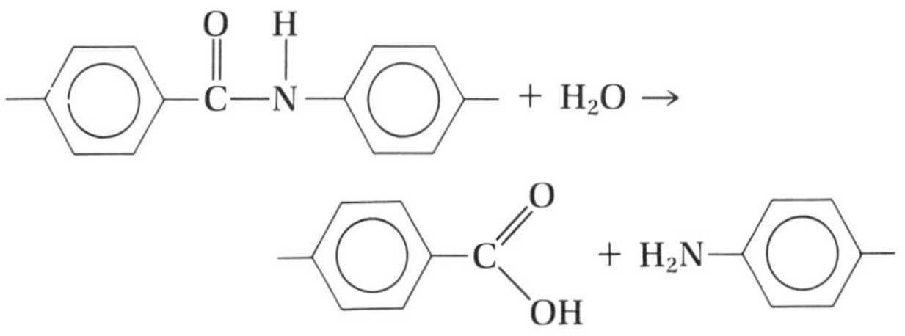

(2)

From the law of mass action, the rate of formation of new macromolecular ends is expressed as

$$\frac{d[NH_2]}{dt} = k[\text{amide}][H_2O] \tag{3}$$

where k is the rate constant. The hydrolysis reaction rate is a constant for fiber strength deterioration because (a) only 4% of the amide linkages have to hydrolyze to produce zero strength [31,82,83] and (b) the H_2O consumed in the reaction will readily be replaced in normal service environment conditions. Smith [74] has reported that the equilibrium moisture content of Kevlar 49 yarns, $[H_2O]$ in Eq. (3), is directly proportional to the RH at 23°C. Hence, from Eq. (3), the rate of hydrolysis will also be directly proportional to RH. The rate of hydrolysis can be accelerated by any H_2SO_4 and NaOH impurities present in the fiber. The H_2SO_4 will act as a true catalyst and will remain chemically unmodified after each hydrolytic reaction, whereas NaOH will accelerate the hydrolytic reaction and then react to form a COO^-Na^+ salt macromolecular end group. From exposing Kevlar 49 yarn to H_2SO_4 and NaOH solutions and monitoring yarn strength degradation, the rates of hydrolysis have been found to be $\sim 10^5$ and $\sim 10^3$ times faster in the presence of 5% H_2SO_4 and NaOH solutions, respectively [84]. These rates were lower at higher acid or alkali concentrations.

TABLE 8
Electrical Properties of Kevlar 49 Fabric (style 120)–Epoxy Composites

Electric Property	Value
Dielectric constant (58% vol % fiber)	
Perpendicular to warp at 9.3×10^9 Hz	3.3
Parallel to warp at 9.3×10^9 Hz	3.7
Loss tangent (58 vol % fiber)	
Perpendicular to warp at 9.3×10^9 Hz	0.010
Parallel to warp at 9.3×10^9 Hz	0.013
Dielectric strength (48 vol % fiber)	24.4 V/mm
Volume resistivity (48 vol % fiber)	5×10^{15} Ω-cm
Surface resistivity (48 vol % fiber)	5×10^{15} Ω
Arc resistance (48 vol % fiber)	125 S

From Ref. 1.

The rate of the hydrolytic degradation has been shown to be related to temperature T by the relationship [31]

$$\ln[d\sigma/dt] = -E/RT + C \tag{4}$$

where E is the activation energy for hydrolytic chain scission of the PPTA amide linkage and C and R are constants. Based on the assumption that applied stress lowers the activation energy for hydrolysis by $B\sigma_{\%}$, where B is a constant and $\sigma_{\%}$ is the applied stress expressed in terms of a percentage of the fiber strength at a specific instant in time t, then the magnitude of fiber strength degradation Δ_t after a time t in a specific environment is related to t, RH, E, $\sigma_{\%}$, and T by

$$\Delta_t = \frac{\mathrm{RH}}{100} Dt \exp - \left(\frac{E - B\sigma_{\%}}{RT}\right) \tag{5}$$

where D is a constant [31].

Hydrolytic degradation of bare Kevlar 49 yarns has been reported as a function of time (Table 9) for exposure conditions of 100% RH in the 65–200°C range. Little spool-to-spool variation was found in the hydrolytic degradation rate at 100°C/100% RH exposure, which indicates that potential H_2SO_4 that could result from fiber fabrication and would catalyze the hydrolytic degradation is essentially absent, immobilized, or of constant concentration in the "as-received" fiber.

TABLE 9
Rates of Hydrolytically Induced Strength Degradation of Kevlar 49 Fiber as a Function of Temperature at 100% RH

Exposure Conditions at 100% RH		Strength Degradation
Temperature (°C)	Time (days)	(%/year)
200	2	12,426
175	4	5,585
175	4	5,466
150	14	1,563
125	32	199
125	40	264
100	162	64
Each Test Performed on a Different Spool		
100	281	77
100	281	77
100	281	76
100	281	78
100	281	86
100	281	83
65	440	3[a]
65	440	1.5[a]
65	440	5[a]
65	440	4[a]

[a] Statistically meaningful data.

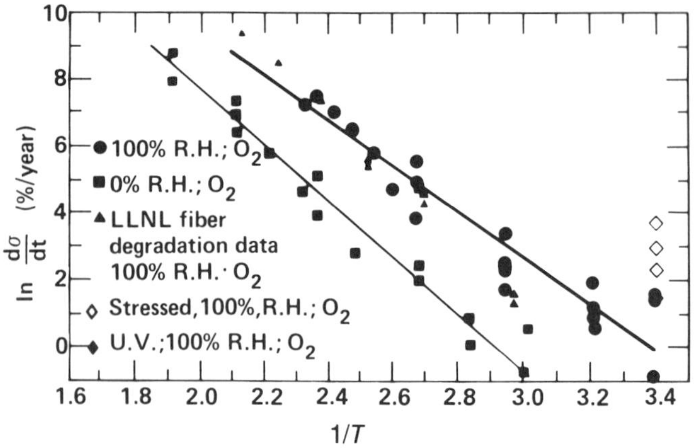

FIGURE 15 **Plot of ln($d\sigma/dt$) versus 1/T for hydrolytic and oxidative degradation of Kevlar fibers.**

Figure 15 plots ln($d\sigma/dt$) versus $1/T$ for this hydrolytic aging data, together with other literature data on hydrolytic and oxidative degradation [74,85–89]. Kevlar 49 fibers degrade about 200 times faster in a 100% RH environment than in a 0% RH environment at the same temperature. The hydrolytic aging data exhibit a discontinuity from linearity at 100°C in the plot of ln($d\sigma/dt$) versus $1/T$ because of the onset of physically induced fiber damage as a result of high steam pressures, which is indicated by higher moisture sorption in such fibers [31]. For estimated rates of fiber degradation in a variety of service environment conditions as determined from extrapolation of the plot of ln($d\sigma/dt$) versus $1/T$ to lower temperatures together with use of Eq. (5), see Table 10. The data indicate that in typical service environment conditions, hydrolytically induced fiber strength degradation is not a serious concern, with rates of degradation of 5% or less a year at 65°C, 100% RH, in the nonstressed state.

Ultraviolet Radiation

Aromatic polyamides, such as Kevlar 49, are susceptible to loss of strength upon exposure to UV radiation [90–

TABLE 10
Estimated Kevlar 49 Strength Degradation as a Result of Hydrolyses

Exposure Temperature (°C)	Imposed Stress Level (%)	Strength Degradation (%/year)
65	0	3.0
	20	4.6
	75	16.3
38	0	0.5
	20	0.8
	75	2.8
23	0	0.2
	20	0.3
	75	1.1

94]. Kevlar 49 absorbs sunlight in the near-UV and part of the visible region in the 300–450 nm range [91]. Only small amounts of this light are present in artificial light sources such as incandescent and fluorescent bulbs, or in sunlight filtered by window glass [91]. Brown and Sandreczki [93] have studied the free radicals that are produced in Kevlar 49 fibers after UV irradiation in vacuum (photodegradative radicals) and in air (photooxidative radicals) by electron paramagnetic resonance (EPR) studies. These UV-induced radical formation mechanisms involve either permanent chain scission or radical recombination to form structural irregularities in the PPTA chains. Brown and Sandreczki [93] also observed that the rate of radical production decreases with irradiation time, which suggests that photolytic products absorb in UV wavelengths, hence act as a light-protective shield. Photooxidative degradation causes a loss of fiber strength more rapid than occurs in the absence of oxygen. In composites, UV-induced fiber aging is not considered serious, because fiber strength deterioration occurs only in the surface layers.

Stress

Fiber failure in tension involves shear-induced microvoid growth along the fiber axis, followed by catastrophic fast macroscopic crack propagation through such microvoids (see the section of this article on deformation and failure modes). Long-time (>5 years) stress–rupture experiments on Kevlar 49–epoxy composite strands revealed no evidence of fiber strength deterioration before catastrophic fiber failure [95]. Hence, the microvoids that grow along the fiber axis do not detectably modify the fiber strength, until catastrophic failure. This "sudden death" failure mechanism suggests that strength retention measurements after service environment exposure are insensitive to stress-induced fiber aging. Techniques that monitor fiber microvoid population changes would be a more direct measure of stress-induced fiber aging.

Thermal

The thermal properties of Kevlar 49 fibers are shown in Table 11.

TABLE 11
Thermal Properties of Kevlar 49 Fibers

Thermal Property	Value
Coefficient of expansion, 10^{-6} cm/cm · °C	
Longitudinal (0–100°C)	−2, −6
Radial (0–100°C)	+59
Specific heat, 23°C (J/g · °C)	1.42
Thermal conductivity at 23°C, J · cm/s · m² · °C	
Parallel to fibers	4.110
Perpendicular to fibers	4.816
Heat of combustion, kJ/g	34.8

From Ref. 1.

The rate of loss of tensile strength of Kevlar fibers as a function of temperature in an oxygen environment was plotted in the $\ln(d\sigma/dt)$ versus $1/T$ plot in Figure 15, above.

Kevlar 49 fibers thermally degrade in vacuum ≥ 375°C with an activation energy of 39.8 ± 3.4 kcal/mol, which was determined from monitoring the evolution of the decomposition products [96]. The principal decomposition products are H_2O, CO, and CO_2. The cessation of evolution of volatile decomposition products occurs after 54 wt % loss [96,97], and the resultant carbonized fiber exhibits relatively good structural integrity. In an oxygen environment, the rate of fiber degradation is faster, ultimately causing a 100% weight loss at 700°C for a heating rate of 10°C/min [97]. The rate of oxidative degradation will vary with the impurity concentrations within the fibers, which catalyze the oxidative degradation reactions.

Composites

Mechanical Properties

The unique chemistry and morphology of Kevlar polyaramid filaments is also manifested in composite behavior. The high thermomechanical anisotropy of the filaments causes a similar diversity in composite thermomechanical properties. In tension, Kevlar composites exhibit one of the highest specific strengths available for structural materials; however, compressive strengths are only one-fifth of tensile strengths, and off-axis strengths are a hundred times lower than in the filament direction. In-plane thermal expansion coefficients near zero can be obtained with Kevlar–epoxy composites because of the negative axial expansion of the filaments. The tendency of the filament rigid-rod morphology to fibrillate and, in so doing, to absorb large quantities of energy, results in high impact resistant composites.

Only the mechanical properties of epoxy impregnated, unidirectionally Kevlar 49 reinforced composites will be considered. The behavior of unidirectional material corresponds to that of a single ply of a more complex layup and is the basic building block of composite structures. The important "off-axis" mechanical properties of Kevlar 49–organic matrix composites are considered in a later section.

The strength of an epoxy matrix impregnated Kevlar 49 strand is affected by many factors—for example, matrix properties, fiber volume fraction, composite fabrication procedure, fiber bundle size, and mechanical property test technique. Both a high matrix modulus and good adhesion to the fiber enhance transfer of stresses, which allows the composite strand to act as a single unit, thus maximizing the failure stress. For tensile-critical applications, a fiber volume fraction of 65–70% is an optimum choice [1]. Some mechanical properties of a Kevlar 49–epoxy unidirectional laminate are shown in Table 12.

The stress–rupture behavior of Kevlar 49–epoxy composite strands has been studied in detail by Chiao et al. [1,99]. In the absence of chemical degradation of the fi-

TABLE 12
Mechanical Properties of Unidirectional Kevlar 49 (1420 denier) Composite of Polyether Triamine, Cured Bisphenol A, and Diglycidyl Ether Epoxy

Mechanical Property	Value		
	Tension	Compression	Shear
Longitudinal strength, MPa	1850 ± 50	235 ± 3	—
Longitudinal ultimate strain, %	2.23 ± 0.06	0.48± 0.3	—
Transverse strength, MPa	7.9 ± 1.1	53 ± 3	—
Transverse ultimate strain, %	0.161 ± 0.023	1.41 ± 0.12	—
Shear stress at 0.2% offset, MPa	—	—	24.4 ± 2.4
Shear strain at 0.2% offset, %	—	—	1.55 ± 0.16
Longitudinal Young's modulus, E_{11}, GPa		81.8 ± 1.5	
Transverse Young's modulus, E_{22}, GPa		5.10 ± 0.10	
Shear modulus, G_{12}, GPa		1.82 ± 0.09	
Major Poisson's ratio, μ^{12}		0.310 ± 0.035	
Minor Poisson's ratio, μ^{21}		0.0193 ± 0.0014	

From Ref. 98.

ber, the median stress–rupture lifetime t_{median} (in hours) is related to absolute temperature T and $\sigma_{\%}$, the applied stress level, as a percentage of fiber strength, by Eq. (6) [100].

$$t_{\text{median}} = 7.9 \times 10^{-11} \exp\left(\frac{18246 - 123\sigma_{\%}}{T}\right) \tag{6}$$

More recent data of these studies, at 23°C, indicate that after 5 years under sustained load, all specimens failed with loads of 72% or more.

Ericksen [101] has studied the creep behavior of unidirectional Kevlar 49–epoxy composites. He found that after 10 hours the composite creep strain varied linearly with log time at all stress levels he tested, and that at long times the creep of the composite is controlled by the creep of the Kevlar 49 fiber. In subsequent studies, Ericksen [63] attributed fiber creep to crystallite reorientation.

Kevlar 49–epoxy unidirectional composites are reported to be superior to S- and E-glass epoxy composites in tension–tension fatigue and, for example, exhibit 10^6 cycles to failure at 23°C and at a stress level of 70% of ultimate, for a ratio of maximum to minimum stress level of 0.1 [26].

Off-Axis Mechanical Properties

Polyaramid composites have low off-axis strengths. The surface treatments and coupling agents developed for glass and graphite reinforcements increase interfacial strengths by 2–3 times over untreated composites [102,103]. Those surface modifications also provide increased hygrothermal stability. Similar improved bonding methods are not readily available for polyaramid-reinforced composite systems. Consequently, interface-sensitive properties are weaker in polyaramid-containing systems than in their glass or graphite counterparts. Representative data for these systems are compared in Table 13.

Transverse tensile strengths of polyaramid–epoxy composites are only 40–50% as strong as those of glass–epoxy or graphite–epoxy composites. Transverse tensile fracture surfaces of polyaramid–epoxy composites are characterized by bare, visually unaltered filaments with little epoxy adhering to them [66,81,104–108]. Thus the interphase rather than the cohesive strength of the filaments or matrix limits the off-axis strength of Kevlar–epoxy composites.

The Kevlar 49–epoxy interphase may also be more sensitive to moisture than the bulk filament and resin. Allred and Lindrose [109] observed that moisture diffuses more rapidly in the filament axial direction. Transverse

TABLE 13
Off-Axis Strengths of Filament-Reinforced Epoxy Composites

Filament	Transverse Tension (MPa)	Interlaminar Shear (MPa)	Ref.
Polyaramid	7–18	21–40	42, 66, 81, 104–108
Glass	17–41	55–97	106
Graphite	26–38	85–110	103, 106

tensile strength is a strong function of composite moisture content [81]. These results indicate that the interphase bonding is weakened by the presence of moisture. The high radial expansion coefficient of the filaments also causes an unfavorable tensile stress state at the interface [110]. When taken together, these mechanical response and environmental sensitivity observations clearly indicate that the performance of Kevlar 49–epoxy composites is limited by weak interfacial adhesion for many applications. However, poor fiber–matrix interface bonding readily allows crack deflection along the interface, which leads to enhanced performance in pressure vessel and ballistic applications.

Filament surface energetics plays a major role in wetting with composite matrix resins and subsequent secondary bond formation. Kevlar surface energetics has been studied by Penn et al. [111,112], and Wesson and Allred [113]. Wesson and Allred [113] showed that the surface energy γ of Kevlar is dominated by dispersion forces with $\gamma^d = 38.0$ mJ/m^2 and the acid/base component $\gamma^{a/b} = 10.5$ mJ/m^2. Liquid epoxy resins exhibit similar surface energetics. For typical bisphenol-A-based epoxies, $\gamma^d = 35.5$ mJ/m^2 and $\gamma^{a/b} = 12.0$ mJ/m^2. Thus, there is a good match of surface energy components between polyaramid filaments and epoxy matrix resins.

The low off-axis properties of polyaramid-reinforced composites have stimulated numerous investigations to improve filament–matrix adhesion. Four types of approach have been taken in these studies: (a) filament surface oxidation or etching, (b) matrix chemical modification, (c) coupling agents, and (d) formation of reactive surface groups on the filaments. The fiber manufacturer tailors the fiber surface for each composite application.

Filament surface oxidation or plasma etching before impregnation improves composite off-axis strengths by approximately 30% [106,114–116]. Gains in off-axis strengths using these methods can be accompanied by large losses in filament tensile strength [114,115]. Strength improvements can be due to removal of the oxide layer from the filament surface (see SURFACE CHEMISTRY, CHEMICAL STRUCTURE) and, perhaps, to improved wetting. When this approach is used on a continuous basis, where the activated surface can be impregnated with a matrix resin before exposure to the atmosphere, 200% improvements in strength are seen [117]. Under these conditions, some covalent bonds appear to form between the polymer and free radicals on the filament surface.

Matrix chemical modification to improve interfacial bonding has not been thoroughly examined for polyaramid–epoxy systems. Chiao and coworkers [98,100] observed that interlaminar shear and transverse tensile strengths are highly dependent on the matrix resins system. It is also known that anhydride-cured resins are more sensitive to adsorbed/absorbed moisture on polyaramid filaments. Increasing matrix toughness by the addition of a dispersed rubber second phase can cause a substantial increase in transverse tensile strength [106].

Surface chemical modification of polyaramid filaments has been studied by various investigators. These approaches have been aimed at improved wetting or providing a filament–matrix coupling. The use of organic sizings designed to function as coupling agents with epoxy resins has been studied by Penn, Bystry, and Marchionni [118]. Although the chemistry of the sizings was selected for compatibility with the polyaramid and the epoxy matrix, none of them resulted in a significant improvement in interfacial bond strength determined by filament pullout or short-beam shear tests. All failures appeared to be interfacial. It was concluded that the bond between Kevlar and epoxy is as good as it can be using only intermolecular forces, and that only chemical, covalent bonds across the interface will strengthen the Kevlar–epoxy bond. If the surface oxide discussed previously (see SURFACE CHEMISTRY) acts as a weak boundary layer, then traditional coupling agents could not function as intended unless it were removed.

A number of investigators have attempted to improve Kevlar–epoxy bonding by creating reactive amine (NH_2) sites on the filament surface. Such sites could react with epoxide groups in the matrix resin and form covalent bonds across the interface.

Hoffman et al. [119] hydrolyzed the amide linkages located near the filament surface with strong acids or bases. The resultant amine groups were reactive with epoxides, but the hydrolysis reaction was accompanied by extensive losses in filament strength.

In what appears to be a more promising approach, Allred, Merrill, and Roylance [42] have incorporated amine sites on polyaramid filaments by exposing them to amine-containing rf glow discharge plasmas. Ammonia plasmas were emphasized in their work. Surface amination does not degrade filament bulk properties. The amine groups are stable in air and water for extended periods of time. When the amines are reacted with epoxy resins in fabricating a composite laminate, the composites exhibit higher interlaminar tensile and peel strengths. Shear strength is also improved [108]. Failure modes are observed to change from interphase dominated to a mixture of filament splitting and matrix cracking as surface amine concentration increases. Moisture absorption occurs by a Fickian bulk diffusion mechanism after amination, whereas absorption occurs by a non-Fickian interfacial wicking mechanism in untreated Kevlar–epoxy. The absorption rate is correspondingly reduced by a factor of 3 after amination.

An extension of the plasma amination approach to a continuous process has produced further improvements in composite properties [117]. Fivefold improvements in peel strength and 250% increases in transverse tensile strength have been observed. Flexural, compressive, and impact properties also exhibit small improvements. These improvements are largely retained in the presence of moisture aging, which demonstrates the value of forming covalent bonds at the interface. The resultant composite properties allow polyaramid-reinforced composites to compete more favorably with glass- and graphite-reinforced composites in structural applications.

In the most recent study, Wu and Tesoro [120] have incorporated amine functional groups into fiber surfaces by bromination followed by ammonolysis and by nitration followed by reduction, without impairment of fiber properties. They report remarkably improved peel strength for the treated fibers.

Roger J. Morgan and Ronald E. Allred

References

1. C. C. Chiao and T. T. Chiao, "Aramid Fibers and Composites," Lawrence Livermore National Laboratory Report UCRL-80400, 1977.
2. P. W. Morgan, *Condensation Polymers: By Interfacial and Solution Methods,* Wiley-Interscience, New York, 1965.
3. G. S. Fielding-Russell, *Textile Res. J., 41,* 861 (1971).
4. D. O. Hummel, H. Siesler, E. Zoschke, I. Vierling, U. Morlock, and T. Stadtlander, *Melliand Textilber., 54,* 1340 (1973).
5. L. H. Gan, P. Blais, D. J. Carlsson, T. Suprunchuk, and D. M. Wiles, *J. Appl. Polym. Sci., 19,* 69 (1975).
6. H. Kakida, Y. Chatani, and H. Tadokoro, *J. Polym. Sci. (Polym. Phys. Ed.), 14,* 427 (1976).
7. K. Tashiro, M. Kobayashi, and H. Tadokoro, *Macromolecules, 10,* 413 (1977).
8. I. Sugiya, S. Kobayashi, S. Iwayanagi, and T. Shibata, *Polym. J., 14,* 43 (1982).
9. S. L. Kwolek, U.S. Patent 3,600,350 (1971).
10. S. L. Kwolek, U.S. Patent 3,671,542 (1972).
11. S. L. Kwokek, U.S. Patent 3,819,587 (1974).
12. G. B. Carter and V. T. J. Schenk, in I. M. Ward, Ed., *The Structure and Properties of Oriented Polymers,* Applied Science, London, 1975, Ch. 13.
13. L. Penn, H. A. Newey, and T. T. Chiao, *J. Mater. Sci., 11,* 190 (1976).
14. L. Konopasek and J. W. S. Hearle, *J. Appl. Polym. Sci., 21,* 2791 (1977).
15. M. G. Dobb, D. J. Johnson, and B. P. Saville, *J. Polym. Sci. (Polym. Symp.), 58,* 237 (1977).
16. M. G. Dobb, D. J. Johnson, and B. P. Saville, *J. Polym. Sci. (Polym. Phys. Ed.), 15,* 2201 (1977).
17. D. J. Carlsson, L. H. Gan, and D. M. Wiles, *J. Polym. Sci. (Polym. Chem. Ed.), 16,* 2353 (1978).
18. K. Haraguchi, T. Kajiyama, and M. Takayanagi, *J. Appl. Polym. Sci., 23,* 903, 915 (1979).
19. T. Takahashi, H. Iwamoto, K. Inoue, and I. Tsujimoto, *J. Polym. Sci. (Polym. Phys. Ed.), 17,* 115 (1979).
20. W. B. Black and J. Preston, *High Modulus Wholly Aromatic Fibers,* Dekker, New York, 1973.
21. J. Preston, *Polym. Eng. Sci., 15,* 199 (1975).
22. A. Ciferri, *Polym. Eng. Sci., 15,* 191 (1975).
23. M. Jaffe, *ACS Prepr. Polym., 19*(1), 355 (1978).
24. K. Matsuda, *ACS Prepr. Polym. 20*(1), 122 (1979).
25. R. J. Morgan, J. Hodson, R. Lim, E. T. Mones, and C. O. Pruneda, Lawrence Livermore National Laboratory Report RM-1055C, 1984.
26. E. I. du Pont, Technical brochure, "Weight Savings for Aircraft Using Kevlar Aramid Fiber," 1983.
27. H. Blades, U.S. Patent 3,767,756 (1973).
28. H. Blades, U.S. Patent 3,869,430 (1975).
29. Y. Onogi, J. L. White, and J. F. Fellers, *J. Polym. Sci. (Polym. Phys. Ed.), 18,* 663 (1980).
30. R. J. Morgan, E. T. Mones, W. J. Steele, and S. B. Deutscher, *ACS Prepr. Polym., 21,* 264 (1980).
31. R. J. Morgan, C. O. Pruneda, N. Butler, F. M. Kong, L. E. Caley, and R. L. Moore, *Proceedings of the 29th Natl. SAMPE Symposium,* 1984, pp. 891–902.
32. V. G. Kulichikhin, N. V. Vasil'eva, L. D. Serova, V. A. Platonov, L. P. Mil'kova, L. N. Andreyeva, A. V. Volokhina, G. I. Kudryavtsev, and S. P. Papkov, *Vysokomol. Soedin., A18,* 590 (1976).
33. E. E. Magat, *Phil. Trans. R. Soc. London, A294,* 55 (1979).
34. J. W. White and J. F. Fellers, *J. Appl. Polym. Sci. (Appl. Polym. Symp.), 33,* 137 (1978).
35. E. S. Gould, *Mechanism and Structure in Organic Chemistry,* Holt, Rinehart, & Winston, New York, 1959, Ch. 11.
36. J. A. Duffy and J. A. Leisten, *J. Chem. Soc. I,* 853 (1960).
37. R. J. Morgan and C. O. Pruneda, *Polymer, 28,* 340 (1987).
38. W. F. Thompson, G. L. Powell, and W. E. Lever, "Characterization of Kevlar 49 Yarn," U.S. Department of Energy Report Y2123, 1978.
39. L. Penn and F. Larsen, *J. Appl. Polym. Sci., 23,* 59 (1979).
40. L. E. Caley, P. K. Ambalal, J. F. Carley, T. S. Ford, R. L. Moore, D. J. Noecker, M. E. Ross, R. J. Sherry, and J. M. Yoshiyama, "Quality Assurance Tests of Five Y-12 Kevlar 49 Spools Used to Fabricate Strands and Reliability Specimens for Stress–Rupture Tests," Lawrence Livermore National Laboratory Report UCID-20078, 1984.
41. M. Wardle, Textile Fibers Department, E. I. du Pont (private communication).
42. R. E. Allred, E. W. Merrill, and D. K. Roylance, in H. Ishida and G. Kumar, Eds., *Molecular Characterization of Composite Interfaces,* Plenum, New York, 1985, pp. 333–375.
43. R. E. Allred, in *Proceedings of the 7th Annual Meeting of the Adhesion Society,* 1984.
44. N. J. Abbott, J. G. Donovan, M. M. Schoppee, and J. Skelton, "Some Mechanical Properties of Kevlar and Other Heat Resistant, Nonflammable Fibers, Yarns and Fabrics," Air Force Materials Laboratory Technical Report AFML-TR-74-65, Part III, 1975.
45. M. G. Dobb and R. M. Robson, "Variation of Mechanical Properties of Aromatic Polyamide Fibers with Microstructure and Conditions of Formation," M.O. Report, M.O.D. Agreement No. D/ERIA/9/4/2062/0104/PERME, England, 1983.
46. M. G. Northolt, *Eur. Polym. J., 10,* 799 (1974).
47. M. G. Northolt and J. J. van Aartsen, *J. Polym. Sci. (Phys. Ed.), 11,* 333 (1973).
48. M. G. Dobb, A. M. Hindeleh, D. J. Johnson, and B. P. Saville, *Nature, 253,* 189 (1975).
49. J. W. Ballou, *ACS Prepr. Polym., 17,* 75 (1976).
50. T. A. Hancock, J. S. Spruiell, and J. L. White, *J. Appl. Polym. Sci., 21,* 1227 (1977).

51. R. J. Morgan, C. O. Pruneda, and W. J. Steele, *J. Polym. Sci. (Polymer Phys. Ed.), 21,* 1757 (1983).

52. J. Hodson, C. O. Pruneda, R. P. Kershaw, and R. J. Morgan, *Compos. Tech. Rev., 5,* 115 (1983).

53. C. O. Pruneda, R. J. Morgan, F. M. Kong, J. A. Hodson, R. P. Kershaw, and A. W. Casey, *Proceedings of the 29th Natl. SAMPE Symposium,* 1984, pp. 1213–1221.

54. M. Panar, P. Avakian, R. C. Blume, K. H. Gardner, T. D. Gierke, and H. H. Yang, *J. Polym. Sci. (Polym. Phys. Ed.), 21,* 1955 (1983).

55. L. I. Slutsker, Z. Yu. Chereiskii, L. E. Utevskii, V. W. Kuz'min, V. D. Kalmykova, T. S. Sokolova, A. V. Volokhina, and G. I. Kudryavtsev, *Vysokomol. Soedin., A17,* 1569 (1975).

56. P. Avakian, R. C. Blume, T. D. Gierke, H. H. Yang, and M. Papar, *ACS Prepr. Polym. 21,* 8 (1980).

57. S. C. Simmens and J. W. S. Hearle, *J. Polym. Sci. (Polym. Phys. Ed.), 18,* 871 (1980).

58. L. I. Slutsker, L. E. Utevskii, Z. Yu. Chereiskii, and K. E. Perepelkin, *J. Polym. Sci. (Polym. Symp.), 58,* 339 (1977).

59. L. I. Slutsker, *Proceedings of the Japanese Int. Symposium on Fiber Science and Technology,* Hakone, Japan, 1985, p. 137.

60. J. F. Fellers and J. S. Lee, "SAXS Studies of Polymeric Materials Used in High Performance Composites," Lawrence Livermore National Laboratory Contract P.O. 2819401, Final Report, June 1985.

61. M. G. Dobb, D. J. Johnson, A. Majeed, and B. P. Saville, *Polymer, 20,* 1284 (1979).

62. C. O. Pruneda, W. J. Steele, R. P. Kershaw, and R. J. Morgan, *Compos. Tech. Rev.,* 3(3), 103 (1981).

63. R. H. Ericksen, *Polymer, 26,* 733 (1985).

64. J. W. Greenwood and P. G. Rose, *J. Mater. Sci., 9,* 1809 (1974).

65. J. W. S. Hearle and B. S. Wong, *J. Mater. Sci., 12,* 2447 (1977).

66. R. J. Morgan, E. T. Mones, W. J. Steele, and S. B. Deutscher, *Proceedings of the 12th Natl. SAMPE Technical Conference,* 1980, pp. 368–379.

67. M. G. Dobb, D. J. Johnson, and B. P. Saville, *Polymer, 22,* 960 (1981).

68. L. T. Drzal, *Proceedings of the 15th Natl. SAMPE Technical Conference,* 1983, pp. 190–201.

69. S. J. Deteresa, S. R. Allen, R. J. Farris, and R. S. Porter, *J. Mater. Sci., 19,* 57 (1984).

70. R. J. Morgan, E. T. Mones, and W. J. Steele, *Compos. Tech. Rev.,* 2(3), 3 (1980).

71. I. M. Brown, T. C. Sandreczki, and R. J. Morgan, *Polymer, 25,* 759 (1984).

72. R. L. Moore, "The Tensile Mechanical Properties of Kevlar 49 Single Filaments," Lawrence Livermore National Laboratory (unpublished).

73. H. D. Wagner, S. L. Phoenix, and P. Schwartz, *J. Compos. Mater., 18,* 312 (1984).

74. W. S. Smith, "Environmental Effects of Aramid Composites," E. I. du Pont, Technical Paper, 1980.

75. R. J. Morgan, F. M. Kong, and J. K. Lepper, "Laser-Induced Damage Mechanisms of Kevlar 49–Epoxy Composites," *J. Compos. Mater. 22,* 1027 (1988).

76. J. M. Augl, "Moisture Sorption and Diffusion in Kevlar 49 Aramid Fiber," Naval Surface Weapons Center Report NSWC/TR-79-51, 1979.

77. M. Moss, "Effect of Water-Bearing Fiber in Kevlar 49–Epoxy Composites," Sandia National Laboratories Report SAND-77-0444, 1978.

78. J. A. Barrie, in J. Crank and G. S. Park, Eds., *Diffusion in Polymers,* Academic Press, New York, 1968.

79. M. E. Roylance, *Polym. Eng. Sci., 22,* 988 (1979).

80. R. E. Allred, *J. Compos. Mater., 15,* 100 (1981).

81. R. E. Allred and D. K. Roylance, *J. Mater. Sci., 18,* 652 (1983).

82. R. S. Jones, E. W. Choe, H. Noether, I. L. Hay, P. K. Sullivan, M. J. Hannon, M. Jaffe, J. Soehngen, and M. Tan, "Structure–Property Relationships of Aramids," Celanese Research Company (unpublished).

83. M. Arpin and C. Strazielle, *Polymer, 18,* 591 (1977).

84. R. J. Morgan and F. M. Kong, Lawrence Livermore National Laboratory (unpublished).

85. L. Penn and R. J. Sherry, Lawrence Livermore Laboratory Report UCRL-79499, 1977.

86. Kevlar 49 Technical Data Report K-2, E. I. du Pont, 1978.

87. W. D. Humphrey, *Proceedings of the 29th Mechanical Failure Composites Meeting,* 1979, pp. 177–197.

88. J. W. Mead, K. E. Mead, I. Auerbach, and R. H. Ericksen, *ACS Prepr. Org. Coat. Plast. Chem., 44,* 336 (1981).

89. H. Golopol, J. Clarkson, R. Moore, and N. W. Hetherington, Lawrence Livermore Laboratory Report UCRL-86034, 1981.

90. W. L. Harper and C. E. McAlister, "The Effect of Radiation on Kevlar 49," U.S. Department of Energy Report Y-DA-6527, 1975.

91. DuPont Technical Report C-29, "The Effect of Ultraviolet Light on Products Based on Fibers of Kevlar 29 and Kevlar 49 Aramid," 1977.

92. D. J. Carlsson, L. H. Gan, and D. M. Wiles, *J. Polym. Sci. (Polym. Chem. Ed.), 16,* 2353 (1978).

93. I. M. Brown and T. C. Sandreczki, "Characterization of Kevlar 49 Fibers by Electron Paramagnetic Resonance," McDonnell Douglas Research Laboratories, Lawrence Livermore National Laboratory Contract P.O. 3115901, Final Report, 1982.

94. J. Cook, PERME, Waltham Abbey, England (unpublished data).

95. C. C. Chiao, R. J. Sherry, and T. T. Chiao, *Composites, 7,* 9 (April 1976).

96. R. G. Garza, C. O. Pruneda, and R. J. Morgan, Lawrence Livermore National Laboratory (unpublished).

97. T. Kunugi, H. Watanabe, and M. Hashimoto, *J. Appl. Polym. Sci., 24,* 1039 (1979).

98. L. L. Clements and R. L. Moore, *SAMPE Q.,* 9(1), 6 (1977).

99. T. T. Chiao and R. L. Moore, *J. Compos. Mater., 7,* 547 (1972).

100. C. C. Chiao, R. J. Sherry, and N. W. Hetherington, *J. Compos. Mater., 11,* 79 (1977).

101. R. H. Ericksen, *Composites,* 7(3), 189 (1976).

102. E. P. Pluddemann, "Mechanism of Adhesion Through Silane Coupling Agents," in E. P. Pluddemann, Ed., *Inter-*

faces in Polymer Matrix Composites, Academic Press, New York, 1974, pp. 173–216.

103. D. M. Riggs, R. J. Shuford, and R. W. Lewis, "Graphite Fibers and Composites," in G. Lubin, Ed., *Handbook of Composites,* Van Nostrand Reinhold, New York, 1982, pp. 196–271.

104. N. B. Moore, P. S. Bruno, and S. C. Browning, *Proceedings of the AIAA/SAE 10th Propulsion Conference,* American Institute of Aeronautics and Astronautics, Paper 74-1208, 1974.

105. R. E. Allred and F. P. Gerstle, Jr., *Proceedings of the 30th Reinforced Plastics/Composites Institute Technical Conference,* Society of the Plastics Industry, 1975, Paper 9-B.

106. R. E. Allred, H. K. Street, and R. J. Martinez, *Proceedings of the 24th Natl. SAMPE Symposium,* 1979, pp. 31–50.

107. E. M. Wu, Washington University, St. Louis, AMMRC Report TR-80-19, 1980.

108. R. E. Allred, *Proceedings of the 29th Natl. SAMPE Symposium,* 1984, pp. 947–957.

109. R. E. Allred and A. M. Lindrose, in *Composite Materials: Testing and Design,* ASTM STP 674, American Society for Testing and Materials, Philadelphia, 1979, pp. 313–323.

110. L. S. Penn, in H. Ishida and G. Kumar, Eds., *Molecular Characterization of Composite Interfaces,* Plenum, New York, 1985.

111. L. S. Penn and B. Miller, *J. Colloid Interfac. Sci., 78,* 238 (1980).

112. L. S. Penn and T. K. Liao, *Compos. Tech. Rev., 6,* 133 (1984).

113. S. P. Wesson and R. E. Allred, *Proceedings of the 7th Annual Meeting of the Adhesion Society,* 1984, p. 27.

114. V. A. Chase, Whittaker Corporation Report, Naval Air Systems Command Contract N00019-74-C-0055, 1975.

115. M. R. Wertheimer and H. P. Schreiber, *J. Appl. Polym. Sci., 26,* 2087 (1981).

116. E. M. Petrie and J. C. Chottiner, *Proceedings of the 40th Annual SPE Technical Conference and Exhibit,* ANTEC 82, 1982, pp. 777–779.

117. R. E. Allred, *Proceedings of the 23rd Natl. SAMPE Symposium,* 1984, pp. 947–957.

118. L. S. Penn, F. A. Bystry, and H. J. Marchionni, *Polym. Compos., 4,* 26 (1983).

119. A. S. Hoffman, T. S. Keller, A. Miyake, B. D. Ratner, and B. J. McElroy, in C. Kim and I. S. Ki, Eds., *Proceedings of the Pacific Chemical Engineering Congress,* Korean Institute of Chemical Engineering, Seoul, 1983.

120. Y. Wu and G. Tesoro, *J. Appl. Polym. Sci., 31,* 1041 (1986).

Aromatic Polyamides

See Aramid Fiber Composites

Articulating Mechanical and Robotic Systems Fabricated from Advanced Composite Materials

Introduction

Intense competition in the international marketplace for high performance machines and robotized equipment in military, aerospace, and advanced manufacturing environments has been responsible for the imposition of more stringent criteria on this class of articulating systems. Typically, high speeds of operation are demanded, power consumption is often restricted, and generally tighter tolerances are imposed on the output characteristics of these high performance systems. While significant progress has been made in the development of viable methodologies and also technologies for the many subsystems of machine and robotic systems (vision, sensing, etc.), the absence of a viable approach for developing lightweight structural components for these systems has been recognized as a major impediment to the evolution of the next generation of machine and robotic systems [1–5].

For instance, the current generation of robotic systems are fabricated with articulating members possessing a ratio of payload weight to arm weight of approximately 1:20 [3]. Consequently, these members are characterized by high stiffnesses, massive weights, and also large moments of inertia, which severely restrict dynamic performance. If superior performance characteristics are to be achieved, such as higher speeds and shorter settling times, then from a mechanical design viewpoint, lightweight structural members must be designed, engineered, and fabricated.

This design field requires that the knowledge base in materials science be integrated with the knowledge base pertaining to the kinematics and dynamics of mechanisms. Among the number of different methodologies that have been proposed for analyzing this class of problems [6], the finite element method is probably the most popular [7–11]. The essential ingredients of these formulations are captured by the simplified finite element equations of motion:

$$[M]\{\ddot{U}\} + [C]\{\dot{U}\} + [K]\{U\} = -[M]\{\ddot{R}\} \qquad (1)$$

for a general high speed machine system in which the body forces and surface tractions have been neglected. In Eq. (1), matrices $[M]$, $[C]$, and $[K]$ represent, respectively, the global mass, damping, and stiffness characteristics of the machine being analyzed, which typically include inertial coupling and Coriolis terms. The discretized deformation field is contained in column vector $\{U\}$, the overdot denotes a time derivative, and $\{\ddot{R}\}$ represents the discretized acceleration field resulting from an analysis of the machine system, assuming all the structural members to be rigid bodies. Equation (1) is then solved for the elastic deflections of the members

$\{U\}$, which not only permits the stresses to be determined using the material constitutive relationship, but in addition, allows the study of the vibrational characteristics of the device, because the deformation field is generally time dependent.

Some global observations may be deduced from an examination of Eq. (1). If all the terms are premultiplied by $[M]^{-1}$, then for a prescribed mode of operation $\{\ddot{R}\}$, the response of the general machine system is governed by the magnitudes of $[M]^{-1}[C]$ and $[M]^{-1}[K]$, namely, the damping properties of the link material and also the stiffness-to-weight ratios of the members. Both these mechanical properties depend on the material from which the links are fabricated. Thus when both these characteristics are maximized, the deflections and stresses will be a minimum, and transient vibrations, such as those resulting from the rapid movement of a robotic manipulator, will be attenuated most rapidly to provide a shorter cycle time in a production environment, for example.

Thus the solution to this class of problems involves fabricating the structural members of machine systems with materials possessing good damping properties and also high stiffness-to-weight ratios. Two generic approaches emerge that satisfy the design criterion for fabricating high stiffness, lightweight structural members. The first approach [12–16], which has been applied to the design of high speed mechanisms, involves the fabrication of members in monolithic materials, such as the commercial metals, and the high stiffness characteristics are achieved by optimizing the geometry of the part. A consequence of this approach is that in spite of the complex cross-sectional geometries and tapers associated with this design philosophy, the desired objectives cannot be achieved because the search for the optimal solution is impeded by a constraint imposed on the material selection process, which requires the member to be fabricated in one of the monolithic materials. The physical properties of these materials are inherently less desirable for these machinery applications than many of the advanced composite materials.

The second approach, which was originally proposed for articulating mechanical systems by the authors [17], requires the articulating members to be fabricated in advanced composite materials. Not only do these modern materials have strength- and stiffness-to-weight ratios superior to the commercial metals, but the designer, to achieve the desired performance characteristics, has the freedom to optimally synthesize both the properties of the link material and the geometry. This philosophy is succinctly presented in Figure 1, for polymeric lami-

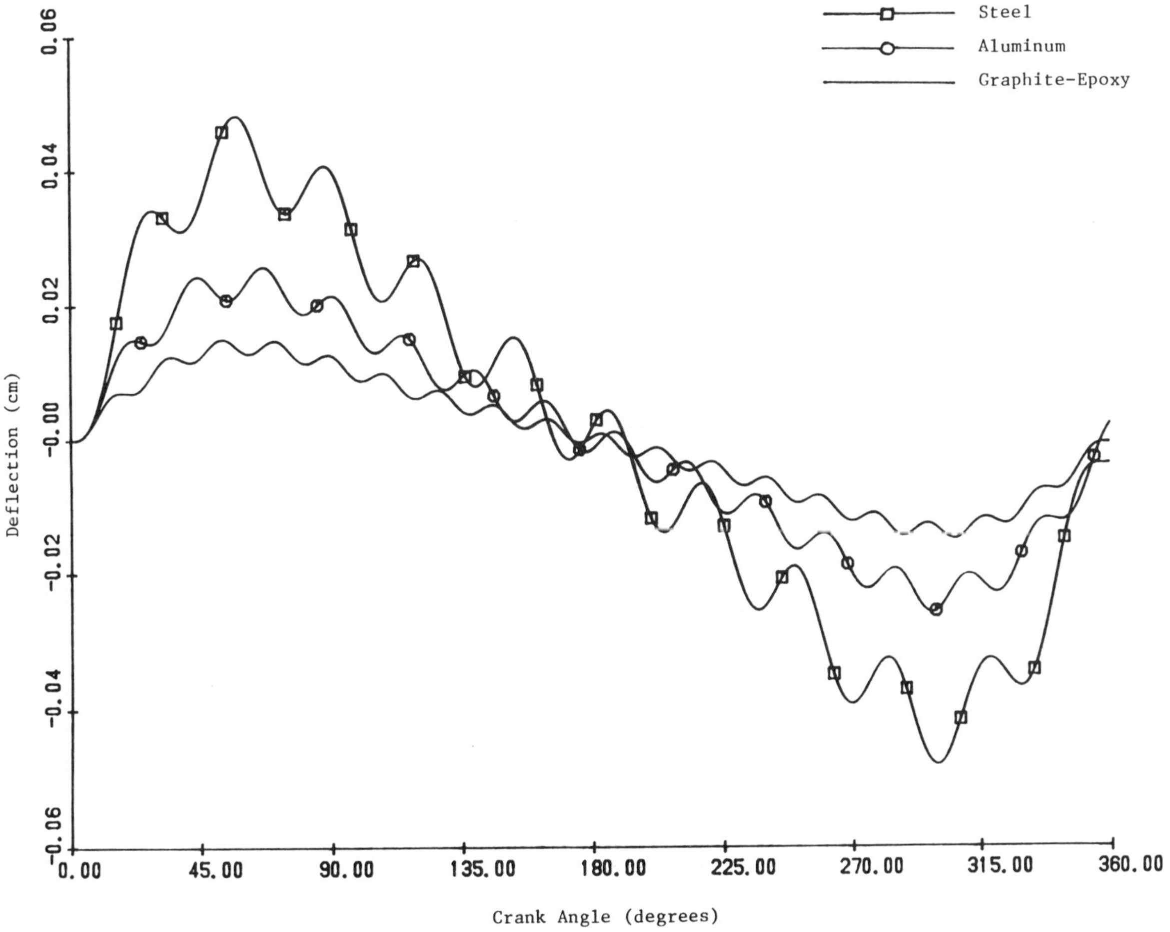

FIGURE 1 **Midspan transverse deflections of connecting rod: simulation results.**

nated composites, and the details of the methodology are presented in Ref. 18.

The development of design methodologies in the field of articulating mechanical systems fabricated in advanced composite materials began with a theoretical publication [17], progressed through experimental studies [19], and also combined analytical and experimental studies [20,21] that validated and refined both the proposed methodology and the associated mathematical models. The elastodynamic analysis of articulating mechanical and robotic systems can be efficiently undertaken by employing finite element methods, and these strategies are summarized in the literature [22–24].

Having established the necessity for fabricating components of articulating mechanical and robotic systems in advanced composite materials, we proceed to summarize the salient features of the following significant research thrusts in this rapidly evolving discipline:

- The elastodynamic response of planar composite-based mechanism systems
- The optimal design of composite-based components of mechanism and robotic systems
- The effect of hygrothermal environments on the elastodynamic response of composite-based articulating systems
- Design-for-manufacture issues pertaining to the design of composite-based mechanism and robotic systems.

Elastodynamic Response of Planar Composite-Based Mechanism Systems

A review of the literature documenting the characteristics of a large variety of engineering materials [25] soon reveals that polymeric composite laminates offer the most promising class of materials with which to fabricate the structural members of articulating mechanical systems. The high specific stiffnesses and damping properties of these modern materials, coupled with the inherent freedom to synthesize different material characteristics through tailoring, offer the design engineer the opportunity to design, fabricate, and engineer mechanical and robotic systems with optimal performance characteristics.

This design philosophy was originally presented in Ref. 17, which was focused on developing the equations governing the elastodynamic behavior of mechanism components fabricated from linear elastic anisotropic materials. The equations of motion for analyzing the behavior of components made from materials with these properties are derived using a mixed variational principle by selecting an appropriate strain energy density function. A two-dimensional theory was developed for studying the elastodynamic motion of bodies with orthotropic properties in which the principal directions of the material are not colinear with the geometrical or body axes of the link. Moreover, the theory also permits the analysis of composite components exhibiting orthotropic properties to be undertaken by using a thickness concept to model the effects of the different stiffnesses of the laminae, and an effective mass density concept is adopted to model the differing densities. In this manner, the mechanical behavior of a multilayered body was represented by a continuum model that is able to describe, in an approximate manner, the characteristic features of the response.

The foregoing mathematical models were incorporated within the variational principle, which was written in a matrix notation suitable for subsequent finite element analysis. As an example of the approach, a simple beam model was developed for studying the motion of a flexible connecting rod of a slider–crank linkage made from graphite–epoxy materials. The finite element equations of motion are integrated numerically using the Newmark method, and several studies are presented comparing the response of links fabricated with composite laminates and the response characteristics of links made from steel and aluminum.

Combined analytical and experimental studies involving the investigation of the elastodynamic response of flexible four-bar linkages and slider crank mechanisms [20,21] followed the authors' pioneering theoretical publication [17]. In parallel with this work on refining the mathematical modeling process, comparative experimental studies were undertaken by fabricating members in aluminum, steel, and two graphite–epoxy layups for the same linkages [19]. Each link was characterized by the same flexural rigidity and the same length, but naturally the mass of each member was different. By sequentially incorporating these links in the two mechanisms, and operating the systems at prescribed speeds to ensure that each specimen was subjected to the same acceleration fields, the experimental results presented in Figure 2 were obtained. This graph shows the elasto-

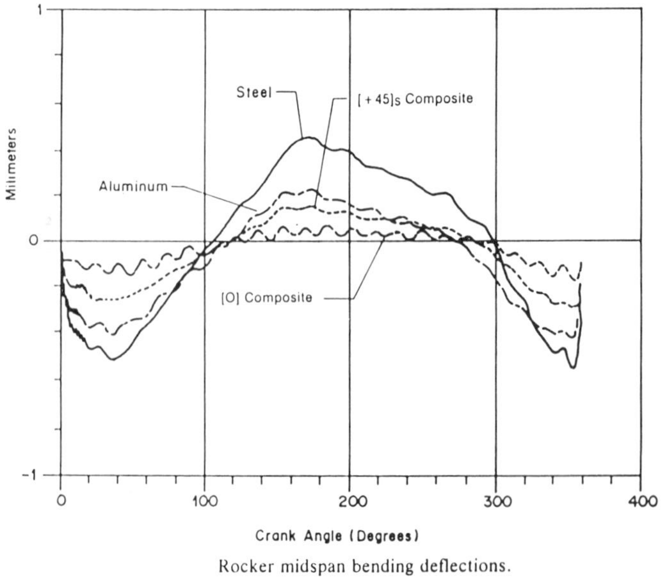

FIGURE 2 **Rocker midspan bending deflections: experimental results.**

dynamic response of the midspan transverse deflection of flexible rocker links of four-bar linkages.

Machinery Noise

Noise pollution from machine systems is receiving increasing attention in design offices because of legislation governing the levels of noise radiation to which employees may be exposed in the workplace. Since the damping capacity of polymeric laminates increases with frequency [26], this phenomenon can be usefully exploited in the design of high speed machinery, which frequently generates high levels of acoustical radiation as a result of accentuated acoustoelastodynamic effects associated with bearing clearances, for example [27,28]. Preliminary experimental work on flexible four-bar linkages with a prescribed radial clearance in the bearing at the joint connecting the coupler link and the rocker link have clearly demonstrated the potential of composite laminates in a noise control regime [29]. Similar results have been achieved on automobile engines featuring graphite–epoxy connecting rods and pistons [30].

Hygrothermal Effects

Rarely in engineering practice will a machine operate under conditions of uniform temperature or constant ambient relative humidity, and these environmental conditions must be incorporated into methodology for designing industrial parts to operate in rolling mills or foundries, or military parts, which must reliably operate both in hot, humid, equatorial climates and in cold, dry, polar regions.

Generally, the diffusion of both heat and moisture will occur simultaneously in the laminated components of the machine, and the situation is further aggravated when the composite members are subjected to mechanical loading. For example, as the relative humidity increases, the stiffness of the composite links of a mechanism will be reduced; in addition, the mass of the members will be increased, resulting in an increased inertial loading and also higher stresses. These stresses can create more favorable conditions for moisture diffusion by developing a more permeable medium through the formation of microcracks, for example, and this causes a further reduction in the stiffness properties of the composite, a further increase in the mass of the member, and the diffusion of even more moisture [31,32]. This coupled phenomenon, which is relevant to both static and dynamic environments, continues until a saturation limit is reached.

The ambient temperature can significantly affect the behavior of a fibrous polymeric material in three ways. First, a differential expansion between the fiber and resin, and also between differently orientated laminae in a laminate, can result in internal stresses. The situation is even more severe with hybrid designs incorporating several different types of fiber [33]. Second, changes in temperature result in changes in the properties of the constituent materials, in particular the matrix. In the absence of externally imposed mechanical loading on the composite, the effect of temperature is usually reversible at ambient temperatures. Changes in the matrix properties affect the properties of the composite that are strongly dependent on the constitutive behavior of the matrix, such as intralaminar shear strengths. Third, the creep resistance is strongly dependent on temperature because of the viscoelastic properties of the matrix, and these phenomena can lead to flow and irreversible microcracking.

Most analytical and experimental investigations address this phenomenon by uncoupling the temperature effects and the moisture effects [34]. The consequences of this simplification are aggravated when dynamic mechanical loading must also be considered. Preliminary analytical results indicate that the stress levels predicted by coupled and uncoupled hygrothermomechanical analyses can differ by 20–80% under conditions of static mechanical loading [35–37]. Clearly the consequences of employing decoupled models for designing composite-based machine products that must reliably operate at high speed could be disastrous.

A potentially viable approach to modeling the diffusion of moisture into the composite and developing an understanding of the effect of the moisture content in the constitutive characteristics of these materials, would be to treat the moisture–composite aggregate as a mixture in the context of the theory of interacting continua [38]. The theory of interacting continua has been used successfully to model the interaction of fluids and polymeric materials undergoing large deformations [39–42]. Motivated by these successes, the theoretical developments have been extended to model the interaction of moisture with polymeric composite materials, and the preliminary results appear to be very promising [43].

Fundamental experimental investigations focused on characterizing the elastic moduli and damping capacities of laminated beamlike structures in hygrothermal environments have been presented in Refs. 44 and 45. The mathematical basis for modeling the hygrothermoelastodynamic response of mechanism and robotic systems has recently been developed [46,47] by extending classical ideas of variational calculus to this challenging area. The importance of employing appropriate nonlinear theories to study the vibrational response of beam- and platelike laminated composite structures has been highlighted in Ref. 48, and it has been clearly demonstrated that hygrothermal effects can significantly modify the elastodynamic response of articulating machine and robotic systems [49,50].

Optimal Design of Structural Members for Mechanism Systems Fabricated in Composite Laminates

The design of a mechanism system to satisfy performance characteristics is a common industrial activity when the system must operate at low speed. This task is not easily undertaken when the mechanism system must operate at high speeds, since the design team must employ elastodynamic equations of the form of Eq. (1) in the design process. The design of linkage members to satisfy often conflicting material constraints associated with requirements for high specific stiffness, damping, impact resistance, toughness, and manufacturability, for example, frequently cannot be achieved by fabricating members in the commercial metals, and designs in composite hybrids offer the only viable strategy [51].

The purpose of hybridization is to tailor the properties of a composite material to comply with the design specification of the specific structural machine element. Generally this tailoring process is dictated by the desire to synthesize specific material properties, or else to comply with cost constraints, by replacing a high cost fiber by a low cost fiber in the less critical regions of the structural member.

Hybridization can enter the design of a structural component at several levels. At the microstructural level, different fibers may be intermingled in a common matrix; at the laminate level, different plies may be employed in the fabrication process; at the microscopic level, subcomponents fabricated with chopped fibers, continuous fibers, or metallic honeycomb structures may be assembled to form a single component. While all these philosophies offer distinct advantages for the designer, they also create a variety of challenging problems for the applied mechanician.

The kernel of these problems is clearly the interfacial modeling between the different constituents whose properties are often quite incompatible. For example, thermal strain effects merit close evaluation with hybrids not only in service but also during manufacture, since there are often wide variations between the anisotropic thermal expansion coefficients and elastic moduli of typical mixtures of fibers like carbon and glass [33].

When hybrid laminates with mixed plies are subjected to flexural loading, which occurs in many machine systems, the flexural modulus has been found to be highly sensitive to the number and dispersion of the plies with apparent deviations from the standard rule-of-mixtures approach [52]. These effects must clearly be accounted for in enhanced models of failure modes. The development of viable models in this area is crucial in the design of machine components fabricated in composites, not only to ensure that appropriate material properties are synthesized, but also to ensure adherence to cost control guidelines.

Composites are more expensive than the commercial metals when compared on a weight basis, but of course less composite material is required in an application governed by the stiffness-to-weight ratio of the structural member. When confronted by a cost constraint on a particular machine element, the designer can invoke a hybridization philosophy, whereby the more expensive high strength fibers are retained in regions of the part presenting high stress levels but are replaced in the lower stress regions by cheaper, less strong fibers such as Kevlar or glass–fiber systems. For example, by using this approach, the material cost per kilogram can be typically reduced to one-third that of an all-graphite design.

Where the performance of a machine is principally governed by the stiffness-to-weight ratio of the parts, the cost penalty associated with fabricating the parts with a composite material is probably warranted by simple economics alone, since the capital cost of a machine is meaningful only when compared to its performance and the associated productivity. Thus the price per kilogram of the material is not the issue; what is relevant to the decision-making process is the ultimate payoff through the creation of a superior product. Typically this scenario occurs in machine systems containing fast-moving components whose speed is limited by their own inertia rather than by the external loads imposed on the components. Typical examples are machines developed to handle light media in the packaging, printing, and textile industries or robotic devices handling small parts and subassemblies for the pharmaceutical, food-processing, and electronics industries.

Composite-Based Robotic Systems

The demand for increased productivity has resulted in many industries incorporating microprocessor-driven robots into production line facilities. The operations performed by these intelligent machines are numerous and varied, but many workstations and flexible manufacturing cells would be greatly enhanced if the robots currently employed in these production environments could operate at higher speeds with more accurate end-point positioning.

These objectives cannot be achieved with existing designs because robots are principally assemblages of beam-shaped members, which are prone to bending deflections that limit the end-point accuracy. In addition, in an attempt to reduce link deformations, the ratio of the weight of the manipulator arm to that of the payload is rarely better than 10:1. Thus the deflections are primarily due to the weight of the mechanical linkage itself. The large mass, hence inertia, of the moving members of the robot naturally inhibits the ability of the device to respond quickly. Hence, if higher operating speeds are to be achieved, these systems should feature structural members with less weight, provided structural stiffnesses are not sacrificed.

The design goals of high speed operation and improved end-effector positioning have received considerable attention in the literature, with the emphasis being on the manipulator control system, the elastodynamic characteristics of the robotic linkage, and also the interaction between these diverse design areas. A variety of techniques have been proposed, involving marriages of established methods in the fields of automatic controls and mechanism design, and these range from finite element analyses of flexible linkages to state-variable feedback methods and frequency domain techniques.

A small body of the research literature has focused on the design of high stiffness, low mass robot arms fabricated in advanced composite materials, rather than the traditional monolithic materials, and the proposed methodologies promise manipulator designs with much smaller deflections at the end effector and also higher speeds of operation using existing actuators.

Composites promise high strength, relatively low cost materials under conditions of high volume production. Thus the fabrication of manipulators in these materials should, in the future, ensure the development of a new generation of low cost robots. Moreover, the manufacturing methods associated with fabricating composite materials will probably be undertaken by robots themselves.

It is true, however, that because of relatively low volume production, composite materials currently cost more than commercial metals when compared on a weight basis, and this factor is often used in arguments for not exploring the fabrication of components in these materials. However, making a part of a machine in a composite material may be justified by simple economics alone. In particular, the capital cost of a machine is meaningful only when compared to its productivity.

Thus the possible increase in the initial cost of fabricating the articulating members of the mechanical linkage of a robot must be carefully weighed against the enhanced performance of the system during its lifetime. This enhanced performance might be quantified in terms of increased payload, higher operational speed with associated higher productivity, lower power consumption, or increased end-point accuracy due to smaller structural deflections. This latter characteristic may then permit the robot to undertake more complicated, more demanding tasks such as the assembly of products, which is the function of less than 5% of all industrial robots in use today.

Moreover, the increased cost of fabricating the mechanical linkage in a composite has probably increased the overall cost of the robot only by a small percentage because the complete robot system includes actuators, software, and a sophisticated control system. Hence the important point here is not the price per kilogram of the material, but rather the material properties that are being purchased for a given sum of money.

A variational formulation for the dynamic viscoelastic finite element analysis of robotic manipulators is presented in Ref. 23, and a comparative study is undertaken to compare the elastodynamic response of manipulators fabricated with steel arms and composite arms of the same flexural rigidity performing identical prescribed maneuvers and carrying the same payload. The simulation results are presented in Figure 3, from which it is evident that the composite design is superior, with its shorter settling times and the smaller peak dynamic deflections, hence stresses.

The foregoing preliminary investigation was followed by a proof-of-concept study, undertaken by designing and fabricating a graphite–epoxy forearm that was retrofitted to a General Electric P50 articulating process robot [53]. This pioneering investigation demonstrated for the first time the significant performance enhancements to be accrued from fabricating arms for commercial robots in advanced composite materials. The hollow box section structural member employed in this study was fabricated in two parts: a channel section and a plate. The individual parts were fabricated by the hand layup technique, then bonded together to complete the manufacture of the hollow segment of the forearm connecting the wrist assembly with the joint at the upper arm. These nonoptimally designed parts were fabricated in Fiberite Hy-E 1034E 6000 filament graphite with 934 epoxy resin. The fiber volume fraction of the prepreg was approximately 60%, the thickness of each ply was 0.005–0.007 in., and each wall thickness comprised 32 plies with the following stacking sequences and layups: $[0/0/0/90]_4$ $[0]_4$ $[90/0/0/0]_3$.

Before the prototype composite are was installed on the P50 robot, the original robot was programmed to execute a series of prescribed maneuvers, and the elastodynamic behavior was monitored and stored on a digital data acquisition system. The maneuvers were then repeated with the graphite–epoxy arm retrofitted to the robot and the two response histories compared. A typical comparison is presented in Figure 4. The most important conclusion from these results was that the amplitude of the transient response of the original aluminum P50 arm was reduced by 70–80% by using the prototype composite arm. Furthermore the experimental results clearly demonstrated the intuitive notions that the higher the velocity of the end effector, the larger the amplitude of the overshoot; the heavier the payload, the larger the amplitude; and the heavier the payload, or the higher the end-effector velocity, the longer the settling time upon completion of a maneuver.

Optimal Design of Robot Arms

The response of a robot arm, for a set of prescribed initial and boundary conditions, is dictated by the ability of the designer to synthesize an appropriate material and also by the geometrical configuration of the arm. Optimal computer-aided design (CAD) tools are essential for this work, not only to synthesize the appropriate laminate properties but also to arrive at the appropriate shape of the structural member and ultimately to ensure its economy of manufacture. A prerequisite to achieving this computer-aided design capability is the development of viable models for predicting a broad range of component characteristics, such as fatigue life, impact resistance, and damping properties; in addition, failure theories must be incorporated into the design process.

While this capability does not now exist, the subject of optimally designing laminates generally for static conditions has recently begun to receive attention from a small body of researchers [54–60]. It is anticipated that these activities will evolve to include dynamical situations and in addition, the results of the purely theoretical works will be validated experimentally, to establish the viability of the proposed methodologies.

These proposed CAD tools should ultimately incorporate an artificial intelligence capability, which would permit the designer to select appropriate materials, configuration of reinforcement, and fabrication techniques in a rational manner. Naturally, this CAD capability implies the need to incorporate an extensive data base of material properties, and it should be able to document relevant manufacturing issues.

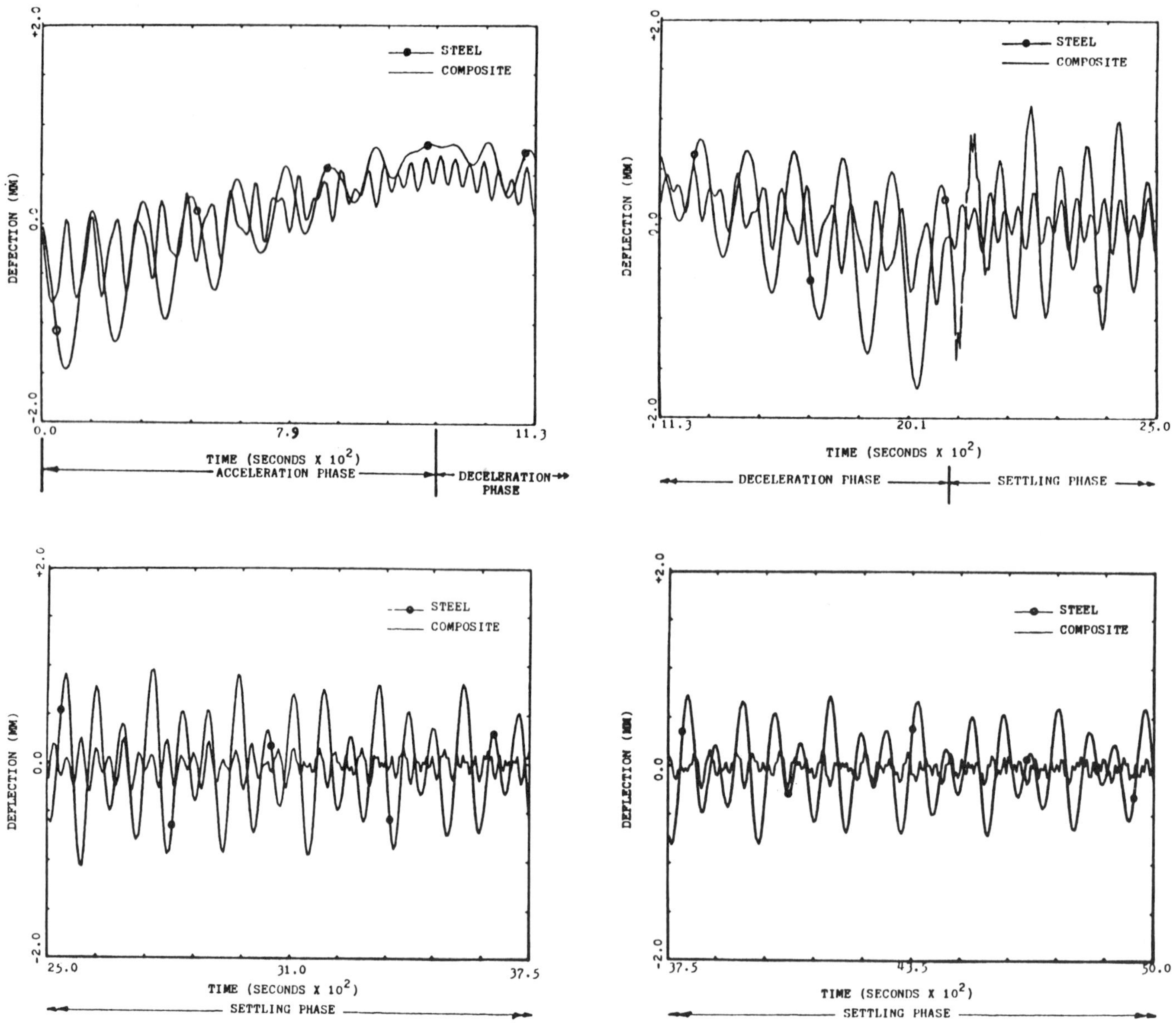

FIGURE 3 Simulation of transverse deflections of end effector.

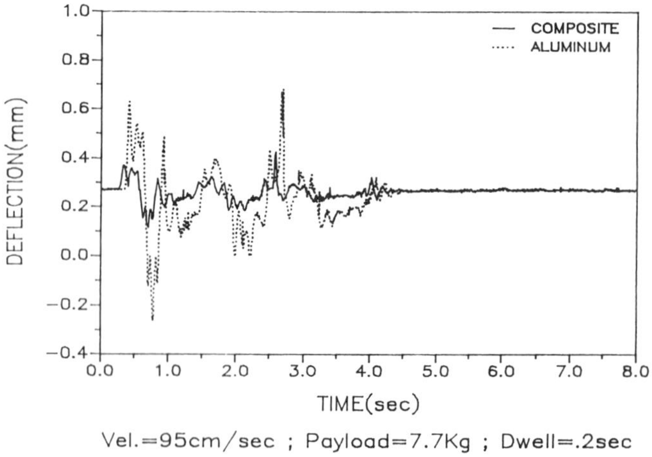

FIGURE 4 Comparison of experimental transverse deflections of end effector.

Design-for-Manufacture Issues

The insatiable demand in the aerospace, military, and industrial marketplaces for high performance articulating mechanical systems has triggered the evolution of a new generation of mechanism and robotic systems featuring state-of-the-art components and subassemblies based on advanced composites. The design implications are presented in Figure 5. Whereas the methodologies for designing composite-based components have been relatively well developed, these methodologies do not explicitly incorporate the constraints and characteristics imposed by the specific manufacturing processes employed to fabricate composite-based parts. Each manufacturing process imparts significantly distinct stiffnesses and damping and mass characteristics to the part, as a result of the distinct fiber volume fractions,

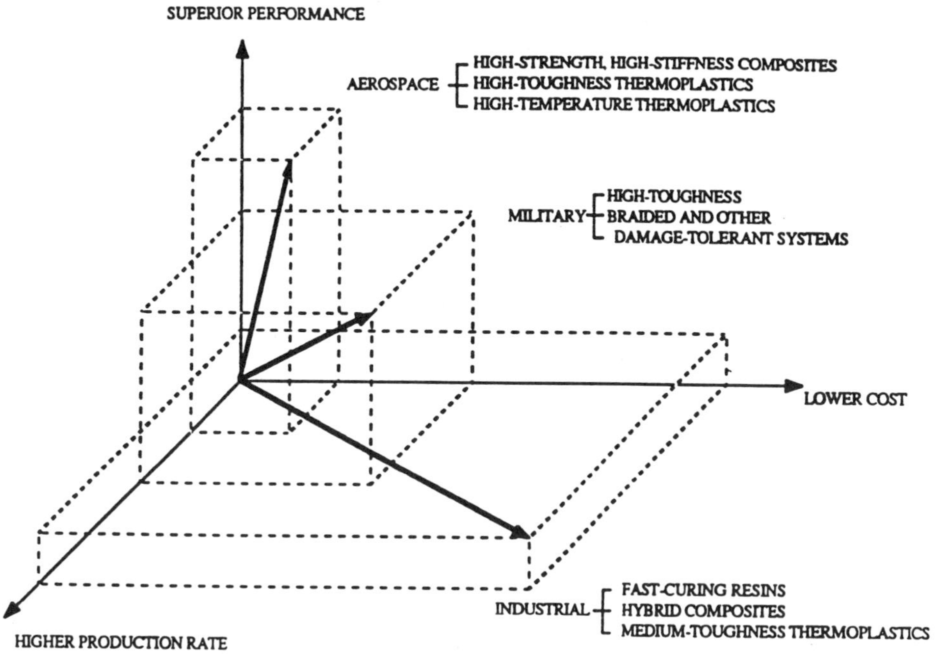

FIGURE 5 **The diverse costs, production rates, and performance characteristics of composite-based parts for aerospace, military, and industrial environments.**

spatial distribution of the fibers, fiber orientations, and layups associated with each process. Thus, the material selection, the part geometry, and the constraints and characteristics of the manufacturing processes are inextricably intertwined, and the appropriate manufacturing processes cannot be selected a posteriori, which is the standard practice in the traditional design of monolithic parts.

Consider, for example, the classical problem of designing the connecting rod of an internal combustion engine in a monolithic material. Given the design specifications, the design process conveniently decomposes into distinct phases of kinematic analysis, machine dynamics, and stress analysis. The stress analysis phase determines the final details of the form-design of the part, and the decision on the appropriate manufacturing processes is subsequently undertaken discretely, in isolation from the design process. Clearly the choice of the appropriate manufacturing processes, such as forging and casting, has impact on the Young's modulus and Poisson's ratio of the material because of the distinct grain structure imparted by each process. These two material characteristics, which emerge from the consideration of manufacturing processes, are subsequently incorporated in the design process to establish the final part geometry.

In sharp contrast to this scenario, the design and fabrication of the connecting rod in an advanced composite material has several distinct advantages over the traditional monolithic design. These advantages include lower weight, increased specific stiffnesses and strengths, and superior damping characteristics, which result not only in overall superior vibrational response characteristics but also in significantly decreased levels of acoustical radiation of the engine. Thus, in addition to the advantages related to the individual component, the properties and characteristics associated with the use of advanced composite materials have a significant impact on the overall system design. These design-cascading effects may permit, for example, elimination of a balance shaft in the engine; or, it may be possible to use simplified engine mounts, or the transmission shafts may require fewer bearings. Finally, of course, the bearing requirements on the connecting rod and the crankshaft are less stringent.

In sharp contrast to a conventional drop-forged steel connecting rod, the design of a composite-based connecting rod cannot be conveniently divorced from the consideration of the manufacturing processes. If the connecting rod is pultruded, filament wound, or braided, the geometry of the part will be constrained by each process, and each manufacturing process will, in turn, impart distinct stiffnesses and damping and mass characteristics to the member, because of the distinct fiber volume fractions, spatial distribution of the fibers, fiber orientations, and layups associated with each process.

There is a dearth of viable process-driven design-for-manufacture (DFM) methodologies for designing and fabricating the wide variety of composite-based parts encountered in the diverse defense, manufacturing, and aerospace industries. This section addresses the design-for-manufacture issues for synthesizing cost-effective, high strength, lightweight, composite-based components for articulating mechanism and robotic systems. The traditional DFM strategies for parts fabricated from

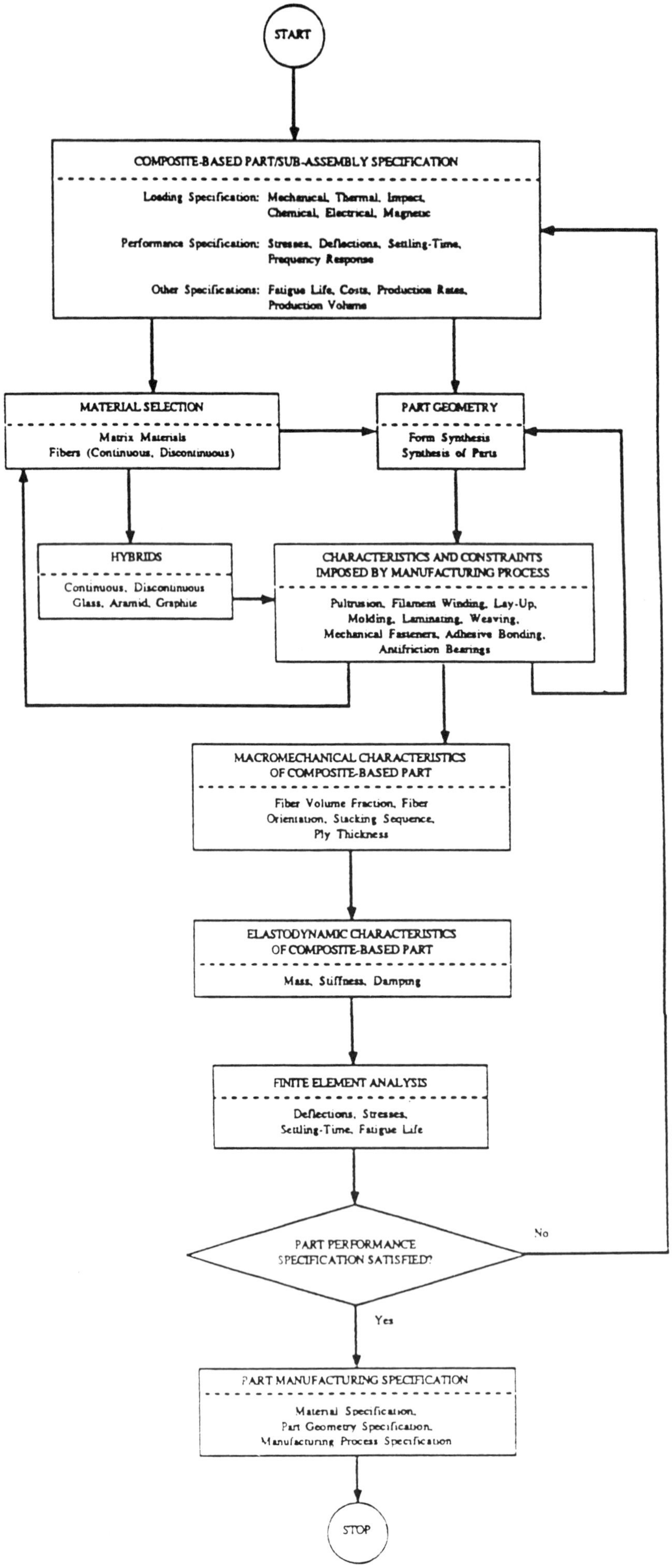
START
COMPOSITE-BASED PART/SUB-ASSEMBLY SPECIFICATION
Loading Specification: Mechanical, Thermal, Impact, Chemical, Electrical, Magnetic
Performance Specification: Stresses, Deflections, Settling-Time, Frequency Response
Other Specifications: Fatigue Life, Costs, Production Rates, Production Volume
MATERIAL SELECTION
Matrix Materials
Fibers (Continuous, Discontinuous)
PART GEOMETRY
Form Synthesis
Synthesis of Parts
HYBRIDS
Continuous, Discontinuous
Glass, Aramid, Graphite
CHARACTERISTICS AND CONSTRAINTS IMPOSED BY MANUFACTURING PROCESS
Pultrusion, Filament Winding, Lay-Up, Molding, Laminating, Weaving, Mechanical Fasteners, Adhesive Bonding, Antifriction Bearings
MACROMECHANICAL CHARACTERISTICS OF COMPOSITE-BASED PART
Fiber Volume Fraction, Fiber Orientation, Stacking Sequence, Ply Thickness
ELASTODYNAMIC CHARACTERISTICS OF COMPOSITE-BASED PART
Mass, Stiffness, Damping
FINITE ELEMENT ANALYSIS
Deflections, Stresses, Settling-Time, Fatigue Life
PART PERFORMANCE SPECIFICATION SATISFIED?
No
Yes
PART MANUFACTURING SPECIFICATION
Material Specification,
Part Geometry Specification,
Manufacturing Process Specification
STOP

monolithic materials are typically focused on manufacturability considerations. It is necessary to address these issues adequately at the design stage to increase cost effectiveness and minimize scrap, for example. Similarly, when considering the assembly of subsystems, traditional DFM approaches focus on minimizing the number of parts, the number of assembly operations, and the use of standard out-sourced parts and components, for example.

In sharp contrast to the classical DFM approaches for monolithic materials, the DFM strategies for composite materials proposed herein not only incorporate the manufacturing processes but are in fact exclusively driven by the manufacturing processes themselves. The proposed DFM strategy will manifest itself at the macromechanical level in the synthesis of the desired material properties and part shapes, while at the subassembly level, this proposed methodology would render obsolete some conventional design-for-assembly (DFA) methodologies for monolithic materials. Parts that are currently fabricated as subassemblies in monolithic materials could be fabricated as a single part in composite materials, which are often cheaper to manufacture. For example, in the traditional approach, a typical subassembly is initiated by the fabrication of parts featuring machined interfaces, to facilitate assembly. This typically labor-intensive activity may account for up to 50% of the total cost of the subassembly. The proposed DFM strategy for composite-based parts and subassemblies can significantly alleviate the costs associated with machining, assembly, and labor by permitting the fabrication of the complete subassembly by filament winding, for example.

A rigorous design-for-manufacture strategy for composite-based parts and subassemblies would require an iterative algorithm that encompassed the materials selection process and issues pertaining to part geometries and manufacturing considerations, for example, as presented in Figure 6. Such a DFM strategy would be initiated by the part or subassembly specification, which would include diverse constraints imposed on the static and dynamic deflections, mass and inertia characteristics, payload requirements, impact resistance, fatigue life, environmental factors such as temperature and humidity, and of course, cost, quantity, and production rate considerations. The specification for the part or subassembly provides the basis for initiating an iterative design-for-manufacture algorithm that integrates the various facets of the DFM strategy by simultaneously considering part geometry, manufacturability, and material selection, for example.

It is clearly evident that the choice of continuous versus discontinuous fibers in the material selection phase of the DFM strategy for the design of robot arms, for example, has a significant impact on the choice of manufacturing processes, which in turn constrains the choice of part geometry. Furthermore, each manufacturing process results in a distinct fiber volume fraction, and also a distinct fiber orientation and spatial distribution of the fibers in the composite arm. This results in distinct anisotropies (e.g., aelotropies and orthotropies), which impart distinct global mechanical characteristics, such as stiffness, damping, and mass, to the robot arm under consideration. For example, in the case of filament winding techniques alone, different combinations of hoop, longitudinal, and helical winding patterns will invariably result in filament-wound arms with different stiffnesses, damping capacities, and mass characteristics, which in turn govern the overall elastodynamic performance of a robot arm under prescribed service conditions.

Future Work

A cursory review of the recent literature focusing on the anticipated scientific and technological developments in the field of robotic systems during the 1990s suggests that composite-based structures will be significant components of the next generation of robots [1–5,61–63]. Not only is this current research thrust fueled by the insatiable demand for higher productivity at reduced costs in the industrial sectors of the global economy; less conventional terrestrial activities have catalyzed new research thrusts, as well. Typical examples are the hazardous environments encountered in the nuclear and chemical industries and in fire fighting, and the unstructured environment associated with deep-sea exploration. Furthermore, the imminent industrialization of space has served to focus attention on the development of viable autonomous and man-controlled robotic systems for deployment in the hostile, unstructured environment of deep space.

The new generation of robots for these diverse industrial, military, and aerospace environments will feature a broad spectrum of "smart" capabilities that will permit the robots to feel, see, and also think in the unstructured environments in which they will be required to operate. According to conventional thinking, these capabilities should reside at the robot end effector; however, the authors are convinced that these "smarts" should also be featured in the materials from which the structural components are manufactured. This capability would permit the elastodynamic response of robotic systems to be optimally tailored in real time, to significantly enhance system performance. Consequently, a pioneering research program has been initiated to develop a new generation of ultra-advanced, intelligent composite materials featuring electrorheological (ER) fluids that can adjust their global properties to the ambient environment in a fairly autonomous sense [64–68].

These ultra-advanced composite materials capitalize on the superior characteristics of advanced composite materials, which are interfaced with dynamically tunable ER fluids contained in voids in the advanced composite structure. Changes in the electrical field imposed on the

FIGURE 6 **Design-for-manufacture strategy for composite-based parts and subassemblies.**

ER fluids dramatically alter the rheological characteristics of the fluids, hence the global mass, stiffness, and dissipative characteristics of the ultra-advanced composite structure. The instantaneous response time of the ER fluids and the inherent ability of these materials to interface with solid state electronics and modern control systems provides designers, for the first time, with a unique capability to synthesize ultra-advanced intelligent composite structures, whose continuum electroelastodynamic response can be actively controlled in real time.

This class of innovative materials derives intelligence from the merger of sensors, embedded in the finite element control segments of the ultra-advanced composite material continuum, microprocessors, and dynamically tunable electrorheological fluids. Several different sensing technologies are available for monitoring the elastodynamic behavior of the ultra-advanced composite structure. While several conventional classes of sensors (e.g., strain gages, thermocouples, accelerometers) could be embedded in this potentially revolutionary class of materials, they entail several inherent disadvantages in these applications, such as the necessity to also embed electrical cables in the material, thereby weakening it; moreover, they are not immune to electromagnetic interference, and they require electrical insulation. Fiber optic sensors and data links, which do not suffer from these disadvantages, are proposed for the new class of ultra-advanced intelligent materials. Moreover, the fibers are sufficiently small and are of high strength; thus they do not significantly reduce the mechanical properties of the material in which they are embedded. The output signals from the sensors are fed to the appropriate microprocessor, which evaluates the signals before determining an appropriate control strategy to achieve the synthesis of the desired elastodynamic response characteristics. This is typically accomplished by controlling the rheological characteristics of the ER fluid domains in the finite element segment associated with the particular sensor. Alternatively, intelligent sensors may be employed by integrating the sensing function and the associated electronic data processing function on a single integrated circuit chip. The change in the rheological characteristics of the ER fluid in a typical finite element control segment in turn alters the global mass, stiffness, and damping characteristics of the ultra-advanced composite structure to achieve the designed vibrational response.

It is anticipated that advanced composite materials and innovative ultra-advanced intelligent composite materials will be crucial ingredients in the evolution of articulating mechanical and robotic systems for the military, aerospace, defense, and advanced manufacturing industries of the future. Significant strides in advanced materials technology would permit a major acceleration in the pace at which humankind unravels new frontiers, both on earth and throughout the universe.

B. S. Thompson and M. V. Gandhi

References

1. J. S. Albus, "Industrial Robot Technology and Productivity Improvement," *Exploratory Workshop on the Social Impacts of Robotics: Summary and Issues,* Congressional Office of Technology Assessment, Report OTA-BP-CIT-11, pp. 62–89, Washington, DC, 1982.
2. *Computerized Manufacturing Automation: Employment, Education, and the Workplace,* Congressional Office of Technology Assessment, Report OTA-CIT-235, pp. 77–80, 316–318, Washington, DC, 1984.
3. K. Sadamato, *Japan Robot News, 1,* 1 (1982).
4. F. Fukuchi and M. Awane, *Hitachi Rev., 34,* 1 (1985).
5. R. M. Christensen, *ASME Appl. Mech. Rev., 38,* 1267 (1985).
6. G. G. Lowen and C. Chassapis, *Mech. Machine Theory, 21*(1), 33 (1986).
7. B. M. Bahgat and K. D. Willmert, *Mech. Machine Theory, 11,* 47 (1976).
8. A. Midha, A. G. Erdman, and D. A. Frohrib, *Mech. Machine Theory, 13,* 603 (1978).
9. C. Bagci and S. Kalaycioglu, *ASME J. Mech. Design, 101*(3), 417 (1979).
10. B. S. Thompson and C. K. Sung, *ASME J. Mech., Transm., Autom. Design, 106*(4), 482 (1984).
11. B. S. Thompson and C. K. Sung, *Mech. Machine Theory, 21*(4), 351 (1986).
12. I. Imam and G. N. Sandor, *ASME J. Eng. Ind., 97,* 609 (1975).
13. M. R. Khan, W. A. Thornton, and K. D. Willmert, *ASME J. Mech. Design, 100,* 319 (1978).
14. W. A. Thornton, K. D. Willmert, and M. R. Khan, *ASME J. Mech. Design, 101,* 392 (1979).
15. W. L. Cleghorn, R. G. Fenton, and B. Tabarrok, *Mech. Machine Theory, 16,* 399 (1981).
16. C. Zhang and H. T. Grandin, *ASME J. Mech., Transm., Autom. Design, 105,* 267 (1983).
17. B. S. Thompson and M. V. Gandhi, "The Finite Element Analysis of Flexible Components of Linkage Mechanisms Made from Fiber-Reinforced Composite Materials," ASME Paper 80-DET-63 (1980).
18. C. K. Sung and B. S. Thompson, *ASME J. Mech., Transm., Autom. Design, 109,* 74 (1987).
19. C. K. Sung, B. S. Thompson, P. Crowley, and J. Cuccio, *Mech. Machine Theory, 21,* 103 (1980).
20. B. S. Thompson, D. Zuccaro, D. Gamache, and M. V. Gandhi, *ASME J. Mech., Transm., Autom. Design, 105,* 528 (1983).
21. B. S. Thompson and C. K. Sung, *J. Sound Vibration, 111*(2), 399 (1986).
22. R. Ramirez, "Design of a High-Speed Graphite Composite Robot Arm," ASME Paper 84-DET-50 (1984).
23. B. S. Thompson and C. K. Sung, *ASME J. Mech., Transm., Autom. Design, 106*(2), 183 (1984).
24. C. K. Sung and B. S. Thompson, *ASME J. Mech., Transm., Autom. Design, 109,* 74 (1987).
25. D. Hull, *Introduction to Composite Materials,* Cambridge University Press, Cambridge, 1981, p. 5.
26. A. B. Schultz and S. W. Tsai, *J. Compos. Mater., 2*(3), 368 (1968).
27. M. Sharif-Bakhtiar and B. S. Thompson, *ASTM J. Compos., Technol. Res., 8*(3), 83 (1986).

28. N. D. Perriera and S. Dubowsky, *J. Acoust. Soc. Am., 67,* 551 (1980).

29. B. S. Thompson, *J. Sound Vibration, 89*(1), 4 (1983).

30. K. W. Kirberg, Institut für Kunststoffverarbeitung an der RWTH Aachen, Aachen, West Germany, personal communication, April 1986.

31. M. Blikstad, P. O. W. Sjoblom, and T. R. Johannesson, *J. Compos. Mater., 18,* 32–46 (1984).

32. Y. A. Tajima, *SAMPE Q., 11*(4), 1 (1980).

33. B. Harris, *Engineering Composite Materials,* Institute of Metals, London, 1986.

34. G. S. Springer, *Environmental Effects of Composite Materials,* Vols. I and II, Technomic, Lancaster, PA, 1984.

35. T. J. Chung and J. L. Prater, *J. Thermal Stresses, 3,* 453 (1980).

36. G. C. Sih, M. T. Shih, and G. C. Chou, *Int. J. Eng. Sci., 18,* 19 (1983).

37. T. J. Chung and R. J. Bradshaw, *J. Compos. Mater., 15,* 502 (1981).

38. R. J. Atkin and R. E. Craine, *Q. J. Mech. Appl. Math., 29*(2), 209 (1976).

39. M. V. Gandhi and M. Usman, "Exact Solutions in Mixture Theory," *Arch. Mech., 3*(2) (1988).

40. M. V. Gandhi, K. R. Rajagopal, and A. S. Wineman, *J. Elasticity, 15*(2), 155 (1985).

41. K. R. Rajagopal, A. S. Wineman, and M. V. Gandhi, *Int. J. Eng. Sci., 24*(8), 1453 (1985).

42. M. V. Gandhi, K. R. Rajagopal, and A. S. Wineman, *Int. J. Eng. Sci., 25*(11/12), 1441 (1987).

43. M. V. Gandhi and M. Usman, *Proceedings of the Third Japan–U.S. Conference on Composite Materials,* 1986, pp. 785–792.

44. C. Y. Lee, B. S. Thompson, and M. V. Gandhi, *ASME J. Eng. Mater. Technol., 110*(2), 174 (1988).

45. C. Y. Lee, M. Pfeifer, B. S. Thompson, and M. V. Gandhi, *J. Compos. Mater.* (submitted).

46. C. K. Sung, B. S. Thompson, and M. V. Gandhi, ASME Paper 87-DAC-60, in *Advances in Design Automation,* Vol. 1, *Design Methods, Computer Graphics and Expert Systems,* DE-Vol. 10-1, American Society of Mechanical Engineers, New York, 1987, p. 231.

47. C. K. Sung, M. V. Gandhi, and B. S. Thompson, *Proceedings of the Applied Robotics and Design Automation Conference,* St. Louis, MO, 1980, VI-2.1.

48. M. V. Gandhi and M. Usman, presentation at the Symposium of Performance Composites in Severe Environments, ASME Winter Annual Meeting, Boston, December 1987: *ASME J. Eng. Mater. Technol., 10*(2), 140 (1988).

49. C. K. Sung, B. S. Thompson, and M. V. Gandhi, *Proceedings of the Seventh World Congress on the Theory of Machines and Mechanisms,* Seville, Spain, Vol. 3, 1987, p. 1627.

50. L. Chao, M. V. Gandhi, and B. S. Thompson, *Proceedings of the Tenth OSU Applied Mechanisms Conference,* New Orleans, Vol. 3, 1987, 7B-III.

51. D. X. Liao, C. K. Sung, and B. S. Thompson, *Proceedings of the Ninth OSU Applied Mechanisms Conference,* Vol. 2, 1985, II-1.

52. H. D. Wagner, I. Roman, and G. Marom, *J. Mater. Sci., 17,* 1359 (1982).

53. S. B. Choi, M. V. Gandhi, B. S. Thompson, and C. Y. Lee, *J. Robotic Syst., 5*(1), 73 (1988).

54. D. X. Liao, C. K. Sung, and B. S. Thompson, *J. Compos. Mater., 20,* 485 (1986).

55. T. R. Tauchert and S. Adibhatla, *J. Compos. Mater., 18,* 58 (1984).

56. T. R. Tauchert and S. Adibhatla, *J. Compos. Mater., 16,* 433 (1982).

57. Y. Hirano, *J. Compos. Mater., 13,* 329 (1979).

58. S. Adali, *ASME J. Mech., Transm., Autom. Design, 105,* 214 (1983).

59. L. A. Schmit and B. Farshi, *Int. J. Numer. Methods Eng., 17,* 519 (1983).

60. C. K. Sung, B. S. Thompson, M. V. Gandhi, and C. Y. Lee, *Int. J. Mech. Sci., 29*(12), 821 (1987).

61. T. M. Halligan, *Production,* July 1986, p. 50.

62. H. L. Stein, *Specifying Eng.,* July 1985, p. 48.

63. A. J. Holzer, *Comput. Ind.,* 1981, p. 199.

64. B. S. Thompson and M. V. Gandhi, presentation at ASM/ESD Advanced Composites Conference and Exposition, in *Advanced Composites III: Expanding the Technology,* 1987, p. 23.

65. M. V. Gandhi, B. S. Thompson, S. B. Choi, and S. Shakir, *ASME J. Mech., Transm., Autom. Design* (in press).

66. M. V. Gandhi, B. S. Thompson, and S. B. Choi, *Proceedings of the Fourth Japan–U.S. Conference on Composite Materials,* 1988, p. 875.

67. M. V. Gandhi, B. S. Thompson, and S. B. Choi, *J. Compos. Mater.* (in press).

68. M. V. Gandhi and B. S. Thompson, "Ultra-Advanced Intelligent Composite Materials for Active Vibration Control Applications," Patent disclosure, Michigan State University, June 1988.

Asbestos and Substitutes

See Fillers

Autoclave Processing

Autoclave processing of composites is an extension of the vacuum bag technique in which pressure in excess of atmospheric is exerted on the part by a gaseous medium during cure. The pressure vessels that contain the compressed gas during the processing of the composite, called autoclaves, are used for the production of high quality, complex, composite hardware for the aerospace and other high technology industries. The method is well suited for large parts and production quantities from one to several thousand parts.

Similar techniques using different pressurizing mediums are hydroclaving (liquid water) and thermoclaving (various temperature-resistant organic liquids such as silicate esters). These techniques tend to be very specialized and have not been widely adopted by industry.

Steam autoclaves have been used for at least the past century for the sterilization of medical instruments and for the production of rubber articles. From about 1930 through 1945, steam autoclaves were used to produce many complex and beautiful structural items from phenolic-impregnated plywood. The most striking example of this is the Hughes flying boat, whose entire load-bearing structure and exterior skin were autoclave molded of phenolic/birch plywood.

Autoclave processing as it is known today probably started during World War II with the manufacture of polyester fiberglass radomes. Steam was abandoned as a pressurizing medium and replaced with air or inert gas. Large, ogival shapes with internal ducts for de-icing were built in one piece using a lost-wax technique. The development of structural epoxies with high viscosities and elevated temperature cures in the late 1940s fitted well with the capabilities of autoclave curing. The aircraft industry, with its requirements for numerous prototypes and limited production of extremely reliable hardware, has since embraced the technique as the basic production method for composite hardware.

Description

Autoclave processing is the method of curing or thermal forming a composite under gas pressure greater than atmospheric. The process is usually accomplished on rigid male or female molds. The composite part is laid up and enclosed in a flexible, leakproof envelope or vacuum bag. Full or partial vacuum is drawn within the bag, and gas pressure greater than atmospheric is exerted on the exterior of the bag. The part temperature is then raised to accomplish cure and to reduce the viscosity of the polymeric matrix so that flow, wetting of the reinforcement, and consolidation of the composite occur.

The technical reasons for the use of augmented pressure are threefold. First, the pressure exerts mechanical forces on the unconsolidated composite, thereby increasing the efficiency of mechanical transport of volatiles to the vacuum ports and causing increased wetting and flow of the resin because of the higher shear forces. Second, the volume of the trapped air and released volatiles is reduced in direct proportion to the absolute pressure at a given temperature; therefore porosity and voids are reduced in size. Third, the transport of volatile materials by solution in the molten polymer and diffusion to the vacuum ports is more efficient at elevated pressure.

An additional very practical reason for augmented pressure when dealing with polymers that suddenly release large volumes of volatile material is to prevent the vacuum bag from inflating like a balloon and the layup from being completely disrupted. The pressure exerted by volatile materials at elevated temperature can be extremely high. For instance, water at the common curing temperature for epoxy resins (180°C) has a vapor pressure of about 1 MPa. Control of sudden volatile release is accomplished by high external pressure, reduced vacuum within the bag, and low rates of temperature rise.

Equipment

The majority of autoclaves for composite fabrication are cylindrical pressure vessels with domed ends, one of which is the door or entrance. The autoclaves are usually mounted horizontally on the factory floor to provide easy access to the interior. Construction is of welded steel conforming to pressure vessel and boiler codes. Diameter is the limiting factor in size, since the hoop stress in a cylindrical pressure vessel is directly proportional to the diameter and approximately twice the longitudinal stress. Large-diameter autoclaves therefore require extremely thick walls and become very expensive. Most research or development autoclaves are about 1 meter in diameter; production autoclaves run from 1 to about 8 meters in diameter (Fig. 1). Lengths of up to 22 meters have been manufactured. All this equipment has massive doors with a self-locking design to prevent opening under pressure. The doors and locking mechanisms in most large autoclaves are mechanically operated.

Modern autoclaves are equipped with internal heating, internal insulation, cooling coils, fans for circulating the gas, vacuum lines, venting lines, lights, gas inlets, and instrumentation. All these facilities penetrate the shell through pressure glands or pressure-safe connections. Doors are pressure sealed with large silicone rubber O-rings. These and other heat-sensitive areas are frequently water cooled. Heating and cooling cycles may be preprogrammed and microprocessor controlled.

Production autoclaves, used for curing polyester, epoxy, and phenolic resin systems at temperatures from 120 to 177°C are usually rated to about 260°C. The newer high temperature resistant composites made from polyimides and engineering thermoplastics may require curing or thermoforming temperatures up to 450°C. Autoclaves for these materials and most research autoclaves therefore have temperatures capabilities from about 450 to 540°C.

Steam and air have been largely abandoned as pressurizing media: steam because of its deteriorating and corrosive effects on parts and equipment, and air because of the potential for autoignition and possible explosion of the contents of the autoclave. Nitrogen and carbon dioxide gas are now generally used. Nitrogen may be supplied by a battery of cylinders of compressed gas or by evaporation and pressurization from a liquid nitrogen source. Carbon dioxide may be obtained from pressurized gas cylinders or by direct volatilization from supercooled liquid. Production autoclaves are rated in the 1–2 MPa range, while research equipment may be rated as high as 7 MPa.

FIGURE 1 **(*Top photo*) Autoclave 18.2 meters long × 7.6 meters wide with control console built on site. Photograph courtesy of Beech Aircraft, Wichita KS. (*Bottom photo*) Autoclave 10.6 meters long × 3.0 meters wide.** (Photograph courtesy of Bonded Technology, Cromwell, CT.)

High capacity, industrial vacuum pumps are connected to the vacuum outlet of the autoclave, which is usually terminated with a multiport vacuum manifold inside the vessel. It is good practice to trap the volatile materials generated during processing with a cold trap to protect the vacuum pumping system. Knowledge of the volume and composition of the condensed volatiles can also be very helpful in process development and control.

Instrumentation consists of multiple thermocouples, pressure-measuring devices for determining pressure and degree of vacuum and, in some cases, dielectric monitoring devices. The data from these sensors are presented in real time at the control panel and usually are permanently recorded.

Materials

Preimpregnated materials have largely replaced wet layups in modern practice. More precise control of resin content, volatile content, and reinforcement distribution and placement is achieved with their use. Furthermore, layup operations are easier, "rubbing out" of excess resin is eliminated, waste is reduced, operations are safer, and cleanup is facilitated. An ideal prepreg for autoclave operations is flexible (has drape), has enough tack to stick to previous layers with slight pressure, conforms easily to contours, and is not noxious. Specifications for these materials are highly developed and are treated elsewhere in this encyclopedia.

Present practice with epoxies, polyesters, and thermoplastics tends toward the use of net resin content, low flow prepregs. With phenolics, bismaleimides, polyimides, and solvated thermoplastics, which eliminate large volumes of volatile materials during processing, considerable resin bleeding occurs at times of low viscosity. In these cases allowance must be made in the prepreg for the anticipated resin loss during processing.

Honeycomb core materials, foam reinforcements, metallic inserts, metal foils, lightning-protecting metallic grids, and many other materials may be included in the layup for autoclave-processed composite parts. However, it is important to remember that the autoclave pressure must not exceed the compressive strength of these materials throughout the temperature range of the process. Metallic inserts are usually bonded into the composite part after curing. In honeycomb structure, however, these inserts may be cocured into the part during the autoclave cycle. In this case it is customary to surround the insert with syntactic foam or "foam in place" materials to increase the compressive and shear strength of the honeycomb in the area of the insert.

Tooling

Tools for autoclave use are usually simple male or female forms. The tool surface is the more dimensionally precise and has a better surface finish than the bag surface in an autoclave molded part. Occasionally, matched male and female tools or tools with complex metallic inserts are used in autoclaves. In these cases the tools usually are completely encased in the vacuum bag to provide uniform pressure on the part.

The choice of materials for tooling is usually based on the temperatures to be encountered during process operations and the production quantities required.

Well-dried plaster and wood masters may be used to produce one or two prototype parts, provided a low temperature cure is employed. For production parts of polyester, epoxy, and phenolic composite, aluminum, steel, and cast epoxy tooling is widely used. Large epoxy tools are reinforced with fiberglass and other reinforcements.

In all elevated-temperature molding, the differential thermal expansion between the tool and the composite part must be considered. Fiberglass composites closely match the thermal expansion of steel but may differ greatly in coefficient of expansion from other tooling materials. Composites reinforced with graphite or Kevlar present special difficulties because of their very low coefficients of expansion in the plane of the laminate. Two corrections must be made for these thermal mismatches. First, desired dimensional accuracy will be achieved only by applying a correct shrinkage factor to the room temperature tool dimensions. Second, care must be taken to prevent parts from locking into or onto male and female features of the tool. In very low thermal expansion composite parts greater than about half a meter in size, fiber disruption, compression buckling, and delamination may occur unless the expansion characteristics of tool and part are closely matched.

The advantages and disadvantages of various tooling materials are briefly presented below.

1. **Cast Epoxy–Glass Reinforced**
   Advantages: complex shapes and curved surfaces easily replicated from plaster masters; lightweight; tough; smooth surface; repairable; low cost.
   Disadvantages: temperature limited to 200°C maximum; high coefficient of thermal expansion (CTE).
2. **Aluminum Alloys**
   Advantages: low cost; machinability; excellent surface finish; high thermal conductivity.
   Disadvantages: high CTE; poor dimensional stability; tendency to warp at +260°C.
3. **Steel, Cast Iron, and Similar Ferrous Materials**
   Advantages: low cost; machinability; stable above 260°C.
   Disadvantages: massive; heavy; high thermal mass; relatively high CTE.
4. **Cast Epoxy–Graphite Reinforced**
   Same as 1, except advantage of very low CTE.
5. **Polyimide–Graphite Reinforced**
   Advantages: temperature resistant to 260–340°C; CTE closely matches advanced composite parts; tough; impact resistant; lightweight; thin tools are easy to heat.
   Disadvantages: expensive; difficult to process; operation above glass transition point of resin may cause dimensional instability and surface deterioration due to oxidation; limited production experience.
6. **Bulk Graphite**
   Advantages: very dimensionally stable at elevated temperatures; low CTE.
   Disadvantages: easily damaged; soft surface; brittle; thick sections required; stock size limitations may require joining of pieces; high thermal mass.
7. **Ceramic and Ceramic Composite**
   Advantages: stable at elevated temperatures; fairly low CTE; hard surface.
   Disadvantages: massive; heavy; difficult to heat; may not be vacuum tight; limited production experience.

Production tools frequently contain built-in heating elements and sensors, vacuum ports around the periphery of the layup area, and provisions for stiffening and handling. Small tools usually are mounted on wheeled steel tables. In very large tools the wheeled support structure may be integral with the tool, with rails provided to move the tool in and out of the autoclave.

Tooling Accessories

Separate mold pieces may be used to exert pressure in selected areas and to better define specific shapes and areas. Shaped metal or silicone rubber forms are placed directly on the finished layup and under the vacuum bag. Rounded edges, fillets, or depressed areas, where the vacuum bag may bridge or exert excessive pressure, are more precisely defined by these devices.

Caul plates, smooth, flexible sheets of metal or silicone rubber, frequently are placed on the layup before bagging to obtain smooth, flat surfaces on the bagged surface.

Layup Procedures

The sequence of operations involved and the materials used in laying up composite parts for autoclave processing are almost identical to those in vacuum bag or oven curing. A typical sequence is outlined.

1. *Tool Preparation.* Clean tool surface and treat with mold release.

2. *Layup.* Place precut composite prepreg pieces on tool in the precise sequence, orientation, and ply count defined in the engineering drawing or shop planning. The operation may be done by skilled technicians or by automatic tape-laying equipment.

3. *Bulk Reduction and Consolidation.* In thick, multiple layups or with materials of high bulk factor (prepreg thickness divided by final molded thickness), it often is necessary to reduce the layup thickness at intervals during the layup procedure. The partial layup is temporarily vacuum bagged and heated. This consolidation is usually performed in an oven at temperatures low enough to preclude excessive advancement of thermosetting resins. The sequence of events in a typical bulk reduction–consolidation cycle is shown in Figure 2.

4. *Post-Layup Operations.* After the layup has been completed, a series of steps may be taken to control resin flow and provide adequate venting of volatiles to the vacuum system. Resin dams may be installed around the periphery of the layup to restrict lateral resin flow. Dams may consist of metal chain, helically wound metal wire, felt strips, and many other porous materials. Peel plies may be placed on the surface of the layup. These plies are sacrificial. They are torn off and discarded after cure to provide a clean, matt surface. A separator ply or plies may then be applied, usually consisting of Teflon-treated glass cloth, followed by several layers of bleeder material. The function of the bleeder material is to absorb excess resin and to provide passage of volatiles to the vacuum ports.

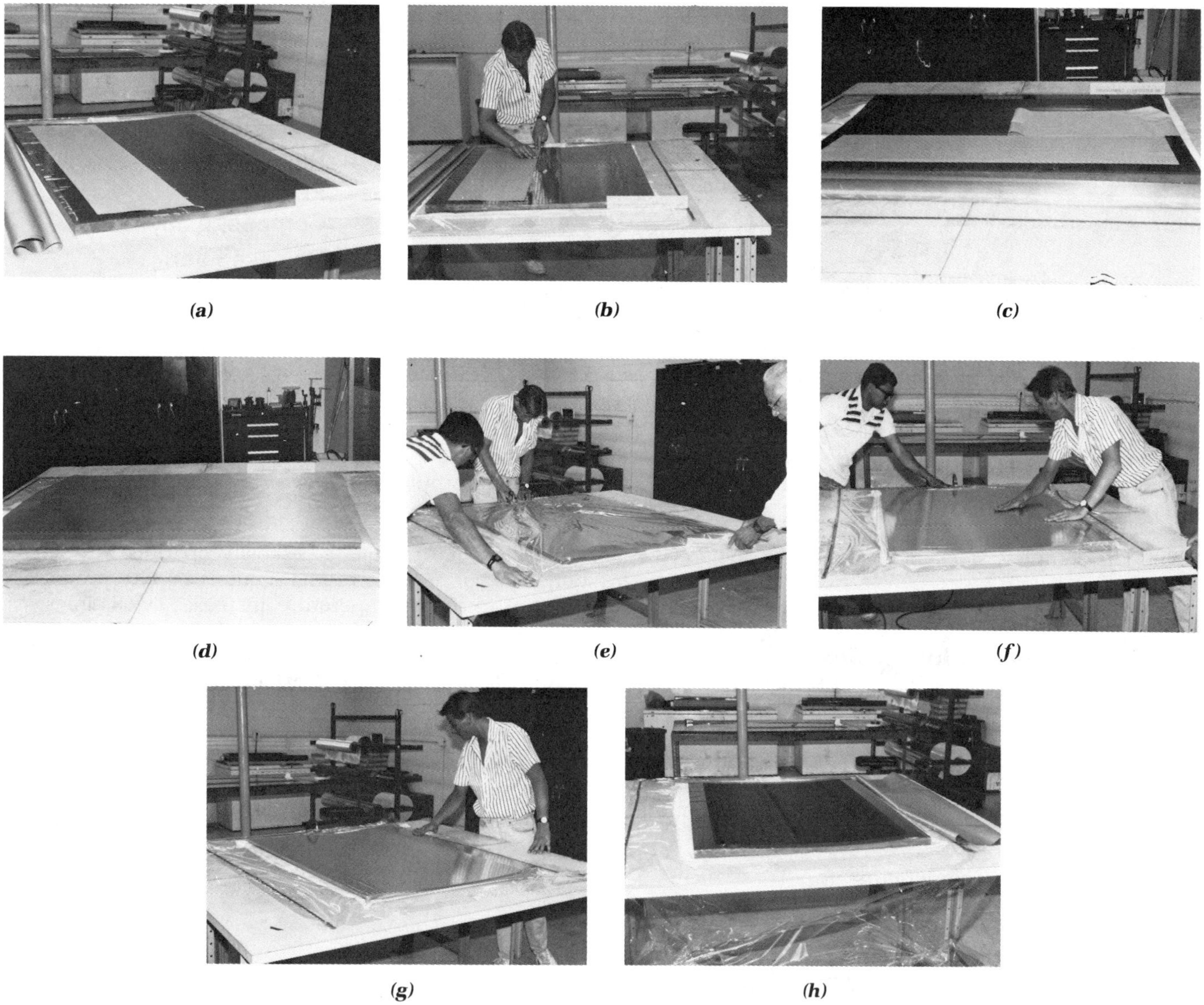

FIGURE 2 Bulk reduction–consolidation sequence. (*a*) Bulk graphite mold with release film and first piece of unidirectional graphite fiber–polyimide (du Pont K111) prepreg in place. Separator paper on prepreg has not been removed. (*b*) Technician continues layup of prepreg by carefully aligning edges of adjoining pieces. (*c*) Completed layup with Kraft separator paper partially removed. (*d*) Completed layup with porous separator film on top. (*e*) Vacuum bag of nylon film being installed. Tool and layup is double-bagged; that is, tool is completely enveloped, top and bottom, by vacuum bag. (*f*) Technicians smooth layup and eliminate wrinkles as vacuum is pulled. (*g*) Technician checks seal of vacuum bag with full vacuum applied. The tool is now placed in an oven at low temperature for a short period. (*h*) Consolidated layup on tool after removal of upper vacuum bag. The consolidation sequence is now repeated until the full number of plies is in place and consolidated.

5. *Vacuum Bagging.* This critical operation consists of enveloping and sealing the layup within an impermeable membrane. Materials used, in approximate ascending order of temperature resistance, are vinyl films, nylon films, Kapton film, silicone rubber sheet, fluorosilicone rubber sheets, aluminum foil, and stainless steel foil. The metal foils and Kapton film have very low elongation and are therefore useful for flat and single-curvature shapes only. In conforming to highly contoured shapes, gores and/or splices may be made in the bagging material.

6. *Sealing.* The completed vacuum bag is sealed, vacuum tight, to the tool, with pressure-sensitive sealant compositions in tape form. Penetrations of the bag for thermocouples, other sensors, gores, and vacuum ports are also sealed.

7. *Final Preparation Before Autoclaving.* All connections are made to sensor jacks, vacuum ports, integral heaters, etc. Vacuum is pulled, and the bag and tool are observed for leaks. Finally, if the tool is integrally heated or if heating blankets are used, thermal insulation is ap-

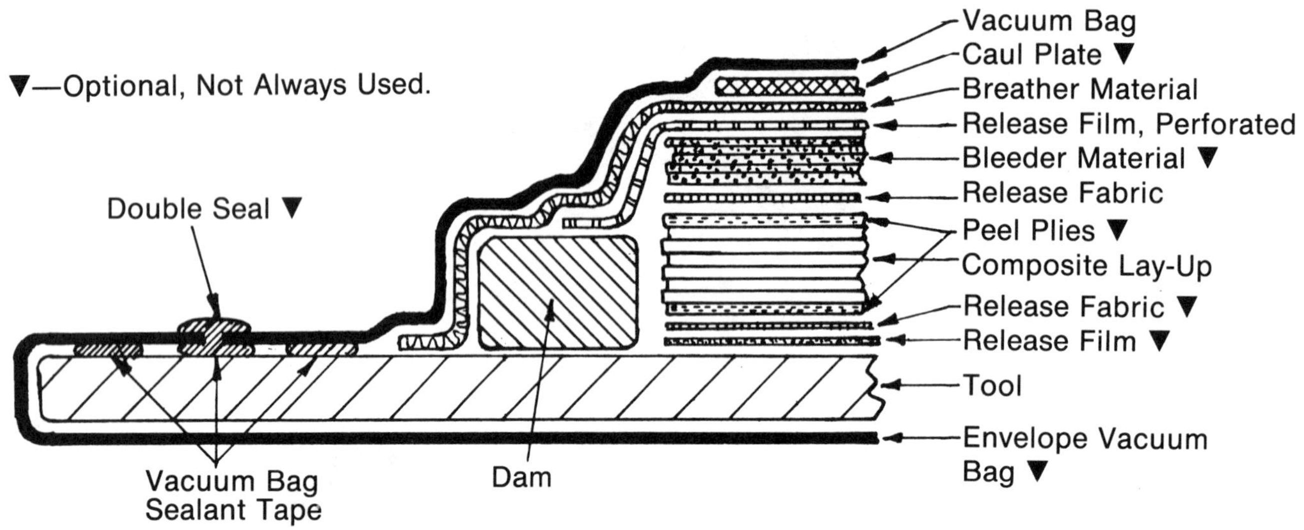

FIGURE 3 **Vacuum bagging stacking sequence for autoclave processing.**

plied. The complete stacking sequence of a layup is diagrammed in Figure 3.

Autoclave Processing Cycles

The autoclave cycle is designed to cause critical process events to occur in the right sequence and to the required extent. In thermosets, these events may include the release of volatiles, de-esterification, polymer chain extension, and such chemical reactions as condensation and imidization, cross-linking, viscosity buildup, and gelation. Control of these events requires precise programming of rates of temperature increase, dwell times at specific temperature plateaus, and the timing and extent of vacuum and autoclave pressure application.

Autoclave cycles for polyesters and epoxies that emit little volatile material during cure are relatively simple. Cycles for phenolics, bismaleimides, and polyimides become increasingly more complex. Some of these are illustrated in Figure 4.

Autoclave cycles for engineering thermoplastics con-

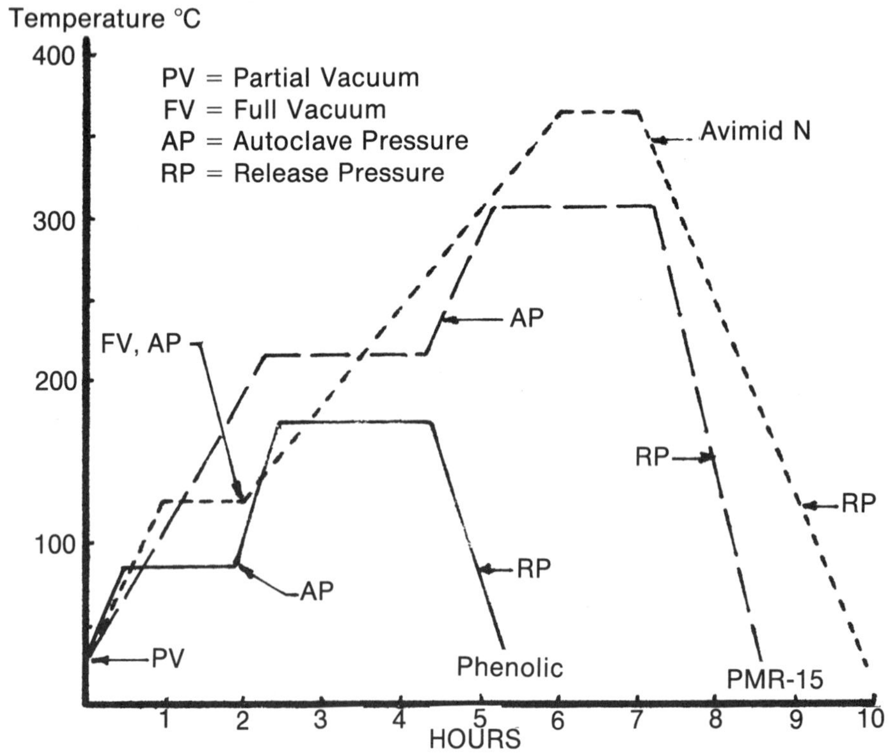

FIGURE 4 **Typical autoclave cycles for thermosetting matrices.**

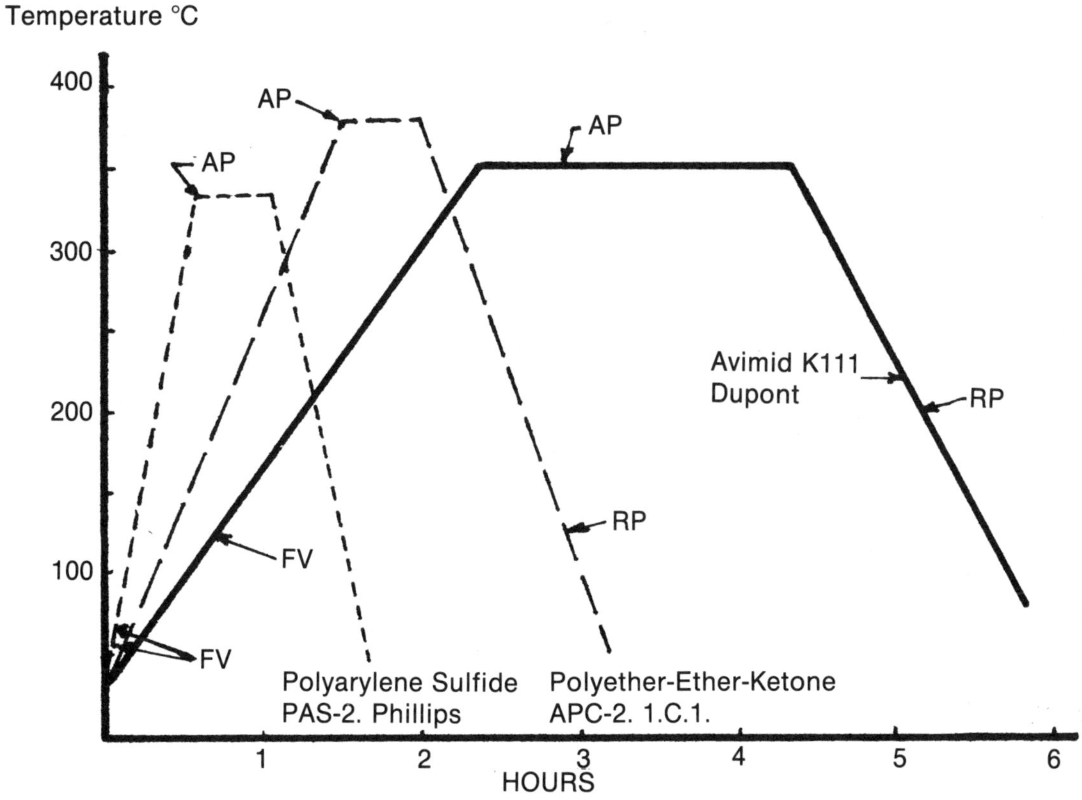

FIGURE 5 Typical autoclave cycles for thermoplastic matrices.

sist of a simple cycle of heating, melting, and consolidation. For crystalline polymers, cooling rates are controlled to achieve the required ratio of crystalline to amorphous phases. High temperatures must be achieved to attain complete melting or sufficiently low viscosity for consolidation to occur. These temperatures range from 325 to 400°C and place great stress on bagging and sealing materials (Fig. 5).

Safety Considerations

All modern autoclaves are constructed and maintained according to strict pressure vessel and boiler codes (in the United States, the ASME code is the basis of regulation). Nevertheless, high pressure operations are inherently dangerous and should be carried out with strict attention to correct procedures.

The primary hazard in current autoclave operation, however, is not explosion, but asphyxiation and exposure to noxious fumes. All organic materials slowly degrade at elevated temperatures to produce toxic gases. These products, combined with an atmosphere containing practically no oxygen, are extremely hazardous to personnel who may enter an autoclave prematurely. After an autoclave cycle, the autoclave atmosphere must be completely vented and substantially replaced with ambient air before personnel are permitted to enter.

Specifications

The quality of autoclave-molded composites is usually governed by material specifications for the finished product. Process specifications are rarely invoked. Shop operations are usually controlled in detailed, step-by-step planning sheets derived from engineering requirements. Inspection points during the process are specified in the planning documents.

Recent Developments

A demonstration of the autoclave molding of a large primary aircraft structure was recently completed by Rockwell International Corporation for the U.S. Air Force. This demonstration of a B1-B wing skin is approximately 15 meters long (Fig. 6). The part was laid up by automatic tape layup equipment from graphite epoxy unitape and cured in one piece [1]. The all composite tool is shown in Figure 7.

Techniques for the autoclave curing or forming of high temperature resistant composites are undergoing rapid development. These materials include advanced polyimides such as du Pont's Avimid N, the Thermid acetylene terminated resins of National Starch Corporation, du Pont's K111 (a particularly interesting polymer that lays up "like an epoxy and, after curing, becomes a

FIGURE 6 **B1-B wing skin after autoclave processing.** (Photograph courtesy of Rockwell International.)

true high temperature thermoplastic [2], polybenzimidazole, and numerous true thermoplastics including polyether ether ketone, polyarylene sulfides, and polyamide imides. Processing of these polymeric composites has been greatly aided by the development of high temperature resistant bagging materials [3].

Other current research probably will culminate in the automatic computer control of processes through the use on in situ dielectric sensors. This development would enhance the yield of high quality composites from difficult-to-process PMR-15 and other high temperature polymers [4].

FIGURE 7 **All-composite tool of graphite fabric–epoxy used for B1-B wing skin.** (Photograph courtesy of Rockwell International.)

Economic Aspects

Autoclave operations consume large amounts of energy and materials, including industrial gases and bagging materials. They are also labor intensive and time-consuming. For large production runs of relatively small composite parts, matched tool molding in presses is usually more economical. Nevertheless, for large parts exceeding the size of available presses, autoclave processing is the only choice for the foreseeable future. Also, for the intermediate production runs required by the aerospace industry, autoclave processing is usually the most economical technique.

L. Brian Keller

References

1. S. Taha, "Manufacturing Technology for Large Aircraft Composite Structure-Wing," End-of-Contract Briefing, Rockwell International Corporation, Jan. 21, 1988.
2. A. R. Wedgewood, K. B. Su, and J. A. Narin, *SAMPE J., 24*(1), 41 (1987).
3. B. Keller, J. Grasty, and A. Castillo, *SAMPE J., 24*(1), 50 (1987).
4. D. R. Day and D. D. Shephard, *Proceedings of the 32nd Int. SAMPE Symp.*, 1987, pp. 1473–1479.

Autoclave Process Modeling

The process of fabricating high performance thick composites currently lacks a firm scientific basis. Based on empiricism and know-how, today's technology is not yielding the part-to-part reliability that will be required of the larger, more complex structures currently on the drawing boards. The major problems in achieving this reliability are posed by the occurrence of voids, resin concentration gradients, and matrix microcracks in the final part. Voids lead to a drastic reduction in the transverse and shear moduli, and in the strength of the composite. The problems due to resin concentration gradients are numerous: some regions have high resin fraction and, therefore, lower than the predicted composite properties, while other regions have low resin fraction, creating fiber-to-fiber interaction.

These phenomena occur during a complex fabrication process involving heat, mass, and momentum transfer with simultaneous chemical reaction in a multiphase system with time-dependent material properties and boundary conditions. The processing science of epoxy resin composites based on fundamental principles has developed in the past few years. It is the result of extensive theoretical modeling and experimental work [1–11]. In this article we review earlier studies on developing science-based cure models for composite processing.

We also present our modeling results and compare them with experimental data. Finally, based on the insight gained by modeling, we propose guidelines for the selection of optimum cure cycle for an autoclave process.

Background

The Autoclave Process

An often-used production procedure for fabricating a high performance structural laminate is the autoclave–vacuum degassing (AC/VD) laminating process. An autoclave is a pressure vessel that operates under controlled temperature and pressure. The vessel, shaped like a shell, usually has two door assemblies, as well as shell, utility, instrumentation penetrations, and vessel support. Vessel internals include insulation, sheet metal shroud, ductwork, blower drive assembly, electrical heater assembly and cooling coils.

In an AC/VD laminating process, individual plies of a preimpregnated material are layered up in a prescribed orientation to form a laminate. The laminate is laid against a smooth tool surface and covered with successive layers of glass bleeder fabric, Mylar or Teflon sheets and, finally, a vacuum bag. A typical entire sandwich is shown schematically in Figure 1 for a unidirectional laminate. The inset depicts a magnified view of the fiber orientation for such a laminate.

The curing process is accomplished after bagging, by moving the entire tool into an autoclave and exposing it to temperature and pressure according to a given cure cycle. Usually, vacuum is initially applied to the bag while the autoclave temperature and pressure are raised. The transfer of autoclave pressure to the resin in the laminate does not occur hydrostatically because the resin is not enclosed in a constant-volume system. Flow can occur initially both vertically (thickness direction) and horizontally. Furthermore, the network of fibers eventually can act as a network of springs to which the vacuum bag and bleeder assembly transfer the load from the autoclave pressure. This load can be transferred through the fiber network to the tool surface with only a slight frictional loss. As the fiber spring network is loaded and compressed, more resin flow occurs; however, at this point, the resin viscosity is increasing, and the permeability is decreasing, making resin flow more difficult.

The quality of the final product is significantly affected by the choice of the cure cycle. Thus, the cure cycle selection for each particular application is important and should be optimized. The processing parameters that need optimization are the autoclave temperature,

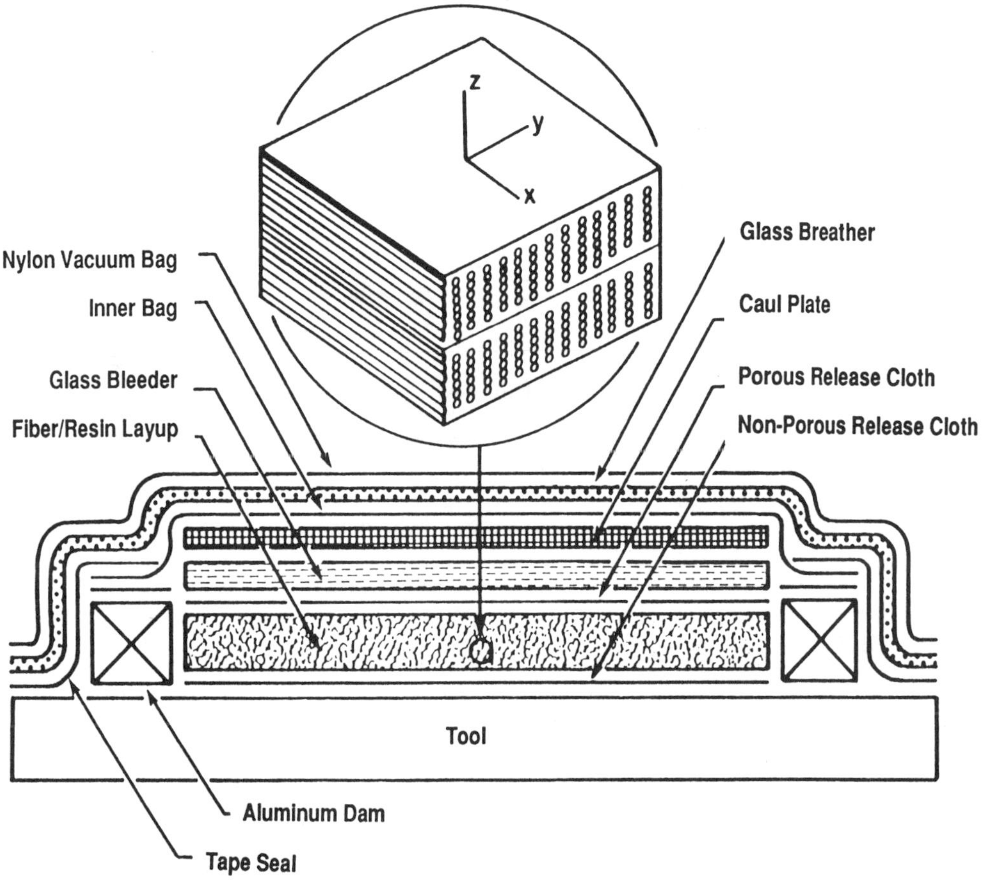

FIGURE 1 Schematic of a laminate layup for autoclave processing. Inset shows magnified view of fiber orientation within the laminate.

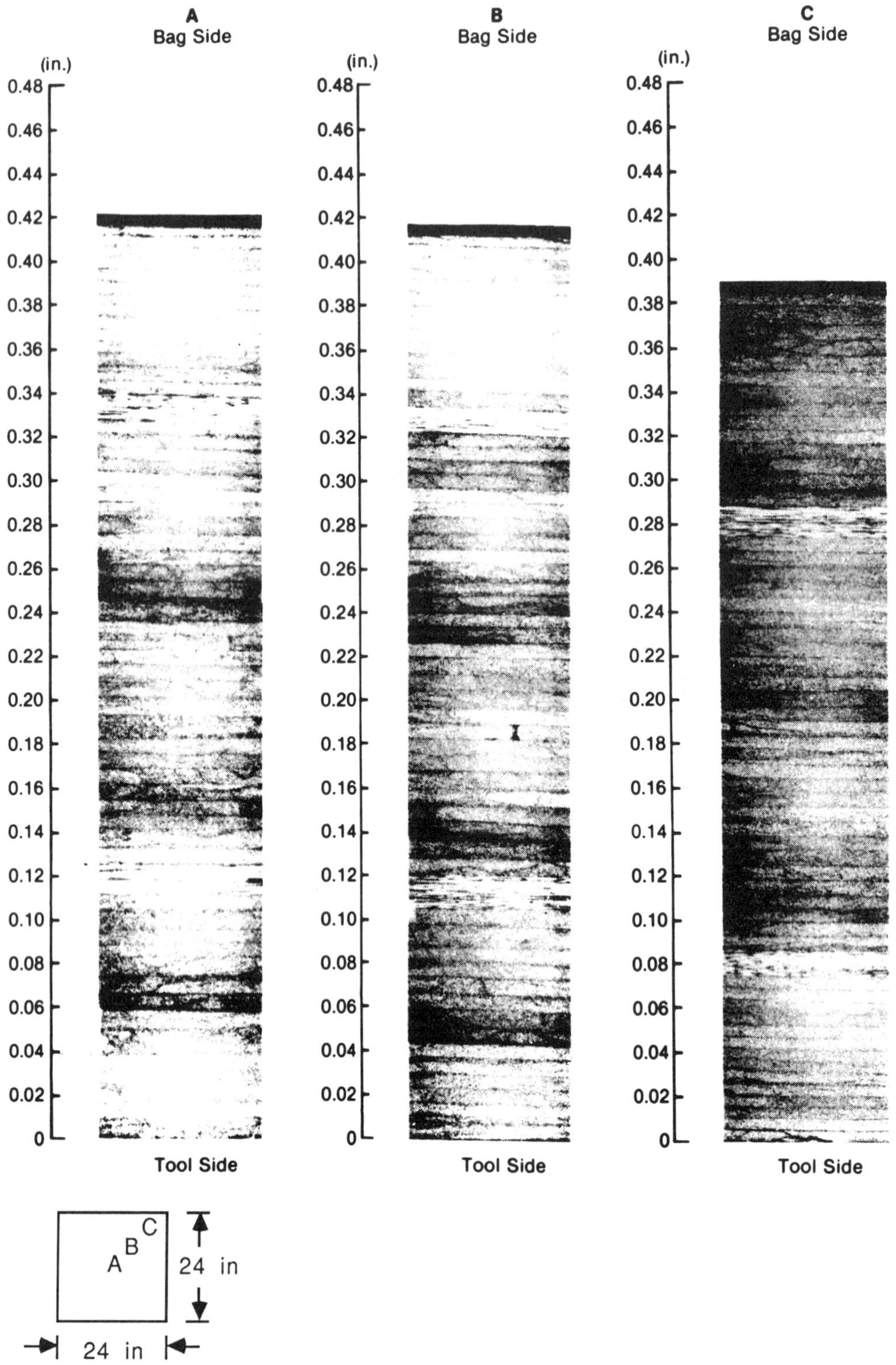

FIGURE 2 **Photomicrographs of a 60-ply laminate.** (From Ref. 12.)

the autoclave pressure, and the bleeder bag pressure. In turn, the variables that need to be controlled are resin temperature and resin pressure.

In general, the choice of the time-at-temperature cycle for a particular fiber matrix system is primarily based on two factors: (a) the time–temperature viscosity dependence of the resin matrix, which in turn requires a realistic model for the curing kinetics, and (b) the desired glass transition temperature T_g of the cured laminate.

In the past, the processing parameters and the cure cycle used in the fabrication of composite structures were derived from experience as a result of iterative experimental testing. Two major drawbacks exist in the empirical methods: (a) extensive experimental testing is necessary to determine the proper cure cycle and (b) a cure cycle found to be suitable for a given material for one application often is inapplicable for a different material or for the same material in a different application.

Clearly, this demonstrates that more fundamental work was needed to fill the gaps in the technology of composites fabrication to allow a more efficient selection of process parameters. This article addresses the problems of void growth, resin flow, and related aspects of consolidation. In particular, we examine the effects of humidity and temperature on void growth, and the effects of viscosity and degree of cure on flow. The resin pressure in a curing laminate affects not only such flow-related phenomena as degree of consolidation but also influences void growth, stability, and transport. Suppression of water vapor void growth during composite processing requires proper understanding and control of the resin pressure within the laminate.

Evidence of Voids, Resin Gradients, and Matrix Microcracking

The occurrence of voids, resin gradients, and matrix microcracking has been thoroughly documented in thick laminates [2,12,13]. The voids are apparently associated with the prepreg surface in almost all cases. Void formation can be due to mechanical entrapment as well as nucleation of stable voids in the resin. Mechanical entrapment could include entrained gas bubbles from the resin mixing operation; bridging voids from large particles or particle clusters (quenched DDS curing agent, airborne particles, or paper release agent); voids from wandering tows, fuzz balls, or broken fibers; air pockets and wrinkles created during lay up; and ply terminations. Nucleation can occur either homogeneously within the resin or heterogeneously at a resin–fiber or a resin–particle interface. If mechanical entrapment occurs, the initial void distribution is dictated by the processing steps before the AC/VD step. In this situation, the initial void distribution must be fed into the model as an experimentally determined parameter.

Once a stable void has formed in a thermosetting resin, its growth is due to diffusion of water from the bulk resin into the void. Grayson and Wolf [14,15] used the technique of precision abrasion mass spectrometry to study the volatile compounds present within a void located in a graphite–epoxy laminate. They detected water as the main component contributing to void growth.

Resin gradient in a cured laminate is evident by the changing thickness from the center to the corner of the laminate (see Fig. 2). The mechanism of microcracking is yet unknown, but in the most general case, the microcracks run continuously through the matrix. Also, matrix microcracking is evident only in some thermosetting resin composites—for example, the bismaleimide and polyimide composites.

The General Model Framework

The model framework for describing the interrelationship between the different phenomena occurring during composite processing appears schematically in Figure 3. The phenomena are, of course, a part of the complete description of the entire processing sequence. As such, they depend on the same material properties and process parameters that govern the entire process. Note that the void and resin flow models are intimately tied to the kinetic, viscosity, and heat transfer models. The overall model, called the master cure model, consists of six interrelated submodels: kinetic, viscosity, heat transfer, resin flow, void growth, and residual stress models. The master cure model can be used for automating the cure cycle in conjunction with a process simulator and a cure optimizer with real-time control.

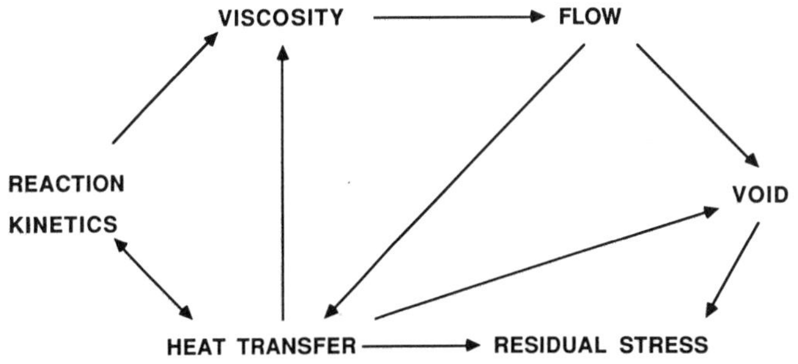

FIGURE 3 **Schematic of overall process model framework.**

Development of the Models

Chemoviscosity

Chemoviscosity indicates the viscosity variations induced by chemical reactions and temperature changes in a reacting system.

Viscosity control during composite processing is particularly critical for thermosetting systems. Viscosity varies with temperature, flow condition and time, and chemical reactions occurring in the liquid state of the resin. In the past, several chemoviscosity and chemorheology models have been proposed [7,8,16].

Most viscosity models are semiempirical and consider the rheological behavior of a *reacting system* as basically governed by two effects: the structural changes caused by the cure reactions (*chemical aspect*), and the variations in molecular mobility induced by the temperature variations (*physical aspect*).

For a reacting system, Nicolais and Apicella [8] have obtained an expression for viscosity that contains both temperature and molecular dependence. They have used the Flory approach to describe the molecular characteristics of the nonlinear polymerization—that is, the chemical aspect—and the Williams–Landel–Ferry (WLF) equation to describe the dependence of viscosity on temperature—that is, the physical aspect. By the Flory approach, on a molecular level, the increase in viscosity corresponds to the increase in molecular weight and to the incipient formation of infinite branched molecules. Growth and branching of the polymer chains occur in the liquid state, where the system is still soluble and fusible. On the other hand, infinite network is developed after the gel point by intermolecular reactions of the branched molecules. Finally, an insoluble cross-linked solid is formed.

Even though the Flory approach is based on fundamental principles, it is often difficult to apply in cases of commercial resin systems whose constituents are not disclosed by the manufacturer. In such situations, a semiempirical equation for viscosity as a function of degree of cure α and temperature T is an alternative approach. Applying the latter approach, Lee, Loos, and Springer's viscosity relationship [16] characterizes the viscosity of Hercules 3501-6 resin.

$$\mu = \mu_\infty \exp\left(\frac{E_\mu}{RT} + K_\mu \alpha\right) \tag{1}$$

where μ (Pa · s) is viscosity, T (K) is temperature, α is the degree of cure, R is a universal gas constant, and μ_∞, E_μ, and K_μ are constants given in Ref. 16. The degree of cure α is predicted from

$$\frac{d\alpha}{dt} = (K_1 + K_2\alpha)(1 - \alpha)(B - \alpha) \qquad \text{for} \quad \alpha \leq 0.3 \tag{2a}$$

$$\frac{d\alpha}{dt} = K_3(1 - \alpha) \qquad \text{for}\, \alpha > 0.3 \tag{2b}$$

where the parameters K_1, K_2, K_3, and B are described in Ref. 16.

Equation (1) is applicable only up to a degree of cure of 0.5 (i.e., before the resin gels). However, even within this limited range, this viscosity relationship is quite useful, since almost no flow occurs after gelation. In the past [4–6], we made extensive use of Springer's viscosity relationship for Hercules 3501-6 as an input to our resin flow and void growth models.

Heat Transfer

Several workers have done heat transfer modeling of composite cure [1,11,17]. This calls for energy balance to predict cure as a function of time and location. By neglecting energy transfer by convection, the energy equation may be expressed as

$$\frac{\partial(\rho C_v T)}{\partial t} = \frac{\partial}{\partial x}\left(K_x \frac{\partial T}{\partial x}\right) + \frac{\partial}{\partial y}\left(K_y \frac{\partial T}{\partial y}\right) + \frac{\partial}{\partial z}\left(K_z \frac{\partial T}{\partial z}\right) + \frac{dH}{dt} \tag{3}$$

where ρ and C_v are the density and specific heat of the composite; K_x, K_y, and K_z are the thermal conductivities in the x, y, and z directions; and T is the temperature of the composite. The rate of heat generation by chemical reaction dH/dt is defined as

$$\frac{dH}{dt} = \frac{d\alpha}{dt} H_r \tag{4}$$

where H_r is the total heat of reaction and α, the degree of cure, is predicted by the chemoviscosity and chemorheology models.

Resin Flow

The main approach suggested in the literature [1,3,9] for predicting resin flow behavior in composites during autoclave processing is to describe the flow in terms of Darcy's law. Darcy's law requires knowledge of the fiber network permeability and porosity, and resin viscosity. Fiber network permeability is a function of fiber diameter, the porosity of the medium, and the shape factor of the fibers. The shape factor is dependent on the pore structure and fiber orientation. Viscosity of a thermoset resin is essentially a function of degree of cure and temperature.

The Loos and Springer model [1] describes two-dimensional resin flow in unidirectional laminates during bleeder ply molding (i.e., autoclave processing). These workers considered flow in both the horizontal (in the plane of the laminate) and vertical (perpendicular to the laminate) directions. However, they decoupled the two flows. This means that profiles calculated for the vertical and horizontal flows could result in different resin pressures existing at the same point in the laminate. Also, Loos and Springer considered that the layers of prepregs that make up a laminate lack deformability. This implies that the applied autoclave pressure on the laminate is borne solely by the resin during cure, contrary to experimental evidence [17].

To overcome the shortcomings of the Loos and Springer model, Davé, Kardos, and Duduković [3] and Gutowski, Morigaki, and Cai [9] have independently developed identical models for resin flow through a saturated porous medium that allows for three-dimensional resin flow and one-dimensional consolidation of the composite. (We shall see later that these flow models differ only in the selection of the coordinate system and in terminology.)

The Davé–Gutowski model is based on a theory of consolidation and flow through a saturated porous medium. It considers that the total force acting on a porous medium is counteracted by the sum of the opposing forces, including the force due to the springlike behavior of the fiber network and the hydrostatic force due to the pressure of the liquid within the porous medium. The flow in the laminate is described in terms of Darcy's law. The Davé–Gutowski model is applicable to transient flow during squeezing of a *saturated* porous medium—for example, autoclave processing. (Saturation of a fluid phase (i) is defined as the ratio of the volume of fluid phase i in a sample to the total accessible pore volume in the sample.)

The assumption that the porous medium is fully saturated has an important ramification. It permits us to neglect the effect of capillarity in the porous medium, that is, surface tension and contact angle. Capillary systems in porous media always involve a solid phase and *at least two* fluid phases.

THE PHYSICAL SITUATION. During the processing of a laminated composite by the autoclave process, at some time t_p ($t_p \geq 0$), autoclave pressure is applied to the composite–bleeder sandwich shown in Figure 1. As a result, resin flows out from the composite in the directions normal and parallel to the tool face. The resin flow rate specifically depends on (a) the magnitude of that portion of the applied autoclave pressure transmitted to the resin, (b) the laminate boundary pressure, (c) the viscosity of the resin, (d) the specific permeability of the fiber network in each direction, (e) the stress–strain behavior in confined compression of the fiber network, and (f) the dimensions of the composite laminate. The flow of resin from the composite laminate during autoclave processing is transient, because there is a continuous depletion of the resin through a fiber composite with time. The flow of resin through a fiber composite laminate is assumed to be a laminar flow having a low Reynolds number; therefore, Darcy's law is valid.

When a porous fibrous bed, completely saturated with incompressible liquid (resin), is subjected to compression during the processing of graphite–epoxy composites, the phenomenon of flow through the fibrous bed due to consolidation is similar to the situation of a water-saturated soil subjected to confined compression, that is, the stress–strain behavior when there is no lateral strain and the axial strain is exactly equal to the volumetric strain. In both cases, the porous medium under consolidation behaves like a nonlinear elastic body.

Consolidation theory considers the porous medium to be a skeleton of solid network-enclosing pore space that may be filled with gas, liquid, or both. The porous medium itself is considered to behave like a spring. Therefore, if a fiber–resin composite is placed under stress in such a way that its volume is decreased, three possible routes by which this decrease might occur are:

1. Compression of the porous medium (like compressing a spring).
2. Compression of the resin within the pores.
3. Escape of the resin across the boundaries of the sample.

For the case of one-dimensional consolidation of a composite, when escape of the resin across the boundaries is allowed, compression of the resin within the pores is negligible. Then the stress during consolidation is given by the following equilibrium equation:

$$\sigma = p + P \tag{5}$$

where

σ = total externally applied stress
p = effective stress (i.e., the stress borne by the springlike skeletal structure of the fiber network)
P = hydraulic pressure of the resin in the composite

The process of consolidation is better understood by the piston-and-spring analogy described by Taylor [18]. In Figure 4, which depicts this analogy, (*a*)–(*e*) show the lengths assumed by a spring under various loads. In (*f*)–(*l*), the same spring is assumed to be immersed in a watertight cylinder filled with some incompressible liquid. In Figure 4g, a frictionless but tightly fitted piston provided with a stopcock is placed in the cylinder and loaded with a total of 20 pounds. Due to incompressibil-

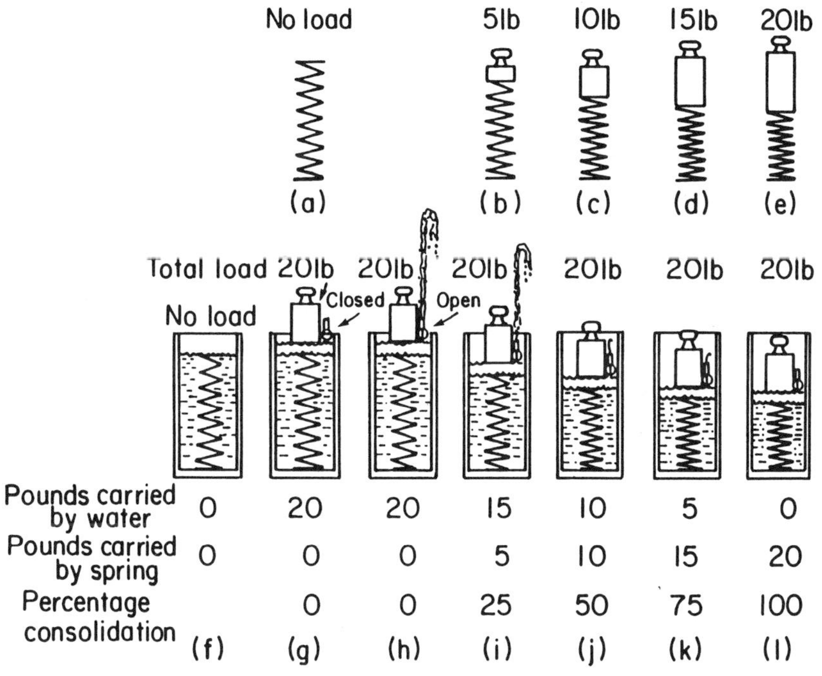

FIGURE 4 The piston-and-spring analogy to consolidation of resin-filled fiber composites. (From Ref. 18.)

ity of the liquid, no consolidation (i.e., compression) of the spring takes place with the stopcock closed. Now at time $t = 0$, the stopcock is opened. (Fig. 4*h*). At the instant $t = 0$, the pressure conditions are unchanged from the conditions at $t < 0$, as shown in Figure 4*h*.

As time passes, more and more liquid escapes and the piston sinks by compressing the spring. However, at any instance, the total force acting down is always counteracted by the sum of the opposing upward forces. The upward forces include the force due to the spring and the hydrostatic force due to the liquid. Since the velocity of the descending load is small, it is assumed that the load and the piston are quasi-static. This means that the sum of the forces acting downward is equal to the sum of the forces acting upward. Finally, at long times (Fig. 4*l*) the spring compresses to the same length as in Figure 4*e*.

In the mechanical analogy presented above, the 20 pound load and piston represent the externally applied force on the laminate due to the autoclave pressure, which is transmitted through the bleeder plies and vacuum bag assembly. The spring represents the compressible fiber bed and the liquid represents the epoxy in the pore space of the fiber bed. The stopcock opening is analogous to the permeability and the compressibility of the spring to the compressibility of the fiber bed. The more compressible the fiber bed, the longer the time required for consolidation; the more permeable the fiber bed, the shorter the time required. In extending the analogy to the flow of resin through a fiber bed during processing, it is assumed that the viscous drag force (hence the shear stress) exerted on the fibers by the resin is negligible. Furthermore, by the findings of Williams et al. [19], it is assumed that the pressure drop due to the capillary forces is negligible compared with the externally applied pressure, permitting us to neglect capillarity effects. Also, the force due to friction between fibers has been considered to be a part of the spring response in this analysis.

MATHEMATICS OF THE FLOW MODEL. The following assumptions underlie the theory of resin flow through a composite due to consolidation.

1. The porous fiber bed is completely saturated with incompressible resin.
2. The individual fibers are incompressible.
3. The resin flow through a porous fiber bed obeys Darcy's law.
4. The velocity and stress–strain increases are small, and the situation is quasi-static.
5. Body forces such as self weight are negligible.

The generalized situation of resin flow through a porous fiber bed during autoclave processing is the case of one-dimensional consolidation with three-dimensional seepage (flow). Figure 5 shows this situation with the appropriate boundary conditions. Note that the boundary conditions P_{BC_1}–P_{BC_5} need not be the same. The governing differential equation satisfying the consolidation of a composite laminate within a given time interval with three-dimensional seepage (flow) and one-dimensional confined compression condition (no boundary motion in x or y directions) is derived below.

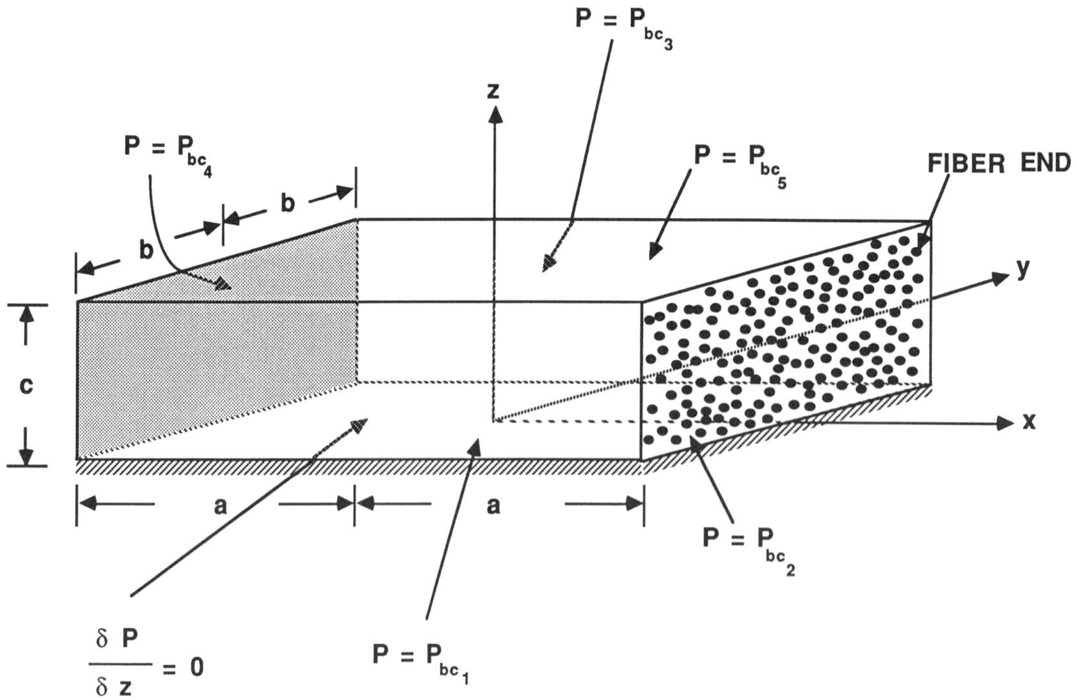

FIGURE 5 Geometry of a flat, unidirectional laminate made of continuous fibers with the appropriate boundary conditions. The boundary conditions P_{BC_1}–P_{BC_5} need not be the same.

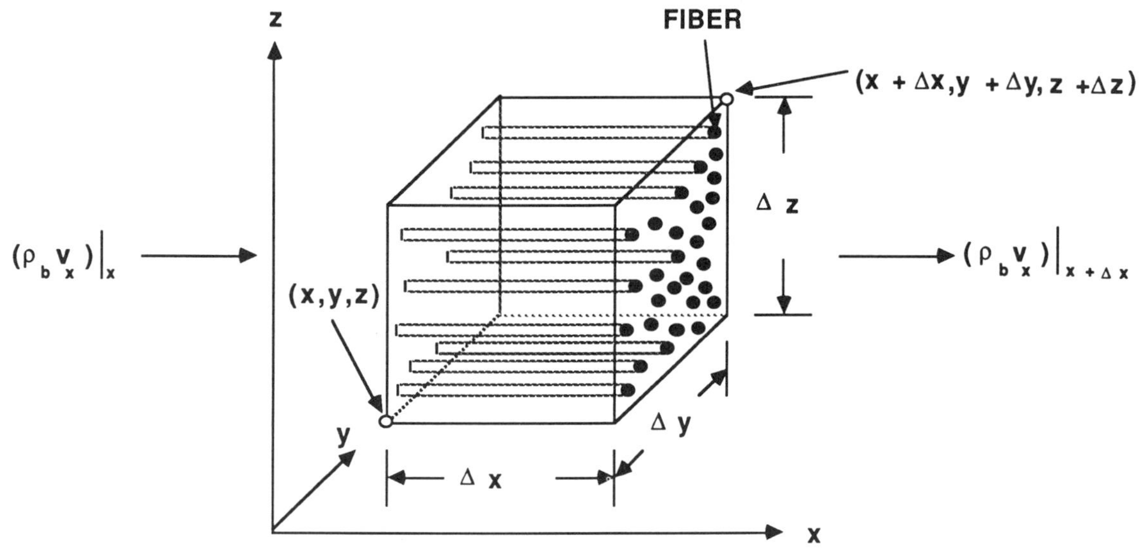

FIGURE 6 **Representative volume element (RVE) of a composite of volume $\Delta x\ \Delta y\ \Delta z$ through which a resin is flowing. [For clarity, resin in only the x direction is shown.]**

Consider a representative volume element (RVE) $\Delta x\ \Delta y\ \Delta z$ of the composite through which the resin is flowing (see Fig. 6). Note that resin flow in the x direction is shown in Figure 6 for clarity only. A representative volume element (RVE) is defined such that no matter where it is placed within the porous medium domain considered, it will contain *both* the solid phase and fluid space (pores). In a composite, the solid phase is the fibers and the fluid phase is the resin.

Let ρ and ε be the resin density and porosity in the RVE. Then the bulk resin density ρ_b is equal to $\rho\varepsilon$.

The equation of continuity for the flow of resin through the composite is developed by writing a resin mass balance over the RVE:

$$\begin{Bmatrix}\text{rate of mass}\\ \text{accumulation}\end{Bmatrix} = \begin{Bmatrix}\text{rate of}\\ \text{mass in}\end{Bmatrix} - \begin{Bmatrix}\text{rate of}\\ \text{mass out}\end{Bmatrix} \tag{6}$$

which becomes

$$\begin{aligned}\Delta x\,\Delta y\,\Delta z\,\frac{\partial \rho_b}{\partial t} &= \Delta y\,\Delta z\,[(\rho_b v_x)|_x - (\rho_b v_x)|_{x+\Delta x}]\\ &\quad + \Delta x\,\Delta z\,[(\rho_b v_y)|_y - (\rho_b v_y)|_{y+\Delta y}]\\ &\quad + \Delta x\,\Delta y\,[(\rho_b v_z)|_z - (\rho_b v_z)|_{z+\Delta z}]\end{aligned} \tag{7}$$

where v_x, v_y, and v_z are the interstitial velocities of the resin within the composite in the x, y, and z directions.

The resin contained within the laminate is not confined to a constant-volume system during autoclave processing. Therefore, resin density ρ can be assumed to be constant even if a slight volume shrinkage of the resin occurs during cure. Under these assumptions, the equation of continuity (Eq. 7) for a time-dependent RVE $\Delta x\ \Delta y\ \Delta z$ becomes

$$\begin{aligned}\frac{\partial}{\partial t}(\varepsilon\Delta x\,\Delta y\,\Delta z) &= -\left[\frac{\partial}{\partial x}(\varepsilon v_x)\right.\\ &\quad \left. + \frac{\partial}{\partial y}(\varepsilon v_y) + \frac{\partial}{\partial z}(\varepsilon v_z)\right]\Delta x\,\Delta y\,\Delta z\end{aligned} \tag{8}$$

Velocities v_x, v_y, and v_z are defined by Darcy's law for fluid flow through an *orthotropic* material whose coordinate axes are aligned with the principal axes of the material.

$$v_x = \frac{q_x}{\varepsilon} = -\frac{k_x}{\varepsilon\mu}\frac{\partial P}{\partial x} \tag{9a}$$

$$v_y = \frac{q_y}{\varepsilon} = -\frac{k_y}{\varepsilon\mu}\frac{\partial P}{\partial y} \tag{9b}$$

$$v_z = \frac{q_z}{\varepsilon} = -\frac{k_z}{\varepsilon\mu}\frac{\partial P}{\partial z} \tag{9c}$$

In Eq. (8), porosity ε is replaced by void ratio e (defined as the volume of pore space per unit volume of solids in the porous medium) using the relationship

$$\varepsilon = \frac{e}{1+e} \tag{10}$$

The volume of solids (fibers) in the time-dependent RVE $\Delta x\ \Delta y\ \Delta z$ is $(\Delta x\ \Delta y\ \Delta z)/(1+e)$. Since this volume is constant, Eq. (8) is reduced to Eq. (11) in the limit when Δx, Δy, and Δz approach zero.

$$\begin{aligned}\frac{1}{1+e}\frac{\partial}{\partial t}(e) &= \left[\frac{\partial}{\partial x}\left(\frac{k_x}{\mu}\frac{\partial P}{\partial x}\right)\right.\\ &\quad \left. + \frac{\partial}{\partial y}\left(\frac{k_y}{\mu}\frac{\partial P}{\partial y}\right) + \frac{\partial}{\partial z}\left(\frac{k_z}{\mu}\frac{\partial P}{\partial z}\right)\right]\end{aligned} \tag{11}$$

Defining a new variable, m_v, the coefficient of volume change in the left-hand side of Eq. (11), where

$$m_v = \frac{-\dfrac{de}{dp}}{(1 + e)} \tag{12}$$

results in

$$-m_v \frac{\partial}{\partial t}(p) = \left[\frac{\partial}{\partial x}\left(\frac{k_x}{\mu}\frac{\partial P}{\partial x}\right) + \frac{\partial}{\partial y}\left(\frac{k_y}{\mu}\frac{\partial P}{\partial y}\right) + \frac{\partial}{\partial z}\left(\frac{k_z}{\mu}\frac{\partial P}{\partial z}\right)\right] \tag{13}$$

The term m_v is used to describe the stress–strain behavior in confined compression; de/dp in the term m_v is obtained from the stress–strain relationship of the porous fiber network [4].

In Eq. (13), p is replaced by $(\sigma - P)$ using Eq. (5), and we assume negligible spatial variations in viscosity μ. This reduces Eq. (13) into the form of the governing differential equation presented previously [3].

$$\frac{\partial P}{\partial t} - \frac{\partial \sigma}{\partial t} = \frac{1}{\mu m_v}\left[\frac{\partial}{\partial x}\left(k_x \frac{\partial P}{\partial x}\right) + \frac{\partial}{\partial y}\left(k_y \frac{\partial P}{\partial y}\right) + \frac{\partial}{\partial z}\left(k_z \frac{\partial P}{\partial z}\right)\right] \tag{14}$$

In Eq. (14), the only unknown is P. The permeability parameters k_x, k_y, and k_z are assumed to obey Kozeny's law. Values for μ and m_v are obtained from the appropriate relationship for the viscosity of the resin and the stress–strain behavior of the porous fiber bed. The applied pressure on the fiber bed σ, either can be constant during the cure cycle or can be allowed to vary according to a prescribed cure cycle. Details on these input parameters are given in our earlier paper [4].

It may be noted that in our numerical analysis for unidirectional graphite–epoxy laminates using the implicit alternating direction finite difference method [4], the motion of the moving boundary layer in the thickness direction of the laminate was accounted for by changing the grid size after each time step during the consolidation process. During each time step, the velocity of the moving boundary and stress–strain increases in the laminate were assumed to be small; that is, the situation was quasi-static. The changing porosity of the laminate as consolidation proceeded was thereby implicit in our model [3,4]. (We shall see later in Fig. 10*c* that the consolidation of the laminate during cure due to resin bleed-off is evident from the decreasing height of the laminate.)

Another method to account for the motion of the moving boundary layer in the thickness direction is by defining a new convected coordinate ξ as Gutowski et al. [9] have done.

$$\xi = \frac{V_0}{V_f} z \tag{15}$$

In Eq. (15), V_0 is the initial volume fraction of the fibers and V_f is the fiber volume fraction at any later time during consolidation.

If Δz is the original thickness of an RVE, then by Eq. (15), the instantaneous thickness at any later time during consolidation is given by

$$\Delta\xi = \left(\frac{V_0}{V_f}\right)\Delta z \tag{16}$$

Substituting for instantaneous thickness $\Delta\xi$ instead of the original thickness Δz in Eq. (7), and assuming that ρ is constant in ε in the RVE, $\Delta x\ \Delta y\ \Delta\xi$, gives the following equation in the limit as Δx, Δy, and $\Delta\xi$ approach zero.

$$\frac{\partial}{\partial t}\left[\varepsilon \frac{\partial \xi}{\partial z}\right] = -\left[\frac{\partial}{\partial x}\left(\varepsilon v_x \frac{\partial \xi}{\partial z}\right) + \frac{\partial}{\partial y}\left(\varepsilon v_y \frac{\partial \xi}{\partial z}\right) + \frac{\partial}{\partial z}(\varepsilon v_z)\right] \tag{17}$$

Applying the appropriate form of Darcy's law for an orthotropic material, which considers the variable thickness dimension, we can write:

$$\frac{\partial}{\partial t}\left[\varepsilon \frac{\partial \xi}{\partial z}\right] = \left[\frac{\partial}{\partial x}\left(\frac{k_x}{\mu}\frac{\partial P}{\partial x}\frac{\partial \xi}{\partial z}\right) + \frac{\partial}{\partial y}\left(\frac{k_y}{\mu}\frac{\partial P}{\partial y}\frac{\partial \xi}{\partial z}\right) + \frac{\partial}{\partial z}\left(\frac{k_z}{\mu}\frac{\partial P}{\partial z}\frac{\partial z}{\partial \xi}\right)\right] \tag{18}$$

Finally, by substituting for $\varepsilon = (1 - V_f)$ and $\partial\xi/\partial z = V_0/V_f$, dividing throughout by V_0, which is a constant, and assuming no spatial variations in μ, we get

$$\frac{\partial}{\partial t}\left(\frac{1 - V_f}{V_f}\right) = \frac{1}{\mu}\left[\frac{\partial}{\partial x}\left(\frac{k_x}{V_f}\frac{\partial P}{\partial x}\right) + \frac{\partial}{\partial y}\left(\frac{k_y}{V_f}\frac{\partial P}{\partial y}\right) + \frac{1}{V_0^2}\frac{\partial}{\partial z}\left(k_z V_f \frac{\partial P}{\partial z}\right)\right] \tag{19}$$

Gutowski et al. [9] have further assumed that k_x, k_y, and V_f are independent of x and y, and depend only on z. Therefore, k_x/V_f and k_y/V_f terms were brought outside the partial derivatives in the first and second terms on the right-hand side of Eq. (19).

It is important to note that Eqs. (14) and (19) are equivalent, except that the moving boundary condition in the thickness direction of the laminate is explicitly incorporated in Eq. (19). On the other hand, Eq. (14) applies for a quasi-static condition, and the motion of the boundary condition must be accounted for during the numerical computation. Also, note the simplifying assumption in Eqs. (14) and (19) that there exist no spatial variations in viscosity. This implies temperature homogeneity. This is a reasonable assumption for thin composite laminates with high thermal conductivity fibers, such as graphite.

If the porous fiber bed is extremely stiff, then the change in the bed thickness upon consolidation is negligible. For this case, closed-form solutions for resin pressure in the laminate are available [3]. Closed-form solutions apply when the permeabilities k_i, the coefficient of volume change m_v, and the applied stress σ are constant during the flow process. Also, the boundary conditions should be either $P = 0$ or grad $P = 0$. Constant coefficient of volume change implies linear stress–strain behavior.

One-dimensional vertical resin flow occurs during autoclave processing when all the four sides are dammed. Two-dimensional resin flow occurs during autoclave processing of a unidirectional laminate when only two sides are dammed to prevent fiber washout.

Under the foregoing conditions, for the laminate shown in Figure 5, the closed-form solution for one-dimensional vertical flow in the z direction is given by

$$P(z,t) = \frac{4\sigma_0}{\pi} \sum_{j=0}^{\infty} \frac{(-1)^j}{2j+1} \cos\left[\frac{(2j+1)\pi z}{2c}\right] \exp\left\{-\left[\frac{(2j+1)^2\pi^2 T_v}{4}\right]\right\} \tag{20}$$

where σ_0 is the constant externally applied stress (autoclave pressure) and

$$T_v = \frac{k_z t}{\mu m_v c^2} \tag{21}$$

and the solution for two-dimensional flow in the x and z directions is given by

$$P(x,z,t) = \frac{16\sigma_0}{\pi^2} \sum_{q=0}^{\infty} \sum_{j=0}^{\infty} \left[\frac{(-1)^{q+j}}{(2q+1)(2j+1)}\right] \left\{\cos\left[\frac{(2q+1)\pi x}{2a}\right] \cos\left[\frac{(2j+1)\pi z}{2c}\right]\right\} \exp\left\{-\frac{1}{\mu m_v}\left[\frac{(2q+1)^2\pi^2}{4a^2/k_x} + \frac{(2j+1)^2\pi^2}{4c^2/k_z}\right] t\right\} \tag{22}$$

Equations (20) and (22) were used for calculating the resin pressure at point X, the midpoint of the laminate (see Fig. 7). These results are compared with the numerical solutions obtained using the finite difference method. Agreement is excellent, there being no perceptible significant differences. Comparisons between the finite difference solutions and closed-form solutions were also made at other points in the laminate. Equally good agreement was obtained. The accuracy of the numerical procedure was thereby confirmed.

Void Growth

This section presents a model for diffusion-controlled void growth and dissolution. The void model predicts the size of a void located in an infinite resin medium as a function of cure time. It also delineates conditions under which voids can be either prevented from growing or made to collapse during the cure cycle.

The rate of growth or dissolution of gas bubbles in epoxy during the processing of graphite–epoxy composites is controlled by the diffusion of a soluble species through the epoxy. Void growth or dissolution is often encountered in everyday life. Even though this phenomenon looks simple, it really is quite complex. Even the simplest problem of mass transport between an isolated, stationary spherical gas bubble and the surrounding isothermal, isobaric liquid of infinite extent is complex. In general, although the governing differential equations are well known, exact solutions are difficult to obtain. This is because of the motion of the gas–liquid interface and the resulting convective transport. Besides, under certain conditions the growth rate is not determined solely by diffusion-controlled mass transfer. In such cases, the effect of liquid inertia becomes important. Surface tension and viscous forces must also be taken into consideration. In the past, many investigators have tackled the problem of void growth and dissolution accounting for various considerations.

The behavior of bubble due to mass transfer can be divided into three regimes. For low values of the driving force, where the role of liquid inertia may be neglected, the growth is diffusion-limited. At the other extreme, for high values of the driving force, the growth is limited by hydrodynamic inertia. While the growth behavior in both these regimes can be described asymptotically, the behavior in the intermediate regime is difficult to understand and requires a numerical solution.

For a complete description of diffusion-limited bubble growth, it is important to consider the movement of the moving boundary and the resulting convective transport. A detailed description of previous investigations of diffusion-controlled bubble growth is presented elsewhere [20].

PROBLEM DEFINITION. The occurrence of voids in a high performance structural composite is a major critical problem. Their genesis and elimination must be understood and this comprehension incorporated in any useful process simulation or on-line control models. The aim of this research, in part, was modeling the stability and growth or dissolution of voids in a resin medium under temperature and pressure conditions similar to those used in actual processing cycles. In this discussion, the terms "void" and "bubble" are synonymous.

Two physically realistic possibilities may be considered for void growth. The first is the growth of a pure water void by diffusion of water from the surrounding liquid resin into a void filled with its vapor. The second involves the same diffusion process except that the water enters a void containing both air and vapor. In either case, based on Grayson and Wolf's experimental results [14,15], it can be concluded that dissolved water is the main component contributing to void growth. Water is dissolved in the resin during prepreg storage or layup.

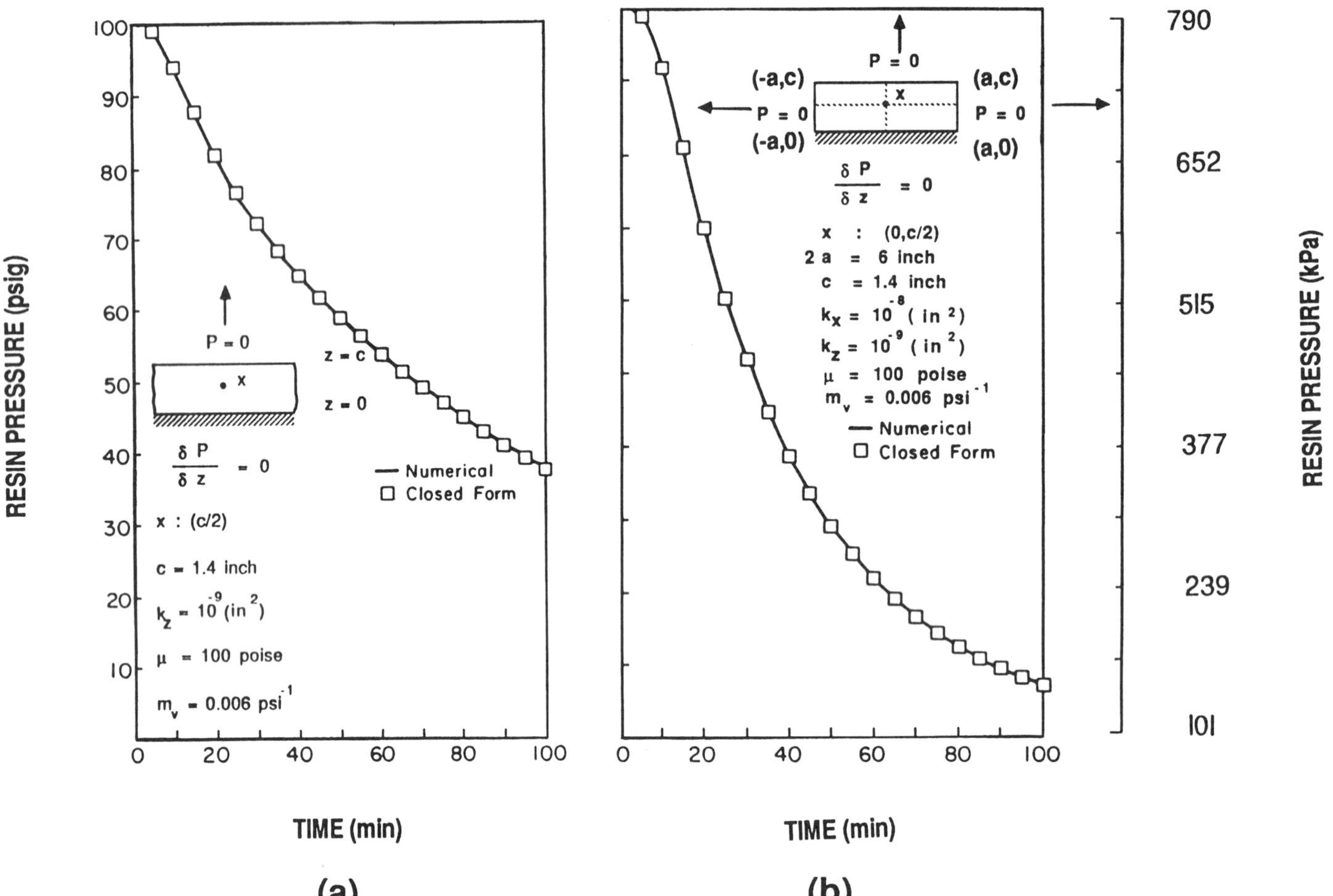

FIGURE 7 Variation of resin pressure with time at the point X for (a) one-dimensional and (b) two-dimensional flow under the conditions shown. (To convert the units in this figure to SI units, see the conversion table at the front of the book.)

There are two possibilities for void nucleation. If air is entrapped during prepreg layup, initial nuclei of finite size always exist. Water dissolved in the resin diffuses into these nuclei whenever conditions are favorable for moisture diffusion. In the absence of air nuclei, water vapor nuclei can form at degrees of supersaturation dictated by the nucleation theory [5]. For low molecular weight materials such as water, critical size nuclei are extremely small. In our void growth model, it is assumed that the critical nucleus size is essentially zero. According to Kingery [21], the number of molecules present in a nucleus of critical size ranges from 50 to 100 molecules.

In Figure 8, which shows the physical situation during void growth, C_∞ is the water concentration in the bulk resin and C_i is the concentration at the bubble–resin interface. As the void grows, the boundary between the resin and void moves; C_∞ will depend on the exposure of the prepreg to water vapor before the autoclave processing. For conservative estimates, the dissolved water in the resin is assumed to be in equilibrium with the humid air at the temperature of prepreg layup. By Raoult's law or Henry's law, or some other rule for nonideal solutions, the water concentration C_i in the resin–water solution at the bubble–resin interface depends on the partial pressure of water vapor in the void and the temperature of the resin. In the case of a pure water void, the partial pressure of the water in the void is equal to the total pressure in the void, which in turn is equal to the resin pressure for slow diffusion-controlled void growth. (If air is initially present in the void, then C_i also depends on the mole fraction of air in the air–water void.)

The exact modeling of void growth would require the solution of many time-dependent, coupled partial differential equations, necessitating excessive computational time. Such an effort does not seem warranted, since many physical properties are not known with sufficient accuracy. Two alternate simplified approaches are possible. One could consider a typical "equivalent" capillary through the plies—a serpentine path that matrix resin and voids might take through connected pores formed by the fibers of the stacked plies—and examine void growth perpendicular to the plies. A knowledge of the pore geometry of the fiber–resin medium is needed for this approach. Also, the surface tension, contact angle, viscosity, and density of the resin as a function of temperature are required as input parameters. The second

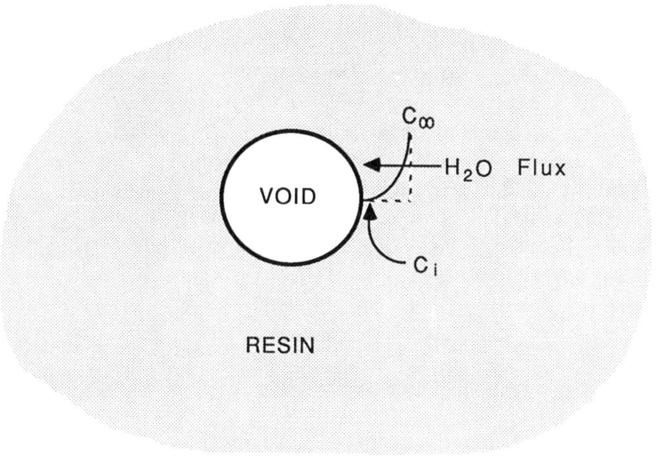

FIGURE 8 **Physical situation of void growth in an infinite resin medium.**

approach—namely, to consider void growth in an isotropic pseudohomogeneous medium—was selected because experimental evidence indicates that voids do not grow preferentially perpendicular to the plies. This approach then becomes equivalent to bubble growth in an infinite isotropic liquid medium. This represents a worst-case scenario of void formation by solvent (moisture) diffusion.

MATHEMATICS OF VOID GROWTH. In developing the void model, the following simplifying assumptions were made.

1. The void is stagnant between two plies and its center is not moving with respect to fixed coordinates in the laminate.

2. The void is approximately spherical and its effective size is calculated based on an equivalent sphere.

3. There is no interaction between voids (no coalescence).

4. The void is considered to be in an infinite isotropic fluid medium.

5. For pure water voids, void nucleation is instantaneous. For air–water voids, air void nucleation or entrapment has already occurred during layup or at the beginning of the cure cycle.

6. At any given time, the temperature and moisture concentration in the bulk resin are uniform.

7. At each temperature, a pseudo-steady state is established with respect to the concentration profile; that is, the profile does not change during diffusion at any particular temperature.

8. The viscous, inertia, and surface tension effects are neglected, since they are significant only during the initial expansion of the original void nucleus [22]. As the void grows, its growth rapidly becomes limited by the rate of arrival of the diffusing species. Also, even in a highly viscous liquid, the viscous effect becomes vanishingly small because the rate of growth da/dt asymptotically approaches zero [23].

Under these assumptions, the conservation equation of the diffusing species may be expressed as:

$$\frac{\partial C}{\partial t} + \frac{da}{dt}\frac{a^2}{r^2}\frac{\partial C}{\partial r} = \frac{D}{r^2}\frac{\partial}{\partial r}r^2\frac{\partial C}{\partial r} \tag{23}$$

with the initial and boundary conditions

$$C(0,r) = C_\infty \tag{24a}$$

$$C(t,\infty) = C_\infty \tag{24b}$$

$$C[t,a(t)] = C_i \tag{24c}$$

In Eq. (23), C is the concentration of the diffusing species, a is the void radius, r is the distance from the center of the void to any point in the resin, D is the diffusivity, and t is the time.

When the density of the gas changes much less rapidly than the radius of the void, the equation of continuity that the void radius $a(t)$ satisfies is

$$\frac{da}{dt} = \frac{1}{\rho_g}D\left(\frac{\partial C}{\partial r}\right)_{r=a} \tag{25}$$

where ρ_g is the gas density.

Under constant temperature and material properties, the following asymptotic solution for large values of the growth constant β was developed by Scriven [22]:

$$a = 4\beta(Dt)^{1/2} \tag{26}$$

where

$$\beta = \frac{C_\infty - C_i}{\rho_g} \tag{27}$$

The change in diameter per unit time (growth rate) can be obtained by differentiation of Eq. (26), which yields

$$\frac{d(a^2)}{dt} = 16D\beta^2 \tag{28}$$

for void growth from critical nucleus size of essentially zero diameter. The initial condition is

$$\text{at } t = 0, a = 0 \tag{29}$$

Equations (28) and (29) may now be used to calculate void growth during the cure cycle. The driving force β is adjusted according to the changing temperature and pressure in the laminate, which affect C_i and ρ_g. This is equivalent to assuming that on the time scale of void growth during autoclave processing, a relatively slow process, all variables have a chance to reach a new equilibrium. Cable and Frade [23] have concluded that the foregoing approximation is valid for nearly the whole course of growth from finite void size (except the initial growth) for low to moderate solubility. This situation is similar to that of water vapor void growth in epoxy resins. Therefore, it can be assumed that the pressure and temperature are uniform throughout the resin at any moment in time but vary according to the prescribed cure cycle. Equations (28) and (29) represent the essence of the model. All that remains is to properly evaluate the model input parameters such as the diffusivity D, the resin water concentration C_∞, the interface water concentration C_i, and the gas density ρ_g, as functions of temperature and pressure during the cure cycle. These input parameters have been presented in our earlier paper [5].

VOID STABILITY EQUATION. While pseudo-steady-state void growth due to slow moisture diffusion is described by Eq. (28), it is desirable that void growth by diffusion of moisture be altogether eliminated during composite processing. Void growth during composite processing is possible if the concentration of the dissolved water in the resin exceeds the water concentration at the void–resin interface. The concentration of the dissolved water in the resin is a function of the relative humidity to which the prepreg is exposed before processing. For slow diffusion-controlled void growth, the concentration of water at the void–resin interface is a function of the resin pressure and temperature, and on the mole fraction of water vapor in the void. Since the driving force for diffusion rises with temperature, the only way to prevent void growth is by maintaining the resin pressure at every point in the laminate greater than the minimum pressure required to prevent void growth by diffusion. This is achieved by maintaining the resin pressure at every point in the laminate greater than P_{min}, the minimum resin pressure required to prevent void growth by diffusion. Given a prepreg equilibrated with moisture at a particular relative humidity, to prevent the potential for pure water void growth by diffusion at all times and temperatures during the curing cycle, the resin pressure at any point within the curing laminate must satisfy the following inequality:

$$P_{min} \geq 4.962 \times 10^3 \exp\left(\frac{-4892}{T}\right) (\mathrm{RH})_0 \qquad (30)$$

where $(\mathrm{RH})_0$ (%) is the relative humidity to which the prepreg is equilibrated before processing, P_{min} (atm) is the minimum resin pressure required to prevent water–vapor void growth by moisture diffusion at any time during the cure cycle, and T (K) is the temperature at any time during the curing cycle.

Equation (30), called the "void stability equation," was derived from the condition of impossibility of void growth by diffusion at any temperature if the growth constant β is zero or less. The void stability equation holds exactly for a pure water–vapor void in any resin when the water solubility relationship of the resin has a power law dependence with relative humidity. It would also hold well for small air–water voids after an initial growth period. This is because the growth pattern of an initially small air–water void is similar to that of a pure water void after an initial growth period (see Fig. 13, below). In case the diffusing species is other than water, the appropriate void stability equation for that diffusing species can be obtained by the same criteria as above; that is, void growth cannot occur if C_i is maintained greater than C_∞ throughout the cure cycle.

A plot of the void stability equation for two relative humidities (50 and 100%) yields a void stability map, shown in Figure 9. This map indicates that vacuum can be applied without encouraging void growth if such an application is coordinated with the temperature of the system.

Input Data

Using the models described above, numerical computation results for flow of Hercules 3501-6 epoxy resin through graphite fibers and water vapor void growth in epoxy resin have been reported [4–6]. It was assumed that the thermal gradients within a curing laminate are

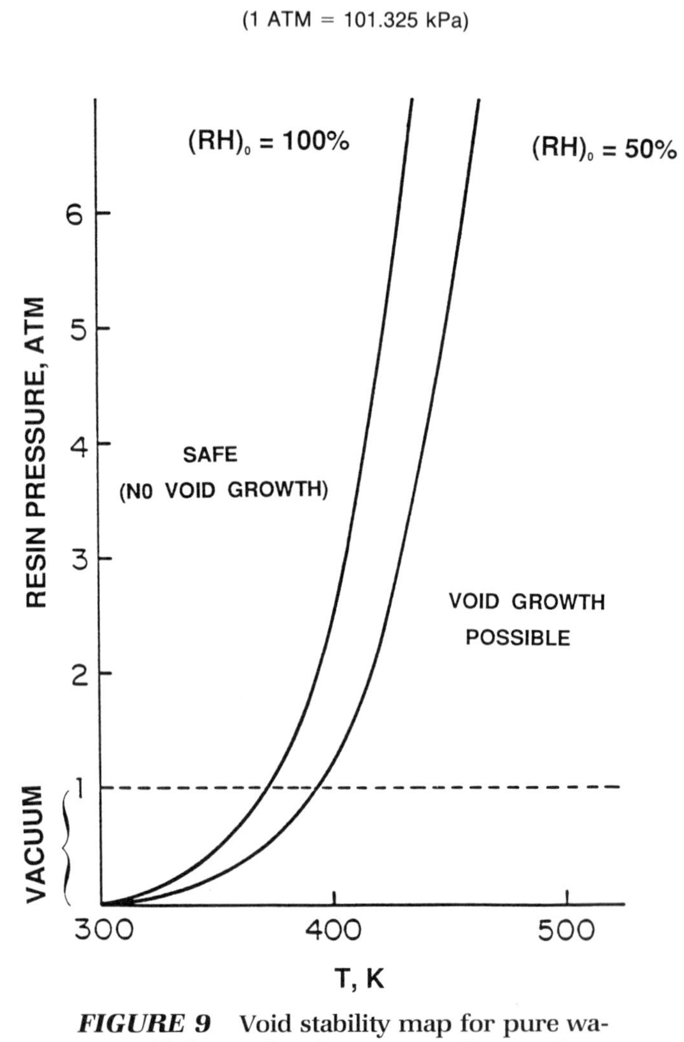

FIGURE 9 Void stability map for pure water void formation in epoxy resin matrices. Note the significant effect of initial relative humidity exposure of the resin.

negligible for heat-up rates up to 3°C/min. This was in accordance with prior experimental results [17]. The existence of negligible thermal gradients within the laminate implies temperature homogeneity. Therefore, for a cure cycle with low heat-up rates, like the one shown in Figure 10*a*, the heat transfer model can be excluded from the overall scheme of Figure 3.

As a result of the foregoing assumption, the required input data for the resin chemoviscosity, flow, and void growth models are:

- Constants in the chemoviscosity model (see Ref. 16)
- Fiber diameter
- Fiber density
- Laminate size
- Resin density
- Resin content
- Ply thickness
- Specific permeability of the fiber bed
- Relationship between effective stress and volumetric strain (i.e., change in void ratio or porosity)
- Vapor pressure as a function of temperature
- Vapor density as a function of temperature
- Solubility and concentration of water in the resin
- Diffusivity of water in the resin
- The initial relative humidity exposure of the resin (or prepreg)

Details on the above input data are available elsewhere [20].

Results and Discussion

This section presents numerical and experimental results of the resin flow and void growth models.

Resin Flow Through a Unidirectional Graphite–Epoxy Laminate

NUMERICAL RESULTS. In the present problem of resin flow within a unidirectional laminate, it is assumed that horizontal flow perpendicular to the fiber direction is negligible compared to the vertical flow or to the horizontal flow parallel to the fiber direction. Horizontal resin flow perpendicular to the fiber direction in a unidirectional laminate can be neglected when the length and the width of the laminate are comparable in size, and the specific permeability in the direction perpendicular to the fibers is much less than that in the direction of the fibers. However, vertical flow cannot be neglected, because the cross-sectional area of the top of the laminate is usually much larger than the side areas of the laminate. As a result, vertical flow is usually of the same order of magnitude as the horizontal flow in the fiber direction.

The cure cycle analyzed and the corresponding viscosity profile of Hercules 3501-6 epoxy resin are shown in Figure 10*a*. In this standard commercial cycle [17], the autoclave pressure was maintained at 790 kPa and the boundary conditions P_{BC_1}–P_{BC_5} (see Fig. 5) were assumed to be zero throughout. The boundary condition at the tool face was no-flow boundary condition. These boundary conditions are called "free-bleeding" conditions in autoclave processing.

Figures 10*b* and 10*c* show resin pressure profiles within a 256-ply laminate for two-dimensional resin flow (i.e., horizontal resin flow along the fiber direction and vertical resin flow). Resin flow behavior was studied in 15.24 cm × 15.24 cm laminates containing 64, 128, and 256 plies of 0.018 cm initial thickness. Figure 10*b* shows the resin pressure profiles in the horizontal midplane of a 256-ply laminate. Note that the pressure profiles are drawn in only half the laminate because of symmetry at the vertical midplane. Figure 10*c* shows the resin pressure profiles along the vertical midplane of the 256-ply laminate. Note that the thickness of the laminate decreases during cure because of laminate consolidation and resin bleed-off. At any time during cure, the thickness of the laminate is given by the point at which the resin pressure profile intersects the x axis in Figure 10*c*. From Figures 10*b* and 10*c*, it is also seen that the resin pressure in the horizontal or vertical direction is not linear, unlike the assumption made in several previous resin flow models. In Figures 10*b* and 10*c*, the resin pressure at 40 minutes after the start of cure, and thereafter, is zero. This is because the resin pressure throughout the laminate has fallen to the bleeder bag pressure (i.e., 0 psig). This suggests that "free bleeding" is not recommended in composite processing because of the likelihood of voids forming at low resin pressures.

In reality, dams surround a curing laminate to prevent fiber washout. Experimental results [24] indicate that dams also prevent "free bleeding" by continuously increasing the resin pressure at the boundary of the laminate.

Figure 11 shows profiles of horizontal, vertical, and total resin mass loss as a function of cure time for 64-, 128-, and 256-ply laminates. In a thin laminate (64-ply), the vertical resin loss is significantly higher than the horizontal loss. In a thick laminate (256-ply), the reverse holds true. However, the total percent resin mass loss from both the thick and the thin laminates is the same. This is because the resin pressure everywhere within the 64-, 128-, or 256-ply laminates falls to the bleeder bag pressure (0 psig) before the resin gels. Another interesting observation from Figure 11 is that the horizontal and vertical resin losses are of the same order of magnitude. Therefore, if the flow equation were decoupled in the two flow directions, the results would be incorrect.

COMPARISON OF FLOW MODEL PREDICTIONS WITH EXPERIMENTAL RESULTS. At this stage it would be worthwhile to compare the predicted results for one-dimensional vertical flow with the experimental results of Campbell, Mallow, and Amuedo [17]. Experiments were done on 15.24 cm × 15.24 cm laminates fabricated from Hercules AS-4/3501-6 prepregs of 10, 25, and 60 plies using the cure cycle shown in Figure 10*a*.

Table 1 presents the numerical and experimental results. The initial resin content in the prepregs used for these experiments was 42% instead of 38% used in this analysis. While the 60-ply results are questionable [17],

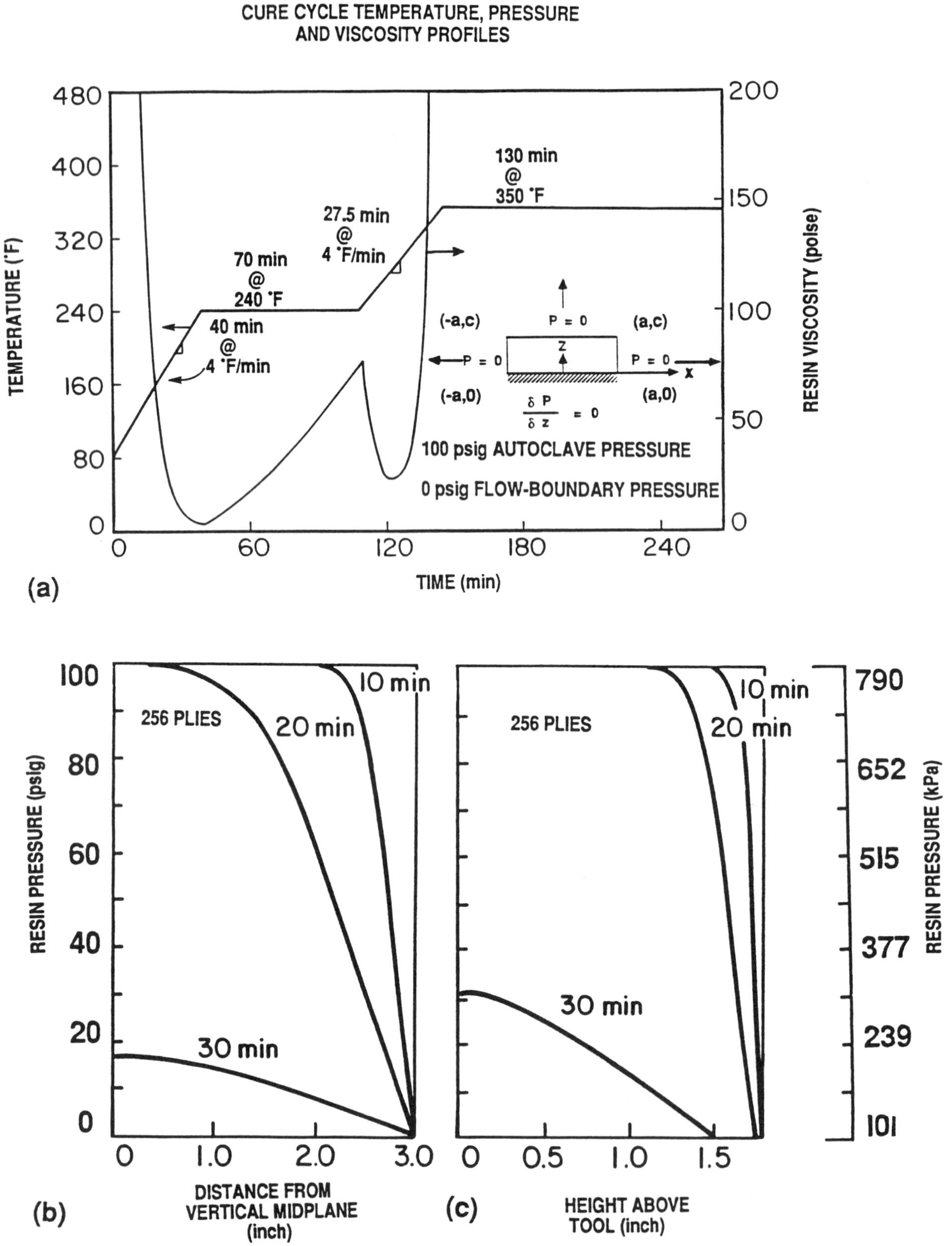

FIGURE 10 (*a*) A typical autoclave cure cycle. (From Ref. 17.) (*b*) Resin pressure profiles in the horizontal midplane of a 256-ply laminate. (*c*) Resin pressure profiles in the vertical midplane of a 256-ply laminate. The conditions for two-dimensional resin flow and the cure cycle are shown in (*a*). [Note that the pressure profiles in the horizontal midplanes are drawn for only half of the laminate due to symmetry at the vertical midplane.] (To convert the units in this figure to SI units, see the conversion table at the front of this book.)

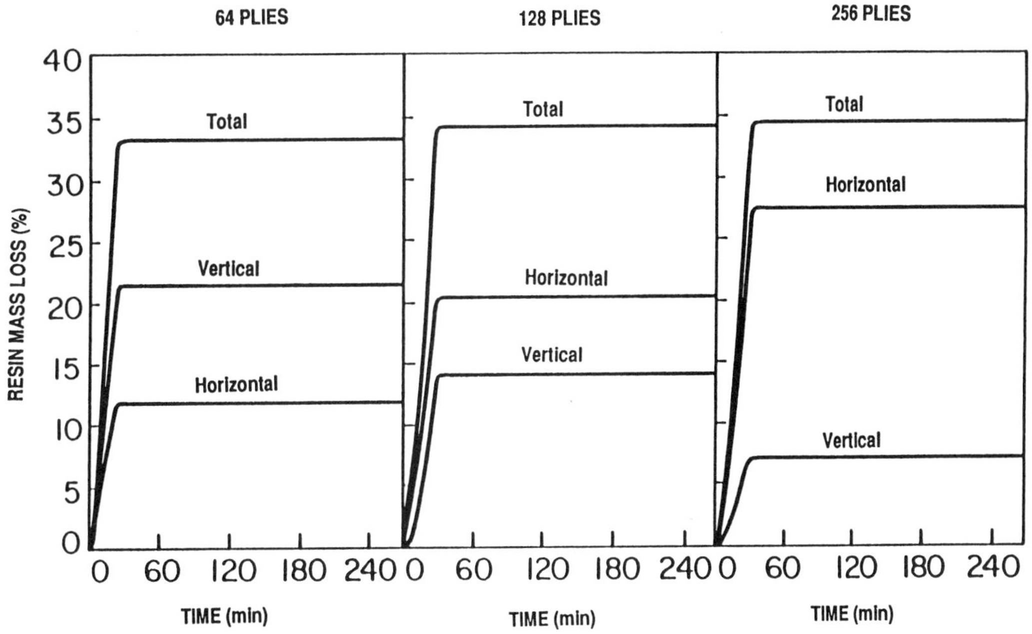

FIGURE 11 Time profiles of resin mass loss (%) during autoclave processing for two-dimensional flow through 64-, 128-, and 256-ply laminates under the flow conditions and cure cycle shown in Figure 10*a*.

the average resin mass loss from the 10- and 25-ply laminates was 42.8%, with 6% standard deviation. The average ply thicknesses before and after cure were 0.0178 and 0.0137 cm respectively. The predicted results, using the same cure cycle and initial thickness per ply, gave a resin mass loss of about 35% for one-dimensional vertical flow from either the 64-, 128-, or 256-ply laminate. The predicted final thickness per ply was 0.0145 cm. Thus, for one-dimensional vertical flow, the model predictions are comparable to the experimentally obtained values.

Comparison of measured and predicted values of the resin pressure for one-dimensional vertical flow during simulated autoclave processing using Eq. (19) was made by Gutowski et al. [9]. Excellent agreement between experimental data and theoretical prediction was observed.

TABLE 1
Comparison Between Experimental and Predicted Results for One-Dimensional Vertical Resin Flow

	Model Prediction	Experimental[a]
Laminate size	15.24 cm × 15.24 cm	15.24 cm × 15.24 cm
Number of plies	64, 128, and 256	10, 25, and 60
Initial resin content	38%	42%
Resin mass loss	35%	42.8%[b]
Initial thickness per ply	0.0178 cm	0.0178 cm
Final thickness per ply	0.0137 cm	0.0145 cm

[a] From Campbell et al. [17].
[b] With 6% standard deviation.

Void Growth

NUMERICAL RESULTS. Void behavior was examined during the various stages of the cure cycle shown in Figure 12. In the past [25], laminates made using this cure cycle had unpredictable quality. Sometimes they had considerable porosity, but often they would be per-

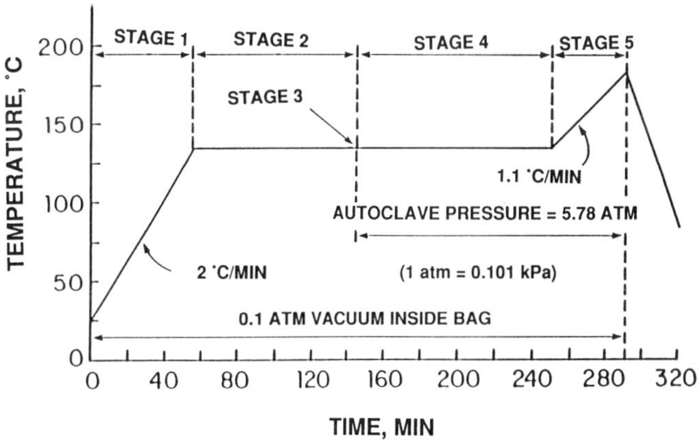

FIGURE 12 Cure cycle for examining void growth behavior. (From Ref. 26.)

fectly acceptable. These fluctuations were found to be related to the environmental conditions at the time of layup and processing.

To simulate void growth in an infinite isotropic resin medium for the cure cycle shown in Figure 12, it was assumed that the resin pressure is 10.1 kPa (0.1 atm) during stages 1 and 2 and 585.7 kPa (5.78 atm) during stages 3–5. In reality, the resin in a curing laminate never experiences the total bleeder vacuum or the total autoclave pressure continuously during the cure cycle (see Fig. 22, below). This is because of the springlike behavior of the fibers, which partake the applied autoclave pressure load. Thus 10.1 kPa (0.1 atm) vacuum and 585.7 kPa (5.78 atm) autoclave pressure represent the lower and upper bounds, respectively.

Figure 13 shows the effect of the processing cycle on the void diameter of a pure water void and an air–water void of 0.1 cm initial diameter. Void growth occurs under the conditions specified in the figure. It is assumed that the air–water void initially consists of pure air, even though there is likely to be a small but finite amount of water vapor present. The void growth profile in Figure 13 can be divided into the various stages of void growth and dissolution and interpreted as follows.

During stage 1, the resin pressure is constant at 0.1 atm but the temperature is increasing (see Fig. 12). This causes the diffusivity of the water in the resin to increase and the interface concentration C_i to decrease. Both these effects favor void growth, and a rapid increase in the void diameter results during this stage.

During stage 2, the resin pressure and temperature are constant at 10.1 kPa (0.1 atm) and 135°C. Hence the diffusivity D, the gas density ρ_g, and the interface concentration C_i are constant. (For an air–water void, ρ_g and C_i are not quite constant, but the variation is slight.) Also, the concentration in the bulk resin C_∞, which is dependent only on the initial humidity exposure of the resin, is constant. Therefore, the growth constant β in Eq. (27) is constant. By Eq. (26), the void growth during stage 2 is a function of the square root of time only.

At the beginning of stage 3, the resin pressure is increased to 585.7 kPa (5.78 atm). The volume decrease of the void is calculated using ideal gas law. Stages 4 and 5 mimic stages 1 and 2, except that there is a negative driving force for void growth. Hence void dissolution occurs rather than void growth. Although the void behavior is calculated through stage 5 using the diffusion-controlled void growth and dissolution model, it is likely that the resin viscosity rises sufficiently during stage 4 to invalidate the model assumptions. In reality, if void dissolution during stages 4 and 5 is possible, it will more likely be controlled by inertia than by diffusion. Furthermore, during void collapse, the surface tension effects gradually become dominant as the void size reaches zero.

The void growth model presented here does not account for the restrictions imposed on void growth by fibers and void breakup. This is unlike the real situation during composite processing. Therefore, the calculated void diameters in Figure 13 represent the worst-case scenerio during composite processing, that is, void growth in an infinite resin medium with no void breakup.

An important conclusion from Figure 13 is that for a small void, it is immaterial whether the initial void contains pure water or an air–water mixture. The diameters at any particular time during the cure cycle are nearly the same when the initial void diameters were identical. An air–water void has a very large driving force during the few minutes of the cure cycle for water diffusion into the void. This causes rapid diffusion into the void of a large amount of water vapor (relative to the original amount of dry air in the void). Consequently the mole fraction of water vapor in the air–water void quickly approaches unity. Thereafter, the rate of diffusion of water vapor across the interface of the air–water void is nearly the same as that for a pure water void.

It is also noteworthy that after the initial period of growth, the higher density of the air–water vapor mixture makes the growth constant β for the air–water void lower than that for a pure water void. Therefore, an air–water void is slightly smaller than a pure water vapor void when both are produced under identical conditions. However, an air–water void cannot completely dissolve during the cure cycle. A pure water void is capable of complete dissolution. (We shall see later that air voids introduced into the laminate before processing simply act as nucleation sites for void growth.)

For the cure cycle presented in Figure 12, Figures 14–16 show the effect of initial void diameter, the initial relative humidity exposure of the resin, and the initial vacuum applied during stages 1 and 2 on the final diameter of a pure water vapor void. The conditions of growth are specified in the figures. In Figure 14 it is seen that for an initial void diameter of zero, the final diameter is

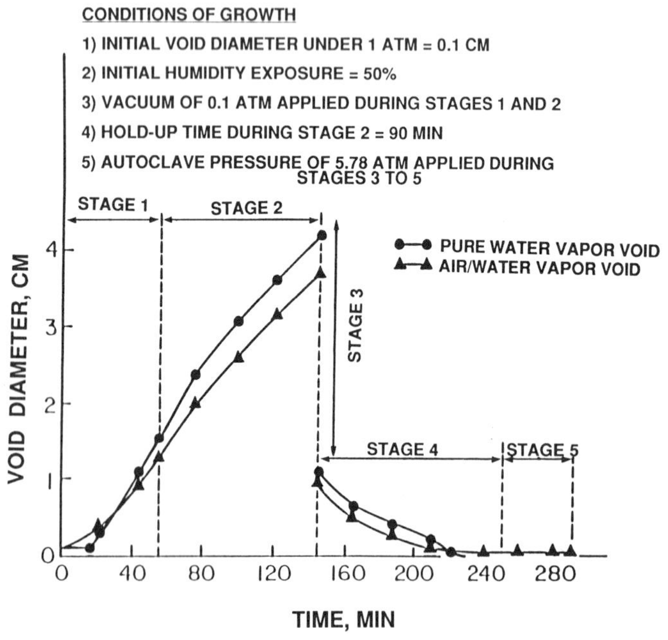

FIGURE 13 Progression of void growth during the cure cycle of Figure 12 for both pure water and air–water voids. Note the change when pressure is applied at the end of stage 2. (0.1 atm = 10.1 kPa.)

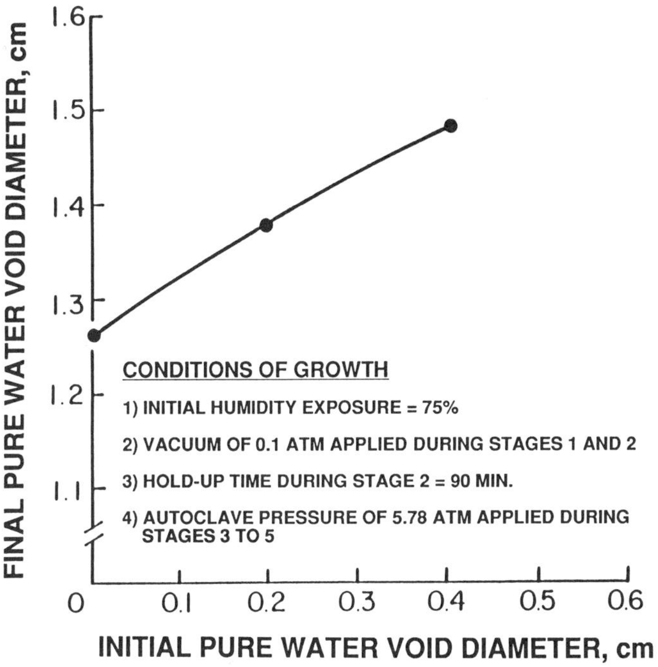

FIGURE 14 **Effect of initial void size on the final void size of a pure water void for the growth conditions specified. (0.1 atm = 10.1 kPa.)**

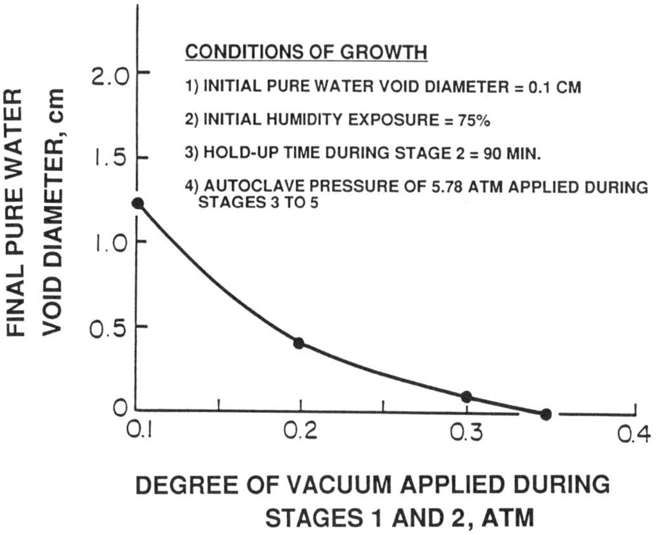

FIGURE 16 **Effect of resin pressure (i.e., degree of vacuum in atm) during stages 1 and 2 on the final void size of a pure water void for the growth conditions specified. (0.1 atm = 10.1 kPa.)**

about 1.25 cm under the specified conditions of growth. This implies that under the specified conditions of growth, initially void-free laminates can have considerable porosity after curing.

The final diameter of a pure water void for different initial relative humidity exposure of the resin is shown in Figure 15. Water solubility data of the resin in the prepreg (Narmco 5208) equilibrated at different relative humidities was best fit by a parabola [5]. That is, the water solubility increased with the square of the initial relative

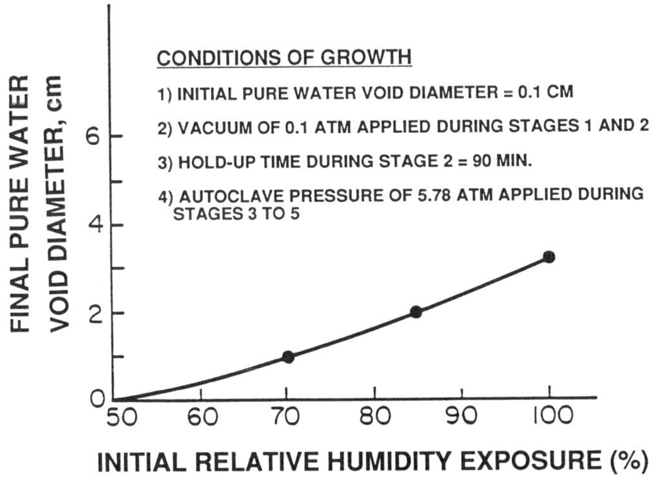

FIGURE 15 **Effect of initial relative humidity exposure of the resin on the final void size of a pure water void for the growth conditions specified. (0.1 atm = 10.1 kPa.)**

humidity exposure. The marked increase in the final void size illustrates the pronounced effect of initial relative humidity exposure on the final void size. This behavior is due to the higher concentration of dissolved water in the bulk resin as an increasing function of initial relative humidity exposure.

Figure 16 shows the final diameter of the void for different degrees of vacuum applied during stage 1. It is apparent that the larger the vacuum applied (lower pressure), the greater the potential for big voids.

Figures 14–16 identify the two most important parameters that affect the final diameter, namely: the initial relative humidity exposure of the prepreg and the applied pressure during the cure cycle. In terms of industrial processing of laminates, these results show that the humidity exposure of the resin before and during processing must be maintained below 50% if an initial vacuum of 10.1 kPa (0.1 atm) is to be applied and void-free laminates are desired. Maintaining this low relative humidity might well be prohibitively expensive, depending on the climate of the production facility.

EXPERIMENTAL VERIFICATION OF THE VOID STABILITY EQUATION. Several publications dealing with various aspects of bubble formation in fluids of low and medium viscosity (up to a few poise) have appeared in the literature. However, there is little experimental information on bubble formation in high viscosity fluids like epoxy resins. The problems of working with a sticky, viscous fluid like an epoxy resin have made such investigations difficult.

In a typical composite processing cure cycle, bubble formation and growth usually occur under conditions of constant autoclave and bleeder pressures. Therefore, to simulate void growth during autoclave processing, it is important that the experiment be conducted under constant resin pressure rather than constant temperature. Releasing the resin pressure under an isothermal temperature, rather than increasing the temperature at a constant resin pressure, could result in inertia-controlled growth instead of diffusion-controlled growth [26]. These two bubble growth behaviors are the extreme asymptotic regimes of the general situation for growth of a gas bubble as a result of mass transfer of one component from the liquid to the gas forming the bubble. Therefore, in any bubble formation and growth studies, the correct simulation of the physical conditions is extremely important.

The work reported in this section was undertaken to experimentally verify the void stability equation (Eq. 30) for water vapor void formation in an epoxy resin. The apparatus was constructed to allow the convenient photographic study of the formation and growth of water vapor bubbles in epoxy resin. Details of the experiment have been published [20]. The bubble formation experiment consisted in raising the temperature of the resin at a constant pressure and recording the temperature at which bubble formation begins. Next the resin pressure was increased to another constant value and the operations were repeated. The resin was equilibrated to 39% relative humidity before the bubble formation experiment.

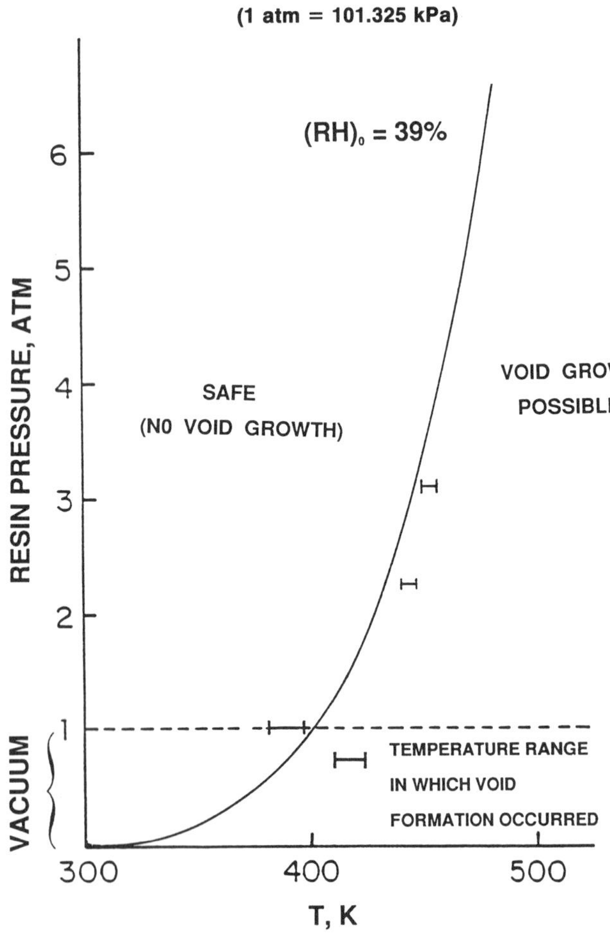

FIGURE 17 Comparison of the experimental results and the theoretical predictions based on the void stability equation.

Figure 17 compares the experimental results and theoretical prediction based on the "void stability equation" (i.e., Eq. 30). In Figure 17, the horizontal bar indicates the temperature range in which bubble formation occurred under a constant resin pressure. Figure 17 shows reasonably good agreement between theoretical prediction and the experimental results. This supports the validity of using the "void stability equation" for predicting the minimum resin pressure required to prevent water vapor voids from forming in a highly viscous medium like an epoxy resin. Further investigations to validate the "void stability equation" are in progress.

Guidelines for Autoclave Cure Cycle Selection

Cure Cycle Selection

The choice of the temperature–time history is primarily based on kinetic and viscosity data of the resin and the desired T_g of the cured laminate. Figure 10*a* showed the profiles of a standard cure cycle and the corresponding viscosity of Hercules 3501-6 epoxy resin; note that the resin viscosity during the initial part of the second temperature ramp is relatively low. This could allow void growth by moisture diffusion.

Figure 18*a* shows the profiles of time-at-temperature and the corresponding minimum resin pressure P_{min} required to prevent void growth by moisture diffusion in resin equilibrated at 50 and 60% relative humidity before processing. Note that the potential for void growth by moisture diffusion does not exist during either the first temperature ramp or the first temperature hold. This is because the resin pressure at the laminate boundary (which is the lowest resin pressure within the laminate) is maintained higher than P_{min}. However, during the second temperature ramp and the initial part of the second hold, diffusion-controlled water vapor void growth is possible because the resin pressure at the laminate boundary has fallen below P_{min}. Finally, during the later part of the second hold, further void growth is unlikely because the resin has already gelled.

One key conclusion can be derived from the foregoing discussion. If the resin pressure everywhere within a curing laminate, including the laminate boundary, is maintained greater than P_{min}, then the potential for void growth by moisture diffusion can be eliminated. In this discussion, the terms "resin pressure at the boundary" and "laminate boundary pressure" are synonymous. However, the laminate boundary pressure is *NOT* the same as the vacuum bag pressure. These two can be equal only for the case of "free bleeding," that is, when there are no dams and the bleeder cloth resistance to fluid flow is negligible.

To prevent void growth by moisture diffusion during the autoclave process, we suggest the following basic guideline for autoclave and laminate boundary pressure selection. *For any given time-at-temperature cycle,*

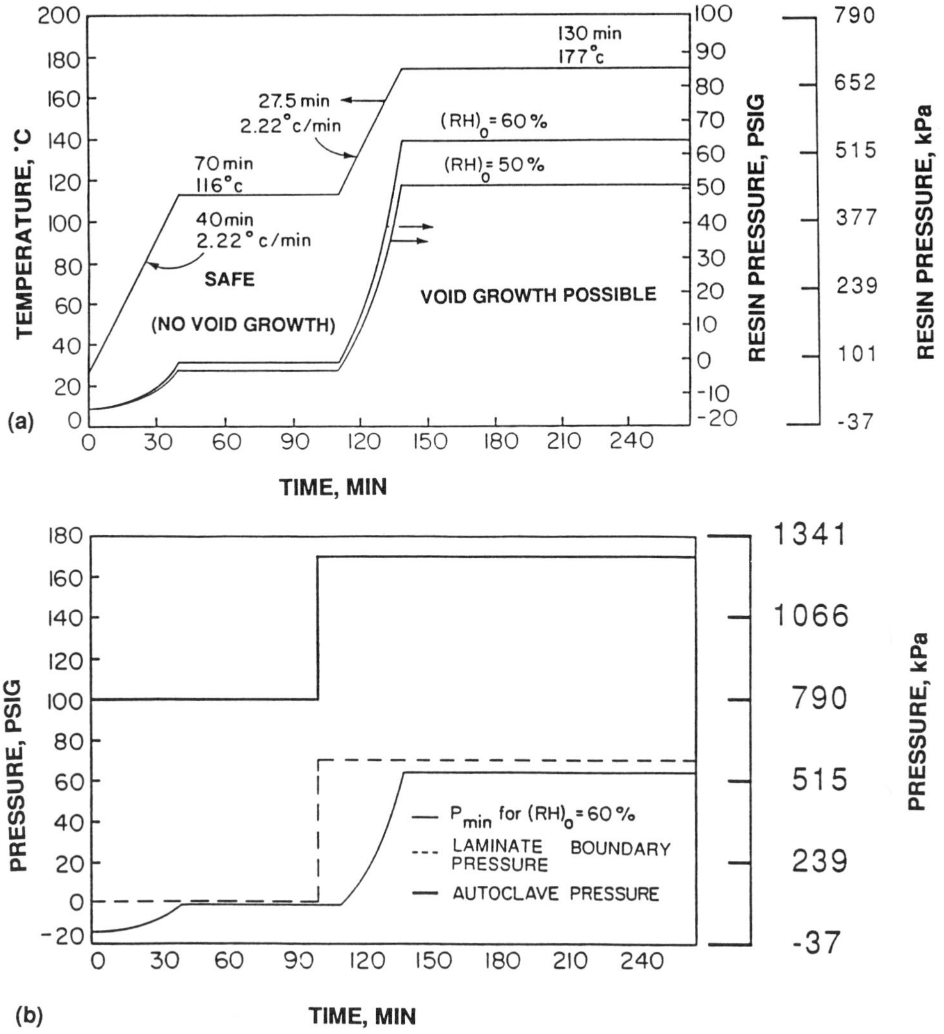

FIGURE 18 **(*a*) Profiles of time-at-temperature and the corresponding P_{min} required to prevent void growth by moisture diffusion. (*b*) Acceptable profiles of autoclave and boundary pressures for void-free laminates. The relative humidity to which the prepreg is exposed before processing is assumed to be 60%.**

choose a laminate boundary pressure cycle such that the boundary pressure at any time during cure is greater than P_{min}. Furthermore, maintain a positive difference between the autoclave and boundary pressures throughout the cure cycle. This prevents resin backflow into the laminate.

Figure 18*b* illustrates the use of this rule in the selection of the laminate boundary and autoclave pressures for the case of "free bleeding." Figure 19 shows the resin pressures at points *R* and *Q* in the laminate (see inset) for the pressure cycles shown in Figure 18*b* and the temperature–time history shown in Figure 18*a*. Note the profile of the minimum resin pressure required to prevent void growth by moisture diffusion in prepregs equilibrated at 60% initial relative humidity in Figure 19. It is clear from Figure 19 that the resin pressure at any point within the curing laminate is higher than P_{min}. Therefore, conditions favoring void growth by moisture diffusion do not exist anywhere within the curing laminate during the cure cycle.

The Guidelines

Based on our analysis of the autoclave process, the following guidelines are suggested for the selection of the cure cycle.

1. Fix the time-at-temperature cycle on the basis of kinetic and viscosity data of the resin and the desired glass transition temperature of the cured laminate.
2. Determine the profile of P_{min} versus cure time for the chosen time-at-temperature cycle by applying the void stability equation.
3. Select a profile for the laminate boundary pressure versus cure time such that the boundary pressure

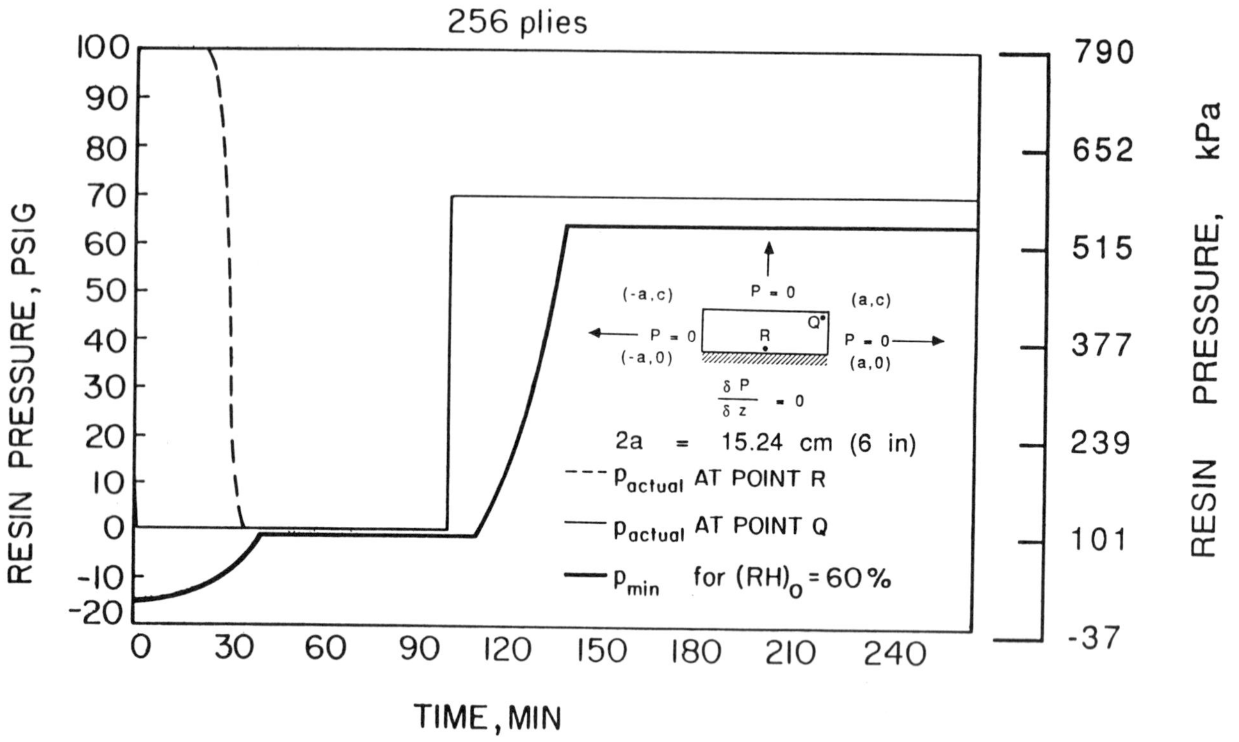

FIGURE 19 **Profiles of resin pressure at points *R* and *Q* in a 256-ply laminate for the time-at-temperature cycle and the pressure cycles shown in Figures 18*a* and 18*b*, respectively.**

is maintained greater than P_{min} throughout the cure cycle.

4. Select and maintain a positive difference between the autoclave and boundary pressures throughout the cure cycle. (This prevents resin backflow into the laminate.)
5. Determine resin mass loss, final laminate thickness, etc., for the chosen cure cycle.
6. If the resin mass loss, final laminate thickness, etc., are different from those desired, iterate between steps 4 and 5.

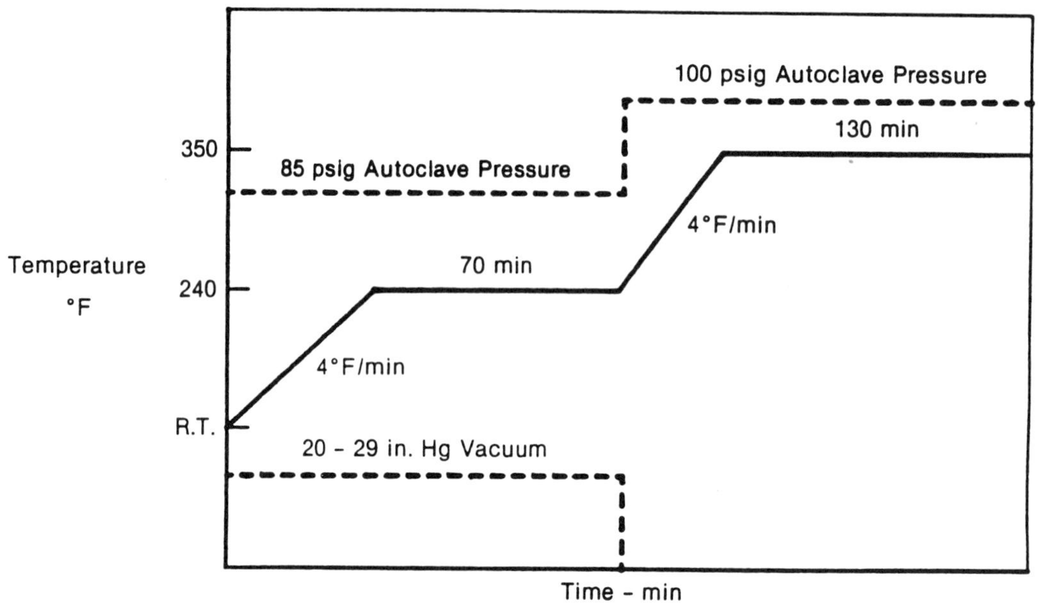

FIGURE 20 **Cure cycle for fabricating dry and wet graphite–epoxy laminates.** (From Ref. 24.) **(To convert the units in this figure to SI units, see the conversion table at the front of the book.)**

Experimental Verification of the Guidelines for Pressure Selection

The proposed guidelines for pressure selection have been verified experimentally [24]. A set of four AS-4/3502 graphite–epoxy laminates was fabricated using the cure cycle shown in Figure 20. These laminates were ultrasonically tested before and after cure. Details on laminate fabrication and nondestructive testing are reported elsewhere [24]. Variables investigated included prepreg moisture content, initial laminate compaction, laminate thickness, and ply drop-off.

The four laminates were processed according to the following conditions:

Laminate 1 had moisturized prepreg and poor compaction.
Laminate 2 had dry prepreg and poor compaction.
Laminate 3 had moisturized prepreg and good compaction.
Laminate 4 had dry prepreg and good compaction.

To produce moisturized prepreg, the plies were desiccant dried and then exposed to a wet environment (32°C,

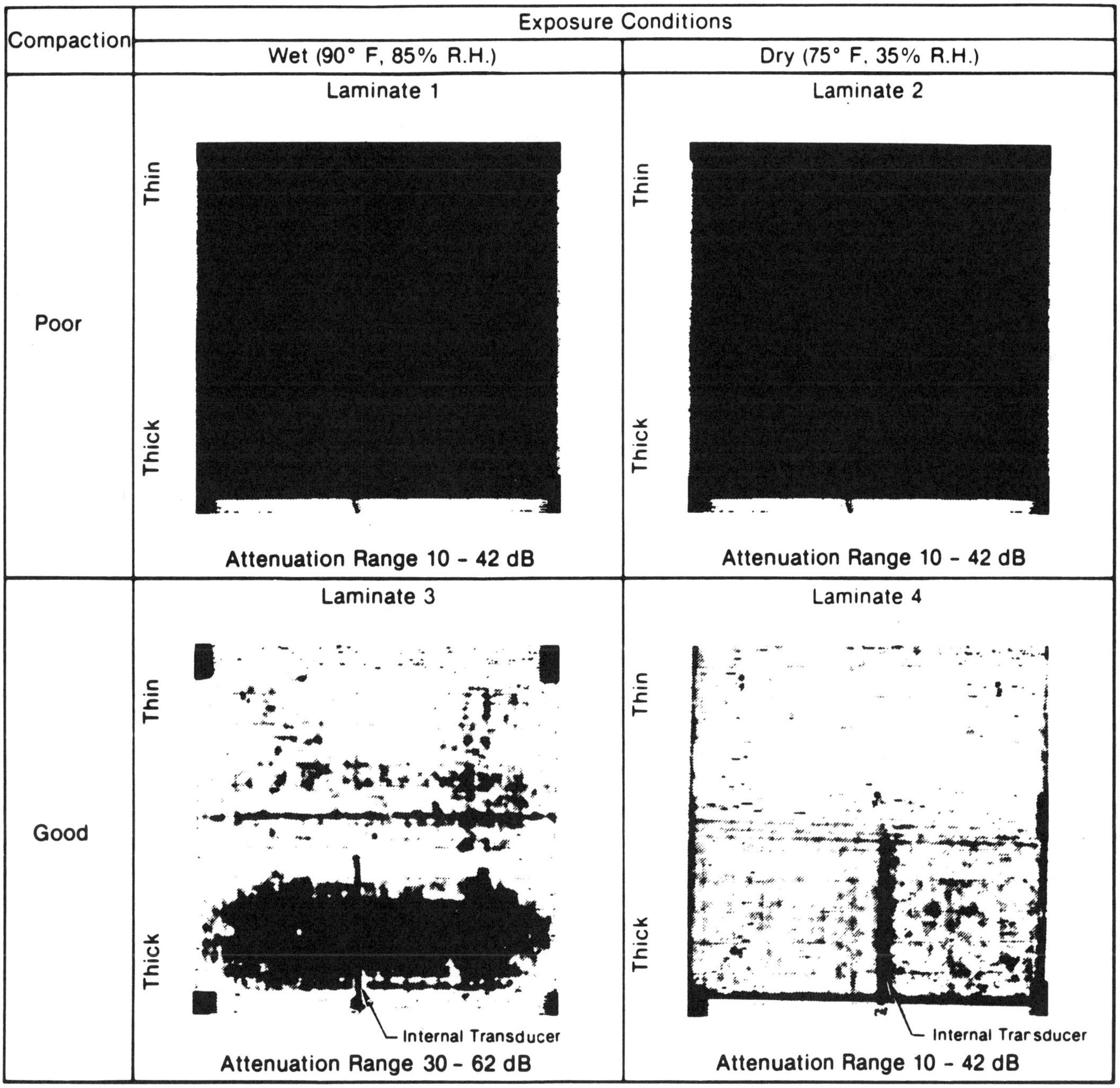

FIGURE 21 Precure nondestructive ultrasonic test results; black areas indicate high porosity. (From Ref. 24.) (To convert the units in this figure to SI units, see the conversion table at the front of the book.)

85% RH) for 72 hours before collation. The dry condition consisted of exposing desiccant-dried plies to 24°C, and 35% RH for 72 hours. Poor compaction consisted of collating the plies without any intermediate vacuum debulks. Good compaction consisted of several intermediate vacuum debulks during ply layup.

All four laminates were tested ultrasonically before cure. The results of the nondestructive ultrasonic test (NDT) are shown in Figure 21. Essentially, no signal was transmitted through the poorly compacted laminates, while the well-compacted laminates transmitted a much larger percentage of the original signal. Note that the ply drop-off and thick areas of the laminates were of much poorer quality than the thin areas.

Profiles of the typical resin pressures monitored by the laminate, tool, and bleeder transducers are shown in Figure 22. The laminate and tool transducers measured resin pressures within the laminate. The bleeder transducer measured the resin pressure at the laminate boundary. Also shown in Figure 22 are the profiles of time-at-temperature and the corresponding P_{min} for 35 and 85% initial relative humidity exposure of the prepreg. From Figure 22, it is evident that the resin pressure within the wet laminate fell below P_{min} before resin gelation. On the other hand, the resin pressure in the dry laminates remained well above P_{min} throughout the cure cycle. Consequently, by the void stability equation, the moisturized laminates should have considerable porosity while the dry laminates should be void-free.

After cure, all four laminates were inspected ultrasonically; the results are presented in Figure 23. As expected, the dry, well-compacted material (laminate 4) gave the best results, while the wet, poorly compacted sample (laminate 1) yielded the worst results. Good initial laminate compaction did significantly reduce the amount of porosity in the wet, well-compacted laminate (laminate 3), but still did not yield a good quality laminate, especially in the thick area of the sample.

A comparison of the precure and aftercure NDT results of the dry laminates (laminates 2 and 4) in Figures 21 and 23 indicates that the extent of compaction did not significantly offset laminate quality; that is, both dry laminates were of good quality after cure. In fact, the aftercure NDT result of the dry, poorly compacted sample (laminate 2) with initially trapped air showed good quality. This confirms that void formation and growth due to moisture is the primary cause of porosity in cured laminates and void transport during composite processing is a feasible mechanism of driving out initially trapped air voids from thin dry laminates. In thick laminates, the role of void transport during processing is not fully understood. The final quality of a thick laminate probably depends on the capillary number, which refers to the ratio of viscous to capillary forces, the initial moisture content, and the cure cycle.

The results thus far verify the proposed guidelines for autoclave and laminate boundary pressure selection and indicate that:

1. Moisture is the major contributing factor to void formation.
2. Initially trapped air voids act as nucleation sites for void growth by moisture diffusion; initial laminate compaction helps to reduce trapped air voids.
3. In dry laminates, initially trapped air voids can be transported out by resin flow.
4. Thick sections are more prone to void formation than thin sections.

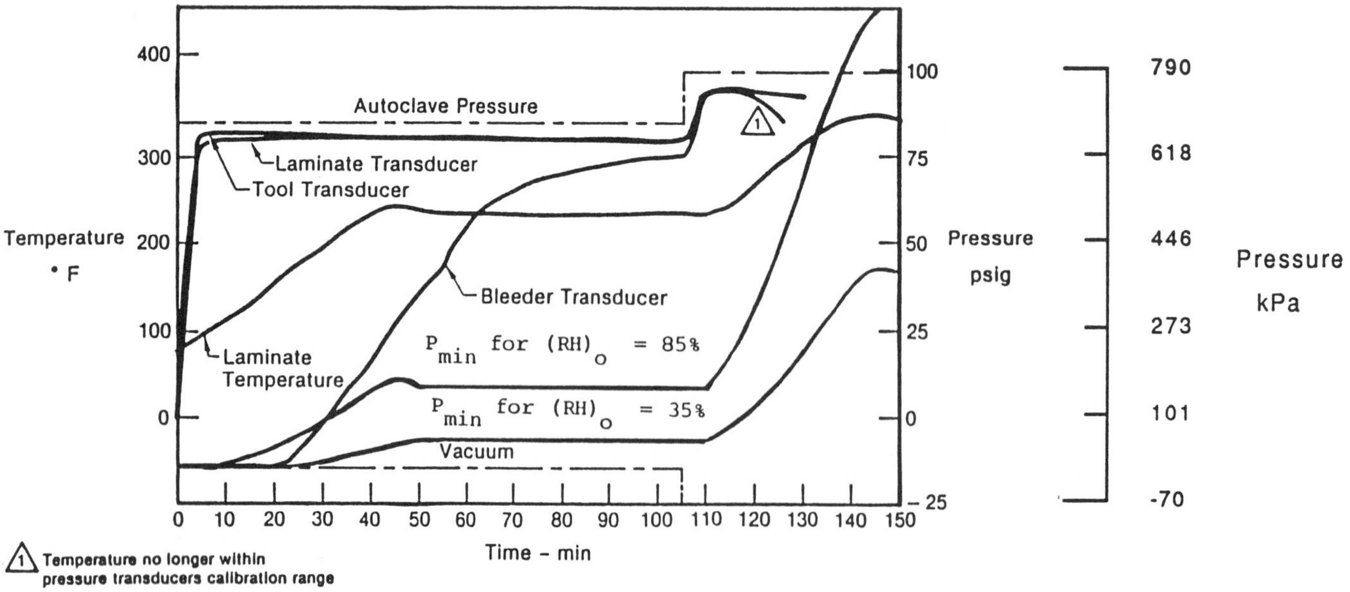

FIGURE 22 **Profiles of typical resin pressures monitored by the laminate, tool, and bleeder transducers. Also shown are the profiles of time-at-temperature and the corresponding P_{min} for 35 and 85% initial relative humidity exposure of the prepreg.** (From Ref. 24.) **(To convert the units in this figure to SI units, see the conversion table at the front of the book.)**

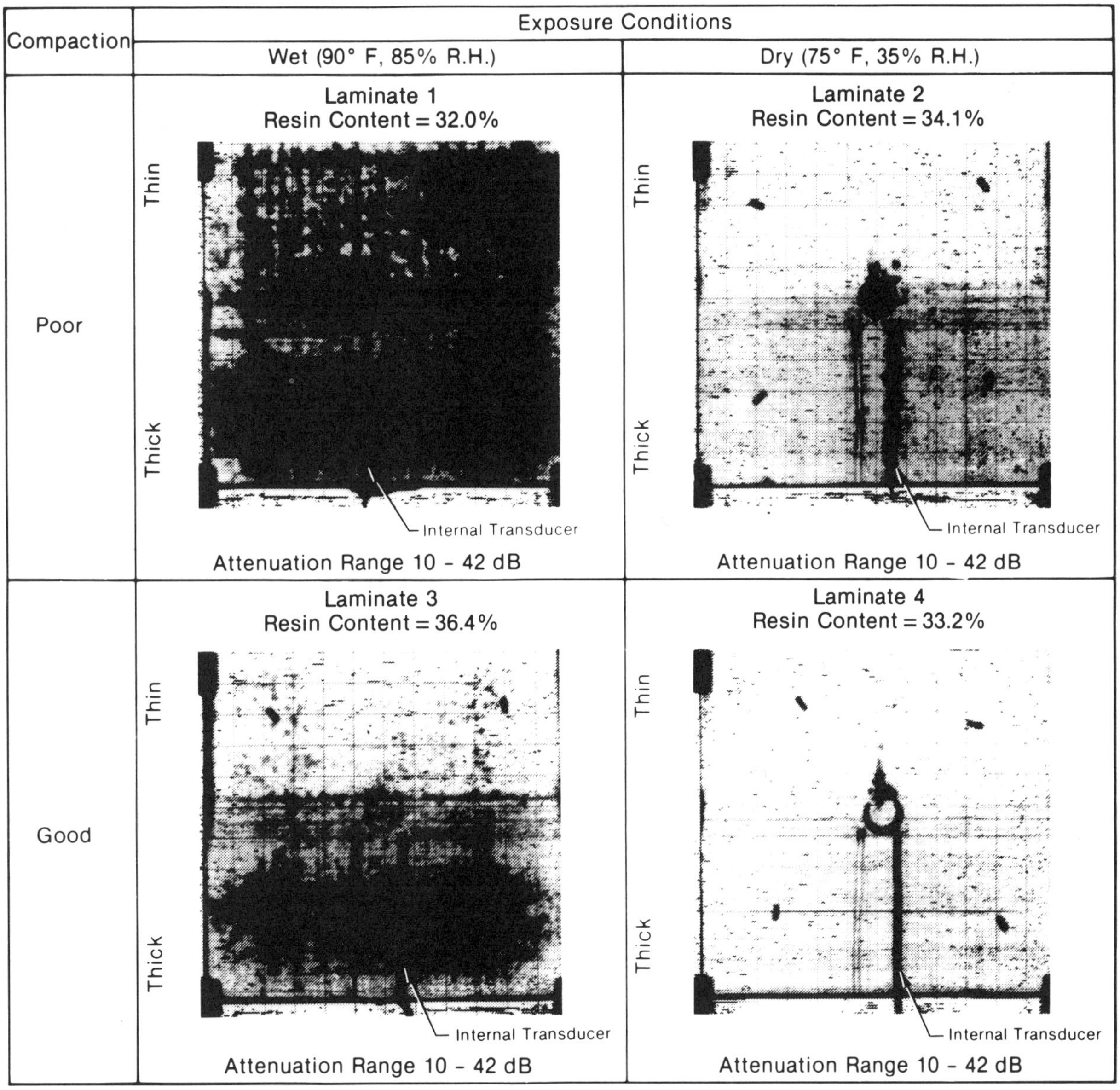

FIGURE 23 **Aftercure nondestructive ultrasonic test results. Note that the dry, well-compacted laminate gave the best results, while the wet, poorly compacted laminate yielded the worst results. (From Ref. 24.) (To convert the units in this figure to SI units, see the conversion table at the front of the book.)**

Conclusions

Any on-line process control model used for computer-aided manufacturing of high performance composite laminates must include a thorough treatment of void stability and growth as well as resin flow. These two key components, along with a heat transfer model and additional chemorheological information on kinetics and material properties, should permit optimized production of void-free, controlled thickness parts. Toward this goal, a number of advances have been made.

A generalized three-dimensional model for resin flow during composite processing has been developed. The model incorporates consolidation theory, which considers that the total force acting on a porous medium is counteracted by the sum of the opposing forces, including the springlike effect of the fiber network and the hydrostatic force due to the pressure of the liquid within the porous medium.

Simulation of resin flow during autoclave processing of fiber-reinforced graphite–epoxy laminates was done. Simulation results show that two-dimensional flow under the condition of "free bleeding" (i.e., without any dams surrounding the laminate and negligible bleeder cloth resistance) would lead to rapid loss in the resin pressure. This in turn could lead to severe problems of void formation. It was also found that in a thin laminate most of the resin flow is vertical, while in a thick laminate most of the flow is horizontal. Reasonable agreement was obtained between predicted and experimentally determined values for resin loss and final thickness per ply.

A model has been constructed to describe time-dependent void growth and stability for any cure cycle. Although the analysis is approximate, it does account for the moving void–resin boundary layer, and its effect on the concentration profile and diffusion of water in the resin. It was found that for the duration of the cure cycle it makes little difference whether the initial small voids contain pure water or mixtures of air and water.

At the end of a cure cycle, the spread in the final void sizes is not as marked as the spread in the initial void sizes. Furthermore, even at zero initial void diameter (i.e., void-free resin), the final void diameter may be relatively large due to rapid nucleation and growth. A particularly significant result is that application of vacuum during the initial stage of the cure cycle has the potential of creating large voids. Moreover, the initial moisture content of the prepreg is important. The void diameter at the end of cure cycle is directly dependent on the initial relative humidity exposure of the prepreg.

A pressure–temperature stability map can be constructed as a function of humidity exposure, which identifies the resin pressure values for each temperature below which void growth is possible and above which voids cannot grow but, rather, tend to collapse. The pressure–temperature void stability equation was experimentally verified for an epoxy resin equilibrated to water vapor under 39% relative humidity.

By combining models for resin flow and void growth during composite processing, it has been shown that it is feasible to select a cure cycle such that conditions favorable for void growth are never allowed at any point within the laminate during the autoclave process. This is achieved by maintaining the resin pressure at every point in the laminate greater than P_{min}, the minimum resin pressure required to prevent void growth by moisture diffusion. Finally, guidelines have been proposed for the selection of an optimum cure cycle.

Experimental results that verify the proposed guidelines have been presented. Experimental results indicate that (a) moisture is the primary cause of void formation, and (b) initially trapped air bubbles act as nucleation sites for void formation.

Nomenclature

Lowercase Letters

a = half-laminate length (Fig. 5) or void radius in Eq. 23
b = half-laminate width (Fig. 5)
c = laminate thickness (Fig. 5)
e = void ratio (Eq. 10)
j, q = integers in the series solution in Eqs. (20) and (22)
k_x, k_y, k_z = specific permeability in the x, y, and z directions (Eq. 9)
m_v = coefficient of volume change (Eq. 12)
p = effective stress due to the springlike behavior of the porous network
q_x, q_y, q_z = superficial resin velocity (Eq. 9)
r = distance from the center of the void to any point in the resin
t = time
v_x, v_y, v_z = interstitial fluid velocity
x, y, z = coordinate axes

Uppercase Letters

B = constant in Eq. (2a)
C = concentration of the diffusing species
C_v = specific heat of the composite
C_i = concentration of the diffusing species at the resin–bubble interface
C_∞ = concentration of the diffusing species in the bulk resin
D = diffusivity of the diffusing species (Eq. 23)
E_μ = constant in Eq. (1)
H_r = total heat of reaction
K_1, K_2, K_3 = constants in Eqs. (2)
K_x, K_y, K_z = thermal conductivities of the composite in the x, y, and z directions
P = resin pressure
P_{BC_1}–P_{BC_5} = laminate boundary pressure during bleeder ply molding (Fig. 5)
P_{min} = minimum resin pressure required to prevent void growth by moisture diffusion (Eq. 30)
R = universal gas constant
$(RH)_0$ = relative humidity to which the resin is equilibrated before processing
T = temperature
T_g = glass transition temperature
T_v = defined in Eq. (21)
V_f = volume fraction of fiber
V_0 = initial volume fraction of fiber

Greek Letters

α = degree of cure
β = growth constant (Eq. 27)
ε = porosity
μ = resin viscosity
μ_∞ = constant in Eq. (1)
ρ = resin density
ρ_0 = initial resin density
ρ_b = bulk resin density
ρ_g = density of gas in the void
σ = external pressure applied to the fiber–resin mass (Eq. 5)
σ_0 = constant external pressure applied to the fiber–resin mass
ξ = moving coordinate defined in Eq. (15)

Acknowledgments

The author expresses his sincere gratitude to Dr. John L. Kardos and Dr. M. P. Duduković of Washington University, St. Louis, for their constant guidance and support. The author also thanks Mr. Andrew Mallow of McDonnell Douglas Corporation for granting permission to reproduce Figures 2, 10*a*, 20, 21, 22, and 23, and some discussion thereof. Finally, help from Mrs. Shirley Matzek in preparing this manuscript is gratefully acknowledged.

Raju S. Davé

References

1. A. C. Loos and G. S. Springer, *J. Compos. Mater., 17,* 135 (1983).
2. J. L. Kardos, M. P. Duduković, E. L. McKague, and M. W. Lehman, *in* C. E. Browning, Ed., *Composite Materials: Quality Assurance and Processing,* ASTM Special Technical Publication 797, American Society of Testing and Materials, Philadelphia, 1983, p. 96.
3. R. S. Davé, J. L. Kardos, and M. P. Duduković, *Polym. Compos., 8,* 29 (1987).
4. R. S. Davé, J. L. Kardos, and M. P. Duduković, *Polym. Compos., 8,* 123 (1987).
5. J. L. Kardos, M. P. Duduković, and R. Davé, *Advances in Polymer Science: Epoxy Resin and Composites,* Vol. 80, Springer Verlag, Berlin, 1986, p. 101.
6. R. S. Davé, J. L. Kardos, and M. P. Duduković, *Proceedings of the American Society for Composites, 1st Technical Conference,* 1986, p. 137.
7. C. L. May, Ed., *Chemorheology of Thermosetting Polymers,* ACS Symposium Series 227, American Chemical Society, Washington, D. C., 1983.
8. L. Nicolais and A. Apicella, *Pure Appl. Chem., 57,* 1701 (1985).
9. T. G. Gutowski, T. Morigaki, and Z. Cai, *J. Compos. Mater., 21,* 172 (1987).
10. T. G. Gutowski, Z. Cai, S. Bauer, D. Boucher, J. Kingery, and S. Wineman, *J. Compos. Mater., 21,* 650 (1987).
11. J. Mijović and H. J. Wang, *Proceedings of the SPE ANTEC Symposium,* 1986, p. 554.
12. F. C. Campbell, A. R. Mallow, and G. A. Blase, "Computer-Aided Curing of Composites, 6th Quarterly Interim Technical Report," Contract F33615-83-C-5088, AFB, OH, AFML, Wright-Patterson, 1985.
13. F. C. Campbell, A. R. Mallow, G. A. Blase, and F. R. Muncaster, "Computer-Aided Curing of Composites, 8th Quarterly Interim Technical Report," Contract F33615-83-C-5088, AFB, OH, AFML, Wright-Patterson, 1986.
14. M. A. Grayson and C. J. Wolf, in J. Koenig, Ed., *Advances in Chemistry Series,* Vol. 174, American Chemical Society, Washington, DC, 1979, p. 53.
15. M. A. Grayson, *ACS Symposium Series,* Vol. 132, American Chemical Society, Washington, DC, 1980, p. 449.
16. W. I, Lee, A. C. Loos, and G. S. Springer, *J. Compos. Mater., 16,* 510 (1982).
17. F. C. Campbell, A. R. Mallow, and K. C. Amuedo, "Computer-Aided Curing of Composites, 4th Quarterly Interim Technical Report," Contract F33615-83-C-5088, AFB, OH, AFML, Wright-Patterson, 1985.
18. D. W. Taylor, *Fundamentals of Soil Mechanics,* Wiley, New York, 1969.
19. J. C. Williams, C. E. M. Morris, and B. C. Ennis, *Polym. Eng. Sci., 14,* 413 (1974).
20. R. S. Davé, *Computer-Aided Processing of Composites,* D.Sc. dissertation, Washington University, St. Louis, 1986.
21. W. D. Kingery, *Introduction to Ceramics,* Wiley, New York, 1960, p. 292.
22. L. E. Scriven, *Chem. Eng. Sci., 10,* 1 (1959).
23. M. Cable and J. R. Frade, *J. Mater. Sci., 22,* 149 (1987).
24. F. C. Campbell, A. R. Mallow, B. L. Boman, and G. A. Blase, "Computer-Aided Curing of Composites, 7th Quarterly Interim Technical Report," Contract F33615-83-C-5088, AFB, OH, AFML, Wright-Patterson, 1986.
25. G. G. Brown and E. L. McKague, "Processing Science of Epoxy Resin Composites, Interim Report," Contract 33615-80-C-5021, General Dynamics, Convair Division, San Diego, CA 1981.
26. J. Szekely and G. P. Martins, *Chem. Eng. Sci., 26,* 147 (1971).

Automotive Composites

In this article, "automotives" refer to land vehicles that are self-propelled by an internal engine; "composites" are defined as organic matrices reinforced with fiber. The geography of automotive vehicles can be divided into various classifications, such as Society of Automotive Engineers (SAE) powerplant, powertrain, and drivetrain systems; braking and steering systems; and transmission and driveline systems. Plastic and composite applications are assigned to four automotive areas: under-the-hood applications, exterior body panels, frame/chassis, and passenger compartment. Tubing, gaskets, mountings, seals, glazing, reflectors and medallions, seat covering and cushioning, headliner padding, and windshield washer bottles do not fall into these areas. Such automotive products are made with neat (unreinforced) plastics and elastomers, especially commodity volume polyolefins (polyethylenes), acrylics, polyvinyl chloride (PVC), styrenics, and cellular polyurethanes.

Applications refer primarily to passenger cars, but overall they can be ascribed to vans, trucks, buses, campers, and recreational vehicle (RVs) such as snowmobiles and golf carts. Specialty automotive off-highway vehicles such as RVs, construction/industrial equipment (bulldozers, dumpers, loaders, earthmovers, graders and scrapers, and cranes), and agricultural vehicles have selective applications according to their performance requirements.

Historical Background

The passenger car developed through the nineteenth century, reaching a watershed in about 1900, when the hallmarks of the modern car were established. The Olds-

FIGURE 1 Oldsmobile curved dash.

mobile "curved dash" (named for its bold design of the curved front: Fig. 1) signaled the end of the car as merely a replacement of the horse-drawn carriage for higher society, and the beginning of a vehicle for the general population. Approximately 5000 curved-dash vehicles were sold each year from 1900 to 1905, which was high volume production for the time. This volume demonstrated that there was a popular market for the car, and Henry Ford implemented the market requirement with the assembly line concept. The original lightweight cars, closely styled to the earlier horse-drawn carriages, evolved to the "muscle cars" of the 1950s with more than 300 horsepower. By the 1980s cars were downsized in dimensions, weight, and horsepower. Composites became recognized as an ideal material to meet the new requirements of the modern car.

Demonstration cars with all-composite bodies have been built since the early 1900s, beginning with Henry Ford's coupe. The Rohm & Haas acrylic Explorer, the Borg-Warner acrylonitrile-butadiene-styrene (ABS) Cycolac Research Vehicle (CRV), and the Mobay polyurethane car were built in the late 1950s and early 1960s. The all-composite-body Pontiac Phoenix Ventura, Chrysler PXL, and Ford Lightweight Vehicle (LWV) (Fig. 2) represented demonstration cars of the 1970s, and Ford's Probe V concept car indicates the all-composite-body car of the 1990s. European concept cars add to the list. Commercial volume all-composite Chevrolet Corvettes have been produced since the late 1950s; and Pontiac Fieros represent the all-composite-body car of the 1980s. A principal difference between these two cars is that the Corvette is a low production volume car with an annual production range between 25,000 and 50,000 units, whereas the annual production range for the Fiero was 100,000–150,000 units.

FIGURE 2 Firebird I, II, and III (which were followed by Firebird IV): General Motors Corporation concept cars of the 1950s and 1960s.

Performance Requirements

Various property requirements for automotive composites, depending on the vehicle, applications, and fabrication process are listed below.

Conductivity: for shielding from electrostatic discharge (ESD), and electromagnetic and radiofrequency interference (EMI/RFI).

Corrosion and chemical resistance (e.g., to chloride salts, alcohols, oil and grease, detergents, waxes, antifreeze, foods, organic solvents).

Design freedom (e.g., for engineering suspension system packaging, styling, aerodynamic shaping); drag coefficient C_d objective for 1990 is 0.30, down from 0.4 to 0.5 in 1979 (1986 Sable had 0.29 and Probe V concept car is 0.137).

Electrical properties (e.g., resistivity, dielectric strength, dielectric constant, frequency range).

Flame retardant

In-place assembly costs: galvanized steel costs about $1.10/kg compared with composite costs of more than $2.20/kg and often much higher. Sheet stamping and metal fabricating cycle times are faster than many composite cycles, but composite fabricating cycles are being reduced. Composite assemblies can often be unitized during fabrication, greatly reducing assembly costs. Nevertheless one sheet stamping plant can supply more than one assembly plant, but one composite body fabricating plant cannot supply even one assembly plant.

Insulation: electrical, thermal.

Matable with metal assemblies (i.e. low coefficient of thermal expansion).

Mechanical properties: including resistance to dynamic fatigue, vibration and noise dampening, torque resistance; impact resistance, strength, and modulus.

Paintability, colorfastness, and platability: on-line versus off-line painting exterior body panels [e.g., heat sag for reaction-injection molded polyurethane (RIM/PUR) with 20–30% fiberglass at 163°C/hour =

0.125 cm for off-line painting]; fascia heat sag specification is 1.25 cm at 121°C/hour.

Surface finish: varies from low visibility quarter panels to high visibility hoods, to frame/chassis parts for which finish is not a requirement.

Temperature resistance: from −40 to 177°C.

Ultraviolet resistance

Automotive composite applications serve as a proving ground for many other potential composite applications. The evaluation of composite performance in automotive vehicles provides information for seating materials, decorative techniques (metallizing, hot stamping, painting, color compounding), consumer-acceptable finishes, corrosion resistance, electrical/electronic applications (electrical properties, flame retardancy, conductivity for EMI/RFI shielding and ESD), mechanical and chemical properties at high temperatures, noise and vibration dampening, torque resistance, ergonomics, optical clarity, consumer appeal, styling, coassembling with metal components, and volume production economics.

Future uses of composites in automotive applications depend on process economics. A major activity for the rest of the twentieth century is to reduce the cost of assembled automotive composite parts. Reducing cycle time is one of the primary targets, but other objectives are equally important. When parts can be molded as a single unitized component, such as exterior body panels, assembly costs are greatly reduced. The selection of processes involves tradeoffs among various economic factors (see PROCESS EQUIPMENT DEVELOPMENT).

Resin Matrices and Reinforcements

Resin Matrices

The following resins are used in automotive composites. Resins preceded by an asterisk are thermosetting; all others are thermoplastic.

ABS and other styrenics such as high impact polystyrene (HIPS), styrene maleic anhydride (SMA), and styrene acrylonitrile (SAN)
Acetal
* Acrylamate
Acrylic, high impact
* Epoxy
Ionomer
Ketone polymers; polyaryl ketone, polyetherketone
Liquid crystal polymer (LCP)
Nylon; amorphous and crystalline nylon 66 and nylon 6
* Phenolic
Polyamide-imide (PAI)
Polyarylate (PAR)
Polycarbonate (PC)
* Polyester, unsaturated (USPE)
Polyester, thermoplastic (see Thermoplastic polyesters PBT and PET)
Polyetherimide (PEI)
Polyethylene; high density polyethylene (HDPE)
* Polyimide (PI)
Polyphenylene ether–polyphenylene oxide (PPE/PPO)
Polyphenylene sulfide (PPS)
Polypropylene (PP)
Polysulfones: polyarylsulfone, polyethersulfone (PSO, PAS, PES)
Polyurea
Polyurethane (PUR)
Thermoplastic polyesters; polybutylene terephthalate (PBT), polyethylene terephthalate (PET)
Polyvinyl chloride (PVC)

Polyimides and polyurethanes can be polymerized as either thermosetting resins or thermoplastics. Several resins are used in automotive composites as alloys, especially polycarbonate with ABS, SMA, PPE, and PPO. PPO is alloyed with nylon, and all commercial PPO grades contain between 40 and 60 Wt % polystyrene. Resins are usually melt-processed by injection molding when reinforced with mineral fillers such as mica and calcium carbonate, or either chopped or milled fiber reinforcements; but sheet molding compounds (SMC) composed of fiberglass mat-reinforced unsaturated polyester resins are compression molded into horizontal body panels, grille opening panels, station wagon tailgates, and tractor trailer fender/hood/cab units. Bulk molding compounds (BMC), liquid resins, and powders are also used to fabricate composites. Specific composite types and grades are continually being developed, evaluated, and challenged for every automotive application. Certain composites are preferred for conventional applications, and these composites are specified by individual automotive companies, with the names of certified vendors. Selected composites and applications, ranging from "workhorse" to high performance composites, are described below.

FIBERGLASS-REINFORCED ACRYLIC. Body panels of fiber-reinforced acrylic replaced aluminum panels in General Motors Truck and Coach Division RTS-2 (Rapid Transit Series) buses in the late 1970s. Although the process economics failed to meet expectations, the panels proved to be roadworthy, and the process was found to be a viable method for the low volume production of large panels. The RTS-2 models used about 14 different types of exterior body panels made from press-reinforced acrylic sheet. The panels provided strength and stiffness from the fiberglass mat, and resistance to UV radiation (weatherability) from the acrylic. Press-reinforced acrylic sheet body panels were used for the AirTrans buses at the Dallas–Fort Worth, Texas, airport and on the Bricklin car from Canada, in the 1970s. Cross-linked acrylic sheet is catalytically cured with methyl ethyl ketone (MEK), and a 0.25 cm fiberglass mat-reinforced polyester preform is pressed into the acrylic sheet at 0.17–0.35 MPa at 93°C. The 0.20 cm thick cross-linked acrylic sheet is vacuum formed to the automotive panel shape before press forming. To increase production rates, catalytic curing time must be reduced. The catalyst, therefore, is a principal factor for improving process economics. The procedure, as observed by the author, was done in random locations in the plant: the preform was prepared in one location, vacuum formed somewhere else, and press formed in another area in the plant. Process economics can be improved with carefully planned job sites, such as in-line stages. After press forming, the panels are transferred to racks, and removed to a finishing station for sandblasting and trimming. Metal reinforcements are added, and pultruded trim is assembled on some of the panels. The panels are ready for shipment to automotive companies.

POLYPROPYLENE. Composites are used under the hood, and in both interior compartment and exterior applications. Principal automotive applications are as follows.

Under the Hood
Air cleaner housing
Air conditioner adapter
Shrouds: fan shroud, extension shroud
Fans: radiator, smog filter, squirrel cage
Heater/air conditioner housing
Clutch fan
Timing belt cover
Transmission underpans

Interior Compartment
Dash pad retainer
Radio support bracket

Exterior Compartment
Fender liner, structural
Housings: backup lamp housing, rear decorative lamp housing
Support panel for front end
Vertical body panels

Forty percent continuous strand fiberglass mat-reinforced polypropylene is used for front and rear bumper beams on certain General Motors and Ford passenger car models made in the United States. Continuous strand mat-reinforced polypropylene is also stamped into station wagon load floors as an alternative to high density polyethylene composites, seat shells, battery trays, gear case covers, oil pans, and rocker panels. Unreinforced polypropylene is used in several automotive applications such as windshield wiper bottles, accelerator pedals, glove box and glove box door, rear package shelf, panels (instrument panel components, inner door panels, kick cowl panels, rear pillar post panels, quarter panels, etc.)

THERMOPLASTIC POLYESTERS: POLYBUTYLENE TEREPHTHALATE (PBT) AND POLYETHYLENE TEREPHTHALATE (PET). We overview PBT and PET in the same series of engineering thermoplastics, then sort them according to applications and properties. Chrysler Engineering Standard/Material Standard, Chrysler Plastic Number (CPN) MS-DB454, "Plastic-Thermoplastic-Polyester-Terephthalate Type," and American Society for Testing and Materials Standard ASTM D 3220-81 demonstrate the generic reference to thermoplastic polyesters and their distinctions. The Chrysler document states: "This standard represents the requirements of unfilled and glass filled, thermoplastic, polyester (terephthalate type) compounds that are injection moldable. These compounds are suitable for use in engine electrical parts, such as distributor rotors." Chrysler CPN MS DB454 and ASTM D 3220 do not include higher performance thermoplastic polyester such as PET containing 55% (weight) fiberglass + mineral (mica) reinforcement, which has a room temperature flexural modulus of 18 GPa.

Thermoplastic polyesters are used in the following automotive applications.

Air conditioner pressure switches
Air deflector
Alternator fans
Body panels and components: body filler panels, grille components, headlamp housings, louvers, mirror housings and parts, quarter-window bezels, side-rail supports for T-bar roof, spoilers, sunroof frame, trim, valence panels, vent grilles
Bumpers
Bumper filler panels
Carburetor parts
Deck lids
Door handles, door hardware
Electrical/electronic parts: housings for computers, connectors, electric motor end bells, fuse panels, relay bases, sockets, switch components; high energy ignition system parts: distributor components (caps, rotor, adaptors), coil bobbins, coil cases, switch, lamp sockets
Emission control system parts
Fuel filter housing (diesel engines)
Fuel pump end caps
Gears
Gearshift fingers
Hinges
Idle speed controls
Instrument cluster
Speedometer frames
Steering column housing components
Suspension spring housings
Transmission components: servo pistons, thrust washers
Windshield wiper structures: water spray nozzle, wiper pivot body, switch superstructure

Other composites are used for several of these applications, as alternative composites or together with thermoplastic polyesters. Thermoplastic polyesters are used for their resistance to automotive chemicals, from antifreeze solutions to lube oils, for dimensional stability such as stiffness, and for resistance to warp and creep over a wide temperature range including paint bake oven temperatures of 177°C for 30 minutes. There are important differences between PBT and PET, which are shown in the property tables for these two resins. PET resins have higher deflection temperatures under load at 1.8 MPa, permitting PET composites to be assembled by vapor phase and wave soldering methods—for example, in the manufacture of electrical connectors. PET resins have higher melt temperatures up to 310°C, versus about 260°C for PBT. PET resins are more sensitive to hydrolysis, which is the degradative reaction of water with the PET polymers (not to be confused with hydroscopicity, which refers only to moisture absorption). PBT and PET differ in their crystallization behavior, which affects their responses to temperature gradients (temperature vs. time), especially during melt processing such as injection molding. Table 1 compares mechanical and electrical properties of commercial PBT and PET, both reinforced with 30 wt % chopped fiberglass.

REINFORCED REACTION-INJECTION MOLDING/ POLYURETHANE (RRIM/PUR). These composites are molded into bumper assemblies, exterior body panels, steering wheels, transmission and firewall deadeners, and lawn tractor engine covers. The composites are also used for lawn tractor valve covers and canopy roofs, for recreational vehicle housings, and for panels, belly pans, and hoods. RRIM/PUR is used for passenger car vertical body panels, especially panels with distinctive contours. High modulus RIM composite [34.5–103.5 MPa] is molded into quarter panels, air dams, spoilers and other aerodynamic add-ons, and fender extensions. These RIM resins are not reinforced. RRIM/PUR is not used for horizontal panels: hood, roof, and trunk lid panels; (where SMC is the composite of choice).

In contrast to add-on body components, exterior fender and door panels are specified for stiffness, dimensional stability, and impact resistance. A panel should be stiff enough to resist the force of airflow up to 161 km/h (100 mph) and to resist excessive yield when leaned on by a person. Horizontal panels require higher modulus and dimensional stability, both during assembly and for road performance, compared with fender panels. An advantage of RRIM/PUR is the composite's impact resilience: a panel shows a deflection of 15–20 cm at −29°C on repeated cycle with no damage, provided there is no internal stress.

Until paint systems with lower cure temperatures are used by automotive companies, composite body panels are exposed to oven temperatures of 163–182°C. Sag values for RRIM with 20–30% fiberglass exposed to 163°C for 30 minutes sag 0.125 cm. The addition of 25% milled fiberglass improves sag resistance during the paint–bake oven cycle. Sag is decreased with higher functionality isocyanates in the polyurethane chemistry system. An isocyanate with a functionality of 2.4 yields lower sag, but it also lower room temperature flexural modulus, than an isocyanate with a functionality of 2.0.

Gloss and paint adhesion to the composite substrate are two principal performance properties of composite panels. These two properties are tested after 1000 hours of exposure in a QUV Weather Tester. Adhesion is tested using the cross-batch method and 610 Scotch brand tape. After exposure to the QUV Weather Tester, RRIM loses gloss, but the gloss is restored after washing and waxing. QUV tests use FS-40 sunlamps at 60°C for 16 hours, followed by 8-hour condensation at 50°C. The following effects of accelerated weathering are observed with a "Weather-Ometer": discoloration, fading, chalking, checking, blistering, 60° gloss, and adhesion.

Isocyanate chemistry has lead to polyurea RIM resins that allow faster cycles, lower clamp force during molding, thinner wall parts [0.25 vs. 0.32 cm for melt-injection molding], and good heat sag resistance at 182°C. The system eliminates ethylene oxide and catalysts (it is autocatalytic), while conventional PUR requires a stannous catalyst and ethylene oxide to speed up the glycol chain extender reaction with the "A" component. Ethylene oxide blocks increase moisture absorption and tend to blend the hard (plastic) blocks with the soft (elastomeric) blocks on the polymer chain. Residual catalyst reduces thermal properties and creates mold release problems. The elimination of ethylene oxide and catalysts removes these problems. The reaction of isocyanate (with diethyltoluene diamine, DETDA) is not only autocatalytic, it is virtually instantaneous and yields 100% polyurea instead of polyurethane linkages. Two types of RIM/polyurea are standard temperature systems and higher temperature systems. The former can be used on modified RIM machines; the higher temperature systems make better use of faster reactive polyurea components, but require faster RIM machines.

Acrylamate RIM systems are under development for the following automotive applications: body panels (inner panels), front body structures, load floors for station wagons, radiator supports, firewalls, door and window frames, floor pans, spare tire covers (commercial), trunk floors, seat structures, frame members (secondary substructures), and heavy truck spoilers (air dams). Acryla-

TABLE 1
Properties at Room Temperature of 30% PBT and 30% PET + Flame Retardant

Properties	PET[a]	PBT[b]
Electrical		
Dielectric strength, 1.6 mm., V/0.0254 mm	650	750
Dielectric constant 10^6 Hz	3.6	3.8
Surface resistivity, Ω^2	10^{14}	
Volume resistivity, Ω-cm	10^{15}	3.4×10^{16}
Arc resistance, s	117	28
Flammability, UL94 (designated thickness)	V-O	V-O
Oxygen index %	33	
Mechanical		
Tensile strength, MPa	150	117
Flexural modulus, GPa	10.3	7.6
Deflection temperature under load at 1.82 MPa, °C	226	204
Izod impact strength, notched J/m	85.4	69.4

[a] Rynite 530 FR, made by E. I. du Pont de Nemours & Company, Inc.
[b] Valox 4205E, made by General Electric Company.

mate RIM composites can be an alternative to thermoformed ABS sheet and compression-molded sheet molding compound. Strength-to-weight ratios can be increased by reducing the acrylamate density with polyvinyl dichloride (PVdC) expandable beads during the reaction injection molding process. High modulus [>31 GPa], high impact strength acrylamate RIM composites are expanding the use of these composites to include whole automotive interior vertical panels and horizontal panels such as trunk lids, recreational vehicle body panels, body panels for agricultural and materials handling vehicles, truck COE (cab-over-engine) panels, and secondary airframe structures. The composites are reinforced with a hybrid system of 35% (weight) graphite fibers plus 10% glass fibers, with 55% resin content. Applications are being developed for performance requirements better than RIM/PUR and below high temperature composites such as reinforced ketone polymers (PAEK, PEEK) and imide polymers. Table 2 compares mechanical and other properties of two grades of acrylamate composites. Structural composites can be fabricated with existing RIM equipment without machinery modification.

Acrylamate is subjected to mat molding/reaction-injection molding (MM/RIM) to produce wheel well covers. MM/RIM is a hybrid process combining resin transfer molding (RTM) and casting, which is a low pressure injection of the reactive liquids into the mold. The mat reinforcement is placed on the open mold before the RTM–casting process begins. Polycarbonate MM/RIM composites are also fabricated into automotive products. The acrylamate MM/RIM resins can have slightly better flexural modulus, while the polycarbonate MM/RIM systems are tougher (higher impact strength), and can have slightly better flow properties.

POLYAMIDE-IMIDE (PAI). Graphite-reinforced PAI composites are distinguished by their high mechanical property values at room temperature and above 260°C, their electrical properties over a wide frequency range to 10^9 Hz, and their inherent ignition resistance without the addition of flame retardants. These characteristics distinguish the high temperature engineering thermoplastics represented by the sulfone polymers, polyphenylene sulfide, the ketone polymers, and such imide polymers

TABLE 2
Properties of Two Acrylamate Composites

Property	Acrylamate: 30% Glass Reinforced	Acrylamate: 40% Glass Reinforced
Mechanical		
Tensile strength, MPa	114	125
Tensile modulus, MPa	7,650	8,680
Tensile elongation, %	2.1	2.1
Flexural strength, MPa	185	231
Flexural modulus, MPa	7,100	7,300
Compressive strength, MPa	165	175
Compressive modulus, MPa	91,600	110,000
Izod impact, J·m		
Notched	600	790
Unnotched	1,000	970
Shear strength, MPa	81	88
Barcol hardness	28	30
Poisson's ratio	0.33	0.34
Thermal		
Deflection temperature at 1.82 MPa, °C	240	240
Coefficient of thermal expansion, mm/mm K °C	52.7×10^{-6}	48.6×10^{-6}
Thermal conductivity, W/m·°C	0.195	
Flammability		
Vertical burn (UL Bulletin 94)	HB[a]	HB[a]
Other		
Specific gravity	1.40	1.46
Mold shrinkage	0.002	0.001
Water absorption, %		
At 24 hours, 23°C	0.025	0.021
At equilibrium, 23°C	0.91	0.80

[a] Fire-retardant versions of ARIMAX resin are available from Ashland Chemical Company.

as polyimide and polyetherimide. Polyamide-imide is being developed for the following engine components: intake valve stems, push rods, tappets, spring retainers, rocker arms, connecting rods, piston skirts, piston rings, wrist pins, and timing gears. Engine temperatures and concurrent mechanical strains are too great for most plastics, but selective use of imide polymers is being achieved. Under-the-hood applications can be divided into three general temperature zones:

1. Inside the engine block
2. Ambient temperatures around the engine block
3. Peripheral ambient temperatures

These three zones can be represented with imide polymers, thermoplastic polyesters, and polyolefins and nylon, respectively. The selective use of composites for engine parts is predicated on the temperature in the immediate vicinity of the combustion chamber, which is above the 260°C recommended service temperature of high temperature engineering thermoplastics. Therefore exhaust valves, cylinder liners, cylinder heads, and top groove rings are not candidates for current high temperature composites.

An important attribute of automotive composites is that they are of lighter weight than metals, as shown by the weight savings with PAI intake valve stems. PAI valves weigh 30 grams compared with 140 grams for metal alloy valves. The lower inertia of composite valves reduces the spring load needed to prevent valve float, resulting in 20% higher engine speed and more efficient combustion. Lighter springs allow lighter retainers and tappets, rocker arms, and push rods.

Lighter weight automotive composite parts, compared with metal parts, contribute to overall reduction in vehicle weight much more than the weight difference between the composite and metal parts. The synergy created within the vehicle's total integrated assembly is shown when metal engine parts are replaced with composite parts. Inertial energy is greatly reduced with composite reciprocating parts, which reduces spring load, among other attributes. Spring load reduction is dramatically shown with composite wheels, which will greatly reduce braking and steering system load requirements, and permit a wider design window for suspension systems.

Another example of the "domino" effect of each lighter weight composite part is that parts supporting lighter weight composites have lower loads imposed on them. Since lighter vehicles mean reduced loads on frame components, composite frame structures become feasible. Composite engine parts reduce the load on other parts of the engine and on the entire powertrain system. Composite moving parts, especially powertrain components, reduce dynamic loads and allow further use of lighter parts. The use of composites breeds more use of composites. PAI composites are being developed for components in the transmission, universal joint, and other powertrain components (i.e., gears, bearings, bushings, and seals), where reduced frictional drag and wear resistance are priority requirements.

POLYETHERIMIDE (PEI). These composites are used under the hood in zone 2 (ambient temperatures around the engine block) for electrical connectors and related electronic devices, such as switches and controls, integrated circuit components, and printed circuit boards. The increasing number of computer controls throughout the modern vehicle can be accomplished only with miniaturization, not only under the hood, but in suspension and braking system packaging, etc., where in situ sensors are installed. Microprocessor programmable controls and closed-loop feedback are spreading throughout the vehicle, from on-board computers to monitor and adjust engine/fuel feed parameters, to modulated braking (antilock braking), suspension systems, and powertrains. PEI has a glass transition temperature T_g of 215°C, and its deflection temperature under load of 1.8 MPa to 186°C. The combined T_g and DTUL/1.8 MPa values, together with PEI's excellent electrical properties, allow PEI composites to be used for engine parts, including graphite fiber reinforced laminates, as well as for electrical hardware, dielectric films, flexible circuitry films, wire insulation, and insulating tape. High temperature composites such as PEI, PAI, PES, and ketone polymers can be assembled by vapor phase welding and wave soldering. Table 3 shows typical mechanical properties Ultem PEI.

TABLE 3
Typical Mechanical Properties at Room Temperature of Ultem PEI

Property[a]	Ultem Grade 1000	Ultem Grade 2300
Tensile strength, yield, N/mm²	105	160
Tensile modulus, 1% secant, N/mm² × 10³	3000	9,000
Tensile elongation, yield %	7–8	
Tensile elongation, ultimate %	60	3
Flexural strength, N/mm²	245	230
Flexural modulus, tangent, N/mm² × 10³	3300	8300
Compressive strength, N/mm²	140	160
Gardner impact strength, N·m (N-m)	36	
Izod impact strength at 3.2 mm, J/m		
Notched	50	100
Unnotched	1300	430
Shear strength, ultimate, psi N/mm²	100	100
Rockwell hardness, M	109	125

[a] Determined according to ASTM testing procedures.

Current Automotive Composite Developments

The six most active areas of automotive composite development are exterior body panels, space frames, bumpers, driveshafts, suspension systems, and instrument panels. Other developmental applications include pulleys (commercial), steering wheel columns, and wheels. These developments cover a broad range of different composites and property requirements. They illustrate further that automotive composite applications serve as a proving ground for all other potential composite applications.

Exterior Body Panels

Exterior body panels are compression molded from SMC; reaction-injection molded from PUR, polyurea, and acrylamates; and melt-injection molded or blow molded from several thermoplastic resins and alloys: namely, ABS alloyed with polycarbonate, nylon, or PET; PPO/nylon 66; amorphous nylon and impact PBT. SMC is compression molded into grille opening panels, tailgates, and horizontal panels (hood, roof, trunk lid). Thermoplastic composites are under development as alternative composites to the sheet-molded variety, but SMC has the advantage of higher flexural modulus (stiffness) required for large thin-walled horizontal panels, and for racking prior to assembly. PPO/nylon 66 alloys are being commercially developed for both vertical panels and horizontal exterior body panels for passenger cars. PPO brings higher creep resistance, higher stiffness at 177°C, and lower moisture absorption to the composite; and nylon imparts increased toughness, ductility, and processibility. The ratios of PPO to nylon can be adjusted according to the given panel application.

Body panels can be press formed or resin transfer molded (RTM) from mat-reinforced composites, or injection molded from glass flake reinforced composites. A sandwich panel construction consisting of 0.38 mm thick epoxy–fiberglass skin bonded to a urethane foam core was installed on a 1986 BMW model. The panels do not show surface porosity, and the sandwich panel construction is gaining attention around the world. However, the higher cost of these panels at this time of development would limit their use to premium priced passenger cars.

Space Frames

Space frames are integrally related to composite body panels. Lightweight composite panels do not impose as great a load on the frame as steel panels; and the space frame design removes load from the panels. With the integrated space frame/body panel design, the panels are not load bearing. Composites with tensile strengths lower than 34.5 MPa can be used to fabricate passenger car body panels with space frames. The critical design considerations for body panels are surface finish, heat sag resistance, coefficient of thermal expansion, designability, impact resistance, and modulus.

The structural assembly of the General Motors Pontiac Fiero sports car is based on a space frame that is similar in design concept to the roll cage of a racing car. The key to the frame is that it provides total structural support to the vehicle, including passenger protection. The frame is made with high strength steel in critical areas of the main structure. The frame was originally constructed from six modular units including one-piece door and window rings, to allow the three major body openings to be framed by a single stamping.

The space frame contributes to the future direction of automotive composites in two ways: (a) the space frame expands the selection of composites for body panels (vertical panels hang on the frame; they have no structural function), and (b) the frames can be expected to be fabricated by automated prepreg tape winding (i.e., a derivative of filament winding) and pultrusion.

The space frame/composite panel design concept introduces the following economies in fabrication and assembly, which will increase the use of composite body panels worldwide: reduced workforce, fewer robots, lower cost material handling, and less assembly with more unitized components. These benefits are primarily due to the lighter weight of composites.

The Chevrolet Division of General Motors is introducing passenger cars with space frame construction. Ford's Probe V concept car demonstrates the space frame that will appear on 1990s models. Matra Murena, and Talbot/France carry SMC panels on galvanized steel space frames; and the Fiat SpA Uno prototype car carries a space frame with composite body panels.

Bumper Systems

Bumper systems are multicomponent designs using a wide variety of unreinforced and reinforced thermoplastics, including elastomeric thermoplastic olefins (TPO), RIM/PUR, and alloys (such as nylon/PPO, and PBT and PET alloyed with polycarbonate). Front and rear bumper backup bars (beams) made of 40% continuous strand fiberglass mat-reinforced polypropylene were introduced on 1987 Ford and General Motors passenger car models. These composites, which are stamped into bumper beams, appeared on 1986 Cadillac Seville and El Dorado models, and on the Chevrolet Corvette. In large sedan models the bumper itself can be stamped from mat-reinforced polypropylene, replacing steel.

Reaction-injection molded (RIM) acrylamates were introduced in the late 1980s as an alternative to stamped polypropylene composites. Melt process injection-molded and blow-molded amorphous nylon, and injection-molded high impact nylon 6 and ABS, are also contenders for bumper beams and other bumper components to be used in 1990s passenger car models. Criteria for the selection of composite/fabricating method (material and fabricating method are decided on together) include in-place cost, surface appearance, and mechanical properties such as Izod impact strength (toughness) and modulus (stiffness).

Composite Driveshafts

Introduced for pickup trucks about 1980, composite driveshafts serve as roadworthy prototypes for vans and passenger cars in the 1990s. Ford began evaluating filament-wound driveshafts in the early 1970s. One early unit was constructed with a seven-layered tube, bonded to conventional steel yokes by adhesive. Epoxy–fiberglass plus graphite hybrid fiber-reinforced composites can be filament wound by "wet" processes or with prepregs, or pultruded. The General Motors 1988 GMT-400 pickup truck carries a composite driveshaft that is pultruded around an aluminum tube. The driveshafts are made in various lengths from 155 to 173 cm and 10 cm diameter, with aluminum wall thickness of 0.2 cm. The driveshaft is 60% lighter than the steel shaft it replaces, weighing 4.06 kg. In addition to weight saving and mechanical dampening, which are characteristics of composites used in frame and chassis, and under-the-hood applications, the composite driveshaft has high torsional strength. The composite driveshaft requires less torsional balancing than a steel shaft, and the composite version shows no temperature or environmental problems after road testing. The driveshaft is one of several components in the drivetrain that are potential composite applications. Other drivetrain candidates are the axle and axle housing, and transmission housing.

Suspension System

Designs for suspension systems can be expected to be radically changed during the 1990s. Whether leaf- or coil-springs are used in the suspension system packages, composites will be used in some designs. The automotive suspension system provides two control functions: control of vibration and control of wheel orientations. In addition to front wheel drive (FWD) and corresponding modifications of the MacPherson strut (which has a unitized strut/shock absorber), Ford introduced the "quadralink" independent rear suspension; General Motors and Volkswagen use a trailing twist axle rear suspension on some models (a semi-independent suspension); Lincoln Continental and Mark VII air suspensions replace steel coil springs; Mazda and Mitsubishi Motors introduced electronic suspension; and stabilizer bars and trailing arms underwent design changes. The filament wound composite leaf spring was introduced as a transverse rear suspension on Chevrolet Corvette models in 1981, and added to the front suspension in 1984; and installed on the General Motors Chevrolet Astro van and GMC Safari van in 1985. The composite leaf spring is a single leaf (monoleaf), because filament winding can distribute pre-designed strength into the leaf, according to locations and directions of loads on the leaf. The steel leaf spring is a multiple leaf (multileaf) because distributed strength is designed with layers of varying leaf dimensions. Composite coil springs require minimal angular rotational deflection, and a design that reduces stiffness requirements for state-of-the art composites. Helically filament-wound torsional parts include coil springs, torsion bars, and driveshafts. Composite springs have been tested to more than a half-million cycles of complete compression on an Instron fatigue test instrument, without fatigue. Boeing has demonstrated the capabilities of composite coil springs with the filament-wound door spring for the 767 aircraft. The spring uses unidirectional graphite fibers in an epoxy matrix. The graphite fibers are wound in a rectangular cross section, looped to form the cylindrical shape. The springs are one-third the weight of steel springs.

Instrument Panel Assemblies

The five main components of an instrument panel assembly are further divided into several subcomponents: the instrument cluster retainer assembly, the trim (instrument cluster trim, steering column trim, accessory trim), the duct opening panels, the main substrate carrier grilles (radio speaker, air conditioner/heater/defroster grilles), and the glove compartment and components.

A miscellaneous group of instrument panel (i.p.) assembly components includes switch mounting plates, defroster window plates, hinges, and mechanical fastening units. The wide variety of i.p. applications results in a correspondingly wide variety of unreinforced, reinforced, and foam plastics and sheets including polycarbonate and polycarbonate/ABS and SMA alloys, polypropylene, PPO and PPE, acrylics, PVC and PUR.

New vehicle styling and occupant safety considerations have caused some rethinking about i.p. design, including material selections. A main concern is to determine predictable failure modes as a function of time and temperature from −40 to 121°C. Upper i.p. panel surface components require temperature resistance to 121°C. Lower i.p.'s not exposed to direct sunlight through the windshield require temperature resistance to 104°C. The overall objective is to achieve managed failure modes through total i.p. assembly.

Various computer-aided design programs have been prepared to facilitate managed failure modes, primarily by major resin producers and automotive companies in the United States. Finished product failure is directly related to resin selection, processing parameters (e.g., injection molding barrel temperatures, pressures, dwell time, and screw and barrel configurations/dimensions), product design, and mold cavity design.

Miscellaneous Automotive Composite

Developments in the miscellaneous category can be represented by phenolic pulleys, steering wheels, and wheels. Phenolic pulleys were introduced on Ford Ranger and Bronco, Taurus, Sable, and Lincoln Town Car models in the 1980s. The pulleys can be used with belt-driven applications such as alternators, air conditioners, and crankshafts. Filament-wound steering wheels using epoxy–fiberglass composites are under development for Audi models.

Vinyl ester–fiberglass wheels have been under evaluation for more than 10 years and were road tested in 1986 on Pontiac sports cars. Composite wheels weigh about 50% less than stamped steel wheels, and a 37.5 cm composite wheel can have nearly 16 kg of unsprung weight reduction. Earlier composite wheels showed fracture at the wheel nut location, and they were damaged by curb-to-tire impact and by crowbars used to remove tires from the wheels.

The Probe V concept car is an indication of new composite developments in the powerplant/powertrain/drivetrain, the braking/steering system, in the frame and chassis, as well as an indicator of composite exterior body panels for the passenger car of the 1990s. Aerodynamics and corrosion resistance are primary design considerations for automotive developments in the 1990s, and composites can meet these two design considerations better than metals for many applications.

Process Equipment Developments

Current process equipment developments are divided into three areas: machinery such as injection molding and compression molding, molds, such as runnerless molds and quick mold change designs, and automation and instrumentation.

Machinery

LARGE INJECTION MACHINES. This equipment allows unitized molding of components for instrument panels, bumpers, and exterior body panels. Large presses typically have 2,721–4,535 Mg clamping force, but a few presses have been custom-made with higher clamp ratings. Three advancements allow the use of larger presses: (a) higher performance composites, which produce stiffness and strength with large parts and thin-wall design, (b) presses that are capable of better shot-to-shot injection precision, and (c) improved clamping accuracy.

Higher clamping force does not necessarily mean that an injection machine has higher shot capacity. A 2,721 Mg machine can have the same shot size (up to 14 kg) and the same standard plasticating rate (based on general-purpose polystyrene) as a 1,361 mg unit. The difference in this case is that the 2,721 mg unit can carry a larger mold, which allows production of larger parts. The larger injection machine has larger platens, more clearance between strain rods, and more daylight space between platens in the open position.

Injection machines are activated by either a hydraulic cylinder, a toggle clamp, or hydromechanical clamps. Hydromechanical automatic clamping systems allow fast mold mounting, which is important for large presses. Automatic mold clamping and tie bar removal can reduce setting up time for large platens from more than an hour to several minutes.

NEW SHEET MOLDING COMPOUNDS (SMC) AND ADVANCED COMPRESSION MOLDING EQUIPMENT. Technological advances contribute to the increasing use of SMC automotive panels. More than 30 different exterior panel components, interior compartment parts, and under-the-hood parts are compression molded from SMC. A short list includes horizontal panels (upper and lower panels, and skirts), quarter (window) panels, tailgates, grille opening panels, headlamp and taillamp housings, wheel covers, garnish moldings, interior trim, instrument panel components, fan shrouds, and heater duct adapters. Compression molding machines are capable of 1 minute cycles, down from nearly 3 minutes in 1981. With this state-of-the-art technology, SMC compression molding machines can take advantage of high productivity (fast cycling) unsaturated polyester thermosetting resin systems.

Compression molding machines now have faster clamping, shorter stroke presses (which permit more precise parallelism), hydraulic press closed-loop leveling equipment, and automation (to achieve precision molding and cycle-to-cycle accuracy at 1 minute cycles). Advanced equipment designs for high volume automotive parts production are found primarily in the hydraulics and in press construction. Models are available that eliminate the pause of platen motion during the transition from rapid traverse speeds to pressing speeds. Elimination of platen motion pause is possible with direct-flow hydraulic circuits, which allow adjustable independent control of press speeds. Direct-flow hydraulic circuits reduce pressure surges and valve delay times during press cycles.

Different compression molding machine models offer distinguishing characteristics. Certain models can be accurate to 0.0025 cm using electronic position controls. Other models features oversize tie rods and long bearings for greater molding accuracy. One series of models features extended bronze guide bushings for the moving crosshead, for better platen parallelism and molding accuracy. Press manufacturers can modify the following specifications to a molder's requirements: platen size, speeds, cylinder strokes, daylights, throat depths, knock-out requirements, and capacities. Productivity on a given machine's pressing capacity can be significantly increased by selective machine modifications.

Compression molding SMC with thermosetting resin matrices uses hot molds with room temperature sheet molding compound; compression molding thermoplastic composites such as polypropylene and thermoplastic polyester alloys, uses heated molding mats preheated by on-line conveying the sheet through an oven before molding.

MAT MOLDING/REACTION INJECTION MOLDING (MM/RIM). The MM/RIM process avoids the loss of impact strength that occurs when fiber-reinforced resins are not in a mat form (e.g., chopped fiber). Mat reinforcements using continuous filament greatly increase the molded product's mechanical strength and modulus. Polyester oligomer–urethane acrylamate and highly cross-linked polycarbonate systems are used as resin matrices in mat molding/RIM composites. The acrylamates are based on proprietary catalyzed acrylesterol–isocyanate chemistry that yields low viscosity, high modulus resins. Wheel covers were one of the earliest commercial automotive applications for MM/RIM acrylamate composites. The highly cross-linked polycarbonate system is composed of 1 part polyol to 1.4 parts polymeric diphenylmethane methane diisocyanate (MDI). (Toluene diisocyanate [TDI] polymers are used in smaller tonnages in PUR.) Mats are made with a variety of fibers and fabrics, including aramid, graphite and E-glass fibers, as well as with woven, knitted, and nonwoven fabrics and chopped fiber mats. Acrylamate resins are also reinforced with 30 and 40% (weight) chopped glass fiber for reinforced reaction-injection molding (RRIM) on conventional equipment without machinery modification. Recommended mold temperatures of 82–104°C are within the conventional RRIM processing range. Mold pressure during injection is typically 0.28–0.48 MPa, permitting the use of lightweight tooling materials. Mild steel is usually used for long production runs, and aluminum, kirksite, or nickel-faced epoxy tooling usually is more economical for shorter runs.

RIM/POLYUREA COMPOSITES. These materials allow faster cycles, lower clamp forces, thinner wall parts, and good heat sag resistance at 182°C for on-line paint ovens, compared with conventional polyurethane resin systems. Polyurea systems do not require a catalyst, which permits a wider selection of internal mold release agents and a higher content of these release agents. Mold release agents that react with catalysts cannot be used with catalytically reactive polyurethanes. Polyurea gel times are 1–2 seconds, compared with about 5–6 seconds with glycol-based polyurethanes. Polyurea in-mold residence times are less than 20 seconds; demold time is 15 seconds; green strength is developed faster than earlier resin systems; and total cycle time (part-to-part time) is about 60 seconds.

The two types of RIM/PUR resin matrix systems are standard temperature systems (which can be used on modified RIM machines) and higher temperature systems (which can make better use of the faster reactivity of polyurea components, and require faster RIM equipment). Faster formation of polyurea linkages results in faster curing and improved process economics. Higher temperature, faster reacting RIM/PUR composites require the following equipment modifications to take full advantage of the benefits of these systems: higher temperature molds 116–132°C, modified mold design (e.g., gate design and locations, automatic ejection pins, venting), higher throughput injection units, from 5.45 to 13.64 kg/s for large body panels [vs. 1.82–4.54 kg] for conventional RIM/PUR composites, and modified hydraulics and materials of construction.

BLOW MOLDING. The blow-molding process is used with neat (unreinforced) resins and composites for aerodynamic add-on panels (air dams, spoilers), fender liners and skirts, air ducts, bumper components, battery cases and covers, fuel tanks, interior trim, radiator fan shrouds and air shields, seat backs and load floors, and wheel covers. Fiber-reinforced composites are difficult to blow mold, especially with controlled dimensions. Carbon black filler has been used with high density polyethylene (HDPE) for spoilers, to improve environmental stress crack resistance; and mica-reinforced polypropylene is blow molded into heating ducts. Rotational molding is an alternative to blow molding for certain applications such as fuel tanks, armrests, and sun visors. Rotomolded resins are used in powder forms rather than pellets, and these resins are not fiber reinforced.

FILAMENT WINDING. Used to fabricate suspension leaf springs, driveshafts, and space frame segments, filament winding can be used for coil springs, as well. Faster curing, cost-effective epoxies with lower curing temperatures are being used for filament-wound automotive applications. On the machinery side, automated filament winding and hybrid combinations of filament winding with tape laying, or resin transfer molding improve process economics.

The principal setpoints for creels in filament winding are wind speed, wind angle, and wind tightness. The number of carousels and the number of creels per carousel can be varied, but not without modifying the equipment. Winders that provide flexibility (fast product changeover) and high speed for high volume automotive production have the following features: winders that control each axis of motion for an "infinite" number through 360 degrees; precise placement of fibers; fast changeover of fiber types, winding angles, and winding patterns; and broad resin capabilities. Glass, graphite, and aramid fibers (and hybrid combinations of these fibers) are filament wound. Thermosetting resins are usually used in filament winding, but engineering thermoplastics, especially nylons, thermoplastic polyesters (PBT and PET), and polypropylenes are being developed for filament winding.

PULTRUSION. Although used primarily in nonautomotive applications, pultrusion is a cost-effective process for low volume composite production of parts for buses and tractor trailers. Bus parts include exterior trim panels for mass transit buses, channels for wiring, hand rails, overhead baggage racks, and window moldings. Trailer truck parts include door panels and flooring beams. Frame components such as crossbars, bumper backup beams, and suspension parts such as antisway bars (antiroll bars), trailing arm components, stabilizer bars, and leaf springs were under development during the 1980s. Pultrusion is an alternate process to filament winding for these products. The principal elements for improved cost effectiveness are faster curing resins, flexible tooling (fast changeover), and instrumentation/automation.

Fiber creels are located furthermost upstream, pulled through guide bars for alignment, through a hot resin bath, and downstream through forming and curing stations. Composites used for pultrusion and filament winding are formulated with resins and fibers that are specifically designed for these processes. Denier and surface treatment are two fiber characteristics to consider when selecting fiber grades for these processes.

Pultrusion–filament winding hybrid systems can be used, with downstream fiber creels for filament winding, and creels for pultrusion located furthermost upstream.

James M. Margolis

Avionics

See Systems Avionics Applications

B

Bagging

See Vacuum Bag Processing

Bending Analysis of Composite Plates and Built-up Sections

Nomenclature

Symbol	Units	Description
σ_i, $\bar{\sigma}_i$ ($i = 1, 2, \ldots, 6$)	MPa	stress components
ε_i, $\bar{\varepsilon}_i$ ($i = 1, 2, \ldots, 6$)	m/m	strain components
C_{ij} ($i, j = 1, 2, \ldots, 6$)	MPa	elastic stiffness matrix
S_{ij} ($i, j = 1, 2, \ldots, 6$)	$(\text{MPa})^{-1}$	elastic compliance matrix ($[S] = [C]^{-1}$)
Q_{ij} ($i, j = 1, 2, \ldots, 6$)	MPa	reduced modulus matrix for a lamina
$[T]$	—	transformation matrix of stress or strain from reference (x, y, s) to lamina (1, 2, 6) coordinates
$\bar{Q}_{ij}$ ($i, j = 1, 2, \ldots, 6$)	MPa	transformed reduced modulus matrix for a lamina $[\bar{Q}] = [T]^{-1} [Q] [T]$
S_{ij} ($i, j = 1, 2, \ldots, 6$) $\bar{S}_{ij}$ ($i, j = 1, 2, \ldots, 6$)	$(\text{MPa})^{-1}$	compliance matrices $[S] = [Q]^{-1}$ and $[\bar{S}] = [\bar{Q}]^{-1}$
E_1 (E_x)	MPa	longitudinal Young's modulus of the lamina along lamina 1 axis (modulus-type engineering constant for

Symbol	*Units*	*Description*
		the laminate along reference x axis)
E_2 (E_y)	MPa	transverse Young's modulus of the lamina along lamina 2 axis (modulus-type engineering constant of the laminate along reference y axis)
ν_{21} (ν_{yx})	—	major (longitudinal) Poisson's ratio for lamina (1, 2) axes (similar constant for the laminate in reference x-y axes)
ν_{12} (ν_{xy})	—	minor (transverse) Poisson's ratio for lamina (1, 2) axes (similar constant for the laminate in reference x-y axes)
G_{12} (G_{xy})	MPa	shear modulus for lamina (1, 2) plane (similar constant for the laminate in reference x-y plane)
m, n	—	cos θ and sin θ, respectively, or summation indices (1, 2, . . .) in the Fourier series
k, N	—	ply index and number of plies, respectively
u_0, v_0, w_0	m	midsurface displacements along x, y, z directions respectively
u, v, w	m	displacements of a point in the plate along x, y, z directions, respectively
w_0, x; w_0, y	rad	rotation of midsurface point along x and y directions, respectively
w_0, xy	rad	twist rotation of midsurface point in the x-y plane
ε_x^0, ε_y^0 ε_1^0, ε_2^0	m/m	midsurface strains along (x,y) and (1, 2) axes, respectively
ε_s^0, ε_6^0	—	engineering sheer strain at midsurface in the x-y plane and 1-2 plane, respectively
ε_x, ε_y ε_1, ε_2	m/m	strains at any point along (x,y) and (1, 2) axes, respectively
ε_s, ε_6	—	engineering shear strain at any point in the x-y plane and 1-2 plane, respectively
γ_{xy}^0, γ_{xy}	m/m	engineering shear strains in the x-y plane at midsurface and arbitrary point, respectively
γ_{xz}, γ_{yz}	m/m	transverse shear strains for the plate

Symbol	*Units*	*Description*
κ_x, κ_y κ_{xy}	m^{-1}	normal and twisting curvatures
N_x, N_y N_{xy}	N/m	in-plane normal and shear loads per unit length, respectively
$Q_x(Vxz)$ $Q_y(Vyz)$	N/m	transverse shear loads per unit length on planes normal to x and y axes, respectively
M_x, M_y M_{xy}	N · m/m	bending and twisting moments per unit length, respectively
A_{ij}, B_{ij}, D_{ij} $(i, j = 1, 2, \ldots, 6)$	N/m, N, N · m	stretching, coupling, and bending moduli of a laminate, respectively
a_{ij}, b_{ij}, d_{ij} $(i, j = 1, 2, \ldots, 6)$	$N^{-1} \cdot m$, N^{-1}, $(N \cdot m)^{-1}$	stretching, coupling, and bending compliance of a laminate, respectively
U	N · m	Airy's stress function in x and y
w	m	deflection of plate in z direction
q	MPa	load intensity on the plate
N_n, M_n	N/m, N · m/m	in-plane and bending loads normal to a plate edge, respectively
Q_n	N/m	transverse shear load per unit length at a plate edge
u_n, u_t	m, m	displacements normal and tangent to a plate edge, respectively
N_{nt}, M_{nt}	N/m, N · m/m	in-plane shear and twisting moments in the n-t coordinates (n-normal, t-tangent) to a plate edge, respectively
$\overline{W}$	(units of work)	work function
A_{mn}	(m)	coefficient in the series solution of deflection
EI	$N \cdot m^2$	flexural stiffness
d_k	m	thickness of kth ply
z_k	m	distance of kth ply
I_k	m^4	moment of inertia of kth ply
h_k	m	distance of kth ply from the neutral axis
z_{na}	m	distance of neutral axis
$EI_{\overline{X}}, EI_{\overline{Y}}$ $EI_{\overline{XY}}$	$N \cdot m^2$	centroidal moments of inertia about $(\bar{x}, \bar{y})$ axes and product of inertia, respectively
EI_X, EI_Y $EI_{XY} = 0$	$N \cdot m^2$	principal moments of inertia about (x, y) axes and product of inertia (= 0), respectively
$\overline{X}, \overline{Y}$	m	centroidal location of built-up section
EI_{X_0}, EI_{Y_0} $EI_{X_0Y_0}$	$N \cdot m^2$	element centroidal moments of inertia about x_0, y_0 axes and product of inertia, respectively
X_i, Y_i	m	location of element centroid of

Symbol	*Units*	*Description*
		*i*th idealized element
ε_z, σ_z	m/m, N/m^2	axial strain and stress at the idealized element centroid (or arbitrary point) of the built-up section, respectively
P	N	axial load or vertical end load on the cantilever beam
Θ_p	rad	rotation of principal axes
L	m	length of beam
δ_x, δ_y	m, m	deflection of beams along x and y axes when bent about centroidal y and x axes, respectively

The successful nonstop, non-refueled flight of the *Voyager* around the world in July 1986 by Dick Rutan with Jeana Yeager, and the ever-increasing amount of dollars spent by the major aerospace corporations in recent years, continue to emphasize that the material of the future is the advanced fiber-reinforced composite. Some key attributes of advanced fiber-reinforced composites that best illustrate their superior performance over conventional materials include low weight, greater strength and modulus-to-weight ratio, tailorable, producible, and survivable/damage-tolerable structures. Typically, these advanced composites include main load-bearing fibers, such as graphite, boron, aramid, and glass in a matrix–resin material. A graphite–epoxy system that is commonly used in the United States contains Thornel T300 fibers and Narmco 5208 resin, abbreviated as T300/5208. In addition to the many graphite fibers available, high performance organic fibers such as du Pont's aramid (Kevlar) and impact-resistant polyethylene fiber series, Spectra 900 and Spectra 1000 by Allied Fibers are also state-of-the-art. Details on commonly available fiber-reinforced composite materials, their constituent materials (e.g., fiber and resin), processing, and their physical, mechanical, and electrical properties may be obtained from handbooks and other references on composite materials [1–3].

A unidirectional laminate made up of single plies of materials in which all the fibers are parallel is used to determine the basic mechanical properties, although other multidirectional laminates are also used to determine the laminate properties. Some of the standard testing procedures and many aspects on which research is being undertaken can be obtained by the conference proceedings regularly published by the American Society for Testing and Materials (ASTM) [4–8] and the Engineering Services Data Unit (ESDU) [9–11].

Micromechanics is a study of mechanical properties of unidirectional composites in terms of the constituent materials (fiber and resin). In contrast, macromechanical studies of fiber-reinforced materials deal with the behavior of a laminate to obtain only averaged apparent mechanical properties, rather than the detailed interactions of the constituents of the composite. The macromechanical properties of a lamina leading to the development of laminated plate theory to obtain the overall mechanical properties of a fiber-reinforced composite plate are presented in Refs. 12–16. Unlike isotropic material, laminated plates exhibit behavior from highly anisotropic to nearly isotropic, depending on the fiber orientations and the stacking sequence of the plies in the construction of the plate. The unique features of fiber composites such as the shear-to-normal coupling in the case of stretching loads, twist-to-bending coupling in the case of bending loads, and load coupling between stretching and bending need to be thoroughly understood for the optimum design of composite structures.

In most of the applications of fiber composites, the laminates are designed to take stretching loads directly and are seldom subjected to direct bending. This is because these materials are much stronger in resisting stretching loads as opposed to bending. Fiber composites effectively react to bending loads as differential tension and to compression loads through a "sandwich" construction. The relative advantages of sandwich construction versus a thick solid laminate in terms of stiffness, strength, and weight are shown in Table 1. The sandwich construction involves splitting the required thickness of fiber composite plate into top and bottom skins bonded to a lightweight honeycomb core material. The facings or skins of a sandwich panel used as a beam act similar to the flanges of an I-beam by taking the bending loads with one facing in compression and the other in tension. The honeycomb core corresponds to the web of the I-beam, resisting the shear loads. The bending stiffness of a solid laminate can be increased several times by adopting the sandwich construction in resisting both stretching and bending loads effectively. The Hexcel Company is a major manufacturer of honeycomb sandwich panels in both metallic and composite materials. A handbook of Hexcel product lines including construction of sandwich honeycomb panels, nomenclatures, data base of physical properties, and simple design of sandwich structures may be obtained from the company. Other references addressing the technical aspects of the design of honeycomb sandwich structures may be obtained from Refs. 3, 7, 12, 17, and 18.

Even though the best use of fiber composites is achieved by tailoring these materials to take directional loads, bending loads are unavoidable for the following reasons. Depending on the load path in a structure, a thick laminate will be subjected to direct bending. A

TABLE 1
Effect of Sandwich Construction Versus a Solid Laminate

PROPERTY \ CONSTRUCTION	SOLID LAMINATE (t)	t/2 – CORE – t/2 (2t)	t/4 – CORE – t/4 (4t)
RELATIVE STIFFNESS (D)	100	700	3700
RELATIVE STRENGTH	100	350	925
RELATIVE WEIGHT	100	103	106

stretching load causes bending in an unbalanced or unsymmetric laminate. When sandwich construction is not feasible, bending loads will be taken by a solid thick laminate of fiber composites. Hence, this article introduces the essentials of the classical laminated plate theory briefly and emphasizes the governing differential equations and solutions pertaining to bending and stretching of anisotropic plates. Finally, the equations for the cylindrical bending of a plate strip are formulated from the general plate equations. The techniques proposed in solution to the plate strip problem are further extended to analyze bending of composite built-up beam sections, which are similar to a strength-of-materials approach in the case of isotropic beam bending theory.

Macromechanical Behavior of a Lamina

The generalized Hooke's law relating stresses to strains can be written in contracted notation as

$$\sigma_i = C_{ij}\,\varepsilon_j \qquad (i,j = 1, \ldots, 6) \tag{1}$$

where σ_i are the stress components, C_{ij} is the stiffness matrix, and ε_j are the strain components. The stiffness matrix C_{ij} has 36 constants in Eq. (1). By consideration of the elastic potential or strain energy [12–16] there are only 21 independent constants. These 21 independent stiffness constants characterize *anisotropic* materials, since there are no planes of symmetry. If there are two orthogonal planes of material property symmetry for a material, symmetry will exist relative to a third mutually orthogonal plane. The stress–strain relations in coordinates aligned with principal material directions characterize an *orthotropic* material. In this case, there is no interaction between normal stresses σ_1, σ_2, σ_3 and the shearing strains ε_4, ε_5, ε_6, such as occurs in anisotropic materials, and there are only nine independent stiffness constants. If at every point of a material there is one plane in which the mechanical properties are equal in all directions, then the material is characterized as *transversely isotropic* and there are only five independent constants. For the case of an infinite number of planes of material symmetry, the foregoing relations simplify to the *isotropic* material case with only two independent elastic stiffness constants. The inverse of Eq. (1) is given by

$$\varepsilon_i = S_{ij}\,\sigma_j \qquad (i,j = 1, \ldots, 6) \tag{2}$$

where S_{ij}, the inverse of C_{ij}, is known as the compliance matrix.

Classical Lamination Theory

A laminate of fiber-reinforced composite material consists of two or more lamina bonded together to act as an integral structural element. The lamina principal material directions are oriented to produce a structural element capable of resisting load in several directions. Classical lamination theory embodies a collection of stress and deformation hypotheses developed to predict the behavior of such a laminated plate [12–16].

Lamina Behavior

The engineering stress–strain relation in principal material coordinates for a lamina of an orthotropic material under plane stresses is given by

$$\sigma_i = Q_{ij}\,\varepsilon_j \qquad (i,j = 1, 2, \ldots, 6) \tag{3}$$

where Q_{ij} is the reduced modulus matrix of elastic constants, with no interaction of shear and normal coupling coefficients. Hence, by the symmetry of these coefficients, there are only four independent constants. The inverse of Q_{ij}, namely S_{ij}, is known as the compliance matrix. The relations between the stiffness Q_{ij} and the engineering constants are given by

$$\begin{aligned} Q_{11} &= \frac{E_1}{1 - \nu_{12}\nu_{21}} \\ Q_{22} &= \frac{E_2}{1 - \nu_{12}\nu_{21}} \\ Q_{12} &= \frac{\nu_{21}E_2}{1 - \nu_{12}\nu_{21}} = \frac{\nu_{12}E_1}{1 - \nu_{12}\nu_{21}} \\ Q_{66} &= G_{12} \end{aligned} \tag{4}$$

where $E_1, E_2, \nu_{12}, \nu_{21}, G_{12}$ are the elastic engineering constants of the lamina in its principal material directions. There are only four independent coefficients, Q_{11}, Q_{22}, Q_{12}, and Q_{66}. Table 2 lists typical values of engineering constants, strength, and hygrothermal properties for some unidirectional composites.

In any other coordinate system, the transformed reduced modulus $\overline{Q}_{ij}$ is given as

$$\overline{\sigma}_i = \overline{Q}_{ij}\,\overline{\varepsilon}_j \qquad (i, j = 1, 2, \ldots, 6) \tag{5}$$

where $\overline{Q}_{ij}$ consists of all nonzero modulus matrices with six independent constants, taking symmetry into account.

The stress tensor transformation matrix $[T]$ from the plate reference coordinates x, y, s (off-axis) to the principal lamina coordinates (on-axis) is given by (see Fig. 1)

$$[T] = \begin{bmatrix} m^2 & n^2 & 2mn \\ n^2 & m^2 & -2mn \\ -mn & mn & m^2 - n^2 \end{bmatrix} \quad m = \cos\theta; n = \sin\theta \tag{6}$$

In other words,

$$\begin{Bmatrix} \sigma_1 \\ \sigma_2 \\ \sigma_6 \end{Bmatrix} = [T] \begin{Bmatrix} \sigma_x \\ \sigma_y \\ \sigma_s \end{Bmatrix} \tag{7a}$$

and the strain transformation is given by

$$\begin{Bmatrix} \varepsilon_1 \\ \varepsilon_2 \\ \frac{1}{2}\varepsilon_6 \end{Bmatrix} = [T] \begin{Bmatrix} \varepsilon_x \\ \varepsilon_y \\ \frac{1}{2}\varepsilon_s \end{Bmatrix} \tag{7b}$$

The transformed reduced stiffness coefficients $\overline{Q}_{ij}$ are given by

$$[\overline{Q}] = [T]^{-1}[Q][T] \tag{8}$$

Performing the matrix operation above results in

$$\begin{array}{c|cccc} & Q_{11} & Q_{22} & Q_{12} & Q_{66} \\ \hline \overline{Q}_{11} & m^4 & n^4 & 2m^2n^2 & 4m^2n^2 \\ \overline{Q}_{22} & n^4 & m^4 & 2m^2n^2 & 4m^2n^2 \\ \overline{Q}_{12} & m^2n^2 & m^2n^2 & m^4 + n^4 & -4m^2n^2 \\ \overline{Q}_{66} & m^2n^2 & m^2n^2 & -2m^2n^2 & (m^2 - n^2)^2 \\ \overline{Q}_{16} & m^3n & -mn^3 & mn^3 - m^3n & 2(mn^3 - m^3n) \\ \overline{Q}_{26} & mn^3 & -m^3n & m^3n - mn^3 & 2(m^3n - mn^3) \end{array} \tag{9}$$

The inverse of the reduced modulus Q_{ij} is known as the compliance matrix S_{ij}, which relates the stress to strain as

TABLE 2
Unidirectional Properties of Fiber-Reinforced Composite Materials

Property			Material				
Description	Symbol	Unit	Graphite–Epoxy (T300–5208)	Boron–Epoxy (BC4–5505)	Graphite–Epoxy (AS/3501)	Glass–Epoxy (Scotch Ply 1002)	Aramid–Epoxy (Kevlar 49–Epoxy)
Longitudinal Young's modulus	E_1	GPa	181	204	138	38.6	76
Transverse Young's modulus	E_2	GPa	10.3	18.5	8.96	8.27	5.5
Longitudinal (major) Poisson's ratio	ν_1	—	0.28	0.23	0.30	0.26	0.34
Shear modulus	G_{12}	GPa	7.17	5.59	7.1	4.14	2.30
Longitudinal coefficient of thermal expansion	α_1	μm/m/K	0.02	6.1	−0.3	8.6	−4.0
Transverse coefficient of thermal expansion	α_2	μm/m/K	22.5	30.5	28.1	22.1	79.0
Longitudinal swelling coefficient	β_1	m/m	0	—	—	—	—
Transverse swelling coefficient	β_2	m/m	0.44	—	—	—	—
Longitudinal tensile strength	X_T	MPa	1500	1260	1447	1062	1400
Longitudinal compressive strength	X_C	MPa	1500	2500	1447	610	235
Transverse tensile strength	Y_T	MPa	40	61	51.7	31	12
Transverse compressive strength	Y_C	MPa	246	202	206	118	53
Shear strength	S		68	67	93	72	34
Thickness of single ply	t_{ply}	μm	125	125	125	125	125
Specific gravity	ρ	—	1.6	2.0	1.6	1.8	1.46

From Ref. 12.

$$\varepsilon_i = S_{ij}\sigma_j \qquad (i,j = 1,2,\ldots,6) \tag{10}$$

The transformed compliance matrix $\bar{S}_{ij}$ in any other coordinate system is given by

$$\bar{\varepsilon}_i = \bar{S}_{ij}\bar{\sigma}_j \qquad (i,j = 1,2,\ldots,6) \tag{11}$$

The relations between the lamina compliance matrix S_{ij} and the engineering constants are given by

$$\begin{aligned} S_{11} &= \frac{1}{E_1} \\ S_{22} &= \frac{1}{E_2} \\ S_{12} &= -\frac{\nu_{21}}{E_1} = -\frac{\nu_{12}}{E_2} \\ S_{66} &= \frac{1}{G_{12}} \end{aligned} \tag{12}$$

Assuming tensorial strain transformation for the strains (as given by Eqs. 6 and 7b), the transformed compliance $\bar{S}_{ij}$ is given by

$$[\bar{S}] = [T]^{-1}[S][T] \tag{13}$$

As before, the matrix operation yields

	S_{11}	S_{22}	S_{12}	S_{66}
$\bar{S}_{11}$	m^4	n^4	$2m^2n^2$	m^2n^2
$\bar{S}_{22}$	n^4	m^4	$2m^2n^2$	m^2n^2
$\bar{S}_{12}$	m^2n^2	m^2n^2	$m^4 + n^4$	$-m^2n^2$
$\bar{S}_{66}$	$4m^2n^2$	$4m^2n^2$	$-8m^2n^2$	$(m^2 - n^2)^2$
$\bar{S}_{16}$	$2m^3n$	$-2mn^3$	$2(mn^3 - m^3n)$	$(mn^3 - m^3n)$
$\bar{S}_{26}$	$2mn^3$	$-2m^3n$	$2(m^3n - mn^3)$	$(m^3n - mn^3)$

(14)

The lamina modulus (Q_{ij}) and the compliance (S_{ij}) are sometimes referred to together as the on-axis "stiffness," and the transformed modulus $(\bar{Q}_{ij})$ and compliance $(\bar{S}_{ij})$ are known as off-axis "stiffness" for the basic ply or lamina.

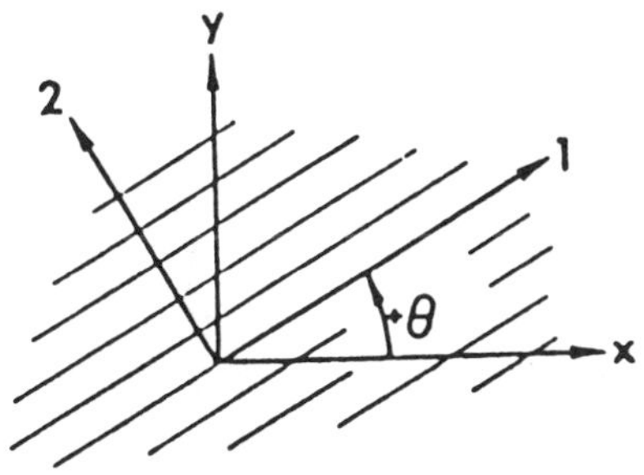

FIGURE 1 Positive rotation of principal material axes (1,2) from arbitrary (*X,Y*) axes.

Multidirectional Anisotropic Laminated Plate Theory

Knowledge of the stress and strain variation through the thickness of a laminate subjected to both the extensional (in-plane stretching) and bending loads is required for the efficient design of a composite part. The geometry of deformation in the *xz* plane of a laminated composite plate is depicted in Figure 2. A line originally straight and perpendicular to the middle surface of the laminate is assumed to remain straight and normal to the midplane after deformation. The thin laminate assumption made as above ensures ignoring the transverse shear strains: that is, $\gamma_{xz} = \gamma_{yz} = 0$, where z is the direction normal to the midplane. In addition, the extension of normal is ignored; that is, $\varepsilon_z = 0$. These assumptions are collectively known as Kirchhoff–Love hypotheses for thin plates and shells. For a layered composite plate, we need to assume in addition that each ply (lamina) is orthotropic, homogeneous, constant in thickness, and linearly elastic until failure. Also, continuity across plies is required to prevent slippage between the plies.

Considering the deformations of points *B* and *C* and letting *u, v, w* correspond to the displacements in the *x, y,* and z directions, respectively, we can write

$$u_c = u_0 - z_c\beta \tag{15}$$

where u_0 is the displacement in the *x* direction of the midsurface point *B* and β is the slope of the deformed midsurface of the laminate, given by

$$\beta = w_{0,x} \tag{16}$$

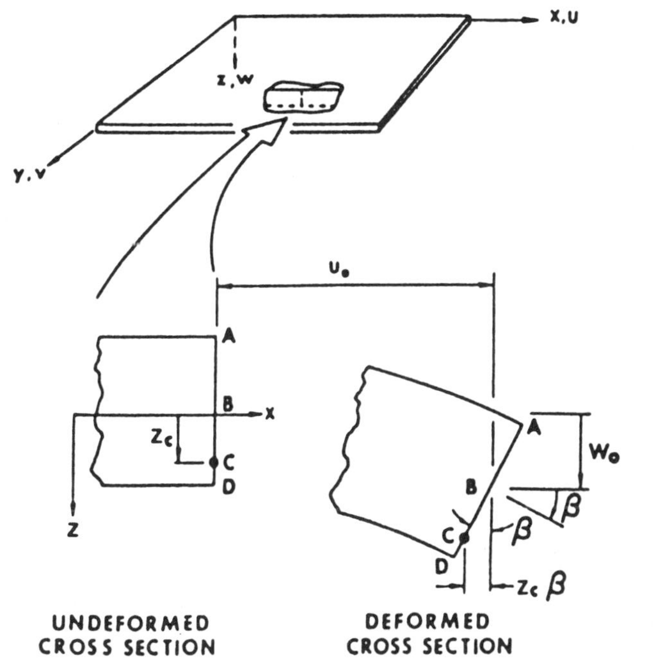

FIGURE 2 Geometry of deformation in the *XZ* plane. (From Ref. 13.)

where the comma in the subscript indicates partial differentiation ($w_{0,x} = \partial w_0/\partial x$). Thus, the displacement u at any point z through the laminate thickness is

$$u = u_0 - zw_{0,x} \tag{17}$$

Similarly,

$$v = v_0 - zw_{0,y} \tag{18}$$

For small strains and using linearized theory of elasticity, the strain–displacement relations are given by

$$\begin{aligned} \varepsilon_x &= u_{,x} \\ \varepsilon_y &= v_{,y} \\ \gamma_{xy} &= u_{,y} + v_{,x} \end{aligned} \tag{19}$$

Substituting Eqs. (17) and (18) in Eq. (19) results in

$$\begin{aligned} \varepsilon_x &= u_{0,x} - zw_{0,xx} \\ E_y &= v_{0,y} - zw_{0,yy} \\ \gamma_{xy} &= u_{0,y} + v_{0,x} - 2zw_{0,xy} \end{aligned} \tag{20}$$

or in matrix notation,

$$\begin{Bmatrix} \varepsilon_x \\ \varepsilon_y \\ \gamma_{xy} \end{Bmatrix} = \begin{Bmatrix} \varepsilon_x^0 \\ \varepsilon_y^0 \\ \gamma_{xy}^0 \end{Bmatrix} + z \begin{Bmatrix} \kappa_x \\ \kappa_y \\ \kappa_{xy} \end{Bmatrix} \tag{21}$$

where the midsurface strains are given by

$$\begin{Bmatrix} \varepsilon_x^0 \\ \varepsilon_y^0 \\ \gamma_{xy}^0 \end{Bmatrix} = \begin{Bmatrix} u_{0,x} \\ v_{0,y} \\ u_{0,y} + v_{0,x} \end{Bmatrix} \tag{22}$$

and the midsurface curvatures by

$$\begin{Bmatrix} \kappa_x \\ \kappa_y \\ \kappa_{xy} \end{Bmatrix} = - \begin{Bmatrix} w_{0,xx} \\ w_{0,yy} \\ 2w_{0,xy} \end{Bmatrix} \tag{23}$$

Recalling Eq. (5), the stress–strain relation for any kth lamina, in reference to the plate x,y coordinates, is given in an expanded form in Eq. (24).

$$\begin{Bmatrix} \sigma_x \\ \sigma_y \\ \tau_{xy} \end{Bmatrix}_{(k)} = \begin{bmatrix} \bar{Q}_{11} & \bar{Q}_{12} & \bar{Q}_{16} \\ \bar{Q}_{12} & \bar{Q}_{22} & \bar{Q}_{26} \\ \bar{Q}_{16} & \bar{Q}_{26} & \bar{Q}_{66} \end{bmatrix}_{(k)} \begin{Bmatrix} \varepsilon_x^0 \\ \varepsilon_y^0 \\ \gamma_{xy}^0 \end{Bmatrix} + z \begin{Bmatrix} k_x \\ \kappa_y \\ \kappa_{xy} \end{Bmatrix} \tag{24}$$

Figure 3 shows the positive sign conventions for the stretching (N_x, N_y, N_{xy}) and bending (M_x, M_y, M_{xy}) loads per unit length. The integration of the stresses given by Eq. (24) should match the externally applied stress resultants (stretching loads) and moments (bending loads) to satisfy equilibrium of the laminates. In reference to Figure 4, and assuming that there are N layers in the laminate, with each layer oriented differently so that the transformed modulus $\bar{Q}_{ij}$ is a discontinuous function of Z, we get

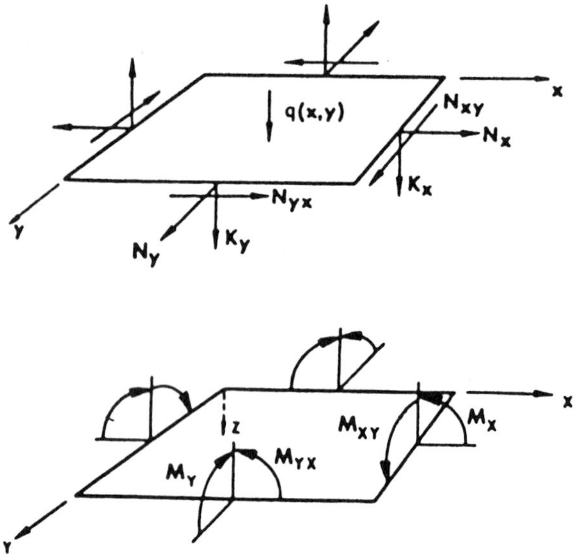

FIGURE 3 (Top) Plate forces (Bottom) Plate moments.

$$\begin{Bmatrix} N_x \\ N_y \\ N_{xy} \end{Bmatrix} = \int_{-h/2}^{+h/2} \begin{Bmatrix} \sigma_x \\ \sigma_y \\ \tau_{xy} \end{Bmatrix} dz = \sum_{k=1}^{N} \int_{h_{k-1}}^{h_k} \begin{Bmatrix} \sigma_x \\ \sigma_y \\ \tau_{xy} \end{Bmatrix} dz \tag{25}$$

and

$$\begin{Bmatrix} M_x \\ M_y \\ M_{xy} \end{Bmatrix} = \int_{-h/2}^{+h/2} \begin{Bmatrix} \sigma_x \\ \sigma_y \\ \tau_{xy} \end{Bmatrix} z\,dz = \sum_{k=1}^{N} \int_{h_{k-1}}^{h_k} \begin{Bmatrix} \sigma_x \\ \sigma_y \\ \tau_{xy} \end{Bmatrix} z\,dz \tag{26}$$

where h_k and h_{k-1} are defined in Figure 4. Note that $h_0 = -h/2$.

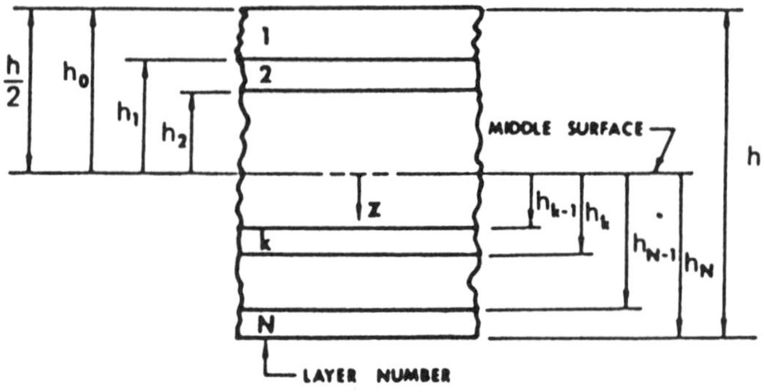

FIGURE 4 Geometry of an N-layered laminate.

Substitution of Eq. (24) in Eqs. (25) and (26) and using Eqs. (22) and (23) gives

$$\begin{Bmatrix} N_x \\ N_y \\ N_{xy} \end{Bmatrix} = \begin{bmatrix} A_{11} & A_{12} & A_{16} \\ A_{12} & A_{22} & A_{26} \\ A_{16} & A_{26} & A_{66} \end{bmatrix} \begin{Bmatrix} \varepsilon_x^0 \\ \varepsilon_y^0 \\ \gamma_{xy}^0 \end{Bmatrix} + \begin{bmatrix} B_{11} & B_{12} & B_{16} \\ B_{12} & B_{22} & B_{26} \\ B_{16} & B_{26} & B_{66} \end{bmatrix} \begin{Bmatrix} \kappa_x \\ \kappa_y \\ \kappa_{xy} \end{Bmatrix} \tag{27}$$

and

$$\begin{Bmatrix} M_x \\ M_y \\ M_{xy} \end{Bmatrix} = \begin{bmatrix} B_{11} & B_{12} & B_{16} \\ B_{12} & B_{22} & B_{26} \\ B_{16} & B_{26} & B_{66} \end{bmatrix} \begin{Bmatrix} \varepsilon_x^0 \\ \varepsilon_y^0 \\ \gamma_{xy}^0 \end{Bmatrix} + \begin{bmatrix} D_{11} & D_{12} & D_{16} \\ D_{12} & D_{22} & D_{26} \\ D_{16} & D_{26} & D_{66} \end{bmatrix} \begin{Bmatrix} \kappa_x \\ \kappa_y \\ \kappa_{xy} \end{Bmatrix} \tag{28}$$

where the matrix of coefficients A_{ij}, B_{ij}, and D_{ij} (i, j, = 1, 2, . . ., 6) are the extensional, coupling, and bending moduli, respectively, for the laminate and are given by

$$A_{ij} = \sum_{k=1}^{N} \overline{Q}_{ij}^{k} (h_k - h_{k-1})$$

$$B_{ij} = \tfrac{1}{2} \sum_{k=1}^{N} \overline{Q}_{ij}^{k} (h_k^2 - h_{k-1}^2) \tag{29}$$

$$D_{ij} = \tfrac{1}{3} \sum_{k=1}^{N} \overline{Q}_{ij}^{k} (h_k^3 - h_{k-1}^3)$$

The laminate constitutive equations (27 and 28) are abbreviated in the form of

$$\begin{Bmatrix} N \\ \hline M \end{Bmatrix} = \begin{bmatrix} A & B \\ \hline B & D \end{bmatrix} \begin{Bmatrix} \varepsilon^0 \\ \hline \kappa \end{Bmatrix} \tag{30}$$

The laminated composite plate is nonhomogeneous and anisotropic and is regarded as a complex structure rather than as a material. Equations (27) and (28) are coupled through $[B]$. Now, solving Eq. (30) for κ and M, we get

$$\{\varepsilon^0\} = [a]\{N\} - [b]\{\kappa\} \tag{31a}$$

$$\{M\} = [b]^T\{N\} + [d]\{\kappa\} \tag{31b}$$

where

$$[a] = [A]^{-1}$$

$$[b] = [a][B] \tag{32}$$

$$[d] = [D] - [b]^T[B]$$

Solving Eq. (31b) for κ and substituting in Eq. (31a) results in

$$\begin{Bmatrix} \varepsilon^0 \\ \hline \kappa \end{Bmatrix} = \begin{bmatrix} a + bd^{-1}b^T & -bd^{-1} \\ \hline -d^{-1}b^T & d^{-1} \end{bmatrix} \begin{Bmatrix} N \\ \hline M \end{Bmatrix} \tag{33}$$

Equation (33) is the complete inversion of Eq. (30). The stresses and strains are determined as follows. Substituting Eq. (21) in Eq. (31a), we get the strains for any kth lamina located at a distance Z from the midplane of the laminate, as

$$\begin{Bmatrix} \varepsilon_x \\ \varepsilon_y \\ \gamma_{xy} \end{Bmatrix}_{(k)} = [a] \begin{Bmatrix} N_x \\ N_y \\ N_{xy} \end{Bmatrix} - [b_{ij} - Z_k\, \delta_{ij}] \begin{Bmatrix} \kappa_x \\ \kappa_y \\ \kappa_{xy} \end{Bmatrix} \tag{34}$$

where δ_{ij} is the (3 × 3) Kronecker delta matrix, containing ones along the diagonal and zeros for other terms. In view of Eqs. (21) and (34), the lamina stresses are obtained from Eq. (24) as

$$\begin{Bmatrix} \sigma_x \\ \sigma_y \\ \tau_{xy} \end{Bmatrix}_{(k)} = [\overline{Q}][a] \begin{Bmatrix} N_x \\ N_y \\ N_{xy} \end{Bmatrix} - ([\overline{Q}][b] - Z_k[\overline{Q}]) \begin{Bmatrix} \kappa_x \\ \kappa_y \\ \kappa_{xy} \end{Bmatrix} \tag{35}$$

Thus, given the in-plane loads and the bending curvatures, the strains and the stresses for any kth lamina at a distance Z_k from the midplane of the laminate can be determined from Eqs. (34) and (35), respectively.

Anisotropic Plate Bending and Stretching Analysis

The constitutive equations for a multidirectional, nonhomogeneous anisotropic plate are given by Eqs. (30) and (33), the latter being the inverse of the former. The partial inversion of Eq. (30) is given by Eq. (31), which is useful for analyzing structures with fixed cylindrical cross sections such as tubes and pressure vessels. However, in the case of a differential plate element governed by the constitutive relation (31), there are six independent quantities (N_x, N_y, N_{xy}, κ_x, κ_y, and κ_{xy}). By using Airy's stress function $U(x,y)$ for the stretching loads and defining the curvatures explicitly in terms of deflection w, we get only two independent quantities, U and w, as follows:

$$\begin{Bmatrix} \varepsilon_1^0 \\ \varepsilon_2^0 \\ \varepsilon_6^0 \end{Bmatrix} = [a] \begin{Bmatrix} U_{,yy} \\ U_{,xx} \\ -U_{,xy} \end{Bmatrix} + [b] \begin{Bmatrix} w_{,xx} \\ w_{,yy} \\ 2w_{,xy} \end{Bmatrix} \tag{36}$$

and

$$\begin{Bmatrix} M_x \\ M_y \\ M_{xy} \end{Bmatrix} = [b]^T \begin{Bmatrix} U_{,yy} \\ U_{,xx} \\ -U_{,xy} \end{Bmatrix} - [d] \begin{Bmatrix} w_{,xx} \\ w_{,yy} \\ 2w_{,xy} \end{Bmatrix} \tag{37}$$

where the subscript comma denotes differentiation.

Detailed derivation of equilibrium equations and the associated boundary conditions for anisotropic plates and corresponding solution techniques may be obtained from references 19 to 23. However, a brief discussion on the governing differential equations, boundary conditions, and simple solutions is presented for the sake of continuity.

To find U and w, we need two equations. One is obtained by considering the equation of equilibrium of Figures 3b and 5, in the Z direction, given by

$$M_{x,xx} + 2M_{xy,xy} + M_{y,yy} = -q \quad (38)$$

where q is the laterally applied load intensity on the plate. Another equation is the compatibility equation for the in-plane strain components, given by

$$\varepsilon^0_{1,yy} + \varepsilon^0_{2,xx} - \varepsilon^0_{6,xy} = 0 \quad (39)$$

Substituting Eq. (36) in (39), and Eq. (37) in (38) results in

$$\begin{aligned} a_{22}U_{,xxxx} &- 2a_{26}U_{,xxxy} + (2a_{12} + a_{66})U_{,xxyy} \\ &- 2a_{16}, U_{,xyyy} + a_{11}U_{,yyyy} + b_{21}w_{,xxxx} \\ &+ (2b_{26} - b_{61})w_{,xxxy} + (b_{11} + b_{22} - 2b_{66})w_{,xxyy} \\ &+ (2b_{16} - b_{62})w_{,xyyy} + b_{12}w_{,yyyy} = 0 \end{aligned} \quad (40)$$

and

$$\begin{aligned} b_{21}U_{,xxxx} &- (2b_{26} - b_{61})U_{,xxxy} + (b_{11} + b_{22} - 2b_{66})U_{,xxyy} \\ &+ (2b_{16} - b_{62})U_{,xyyy} + b_{12}U_{,yyyy} + d_{11}w_{,xxxx} \\ &+ 4d_{16}w_{,xxxy} + 2(d_{12} + 2d_{66})w_{,xxyy} \\ &+ 4d_{26}w_{,xyyy} + d_{22}w_{,yyyy} = -q \end{aligned} \quad (41)$$

where $a_{ij}, b_{ij}, d_{ij} (i, j = 1, 2, 6)$ are the coefficients in the respective matrices of $[a]$, $[b]$ and $[d]$. Equations (40) and (41) constitute the classical theory of laminated plates. The complexity of the governing equations causes difficulty in obtaining exact solutions. However, the numerical method of finite difference [20] may be used to solve such complex problems. Also, finite element modeling techniques using NASTRAN CQUAD4/CQUAD8 elements along with PCOMP/PSHELL cards are traditionally followed to solve the coupled anisotropic plate problem [24]. The presence of both dependent variables in both equations indicates the coupling between stretching and bending through the coupling matrix $[b]$. In the case of symmetrical layups about the midplane of the plate, the coupling matrix $[B]$ becomes zero (hence $[b] = 0$); then Eqs. (40) and (41) reduce to

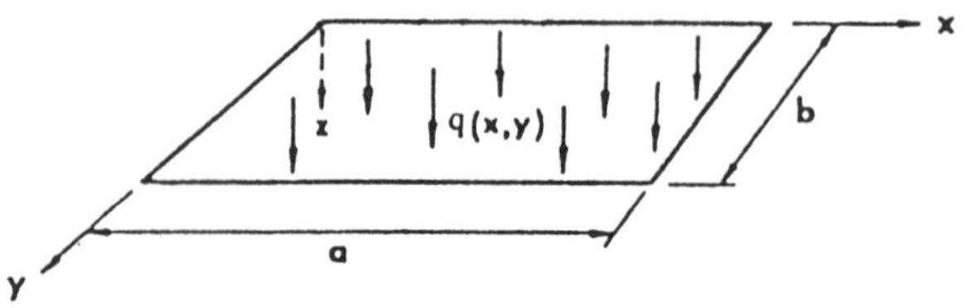

FIGURE 5 Simply supported laminated rectangular plate under distributed lateral load.

$$\begin{aligned} a_{22}U_{,xxxx} &- 2a_{26}U_{,xxxy} + (2a_{12} + a_{66})U_{,xxyy} \\ &- 2a_{16}U_{,xyyy} + a_{11}U_{,yyyy} = 0 \end{aligned} \quad (42)$$

and

$$\begin{aligned} D_{11}w_{,xxxx} &+ 4D_{16}w_{,xxxy} + 2(D_{12} + 2D_{66})w_{,xxyy} \\ &+ 4D_{26}w_{,xyyy} + D_{22}w_{,yyyy} = q \end{aligned} \quad (43)$$

Thus each equation contains only one dependent variable; hence there is no coupling between stretching and bending.

However, there are six nonzero coefficients in each of these equations, indicating that the laminate is homogeneous ($D_{ij} = A_{ij}h^2/12$) but still *anisotropic*, both in the in-plane mode and in bending. Equations (42) and (43) would correspond to unidirectional laminates, where the lamina coordinates make an angle to the plate reference coordinates. In the case of cross-ply laminates, where the laminae principal material directions correspond to the plate x and y coordinates, Eqs. (42) and (43) further reduce to

$$a_{22}U_{,xxxx} + (2a_{12} + a_{66})U_{,xxyy} + a_{11}U_{,yyyy} = 0 \quad (44)$$

and

$$D_{11}w_{,xxxx} + 2(D_{12} + 2D_{66})w_{,xxyy} + D_{22}w_{,yyyy} = q \quad (45)$$

Each equation contains four coefficients, suggesting that the plate is *orthotropic* in both in-plane and bending effects.

Boundary Conditions

To get unique solutions to the pair of anisotropic plate governing differential equations (42 and 43) and (44 and 45), the solutions should also satisfy the necessary geometric (u_n, u_t, w, $w_{,n}$) and natural (N_n, N_{nt}, M_n, M_{nt}) boundary conditions. Thus, one member of each pair of the following four quantities must be prescribed along the boundary [25]:

$$\begin{aligned} &u_n; N_n \\ &u_t; N_{nt} \\ &w_{,n}; M_n \\ &w; M_{nt,t} + Q_n \end{aligned} \quad (46a)$$

where the quantity $(M_{nt,t} + Q_n)$ is the well-known Kirchhoff condition. In particular, for a rectangular plate, one of the following combinations of boundary conditions should be satisfied [13].

$$\begin{array}{lllll} S1 & w = 0 & M_n = 0 & u_n = u_n & u_t = u_t \\ S2 & w = 0 & M_n = 0 & N_n = N_n & u_t = u_t \\ S3 & w = 0 & M_n = 0 & u_n = u_n & N_{nt} = N_{nt} \\ S4 & w = 0 & M_n = 0 & N_n = N_n & N_{nt} = N_{nt} \end{array} \qquad (46b)$$

$$\begin{array}{lllll} C1 & w = 0 & w_{,n} = 0 & u_n = \bar{u}_n & u_t = \bar{u}_t \\ C2 & w = 0 & w_{,n} = 0 & N_n = \bar{N}_n & u_t = \bar{u}_t \\ C3 & w = 0 & w_{,n} = 0 & u_n = \bar{u}_n & N_{nt} = \bar{N}_{nt} \\ C4 & w = 0 & w_{,n} = 0 & N_n = \bar{N}_n & N_{nt} = \bar{N}_{nt} \end{array} \qquad (46c)$$

where n is normal and t is tangent to a plate edge. Also, simply supported conditions are denoted by S; and clamped or fixed conditions are denoted by C. The bared quantities can be either zero or some specified values. The various governing differential equations and their solutions to some practical plate problems are illustrated in Ref. 13.

Effect of In-Plane Loads on the Plate Deflection

In deriving the equation of equilibrium (Eq. 39), the contribution due to the in-plane loads in the z direction was not considered. Considering Figures 6 and 7, in the deflected position of the plate, yet within the linear elastic limit of the material of the plate, the summation of the forces (neglecting higher order terms) in the z direction will be added to the right-hand side of Eq. (38) as

$$M_{x,xx} + 2M_{xy,xy} + M_{y,yy} = -(q + N_x w_{,xx} + N_y w_{,yy} + 2N_{xy} w_{,xy}) \qquad (47a)$$

Substituting for the moments from Eq. (28), where $B_{ij} = 0$, gives

$$\begin{aligned} D_{11} w_{,xxxx} &+ 4D_{16} w_{,xxxy} + 2(D_{12} + 2D_{66}) w_{,xxyy} \\ &+ 4D_{26} w_{,xyyy} + D_{22} w_{,yyyy} \\ &= q + N_x w_{,xx} + N_y w_{,yy} + 2N_{xy} w_{,xy} \end{aligned} \qquad (47b)$$

Again, closed-form solutions are difficult and numerical methods using finite difference techniques should be sought. Equation (47b) may be used to calculate buckling loads for the plate.

Solution Techniques for Certain Small-Deflection Problems

The plate is considered thin when its thickness is small compared with its length (a) or width (b). The deflection of a thin plate is also small compared with its thickness. Hence, the small-deflection plate theory satisfies Kirchhoff–Love hypotheses in addition to the following conditions (from Refs. 26 and 27, respectively) on its thickness h:

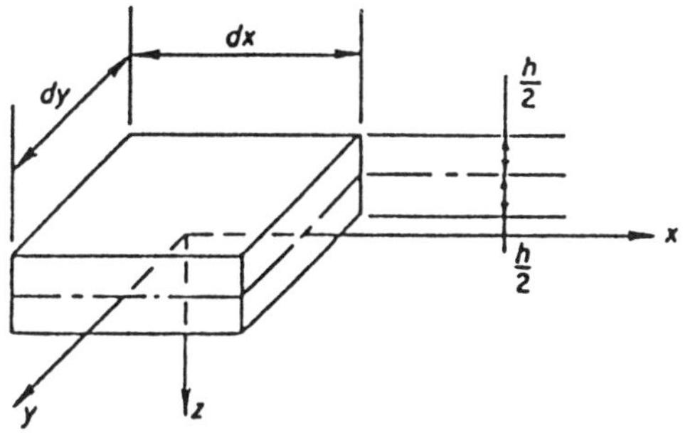

(a)

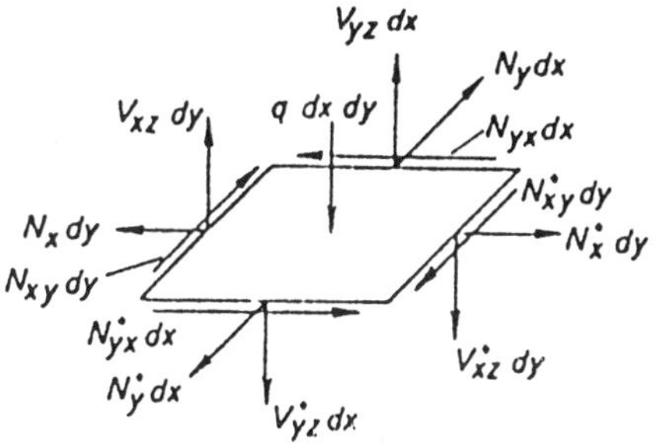

$N_x^* = N_x + N_{x,x}\,dx$
$V_{yz}^* = V_{yz} + V_{yz,y}\,dy$
(similar expressions for N_y^*, N_{xy}^*, N_{yx}^*, and V_{xz}^*)

(b)

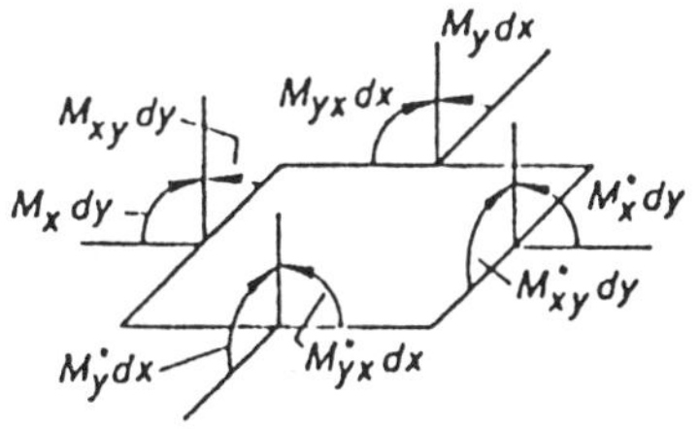

$M_x^* = M_x + M_{x,x}\,dx$
$M_{yx}^* = M_{yx} + M_{yx,y}\,dy$
(similar expressions for M_y^*, M_{xy}^*)

(c)

FIGURE 6 Variation of plate forces and moments. (From Ref. 15.)

1. $h \leq 0.25$ (a or b, whichever is smaller)
   $h/2 \geq$ maximum deflection
2. $h \leq 0.05$ (a or b, whichever is smaller)
   $h \geq$ maximum deflection

Also, the out-of-plane stresses σ_z, τ_{xz}, τ_{yz} are zero.

The out-of-plane displacement and moment resultants of composite plates depend on the elastic stiffness matrices of these plates. Elasticity solutions with the Navier method, the Levy method, and the Ritz method are some of the techniques employed to get closed-form solutions to the governing differential equations. The un-

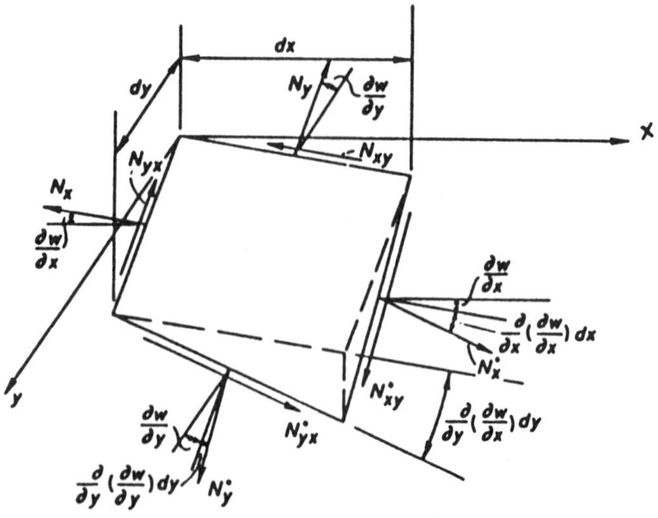

FIGURE 7 **Effect of in-plane loads on the deflection of plate.**

symmetric laminate characterized by Eqs. (40) and (41) ($B_{ij} \neq 0$) is more complex to analyze except perhaps by the finite difference method. Whitney [28] has given closed-form solutions to special cases of cross-ply and angle-ply laminated plates, considering the bending–extensional coupling under transverse loads. Approximate solutions to coupling problems by reducing to equivalent orthotropic laminates may be obtained from Smith [29]. Since the discussions on the coupling problems are beyond the scope of this review, readers should consult Refs. 28 and 29 for further information. The bending of an orthotropic plate ($A_{16} = A_{26} = D_{16} = D_{26} = B_{ij} = 0$) characterized by Eq. (45) can be analyzed using an elasticity approach known as the Navier method, as outlined below.

Consider a simply supported rectangular orthotropic plate of dimensions $a \times b$, subject to an arbitrarily distributed normal load intensity of $q(x,y)$ as shown in Figure 5. The Navier method consists of solving Eq. (45) using double Fourier series for the deflection $w(x,y)$ and the load intensity $q(x,y)$. The simple support boundary conditions are

$$\begin{aligned} x &= 0,a \quad w = 0 \quad M_x = -D_{11}w_{,xx} - D_{12}w_{,yy} = 0 \\ y &= 0,b \quad w = 0 \quad M_y = -D_{12}w_{,xx} - D_{22}w_{,yy} = 0 \end{aligned} \tag{48}$$

Assuming

$$q(x,y) = \sum_{m=1}^{\infty}\sum_{n=1}^{\infty} q_{mn} \sin\frac{m\pi x}{a} \sin\frac{n\pi y}{b} \tag{49}$$

For a uniformly distributed load intensity q_0, Eq. 49 becomes

$$q(x,y) = \sum_{m=1,3.}^{\infty}\sum_{n=1,3.}^{\infty} \frac{16q_0}{\pi^2}\frac{1}{mn} \sin\frac{m\pi x}{a} \sin\frac{n\pi y}{b} \tag{50}$$

assuming

$$w(x,y) = \sum_{m=1}^{\infty}\sum_{n=1}^{\infty} a_{mn} \sin\frac{m\pi x}{a} \sin\frac{n\pi y}{b} \tag{51}$$

Substituting Eqs. (49) and (51) in Eq. (45) results in

$$a_{mn} = \frac{\left(\frac{q_{mn}}{\pi^4}\right)}{D_{11}\left(\frac{m}{a}\right)^4 + 2(D_{12} + 2D_{66})\left(\frac{m}{a}\right)^2\left(\frac{n}{b}\right)^2 + D_{22}\left(\frac{n}{b}\right)^4} \tag{52}$$

Hence, comparing (49) and (50) for a uniformly distributed load q_0 the deflection $w(x,y)$ is given by

$$w = \frac{16q_0}{\pi^6} \sum_{m=1,3,\ldots}^{\infty}\sum_{n=1,3,\ldots}^{\infty} \frac{\left(\frac{1}{mn}\right)\sin\frac{m\pi x}{a}\sin\frac{n\pi y}{b}}{D_{11}\left(\frac{m}{a}\right)^4 + 2(D_{12} + D_{66})\left(\frac{m}{a}\right)^2\left(\frac{n}{b}\right)^2 + D_{22}\left(\frac{n}{b}\right)^4} \tag{53}$$

Using Eq. (53), the lamina strains ε_i^k from Eq. (21) and the lamina stresses σ_i^k from Eq. (24) are calculated neglecting the midsurface extensional strains ε_i^0. The stress couples (moments M_i) can also be calculated from Eq. (28) after setting $B_{ij} = 0$.

NUMERICAL EXAMPLE. A simply supported rectangular plate of size $a = 0.216$ m, $b = 0.508$ m and thickness $h = 2.794 \times 10^{-3}$ m, is made of layered composite materials. The bending modulus D_{ij} is given as

$$D_{ij} = \begin{bmatrix} 32.228 & 12.837 & 0 \\ & 32.228 & 0 \\ \text{sym} & & 14.634 \end{bmatrix} \text{N} \cdot \text{m}$$

The deflection at the center of the plate for a uniform load intensity of $q_0 = 3.5$ kPa is estimated as $w_{max} = 2.413 \times 10^{-3}$ m $< h(2.794 \times 10^{-3}$ m, satisfying the small-deflection assumption made earlier.

The plate moments at the center of the plate are

$$\begin{aligned} M_x &= 16.195 \text{ N} \cdot \text{m/m} \\ M_y &= 7.59 \text{ N} \cdot \text{m/m} \\ M_{xy} &= -5.43 \text{ N} \cdot \text{m/m} \end{aligned}$$

where the summation indices for the series solutions are taken up to $m = 1,3$ and $n = 1,3$.

Now, consider the rectangular plate shown in Figure 5, but with the elastic stiffnesses satisfying anisotropic conditions ($B_{ij} = 0$, $A_{16} = A_{26} = 0$; $D_{16} \neq D_{26} \neq 0$). The Navier method presented earlier for the orthotropic plate will be difficult because of the nonzero D_{16} and D_{26} terms. However, an elasticity solution based on the Ritz method (outlined below) is tedious. Finite element modeling techniques such as NASTRAN [24] programs may be used to solve coupled anisotropic problems where $B_{ij} \neq 0$.

The Ritz method is based on minimizing the work function $\overline{W}$ given below, for an anisotropic simply supported plate subject to uniformly distributed load intensity q_0 as

$$\overline{W} = \int_A q_0 w \, dx \, dy - \tfrac{1}{2} \int_A [D_{11}(w_{,xx})^2 + 2D_{12}(w_{,xx})(w_{,yy}) + D_{22}(w_{,yy})^2 + 4D_{66}(w_{,xy})^2 + 4D_{16}(w_{,xx})(w_{,yy}) + 4D_{26}(w_{,yy})(w_{,xx})] \, dx \, dy \quad (54)$$

The out-of-plane displacement $w(x,y)$ is assumed as a double Fourier series (satisfying boundary conditions of Eq. 48), as

$$w(x,y) = \sum_{m=1}^{\infty} \sum_{n=1}^{\infty} A_{mn} \sin \frac{m\pi x}{a} \sin \frac{n\pi y}{b} \quad (55)$$

Partial derivatives in Eq. (54) are calculated using Eq. (55) and substituted back in Eq. (54). Now, Eq. (54) will contain terms involving A_{mn}. The stationary value of the work function $\overline{W}$ should satisfy

$$\frac{\partial \overline{W}}{\partial A_{mn}} = 0 \quad (56)$$

Equation (56) gives set of simultaneous equations in A_{mn} depending on the number of terms assumed. Knowing A_{mn}, $w(x,y)$ can be found to a reasonable accuracy.

For a detailed description of the Ritz method, Ref. 30 may be consulted, while Ref. 31 might be helpful in providing techniques to carry out some of the integrations typically involved in this procedure.

Anisotropic Plate Strip Theory

Considering a narrow rectangular plate with a large aspect ratio, subjected to load intensity varying only as a function of x, that is, $q(x)$, the governing differential equations and the solutions become independent of "y." Hence a two-dimensional plate becomes a one-dimensional plate strip problem. The plate strip and rectangular beam theories of laminated sections are covered in detail in references 15 and 25. A brief discussion on the one-dimensional plate strip cylindrical bending theory is given here for the sake of continuity and in view of later reference in the section on bending analysis of layered composite built-up sections.

Cylindrical Bending

One dimensional analysis for an infinite plate strip of a most general laminate subjected to cylindrical bending is given in detail by Whitney [25]. However, most commonly used thin laminates are essentially either balanced and symmetric ($A_{16} = 0$ and $B_{11} = B_{16} = 0$ respectively), or nearly symmetric ($B_{11} = B_{16} \approx 0$) laminates in view of manufacturing considerations. Hence, consider a practical laminate plate strip of finite width 'b' so that $\frac{L}{b} >> 1$, as shown in Figure 8. Assuming cylindrical bending (displacement along y-axis, $v = 0$, see Ref. 15) of such finite width but long laminate plate strip, leads to the following strain-displacement relations given by Equations (57). In addition, small deformation of the plate strip satisfying Kirchhoff–Love hypotheses described earlier is assumed in this case. Hence, the following treatment on the cylindrical bending of a finite width but long plate strip is a special case of a most general analysis given in [25] on this subject, by considering only laminates of practical interests as outlined above.

$$\varepsilon_x = u_{0,x} - Zw_{,xx}$$
$$\varepsilon_y = 0 \quad (57)$$
$$\gamma_{xy} = 0$$

Equations (27) and (28) reduce to

$$N_x = A_{11}u_{0,x} - B_{11}w_{,xx}$$
$$N_y = A_{12}u_{0,x} - B_{12}w_{,xx} \quad (58)$$
$$N_{xy} = A_{16}u_{0,x} - B_{16}w_{,xx}$$

and

$$M_x = B_{11}u_{0,x} - D_{11}w_{,xx}$$
$$M_y = B_{12}u_{0,x} - D_{12}w_{,xx} \quad (59)$$
$$M_{xy} = B_{16}u_{0,x} - D_{16}w_{,xx}$$

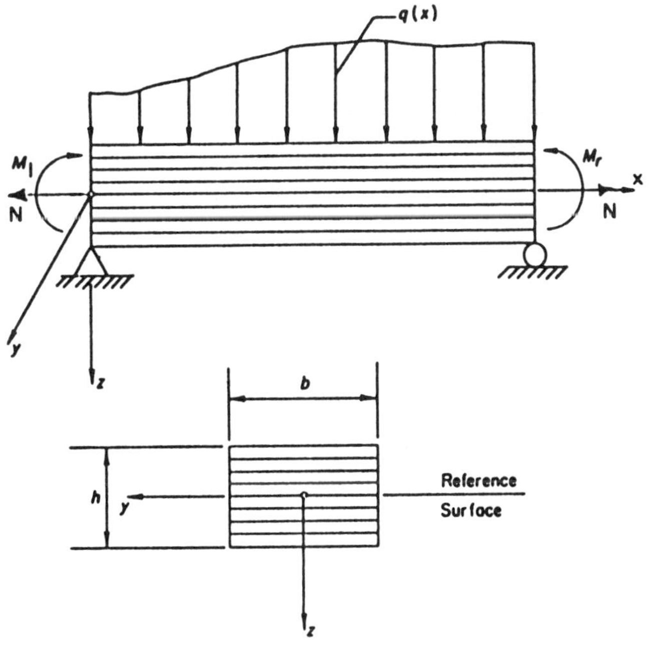

FIGURE 8 Laminated plate strip. (From Ref. 15.)

Considering only moment M_x ($N_x = 0$),

$$u_{0,x} = \frac{B_{11}}{A_{11}} w_{,xx} \tag{60}$$

$$w_{,xx} = \left(\frac{A_{11}}{B_{11}^2 - A_{11}D_{11}}\right) M_x(x) \tag{61}$$

and

$$u_0(x) = \frac{B_{11}}{B_{11}^2 - A_{11}D_{11}} \int_0^x M(\xi)\, d\xi + k \tag{62}$$

Assuming the constant $k = u_0(0) = 0$ at $x = 0$, the first equation of Eq. (57) becomes

$$\varepsilon_x^{(k)} = \frac{A_{11}}{B_{11}^2 - A_{11}D_{11}} \left(\frac{B_{11}}{A_{11}} - z\right) M_x(x) \tag{63}$$

and

$$\sigma_x^{(k)} = \overline{Q}_{11}^k \left(\frac{A_{11}}{B_{11}^2 - A_{11}D_{11}}\right)\left(\frac{B_{11}}{A_{11}} - z\right) M_x(x) \tag{64}$$

Equations (63) and (64) give the lamina strains and stresses in the axial directions. Now considering N_x ($M_x = 0$), it can be shown [15] that the lamina axial strain is given by

$$\varepsilon_x^{(k)} = \frac{B_{11}}{A_{11}D_{11} - B_{11}^2} \left(\frac{D_{11}}{B_{11}} - z\right) N \tag{65}$$

and the lamina axial stress by

$$\sigma_x^{(k)} = \overline{Q}_{11}^k \left(\frac{B_{11}}{A_{11}D_{11} - B_{11}^2}\right)\left(\frac{D_{11}}{B_{11}} - z\right) N \tag{66}$$

For both load cases M_x and N, the lamina stresses $\sigma_y^{(k)}$ and $\tau_{xy}^{(k)}$ can also be calculated using the appropriate modulus components.

Simple Beam Solution

Dietz [32] has shown that a simple mechanics-of-materials approach for the laminated plate strip yielded similar results for the lamina stresses obtained as above in Eq. (64). Briefly, modulus-weighted areas and moments of inertia of individual lamina are used to find the neutral axis and flexural stiffness of the laminated strip. In other words, the distance of the neutral axis (z_{na}), flexural rigidity (EI), and bending stress ($\sigma_x^{(k)}$) are given as

$$z_{na} = \frac{\sum E_k A_k z_k}{\sum E_k A_k} \qquad EI = \sum E_k I_k \tag{67a}$$

and

$$\sigma_x^{(k)} = \frac{M_x(x) E_k z}{EI} \tag{67b}$$

where E_k, I_k, and z_k are, respectively, the modulus, moment of inertia, and distance of individual lamina from some reference surface. In reference to Figure 4 for the lamina stacking sequence, the depth, area, and distance to the center of a lamina, respectively, are given as

$$\begin{aligned} d_k &= h_k - h_{k-1} \\ A_k &= b(h_k - h_{k-1}) \\ z_k &= \tfrac{1}{2}(h_k + h_{k-1}) \end{aligned} \tag{68}$$

Substituting Eq. (68) in Eq. (67) gives

$$z_{na} = \frac{b \sum \overline{Q}_{11}^k (h_k - h_{k-1}) \frac{1}{2}(h_k + h_{k-1})}{b \sum \overline{Q}_{11}^k (h_k - h_{k-1})} \tag{69}$$

Using Eq. (29), Eq. (69) becomes

$$z_{na} = \frac{B_{11}}{A_{11}} \tag{70}$$

For a unit width of beam, moment of inertia of a lamina is given by

$$I_k = \left[(z_k - z_{na})^2 + \frac{d_k^2}{12}\right] d_k \tag{71}$$

Substituting for the variables, we get

$$I_k = \frac{h_k^3 - h_{k-1}^3}{3} - (h_k^2 - h_{k-1}^2)\frac{B_{11}}{A_{11}} + (h_k - h_{k-1})\left(\frac{B_{11}}{A_{11}}\right)^2 \tag{72}$$

Hence, the flexural rigidity EI in Eq. (67), in view of Eq. (29), becomes

$$EI = D_{11} - \frac{B_{11}^2}{A_{11}} \tag{73}$$

Equation (73) is the equivalent flexural rigidity EI for the plate strip consistent with Eq. (61) except for a change in sign. The distance z from the neutral axis of the plate strip is

$$z = z_{na} - z_k \tag{74}$$

Equation (67b) for the lamina stress $\sigma_x^{(k)}$ is obtained knowing the equivalence of E_k with $\overline{Q}_{11}^k$ and using Eq. (70), as

$$\sigma_x^{(k)} = \overline{Q}_{11}^k \left(\frac{A_{11}}{A_{11}D_{11} - B_{11}^2}\right)\left(\frac{B_{11}}{A_{11}} - z\right) M_x(x) \tag{75}$$

Equation (75) is identical to Eq. (64) for the lamina stress in a plate strip subjected to M_x.

Bending Analysis of Laminated Composite Built-up Sections

Primary composite load-bearing structures are used not only in the form of membrane (tension or compression) skins and bending plates but also as built-up sections (I-beams, C-channels, T-beams, etc.), and in a variety of complex shapes depending on need. Even though these built-up sections are mainly one-dimensional beam problems, they are difficult to analyze because of anisotropic properties of the basic composite material used to build the structures. A rigorous analysis would involve complex mathematical finite element modeling using NASTRAN programs [24] to solve a simple beam-bending problem. However, recently [33] the author proposed a simple analysis tool for the bending of composite built-up sections similar to the mechanics of materials approach by Dietz [32] for the laminated plate strip analysis. A brief discussion is given here, with highlights on certain new techniques used in the bending analysis of built-up section.

In reference to Figure 9*a*, a T-section with a bent web, made of laminated composite material, the objective is to find the stress and strain at any point on the cross section due to bending and axial load. A simple mechanics-of-materials approach as used in the isotropic beam bending analysis [34,35] would require such basic geometrical properties as the location of the centroid $(\overline{X},\overline{Y})$ and axial $(\overline{AE})$ and flexural $(\overline{EI})$ stiffnesses associated with the cross section. Assuming that the shear load passes through the shear center so that pure bending occurs, and no twisting, along with the usual Kirchhoff–Love assumptions for small-deformation theory, an analytical tool is developed. Both classical lamination theory and linear elastic isotropic theory in summing up the effect of idealized rectangular sections of the built-up cross section are used in determining the overall properties for the entire cross section. A set of reference axes $(x_{\text{ref}}, y_{\text{ref}})$ is chosen as shown in Figure 9*b*. The built-up section is idealized into several rectangular sections of constant thicknesses using median line lengths and ignoring the radii. The areas and centroids of each of these idealized elements should be known using moduli-weighted properties, as outlined later. The planar orientation of each element with reference to the chosen coordinates must also be known.

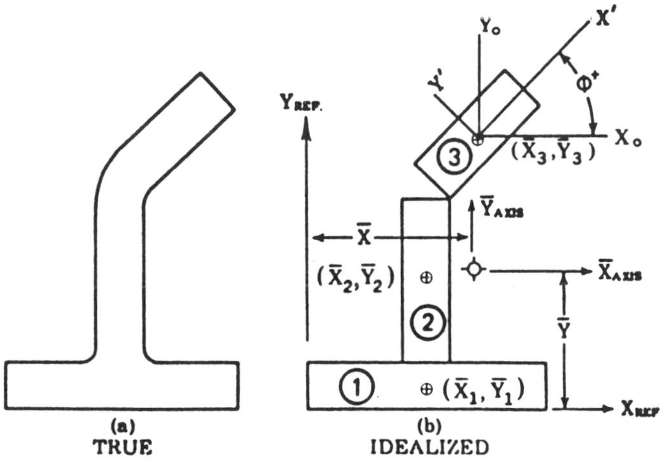

FIGURE 9 Section geometry.

In the case of a laminated plate, of thickness h, the equivalent engineering constants are calculated performing uniaxial tensile, compressive, and shear tests on the laminate coupons. For a general laminate, the engineering constants are extracted from the stretching part of the compliance matrix of Eq. (33), as

$$[A'] = [a] + [b][d^{-1}][b^{T}] \tag{76}$$

The engineering constants are explicitly given as

$$\begin{aligned} E_x &= \frac{1}{a_{11}h} \\ E_y &= \frac{1}{a_{22}h} \\ \nu_{yx} &= -\frac{a_{21}}{a_{11}} \\ \nu_{xy} &= -\frac{a_{12}}{a_{22}} \\ G_{xy} &= \frac{1}{a_{66}h} \end{aligned} \tag{77}$$

where

$$[A'] = \begin{bmatrix} a_{11} & a_{12} & a_{16} \\ a_{12} & a_{22} & a_{26} \\ a_{16} & a_{26} & a_{66} \end{bmatrix} \tag{78}$$

The engineering constants associated with the flexural part of the compliance matrix of Eq. (33) are given by

$$[D'] = [(D - b^{T}B)^{-1}] \tag{79}$$

or

$$\begin{aligned} (EI)_x &= b/d_{11} \\ (EI)_y &= b/d_{22} \\ GJ &= b/d_{66} \end{aligned} \tag{80}$$

where b is the laminate width and

$$[D'] = \begin{bmatrix} d_{11} & d_{12} & d_{16} \\ d_{12} & d_{22} & d_{26} \\ d_{16} & d_{26} & d_{66} \end{bmatrix} \tag{81}$$

Assuming symmetric layups for as many as n idealized sections ($n = 3$ in Fig. 9*b*), $[A'] = [A]^{-1}$ and $[D'] = [D]^{-1}$, since $B_{ij} = 0$. The modulus-weighted centroidal distances $(\overline{X},\overline{Y})$ and flexural stiffnesses $(EI_{\overline{X}}, EI_{\overline{Y}})$ are defined as

$$\bar{X} = \frac{\sum_{i=1}^{n} (AE)_i x_i}{\sum_{i=1}^{n} (AE)_i} \quad (82)$$

$$\bar{Y} = \frac{\sum_{i=1}^{n} (AE)_i y_i}{\sum_{i=1}^{n} (AE)_i} \quad (83)$$

where

$$E_i = \frac{1}{(a_{11}h)_i} \quad (84)$$

and

$$EI_{\bar{X}} = \sum_{i=1}^{n} (AE)_i (y_i)^2 + \sum_{i=1}^{n} (EI_{x_0})_i - (\bar{Y})^2 \sum_{i=1}^{n} (AE)_i \quad (85)$$

$$EI_{\bar{Y}} = \sum_{i=1}^{n} (AE)_i (x_i)^2 + \sum_{i=1}^{n} (EI_{y_0})_i - (\bar{X})^2 \sum_{i=1}^{n} (AE)_i \quad (86)$$

where

$$EI_{x_0} = EI_{x'} = \frac{b}{d_{11}} \quad (87)$$

$$EI_{y_0} = EI_{y'} = \frac{b^3}{12a_{11}} \quad (88)$$

The difference between the element centroidal flexural stiffnesses $I_{x'}$ and $I_{y'}$ is explained in Figure 10. Except for the top and bottom layers of 0° orientation in Figure 10*a*, switched to ± 45° in Figure 10*b*, the total numbers of layers and orientations are same in both cases. Hence, there is a difference between the magnitudes of flexural stiffnesses, $EI_{x'}$ and $EI_{y'}$ of the section about x' and y' axes, as indicated in Figure 10. In the case of an inclined idealized section, the element centroidal stiffnesses ($EI_{x'}$, $EI_{y'}$) have to be rotated through (x_0,y_0), which are parallel to the reference coordinates (X_{ref},Y_{ref}) (refer to Fig. 9*b*) as

$$EI_{x_0} = EI_{x'} \cos^2\phi - EI_{x'y'} \sin 2\phi + EI_{y'} \sin^2\phi \quad (89)$$

$$EI_{y_0} = EI_{x'} \sin^2\phi + EI_{x'y'} \sin 2\phi + EI_{y'} \cos^2\phi \quad (90)$$

where $EI_{x'}$, $EI_{y'}$ are obtained through Eqs. (87) and (88), respectively. While the product of inertia $EI_{x'y'}$ is zero for a symmetrically layered section about the element centroidal axes $x'y'$, the rotated $EI_{x_0y_0}$ is given by

$$EI_{x_0y_0} = \tfrac{1}{2} \sin 2\phi \, (EI_y' - EI_{x'}) \quad (91)$$

Transferring to the gross section centroidal axes, we have

$$EI_{\bar{X}\bar{Y}} = \sum_{i=1}^{n} (AE)_i x_i y_i + \sum_{i=1}^{n} (EI_{x_0y_0})_i - \bar{X}\,\bar{Y} \sum_{i=1}^{n} (AE)_i \quad (92)$$

Equations (85), (86), and (92) give the centroidal flexural stiffnesses of the built-up cross section. The parallel axes theorem has been used in arriving at these equations.

In reference to Figure 11, the axial strain ε_z at any point $p(x,y)$ in the cross section of the composite beam subject to bending along two centroidal axes (M_X,M_Y) and an axial load (P) is given analogous to other sources [34,35], as

$$\varepsilon_z = \frac{M_{\bar{X}}(EI_{\bar{X}\bar{Y}}) - M_{\bar{Y}}(EI_{\bar{X}})}{(EI_{\bar{X}})(EI_{\bar{Y}}) - (EI_{\bar{X}\bar{Y}})^2} x + \frac{M_{\bar{Y}}(EI_{\bar{X}\bar{Y}}) - M_{\bar{X}}(EI_{\bar{Y}})}{(EI_{\bar{X}})(EI_{\bar{Y}}) - (EI_{\bar{X}\bar{Y}})^2} y + \frac{P}{\overline{AE}} \quad (93)$$

and the average axial stress σ_z at that point is given by

$$\sigma_z = E\varepsilon_z \quad (94)$$

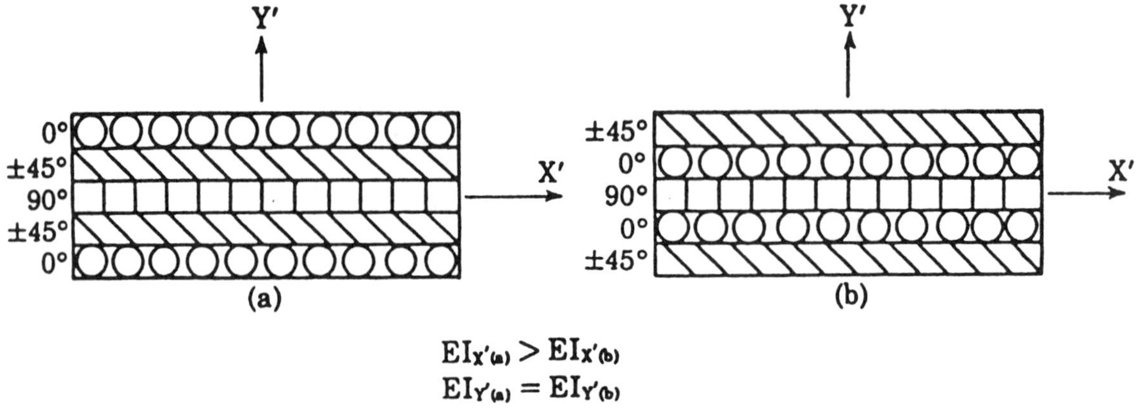

FIGURE 10 Composite element comparison.

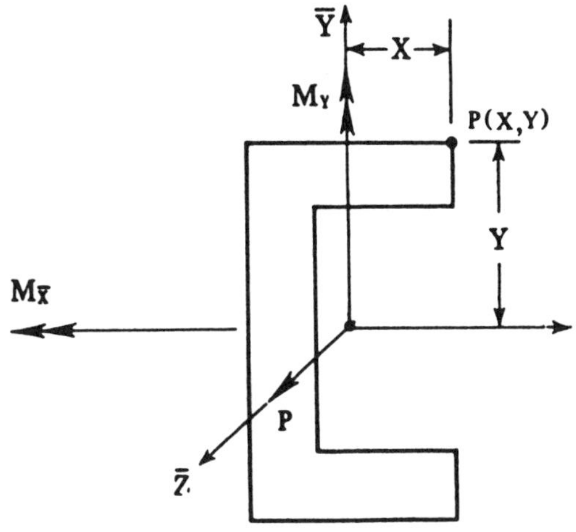

FIGURE 11 Complex bending geometry.

where

$$\overline{AE} = \sum_{i=1}^{n} (AE)_i \tag{95}$$

The accuracy of this analysis increases as the number of idealized sections increase; hence the computations may be computerized. The program COMPSEC was written in FORTRAN to facilitate computational efforts. While any arbitrary point $p(x,y)$ on the cross section may be selected to compute the strain in Eq. (93), the centroidal location (X_i,y_i) of an idealized element is more appropriate (refer to Fig. 9*b*). Having obtained ε_z at the centroid of an idealized element, the ply-stress and ply-strain analysis will be done to ensure the safety of the overall section, as follows [12].

The laminate effective Poisson's ratio for any idealized laminate element is calculated as

$$\nu_{21} = -\frac{a_{21}}{a_{11}} \tag{96}$$

The laminate effective shear coefficient is given by

$$\nu_{61} = \frac{a_{61}}{a_{11}} \tag{97}$$

Then, the Poisson's strain in the lateral direction (ε_x) and the shear strain (ε_s) are given by

$$\varepsilon_x = -\nu_{21}\varepsilon_z \tag{98}$$

$$\varepsilon_s = \nu_{61}\varepsilon_z \tag{99}$$

If the orientation of any *k*th lamina is known within the idealized laminate element, then the principal (on-axis) lamina (ply) strains and stresses are given, respectively, by

$$\{\varepsilon_1,\varepsilon_2,\tfrac{1}{2}\varepsilon_6\}_{(k)} = [T]_{(k)}\{\varepsilon_z,\varepsilon_x,\tfrac{1}{2}\varepsilon_s\} \tag{100}$$

and

$$\{\sigma_1,\sigma_2,\sigma_6\}_{(k)} = [Q]_{(k)}\{\varepsilon_1,\varepsilon_2,\varepsilon_6\}_{(k)} \tag{101}$$

where $[T]$ is the tensor transformation matrix given by Eq. (6) and $[Q]$ is the ply modulus matrix defined in Eq. (4).

While Eqs. (93) through (101) may be calculated for each of the idealized laminate elements, time and effort may be saved by physical interpretation of the actual problem at hand, by locating some of the critically stressed idealized elements and performing the recurring calculations (93)–(101) to those elements only. For example, axial pulling causes uniform stress or strain in a cross section, while a bending produces maximum stress or strain at the extreme fiber distance from the axis of bending. Also, the calculation of strain (ε_z given by Eq. 93) may be simplified if the centroidal axes happen to be principal axes (X,Y), since the product of inertia (EI_{XY} becomes zero). For cross sections with at least one axis of symmetry, the product of inertia will be zero (e.g., T- or channel sections), and the expression for ε_z in Eq. (93) becomes simple. However, for a nonsymmetric section, like a z section, the principal axes for the moment of inertia may be calculated (see Fig. 12) as

$$\tan 2\Theta_p = -\frac{2EI_{\bar{X}\bar{Y}}}{EI_{\bar{X}} - EI_{\bar{Y}}} \tag{102}$$

$$EI_X = \frac{EI_{\bar{X}} + EI_{\bar{Y}}}{2} + \left[\left(\frac{EI_{\bar{X}} + EI_{\bar{Y}}}{2}\right)^2 + (EI_{\bar{X}\bar{Y}})^2\right]^{1/2} \tag{103}$$

$$EI_Y = \frac{EI_{\bar{X}} + EI_{\bar{Y}}}{2} - \left[\left(\frac{EI_{\bar{X}} + EI_{\bar{Y}}}{2}\right)^2 + (EI_{\bar{X}\bar{Y}})^2\right]^{1/2} \tag{104}$$

and the strain ε_z is given as

$$\varepsilon_z = -\frac{M_Y}{EI_Y}X - \frac{M_X}{EI_X}Y + \frac{P}{\overline{AE}} \tag{105}$$

To verify the accuracy of the theoretical development, cantilever bending tests about two centroidal axes of a

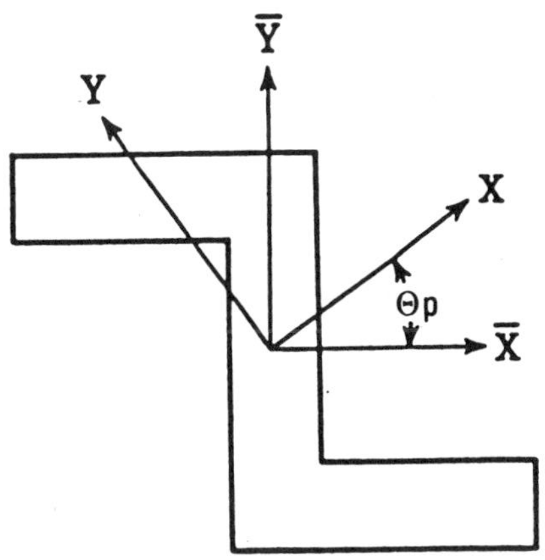

FIGURE 12 Principal axes definition.

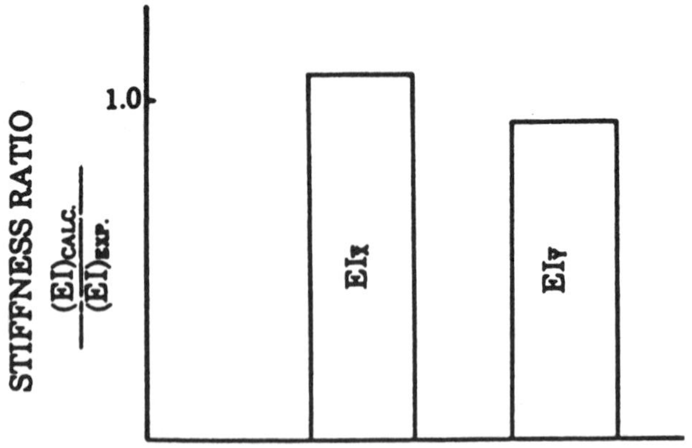

FIGURE 13 Section stiffness comparison.

graphite–epoxy channel section were conducted. The flexural stiffnesses ($EI_{\bar{X}}$, $EI_{\bar{Y}}$) were calculated by noting the end deflections (δ_Y and δ_X) as

$$EI_{\bar{X}} = \frac{PL^3}{3\delta_y} \quad (106)$$

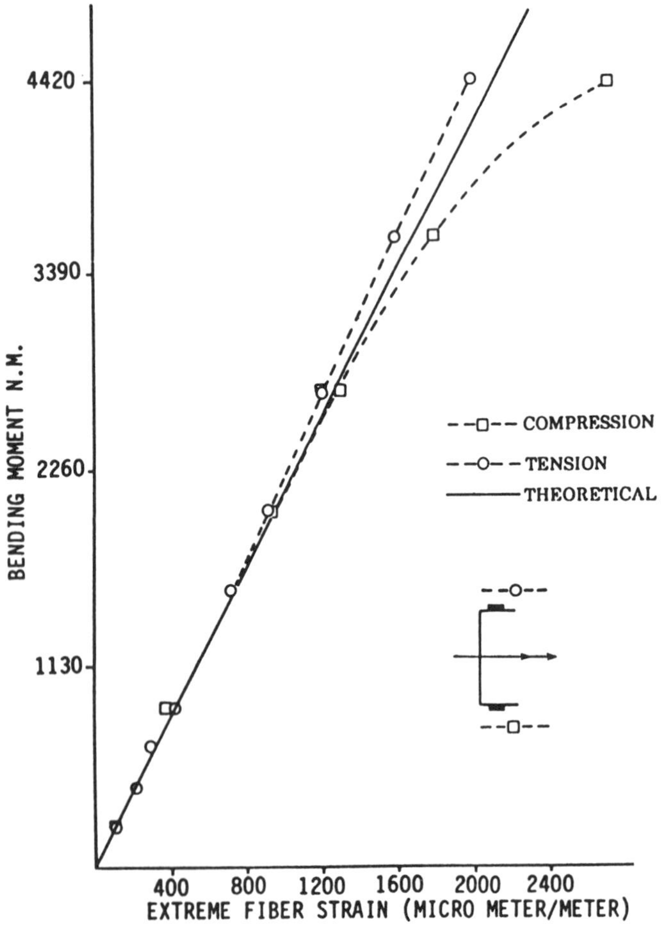

FIGURE 14 Calculated and measured strains for the channel.

and

$$EI_{\bar{Y}} = \frac{PL^3}{3\delta_x} \quad (107)$$

where P is the end vertical load and L is the length of the beam. The extreme fiber strains (ε_z) on the top and bottom of the channel were also measured. The analytical bending stiffnesses given by Eqs. (85) and (86) were compared with those measured and the results are shown in Figure 13. The correlation was good, as evidenced in the figure, with no more than 8% difference. Also, the measured strains and the calculated strains through Eq. (93) were plotted as shown in Figure 14. Excellent agreement in the lower range of the load is apparent from the graph. But at the higher load, stability of the section became a factor and nonagreement with linear theory was noted. The channel section finally failed in instability.

Conclusions

This discussion of the development of fiber-reinforced composite materials for both primary and secondary aerospace structures emphasized the necessity for bending analysis of laminated composite plates, in spite of the fact that the best use of fiber composite is achieved by tailoring and putting the required amount of material in the direction of loads. The constitutive (stress–strain) relation of a lamina of orthotropic material was reviewed. An overview of the classical lamination theory was developed to arrive at the constitutive relation for a multiple layered composite anisotropic plate. The governing differential equations for bending and stretching of nonhomogeneous, unsymmetric, and coupled anisotropic plates were derived. These general differential equations were reduced to anisotropic, symmetric laminates and also to orthotropic laminates. Considering a rectangular plate, the required geometric and natural boundary conditions were stated to find unique solutions satisfying both the governing differential equations and the boundary conditions. The difficulty of obtaining closed-form solutions to the coupled anisotropic plate was realized, and alternate numerical finite difference and NASTRAN finite element modeling techniques were emphasized to solve such problems. The Navier and Ritz methods of solution for orthotropic and anisotropic bending plates, respectively, were reviewed. The two-dimensional anisotropic plate equations were reduced for the one-dimensional cylindrical bending of a plate strip problem, in which case a simple mechanics-of-materials approach was successfully demonstrated to yield the same results predicted by the plate strip theory. Finally, a simple analytical tool was developed using the classical plate theory in conjunction with the isotropic beam bending theory, for the bending analysis of composite built-up sections. Analytical bending stiffnesses and strains were in excellent agreement with the experimental results, substantiating the validity of the analytical tool developed for the built-up section.

Recommendations

The bending analysis of composite structures is quite an extensive subject, although only small-deflection linear theory of anisotropic plate bending and stretching was reviewed. For large deflection of plates, membrane forces become predominant and nonlinear effects will be present [19]. For highly anisotropic thick plates, transverse shear deformation [36] and interlaminar stresses [13] will be predominant. The reader should consult the foregoing references and current journal articles in these areas for better understanding and for awareness of the limitations on the linear small-deflection analysis of thin anisotropic plates as discussed here.

Acknowledgments

The author is grateful to John G. Avery and Gary C. Cassatt of Advanced Programs, Boeing Military Airplanes (BMA), Wichita, for their encouragements and to BMA, in the preparation of this manuscript. The coordination efforts of Mrs. Lou G. DeHaven for the completion of this manuscript in the short time are greatly appreciated.

T. P. Sivam

References

1. G. Lubin, Ed., *Handbook of Fiberglass and Advanced Plastics Composites*, Van Nostrand Reinhold, New York, 1969.
2. R. T. Schwartz, and H. S. Schwartz, Eds., *Fundamental Aspects of Fiber Reinforced Plastic Components*, Wiley-Interscience, New York, 1968.
3. Plastics for Aerospace Vehicles, Part 1, "Reinforced Plastics," MIL-HDBK-17A, U.S. Department of Defense, Washington, DC, 1971.
4. *Composite Materials*: *Testing and Design*, ASTM STP 460, 1969.
5. *Composite Materials*: *Testing and Design* (Second Conference), ASTM STP 497, 1972.
6. *Composite Materials*: *Testing and Design* (Third Conference), ASTM STP 546, 1974.
7. *Composite Materials*: *Testing and Design* (Fourth Conference), ASTM STP 617, 1977.
8. S. W. Tsai, Ed., *Composite Materials*: *Testing and Design* (Fifth Conference), ASTM STP 674, 1979.
9. *Buckling of Sandwich Panels*, Vol 3, ESDU International Limited, London, June 1980.
10. *Buckling of Laminated Composite Plates*, Vol 13, ESDU International Limited, London, December 1986.
11. *Laminated Composites*, Vol 14, ESDU International Limited, London, October 1986.
12. S. W. Tsai, and H. T. Hahn, *Introduction to Composite Materials*, Technomic, Lancaster, PA, 1981.
13. R. M. Jones, *Mechanics of Composite Materials*, Scripta, Washington, DC, 1975
14. J. E., Ashton, J. C. Halpin, and P. H. Petti, *Primer on Composite Materials; Analysis*, Technomic, Lancaster, PA., 1969, 2nd ed., 1984.
15. L. R. Calcote, *Analysis of Laminated Composite Structures*, Van Nostrand Reinhold, London, 1969.
16. B. C. Haskin, and A. A. Baker, Eds., *Composite Materials for Aircraft Structures*, AIAA Education Series, AIAA Inc., New York, 1986.
17. "Structural Sandwich Composites," MIL-HDBK-23A, U.S. Department of Defense, Washington, DC, 1968.
18. C. C. Chamis, Ed., *Structural Design and Analysis—Composite Materials*, Volumes 7 and 8, Academic Press, New York, 1974.
19. S. G. Lekhnitski, *Anisotropic Plates*, 2nd ed., Gordon & Breach, New York, 1968.
20. S. B. Dong, R. B. Matthiesen, K. S. Pister, and R. L. Taylor, "Analysis of Structural Laminates," Air Force Report, ARL-76 1961.
21. H. L. Langhaar, *Energy Methods in Applied Mechanics*, Wiley, New York, 1962.
22. J. E. Ashton, *J. Compos. Mater.* 4, 162 (1970).
23. J. M. Whitney, *J. Compos. Mater.* 7 (1970), 192.
24. R. H. MacNeal, Ed., *MSC/NASTRAN, Handbook for Linear Static Analysis*, Version 61, MacNeal-Schwendler Corporation, December 1981, [or MSC/NASTRAN User's Manual, Vol. I].
25. J. M. Whitney, *Structural Analysis of Laminated Anisotropic Plates*, Technomic, Lancaster, PA, 1987.
26. R. J. Roark and W. C. Young, *Formulas for Stress and Strain*, McGraw-Hill, New York, 1975.
27. D. McFarland et al., *Analysis of Plates*, Spartan Books, New York, 1972.
28. J. M. Whitney, *J. Compos. Mater.* 3, 20 (1969).
29. B. L. Smith, "Analysis of Thin Laminated Composite Plates," Parts I and II, University of Alabama, College of Engineering, BER Report 202-94, September 1975 and September 1976.
30. J. E. Ashton and M. E. Waddoups, *J. Compos. Mater.* 3, 148 (1969).
31. D. Young and R. P. Felgar, *Tables of Characteristic Functions Representing the Normal Modes of Vibration of a Beam*, University of Texas, Publication 4913, 1949.
32. A. G. H. Dietz, "Design Theory of Reinforced Plastics," in R. H. Sonneborn, Ed., *Fiberglass Reinforced Plastics*, Reinhold, New York, 1954, Chapter 9.
33. T. Paramasivam and J. Pilla, *Proceedings of the 19th Int. SAMPE Tech. Conf.*, 1987, pp. 198–208.
34. E. F. Bruhn, *Analysis and Design of Flight Vehicle Structures*, Jacobs, Indiana (1973).
35. D. J. Perry, *Aircraft Structures*, McGraw-Hill, New York, 1982.
36. J. M. Whitney, *J. Compos. Mater.* 3, 534 (1969).

Biological Composites

See Natural Composites, Fiber Modification

Biomaterials

See Orthopedic Devices

Bismaleimides

See Polyimides

Blends and Alloys

See Polymer Blends and Alloys

Bolted Joints

See Joining, Mechanical Fastening

Bonded Joints

See Joining, Adhesive; Joining, High-Temperature Adhesives

Bonding

See Joining, Adhesives

Bond Interface

See Interface in Fiber Reinforced Composites; Interfacial Analysis

Bond Strength

See Joining, Adhesives; Joining, High-Temperature Adhesives; Joining, Polymer Composites, Fusion; Joining, Metal Matrix Composites, Fusion; Joining, Metal Matrix Composites, Nonfusion

Boron Fibers

See Fiber Reinforcement

Braiding

Introduction

Braiding is one of the traditional methods of interconnecting strands of fibers into fabrics, decorative designs, ropes, and reinforcements. In this process, the strands (yarns) are usually fed in the direction of braid formation, and they interlace each other to form tubes, narrow flat strips, contoured shapes, or solid three-dimensional (3D) structures.

In other fabric-forming processes, yarns are manipulated in many different ways. In weaving, yarns are fed in the direction of fabric formation and interlace with additional sets of yarns fed at 90°. In knitting, yarns are interlooped with each other to form a sheet or tube. Additional straight yarns may be added to knits and held in place by the loops. Several other methods of forming fibrous structures have been developed by the textile industry: entangling (needled fabrics), bonding (nonwovens), knotting (nets) and twisting (lace fabrics, rope), tufting (carpets), and stitching.

Although most fabrics are made as sheets, braids are typically made as tubes or solid ropes. They vary greatly in size and end-use application, ranging from microscopic eye sutures to shoe laces and high strength marine ropes for towing ships.

Circular braids have been used to reinforce flexible pressure hoses for some time, and their selection for use in reinforcing composites was a natural extension of this application. Tubular shapes are most commonly used, but braids can be made in a wide range of shapes, and the technology is expanding rapidly to include complex contours and even solid forms.

Unique Features of Braids

The principal feature of braids that makes them desirable for reinforcing composites is the ability to form a wide range of geometric shapes. In addition, braiding offers a cost-effective method of arranging fibers and, in some cases, applying a matrix. Braids cannot be made into wide sheets that can readily be woven or knit; but they can form more complex structures than materials made by any other textile process.

Braiding can produce complex shapes because of several unique features. First, the process, unlike others, includes a relatively large "convergence zone" in which the interlaced strands are pulled tight into the final pattern (Fig. 1). During this tightening, the fabric can be formed over a complex shaped mandrel—which may be asymmetrical. Second, in braiding, all yarns feed "on demand" rather than at a constant or predetermined rate. This feature permits more material to be fed where it is needed—such as at the outside of a bend. Third, braiding machines can be designed to provide for readily inserting strands or other elements into the core or even the wall of the fabric. This makes it possible to overbraid mechanical elements such as rigid connectors that remain in the fabric structure. Contoured mandrels or items such as structural foam can also be inserted into the core and overbraided. The fourth feature that contributes to geometric flexibility is the opportunity to change the sequence of interlacing (i.e., the braiding pattern). This makes it possible to create intricate lace fabrics or complex 3D parts.

This article describes braiding by reviewing the process, the structures formed, the conversion to composites, and the mechanical properties of braids. The process description includes machinery currently used for composites applications as well as equipment that may eventually be applied to composites but is now used only for textiles. Braided structures are described in terms of their micro- and macrogeometric variations. The emerging field of 3D braids is covered in a separate section.

FIGURE 1 **Circular braiding of complex shape.** (Photograph courtesy of Fibrous Materials Research Laboratory, Drexel University.)

The Braiding Process

In the basic braiding process, a number of strands are held together at one point and then interlaced. The simplest example is a three-stranded braid (or plait) of hair in a child's "pigtail." The process has been developed into a continuous high speed operation by a number of ingenious inventions.

A generalized schematic of a braiding process appears in Figure 2, and the key components and their functions are listed in Table 1. Braiding yarns interlace each other in a prescribed pattern. Optional additions include a mandrel for shape, axial yarns, and core yarns.

The braiding yarns form a self-supporting interconnected network. In a typical structure they run in right- and left-handed interlaced helices. Multiple layers may be formed by overbraiding with multiple passes or multiple machines. Additional yarns or even rigid components may be added to this basic structure—in the core, between layers, or within one layer.

The machine action to interlace the braiders operates as follows. Braiding yarns are wound on bobbins, typically plastic or paper spools, each of which is placed on a carrier assembly. The carriers "float" relative to the machine and are not fixed to any part. During operation, they are passed from one position to another and trace out intersecting paths in the braiding plane in a specified braiding pattern. This action moves the "free" ends of the braiding yarns relative to their "fixed" ends at the braid formation point. The possible carrier paths and the carrier type are determined by the type of braiding machine.

In a simple circular braider, the carriers move in serpentine manner around a circle—half in each direction. They cross paths frequently to give interlacing patterns such as one-over-one or two-over-two as they form a tube.

Many other shapes can be formed by varying the pattern by which the carriers move in the braiding plane. A "braided shape" differs from one obtained by overbraiding a shaped mandrel because it can be solid. Commercial machines are available to braid simple shapes such as solid cylinders and solid squares, and new 3D equipment is being developed to braid complex shapes.

Braiding Machine Mechanisms

The traditional means [1] for moving carriers in a particular pattern is a "horn gear" mechanism, as represented in Figure 3: a carrier rides in a notch of a horn gear (actually a notched disc); as it approaches another horn gear rotating in the opposite direction, the carrier switches to the new gear. This transfer involves a contacting element (not shown), which acts like a railroad "switch-track," and cam surfaces (also not shown), which hold the carriers against the horn gears.

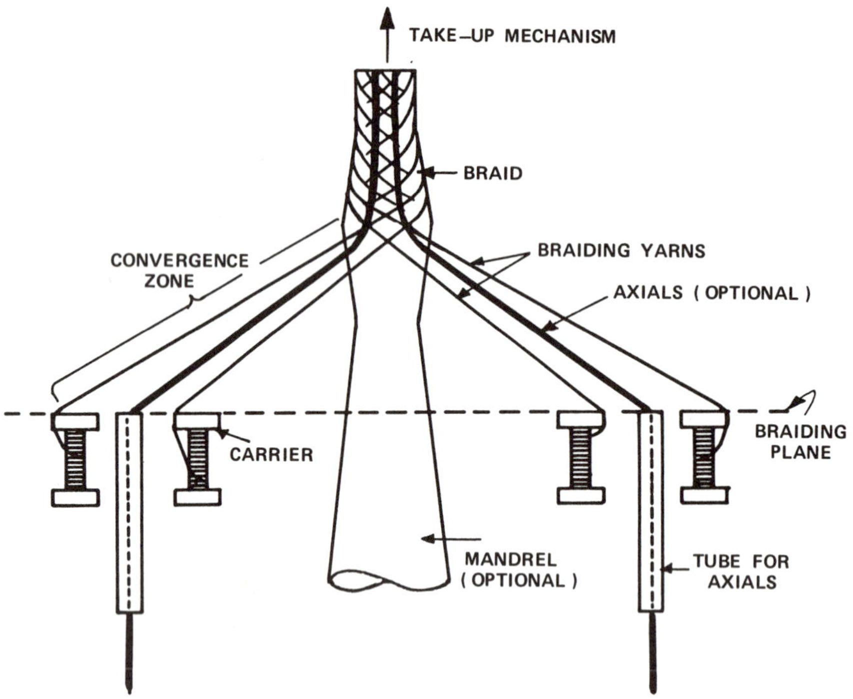

FIGURE 2 Schematic of circular braider.

The carrier mechanism also serves as a key element of braiding. It must maintain constant yarn tension and permit the yarns to either feed or retract. During braiding, the carriers change their distance to the convergence point as they move in the braiding plane. When they move closer to the convergence point, the carriers must retract the previously fed length to prevent the yarns from becoming slack.

A traditional carrier mechanism for feeding and retracting yarns is shown schematically in Figure 4. The yarn passes from its bobbin over a set of spring-loaded pulleys to form a length reservoir and then through an exit guide. The bobbin is prevented from rotating by a pawl and ratchet. Yarn feeds from the reservoir or retracts into it by deflecting a spring-loaded pulley. When the reservoir length is depleted, yarn tension rises momentarily and trips the pawl to release the ratchet. New length is fed into the reservoir, and the action repeats.

The required retraction length depends on the type of braider and the variation of distance between the carriers and the convergence point. Typical retraction lengths range from 5 to 36 cm with this mechanism.

Process Limitations

Several factors limit the braiding process: the entire supply of braiding yarns must be moved; the machine size is large relative to the output product; only moderate aspect ratio parts can be made; the fiber orientation angle cannot be varied arbitrarily; and different machines are usually required to vary the braiding pattern. We discuss these limitations in turn.

Since a "free end" of each braider must be moved to create an interlaced pattern, the entire supply of each braider must be transported on a "floating" carrier that can interlace others. Moving a supply of yarn causes a

TABLE 1
Components of a Braid

Component	Geometry	Function
Braiding yarns	Angled to braid axis Interlaced with each other and axials	Interconnect structure Reinforce at braid angle Immobilize structure
Axial yarns	Parallel to braid axis Included within a braided layer	Reinforce at 0°
Core yarns	Parallel to braid axis In core or between braided layers	Reinforce at 0° Fill core
Core components	Mandrels (usually removable) Fillers (e.g., foam) Rigid parts (e.g., mechanical connectors)	Establish shape Stiffen structure

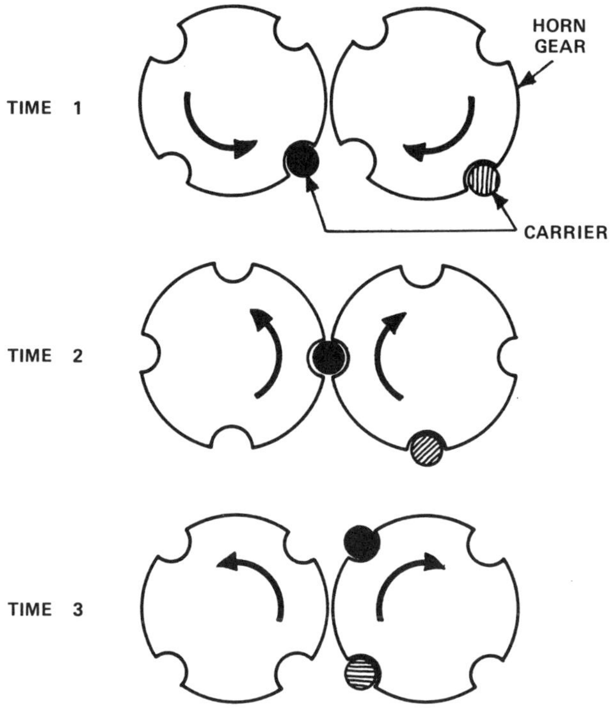

FIGURE 3 Schematic of carrier transfer action in conventional horn gear machine.

problem because a relatively large mass must be moved throughout the braiding plane. Very small bobbins and carriers would be desirable to operate at high speed with small machines, but they would have to be replenished too frequently.

Braiding machines tend to be large because the carriers move the braiding yarns through the braiding plane.

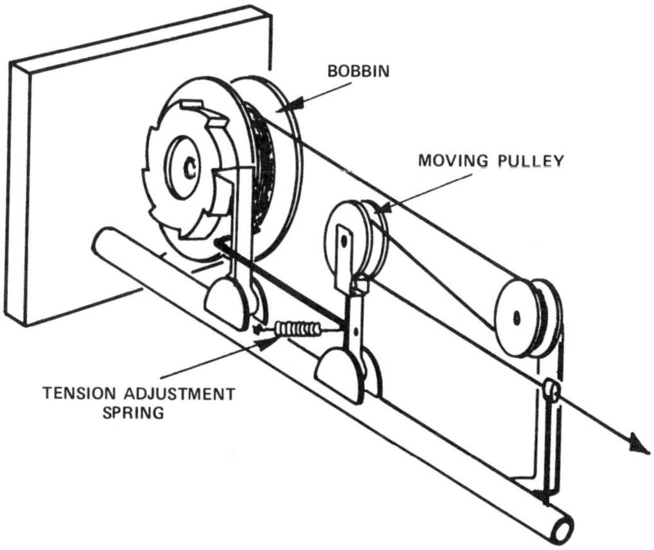

FIGURE 4 Schematic of carrier mechanism for feeding and retracting yarn at constant tension.

The yarns converge to form a fabric, and the machine is much larger than the product formed (the ratio of machine to part diameter may be as great at 50:1). By contrast, in weaving, the machine width is approximately equal to the sheet formed because the yarns feed through the loom in a parallel array.

Wide sheets cannot be braided readily because the yarns converge in all directions. The process is most suited to shapes with a moderate ratio of width or circumference to thickness.

The angle of the component yarns can be varied through a wide range. Braiding yarns can be inclined between about 5–85° from the axis, and both core and axial yarns can be inserted at 0°. However, in any given braid, the attainable angles depend on process parameters such as yarn width, braid diameter, and type of braid. This is discussed later.

The braiding patterns on most braiding machines can be varied only to a limited extent. For example, many circular machines can braid "2 up, 2 down" or "1 up, 1 down," circular of flat braids, and braids with one-quarter, one-half, or the maximum number of carriers. Different machines are needed to vary the array size and part dimensions.

Types of Braiding Machine

Braiding machines of a number of types have been developed over many years for textile and wire applications. A limited number of them have been adapted to make fabrics to reinforce composites. In addition, a number of entirely new machines, particularly ones for 3D braids, are being developed exclusively for composites. Table 2 describes the basic braiding machines and their key features. As in other textile processes, each type of braider can produce only a small fraction of the possible structures. The key machine characteristics are the number of carriers, the shape of the braiding array, the braiding patterns, and the accessibility for inserting axial yarns and core components.

Circular machines are most commonly used for composites applications because of their availability and versatility. They range in size from 16 to 144 carriers and produce fabrics from about 1 cm to 1 meter in diameter. The maximum production rate is about 1.5 meters/minute. Often operated in a horizontal mode (Fig. 5), the machines include a mechanism to drive a mandrel, which is overbraided in both directions.

Grouped carrier braiders, sometimes referred to as Wardwell or wire braiders, do not use horn gears and operate on a different principle. Rather, two groups of carriers rotate in opposite directions (Fig. 6). Yarns from the lower group are flipped over and under the upper carriers to form the braid. The relative simplicity of this process permits production rates about three times that of conventional circular braiders. This type of machinery is starting to be used for composites.

Horn gear machines have been made with extremely complicated arrays to form intricate braiding patterns. In Jacquard machines some of these patterns can be obtained by controlling the transfer of carriers between

TABLE 2
Types of Braiding Machines

Type and Description	Structures Formed	Application to Composites
Circular		
Carriers move in circular array in opposite directions Typically 144 carriers, maximum Maximum operating speed, 400 picks/minute Number of axials: half the maximum number of carriers	Tubes, flat strips, hollow contoured shapes Single or multiple layers Biaxial or triaxial Cores may be filled Typical range of yarn (18–76 picks/cm) Typical part of diameter range (1–100 cm)	Most commonly used equipment
Grouped Carrier		
Two groups of carriers, axially spaced, rotate in opposite directions; yarns from one group flip around others Maximum operating speed, typically 1200 picks/minute	Tubes	Starting to be used because of high operating speed
Jacquard		
Circular braider with programmable control of carrier switching between horn gears	Tubes Flat strips with complex lace patterns	None
Special Pattern Machines		
Machines with horn gears arranged in any of a number of complex arrays Gears may have different diameters Variable switching may be included	Wide range of interlacing patterns including braided holes	None
Solid Rope		
Carriers move in circular "loop the loop" pattern All carriers move in same rotational direction Typically 24 carriers, maximum	Cylindrical, solid ropes Core yarns can be added. Structure is not torsionally balanced since braiders rotate in same direction	None
Packing Braider		
Carriers move around square array of axials Maximum size typically 5 × 5 cm Inverted braiding into impregnation bath and forming die Maximum size array 4 × 4 Maximum number of carriers 52 Maximum number of axials 25 Maximum speed 230 picks/minute	Solid square cross section	Starting to be used for composites
3D		
Braiding yarns move relative to each other and axials to form integrated shape Several types of processes under development Rectangular and circular machines Maximum number of braiders: > 2000, circular; > 12,000, rectangular	Wide range of complex shapes including thick-walled beams, contoured tubes, bifurcations, and shapes with extensions	Developed exclusively for reinforcing composites
Hinged		
Circular braider in which output is fed back into core overbraided; machine is hinged to remove completed product	Endless belt	None

FIGURE 5 Horizontally mounted 144 carrier braider with mandrel advance system. (Photograph courtesy of Wardwell Company.)

gears. Machines have been made with on-line computer control for this purpose, but to date the technology has not been applied to composites.

Another type of machine not yet applied to composites is the hinged braider, which forms endless belts by refeeding the output into the core of the braid.

Many variations and combinations of machines have also been used, such as operating two machines in series. For composites applications, braiding machines can also be combined with other fiber placement processes such as pultrusion or filament winding.

FIGURE 6 Grouped carrier circular braider. (Photograph courtesy of Wardwell Company.)

Structural Features of Braided Composites

Braiding offers a major feature for composites reinforcement by its ability to arrange fibers in many different ways. The process can vary the microgeometry by orienting fibers at any angle in the composite plane and in some cases through its thickness. Braiding can also vary the macrogeometry and produce very varied parts, ranging from thin-walled tubes to solid shaped beams. Some important geometric restrictions do exist in current processes.

Microgeometry

The microgeometry of braided composites is characterized primarily by the angular orientation of the fibers and secondarily by their crimp. As Table 3 indicates, many microstructural variations can be achieved.

The biaxial braid, which is the most common, consists of two sets of yarns forming a symmetrical array with the yarns oriented at a fixed angle from the braid axis. Figure 7 shows an example of a commonly used "2 up, 2 down" pattern. This "Chinese finger puzzle" structure can be quite mobile and can be rearranged to map out many shapes. The yarns of this type of braid may develop substantial crimp, since each yarn is interlaced at many points. The level of crimp approaches zero if the yarn

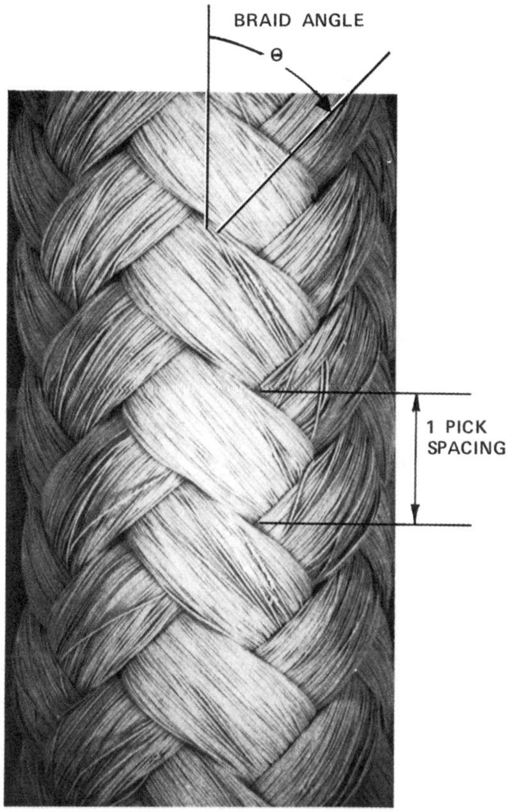

FIGURE 7 Microgeometry of conventional two-over-two biaxial circular braid.

TABLE 3
Microgeometry Variations

Property	Description	
Fiber orientation	Biaxial braids:	two angles ($\pm\theta$)
	Triaxial braids:	three angles ($\pm\theta$, 0) fully interconnected
	Core braids:	0° yarns added inside a braided layer
Multiple layers	Previously formed layers can be overbraided in forward or reverse direction	
Through-thickness reinforcement	3D braids include yarns that reinforce structure in all directions	
Hybrids	Different fibers can be used as braiders, axials, and core yarns	
Crimp	Fiber crimp can be adjusted by changing yarn cross section and braiding tension	

cross section is made very flat. In a triaxial braid, which consists of a biaxial braid with added axials, the yarns are locked together and geometric rearrangement is not possible after braiding. The axials lie within the fabric formed by the biaxial braiding yarns. They generally have little crimp and provide significant reinforcement to the final structure because they lie in the 0° direction.

To form a hybrid fabric, the axials can be made of fibers different from the braiders. Core yarns running in the axial direction can also be added to the structure for 0° reinforcement, but they will not be interlaced between the braiding yarns.

Multilayer fabrics can be formed by simply braiding back and forth or overlapping in the same direction. Each layer can be biaxial or triaxial. The fiber type and the braid angle can be varied as needed.

The braid angle can range from about 5° in an almost parallel yarn braid to approximately 85° in a "hoop" yarn braid. However, because of geometric limitations of yarn jamming and considerations of the fabric "cover," the angle that can be achieved in any case depends on the braid dimensions and the process parameters.

In biaxial braiding, the allowable angle range depends on the following factors: the number of carriers N_c, the braiding yarn width w, the diameter of the layer being braided D, and the cover factor F—which is defined as the fraction of the projected area covered by yarns. By simple geometric analysis of a repeating braid unit, the braid angle θ and the yarn volume fraction V_y can be computed and shown to be approximately:

$$V_y = 1 - \sqrt{1 - F}$$

$$\cos\theta = \frac{wN_c}{2\pi V_y D}$$

This relation, plotted in Figure 8, shows the braid angle versus the braided tube diameter, normalized by the

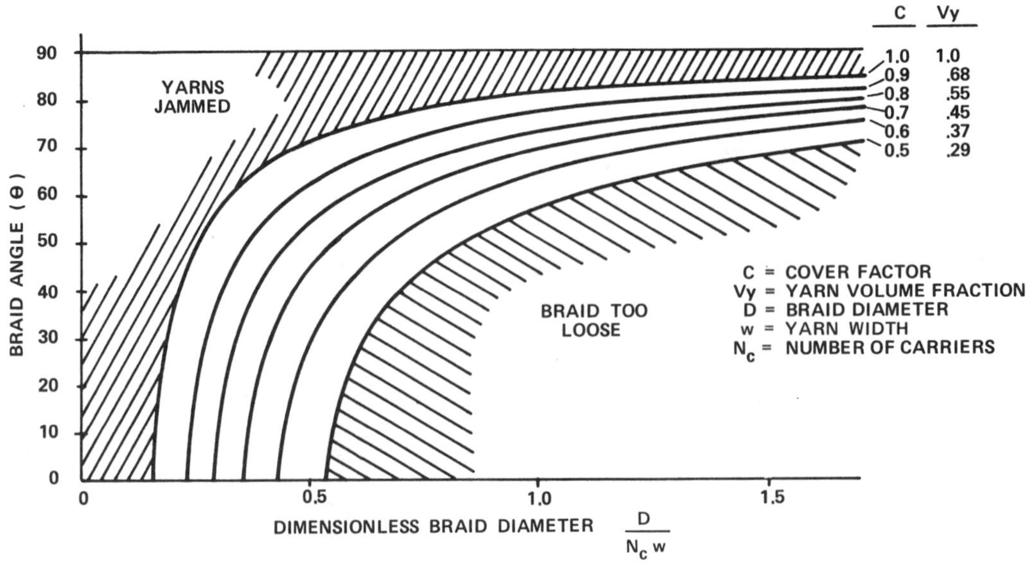

FIGURE 8 **Operating range of circular biaxial braider: braid angle versus diameter and other process parameters.**

yarn width and the number of carriers. Graphs of varying cover factor (0.5–1.0) are plotted.

The yarns jam against each other at a cover factor of 1.0 and the braid cannot be tightened beyond this point. This limit is indicated by the upper hatched region on the graph. At very low cover factors ($< \sim 50\%$), the braids become unstable and the fiber volume fraction drops to very low levels.

This result shows that only limited combinations of braid angle and diameter can be made on a given machine with a given yarn. For example, large diameter braids generally require the yarns to lie at high angles; otherwise the braid will be very open, with a low cover factor. If lower braid angles were required, wider yarns or a machine with more carriers could be used.

In parts with varying diameter along the length, it is often desirable to maintain constant cover factor. This can be accomplished by varying the braid angle with diameter according to Figure 8. Note that the microgeometry equations can be extended to noncircular braids, triaxial braids, 3D braids [30], or cases in which the yarn width varies.

Hybrid structures can be braided in several ways as shown earlier (Table 3). Yarns of different size or type fiber can readily be positioned in the varying braid components and used as braiders, axials, or core yarns. This makes it possible to adjust the properties or size of fibers at different angles.

Macrogeometry

In addition to microgeometric variations, braids can be made with a range of macrogeometric shapes. Braiding produces more varied structural shapes than any other preform process. Table 4 lists the principal variations possible. These near-net-shape preforms that can be braided greatly simplify composite parts manufacturing insofar as a single braided structure often replaces a layup of cut layers.

Tubes (Table 4) are the simplest and most common form of braid. They are typically collected in a flattened form and expanded during parts manufacture.

Contoured shapes are formed over mandrels, which may be removed after matrix application. In some cases, mandrels are made of structural foam and left inside the braid in the final part. In the example of a contoured shape being braided shown previously (Fig. 1), the braiding yarns easily cover a shaped mandrel as the braid converges. Asymmetric shapes can also be braided in the same way.

Bends can be braided by advancing the output material in a rotary manner and braiding more material on the outside of the bend than on the inside. Figure 9 shows an example of such a structure. The ability of the carriers to feed yarn "on demand" facilitates this action. Axial yarns added to a braided bend stabilize the structure and prevent it from being straightened.

Flat slabs or relatively narrow strips can be braided by circular or 3D methods. It is not possible to braid wide sheets with current processes because the yarns con-

TABLE 4
Macrogeometric Variations

Shape	Equipment or Method
Thin-Walled Structures	
Tubes	Circular braider
Flat strip	Modified circular braider
Contours	Circular braid over shaped mandrel
Thick-Walled Structures	
Solid rods	Solid rope braider; circular braid over core
Solid squares	Packing braiders
Solid shapes	3D braiders (various types)
Special Structure Features	
Bends	Pull forming braid along curved path
Holes	Overbraid mandrel with protruding pins to deflect yarns
Structural voids	Include removable components (such as elastomers) in core and/or axial positions
Bifurcations	In circular braid: pass one leg of structure through convergence zone In 3D braids: change braid pattern from single to multiple regions
Mobile structures	Braids that are not jammed and do not include axials can be re-formed to other shapes
Endless belt	Pass output into core and overbraid multiple layers
Included rigid components	Various mechanical elements can be added to core and overbraided (e.g., connectors, structural foam)

FIGURE 9 Braiding of bent shape for rocket motor component using prepreg yarns. (Photograph courtesy of Albany International Research Company.)

verge in all directions during formation. The maximum possible aspect ratio (width to thickness) is about 200 : 1.

Holes can be braided [2,3] into a structure as shown in Figure 10. Rounded pins inserted into a mandrel or through the forming braid deflect the braiding yarns and rearrange them to surround the hole. This procedure eliminates the need to drill a hole and cut fibers. In some cases, this offers significant property advantages. Holes can also be braided with special horn gear arrays and mechanical switches that reverse the carriers at the hole location.

Structural voids [4] can be formed by braiding materials such as elastomers into the core or axial positions of the structure (see Fig. 20). These components can be removed after the composite has been made rigid. The remaining voids offer improved specific properties by removing weight with only a moderate reduction in properties. In 3D fabrics, these structural voids are reinforced by the braiding yarns.

Bifurcations, or branched structures, can be braided on circular or 3D braiders. In circular braiders, a bifurcated mandrel is used. Individual branches of a bifurcated shape can be braided by passing the unused part of the mandrel through the convergence zone. It is then overbraided in a subsequent layer. Figure 11 shows examples of this type of structure. Bifurcations can also be formed by changing the pattern in 3D braiding from a single region to two unconnected ones.

Mobile structures result when 2D or 3D braids are made without axial yarns. These structures can readily bend to conform to varied shapes. Biaxial tubes, the simplest such structure, can also change cross-sectional shape and dimensions and even be folded back on themselves, as in Figure 12. This feature of braids permits bent shapes such as bicycle frames and tennis racquets to be formed from simple tubes.

Mechanical elements [5] such as metallic connectors can be overbraided, remaining in the structure after the matrix is added and consolidated. Windmill rotors offer an example of how interbraided connectors simplify part manufacture and form efficient connections.

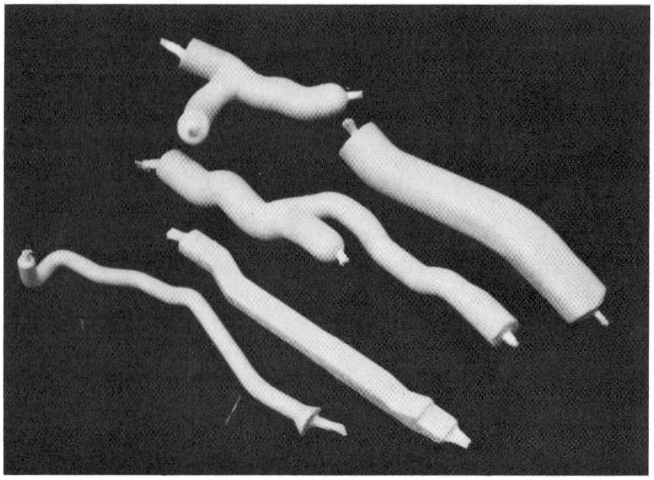

FIGURE 11 **Complex shaped parts with braided bends and bifurcations.** (Photograph courtesy of Fiber Innovations Inc.)

Endless belt braids are made by simply feeding the output of a braiding machine back into the core so that it can be overbraided. The length and number of layers can be adjusted as needed. (The end of the last layer does have to be secured.) To complete fabricating this structure, the braiding machine must open by a hinge mechanism to permit removal of the endless belt.

3D Braids

The newest type of braid designed especially for composites applications can be made as thick-walled complex shapes in which the braiding yarns interlock the entire structure. The 3D braids provide through-thickness reinforcement, and they form complex geometric shapes such as I-beams. In addition, the fibers of 3D braids are adequately interconnected, ensuring that the

FIGURE 10 **Direct braiding of holes using mandrel with rounded pins.** (Photograph courtesy of Albany International Research Company.)

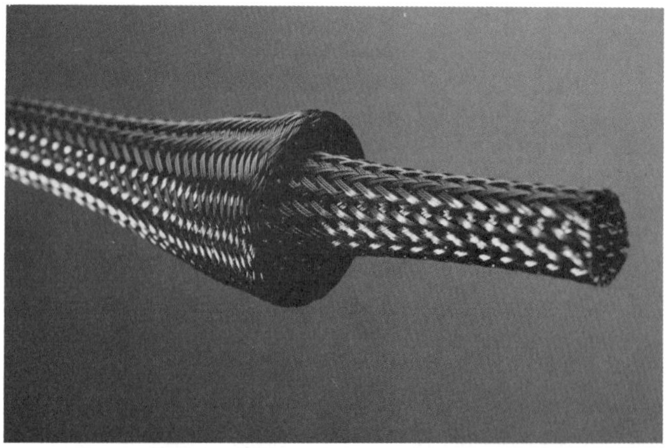

FIGURE 12 **Example of structural mobility: biaxial tube deformed into complex shape.**

structure is stable without any matrix. Other near-net-shape processes such as filament winding and pultrusion depend on the matrix to provide structural integrity.

Although solid braids of circular and square cross section have been produced for some time on commercially available machines, 3D braids in complex shapes are currently being made only on developmental or proprietary equipment. Much of this equipment relies on hand operations for some portion of the process, but great advances are being made toward full automation.

The generalized solid braiding machine shown in Figure 13 illustrates the formation of 3D braids. Two sets of yarns feed the process—axials (optional) and braiders. The axials pass through the array of braiding yarns. In the final structure, they lie inside the braid parallel to its axis. The braiding yarns move relative to each other and to the axials to interconnect the structure. Their final orientation is angled relative to the braid axis as in the case of biaxial braids, but the yarns also are angled relative to the thickness direction as they pass through the fabric.

A number of braiding patterns have been developed that form stable 3D structures (see Table 2). Some of these patterns can be made on automatic machinery with conventional horn gear and carrier mechanisms arranged in special arrays. However, most 3D patterns require entirely new mechanisms to drive the carriers.

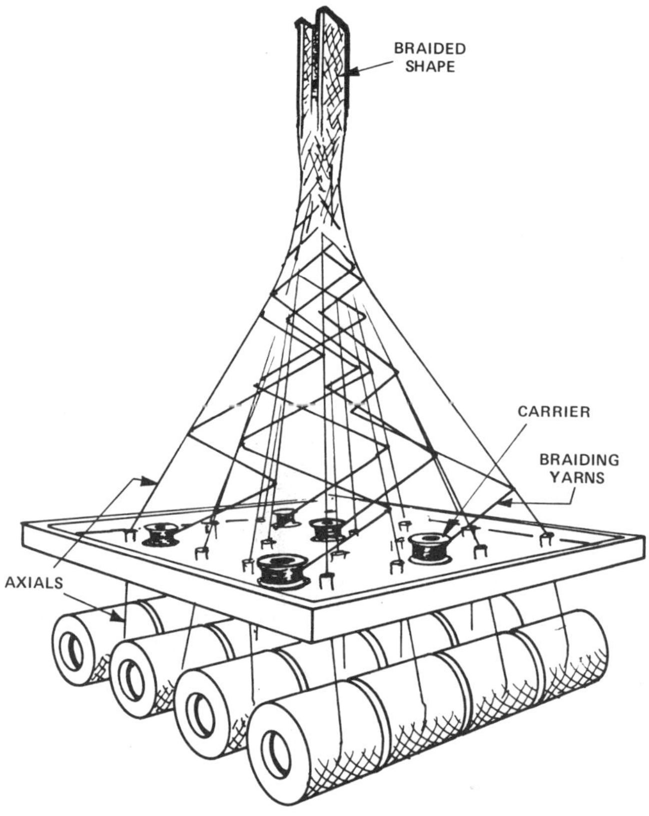

FIGURE 13 Schematic of generalized 3D braiding machine for making shaped part.

Traditional Solid Braids: Horn Gear Machines

Commercial horn gear machines are available to form solid braids with fully interconnected layers. Only circular and square cross sections can be braided. These machines use carriers with relatively long reservoir lengths to accommodate large differences in braider path lengths in the convergence zone.

Solid ropes are made on circular machines, which drive all the braiders in the same rotational direction. The braiders are interlaced by looping them around each other at one or more locations. This is done by moving the carriers around a horn gear positioned inside a circular array of gears. The resulting structure is highly interlaced and has a smooth surface. Its cross-sectional shape is very stable. Since all the carriers move in the same direction, this braid is not torsionally balanced when tensioned (most ropes are not).

Packing braids have solid square cross sections and are currently made on machines with square arrays of up to 16 horn gears. Carriers are passed from one horn gear to the other to move the braiders diagonally through the cross section. Axial yarns are fed into all available locations within and between the horn gears (up to 25). Their mass distribution is usually arranged to create a square cross section with well-defined corners.

Packing braiders, such as the one shown in Figure 14, braid downward; thus the yarns can be impregnated in the convergence zone by passing them through a bath and then a forming die—which may be heated. In use, these packings often come in contact with rotating shafts, which wear through the braid. To maximize life, they are typically constructed with a 70 : 30 weight ratio of braiders to axials.

3D Braids for Complex Shapes: Machines Without Horn Gears

A number of unique braiding [4,5] patterns have been developed to create complex-shaped preforms in which the braiders provide through-thickness reinforcement. In some cases large numbers of braiders are used—up to 12,000 reported [13]. Totally new braiding equipment developed for this purpose does not use horn gears or conventional carriers.

Of the 3D braids discussed in the technical and patent literature, two are described here: the row-and-column and the two-step methods. Other 3D braids include the AYPEX process [6] and the Interlock Twiner [7]. In the AYPEX process, braiders are transported through the section by a mechanism that exchanges bobbins on two adjacent locations. In the Interlock Twiner braider, a grouped carrier process forms a braid in which yarn bundles are entwined by braiders to form a specified shape.

Row-and-Column Braids

The row-and-column type of 3D braid [10–15] has been referred to by several names: Omniweave, Magnaweave, through-the-thickness braids, and four-step braids. It

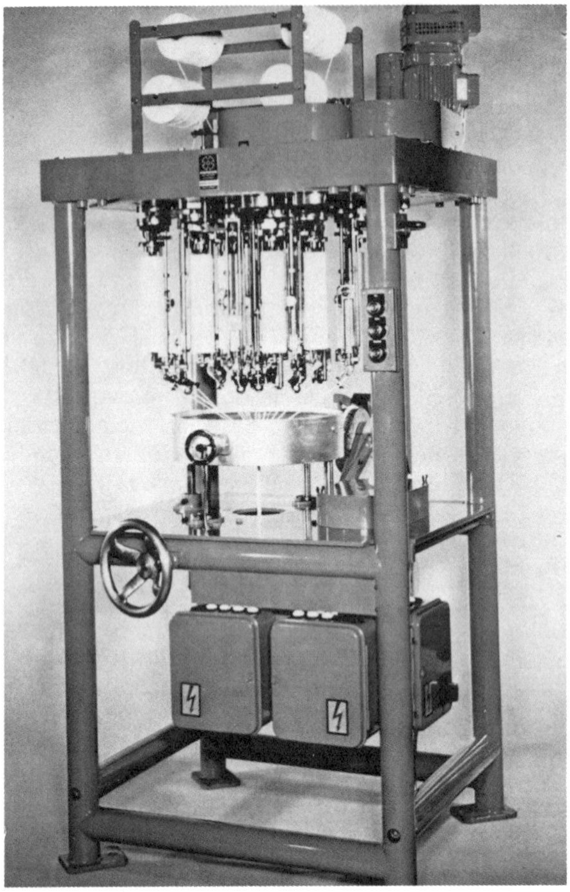

FIGURE 14 **Packing braider for braiding solid squares. Note inverted braiding, axial yarn supply, impregnation bath.** (Photograph courtesy of Herzog Company.)

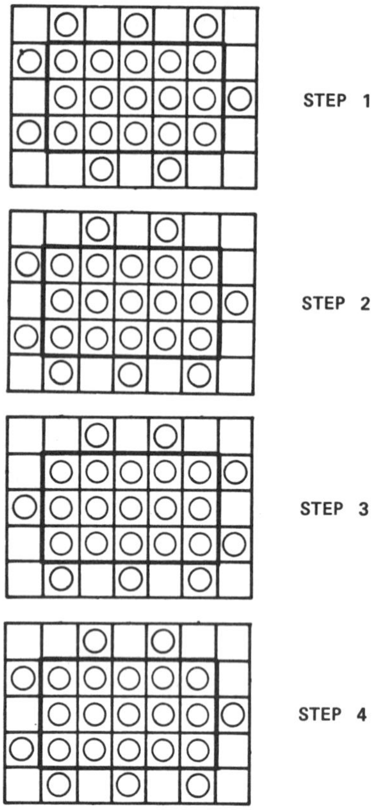

FIGURE 15 **Braiding pattern for 3D shapes using row-and-column process.**

has been under development longer than any other 3D braid and is being manufactured on proprietary and developmental equipment.

The structure can be described by referring to Figure 15, which shows the braiding pattern for a rectangular slab. The braiders are arranged in rows and columns to form the required shape; then additional braiders are added to the outside of the array in alternating locations. The exact location and quantity of braiders lying outside the main array depends on the exact shape and structure required [16,17].

Four steps of motion are then imposed on the braiders by the machine as shown. In the first, the columns are moved vertically in alternating directions (odd columns down, even ones up). In the second, the rows are moved in alternating directions. The remaining two steps repeat the vertical and horizontal moves in reverse. The yarns are mechanically "beat" into the structure to tighten the fabric between each step (as in weaving). As these steps are continued, each braider moves throughout the cross section in a complex path. This basic row-and-column braid is referred to as a 1 × 1 × 1.

The braiding pattern may be altered in various ways to produce structural shapes. For example, a complex shape can be braided by arranging the braiders in the shape of an I-beam. The basic four steps are repeated, but selected rows or columns are not moved. Special programs have been written [16] to determine the specific braid pattern needed for a given shape.

Axial yarns can be included in the braid by limiting the motion of selected braiders. They do not move through the cross section but remain fixed in one location. In another variation, additional braiders are added outside the basic array, and the rows and columns are moved by more than one unit.

Thick-walled tubular structures can also be made by this braiding method. The rows and columns are replaced by an array of radial and circumferential braiders. The braiding pattern is similar to that of rectangular arrays, but in some cases, fewer radial rows are used in the low radius regions to render more uniform the number of yarns per area.

As an example of the complex shapes possible [12–14], circular and rectangular braiding were combined to form a ship propeller by R. Brown at Atlantic Research Corporation. A thick cylindrical hub was braided with some yarns protruding from the surface. These yarns were later braided by a rectangular-type pattern to form the blades.

Machinery is being developed to make these braids. Figure 16 shows a rectangular braider making a thick Kevlar slab. This machine, although not yet fully automated, can braid a wide range of shapes such as beams

FIGURE 16 3D Braider for rectangular shapes. (Photograph courtesy of Atlantic Research Corporation.)

of arbitrary cross section. Lengthwise variations can be introduced by changing the braiding pattern or the number of braiders. Bobbins are not used to supply yarn. Rather, a fixed length of braiding yarns is strung up and braided. Elastomeric strings attach each yarn to small cylinders, which are moved by the machine. The elastomeric strings perform the carrier function of taking up yarn slack when needed. New carriers suitable for this process are being developed.

Figure 17 shows an automatic [14] circular version of this process developed by Atlantic Research Corporation. This configuration reduces machine size by moving the carriers on a surface parallel to the braiding direction. The carriers move on the inside of a cylindrical surface, rather than in a flat braiding plane. They move circumferentially and longitudinally on the cylindrical surface rather than circumferentially and radially in the plane.

Two-Step Braids

The two-step method of 3D braiding [4,18] differs from the row-and-column process in several ways. The interlacing pattern is different, and the structure includes a larger number of axials for efficient reinforcement and a smaller number of braiders for ease of manufacture.

The braiding pattern described in Figure 18 shows the relative position of all yarns at one location along the braid. The axials are indicated by solid circles and the braiders by open circles. The braiders are distributed in alternating spaces around the perimeter of an array of axial yarns as shown. The basic process consists of two steps in which the braiders move completely through the structure between the axials. To understand direction and extent of motion of each carrier, refer to the figure. As the two machine steps are repeated, the braiders completely intercinch the axial yarns after several cycles.

Surprisingly, this relatively simple sequence of braiding moves can form parts of essentially any shape, including hollow and circular. In each case, the axial array fills the required cross section and the braiders follow the same two steps. Figures 19 and 20 show examples of braids made by this process—a solid T-beam preform and a T-beam with structural voids.

The relatively simple geometry of two-step braids permits an important process simplification. The braiding yarns pass straight through the structure in diagonal

FIGURE 17 Automatic 3D braider for circular shapes. Note that carriers move parallel to braid direction. (Photograph courtesy of Atlantic Research Corporation.)

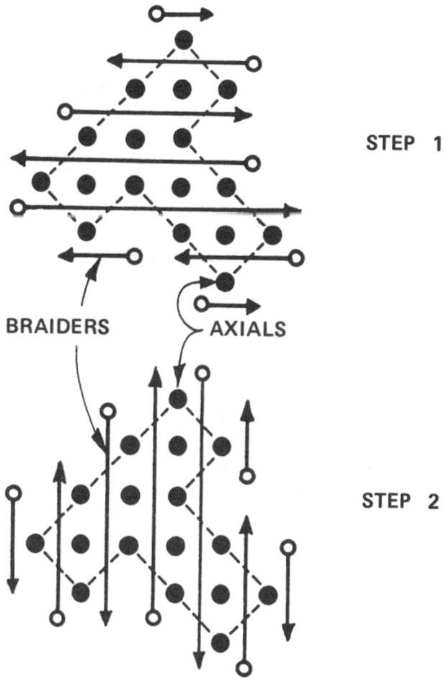

FIGURE 18 Braiding pattern for 3D shape using two-step process.

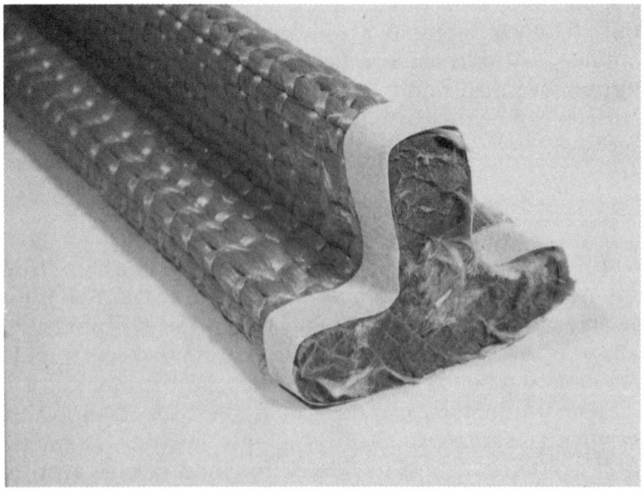

FIGURE 19 Braided T-beam preform made by two-step process. (Photograph courtesy of E. I. du Pont de Nemours and Company, Inc.)

paths and lie completely outside the axial array after each step. This permits the braid to be pulled tight by the yarn tension alone [19]. A "beat up" action, which would require mechanical elements to pack in the yarns, can be avoided. Having the braiders simultaneously outside the array also creates the possibility of adding various inserts to the structure and rearranging the axial array geometry.

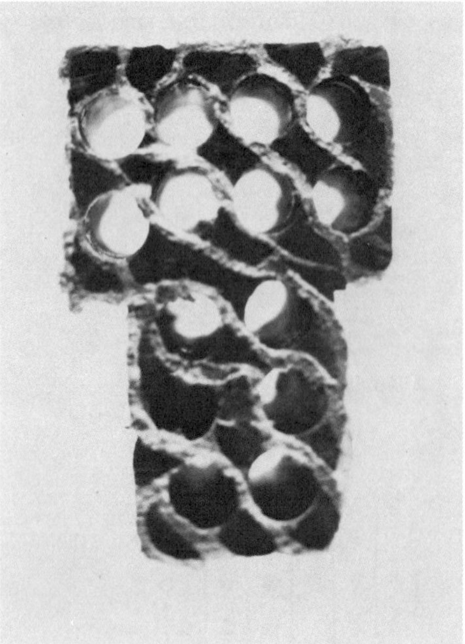

FIGURE 20 Section of braided T-beam with structural voids made by two-step process. (Photograph courtesy of E. I. du Pont de Nemours and Company, Inc.)

The microgeometries of the 3D braiding processes [17,19,30] discussed differ in fiber orientation angles. In typical two-step patterns, the braiders lie between 40 and 70° from the axials and pass through the thickness at about 45° from the surface. Row-and-column braids typically have corresponding angles of 20 and 5°.

Methods of Forming Composites from Braids

Traditional methods of consolidating composites can be applied to braided structures. The fibers must be impregnated or combined with a matrix material and then consolidated into a rigid structure. A number of such methods have been developed using thermosetting and thermoplastic polymers, metals, and ceramics.

In addition to conventional methods, several new systems have evolved to combine the matrix addition or consolidation process with braiding. Matrix can be added to the fibers during braiding, and the structure can be consolidated or made rigid in-line. A nearly completed part can be formed by this coupled process with both economic and technical benefits.

Application of Thermosetting Matrix

The simplest method of forming a braided composite with a thermosetting matrix involves the use of prepreg yarns (yarns impregnated with a partially cured polymer). The prepreg must be made with a nontacky surface, to permit the yarns to slide over each other in the convergence zone of the braiding machine. Prepregs for other applications such as laminates are made tacky, so that the structure holds together during assembly.

After braiding, the structure is placed over appropriate tooling to combine different components and form the required part geometry. Mobile biaxial braids are often bent or reshaped at this point. Heat and pressure may be applied by various methods to consolidate the structure: molds that apply internal pressure through a bladder, matched molds, or autoclaves. The matrix flows throughout the fibers in the braid and cures by the imposed heat and pressure. The bent tube shown earlier (Fig. 9) is being braided with prepreg yarns.

Resin transfer molding [5,20] can also be used to add a thermosetting matrix to a "bare fiber" braided preform. An assembly of these preforms and other components, such as metal connectors, is placed in a mold and resin is pumped in to fill the fiber interstices. Various techniques, including vacuum, are used to enable the resin to fully surround the fibers. The mold is generally heated to cure the matrix.

Thermosetting resins can also be applied during braiding in several ways. Using an inverted braider, such as a packing braider, the resin can be placed in a bath that covers the last portion of the convergence zone. The impregnated structure then passes through a heated die. This system is not now being used commercially for composites.

Another method of applying thermosetting resin during braiding was developed by U.S. Composites Corporation [21,22] (Fig. 21). In this process, the yarns pass over a contoured surface located in the convergence zone that exudes resin at a metered rate. The yarns pick up resin as they wipe this surface and slide over each other. The braid is formed wet, and the resin flows throughout the fibers to fill the voids.

Application of Thermoplastic Resin

Thermoplastic resins are usually combined with the fibers before braiding. Resin transfer techniques have not been developed because of the high melt viscosity. After braiding, or in some cases during braiding, the matrix is heated and flows between the fibers.

One method of combining the thermoplastic matrix with the reinforcing fibers before braiding uses "commingled" thermoplastic fibers [29]. These yarns can be braided into required shapes and then heated under pressure in a mold or high pressure autoclave. The thermoplastic fibers melt and flow around the reinforcing fibers to form a composite. Melt-impregnated and powder-impregnated yarns can also be processed in this way.

Yarns that contain thermoplastic matrices can be consolidated during braiding. By heating the yarns in the convergence zone, the matrix can be forced to flow between the fibers. Yarn tension develops pressure between the fibers and assists the resin flow. This manufacturing method, done on a limited basis to date, offers a means of forming completed parts on a braiding machine.

FIGURE 21 **On-line application of thermosetting resin during braiding.** (Photograph courtesy of U.S. Composites Corporation.)

Mechanical Properties of Braided Composites

Braided composites provide two major mechanical features. They resist delamination, and they can be used to reinforce shaped parts efficiently. By forming shapes directly, braids eliminate the need for cutting fibers in joints, splices, or overlaps and the associated local strength loss.

The in-plane mechanical properties of most braided composites are approximately equal to those of laminates with the same angle distribution and fiber volume fraction. For example, the in-plane properties of a biaxial braid are close to those of a similarly constructed laminate of undirectional fibers. Fiber crimp due to interlacing may reduce properties somewhat, as it does in woven fabric reinforcements.

The data base of braided composite properties is far from complete, particularly with regard to specially shaped structures. An excellent review of available information has been prepared by Ko [23], some of whose data are included here.

In-Plane Properties

A typical circular braid includes multiple layers of biaxial or triaxial material. Each braided layer consists of fibers oriented at either two or three different angles. The properties of the assembly can be computed by conventional laminate theory, with each fiber angle modeled as a separate layer. That is, a biaxial layer is modeled by two angled layers and a triaxial one by two angled layers and one 0° layer.

Table 5 gives the key properties of circular braids made of different materials as reported by several investigators. In the first example, the measured properties are compared with values computed by laminate theory. The relatively good agreement indicates that braids can be engineered as laminates to achieve required properties.

The data also show that properties can be varied through a wide range by using different fibers and orientations. Biaxial tubes with high braid angles are more than 10 times stronger in the hoop than in the longitudinal direction; and triaxial tubes with moderate angles show the reverse anisotropy. In most cases, compressive strength is lower than tensile strength, as in laminates.

Table 6 gives the in-plane properties of 3D braids. As in Table 5, the 3D braids perform as equal angle laminates. Additional data included in the study by Gausse and Alper [27] but not appearing in these tables show that in both tensile and compressive strength, 3D braids are the same as equal-angle laminates.

An important limiting factor in both circular and 3D braids is the restricted angle distribution described earlier. Although the properties of high performance lami-

TABLE 5
Properties of Circular Braid Composites

Material	Structure	Fiber Volume Fraction (%)	Braid Angle	Strength (MPa)[a] Longitudinal	Strength (MPa)[a] Loop	Strength (MPa)[a] In-Plane Shear	Modulus (GPa)[a] Longitudinal	Modulus (GPa)[a] Hoop	Modulus (GPa)[a] In-Plane Shear	Ref.
Kevlar–epoxy	Biaxial braid	55	(46/46/15/15/46/46)	347.8	96.6	198.0	21.4	11.0	13.8	24
	Laminate (computed)	55	(46/46/15/15/46/46)	276.0		95.9	20.7	11.7	14.5	
S-2 Fiberglass–epoxy	Biaxial braid	75	78		731/276					25
		75	82/50		1028/331					
		75	86/75	82.8/96.6	1256/379					
		75	89	20.7/221	1325/704					
Graphite–epoxy	Triaxial braid	33.8	0/45				61.4/62.8	6.76/−		2
		29.3	0/63				49.0/49.7	15.2/−		
		56.3	0/80				−/52.4	43.6/−		

[a] Tension/compression values.

TABLE 6
In-Plane Properties of 3D Braided Composites

Material	Structure	Fiber Volume Fraction (%)	Braid Angle	Strength (MPa)[a] Longitudinal	Strength (MPa)[a] Transverse	Modulus (GPa)[a] Longitudinal	Modulus (GPa)[a] Transverse	Modulus (GPa)[a] In-Plane Shear	Poisson's Ratio	Ref.
Graphite–epoxy	Row-and-column braid (1 × 1 × 1)	59	± 18	980–428	34.5	90.3/75.8	10.3	12.9	1.06	27
	Laminate (computed)	59	± 18			85.5	11.0	16.5	1.15	
	Row-and-column braid	58	± 24/0	750/473	22.8	106/93.1	9.7	12.7	0.81	
	(1 × 1) ½ fixed laminate (computed)	58	± 24/0			108.9	11.0	10.3	0.73	
	Control laminate	60	$(\pm45/0_2/\pm45/0_2\pm45/0/90)_S$	910/420	416	65.5/60.7	31.0	21.0	0.42	
	Laminate (computed)	60	$(\pm45/0_2/\pm45/0_2\pm45/0/90)_S$			66.2	31.7	17.9	0.48	
Graphite–Kevlar–epoxy	Two-step	69 (40 axials 29 braiders)	± 55/0	533		32				28
		69 (60 axials 9 braiders)	± 55/0	640		76			0.54	

[a] Tension/compression values.

TABLE 7
Impact Resistance of 3D Braided Composites

Material	Structure	Fiber Volume Fraction (%)	Braid Angle	Energy Absorbed (Js) First Damage	Energy Absorbed (Js) Maximum-Load	Energy Absorbed (Js) Through Penetration	Damaged Area (mm^2) At 13.6 J	Damaged Area (mm^2) At 150 J	Damaged Area (mm^2) At 381 J	Ref.
Graphite–epoxy	Row-and-column braid (1 × 1 × 1)	59	± 18	2.5	15.1	65.4	1290	2516		27
	Row-and-column braid (1 × 1) ½ fixed	58	± 24/0	4.4	12.0	51.1	2129	2322		
	Control laminate	60	$(\pm45/0_2/\pm45/0_2\pm45/0/90)_S$	2.3	17.8	72.6	3225	5547		
Graphite–PEEK	Row-and-column braid (1 × 1 × 1)		± 20					65	645	29
	Laminate							1032	6128	
FP(Alumina)Al-Li	Row-and-column braid (1 × 1 × 1)	17	± 20	66.5	266.6					
	Laminate	34	0	14.3	127.2					15

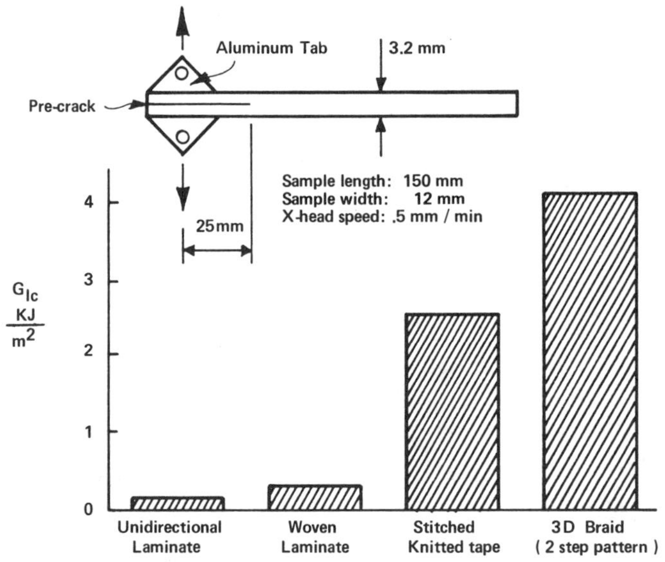

FIGURE 22 Effect of 3D reinforcement on mode I fracture toughness.

nates can be achieved, the range of attainable braid angles depends on the processing equipment and the braiding pattern.

Delamination Resistance

Circular braids resist delamination within a single layer because the braiding and axial yarns are interlaced. Although a braided layer may include up to three "sublayers" of fibers at different angles, it is almost impossible to separate them. Multiple layers, however, perform the same as conventional laminates. On the other hand, 3D braids are fully interconnected by braiding yarns, which reinforce the structure in the thickness direction. Even small amounts of this reinforcement increase delamination resistance dramatically by limiting cracks to propagation through the matrix.

Delamination resistance can be measured by a number of loading modes that separate the layers. The performance can be measured by the loads required to cause the delamination or by the extent of damage caused by a given loading.

Table 7 shows the results of impact tests on 3D braids and laminate controls, of very different types, as measured by various investigators. This table includes graphite with both thermosetting and thermoplastic matrices and ceramic fibers with metal matrices. The results compare energy levels and damage areas.

For the thermosetting matrix (epoxy), approximately equal amounts of energy are required to damage or fail braids and laminates, but the braid damage is confined to much smaller areas. In a thermoplastic matrix [29] study using polyether ether ketone (PEEK), the braids absorbed more energy and limited the damage area by a factor of 10. In a metal matrix study, the braids were able to absorb considerably more energy than unidirectional laminates. The available impact data strongly suggest that through-thickness reinforcement of 3D braids limits internal delaminations.

Figure 22 shows the results of mode I delamination tests [26] for 3D braids and related structures. The loading mode, shown schematically, directly tests the ability of through-thickness fibers to retard crack growth. The fracture toughness reported was determined by means of conventional techniques and equals the energy required to fail a given surface area.

The 3D braid, made by the two-step process, clearly resists this delamination better than laminates made of unidirectional tapes or fabrics. It also performs better than a stitched laminate, possibly because sewing damages fibers.

The level of improvement in delamination resistance offered by 3D depends on the matrix properties. The benefit appears to increase as the fracture toughness of the matrix decreases.

Special Properties

The difference in the properties of braids and other composites may be more pronounced in shaped parts than in test coupons. For example, the properties of holes in

TABLE 8
Effect of Holes on Properties of Braided Composites

Material	Structure	Fiber Volume Fraction (%)	Braid Angle	Open Hole Strength (MPa)[a]	Bearing Stress (MPa)[a]	Strain Ratio at Hole (Hoop to Average) 0°	45°	90°	Ref.
Graphite–epoxy	Biaxial braid, braided hole	50	± 45		447/–	5.3	4.1	3.2	3
	Biaxial braid, machined hole		± 45		252/–	7.1	5.8	11.8	
Graphite–epoxy	Row-and-column braid (1 × 1 × 1)	59	± 18						27
	Machined hole control (no hole)			881/418 980/428	245/373 980/428				
	Row-and-column braid (1 × 1) ½ fixed	58	± 24/0						
	Machined hole control (no hole)			863/422 750/473	274/396 750/473				
	Control laminate machined hole control (no hole)	60	$(\pm45/02/\pm45/0_2/\pm45/0/90)_s$	593/536 910/420	677/577 910/420				

[a] Tensile/compressive strength.

circular and 3D braided composites indicate that braids minimize the structural weakening usually associated with this geometric feature.

Table 8 shows the difference between braided and machined holes in biaxial braids. Brookstein [3] found that the bearing stress increases by a large factor when the holes are braided rather than machined. This probably occurs because the braiding fibers provide extra support around the hole and reduce the hoop strain as shown.

Measurements on machined holes by Gausse and Alper [27] indicate that 3D braids have lower bearing strength than laminates, but they are more resistant to open holes.

Applications and Future Trends

Current applications for braided composites include an increasing number of parts for a variety of markets: recreation, industrial, medical, automotive, military, and aerospace. In most cases the braid enables the inclusion of some geometric feature that could not be obtained with equal performance or cost effectiveness by another composites manufacturing process. Table 9 gives representative examples of braiding applications currently being manufactured or developed.

The braiding machinery now used for composites includes conventional circular braiders (often horizontally mounted) and several types of proprietary 3D braider. As discussed previously, braiders are likely to see heavier use in noncomposites applications. For example, new "grouped carrier" machines are being developed for high speed braiding. Machines for 3D braiding, often operated with extensive manual assistance, are being automated and equipped with special high retraction carrier systems. This equipment is being developed exclusively for composites applications. It offers the potential of fully automated production of a wide range of thick- or thin-walled complex shapes for sophisticated applications such as rocket nozzles, orthopedic devices, and shaped beams.

Many braiding machines are now being operated under computer control. The rate of mandrel advance can be programmed and electronically linked to the braiding rate. This permits a shaped structure to be braided with a predetermined angle or cover factor. Multiple layers can be made with automatic reversals in braiding direction.

The trend of combining braiding with other fiber placement and matrix-related operations is likely to continue. For example, combined braiding and pultrusion, now being researched, offers the possibility of making rigid rods and beams with specified angle distributions in a single process. Combined braiding and filament winding can result in shaped parts that cannot be made by either process alone. Braiding provides a means of making complex shapes, and filament winding provides a convenient means of adding circumferential fibers. Combinations of matrix addition or matrix consolidation with braiding, as discussed previously, can produce fully rigid, shaped parts in a single operation.

Peter Popper

TABLE 9
Examples of Applications in Use or Under Development for Braided Composites

Aerospace	**Industrial**
Ducting	Fasteners
Rocket nozzles	Connecting rod linkages
Rocket motor components	Windmill rotors
Missile rings	Electrical components
Helicopter rotor spars	Pressure vessels
	Driveshafts
Recreational	Structural members
Baseball bats	
Golf clubs	**Medical**
Sailboard masts	Prosthetic joints
Tennis racquets	Artificial limbs
Bicycles	
Skis	**Automotive**
Racing sculls	Racing car bodies
	Connecting rods

References

1. H. Reiligmann, *Textile Technology—Machine Braids,* All-Textile Employers Association, Frankfort, KY, 1981.
2. T. Tsiang, D. S. Brookstein, and J. Dent, *Proceedings of the 29th SAMPE Natl. Symposium,* 1984, pp. 880–890.
3. D. S. Brookstein, *Compos. Struct., 6,* 87–94 (1986).
4. P. Popper and R. McConnell, *Proceedings of the 32nd International SAMPE Symposium,* 1987, pp. 92–103.
5. R. J. Post, *Proceedings of the 22nd SAMPE Natl. Symposium,* 1977, pp. 486–503.
6. R. D. Weller, "AYPEX: A New Method of Composite Reinforcement Braiding," 3-D Composite Materials, NASA Conference Publication 2420, November 1985.
7. P. M. Cole, "Three-Dimensional Structures of Interlocked Strands," U.S. Patent 4,737,399.
8. R. M. Blunk, "High Speed Bias Weaving," U.S. Patent 3,426,804 (February 1969).
9. M. A. Maistre, *AIAA/SAE 11th Propulsion Conference Proceedings,* 1975.
10. R. Florentine, "Apparatus for Weaving a Three-Dimensional Article," U.S. Patent 4,312,261 (January 1982).
11. E. R. Stover, W. C. Mark, I. Markowitz, and W. Mueller, "Preparation of an Ominweave-Reinforced Carbon–Carbon Cylinder as a Candidate for Evaluation in the Advanced Heat Shield Screening Program," AFML-TR-70-283, March 1971.
12. R. T. Brown, *Proceedings of the 30th SAMPE Symposium,* 1985, pp. 1509–1518.
13. R. T. Brown and M. E. Harman, "Advanced Textile Braiding Techniques," paper presented at Conference on Advances in High Performance Composite Technology, Clemson University, Clemson, SC, 1987.

14. R. T. Brown, "Automation of 3D Braiding Devices," FIBER-TEX 87, Greenville, SC, November 1987.

15. F. K. Ko, D. W. Whyte, and C. M. Pastore, "Control of Fiber Architecture for Tough Net-Shaped Structural Composites," MiCon 86, in ASTM STP 979, 1988, pp. 290–298.

16. C. M. Pastore, "A Processing Science Model for Three-Dimensional Braiding," Ph.D. thesis, Drexel University, Philadelphia, 1988.

17. W. Li, T. Kang, and A. El Shiekh, "Structural Mechanics of 3D Braided Preforms for Composites. Part 1. Geometry of Fabric Produced by 4-Step Process," NASA Conference Publication 3001, FIBER-TEX 87, Greenville SC, November 1987.

18. P. Popper and R. F. McConnell, "Complex Shaped Braided Structures," U.S. Patent 4,719,837 (January 1988).

19. W. Li and A. El Shiekh, *SAMPE Q., 19*(4), 22 (1988).

20. G. C. Sharpless, *Proceedings of the Third Textile Structural Composites Symposium,* Fibrous Materials Laboratory, Drexel University, June 1988.

21. A. H. Kruesi, "On-Line Resin Impregnation," U.S. Patent 4,494,436.

22. A. H. Kruesi and G. H. Hasko, *Proceedings of the 32nd SAMPE Int. Symposium,* 1987, pp. 309–317.

23. F. K. Ko, "Braiding," in *Engineered Materials Handbook,* ASM International, Metals Park, OH, 1987, pp. 519–528.

24. M. L. White, *Proceedings of the Rotor System Design Conference,* American Helicopter Society, 1980.

25. D. E. Flinchbaugh, *Proceedings of the Composites Materials Conference,* August 1985.

26. H. B. Soebroto, *Proceedings of the Third Textile Structural Composites Symposium,* Fibrous Materials Lab, Drexel University, June 1988.

27. L. W. Gausse and J. M. Alper, *J. Compos. Techn. Res.,* 9(4), 141 (1987).

28. F. K. Ko, H. B. Soebroto, and C. Lei, *Proceedings of the 33rd SAMPE Int. Symposium,* 1988, pp. 912–921.

29. F. K. Ko, H. Chou, and E. Ying, *Proceedings of the Advanced Composites Conference,* 1986.

30. G. W. Du, T. W. Chou, and P. Popper, *Proceedings of the FIBER-TEX 88,* 1988.

C

CAD/CAM/CAE

Computer-aided design and computer-aided manufacturing (CAD/CAM) are a family of related technologies. A few years ago, they were seen as a means to increase efficiency in design and manufacturing activities. System purchases were often based on predictions of improvements in efficiency. As the technologies grew, efficiency became secondary in many organizations. Expansions of successful system installations tended to be justified on the basis of increased effectiveness. Improvements in quality, product cycles, and manufacturing unit costs were pursued. Often, efficiencies at the design stage were consciously sacrificed in favor of the larger goals. And it is at the earliest moments of the design stage that maximum leverage can be gained through critical design selection. Key design directions are made at that early phase in the design process which determine final reliability, cost, and competitive factors.

Thus while efficiencies were made in the drafting room through increasingly effective drawing production methods, similar computer-aided analytical tools were being developed to assist the engineer in design selection and optimization. This technology became known collectively as computer-aided engineering (CAE). Its development and proliferation have been, until recently, relatively slow because of the requirement for exceptionally fast hardware systems and for sophisticated ease-of-use software designs.

Now, as the technologies mature further, the idea of using them to change the ways we do business is taking hold. CAD/CAM/CAE is emerging as a means to make businesses more competitive, often in ways more complex and profound than improvements in efficiency and effectiveness. The meaning of the term is also changing as overlaps and linkages among technologies come into play. This article explores the application of CAD/CAM/CAE technologies for achieving efficiency, effectiveness, and competitive advantage, as well as the shifting definition of the term, CAD/CAM/CAE.

Terminology

The "CAD" in "CAD/CAM" is sometimes understood as "computer-aided drafting" and sometimes "computer-aided design." Increasingly, CAD is heard as a word, somewhat disassociated from being an acronym. It then tends to encompass the broad use of computers to improve the design engineering process, from concept to delivery. However, the two phrases mix terms of very different scope. "Drafting" is the producing of drawings; "design" invokes creativity during the ordering of information. Drafting is the most common way of giving form to designs, in the same manner as typing is the most common way of giving form to writing. Drafting systems for most engineering disciplines are similar; they must all permit the rapid production of drawings. Electronic, mechanical, and architectural forms of drafting demand lines and curves, text and symbols, panning and zooming, and so on.

Computer-aided "design" systems are application-oriented. They tend to be available to designers on an "as needed" basis. These systems combine the rapid graphic capture of concepts in a way that permits a variety of analytical techniques to be readily applied. Their forte is early optimization of design. Their output either is used directly or is passed on to separate drafting systems for drawing production.

The "CAM" of CAD/CAM is sometimes narrowly defined to mean "computer-aided machining," the link between a design computer and a numerically controlled cutting machine. More often, it is more broadly defined as "computer-aided manufacturing," the automation of the factory floor. The boundary between computer-aided manufacturing (CAM) and computer-integrated manufacturing (CIM) is not clear. It is generally accepted that CAM is an element of CIM, the link between modern design engineering and manufacturing.

The term "CAE" lacks a clear lineage, as do the other terms. To some, it applies to sophisticated analytical techniques directed toward a specific part being designed. Originating in the electronic (printed circuit board and integrated circuit chip) design field and recently offering applications in mechanical design, this category splits into ECAE and MCAE for electronic and mechanical design systems, respectively. Systems currently available on the market marry this technology to the "design"-oriented CAD system, thereby providing a powerful modeling and analysis tool to the engineer/designer, who can employ it at the inception of the design process.

A larger, broader interpretation of the term gathers all automated functions occurring within engineering. In this sense, "CAE" includes all three subcategories: drafting, design, and analysis . . . and goes on to include artificial intelligence, software development, graphic visualization of scientific processes, etc.

These vague semantic differences take on importance when participants in discussions assign different levels of meaning to the terms in their thinking. Here the broader meanings are assumed, except when stated otherwise.

Historical Perspective

Despite the wide variety of interpretations of the CAD/CAE acronym, it is generally acknowledged that many CAD systems in use today, regardless of industry of application (mechanical, electronic, or architectural), began operation with the expectation that they would be dedicated to a drafting/documentation function. These systems have been envisioned as an electronic pencil replacing the drafting board, and their nature has been seen as replacing templates, triangles, and ruler. Thus their justification has been in terms of labor savings.

Design functions performed on CAD systems have been, and still are largely in the area of layout. Companies have coined the term "designer/drafter" in response to the increasingly blurred distinction between the two job classifications. In fact, automation-supported detailing has caused the detail-drafter to become extinct in many companies. Economies of production have become apparent through blink-of-eye copying of existing designs, which can be edited for variants. By entering a few nominal dimensional values, an engineer could automatically generate a fully annotated engineering drawing of a new family member. Generally, these cases have been the ones cited for high productivity multiples (e.g., 25 : 1). Thus, historically, "efficiency" has been used to justify the purchase of CAD systems.

Historically, too, CAD systems frequently were positioned deep within drafting departments, where regardless of local productivity improvements, they had little impact on the financial fundamentals of the business. Often, they did offer some means of communicating from the drafting systems to other applications such as CAE analysis packages, to CAM manufacturing processes, and to inventory control programs. These parallel applications most often ran on different systems and were used by different operators.

This gave rise to an inevitable evolution of facility and, possibly, of personnel organization. As the CAD/CAM/CAE tools became increasingly sophisticated—which meant that they did much, much more while becoming easier to operate—the classic drafting room, formerly loaded with drawing boards, could not simply have computer terminals occupied by former board users.

Instead, companies today are finding that former drafting rooms are greatly reduced in size or completely obsolete. The computer-aided systems are being occupied by various combinations of individuals. In some cases, engineers operate CAD terminals; in other cases, drafters or designers work at terminals with engineers at their elbows. In some companies, engineers have their own access to the CAD system from their offices and edit designs as their drafter/designer team members file them for review. Frequently, such access is through relatively low cost personal computers, which lack the full functionality of a dedicated design workstation. However such access is sufficient for viewing and editing.

These scenarios are neither idealistic nor complex to manage. They do however require a possible alteration in drawing production philosophy. The scenario just described assumes a "team" approach wherein the design is optimized "on the fly" by both the engineer and the designer/drafter. While the engineer continues to audit, analyze, and manage the project, the response time for the communication loop has been vastly reduced.

The design/documentation process becomes less "serial," as was the case in the traditional drafting board approach of engineers and designers passing sketches to the drafter, who would compose them into proper format. Despite the speed of the drafter, it frequently happened that because of the serial nature of that process, the engineer would be modifying the design while the first edition of the drawing was still being formed. The engineer would contact the drafter, who would then edit or start over. This process tended to relegate the drafting organization to a pure documentation role whose contribution related to format consistency and to manufacturability/inventory considerations. Direct involvement in the design was hampered by logistics because the loop was not only serial but, often, long and slow.

In contrast, computer-aided design enables this serial process to be tightened considerably. While the engineer retains the primary design responsibility, the drafter/designer becomes engaged in the design process in real time. And often, the engineer lacks some of the skill sets supplied by people who are closer to manufacturing issues such as inventory control or manufacturability considerations. In this way, the design evolves with all such factors entering along the way.

The key is to fit the system to the organization's needs, definitely not the other way around. Some engineers are reluctant to personally operate CAD terminals, and others are filling project management roles with designers supplying the design optimization effort. An organization must consider whether to force uncomfortable fits and possibly sustain low productivity, or to match equipment with aptitude and take a longer view. Many companies have found that as long as most of the higher performance employees use the new system, the resulting culture change will sweep the rest along. Each installation of CAD equipment is a customized one, in the sense that the personnel and practices must adjust and accommodate the nuances of an automated system. Initially, it can be a benefit for the drafter/designers to become proficient and then attract key engineers toward the technology. Such an evolutionary approach minimizes the disruption, provides time for adjusting the firm's fundamental practices and procedures, and avoids ill feeling.

There is no room for caste-system thinking, where drafters are not permitted to enter into the design process. The very nature of the CAD system is that it promotes rapid iterative consideration of design alternatives. It is precisely the computerized speed and the network communication ability that provides the time for extra (early) consideration of design alternatives, as well as the opportunity to elevate the level of contribution of all concerned. This yields greater effectiveness.

For the foreseeable future, most organizations using CAD systems will undergo a long-term evolution. This is inevitable as their personnel pursue the design process with computer-aided tools, and because the tools themselves will evolve and prompt procedural changes. To envision only tool changes, with the design process remaining the same, is to fail to see the opportunity to move beyond efficiency improvements toward increased effectiveness.

Present Perspective

Competitive Advantage: Automating the Entire Product Design Process

In new product development, tasks ranging from labor-intensive mechanical drafting to control of manufacturing machinery have been streamlined through the application of computer technologies. Automation of the later stages of product development (i.e., drafting, detailed analysis, preproduction engineering, and manufacturing) has rapidly accelerated in the past few years.

Visits to numerous companies reveal that the environment in which today's engineers work has changed little in decades. While functions surrounding them have been automated, engineers who work in the earliest, highly leveraged portion of the design process still work largely in a manual environment. And, as one engineer put it, they tend to work in islands of nonautomation.

While it is true that some automated analysis (CAE) packages have been available over the past decade, until the advent of minicomputers and high-end personal computers (at reasonable cost), such packages ran on large mainframe computers. In addition, users of those packages required considerable training and experience just in the nuances of the software itself, to say nothing of the complexities of the analytical technology. This combination was overwhelming to the design engineer, whose priorities were structured in broader scope and who could ill afford to acquire such esoteric expertise. The net result was a tendency to overdesign for safety and reliability. A far worse effect, caused by reluctance, inability, or insufficient time, was a failure to develop innovative design that could provide lower cost, improved value, or possibly functional innovation.

This isolation has occurred because the computer automation tools used by mechanical drafting departments have not, until recently, been well suited to the engineer's conceptual development environment. CAD/CAM systems used for mechanical drafting, by virtue of their architecture, could not easily accommodate the information flow necessary for engineers to practice traditional top-down design methods.

The Hierarchical Approach

Engineers and designers use hierarchy to manage complexity. This is particularly true when approaching the design of assemblies. The initial concept is abstract and lacks the detailed features of a final design. It is subdivided into component parts, which are considered individually or are given to others who provide special knowledge. The component breakdown can be either functional or structural.

Once the component breakdown has been made, further refinement can be made to a large extent in parallel. Each refinement step is accompanied by decisions that reflect constraints from earlier steps and further constrain future design. In addition, interrelationships are set up between components which add to overall design constraints.

Such a process is called "hierarchical" and can be described as a top-down process. Starting with a final, most likely imperfect vision of the complete product, each of the constituent parts is subjected to an organized, well-connected series of design iterations. Each comes into focus, while establishing design relationships with companion components. All must march in step toward final design as their relationships become increasingly clear. As this process proceeds, there is a diminution in the variability of design or degrees of freedom available to the engineer and designer.

Automating Hierarchical Design

The process just described is one of the conceptualization followed by optimization and audit. It is the elemental part of the design process. By its position in that process, it exacts great leverage on the competitive aspects of the product in terms of cost, quality, and performance characteristics.

The technology to automate this segment of the design process is available and is known as mechanical or electrical computer-aided engineering, or CAE in general. Although CAE systems are still in their infancy, some powerful new trends have already emerged.

The early 1980s brought a virtual revolution in the quality of computer-aided design tools, beginning with the emergence of powerful digital circuit design tools for electronic engineers. Such tools adopted leading-edge artificial intelligence techniques, used in connection with silicon compilers and similar sophisticated tools, as well as in connection with the design of integrated circuits. These tools are classified as electrical CAE.

It has taken considerably longer to provide meaningful assistance to the mechanical engineer because of the much greater diversity of tasks and kinds of analytical technique employed. With the advent of powerful but inexpensive microprocessors and the resulting new generation of engineering workstations with standard operating systems, graphics standards, and data standards, it has become possible to develop systems that provide meaningful assistance to mechanical engineers and designers in the conceptual design phase of product design.

In the late 1980s, the emergence of integrated mechanical CAE systems began to be appreciated by the industry. Such systems provided the capability for modeling the part or assembly with considerable ease and then applying changes to material, geometry, loads, speeds, temperature, or other dynamic attributes. Such changes are reflected both in the geometry of the physical design and in the analytical model, which predict the performance of the physical design under a variety of operating conditions. With this combination of hardware and software, the automation of the hierarchical design process is possible.

The software developments of the late 1980s enable the entire hierarchical design process to be computerized. Of primary importance is the evolution of three-dimensional, solid modeling capability of desktop systems available to the designer and engineer. Such systems, by their fundamental architecture, provide for the top-down approach to design. Rather than beginning with the point/line construction approach of individual (presumably correctly designed) components, one develops solid objects representing the product's overall concept and subsequently refines their appearance. The images produced appear to be "solid"; they are colored, shaded, three-dimensional objects, which when modified reveal holes, slots, etc. They can be rotated and manipulated on the screen for ease of design. Moving from the initial concept of the complete product, subcomponents are then separated out and developed, using the constraints recently developed from the top-down design.

As the process continues, an assembly of components is developed. If the product consists of only one component, the process is that much simpler. In this case, the "assembly" is the fit of the single-piece product with its environment, and the constraints imposed on it are defined by that environment. Nevertheless, once a data base of three-dimensional shapes has been developed, and individual component parts have been identified, several opportunities not available to previous generations of engineering departments become available.

At this point, the automated design environment provides the engineer and designer with a delightful opportunity. Three key elements come together. The geometry is depicted in an elegant manner; the supporting data base contains material properties and is flexible; and there is easy access to analytical functions with which the design can be tested. And best of all, all three of these tools can be in one system that is easily manipulated by a single individual.

With such a system, it is very easy to develop standard mechanical attributes such as volume, mass, center of gravity, moments of inertia, and radius of gyration. Furthermore, with specific software tools, one can identify points or fields of pressure and restraint and develop standard stress–strain evaluations through highly automated finite element analysis routines. CAE systems of today provide these tools with vastly improved convenience of use over those available in past years.

Approximate analytical models, formulas, and engineering approximations have for too long been buried inconveniently in a variety of handbooks. The systems described here retain these in a format that permits engineers who use them infrequently to pick them up without digressing into a minor study of their nature. In this way, the engineer is freed from the mathematical tedium of solving complex equations or the alternative of writing custom computer programs to solve them. The design system provides this background on demand.

Finally, all this must be readily applied to the design under development. Through straightforward commands, the system will permit the design to be segmented into elements and those elements associated with the analytical techniques. The results are displayed in a variety of formats, and the engineer is free to iterate the analysis while changing parameters.

A comment about the application of this "modeling" capability combined with the technology of "finite element analysis": suppliers of CAD/CAE systems that provide this combination emphasize that these systems do not supplant the highly trained finite element analysis engineers who bring critical expertise to the analytical problem. These professionals employ skill and experience to discern possible weaknesses in the modeling approach that can lead to impressive but invalid results. However, the wide availability of desktop automation permits the design engineer to perform fairly sophisticated evaluations of the design on the spot. Such evaluations would rarely go to the professional analyst, and when their skills are critical to ensuring failure-proof functionality, they receive a reasonably elegant "preanalyzed" design to evaluate.

In many applications mechanical CAE analysis is not appropriate, yet the need for rapid, easily used modeling is essential. Engineers and designers can be highly productive on pure drafting-oriented CAD systems as well as on solid modeling design-oriented systems. The availability of these systems in the organization may dictate the type used; or the preferences of the user may come into play. In either case, the benefits of the automated process are essentially the same, and these systems can be highly effective in the early stages of the design.

The Design Engineer's Role

Integrated mechanical CAD/CAE tools now make possible the resurgence of the classical design engineer—someone who knows not only how to design to meet a need, but also how to do simple and yet effective performance analysis at the point of maximum efficiency. An engineer or designer using computerized tools can try many variations of a selected solution to arrive at the best solution. Then, having picked the best solution, one can optimize the performance of the selected approach either in a rigorous mathematical sense or by a comparison process that uses an end-use-driven figure or a subjective approximation. This process can now be undertaken using a tool that allows one to try to design variations at a rate generally 100 to 1000 times faster than is possible with conventional paper, pencil, and eraser methods followed by premature prototyping.

The computer-aided systems described here emphasize the blend of the concepts of computer-aided design with those of computer-aided engineering. This blend is important to the engineer and designer because these systems are designed to operate as part of the initial design conception/optimization thinking process. The engineer or designer is not forced to "think out of sequence" by conceiving detail components first. They do not force one to "draft" the design with documentation-styled procedures. These systems capitalize on the colored, shaded, three-dimensional representation of the object by affording sophisticated desktop analysis of mechanical properties. And finally, the completed design can be handed to a detail drafter for proper documentation. That individual can use the compatible system to automatically reformat the solid model into a mechanical drawing format and then edit it further.

Considering a CAD/CAM/CAE System

Turnkey Systems

A turnkey system is a combination of hardware and software acquired from a single supplier that requires no additional purchases before it can be put to its intended use. People buy turnkey systems because they have an understandable fear of confronting the problems of interfacing pieces of hardware and software on their own.

Turnkey system suppliers take the responsibility of making the system's components work well together both before system installation and (usually) for the life of the product. These vendors are successful and make a reasonable profit by providing field service. They interact with the developers of the individual components and influence their evolution. This association tends to maintain compatibility among these components as independent development efforts enhance their performance.

However, there is a downside to using a turnkey system. As the system design ages and sales fall, the vendor sometimes makes radical changes and fails to provide continuity between old and new versions. All too often, old data die with the old system. There is no easy precaution against this, except to discuss the system evolution with long-term users and judge the "development style" employed by the vendor. However, this approach is not foolproof, since times change, as do marketing conditions and management philosophy.

There are two highly successful variations on the turnkey approach. The first is hardware-based: that is, one relies on a major computer company to install both its own hardware and a set of recommended software modules, and contracts with the company to "make it work." The second approach is the opposite: to select a major software company and ask its experts to identify and install their software on a platform of their choice. These two options are represented by several of the largest computer and software manufacturers in the world.

User-Selected System

The user-selected system is a valid approach only for very specialized and unusual applications, or for users who must be dangerously close to leading-edge technology. Users can select a software package and then purchase a computer to run it on. With a careful eye on compatibility, they can choose graphics workstations and output devices. With multiple users, a local area network (LAN) can be added. This strategy allows a firm to take advantage of the highly competitive pricing in high technology markets. Care must be taken, because despite assurances by independent vendors of hardware and software that their products will work together, there are subtle and often complex problems, which the user must resolve. Both software and hardware change, and unforeseen interface problems tend to develop. As the business dynamics of the various companies change, products often cease to be compatible. The user must then either ignore some highly valuable software improvements or devote an unreasonable amount of time and pay a lot for the expertise required to maintain compatibility among the latest versions of the products.

System Integrators

The incompatibility problem can be minimized by employing a system integrator. Such a service firm will help select the equipment and software, negotiate with the suppliers, perform the integration, and leave the user with the apparent equivalent of a customized turnkey system. Such a service supplier might also leave the user with the seeds of a potential disaster.

While the integrator can successfully compose a workable combination, and in fact may produce a combination far more clever than the owner could conceive, long-term reliability is no greater than it would be for a self-assembled system. It is true that some system integrator companies have strong computer departments that can undertake support of a multivendor system; most do not. Extreme care must be taken to evaluate thoroughly the long-term relationship that is essential with this type of service supplier.

Some Key Considerations

Limits on Number of Users

Since most turnkey systems are configured as a computer with semi-independent workstations, growing beyond the number of workstations that the computer can support with adequate responsiveness generally forces the user to acquire another computer. Since sharing a large engineering data base under a multicomputer arrangement is complicated, some organizations elect to employ independent workstations or personal computers. This approach takes two forms: either the computers are completely isolated, or they are networked to the larger system for file access. In either case, they possess all the necessary CAD/CAM/CAE software necessary to manipulate the design. The main computer functions as a "file server." This approach has gained favor throughout the industry and continues to become even more popular.

The demands of users for highly responsive CAD/CAM/CAE workstations, coupled with the availability of high power computers, will lead to a proliferation of networks of powerful independent workstations. And as the networking activity itself often adds to the computational overhead of the system, thus impairing responsiveness, tightly integrated "networks on chips" will be developed by the industry to alleviate the strain. The extent of the impact of networking on the system is tied to the computer architecture employed; some systems handle this load with no difficulty.

The Selection Process

As an organization approaches the acquisition or expansion of CAD/CAM/CAE facilities, certain questions arise.

1. What are the business objectives CAD/CAM/CAE is to meet?
2. With whom must CAD/CAM/CAE compatibility be maintained?
3. Are any special capabilities needed that would lead to the selection of a particular system?
4. How important is CAD/CAM/CAE system cost in the selection process?.
5. How can one evaluate both the vendor and the likelihood that the vendor relationship will be a positive and lasting one?
6. What are the key elements of the implementation of a CAD/CAM/CAE system?
7. What are the realities of maintaining the system after it has been installed?
8. What are the training costs associated with start-up?

Supporting Business Objectives

Consideration must be given to the role the automated system plays in the larger business scheme. As described earlier, reasons of efficiency or effectiveness are real and are justified by the performance of the technology. However, a realistic appreciation of the consequent effect on business earnings is in order. The use of automated systems at the point of sale or closely tied to it has a more direct relation to increased sales and, hopefully, earnings.

System Compatibility

Compatibility needs may start within the organization. If the new acquisition is an expansion, there usually must be some compelling reason to bring in a different technology. If it is a new installation, there may be a need to maintain compatibility with CAD/CAM/CAE systems elsewhere in the parent organization.

Compatibility with customers may be needed. If an organization is designing as a service to customers, the value of the service is greatly diminished if the customer cannot make full, easy use of the results. Sometimes a rational expedient consists of installing a CAD/CAM/CAE system that is a "clone" of the system type used by the most important customers. The advantage of this can be great enough to justify acquiring a system that is mediocre in performance and/or capabilities. This strategy must be approached with caution and with an eye to the future. It should be assured that the system thus chosen is adequate for its mission and that the customers are not about to abandon the technology.

It is important to add that compatibility between systems of different manufacture has been addressed on an ongoing basis through the identification of "interchange standards." The difficulty in defining these standards is that system designers provide unusual features in an attempt to achieve product identity. These features rarely translate across to another system. However, the quality of interchange standards is improving and could greatly simplify compatibility considerations. Standards are not a panacea, however.

Special Capabilities

CAD/CAM/CAE technology has matured greatly during the past five years. For most applications, it can be considered to be a commodity. More or less equivalent products can be bought from any of several vendors. There are, however, a few areas of capability that are not near commodity status yet. One such area is in the mechanical CAE field and involves analytical computations with nonlinear and/or discontinuous material property curves. This can be an important consideration in design with composite materials.

In considering special capabilities, one must again use care. Is the technology containing the desired feature complete in other areas? Is the vendor likely to survive against competition? If the answer to either question is "no" or "doubtful," the user must ask whether sufficient value will be obtained from the product anyway. The alternative is to wait for a more stable and/or complete product to emerge.

The issue of special capabilities is usually confused by vendor claims of unique advantage. Many of the features thus claimed are not truly special capabilities but arbitrarily imposed uniqueness, adding little or no real value.

System Cost in the Purchase Equation

Often, large organizations can make acquisitions on a cost-benefit basis; that is, as long as the return on investment is good, the absolute cost of the system is not terribly important. In other cases, especially in smaller organizations, keeping the system cost under some modest limit can be of overriding importance. The personal computer implementations of CAD/CAM/CAE have a special appeal for financially limited installations. The capabilities of these low cost computers have grown to the point of often being the chosen platform, even in some large corporate applications. Some very sophisticated software now exists for these machines, both in two- and three-dimensional design graphics (CAD) and in related areas such as stress analysis and mold design (CAE). A pitfall exists in that there are also some very poor software products being sold for personal computers. The buyer must be very sure to know the difference.

On the other side of the cost issue is the large, affluent organization that can buy whatever seems appropriate and promises a good return. These selections tend toward the larger machines, and the more established CAD/CAM/CAE vendors.

Vendor Evaluation: An Ongoing Relationship

Acquiring a CAD/CAM/CAE system is not a simple commercial transaction; it is the beginning of a relationship between the vendor and the purchaser. It is important that the purchaser examine the character of the vendor thoroughly before agreeing to buy. Here are some points on which to evaluate both the vendor and the likelihood of the relationship having a positive and lasting value.

STABILITY. Take the time to determine the likelihood of the vendor's continued existence. Computer-aided systems are part of a rapidly moving technology, and survival of the players in this market is constantly under challenge. Use business analysts' ratings, study annual reports, and ask for the vendor's business plan. Most important of all, seek other users and discuss vendor's performance with them. Ask the vendor for references and ask them for other users.

INTEGRITY. The extreme complexity of these systems can result in "bugs," or defects in system performance. Ask current users about the vendor's performance in resolving such problems as promised, and in a timely way. Ask how the vendor responds to suggestions for improvement, how system-related questions are handled, and how the vendor responds to references to the competition. A good many system deficiencies can be overlooked if the vendor is faithful in day-to-day business dealings.

TRAINING. It is true the vendor supplies the training required to operate the system. However, this usually amounts to 5 or 10% of the training ultimately required. The remainder relates to the work environment. System training must be integrated with work skills. Procedures and standards must be developed for the new mode of operation. These issues are intimate to your organization's needs and practices; no vendor can anticipate them.

Consideration must be given to the location of the training. There may be travel and living expenses involved for your operators. With small systems, training typically takes place on your site.

STRATEGY. Determine the technical direction in which the vendor is heading. Is there a move under way to shift to a different hardware platform or to a different operating system? Is the vendor leaning toward developing an application strength that departs from your area of interest? Are you about to buy technology that will suffer from lack of support, or from modest, low priority support. Vendors usually cannot afford to apply their development forces to old and new technologies with equal emphasis.

USER GROUP. The user group is an important forum for exchanging information on operating techniques. Such a group is a very important aspect of your relationship with the vendor. Since fellow users meet to share information about using the system, they can join forces to obtain software improvements.

SCREENING VENDORS. Develop a list of "required" and "desired" features. Take the time to spell out a clear statement, in functional terms, of what you want the system to do; do not stray into "how" the system does it, early in your investigation. Use input from your anticipated users; your lists will have practical validity, and the users will benefit from contributing to the selection process.

As you investigate various systems, continue to reassess the importance of the "required" versus the "desired," and consider ranking the items by importance. This approach will lead a basically subjective process in a quantitative direction. Some people develop a "weighting scheme" associated with these lists and drive the quantitative approach further. The key here is to ensure that the relative importance of features is clear, so that WHAT systems do remains more important than HOW MANY "checklist" features are available.

It is important to avoid having a short list of deficiencies or weaknesses negate a long list of capabilities and impressive benchmarks. Careful consideration must be given to the "weight" of negative findings.

When the finalists are identified, have a "performance test" run in your presence. Attempt to discern between "operator skill" and true "system performance." Again, testimony of existing users is extremely valuable.

Implementation

One good plan is to use the knowledge provided by the vendor to arrange the sequence of events. Because vendors have a vested interest in your successful start-up, most are extremely helpful. They have the experience of other installations and can highlight key issues that must be attended to, early on. Independently seek out and tap other users to see how they implemented their systems. Involve your own users as much as possible. Early participation from them builds "ownership" on their part. Since they will be "inheriting" the system and will have to deal with it daily, their input tends to be realistic and reflects the pragmatic aspects of integrating the system into existing procedures.

A note of caution is appropriate. Because of rapid growth and personnel turnover, CAD vendors often have unqualified and inexperienced people in customer engineering positions. The new users should be prepared for this, and should insist that the vendor home office send in qualified people, if problems develop.

System Manager

Identify a system manager and involve that individual from the start, if possible. Even if the system is relatively small, the "system manager" still exists and takes the form of a "champion" of the technology. In both cases, this individual is the focal point for the technology and is the innovator/driver of the implementation. This individual usually becomes the primary contact with the CAD vendor.

Part of the system manager's position is the responsibility to assist the local user community. By doing that, this person maintains an awareness of user practices, preferences, and needs. In most organizations, this awareness is reflected in a "user manual" or similar procedural format. Given the dynamics of all automated systems, however, the needs of the users are never ending. Both increased proficiency of users and improvements to the system drive changes to practices and procedures. As discussed earlier, these changes prompt evolution of not only the "way" the group does its job, but also "what" they do and "with whom" they relate.

SYSTEM CUSTOMIZATION. "Customization" refers to the establishment of "user conventions." While systems function upon delivery, they are designed to permit the user to establish local conventions and practices as part of the daily routine. Even in the case of a single-user situation, the schemes for file naming, directory use, layer conventions, general backup, and archiving procedures must all be decided on. With multiple users and when documentation will survive for years, the need for formal organization is critical.

TRAINING. While all large and most small systems come with provisions for training, the training received is strictly system-technology related; it cannot include the procedures associated with the integration of the system into current practices. Months of training mixed with practice are typical, but it is the larger integration problem that taxes most organizations. Unless the users are CAD literate with an earlier system or different application, large organizations require a full year to become smoothly functional in the best case. They require the second year to improve their efficiency, and typically can address effectiveness issues in their third year. This timing is rough and can easily be stretched by years.

System "Personality"

Conversations with the users of any computer-aided system indicate a perception of a "personality" associated with the setup. This is the overall system character as perceived by the users. It consists of factors such as predictability, performance/responsiveness, technical depth, compatibility (with other systems), and ease of use. In the case of the more sophisticated offerings, the extent of ongoing development is important and is evidenced by orderly software upgrades.

Another element of "personality" is the extent to which the system "helps" or "anticipates" the user, from the basic "help function" delivered on-line to get one out of jams, to the more sophisticated "cues" shown on the screen, which indicate the most probable or logical next step.

And of course, the capacity of the system manager to provide timely assistance, the productivity of vendor relations, the degree of comfort in the physical environment, and the quality of the local user group all add to the overall perception of the system.

Building Composites with Computers

The characterization of composites demands the computer analysis of both the material and the subsequent product design. The collection of finite element modeling and analysis programs that perform this task are categorized here as part of mechanical computer-aided engineering (CAE) technology.

Recent developments in high temperature advanced composite materials have accentuated the inadequacies of most finite element analysis programs. It comes as no surprise to find that these programs began as liner analysis of geometries made from isotropic materials. Today, there is a shortage of well-crafted analytical tools to accurately evaluate composite design and performance.

Consequently, there have been special-purpose programs developed to facilitate design of and with composites. As is typical of "home brew" analytical routines, however, these programs are largely difficult to use and lack reasonable user interfaces.

On the positive side, many vendors of finite element analysis software are actively pursuing serious advancements in this technology. Efforts are under way to model fiber orientation, fiber degradation in the matrix, and varieties of nonlinear behavior. Almost of equal importance are the efforts to provide "friendly access" to these routines, so that their use is less arduous than in the past. Several serious commercial applications are available.

The scope of this writing provides a sense for the issues involved; a detailed examination is best made in conjunction with a particular CAE finite analysis software package in hand. In the process of evaluation, one can determine whether the strengths and weakness of a specific package suit the application.

The added complexity of composite layering demands the calculation of the material properties of both the fiber and the matrix (when present) in each ply, and then synthesize the properties of the laminate as a whole. In addition, the engineer would expect the analysis to reveal which ply might exhibit early failure and to pinpoint the geometric location of the failure relative to the part design. These needs and others point to the need for computer-aided modeling and analysis.

Composite design software can be divided into five basic categories:

- *Lamination property programs:* determine data such as the mechanical and thermal properties and hygral (moisture) effects on composite materials.
- *Joint programs:* determine the effects of loads on bolted and bonded joints in composites.
- *Structural analysis programs:* analyze simple shapes such as beams, plates, and tubes.
- *Optimization programs:* choose the best combination of matrix and fiber to meet certain engineering criteria.
- *Fatigue behavior programs:* evaluate behavior such as delamination, fiber breakage, matrix cracking, and matrix failure.

The "Elements"

Many finite element analysis programs treat structures made of composites with a combination of two-dimensional shell elements and three-dimensional solid elements. With them, three types of modeling element are used. The simplest is the "2D laminated composite shell," which produces plane stresses and strains. Additional complexity is accommodated by the "sandwich shell" and the "3D solid overlay shell." The use of three-dimensional solid elements accommodates three angles of fiber direction and is especially useful for modeling areas of high stress or strain, such as free edges. Three-dimensional elements can be used to determine the normal shear or interlaminar shear at any element location.

The layered composite shell can have many layers of different thicknesses, with different fiber orientations and material properties allowed in each ply. The sandwich shell element includes two face sheets that resist extensional and in-plane shear deformation. Sandwiched between the face sheets is a metal, foam, metal composite, or honeycomb composite. Each ply in the solid overlay element can have different properties or lamination angles, though each group of plies is considered one element.

The Results

Computer-aided systems today provide elegant output. While reams of data are always available, outstanding graphic representations are both time-saving and elucidating. Repeatedly, investigators identify critical aspects of a design from color-keyed stress–strain patterns. Achieving similar results through tabular data would be difficult and certainly time-consuming. Moreover, with iterative methods, one can cycle through successive passes at an analysis: viewing the pattern, adjusting material or geometry, and rerunning the analysis.

The technology for this type of display and subsequent video output and high quality color copy on paper is rapidly advancing. Driven by other parts of the computer industry, computer graphics have become a dynamic technology. Even traditional business applications are becoming graphics-oriented.

Summary

The use of computer-aided techniques is essential to the study of composites. Regardless of the materials studied, the selection and implementation of a computer-aided design system is neither trivial nor an event of short duration. This article provides an updated introduction to those who anticipate the use of computers as design tools. It informs the new user of larger issues surrounding the acquisition and use of computers in design, particularly in a corporate organization, offering thoughts, cautions, and encouragement based on experience by addressing people issues as well as system issues. The ideas naturally reduce to the single individual, or small installation, also.

Computers and system installations vary from the most modest installation to extraordinarily large, complex arrays of hardware, networks, applications, and users. However, today the individual must look beyond issues of "efficiency" and "effectiveness" and, instead, identify "what can I do that is substantially different from before, in a way that allows me to use computer-aided design systems to leap over my competitor." And the competitor may be an academic peer or a commercial adversary.

Examples such as the evolution of the classic computer-aided "drafting" tool toward the computer-aided "design" tool, and the rapidly advancing CAE finite element analysis tools, are exactly that: "tools." With them, one can do things completely differently, not just faster.

Their substantial technical depth combined with the daily improvement in their usability arm the investigator or designer with the opportunity to soar.

Joseph D. Halford and Robert N. Wieland

Bibliography

Brody, Herb, *High Technology Magazine*, 1987, pp. 12–18.

Miller, R., *IEEE Spectrum*, May 1987.

Mills, R., *CAE: Computer Aided Engineering*, July 1987.

Norwood, D., *CAE: Computer-Aided Engineering*, November 1986.

Villers, Philippe, *CAE Magazine*, 1986, pp. 67–69.

Wedeking, Edwina L., *Machine Design*, Jan. 22, 1987.

Carbon–Carbon Composites

A carbon–carbon composite is a combination of carbon or graphite fiber reinforcement in a carbon or graphite matrix. The first carbon–carbon composite was prepared in 1958 at the Chance Vought Aircraft Company laboratory [1], quite by accident. A chemical analysis was being performed to determine the quantity of reinforcement/matrix in an organic matrix composite. Part of the process was to expose the composite to the air, but the lid to the crucible was left on inadvertently and, instead of oxidizing, the organic matrix composite pyrolized. The resulting charred composite or "burnt toast" was noted to exhibit structural characteristics, hence the birth of carbon–carbon. However, the advancement of carbon–carbon technology was slow for several years. By the late 1960s, with the advent of carbon fiber and weaving technology, structural carbon–carbon composites with high temperature strength, stiffness, and toughness began to emerge as a new class of engineering materials.

However, besides the technology borrowed from the graphite industry [2], the practices of other industries and technologies have been drawn upon in establishing carbon–carbon processing techniques. For example, the use of reinforced plastic composites that can be processed into carbon–carbon composites represents an extension of the technology used in the plastics industry. By taking advantage of pressure carbonization as an effective, efficient method to densify carbon–carbon composites, the equipment used was modified hot isostatic press (HIP), originally developed for metal processing. Similarly, chemical vapor deposition (CVD) techniques originally developed for producing pyrolytic graphite structures and coatings are used to infiltrate and densify carbon–carbon composites.

Typical characteristics of carbon–carbon composites are shown in Table 1 and Figures 1–5.

The light weight and low density (1400–2000 kg/m^3) of carbon–carbon composite (vs. aluminum, titanium, and niobium densities of 2700, 4500, and 8400 kg/m^3, respectively) make this material extremely attractive in high temperature aerospace applications. Its high strength and stiffness, complemented with low thermal expansion, make carbon–carbon composite one of the materials that is most resistant to thermal shock. Carbon–carbon also exhibits a negative coefficient of thermal expansion until the temperature reaches about 500°C. This unique feature parallels the volume expansion of water while temperature decreases. Another distinctive feature of carbon–carbon is that it actually gains and

TABLE 1
Typical Material Characteristics of Carbon–Carbon Composites

Light weight and low density
High strength and stiffness
Low thermal expansion
High thermal conductivity, decreasing with increase in temperature
High thermal shock resistance
High fracture toughness
Pseudoplastic behavior
Good fatigue and creep resistance

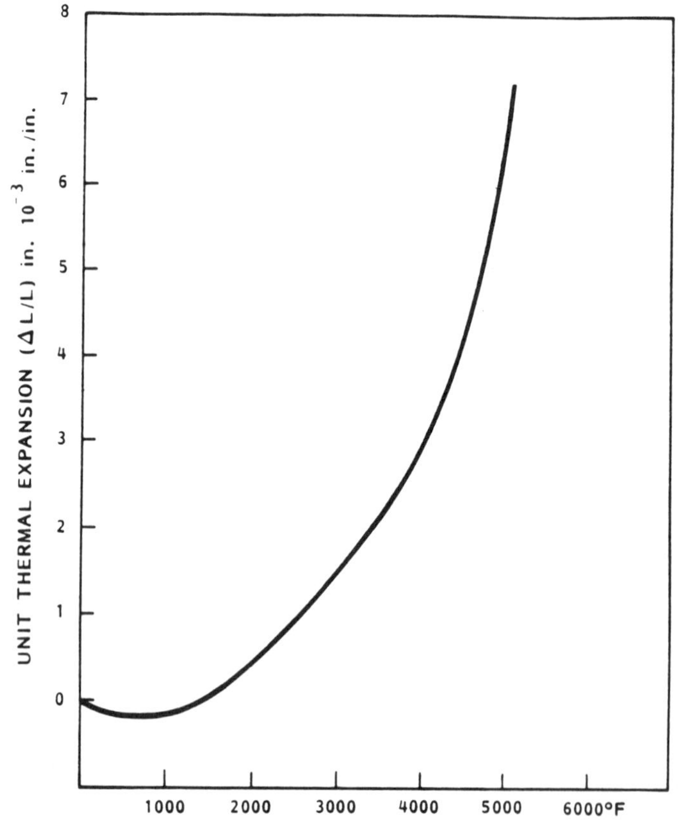

FIGURE 1 Thermal expansion of carbon–carbon as a function of temperature: near zero at room temperature; exhibits a negative expansion until about 50°C. (From Ref. 3.) To convert the units in this figure to SI units, see the conversion table at the front of this book.

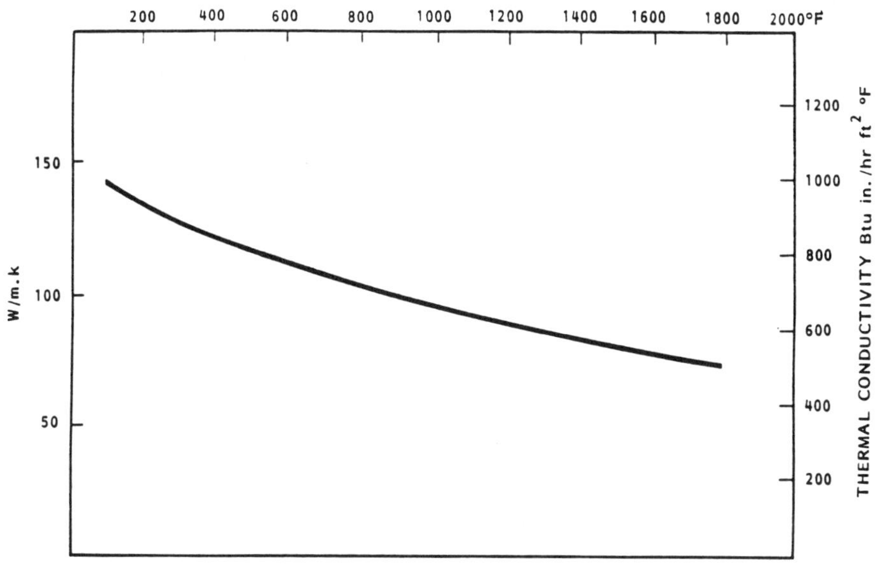

FIGURE 2 Thermal conductivity of carbon–carbon decreases as temperature increases. (From Ref. 3.) To convert the units in this figure to SI units, see the conversion table at the front of this book.

retains strength while temperature increases, as evidenced in Figure 3. Aluminum softens at 190°C and melts at 580°C, steel softens at 635°C and melts at 1400°C, while carbon–carbon strength and stiffness increase from 20°C to 2000°C. Contrary to its brittle appearance, carbon–carbon is quite fracture "tough": indeed, tails have been driven into without fracturing the composite. Its pseudoplastic and creep resistance behavior can be observed in Figures 4 and 5.

Constituents of Carbon–Carbon Composites

Carbon Fibers

For reference, various basic carbon crystal forms are illustrated in Figure 6. The elastic constants of the graphite hexagonal lattice (Fig. 7) layers are 1060 GPa for the layer planes and 36.5 GPa perpendicular to these planes.

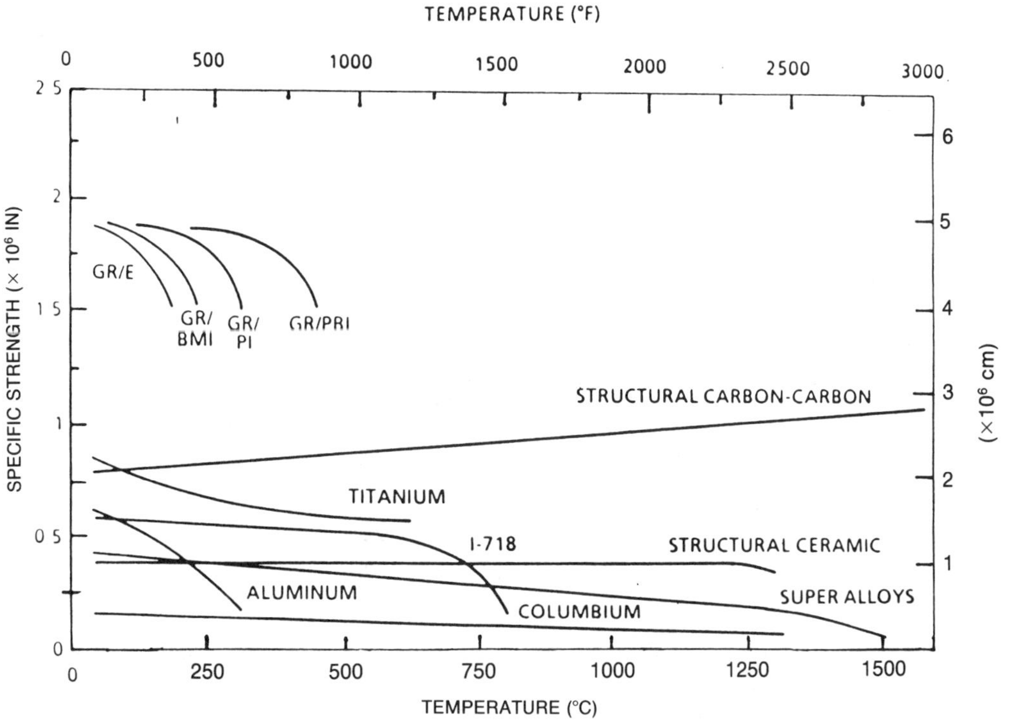

FIGURE 3 Carbon–carbon actually gains and retains strength when temperature increases (whereas other materials deteriorate). (From Ref. 4.) To convert the units in this figure to SI units, see the conversion table at the front of this book.

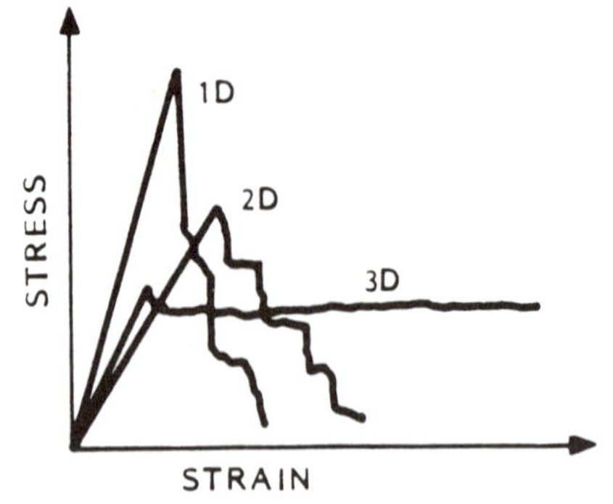

FIGURE 4 Pseudoplastic fracture behavior of carbon–carbon composites; this effect is especially enhanced in 3D. (From Ref. 5.)

Ideally, carbon fibers may be fabricated to have in-plane tensile modulus or Young's modulus approaching 1060 GPa, the modulus of a single graphite crystal. With carefully controlled and developed fiber microstructure—that is, lining up of the crystallographic basal planes along the longitudinal fiber axis (Fig. 8)—researchers were able to commercialize a fiber with a nominal modulus of 828 GPa and to develop experimentally a fiber having a modulus of 966 GPa, approaching the modulus of a single graphite crystal [8].

The experimental data of Figure 9 indicate that the preferred graphite basal plane orientation is determined by fiber processing temperatures and precursor type. The fiber mechanical properties and electrical resistivities are determined by the preferred orientation of its graphite basal plane and by fiber processing temperature (Figs. 10 and 11). As the fiber processing temperature increases, both the resistivity and the basal plane preferred angle θ decrease, aligning closer to the fiber axis, but the tensile modulus and the electrical conductivity of the carbon fiber increase. In general, fiber mechanical properties, thermal properties, and preferred basal plane orientation are interrelated, as evidenced by Figures 10–14. Table 2 gives a rule of thumb for fiber processing temperature in relation to mechanical properties and basal plane orientation.

There are three common types of carbon fiber precursors: rayon, polyacrylonitrile (PAN), and petroleum pitch. The fiber microstructural morphology of each is shown in Figure 15. Each type of precursor fiber has its own distinctive form, which may affect the fiber–matrix interaction and matrix wetting characteristics. The unusual morphology of special pitch precursor fibers may help to improve fiber packing efficiency by permitting closer fiber packing in the composite, hence higher loadings than are possible with round fibers. Figure 16 shows circular pitch-based P-55-2K fibers with a peculiar sizing phenomenon. The sizing of the fiber provides ease of handling the fiber and promotes weavability.

PAN-based carbon fibers of noncircular cross section (Figs. 17 and 18) have been manufactured using unique melt-spun precursor technology. These fibers are currently in the developmental stage, and researchers believe that by varying the carbon fiber shape, significant improvement in the overall balance of properties of the composite may be obtained. The basic manufacturing processes for the three common carbon fibers are illustrated in Figures 19–21. Some typical properties of commercial high modulus carbon fibers with different precursors are shown in Table 3. A general classification of carbon fibers according to their strength and modulus is presented in Figure 22.

PAST AND PROJECTED TRENDS IN CARBON FIBER CONSUMPTION AND PRODUCTION. Extensive progress has been made in the improvement of fiber strength. Figure 23 shows the growth of carbon fiber strength and the trend of progress. Joint efforts, global involvement in business and technical relations, and the production capacity of major fiber and prepreg manufacturers are shown in Figures 24 and 25, and Table 4. Table 5 reveals past and projected consumption data for PAN-based fiber.

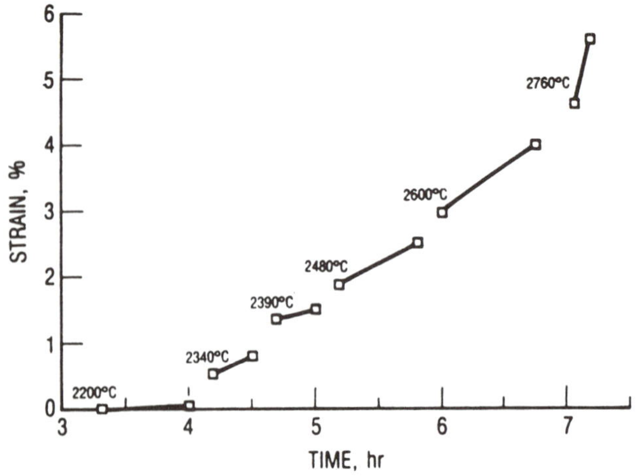

FIGURE 5 Creep behavior of unidirectional carbon–carbon composite 2000 and 3000°C. (From Ref. 6.)

TABLE 2
Guidelines for Carbon Graphite Fiber Mechanical Properties and θ as Related to Processing Temperatures

	Processing Temperature (°C)		
Property	1000	1500	2000–3000
Type of fiber	Carbon	Intermediate	Graphite
Crystal orientation to fiber axis, θ	30°	12–15°	7–8°
Tensile strength, kPa	< 690	965–1100	1725–2410
Tensile modulus, MPa	< 65	170–240	515–825

From Ref. 15.

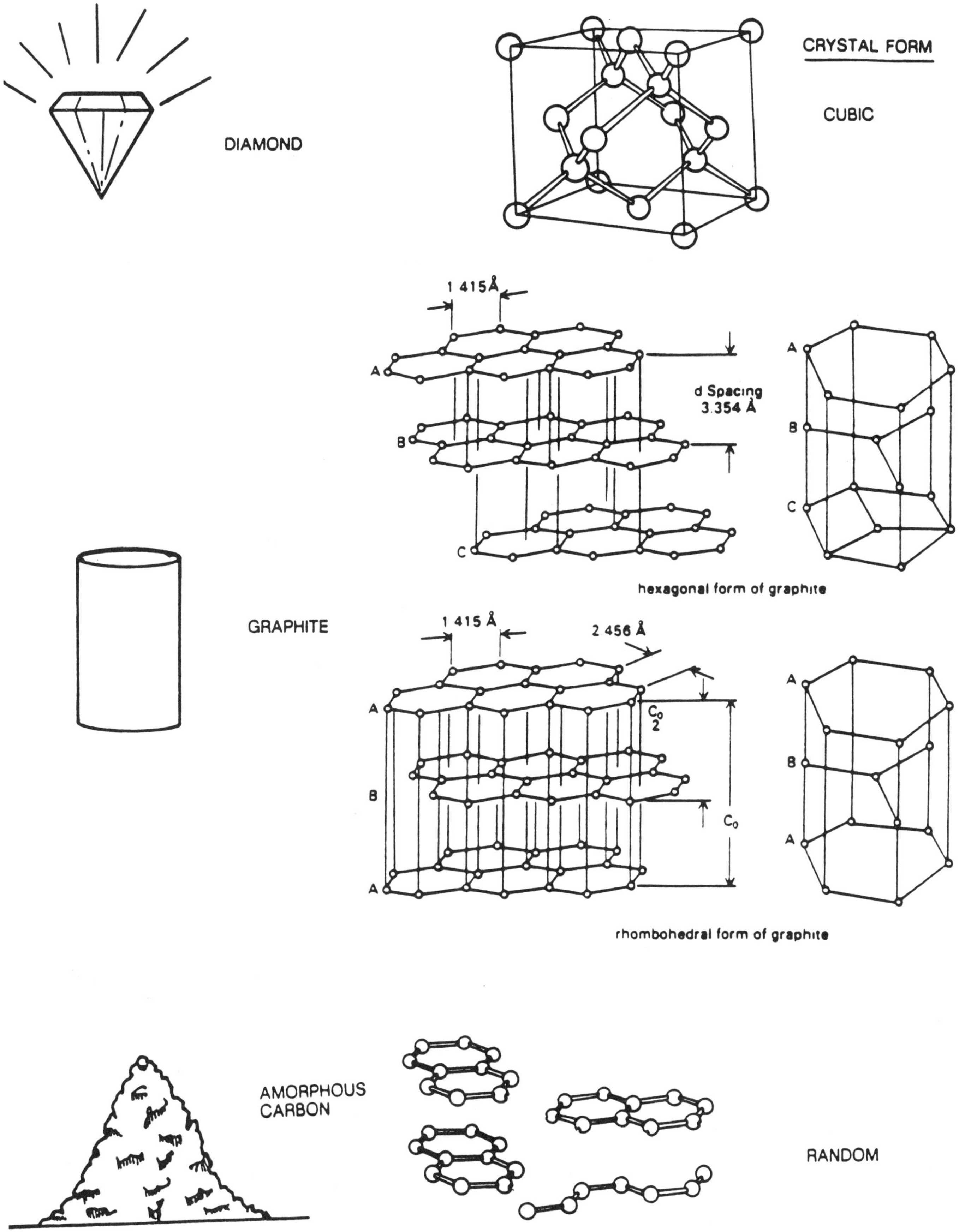

FIGURE 6 Basic carbon crystal forms.

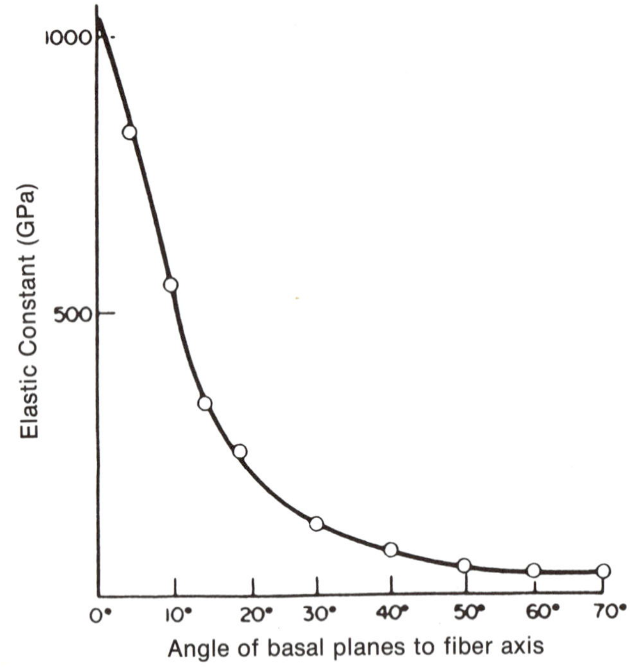

FIGURE 7 Elastic constants of the graphite hexagonal lattice as a function of angle of the basal plane θ. (From Ref. 7.)

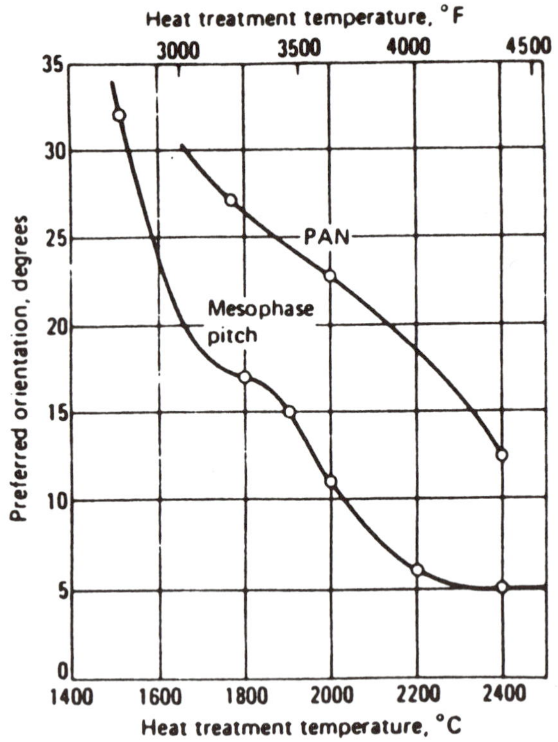

FIGURE 9 Preferred angle of the basal phase θ, as determined by fiber processing temperature and type of precursors. (From Refs. 9,10.)

PAST AND PROJECTED TRENDS IN CARBON FIBER COST HISTORY. The price of carbon fiber has dropped dramatically in the past decade despite two distinct periods of rampant inflation. During the 1980s, metal material prices were pushed up while graphite prices dropped. As the use of carbon fiber increases, price will continue to decrease. Thanks to modern technology, the carbon fiber cost trend has been declining since the late 1960s. Some carbon fibers now cost less than $50/kg (Fig. 26).

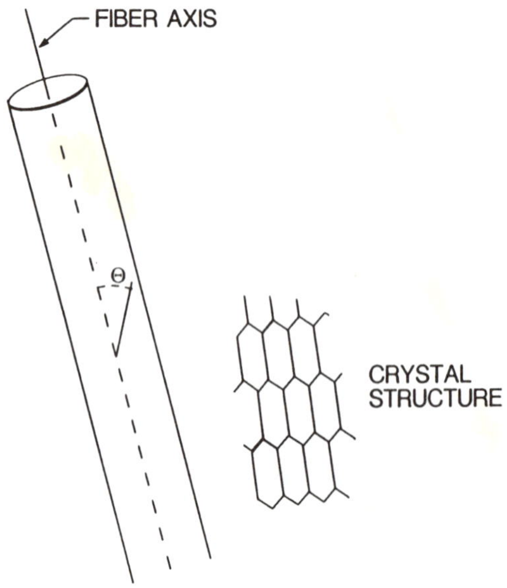

FIGURE 8 Angle of the basal plane θ relative to fiber axis.

Matrix Materials

There are three major categories of matrix precursors used for carbon–carbon processing, as shown in Table 6:

Thermosetting resins
Thermoplastic pitches
Carbon-rich vapors (hydrocarbons)

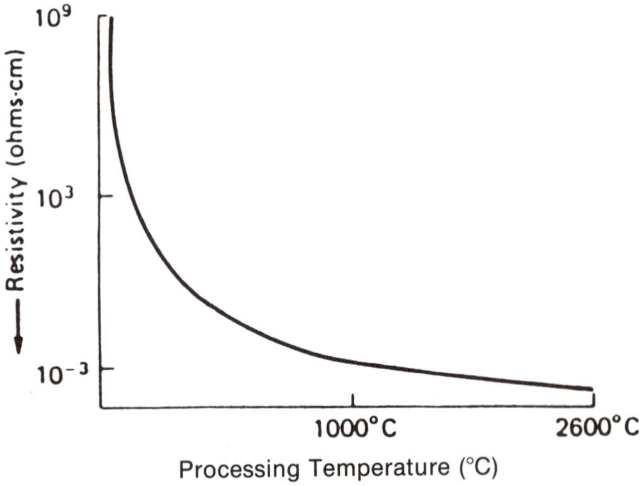

FIGURE 10 Resistivity versus fiber processing temperature. (From Ref. 11.)

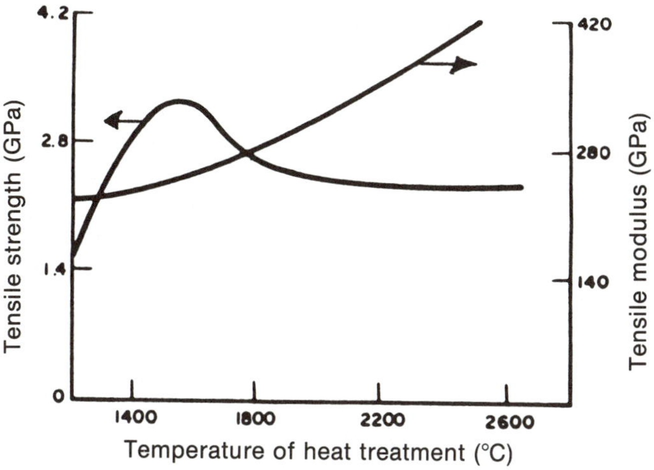

FIGURE 11 Tensile strength and modulus versus fiber processing temperature. (From Ref. 12.)

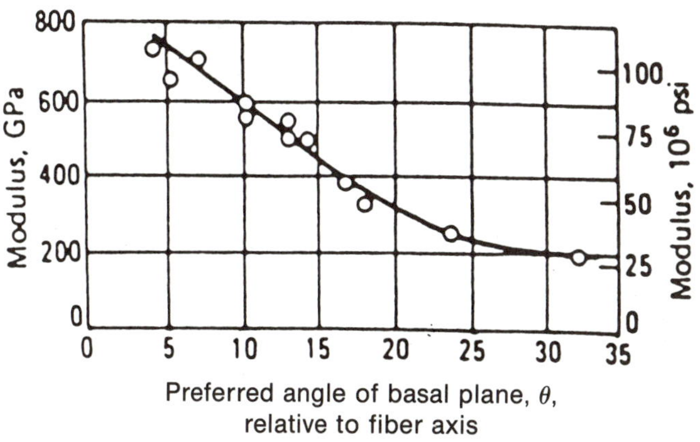

FIGURE 13 Tensile modulus of pitch-based carbon fiber versus preferred angle of basal plane θ, relative to fiber axis. (From Ref. 14.) Note the similarity to Figure 7.

By varying the combination of matrix precursors as well as their sequencing during processing, one is able to provide a wide variety of tailor-made carbon–carbon composites for specific design usages. Thermosetting resin matrices are popular in plastic composite, filament-wound, laminated, and chopped fiber materials. They can be converted and cross-linked, polymerized, to form insoluble or infusible solids. Upon pyrolysis, these resins form glassy (amorphous) carbon.

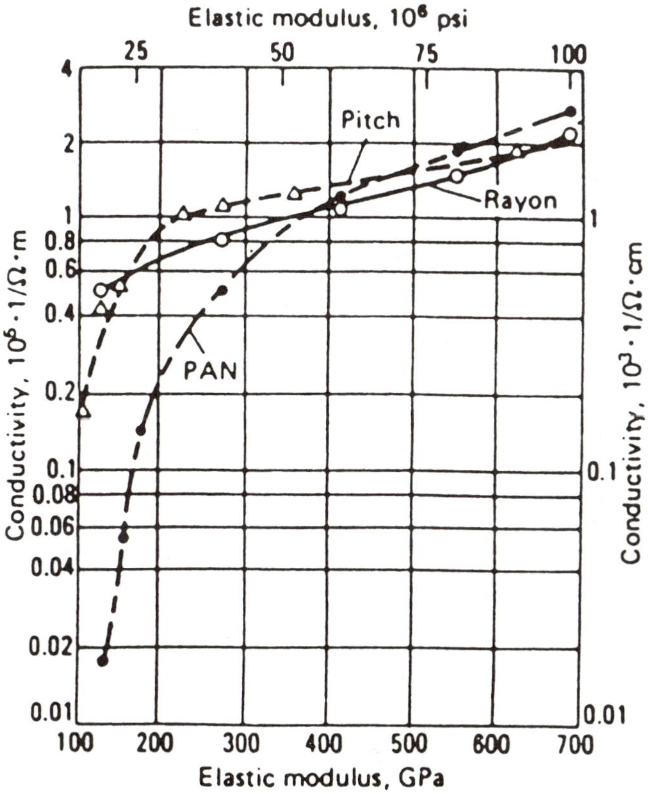

FIGURE 12 The conductivity of carbon fibers increases with fiber tensile modulus (or fiber processing temperature). (From Ref. 13.)

The thermoplastic matrix precursors such as coal tar and petroleum-based pitches are mixtures of polynuclear aromatic hydrocarbons. These materials are capable of experiencing physical changes only from low to high conditions of temperature and pressure while remaining relatively constant chemically. This characteristic is evidenced in the microstructure of the coke. After the carbonization of pitch, coke is formed, and its microstructure is influenced by the pressure and temperature of processing. Figure 27 shows the coke microstructure formed after atmospheric carbonization at 800°C. The microcracks are due to shrinkage during process cooling. Otherwise, the general appearance of the micro-

TABLE 3
Typical Properties of Commercial High Modulus Carbon Fibers

	Precursors		
Property	Rayon	PAN	Pitch
Tensile strength, MPa	2200	2400	2100
Tensile modulus, GPa	393	390	380
Density, × 10^3 kg/m^3	1.66	1.81	2.0
Filament diameter, m	6.5	6.5	10
Elongation at break, %	0.6	0.6	0.5
Elastic recovery, %	100	100	100
Carbon assay, %	99.9	99+	99+
Surface area, m^2/g	1	1	1
Thermal conductivity, W/m·K	122	70	100
Electrical resistivity, $\mu\Omega$·m		9.5	7.5
Longitudinal thermal expansion at 21°C, × 10^6/K		−0.7	−0.9
Specific heat at 21°C, J/kg/K		925	925

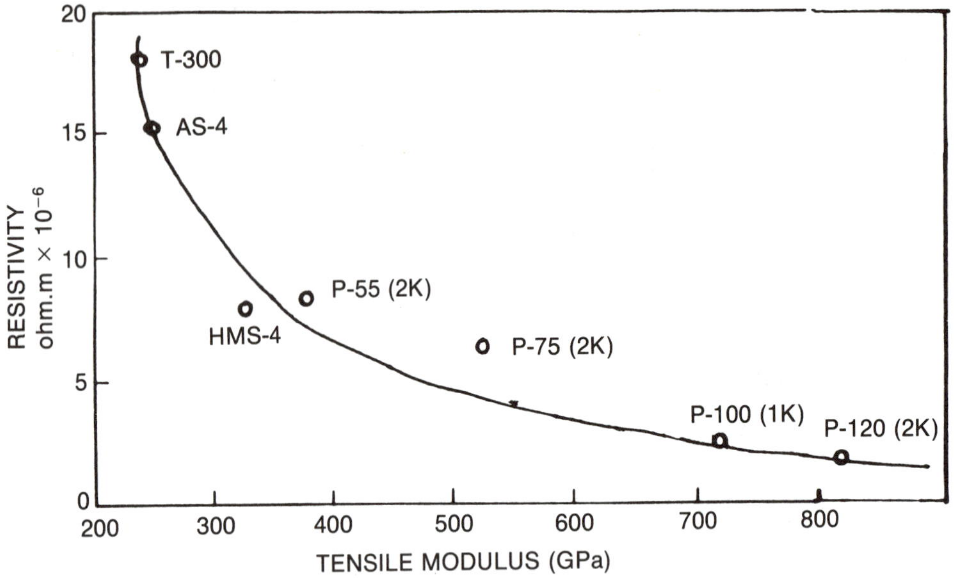

FIGURE 14 Resistivity of commercially available carbon fiber versus tensile modulus.

TABLE 4
Major Carbon Fiber Manufacturers and Their Production Capacities as of October 1986

Country	Manufacturers	Trade Name	Capacity (tons/yr)	Remarks
Japan	Toray	Torayca	1500	
	Toho Rayon	Besfight	1380	
	Asahi-Nippon	Hi-Carbolon	300	
	Mitsubishi Rayon	Pyrofil	120	
	Sumika-Hercules	Magnamite		Importation-900 tons/yr (1987)
	Kureha Chemical	Kureca	400	
	Nippon Carbon	Carbolon	60	Isotropic pitch
United States	Hercules	Magnamite	1050	
	Amoco Performance	Thornel	360	
	Amoco Performance	Thornel P	230	Mesophase pitch
	Celien C.F./BASF	Celien	340	
	Great Lakes Carbon	Fortafil	300	
	Stackpole	Penex	115	
	Hysol-Grafil	Grafil	150	
	Ashland Chemical	Carboflex	20	Isotropic pitch
United Kingdom	Hysol-Grafil	Grafil	350	
	R. K. Textiles	R. K. Carbon	20	
France	Soficar	Filkar	300	
West Germany	Sigri Elektrographit	Sigra-fil	230	
	Enka	Tenax	350	
Other	Afikim (Israel)	ACF	115	
	Taiwan Plastics	Tairyfil	150	
	Korean Steel & Chemical			
Total			7150	PAN base
			710	Pitch base

From Ref. 18.

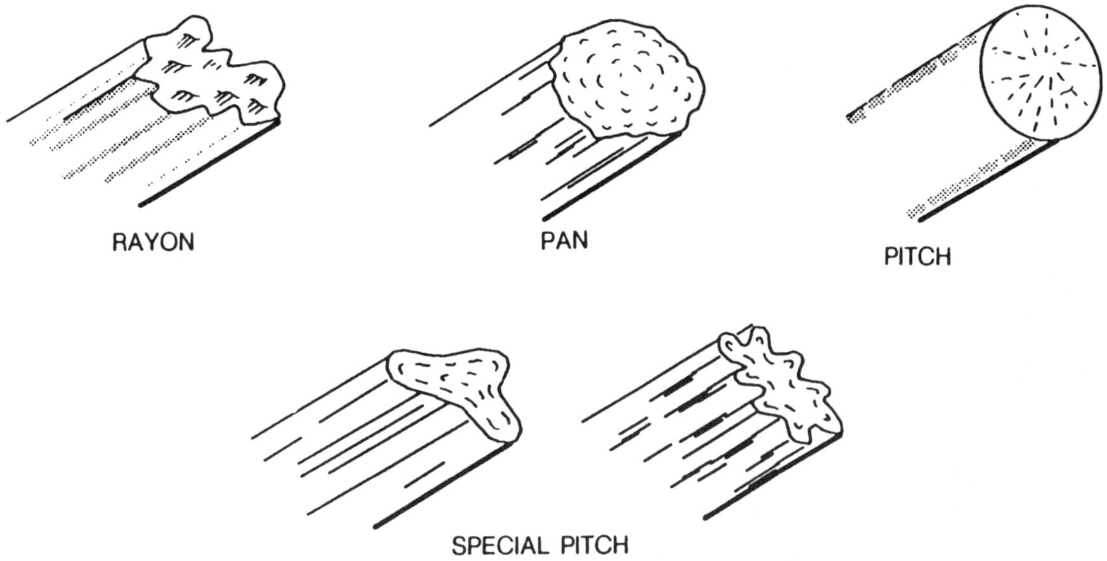

FIGURE 15 Typical carbon fiber microstructural morphology. (From Ref. 15.)

structure is uniform. One theory attributes the uniform and needlelike coke surface to the deformation of mesophase spheres. After heat treatment at 2500°C, more voids or porosity become visible (Fig. 28), apparently because of shrinkage during cooling and the graphite structure of coke after heat treatment.

Figure 29 reveals that the microstructure of coke formed at 1033 MPa and 650°C is rather isotropic and has matrix pocket formation. One theory indicates the possibility of "capturing" low volatiles that escape at the high pressure used during carbonization. Figure 30 indicates the petroleum coke microstructure formed at 1033 MPa and 650°C after heat treatment to 2500°C. The graphite structure of coke after heat treatment enhances the underlying matrix pocket formation.

FIGURE 16 Circular P-55 (2K) pitch-based carbon fibers with peculiar sizing phenomenon. (Photomicrograph courtesy of Lockheed Palo Alto Research Laboratory.)

FIGURE 17 PAN-based carbon fibers with rectangular or "ribbon"-shaped morphology. (Photomicrograph courtesy of BASF Structural Materials, Inc., Celion Carbon Fibers.)

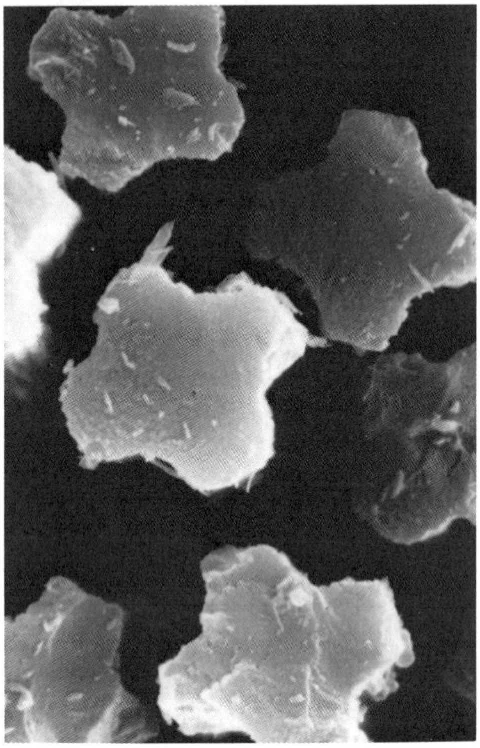

FIGURE 18 PAN-based carbon fibers with tetralobal or "cross"-shaped morphology. (Photomicrograph courtesy BASF Structural Materials, Inc., Celion Carbon Fibers.)

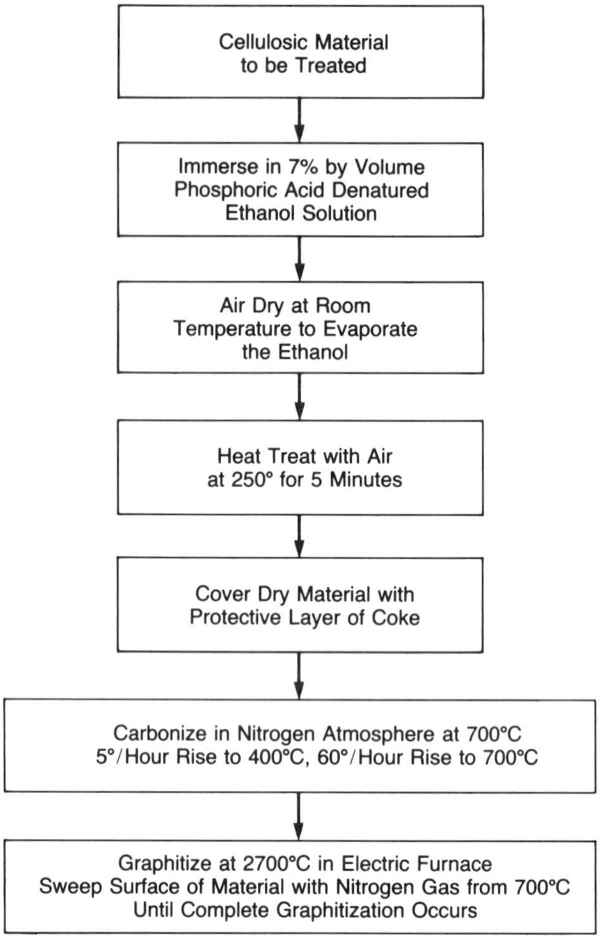

FIGURE 19 Fabrication of rayon precursor carbon fibers. (From Ref. 16.)

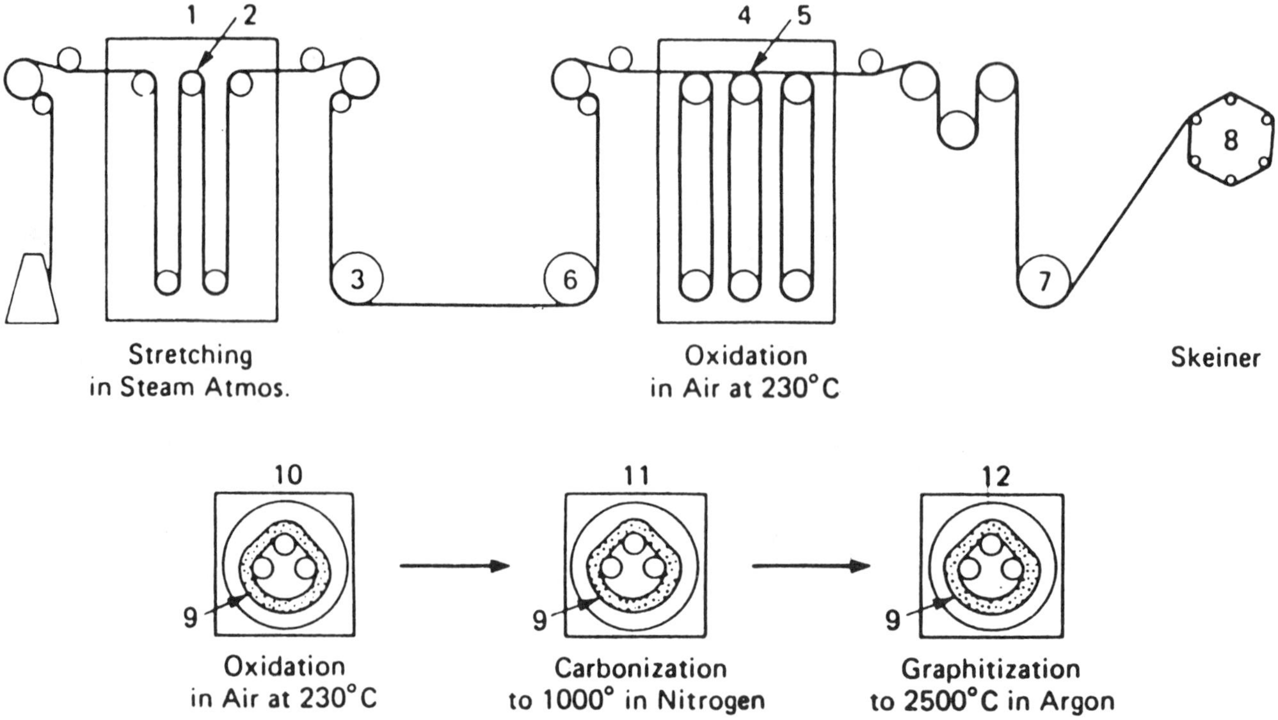

FIGURE 20 Fabrication of PAN precursor carbon fibers. (From Ref. 17.)

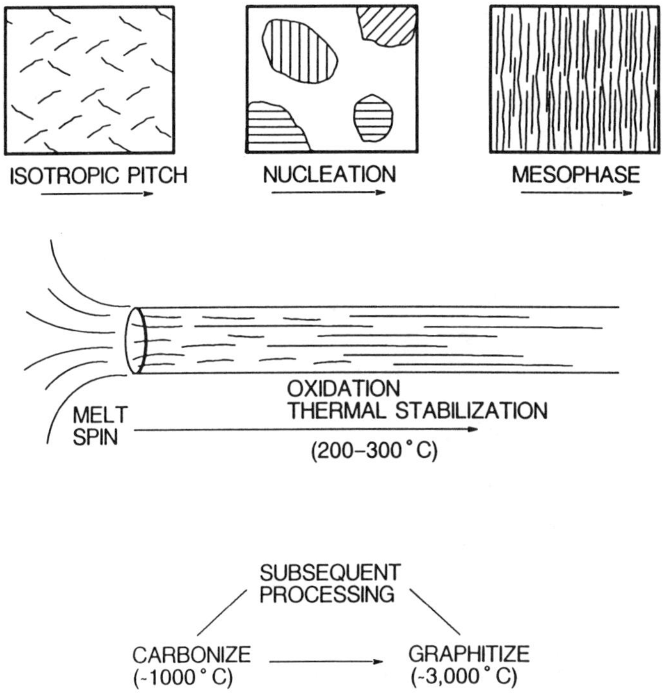

FIGURE 21 **Fabrication of pitch precursor carbon fibers.** (From Ref. 15.)

Contrary to common belief, pitch is a highly process-controlled petroleum by-product. There are three common petroleum pitch processes: thermal, oxidation, and thermal oxidation. The process illustrated in Figure 31 is thermal.

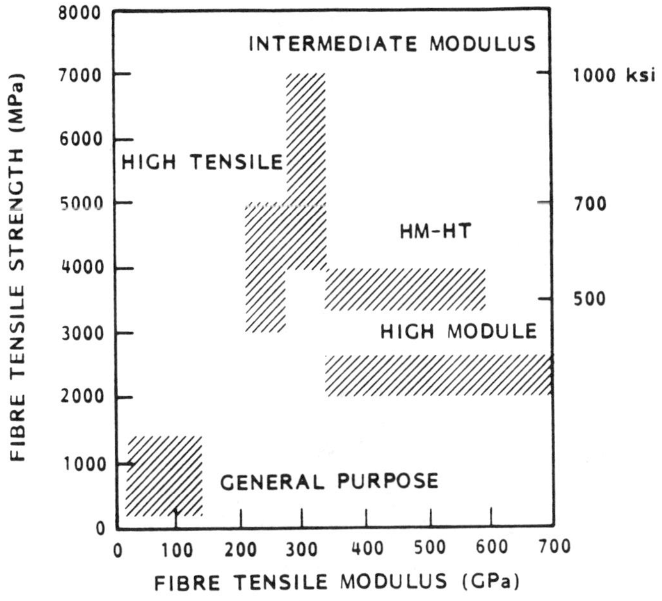

FIGURE 22 **Classification of carbons fibers by tensile strength and modulus.** (From Ref. 18.)

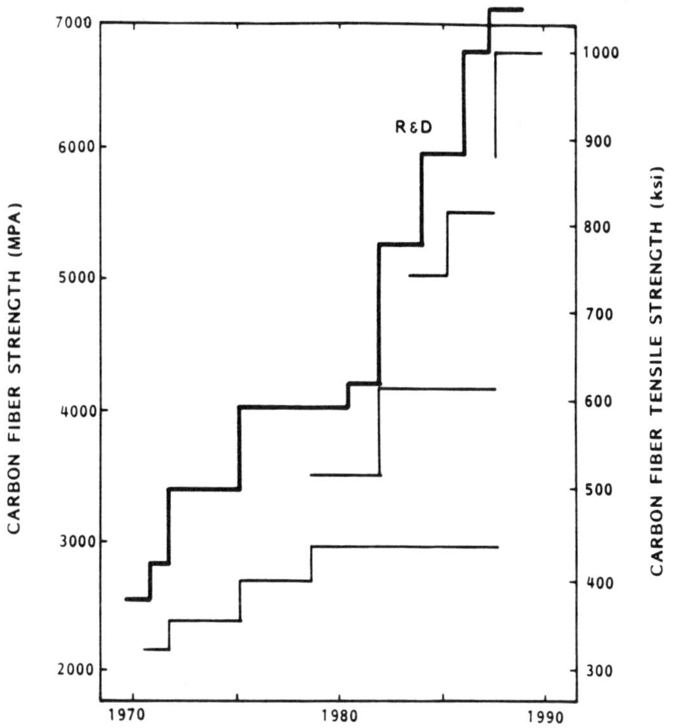

FIGURE 23 **Carbon fiber progress in tensile strength.** (From Ref. 18.)

Hydrocarbon gas is another form of matrix precursor. The hydrocarbon gas commonly used for the chemical vapor deposition/infiltration (CVD/CVI) operation is methane gas. Methane on the preform substrate is transformed into carbon by pyrolysis. Typical properties of the pitches are shown in Table 7, while other matrix precursors properties are available in standard references.

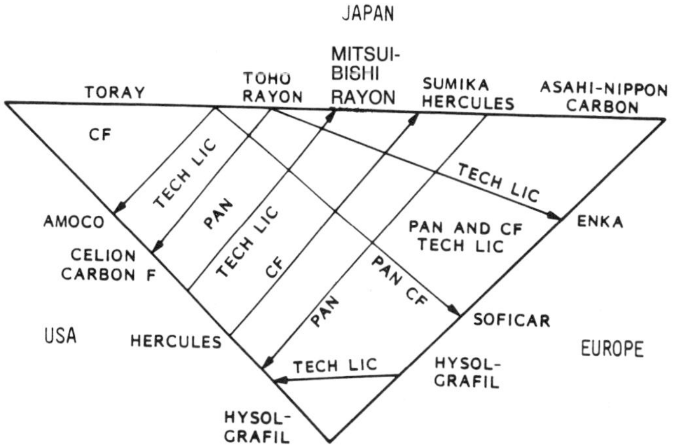

FIGURE 24 **International technical and business relationship in the manufacture of carbon fiber manufacture (see also Table 4).** (From Ref. 18.)

TABLE 5
PAN-Based Carbon Fiber Consumption 1983–1987 and Projections 1988–1995 (tons/year)

Area	1983	1984	1985	1986	1987	1988	1989	1990	1995
United States	1,100	1,320	1,580	1,900	2,280	2,740	3,290	3,950	5,400
Japan	500	570	600	660	730	800	800	970	1,500
Europe	300	450	580	700	850	1,010	1,200	1,750	2,800
Others	340	450	400	450	500	550	600	660	1,000
Total	2,240	2,790	3,160	3,710	4,360	5,100	5,970	7,330	10,700

From Ref. 18.

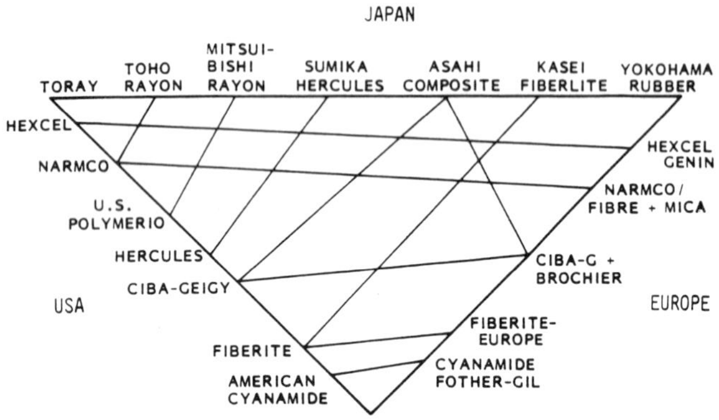

FIGURE 25 International technical and business relationships for the manufacture of prepreg. (From Ref. 18.)

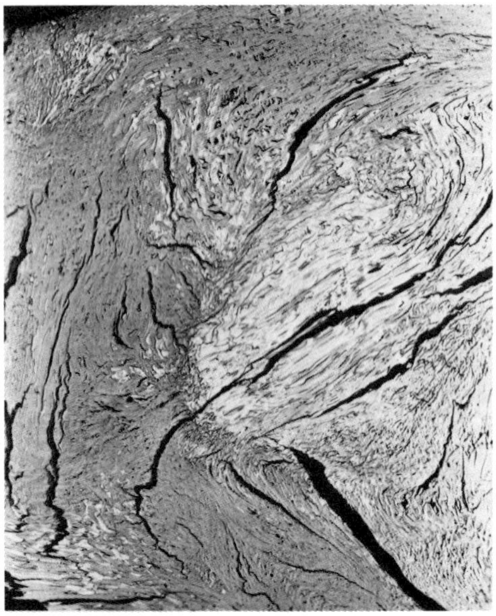

FIGURE 27 Microstructure of petroleum pitch A-240 being carbonized at atmospheric pressure at 800°C, before heat treatment. (Photomicrograph courtesy of Lockheed Palo Alto Research Laboratory.)

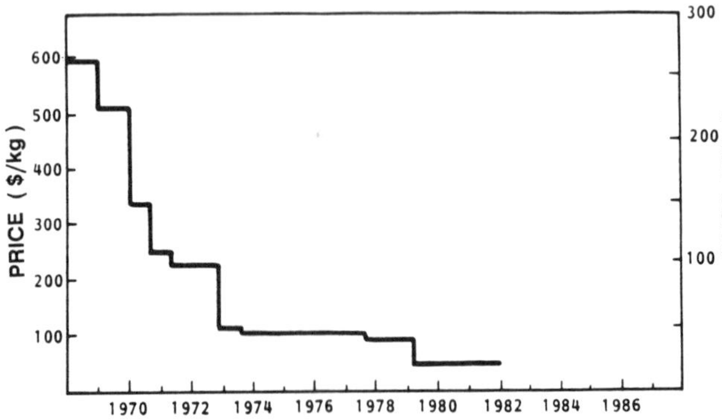

FIGURE 26 Carbon fiber price history. (From Ref. 4.)

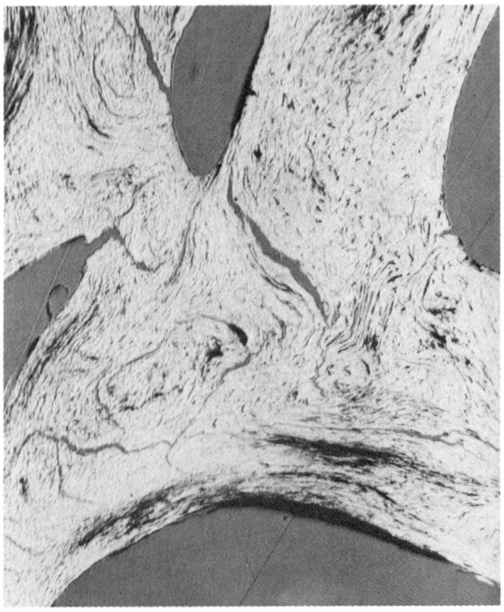

FIGURE 28 Microstructure of petroleum pitch A-240 being carbonized at atmospheric pressure at 800°C, after heat treatment at 2500°C. (Photomicrograph courtesy of Lockheed Palo Alto Research Laboratory.)

TABLE 6
Three Common Carbon Matrix Precursors

Thermosetting Resin	Thermoplastic Pitch	Hydrocarbon
Furfurals	Coal tar based	Methane Propane
Phenolics	Petroleum based	Acetylene Benzene

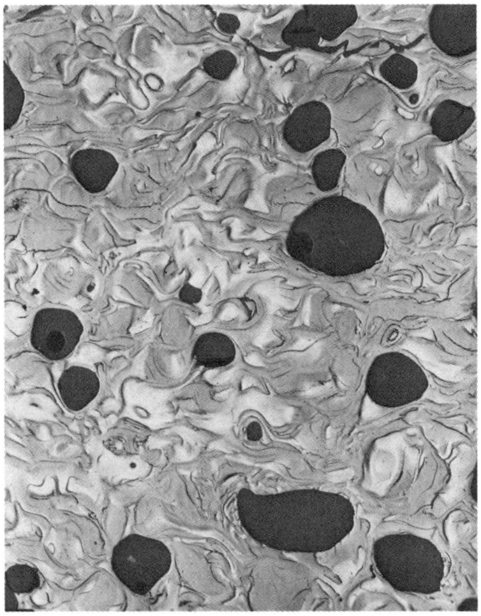

FIGURE 29 Microstructure of petroleum pitch A-240 being carbonized at 1033 MPa at 650°C, before heat treatment. (Photomicrograph courtesy of Lockheed Palo Alto Research Laboratory.)

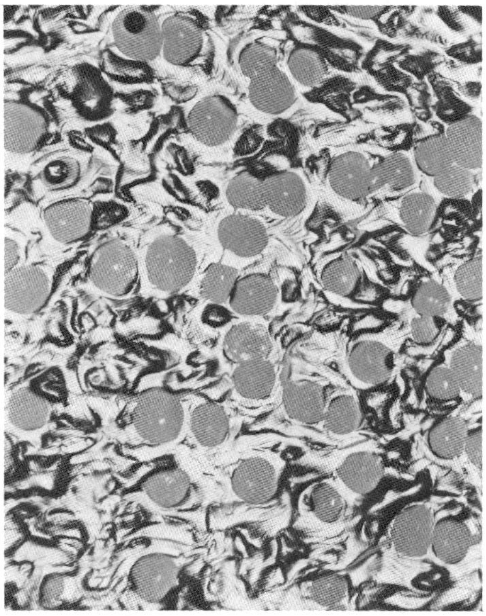

FIGURE 30 Microstructure of petroleum pitch A-240 being carbonized at 1033 MPa at 650°C, after heat treatment to 2500°C. (Photomicrograph courtesy of Lockheed Palo Alto Research Laboratory.)

TABLE 7
Typical Properties of Various Thermoplastic Carbon Matrices: Pitches

	Carbon Matrix Type			
Property	**Petroleum Based**		**Coal Tar Based**	
Softening point, °C	121		94–107	
Density, kg/m^3	1220		1280–1310	
Viscosity, Pa·s	~ 0.03	at 250°C	0.03–0.05	at 250°C
Benzene insolubles, wt %	8		24–28	
Quinoline insolubles, wt %	Negligible		2–7	
Sulfur, wt %	2.5		0.1–0.6	
Ash, wt %	0.05		0.2–0.5	
Carbon, wt %	50–55		52–62	

Common Carbon–Carbon Fabric and Preform Constructions

To produce a carbon–carbon composite meeting the desired design requirements demands hand-in-glove cooperation between the textile and material design engineers. Weave design, such as style (plain or satin), fiber selection (denier), and fiber bundle spacing, must be carefully selected by the textile engineer to give the optimum fabric or preform. Table 8 summarizes the types of fabric and preform construction.

For a design to be acceptable in quality, there must be minimal fiber damage during the fiber placement operation. Fiber position as designed must be achieved accurately, and maximum conformity and retainment of the product shape must also be realized.

Translation of the physical and mechanical properties of the fabric or preform into structural and thermal properties falls in the hands of the material design engineer. With this understanding, let's examine some common weave geometries and analyze the advantages and disadvantages of each pattern, to see how this factor can play an important role in carbon–carbon composites.

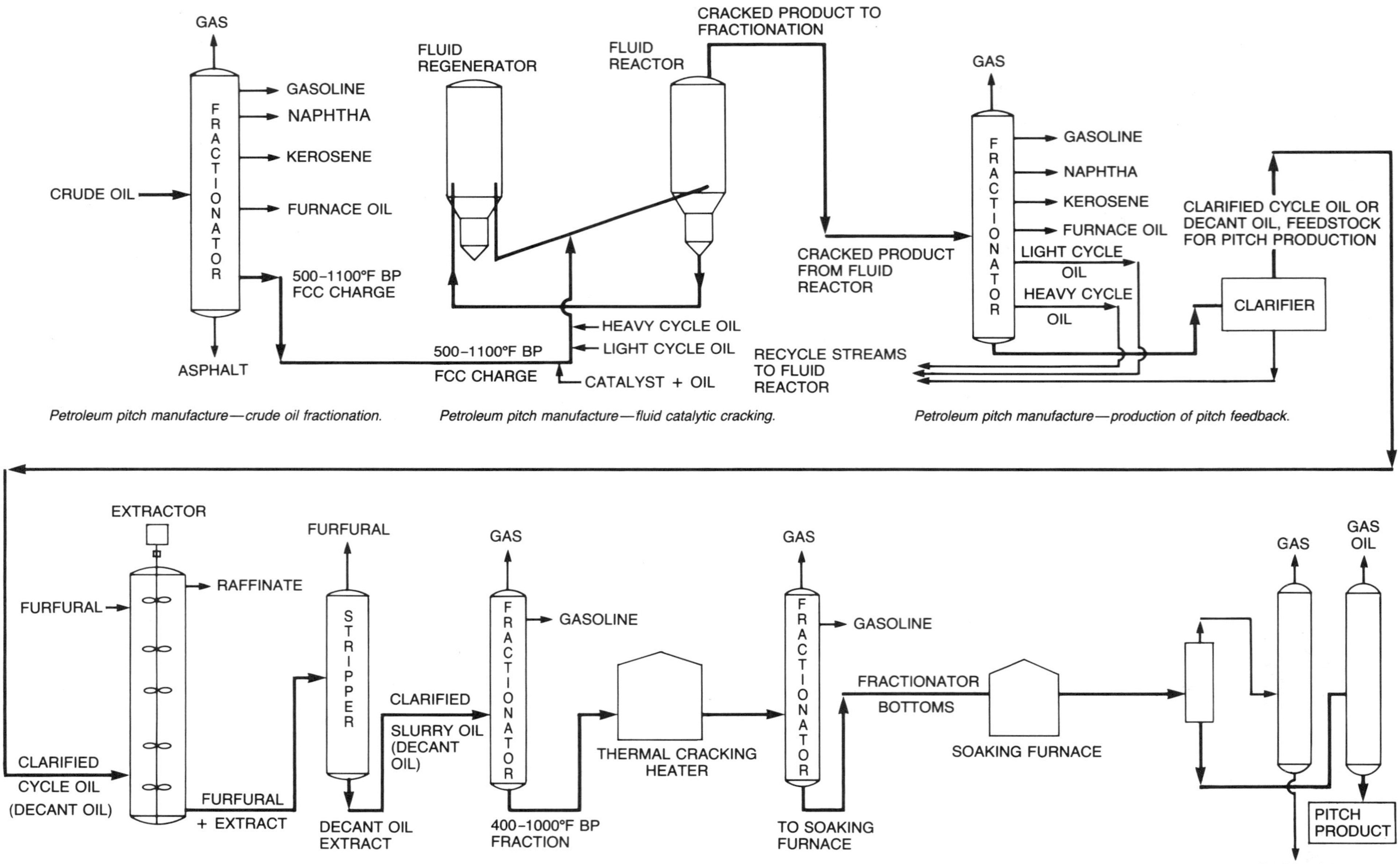

Petroleum pitch manufacture—crude oil fractionation.

Petroleum pitch manufacture—fluid catalytic cracking.

Petroleum pitch manufacture—production of pitch feedback.

Petroleum pitch manufacture—concentration of aromatics by solvent extraction.

Petroleum pitch manufacture—thermal process, initial step.

Petroleum pitch manufacture—thermal process, final step.

FIGURE 31 Petroleum-based pitch precursor matrix manufactured by thermal process. (From Ref. 19.)

TABLE 8
Types of Fabric and Preform Construction

Type	Variations	Construction
Uniaxial/unidirectional (1D)	Uniaxial	Uniaxial tape
		Laminate
		Warp fibers stitched together
		Uniaxial filament winding
Biaxial/two-directional (2D)	Biaxial 2D	Fabric with warp and fill interlock
		Biaxial 2D braiding
		Filament winding
	Biaxial 3D	Fabric with layers of warp angle interlock
		Filament winding with layer angle interlock
Triaxial/three-directional (3D)	Triaxial 3D	Triaxial 3D braiding
		Fabric with layers of warp angle interlock with stuffers in warp direction
	Cartesian 3D (orthogonal)	x, y, z axes orthogonal to each other
	Polar 3D	Axes of fiber oriented in polar coordinates
Multiaxial/multidirectional	Tetraxial 3D/4D	x and y fibers at 45° to each other along z axis
	Pentaxial 3D/5D	± 45° in-plane reinforcement with respect to x–y along z axis
	Heptaxial 3D/7D	Additional face and diagonal fibers
	Undecaxial 3D/11D	Additional face and diagonal fibers

One-Directional (1D, Unidirectional) Fabric Construction

A unidirectional material has the most "unbalanced" weave, and such weaves usually are reserved for applications that require axial symmetry along the fiber axis. Thus the term "uniaxial 1D" is also used. The warp fibers are commonly held together in the laminate form or by being stitched together. Warp/end fibers are yarns running lengthwise in a woven fabric roll. Fill/pick fibers are oriented perpendicular to the warp fibers. Traditionally, filament winding is a composite fabrication process in which filamentary fiber is wound continuously in nonwoven, parallel, and uniaxial arrays. Filament winding provides great control over fiber placement and uniformity thickness of structure, such as rocket motor casings and pressure vessels. Nowadays this technology has been expanded successfully to apply to the fabrication of noncylindrical, complex geometrical structures.

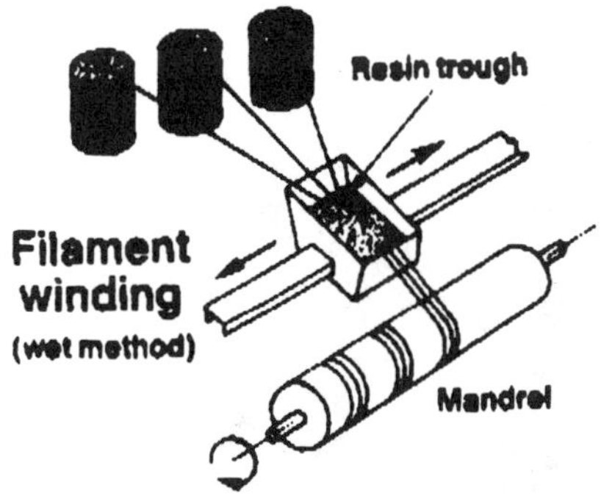

FIGURE 32 Filament winding (wet method).

The dry process involves winding resin-impregnated carbon fibers on a rotating mandrel in predetermined patterns. In the wet method (Fig. 32), the fibers pick up the matrix by passing through either a trough or a metered system. After the fibers are wound, the part is densified and removed from the mandrel.

Two-Directional (2D) Fabric Construction

Biaxial 2D construction may provide fibers in the z direction by varying angles of the "through the thickness" warp fibers (Fig. 33). Conventional 2D fabrics are shown schematically and in photos in Figures 34 and 35. Plain weave, which is the most highly interlaced, is the tightest basic fabric design and the most resistant to in-plane shear movement; hence, sometimes it is difficult to wet out during densification. However, plain weave can be relied on to give reproducible laminate thickness.

Basket weave is a variation of plain weave and has two warp and two fill yarns.

Satin weave represents a family of constructions with a minimum of interlacing. It can be woven as standard four- (crowfoot), five-, or eight-harness fiber. In four-end

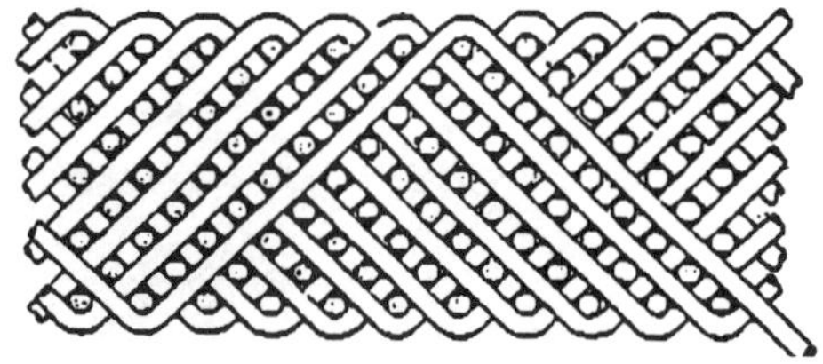

FIGURE 33 Biaxial 2D weaving with warp angle interlock. (From Ref. 20.)

satin weave, warp yarn is skipped over three fill yarns and goes under one fill. The yarn skip pattern repeats. For five-end and eight-end satin weave, the warp yarn skips over four and seven fill yarns, respectively, and goes under one. The satin weave construction offers the floating yarn advantage, which contributes more strength to a composite because of its straight length.

Satin weave has low resistance to shear distortion: as the number of harnesses increases, so does the degree of drapability (ability to conform to complex contour shapes). Smooth surface, minimum thickness, and high tensile and flexural strength are just a few reasons for satin weave to be preferred for many aerospace applications.

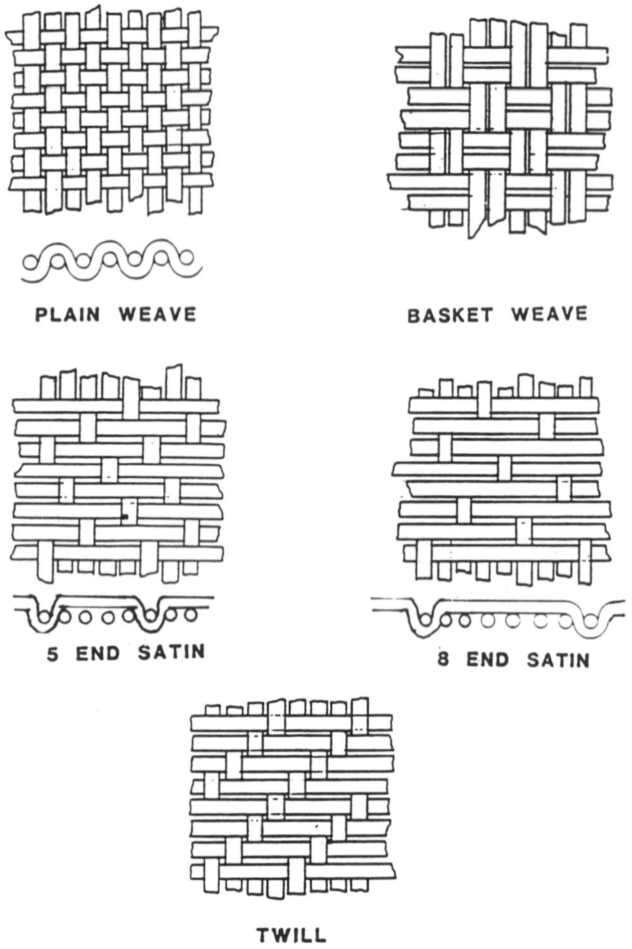

FIGURE 34 Schematics of conventional 2D fabric. (From Ref. 21.)

Twill weave warps and fills interlace regularly, forming a distinct diagonal line on the fabric surface. Twill will follow contours more easily than plain fabric and is much easier to wet out.

Biaxial braid, a variation of 2D weaving [22], is the most common because it conforms to various shapes, contours, and diameters. The braiding angle can be varied from 10° to 85°. Small angles (~ 10°) are best for tensile strength; 45° is best for torsional strength; and large angles (~ 85°) are best for hoop strength. The most common angle is 45°, which gives a balance and isotropic properties.

Biaxial 2D filament winding laminates consist of unidirectional laminates on plies with angles that vary from nearly 0° to 90° to the axis of the mandrel.

Three-Directional (3D) Preform Construction

BIAXIAL 3D. In biaxial 3D weaving, the warp fiber may pass completely "through the thickness" or it may interlock only the adjacent layers or any number of the

FIGURE 35 Photos of conventional 2D fabric up to 24″ wide. (Photo courtesy of Techniweave, Inc.)

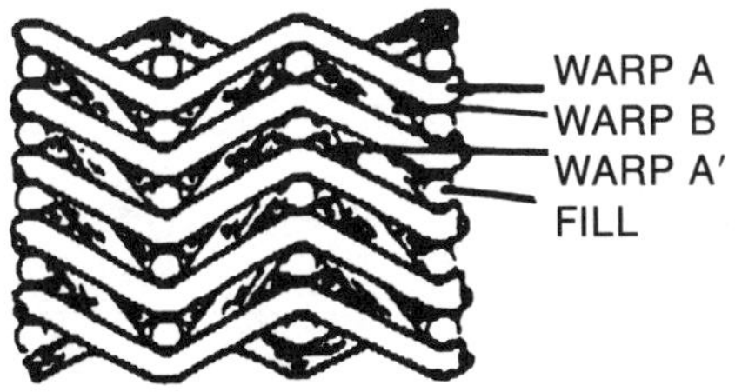

FIGURE 36 Biaxial 3D weaving with layer angle interlock. (From Ref. 20.)

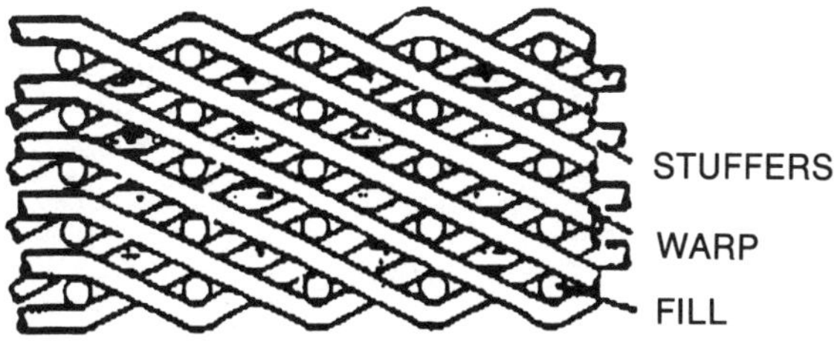

FIGURE 37 Triaxial 3D weaving with stuffers in warp direction. (From Ref. 20.)

adjacent layers (Fig. 36). Filament-wound biaxial 3D weaving results in a fabric with layer interlock in which the warp and fill fibers are not normal to each other.

TRIAXIAL 3D. Triaxial braiding is a modified 3D weaving technique. Triaxial braid is constructed by introducing a third fiber, parallel to the braid axis. The parallel, or 0°, fibers add tensile strength and stiffness while resisting bending. However, triaxials are locked—they cannot conform or be molded to other shapes. Because of these restrictions, a triaxial fiber must be braided into a mandrel, then cured on the mandrel. The mandrel must also be removed after cure. A good description of using braid for composite preforms for manufacture of structural shape appears in Ref. 23.

Triaxial 3D weave is similar to biaxial 3D except that stuffers or fills are added for the additional axis to improve properties in the warp direction (Fig. 37).

3D CARTESIAN COORDINATE WEAVE. Perhaps one of the most widely used and simplest types of multidirectional preform among carbon–carbon composite applications is the 3D orthogonal weave construction or block preform (Fig. 38). This type of preform offers the flexibility to orient selected fiber amounts, types, and spacing between adjacent fiber bundles to achieve optimum material design and composite structural loads.

Typical Cartesian orthogonal block weave characteristics are shown in Table 9. Two common 3D Cartesian coordinate weave processes are represented in Figures 39 and 40. End product configurations of this type of construction are usually accomplished by machining after preform densification.

3D POLAR COORDINATE WEAVE. Another popular type of weave construction for cylindrical shapes or other shapes of revolving objects is the polar coordinate 3D weave (Fig. 41). By optimizing fiber amount, types, and spacing (as in the Cartesian coordinate weave) in the axial, circumferential, and radial directions, optimum composite design may be achieved. Typical characteristics of the 3D polar coordinate preform are shown in Table 10.

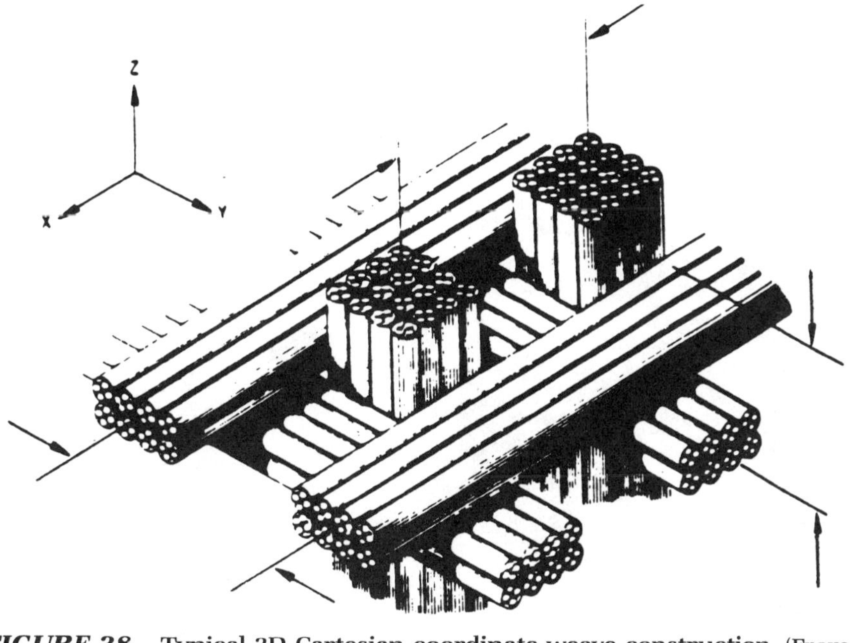

FIGURE 38 Typical 3D Cartesian coordinate weave construction. (From Ref. 24.)

TABLE 9
Characteristics of 3D Cartesian Coordinate Preforms

Material	Bulk Density (kg/m^3)	Number of Yarn Bundles			Center-to-Center Bundle Spacing (mm)		Fiber (vol %)[a]		
		x	y	z	x, y	z	x	y	z
Thornel 50[b]	640	1	1	1	0.56	0.58	0.14	0.14	0.13
	750	1	1	2	0.71	0.58	0.11	0.11	0.23
	680	2	2	1	1.02	0.58	0.14	0.14	0.12
	800	2	2	6	0.69	1.02	0.12	0.12	0.24
Thornel 75[b]	700	1	1	2	0.56	0.58	0.09	0.09	0.17
	650	2	2	1	0.84	0.58	0.12	0.12	0.09
	720	2	2	2	1.07	0.58	0.09	0.09	0.18

[a] Volume fraction of total preform volume occupied by fiber in each orthogonal direction.
[b] Center-to-center spacing.
From Ref. 24.

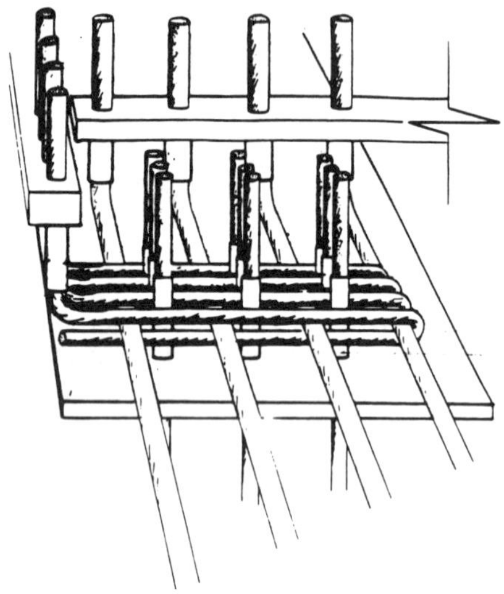

FIGURE 39 3D Cartesian coordinate orthogonal block weave process. (From Ref. 25.)

Multidirectional Preform Constructions

In general, in designing carbon–carbon composites, mechanical properties can be improved by increasing the composite fiber volume ratio. Fibers can be woven or placed in the direction in which strength is required. Balanced weave may be the most desirable where isotropic composite properties are required. To impart additional shear strength to a carbon–carbon, composite, more diagonal fibers may be placed as reinforcement.

TETRAXIAL 3D OR 4D. Four-directional (4D) design (Fig. 42*b*) represents the basic 3D orthogonal structure (Fig. 42*a*) with x- and y-direction fibers at 45° axis.

PENTAXIAL 3D or 5D. Five-directional (5D) design (Fig. 42*c*) resembles the 4D, but the weave configuration is ± 45° in-plane reinforcement with respect to the x-y fibers along the z axis.

HEPTAXIAL 3D OR 7D. Figure 43*a* shows diagonal reinforcement across the four corners, T, U, V, and W, when combining with the basic 3D x, y, and z fibers, producing one type of 7D design. Another type of 7D

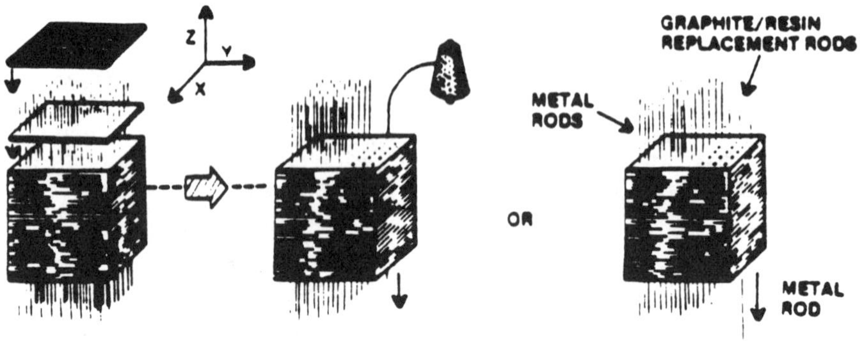

FIGURE 40 3D Cartesian coordinate pierced fabric block weave process. (From Ref. 26.)

TABLE 10
Typical Characteristics[a] of 3D Polar Coordinate Preform

Diameter (mm)	Weave Spacings (mm)			Fiber Volume (%)			
	z	θ	r	z	θ	r	Total
99	1.5	1.8	2.8	15	23.2	9.3	47.5
286	1.6	2.3	2.7	11.8	22.1	13.2	47.1
500	1.4	8.7	1.4	6.0	30.0	3.9	39.9
1156	1	2	1.7	11	23	11	45

[a] z = axial direction, θ = circumferential direction, r = radial direction.
From Ref. 4.

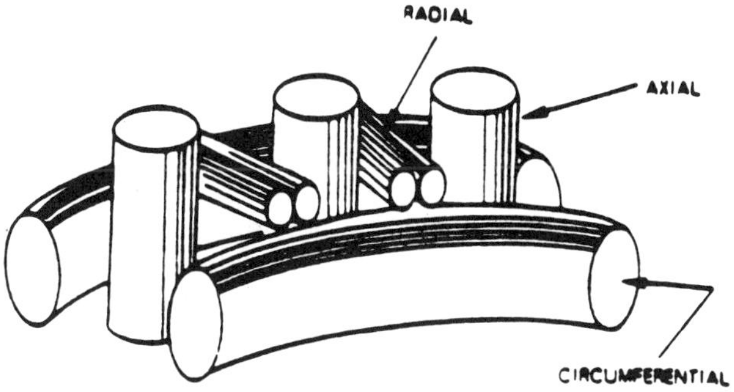

FIGURE 41 Typical 3D polar coordinate weave construction. (From Ref. 24.)

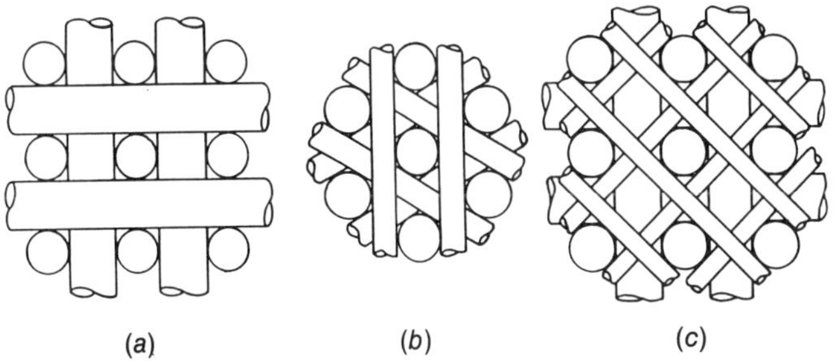

FIGURE 42 Multidirectional perform constructions: (*a*) 3D orthogonal, (*b*) 4D hexagonal, and (*c*) 5D isotropic.

structure contains diagonal yarns across the face of the preform (Fig. 43*b*). Both the diagonal across-the-corner or across-the-face constructions can be used to produce 4D constructions by elimination of the baseline 3D orthogonal portion of the structure.

UNDECAXIAL 3D OR 11D. A combination of the 3D base structure, diagonal across-the-corners, and diagonal across-the-face will produce a third type of isotropic woven structure with 11 directions of reinforcement (Fig. 43*c*).

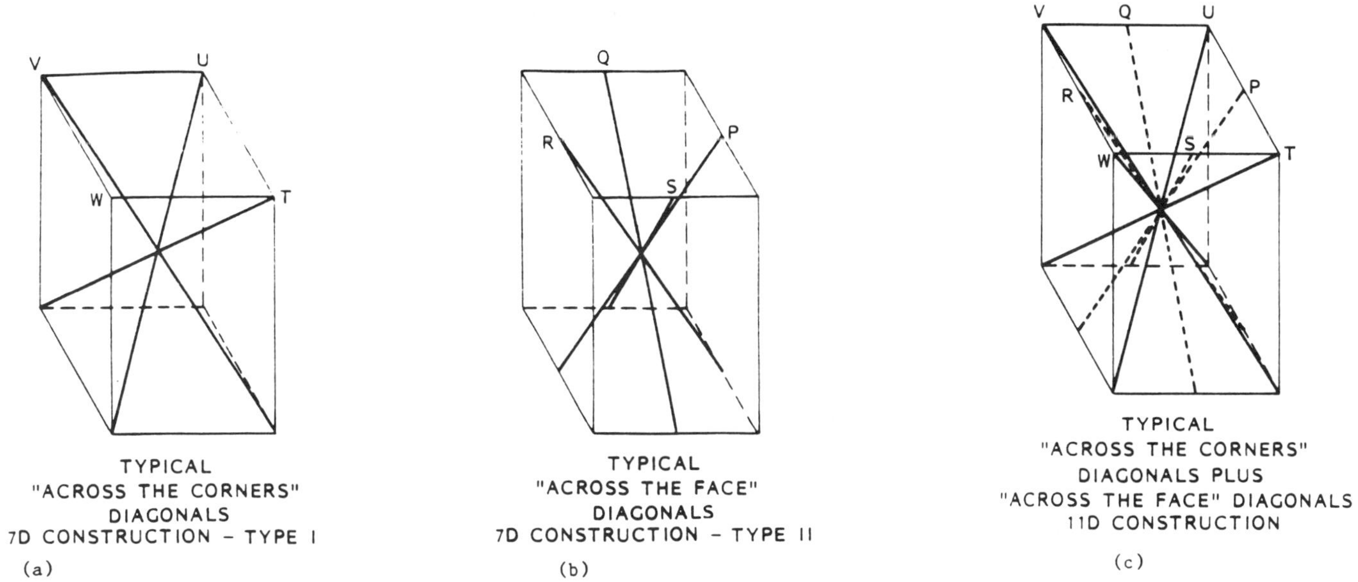

FIGURE 43 Multidirectional preform construction. (From Ref. 24.)

Carbon–Carbon Composite Processing

Three methods of processing carbon–carbon composites, common to the aerospace industry (Fig. 44), involve solid, liquid, and vapor phase pyrolysis and heat treatment (graphitization).

Liquid phase impregnation makes use of the knowledge that most thermosetting resins polymerize at low temperature (< 250°C) to form a highly cross-linked, thermosetting, nonmelting, amorphous solid. Upon pyrolysis, these resins form a glassy carbon that does not graphitize at temperatures up to 3000°C [27,28].

Thermoplastic pitch at the liquid impregnation phase (at 400°C) undergoes various changes, including volatilization of low molecular weight compounds, decrease in viscosity, polymerization, cleavage and rearrangement of molecular structures, and forming of mesophase pitch.

Impregnation is usually achieved under an inert atmosphere to minimize oxidation of carbon. It is also usually done under vacuum or slight pressure, to ensure complete impregnation of the preform with the impregnant.

Low and High Pressure Carbonization

Carbonized thermoplastic pitch is usually processed at a controlled heating rate to the temperature range of 800–1500°C in about 80 hours. During that period, the process is operated under an inert atmosphere and the pitch is carbonized to coke. The efficiency of the carbonization step is improved if high pressure, in some cases up to 1033 MPa, is used. The pressure is increased gradually while the temperature is increased at a controlled rate. Carbonization processing time may be cut into half by improving efficiency.

Graphitization

The last step in the processing of carbon–carbon composites is usually heat treatment at temperatures between 1500 and 2750°C. This step transforms coke into a more graphitic structure, thus opening up more porosity for reimpregnation. The entire process may be repeated until the desired density is reached.

Gas Phase Impregnation/Chemical Vapor Deposition/Chemical Vapor Infusion

The CVD processes operate on isothermal, thermal gradient, or differential pressure principles. Hydrocarbon and carrier gases are introduced and allowed to infuse/infiltrate into the hot substrate. The isothermal process, which is usually conducted at reduced pressure, generally produces a uniform deposit through the substrate. The crust on the substrate preform must be machined off between cycles to achieve a high density composite.

In the thermal gradient process, the induction coil and susceptor must be tailored to the shape of the preform substrate. The hydrocarbon and carrier gases are driven across the substrate preform to the outer region of the temperature gradient of the substrate. The thermal gradient process offers shorter processing time but limits its operation to one item per batch, whereas the isothermal process can operate more than one item at a time.

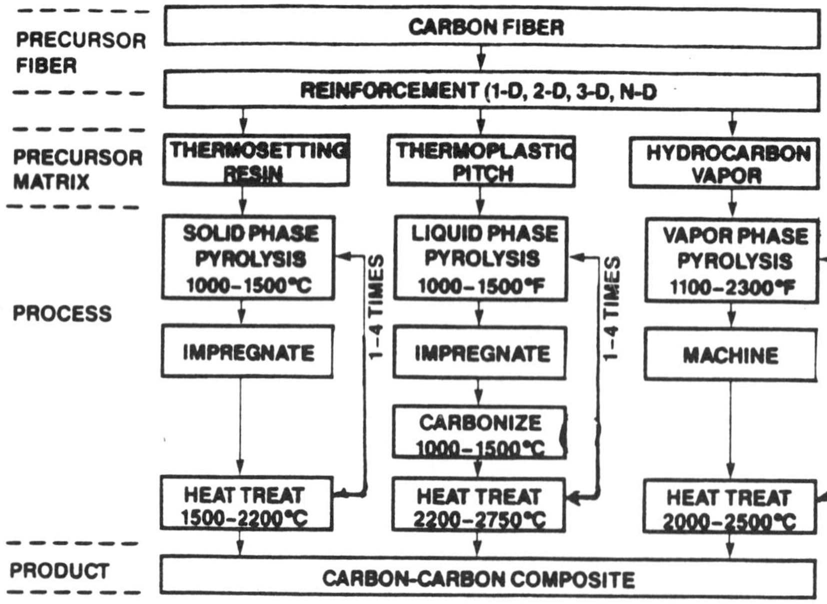

FIGURE 44 Three common methods of carbon–carbon composite processing in the aerospace industry. For thermoplastic pitch, carbonization may be pressurized between atmosphere and 1033 MPa to improve process efficiency; for hydrocarbon vapor, vapor phase pyrolysis is slightly pressurized to 1300–1400 Pa. (From Ref. 1.)

TABLE 11
Typical Properties of Unidirectional Carbon–Carbon Composites[a]

Property	Parallel		Perpendicular	
	HTU	HMS	HTU	HMS
Tensile strength, MPa	600	575	4	5
Tensile modulus, GPa	125	220		
Compressive strength, MPa	285	380	25	50
Compressive modulus, GPa	10	250	7.5	
Bend strength, MPa	1250–1600	825–1000	20	30
Shear strength, MPa	20	28		
Fracture toughness, kJ/m^2	70	20	0.4	0.8

[a] HTU = high tensile, untreated; HMS = high modulus, surface treated.
From Ref. 29.

TABLE 12
Typical Properties of 2D Carbon–Carbon Composites

Property	Type of Composite	
	RCC[a]	ACC[b]
Tensile strength, MPa	62	348
Tensile modulus, GPa	14.5	106
Compressive strength, MPa	185	160
Compressive modulus, GPa	31	137
Shear strength, MPa		
In-plane	38	37
Interlaminar	10.3	8.3
Tensile strength (interlaminar), MPa	5.3	4.6
Coefficient of thermal expansion at 1093°C, 10^{-6}/K	2.6	1.8
Thermal conductivity at 1093°C, (W/m·K)	12	36
Density, kg/m^3	1578	1771
Nominal design thickness, mm	5.9	5.8
Number of plies	19	19

[a] Space shuttle nose cone and leading edge, rayon precursor uncoated carbon–carbon material.
[b] Advanced carbon–carbon materials, PAN precursor.
From Ref. 30.

Another type of CVD process is based on differential pressure, or the pressure gradient technique. The positive pressure forces the hydrocarbon and carrier gas to go through the preform substrate, thus carbonizing the preform substrate. The CVD process offers certain property advantages over other densification methods. For example, it is believed that CVD can improve fiber–matrix bonding. Typical properties of various directional carbon–carbon composites are shown in Tables 11–14.

Carbon–Carbon Design Considerations, Selection Criteria, and Case Studies

It is indisputable that the properties of carbon–carbon composites depend on the following parameters.

- Type of carbon fiber
- Carbon fiber distribution
- Weaving characteristics/construction
- Matrix precursor selection
- Processing conditions and sequence

TABLE 13
Typical Properties of 3D Polar Weave Carbon–Carbon Composites in Axial (z), Circumferential (θ), and Radial (r) Directions

Property	z	θ	r	z	θ	r
Tensile strength, MPa		83–172				138–241
Tensile modulus, GPa		41–62			69–138	
Compressive strength, MPa	103–172	103–138		103–138	103–138	
Compressive modulus, GPa	34–55	48–69		41–48	41–48	
Shear strength, MPa			14–28			7–28
Density		1400–1600			1900	

From Ref. 4.

TABLE 14
Typical Properties of 3D Cartesian Weave Carbon–Carbon Composites

Property	3D Orthogonal Weave[a] z	3D Orthogonal Weave[a] x–y	3D Pierced Fabric[b] z	3D Pierced Fabric[b] x–y
Density, kg/m^3	1900	1900	1650	1650
Tensile strength, MPa				
At room temperature	310	103	103.4	34.5
At 1900 K/2758 K	400	124	68.9	64.6
Tensile modulus, GPa				
At room temperature	152	62	41.3	11.0
At 1900 K/2758 K	159	83	10.3	6.2
Compressive strength, MPa				
At room temperature	159	117	82.7	62
At 1900 K/2758 K	196	166	158.5	109.6
Compressive modulus, GPa				
At room temperature	131	69	22.7	11.0
At 1900 K/2758 K	110	62	12.4	6.2
Thermal conductivity, W/m·K				
At room temperature	246	149	55.4	83
At 1900 K/2758 K	60	44	24.2	27.7
Coefficient of thermal expansion 10^{-6}/K				
At room temperature	0	0		
At 1900 K/811 K	3	4	0.2	1.0
At 3000 K/2758 K	8	11	4.7	9.4

[a] High modulus fiber (Thornel 50); Coal tar pitch as matrix; HIP process.
[b] Low modulus fabric (WCA) in x–y and high modulus fiber (Thornel 50) in z; phenolic resin as matrix; HIP process.
From Refs. 31, 32.

Type of Carbon Fiber

That the use of different types of carbon fiber influences the properties of a composite product is intuitively clear. In general, where structural properties are important, high strength, high modulus fibers are usually selected. The highest modulus fiber will provide the highest thermal conductivity and the lowest thermal expansion. Thus in applications calling for low thermal conductivity, a low modulus fiber usually is preferred. One drawback of the high modulus fiber is its relatively poor wetting characteristics, which perhaps are due to the highly crystalline surface that is induced by the high processing temperature, with effects on the degree of graphitization.

Another rule of thumb in selecting a fiber is that a design factor of 2 should be employed; that is, the fiber modulus should be twice the desired composite modulus. However, this design factor depends highly on other criteria, such as matrix type, weave geometry, and processing conditions.

Carbon Fiber Distribution

Experience has shown the mechanical properties of a carbon–carbon composite usually improve when fiber volume increases. Different types of tape, fabric, and weave geometry affect the fiber volume of the composite. Typically unidirectional tape can build up a fiber volume of 65%. Two-directional weave typically has a higher fiber volume of 50–60%, versus 40–55% for three-directional weave. Higher composite fiber volume by no means categorically indicates that the composite is stronger.

As a matter of fact, the more balanced, three-directional orthogonal weave, which has less fiber in volume, represents a higher shear strength than two-directional carbon–carbon in general. Ease of fabrication in two-directional carbon–carbon compared with three-directional carbon–carbon compensates for the low shear properties and high risk of delamination during processing of 2D. Theoretically, the variations in tape lay ups, fabric types, and weave geometries are unlimited. The task of selecting the appropriate fiber distribution and arrangement in carbon–carbon becomes an art. The variations of 3D orthogonal weave shown in Figure 45 range from a balanced isotropy of 1,1,1 to the highly z-oriented strength and improved shear properties of 2,2,6.

It is a peculiarity of the cylindrical weave geometry (Fig. 46) that to maintain uniform density structure, axial fiber sizes can be increased from the inner to the outer diameter of the cylinder, or to maintain uniform strength, partial radial fibers can be inserted in a pattern, as also shown. Hybrid weaving can tailor strength and

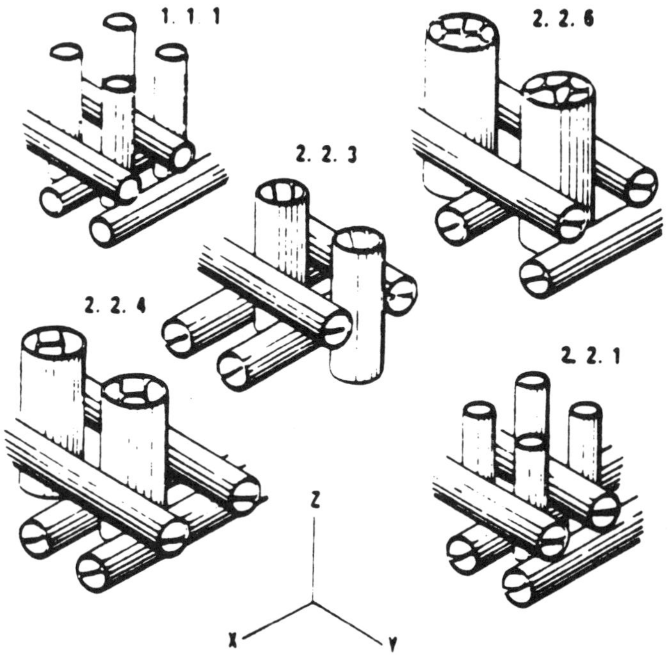

FIGURE 45 Types of 3D orthogonal weave. (From Ref. 24.)

take advantage of various fiber types by combining them in a fashion that will meet design requirements.

Weaving Characteristics and Construction

Carbon–carbon composites manufactured using the fine weave orthogonal and the pierced-fabric processes were compared. Both these preforms were prepared from the same high modulus carbon fiber, but they differed significantly in fiber volume and fiber distribution (Table 15). It is evident that weave construction affects the effectiveness of the translation of fiber properties to the composite.

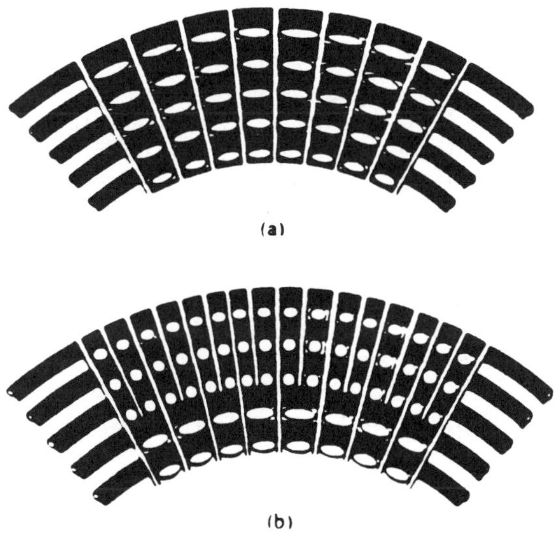

FIGURE 46 Axial and radial yarn compensation. (*a*) Axial yarn compensation from i.d. to o.d. (axial yarn diameter varies). (*b*) Radial yarn compensation from i.d. to o.d. (axial yarn diameter constant). (From Ref. 24.)

Matrix Precursor Selection

In the selection of a matrix precursor for carbon–carbon processing [2], the following characteristics must be considered.

- Viscosity
- Coke yield
- Coke microstructure
- Coke crystal structure

Generally, it is easier to graphitize a thermoplastic pitch than a thermosetting resin. Thermosetting resin remains glassy and is not considered to be graphitizable unless it is under stress.

Five principal factors should be considered when selecting a thermosetting resin for densification processing of multidirectional carbon–carbon composites. They are as follows.

1. Coke yields are in the range of 50–70 wt %; limited data indicate that coke yields are not increased by carbonizing under pressure (unlike thermoplastic pitch).
2. Coke structures are glassy and do not graphitize up to 3000°C.

TABLE 15
Weaving Characteristics of Pierced Fabric Versus 3D Orthogonal Weaving in Relation to Carbon–Carbon Composite x–y Properties

Property	Pierced Fabric	Fine Weave, Orthogonal
Preform density, kg/m^3	900	800
Fiber volume, %		
x–y direction	50	32
z direction	9	13
Tensile strength, MPa	104.7	99.2
Tensile modulus, GPa	57.9	55.8
Tensile strain to failure, %	0.2	0.2
Compressive strength, MPa	90.9	68.9
Compressive modulus, GPa	70.3	42.7
Compressive strain to failure, %	0.2	0.2
Coefficient of thermal expansion, 10^{-6}/K		
0–908 K	0	0
908–3019 K	2.0	2.0
Thermal conductivity at 519 K, W/m·k	126.3	114.2
Composite density	1.8	1.8

From Ref. 2.

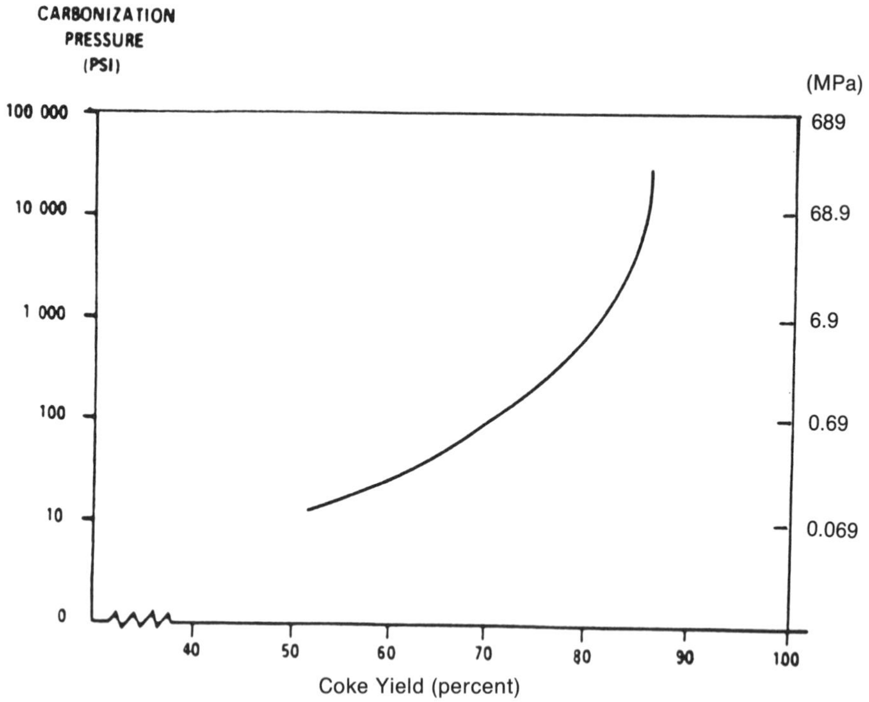

FIGURE 47 Effect on coke yield for petroleum pitch of carbonization under pressure. (From Ref. 24.)

3. Coke density is very low (< 1500 kg/m^3).
4. High shrinkage occurs during carbonization.
5. Stresses applied (or induced) during heat treatment can lead to a graphitic coke structure.

The factors that follow should be considered when selecting a pitch for densification processing of multidirectional carbon–carbon composites.

1. Coke yields at atmospheric pressure are about 50 wt %; however, by carbonizing the pitch under pressure of 10 MPa or greater, coke yields will approach 90% for some pitches (Fig. 47).
2. Coke structures are graphitic.
3. Coke density is high (~ 2000 kg/m^3).
4. Pressure applied during carbonization affects coke microstructure.

Case Studies

According to one case study [33], altering matrix precursors may have significant effects on the properties of a carbon–carbon composite. Two different pitch precursors were used on three similar preforms, but all other processing parameters were the same. The billets were processed to the desired density. The results of the experiment are displayed in Table 16. Apparently, variations of thermal expansion can be obtained by using different precursors, increasing the density of the composite, or varying the number of yarns in each direction. Since the three billets were cut from the same preform, fiber type and distribution are similar before processing. Therefore, variations in properties are primarily attributed to differences within the matrix microstructure and distribution.

Another case study [34] that used various matrix precursors and combinations of CVD and low pressure densification processes had demonstrated matrix precursor influence on the cylindrical properties of 3D carbon–carbon composites. The results are shown in Table 17.

Processing Condition and Sequence

In general, high pressure densification is more efficient than atmospheric densification. The high pressure assists impregnant infiltration into the preform and also keeps the volatile from escaping, thus increasing coke yield. For thick wall structures, such as block, high pressure densification is much preferred over CVD because of its infiltration efficiency. For thin wall structures (< 5 cm), high pressure densification is not recommended. Figure 48 shows that the high pressure densification process on pressure impregnation carbonization (PIC) can produce higher density composite than the same number of atmospheric carbonization cycles.

Economics and Cost

Carbon–carbon composite quality is a function of carbon fiber and matrix properties, weave/fiber placement consistency, and process conditions. Needless to say, the cost of the carbon–carbon depends not only on these factors but also on the end-product configuration,

TABLE 16
Matrix Precursor Effects, Using HIP Process on 3D Orthogonal Carbon–Carbon Composite Properties[a]

Property	Impregnant		
	Allied 15V	Koppers A	Allied 15V
Bulk density, kg/m^3			
By weight/volume	1960	1960	1870
By porosimetry	1960	1970	1890
Apparent density by porosimetry, (kg/m^3)	2160	2170	2110
Open pore volume, %	8.9	9.8	10.1
Coefficients of thermal expansion at 20–1400°C, 10^{-3}/K			
x-direction	2.3	3.0	4.1
Fiber volume, %	14	11	11
z direction	1.8	1.7	3.9
Fiber volume, %	26	27	26

[a] 1, 1, 2 Preform; high modulus fiber (P.55); HIP process.
From Ref. 33.

hence, ease of manufacturing. Therefore, unlike the well-standardized conventional metals industry, uniform price quotations for carbon–carbon composites are not possible at this time. However, for a general idea, products range in price from a simple block of approximately $500/kg using an automated weaving process to $3000/kg using the labor-intensive weaving process. For a complex developmental structure, the cost per kilogram is between $9000 and $11,000.

Applications

Carbon–carbon materials have been used in aerospace and military applications since mid-1960s. Components such as the reentry nosetip (Fig. 49), heat shield, rocket nozzle throat (Fig. 50), and exit cone (Fig. 51) have long been employing carbon–carbon composites, mainly for their controllable and predictable ablation, erosion, and recession characteristics. Most notable are the space

TABLE 17
Matrix Precursor Effects, Using CVD/Low Pressure Densification Process, on 3D Polar Carbon–Carbon Composite Properties

Matrix Precursor	Density (kg/m^3)	Strength (MPa)		Modulus (GPa)		Strain to Failure (%)	
		Circumferential, Tensile	Axial, Compressive	Circumferential, Tensile	Axial, Compressive	Circumferential	Axial
Phenolic	1620	118.5	52.9	70.3	20.2	0.18	0.59
High solid phenolic	650	106.8	59.9	64.8	20.7	0.17	0.50
High melt pitch	1640	94.4		106.1		0.08	
Low melt pitch	1650	128.2		64.1		0.05	
Isothermal CVD (not graphitized)	1590	113.7	103.1	77.2	33.8	0.15	0.48
Isothermal CVD/Phenolic	1730	106.8	73.9	77.9	27.6	0.13	0.4
Differential pressure CVD (not graphitized)	1350	136.4	115.1	68.2	23.4	0.20	0.72
Differential pressure CVD (graphitized)	1280	130.2		61.3		0.20	
Differential pressure CVD/phenolic	1580	128.2	82.7	64.1	48.6	0.20	0.65

From Ref. 34.

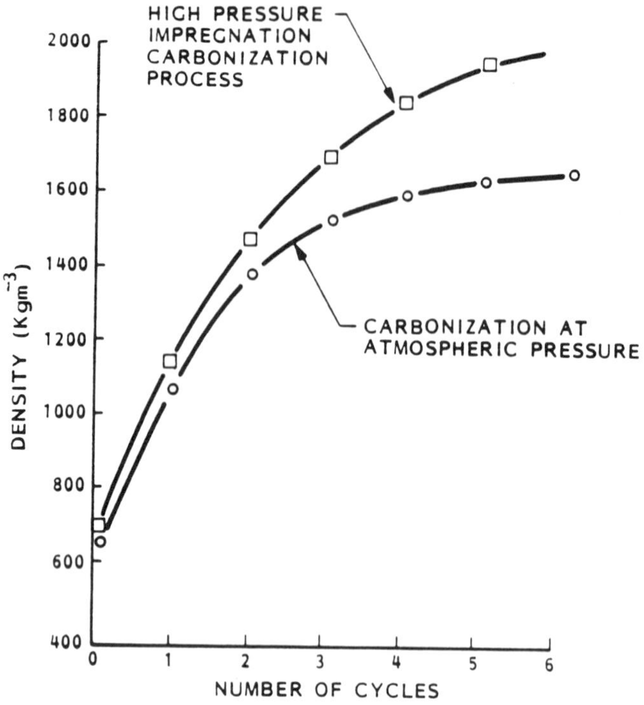

FIGURE 48 **High pressure densification process produces higher density composite than is obtained in the same number of atmospheric carbonization cycles.** (From Ref. 24.)

FIGURE 50 **Carbon–carbon composites are used in this rocket nozzle throat.** (Photo courtesy of Textron Specialty Materials.)

shuttle nose cone and leading edge (Fig. 52), which serve as the thermal protection systems for the most severely heated parts of the shuttle. These components are made of coated 2D carbon–carbon material. Figure 53 shows the nose cone under reentry test at about 1300°C. Figure 54 shows a model of carbon–carbon leading edge glowing hot.

The inherently excellent wear rate of carbon–carbon composites has suggested another area of application, namely aircraft parts requiring a high coefficient of friction. Various carbon–carbon materials' coefficients of friction increase from 0.2 to 1.1 while temperatures increase from 50 to 800° [35]; hence their use for advanced aircraft such as Airbus, Concorde, Delta 2000, and Falcon

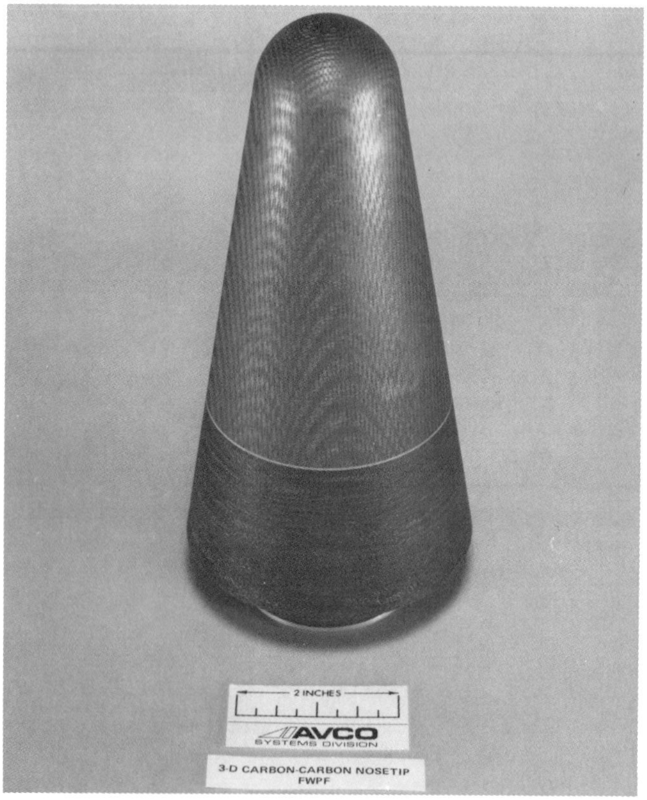

FIGURE 49 3-D carbon–carbon nosetip using fine-weave pierced fabric process. (Photo courtesy of Textron Specialty Materials.)

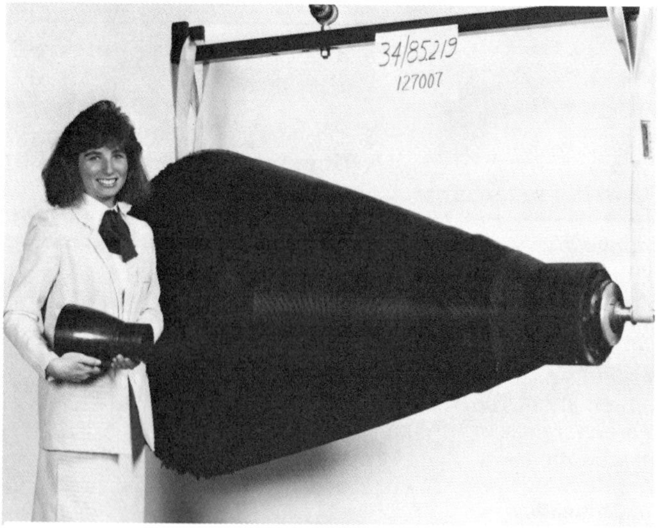

FIGURE 51 An exit cone made from carbon–carbon composites. (Photo courtesy of Textron Specialty Materials.)

FIGURE 52 Carbon–carbon composites are especially appropriate for the space shuttle nose cone and leading edge because they serve as thermal protection components. (Photo courtesy of Lockheed Missiles & Space Company, Inc.)

900. The stopping power is four times more than that of metal brakes. Carbon–carbon has also been used on the Lotus Formula 1 race car clutch. Other recent uses include thermal protection for shipboard vertical launchers, hot turbine engine wheels and rotors, and F-100 jet engine afterburners.

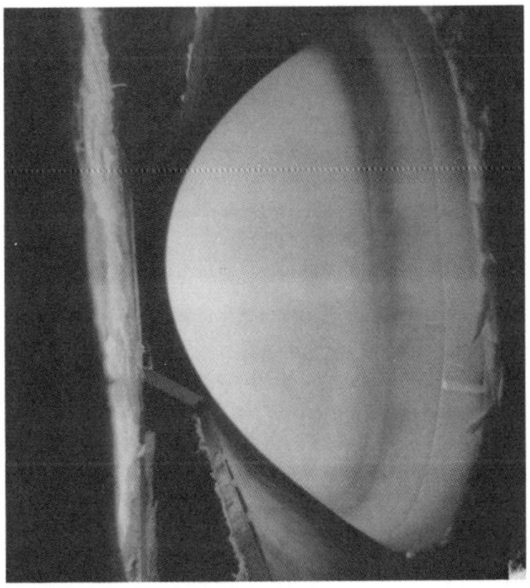

FIGURE 53 A nose cone under reentry test at 1300°C. (Photo courtesy of LTV Missiles and Electronic Group.)

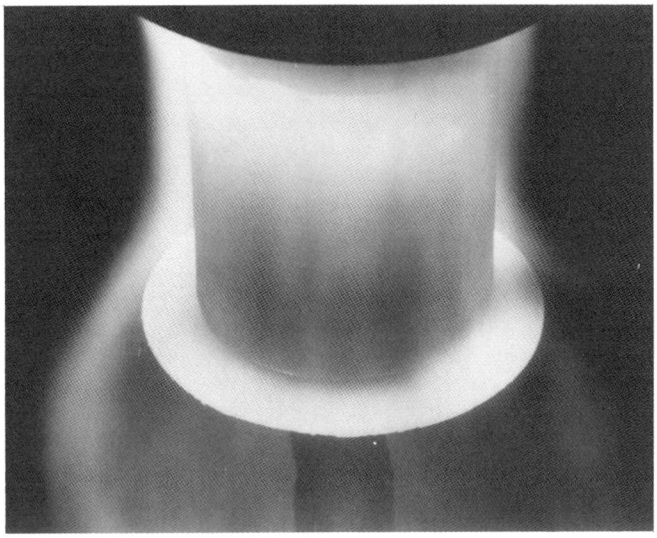

FIGURE 54 A model of carbon–carbon leading edge that is glowing hot. (Photo courtesy of LTV Missiles and Electronic Group.)

Carbon–carbon composites have also been fabricated into space structural components, mirror substrates, and various Strategic Defense Initiative applications. A typical space system has a temperature range of −100 to +100°C. With solar radiation on one side and cold space on the other, thermal expansion of the space material is a significant design factor. Since the coefficient of thermal expansion of carbon–carbon is negligible between these temperatures, it is a very desirable space structural material.

In the biomedical field, the biocompatibility, low density, and good fatigue and fracture toughness behavior of carbon–carbon offer unique advantages in prosthetics. Carbon–carbon prosthetic devices have been shown to be compatible with the body, even serving as seeds for bone and tissue ingrowth [36]. Carbon–carbon also has been studied for potential use in simulated ligament regeneration [37].

Current Developments

Oxidation Protection Coatings

The successful reuse of carbon–carbon primary structural material in high temperature applications strictly depends on the quality of its oxidation protection coatings. With the coating technology borrowed from the graphite industry, researchers have identified several criteria for good oxidation protection. Such a coating must be:

- Oxidation resistant
- Low in oxygen permeability
- Chemically compatible with carbon–carbon, both closely matched with respect to coefficient of thermal expansion and having good adhesion ability

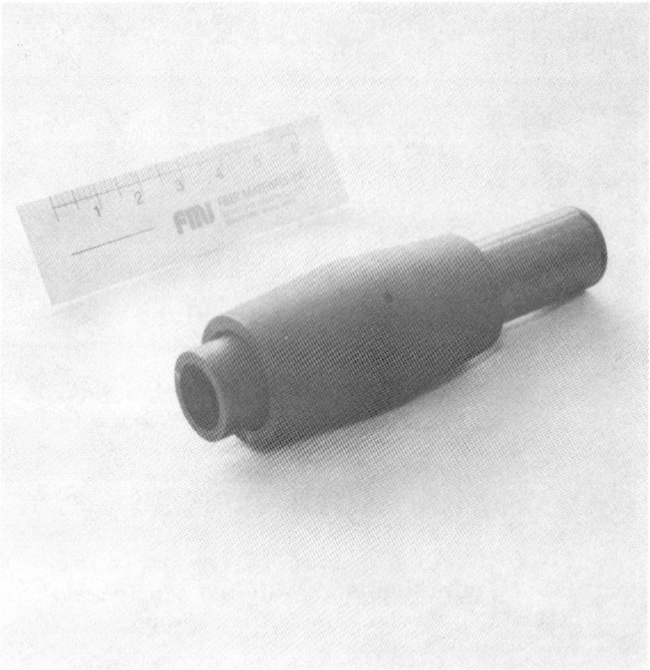

FIGURE 55 Triaxially braided tubes with joints has shown promising results. (Photo courtesy of Fiber Materials, Inc.)

FIGURE 56 Fabrication of I-beams is possible with the advent of carbon fiber weaving technology. (Photo courtesy of Techniweave, Inc.)

Carbon–carbon oxidation protection coating has proved its worth in the space shuttle nose cone and leading edge, in which the pack cementation process is used to convert several outer layers of the carbon–carbon into an effective silicon carbide oxidation-resistant coating. The microcracks and pores of the coating are filled with silicon oxide from the reaction with tetraethylorthosilicate.

Other coating materials include compounds of various noble metals, and these, together with combinations of pack cementation, "brush on," or chemical vapor deposition processes, have shown very promising oxidation protection results at high temperature. Research efforts have been concentrated on improving coatings, over-coatings, and sealants, as well as use of oxidation inhibitors dispersed in the carbon matrix or carbon fiber, and proper selection of the sequence of coating processes. Perhaps it is a matter of time before researchers pin down the formula and process that will provide extended oxidation protection coating for structural carbon–carbon composites.

Joints and Attachments

With the increased use of carbon–carbon composites in various applications, come newly discovered problems, as well. One problem area is the technology of carbon–carbon joining and attaching, where only limited expertise has been developed since the mid-1960s. Various levels of carbon–carbon joining processes have been investigated. At the preform level, processes like "through the thickness" stitching, sewing, and prerigidized fiber rod (dowel) are being evaluated and have shown very encouraging results. At the composite level, cocuring, high temperature bonding, molybdenum brazing (Fig. 55), and mechanical joining using carbon–carbon fasteners and bolts have all shown promising results. With some creative joining design approaches and combinations of the foregoing processes, standard carbon–carbon joining and attaching processes may be developed in the very near future.

Near-Net-Shape Manufacturing

Conventional carbon–carbon component manufacturing has relied on machining for final product configuration. This approach is not only costly but very time-consuming. Effort has been expended to fabricate near-net-shape carbon–carbon components to improve manufacturing efficiency. With the advent of carbon fiber weaving technology, fabrication of integral woven flanges, I-beams (Fig. 56), rib stiffeners (Figs. 57–60) for structural supports, thin-walled exit cones (Fig. 61), and thin-walled tubings (Fig. 62), has been demonstrated. However, the challenge of weaving near-net-shape large complex structures remains.

Thomas Ngai

FIGURE 57 (*Left*) Integrally woven rib-stiffened preform. (*Right*) Rib-stiffened plate. (Photos courtesy of Techniweave, Inc.)

FIGURE 58 (*Left*) 3D carbon-carbon T-beam. (*Right*) Support structures. (Photos courtesy of Fiber Materials, Inc.)

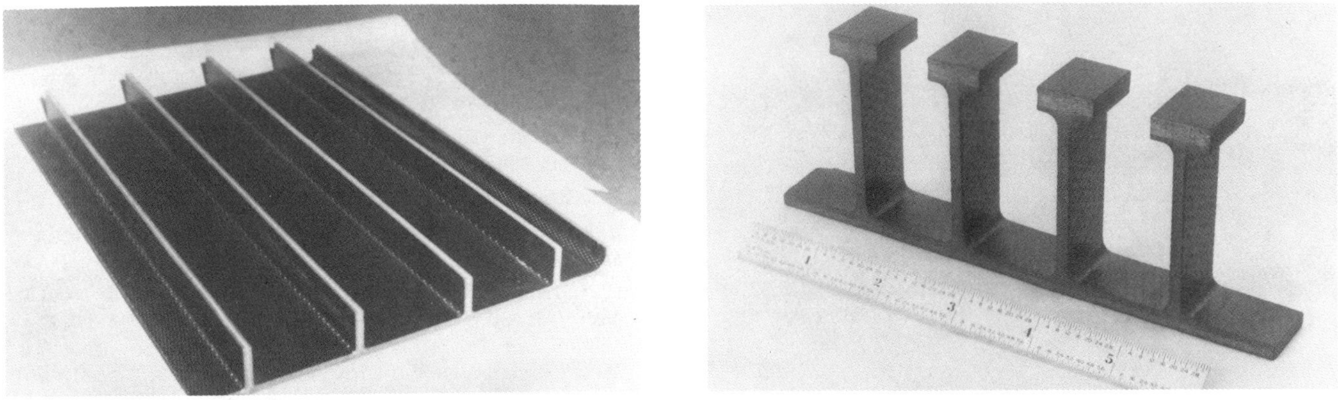

FIGURE 59 (*Left*) 2D carbon-carbon stiffened skin structure. (*Right*) 2D carbon-carbon stiffened panel subelement. (Photos courtesy of LTV.)

FIGURE 60 Fabric structures: continuous length structural components in a variety of shapes. (Photo courtesy of Fiber Materials, Inc.)

FIGURE 61 **This thin-walled exit cone is also possible because of carbon fiber weaving technology.** (Photo courtesy of Fiber Materials, Inc.)

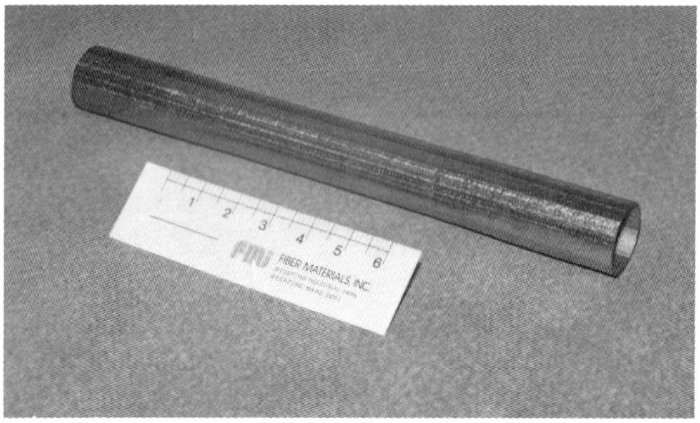

FIGURE 62 **Thin-walled tubings using carbon fiber weaving technology.** (Photo courtesy of Fiber Materials, Inc.)

References

1. J. H. Brahney, *Aeros. Eng.*, *7*(6), 15 (1987).
2. L. E. McAllister and W. L. Lachman, in A. Kelley and S. T. Mileiko, Eds., *Handbook of Composites*, Vol. IV, North-Holland, Amsterdam, 1983.
3. D. B. Powell, Southern Research Institute, personal communication, Nov. 25, 1987.
4. T. S. Lewis, Hercules Aerospace Corporation, personal communication, Nov. 24, 1987.
5. E. Fitzer, *Carbon*, *25*(2), 178 (1987).
6. L. A. Feldman, Aerospace Report TDR-0086 (6728-02)-2, Aerospace Corporation, El Segundo, CA, September 1987.
7. W. Watt, in M. B. Bever, Ed., *Encyclopedia of Materials Science and Engineering*, Vol. 1, Pergamon/MIT Press, Cambridge, MA, 1986, p. 509.
8. D. A. Schulz, *SAMPE J*, *23*(2), 27 (1987).
9. G. D. D'Abate and R. J. Diefendorf, *Proceedings of the 17th Biennial Conference on Carbon*, American Carbon Society, 1985, p. 390.
10. C. W. LeMaistre and R. J. Diefendorf, *SAMPE Q*, *4*(4), 1 (1973).
11. I. Litant, U.S. Patent 3,406,126 (Oct. 15, 1968), to AVCO, Inc.
12. R. Moreton, W. Watt, and W. Johnson, *Nature*, *213*, 690 (1967).
13. A. A. Bright and L. S. Singer, *Proceedings of the 13th Biennial Conference on Carbon*, American Carbon Society, 1977, p. 100.
14. W. Ruland, *ACS Polym. Prep.*, *9*(2), 1368 (1968).
15. A. C. Johnson, Lockheed Technical Excellence Seminar Series, December 1987.
16. R. Bacon, R. O. Moyer, and G. Cranch, U.S. Patent 3,305,315 (Feb. 21, 1967), to Union Carbide Corporation.
17. R. Prescott and I. W. Orr, U.S. Patent 3,533,743 (Oct. 13, 1970), to Great Lakes Carbon Corporation.
18. M. T. Katsumoto, in *Proceedings of the SAMPE 13th Annual Composite Workshop*, San Jose, CA, Jan. 28, 1988.
19. J. W. Newman, in M. L. Deviney and T. M. O'Grady, Eds., *Petroleum-Derived Carbons, ACS Symposium Series 21*, American Chemical Society, 1976, Washington, DC, p. 53.
20. G. P. Williams, Techniweave, Inc., personal communication, Feb. 20, 1988.
21. P. L. Martin, Fiber Materials Inc., personal communication, Nov. 19, 1988.
22. A. J. Klein, *Adv. Compos.*, *2*(5), 37 (1987).
23. R. J. Post, *Proceedings of the 22nd Natl. SAMPE Symposium and Exhibition*, 1977, pp. 486–503.
24. W. L. Lachman, J. A. Crawford, and L. E. McAllister, in B. Noton, R. Signorelli, K. Street, and L. Phillips, Eds., *Proceedings of the Intl. Conference on Composite Materials*, Metallurgical Society of the American Institute of Mining, Metallurgical, and Petroleum Engineers, 1978, pp. 1302–1319.
25. R. S. Barton, *SPE J.*, *24*, 31 (1968).
26. L. E. McAllister and A. R. Taverna, in E. Scala, E. Anderson, I. Toth, and B. Noton Eds., *Proceedings of Intl. Conference on Composite Materials*, Vol·I, Metallurgical Society of the American Institute of Mining, Metallurgical, and Petroleum Engineers, 1976, p. 307.
27. K. Kobayshi, S. Sugawara, S. Toyoda, and H. Honda, *Carbon*, *6*, 359 (1968).
28. H. Honda, K. Kobayshi, and S. Sugawara, *Carbon*, *6*, 517 (1968).
29. C. R. Thomas and E. J. Walker, *High temp. High pressure*, *10*, 79 (1979).
30. C. M. Standard, LTV Missiles and Electronic Group, personal communication, Dec. 14, 1987.
31. A. Levine, *Proceedings of the 12th Biennial Conference on Carbon*, Part II, Testing, 1975.
32. D. L. Schmidt, *SAMPE J.*, *8*(3), 9 (1972).
33. R. A. Meyer and S. R. Gryetvay, in J. D. Bacha, J. W. Newman, and J. L. White, Eds., *Petroleum-Derived Carbons*, ACS Sym-

posium Series 25, American Chemical Society, Washington, DC, 1986, p. 380.

34. C. K. Mullen, and P. J. Roy, *Proceedings of the 17th Natl. SAMPE Symposium*, IIIA, 1972, Paper 2, pp. 1–8.
35. E. Fitzer, A. Gkogkidis, and Morgenthaler, presentation at the International Carbon Conference, Baden-Baden, Germany.
36. D. Adams and D. F. Williams, *Biomaterials, 5,* 59 (1984).
37. J. M. Moran, *SAMPE J. 21*(3), 10 (1985).

Carbon Fiber, Evolving Technology

During the past 5 years, dramatic improvements in carbon fiber performance characteristics have been achieved. In large part, these improvements were precipitated by the recognition of the aircraft industry that tougher composite structures would require higher strain-to-failure fibers. Without such fibers and their contribution to tougher, more structurally effective composites, significant penetration of the large market for commercial aircraft structure was not likely to occur in the near future. At the same time, higher modulus, (~ 276 GPa) high strength fibers were sought for military aircraft, to simultaneously meet increased range, payload, and maneuverability requirements.

These higher strain and stiffness fibers are being coupled with programs to develop tougher matrix materials, and together they are fueling the evolution of damage-resistant composite systems capable of meeting primary, load-bearing commercial aircraft structural requirements. In addition, performance envelopes of military aircraft can be greatly expanded with the use of such composite systems.

Carbon Fiber Status and Directions

PAN-Based Fiber

As pointed out by Riggs [1], PAN-based carbon fibers, during their first 15 years of commercialization, fell into three classes: (a) high strength and (b) high modulus, both based on wet-spun precursor (~ 95% PAN), and (c) ultrahigh modulus (Celion GY-70), as derived from dry-spun PAN homopolymer (see Table 1). Late in the 1970s, it was discovered by Pepper and Harris [2] that further drawing (orientation) of wet-spun PAN precursor before processing resulted in finer diameter fibers with significantly higher strength and modulus than the high strength fiber class shown in Table 1. In fact, strain-to-failure in excess of 2% was reported for a fiber with strength and modulus of 5200 MPa and 255 GPa, respectively.

Five years later, in response to aircraft industry requests for a high strength fiber with a modulus above 275 GPa, commercial carbon fiber suppliers began the development of higher modulus, high strength fiber following the fine diameter fiber lead of Pepper and Harris. The resulting fibers are now referred to as "intermediate modulus" (IM), as indicated in Table 1. These evolutionary improvements are most likely based on the cumulative knowledge of the role played by such parameters as precursor orientation and retention during process, filament diameter effect, purity of precursors, and freedom from in-process damage and/or contamination. Several variants of the evolving high strain IM fiber are now commercial.

Representative properties of the four major classes of PAN-based carbon fiber are summarized in Table 2. Individual properties of the leading intermediate modulus fibers offered by major carbon fiber suppliers are listed in Table 3. T-1000, along with its performance characteristics, is also listed in Table 3 to illustrate the dynamics of current carbon fiber (CF) development efforts. This fiber, with its 40% increase in strength and strain, is generating a new surge of interest by composite materials engineers and designers.

Finer denier, PAN-based carbon fiber can be prepared by two different approaches. First, Pepper and Harris [2] demonstrated improvement by postdrawing an already formed PAN fiber followed by thermal stabilization and carbonization steps identical to standard processing. Second, instead of allowing a normal 10–15% shrinkage during stabilization, it is possible to significantly reduce shrinkage and achieve a finer diameter stabilized fiber before carbonization. Both processes could produce two results: (a) greater orientation of the stacked, condensed-ring structure parallel to the fiber direction, and (b) a more uniform structure with less structural distinction between skin and core, hence, fewer incipient flaws.

TABLE 1
Broad Categories of Carbon Fibers

Carbon (%)	Fiber Type	Heat Treatment Temperature (°C)[a]	Modulus (GPA)
92–96	High strength	1200–1400	228–241
92–96	Intermediate modulus	~ 1400	290–310
99	High modulus	1800–2500	276–380
99^+	Ultrahigh modulus	$2800–3000^+$	$483–690^+$

[a] Principally PAN-based carbon fibers.

TABLE 2
Properties of PAN-Based Carbon Fiber[a]

Property	High Strength[b]	Intermediate Modulus[c]	High Modulus[d]	Ultrahigh Modulus[e]
Filament diameter, μm	5.5–8.0	5.0–5.2	5.4–7.0	8.4
Density, g/cm³	1.75–1.80	1.73–1.80	1.78–1.81	1.96
Carbon content, wt %	92–95	~ 94	99–99⁺	99⁺
Tensile strength, MPa	3105–4555	5000–5200	2415–2555	1865
Tensile modulus, GPa	228–262	290–310	359–393	517
Elongation at break, %	1.3–1.8	1.7	0.6–0.7	0.38
Electrical resistivity, μΩ · m	15–18	11–13	9–10	6.5
Thermal conductivity, W/(m · K)	8.1–9.3	~ 38	64–70	~ 120

[a] Strand data.
[b] Thornel T-300, T-500, T-600, T-700, Amoco Performance Products; Celion G30-500, BASF; AS2/AS4/AS6, Hercules Corporation.
[c] Magnamite IM8, Celion G40-700, Thornel, T650-42.
[d] Thornel T-50, Celion G-50, Magnamite HMS4.
[e] Celion GY-70.

These factors would give rise to higher modulus (orientation) and higher strength (reduced flaws), respectively.

Pitch-Based Fiber

In the discussion on pitch-based fiber in reference 1, the differences between the physical and mechanical properties of such materials and those of PAN-based fiber were identified. To summarize: significantly higher modulus levels are achievable with pitch (up to 900 GPa); maximum strengths at lower moduli (240–300 GPa) are less than half the current levels for leading PAN-based fiber; densities are 20% higher for pitch fibers; and, pitch-derived fiber thermal and electrical conductivities are higher than PAN-based high modulus equivalents by factors of 1.5 to 2 and 3, respectively (Tables 2 and 4).

At the outset of pitch-based carbon fiber development, the economic drivers appeared to favor the pitch-based variant: petroleum pitch cost relatively little and was abundant; conversion of precursor to carbon fiber was higher; and no lengthy stabilization process was required for the precursor fiber. However, reproducible high strength fiber was not achieved with pitch precursor in a timely manner. This failure, accompanied by the development and significant growth of PAN-based fiber markets, resulted in a dominant position for the latter carbon fiber.

Properties of mesophase pitch-based carbon fibers currently available from Amoco are listed in Table 4. High modulus, combined with the excellent stability of high carbon content, make these materials attractive for high stiffness, lightly loaded space structures. High carbon content and attendant thermal stability are features attractive in carbon–carbon composite manufacturing, where repeated carbonization cycles at temperatures exceeding 2000°C are used following a series of pitch or phenolic resin reimpregnation steps.

Composite Properties

Epoxy composite properties of the three most commonly used fibers in the United States are represented in Table 5, as extracted from vendor data sheets. It is noteworthy that the compression strength (and, therefore, strain) is significantly lower than tensile strength for all three fiber types. Since it is compression strain after im-

TABLE 3
Properties of Intermediate Modulus Carbon Fibers

Property	Magnamite IM8[a]	Celion G40-700[b]	Thornel T-650-42[c]	Thornel T-1000[c,d]
Filament diameter, μm	5	5.0	5.1	5.2
Density, g/cm³	1.80	1.73	1.78	1.82
Tensile strength, MPa	5200	5000	5000	7000
Tensile modulus, GPa	310	300	290	300
Elongation at break, %	1.66	1.66	1.7	2.3

[a] Hercules.
[b] BASF/CCF.
[c] Amoco.
[d] Toray.

TABLE 4
Properties of Mesophase Pitch-Based Carbon Fiber

Property	Thornel P25	Thornel P55S[a]	Thornel P75S[a]	Thornel P100S[a]
Filament diameter, μm	11	10	10	10
Density, g/cm³	1.9	2.0	2.0	2.15
Carbon content, wt %	97⁺	99	99	99⁺
Tensile strength, MPa[b]	1400	1900	2100	2200
Tensile modulus, GPa[b]	160	380	520	720
Elongation at break, %	0.9	0.5	0.4	0.3
Electrical resistivity, μΩ · m	1.3	8.5	7	2.5
Thermal conductivity, W/(m · K)		110	185	520

[a] Thornel is a trade name of Amoco Performance Products; S indicates surface treatment.
[b] Mechanical properties measured on impregnated strands.

TABLE 5
Properties of High Strength, High Modulus, and Intermediate Modulus Fibers in Epoxy Composites

Property	High Strength[a]	Intermediate Modulus[b]	High Modulus[c]
Tensile strength, MPa	2300	2600	1420
Tensile modulus, GPa	145	180	210–250
Compression strength, MPa	1600	1800	900
Compression modulus, GPa	135	150	190–230
Short beam shear strength, MPa	120	120	80

[a] Magnamite AS4, Celion G30-500, Thornel T-650-35.
[b] Magnamite IM8, Celion G40-700, Thornel T-650-42.
[c] Magnamite HMS4, Celion G50-300, Thornel T-50.

pact (CAI) [3,4] that is limiting aircraft structural performance, both the extent of damage and its effect on residual compression strength are critical. In addition to acceptable initial strain levels, toughness is imperative to prevent serious fiber damage in the impact event. Significant fiber (and composite) damage would permit local fiber buckling in the damage zone, which would precipitate compression failure at strain levels below structural limits. The high level of interest in tough thermoplastic matrix materials by the aircraft industry is then quite clear. The fiber–matrix interface and the effects of fiber sizes on this interface constitute yet another parameter that can and must be used to optimize toughness and damage tolerance of composite components, as discussed next.

A principal use for pitch-based carbon fiber–epoxy composites is in space structures, where structural rigidity is the key requirement and loads are modest. The mechanical properties of epoxy composites, shown in Table 6, include very high tensile and compressive moduli, which are useful in rigid supports for transmitters and antennae for directionally accurate signal transmission and reception. The higher thermal and electrical conductivities of pitch-based fiber composites (vs. PAN-

TABLE 6
Properties of Pitch-Based Thornel Fibers in Epoxy Composites

Property	P55S	P75S	P100S	P120S
0° Tensile strength, MPa	940	790	1250	1440
0° Tensile modulus, GPa	246	340	500	520
Elongation at break, %	0.38	0.24	0.25	0.28
0° Compression strength, MPa	510	460	290	
0° Compression modulus, GPa	200	310	520	
Elongation at break, %	0.26	0.22	0.06	
Short beam shear strength, MPa	64	55	39	30

based equivalents) are expected to be valuable properties in providing more rapid dissipation of both heat and electrical charges, particularly in structures to be used in space.

Damage-Tolerant Structure

As a result of the NASA Aircraft Energy Efficiency (ACEE) program, which spanned the mid-1970s into the 1980s, the application of advanced composites in secondary and non-load-bearing primary structure for commercial transport aircraft was demonstrated. Up to 5% of the structure is being converted to structural composites in newer aircraft (such as Boeing's 757 and 767 and the Airbus A320), and much of this is carbon fiber reinforced. However, it has not yet been possible to demonstrate the same feasibility for primary load-bearing structure for commercial aircraft because of the lack of damage tolerance of the carbon fiber composites. Nevertheless, the NASA-led efforts defined both the performance requirements and the materials tests needed to screen and qualify potential composite candidates. These have resulted from funded development and verification programs at Boeing, Lockheed, and Douglas Aircraft and are described in NASA Reference Publication 1092. (For standard test descriptions, see Table 7.)

The most critical of the five tests defined is the compression-after-impact test; here compression strain of 0.6% must be sustained after the laminate has been subjected to a fully specified impact event. Both the fiber and the matrix are important; however, current consensus is that available high strain carbon fiber variants with elongations at failure of 1.7–2% should be sufficient to meet the needs.

Protection of the reinforcing carbon fibers (and the overall composite) by utilization of a suitably tough matrix has been the subject of extensive effort for the past several years. The high elongation and the attendant toughness of the thermoplastic polymers [5] have been the focus of attention [6].

A review of some of the leading thermoplastic resin candidates has been given by Johnston [3] along with key physical and mechanical characteristics. Johnston and coworkers at Langley [3,7] have also shown the relationship between neat resin tensile modulus and composite toughness and have pointed out that certain thermoplastics such as PEEK and Torlon possess both high modulus and toughness. This is most significant, since it was shown by Hahn and Williams [7] that composite compression strength is directly proportional to neat resin modulus; that is, the more rigid the matrix, the greater will be its resistance to fiber buckling and the onset of compression failure.

Since thermoplastic matrix polymers are solids at ambient conditions, it is particularly important that matrix and fiber be brought together in a processible form to enable component fabrication. Evolving forms include thermoplastic powder-impregnated carbon fiber yarns [8], commingled thermoplastic and carbon fibers [9], as well as melt-coated sheet and ribbon variants [10]. Wetting and compatibility of the reinforcing carbon fiber and the matrix is extremely important to enable the formation of low void composites. The required wetting and

TABLE 7
Standard Tests for Toughened Resin Composites

Test Designator	Test Type	Ply Orientation	Nominal Thickness	Test Specimen Size (cm)	Principal Contributor
ST-1	Compression after impact	$[+45/0/-45/90]_{ns}$	0.64 cm	18 × 32 (before impact) 13 × 32 (after impact)	Charles F. Griffin Lockheed–California Company Burbank, California
ST-2	Edge delamination tension	$[\pm 30/\pm 30/90/\overline{90}]_s$ $[\pm 35/0/90]_s$	11 plies 8 plies	4 × 25	T. Kevin O'Brien Structures Laboratory AVRADCOM Research and Technology Laboratories NASA Langley Research Center Hampton, Virginia
ST-3	Open-hole tension	$[+45/0/-45/90]_{ns}$	0.64 cm	5 × 31	D. J. Watts Douglas Aircraft Company Long Beach, California
ST-4	Open-hole compression	$[+45/0/-45/90]_{ns}$	0.64 cm	13 × 32	Charles F. Griffin Lockheed–California Company Burbank, California
ST-5	Hinged double-cantilever beam	$[0]_n$	0.31 cm	4 × 23	Motoaki Ashizawa Douglas Aircraft Company Long Beach, California

From NASA Reference Publication 1092, July 1983.

compatibility can be greatly facilitated through the use of an appropriate fiber finish; and the finish also enables fiber handling operations (fiber packaging, prepregging, weaving, filament winding, etc.) to be carried out with a minimum of fiber damage—that is, property degradation. Finally, with the growing utilization of thermoplastic matrices, the combining and shaping of prepregs or preforms into structural assemblies must be carried out using time–temperature–pressure cycles that do not damage fiber during shaping nor during cool-down from the very high temperatures (300–400°C) employed in the final fabrication steps. Proper fiber finishes should be ductile, high elongation materials capable of damping the effects of mismatches in coefficient of thermal expansion (CTE) between fiber and matrix.

In the early phases of composite development, proprietary epoxy-based finishes were formulated and applied to the carbon fibers. Such finishes were compatible with the epoxy matrix materials used at the time. However, with the advent of higher temperature matrix resins such as polyimides, improved finishes were sought to replace the epoxy-based systems, which undergo thermal degradation above 200°C. As a result, carbon fiber manufacturers are continuing to develop for their products proprietary finishes that both facilitate handling and provide compatibility with the new generation of matrix materials.

Efforts are continuing to develop compatible fiber capable of handling the high temperature fabrication conditions that accompany the leading thermoplastics (~ 400°C). The CCF Division of BASF offers a no-size fiber labeled "entangled" to permit fiber handling without damage [11]. Others such as Courtaulds seek new finishes based on in-house and external interphase characterization efforts [12].

The role of the finish on carbon fiber has several dimensions, and each has a direct bearing on the performance of composites in which it is a constituent. The functions of the finish and its potential roles are as follows.

1. Provides assistance in the handling of yarns without breakage of fibers or buildup of fuzz balls at guides. Finishes are essential for economical weaving of carbon fiber fabrics. In addition, much higher winding speeds are readily employed, which is important in manufacturing, filament winding, etc.

2. Facilitates wetting of the fiber by the matrix. A compatible finish, which is solution or emulsion applied, can provide a good, uniform fiber coating under the proper conditions, hence can prevent or minimize the trapping of air and creation of voids at the interface.

3. Forms between fiber and matrix an interphase that can be more ductile than either constituent and spreads the load transferred to a longer length of fiber. (In the instance of a failed fiber, this capability can often arrest failure of adjacent fibers.)

4. Renders the fiber and the matrix compatible by the formation of an interphase that dampens the effects of expansion/contraction of a brittle matrix having a CTE significantly different from the reinforcing fiber.

5. Minimizes the effects of the interphase by proper selection of a finish that does one or more of the following:

Forms non-moisture-sensitive bonds to both matrix and fiber.

Possesses a low coefficient of moisture expansion, which minimizes swelling.

Retards diffusion of moisture along the interphase, which can be the preferred route because of porosity and/or ductility.

Emerging New Fibers

Although the intermediate modulus carbon fibers were described previously and T-1000 fiber properties were listed in Table 3, it is also necessary to include T-1000 fibers in the "emerging" category by virtue of the evolutionary dynamics still being illustrated. These fibers are new, and extensive evaluations have not yet been reported. Initial composite properties published by Toray [13], however, show that new levels of tensile strength and modulus are indeed realized (Table 8). Strengths and failure strains of 3.5 GPa and 2.0%, respectively, have been measured. However, it should be noted that composite compression and shear strengths are no better than the corresponding properties of high strength and IM fiber composites (Table 5). Implications are that sophisticated design will be required to locate these fibers in structures where the outstanding tensile properties can be utilized without exposure to excessive compression loads.

In addition to the evolutionary development of such PAN-based fibers, considerable progress is occurring in other areas. First, a melt-spun precursor (MSP) technology [14], developed by American Cyanamid, has been sold to BASF following the codevelopment of a novel route to high strength carbon fiber precursors. This route may prove to be more economical than conventional solution spinning, since fiber coagulation baths and/or solvent removal steps can be eliminated. The physical properties such as graphite layer spacings and orientation, as well as the fiber and composite mechanical properties of G30-500 made by the MSP process, have been shown to be equivalent to properties for standard high strength Celion G30-500 based on wet-spun precur-

TABLE 8
Thornel T-1000 Composite Properties

Property	Resin 3620[a]	Resin 3632[a]
0° Tensile strength, MPa	3490	3550
0° Tensile modulus, GPa	158	158
Elongation at break, %	2.0	2.0
0° Compression strength, MPa	1700	1650
Interlaminar shear strength, MPa	110	95

[a] Toray epoxy resins with high elongation: 3620 = 4.0%; 3632 = 4.9%.

sor. Strand tensile strength and modulus properties measured on carbon fiber derived from the two processes are reported as follows.

	MSP-based	Wet-spun Control
Tensile strength, MPa	4070	4000
Tensile modulus, GPa	237	234
Elongation to break, %	1.72	1.71

Comparable results have also been reported for composite tensile, compression, and shear tests [14].

When "melt spinning" (or pseudo-melt-spinning) processes are used in fiber spinning, the extruded fiber assumes the shape of the hole in the spinneret (extrusion face). With the MSP process, the CCF unit of BASF has been able to make fibers with controlled cross-sectional shapes (e.g., rectangular and tetralobal). Such a capability could prove valuable in improving composite design and performance.

For example, rectangular-shaped fibers could be induced to pack more efficiently and provide composites with higher fiber volume fractions, hence, higher mechanical properties. It is also possible to increase interfacial surface areas between matrix and fiber. This could be an important approach to improved carbon–carbon composite performance where interface forces are extremely weak. Increased interfacial area could provide improved intracomposite cohesion.

In another area, Du Pont has announced a new pitch-based carbon fiber labeled PRD-172 [15]. This fiber is reported by Du Pont to be derived from a special liquid crystal based pitch fragment developed by their Conoco subsidiary. Du Pont's knowledge and experience gained in the processing of the liquid-crystalline Kevlar polymer as a fiber has most likely been invaluable. Initial properties reported by Du Pont in product bulletins and in the literature are shown in Table 9. It should be noted that the reported data were obtained from single-fiber tests as opposed to impregnated strands or composite tensile specimens currently used by the composites industry. Such data will reportedly follow.

Data on Amoco's pitch-based carbon fiber are included (see Table 4) for direct comparison, but such mechanical property comparisons are difficult because of test method differences (single filament by Du Pont vs. impregnated strand by Amoco). It does appear that tensile strengths of the Du Pont materials may be higher; however, there is no indication that a surface treatment has been carried out. Surface treatments are oxidative, and the severity of required treatment conditions increases as carbon content, modulus, and accompanying inertness increase, and oxidation is known to deteriorate the tensile strength of carbon fibers [16]. Comparative judgment should be deferred until impregnated strand and composite laminate data are made available by Du Pont.

Finally, research efforts by General Motors in the past few years have expanded the possibilities of catalytically grown carbon fiber [17]. Discontinuous fibers of meaningful lengths and diameters (centimeters and 7–30 μm, respectively) have been grown from methane (also benzene) and H_2 on microscopic Fe or Fe–alloy particles, which function as a catalyst. Such fibers are referred to as vapor-grown carbon fibers (VGCF). In the process, the submicrometer nucleating particle grows away from the supporting surface at the end of a precipitating carbon filament and the filaments thicken as more carbon is deposited. Formation takes place at temperatures in the 900–1100°C range.

Fibers formed are distributed over a wide range of dimensions with lengths extending from submicrometer to a few centimeters and diameters from submicrometer to the interesting range of 7–30 μm. Strengths and moduli of the best fibers have been measured at 2.9 and 237 GPa, respectively. Recent progress is showing that carbon filament growth in the range of millimeters per minute is possible. If progress in making fibers with both uniform properties and dimensions occurs, such materials may be able to penetrate developing markets with lower cost reinforcement to achieve metallike stiffness.

Expanding Applications and Market Projections

Applications

AIRCRAFT/AEROSPACE. Aircraft applications are projected to remain the dominant market for carbon fibers well into the 1990s [18–20]. This is the direct result

TABLE 9
Du Pont PRD-172 Pitch-Based Carbon Fiber Properties

Property[a]	E-35[b]	E-75	E105	E130
Filament diameter, μm	9.6	9.4	9.3	9.2
Density, g/cm^3	2.1	2.16	2.17	2.19
Tensile strength, MPa	2800	3100	3300	3900
Tensile modulus, GPa	240	520	720	890
Elongation at break, %	1.0	0.56	0.55	0.55

[a] Data obtained on single-filament tests.
[b] Other modulus variants also available; E-55, E-120.

TABLE 10
Economic Rationale for Composites in Transportation: Fuel Cost Sensitivities

Base factor: 1 liter of fossil fuel moves 1 kg a distance of 18,000 km		
Transportation Mode[a]	**Lifetime Kilometers**	**Value of 1 kg Weight Reduction at 4 l/$ Fuel Cost**[b]
Commercial aircraft[c]	$(50–100) \times 10^6$ km	$750–$1500
Trucks/trailers	$(2–10) \times 10^6$ km	$30–150
Autos	0.15×10^6 km	$2

[a] Reusable cargo containers can have the value as shown, depending on container lifetime/durability; ships included.
[b] Regional differences in fuel costs change the value of weight reduction: fuel costs in Europe at $0.6–0.9/l increase weight reduction value to $4–$7/kg.
[c] Note that although aircraft may operate more efficiently at high altitude cruise during long distance flights, they must also transport the weight of fuel on board (for 747 SPs, this exceeds 100×10^3 kg at takeoff).

of the value of weight saved in commercial aircraft combined with the positive results of composite structure demonstrations funded under the NASA ACEE program. The value of the weight saved is derived directly from the lifetime distance that a commercial aircraft flies ($\sim 50 \times 10^6$ to 100×10^6 km) combined with the author's rule of thumb that one liter of fossil fuel will move one kilogram a distance of 18,000 km (see Table 10). When this is compared with the average lifetime miles of an auto ($< 150 \times 10^3$ km) or a large truck (2×10^6 to 10×10^6 km), it is at once apparent that the value of a kilo saved over the lifetime of an aircraft is more than 2 orders of magnitude larger than the value of a kilo saved in an automobile. Weight savings for trucks and trailers fall at an intermediate value. Hence, the priority order of transportation applications of composites (based solely on fuel efficiency) should be aircraft, trucks and trailers, and autos. This is, indeed, the case.

Examining the aircraft segment, an exponential increase in use of carbon fiber composites has taken place through the 1980s (see Fig. 1 with the reasonable assumption that approximately half the reinforcement is

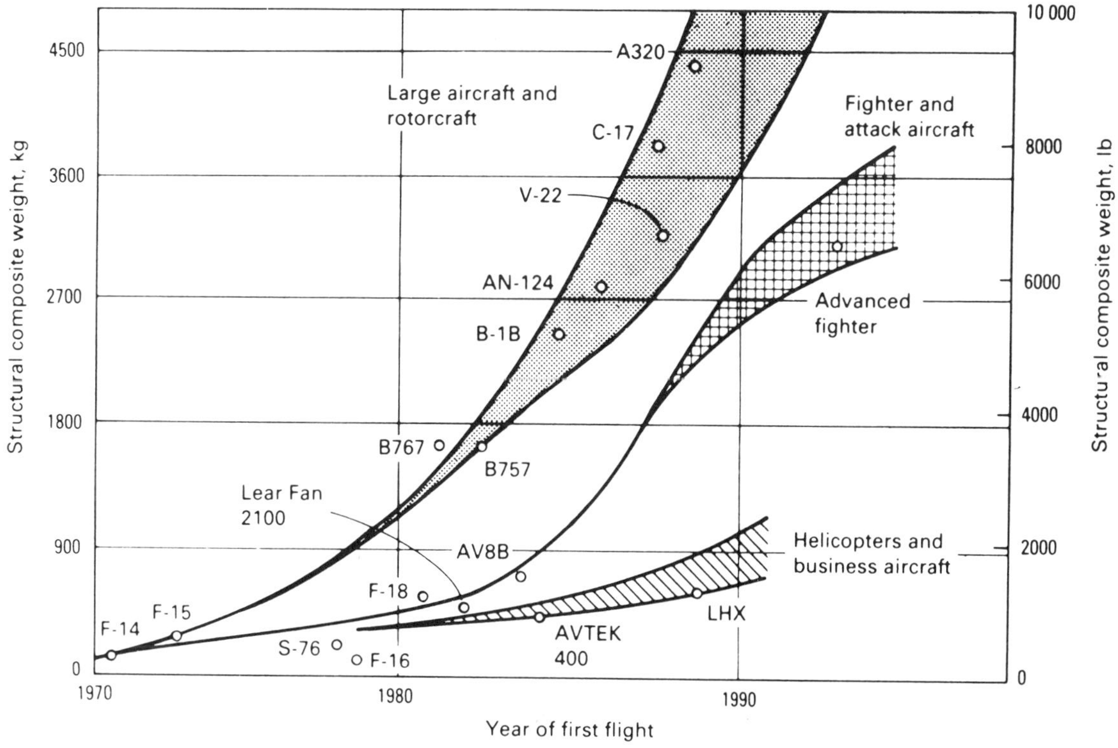

FIGURE 1 **Composite aircraft structure by weight.** Reproduced with permission from Engineered Material Handbook, Vol. 1, Composites, ASM International, Metals Park, OH, 1987.

CF). Military aircraft usage is driven not by the economics of fuel consumption, but by weight reduction, to provide increased performance—increased range, payload, armament, and/or maneuverability. Primary load-bearing composite structure is already being used on military aircraft (largely fighters, which must offer a pilot escape options). Usage will increase in transports, as in the announced projections for the C-17 [21]. Key carbon fiber composite applications on military aircraft that have been instrumental in establishing their viability are listed below, in sequential order of development.

F-15	Speedbrake
F-16	Horizontal and vertical tailskins
F-18	Doors, tailskins, wing-box
B-1	Doors, tailskins, flaps
AV-8B	Doors, rudder, tails, ailerons, flaps, wing-box, body elements
V-22	Fuselage, wing-box, wing-skins, full tail, engine housings (3100 kg of CF/epoxy)

In addition, the C-17 military transport is now in full-scale development, with the first flight scheduled for 1990 and delivery of the first 12 aircraft in 1992. Latest Douglas projections are for 2728 kg of carbon fiber composite per plane, with a total of 210 units scheduled for delivery by 1998.

ENGINES. Carbon fiber–polyimide engine applications are evolving in spite of considerable difficulty at present with consistent fabrication of PMR-15 composites. Committed applications are to external nozzle flaps and interfacing rings. Augmenter and bypass ducts along with compressor shrouds are in development. Solution to long-term stability of carbon fiber–polyimide structures at 260–300°C will lead to extensive applications in engines, frames, and heating ducts for commercial aircraft.

SPACE AND MISSILE SYSTEMS. The major structural components used in space require rigidity and thermal stability, as well as carefully designed and tailored coefficients of thermal expansion for minimum changes with temperatures (light-to-dark cycles). High modulus carbon fibers, including pitch-based materials, are particularly attractive because of their high rigidity and stability.

CARBON–CARBON COMPOSITES. These materials were developed as nose cones to withstand the severe conditions of reentry from space and as nozzles for rocket motors. Carbon–carbon composites are typically formed from carbon fiber–phenolic prepregs which, after curing, are exposed to temperatures above 2000°C for carbonization. Successive impregnations with high carbon content phenolic or pitch resins are followed by high temperature carbonization cycles—normally three or four cycles to achieve high density, low void structures. Because of the repeated exposure to temperatures in excess of 2000°C, high modulus fibers, having already been treated at these temperatures, are preferred. The material of choice is rayon-based fiber because of its higher interface area, provided by a heavily crenulated surface, and the extensive processing experience and data base available. Both pitch and high modulus PAN-based carbon fibers are in use and in continuing evaluation to replace rayon-based fiber. Because of lack of a clear-cut alternative, the AF has maintained Avtex, the sole supplier of rayon precursor.

Automotive/Transportation

The principal impediment to the use of carbon fiber composites in automotive applications is cost. In comparison to aircraft applications, the rationalization and payback for more expensive materials is lacking. As pointed out in the preceding section, the primary driver for aircraft use is reduced lifetime operating cost—based on 50 million to 100 million lifetime kilometers for commercial carriers. Since the average auto is driven little more than 150,000 kms, the cost of fuel used in moving the vehicle itself is relatively insignificant at approximately $2/kg over the lifetime distance of 150,000 km. In a realistic example of vehicle weight reduction by 100 kg on a 1000 kg vehicle, the in-use saving is only $200. This is not a significant incentive.

For use in autos then, cost/performance advantages must significantly exceed minor efficiency gains. One example of such an application is modification of a truck driveshaft from a two-piece to a one-piece unit, providing significantly improved performance, as well as reduced cost, by elimination of the center bearing. The modification is a combination of carbon fiber and glass fiber composite overwraps to reduce weight and increase stiffness, thereby moving fundamental vibrations to higher frequencies. Furthermore, filament-wound leaf springs are already in production with glass fiber reinforcement; for heavier commercial vehicles where stiffness, weight reduction, and costs of downtime are factors, some selected carbon fiber reinforcement can be expected. In addition, if filament-wound "space frames" evolve for automobiles, it is possible that they will utilize some carbon fiber for stiffening, to provide an optimal balance of performance and weight.

The applicability of filament-wound composite technology illustrates that successful automotive applications for carbon fiber are possible when high speed, low cost processing routes are employed. Pultrusion of reinforced shapes is a similar established, volume production route that can incorporate carbon fiber composites for stiffness. However, additional high speed fabrication processes and/or novel intermediate product forms are still needed to facilitate the design and low cost production of complex automotive components. To be cost-effective, such processes and product forms will enable significant part consolidation and, in many cases, will utilize the designed-in selective stiffness made possible by carbon fiber inclusion. One conceptual approach would be the use of carbon fiber in inserts or preforms incorporated into resin transfer molding (RTM) processes.

Other Applications

Without doubt, the most significant applications for carbon fiber composites other than aircraft/aerospace is in sporting goods. These applications started early on and, by the mid-1970s, before significant aerospace qualifications and applications existed, they were the largest market. Heavy utilization for recreational products continues, with current consumption at about one-third of available carbon fiber. Since testing requirements are minimal and qualification is not necessary, fiber sells into these applications at approximately 25% less than into aircraft usage. Typical applications continue to be tennis rackets, fishing rods, golf club shafts, skis and poles, sailing masts, and bicycles.

Opportunities exist for carbon fiber composites in markets other than aircraft and recreational; however, the rationale for their utilization is more diffuse and difficult to correlate. Three commonly recognized areas of opportunity are appliances, industry, and construction. Appliances are not expected to utilize much carbon fiber, since requirements and CF cost will favor glass fiber. However, industrial markets will clearly evolve. Products will most likely take on the form of filament-wound and pultruded tubes and beams. Beams would be used for heavy equipment and catwalks, particularly in plants, where the environment and the service call for resistance to chemicals and creep. Stiffness and vibrational damping provided by carbon fiber composites will be important in moving equipment (e.g., cranes and booms). Longer term applications, such as construction materials in bridges, buildings, etc., will evolve where long-term durability and reliability are required; however, reliability must be established and 10–20 years will be needed.

Market Growth and Forecast

The aircraft/aerospace (A/A) market for carbon fibers has been dominant in the United States throughout the 1980s. In market information recently released by BASF Structural Materials [18] and by the Freedonia Group [19], the A/A segment became dominant worldwide in the 1987–1988 time frame. Data extracted from the BASF release are shown in Table 11. As discussed in the preceding section, this trend is a result of several events: (a) completion of A/A qualification programs; (b) expanded use in military aircraft structure including primary structure on the AV-8B, which uses 3100 kg of carbon fiber composite (see Fig. 2); and (c) the move to CF composites in secondary and non-load-bearing primary commercial aircraft structure.

Military aircraft use of carbon fiber composites will not decrease, but will simply approach a stable level within the next decade. U.S. military programs are projected to account for 70% of the increased usage forecast between 1988 and 1993. This is the rationale for the domestic precursor requirement for carbon fiber of the Department of Defense (DOD), which was passed into law in 1987. The law requires that 50% of DOD requirements be supplied from carbon fiber based on domestic precursor by 1992. This will require fiber requalification efforts for both Hercules and the CCF Division of BASF Structural Materials. As reported earlier, CCF has developed a domestic precursor (MSP) and has announced plans for a 0.91×10^6 kg CF plant to be on-stream in 1990.

The BASF Structural Materials [18] presentation also revealed statistics on carbon fiber capacity and demand for 1983 and 1988 and the demand forecast for 1993 (Fig. 3). In the United States, demand in 1988 slightly exceeded domestic capacity and, with the growth projected for 1993, approximately 3.2×10^6 kg of additional capacity will be needed.

In the analysis of price trends between 1977 and 1988, BASF figures show an in-year average carbon fiber price index decline for 235 GPa modulus fiber of 40%. (The 235 GPa product was the standard material at the beginning of the period and is the product for which most of the qualification data exist.) When the U.S. gross national product deflator is used to adjust the actual price index to constant dollars, the average price decrease in constant dollars is shown to be 67% (Fig. 4).

TABLE 11
Worldwide Carbon Fiber Demand (metric tons × 10^3) by Market

	United States	Europe	Rest of World	Worldwide Total
1988				
Aerospace	1.77	0.50		2.27
Industrial	0.32	0.14	0.36	0.91
Recreation	0.27	0.18	1.41	1.77
	2.36	0.82	1.77	4.95
1993				
Aerospace	3.64	0.95	0.05	4.64
Industrial	0.64	0.41	0.73	1.82
Recreation	0.27	0.23	1.95	2.41
	4.55	1.59	2.73	8.87

From Ref. 18.

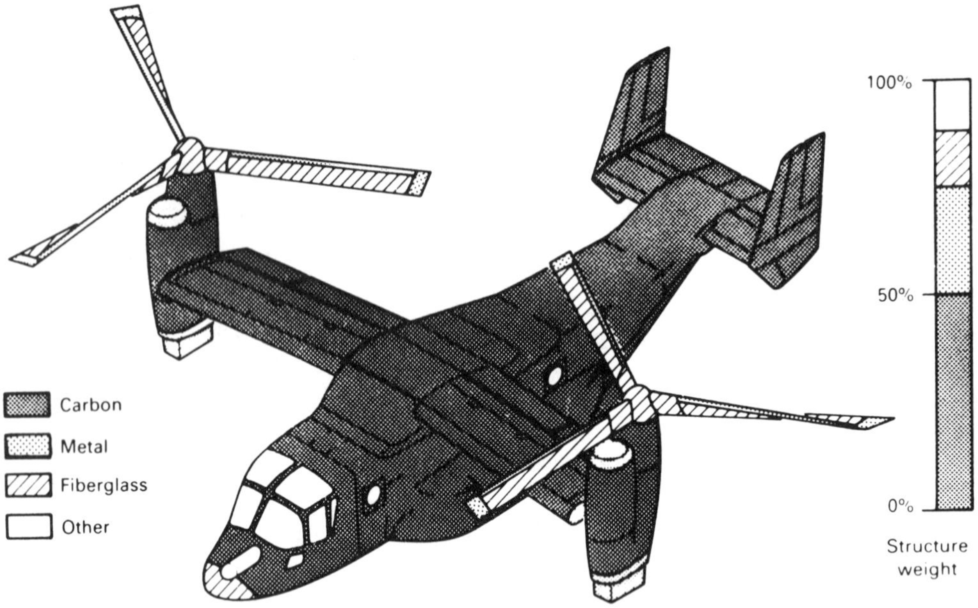

FIGURE 2 V-22 material applications. Structure (wing, fuselage, empennage, nacelle, rotor): 6120 kg; carbon-epoxy: 3100 kg. (Reproduced with permission from Engineered Material Handbook, Vol. 1, Composites, ASM International, Metals Park, OH, 1987.)

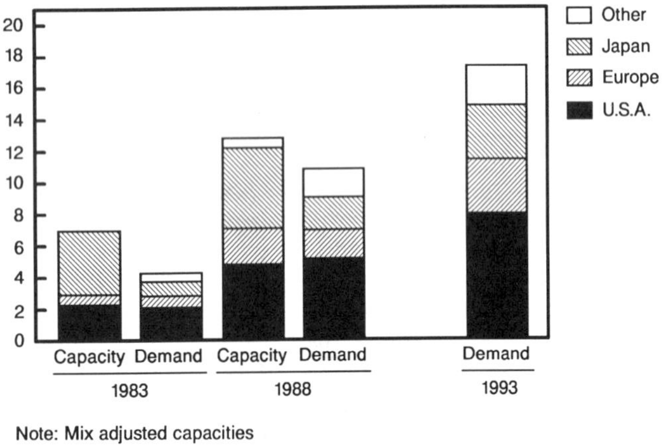

FIGURE 3 Worldwide carbon fiber supply/demand (millions of pounds). (Figure courtesy of K. Bloomer, BASF Structural Materials Inc., Celion Carbon Fibers Div.)

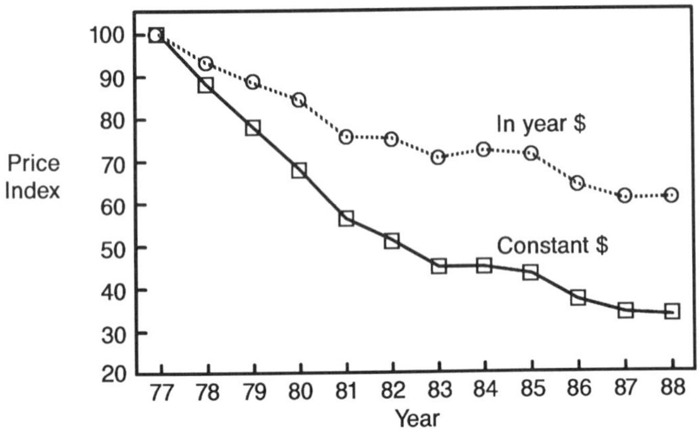

FIGURE 4 Average carbon fiber prices nominal 235 GPa modulus fibers (1977 = 100). (Figure courtesy of K. Bloomer, BASF Structural Materials Inc., Celion Carbon Fibers Div.)

It is anticipated that price trends will level off as demand and supply approach a pseudoequilibrium of periodic shortages and excesses. The "standard" 235 GPa product should level off in price—it has followed the BCG learning curve* for the past decade. Continuing investment is expected to be used to install the needed capacity increases and to continue significant research and development spending to complete development of the "emerging products" discussed above. It is the stated belief of carbon fiber manufacturers and CF composite structure end users that the best route to lower cost composites will be via both evolution and breakthrough developments of improved fabrication techniques. The cost of the carbon fiber used in a CF composite fabricated part represents only 10% of the cost of the part. Heavy DOD funding of the development of thermoplastic

* Boston Consulting Group (BCG) model proposes that evolving products decrease 20% in selling price for each doubling of cumulative volume produced during the manufacturing "learning" period.

composites, which is accompanied by significant matching of expenditures by thermoplastics manufacturers, is being carried out in the belief that significantly lower cost fabrication coupled with higher performance will evolve with their emergence.

P. E. McMahon

References

1. J. P. Riggs, "Carbon Fibers," this volume.
2. R. T. Pepper and F. L. Harris, *Proceedings of the Sixth Conference on Fibrous Composites and Structural Design,* 1983.
3. N. J. Johnston, *Proceedings of the Sixth Intl. Conference on Composite Materials,* 1987.
4. NASA Reference Publication 1092, "Standard Tests for Toughened Resin Composites," rev. ed., July 1983.
5. N. J. Johnston and P. M. Hergenrother, *Proceedings of the 32nd Intl. SAMPE Symposium,* 1987, pp. 1400–1412.
6. L. K. English, *Mater. Eng.,* pp. 43–46, January 1989.
7. H. T. Hahn and J. G. Williams, NASA TM-85834, 1984.
8. T. Hartness, *J. Thermoplas. Compos. Mater., 1,* 210 (1988).
9. M. E. Ketterer, NASA Contract Report 3839, November 1984.
10. (a) P. E. McMahon and M. Maximovitch, at Third International Conference on Composite Materials, Paris, August 1980. (b) C. C. Martin, J. O'Connor, and A. Y. Low, *Proceedings of the 29th Intl. SAMPE Symposium,* 1984, pp. 753–760. (c) T. Hartness, *SAMPE J., 20*(5), 26 (1984). (d) D. M. Lee, D. F. Register, M. L. Lindstrom, and R. W. Campbell, *J. Thermoplast. Compos. Mater., 1,* 161 (1988). (e) E. M. Silverman and R. J. Jones, *SAMPE J., 24*(4), 33 (1988).
11. BASF Structural Materials, Celion product brochures.
12. C. Megerdigian, R. Robinson, and S. Lehman, *Proceedings of the 32nd Intl. SAMPE Symposium,* 1988, pp. 571–582.
13. S. Yamane, T. Hiramatsu, and T. Higuchi, *Proceedings of the 32nd Intl. SAMPE Symposium,* 1987, pp. 928–937.
14. G. P. Daumit, Y. S. Ko, C. R. Slater, J. G. Venner, D. W. Wilson, C. C. Young, and H. H. Zabaleta, *Proceedings of the 20th SAMPE Tech. Conf.,* 1988, pp. 414–422.
15. Du Pont product brochures; *Adv. Compos., 3,* 64 (November–December 1988).
16. (a) P. E. McMahon, *SAMPE Q., 6*(1), 7 (1974). (b) L. Manocha, E. Yasuda, Y. Tanabe, and S. Kimura, *Carbon, 26,* 333 (1988).
17. (a) G. G. Tibbetts, M. Endo, and C. D. Beetz, Jr., *SAMPE J., 22*(5), 30 (1986). (b) G. G. Tibbetts and C. D. Beetz, Jr., *J. Phys. D: Appl. Phys., 20,* 292 (1987). (c) G. G. Tibbetts, M. G. Devour, and E. J. Rodda, *Carbon, 25,* 367 (1987).
18. K. G. Bloomer, at Fiber Producer Conference, October 1988.
19. Freedonia Group, *Plast Design Forum,* November–December 1988, p. 75.
20. SACMA News Release, *Adv. Compos.,* January–February 1989, pp. 22, 24.
21. *Adv. Compos., 3,* 53 (1988).

Carbon Fibers*

Over the past 15 years, carbon fibers have emerged as the main reinforcement fiber for high performance composite materials. The development of strong and stiff carbon fibers for a wide range of applications and their use in lightweight structural parts are among the principal technological achievements of this period. These fibers demonstrate a unique combination of static and dynamic performance advantages in demanding uses [1–3]. High strength and stiffness, light weight, improved fatigue resistance and vibration damping, corrosion resistance, good friction and wear qualities, low thermal expansion, and thermal and electrical conductivity combine to make carbon fiber composites an attractive substitute for various metals, special alloys, and other materials. These performance characteristics, in concert with composite structure, offer a degree of versatility in design and fabrication that is unavailable in other materials.

Carbon fibers are manufactured by the controlled pyrolysis of a precursor material (in fiber form)—for example, rayon, various pitches (petroleum or coal tar), and polyacrylonitrile (PAN), of which the latter is dominant. The utility of carbon fiber principally depends on its use as a reinforcement in various matrices, including thermoset synthetic resins such as epoxies, bismaleimides, polyimides, vinyl esters, and to a growing extent in a number of thermoplastics; it is used to a lesser degree in carbon, metal, ceramic, and glass matrices.

The term "carbon fiber" generally refers to materials that have been heat-treated at temperatures of 1000–3000°C and have markedly different properties and structure; they contain 92–99.99$^+$% carbon. In practice, the term sometimes refers to materials treated in the range of 1000–2000°C. The term "graphite fiber" refers to fibers processed at or in excess of 2500°C, although technically, these fibers lack the three-dimensional crystalline structure of graphite, except for the highest modulus, pitch-based materials. In this article, the term "carbon fiber" is used in its broadest sense.

Carbon fibers have been broadly categorized, as listed in Table 1, principally according to tensile or Young's modulus (or stiffness) and heat treatment temperature.

Properties

Carbon fiber properties are highly anisotropic; the properties of interest and advantage are in the longitudinal (fiber axis, 0°) direction, and the properties transverse (or 90°) to the fiber direction are distinctly different. Although carbon fibers and their composites are often characterized by unidirectional properties, in practice,

* Reprinted from *Encyclopedia of Polymer Science and Engineering*, Vol. 2, 2nd ed.

TABLE 1
Broad Categories of Carbon Fibers

Carbon %	Fiber Type[a]	Heat Treatment Temperature (°C)[b]	Modulus (GPa)[b]
92–96	High strength	1200–1400	228–241
99	High modulus	1800–2500	276–380
99^+	Ultrahigh modulus	$2800–3000^+$	$483–690^+$

[a] In the UK, high strength fibers are called Type II and high modulus, Type I.
[b] Principally PAN-based carbon fibers.

multiplied, multidirectional composites are usually required for optimum performance and structural integrity. These are built up with the reinforcing fiber employed in multiangular configurations; this subjects the fiber to loads in directions other than directly along the fiber axis and results in reduced properties.

Representative unidirectional properties of PAN and mesophase pitch-based carbon fibers are given in Tables 2 and 3, respectively. As tensile modulus increases (increasing degree of orientation along the fiber axis), carbon content, density, electrical conductivity (inverse of resistivity), and thermal conductivity increase; tensile strength and elongation decrease. As shown in Figure 1, carbon fibers are basically elastic, exhibiting linear stress–strain characteristics. In comparing the pitch and PAN-based data, at a given modulus, the pitch fibers are generally higher in density, lower in tensile strength and elongation, and higher in electrical and thermal conductivity. There is no pitch-based equivalent to the PAN high strength fibers because of fundamental differences in precursor composition, structure, and orientation and resultant differences in the strength–modulus profile developed as a function of processing temperature. However, recent experimental developments in Japan have shown substantial improvements in the strength of mesophase pitch fibers at lower moduli [4]. The very high specific mechanical properties (strength and modulus divided by specific gravity) of carbon fibers and their composites obtainable at low weight are competitive with those of steel, but at weights reduced by as much as a factor of 5. These specific, unidirectional properties of various types of carbon fibers are compared with other reinforcing fibers and various metals in Figure 2.

Recent developments are resulting in higher fiber strengths at higher modulus, and this combined with the excellent compressive properties, exceptional fatigue characteristics, and cost reduction potential of carbon fiber composites assures their further growth as a structural material. In hybrid composite constructions, different fibers are combined in a given system (e.g., carbon–glass, carbon–Kevlar, glass–Kevlar) [5–8], providing cost/performance advantages.

Composites

Structural composites are defined as products that utilize fiber reinforcements (50–70 wt %) of very high strength and stiffness in combination with polymers, metals, carbon, glass, or ceramics to form structural materials with unique property-to-weight characteristics. The matrix binds the reinforcing fibers together, forming a coherent structure, and provides a medium by which

TABLE 2
Properties of PAN-Based Carbon Fiber[a]

Property	High Strength[b]	High Modulus[c]	Ultrahigh Modulus[d]
Filament diameter, μm	5.5–8.0	5.4–7.0	8.4
Density, g/cm^3	1.75–1.80	1.78–1.81	1.96
Carbon content, wt %	92–95	$99–99^+$	99^+
Tensile strength, MPa	3105–4555	2415–2555	1865
Tensile modulus, GPa	228–262	359–393	517
Elongation at break, %	1.3–1.8	0.6–0.7	0.38
Electrical resistivity, $\mu\Omega$·m	15–18	9–10	6.5
Thermal conductivity, W/(m·K)	8.1–9.3	64–70	~120

[a] Strand data.
[b] Thornel T-300, T-500, T-600, T-700, Union Carbide Corporation; Celion 3000, 6000, 12000, Celanese Corporation; AS2/AS4/AS6/IM6, Hercules Corporation.
[c] Thornel T-50, Celion G-50, HMS, Hercules Corporation.
[d] Celion GY-70.

TABLE 3
Properties of Mesophase Pitch-Based Carbon Fiber

Property	Thornel P55S[a]	Thornel P75S[a]	Thornel P100[a]
Filament diameter, μm	10	10	10
Density, g/cm^3	2.02	2.06	2.15
Carbon content, wt %	99	99	99$^+$
Tensile strength, MPa	1895	2070	2240
Tensile modulus, GPa	380	517	690
Elongation at break, %	0.5	0.4	0.3
Electrical resistivity, $\mu\Omega{\cdot}$m	7.5	4.6	2.5
Thermal conductivity, W/(m·K)	110	185	515

[a] Thornel is a trade name of Union Carbide Corporation.

to transfer applied stresses from one filament to the next. The matrix material affects high temperature mechanical properties, transverse strength (i.e., nonunidirectional), and moisture resistance, as well as other properties, and is a key factor in toughness, shear strength, and oxidation and radiation resistance. The matrix system also strongly influences the fabrication process and associated parameters for forming composite materials into intermediate and final products.

Most carbon fiber composites are based mainly on thermosetting epoxy matrices; they offer ease of fabrication, excellent adhesion, low shrinkage during cure, a good balance of physical and electrical properties, and good heat resistance and stability under a wide range of environmental conditions. Other thermoset and thermoplastic matrix systems are growing in significance.

Typical properties of composites are given in Table 4. Fiber loading is 60–65 vol % (65–70 wt %); unidirectional fiber tensile modulus in the unidirectional laminate is directly proportional to the fiber volume loading (rule of mixtures); fiber tensile strength translates at 85–95% efficiency. The marked differences in properties between unidirectional (0°), transverse (90°), and quasi-isotropic (0°, ±45°/90°) fiber orientations can be seen. The high modulus fiber composite data reflect the lower strength, higher modulus properties of the fiber and the lower shear strengths inherent with higher modulus fibers.

Carbon fibers (longitudinally) have a coefficient of thermal expansion (CTE) that is generally slightly negative or near zero. Composite CTE values depend on fiber orientation, fiber type, and layup sequence. Production of composites with a fairly broad range of CTE values

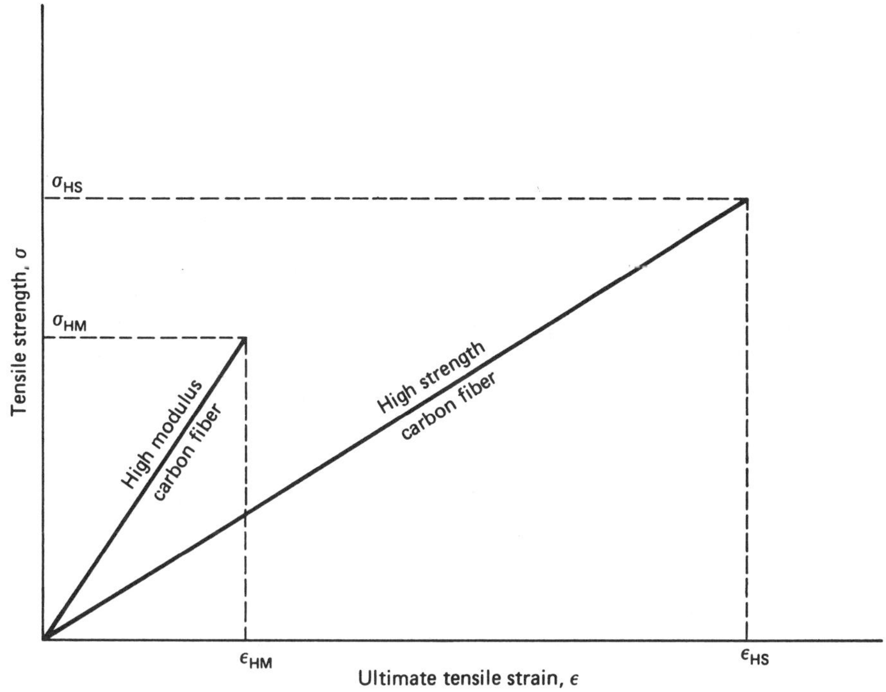

FIGURE 1 **Schematic of tensile stress–strain characteristics of carbon fiber illustrates basically elastic behavior. Tensile modulus $E = \sigma/\varepsilon$.**

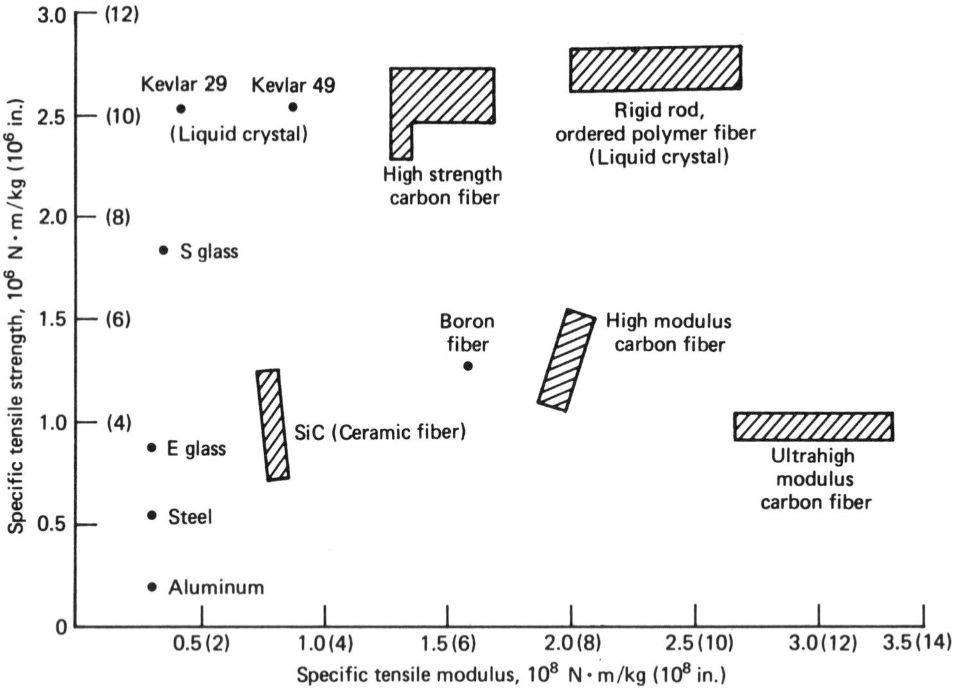

FIGURE 2 **Comparative specific strength vs specific modulus for various materials and reinforcing fibers.**

permits manufacture of components with a zero CTE. This feature is widely used in aerospace applications, for example, for maintaining critical instrument positioning. In addition, a CTE can be specifically matched to other types of materials to which the carbon fiber composite component might be attached.

Carbon fiber composites are compared with steel and aluminum in Figure 3. Compared with aluminum, titanium, steel, and glass fiber, carbon fiber composites exhibit unusual fatigue endurance characteristics. Therefore, carbon fiber composites are candidates for the replacement of metals in the most critical applications (e.g., commercial aircraft).

Although most carbon fiber is used in continuous filament form, thus maximizing static and dynamic performance characteristics, it is increasingly applied in

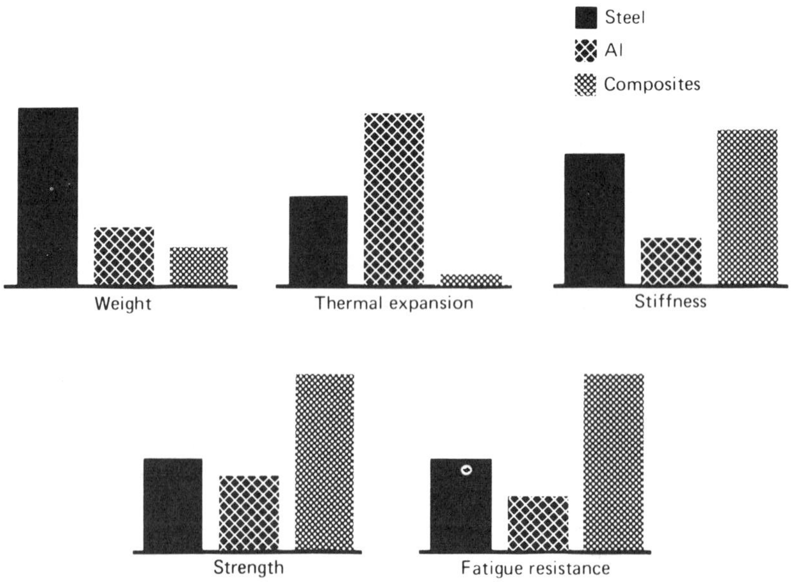

FIGURE 3 **Properties of carbon fiber composites compared with those of steel and aluminum [9].** (Courtesy of Van Nostrand Reinhold Company.)

TABLE 4
Properties of High Strength and High Modulus Carbon Fiber Composites in Epoxy Matrices

Property	Symbol	ASTM Test Method	High Strength: Celion 3000[a,b] 23°C	High Strength: Celion 3000[a,b] 82°C	High Strength: Celion 6000[a,c] 23°C	High Strength: Celion 6000[a,c] 177°C	High Modulus Celion G-50[d] 23°C
Unidirectional Laminate							
Longitudinal, 0° tensile strength, MPa	σ_{11}	D 3039	1647	1633	1785	1757	1165[f]
Modulus, GPa	E_{11}		151	143	145	147	212[f]
Ultimate strain, %	ε_{11}		1.1	1.1	1.2	1.2	0.55
Compression strength, MPa	σ_{11}^{c}	D 3410	1470	1525	1221	1118	841[e]
Modulus, GPa	E_{11}^{c}		141	151	138	133	189[e]
Ultimate strain, %	ε_{11}^{c}		1.7	1.1	1.1	1.4	0.45
Flexure strength, 4 pt MPa	σ_{11}^{f}	D 790	1750	2012	1997	1994	1338[e]
Modulus, GPa	E_{11}^{f}		135		133		192[e]
Interlaminar shear strength, short beam, MPa	σ_{ILS}	D 2344	125	79.3	97.0	76.0	79
Transverse, 90°							
Tensile strength, MPa	σ_{22}	D 3039	69.6	30.4	48.0	45.0	35.8
Modulus, GPa	E_{22}		9.3	9.0	9.6	9.6	7.24
Ultimate strain, %	ε_{22}		0.78	0.33	0.52	0.56	0.49
Other							
Density, g/cm^3	ρ	D 792	1.59	1.59	1.55	1.55	1.61
Shear strength, in-plane, MPa	σ_{12}	D 3518	76.6	58.0	72.0	71.0	59.3
Shear Modulus, in-plane, GFa[f]	G_{12}		6.2	4.1	4.8	4.1	4.07
Poisson ratio 0° coupon	ν_{12}		0.32	0.32	0.3	0.3	0.24
Coefficient at thermal expansion, 10^{-6} mm/mm K							
At 0°C	α_{11}		−0.22	−0.4	−0.31	−0.56	
At 90°C	α_{22}		15.0	27.1	19.8	35.8	
Quasi-isotropic Laminate, (0° ± 45°/90°)							
Tensile strength, MPa	σ_x	D 3039	494	444	537	528	305
Tensile modulus, GPa	E_x		53.1	48.0	49.7	50.0	73
Ultimate strain, %	ε_x		1.1	0.9	1.2	1.2	0.42

[a] Properties were obtained on autoclave-cured laminates using Rigidite 5208-Celion 3000 and Rigidite 5213-Celion 6000 (Celanese Corp.), respectively; prepreg manufactured by Narmco Materials, Inc. All unidirectional tension and flexural values normalized to 62 vol % fiber. Nominal fiber loading for all other properties, 60–65%.
[b] High strength carbon fiber in a high temperature epoxy matrix of the type used in aircraft and aerospace applications (5208); 3000 filaments per strand.
[c] High strength carbon fiber in a general epoxy matrix used in recreational and industrial applications (5213); 6000 filaments per strand.
[d] Properties developed with Celion G-50, 12,000 filament high modulus carbon fiber in Fiberite 934 high temperature epoxy; prepreg manufactured by Fiberite Corporation.
[e] Normalized to 62 vol % fiber; all other values in this column are fiber loading of 60–65%.
[f] In-plane shear properties determined by tension test of ±45° balanced laminate in 0° direction.

chopped fiber form (randomly dispersed and oriented), particularly in engineering thermoplastic injection molding compounds. The effect of carbon fiber reinforcement of selected thermoplastic resins is compared to unreinforced resin in Table 5. Carbon fiber offers improved mechanical properties, lower coefficients of thermal expansion, reduced creep, enhanced fatigue resistance, higher thermal and electrical conductivity, lower coefficient of friction and improved wear, and reduced mold shrinkage and abrasion. The improved electrical and thermal conductivity and wear and friction properties are the main advantages.

Property Development

Tensile, or Young's modulus, an intrinsic material property, is correlated to preferred orientation; that is, the more highly oriented the layer planes in carbon fiber, the higher the tensile modulus. This relationship is independent of the carbon fiber precursor [10,11]. Young's modulus can be increased by higher temperature and higher stress (stretch) processing. The theoretical modulus of carbon fiber is about 1000 GPa. Products with values approaching 550–860 GPa are commercially available, albeit at very high cost.

TABLE 5
Thermoplastic Molding Compounds Reinforced with Chopped Carbon Fiber[a]

Base Resin	PAN Carbon Fiber Content (%)	Flexural Strength, ASTM D 790 (MPa)	Flexural Modulus, ASTM D 790, (GPa)	Tensile Strength, ASTM D 638, (MPa)	Surface Resistivity ASTM D 257, (Ω/sq)
Nylon 6,6	0		2.83	81.4	>10,000,000
	30	351.9	20.0	241.5	150
Polycarbonate	0		2.28	62.1	
	30	248.4	13.1	165.6	3,500
Polyester[b]	0		2.35	58.6	
	30	200.1	15.9	151.8	500
Acetal	0		2.59	60.7	
	20	94.5	9.32	81.4	2,000
Polyphenylene sulfide	0		4.14	74.5	
	30	234.6	16.90	186.3	250

[a] From product literature, LNP Corporation.
[b] Polybutylene terephthalate.

The development of tensile strength in carbon fibers is not as straightforward nor as well understood and involves much more complex process, property, and structure interactions. Carbon fiber strength (as expected for a brittle, elastic material) is defect-controlled (or flaw-controlled), and the strength depends on the gage length [12,13] at which it is measured. The types of flaw present have been broadly characterized as extrinsic (i.e., internal and surface) [14,15] or intrinsic (i.e., resulting from an integral part of the fiber microstructure).

Better control of defect structure has significantly improved the strength of commercial carbon fibers in recent years (from ~ 2800 MPa to > 4800 MPa), and graphite whiskers, almost perfect forms of carbon fiber, have been made for many years with strengths of 20 GPa at moduli on the order of 700 GPa [16]. However, both the absolute strength and the strength–modulus relationship for carbon fibers are far from theoretical expectations. The strength for many materials ranges from 10 to 20% of the modulus. The highest strength carbon fibers and graphite whiskers have strengths on the order of 1–2% that of the modulus. Furthermore, theoretically, the strength of a material should increase as the tensile modulus increases. However, early in the development of PAN-based carbon fibers, this was observed not to be the case [17]. In general, the strength goes through a maximum in the 1300–1500°C range at a modulus of 230–250 GPa with increasing processing temperature and increasing modulus, although recent work is broadening this maximum to higher levels (270–310 GPa); deviations have been reported for PAN carbon fibers resulting from stress graphitization at high temperature, effects of fiber cross section shape, and minimization of impurities in the fiber [18–20].

For mesophase pitch-based carbon fiber, the relation of strength to temperature and modulus development is quite different [21]. These products generally increase in strength monotonically with temperature (and reach higher moduli at lower temperature than PAN fiber), reaching a plateau or showing a slight decrease in the 345–550 GPa range, and increasing further with higher heat treatment temperature (and moduli).

Structure of PAN Carbon Fiber and Property Development

A key characteristic of carbon fibers is high structural anisotropy, which can be achieved in a filament built up of two-dimensional, highly anisotropic graphitic crystallites arrayed in undulating sheetlike assemblies oriented along the fiber axis [10,22,23]. In a medium modulus fiber, sheets are smaller, less oriented, and more cross-linked than in high modulus materials, where they are wider, thicker, and more axially oriented. These sheets have a small amplitude of undulation near the surface and are oriented parallel to the fiber axis; toward the center, the amplitude increases, resulting in a larger *c*-axis component oriented parallel to the axis. The gradient in preferred orientation produces gradients in the local modulus and in the local coefficient of thermal expansion. This structure is thought to be a result of the stabilization process, whereby surface layers are efficiently cross-linked, but the fiber interior relaxes owing to the exotherms or the high temperatures encountered in stabilization. For the higher modulus fiber (> ~ 310 GPa), a distinct skin-core heterogeneity develops, varying in extent, but averaging about 150 nm in thickness [22,24,25]. The crystallites in the skin are larger and have a better preferred orientation than those in the core. Many large, misoriented crystallites have been found in this skin region; in the absence of microscopic flaws, these are likely to reduce strength. In lower modulus fibers (lower final heat treatment temperature), the first few layer planes forming the surface are of better orientation than those closer to the interior. Among the structural models proposed [26], the one depicted in Figure 4 [27] represents one of the better overall studies of carbon fiber structure. It is evident that most of the planes are

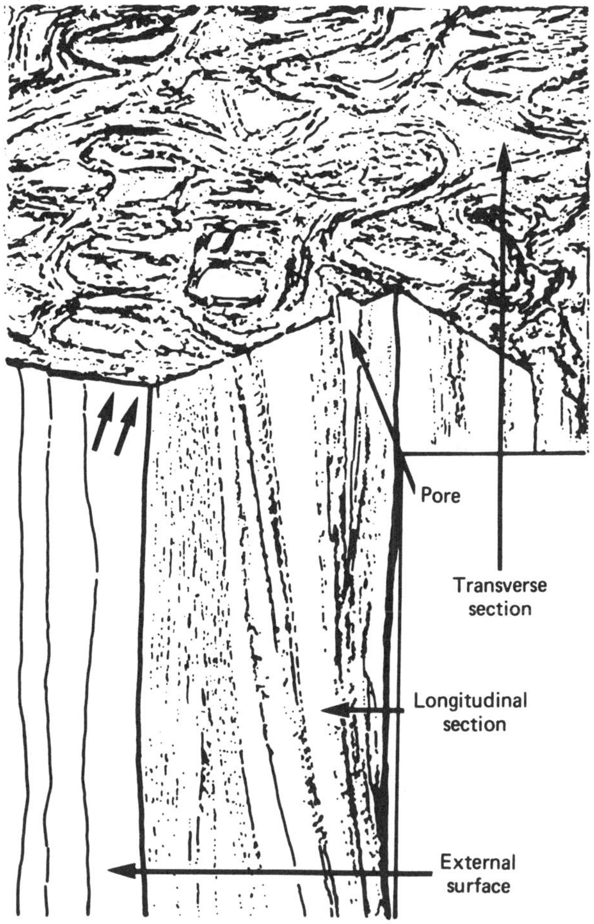

FIGURE 4 **Schematic of high modulus carbon fiber structure; double arrow indicates surface [27].** (Courtesy of ***Revue de Chimie Minérale.***)

oriented parallel to the surface (indicated by the double arrow in Fig. 4), although all orientations are statistically present in the middle of the fiber.

In general, wide angle X-ray diffractions reveal substantial structural changes that occur as a function of increasing temperature [28]. At lower moduli (240 GPa), the structure is turbostratic, with the hexagonal basal planes lacking interplane registration (i.e., three-dimensional order). With increasing temperature, stacking of the basal planes along the *c*-axis perpendicular to the fiber axis increases and results in higher long-range order, smaller average distance between planes, and improved angular orientation. At high modulus (> 517 GPa), the appearance of various reflections demonstrates the formation of a more three-dimensional structure. The development of both interbasal plane registration and improved degree of orientation contributes to higher modulus.

The principal limitation of carbon fiber is the relatively low strain-to-failure, particularly in higher modulus materials. Improvements in strength–strain have reduced extrinsic defects by the following process modifications.

1. Production of precursor to minimize incipient flaw damage caused by inclusion of foreign bodies, surface damage, and interfilament coalescence during the spinning [20,29].
2. Choice of precursor, reducing orientation loss in stabilization owing to comonomer composition.
3. Reduction and elimination of coalescence of filaments in stabilization by applying protective coatings on the precursor and proper choice of stabilization conditions [30–33].
4. Control of the stabilization process to prevent formation of a pronounced skin–core structure.
5. Rigorous exclusion of all oxidizing species from the carbonization process by improved furnace design and process control.
6. Improvement of surface treatments that minimize strength degradation.

The control of fine structure in spinning [34,35] and resolution of the effects of heterogeneity development, crystallite size, and misorientation are important [36,37]. As illustrated in Figure 5, the former entails a better knowledge of the implications of the Griffith failure criteria. For carbon fibers, the surface energy expected from bond breaking alone has been estimated at 5–10 J/m^2. However, it is possible to create and measure apparent

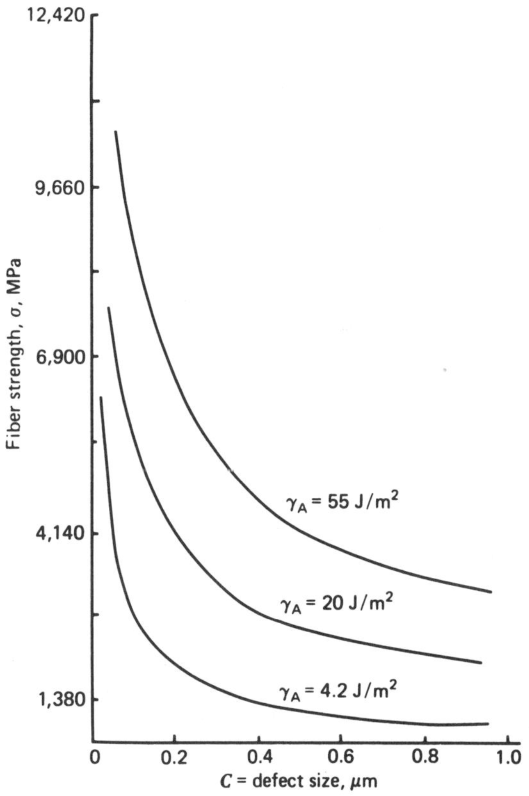

FIGURE 5 **Effect of fiber surface energy on carbon fiber strength as determined by defect size, based on Griffith failure criteria. Griffith relation: $\sigma^2 = 2E\gamma/\pi C$, where E = 240 GPa; γ_A = apparent surface energy [34].**

surface energies from 14 to 82 J/m^2, providing a mechanism to increase the energy consumed in crack propagation [34]. In the latter case, the presence of misoriented crystallites, rather than the flaw itself, determines the onset of tensile failure, and defect size itself may not be the controlling factor. Recent significant increases in the strength of commercial carbon fibers in the 240–275 GPa modulus range are also apparently related to a reduction in filament diameter from 7 to about 5 μm. The smaller diameter may obviate the detrimental effects of the latent heterogeneity and permit improved perfection of orientation in the bulk (i.e., core) of the fiber.

Manufacture and Processing

Carbon fibers are prepared by the controlled pyrolysis of an organic precursor in fibrous form. Commercial products are based on rayon (ore regenerated cellulose), pitch (petroleum and coal tar based), and PAN. Although now of minor importance, rayon-based products were developed first, followed by PAN-based material, and the pitch products (with mesophase pitch of prime importance). Characteristics of these precursors are given in Table 6.

Precursor weight loss is a central issue in the processing and economics of carbon fiber technology. Many materials have been evaluated as potential precursors with the objective of increasing carbonization yields and reducing fiber costs. These materials include such non-heterocyclic aromatic compounds as phenolics [38,39], polyamides [40,41], and polyphenylene [42]; heterocyclic aromatic compounds such as polyimides [43,44] and polybenzimidazole [40,43]; and aliphatic compounds such as polyethylene and polypropylene [45], polyvinyl chloride, including pitch material [46–48], polymethyl vinyl ketone [49], and polyvinyl alcohol and polyvinyl acetate [50,51]. Although a number of these, particularly the aromatic and heterocyclic polymers, give high yield and good properties, none has proved to be competitive to PAN and pitch.

Although the carbon fibers based on rayon, PAN, or pitch differ in properties, uses, economics, and process chemistry and technology, processing involves a number of common steps: extrusion of a polymer solution or melt into fine filaments; stabilization (preoxidation or thermosetting) carried out at relatively low temperatures (200–450°C), usually in air, which converts the precursor to an infusible state, stable to subsequent higher temperature processing; carbonization in an inert atmosphere (usually pure nitrogen) at 1000–2000°C, with an increase in carbon content to 85–99%; graphitization in an inert environment (Ar or N_2), at temperatures in excess of 2500°C, with an increase in carbon content to greater than 99%; and surface treatment to promote adhesion to a matrix (usually a polymeric resin). The term "graphitization" is used in the process description to separate the highest temperature processing stage from carbonization; the product of the graphitization step does not generally have the true three-dimensional order characteristic of polycrystalline graphite.

TABLE 6
Characteristics of Precursors

Precursor	Structure	Weight Retention Based on Precursor (%)
Rayon	$—(C_6H_{10}O_5)_n—$	20–25
PAN	$—(CH_2—CH(CN))_n—$	45–50^{+a}
Mesophase pitch	Polynuclear aromatics	75–85^b

^{a} Depending on modulus level (processing temperature).
^{b} Does not include losses in the initial mesophase conversion step before precursor fiber preparation.

Rayon-Based Carbon Fibers

Rayon-based carbon fibers gained commercial significance in the late 1950s and were dominant during the 1960s and early 1970s [52]. The early low strength, low modulus fibers were followed by a substantially improved high strength, high modulus yarn [53,54], but at much higher processing cost. It is of historical interest to note that the first carbon fibers, based on cellulose, were made by Edison [55] before 1880 for incandescent lamp filaments.

Rayon is manufactured in three types [56]: viscose, cuprammonium, and saponified acetate. For carbon fiber production, a special grade of viscose rayon is normally used because it tends to have relatively few defects. The conversion to carbon fiber consists of a low temperature (200–350°C) heat treatment, usually carried out in air, which gives a structure stable to higher temperature processing; carbonization in an inert environment at 1000–2000°C; and graphitization in an inert atmosphere (with or without stretch) at temperatures up to 3000°C [56,57]. In the first stage, extensive decomposition occurs and 50–60% of the initial precursor weight is lost. The excessive weight loss is related to thermal cleavage of carbon–oxygen and carbon–carbon bonds at 210–320°C, resulting in the evolution of H_2O, CO, CO_2, and tar (which results from the production and subsequent decomposition of levoglucosan) [58]. This stage is normally carried out in a reactive atmosphere such as air, chlorine, or HCl to inhibit tar formation and improve yield; Lewis acid chemical impregnants are also used [59]. Chain fragmentation eliminates any orientation present in the precursor, necessitating application of stress in later steps to improve the properties; the resulting product is an amorphous char. In the carbonization step, there is further weight loss with the early development of a carbon layer structure. This material can then be processed to higher temperatures.

Overall yield for the process is 20–25% and the properties resulting from the above sequence are relatively poor: tensile strengths of 345–690 MPa and tensile moduli of 20–55 GPa, at densities of 1.43–1.70 g/cm^3.

Properties can be improved by stretch graphitization at 2800–3000°C. At this temperature, plasticity in the yarn is sufficient to effect a high degree of stretch to promote axial orientation of the carbon layers. The high temperature increases the size and perfection of the layers. Fibers with moduli as high as 620 GPa and strengths of 3.1 GPa have been prepared by this route [57,60]. This hot-stretching process is, however, inefficient and very expensive.

Pitch-Based Carbon Fibers

The principal pitch materials used as starting materials for carbon fiber manufacture are based on petroleum asphaltenes and coal tar, composed of a complex mixture of organic compounds [61,62]; representative structures appear below.

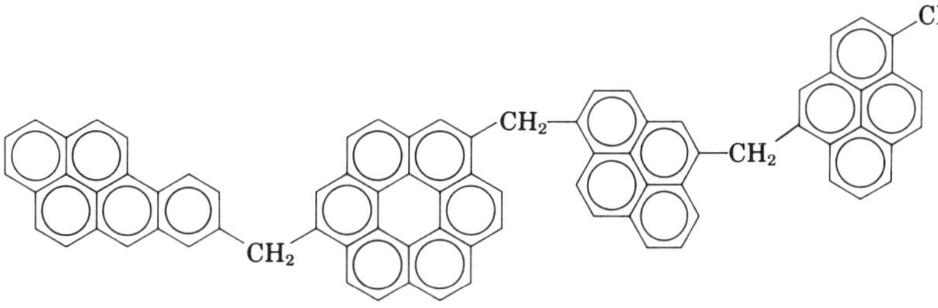

$C_{79}H_{46}$
coal tar pitch

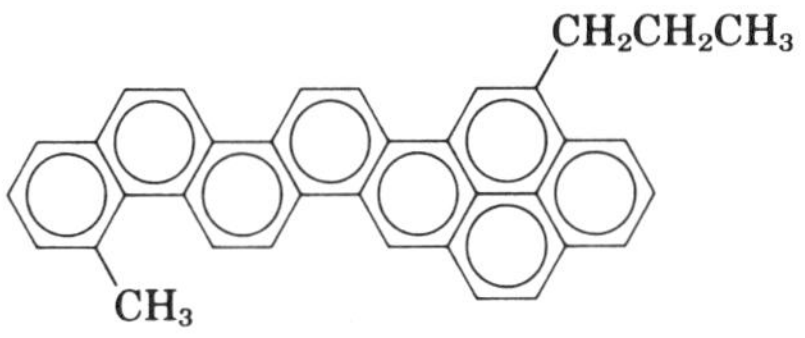

$C_{36}H_{23}$
petroleum pitch

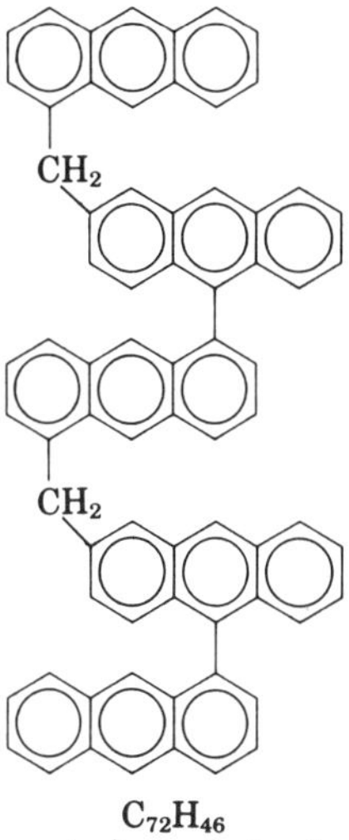

$C_{72}H_{46}$
petroleum-derived mesophase pitch

In general, the suitability of a pitch material for conversion to carbon fiber [63] depends on high carbon content (> 90%), high aromatic content (> 50%), molecular weight and molecular weight distribution, product consistency and low impurities, viscosity and rheological characteristics consistent with ease of spinning, and ready conversion to a nonfusible state for subsequent carbonization.

Compared with petroleum pitches, coal tar has a number of advantages [63,64], such as higher aromaticity and aromatic-to-total carbon ratio, higher specific gravity, lower sulfur content, higher viscosity and better rheological properties, and lower ash level. However, the high content of quinoline insolubles (QI) in coal tar (10–20% vs. essentially none in petroleum pitch) reduces its usefulness as a carbon fiber precursor; the QI particles cause spinning instability and lead to property-limiting defects in the carbon fiber. These fine, infusible particles consist of coke and carbon resulting from thermal decomposition during the coking process. Additional QI can be formed during the initial heat treatment of pitch.

Raw pitches (coal- or petroleum-based) are isotropic, and early pitch-based carbon fibers were produced from this material. However, the physical properties were poor (tensile strength and modulus on the order of 1035 MPa and 140 GPa, respectively). As with carbon fiber produced from rayon, properties can be substantially enhanced by hot-stretching [65] at high temperatures, but this procedure is difficult and costly.

In the early 1960s it was found that pitches, while still in a liquid state, could form an intermediate, highly oriented, optically anisotropic liquid crystal phase (mesophase) when heated for a certain period of time above 350°C [66,67]. The heating promotes dehydrogenation and polycondensation, resulting in more and larger molecules, which coalesce into a nematic liquid-crystalline phase [68]. In addition, it was discovered that mesophase pitch could be melt-spun into a carbon fiber precursor with a high degree of preferred axial orientation, that is, shear and elongational forces exerted during extrusion preferentially orienting the liquid crystal domains in the pitch melt [69]. This orientation could be maintained and further developed through carbonization and graphitization without application of hot stretch. Modulus develops quite readily as a function of processing temperature, and very high values (830 GPa) can be attained without application of tension. This is a primary advantage of the pitch process over those based on rayon or PAN precursors [70].

In the United States, the most commonly used petroleum pitch has been Ashland 240, manufactured by Ashland Petroleum Company [71].

In the formation of a typical petroleum, mesophase pitch-based carbon fiber [63,69–76], the isotropic pitch is heated for prolonged periods to form the mesophase before spinning. This material is melt-spun at high drawdowns to effect a precursor with high preferred orientation and relatively small filament diameter. The precursor fiber is stabilized in air to form an infusible product, capable of rapid carbonization, which is treated at higher temperatures to produce a range of carbon fiber products. A final step can be an oxidative (liquid or gaseous) surface treatment to effect adhesion in a resin matrix. A series of other process variations has been developed by different organizations, as described in the patent literature [77–82].

Mesophase pitch products generally have lower tensile strength and strain and higher modulus than PAN-based fibers, and this has restricted their range of application. Part of this problem has been related to inherent difficulties in spinning mesophase pitch. The spun products are brittle and difficult to handle; in addition, the polycondensation reactions continue during spinning, evolving gases, and, with the continued coalescence of liquid crystal domains, there is a tendency to coking. This leads to spinning instabilities, strength-reducing flaws, and high variability in properties.

A recent development in pitch technology shows promise for improving the strength–modulus profile of pitch-based (coal or petroleum) carbon fiber products [4]. A premesophase process developed by the Japanese Government Industrial Research Institute, Kyushu, of the Agency of Industrial Science and Technology, yields carbon fiber with much higher strengths and elongations at lower modulus levels (comparable to some PAN products) with claims for very low cost. Reported properties are tensile strengths of 2.94 GPa at a modulus of 195 GPa (1.5% strain). The spun material is mainly optically isotropic, but converts readily to an anisotropic material upon carbonization above 600°C.

In the proposed process, petroleum pitch or a coal tar depolymerizate is treated with tetrahydroquinoline under hydrogen pressure at 340–450°C. This reduces molecular weight and viscosity, permitting lower spinning temperatures and suppression of excess condensation of pitch during spinning. The initial step is followed by a second-stage, high temperature (450–500°C) stripping at atmospheric or reduced pressures. A mesophase fraction (40–60%) develops which, unlike that obtained in earlier processes, is completely soluble in the matrix pitch, forming an isotropic, homogeneous mixture in the molten state; during carbonization, this is converted to an anisotropic mesophase. Substantially more rapid, more stable spinning can be effected with fewer defects. It is reported that more than 14 Japanese companies have taken an option on this technology and are working to develop the process further.

PAN-Based Carbon Fibers

PAN-based carbon fibers represent by far the most important carbon fibers used today and are expected to retain this position into the foreseeable future, although premesophase pitch-based carbon fibers seem to have important long-term potential. PAN-based carbon fibers offer a broad range of excellent structural properties, ready compatibility with matrix systems, and well-established manufacturing capabilities and positions in expanding applications.

The potential for PAN as a preferred precursor for carbon fiber was first recognized in Japan in the late 1950s [83]; this was followed by demonstrations in the UK in the early 1960s that fibers with attractive properties could be prepared [84]. This early British work demonstrated that the strength and modulus of PAN-based carbon fibers could be significantly enhanced by maintenance of stress during the initial air stabilization stage [84,85]. Development and application of carbon fibers were aggressively pursued in the UK, with strong government sponsorship.

Significant effort in this area got underway in the United States in the 1960s under sponsorship of the Air Force Materials Laboratory (AFML). Sustained support by the AFML and NASA has been extremely important in the development and application of the technology.

The overall process for the conversion of PAN to carbon fiber involves the following steps, which are discussed separately below: precursor preparation, stabilization (preoxidation), carbonization (graphitization), and surface treatment. Starting with a linear polymer containing about 68% carbon, a controlled transformation is effected through a series of heat treatment steps in different environments at progressively increasing temperatures. Although millions of tons of conventional acrylic fibers are manufactured worldwide for textile and other industrial applications, large-scale economics cannot be transferred to the manufacture of PAN precursor. The basic process technology used for the conventional and precursor fibers (i.e., polymerization and spinning) is essentially the same; however, the precursor fibers differ importantly in comonomer content (lower) and composition, number of filaments per tow (lower), and decitex per filament (lower); they are thus very much specialty variants.

PRECURSOR PREPARATION

Composition. The precursor fiber can be spun as a homopolymer or, more commonly, a co- or terpolymer composition. Typical of these are methyl acrylate [86–88], methyl acrylate or methacrylate and itaconic acid [89,90], methacrylic acid [91–93], methyl acrylate and methacrylic acid [94,95], and hydroxyethylacrylonitrile and sodium alkyl sulfonate [96,97]. The use of relatively low (0.5–5.0%, mainly 0.5–2.0%) concentrations of comonomers improves the dissolution and stability of the polymer solution and facilitates spinning. More importantly, these low concentrations contribute to higher processing rates in stabilization, permitting higher temperatures and improved properties through minimization of chain rupture and cleavage; thus, fiber orientation is retained. In addition, the copolymers can be chosen to modify the nature of the exothermic reaction in stabilization and catalyze the cyclization rate in this step to reduce stabilization time. This is particularly true of acidic components.

Orientation. The fiber is highly drawn in the spinning process to align the molecular chains and impart a preferred orientation to the carbon fiber.

Filament diameter. Precursors are commonly drawn to relatively small filament diameters of 8–15 μm; this increases the stabilization rate (diffusion-controlled) and strength development. Efforts to draw the filaments to excessively small diameters causes process instabilities, broken filaments, the introduction of voids, and other strength-reducing defects.

Internal structure and impurities. Geometric and morphological characteristics of the precursor surface and internal structure are faithfully transmitted to the carbonized product, and this can have significant impact on property development. For example, modification of spinning conditions in wet spinning can alter the precursor internal structure (microporosity and fibrillarity) in such a manner as to increase the energy consumption in fiber failure and, consequently, final strength [34,35].

Contaminants in the precursor can lead to surface defects, gross voids, or inclusions that seriously reduce the strength of the carbon fiber. Spinning under dust-free conditions can result in a significant increase in carbon fiber strength [20,29].

Fibers are formed from PAN-based polymer by different spinning processes (e.g., wet, dry, dry-jet/wet, and semimelt). Dry-spun homopolymer PAN tends to yield higher modulus, whereas wet-spun copolymers give comparatively higher strength. The principal commercial processes are wet and dry-jet/wet spinning.

STABILIZATION. The stabilization step converts the precursor to a thermally stable structure capable of withstanding high temperature processing at high rates with good yield. In this step, carried out at 200–300°C in air, the structure is changed from a linear polymer system to a highly condensed cyclized (ladder) structure of stacked, partially hydrogenated naphthyridine rings with conjugated $>C{=}N$ bonds (see Fig. 6). From a processing engineering viewpoint, the principal feature of the overall reaction is the large exotherm, mainly associated with cyclization of the nitrile groups [99]. Above 220°C, cyclization is spontaneous, and the potential exists for the development of a rapid, uncontrolled release of heat which, on a large scale, can lead to catastrophic results. For this reason, and to ensure optimum property development, the process is usually run in a manner to promote a slow, controlled release of heat and thus prevent excessive fragmentation of the polymer chain and the development of an incipient, property-degradative structure owing to interfilament coalescence or formation of a highly heterogeneous, skin–core structure in the stabilized fiber. As stabilization proceeds, the reactants responsible for the heat generation are consumed and the rate of heat generation is diminished. In general, at some point the thermally treated PAN is said to be stabilized. The effects of this process are illustrated in Figure 7, where the differences in thermal and weight loss characteristics with and without stabilization are depicted; the exotherm progressively decreases with time. Although there is broad agreement on the kinetic parameters for the overall cyclization [101,102], there is no agreement on the mechanism.

Although the stabilization chemistry of PAN is complex, it is clear that in the presence of oxygen, chain scission, cross-linking, dehydrogenation, and cyclization take place. Cyclization can be effected purely thermally, but when PAN is heated in an oxidizing environment, the rate of cyclization is higher and the carbon fiber properties are improved [101,103,104]. The kinetics and mechanism for the conversion of polyacrylonitrile to a cyclized ladder structure have been investigated thoroughly [57,99,101–118]. Although there is some controversy, the following statements can be made:

1. Oxygen promotes the initiation of cyclization (exotherm starts at lower temperature), and this initiation is involved with chain scission and the formation of initiator sites.
2. As the temperature is raised, chain scission, the proportion of initiator sites, and the overall kinetics of the stabilization process are increased.
3. Oxygen participates in a number of cross-linking reactions (at the chain end, after scission, and on the polymer backbone).
4. Dehydrogenation reactions, promoted by an oxidative environment, lead to a conjugated and planar backbone structure, which, combined with cyclization, yields a prestructure of sheetlike character capable of being stacked, a preferred arrangement for subsequent carbonization.
5. The combined effects of chain scission and the tacticity of PAN (random atactic) preclude long segments of cyclized chains from being formed.

Recent work has substantially furthered the development of a kinetic model to describe and simulate heat evolution (see Fig. 8) from a PAN copolymer based on a consecutive first-order process with an exothermic second step and an autoaccelerated process [100]. Catalytic modification of the kinetics and heat flow characteristics of stabilization, in an effort to decrease reaction times and provide a more controllable reaction, has also received considerable attention. In addition to the comonomers noted above, approaches include impregnation of

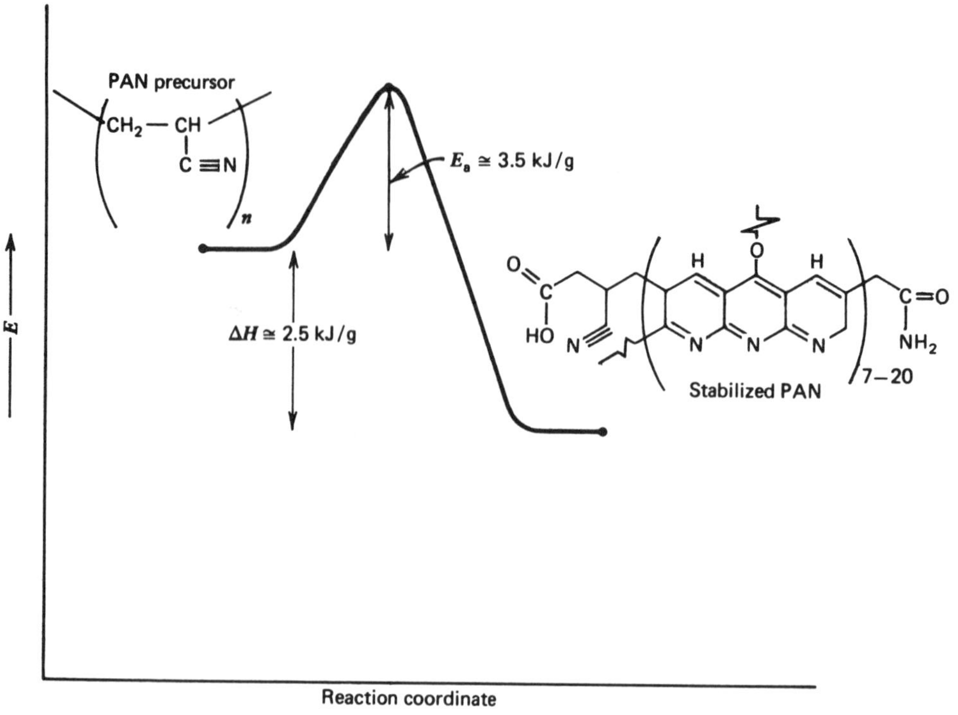

Element	Composition (%)	
	Precursor	Stabilized
Carbon	67.9	63.9
Hydrogen	5.7	3.9
Nitrogen	26.4	21.1
Oxygen	0	11.1

FIGURE 6 Kinetic, thermal, and structural changes in the stabilization of PAN precursor [98].

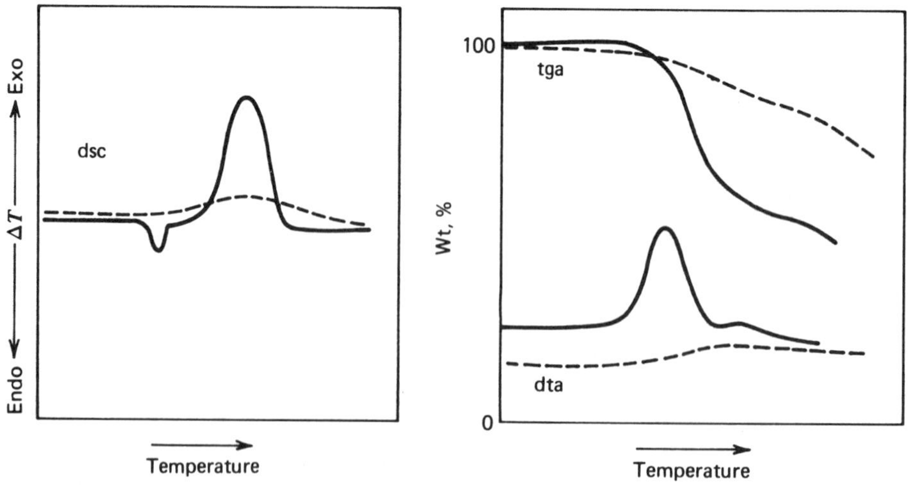

FIGURE 7 The effects of PAN stabilization on thermal and gravimetric responses to temperature: solid lines indicate stabilized; dashed line, not stabilized; tga = thermal gravimetric analysis; dta = differential thermal analysis; dsc = differential scanning calorimetry.

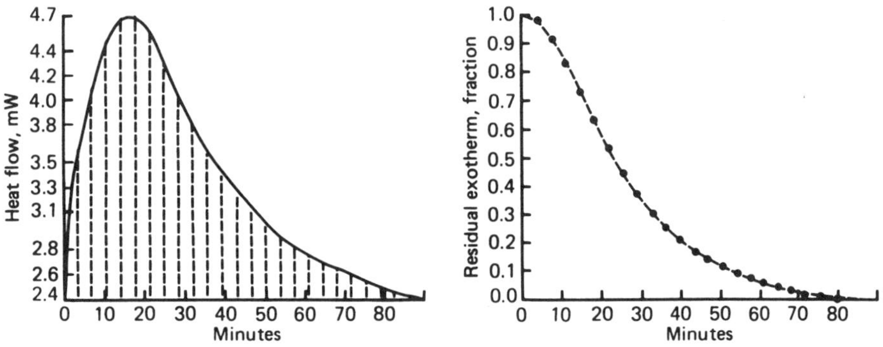

FIGURE 8 **Heat flow and residual exotherm as a function of time with PAN stabilization in air at 265°C [100].**

the fiber with various Lewis acids [119,120], amines [118,121,122],. and phosphonium bromides [100].

The stabilization step is highly temperature-dependent, whereas the oxidation process is diffusion-controlled. In the usual process, where stabilization is separated from spinning and conducted continuously with the subsequent stages in a hot air circulating oven, these factors are taken into account by a thermally staged heating sequence, beginning at low temperatures and finishing at the highest temperature in the schedule [123,124]; process rate and product quality are maximized. Over the past 15 years, optimization of this process has reduced stabilization time by 95%. In a newer method, stabilization is conducted in-line with the spinning process at high speeds, and very short total residence times (5–10 min) are claimed by contacting the fiber with a series of hot rolls at increasing temperatures; the process is interrupted at this point [97].

Current standard processing technology requires 1–2 hours for adequate stabilization. To preserve and improve the molecular orientation, fiber stress must be maintained during the stabilization process by limiting shrinkage to less than 15% [85,88]; natural shrinkage can be as high as 75%. The diffusion control of oxidation leads to the formation in the stabilized fiber of a skin–core structure that influences subsequent property development. In the extreme case, an overstabilized skin and understabilized core combination develops (improper stabilization conditions), which, upon carbonization, results in gross internal voids. With an optimum stabilization process, a latent skin–core structure may influence tensile property development in higher modulus products [24,25].

During stabilization, an overall net weight loss of 5–8% occurs because of dehydrogenation reactions that evolve H_2O, decarboxylation reactions that evolve CO_2, and the fact that some nitrile groups do not enter into cyclization, but are evolved as HCN. Carbon content slightly decreases from about 68 to 62–65%; optimal oxygen content is 8–10%.

The stabilized product resembles textiles in its physical properties (see Table 7) and handleability, which, combined with high thermal stability and flame resistance, has led to the development of applications, particularly in the area of asbestos replacement.

CARBONIZATION. This step involves a rapid thermal pyrolysis in an inert environment (usually high purity nitrogen) at 1200–2000°C (1200–1400°C for high strength products). In the 350–1300°C temperature region, a thermally stable pyridinoid structure is formed which subsequently collapses into a stacked-ring carbon structure to form carbon fibers; the yield is about 50% based on the precursor. Almost all other noncarbon elements are eliminated as volatiles; only low nitrogen concentrations remain. These off-gases include substantial amounts of HCN, NH_3, N_2, and H_2O with lesser amounts of low molecular weight nitriles, CO_2, CO, H_2, and CH_4 [98,125]. The off-gas stream is toxic and must be treated by liquid phase scrubbing or catalytic combustion before venting.

Various mechanisms have been proposed [57,98] for this condensation, which is depicted in Figure 9 for the two principal temperature ranges [57,98]. In the 350–600°C region, the main reactions most likely involve those functional groups (chain end) that do not fit into the condensed ring structure. This leads to some intermolecular cross-linking, which promotes reorganization and coalescence of the cyclized sections and results in evolution of the indicated compounds. In the 700–1000°C range, the main reaction is the intermolecular collapse of the pyridinoid structure into the ring structure. The principal off-gas emission is HCN, with smaller amounts of other nitriles and NH_3. When hydrogen is virtually depleted from the structure (i.e., at ~900°C), elemental nitrogen also becomes an important off-gas product.

TABLE 7
Physical Properties of Stabilized PAN[a]

Property	Value
Density, g/cm^3	1.3–1.4
Moisture regain at 65% RH, %	8–10
Electrical resistivity, Ω·cm	>10^{10}
Tensile strength, MPa	207
Tensile modulus, GPa	11.0
Elongation, %	8–10
Color	Black

[a] Preoxidized

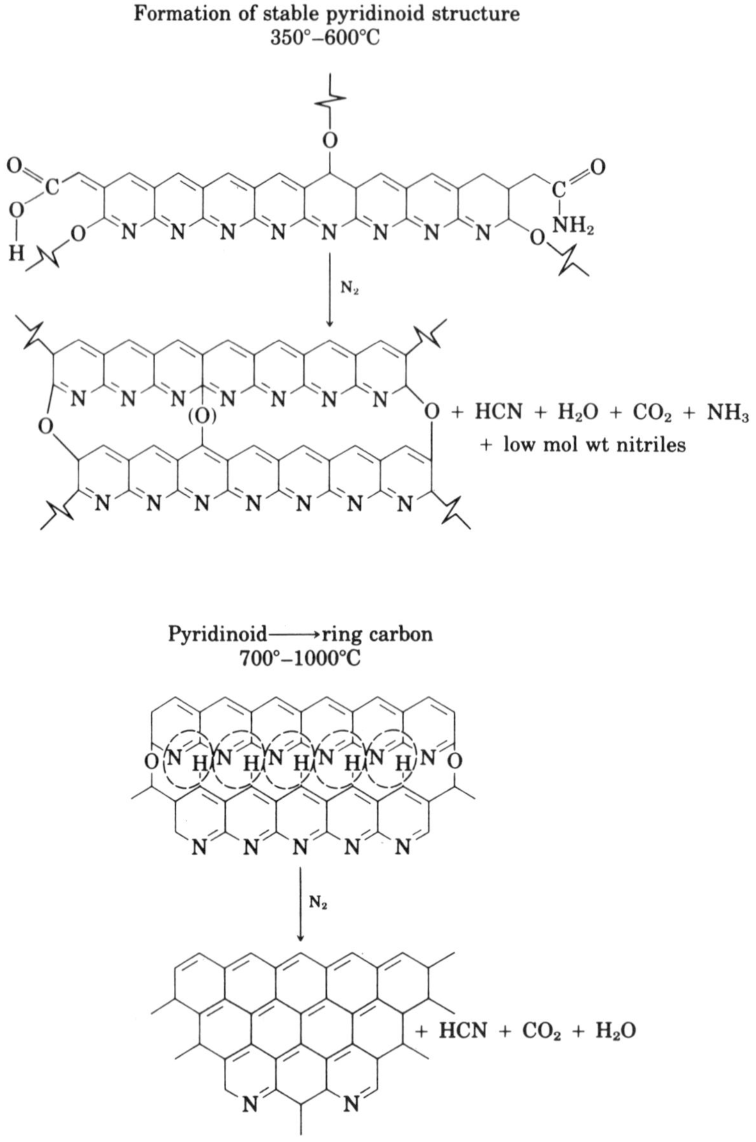

FIGURE 9 **Condensation reactions occurring in the carbonization of PAN [98]. Circles of dashes indicate HCN loss.**

With increasing process temperature, thermal and electrical conductivity, density, strength, and modulus increase progressively, reflecting increased preferred layer orientation and crystal growth. As can be seen in Figure 10, the tensile strength tends to go through a maximum at 1200–1500°C, whereas the modulus continues to increase monotonically. A number of the offgases are corrosive, and failure to exclude these and adventitious oxygen from contact with the fiber at high temperatures leads to flaw formation and reduction in tensile strength. The patent literature provides examples of carbonization furnaces designed for these requirements [126,127].

With the massive rearrangements that occur in carbonization at high rates with relatively large amounts of volatiles evolved, it is quite remarkable that a very high performance fiber is readily produced.

GRAPHITIZATION. Additional heat treatment, normally at temperatures above 2500°C for short periods of time in argon or nitrogen, further improves the preferred crystallite orientation and order and can lead to ultrahigh modulus products (i.e., $>$ 517 GPa) with processing temperature of about 3000°C, without stretch in graphitization, although some tension is maintained. The high temperature processing, typically in an induction furnace [128–130], is expensive, and catalysts (e.g., boron compounds) have been investigated to reduce the temperatures and time required to develop a certain modulus [131–134]. Boron catalysts have also been claimed to increase strength at higher modulus [134–136].

SURFACE TREATMENT. Since pristine carbon fibers show only a modest degree of adhesion to various matrices, surface treatments are normally applied immedi-

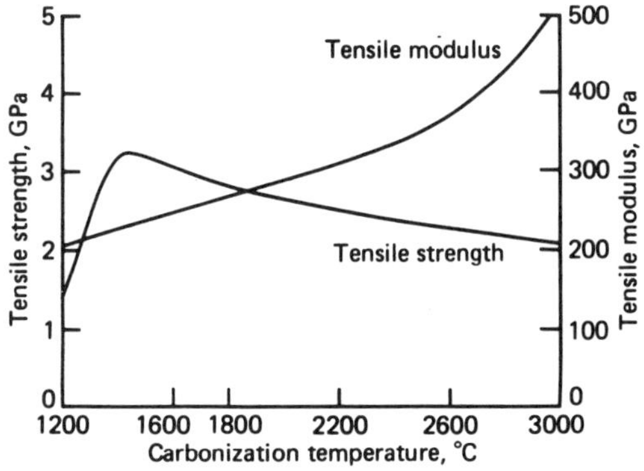

FIGURE 10 **General effects of carbonization temperature on tensile strength and modulus for PAN-based carbon fibers [17]. (Courtesy of Macmillan Journals, Ltd.)**

ately after carbonization (or graphitization). These treatments most often consist of controlled oxidation of the carbon fiber and are applied from the gaseous or liquid phase over a wide range of conditions, depending on the method. The performance of carbon fiber composite depends strongly on the degree of adhesion between the fiber and the matrix, and that, in turn, depends on the state of the carbon fiber surface. Thus, the need to modify the surface for adequate performance was recognized in the earliest work [137]. The objective of these treatments is to etch the surface, removing any weakly bonded carbon debris as well as condensed furnace vapors and other impurities; leave polar hyrophilic oxygen-containing groups on the surface, increasing its wettability and bondability to the resin matrix; and permit some control over the extent of specific surface area and the rugosity change.

The adhesion between the carbon fiber and a given matrix is related to the surface structure, morphology, anisotropy, heterogeneity (edge vs. basal plane configurations), and, to a smaller extent, specific surface area; the chemical composition and functionality of the polar surface groups; the surface energetics that permit a quantitative characterization of the foregoing effects; and the nature and properties of the interphase layer between the fiber and the bulk resin.

Oxidative gas phase treatments are typically carried out by heating the carbon fiber in air (or in O_2–N_2 mixtures with varying oxygen concentrations) at 350–1100°C [14,138]. At temperatures below 1800°C, other oxidizing gas environments, with or without specific oxygen addition, have been used (e.g., CO_2 [139], Cl_2 [140], NO_2–NO [141], NH_3 [142], and plasma-ionized inert gases [143,144]). Liquid phase oxidations have been extensively investigated, including both direct wet chemical oxidation and electrochemical (anodic) treatments. The former include aqueous solutions of nitric acid [145], hypochlorite [146,147] and chlorate [148], and dichromates in sulfuric acid [149]. Aqueous electrolytic treatments employ electrolytes such as hypochlorite [137,150], ammonium hydroxide [151], sodium hydroxide [152,153], and ammonium sulfate [154].

The various surface treatments differ substantially in severity and time and temperature conditions. Extensive oxidation can generate surface flaws and reduce fiber strength. Various processes have been evaluated and efforts made to maximize overall composite performance without loss of tensile properties [155–157]. High temperature gas phase treatments, in particular, may be degradative, and the increased kinetics and process rates they afford are offset by difficulties in control. The nature of the gas media can also significantly influence surface etching, the specific surface area, and surface pore sizes [141].

The heterogeneous surface structure of a carbon fiber affects interfacial adhesion and a number of attendant mechanical and environmental properties of the composite. The surface, in addition to being the most highly oriented part of the fiber, contains exposed, active edge planes as well as basal planes. The carbons in these high energy edge planes have a free bond and form strong carbon–oxygen complexes [158], which can further react and bond with the matrix resin. The lower energy and predominantly basal surface planes interact with the matrix to form a number of different, weaker bonds [159,160]. The increasing basal plane character with increasing modulus (and consequent reduction in the fraction of higher energy surface), combined with a developing microstructural heterogeneity, result in a surface that is hard to wet and bond. This manifests itself in the lower interlaminar shear strengths of high modulus fiber composites.

Oxidation increases the surface oxygen content of ultrahigh modulus [161] and high strength [162,163] carbon fibers, and this increase correlates with increased shear strength and an increase in the ratio of polar to dispersive components of the fiber surface free energy [162–165]. Oxidation does not increase the surface area of high strength fibers; subsequent high temperature (300–750°C) vacuum treatment reduces surface oxygen content below that observed before surface treatment; shear strength falls only slightly. Oxidative treatment evidently increases shear loading capability by introducing reactive surface groups, changing surface morphology, and reducing contamination.

In composites, surface treatment dramatically increases interlaminar shear strength (by a factor of 2–4). About 30% of this increase may be owing to removal of a weak layer in the as-manufactured surface and change in the surface structure. The remainder is caused by the bonding of oxygenated species on the fiber surface to the matrix resin and by increased surface area. For lower modulus (high strength) fibers with a fairly high proportion of active-edge carbons, increased surface area is not a factor; for higher modulus products, the treatments producing a satisfactory increase in shear strength usually result in increased surface area.

A final surface modification step consists of the application of a thin layer of finish, or size. Most of the time, this material is the same as the matrix resin (e.g., the principal component of the matrix without the curing agent). This coating is applied primarily to improve wetting, provide protection against surface damage, and facilitate handling in subsequent processing. However, recent work shows that tailoring of the finish composition can significantly affect important properties such as impact strength [166].

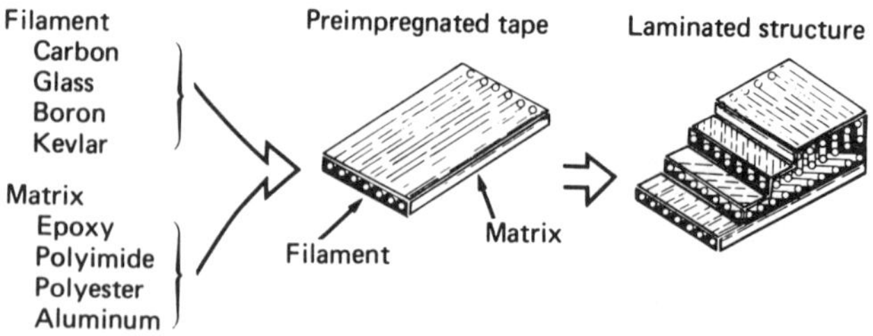

FIGURE 11 **Schematic of production of composite laminate structure.**

Incorporation into Composite Materials and Structures

The utility of carbon fiber, from the standpoints of both performance and cost, is determined by its use in a composite system (combination of fiber and matrix). The variety of commercially available fiber and matrix systems provides a wide spectrum of cost and performance. By changing the orientations of the reinforcing fibers and the strength–modulus characteristics, the structural performance of a composite system can be tailored to meet specific requirements. Similarly, the choice of the matrix material influences service environment and fabricability. In addition, different fibers can be used in the same composite system (hybrid composites) to achieve other performance advantages [5–8]. These systems combine the best attributes of the fiber components; notable examples include the combination of carbon fiber and glass (increased toughness, reduced cost) and carbon fiber and Kevlar (toughness, cost, enhanced specific strength). Matrix systems have been mostly organic thermoset resins (e.g., epoxies, bismaleimides, polyimides, phenolics, polyesters); epoxies are the most widely used because of overall performance characteristics, cost, and ease of fabrication. Other matrices include carbon and metals; thermoplastic resins are being developed for high performance uses. Possible combinations and a laminated composite structure are shown in Figure 11.

Product Forms and Fabrication

Carbon fibers are incorporated into various resin-based composite structures by using a variety of fiber–matrix forms and fabrication procedures.

1. *Unidirectional Prepregs.* Continuous aligned filaments in a resin matrix are used as a unidirectional, partially cured (B-staged) tape, 7.6–122 cm wide, which is laid up in different orientations and cured to form a laminated composite structure; costs are high, but properties are optimum.

2. *Fabric Prepregs.* Carbon fibers in a variety of fabric constructions are impregnated with a B-staged resin and fabricated; both prepreg costs and fabrication costs are lower. Properties are reduced. These prepregs are provided in widths up to 152 cm.

3. *Filament Winding.* Resin-impregnated, continuous carbon fiber tows are successively wound over a rigid form or mandrel (which may or may not be subsequently removed) to form the final shape, which is cured. Advances in computer-controlled winding technology permit high precision in fiber placement, and variations of orientation allow buildup of complex, three-dimensional laminate structures at high productivity. This technology can also be applied to placement of fiber in a mold.

4. *Pultrusion.* Continuous, resin-impregnated carbon fiber tow is fed through a heated die. The emerging cured section is pulled at a constant rate. This technique is used for making complex shapes, but is restricted to constant cross-section shapes; the rate is limited by the cure time of the resin. Articulating dies that would alleviate the cross-sectional constraint are being developed.

5. *Molding Compounds.* Chopped fibers of varying lengths are randomly distributed in a thermoplastic or thermoset resin through conventional compounding technique, followed by injection or compression molding. Processing and fabrication cost advantages are achieved at the expense of efficient use of fiber mechanical properties.

6. *Honeycomb Constructions.* A metal or polymer honeycomb core is covered with carbon–epoxy skins (prepreg layups). This highly efficient structural construction is widely used in the aircraft and aerospace industries.

Current fabrication methods use mainly prepreg materials, and, in the aircraft and aerospace industries, matrix materials are usually high temperature curing (177°C) epoxies. Progress in manufacturing has been substantial. Complex parts can be fabricated in a single operation, reducing the need for additional assembly operations (up to 60% reduction for various aircraft components), fabrics reduce prepreg layup time, and automated computer-controlled prepreg cutting and tape-laying operations have been developed. Although less material is being wasted than in some metal operations, composite materials remain expensive, with typical carbon prepreg final costs of $110–330/kg [3]. Production rates and scrap utilization require improvement.

Composite Design

The variations possible with carbon fiber and other composite systems enable design and fabrication engineers to develop a component's shape, sizing, and material configuration to attain the highest strength in highly stressed locations and directions. Unlike isotropic metals, these materials can be designed to allow stiffness in selected directions while maintaining desired flexibility in other directions.

HYBRID SYSTEMS. Proper use of carbon fiber in hybrid constructions with other fibers minimizes the

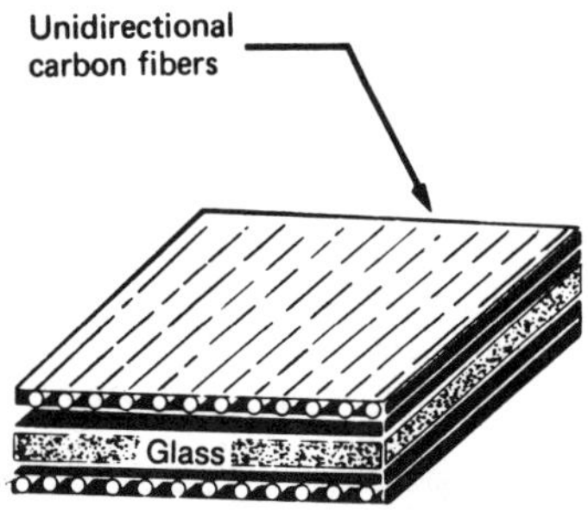

FIGURE 12 **Sandwich construction of carbon fiberglass hybrid composite. Continuous carbon fibers as face sheets, chopped glass core: carbon fiber 8–12%, average weight ratio 1.0; glass fiber 55–60%, average weight ratio 5.4; resin 30%, average weight ratio 2.7.**

amount of carbon fiber required with retention of significant performance advantages; cost/performance effectiveness is thus increased. The hybrid sandwich construction shown in Figure 12 meets the objectives of reducing cost by minimizing the carbon fiber content, maximizing performance by optimal placement and orientation of the fiber, and creating a product form amenable to high speed fabrication. The materials cost of this type of construction is similar to that of other high performance polymeric materials, but these relatively small amounts of carbon fiber result in considerable increases in performance characteristics, such as tension, flexural stiffness, and fatigue endurance (see Fig. 13).

HIGH TEMPERATURE COMPOSITES. In extending the applicability of high performance, structural carbon fiber composites to the aircraft and aerospace industries, materials for higher temperature use and increased toughness are required. Work in this area has been led by NASA, AFML, and materials suppliers or hardware manufacturers. Emphasis has been on the synthesis, characterization, and evaluation of new and improved polymers for use as composite matrices, adhesives, and films.

Materials under investigation for significantly higher temperature composite use include the following in approximate order of increasing use temperature (167):

Material	Use Temperature (°C)
Bismaleimides and maleimide–epoxy blends (BMI)	205–245
Phenolics	
Polyimides (PI)	260–315
Polyquinoxalines (PQ) and polyphenylquinoxalines (PPQ)	315–370
Polybenzimidazoles (PBI)	315–370

The phenolics have not, to date, generally produced widespread interest within the aerospace industry as polymeric binders for structural reinforcements; however, they are replacing some epoxies for interior applications because of better flame resistance. The PPQ, PQ, and PBI systems, offering the highest potential for thermal and oxidative stability, are still in the experimental stage. Processing requires removal of volatiles, high cure temperatures and pressures, and long postcures at high temperature. The properties of the PQ and PPQ systems are not attractive enough at this point to justify extensive development. However, PBI is of more interest, and with increasing raw material availability, warrants further work. As a composite matrix, PBI is formed in situ from

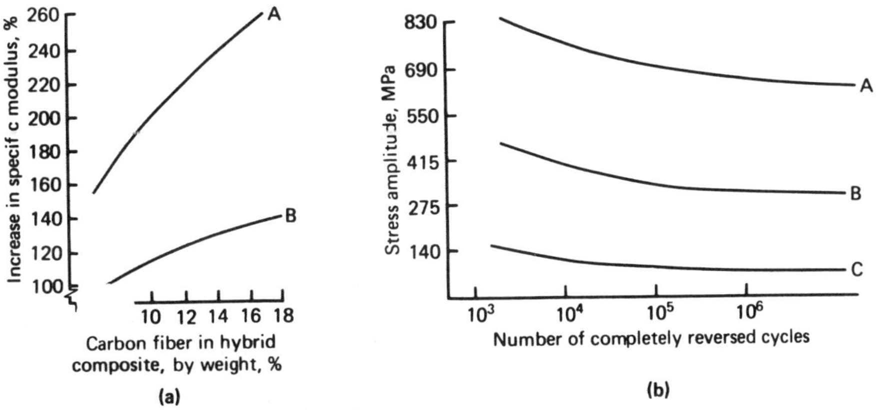

FIGURE 13 **Increases in static and dynamic performance attainable with small amounts of carbon fiber in hybrid composites (see Fig. 12). (*a*) Curve A, increase in specific flexural modulus (%); curve B, increase in specific tensile modulus (%). (*b*) Flexural fatigue. Curve A, all unidirectional high strength carbon fiber in polyester resin; curve B, unidirectional high strength carbon fiber faces over chopped glass core, polyester resin carbon fiber = 8 wt (%); curve C, chopped glass in polyester resin.**

oligomeric (low molecular weight) intermediates, and the high temperature performance of the resulting composites (e.g., survivability for 1 min at 650°C) makes them attractive for severe requirements such as in tactical missiles.

The systems of main interest are currently based on BMI and PI. The former are commercially available in various modifications. The BMI systems are readily modified to process like epoxies with acceptable prepreg tack and drape. An early application will be in the F-16XL fighter aircraft. In addition, they are much tougher than epoxies with potential applicability in primary aircraft structure.

Today polyimides are the matrix materials of choice for substantially higher temperature application with demonstrated performance and producibility. A preferred system is designated LARC-160, which was developed at the NASA Langley Research Center under the Composites for Advanced Space Transportation Systems (CASTS) project.

In the development of a thermally stable, high temperature composite system, the matrix is of prime significance, but two key contributions to this thermal stability arise from the reinforcing carbon fiber. These are the inherent oxidative resistance of the fiber and the thermal stability of the finish applied to the fiber as a handling aid [168]. As shown in Figure 14, the oxidative resistance of carbon fiber increases with modulus, which, in turn, correlates with processing temperature and carbon content for PAN-based fibers. The effect of finishes and different carbon fiber manufacturing bases at a constant modulus of 240 GPa on fiber oxidative stability is shown in Figure 15. The effect of different fiber bases is principally related to different alkali metal contents in the fiber [90]; the relative effects of an epoxy size versus a polyimide compatible size are also evident. The combination of LARC-160 with a thermally stable, high strength, 240 GPa fiber shows no property deterioration of consequence after 1000 hours at 260°C in air.

The NASA–Rockwell program has demonstrated the effect of polyimide–carbon composites on advanced materials for the space shuttle orbiter [169]. Results show the potential for an increase in weight savings from the baseline aluminum–composite design of 1230 kg to over 6360 kg through further materials replacement in structural components and the thermal protection system.

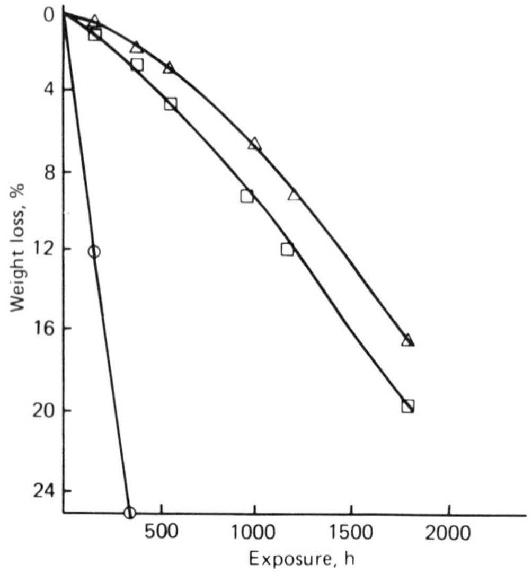

FIGURE 15 **Weight loss of high strength carbon fibers at a constant modulus of 240 GPa in 315°C air as a function of time. Low alkali metal content (i.e., < 100 ppm) is represented by polyimide size (triangles) and epoxy size (squares); circles represent high alkali metal content (i.e., > 1000 ppm) [168].**

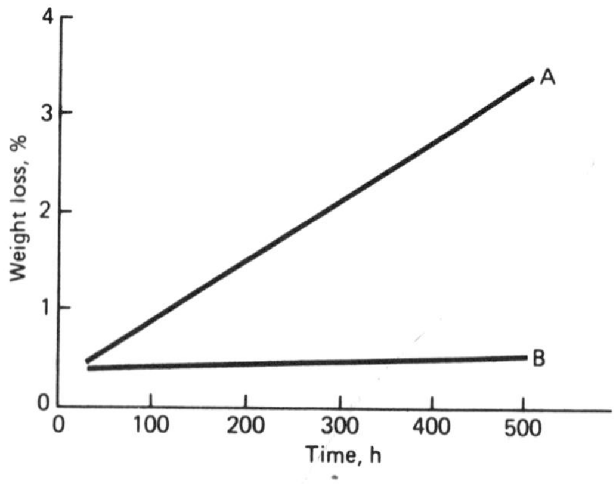

FIGURE 14 **Comparative thermal–oxidative stability of carbon fibers with different moduli; weight loss at 315°C in air. Curve A, Celion 3000, 240 GPa modulus; curve B, Celion G-50, 345 GPa modulus [168].**

TOUGH COMPOSITES. Development of tough (impact-resistant, damage-tolerant) carbon fiber composites is currently receiving the highest priority, and comprehensive programs on both matrix and fiber improvements are in progress under NASA and AFML [170] and independent industrial sponsorship. The potential impact on commercial aircraft is very significant. The present use of approximately 2.5% high performance composite materials on the Boeing 767 and 757 could be increased to 25–50% in the primary structure of future jetliners [171].

Development of improved toughness in carbon fiber composites is a complex problem and involves further development and basic understanding of the effects and interaction of matrix characteristics, fiber properties, and the interphase between fiber and matrix as well as the micromechanics of fracture. The principal approaches to such improvements include modified epoxies; nonepoxy systems (e.g., thermosets and thermoplastics), and fundamental studies on the nature and effect of the interphase. The most promising materials currently are nonepoxy, modified bismaleimides, and a range of new, significantly improved thermoplastic materials [170,172–174].

In efforts to increase toughness while retaining important mechanical properties (particularly under hot/wet conditions), tradeoffs have been difficult to reconcile. As Figure 16 indicates, current high performance epoxies

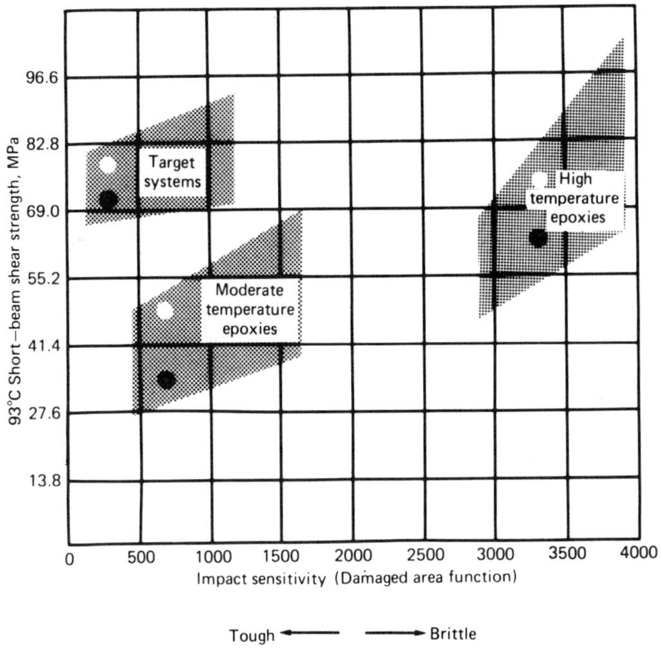

FIGURE 16 **Generalized effect of resin on carbon fiber composite shear strength and impact performance: open circles, dry; solid circles, wet [175].**

provide mechanical and environmental performance at the expense of impact resistance; more ductile epoxies give improved impact, but marginal mechanical performance, and the desired performance must combine the best of both.

Toughness has been improved significantly with modified bismaleimides combined with high strain carbon fiber [175]. This matrix resin offers easy processibility (low flow characteristics), high performance at 130–150°C (wet), no reaction volatiles, high impact resistance, excellent solvent resistance, and good shelf-life characteristics.

To illustrate the combined influence of matrix and higher strain fiber on performance, we show the effect of higher strain and strength fibers on conversion (Fig. 17), and the dramatic effect of both fiber and matrix on impact (Fig. 18). Because of its overall performance, the modified bismaleimide–high strain carbon fiber system is clearly a contender for use in the more demanding primary structure areas.

The second group of new resins gaining increasing attention for demanding composite applications are a series of high performance thermoplastics:

Material	Use Temperature (°C)
Polyamide–imide (PAI)	260
Polyphenylene sulfide (PPS)	200–230
Polysulfone (PS)	150
Polyethersulfone (PES)	175
Polyetherimide (PEI)	200
Polyetheretherketone (PEEK)	150

These materials, as well as a number of lower performance thermoplastics, are being used as chopped fiber (principally glass and carbon), reinforced molding compounds; a few are being developed as matrices for continuous fiber reinforcement.

Thermoplastic composites reinforced with continuous carbon fiber have a number of potential advantages for aircraft [177], including lower fabrication costs, better damage tolerance, repairability and reclaimability, reduced moisture absorption and degradation, and indefinite shelf life. Disadvantages have included lack of chemical and solvent resistance and adequate thermal stability; only crystalline thermoplastics provide the former [178]. Incorporation of comonomers (end or pendent groups) capable of cross-linking to provide chemical resistance has shown some success with norbornene end-group capping of polysulfone [179] and ethynyl-terminated polysulfone [170]. Similarly, work is underway to reduce the processing temperature of linear, aromatic polyimides by diluting the imide content and increasing chain flexibility [170].

A thermoplastic with high promise is polyetheretherketone (PEEK) [178,180,181]. This aromatic polyether with considerable crystalline character offers exceptional toughness as well as chemical resistance. Typical unidirectional carbon fiber composite properties are equivalent to high performance epoxies, and good high temperature performance is achieved. Equilibrium moisture gain is about 0.4%; these materials have excellent hot/wet strength retention—superior to that of epoxies. Comparison of the toughness of PEEK and epoxy laminates shows the PEEK-based materials to be 10 times tougher.

Another area of growing importance is the interphase between matrix and fiber. A much sounder understanding is needed of its effect on composite response to mechanical, chemical, and thermal environments [163, 170]. Recent results indicate that the fiber finish can significantly improve strength and toughness without reducing adhesion or environmental resistance [166]; flex strength and impact resistance can be increased by finish modification.

New Processes: Effects on Mechanical and Electrical Properties

Based on present PAN technology, a number of new fiber variants are being developed with the objectives of improved mechanical performance, modification and control of electrical and thermal conductivity characteristics, improved thermal–oxidative resistance, and enhanced cost effectiveness. High strain fibers have broad significant use implications, and it is expected that intermediate modulus (275–310 GPa) products with breaking strains exceeding 2% (5520–6900 MPa tensile strength) will be developed; high modulus (345–380 GPa) fibers with strains of 1% (3450–3800 MPa strength) are feasible; and ultrahigh modulus (970 GPa) products with strain levels on the order of 0.4% (4140 MPa strength) can be produced. Current commercial high strength (240

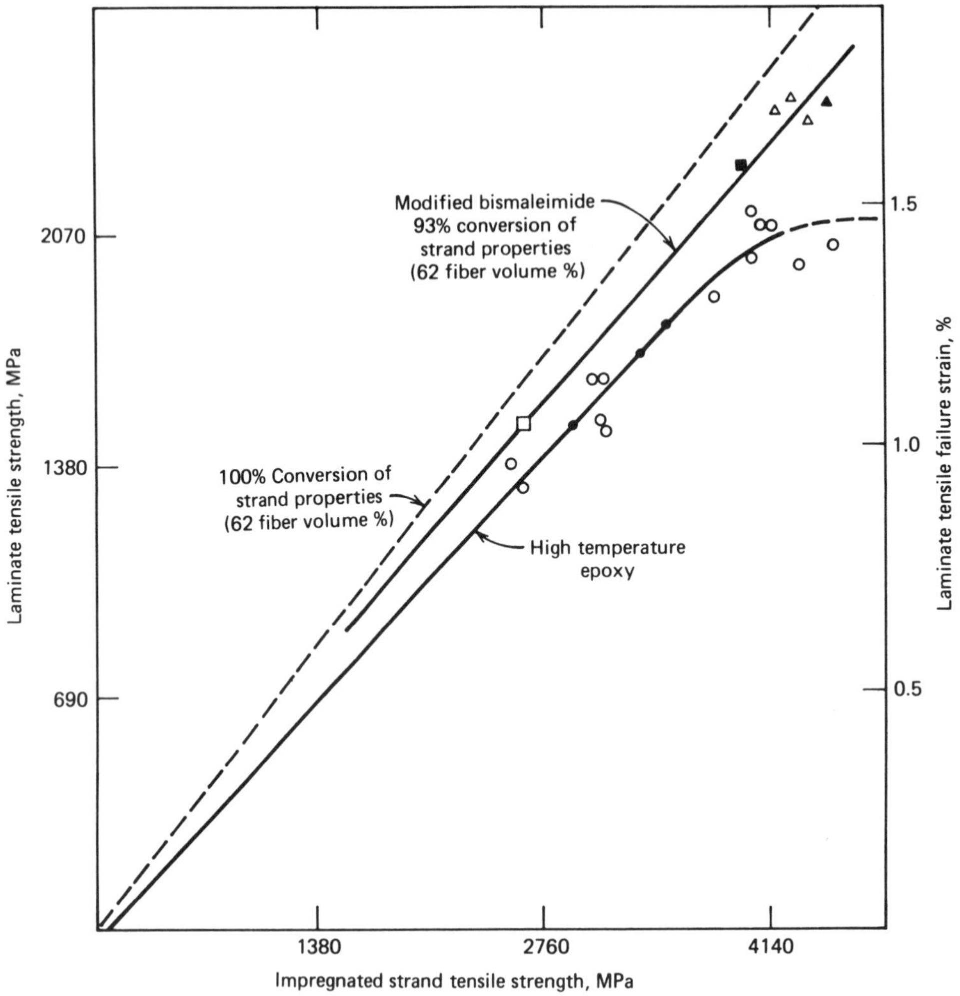

FIGURE 17 **Carbon fiber laminate strength translation in different matrices [176].**

GPa modulus) carbon fibers are becoming increasingly available at higher strain levels of 1.8–1.9% (4480 MPa strength) and intermediate modulus products have recently been introduced with strains to failure of 1.5–1.7% (4140–4830 MPa strength).

Work is also in progress to make more effective use of the electrical characteristics of carbon fibers, particularly in conjunction with reinforcement and structural applications. The inherently high aspects ratios of carbon fiber, coupled with an electrical conductivity more closely approaching that of bulk metals, would make such fibers more attractive for use in electromagnetic interference (EMI) shielding applications. They could also be uniquely suited, for example, for providing lightning protection for aircraft components fabricated from carbon fiber composites while providing a structural load-bearing function. This has been demonstrated [182] by the development of thin, uniformly Ni-coated carbon fibers, where the mechanical properties of the coated fiber are only marginally reduced because of the relatively high properties of the Ni coating. The high electrical conductivity, combined with the fiber aspect ratio, contributes to improved EMI shielding effectiveness at very low (~ 3%) fiber-weight loadings in a resin matrix.

Conductivity may also be increased through modification by chemical dopants (intercalation) [183–187]. A very high specific strength and stiffness material with electrical conductivity equivalent to that of transition metals, such as iron, cobalt, or nickel, and enhanced corrosion resistance would be attractive for applications such as antennae, guy wires, and power transmission lines with much reduced weight. In a sea environment, buoying effects would multiply this advantage, making the effective weight of a carbon fiber conductor about one-eighth that of transition metals. Intercalated, specific conductivities equivalent to that of copper have been demonstrated on an exploratory scale. Although the likely cost of such a material is very high, it would combine a number of other advantages (e.g., very low temperature dependence of the conductivity, stable continuous electrical conduction at elevated temperatures, an ability to withstand high current surges, and enhanced ac conductivity at high frequencies). Further work in modifying the base carbon fiber has shown that carbonization at

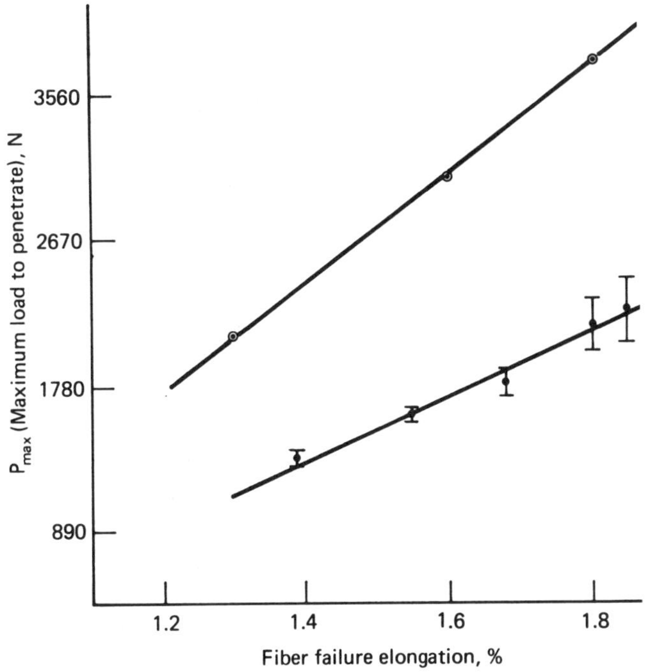

FIGURE 18 **Carbon fiber composite dart-drop impact resistance as a function of fiber strain and matrix system: data points on upper curve, modified bismaleimide; lower curve, high temperature epoxy [176].**

low temperatures (up to ~ 950°C) in varying atmospheres (e.g., acetylene) can alter the magnitude and stability of electrical conductivity and improve the oxidation resistance [188].

New processes of interest include that based on pre-mesophase pitch (discussed above) and those based on the thermal decomposition of hydrocarbon vapors [189]. Recent work by both Japanese [190–192] and U.S. [193] investigators indicates promise for developing materials of some utility. The former have shown that decomposition of benzene over extremely fine (~ 10 nm) particles of iron at 1200°C leads to the initiation of fiber growth, presumably through the catalytic action of iron carbides. Typical fiber diameters range from 10 μm to several hundred micrometers, with lengths up to 25 cm. With post-treatment at 3000°C, fibers with high strengths (4480 MPa) and high conductivity have been obtained and used in intercalation studies [184]. Analogously, it was found that pyrolysis of natural gas (~ 97% methane) in a stainless steel tube leads to the radial growth of fibers (5–1000 μm in diameter and up to 12 cm long) from the tube wall. Here again, fine particles of iron or chromium carbides are believed to provide the catalytic sites for initiation of fiber formation and growth.

Economic Aspects

Carbon fiber cost is principally determined, independent of the usual volume–price relationships, by precursor cost and related mass yield in conversion and capital and energy intensity (the latter putting a high premium on process mass throughput). Pricing is differentiated by precursor type (pitch vs. PAN), product form (mat and paper, chopped, tape, continuous filament), carbon fiber two size (filament count), and modulus level, and ranges from on the order of \$15/kg (nonwoven, low performance chopped fiber pitch mat) to more than \$2500/kg for ultrahigh modulus, continuous filament tows. Manufacturers' literature (e.g., from Celanese, Hercules, or Union Carbide) should be consulted for specific details. In general, prices are higher for continuous filament products, reduced filament count, and increased modulus (see Table 8). Certain pitch products, such as chopped fiber and mat and high and ultrahigh modulus products in the range 380–517 GPa, tend to be less expensive than PAN-based materials. Over the past 10 years, prices for the highest volume, high strength products have decreased from \$120–\$155/kg to \$30–\$65/kg.

The carbon fiber business has developed into a complex and unusual multinational network of interlocking corporate arrangements dominated in manufacturing capacity by Japan and in consumption by the United States (see Fig. 19). Past and projected U.S. consumption are given in Table 9; worldwide consumption is approximately twice that of the United States. Projected and 1983 worldwide capacities are given in Table 10. In contrast, consumption in 1983 was 1770 tons and is projected at 3045 tons for 1985 to 1986. The imbalance between world capacities and consumption is causing delays in European expansions and is very likely to delay projected U.S. and Japanese expansions.

The relatively high raw material cost base of carbon fiber composites has been readily justified in military and commercial aircraft applications and in aerospace uses. The combination of unusual static and dynamic physical properties with low weight and design and manufacturing flexibility has produced performance characteristics unobtainable with other materials. The substantial manufacturing and operational cost advantages are illustrated in Table 11, where the ability to reduce weight and the number of parts of various commercial aircraft components substantially is shown. In Figure 20, the impact of aircraft weight reduction on operating costs is indicated. For an overall reduction in aircraft weight of little more than 10%, the projected fleet

TABLE 8
Prices of Carbon Fiber Continuous Filament

Fiber Type[a]	Price Range (\$/kg)	Differentiation
High strength	30–230	Tow size, strength, market
High modulus	65–130	Precursor base, tow size, modulus
Ultrahigh modulus	330–2800	Modulus, tow size, precursor base

[a] High strength products constitute the largest market, ultrahigh modulus the smallest.

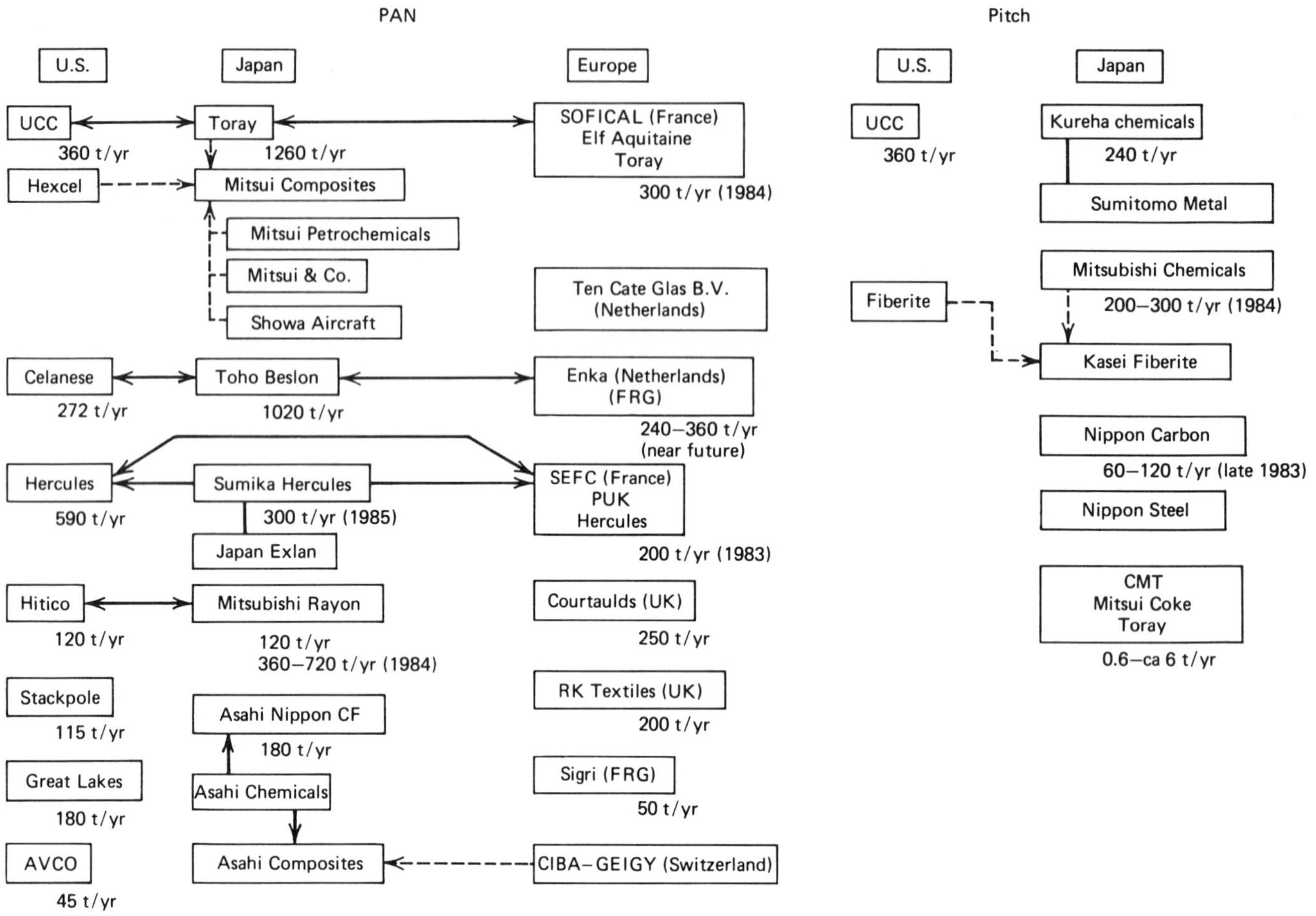

FIGURE 19 **Worldwide network of carbon fiber manufacturers: solid lines, carbon fiber; dashed lines, prepreg and composite, SOFICAL and SEFC announced delays in late 1983 [194].**

lifetime operating savings are well over $10 billion. To achieve this, however, will necessitate extensive use of carbon fiber composites in primary structure components such as the wing and fuselage, and this, in turn, is driving the development of improved composite materials (higher strain fibers, tougher matrices). In addition, the energy requirements associated with manufacture from composite materials are much lower than those for aluminum, steel, or titanium [3]. For example, the energy advantage with composites ranges from 2.5 times better than with steel to 10 times better than with titanium. Further manufacturing cost benefits are obtained with thermoplastic matrix materials and faster, automated fabrication methods.

In automotive and other industrial applications, the performance characteristics of carbon fiber composites remain attractive, but raw material and manufacturing cost premiums have prevented extensive application. However, the manufacturing cost benefits associated with parts consolidation and lower energy requirements [195] combined with further development of hybrid materials forms, optimized composite component design,

TABLE 9
U.S. Carbon Fiber Consumption (tons)

Application	1981	1983	1986[a]	1990[a]
Aircraft and aerospace	370	500	1250	2700
Sporting goods, recreation, marine	95	250	325	410
General industrial, transportation	50	205	275	500
Total	515	955	1850	3610

[a] Projected estimates.

TABLE 10
Carbon Fiber Manufacturing Capacities (tons/year)

Area	Base	1983	1985[a]
Japan	PAN	2580	3300
	Pitch	330	580
		2910	3880
United States	PAN	1680	2840
	Pitch	360	360
		2040	3200
Western Europe	PAN	500	1000
	Pitch		
		500	1000
Total	PAN	4760	7140
	Pitch	690	940
		5450	8080

[a] Projected estimates.

TABLE 11
Savings Afforded by Carbon Fiber Composites[a]

	Reduction (%)		
Component	Number of Parts	Number of Fasteners	Weight
727 elevator	40	~ 60	26
737 stabilizer	55	~ 60	23
747 aileron	47	~ 60	23

[a] From NASA and Boeing data.

and manufacturing flexibility can be expected to lead to wider use.

In all applications, the recent development of multiaxis, computer-controlled filament winding may be particularly significant. The original simple winding concept has evolved into a process for fabricating highly complex composite structures; precise, automated placement of fibers over complex mandrels and into cavities is now practical, permitting rapid, highly reproducible fabrication of a broad range of parts, including nonlinear shapes.

Specifications and Standards

In addition to the various levels of strength and modulus, carbon fiber materials are available in a wide variety of product forms (yarns and tows, fabrics, mat, chopped fiber, paper, roving, molding pellets, braid, various prepreg products, etc.). Most of these are further classified; for example, continuous filaments are produced in counts ranging from 400 (fine yarn) to 320,000 (large tow). Carbon fibers are often purchased in an intermediate product form, such as different types of prepreg (e.g., preimpregnated with a matrix resin). These can be unidirectional or cross-plied tapes, fabrics, or yarns (or rovings). Tape and fabric specifications include resin type and content, width, and thickness per ply.

Minimum specifications are yield (mass per unit length), strength, and modulus, and for more demanding applications, these specifications are accompanied by stringent statistical variability limits. Almost all specifications and performance standards are primarily determined by in-composite performance; typical values are given in Tables 4 and 5. For the higher performance applications, hot-wet, dynamic, and impact values are included. Many of these are specific and proprietary to various companies.

In the aircraft and aerospace industries, testing is extensive and there are long periods of qualification to es-

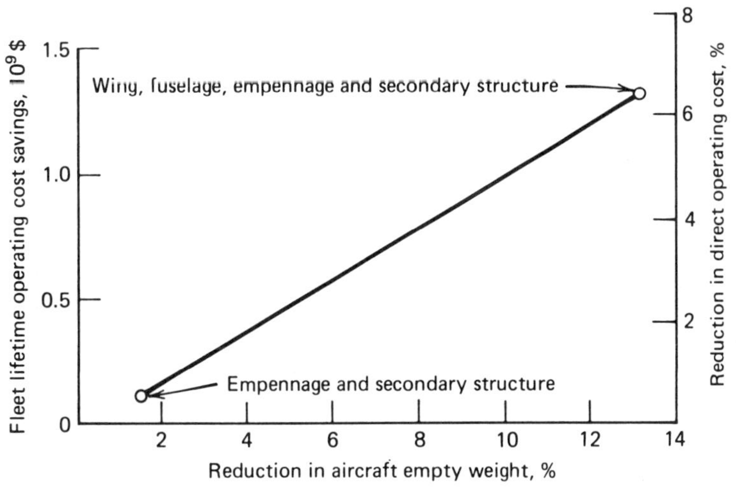

FIGURE 20 **Impact of carbon fiber composites use on weight and cost savings in commercial aircraft. Data from NASA.**

tablish a fiber or composite application. This is accompanied by process and product audit and certification procedures for suppliers.

Analytical and Test Methods

Complete testing and characterization of carbon fiber and carbon fiber composites entail the development of a broad range of information encompassing elementary fiber testing through composite static and dynamic mechanical characterization and other evaluations such as density, fiber, and resin content, void content, moisture absorption, and various thermal, electrical, and wear characterizations. Test methods employed and the extent of the testing are largely determined by use. For the most demanding applications in the aircraft and aerospace industries, composite strength and stiffness are determined in various failure modes and laminate configurations; damage tolerance and fatigue life are tested under various environmental conditions, and failure criteria, inspection requirements, and design bases are established. Statistical allowables are defined as an A-, B- and S-basis, referring, respectively, to the mechanical property value above which at least 99% of values is expected to fall with a confidence of 95%, the value above which at least 90% of the values is expected to fall with a confidence of 95%, and the minimum value acceptable under the governing specification. For any well-defined component, the application specifies a medium service stress, a lifetime requirement in terms of strength and stiffness, residual strength requirements for damage, and an acceptable failure probability.

Fibers

Single filaments (ASTM D 3379-75) and impregnated strand (or yarn, roving, or tow) (ASTM D 2343-67) are tested for longitudinal tensile strength and modulus. Single-filament testing has limited in-process utility and is used primarily on a research scale (i.e., new fiber development, flaw analysis, stress transfer, and interface studies). In the strand test, a yarn containing a large number of filaments is impregnated with resin, the assembly cured, and specimens of a given length (usually 12.7–25.4 cm) cut for testing. The test gives a relatively rapid and realistic assessment of fiber properties in a given matrix and can be used for control and specification purposes as well as research and development characterization.

Composites

The principal and most valid means for characterizing fiber performance is in a composite laminate [196–200]. Typical properties are given in Tables 4 and 5, together with the relevant ASTM references.

Other characterizations of particular importance are fatigue resistance (ASTM D 3479-76, D 671-71 [201,202]) damage tolerance [203], and the use of various nondestructive evaluation techniques [204] alone or in conjunction with other tests. Damage tolerance (i.e., toughness, the resistance to impact or cracking) can be evaluated in a number of ways, particularly by controlled assessment of crack initiation and growth and the effects of a specified impact. Fracture toughness parameters related to the former include definition of a critical stress-intensity factor, which is a measure of the stress intensity near the crack tip (ASTM D 813-81 [205]), and a critical strain energy release (ASTM D 3433-75 [206]), which relates to the energy per unit area of crack surface that is available for crack growth. These factors are uniquely related to each other and to various intrinsic material variables such as fiber volume fraction, orientation, laminate geometry, fiber–matrix interfacial bond strength, fiber strength and stiffness, and strength and toughness of the matrix.

Impact testing and characterization is a difficult, but critical evaluation for composite materials, particularly with respect to establishing aircraft component performance criteria. Historically, standard characterization techniques such as Izod and Charpy impact tests have been used to evaluate toughness (ASTM D 256-81 [207,208]). Because of the complexity of the impact failure mechanisms, however, results of these tests are not useful in predicting in-service performance of carbon fiber composites. Current studies attempt to simulate actual impact conditions to identify the operative failure mechanisms and provide design data [209,210]. These tests involve ballistic or drop-weight impact on laminated composite panels. In addition, many recent studies include microprocessor-sensing capability, where the complete impact event history is recorded and analyzed [176,210–213]. Typical output includes the loads corresponding to incipient damage, the maximum load required for penetration (see Fig. 18 for an example of the latter), and the total energy absorbed at each stage.

Alternatively, the impact resistance of the composite can be determined in a series of low energy (nonpenetration) tests. The amount of damage and delamination incurred is measured using ultrasonic C-scan or visual inspection (see Fig. 16). Comparison of the damage size resulting from equivalent impacts on different materials gives a relative measure of the damage tolerance, and it is possible to identify the onset of damage by variation of the load or energy parameters.

Laminates

To relate carbon fiber composite properties to component performance [196], laminate mechanics provide a means of determining the engineering properties (e.g., stiffnesses) of a composite that consists of various fibers and fiber orientations. These properties can be used in conjunction with structural analysis methods to determine the response (e.g., deflections, force distributions) of a composite component under applied loads. Once the force distributions have been determined, laminate mechanics can again be used to determine the distribution of stresses.

Stress analysis and structural design of composite components, using traditional structural analysis and special laminate mechanics methods, can be applied successfully in estimating the stiffness, buckling strength, and vibration characteristics of composite components. The distribution of stress and strain can be predicted throughout a composite laminate under applied mechanical, thermal, and moisture loads. The prediction of laminate failure and the micromechanics prediction of lamina properties from the properties of the fiber and matrix materials are being studied.

Short fiber composites, such as injection molding and sheet molding materials, require somewhat different methods. The properties in an injection-molded part depend on the orientations of the reinforcing fiber. Methods for estimating such orientations are being developed and evaluated. When combined with methods developed for predicting the bulk properties of such materials as a function of fiber orientation, the performance of injection-molded components can be determined by much the same methods as those used for continuous fiber reinforcement.

Health and Safety Factors

Manufacturers' materials safety information should be consulted for specifics on health and safety factors. Potential hazards include respiratory problems owing to inhalation of dust or fine fibrous particles, and electrical hazards caused by fiber, dust, and fibrous particles; explosive shorting of high voltage systems is possible.

Minimum precautions include good ventilation with localized exhaust at points where severe fragment and dust generation can occur. Protective masks should be available. Periodic monitoring of the particulate content of the working environment is recommended. Conditions where fiber dust generation can occur are particularly conducive to electrical hazards, and filtering the air and good housekeeping are therefore essential. In addition, electrical equipment, especially high voltage panels, must be isolated from the dust. Complete sealing of electrical components and provision for filtered ventilation to maintain a positive air pressure on components is usually indicated. Electrical panels should be inspected and cleaned periodically.

Dermatitis and other skin-related irritations can be caused by broken filaments embedded in the skin. Resin reactions and allergies may occur from exposure to the finish on the fiber or to resin in composite preparation or fabrication. Gloves are recommended for handling these materials.

Carbon fibers should not be disposed of by incineration. Waste fibers should be packaged and disposed of in a landfill intended for the disposal of such wastes or as otherwise required by law.

Uses

Carbon fibers are, by any standard, high performance, relatively expensive materials. Their development was initiated and furthered by the Department of Defense (DOD) and NASA for performance-critical military aircraft and aerospace applications. Their excellent characteristics—that is, low density, high stiffness (modulus) and strength, low axial thermal expansion, high electrical and thermal conductivity, good vibration damping, biocompatibility, resistance to creep, fatigue, corrosion, and wear—have also led to other applications (see also Table 9) [214].

Aircraft and Aerospace Applications

In 1983 the aircraft/aerospace sector constituted about 50% of carbon fiber use; it is projected to grow to 75% by 1990. Over the past 15 years, extensive application development has produced high confidence in carbon fiber composites with demonstrated materials performance, reliability, and predictability, accompanied by weight and fuel savings and manufacturing and operational cost reductions. Military aircraft, commercial aircraft, helicopters, missiles, spacecraft, and general aviation craft all contain carbon fiber composite components. From the first single-component production in 1969 of boron fiber reinforced epoxy skins for the horizontal stabilizer box on the F-14, military applications have progressed to the AV-8B (1982), which has about 26% of the structural weight in carbon fiber–epoxy composites (590 kg). The F-16, in 1977, marked the beginning of predominant use of carbon fiber–epoxy (over boron–epoxy), whereas the F-16XL in 1983, was the first important aircraft using a nonepoxy matrix, namely, carbon fiber–bismaleimide. Development of commercial and general aviation applications was stimulated by two NASA programs: Flight Service Evaluation and Aircraft Energy Efficiency (ACEE). Increased amounts of advanced composites are being used in the newest commercial aircraft, the Boeing 757 and 767 and the McDonnell-Douglas Super 80 and DC-10 [215]. Composite structures are also used in the Lear Avia Lear Fan 2100, a business jet, where carbon fiber–epoxy is used for almost the entire structure; this aircraft is currently in the process of FAA certification.

In spacecraft applications (e.g., antennae, booms, struts, and support trusses), composite materials offer the high dimensional stability (low-to-zero coefficient of thermal expansion) and ultrahigh stiffness achievable with variants of carbon fiber–epoxy materials. The space shuttle orbiter (NASA–Rockwell) represents the first production application of metal matrix composites (boron–aluminum). Its payload bay doors are the largest carbon fiber–epoxy structure built to date, and the first important use of a polyimide matrix with carbon fiber [169,216].

Other Applications

Sporting goods applications include: golf club shafts that weigh 40% less than metal shafts and permit heavier head design; tennis rackets with reduced head torque, improved vibration damping, and larger sweet spots; lightweight fishing rods with rapid vibration damping; and unstayed carbon fiber masts, which allow for novel wraparound sail designs that maximize aerodynamic efficiency.

Various rotating or reciprocating components of industrial machinery take advantage of high specific stiffness and fatigue resistance, vibration damping, friction and wear characteristics, and corrosion resistance, thus enabling higher operating speeds at reduced noise levels and longer lifetimes.

Business machine and computer applications use carbon fiber reinforced thermoplastics for a combination of mechanical and electrical characteristics, the latter providing static dissipation or EMI shielding.

Medical applications are of interest because of the compatibility of carbon fiber with human tissues, in addition to high strength, stiffness, and low degree of X-ray absorption; examples include X-ray film cassettes and holders, prosthetic devices, and artificial limbs and surgical implants.

High modulus carbon fiber is used in high quality loudspeakers because the speed of sound on the surface of a molded carbon fiber filled resin speaker cone is several times higher than on the surface of other speakers.

Carbon fiber composites did not meet early expectations for large, broad applicability in the automotive industry, where carbon fiber composites were thoroughly investigated. The Ford Motor Company built an experimental automobile using mainly carbon fiber composites while maintaining the overall size and appearance of an intermediate-size car (e.g., Ford Granada). Individual component weight reduction ranged from 20 to 60% over steel parts, and the overall effect was a 30% reduction in inertial weight and a 35% increase in fuel economy. However, other routes (e.g., size reduction, improved engine efficiency, and better aerodynamics) proved to be more cost-effective. Carbon fiber composites have proved to be an advantage in racing cars [217]. Mazda (Toyo Kogyo Company) is developing an experimental vehicle that includes a carbon fiber reinforced plastic body. A joint program between Polimotor Research and Ford [195] is focusing on the development of an all-composite engine, much of which is based on chopped or continuous carbon fiber reinforcement of epoxy and higher temperature matrices (i.e., polyimides, amideimides, bismaleimides). Filament-wound carbon fiber light truck driveshafts are nearing production, and hybrid carbon–glass leaf springs appear to be promising.

John P. Riggs

References

1. J. Riggs, *Emerging Non-Metallic Structural Materials Used for Airframes and Other Demanding Applications, Workshop on Substituting Non-Metallic Materials for Vulnerable Minerals,* National Science Foundation, Division of Policy Research and Analysis, Advanced Technologies and Resources Policy Group, Washington, DC, June, 1983, as reported in *Mater. Soc., 8*(2), 351 (1984).
2. W. Winters, *Use of Composites in High Performance Structural Applications,* in Ref. 1, p. 313.
3. R. Hadcock, "Role of Composites in Substitution, Conservation and Displacement of Materials and Potential for Substitution in Military and Commercial Aerospace," *Proceedings from Workshop on Conservation and Substitution Technology for Critical Materials,* NBSIR-82-2495, Vol. II, April 1982, p. 42-1.
4. *Jpn. Econ. J.,* Oct. 18, 1983.
5. H. Kliger, *Proceedings of the 31st Annual Technical Conference, Reinforced Plastics/Composites Institute,* Society of the Plastics Industry, 1976.
6. H. Kliger, *Mod. Plast., 55,* 82 (April 1978).
7. H. Kliger, *Mach. Design, 51,* 150 (Dec. 6, 1979).
8. F. Tortolano, *Design News, 37* (Sept. 21, 1981).
9. J. Delmonte, *Technology of Carbon and Graphite Fiber Composites,* Van Nostrand Reinhold, New York, 1981. p. 2.
10. A. Fourdeux, R. Perret, and W. Ruland, *Proceedings of the First Intl. Conference on Carbon Fibers,* Plastics Institute, 1971, p. 57.
11. D. Johnson, in *Sixth International Conference on Carbon and Graphite,* Society of the Chemical Industry, 1982.
12. R. Moreton, *Fibre Sci. Technol., 1,* 273 (1969).
13. P. McMahon, *SAMPE Q., 6* (October 1974).
14. J. Johnson, *Appl. Polym. Symp., 9,* 229 (1969).
15. J. Johnson and D. Thorne, *Carbon, 7,* 659 (1969).
16. R. Bacon, *J. Appl. Phys., 31,* 283 (1960).
17. R. Moreton, W. Watt, and W. Johnson, *Nature, 213,* 690 (1967).
18. J. Johnson, J. Marjoram, and P. Rose, *Nature, 221,* 357 (1969).
19. C. LeMaistre and R. Diefendorf, *Proceedings of the 10th Biennial Conference on Carbon,* American Carbon Society and Lehigh University, 1971, p. 163.
20. R. Moreton and W. Watt, *Nature, 247,* 360 (1974).
21. A. Bright and L. Singer, *Carbon, 17,* 59 (1979).
22. R. Diefendorf and E. Tokarsky, *Polym. Eng. Sci., 15,* 150 (1975).
23. K. Chen and R. Diefendorf, in T. Hayashi, K. Kawata, and S. Umekawa, Eds., *Progress in Science and Engineering of Composites, ICCM-IV,* 97 (1982).
24. B. Wicks and R. Coyle, *J. Mater. Sci., 11,* 376 (1976).
25. S. Bennett and D. Johnson, *Carbon, 17,* 25 (1979).
26. B. Wicks, *J. Nucl. Mater., 56,* 287 (1975).
27. M. Oberlin and A. Oberlin, *Rev. Chim. Miner., 18,* 442 (1981).
28. J. Stamatoff, H. Goldberg, and I. Kalnin, *Mater. Res. Soc. Symp. Proc., 20,* 57 (1983).
29. R. Moreton et al., *Carbon, 12,* 543 (1974).
30. R. Rulison, U.S. Patent 3,508,874 (Aug. 28, 1970), to Celanese Corporation.
31. A. DiEdwardo et al., U.S. Patent 4,112,059 (Sept. 5, 1978), to Celanese Corporation.
32. British Patent 2,059,407 (April 23, 1981), to Sumitomo Chemical.
33. H. Ogawa et al., British Patent 2,090,302 (July 7, 1982), to Toho Beslon.
34. W. Whitney and R. Kimmel, *Nature London Phys. Sci., 237,* 93 (June 5, 1972).
35. M. Ram and J. Riggs, U.S. Patent 3,841,079 (Oct. 15, 1974), to Celanese Corporation.
36. W. Reynold and J. Sharp, *Carbon, 12,* 103 (1974).

37. S. Bennett, D. Johnson, and W. Johnson, *J. Mater. Sci., 18,* 3337 (1983).
38. R. Lin and J. Economy, *ACS Div. Org. Coat. Repr., 31,* 403 (1971).
39. K. Kawamura and G. Jenkins, *J. Mater. Sci., 7,* 1099 (1972).
40. H. Ezekiel, *ACS Div. Org. Coat. Repr., 31,* 415 (1971).
41. Sano et al., U.S. Patent 3,547,584 (Dec. 15, 1970), to Celanese Corporation.
42. J. Economy and R. Lin, *Appl. Polym. Symp., 29,* 199 (1976).
43. H. Ezekiel and R. Spain, *J. Polym. Sci. C, 19,* 249 (1967).
44. R. Bourdeau and F. Galasso, *Am. Ceram. Soc. Bull., 55,* 785 (1976).
45. British Patent 1,283,714 (Aug. 2, 1982), to ICI, Limited.
46. E. Boucher, R. Cooper, and D. Everett, *Carbon, 8,* 597 (1970).
47. S. Otani, *Carbon, 3,* 31 (1965).
48. S. Otani, *Carbon, 5,* 219 (1967).
49. British Patent 1,293,900 (Oct. 25, 1972).
50. W. Fischer and J. Heckmaier, U.S. Patent 3,732,610 (Mar. 27, 1973), to Wacker Chemie GmbH.
51. German Offen. 1,246,510 (Aug. 3, 1967).
52. C. Ford and C. Mitchell, U.S. Patent 3,107,152 (Sept. 12, 1960), to Union Carbide Corporation.
53. W. Schalamon et al., U.S. Patent 3,716,331 (Feb. 13, 1973), to Union Carbide Corporation.
54. R. Bacon, A. Pallozzi, and S. Slosarik, *Proceedings of the 21st Annual Reinforced Plastics Conference,* Society of the Plastics Industry, 1966, Sect., 8E.
55. T. Edison, U.S. Patent 223,898 (Jan. 27, 1880).
56. R. Bacon, *Chem. Phys. Carbon, 9,* 1 (1973).
57. P. Goodhew, A. Clarke, and J. Bailey, *Mater. Sci. Eng., 17,* 3 (1975).
58. R. Bacon and M. Tang, *Carbon, 2,* 221 (1964).
59. H. Ezekiel, U.S. Air Force Materials Laboratory Report AFML-TR-70-1200, 1971.
60. R. Bacon and W. Schalamon, in Eighth Biennial Conference on Carbon, American Carbon Society, 1967.
61. T. Araki et al., *Appl. Polym. Symp., 9,* 331 (1969).
62. K. Okuda, *Kagaku Keizi, 28,* 20 (1981).
63. L. Singer, *Fuel, 60,* 839 (1981).
64. H. A. Kremer in *Proceedings of the Sixth Intl. Conference on Carbon and Graphite,* Society of the Chemical Industry, 1982.
65. H. Hawthorne et al., *Nature, 227,* 946 (1970).
66. J. Brooks and G. Taylor, *Carbon, 3,* 185 (1965).
67. J. Brooks and G. Taylor, *Adv. Chem. Ser., 55,* 549 (1966).
68. J. Barr et al., *Appl. Polym. Symp., 29,* 161 (1976).
69. L. Singer, U.S. Patent 4,005,183 (Jan. 25, 1977), to Union Carbide Corporation.
70. R. Bacon, *Phil. Trans. R. Soc. Ser. A, 294,* 437 (1979).
71. W. Smith et al., *Y-1921 AEC Report, Contract W-7405-Eng.-26,* Oak Ridge Y-12 Plant, 1974.
72. L. Singer, U.S. Patent 3,919,387 (Nov. 11, 1975), to Union Carbide Corporation.
73. I. Lewis, U.S. Patent 4,032,430 (June 28, 1977), to Union Carbide Corporation.
74. S. Chwastiak, U.S. Patent 4,209,500 (June 24, 1980), to Union Carbide Corporation.
75. J. Barr et al., U.S. Patent 4,276,278 (June 30, 1981), to Union Carbide Corporation.
76. I. Lewis et al., U.S. Patent 4,303,631 (Dec. 1, 1981), to Union Carbide Corporation.
77. T. Araki et al., U.S. Patent 3,718,574 (Feb. 27, 1973), to Kureha Kagaku Kogyo.
78. S. Otani et al., U.S. Patent 4,016,247 (April 5, 1977), to Kureha Kagaku Kogyo.
79. S. Otani et al., U.S. Patent 4,115,527 (Sept. 19, 1978), to Kureha Kagaku Kogyo and Toyo Boseki.
80. D. Angier et al., U.S. Patent 4,184,942 (Jan. 22, 1980), to Exxon Research and Engineering.
81. G. Dickakian, U.S. Patent 4,219,404 (Aug. 26, 1980), to Exxon Research and Engineering.
82. G. Dickakian, U.S. Patent 4,271,006 (June 2, 1981), to Exxon Research and Engineering.
83. A. Shindo, Report from the Osaka Industrial Research Institute, No. 317, Osaka, Japan, 1961.
84. W. Watt, L. Philips, and W. Johnson, *Engineering (London), 221,* 815 (1966).
85. W. Johnson et al., British Patent 1,110,791 (April 24, 1968), and U.S. Patent 3,412,062 (Nov. 19, 1968), to National Research Development.
86. M. Ram et al., U.S. Patent 3,657,409 (April 18, 1972), to Celanese Corporation.
87. British Patent 1,343,504 (Jan. 9, 1974), to Toho Beslon.
88. K. Saito et al., U.S. Patent 4,069,297 (Jan. 17, 1978), to Toho Beslon.
89. V. Furness et al., British Patent 1,196,599 (July 1, 1970), to Courtaulds.
90. T. Hiramatsu et al., U.S. Patent 4,349,523 (Sept. 14, 1982), to Toray Industries.
91. S. Kisbimoto et al., U.S. Patent 4,154,807 (May 15, 1979), to Japan Exlan.
92. M. Ozaki et al., U.S. Patent 4,080,494 (March 21, 1978), to Japan Exlan.
93. K. Fukushima, U.S. Patent 4,113,847 (Sept. 12, 1978), to Japan Exlan.
94. Y. Matsumura et al., U.S. Patent 4,001,382 (Jan. 4, 1977), to Japan Exlan.
95. Japanese Kokai 57-5922 (Jan. 12, 1982) to Mitsubishi Rayon.
96. K. Morita et al., U.S. Patent 3,632,798 (Jan. 4, 1972), to Toray Industries.
97. Y. Kinoshita, U.S. Patent 4,065,549 (Dec. 17, 1977), to Toray Industries.
98. A. DiEdwardo, *Organic Chemistry and Plastic Chemistry,* Vol. 38 (175th American Chemical Society Meeting), ACS, Washington, DC, 1976, p. 692.
99. N. Grassie and R. McGuchan, *Eur. Polym. J., 6,* 1277 (1970).
100. G. Collins, N. Thomas, and G. Williams, *Fibre Sci. Technol., 20,* (1984); *16th Biennial Conference on Carbon,* American Carbon Society, 1983, Extended Abstracts, p. 519.
101. E. Fitzer and D. Muller, *Carbon, 13,* 63 (1975).
102. J. Juron and J. Meybeck, *Eur. Polym. J., 13,* 553 (1977).
103. W. Watt and W. Johnson, *Proceedings of the Third Intl.*

Conference on Carbon and Graphite, Society of the Chemical Industry, 1971, p. 417.

104. W. Att, *Proceedings of the Third Intl. Conference on Carbon and Graphite,* Society of the American Chemical Industry, 1971, p. 431.
105. R. Houtz, *Text. Res. J., 20,* 786 (1950).
106. W. Burlant and J. Parsons, *J. Polym. Sci., 22,* 249 (1956).
107. J. Kennedy and S. Fontana, *J. Polym. Sci., 39,* 501 (1959).
108. W. Vosburgh, *Text. Res. J., 30,* 882 (1960).
109. N. Grassie and J. Hay, *J. Polym. Sci., 56,* 189 (1962).
110. J. Kirby, J. Brandrup, and L. Peebles, Jr., *Macromolecules, I,* 53 (1968).
111. H. Friedlander et al., *Macromolecules, I,* 79 (1968).
112. W. Turner, *J. Appl. Polym. Sci., 9,* 1245 (1968).
113. J. Hay, *J. Polym. Sci. A-1, 6,* 2127 (1968).
114. A. Clarke and J. Bailey, *Nature, 234,* 529 (1971).
115. A. Standate and R. Matkowsky, *Eur. Polym. J., 7,* 775 (1971).
116. N. Grassie and R. McGuchan, *Eur. Polym. J., 8,* 243, 257 (1972).
117. M. Coleman and G. Sivy, *Carbon, 19,* 123, 127, 133, 137 (1981).
118. A. Endrey, *J. Polym. Sci., Polym. Chem. Ed., 20,* 2105 (1982).
119. K. Gump et al., U.S. Patent 3,647,770 (March 7, 1972), to Celanese Corporation.
120. A. DiEdwardo et al., U.S. Patent 3,820,951 (June 18, 1974), to Celanese Corporation.
121. H. Sato et al., U.S. Patent 3,917,776 (Nov. 4, 1975), to Mitsubishi Rayon.
122. I. Kalnin, U.S. Patent 4,364,916 (Dec. 21, 1982), to Celanese Corporation.
123. M. Ram, U.S. Patent 3,539,295 (Nov. 19, 1970), to Celanese Corporation.
124. Y. Matsumura et al., U.S. Patent 4,009,991 (March 1, 1977), to Japan Exlan.
125. A. Fielder et al., *Carbon, 11,* 426 (1973).
126. R. Dix et al., U.S. Patent 4,020,273 (April 26, 1977), to Celanese Corporation.
127. Y. Saji et al., U.S. Patent 4,073,870 (Feb. 14, 1978), to Toho Beslon.
128. F. Douglas et al., U.S. Patent 3,607,063 (Sept. 21, 1971), to United Aircraft Company.
129. F. Germent, U.S. Patent 3,656,910 (April 18, 1972), to Celanese Corporation.
130. British Patent 2,059,406 (April 23, 1981), to Toho Beslon.
131. G. Cooper et al., British Patent 1,267,201 (March 15, 1972), to National Research Development.
132. M. Ram, U.S. Patent 3,656,904 (April 18, 1972), to Celanese Corporation.
133. H. Ezekiel, British Patent 1,295,289 (Nov. 8, 1972).
134. M. Ram, U.S. Patent 3,723,605 (March 27, 1973), to Celanese Corporation.
135. British Patent 1,320,890 (June 20, 1973), to National Research Development.
136. E. Brewster and D. Nelson, *28th Natl. SAMPE Symposium, 65,* 1983.
137. H. Wells et al., U.S. Patent 3,657,082 (April 18, 1972), to United Kingdom Atomic Energy Authority.
138. M. Druin et al., U.S. Patent 3,754,957 (Aug. 28, 1973), to Celanese Corporation.
139. M. Druin, U.S. Patent 3,723,150 (March 27, 1973), to Celanese Corporation.
140. British Patent 1,341,161 (Dec. 19, 1973), to Union Carbide Corporation.
141. S. Kim et al., U.S. Patent 4,374,114 (Feb. 15, 1983), to Celanese Corporation.
142. S. Steingiser, U.S. Patent 3,627,466 (Dec. 14, 1971), to Monsanto Company.
143. K. Hou, U.S. Patent 3,767,774 (Oct. 23, 1973), to Celanese Corporation.
144. A. Boom, U.S. Patent 3,780,255 (Dec. 18, 1973), to Celanese Corporation.
145. A. Scola et al., U.S. Patent 3,660,140 (May 2, 1972), to United Aircraft Company.
146. I. Aitken, U.S. Patent 3,746,506 (July 17, 1973), to National Research Development.
147. M. Druin et al., U.S. Patent 3,894,884 (July 15, 1975), to Celanese Corporation.
148. J. Goan et al., U.S. Patent 3,746,560 (July 17, 1973), to Great Lakes Carbon Company.
149. L. Loo et al., U.S. Patent 3,989,802 (Nov. 2, 1976), to Great Lakes Carbon Company.
150. M. Druin et al., U.S. Patent 3,859,187 (Jan. 7, 1975), to Celanese Corporation.
151. J. Paul, Jr., U.S. Patent 3,832,297 (Aug. 27, 1974), to Hercules Corporation.
152. J. Ray, U.S. Patent 3,671,411 (June 20, 1972), to United States of America, Secretary of the Air Force.
153. P. White, British Patent 1,371,621 (Oct. 23, 1974), to Courtaulds.
154. British Patent 2,071,702 (Sept. 23, 1981) to Toho Beslon.
155. J. Goan, L. Loo, and G. Sharpe, *Proceedings of the 27th Annual Technical Conference of SPI,* Society of the Plastics Industry, 1972, p. 21-E.
156. P. Ehrburger, J. Herque, and J. Donnet, *Proceedings of the Fourth Intl. Carbon and Graphite Conference,* Society of the Chemical Industry, 1976, p. 201.
157. F. Molleyre and M. Bostick, *Proceedings of the Fourth Intl. Carbon and Graphite Conference,* Society of the Chemical Industry, 1976, p. 190.
158. N. Laine, F. Vastola, and P. Walker, *J. Phys. Chem., 67,* 2030 (1963).
159. B. Rand and R. Robinson, *Carbon, 15,* 257, 311 (1977).
160. F. Fowkes, *J. Adhes. 4,* 155 (1972).
161. D. Steutz, U.S. Air Force Materials Laboratory Report AFML-TR-77-214, 1977.
162. L. Drzal and G. Hammer, *Appl. Surf. Sci., 4,* 340 (1980).
163. L. Drzal, *SAMPE J., 19*(5), 7 (1983).
164. D. Kaelble, P. Dynes, and E. Cirlin, *J. Adhes., 6,* 23, 239 (1974); *8,* 121 (1976).
165. D. Kaelble, *Polym. Eng. Sci. 17,* 474 (1977).
166. P. McMahon and L. Ying, NASA Contract. Report 3607, Contract NASI-15749, (1982); *SAMPE Q., 14,* 26 (1983).
167. J. Rhodehamel, Ed., *Proceedings of the Workshop on High Temperature Polymeric Laminates: Processability/Produc-*

ibility, AFWAL-TR-83-4026, 1983; Chemical Propulsion Information Agency Publication 358, 1982.

168. P. McMahon, *Celion® Thermal Stability and its Impact on Performance of High Temperature Composites,* in Ref. 167.

169. J. Jones, *SAMPE J., 19*(2), 10 (1983).

170. See, for example, *Tough Composite Materials Workship,* NASA-Langley Research Center, May 1983, preprints.

171. Various Boeing publications.

172. R. Roylance, W. Houghton, and S. Ghiorse, *Proceedings of the 28th Natl. SAMPE Symposium,* 309, 1983, pp. 309–317.

173. R. Milligan and C. Carpenter, *Proceedings of the 28th Natl. SAMPE Symposium,* 318, 1983, pp. 318–333.

174. C. Cobuzzi, J. King, and R. Castonguay, *Proceedings of the 28th Natl. SAMPE Symposium,* 359, 1983, pp. 359–366.

175. L. Hopper et al., in T. Hayashi, K. Kawata, and S. Umekawa, Eds., *Progress in Science and Engineering of Composites,* ICCM-IV, 153 (1982).

176. P. McMahon and D. Taggart, in T. Hayashi, K. Kawata, and S. Umekawa, Eds., *Progress in Science and Engineering of Composites,* ICCM-IV, 529 (1982).

177. J. Laakso and J. Hoggatt, Naval Air Development Center Final Report, N62269-75-C-0368, 1968.

178. J. Hartness, *SAMPE Q., 14*(2), 33 (1983).

179. C. Sheppard, E. House, and M. Stander, 36th Annual Technical Conference, Reinforced Plastics/Composites Institute, Society of the Plastics Industry, Inc., 1981.

180. F. Cogswell, *Proceedings of the 28th Natl. SAMPE Symposium,* 528, 1983, pp. 528–534.

181. J. Hartness and R. Kim, *Proceedings of the 28th Natl. SAMPE Symposium,* 535, 1983, pp. 535–544.

182. C. Beetz, Jr., and R. Schmitt, *SAMPE J., 19*(3), 6 (1983).

183. M. Dresselhaus and G. Dresselhaus, *Adv. Phys., 30,* 139 (1981).

184. M. Endo, T. Koyama, and M. Inagaki, *Synth. Metals, 3,* 177 (1981).

185. I. Kalnin and H. Goldberg, *Synth. Metals, 3,* 159 (1981).

186. I. Kalnin et al., *Proceedings of the 16th Biennial Conference on Carbon,* American Carbon Society, 1983, Extended Abstracts, p. 274.

187. J. Murday et al., Naval Research Laboratory Memorandum 4993, AS-A122919, January 1983.

188. D. Cagliostro, *Text. Res. J., 53,* 419 (1983).

189. T. Koyama, M. Endo, and Y. Onuma, *Jpn. J. Appl. Phys., 11,* 445 (1972).

190. A. Oberlin and M. Endo, *J. Cryst. Growth, 32,* 335 (1976).

191. M. Endo, A. Oberlin, and T. Koyama, *Jpn. J. Appl. Phys., 16,* 1519 (1977).

192. M. Endo and T. Koyama, *Elect. Eng. Jpn.* (English transl. of *Denki gakkai Zasshi*), *100,* 9 (1980).

193. *Chem. Eng. News,* Aug. 8, 1983, p. 27.

194. *Jpn. Chem. Week,* March 3, 1983.

195. S. Kulkarni, Lawrence Livermore Laboratory Report, DOE Contract W-7405-Eng-48, UCRL-53351, December 1982, Lawrence Livermore Laboratory.

196. J. Whitney, I. Daniel, and R. Pipes, *Experimental Mechanics of Fiber Reinforced Plastics,* Society for Experimental Stress Analyses, Brookfield Center, CT, 1982; C. Chamis, *SAMPE Q., 15*(4), 41 (1984).

197. J. Vinson, Ed., *Advanced Composite Materials—Environmental Effects,* ASTM STP 658, American Society for Testing and Materials, Philadelphia, 1978.

198. C. Chamis, Ed., *Test Methods and Design Allowables for Fibrous Composites,* ASTM STP 734, American Society for Testing and Materials, Philadelphia, 1981.

199. I. Daniel, Ed., *Composite Materials: Testing and Design (6th Conference),* ASTM STP 787, American Society for Testing and Materials, Philadelphia, 1982.

200. C. Browning, Ed., *Composite Materials: Quality Assurance and Processing,* ASTM STP 797, American Society for Testing and Materials, Philadelphia, 1983.

201. R. Evans, Ed., *Fagitue of Filamentary Composite Materials,* ASTM STP 636, American Society for Testing and Materials, Philadelphia, 1977.

202. T. O'Brien, Ed., *Long-term Behavior of Composites,* ASTM STP 813, American Society for Testing and Materials, Philadelphia, 1983.

203. K. Reifsnider, Ed., *Damage in Composite Materials,* ASTM STP 775, American Society for Testing and Materials, Philadelphia, 1982.

204. R. Pipes, Ed., *Non-destructive Evaluation and Flaw Critically for Composite Materials,* ASTM STP 696, American Society for Testing and Materials, Philadelphia, 1979.

205. P. Paris and G. Sih, ASTM STP 381, 1965.

206. G. Irwin, "Fracture", in *Handbuch der Physik,* Vol. VI, Springer-Verlag, Berlin, 1958.

207. P. Mallick and L. Broutman, *J. Test. Eval., 5,* 190 (1977).

208. B. Agarwal and L. Broutman, *Analysis and Performance of Fiber Composites,* Wiley-Interscience, New York, 1980, pp. 246–268.

209. K. Rogers, G. Sidney, and D. Kingston-Lee, Royal Aircraft Establishment Technical Report 70166, UK, 1970.

210. M. Wardle and E. Tokarsky, *Compos. Technol. Rev., 5*(1), 4 (1983).

211. J. Aleska, *J. Test Eval., 6,* 202 (1978).

212. D. Ireland, Instrumented Impact Testing for Evaluating End Use Performance, TR-80-63, Effects Technology, Inc., 1980.

213. A. Miller, P. Hertzberg, and V. Rantala, *Proceedings of the 12th Natl. SAMPE Techn. Conf.,* 1980, pp. 279–293.

214. A. Watts, Ed., *Commercial Opportunities for Advanced Composites,* ASTM STP 704, American Society for Testing and Materials, Philadelphia, 1980.

215. C. Kam and J. Gaidulis, *SAMPE Q., 15*(1), 36 (1983).

216. *Polym. News, 8,* 278 (1982).

217. *Polym. News, 8,* 279 (1982).

Bibliography

Delmonte, J., Ed. *Technology of Carbon and Graphite Fiber Composites,* Van Nostrand Reinhold, New York, 1981, is a good general reference.

Langley, M., Ed. *Carbon Fibers in Engineering,* McGraw-Hill, New York, 1973.

Sittig, M., Ed., *Carbon and Graphite Fibers: Manufacturing and Applications,* Noyes Data Corp., Park Ridge, NJ, 1980.

Carbon Fibers: Property Testing and Analysis

Carbon fibers produced from polyacrylonitrile (PAN) precursor are finding increased usage in fiber-reinforced plastics [1]. Despite their present relatively high cost, carbon fibers are used where weight savings are more important than cost considerations. As new applications for carbon fiber composites are being found, new demands of the composite mechanical performance are occurring. These demands are resulting in a wider choice of carbon fiber and polymeric matrix mechanical properties.

The first generation of composites using carbon fibers was made with thermosetting resins such as epoxies [2]. It is now becoming apparent that these composites are too brittle for many current design applications [3–5]. Recent trends in composite development are toward composites that can withstand impact loads and still function properly. These demands in composite performance are being met by improving the toughness of the matrix resin.

Methods used to increase matrix toughness have included modifying existing epoxy formulations by adding a second phase (such as rubber or a thermoplastic resin) that can absorb energy [3–7], using thermoplastic resins [8–11], and depositing a ductile material on the fiber surface [12–14].

Although the newer materials being used as matrix materials have increased toughness, it is difficult to predict the mechanical properties of a composite based on the properties of the individual components alone [11,15]. The main reason for this is that the interaction of the fiber with the matrix also has an important effect on the mechanical properties of composites [15–17]. This interaction between fiber and matrix includes adhesion and wetting as well as the effect of the fiber on the morphological characteristics of the polymer.

Adhesion between fiber and matrix can be altered by surface treating the fiber [16,17]. Surface treatments of carbon fibers have been developed for epoxy systems, but the optimum surface treatment for epoxy systems may be inadequate for newer resin systems. It may also be possible that by tailoring the interface between fiber and matrix, the mechanical properties of the composite can be controlled. To tailor the interface, it is necessary to understand the nature of carbon fiber surfaces and their reactions when surface treated.

This article reviews carbon fiber preparation and discusses physical properties of carbon fibers. Emphasis is placed on functional groups present on carbon fiber surfaces and the relationship of these functional groups to adhesion in polymeric matrices.

Carbon Fiber Preparation

Most of the presently available carbon fibers are synthesized from polyacrylonitrile starting material. Although several other precursors do exist such as rayon and pitch [18], PAN precursor fibers have the best mechanical properties for structural applications. The technology of carbon fiber synthesis is protected very strongly by carbon fiber producers. However, the basic chemistry of carbon fiber synthesis is known. A brief review is included here.

Thermal Treatments for PAN-Based Carbon Fiber Synthesis

The processes involved in the synthesis of carbon fibers from PAN have been outlined by Deifendorf and coworkers [19,20]. These processes include spinning of PAN into fiber form, oxidizing the fiber at 200–300°C and carbonizing the fiber at 1000–2500°C in an inert atmosphere, surface treating, and sizing. The strength, modulus, and structure of the fiber can be controlled by stretching the fiber during the process as well as by changing the heating rates, the extent of oxidation, and the final carbonization temperature.

The chemical changes occurring during carbon fiber formation from PAN have been reviewed by Watt [21] and by Goodhew et al. [22]. Coleman and coworkers [23,24] have proposed a scheme of the chemical changes that occur during oxidation of PAN at 200°C (Fig. 1). The first step is cyclization within the polymer backbone to form a ladder structure. This ladder structure stabilizes the polymer for heating to higher temperatures. The polymer is stretched during cyclization to maintain alignment of the polymer molecules in the fiber direction. The ladder structure is then oxidized.

The chemical changes occurring during carbonization of the fiber are shown in Figure 2. Although the fiber is carbonized between 1000 and 2500°C, reactions begin to occur at much lower temperatures as the fibers are being heated to the carbonization temperature. At 400–600°C, the cyclic molecules begin to link together, resulting in loss of hydrogen and probably oxygen. This is followed by nitrogen loss and further linking at 600–1300°C to form graphitic sheets.

Carbon Fiber Structure

After carbonization, the carbon is in a sheet form as shown in Figure 3. The carbon is in an sp^2 hybrid state. There is an unbonded electron in an orbital perpendicular to the graphite plane. The unbonded electron coupled with unbonded electrons from adjacent carbon atoms will cause the formation of a conduction band of electrons between the carbon layers. The structure shown in Figure 3*a* is an idealized model for the molecular structure of graphitic carbon [25]. However, the carbon in a carbon fiber will contain some discontinuities [25]. Figure 3*b* represents an imperfect graphite sheet,

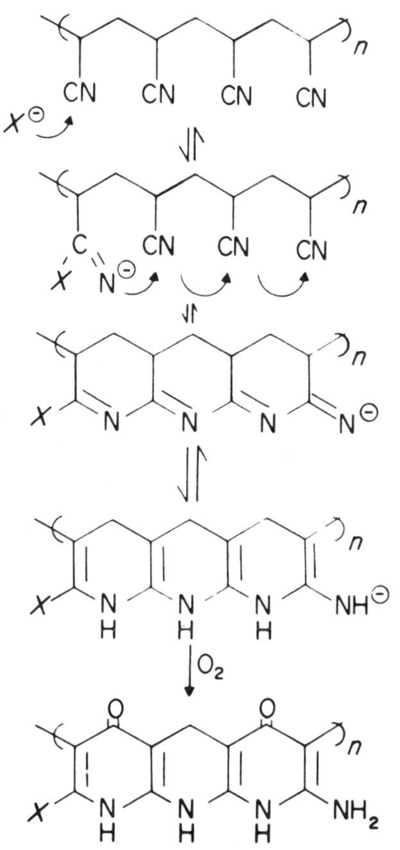

FIGURE 1 **Presently accepted mechanism for cyclization and oxidation of polyacrylonitrile.** (Reprinted with permission from Ref. 23. Copyright 1981 Pergamon Press.)

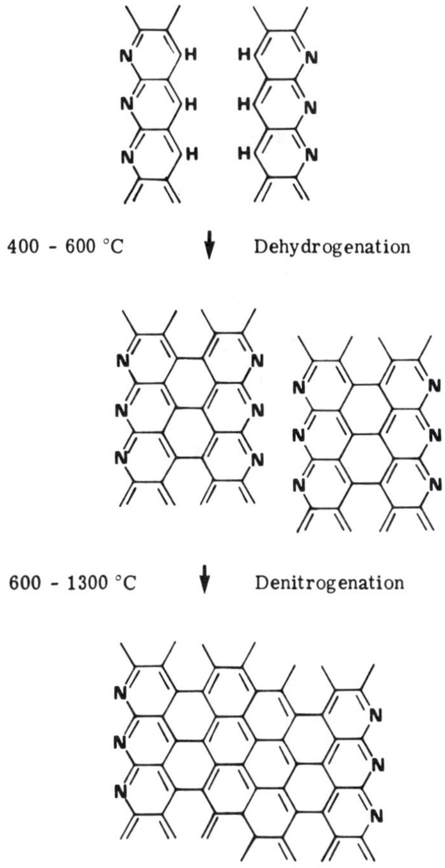

FIGURE 2 **Proposed mechanism for carbonization of cyclicized PAN fiber into aromatic carbon sheets.** (Reprinted with permission from Ref. 22. Copyright 1975 Elsevier.)

which is probably more typical of the structure of carbon fibers. The carbon fiber is made up of many sheets that will coalesce to form aggregates similar in structure to the graphite unit cell. Since the carbon sheet is imperfect, the ideal graphite structure shown in Figure 4*a* cannot be formed. Instead, we find a disordered crystal structure, the turbostratic structure, as shown in Figure 4*b* [26]. These turbostratic crystals can form aggregates of 0.03 μm (several hundred angstroms) in size. The layers of carbon twist and undulate along the length of the fiber as shown in Figure 5 [27]. Between the ordered areas there are areas of amorphous carbon [28].

The structure from the core to the surface of the fiber varies also. A model for the PAN-based carbon fiber structure proposed by Diefendorf and reported by Drzal [29] is shown in Figure 6. In this model, the carbon layers are highly oriented at the fiber surface. The carbon layers in the core are less ordered. At the fiber surface, graphitic basal planes are oriented perpendicular to the outer fiber surface. In the fiber core, the graphitic basal planes are oriented radially from the center of the fiber outward. This model for carbon fiber morphology is referred to as an onion skin structure.

The degree of order of the fiber surface was shown by Bennet and others [30,31] to depend on the carbonization temperature. The sketches of Figure 7 represent longitudinal sections of carbon fibers examined by Bennet with the transmission electron microscope (TEM). The structural order of the fiber surface increases with increasing carbonization temperature. Thus fibers formed at higher temperatures adhere to polymeric resins with difficulty.

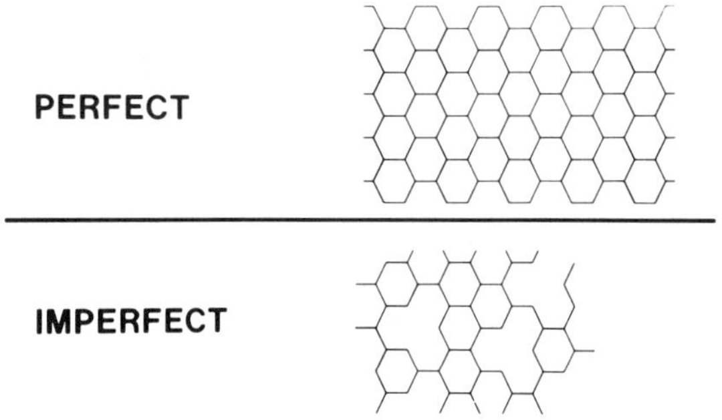

FIGURE 3 **Schematic diagrams of (*a*) perfect graphite sheets and (*b*) imperfect sheets, which are more indicative of the structure in carbon fibers [25].**

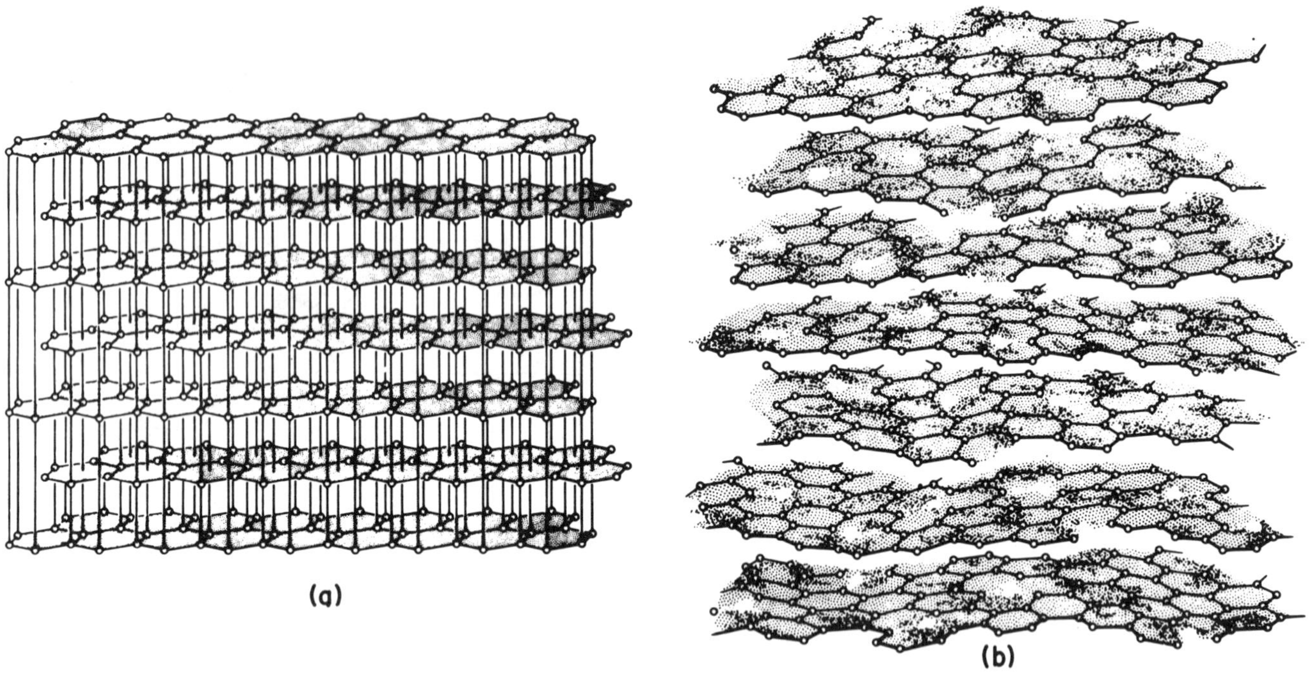

FIGURE 4 Structural models of graphite and carbon structure: (*a*)perfect graphite crystal and (*b*) turbostratic model. (Reprinted with permission from Ref. 26. Copyright 1969 Marcel Dekker.)

Carbon Fiber Surface Treatment

After the carbon fibers come out of the carbonization furnace, they are surface treated. This surface treatment serves several purposes: (a) to remove the outer layer of the carbon fiber surface, which is believed to be disordered carbon and of low shear strength, and (b) to oxidize the fiber surface, thus fixing on the fiber surface functional groups that will promote adhesion to the polymer matrix used for making composites.

Possible surface treatment mechanisms include anodization [32–34], plasma and flame treatment [35], solution oxidation [36,37], gas phase oxidation, and high temperature oxidation. Some of these treatments have been reviewed by Donnet and coworkers [18,38]. The most practical surface treatment for commercial production of carbon fibers is anodization. This is because anodization (electrolytic oxidation) can be performed continuously on carbon fibers. Typical anodizations have been performed in aqueous acidic or basic solutions. Electrolytes include sodium hydroxide, potassium hydroxide, sulfuric acid, nitric acid, and solutions of amine salts. Amine salts have an added advantage in that, after treatment, excess electrolyte can be removed simply by heating the fiber to high temperatures (250°C).

Carbon Fiber Sizing

After surface treatment, the fibers are heated to remove from the fiber surface volatile materials that would otherwise create voids in the composite during high temperature processing. The fibers are sized to protect the fiber surface from surface damage during handling and to protect the surface chemistry created by the surface treatment [39].

Physical Properties of Carbon Fibers

Tensile Breaking Strength and Modulus of Carbon Fibers

The tensile modulus E and strength σ_B of carbon fibers are shown as a function of carbonization temperature in Figure 8 [21]. The carbon fiber modulus increases with increasing carbonization temperature. This increase in modulus is caused by increased graphitization of the carbon at higher temperatures, since the more perfect graphite has a higher modulus than the less ordered carbon sheets.

Carbon fibers formed at higher temperatures (> 2000°C) are referred to in the literature as high modulus or type I fibers. Fibers formed at lower temperatures (1000–1600°C) are referred to as low modulus or type II fibers. Recent developments in carbon fiber synthesis have resulted in carbon fibers with a tensile modulus intermediate between type I and II but with a tensile strength similar to type II [25]. These newer fibers have been called "intermediate modulus." Many improvements in the processing of PAN fibers are being made. A wide range of mechanical properties is available for specific design applications.

The tensile breaking strength of carbon fibers is dominated by flaws within the fiber and on its surface. Some of the flaws that can affect the fiber strength include:

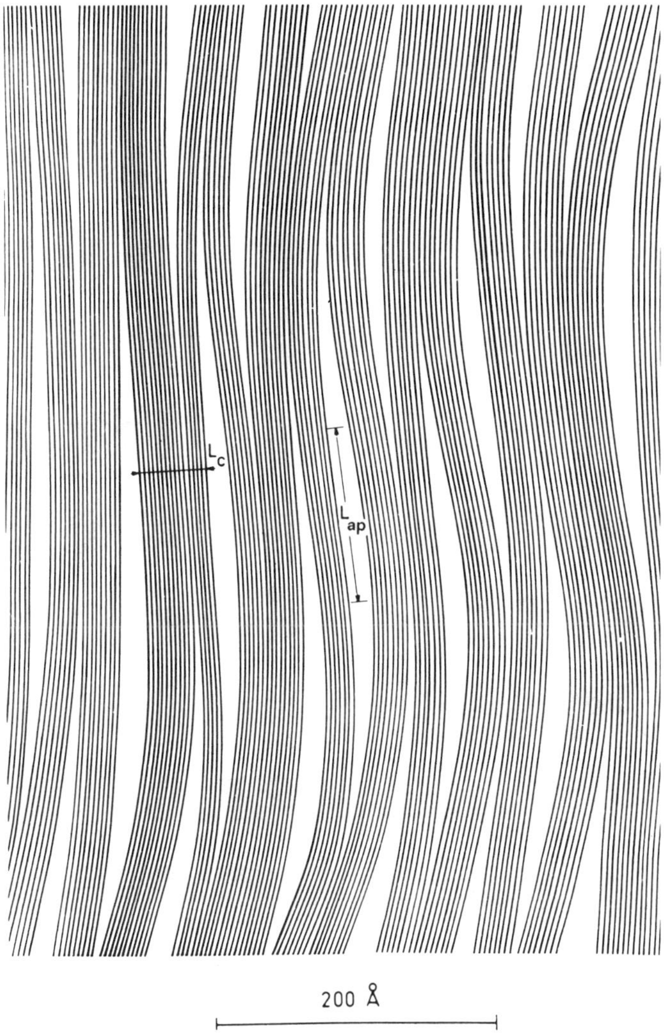

FIGURE 5 **Proposed structure of PAN-based carbon fibers showing layers undulating in and out of crystalline regions: L_c is the width of the turbostratic crystals and L_{ap} is the length of the crystal.** (Reprinted with permission from Ref. 27. Copyright 1970 International Union of Crystallography.)

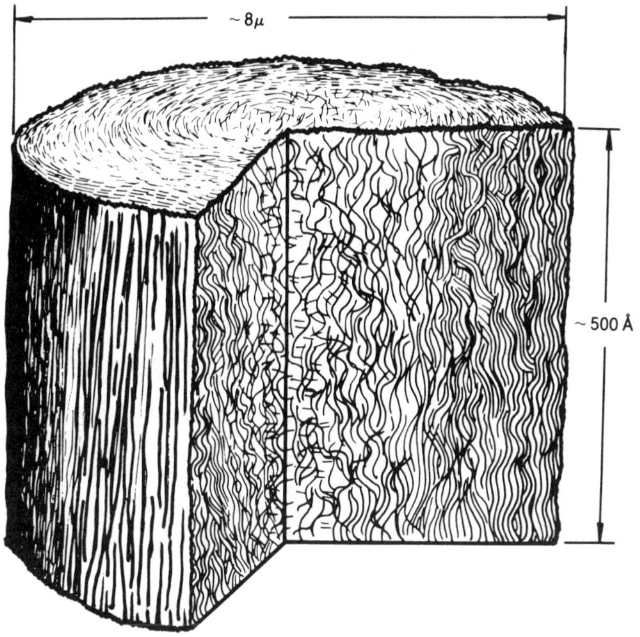

FIGURE 6 **Proposed structure of carbon fiber morphology showing higher order at the fiber surface. Outer layers have basal planes oriented normal to the fiber surface.** (Reprinted with permission from Ref. 29. Copyright 1977 Pergamon Press.)

Discontinuities in the crystal or fibrillar structure of the fiber.
Variations in the thickness of the onion skin layer.
Variations in the overall thickness of the fiber.
Microscopic impurities in the precursor.
Surface defects due to handling and processing.

Since the breaking strength of the fiber is controlled by the presence of flaws, it is expected that as the length of the fiber decreases, the breaking strength will increase. The dependence of strength on length is expected, since there is a lower probability of encountering a defect in the shorter fiber.

Surface treatment has been observed to change the breaking strength of carbon fibers. Bahl et al. [40] and

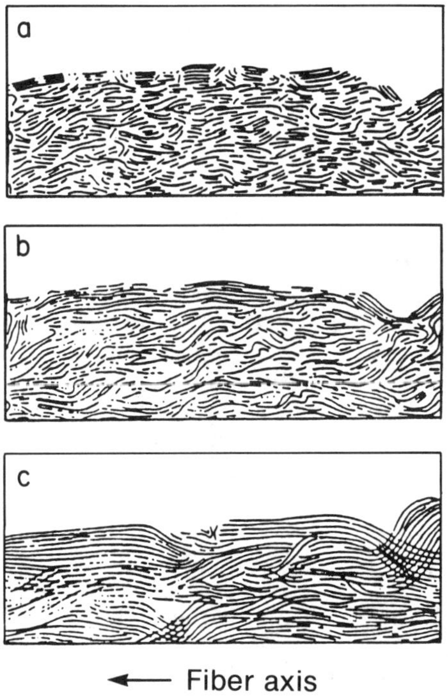

FIGURE 7 **Sketches of TEM images of carbon fiber surface after carbonization at (*a*) 1000°C, (*b*) 1500°C, and (*c*) 2500°C. The crystalline order increases with carbonization temperature [30].**

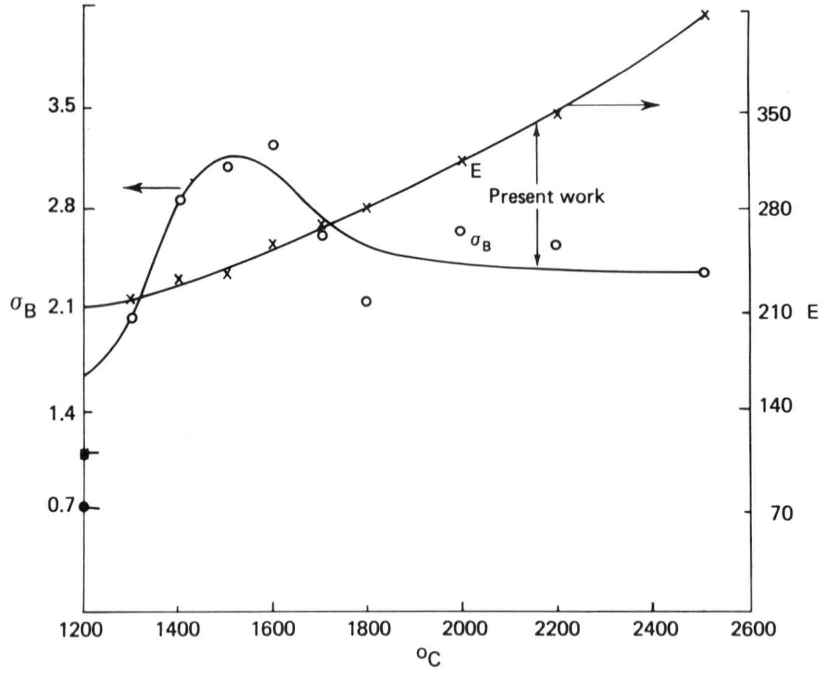

FIGURE 8 **Dependence of PAN-based carbon fiber tensile strength σ_B and modulus *E* on carbonization temperature.** (Reprinted with permission from Ref. 21. Copyright 1985 Elsevier.)

Fitzer and Weiss [41] have observed that treatment of carbon fibers in nitric acid initially increases the fiber tensile strength. Continued anodization results in a loss in strength caused by fiber damage. This initial increase in strength can be explained by removal from the fiber surface of defects, which can initiate fracture.

Models for Fiber Breaking Strength as a Function of Length

The mathematical models used to predict fiber strength as a function of length considers the fiber to be a chain [42–45]. When the chain is pulled in tension, it will break at the weakest link. The probability of a fiber having a specific breaking strength is equal to the probability that each link in the chain will have a breaking strength larger than the fiber breaking strength (i.e., the reliability). This is shown in Eq. (1):

$$g(\sigma) = [R_0(\sigma)]^n \tag{1}$$

where σ is the stress, $g(\sigma)$ is the probability distribution for fiber strength, R_0 is the reliability of a link surviving to stress σ, and n is the number of links in the fiber.

Taking the logarithm of both sides of this equation gives:

$$\ln[g(\sigma)] = \sum_{1}^{n} \ln[R_0(\sigma)] \tag{2}$$

Assuming an infinite number of links, the summation can be replaced with an integral over the length L of the fiber, as shown in Eq. (3).

$$\ln[g(\sigma)] = \int_0^L \ln[R_0(\sigma)\, dx] \tag{3}$$

The reliability function has been assumed [42] to take the form of the Weibull distribution [43] shown in Eq. (4), where α and β are shape and scale parameters, respectively:

$$R_0 = \exp - \left[\frac{\sigma}{\beta}\right]^\alpha \tag{4}$$

Substituting Eq. (4) into Eq. (3), integrating, and taking the exponent of the integral, gives:

$$g(\sigma) = \exp - \left[L\left[\frac{\sigma}{\beta}\right]^\alpha\right] \tag{5}$$

Equation (5) gives the dependence of fiber strength on fiber length. The parameter α is a shape parameter and is indicative of the distribution of flaws in the surface of the fiber [44,45]. A high value of α indicates a high flaw density and thus little dependence of strength on length. A low value of α indicates few flaws on the fiber surface and thus a strong dependence of strength on length. Methods for estimating the parameters α and β are given elsewhere [46].

Surface Energy Measurement

There are many techniques for probing the chemical and physical properties of a solid surface to predict the bonding of organic polymers to solid surfaces. The electronic structure of solid surfaces has been studied by measuring the thermodynamic interaction of the solid surface with simple liquids of known molecular structure. Experimental techniques for measuring the thermodynamic interaction between solid and liquid include contact angle measurement, calorimetry, and gas chromatography. Some of these techniques are discussed below. Specific techniques related to characterization of carbon fiber surfaces are also discussed.

CONTACT ANGLE MEASUREMENT. When a liquid drop is placed on a solid surface, the liquid will either spread on the surface or form a drop. Between itself and the solid, this drop will have an angle that is indicative of the interaction between the two materials [47]. In addition, the liquid will have a vapor pressure with which the solid surface will be in equilibrium. The forces in the drop are balanced as shown in Figure 9 [48]. These forces include the tendency of the drop to minimize its surface area by forming a sphere, and the tendency to spread on the solid surface and thus increase the extent of interfacial contact. This balance of forces has been described by the Young equation:

$$\gamma_{sl} - \gamma_{sv} + \gamma_{lv}\cos(\theta) = 0 \tag{6}$$

where γ_{sl} is the surface energy between solid and liquid, γ_{sv} is the surface energy between solid and vapor, γ_{lv} is the surface energy between liquid and vapor, and θ is the angle of the drop between solid and liquid. By measuring the angle between the liquid drop and the solid surface, the interaction between solid and liquid (γ_{sl}) can be estimated.

Adhesion is defined thermodynamically by the change in free energy when two materials come into contact. The work of adhesion in the contact angle experiment has been defined [47] by Eq. (7).

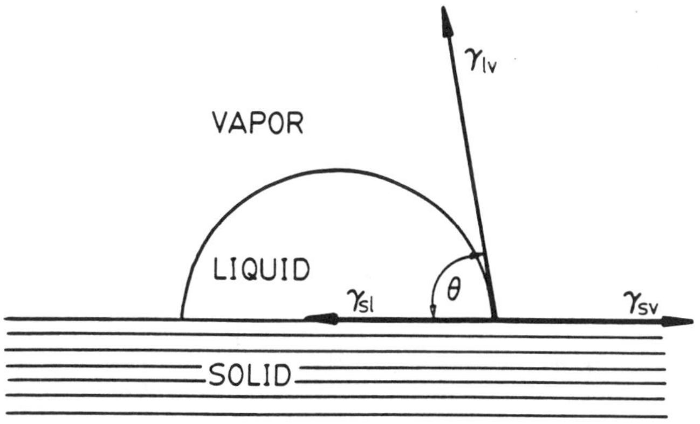

FIGURE 9 Diagram of contact angle experiment [48]: γ_{lv} = surface energy of the liquid–vapor interface, γ_{sl} = surface energy of the solid–liquid interface, and γ_{sv} = surface energy of the solid–vapor interface.

$$W_a^T = \gamma_{lv}[1 + \cos(\theta)] \tag{7}$$

where W_a^T is the total work of adhesion.

Girifalco and Good [49] assumed the interaction between a solid and a liquid could be quantified by an interaction parameter (ϕ) times the geometric mean of the surface tension of the solid and the liquid resulting in Equation (8).

$$\gamma_{sl} = \gamma_{sv} + \gamma_{lv} - 2\Phi\,(\gamma_{sv}\gamma_{lv})^{1/2} \tag{8}$$

Fowkes [50] later postulated that the interaction energy due to wetting of solids by liquids with dispersion force interactions only, could be described by a geometric mean equation as shown in Eq. (9):

$$\gamma_{sl} = \gamma_{sv} + \gamma_{lv} + 2(\gamma_{sv}^d\gamma_{lv}^d)^{1/2} \tag{9}$$

where γ_{sv}^d is the dispersion surface energy of the solid and γ_{lv}^d is the dispersion surface energy of the liquid.

The interaction between solid and liquid due to polar groups has been considered by Fowkes [51] to be defined by acid–base interactions. In this model, Fowkes assumes that the interaction between two materials can be described by a component due to dispersion interactions in the form of a geometric mean relationship plus a component due to acid–base interaction. The acid–base interaction indicates the ability of a polar group on one surface to donate or accept electrons from polar groups on the other surface. The work of adhesion is then described by Eq. (10),

$$W_a^T = W_a^d + W_a^{ab} \tag{10}$$

where W_a^d is the work due to dispersion forces, and W_a^{ab} is the work due to acid–base interactions.

Kaelble et al. [52,53] have developed a technique to determine the polar and dispersion components of the surface energy of carbon fibers and other solid surfaces. In this technique, the contact angle of the fibers in several liquids of varying polar and dispersion components is measured. The work of adhesion W_a is assumed to be equal to the sum of the geometric mean of the polar components of the surface energies plus a geometric mean for the dispersion surface energy components of the liquid and solid surface energies as shown in Eq. (11).

$$W_a = 2(\gamma_{sv}^d\gamma_{lv}^d)^{1/2} + 2(\gamma_{sv}^p\gamma_{lv}^p)^{1/2} = \gamma_{lv}[1 + \cos(\theta)] \tag{11}$$

where γ_{sv}^p is the polar component of the solid surface energy and γ_{lv}^p is the polar component of the liquid surface energy.

The polar and dispersion components of the surface energy were calculated by dividing both sides of Eq. (11) by the square root of $2\gamma_{lv}^d$ as shown in Eq. (12).

$$\frac{\gamma_{lv}[1 + \cos(\theta)]}{2(\gamma_{lv}^d)^{1/2}} = (\gamma_{sv}^d)^{1/2} + \left(\frac{\gamma_{lv}^p}{\gamma_{lv}^d}\right)^{1/2} * (\gamma_{sv}^p)^{1/2} \tag{12}$$

The contact angle is measured in a series of liquids with varying polar and dispersion surface energy components. The components of the fiber surface energy can be determined by plotting the left-hand side of Eq. (12) as a function of $(\gamma_{lv}^{p}/\gamma_{lv}^{d})^{1/2}$ of the liquid. The slope of this plot will be equal to $(\gamma_{sv}^{p})^{1/2}$ of the solid. The intercept will be equal to $(\gamma_{sv}^{d})^{1/2}$ of the solid.

If the surface energy is estimated for a polymer and for a solid using Kaelble's method, the work of adhesion between polymer and solid can be calculated using Eq. (11). However, it should again be noted that Fowkes [51] has argued that the geometric mean relationship to describe the polar group interaction between two materials may better be described by acid–base interactions.

CONTACT ANGLE MEASUREMENT ON SMALL-DIAMETER FIBERS. Since carbon fibers are so small, it is very difficult to measure the contact angle of a drop on a fiber. Nevertheless, several techniques have been developed to measure the contact angle of a drop on a small fiber under a microscope [54]. A simpler method for contact angle determination is to measure the wetting force of the fiber when it comes into contact with a liquid [52,53,55,56]. If the surface energy of the liquid is known, the contact angle of the liquid on the fiber can be calculated by:

$$F = \pi d \gamma_{lv} \cos(\theta) \tag{13}$$

where F is the wetting force and d is the fiber diameter.

Equation (13) assumes that the cross section of the fiber is circular. If the cross section is not circular, the πd term is replaced by the actual circumference of the fiber. The circumference may be determined by measuring the wetting force of the fiber in a liquid that completely wets the fiber [$\cos(\theta) = 0$] as described by Herb et al. [57].

INVERSE GAS CHROMATOGRAPHY FOR MEASUREMENT OF SOLID–VAPOR INTERACTION. The idea of putting carbon fibers in a gas chromatography column and passing probe molecules through the column to measure the fiber–liquid interaction was first used by Brooks and Scola [58]. Initial investigations using this technique, known as inverse gas chromatography (IGC), were inconclusive [59]. However, Schultz et al. [60] used IGC to show that the surface of carbon fibers obtained from Hercules, Inc., was acidic.

IGC measures the retention times of the probe molecules in the column. Molecules with a high adsorption enthalpy will take longer to pass through the column than molecules with a low adsorption enthalpy. If probe molecules of varying acid–base character are used, the acid–base properties of the fiber can be determined. This technique has also been used to determine the surface area of carbon fibers by using nonpolar probe molecules [61].

CALORIMETRIC MEASUREMENT OF SOLID–LIQUID INTERACTION. The thermodynamic interaction between a liquid and a solid can be measured using calorimetry. In this technique, a solid and a liquid are brought into contact in a cell with a sensitive heat detector. As the liquid wets the solid, heat is generated, which is detected by the sensitive device. This technique has been used by Rand and Robinson [62] to measure the heats of wetting of carbon fibers in acidic and basic liquids. It was found that basic probes gave a much higher heat of wetting, indicating an acidic fiber surface. Since the surface areas of carbon fibers are low, the amount of heat generated is low, and precise measurement is difficult.

Detection of Functional Groups on Carbon Fiber Surfaces

Some of the functional groups expected on carbon surfaces are carboxylic acids, phenols, quinones, lactones, ethers, peroxides, and esters (Fig. 10). Functional groups on carbon surfaces have been detected by such methods as polarography [63], titration [63–68], X-ray photoelectron spectroscopy [34], radioisotope labeling [34], and infrared spectroscopy [63]. Some of these techniques are discussed below.

Titrimetric Methods for Carbon Fiber Surface Functionality Determination

Functional groups on carbon surfaces have been identified by reacting the material with reagents that will react with specific functional groups on the carbon surface. The amount of reagent reacted is determined by titration. Several reviews of these techniques for analysis of carbon surfaces are available [63–68]. For these methods to be sensitive enough to detect the functional groups, either very large surface areas or large quantities of material must be available.

X-Ray Photoelectron Spectroscopy of Carbon Fiber Surfaces

X-Ray photoelectron spectroscopy, commonly referred to as XPS or ESCA (electron spectroscopy for chemical analysis), uses the photoelectric effect to analyze the chemistry of solid surfaces [69]. In this technique (Fig. 11), a solid surface is exposed to nearly monochromatic X-rays. This exposure causes electrons to be ejected from the solid surface. These ejected electrons will have a spectrum of kinetic energies as they come from within the structure of the solid. At certain energies, the number of ejected electrons will increase. This peak in intensity is caused by ejection of electrons from atoms in the surface of the solid.

The energy of the electrons at the peak is indicative of the element present. Since the X-rays are nearly monochromatic, the kinetic energy distribution of these electrons will be narrow. The kinetic energy KE will be equal to the photon energy of the X-rays $h\nu$ minus the binding energy of the electron in the atomic structure of the element BE minus a work function Ω:

$$KE = h\nu - BE - \Omega \tag{14}$$

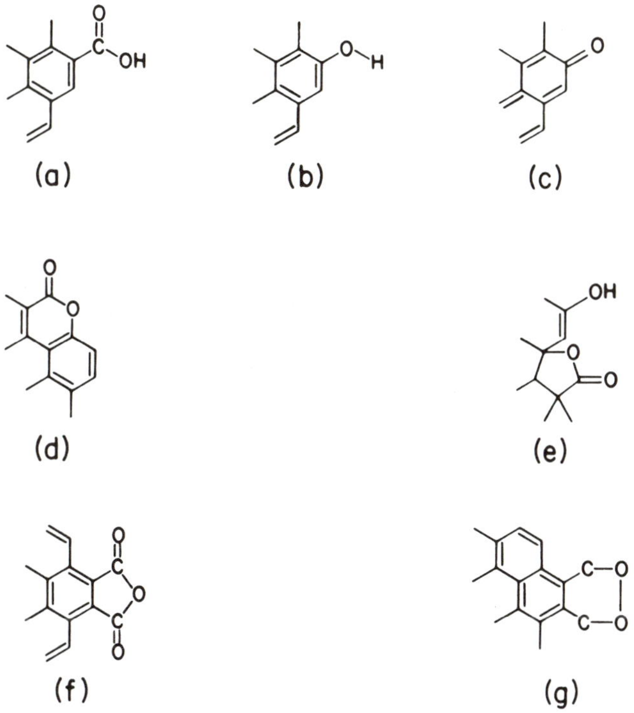

FIGURE 10 **Predicted oxygen functional groups on carbon surfaces: (*a*) carboxyl, (*b*) phenolic hydroxyl, (*c*) quinone, (*d*) lactone, (*e*) fluoresceinlike lactone, (*f*) carboxylic acid anhydride, and (*g*) cyclic peroxide.** (Reprinted with permission from Ref. 63. Copyright 1976 Marcel Dekker.)

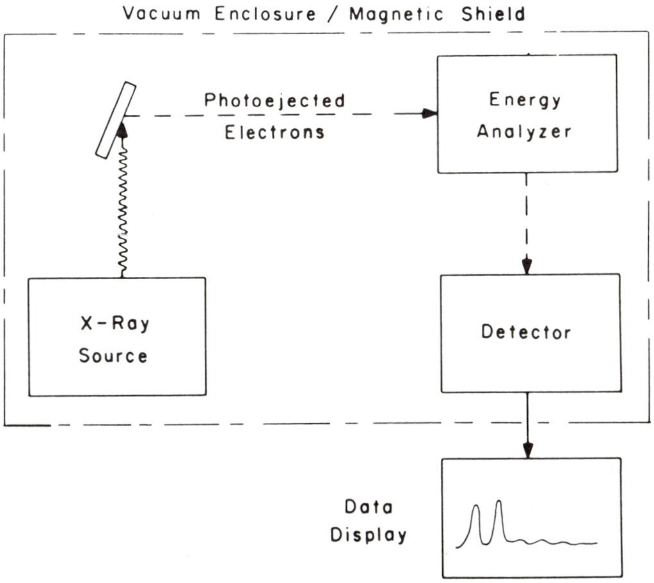

FIGURE 11 **Schematic diagram of X-ray photoelectron spectroscopy experiment.** (Reprinted with permission from Ref. 69. Copyright 1975 Elsevier.)

The intensity of the photoelectron signal for element i is proportional to the number of atoms on the solid surface N_i, the cross section of the atom to X-rays μ, the mean free path λ of the electrons in the solid (which is typically 5 nm), the X-ray energy flux F, and the geometric arrangement of the spectrometer [70]. The relationship for the photoelectron peak intensity I_i for element i is:

$$I_i = \int_0^\infty F\,k\,N_i(x)\,\mu\,\exp - \left[\frac{d}{\lambda}\right] dx \qquad (15)$$

where k is a constant specific to spectrometer, x is the perpendicular distance from the surface into the sample, and d is the distance the electrons travel through the solid before exiting.

This relationship allows determination of the relative percentage of a given element on a surface from the relative peak intensities. Values for the atomic cross sections have been calculated by Scofield [71]. Empirical equations for calculating the electron mean free path have been developed by Cadman et al. [72]. Wagner [73] has determined sensitivity factors for each element to relate peak intensities to atomic concentration.

ANGULAR-DEPENDENT DEPTH PROFILING. Many materials will have a variation of chemical groups present from the surface into the bulk of the material. The surface composition profile can be studied by varying the angle at which the electrons ejected from the solid surface are detected.

The calculated escape depth for photoejected electrons is about 5 nm. As shown schematically in Figure 12, ejected electrons measured at 90° to the surface will come from 0 to 5 nm within the surface; ejected electrons measured at less than 90° will come from less than 5 nm. Thus by varying the angle between the surface and analyzer, a depth profile can be obtained. The escape depth normal to the sample x will be equal to the electron escape depth d times the sine of the takeoff angle θ as shown in Figure 12 and described in Eq. (16).

$$x = d\,[\sin(\theta)] \qquad (16)$$

By decreasing the takeoff angle in the XPS experiment, a higher percentage of atoms from the top few atomic layers of a solid surface can be analyzed. The change in relative amounts of elements detected as the take off angle is changed will give an indication of the distribution of elements in the top surface of the material.

PEAK SHAPE ANALYSIS. If all the elements present on a solid surface were in the same bonding environment, we would expect the kinetic energy of the photoelectrons emitted from the solid surface to have a very narrow distribution. However, the actual width of the photoelectron peak is influenced by interaction of the ejected electrons with the sample as well as by the spectrometer itself.

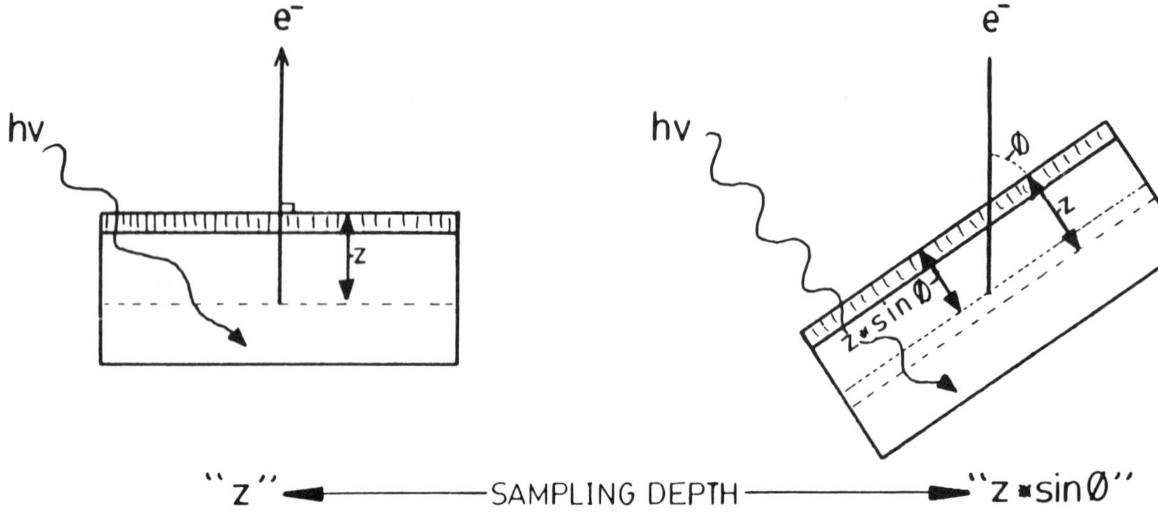

FIGURE 12 **Schematic diagram for angular-dependent depth profiling using XPS [48].**

The elements in most materials are not in just one binding state. The binding energy of electrons in the atom is influenced by the valence state of the atom. Thus the functionality of a solid surface often can be determined by observing shifts in the XPS peaks.

XPS photopeaks are typically fitted with Gaussian-shaped peaks. The peaks are assigned a width typical of the spectrometer being used and of the element being studied. The peaks are shifted in binding energy to represent the chemical environment of the element. The intensity of the fitted peak is proportional to the amount of that functional group present.

For a particular element, the number of peaks used to fit the photopeak is equal to the number of different functionalities expected for that element. Sometimes, the expected functionalities will have similar or overlapping binding energies. In this case, either several overlapping peaks can be used to fit the overall peak, or the width of the fitted peak can be increased.

Much work has been done to determine the binding energy shift caused by specific binding states [74]. The shifts in binding energy expected for carbon, oxygen, and nitrogen in organic materials are summarized below.

Carbon 1s Peak Shape Analysis. The photopeak from the carbon 1*s* atomic orbital has been widely analyzed. The peak due to C—C bonding occurs at a binding energy of 285.0 eV and is often used for instrument calibration. Many standard materials have been examined and the corresponding shift due to functional groups determined. Clark and coworkers have done numerous XPS analyses of polymer surfaces and have reported the carbon 1*s* binding energy shifts of many functional groups. Clark's results have been reviewed briefly by Briggs and Seah [75]. A basic trend [75] is that the R—O—C type of bond will shift the carbon 1*s* photopeak about + 1.5 eV, R—C═O bonds will cause about a + 3 eV binding energy shift, and R—C(═O)—O— bonds will cause about a + 4.5 eV binding energy shift.

Proctor and Sherwood [76] have studied the carbon 1*s* spectrum of carbon fiber and graphitic surfaces. They have shown that in addition to peak shifts due to functional groups, there is a peak at about + 6 eV from the main C—C peak due to interaction of ejected core electrons with plasmons from the conduction electrons in graphite. They also pointed out that certain functional groups on graphitic-type surfaces may occur at different binding energies due to aromaticity of the carbon structure to which the functional group is attached.

Oxygen 1s Peak Shape Analysis. The binding energy shift for oxygen bound to carbon is less well defined than the shift of carbon. Most of the oxygen peaks fall in a narrow (2 eV) range centered around 533 eV. Oxygen doubly bound to carbon tends to have a lower binding energy than oxygen singly bound to carbon.

Nitrogen 1s Peak Shape Analysis. Most nitrogen associated with carbon also falls in a narrow region between 399 and 401 eV. Oxidized nitrogen shifts (6–8 eV) to higher binding energy. Clark and Harrison [75] have shown for polymers that a nitrogen binding energy of about 400 eV is due to amine groups, whereas a binding energy of about 401.5 eV is due to nitrogen bound to oxygen and/or nitrogen bound to carbon containing carbonyl groups.

DERIVATIZATION. Everhart and Reilley [77] have developed a systematic approach to the identification of functional groups on oxidized polymer surfaces. Here, polymer surfaces were exposed to a series of reagents that reacted only with specific functional groups. The reagents chosen also contained an element that could easily be detected with XPS. Most of the reagents used by Everhart and Reilley were fluorine-containing compounds.

This technique does have several drawbacks involving, for example, the specificity of the derivatizing reagent, the determination of the extent of the reaction, the determination of the number of functional groups that have reacted, and the stability of the reagent to X-rays. The question of whether reactions that occur in solution can be extended to a two-dimensional surface remains unanswered.

Secondary-Ion Mass Spectrometry

In secondary-ion mass spectrometry (SIMS), a solid surface is bombarded with ions, causing ejection of ion fragments from the solid surface. The mass-to-charge ratios of the ejected ion fragments are analyzed in a mass spectrometer, permitting the investigator to infer the molecular and atomic structure of the solid. Trace elements also can be detected.

With high primary-ion currents, material is removed from the solid surface rapidly and a depth profile is obtained. At low ion currents, the top few atomic layers of the surface are removed. Detailed information about the molecular structure of the solid surface is thus obtained.

This technique has been applied to the analysis of polymer surfaces by Briggs [78,79] and Brown and Vickerman [80]. Spectra obtained at low current observation led to detailed fingerprint spectra of polymer surfaces. Specific fragments could be assigned to either aliphatic or aromatic compounds.

One problem with analyzing polymer surfaces with SIMS is static charging caused by the primary-ion current. In Briggs's work [78,79], the ion current was neutralized with an electron gun. Recent developments in the analysis of polymer surfaces by mass spectrometry have included a gun that will bombard the solid surface with neutral atoms such as argon [80]. This analysis by neutral atom bombardment, referred to as fast atom bombardment mass spectrometry (FABMS), greatly reduces the static charging problem encountered in the normal SIMS experiment.

Adhesion Principles

Adhesion commonly refers to the potential for stress transfer across an interface between two materials [81]. In a fiber-reinforced composite, adhesion will result in stress transfer between fiber and matrix. The matrix thus acts to transfer stress between adjacent fibers. The adhesion between fiber and matrix will affect shear stress transfer in a composite. In addition, stress will be transferred from the ends of broken fibers to adjacent fibers through the interface and the matrix.

Theories of Adhesion

To form an adhesive bond between two materials, it is necessary that they come into close molecular contact. One of the materials must be capable of flowing, wetting the other material, and solidifying [81–83].

The mechanisms that cause two materials to adhere to one another are not well understood. For bonding of a polymeric material to a solid surface, two mechanisms for stress transfer across the interface are possible: namely, mechanical interlocking and electronic attraction.

The mechanical interlocking theory assumes that adhesion is due to irregularities on the surface into which the liquid material can penetrate. Upon solidification, the solid adhesive material is held in place by the geometry on the adsorbed layer. Mechanical interlocking is thus enhanced by increasing the surface roughness or porosity of the solid material.

The electronic attraction theory assumes adhesion to be caused by the electronic attraction between the atoms in the two materials being bonded. These forces of attraction will result from interaction of specific functional groups on the two surfaces as well as from nonlocalized electronic interactions due to the molecular structure of the materials being bonded.

For either of these mechanisms to be valid, the polymer used as the adhesive must form close contact with the solid. Huntsberger [82,83] has shown that the adhesive bond strength of polymethyl methacrylate to aluminum adherends depends on the temperature at which the bond was formed. This result was thought to be caused by inadequate molecular contact between adhesive and adherend at lower temperatures.

Forces of Attraction Across an Interface

The basic electronic forces that hold "homogeneous" materials together [84,85] include ionic bonding, dipolar interactions, covalent bonding, dispersion forces, metallic bonding, and hydrogen bonding. Ionic bonding results from the electrostatic attraction between oppositely charged ions. Dipolar bonding results from interaction of permanent dipoles within the material. Covalent bonding results from the formation of chemical bonds within the material. Dispersion force bonding results from attraction between local electron density fluctuations in the material caused by electron mobility. Metallic bonding results from attraction of metal ions to a sea of electrons. Hydrogen bonding is similar to ionic bonding and results from sharing of an adjacent hydrogen atom by two other atoms.

When two dissimilar materials are brought into contact, as is the case in an adhesive bond, the resulting electronic attraction can be caused by any combination of the interactions listed above. The attractive forces across the interface have been classified into two broad categories: dispersion and polar. These forces have been discussed by Atkins [86] and Wake [87]. The polar component results from electric dipoles associated with specific atom pairs or functional groups on the materials surface. The dispersion component results from loosely bound electrons such as those in the conduction band of metals or simply from electrons in the atoms or molecules in the material.

If the possibility of interdiffusion between the two materials does not exist, the interaction can be simplified. The attraction across an interface can thus result from dispersion interaction or dipole–dipole interactions. In addition, dipolar groups in one material can induce dipoles and thus create a dipole–induced dipole attractive force.

To predict and understand adhesion between two materials, it is necessary to understand the chemical and physical structure of the materials being bonded.

Tests of Adhesion of Carbon Fibers to Polymeric Matrices

Tests for adhesion of polymeric matrices to carbon fibers can be classified into two general categories: single-fiber tests and multiple-fiber or composite tests. Single-fiber tests normally are easy to perform and can be done on a small quantity of fiber. Sample preparation is generally inexpensive. However, the single-fiber test is possibly not indicative of the performance of an actual composite. Multiple-fiber or composite tests, on the other hand, will give a very good indication of the expected composite performance. These tests have the disadvantages of calling for expensive and time-consuming sample preparation, requiring large quantities of material.

Single-Fiber Adhesion Tests

Single-fiber adhesion tests are performed by embedding a single fiber in a matrix, then measuring the adhesive strength by applying a force and observing the failure of the bond. Two common single-fiber tests are the fiber pullout test and the fiber critical length test. The former is a direct test and the latter, an indirect test.

FIBER PULLOUT. In the fiber pullout test, a fiber is embedded in a very thin film (normally ~ 0.5 mm) [88] or bead [89] of polymer, and the specimen is placed in a tensile testing machine. The fiber is pulled out of the polymer film. The force required to remove the fiber is measured with a load cell. The adhesive strength of the joint is calculated by dividing the measured load by the area of contact of the fiber with the polymer. The area of contact is normally measured in a scanning electron microscope. This technique is very tedious and time-consuming. The small diameter of the fiber requires that the area of interfacial contact be small, for otherwise the fiber may break rather than pulling out, thus altering the data analysis.

FIBER CRITICAL LENGTH EXPERIMENT. In the fiber critical length (FCL) experiment, a single fiber is embedded in a polymeric matrix [90–96]. A typical specimen is shown in Figure 13. The specimen is then pulled in tension. As the strain is increased, the fiber will break. When the fiber first breaks, the matrix restrains the fiber fragments from returning to their unstressed dimension; stress is transferred to the fiber by the matrix. However the axial stress in the fiber at a distance from the break increases until it reaches the original stress in the fiber before the first break. Further increase in the specimen strain will result in another break in the fiber. This fiber breakage will continue with increased strain until the fiber fragments become so small that the matrix can no longer transfer stress over a long enough distance to break the fiber. The length of the broken fiber fragments is referred to as the fiber critical length l_c. The fiber critical length is an indication of the ability of the polymeric matrix to transfer stress to the fiber.

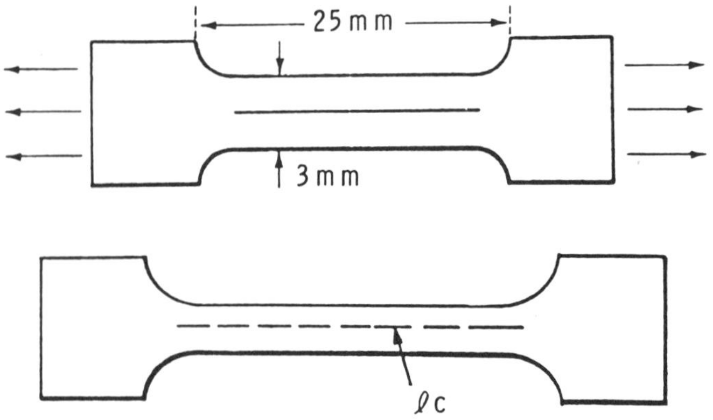

FIGURE 13 Diagram of fiber critical length experiment. (Reprinted with permission from Ref. 90. Copyright 1986 Gordon and Breach Science Publications.)

If the stress transferred across the interface τ acts over a length l_c, then a force balance can be obtained by setting the total force transferred across the interface equal to the breaking strength of the fiber σ_f times the area of the fiber:

$$\int_0^1 2\pi r\tau(l)\,dl = \pi r^2\sigma_f \tag{17}$$

where r is the fiber radius.

The shear stress is assumed to be some function of the distance l along the length of the fiber. If we assume that the shear stress is a maximum at the fiber tip τ_{max} and decays linearly along the length of the fiber fragment l_c, we can write:

$$\int_0^{l_c} 2\pi r\tau_{max}\left[\frac{l_c - l}{l_c}\right] dl = \pi r^2\sigma_f \tag{18}$$

Integrating and rearranging to solve for τ_{max} gives:

$$\tau_{max} = \frac{\sigma_f r}{l_c} \tag{19}$$

Assuming that the average shear stress transferred is equal to half the maximum shear stress, which will occur if the stress decays linearly, Eq. (20) results.

$$\tau_{avg} = \frac{\sigma_f r}{2l_c} \tag{20}$$

Calculation of τ_{avg} gives an indication of the ability of the matrix to transfer stress to the fiber. Since there is normally no interfacial failure between fiber and matrix, it is incorrect to call τ the interfacial shear strength. Rather, τ is a measure of the stress transfer across the interface, and it should be referred to as the stress transfer or stress transfer coefficient.

In an actual experiment, given the statistical nature of fiber fracture, the fiber fragments will range from the critical length to twice the critical length. Ohsawa et al. [91] have used a simple average to calculate l_c from the average fiber length l_a:

$$l_a = \frac{l_c + 2l_c}{2} \quad \text{and} \quad l_c = \tfrac{2}{3}l_a \tag{21}$$

A more sophisticated model by Drzal et al. [92] uses a Weibull distribution to describe the fragment lengths.

To calculate the interfacial shear stress in the FCL experiment, the strength of the fiber at the critical length must be known. However, the strength of the fiber depends on the flaw distribution of the fiber. Typical fragment lengths in a fiber critical length experiment for carbon fibers are about 0.5 mm. Rich and Drzal [93] have measured the strength of carbon fibers at these short lengths. This process is very tedious.

Estimates of the fiber strength at shorter lengths have been made by extrapolation using Eq. (5) and breaking strengths at longer lengths. Correlation between predicted strengths and measured strengths are not good, however [44]. This is because at shorter lengths, the fiber strength approaches the ultimate strength of the material. That is, the flaw-dominated strength, which is what is measured at the longer lengths, is no longer the variable being measured. Hence use of this extrapolation for calculating interfacial stress in the FCL experiment may lead to incorrect conclusions.

Photoelastic Stress Transfer Observation in the Fiber Critical Length Experiment

Additional information can be obtained by observation of the stress transferred between fiber and matrix at the broken end of a fiber fragment. The stress in the matrix at the tip of the broken fiber will rise sharply as shown in Figure 14 [94]. Failure modes at the fiber tip have been reviewed by Mullin and coworkers [95,96]. If the fiber matrix adhesion is poor, failure will occur at the interface (Fig. 15*b*). If the matrix is brittle, matrix cracking will occur and the specimen will fail after one fiber break (Fig. 15*a*). If a ductile matrix is used and the adhesion between fiber and matrix is high, the matrix will fail by shear (Fig. 15*c*). These failure modes can be observed under a microscope with crossed polarizers [90,92,93] to enhance the information obtained from measurement of the fiber critical length.

Surface Properties of Carbon Fibers

Carbon fiber surfaces are treated chemically to enhance bonding of the fiber to the resin in a composite. In sur-

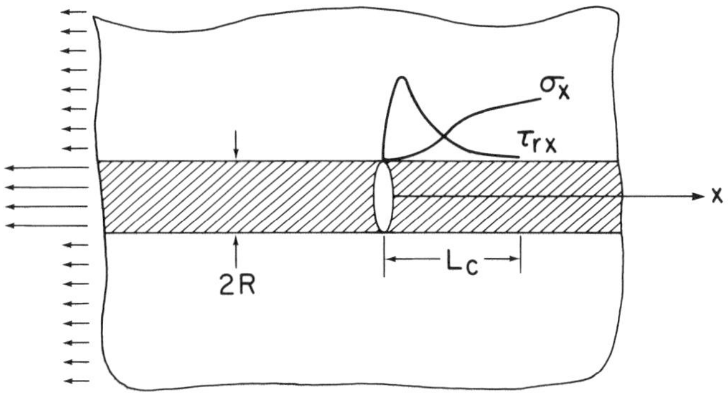

FIGURE 14 **Theoretical stress distribution at the tip of a broken fiber in the fiber critical length experiment [92].**

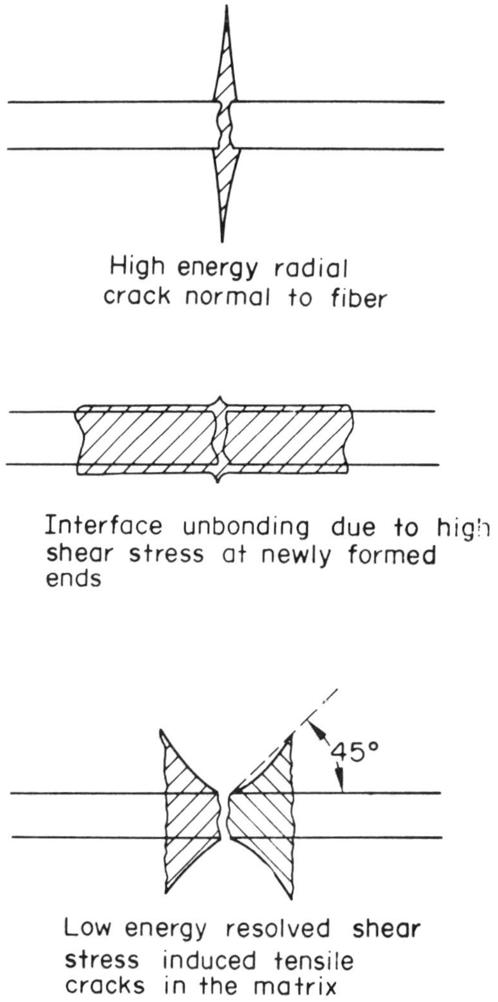

FIGURE 15 **Diagram of possible failure modes in the fiber critical length experiment: (*a*) matrix cracking, (*b*) frictional stress transfer, and (*c*) shear stress transfer.** (Reprinted with permission from Ref. 95. Copyright 1972 Pergamon Press.)

face treating, functional groups are created on the fiber surface. Many studies have been conducted to observe the effect of functional groups on carbon fiber–epoxy matrix adhesion [97–102].

STUDIES OF CARBON FIBER ADHESION TO POLYMERIC MATRICES. Donnet and coworkers [97–99] have studied adhesion between epoxy resins and carbon fibers after anodization in sodium hydroxide or nitric acid. The number of acid groups on the fiber surface was determined by titration. Mild sodium hydroxide was used to neutralize strong acidic groups such as carboxyl and phenol. Sodium ethoxide was used to neutralize weaker acidic groups such as hydroxyl and carbonyl. A direct correlation was found between the number of carboxylic acid groups and the interlaminar shear strength of the composite (Figure 16).

Fitzer et al. [100] have studied the surface treatment of carbon fibers by boiling in nitric acid. By chemically blocking specific functional groups, they were able to determine which groups were most responsible for adhesion. Blocking of strong and weak acidic oxides resulted in a significant reduction of composite shear strength, leading to the conclusion that adhesion is caused by chemical bonding of the epoxy to acidic groups on the fiber surface.

In addition to functional groups being created by surface treatment, it is possible that the surface treatment will affect the molecular and morphological structure of the carbon fiber surface. Pores may be created which can enhance mechanical bonding. This has been best demonstrated by Drzal and coworkers [101,102], whose results are shown in Figure 17, where the interfacial shear strength of a carbon fiber–epoxy matrix bond is plotted against the surface oxygen content of the carbon fiber. The oxygen content is assumed to be indicative of the amount of functional groups on the surface. The AU fiber had no surface treatment and thus a low oxygen content and a low interfacial shear strength. The AS fiber

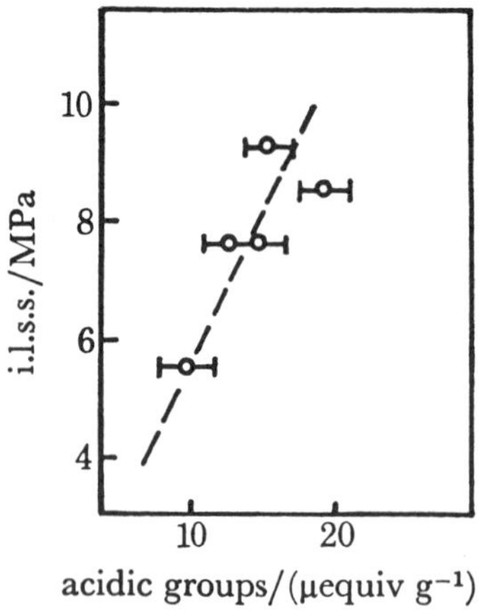

FIGURE 16 **Effect of surface acidic groups on the interlaminar shear strength (i.l.s.s.) of carbon fiber composites.** (Reprinted with permission from Ref. 97. Copyright 1980 Royal Society.)

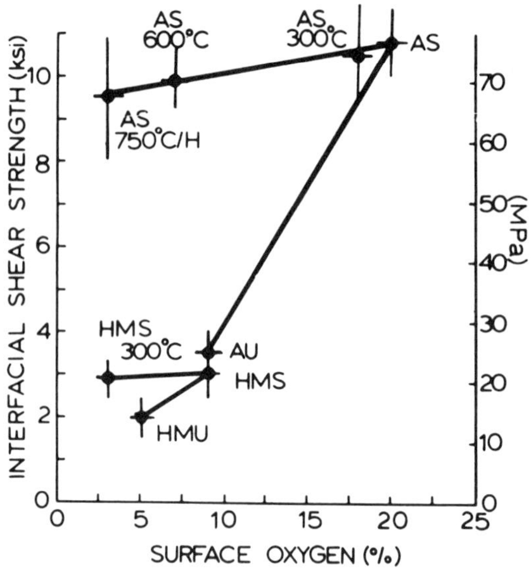

FIGURE 17 **Effect of surface oxygen content on interfacial shear strength measured on single fiber: AU and HMU indicate no commercial surface treatment; AS and HMS indicate commercial surface treatment. Temperatures indicate heat treatment for removal of oxygen.** (Reprinted with permission from Ref. 102. Copyright 1982 Gordon and Breach Science Publications.)

had been surface treated commercially and thus had a high oxygen content and a high interfacial shear strength. Upon removal of the functional groups from the AS fiber, the oxygen content is greatly reduced, indicating loss of functional groups. However, the interfacial shear strength remains high. This indicates that adhesion promotion is not due solely to the addition of functional groups on the fiber surface, but also to the crystallinity, molecular structure, and/or possibly of the fiber surface.

The crystallinity of carbon fiber surfaces has been studied with Raman spectroscopy [103–105]. The Raman spectrum of carbon fibers gives two peaks. One peak due to the graphitic nature of carbon fibers occurs at 1575 cm^{-1}. The other peak at 1355 cm^{-1} has an intensity inversely proportional to the graphite crystal size. By comparing the ratio of the 1355 cm^{-1} peak to the 1575 cm^{-1} peak, the crystal size at the surface of carbon fibers can be estimated. Tuinstra and Koenig [104] have observed that as the 1355 cm^{-1} peak decreases in intensity, the interlaminar shear strength of the composite also decreases. This indicates that bonding is affected by the crystal structure of the carbon fiber surface. Further evidence supporting this idea has been expressed by Brelant [106], who showed an inverse correlation between the thermal conductivity of the composite and its interlaminar shear strength.

EFFECT OF SURFACE TREATMENT ON CARBON FIBER SURFACE PROPERTIES. To understand how the surface treatments enhance adhesion, it is necessary to understand how specific surface treatments alter the surface properties of carbon fibers. Considering the turbostratic carbon structure shown above (Fig. 4*b*), there are three possible locations for oxidation. Functional groups can be created at the crystal edges, between layers, or at the basal planes.

The surface treatments used by commercial producers of carbon fibers are proprietary. Indeed, much of the work on characterization of carbon fiber surfaces has been performed on fibers with proprietary surface treatments [101,102,107–111]. Although these studies show that oxygen and nitrogen functionalities are being added to the fiber surface by surface treatment, they do not give much insight into the reactions occurring during surface treatment. Nor do these studies begin to set a standard for surface treatment.

The most extensive study of carbon fiber surface treatments has been done by Sherwood and coworkers [112–116]. Some of their conclusions are outlined here. The functional groups created by anodization in sulfuric acid and in ammonium bicarbonate were dependent on the anodization potential [112]. Carbonyl (R—C=O) bonds were formed at low potential, while R—O—C bonds were formed at higher potentials [112]. Anodization in nitric acid yielded mostly $>$C=O functionality [114]. Anodization at high pH attacked primarily the edges of the graphitic planes, while anodization at neutral and low pH attacked between the carbon layers, creating a graphite oxide [115]. Nitrogen functionality can be created on the carbon fiber surface by anodization in ammonia-saturated ammonium bicarbonate solution [116].

Acknowledgments

The support of NASA Langley Research Center of the work on which this review is based is appreciated.

James P. Wightman, T. A. DeVilbiss, and John G. Dillard

References

1. M. S. Reisch, *Chem. Eng. News.*, Feb. 2, 1987, p. 9.
2. J. Delmonte, *Technology of Carbon and Graphite Fiber Composites*, Van Nostrand, New York, 1981.
3. J. Diamont and R. J. Moulton, *SAMPE Q.*, *16*(1), 13 (1984).
4. G. D. M. DiSalvo and S. M. Lee, *SAMPE Q.*, *14*(2), 14 (1983).
5. J. Brandt and J. Warnecke, *Proceedings of the 7th Int. Conference of SAMPE, European chapter*, in K. Brunsch, H. D. Golden, and C. M. Herkert, Eds., *High Tech—The Way into the Nineties*, Elsevier, Amsterdam, 1986, pp. 251–260.
6. A. F. Yee and R. A. Pearson, *J. Mater. Sci.*, *21*, 2462 (1986).
7. A. F. Yee and R. A. Pearson, *J. Mater. Sci.*, *21*, 2475 (1986).
8. D. J. Willats, *SAMPE J.*, *20*(5), 6 (1984).
9. J. L. Kardos, *J. Adhes.*, *5*, 119 (1973).
10. T. W. Johnson and C. L. Ryan, *Proceedings of the 31st SAMPE Int. Symposium*, 1986, pp. 1537–1548.
11. D. L. Hunston, *Compos. Technol. Rev.*, *6*, 176 (Winter 1984).
12. J. H. Williams and P. N. Kousiounelos, *Fibre Sci Technol.*, *11*, 83 (1978).
13. R. V. Subramanian, V. Sundram, and A. K. Patel, *Proceedings of the 33rd Annu. Tech. Conf. SPI Reinf. Plast./Compos. Inst.*, 1978, *20F*, 1.
14. R. V. Subramanian, J. J. Jakubowski, and F. D. Williams, *J. Adhes.*, *9*, 185 (1978).
15. L. Ying, *SAMPE Q.*, *14*(3), 26 (1983).
16. S. Lehmann, C. Megerdigian, and R. Papalia, *SAMPE Qt.*, *16*(3), 7 (1985).
17. R. Robinson, S. Lehmann, G. Askew, D. Wilford, C. Megerdigian, and R. Papalia, *Proceedings of the 7th Int. Conference of SAMPE, European chapter*, in K. Brunsch, H. D. Golden, and C. M. Herkert, Eds., *High Tech—The Way into the Nineties*, Elsevier, Amsterdam, 1986, pp. 299–310.
18. J. B. Donnet and R. C. Bansal, *Carbon Fibers*, Dekker, New York, 1984, Ch. 1.
19. R. J. Diefendorf, W. C. Stevens, and S. H. Chen, *Fiber Producer*, *7*(6), 16 (1979).
20. R. J. Diefendorf and E. Tokarsky, *Polym. Eng. Sci.*, *15*, 150 (1975).
21. W. Watt, in W. Watt and B. V. Perov, Eds., *Strong Fibers*, Elsevier, Amsterdam, 1985.
22. P. J. Goodhew, A. J. Clarke, and J. E. Bailey, *Mater. Sci. Eng.*, *17*, 3, (1975).
23. M. M. Coleman and G. T. Sivy, *Carbon*, *19*, 123 (1981).
24. M. M. Coleman and R. J. Petcavich, *J. Polym. Sci., Polym. Phys. Ed.*, *16*, 821 (1978).
25. R. Bacon, in L. F. Vosteen, N. J. Johnston, and L. A. Teichman, Eds., *Tough Composite Materials*, NASA CP-2334, 1983, p. 245.
26. J. C. Bokros, in P. L. Walker, Ed., *Chemistry and Physics of Carbon*, Vol. 5, Dekker, New York, 1969, p. 1.
27. R. Perret and W. Ruland, *J. Appl. Crystallogr.*, *3*, 525 (1970).
28. M. Guigon, A. Oberlin, and G. Desarmot, *Fibre Sci. Technol.*, *20*, 55 (1984).
29. L. T. Drzal, *Carbon*, *15*, 129 (1977).
30. S. C. Bennet, "Strength Structure Relationships in Carbon Fibres," Ph.D. thesis, University of Leeds, 1976.
31. R. J. Diefendorf, in L. F. Vosteen, N. J. Johnston, and L. A. Teichman, Eds., *Tough Composite Materials*, NASA CP-2334, 1983, p. 209.
32. J. T. Paul, British Patent 1,433,712 (1976).
33. K. Saito, H. Ogawa, and T. Shigei, U.S. Patent 4,401,533 (1983).
34. G. Gynn, R. N. King, S. F. Chappell, and M. L. Deviney, "Improved Graphite Fiber Adhesion," AFWAL-TR-81-4096, 1981.
35. D. J. Pinchin and R. T. Woodhams, *J. Mater. Sci.*, *9*, 300 (1974).
36. D. M. Brewis, J. Comyn, J. R. Folwer, D. Briggs, and V. A. Gibson, *Fibre Sci. Technol.*, *12*, 41 (1979).
37. J. C. Goan, L. A. Joo, and G. E. Sharpe, *Proceedings of the 27th Annu. Tech. Conf. SPI Reinf. Plast./Compos. Inst. 21E*, 1 (1972).
38. P. Ehrburger and J. B. Donnet, in W. Watt and B. V. Perov, Eds., *Strong Fibers*, Elsevier, Amsterdam, 1985, p. 577.
39. L. T. Drzal, M. J. Rich, M. F. Koenig, and P. F. Lloyd, *J. Adhes.*, *16*, 133 (1983).

40. O. P. Bahl, R. B. Mathur, and T. L. Dhami, *Polym. Eng. Sci., 24*, 455 (1984).
41. E. Fitzer and R. Weiss, *Proceedings of the 16th Biennial Conference on Carbon*, July 1983, p. 473.
42. S. W. Tsai and H. T. Hahn, *Introduction to Composite Materials*, Technomic, Lancaster, PA, 1980, Ch. 9.
43. W. Weibull, *J. Appl. Mech., 18*, 293 (1951).
44. J. W. Hitchon and D. C. Phillips, *Fibre Sci. Technol., 12*, 217 (1979).
45. P. W. Manders and T. W. Chou, *J. Reinf. Plast. Compos., 2*, 43 (1983).
46. T. A. DeVilbiss, Ph.D. dissertation, Virginia Polytechnic Institute and State University, Blacksburg, VA, 1987.
47. B. O. Bateup, *Int. J. Adhes. Adhes., 1*(5), 233 (1981).
48. H. F. Webster, M.S. thesis, Virginia Polytechnic Institute and State University, Blacksburg, VA, 1985.
49. L. A. Girifalco and R. J. Good, *J. Phys. Chem., 61*, 904 (1957).
50. F. M. Fowkes, *Ind. Eng. Chem., 56*, 40 (1964).
51. F. M. Fowkes in K. L. Mittal, Ed., *Physicochemical Aspects of Polymer Surfaces*, Vol. 2, Plenum Press, New York, 1983, p. 583.
52. D. H. Kaelble, P. J. Dynes, and E. H. Cirlin, *J. Adhes., 6*, 23 (1974).
53. D. H. Kaelble, P. J. Dynes, and L. Maus, *J. Adhes., 6*, 239 (1974).
54. W. C. Jones and M. C. Porter, *J. Colloid Interface Sci., 24*, 1 (1967).
55. G. E. Hammer and L. T. Drzal, *Appl. Surf. Sci., 4*, 340 (1980).
56. J. Schultz, C. Cazeneuve, M. E. R. Shanahan, and J. B. Donnet, *J. Adhes., 12*, 221 (1981).
57. C. A. Herb, J. L. Buckner, and J. R. Overton, *J. Colloid Interface Sci., 94*, 14 (1983).
58. C. S. Brooks and D. A. Scola, *J. Colloid Interface Sci., 32*, 561 (1970).
59. J. V. Larsen, T. G. Smith, and P. W. Erickson, "Carbon Fiber Surface Treatments," NOLTR 71-165, 1971.
60. J. Schultz, L. Lavielle, and H. Simon, *Proceedings of the Intl. Symposium on Science and New Applications of Carbon Fibers*, 1984, p. 125.
61. A. S. Gozdz and H. D. Weigmann, *J. Appl. Polym. Sci., 29*, 3965 (1984).
62. B. Rand and R. Robinson, *Carbon, 15*, 311 (1977).
63. J. P. Randin, in A. J. Bard, Ed., *Encyclopedia of Electrochemistry of the Elements*, Dekker, New York, 1976, p. 1.
64. J. B. Donnet, *Carbon, 6*, 161 (1968).
65. H. P. Boehm, *Angew. Chem., 5*, 533 (1966).
66. D. Rivin, *Rubber Chem. Technol., 44*, 307 (1971).
67. D. Rivin, *Rubber Chem. Technol., 36*, 729 (1963).
68. D. W. McKee and V. J. Mimeault, in P. L. Walker and P. A. Thrower, Eds., *Chemistry and Physics of Carbon*, Vol. 8, Dekker, New York, 1973, p. 151.
69. W. M. Riggs and M. J. Parker, in S. P. Wolsky and A. W. Czanderna, Eds., *Methods of Surface Analysis*, Elsevier, Amsterdam, 1975, p. 103.
70. C. S. Fadley, *Prog. Solid State Chem., 11*, 265 (1976).
71. J. H. Scofield, *J. Electron Spectrosc. Relat. Phenom., 8*, 129 (1976).
72. P. Cadman, G. Gossedge, and J. D. Scott, *J. Electron Spectrosc. Relat. Phenom., 13*, 1 (1978).
73. C. D. Wagner, *J. Electron Spectrosc. Relat. Phenom., 32*, 99 (1983).
74. D. Briggs and M. P. Seah, *Practical Surface Analysis*, Wiley, New York, 1983, p. 363.
75. D. T. Clark and A. Harrison, *J. Polym. Sci., Polym. Sci. Ed., 19*, 1945 (1981).
76. A. Proctor and P. M. A. Sherwood, *J. Electron Spectrosc. Relat. Phenom., 27*, 39 (1982).
77. D. S. Everhart and C. N. Reilley, *Anal. Chem., 53*, 665 (1981).
78. D. Briggs, *Surf. Interface Anal., 4*, 151 (1982).
79. D. Briggs, *Surf. Interface Anal., 9*, 391 (1986).
80. A. Brown and J. C. Vickerman, *Surf. Interface Anal., 8*, 75 (1986).
81. L. H. Sharpe, in Notes from Adhesion Society, short course, Savannah, Georgia, February 1985.
82. J. R. Huntsberger, in R. L. Patrick, Ed., *Treatise on Adhesion and Adhesives*, Vol. 1, Dekker, New York, 1967, p. 119.
83. J. R. Huntsberger, *J. Paint Technol., 30*, 199 (1967).
84. C. Kittel, *Introduction to Solid State Physics*, 5th ed., Wiley, New York, 1976, p. 71.
85. R. J. Good, in R. L. Patrick, Ed., *Treatise on Adhesion and Adhesives*, Vol. 1, Dekker, New York, 1967, p. 9.
86. P. W. Atkins, *Physical Chemistry*, 3rd ed., Freeman, New York, 1986, p. 584.
87. W. C. Wake, *Adhesion and the Formulation of Adhesives*, 2nd ed., Applied Sciences Publishers, Barking, England, 1982, p. 8.
88. L. Penn, F. Bystry, W. Karp, and S. Lee, in H. Ishida and G. Kumar, Eds., *Characterization of Composite Interfaces*, Plenum Press, New York, 1985, p. 93.
89. B. Miller and M. A. Tallent, *Proceedings of the 9th Annual Meeting of the Adhesion Society*, 1986.
90. W. D. Bascom and R. M. Jensen, *J. Adhes., 19*, 219 (1986).
91. T. Ohsawa, A. Nakayama, M. Miwa, and A. Hasegawa, *J. Appl. Polym. Sci., 22*, 3203 (1978).
92. L. T. Drzal, M. J. Rich, J. D. Camping, and W. J. Park, *Proc. 35th Annu. Tech. Conf. SPI Reinf. Plast./Compos. Inst., 20-C*, 1980, p. 1.
93. M. J. Rich and L. T. Drzal, *Proc. 41st Annu. Tech. Conf. SPI Reinf. Plast./Compos. Inst., 2-F*, 1986, p. 1.
94. J. M. Whitney and L. T. Drzal, in N. J. Johnston, Ed., *Toughened Composites*. ASTM STP 937, Am. Soc. Testing and Materials, Philadelphia, 1987, p. 179.
95. J. V. Mullin and V. F. Mazzio, *J. Mech. Phys. Solids, 20*, 391 (1972).
96. J. Mullin, J. M. Berry, and A. Gatti, *J. Compos. Mater., 2*, 82 (1968).
97. P. Ehrburger and J. B. Donnet, *Phil. Trans. R. Soc. London, A295*, 495 (1980).
98. P. Ehrburger, J. J. Herque, and J. B. Donnet, *Proceedings of the Fifth London International Carbon Graphite Conference*, 1978, p. 398.
99. J. Lahaye, J. J. Herque, and J. B. Donnet, *Proceedings of the Fourth London International Carbon Graphite Conference*, 1976, p. 201.
100. E. Fitzer, K.-H. Geigl, and W. Huttner, *Carbon, 18*, 389 (1980).

101. L. T. Drzal, *SAMPE J., 19*(5), 7 (1983).
102. L. T. Drzal, M. J. Rich, and P. F. Lloyd, *J. Adhes., 16,* 1 (1982).
103. F. Tuinstra and J. L. Koenig, *J. Chem. Phys., 53,* 1126 (1970).
104. F. Tuinstra and J. L. Koenig, *J. Compos. Mater., 4,* 492 (1970).
105. A. Ishitani, H. Ishida, G. Katagiri, and S. Tomita, in H. Ishida and J. L. Koenig, Eds., *Composite Interfaces,* North Holland, New York, 1986, p. 195.
106. S. Brelant, *SAMPE J., 21*(5), 6 (1985).
107. F. Hopfgarten, *Fibre Sci. Technol., 11,* 67 (1978).
108. F. Hopfgarten, *Fibre Sci. Technol., 12,* 283 (1979).
109. A. Ishitani, *Carbon, 19,* 269 (1981).
110. T. Takahagi and A. Ishitani, *Carbon, 22,* 43 (1984).
111. K. Waltersson, *Fibre Sci. Technol., 17,* 289 (1982).
112. A. Proctor and P. M. A. Sherwood, *Carbon, 21,* 53 (1983).
113. A. Proctor and P. M. A. Sherwood, *Surf. Interface Anal., 4,* 212 (1982).
114. A. Proctor and P. M. A. Sherwood, *J. Chem. Soc., Faraday Trans. 1, 80,* 2099 (1984).
115. C. Kozlowski and P. M. A. Sherwood, *J. Chem. Soc., Faraday Trans. 1, 8,* 2745 (1984).
116. C. Kozlowski and P. M. A. Sherwood, *Carbon, 24,* 357 (1986).

Carbon Fiber Surface Properties

It is fundamental to composite materials technology that intimate contact—preferably chemical bonding—must exist between matrix and reinforcement for the load to be transferred to the reinforcement. Inasmuch as the reinforcement's surface properties determine the nature of the matrix–reinforcement interactions, its surface is one of the more critical aspects of a composite material. In the case of carbon fiber reinforced composites, the unique properties of the fiber surface have made it particularly difficult to achieve strong matrix–reinforcement interfaces. Problems originating with fiber surface properties date from the first efforts toward a carbon fiber composite technology when it was found that the fibers were resistant to wetting and infiltration by matrix material (Fig. 1). With the development of methods for obtaining reliable infiltration, difficulties associated with the carbon fiber surface have persisted in the form of weak fiber–matrix interfaces and correspondingly low interlaminar shear strengths or transverse tensile strengths in the composites. As a consequence, interface problems related to carbon fiber surface properties continue to be a major focus of research and development efforts in both polymer matrix and metal matrix composites.

Much of the material in this article has been covered, in more detail than is possible here, in excellent review articles by Reynolds [1] and by McKee and Mimeault [2], which are highly recommended to the reader. Although

FIGURE 1 **"Canning" of graphite fiber bundle by aluminum matrix due to failure of the molten aluminum to infiltrate the fibers.** (From Ref. 74.)

more than 15 years old, these sources present incisive critical commentary on points of view that continue to determine the directions of carbon fiber composite research and development. In the intervening years, additional data have made it possible to resolve some of the significant issues about the nature of the carbon fiber surface and its relationship to composite properties. New material considered here includes aspects of graphite chemistry, not previously emphasized in the composite literature, which have important implications for solving the problems of carbon–fiber composite interfaces.

Physical Properties

Fiber Structure

The advantages of carbon fibers as a reinforcement material are almost entirely due to the strength and stiffness of the graphitic basal plane. As we shall see, the fiber surface properties are also intimately connected with the basal plane character of the fiber and are best described in terms of the fiber structure and the properties of graphite. As shown schematically in Figure 2, graphite is a laminar material whose basal planes consist of open hexagons with a 0.1415 nm carbon–carbon spacing [3]. The hexagonal structure of the basal plane is due to the filling of coplanar sp^2 bonding states by three electrons per atom. The remaining electron outside the 1*s* carbon core occupies band states associated with mixed coplanar and interplanar bonding and is largely responsible for the electrical properties of graphite [4].

The planar hexagonal unit of the graphitic basal plane is the elementary structural component of carbon blacks and many other solid carbon materials [5,6]. However, the term "graphite" is reserved exclusively for single-crystal material in which the basal planes are stacked in

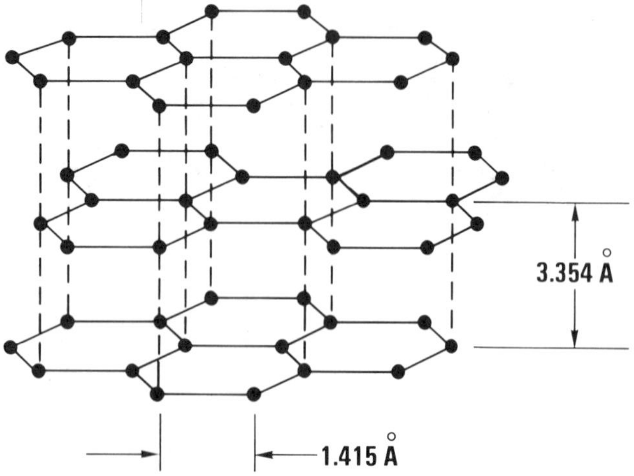

FIGURE 2 **Schematic of the graphite crystal structure. Atomic separation within a basal plane is 0.1415 nm and the separation of basal planes is 0.3354 nm.**

A-B-A-B registry with an interplanar spacing of 0.33539 nm.* The nongraphitic carbons constitute the class of solid carbon materials composed of graphitic basal planes with little or no measurable ordering between planes other than a parallel stacking. The interplanar distance increases with increasing disorder in the stacking of the basal planes, reaching a value of 0.344 nm for turbostratic carbon, in which the stacking sequence is completely random [7]. Since the nongraphitic carbons retain a parallel graphitic basal plane structure, they share with graphite many of the properties associated with the basal plane.

The hexagonal basal plane produced by the coplanar sp^2 bonds is, with an in-plane Young's modulus of 1.0 TPa, one of the strongest structures in nature [8]. However, in single-crystal graphite, the modulus along the *c* axis (perpendicular to the basal plane) is only 36.4 GPa, while the shear modulus is so low (180 MPa) [9] that one of the best-known applications of graphite is as a lubricant. In general, we cannot take advantage of the very high load-bearing capabilities of the graphitic basal plane with solid carbon structural materials because the very weak *c*-axis strength prevents loads from being transferred to the graphitic basal planes.

Early work on carbon whiskers explored the physical possibilities of solid carbon reinforcements and demonstrated properties close to those of single-crystal graphite [10]. However, the important commercial carbon fibers for composite technology have thus far been limited to those prepared by the carbonization of rayon, polyacrylonitrile (PAN), or mesophase petroleum pitch. Rayon-based carbon fibers [11,12] were made obsolete by the development of methods for producing high strength carbon fibers from PAN fibers [13–17]. Currently, PAN fibers are used primarily with polymer matrices, while mesophase pitch is the precursor of choice for metal–matrix composites. In the case of pitch fibers, it should be noted that the initial attempts to produce carbon fibers from isotropic pitches resulted in fibers with very low strength and modulus. Pre-1973 literature on pitch-based fibers refers exclusively to these inferior materials and should not be confused with reports on the very high quality fibers currently manufactured from mesophase pitch [18,19]. Very recently, the range of commercially available carbon fibers has been expanded to include high strength fibers prepared from highly purified mesophase pitch [20].

Regardless of the precursor material, the basic goal in the development of a carbon reinforcement fiber is to produce a high density of graphitic basal planes that are accurately oriented parallel to the fiber axis. The tensile modulus of graphite falls off very rapidly with angular displacement out of the basal plane, and the longitudinal tensile modulus of the carbon fiber is therefore a sensitive function of the average basal plane alignment with the fiber axis. The limits on the basal plane alignment that can be achieved and the mechanisms for obtaining alignment depend on the nature of the precursor material. For both rayon and PAN precursors, the nature of the parent polymeric structure is such that carbonization produces undulating ribbons or fibrils of turbostratic carbon [21–24]. These synthetic precursor materials are nongraphitizable (i.e., the turbostratic domains cannot be converted to graphite), and the resultant fibers are *not* graphite. The turbostratic crystallite size in PAN fibers increases and crystal order and alignment with the fiber axis improve with high temperature treatment above 2000°C. With few exceptions, fibers having a tensile modulus above 250 GPa have been annealed between 2000 and 3000°C. Crystallite dimensions perpendicular to the basal planes in PAN precursor fibers vary from 1.5–2.5 nm after 1000°C carbonization to more than 6.0 nm after heat treatment above 2600°C. The correlation of crystallite size with fiber modulus is of particular significance. A PAN fiber with a longitudinal tensile modulus of 280 GPa has typical crystallite ribbons 13 layers thick and 4.0 nm wide, while a fiber with a 700 GPa modulus is characterized by crystallites 30 layers thick and 9.0 nm wide. Kwizera et al. have reported the direct observation of microfibrils with 100 nm diameters in a high modulus PAN fiber [25].

* It is important for the reader to be aware that the complexity of behavior of solid carbon materials has resulted in a serious problem of nomenclature in carbon technology. The physical and chemical properties of a solid carbon are critically dependent on its detailed structural form. Consequently, less than precise terminology may imply properties that are qualitatively different from those that are intended. In recognition of the unusually serious consequences of casual or imprecise terminology that have been experienced in the carbon community, the International Committee for Characterization and Terminology of Carbon (ICCTC) has formulated conventions and definitions to be used in discussions of solid carbon materials [K.-H. Kochling et al., *Carbon, 20,* 445 (1982)]. Unless otherwise noted, these conventions and definitions are used here. The reader is cautioned that the composite materials literature frequently embraces usage specifically cited by the ICCTC as particularly inappropriate and misleading.

Conversion of the aromatic molecules in mesophase pitch produces graphitic basal plane fibrils that are automatically aligned with the fiber axis [19,26–29]. After carbonization and heat treatment at 3000°C, the full width at half-maximum (fwhm) of the preferred orientation distribution in a mesophase pitch fiber is about 5°, as compared with 30° for a typical PAN fiber with a 200 GPa tensile modulus. Mesophase pitch fibers with a 690 GPa tensile modulus exhibit three-dimensional diffraction patterns and an average crystallite size of 20 nm. Research quantities of mesophase pitch fibers have been produced with a tensile modulus of the order of 990 GPa, which is very close to the value for single-crystal graphite and represents a nearly perfect alignment of basal planes parallel to the fiber axis. As might be expected from its aromatic nature, mesophase pitch is graphitizable and carbon fibers produced from this precursor material are unique in being true graphite fibers. (The ICCTC's prohibition [30] of references to pitch fibers as "graphite" alludes to the older fibers made from isotropic pitch and is not applicable to mesophase pitch fibers.)

Diefendorf and Tokarsky have presented very detailed models of the macrostructures of several types of synthetic precursor fibers with moduli ranging from 40 to 700 GPa [23,24]. Although there are differences associated with the type of precursor, there is a clear correlation between high modulus and the development of an oriented ribbon structure. In all fibers, fibril development and orientation are much better close to the surface than in the interior. Three general macrostructures are observed within carbon fibers: fibril ribbons arranged in a radial spokelike structure, fibrils forming coaxial "onion-skin" layers around the fiber axis, and a random distribution of fibril ribbons. All three macrostructures have been reported for mesophase pitch fibers as well as for fibers from synthetic precursors.

Carbon fibers usually exhibit surface striations or folds running parallel to the fiber axis. The phenomenon was quite pronounced in some of the early PAN fibers and rayon fibers, producing "dog bone" and irregular cross sections [24]. In extreme cases, the folding of the surface skin can be great enough to form closed-off tubes at the surface. Diefendorf and Tokarsky [24] have pointed out that buckling of the surface skin is the result of a higher coefficient of thermal expansion in the poorly ordered interior of the fiber. The differential contraction upon cool-down from processing temperatures places the outer onion skin in compression and causes the buckling. Distortion of the cross section has been minimized in modern carbon fibers. As illustrated in Figure 3, typical surface striations consist of small ridges running parallel to the fiber axis.

FIGURE 3 **Mesophase pitch fibers (Thornel P55). The striations on the 10 μm diameter fibers are characteristic of high modulus fibers. The fractured surfaces reveal the round cross section and modified radial structure of the fibers. (Photomicrograph courtesy of M. K. Norr, Naval Surface Warfare Center.)**

Surface Structure

In spite of the major qualitative differences in the internal structures of fibers produced from different precursors, all high modulus carbon fibers exhibit the same basic surface structure. As early as 1973, McKee and Mimeault [31] were able to conclude that ". . . there seems little doubt that the surfaces of fibers produced at 2500°C and higher temperatures are essentially graphitic in nature and almost pure basal plane in the case of the strongest fibers."

This conclusion was based on observations reported for both rayon and PAN fibers and can now be extended to the surfaces of mesophase pitch fibers. Graphitic surface structure was directly verified for rayon-based fibers by transmission electron micrographs [32] showing basal planes of graphite parallel to the surface. These studies showed that the basal planes were continuous over long distances in spite of sharp bends in individual packets of planes. Diefendorf and Tokarsky have concluded from extensive electron microscopy of carbon fibers that all high modulus fibers, regardless of precursor or internal structure, form an "onion skin" structure on the surface, with the orientation of the basal planes in the surface skin accurately perpendicular to the fiber radius in high modulus fibers [24]. In most cases, the onion-skin sheath is too thin to be seen by standard scanning electron microscope (SEM) techniques. The general picture of fibers having an onion-skin surface is consistent with sheath–core models that have been proposed for fiber macrostructure [33–35] and has been confirmed by experiments in which the macrostructure is developed by etching with oxidizers [36,37] and with rf plasmas [25,38]. Although no quantitative relationship has been established, the reported observations show a correlation between the degree of surface crystallinity and the modulus of the fiber. The surface onion-skin layer has also been confirmed on mesophase pitch fibers, where it becomes so well developed on some of the radial-spoked fibers that it can be observed directly with a scanning electron microscope [39]. The evidence leaves little room for doubt that every carbon fiber, regardless of precursor or internal structure, has developed a graphitic onion-skin outer sheath by the time it has received sufficient heat treatment to provide it with a useful longitudinal tensile modulus.

Inasmuch as fiber–matrix interactions and bonding must take place in the top few atomic layers on the fiber, a critical aspect of the fiber surface structure is whether the graphitic basal planes of the onion skin constitute the *exposed* surface of the fiber or simply provide a substrate for a coating of contaminant, debris, or disorganized carbonaceous material. The evidence just presented for the existence of an onion-skin outer sheath on carbon fibers is not very conclusive on this point. However, the graphitic basal plane surface has chemical properties that are quite unique and can be used for positive identification of a fiber surface that is principally graphitic basal plane. The identification criteria are based on a combination of established structural and chemical properties of the basal plane, which we will review briefly.

During the 1960s, Hennig studied the defect structure and chemistry of single-crystal graphite by using metal decoration and transmission electron microscopy (TEM) to follow the enlargement of defect sites by chemical reactions [40]. This work showed that defect concentrations in the graphitic basal plane are reduced to less than 10^{-10} per carbon atom by a few seconds anneal at 2000°C for the central regions or at 2300°C for the edges of a single-crystal domain [41]. Except for a few low modulus fibers, all carbon fibers are subjected to extended anneal at these, or much higher, temperatures. Thus, Hennig's result tells us that any exposed basal planes on a carbon fiber surface will consist of single-crystal domains whose atomic lattices are of extreme perfection with no significant atomic defect concentration whatsoever. Any defects in such a fiber surface will be located solely at the boundaries between single-crystal domains or within non-basal-plane regions of the surface. The significance of the defect structure derives from the fact that the perfect graphitic basal plane is almost completely inert. As discussed below in more detail, chemical reactivity of the basal plane toward normal atmospheric gases and vapors is limited to defect sites [40]. All other carbonaceous surfaces, including the pyramidal surface of single-crystal graphite, are very active and chemisorb high levels of gas when exposed to air. Thus, the criteria for identification of an exposed basal plane surface are straightforward. If the exposed fiber surface consists entirely of the basal planes of the onion-skin sheath, surface analysis of the air-exposed surface will show only a very low coverage of adsorbed gas (corresponding to the chemically active areas at the boundaries between fibrils) and will exhibit the characteristics of pure graphitic carbon. Residence on the fiber surface of any significant areas of edge planes, disordered carbon, or carbonaceous debris will be reflected by a proportionate amount of adsorbed gases.

The investigation of carbon fiber surface chemistry has been most effectively carried out using X-ray photoelectron spectroscopy (XPS), also known as electron spectroscopy for chemical analysis (ESCA). This spectroscopy is sensitive to only the top few atomic layers of the sample and, in addition to providing compositional analysis of the surface, is capable of determining the chemical states of the constituent atoms by detecting small deviations in the atomic core level binding energies. As shown in the X-ray photoelectron spectra of Figure 4, bare, as-received, untreated carbon fiber surfaces exhibit only a very small adsorption of oxygen after exposure to air. By the criteria we have just outlined, spectra such as these clearly identify the exposed fiber surfaces as almost pure basal plane. Noncarbonaceous contaminants, other than the very small amount of adsorbed oxygen, are virtually nonexistent on these surfaces, and we can conclude that there is no characteristic contamination or debris on the fiber surface. XPS spectra of air-exposed pitch fibers have been found to be nearly identical to spectra from the basal planes of high quality single-crystal Ticonderoga graphite [42]. A 0.9% at. level of oxygen was found on the single crystal graphite as a result of chemical activity at crystal domain boundaries. By comparison, a P55 pitch fiber (379 MPa modulus) surface had 1.4% at. adsorbed oxygen and a P100 fiber (688 MPa modulus) exhibited only 0.7% at. adsorbed oxygen—slightly less than the single crystal. A more definitive verification of the basal plane nature of the fiber surfaces is presented in Figure 5, which shows that the characteristic carbon 1*s* XPS line shape of the fiber is identical to that of a single-crystal graphite basal plane. The shape of this very narrow carbon 1*s* XPS line, which has the asymmetrical shape characteristic of pure carbon, is very sensitive to the chemical state of the carbon [43]. That these results are characteristic of carbon fibers is indicated by a survey of XPS fiber studies [44], as well as by direct comparison of PAN fibers with single-crystal graphite [45]. In general, the amount of oxygen adsorbed on untreated fibers is found to decrease with the fiber modulus. This correlates with the decrease in area of chemically active boundaries which must accompany the coalescence and growth of single-crystal domains (fibrils) at the surface. Thus, we may conclude with some confidence that the *exposed* surfaces of carbon fibers are, in general, almost completely basal plane and that the number of defect sites on the surface decreases with modulus.

While, on the atomic scale, high temperature heat treatments remove defects from single-crystal domains on the fiber surface, there is evidence that cool-down from these temperatures can introduce defects on a much larger macroscopic scale within the graphitic onion-skin sheath. Diefendorf and coworkers [23, 24] have pointed out that the rapid decrease of the tensile modulus of graphite with angular displacement out of the basal plane implies that the longitudinal tensile modulus

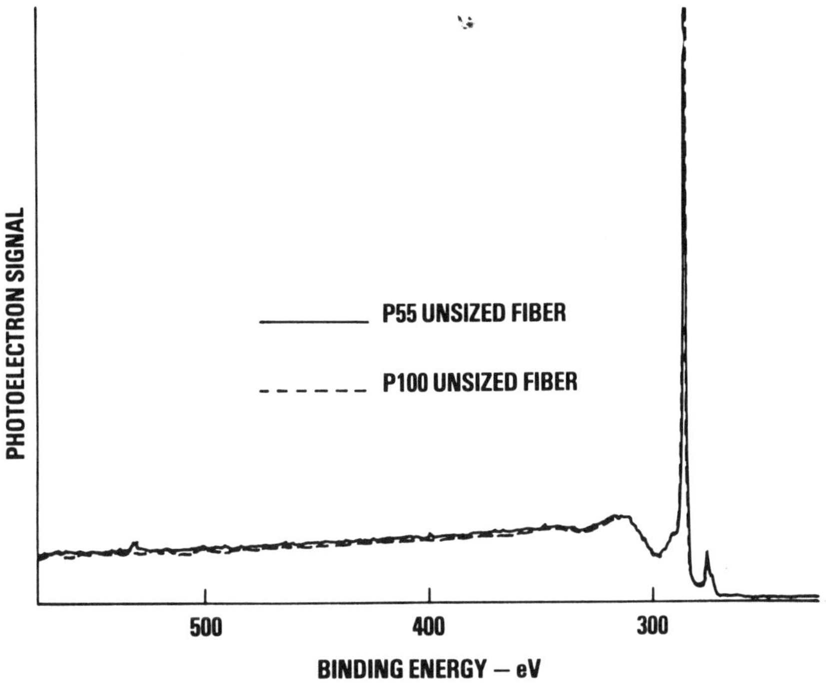

FIGURE 4 X-Ray photoelectron spectra of as-received P55 and P100 mesophase pitch fiber. Carbon and oxygen are the only elements detectable on these fibers, with 1.4 and 0.7% at. oxygen concentrations on the P55 and P100 fibers, respectively. (From Ref. 42.)

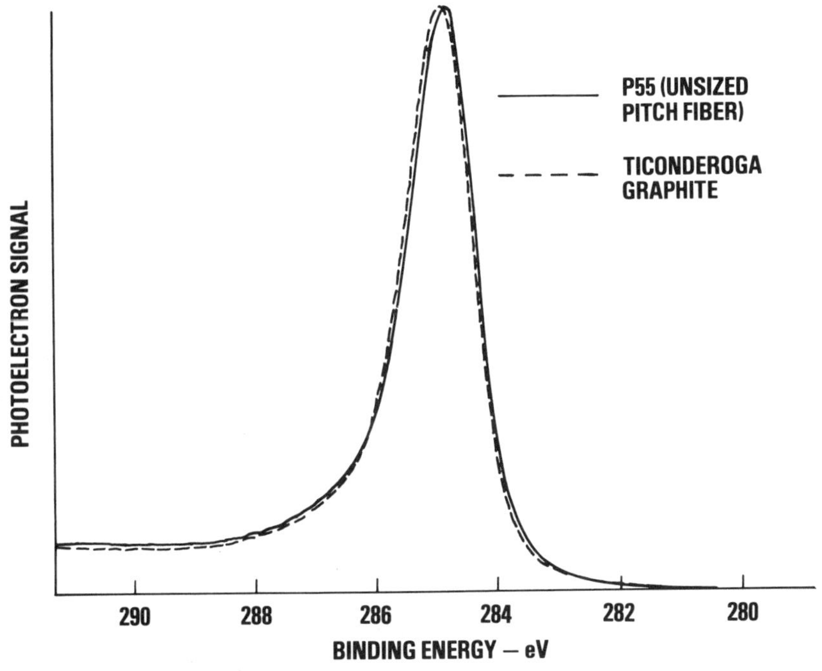

FIGURE 5 Carbon 1*s* X-ray photoelectron spectra of air-exposed P55 mesophase pitch fiber and cleaved single-crystal Ticonderoga graphite. The near-identical line shapes identify the carbon on the two surfaces as being in the same pure graphite chemical state. (From Ref. 42.)

will be smaller in the regions within the fiber where the basal plane ribbons have poor axial orientation and larger where highly developed graphitic planes are well oriented. Measurements for high modulus, surface-treated (HMS) PAN fibers indicate that the onion-skin surface layer has a modulus twice that of the average fiber modulus, while the modulus in the interior is only about half the fiber average. In effect, a PAN fiber is itself a minicomposite material in which the surface onion skin carries a large fraction of the load. Aside from the question of distribution of the load within the fiber, the anisotropies of the fiber can lead to thermal degradation. In particular, the onion-skin layers may fracture under the residual stresses of cool-down from processing temperatures. LeMaistre has analyzed an idealized cylindrical fiber in which residual stresses are created by the large difference in thermal coefficients of expansion of graphite in the "*c*" direction (normal to the basal plane) and the "*a*" directions (lying in the basal plane) [46]. Using an ultimate tensile strength of 5.5 MPa for the "*c*" axis, he calculated that cooling of an 8 μm diameter fiber from a 2500°C processing temperature would produce a circumferential crack in the onion-skin sheath 80 nm below the surface of the fiber. Circumferential cracks of this type have, in fact, been observed by Diefendorf and Tokarsky [24].

LeMaistre's calculations raise the general point that cool-down from processing temperatures introduces residual stresses that can easily exceed the strength of the highly anisotropic onion-skin sheath of the fiber. While circumferential cracking has been observed directly on fibers with relatively thick onion-skin surfaces, there is also substantial indirect evidence that the structural integrity of the basal plane sheath may be compromised by a network of microcracks. It was reported by Diefendorf and Tokarsky [24] that high modulus fiber surfaces are roughened by an unspecified abrasion process referred to as "wiggle-bugging." Examination showed that the roughness was due to 0.2 μm diameter particles, probably single crystals, being pulled from the surface. The loss of small particles from the surface has also been reported in connection with electrochemical treatments [44], where a dark "ink" commonly appeared at the fiber–solution interface when voltages were applied. Subsequent SEM examinations showed a significant increase in surface roughness. On the other hand, many of the results on surface structure have come from the examination of thin macroscopic flakes of graphitic carbon, which had delaminated from the fiber surface [24,47]. Such delamination is consistent with the expected low shear strength between the basal planes of graphite and may simply be the result of cleavage of well-developed single crystals in the onion-skin sheath. Thus, it remains a matter of speculation whether the small particles that separate from the fiber are defined by a microcrack network produced during cool-down or by the size of the surface fibril structure (i.e., single-crystal domains) in the onion-skin sheath.

The tendency of the onion-skin sheath to break up into small particles that are easily pulled out of the fiber surface appears to correlate with careful analysis of failure at the fiber–matrix interface. In carbon fiber composites, failure under transverse tensile loads invariably produces fracture surfaces that follow the contours of the fibers. That is, fracture surfaces parallel to the fibers always propagate around a fiber rather than through it, and fibers on the fracture surface are never split. Electron micrographs such as the one in Figure 6 illustrating this phenomenon, often have been presented as prima facie evidence for the conclusion that only mechanical bonding exists between the fiber and matrix. In fact, more detailed examinations of fracture surfaces indicate that this "obvious" conclusion is seldom correct. Scanning Auger microprobe (SAM) analysis of transverse fracture surfaces in carbon–aluminum composites [48], for example, provides conclusive evidence for strong fiber–matrix bonding in spite of the lack of fiber splitting. In these studies, the composite samples are fractured in the ultrahigh vacuum environment of the SAM instrument, and Auger analysis is carried out immediately thereafter to eliminate the possibility of surface contamination. The fracture surface shown in the electron micrograph of Figure 6 is typical of those obtained from composites with relatively good mechanical properties. There is no fiber splitting, and the impression that fracture occurs precisely at the fiber–matrix boundary is all the more credible for the fact that even the longitudinal striations characteristic of the fiber surface are replicated in the grooves left by the fibers on the matrix side of the fracture. However, Auger spectra from the fiber imprints on the matrix show the presence of a layer of graphitic carbon that could have come only from the fiber. At the maximum resolution (~ 100 nm) of Auger elemental mapping, the delaminated layer of graphite transferred

FIGURE 6 **Transverse fracture surface of a graphite–aluminum composite showing that the fracture path follows the fiber surface. The 10 μm diameter imprint of the fiber on the aluminum matrix (center trough) replicates the striations on the fiber.** (From Ref. 74.)

to the matrix side of the fracture always appears to be continuous, and there are no topographical features that are recognizable as flakes or particles of carbon on the matrix. Thus, only a very thin layer of carbon is transferred from the fiber, which is consistent with the fact that the onion-skin fiber sheath itself is usually too thin to be seen in a conventional scanning electron micrograph. Evidence that failure has occurred within the fiber surface is almost invariably found on the fracture surfaces of carbon–aluminum composites. The only exceptions that have been encountered are in inferior composites where failure occurs entirely within a weak, excessively thick brittle oxide interphase surrounding the fiber [48] and in composites with fiber–matrix bonding that is purely mechanical due to the method of manufacture [49]. Thus, at least for the mesophase pitch fibers used in metal matrix composites, the evidence is reasonably conclusive that the fiber–matrix bonds are usually stronger than the fiber surface structure itself and that interface failure generally occurs through failure in the onion-skin sheath of the fiber.

Inasmuch as PAN fibers also have an onion-skin sheath, one might expect that interface failure by fiber surface delamination also should occur in the polymer–matrix composites that employ this type of fiber. This expectation is borne out by Drzal's studies [50] of fiber adhesion in epoxy matrices. In this work, interfaces that had failed under longitudinal tension were ultramicrotomed and examined by transmission electron microscopy. The matrix side of fractures at interfaces with type HM fibers (heat treated at 2600°C for a tensile modulus of 357 GPa) were found to be covered with a relatively high density of small particles pulled from the fiber surface. This characteristic was observed both for surface-treated fibers and for untreated (HMU) fibers. With type A fibers (heat treated at 1500°C for a tensile modulus of 245 GPa), fiber particles were also observed on the epoxy side of the fracture for untreated (AU) fibers. However, at interfaces with type A fibers that had been subjected to proprietary oxidative surface treatment (AS) the fracture was strictly interfacial, with no transfer of fiber material to the epoxy side. One interpretation of the results on type A fibers is that only a rudimentary onion skin is formed during the relatively low temperature heat treatment given these fibers and that the onion skin is removed by surface treatments. While there are unresolved issues here concerning the function of surface treatments, the evidence is persuasive that the graphitic onion-skin sheath on carbon fibers is usually the weakest link at a well-formulated fiber–matrix interface, regardless of the fiber precursor or the nature of the matrix. (Early experiments by Donnet [47] support this conclusion for rayon fibers, as well.)

The picture of the fiber surface structure that has emerged from the available data is summarized in the schematic of Figure 7. Regardless of internal structure, all carbon fibers have at least a rudimentary onion-skin sheath which is composed of single-crystal fibrils whose graphitic basal planes are accurately aligned with the fiber axis. The exposed surfaces of the fibrils are basal planes of very high perfection, with little or no coverage

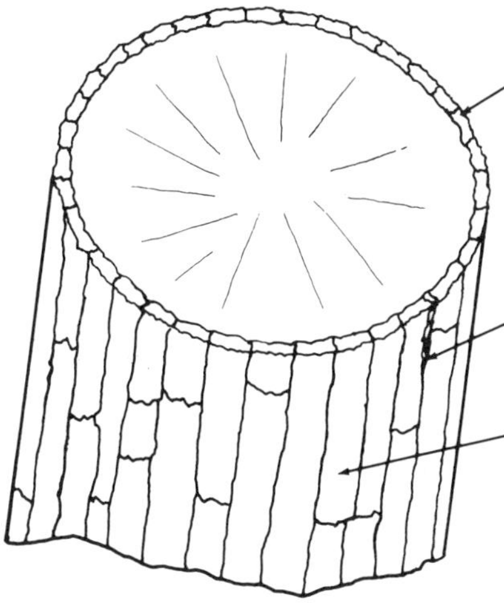

FIGURE 7 **Surface structure of a high modulus carbon fiber. The size of the basal plane domains increases with increasing modulus.**

by contamination or carbonaceous debris, while the chemically active areas of the fiber surface are limited to the boundary regions between fibrils. As a consequence of the growth and coalescence of fibrils during the heat treatments required to increase fiber modulus, the size of the single-crystal basal plane domains increases with fiber modulus and the relative area of the active regions decreases. Fibers processed at temperatures below 2000°C have onion-skin surfaces less perfect than is suggested by Figure 7. The surface fibril domains are much smaller on such fibers, and the disordered regions between fibrils constitute a larger fraction of the surface area. Moreover, the degree of crystalline perfection of the exposed basal planes is uncertain for fibers that have been annealed at temperatures substantially less than 2000°C.

Chemical Properties

The graphitic basal plane character of carbon fiber surfaces implies that it should be possible to understand the chemical behavior of the fibers in terms of the rather extensive literature on graphite chemistry. The strong similarity between the chemical reactivity of carbon fibers and that of single crystal graphite has been directly confirmed by studies of reactions with both aqueous and gaseous reagents [47,51–54]. In particular, the high yield of lamellar graphitic oxide from the reaction of fibers with Hummer's reagent (potassium permanganate in sulfuric acid) is convincing chemical evidence of the graphitic basal plane character of the fiber surfaces. (Graphitic oxide is a lamellar compound in which the carbon atoms retain the hexagonal planar arrangement of graphite in layers that are puckered by the sp^3 hybrid carbon–oxygen bond [55].) More reactive solid carbons produce carbon dioxide when treated with this reagent, and the yield of graphitic oxide correlates directly with the degree of perfection of a graphitic material.

One of the more prominent aspects of graphite chemistry is the anisotropy of chemical reactivity between the basal plane surface and the edge planes or pyramidal surface. While it was recognized at a very early stage that the basal plane surface is less reactive than the edge plane, decades of research on oxidation reaction rates were plagued by inconsistent and conflicting results. During the 1930s, values published for the reactivity anisotropy ratio (pyramidal surface to basal plane reaction rate) ranged from 2 to 25 [56–58]. Subsequent experimental values did not converge in this range and, by the early 1960s, reported values for the anisotropy were as high as 1000 [59]. The puzzle was solved by Hennig in an elegant series of experiments in which he demonstrated that the reactivity of the graphitic basal plane was almost completely determined by its defect concentration [40]. According to these results, defects at the atomic level are the chemically active sites on the basal plane surface and, in the absence of defects, the basal plane is almost completely inert. Hennig's methods allowed him to measure separately the rates for reactions that remove atoms from edges of basal planes and progress parallel to the basal plane ("parallel reactions") and reactions that remove atoms from the surfaces of perfect basal planes and progress perpendicular to the basal planes along the *c* axis ("perpendicular reactions"). The activation energy E_a and coefficient K_0 in the rate constant expression

$$K = K_0 \exp\left[-\frac{E_a}{RT}\right] \tag{1}$$

were measured for both the parallel and perpendicular reaction rates of a number of reagents. The values obtained for oxidizing gases are summarized in Table 1. These measurements show that anisotropy ratios of 10^{12} are typical of common oxidants and, for this type of oxidation reaction, the defect-free basal plane is one of the most inert substances known. The reader will note from these results that if the surface defect density is not measured and controlled at the atomic level, quantitative discrepancies as large as 10 orders of magnitude can easily occur between experiments purporting to measure the chemical properties of the graphitic basal plane.

According to Donnet [47], oxidative reagents can be classified in one of two categories depending on whether they attack disorganized regions of solid carbon and produce CO_2 or react with graphitic carbon to produce graphitic oxide. The reagents of Table 1, which are nonreactive with the basal plane, belong to the first class. Measurements have also been made [40] of the parallel and perpendicular reaction rates for oxidative solutions that are members of the class of graphitic oxide producers. The results differ dramatically from those for the CO_2 producers. Perpendicular reaction rates of 10^{-9} to 10^{-7} min^{-1} were measured at 100°C in these solutions, but parallel reactions could not be detected during a 1 hour reaction time.

The XPS spectra of bare carbon fibers were discussed above in the context of the finding that exposed carbon fiber surfaces consist almost entirely of graphitic basal planes. Gynn and King [44] have summarized the XPS studies of untreated and treated type A, HT, and HM fibers reported by a number of laboratories. Typically, up to 10% at. nitrogen and 5% at. sodium (due to incomplete reaction of the PAN) are found on the surfaces of type A fibers. The type A fibers also exhibit up to 8% at. oxygen on untreated fibers, whereas the untreated higher modulus type HM fibers have less than 5% at. oxygen and show little or no nitrogen or sodium. As we have discussed above, very high modulus pitch fibers exhibit only about 1% at. surface oxygen.

Interpretation of these results can be extended to some firm conclusions that should prove to be of significance in composite technology. The evidence that the exposed fiber surfaces consist of single-crystal basal planes is quite conclusive and, by virtue of Hennig's results and the thermal history of the fibers, the central areas of the basal plane domains on the surface are chemically inert toward most oxidants. Thus, a significant density of chemically active sites on the surface of a carbon fiber should occur only at edge planes or disorganized regions in the boundary zones between the single-crystal fibrils of the onion-skin sheath. We are therefore led to the conclusion that the attachment of functional groups to a fiber must be highly localized at the boundaries of the basal plane domains. By this reasoning, XPS measurements of the amount of reacted material on a fiber constitute measurements of the relative

TABLE 1
Oxidation Reaction Rates of Graphite

		Parallel Reaction			Perpendicular Reaction		
Reactant	T (°C)	K_0 (min^{-1})	E_a (kcal)	K (min^{-1})	K_0 (min^{-1})	E_a (kcal)	K (min^{-1})
Air	900	4×10^{17}	77	2.2×10^{3}	5.6×10^{6}	84	5.6×10^{-9}
Cl_2/O_2	723	7.1×10^{6}	16.7	1.7×10^{3}	2.5	46	2.5×10^{-10}
O_3	40	6×10^{18}	31.6	6.1×10^{-4}	5.1×10^{6}	18	1.4×10^{-6}
CO_2	1100	2.1×10^{16}	94	25		No reaction	

Compiled from Ref. 45.

areas of the boundary zones and the single crystal basal plane domains on the fiber surface. This view is strongly supported by the XPS result that the amount of oxygen on the (untreated) fiber surface decreases with increasing modulus. In this interpretation, the observed dependence of surface coverage on modulus is a direct consequence of the fact that the surface fibrils coalesce and grow, thereby decreasing the boundary-zone/basal-plane area ratio, during the heat treatments required to increase the modulus.

It is a generally held view that the chemical behavior of solid carbon surfaces is determined by the surface functional groups formed by chemical reaction of adsorbed atoms with the unsaturated covalent bonds of the carbon surface [55,60]. For this reason, the functional groups on carbon fibers have received a great deal of attention [43,61–65]. The identification of the surface groups produced by various reagents on solid carbon surfaces has proven to be very complex and is treated in detail in other articles of this encyclopedia. Specific surface groups that have been identified on solid carbons include carboxyl, carbonyl, lactone, quinonic, and hydroxyl groups [55,66–68]. Surface groups containing oxygen, which are the most important and most widely studied, can be either basic or acidic depending on the reaction temperature [69–72]. In view of Hennig's work, the chemistry of surface groups on graphitic carbons should clearly depend on the perfection of the basal planes as well as on the basal-plane/edge-plane ratio on the surface. Although there is great variability in these determining factors, there are very few experiments in which they are either controlled or measured.

Very little is known about the fiber chemistry relevant to metal–matrix composites. Certainly, the functional groups of concern to the polymer matrix community could not survive liquid metal infiltration and would have no direct effect on the interface bonding. At the same time, the bonding of metal to fiber via the carbide bond is undesirable for the carbide formers of interest [73]. While the carbide bond generally has the requisite strength, the carbides tend to hydrolyze and produce unstable composites. However, metals do not usually react directly with graphite on either the pyramidal or basal planes below the carbide formation temperature [40], so it has been necessary to use an intermediary in the establishment of the fiber–metal bond. Fortunately, the oxygen in the fiber sizing has turned out to be an appropriate intermediary [74], which provides bonding via an oxide interphase [48]. While oxygen is an adequate intermediary for aluminum and magnesium, it is not entirely appropriate for other metals [75]. A variety of fiber coatings have been tested in an attempt to improve interphase properties, with some success reported for a pitch coating [76]. Interpretation of fiber coating experiments has generally been uncertain, however, due to interference by extraneous oxygen [42].

The difficulties of directly investigating metallic bonding to mesophase pitch graphite fibers are unusually challenging. While there have been significant problems of reproducibility related to unannounced changes in fiber sizing formulations and quality control [74], the greatest difficulties are related to the lack of experimental accessibility of the interface bond. The only facts about the bond to the fiber surface that seem to have been firmly established are that bonding occurs if a sufficiently well-developed oxide is present at the fiber surface [77] and that there is no reactive dissolution of the fiber into the interphase in the absence of significant carbide formation. Model experiments that leave the graphite–oxide interface accessible for analysis have been difficult to formulate. The obvious approach would be to study the reactions of oxygen with a low coverage of metal particles on a graphite surface. However, it is characteristic of active metal particles on graphite that, in the presence of oxygen, they will simply catalyze the formation of CO_2 at edge plane sites and burn characteristic tracks across the basal planes [40]. No enduring graphite–oxygen bonds are formed in such model experiments, nor have catalytic tracks been observed on fibers at metal matrix composite interfaces. Thus far, graphite–oxygen–metal bonding has been observed only at buried interfaces, where the chemistry is difficult to analyze.

Surface Modifications

In the early phases of carbon fiber composite development, it was found that the attainable shear strengths of carbon fiber composites were only of the order of 30 MPa and decreased with modulus of the fiber [78]. Empirical solutions to the problems with polymer matrices were found in oxidative treatment of the fibers, but the results have not been entirely satisfactory and the process is still not understood. Initially, the two principal schools of thought regarding the function of oxidative surface treatments were (a) that etching of the fiber surface improved fiber–matrix adhesion by increasing the area for fiber–matrix bonding and (b) that the functional groups deposited on the surface by the oxidizing solution increased the strength of the chemical bonds to the matrix [79]. McKee and Mimeault [80] examined the evidence for these concepts and, finding little or no correlation with composite properties to support either rationale, they speculated:

> It is possible that the improved shear properties result not so much from increasing the interaction between matrix and reinforcement but from removing from the surface of the reinforcement "defects" that act as stress risers, causing failure to occur at less than optimum values.

We are now aware that interface failure is usually a cohesive failure in the fiber surface rather than an adhesive failure between the matrix and fiber, and it is not surprising that neither surface area nor the surface functional groups correlate very well with composite strength. However, the role of "defects" is a continuing issue. Chwastiak and Bacon [81], for example, have argued that surface functionalities comprise only a few percent of the surface on high modulus fibers and are so dilute that they are unlikely to have significant influence on the fiber–matrix adhesion. They further point out that fiber wettability is determined by the fiber sizing or "finish" and cannot, in principle, be affected by the polar free energy of the surface groups. They suggested, as an alternative, that the benefits of oxidative surface treatment are entirely due to the removal of a weakly bonded layer of graphite from the fiber surface. Also, in their work on the adhesion of PAN fibers to epoxy matrices, Drzal et al. [50] concluded that surface groups had a secondary effect but that some other mechanism was responsible for the permanent benefits of oxidation. They speculated that PAN fiber surfaces are very "defect laden" and proposed that the major benefits realized from surface treatment were due to removal of the defected surface layer.

While the removal of surface defects by the oxidative treatment has been a recurrent theme, the nature of the purported characteristic defects has not been made very clear. However, examination of the known properties of the fiber surface narrows the possibilities. To explain the observed inverse relationship between shear strength and modulus, it would seem to be necessary for the defect density to increase with increasing modulus. However, in terms of the atomic perfection of the basal planes and the macrostructure of the coalescing fibrils, we know that the fiber surface becomes more, rather than less, perfect with increasing modulus. The only remaining "defect" candidates would seem to be the macroscopic defects associated with microcracking of the onion-skin sheath or, as Drzal et al. and Chwastiak and Bacon have suggested, the weak interplanar bonding of the surface basal planes themselves. Either of these possibilities presents the problem of whether it is possible to remove such defects by an oxidative treatment without severely damaging the fiber. The oxidative reagents will attack the boundary regions much more rapidly than the basal planes, which can be removed from the fiber only by parallel reactions. Thus, we would expect the reaction to deeply notch the fiber at the boundary regions well before a significant number of basal planes have been removed from the single-crystal domains of the fibril surfaces. In high modulus PAN fibers, especially, a large fraction of the load is carried by the onion-skin sheath, and the conclusion seems unavoidable that, even if notching did not take place, removal of the basal planes (i.e., the onion skin) would seriously compromise the mechanical properties of the fiber.

The author has recently proposed an alternative mechanism, which eliminates the necessity of removing defects and appears to explain many of the heretofore mysterious characteristics of oxidative surface treatments [49,74]. The concept is based on the conclusion that the only places on the fiber surface at which bonding to the matrix can occur are at the chemically active boundary zones between the exposed single-crystal basal planes of the fibrils. With increasing modulus, not only does the number of such sites decrease, but the sites become farther apart. Thus, the structure of a high modulus fiber surface does not allow a uniform distribution of stress over the surface. It forces the load to transfer from the matrix in a stress concentration pattern that becomes more severe with increasing modulus because of the increase of fibril size with modulus. Moreover, regardless of whether the dominant "defects" are microcracks or the basal planes themselves, the stress concentrations will necessarily lie at boundaries where the surface structure is the weakest. If this picture is correct, the beneficial function of an oxidative surface treatment must be to relieve the stress concentration by moving active (edge) bonding sites into the interior of the basal plane domains. The chemical properties of the fiber surface imply that this is exactly what should happen. Parallel reactions should erode the edges of the topmost basal planes and create chemically active steps on previously inert areas of the basal plane domains. By this means, the oxidation reaction should both increase the number of bonding sites (as is observed in XPS studies) and distribute the load more uniformly. However, with very high modulus fibers, in particular, fiber notching by the more rapid simultaneous oxidation at the boundaries will limit the benefits that can be achieved.

Conclusion

Carbon fibers are unique among composite reinforcement materials in that their behavior is dominated by an extreme anisotropy in both the mechanical and chemical properties of the most basic structural unit—the graphitic basal plane. This distinction is even more pronounced because of the onion-skin sheath that forms on all carbon fibers and presents a graphitic basal plane surface to the outside world. By virtue of the thermal history of the fiber and the properties of the basal plane, the onion-skin surface imposes a degree of heterogeneity and lack of chemical reactivity that is unparalleled in other reinforcing fibers. While the evidence for the onion-skin fiber sheath and the basal plane nature of the exposed surface of the fiber is conclusive, the perfection of the basal plane domains on a high modulus carbon fiber surface has not been directly demonstrated, nor has it been directly verified that chemical bonding to the fiber occurs only at domain edges. However, it is a logical consequence of well-established properties of graphite that carbon fiber surfaces should have these characteristics. Moreover, highly localized bonding at domain boundaries would systematically create the stress risers that McKee and Mimeault attributed to "defects" and would provide an explanation for many of the mysteries of carbon fiber surface treatments and the relationships between fiber modulus and composite interface behavior.

The model of the carbon fiber surface as a mosaic of inert basal plane domains separated by chemically active boundary regions raises a number of interesting questions about the validity of applying conventional analysis to composites reinforced with carbon fibers. Virtually all the mechanical theories of composites assume that the reinforcement is homogeneous and that bonding between matrix and reinforcement is continuous. While such assumptions may be appropriate to carbon fibers such as type AS, which appear to have little onion-skin character, there should be some stage of development of the surface domain structure at which discrete bonding

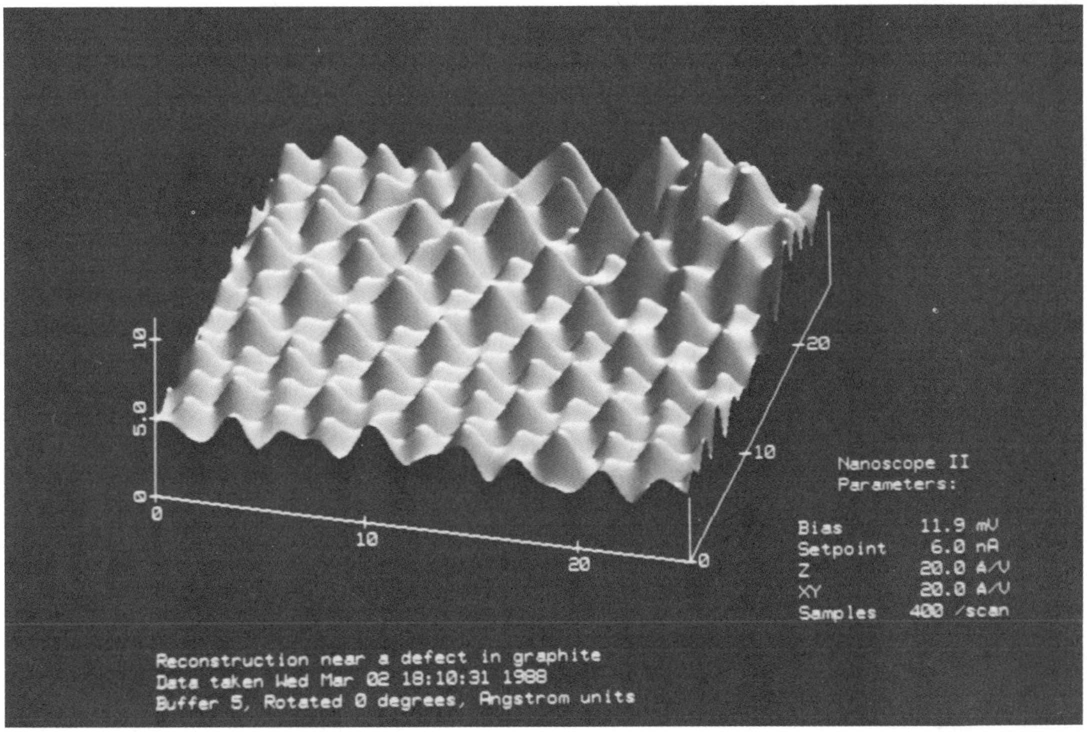

FIGURE 8 Scanning tunneling microscope image of a defect in a graphite lattice. (NanoScope II image courtesy of Dr. Virgil Elings, Digital Instruments, Inc., Santa Barbara, CA.)

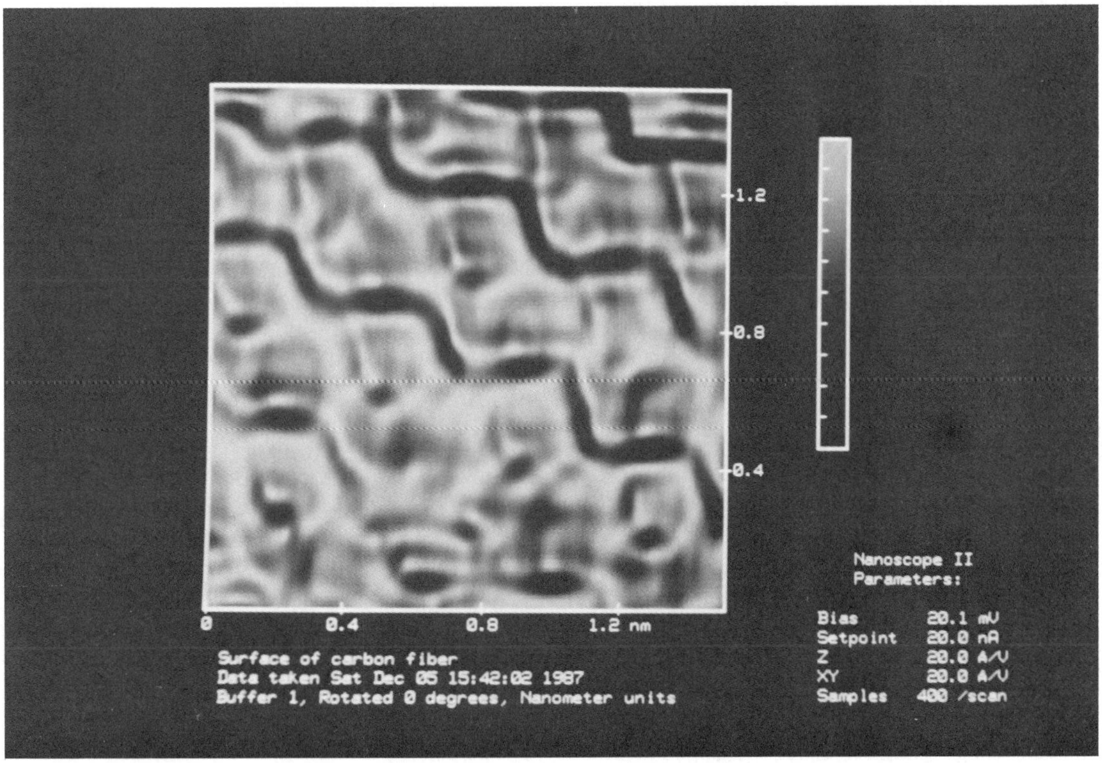

FIGURE 9 Scanning tunneling microscope image of a Thornel P55 carbon fiber surface. (NanoScope II image courtesy of Dr. Virgil Elings, Digital Instruments, Inc., Santa Barbara, CA.)

at stress concentrations begins to dominate interface behavior and the continuous-bonding interface model is no longer qualitatively correct. The meaning of such basic concepts as the "critical load transfer length" [82], for example, may have to be reexamined when applied to composites employing high modulus fibers. By the same token, finite element analyses of composites normally employ meshes and parameters appropriate to continuous bonding of the matrix to a uniform reinforcement and may not be qualitatively correct when applied to high modulus carbon fiber composites. Extending the scope of finite element calculations to address the possibility of load transfer occurring only at stress concentrations along the boundaries of onion-skin domains would be a significant challenge. Aside from the technical difficulties of carrying out such a three-dimensional calculation, there may also be a question of the limits of applicability of continuum mechanics if the weakest element of the surface turns out to be the interplanar bonding between surface basal planes at the fiber surface.

Although there have been significant advances in the understanding of carbon fiber surfaces since McKee and Mimeault speculated about the existence of stress risers on the surface, our understanding of the fiber surface still falls short of being fully adequate to control and optimize interfaces. The critical gaps in knowledge are primarily related to the detailed defect structure of the surface and its implications with respect to chemical and mechanical behavior. To develop valid theories for carbon fiber interfaces, it will be necessary to identify the weak element of the onion-skin sheath associated with the observed cohesive failure of the fiber surface and to determine whether bonding is indeed localized at the edges of surface single-crystal domains, as the indirect evidence so strongly indicates. The delicacy and dimensional scale of these aspects of the fiber surface make them very difficult to address experimentally. One of the more promising experimental approaches to these problems may be through the use of the scanning tunneling microscope (STM) [83], which has been widely applied to atomic imaging of graphite [84] (Fig. 8). STM images of carbon fiber surfaces have been obtained (Fig. 9) that have the symmetry of the graphitic basal plane, but systematic studies of carbon fiber surface structure by STM have yet to be reported. Commercial STM instruments now available are capable of producing images ranging from atomic resolution up to areas of square micrometers. Sophisticated tools such as the STM should make it possible, in the near future, to obtain the information about surface defect structure and chemistry required to formulate valid quantitative theories of carbon fiber interfaces.

Ronald N. Lee

References

1. W. N. Reynolds, *Chem. Phys. Carbon, 11,* 1 (1973).
2. D. W. McKee and V. J. Mimeault, *Chem. Phys. Carbon, 8,* 151 (1973).
3. J. D. Bernal, *Proc. R. Soc. (London) Ser. A, 106,* 749 (1924).
4. I. L. Spain, *Chem. Phys. Carbon, 8,* 1 (1973).
5. J. B. Donnet and A. Voet, *Carbon Black,* Dekker, New York, 1976.
6. G. M. Jenkins and K. Kawamura, *Polymeric Carbons—Carbon Fibre, Glass and Char,* Cambridge University Press, New York, 1976.
7. W. N. Reynolds, *Physical Properties of Graphite,* Elsevier, New York, 1968.
8. E. L. Piper, Society of Mining Engineers, AIME, Preprint 73-H-14, 1973.
9. O. L. Blakslee, D. G. Proctor, E. J. Seldin, G. B. Spence, and T. Weng, *J. Appl. Phys., 41,* 3373 (1970).
10. R. Bacon, *J. Appl. Phys., 31,* 283 (1960).
11. R. Bacon and M. M. Tang, *Carbon, 2,* 211 (1964).
12. R. Bacon and M. M. Tang, *Carbon, 2,* 220 (1964).
13. A. Shindo, *Osaka Kogyo Gijitsu Shikuko Kiko, 12,* 110 (1961).
14. A. Shindo, *Carbon, 1,* 391 (1964).
15. W. Johnson, L. N. Phillips, and W. Watt, British Patent 1,110,791 (April 24, 1968), to National Research Corp.
16. W. Watt, L. N. Phillips, and J. Johnson, *Engineer, 221,* 815 (1966).
17. A. E. Standage and R. Prescott, *Nature, 211,* 169 (1966).
18. L. S. Singer, "High Modulus Carbon Fiber from Pitch," Belgian Patent 797,543 (1973).
19. S. Chwastiak, J. B. Barr, and R. Didchenko, *Carbon, 17,* 49 (1979).
20. The availability of commercial fibers produced from highly purified mesophase pitch has been announced by the Textile Fibers Department of E. I. du Pont de Nemours & Company, Wilmington, Delaware, and by the Carbon Fiber Project Group of Toa Nenryo Kogyo K. K. (Tonen), Tokyo, Japan.
21. A. Fourdeux, R. Perret, and W. Ruland, Paper 9 at International Conference on Carbon Fibers, Plastics Industry, London, 1971.
22. A. Fourdeux, C. Herinckx, R. Perret, and W. Ruland, *C. R. Acad. Sci., 269,* 1597 (1969).
23. R. J. Diefendorf and E. W. Tokarsky, *Polym. Eng. Sci., 15,* 150 (1975).
24. R. J. Diefendorf and E. W. Tokarsky, "The Relationship of Structure to Properties in Graphite Fibers," Air Force Materials Laboratory Report AFML-TR-72-133, Part III, November 1975.
25. P. Kwizera, M. Dresselhaus, D. R. Uhlman, J. S. Perkins, and C. R. Desper, *Carbon, 20,* 387 (1982).
26. I. C. Lewis and L. S. Singer, *Proceedings of the 12th Biennial Conference on Carbon,* 1975, p. 265.
27. H. Honda, M. Kimura, Y. Sanada, S. Sugawara, and T. Furuta, *Carbon, 8,* 181 (1970).
28. S. Chwastiak and I. C. Lewis, *Carbon, 16,* 156 (1978).
29. S. Chwastiak, R. T. Lewis, and J. D. Ruggiero, *Carbon, 19,* 357 (1981).
30. K.-H. Kochling, B. McEnaney, F. Rozploch, and E. Fitzer, *Carbon, 20,* 445 (1982).
31. D. W. McKee and V. J. Mimeault, *Chem. Phys. Carbon, 8,* 179 (1973).

32. J. A. Hugo, V. A. Phillips, and B. W. Roberts, *Nature, 226,* 144 (1970).

33. R. H. Knibbs, *J. Microsc. 94,* 273 (1972).

34. J. V. Larsen, "Carbon Fiber Structure, Surface, and Surface Treatments," Ph.D. dissertation, University of Maryland, College Park, July 21, 1971.

35. G. K. Layden, *Carbon, 10,* 59 (1972).

36. P. Ehrburger, C. Bourgeois, and J. Lahaye, *Carbon, 19,* 1 (1981).

37. P. Ehrburger, C. Bourgeois, and J. Lahaye, *Carbon, 19,* 7 (1981).

38. F. R. Barnet and M. K. Norr, *Carbon, 11,* 281 (1973).

39. J. B. Barr, S. Chwastiak, R. Didchenko, I. C. Lewis, R. T. Lewis, and L. S. Singer, *Appl. Polym. Symp., 29,* 161 (1976).

40. G. R. Hennig, *Chem. Phys. Carbon, 2,* 1 (1966).

41. G. R. Hennig, *Appl. Phys. Lett., 4,* 55 (1964).

42. F. Santiago, A. N. Mansour, and R. N. Lee, *J. Surf. Interface Anal., 10,* 17 (1987).

43. A. Proctor and P. M. A. Sherwood, *J. Electr. Spectr. Rel. Phenom., 27,* 39 (1982).

44. G. Gynn and R. N. King, "Improved Graphite Fiber Adhesion," Ashland Chemical Company Technical Report, Final Report of Air Force Contract F33615-79-C-5060, Air Force Wright Aeronautical Laboratories (AFSC), July 7, 1981.

45. I. A. Ishitani, *Carbon, 19,* 269 (1981).

46. C. W. LeMaistre, "The Origin of Structure in Carbon Fibers," Ph.D. dissertation, Rensselaer Polytechnic Institute, Troy, NY, September 1971.

47. J. B. Donnet, *Carbon, 20,* 266 (1982).

48. H. L. Marcus, D. L. Dull, and M. F. Amateau, "Scanning Auger Analysis of Fracture Surfaces in Gr–Al Composites," Aerospace Report TOR-0078(3762-03)-2, The Aerospace Corporation, El Segundo, CA, Aug. 25, 1978.

49. R. N. Lee, *Proceedings of the Nondestructive Testing and Evaluation of Advanced Materials and Composites Conference,* U.S. Air Force Academy, Colorado Springs, CO, Aug. 19–21, 1986, p. 157.

50. L.T. Drzal, M. J. Rich, and P. F. Lloyd, *J. Adhes., 16,* 1 (1982).

51. P. Ehrburger, J. J. Herque, and J. B. Donnet, in M. L. Deviney and T. M. O'Grady, Eds., *Petroleum-Derived Carbons,* ACS Symposium Series 21, American Chemical Society, Washington, DC, 1976.

52. J. B. Donnet and P. Ehrburger, *Proceedings of the 4th Intl. Carbon and Graphite Conference,* Society of the Chemical Industry, 1976, p. 384.

53. P. Ehrburger and J. B. Donnet, *Carbon, 15,* 143 (1977).

54. P. Ehrburger and J. B. Donnet, *Phil. Trans., R. Soc. London, A294,* 495 (1980).

55. H. P. Boehm, "Chemical Identification of Surface Groups," in D. D. Eley, H. Pines, and P. B. Weisz, Eds., *Advances in Catalysis,* Vol. 16, Academic Press, New York, 1966.

56. E. N. Green and B. Topley, *Nature, 129,* 904 (1932).

57. U. Hofmann, *Chem. Ber., 65B,* 1821 (1932).

58. U. Hofmann, *Z. Elektrochem., 42,* 504 (1936).

59. G. R. Hennig, *Z. Elektrochem., 66,* 629 (1962).

60. J. B. Donnet, *Carbon, 6,* 161 (1968).

61. C. R. Thomas and E. J. Walker, *Proceedings of the 5th London Carbon Graphite Conference,* Society of the Chemical Industry, 1978, p. 520.

62. P. Ehrburger, J. J. Herque, and J. B. Donnet, *Proceedings of the 5th London Carbon Graphite Conference,* Society of the Chemical Industry, 1978, p. 398.

63. E. Fitzer, K. H. Geigl, W. Huttner, and R. Weiss, *Carbon, 18,* 389 (1980).

64. A. Proctor and P. M. A. Sherwood, *Surf. Interface Anal., 4,* 212 (1982).

65. J. Harvey, C. Kozlowski, and P. M. A. Sherwood, *J. Mater. Sci., 22,* 1585 (1987).

66. J. B. Donnet and C. Reitzer, *Cah. Rheol., 6,* 5 (1962).

67. V. A. Garten and D. E. Weiss, *Rev. Pure Appl. Chem., 7,* 69 (1957).

68. M. L. Studebaker, E. W. D. Huffman, A. C. Wolfe, and L. G. Nabors, *Ind. Eng. Chem., 48,* 162 (1956).

69. I. Ogawa, *Biochem. Z., 172,* 249 (1926).

70. H. R. Kruyt and G. S. de Kadt, *Kolloid-Z., 47,* 44 (1929).

71. I. M. Kolthoff, *J. Am. Chem. Soc., 54,* 4473 (1932).

72. A. King, *J. Chem. Soc.,* 1498 (1937).

73. R. N. Lee, Ed., *Proceedings of Workshop on Transverse Strength in Carbon-Fiber/Aluminum Composites,* NSWC Miscellaneous Publication NSWC-MP-84-258, Naval Surface Weapons Center, Silver Spring, MD, April 1984.

74. R. N. Lee, M. K. Norr, and F. Santiago, *Gr/Al Interface Studies: MMC MANTECH Task Interim Report,* NSWC Technical Report, TR 82-314, April 15, 1982, Naval Surface Weapons Center, Silver Spring, MD.

75. H. A. Katzman, "A New Fabrication Technique for Graphite–Copper Composites," The Aerospace Report TOR-0086(6726-02)-1, 1985.

76. H. A. Katzman and G. L. Steckel, "Transverse Strength Properties of Graphite–Aluminum Composites," Aerospace Report TOR-0083(3726-01)-1, The Aerospace Corporation, El Segundo, CA, Oct. 10, 1982.

77. S. D. Tsai, M. Schmerling, and H. L. Marcus, "The Interface Structure in Graphite/Aluminum Composites," University of Texas, Austin, Report, Order AD-A096627, 1980.

78. J. C. Goan and S. P. Prosen, in A. G. Metcalfe, Ed., *Interfaces in Composites,* ASTM STP 452, American Society for Testing and Materials, Philadelphia, 1968.

79. J. W. Herrick, P. E. Gruber, Jr., and F. T. Mansur, "Surface Treatments for Fibrous Carbon Reinforcements," AFML-TR-66-178, Part I, Air Force Materials Laboratory, July 1966.

80. D. W. McKee and V. J. Mimeault, *Chem. Phys. Carbon, 8,* 202 (1973).

81. S. Chwastiak and R. Bacon, "Carbon Fibers for Advanced Composites," Symposium on Advances in Composite Materials, Polymer Division, 182nd ACS Meeting, 1981.

82. G. A. Cooper and A. Kelly, "Role of the Interface in the Fracture of Fiber–Composite Materials," in A. G. Metcalfe, Ed., *Interfaces in Composites,* ASTM STP 452, American Society for Testing and Materials, Philadelphia, 1968, p. 90.

83. G. Binnig, H. Rohrer, C. Gerber, and E. Weible, *Appl. Phys. Lett., 40,* 178 (1982).

84. See, e.g., many papers in R. Feenstra, Ed., *Proceedings of the 2nd Intl. Conference on Scanning Tunneling Microscopy,* July 20–24, 1987, American Vacuum Society, New York, 1988.

Casting

See Metal Matrix Composites, Casting

Cellular Materials

See Core Composite and Sandwich Structures

Centrifugal Casting

See Metal Matrix Composites, Casting

Ceramic Fiber Tiles*

Ceramic reusable surface insulation materials and borosilicate coatings are used in the fabrication of the space shuttle orbiters. When NASA's space shuttle orbiters reenter earth's atmosphere from space, they encounter temperatures as high as 1260°C. To protect the orbiters and the astronauts from such intense heat, an ultrapure silica fiber insulation has been developed: this material transfers heat so slowly that a piece of it can be held by the edges with a bare hand only seconds after removal from an extremely hot oven (Fig. 1). In those few seconds, heat has rapidly escaped from the surface of the material, but fiery interior heat has not yet moved to the outside to replace it.

Conversely, the black reflective coating applied to silica "tiles" allows 90% of the searing heat generated by reentry to be radiated back into the atmosphere. Several hours are needed for the heat trapped inside the material to work its way out through the silica fibers. Consequently, temperatures on the shuttle's aluminum skin never exceed the structural design limit of 177°C.

The shuttle is far from the solid, rigid craft it appears to be in Figure 2. The aerodynamic pressures and vibrations experienced during launch cause the orbiter to bend and twist substantially. In addition, the contraction experienced in the cold temperatures of space and the subsequent expansion in the heat of reentry cause significant flexing of the vehicle. For these reasons, designers decided to use tiles instead of larger panels, which would have been more rigid and more likely to crack. To isolate the tiles from the orbiter's movements, they are bonded to a synthetic felt called a strain isolation pad (SIP), which in turn is bonded to the shuttle's skin. Each orbiter requires nearly 31,000 tiles—each a different size, shape, and thickness.

The basic raw material for the reusable insulation is a 99.7% pure silica fiber that is derived from high quality sand. The insulation made from the fiber was selected

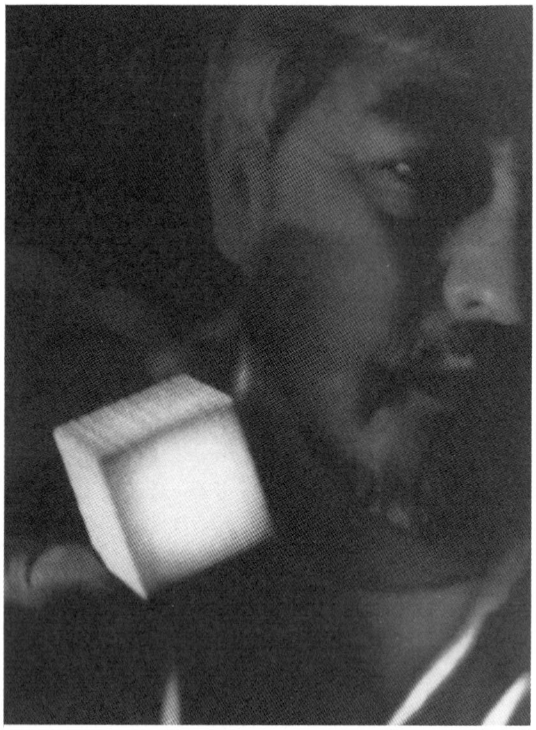

FIGURE 1 White hot at 2268°C, the glow from a cube of fiber ceramic tile insulation held in the technician's bare hand provides the only illumination in this photo.

FIGURE 2 The space shuttle Columbia with reusable surface insulation after mission.

* The information presented was taken with permission from a paper by R. P. Banas, D. R. Elgin, E. R. Cordia, K. N. Nickel, E. R. Gzowski, and L. Aguilar [1] from a fact sheet published by Lockheed Missiles & Space Company [2].

for the shuttle because of its light weight, low coefficient of thermal expansion, and high temperature stability.

The tiles accomplish three major tasks:

1. They dissipate 90% of the reentry heat energy in reradiation back to the earth's atmosphere and deep space from the reaction-cured glass (RCG) coating, which is only 0.30 mm thick. The energy is transformed into surface heating at incandescent temperatures and is dissipated because of the high emissivity inherent in the coating's optical properties. The RCG coating forms the aerodynamic skin of the orbiter vehicle and rapidly heats to 1093°C for 10 minutes during reentry. The peak heating rate typically is 283,913 J/m^2 during this critical period.

2. They provide the unique fused silica fiber insulation structure, which allows a rapid surface temperature response and supports the glass aerodynamic skin of the vehicle. The structural insulation delays in-depth temperature response and takes the thermal shock and acoustic impulses. Each fused silica tile transmits the aerodynamic loads into the orbiter airframe structure through a strain isolation pad. About 5% of the incident energy is dissipated by convection back to the cooler atmosphere during letdown for a landing.

3. They delay the remaining 5% incident energy from reaching the aluminum orbiter airframe structure until after the landing is completed. The low thermal diffusivity, a unique product of density, specific heat, and thermal conductivity, delays this heat pulse typically for 30 minutes, while the temperature of the aluminum structure never exceeds 177°C.

The unique material is manufactured in two forms, LI-900, 0.145 g/cm^3 (Lockheed Insulation) and LI-2200, 0.352 g/cm^3. Since LI-2200 is denser and stronger than LI-900, the former is used to cover access doors, the main landing gear door, and other areas of the orbiters that require tougher insulating material. Together, the two types of material cover 70% of the exterior of Columbia and Challenger, the first two vehicles of the shuttle orbiter fleet.

Beginning with Discovery, the third shuttle, LI-2200 was replaced with a new material that reduces each orbiter's weight by about 454 kg. The fibrous refractory composite insulation material called FRCI-12 weights 0.192 g/cm^3 and is just as strong as the heavier LI-2200. FRCI-12 is composed of 80% pure silica fibers and 20% of a 3M Company fiber called Nextel, which contains a small amount of boron that welds the pure silica fibers and the Nextel fibers into a rigid structure during high temperature sintering in a furnace. The result is a lighter material with strength equal to the denser and more loosely structured LI-2200.

Background

The space shuttle thermal protection system (TPS) evolved over many years and resulted from the contributions and dedication of many individuals and organizations, including Rockwell International Corporation, the NASA research centers, and Lockheed Missiles & Space Company, Inc. (LMSC).

During the early 1960s, LMSC conducted research and development of specially designed and integrated ceramic composite structures based on inorganically bonded ceramic fibers. These structures included compatible inorganic binders for both continuous-filament-wound fibers and cast staple fibers. Centrifugal slurry-casting techniques were demonstrated as well as conventional slurry-casting tower processes for the fibrous class of materials.

As early as 1962 a conical Apollo radome was built of fused silica composites to withstand the reentry environment (Fig. 3). The external shell of this radome was filament-wound with pure silica fiber. A lightweight internal insulation was centrifugally cast from short staple silica fiber, and the two sections were densified with colloidal silica and sintered into a composite structure. This early fiber ceramic reinforced composite structure was 100% inorganic and nonablative. Initial attempts at inorganic ceramic composites designed specifically for ballistic and lifting entry thermal protection were limited to high density solid refractories such as zirconia used in Dyna-soar.

An early LMSC development designated Lockheat was based on the creation of a low density, inorganically bonded refractory fiber skeleton structure having controlled porosity and microstructure. These structures were impregnated with noncharring organic coolants to achieve a transpiration cooling function at extreme entry heat flux rates. A large matrix of alternative fiber and binder materials was explored in the early years, including metallic and inorganic metal oxide fibers and a variety of metal oxide binders. Silica, alumina, and boria re-

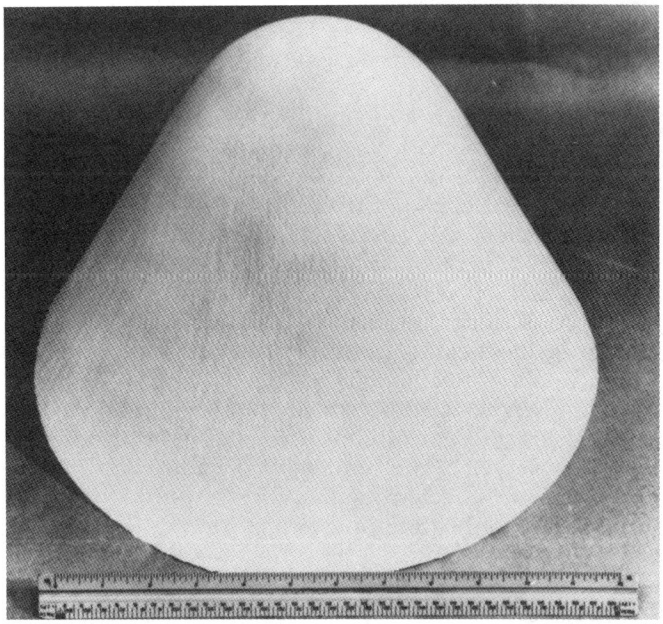

FIGURE 3 **This conical Apollo radome was built of fused silica composites in the early 1960s. This early ceramic structure was 100% inorganic and nonablative.**

FIGURE 4 **During tests of various thermal protective materials on NASA's Pacemaker reentry vehicle, Lockheed LI-1500 (lower portion of body to right of centerline) was the only one to survive 1260°C reentry temperatures unscathed.**

mained the principal components carried over into today's materials. Multilayered composites of pure silica filament-wound and staple fiber materials were produced with coolant impregnants tailored to meet various specific weights in the range of 12.2–24.4 kg/m^2. Very acceptable backface temperature response was demonstrated during the early 1960s in reentry simulation tests of typical composite heat shield designs.

In 1965 a specific formulation of silica-bonded staple fibers was selected for development to meet the requirements for lifting entry applications. This silica formulation was LI-1500 at a density of 0.24 g/cm^3. Initially, the 89% porous silica composite was impregnated for protection against heating from above the "melting point" of silica. "Cooling" resulted at high heating rates because of the endothermic decomposition and transpiration of the selected organic constituent. After the organic component was depleted, the remaining silica fiber insulation provided continued protection for the duration of the more moderate heating phases (Fig. 4).

Materials

LI-900 Tiles

The major goal was to obtain a material that resists devitrification at elevated temperatures. The principal component in LI-900 consists of all-amorphous silica fibers (Q-fiber) with an average diameter of 1.2–1.4 μm, lengths to 0.64 cm, and greater than 99.7% purity. These fibers retain their amorphous structure when exposed to a temperature environment of 1357°C for extended periods. The LI-900 system contains a colloidal silica binder that requires extensive treatment to obtain the purity required for high temperature morphological stability (Fig. 5).

Maintaining uniform shrinkage characteristics was realized by heat treating the fiber before processing it into billets. In addition, unfiberized glass called "shot," if not removed, caused high density, devitrified inclusions in the sintered material. To eliminate the "shot," the fiber was slurried with deionized water and passed through a hydrocyclone cleaner. The cleaned fiber slurry is transferred into a centrifugal extractor to remove excess water and to form a fiber "cake," in preparation for final drying.

LI-2000 is composed of amorphous silica (Q-fiber) and a small amount of silicon carbide powder, which provides additional thermal protection if the material is exposed to excessive temperatures as a result of coating loss at the tile outer surface. With LI-2200, the fiber heat treatment is omitted, and only the hydrocyclone cleaning is performed.

FRCI-12 is a composite fiber material containing amorphous silica (Q-fiber) and aluminum borosilicate fibers (Nextel 312) in a fused fiber matrix. Silicon carbide powder is added for additional thermal protection as it is in LI-2200. The bulk silica and Nextel fibers are heat treated at 1204°C and 1093°C, respectively, to stabilize and standardize fiber properties. Hydrocyclone cleaning is performed on the silica fibers to remove particulate contaminants, followed by drying (see Fig. 6).

CLASS 1. The optical property requirements for a class 1 (white) coating include a ratio of solar absorptance to total hemispheric emittance between 0.2 and 0.4, to achieve low temperatures while on orbit, and an emittance of 0.7 at 649°C to allow maximum reradiation of the convective heating energy during reentry. A dual layer coating (0036B) consisted of a fused, water-impervious top coat of clear glass plus zinc oxide, over a porous subcoat that contained aluminum oxide for high reflectance and silicon carbide for high emittance. The subcoat required drying at 704°C before spraying the top coat. After both layers were applied, glazing at 1148°C was required to produce a water-impervious, dual-layer coating (Fig. 7).

A single-layer coating (0036C) was successfully developed, qualified, and implemented into production in early 1978. During this period, the maximum coating weight requirement was increased from 0.439 to 0.586 kg/m^2 to alleviate a coating cracking problem that was unavoidable with a 0.02 cm thick coating.

CLASS 2. The class 2 (gray) coating (0050) was a dual-layer system that contained Corning 7930 frit with a boria content of 8% and a silicon carbide emittance agent. Because of a high coating residual tensile strain, values of 200–300 μin./in., crack propagation was not inhibited. This coating was replaced with a single-layer system that met the optical property requirements and also with a

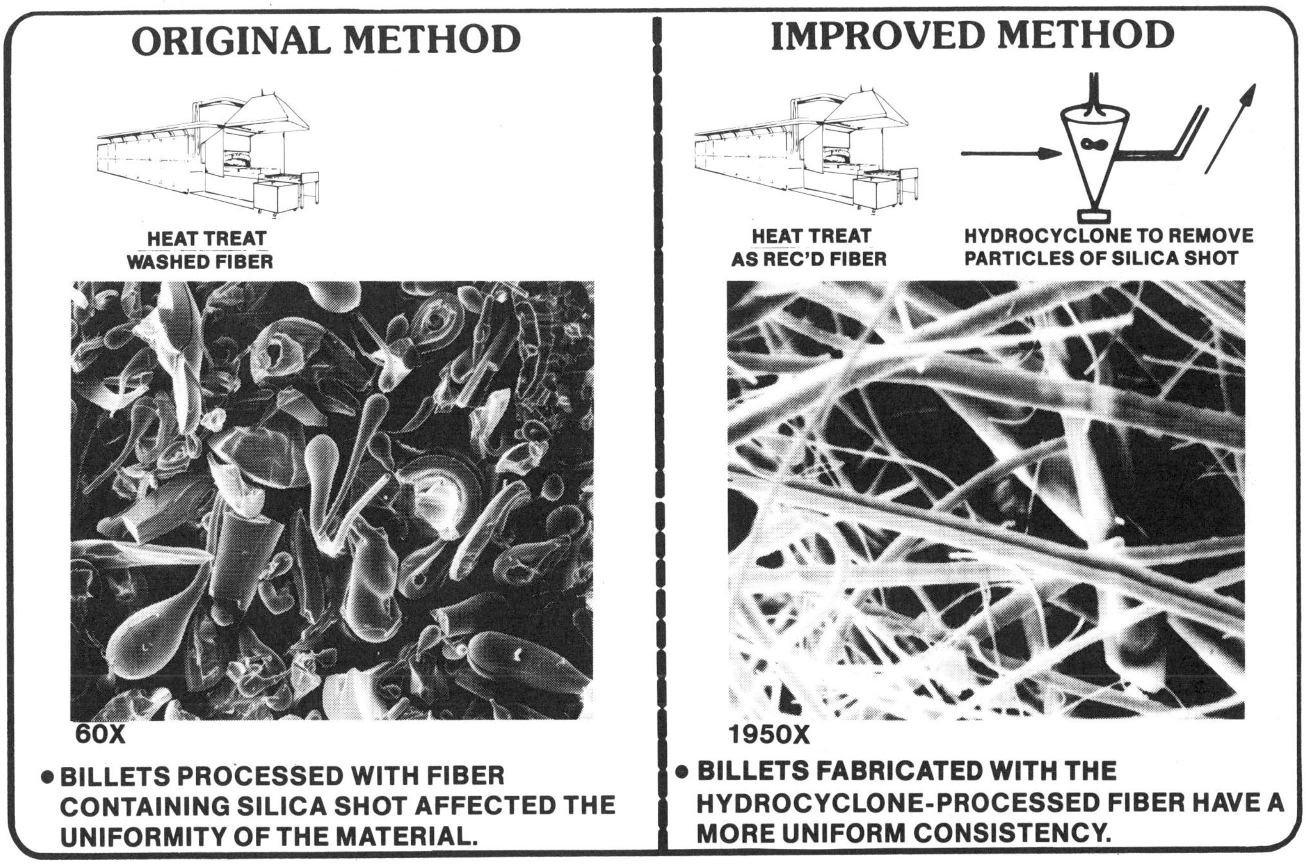

FIGURE 5 LI-900 fiber pretreatment. (Courtesy of Lockheed Missiles & Space Company, Inc.)

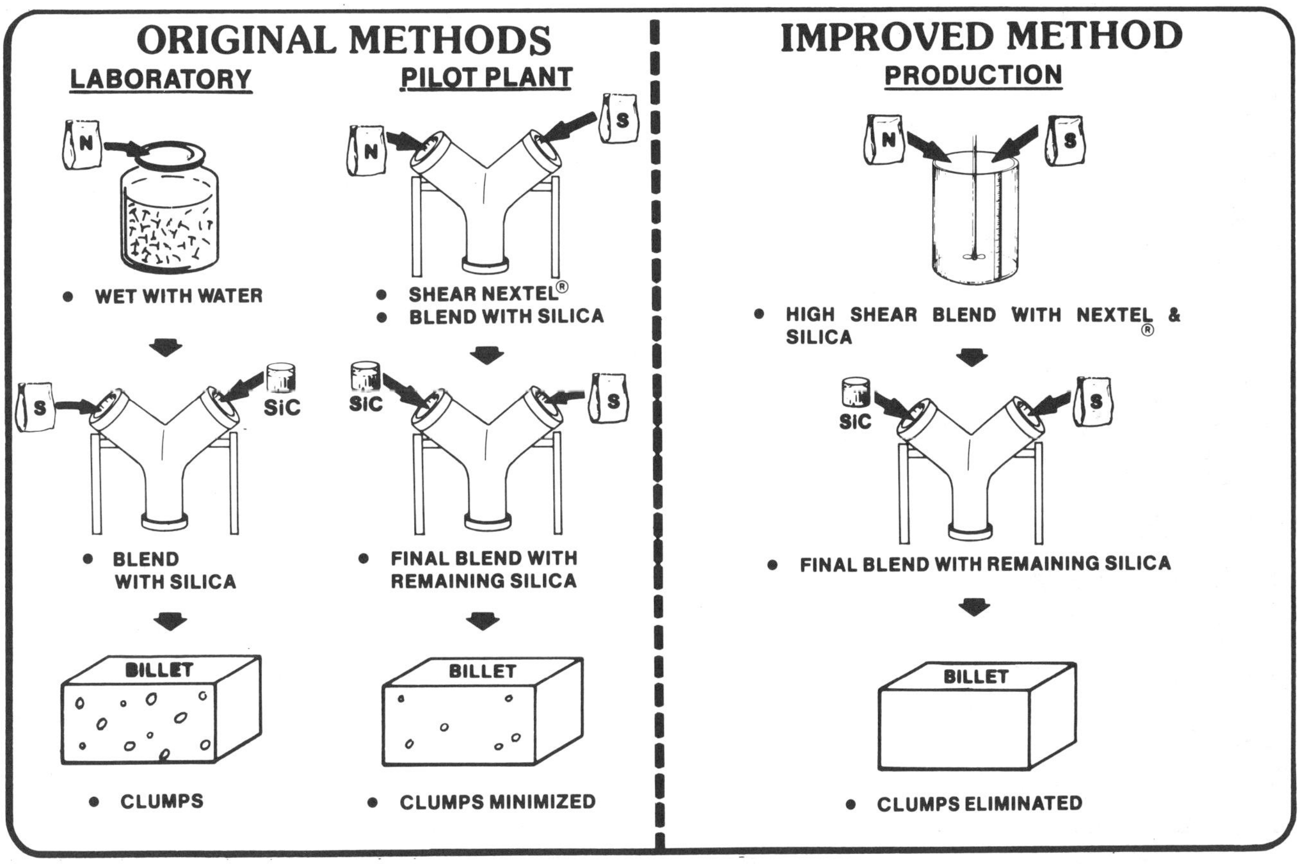

FIGURE 6 Nextel methods for FRCI-12. (Courtesy of Lockheed Missiles & Space Company, Inc.)

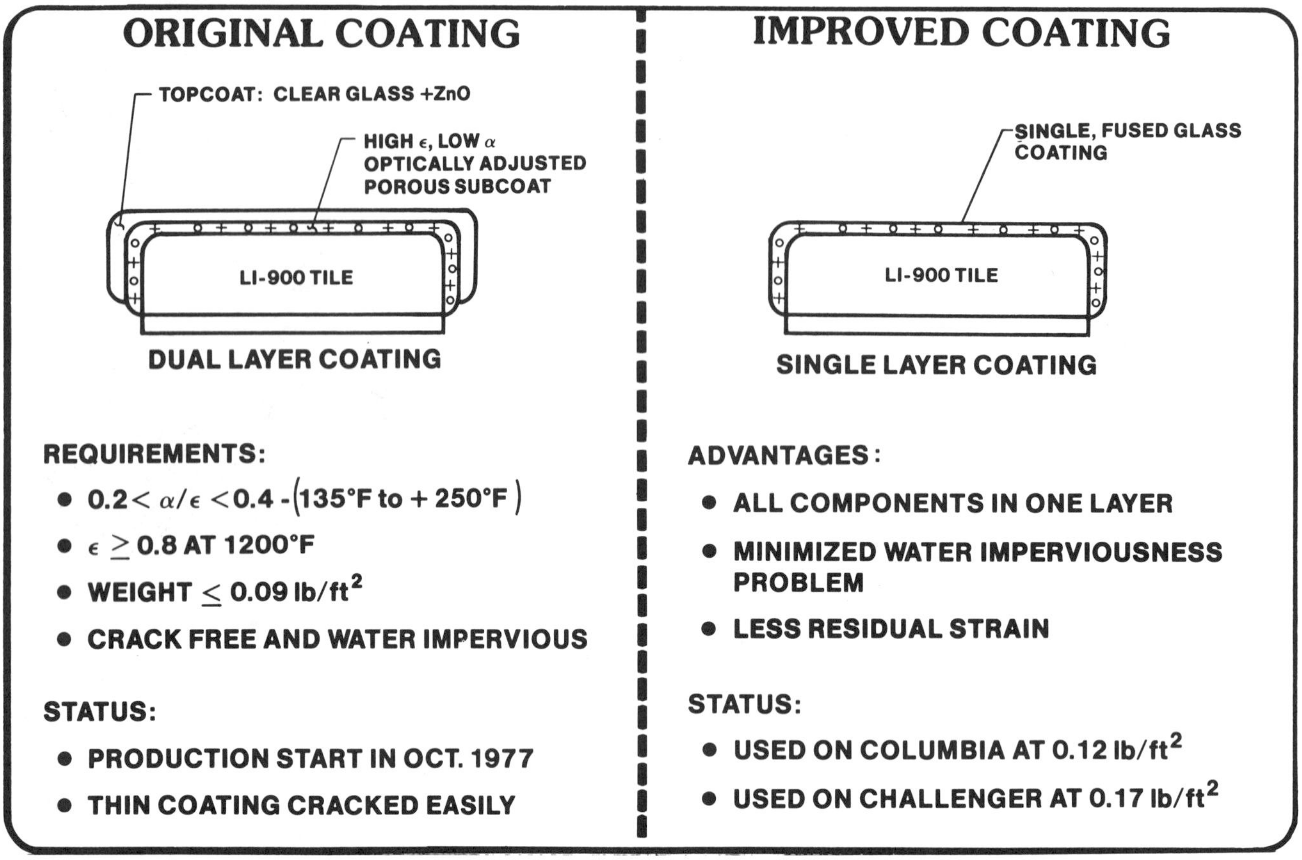

FIGURE 7 Class 1 coating optimization. (Courtesy of Lockheed Missiles & Space Company, Inc.)

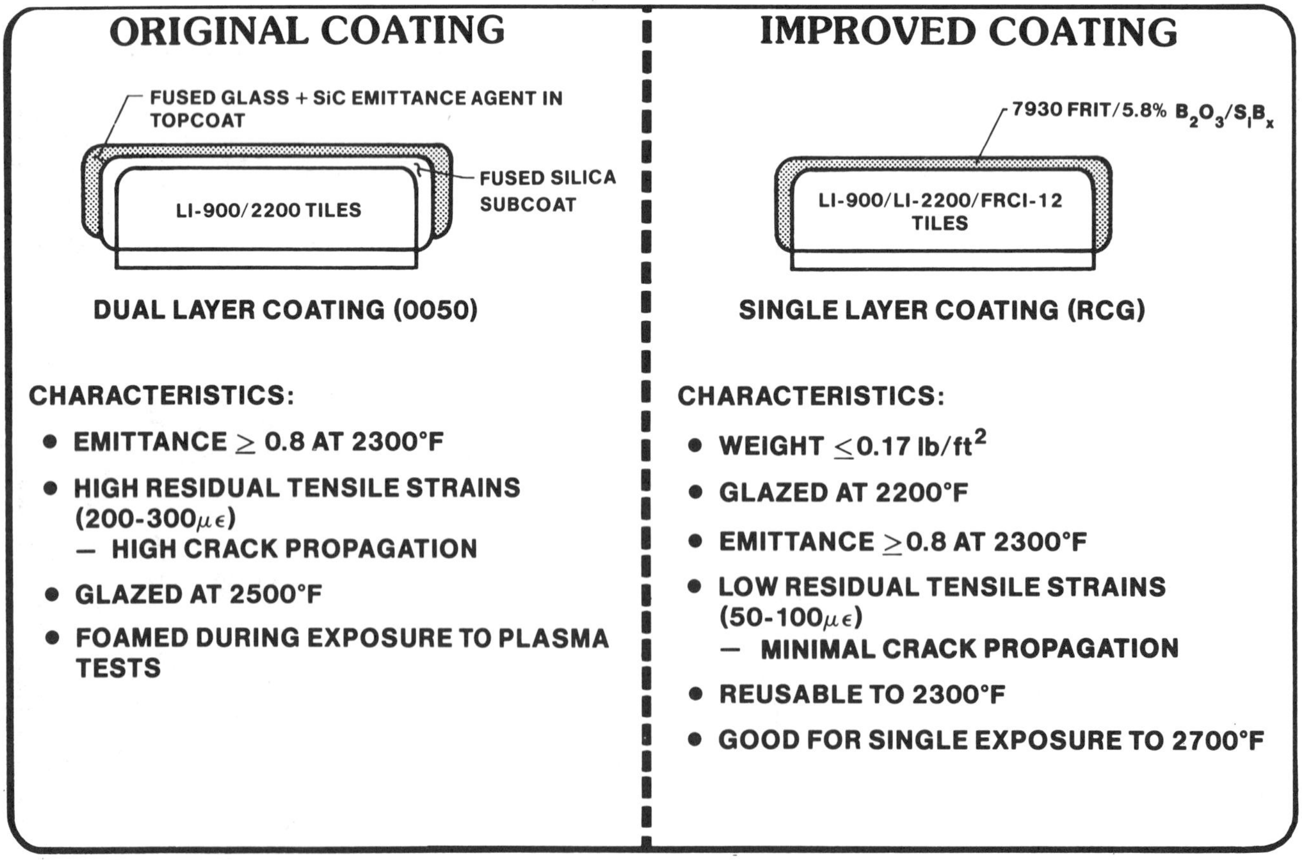

FIGURE 8 Class 2 coating optimization. (Courtesy of Lockheed Missiles & Space Company, Inc.)

lower coating residual tensile strain (Fig. 8). However, the coating was shown to be susceptible to both coating impact damage and crack propagation. This susceptibility is probably common to any thin glass coating, but it proved to be repairable when damage occurred.

UNIVERSAL PATCH. A universal patch material was developed that has a density approximately equal to that of the tile. It was applicable to both silica and FRCI tiles, compatible with the coating–glazing cycle, capable of repairing tile edges and corners, and easily and rapidly applied. Existing, approved shuttle materials used to compound the patch material are silica fibers, colloidal silica, acrylate solution, deionized water, and a combination of Fuchsin and methylene blue dyes. The slurry was placed into a void at twice the void volume, flattened with a Teflon spatula, and dried at 649°C for 8 minutes. It is ready for coating with full patch cure occurring during coating/glazing. Excellent bonding of patch to tile was demonstrated, and no crystallization occurred from exposure to a temperature of 1260°C for 15 hours.

Fabrication

LI-900 Process

The LI-900 process flow is shown in Figures 9–14. LI-900 billets are cast in two sizes, 38.1 × 38.1 × 16.5 cm and 25.4 × 50.8 × 18.5 cm using an automated casting line. Preweighed amounts of fiber are loaded into 26 hoppers on a carousel that automatically positions and empties the hoppers sequentially. The fiber and a predetermined quantity of water are combined in a tank containing a low shear mixer that uniformly disperses the fiber with minimum chopping. At the conclusion of the timed mix cycle, the slurry is automatically transferred into a casting mold positioned directly below the mixing tank.

Entrapped air bubbles are removed from the slurry before the billet is compressed to its final cast size. This is accomplished by a combination of vibration and stirring. Water is removed and the casting is compressed to a specified height in the volume adjustment operation. Concurrently a vacuum is applied to the bottom of the mold to remove a specified quantity of water. At this stage, the standard billet contains 2.36 kg of fiber and approximately 10.9 kg of water. The next step in the casting operation is the dispersion of a colloidal silica binder in the compressed casting at the weight adjustment station. The binder solution components are automatically mixed and dispensed through metering pumps. The solution is dumped on top of the compressed billet, and a vacuum is applied to the bottom of the casting mold. The residual water in the billet is displaced by the binder, which is dispersed throughout the casting to a specified solids concentration. Upon removal from the mold, the wet billet is weighed to provide a check that all steps of the casting operation were performed correctly. A gelling agent is used to set the binder and prevent it from mi-

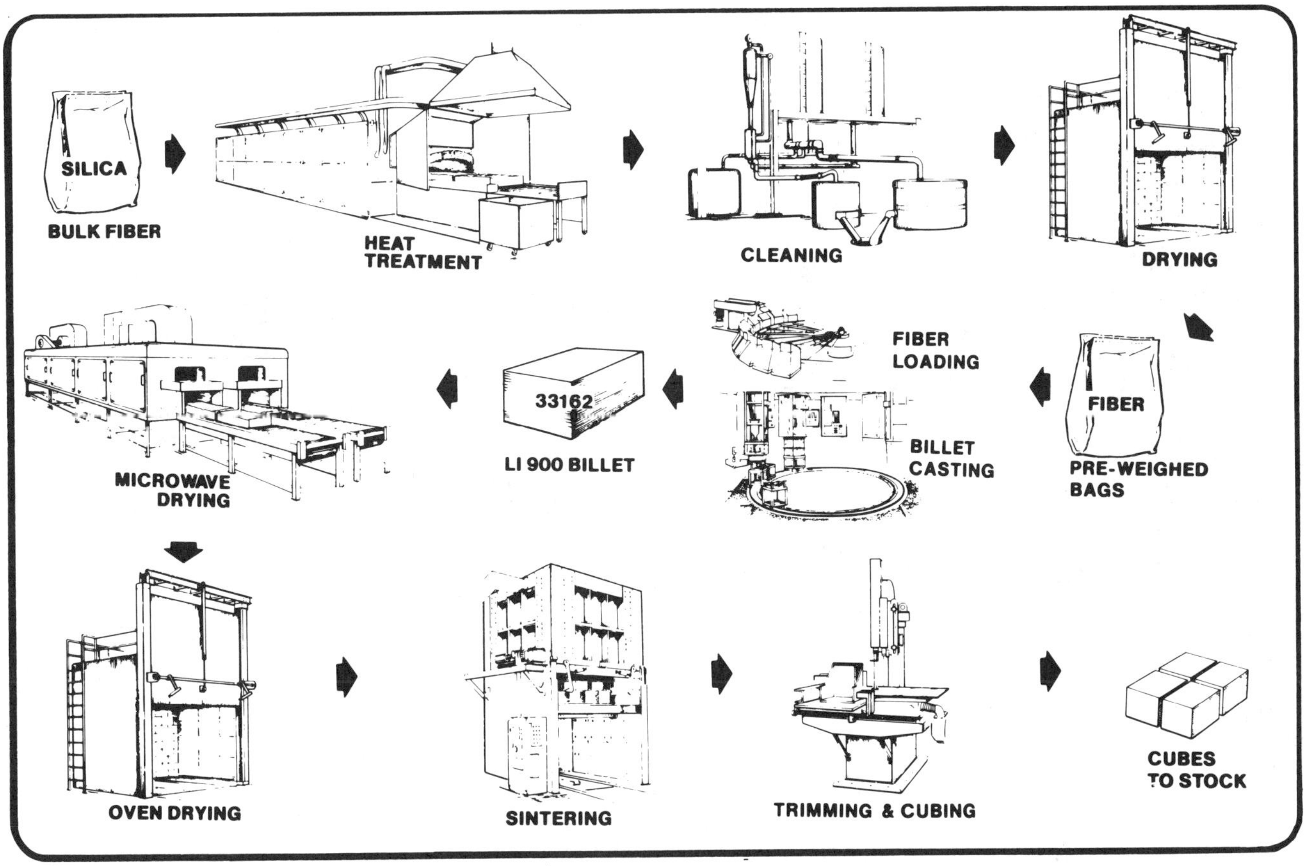

FIGURE 9 LI-900 process flow. (Courtesy of Lockheed Missiles & Space Company, Inc.)

Figure 10

Figure 11

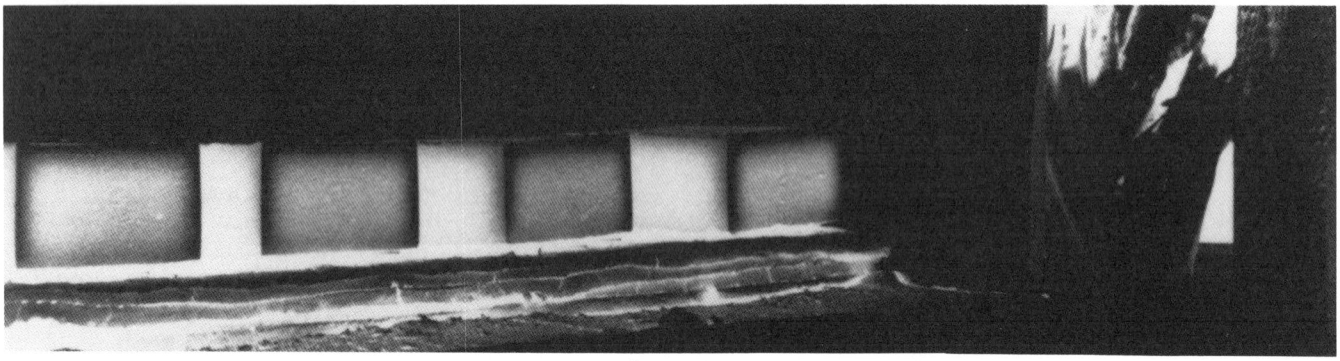

Figure 12

Figure 13

Figure 14

FIGURES 10–14 The basic raw material for LI-900 and LI-2200 is a short staple 99.9 percent pure amorphous silica fiber derived from high-quality sand. In Figure 10, premeasured silica fiber is loaded into a carousel hopper. In Figure 11, the silica fiber contents of one of the hoppers is dumped into a chamber where it is mixed with deionized water prior to pouring into a clear plastic mold and forming into a block. In Figure 12, the blocks are sintered in furnaces at temperatures up to 1315.5°C. In Figure 13, after each block is trimmed and cut into quarters, each quarter is individually machined to precise dimensions to produce a specific tile. After coating and water proofing, the tiles are gathered in arrays that contain an average of 22 single tiles. In Figure 14, the arrayed tiles are held in place with their coated surface down against the outer mold line of the array frame in the same position they will later assume on the Shuttle. They are then machined as a group along their inner mold line.

grating during drying. The billets are dried using either a conventional oven or a microwave dryer and are weighed after drying to assure that the specified amount of water has been removed before sintering is begun.

Beginning in 1975, the dried castings were sintered in specially designed three-zone tunnel kilns at a peak temperature of 1288°C. Early in 1982, however, the sintering operations were transferred to elevator kilns that use six-side heating (vs. five-side heating in the original kilns). This improves the strength distribution within the billets. The sintering schedule is adjusted to produce billets with an average density of 0.141 g/cm^3 by adjusting the sintering time to accommodate the PTX lot shrinkage variations.

LT-2200 Process

The LI-2200 process flow is shown in Figure 15. The billets are cast in a specially designed, manually operated casting tower in a 36.6 × 36.6 × 20.32 cm size. The mixing process differs from LI-900 in that the preweighed fibers are combined with water, SiC, and ammonium hydroxide into a V-blender equipped with an intensifier bar. Since LI-2200 requires a significantly higher casting density than LI-900, the slurry requires some chopping action to obtain the necessary fiber packing. After the blended slurry has been transferred into the casting tower and sealed, void elimination is accomplished by applying a high vacuum to the slurry before billet formation. The billet is formed by removing part of the water by gravity drain, compressing the slurry to a final thickness, and then extracting additional water with vacuum.

The billets are dried in a batch oven at 232°C for 16 hours. The dry density of the LI-2200 is approximately 0.21 g/cm^3. Originally, this was the final operation before sintering. However, the billets sometimes exhibited cracks after sintering. An additional drying at 538°C for 12 hours was developed to eliminate this problem.

The LI-2200 is sintered in elevator kilns. The sintering schedule is similar to that used for LI-900, except that the peak temperature is 1327°C. The soak time at peak temperature is adjusted to maintain final densities within a 0.352 ± 0.032 g/cm^3 density range.

FRCI-12 Process

The FRCI-12 process flow is shown in Figure 16. Silica and Nextel fibers are intermixed and cast into billets using a multistage, wet-slurry blending process and automated casting equipment. Castings are dried using a combination of microwave and convection–air ovens to achieve optimum drying rates. Dry castings are sintered in elevator kilns using microprocessor controllers to achieve uniform, repeatable heating to the optimum sin-

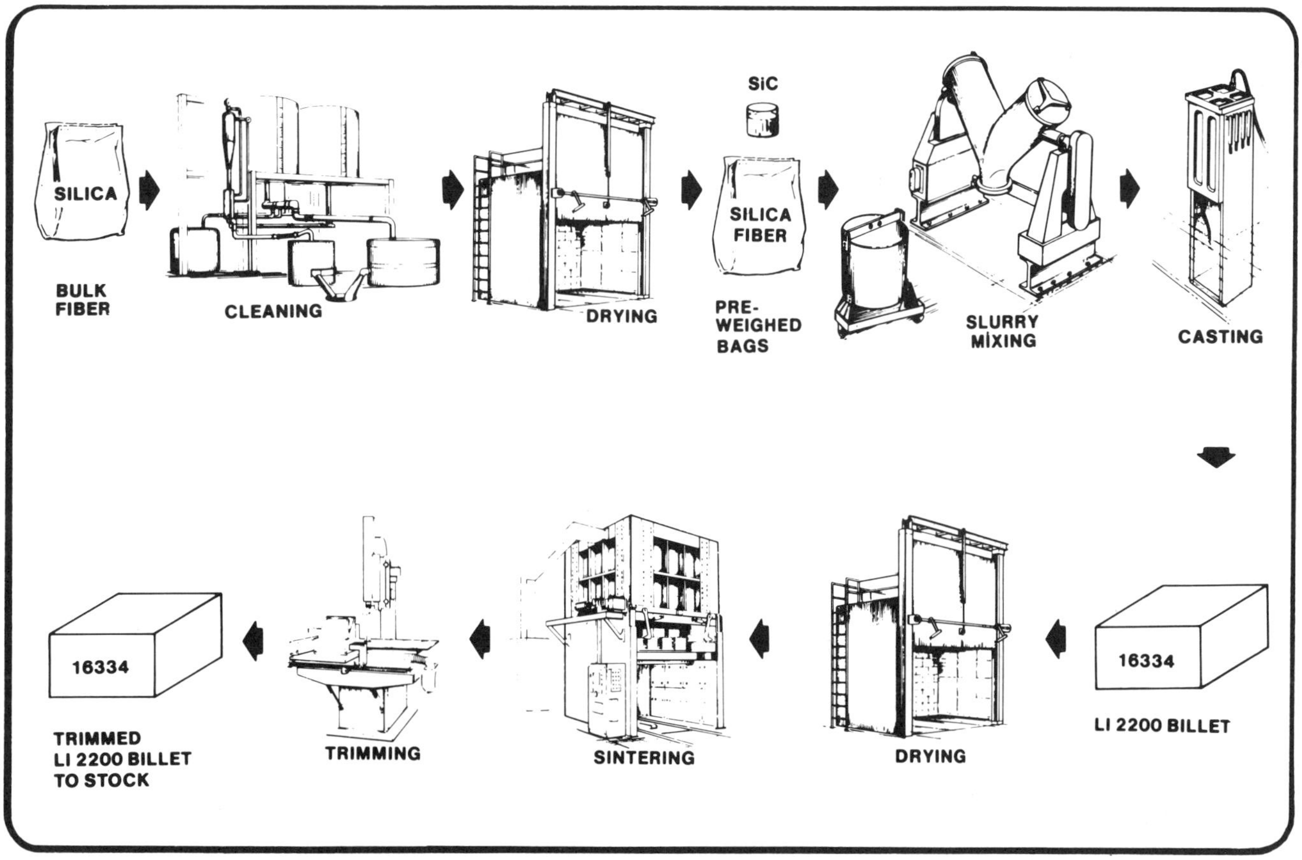

FIGURE 15 LI-2200 process flow. (Courtesy of Lockheed Missiles & Space Company, Inc.)

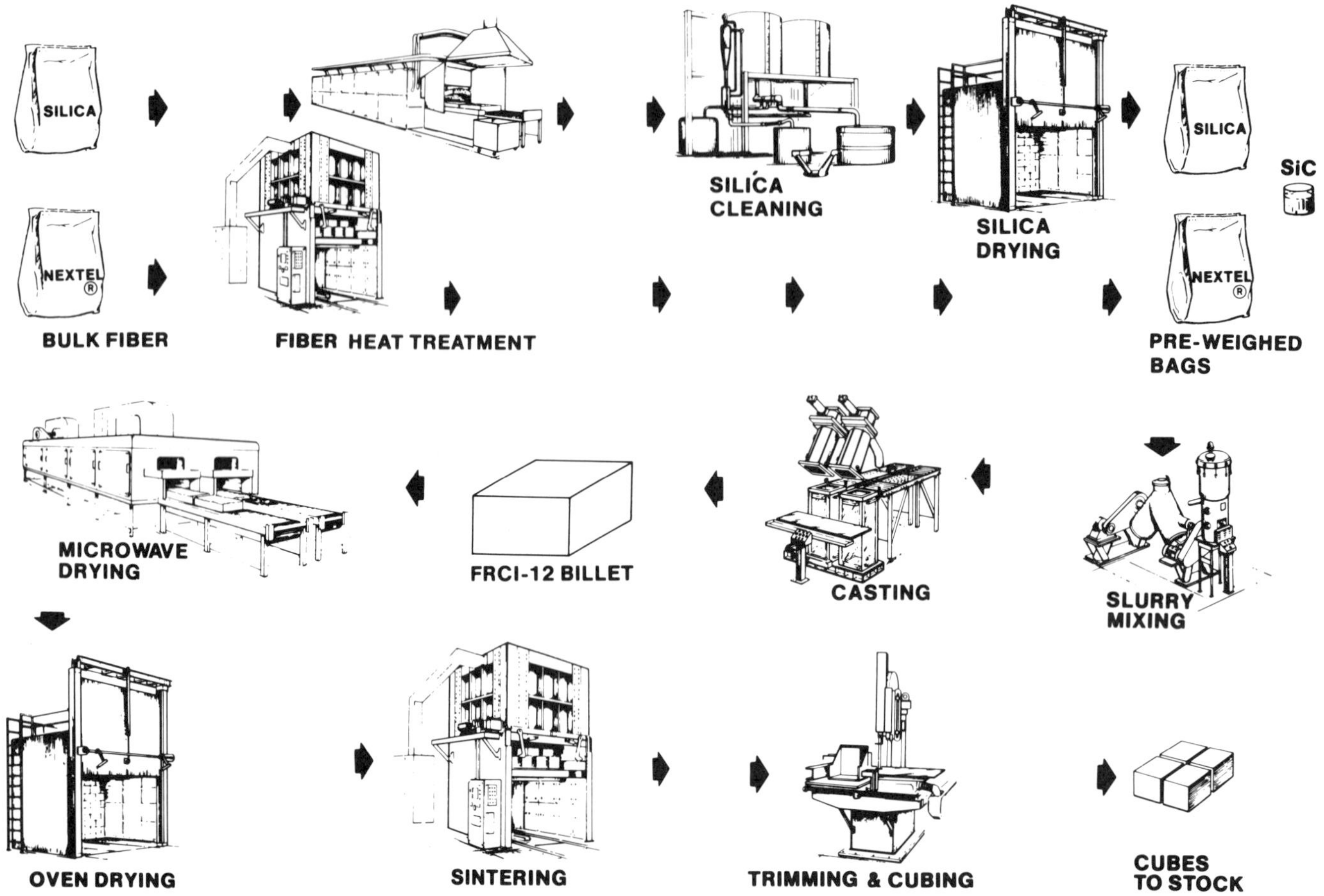

FIGURE 16 FRCI-12 process flow. (Courtesy of Lockheed Missiles & Space Company, Inc.)

tering temperature of approximately 1200°C. The optimum sintering temperature varies as a function of FRCI composition and desired final density.

FRCI can be fabricated with a range of compositions and densities to allow tailoring the material to a specific application. An FRCI formulation with a density of 0.192 g/cm^3 and a silica-to-Nextel fiber ratio of 78/22, identified as FRCI-12, was developed to replace LI-2200. Other FRCI materials with densities of 0.0128–0.320 g/cm^3 have been produced. FRCI-12 tiles were substituted for approximately 2764 LI-2200 tiles on the third orbiter, Discovery. Substituting the lower density FRCI-12 for LI-2200 saves approximately 395 kg of weight per orbiter. Also the tensile strength design value is increased by 50%, and the susceptibility to coating impact damage is reduced by eliminating the residual tensile strain in the coating, as a result of a better match in coefficient of thermal expansion between FRCI and tile coating materials. A high-speed, high shear blender was used for preblending operations in the final production process as shown above (Fig. 10).

Tile Waterproofing

Initially, the tiles were rendered hydrophobic (waterproof) by immersion in a hexamethyl disilazane silicone/Freon solution. Tile weight gain was about 1.0%. Several hundred grams of Freon solvent evaporated from each tile during thermal exposure. The process provided good water imperviousness, but left a dark, carbonaceous residue when heated to between 427 and 538°C resulting in an increase in the ratio of solar absorptance to hemispherical emittance to above the 0.4 specification limit.

The present method involves release of a trimethoxysilane vapor into a vacuum chamber containing the tiles. The tile weight gain is only 0.1%. Also, the material sublimes in a char-free condition, and no change in optical properties is encountered.

Tile Coating

Class 1 and 2 borosilicate glass coatings are applied to tile blanks by spray application of a slurry (Fig. 17). Tile blanks are set up on holding fixtures, masked to provide a breather space near the inner mold line (IML) surface and patched as necessary to cover surface deformities. Class 1 tiles are seal coated with a suspension of colloidal silica in water and dried before application of the coating. Class 2 tiles are wetted with alcohol before application of the coating.

The amount of slurry applied to each tile is controlled by maintaining within predetermined limits slurry viscosity, line pressure, and number of coats (Fig. 18). The coating weight is determined "wet," and a conversion

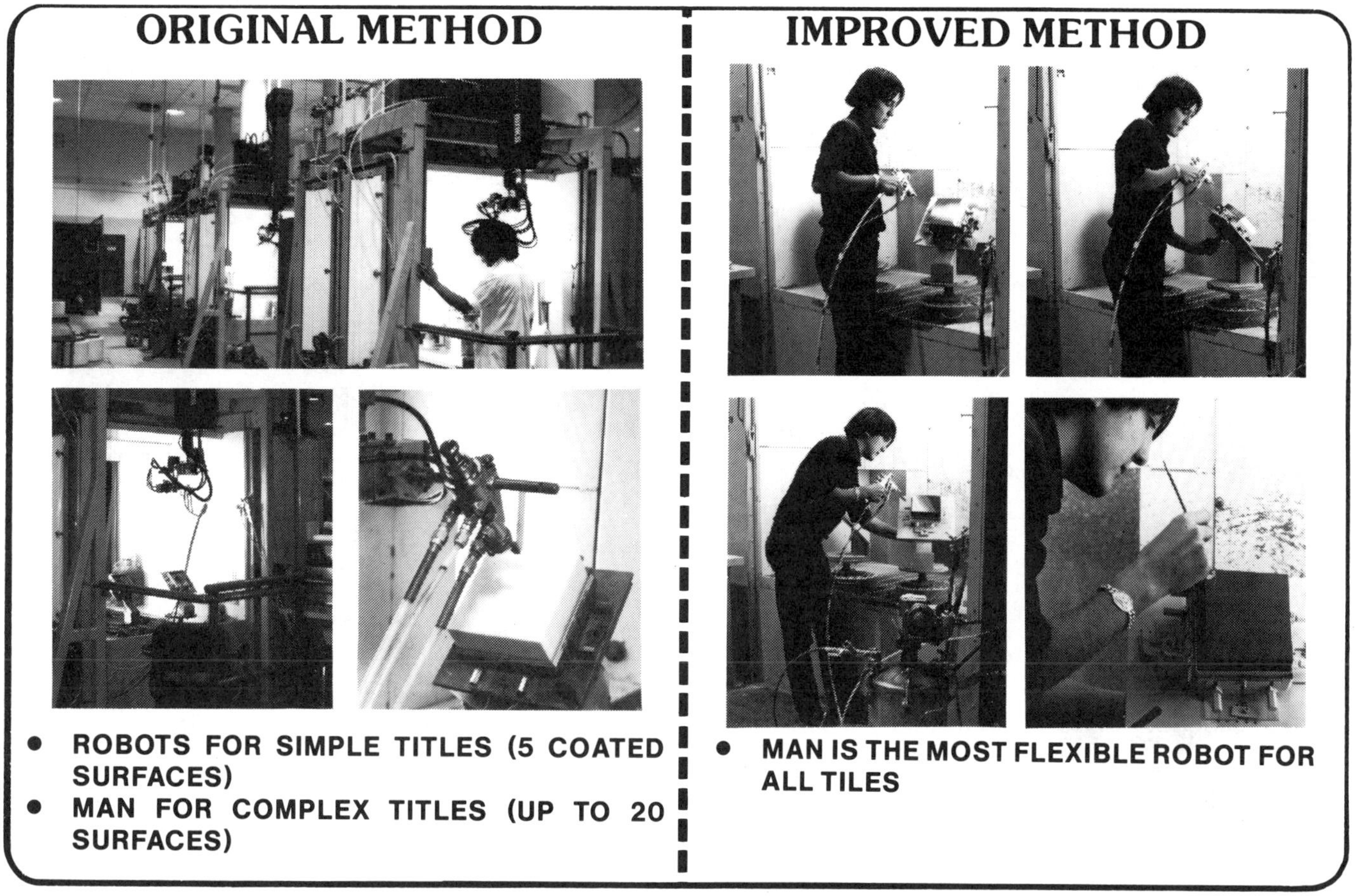

FIGURE 17 Coating spraying methods. (Courtesy of Lockheed Missiles & Space Company, Inc.)

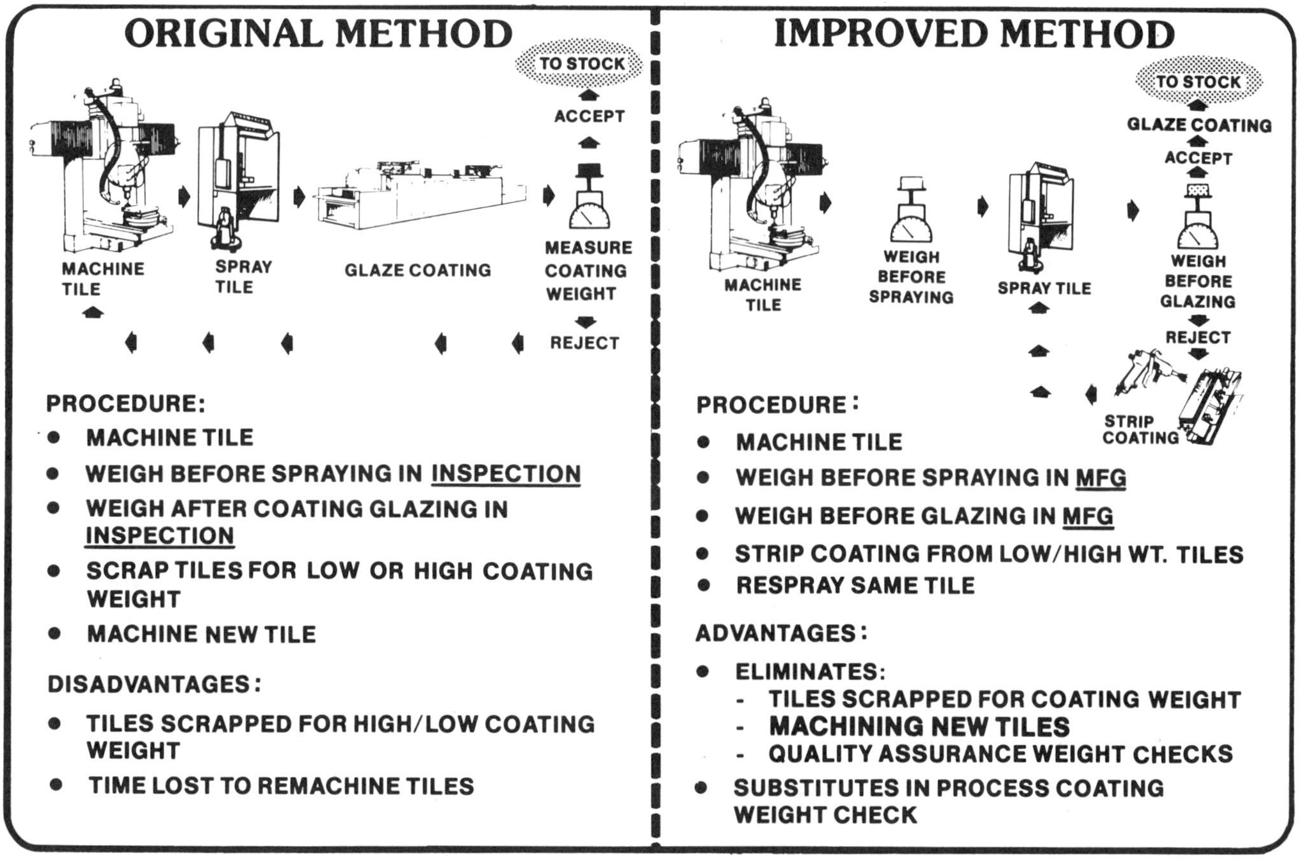

FIGURE 18 Control of coating weight.

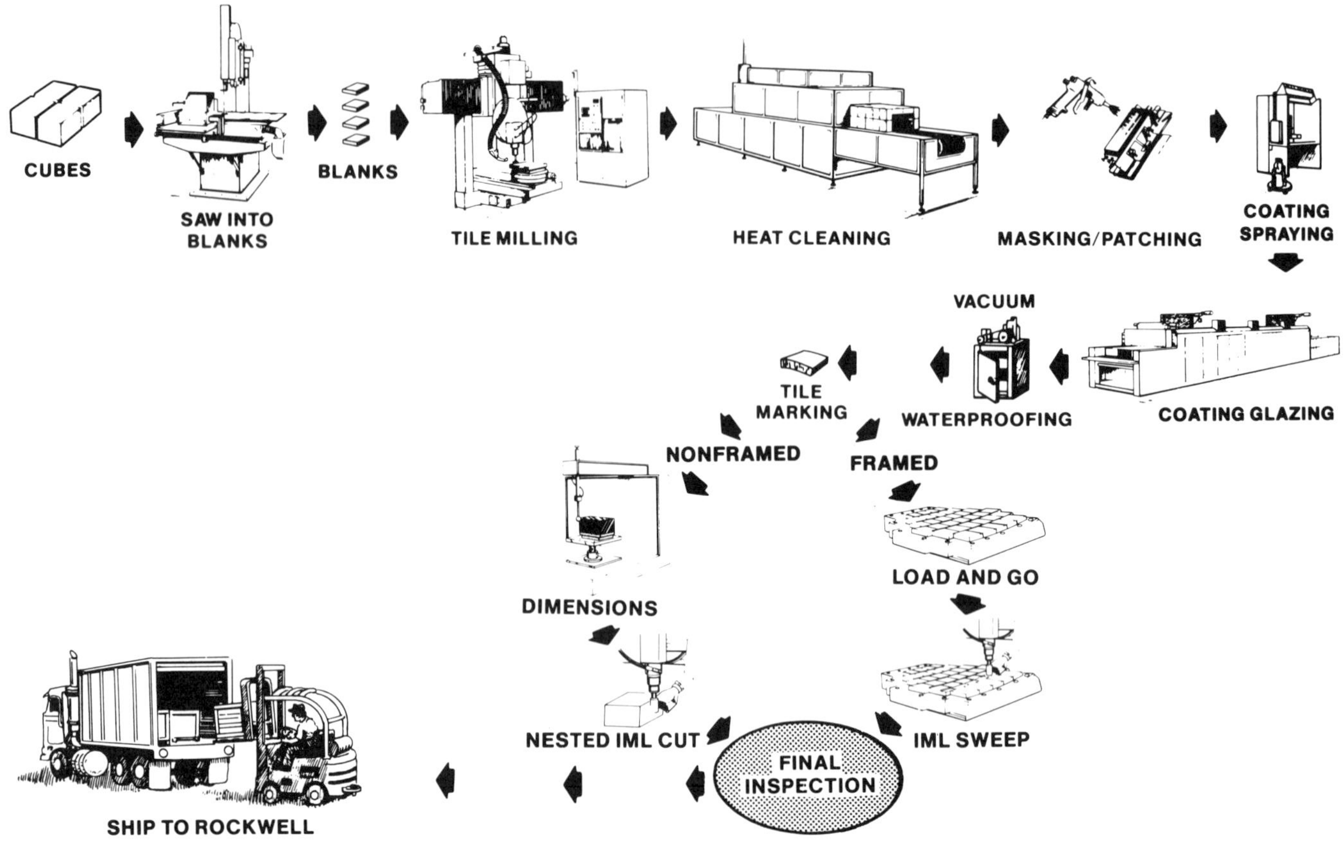

FIGURE 19 Tile process flow. (Courtesy of Lockheed Missiles & Space Company, Inc.)

factor is applied to calculate "dry" weight. Coating weights are between 0.0011 and 0.0027 g/cm^3 for class 1 tiles and between 0.00013 and 0.0027 g/cm^3 for class 2 tiles. The corresponding coating thicknesses are 0.015–0.038 cm for class 1 tiles and 0.20–0.38 mm for class 2 tiles.

The Process Flow

The sequence of operations performed after the insulation material is cut into cubes is shown in Figure 19. When the tiles have been machined, they are heat cleaned to remove organic contaminants, masked to allow an uncoated breather space area along the lower perimeter adjacent to the IML, sprayed with class 1 or class 2 coating, and sintered at 1148–1232°C in an Ipsen tunnel-hearth roller kiln. After vacuum waterproofing with a methyl trimethoxysilane (Dow Corning DC 6070), the tile identification number is painted on the outer mold line (OML) and the tiles are checked dimensionally as required before the IML cuts are made.

The tile IML cut is performed either individually, which is called a "nested" tile, or on a group of tiles simultaneously in an array frame, which is known as an ATA.

The dimensions of a nested tile are checked as required before the IML cut. The dimensions of the ATA tiles are checked by their ability to fit into a premeasured polyurethane array frame with the required tile-to-tile gaps, which are generally 0.11 ± 0.04 cm on the lower wings and fuselage and 0.14 ± 0.04 cm on the upper wings, fuselage, and vertical fin.

Tile Shrinkage and Measurement

A series of test programs led to the realization that tile shrinkage could be correlated to a single factor, tile thickness, if all other parameters were held constant. This knowledge was aided by the use of precision Bendix Cordax machines, which provide accurate, repeatable measurement data for each tile.

The initial results showed that tile length or width changes were related to the glazed tile thickness. A best-fit, least-squares logarithmic equation was developed using the available data. Reference 3 describes the details of this activity.

TPS tiles were originally machined with planar vertical sides. Subsequent shrinkage investigations revealed that special offsets were necessary to control shrinkage during coating glazing. These special offsets are classified into three categories: planar, which is the type of shrinkage discussed above, side-slope, and radius-type compensation. The planar type of shrinkage, which is the largest of the three, accommodates platform shrinkage during the glazing cycle. The side-slope and radius compensations are smaller in magnitude and accommodate

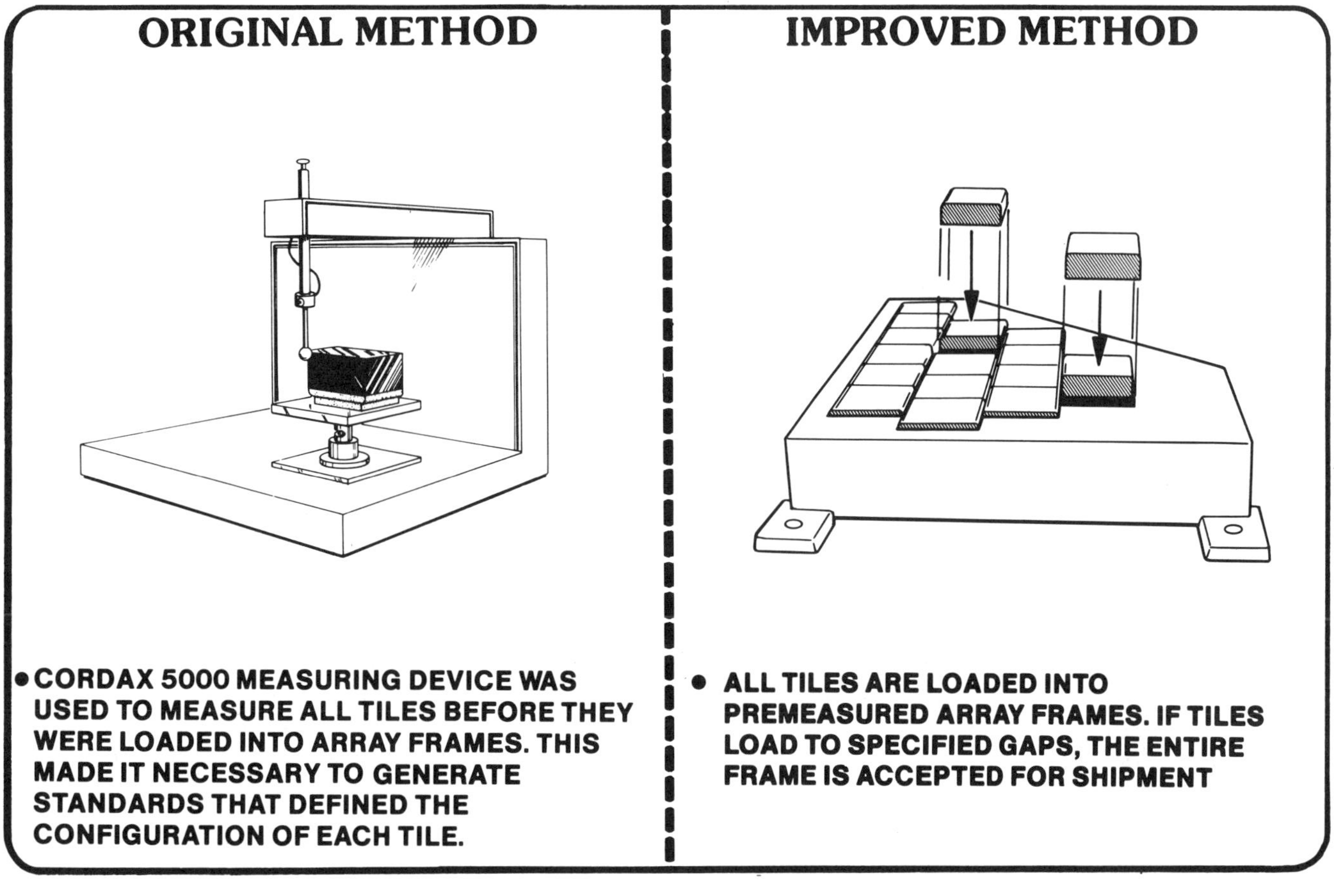

FIGURE 20 Load and go concept. (Courtesy of Lockheed Missiles & Space Company, Inc.)

distortion shrinkages. The side-slope distortion occurs because the OML edge shrinks more than the base of the tile. Radius compensation is necessary because the corners shrink more than the middle side of the tile. All adjustments are made in an equal and opposite sense and constitute typical adjustments made to NC machine part programs.

During the early stages of Columbia tile delivery, simple (square, flat) tiles were manufactured. When more complicated tiles were fabricated, dimensional problems occurred. Tiles whose sides were not parallel or normal to each other (i.e., tiles with a wraparound OML) had a high dimensional rejection rate. An investigation revealed that the shrinkage normal to the in-plane direction, which is defined as the through-the-thickness direction, is about three times larger than the in-plane shrinkage. The in-plane direction usually denoted a plane parallel to the orbiter surface. Shrinkage and distortion of these surfaces follow a complex relationship.

Tiles were measured and verified for platform dimensional conformity before being loaded into array frames to machine the IML (Fig. 20). An automated system to measure tiles was implemented. Two Cordax measuring devices were programmed to automatically summon the inspection standards from a host computer, locate the tile on the machine bed, and automatically determine the acceptability of the tile by using a series of 6–12 predetermined touch points on the tile sides, and 5 touch points on the OML surface.

The average time to measure a tile using a Cordax machine was 5–15 minutes. Another device, the maxi-measure (Fig. 20) was developed to reduce the load on the Cordax measurement machines. This device consists of two parallel plates that measure the maximum dimension of tiles with parallel sides. The average measurement time for the "maxi-measure" device is less than one minute.

References

1. R. P. Banas, *et al.*, "Lessons Learned from the Development and Manufacture of Ceramic Reusable Surface Insulation (RSI) Materials for Space Shuttle Orbiters," NASA/LARC Conference "Shuttle Performance: Lessons Learned," Langley Research Center, March 8–10, 1983.
2. HRSI Fact Sheet, Lockheed Missiles & Space Company, March 1984.
3. B. T. Fitchett, "Dimensional Control of Space Shuttle Tile During Manufacture," *Proceedings of the Seventh Conference on Composites and Advanced Materials,* Ceramic-Metal Systems Division, American Ceramic Society, 1983.

Bibliography

Banas, R. P., and Cordia, E. R., *Proceedings of the Fourth Annual Conference On Composites and Advanced Materials, Ceramic–Metal Systems Division, American Ceramic Society,* 1980.

Banas, R. P., and Cunnington, G. R., Jr., *Proceedings of the AIAA Thermophysics and Heat Transfer Conference,* 1974, Paper 74-730.

Banas, R. P., Gzowski, E. R., and Larsen, W. T., *Proceedings of the Seventh Conference on Composites and Advanced Materials,* Ceramic-Metal Systems Division, American Ceramic Society, 1983.

Banas, R. P., Kural, M. H., Deruntz, J. A., Burns, A. B., Chinn, A. J., Lambert, R., Ritz, J. R., Vanwest, B., and Woneis, J. T., "Space Shuttle Thermal Protection System Development," Lockheed Report LMSC-D152738, Lockheed Missiles & Space Company, Space Systems Division, Jan. 17, 1972.

Banas, R. P., Dolton, T. A., Housten, S. J., and Wilson, R. G., *Proceedings of Thermodynamics and Thermophysics of Space Flight,* 1970, pp. 239–276.

Beasley, R. M., and Clapper, R. B., *Proceedings of the 67th Annual Meeting of the American Society for Testing and Materials,* 1964.

Beasley, R. M., and Izu, Y. D., *Proceedings of the OSU-RTS Symposium on Electromagnetic Windows,* 1964.

Beasley, R. M., Izu, Y. D., et al., "Fabrication and Improvement of LMSC's All-Silica RSI," in *Symposium on Reusable Surface Insulation for Space Shuttle,* Vol. 1, Technical Report NASA TMX-2719, November 1972.

Burns, A. Bruce, and McCarter, C. R., *Proceedings of the SAE Aerospace Congress & Exposition,* 1980.

Creedon, J. F., Banas, R. P., and Stern, P., *Proceedings of the 29th Pacific Coast Meeting, Ceramic-Metal Systems Division, American Ceramic Society,* 1976.

"Development and Design Application of Rigidized Surface Insulation Thermal Protection Systems," Lockheed Missiles & Space Company Report LMSC-D282673, Dec. 30, 1972.

Dotts, R. L., Smith, J. A., and Tillian, D. J., "Space Shuttle Orbiter Reusable Surface Insulation Flight Results. Shuttle Performance: Lessons Learned," NASA CP-2283, Part 2, 1983, pp. 949–966.

Dotts, R. L., Battley, H. H., Hughes, J. T., and Neuenschwander, W. E., *Proceedings of the AIAA* 1982, paper 82-0788-CP.

Elgin, D., and Schirle, J., *Proceedings of the Fifth Annual Conference on Composites and Advanced Materials,* American Ceramic Society. 1981.

Forsberg, K., *Proceedings of the Intl. Aeronautic Congress,* 1979.

Goldstein, H. E., Leiser, D. B., and Katvala, V. W., "Reaction Cured Glass and Glass Coatings," U.S. Patent 4,093,771 (June 6, 1978).

Goldstein, H. E., Smith, Marnell, and Leiser, Daniel, "Silica Reusable Surface Insulation," U.S. Patent 3,952,083 (April 20, 1976).

Hammitt, R. L., "Lightweight Insulation, LI-15, Test Summary," Lockheed Missiles & Space Company, LMSC-685434, Code 9999, April 26, 1968.

Holmquist, G. R., and Cordia, E. R., *Proceedings of the 12th Natl. SAMPE Technical Conference,* 1980, pp. 111–120.

Holmquist, G. R., Cordia, E. R., and Tomer, R. S., *Ceramic Engineering and Science Proceedings: Fifth Annual Conference on Composites and Advanced Materials,* Vol. 2, American Ceramic Society, 1981.

"Improvement of Reusable Surface Insulation Material, Final Report for NAS9-12137, "Lockheed Missiles & Space Company Report LMSC-D266204, March 1, 1972.

Korb, L. J., and Clancy, H. M., *Proceedings of the 26th Natl. SAMPE Symposium,* 1981, pp. 232–249.

Korb, L. J., and Morant, C. A., et al., *Ceram. Bull. 60* 11, (1981).

Leiser, Daniel B., Goldstein, Howard E., and Smith, Marnell, "Fibrous Refractory Composite Insulation (FRCI-12)," U.S. Patent 4,148 962 (April 10, 1979).

Leiser, D. L., Smith, M., Stewart, D., and Goldstein, H., *Proceedings of the Seventh Conference on Composites and Advanced Materials,* Ceramic–Metal Systems Division, American Ceramic Society, 1983.

"LI-1500 Heat Shield, Independent Development Program C548," Lockheed Missiles and Space Company, Report LMSC-D153908, Dec. 24, 1971.

Nakano, H. N., and Izu, Y. D., *Proceedings of the 12th Annual Meeting of the Fine Particle Society, Powder and Bulk Solids Conference,* 1981.

"A Proposal for Space Shuttle Orbiter High-Temperature Reusable Surface Insulation," Lockheed Missiles & Space Company Report LMSC-D336595, March 14, 1973.

Rusert, E. L., "Development of a Rigidized Surface Insulative Thermal Protection System for Shuttle Orbiter, Final Report," NASA CR-114973, Feb. 15, 1971.

"Space Shuttle Concepts. Final Report for NAS8-26362," Lockheed Missiles & Space Company, Report LMSC-D153024, March 15, 1972.

Symposium on Reusable Surface Insulation for Space Shuttle, Vol. 2, Technical Report NASA TMX-2720, November 1972.

Tanabe, T. M., and Izu, Y. D., "Influence of Particle Size Distribution on Ceramic Coating Characteristics," paper presented at the 32nd Pacific Coast Regional Meeting of American Ceramic Society, Seattle, Oct. 24–26, 1979.

"Thermal Protection Concepts for Lifting Entry Vehicles," Lockheed Missiles & Space Company, Report 6-62-64-5, May 5, 1964.

Tomer, R. S., and Cordia, E. R., *Ceramic Engineering and Science Proceedings, Sixth Annual Conference on Composites and Advanced Materials,* Vol. 3, American Ceramic Society, 1982, pp. 601–611.

Wheeler, W. H., Garofalini, S. H., and Beasley, R. M., *Proceedings of the American Ceramic Society Meeting,* 1976.

Williams, S. D., and Curry, D. M., "Nonlinear Least Squares—An Aid to Thermal Property Determination," NASA TM X-58092, June 1972.

Witte, W. G., "Flight Test of High-Density Phenolic–Nylon on a Spacecraft Launched by the Pacemaker Vehicle System," Technical Report NASA TMS-1910, Langley Research Center, November 1969.

Ceramic Laminates

See Laminates, Ceramic

Ceramic Matrix Composites

The recent interest in ceramic matrix composites derives from the opportunity to tailor the various properties of these materials. Ceramic matrix composites here include a great variety of materials. Typically, a matrix (polycrystalline ceramic, glass, or mixture thereof) is modified by the addition of particles, flakes, whiskers, fibers, etc., or even voids. The great flexibility in designing and tailoring these multicomponent systems permits controlling and enhancing many significant properties. Perhaps the most important, in the context of present trends, is the ability to obtain high strength and toughness, especially at high temperatures. Toughness here refers to resistance to the initiation or propagation of fracture. High strength and toughness are not typical of conventional ceramics, where intrinsic material factors limit certain properties. These intrinsic factors have both beneficial and detrimental effects. Ceramics have historically been employed because of the benefits of certain of these intrinsic material properties.

Typically, ceramics are composed of low atomic number constituents with strong, directional ionic–covalent bonding, and highly localized electrons. These characteristics result in ceramics exhibiting the desirable properties of low density, high elastic constants, high Peierl's or yield stress, high melting temperature, chemical stability, electromagnetic transparency, low thermal expansion, and low thermal conductivity. Unfortunately, the same intrinsic features also have detrimental effects. The ionic–covalent bonding produces high Peierl's stresses and thus high yield stresses. This type of bonding also produces noncubic crystal structures with limited numbers of slip systems.

The combination of high Peierl's stress and limited slip systems then results in *very* limited plasticity, thus little or no crack blunting. As a result, the mechanical properties of conventional ceramics suffer severe limitations. For planar cracks, the fracture energy approaches the low value of the true surface energy. The strength then becomes flaw-size dependent, with the fracture stress inversely proportional to the square root of the flaw size (e.g., $\sigma_f \propto c^{-1/2}$). The failure is catastrophic, since the fracture energy γ_f is relatively independent of crack length, and the stress intensity K increases with increasing crack length (e.g., $K_I \propto c^{1/2}$). The strength then becomes quite variable, being sensitive to the distribution of both size and severity (geometry, orientation) of the flaw population.

As a result of these intrinsic limitations on their mechanical properties, ceramics have a somewhat undeserved reputation as being mechanically unreliable. In some cases coupling of other properties to elastic or mechanical effects can cause similar variability to electrical or magnetic loads. This would be true in the case of materials exhibiting piezoelastic, electrostrictive, or magnetostrictive effects. The use of ceramics in electro-optical windows (radomes, IRdomes, laser windows, microwave windows) is also limited by their mechanical unreliability. In many of these applications, significant thermal stresses are present and limit system performance [1]. In applications such as bearings, seals, and cutting tools, mechanical unreliability can also be a severe problem. The ceramic cutting or bearing material is typically subjected to very high mechanical loads, impact loads, and high stresses resulting from localized and rapid heating.

Because of the limitations on the strength and toughness of ceramics, significant difficulties are encountered in the use of ceramics in structural applications or where stresses are significant. Typically, the results are (a) the requirement of an extremely low design strength, far below the mean strength of the material, or the use of very large safety factors, and (b) an extreme reluctance to employ ceramics in critical or man-rated applications where failure would be extremely costly or could result in loss of life. Even in less critical situations, the unpredicted mechanical failure of a ceramic component can be very serious. For example, the failure of a relatively inexpensive ceramic cutting tool in an automated machining system can lead to very costly downtime, with the costs in maintenance and loss of production far exceeding the value of the ceramic part. Such costs may preclude the use of unreliable cutting tool materials, despite the other advantages they may possess.

Approaches to the Improvement of Mechanical Reliability of Ceramics

Despite significant progress in improving the mechanical reliability of conventional ceramics by various techniques, the potential here is severely limited. Typically, conventional (noncomposite) ceramics can be significantly toughened to improve their reliability in only a few ways. One entails the use of various means for producing a very rough fracture surface. Here, the ceramic derives increased toughness from several related effects. These include increased fracture surface area, fracture mode change (e.g., mode I to II or III), and possible hindrance of crack opening by mechanical interlocking of the rough fracture surfaces (Fig. 1). Other techniques include various irreversible processes at or near the crack tip, which consume energy or otherwise limit crack growth (Fig. 2).

Very limited plasticity or viscous flow at the crack tip may result in crack blunting and reduction of crack tip stress intensity K. Viscous flow would normally be associated with glassy phases and high temperatures. Limited plasticity would be present in cubic crystal structures or in other crystal types at high temperatures. Some of these energy-absorbing irreversible processes, such as the stress-induced martensitic transformation in ZrO_2 or stress-induced microcracking, may absorb energy in a "process zone" adjacent to the crack tip. The size and shape of this "process zone" have significant effects on the resultant increases in toughness.

Techniques for producing a rough fracture surface can lead to toughness increases of only 50–100% over the base values (e.g., to values of perhaps 5–10

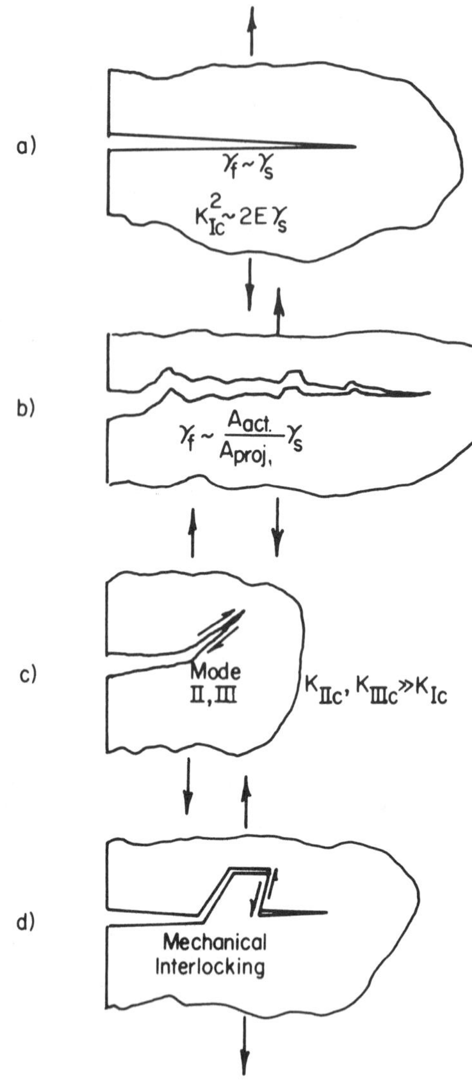

FIGURE 1 **Fracture surface roughness contributions to increased fracture toughness. (*a*) Planar fracture surface with fracture energy equal to surface energy. (*b*) Rough fracture surface with fracture energy increased by increase in true surface area. (*c*) Fracture mode change from easy mode I to more difficult mode II or III. (*d*) Mechanical interlocking behind crack tip (requiring acicular microstructure) hindering crack opening.**

MPa · $m^{1/2}$). The second type of toughening, through irreversible processes, can produce increases of 2–10 times, to K_{I_c} values of 5–20 MPa · $m^{1/2}$. While the maximum values here are at useful levels, there are many limitations on the actual use of these toughening techniques. Producing a rough fracture surface is most readily done via ceramic composites. Otherwise production of a rough fracture surface generally requires the use of highly anisotropic materials, such as aluminum titanate. Such materials, as fabricated, typically contain significant residual microstresses, which may pose significant limitations. In the case of aluminum titanate, the significant limitation is on processing conditions, as noted below.

Materials that exhibit the significant residual microstresses from crystalline anisotropy, or mismatches between different phases, may also benefit from another type of toughening. These materials are also likely to benefit from the toughening afforded by stress-induced microcracking. However, these residual stresses diminish with increasing temperature, and they would disappear entirely at approximately the processing temperature for the material. Thus the toughness derived from these residual stresses must decrease with increasing temperature as well.

Of more importance to most current applications of ceramics, the residual stresses can be high enough to cause material failure through extensive spontaneous microcracking. It is typically necessary to control the microstructure of these materials very closely, especially in regard to maintaining a very small grain size (submicrometer in the case of the aluminum titanate). The toughening possible through martensitic transformation also depends critically on microstructure, requiring precise processing control and precise postfabrication heat treatments. Here also the effect is temperature dependent, decreasing with increasing temperature, and disappearing altogether at the martensitic transformation temperature (500–1200°C in various zirconias). The effects of crack tip blunting through microplasticity are very limited in the case of typical ceramics by the high Peierl's stresses and limited slip systems available in noncubic materials. Effects of viscous flow by glassy phases, which are limited to high temperatures, may indeed be the *source* of failure, through gross viscous flow, at slow loading rates or for static loads of long duration.

Various composite approaches offer much more potential, both in obtaining large increases in fracture toughness, and in providing toughness increases over large temperature ranges, and can provide high toughness in conjunction with high strength as well. It is also frequently possible, through composite approaches, to combine increased toughness and strength with other properties desirable for particular applications. Following is a brief overview of various composite approaches for toughening and strengthening ceramic materials, and possible opportunities for advancements and synergisms in these approaches.

Toughening Mechanisms in Ceramic Composites

Composite material approaches permit the increase in apparent fracture toughness or fracture energy of ceramic materials to useful levels. The mechanisms by which these improvements occur fall into several overlapping categories. Note that there are two *different* types of effect that may be useful: increasing the stress intensity necessary for fracture initiation, and increasing the energy required to continue propagation of a crack.

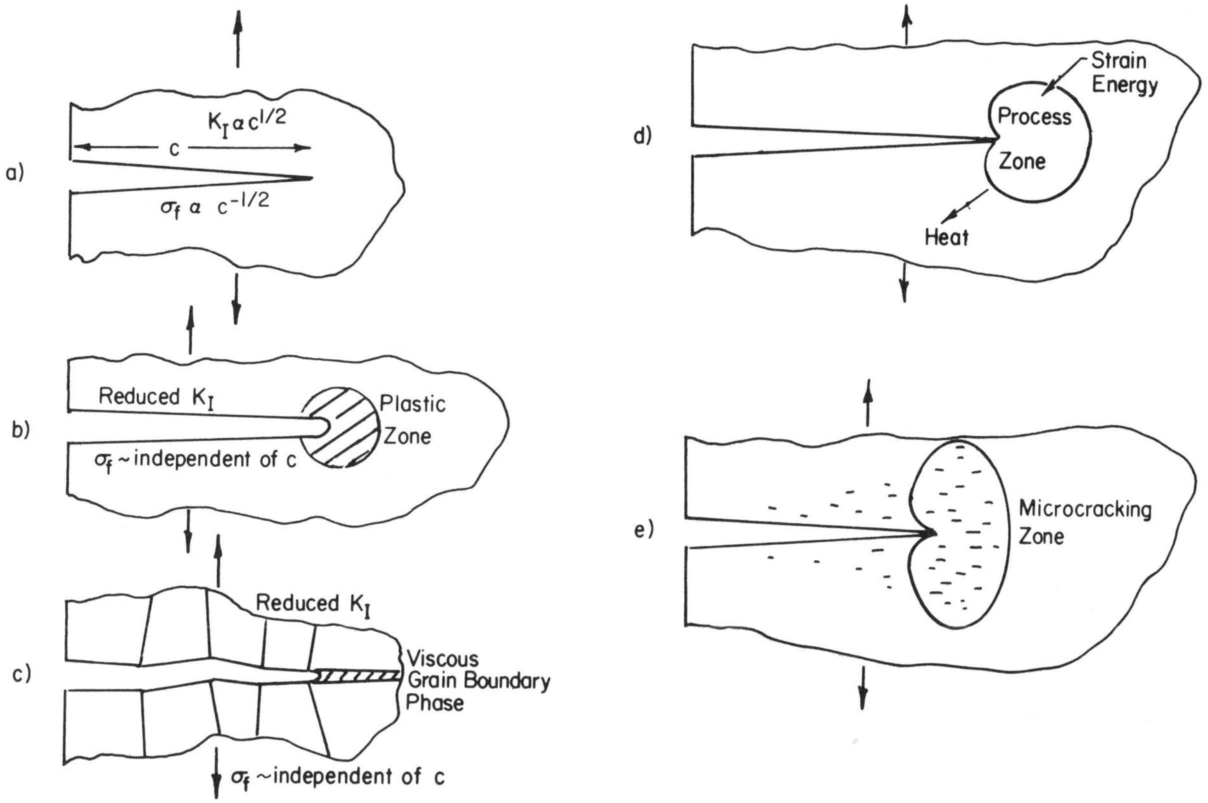

FIGURE 2 **Irreversible processes leading to increased fracture energy and fracture toughness. (*a*) Sharp crack with *K* proportional to $c^{1/2}$. (*b*) Plastic zone at crack tip leading to crack blunting and reduction in crack tip stresses. (*c*) Viscous flow in glassy grain boundary flow leading to same result. (*d*) Process zone adjacent to crack tip absorbing energy during fracture process. (*e*) Microcracking in process zone adjacent to crack tip.**

The first effect controls the strength for materials that fail in a catastrophic manner. The second controls the ability of a material to absorb energy during fracture, or to operate for a long time with the presence of large flaws.

The general categories of mechanisms for toughening ceramics via composite approaches are as follows.

1. Increased fracture surface area.
2. Energy absorption by irreversible processes in a zone near the crack tip.
3. Change in fracture mode (e.g., mode I → mode II or III).
4. Reduction in the stress intensity at the crack tip.

Note that these effects may be *highly* interactive and may be produced by similar means. Thus, it may be extremely difficult, either analytically or experimentally, to sort out the individual contributions. Much of our discussion is therefore general and rather unquantitative. However, the guidelines and interactions noted should be useful in material design and improvement.

Increased Fracture Surface Area

The basis for the fracture surface area technique is the increase in area of a fracture surface over that of a planar crack (cf. Fig. 1*a*, 1*b*). Then,

$$\gamma_f = (A_{actual}/A_{proj})\,\gamma_t \tag{1}$$

where γ_f is the apparent fracture energy, γ_t is the true surface energy, and A_{actual} and A_{proj} are the actual and projected areas of the fracture surface. The maximum effects on fracture energy that can be achieved here depend on the geometry of the fracture surface roughness, which in turn must relate to the heterogeneities and residual stresses in the material; only limited effects will arise from random fluctuations in the crack path. If the fracture surface is schematically represented in various ways with simple geometrical shapes, the area ratios (Eq. 1) range from 2 for hemispheric surfaces to 3 and higher for cylinders. These area ratios are significantly less than those implied by experimental fracture energy values, suggesting that other mechanisms are operative as well.

The limitations on increasing fracture energy by this technique are actually even more limited than suggested by the area ratios above, since only a fraction of the fracture surface typically exhibits this roughness. Thus, for example, in a brittle material containing a dispersion of spherical second-phase particles for toughening, with volume fraction v_p, the area ratios would be further reduced by the factor (approx.) of:

$$A_{actual}/A_{proj} = 1 + 0.77 v_p^{2/3} \tag{2}$$

For a plausible $v_p = 0.20$, we obtain an area ratio of approximately 1.16, a *very* modest increase in fracture energy.

APPROACHES AND MECHANISMS. The methods for producing a rough fracture surface in ceramic composites, thereby increasing the apparent fracture energy, include various composite approaches—particulate, whisker, fiber, and other multiphase materials—in which crack interaction with the microstructure produces a rough fracture surface. The effects in a composite that are required to ensure crack–microstructure interaction include residual stresses (from thermal expansion anisotropy and thermal expansion mismatches); stress-induced microcracking at second-phase particles; elastic mismatches between phases, and weak interfaces between phases (Fig. 3). In each case here, the requirement is for some interaction between a propagating crack and the microstructure that causes the crack to deviate from a planar mode. This can be done by changing the local resistance to fracture—for example, by selecting tough second phases, which the crack must bypass, or weaker paths (lower fracture energy), which the crack will preferentially follow. In the latter case, it is necessary also to ensure that the increased fracture surface area is not offset by the reduced fracture energy on the weaker directions.

A similar problem arises with preexisting microcracks, which produce a very rough fracture surface although the *new* fracture surface area created by the advancing crack may be very small. Another more effective means of producing the desired crack–microstructure interaction is through perturbations in the crack tip stress field. These can be produced by interactions of the external stresses and the crack tip stress field with the microstructure (i.e., with the elastic heterogeneities or anisotropies), but these typically are small perturbations. Much larger effects would result from the presence of the residual stresses that would arise from thermal expansion anisotropy and thermal expansion mismatches between matrix and second phases. For example, in an alumina–boron nitride composite, where the expansion mismatch between the *c* axis in BN and the alumina is about 20×10^{-6} mm/mm K, one would expect residual stresses on the order of 5–10 GPa. In this case, the residual stresses are probably sufficient to cause microcracking, even in the absence of applied loads. Clearly, very high residual stress fields can be generated by thermal expansion mismatches, stress fields that are comparable in magnitude to those in the vicinity of the crack tip.

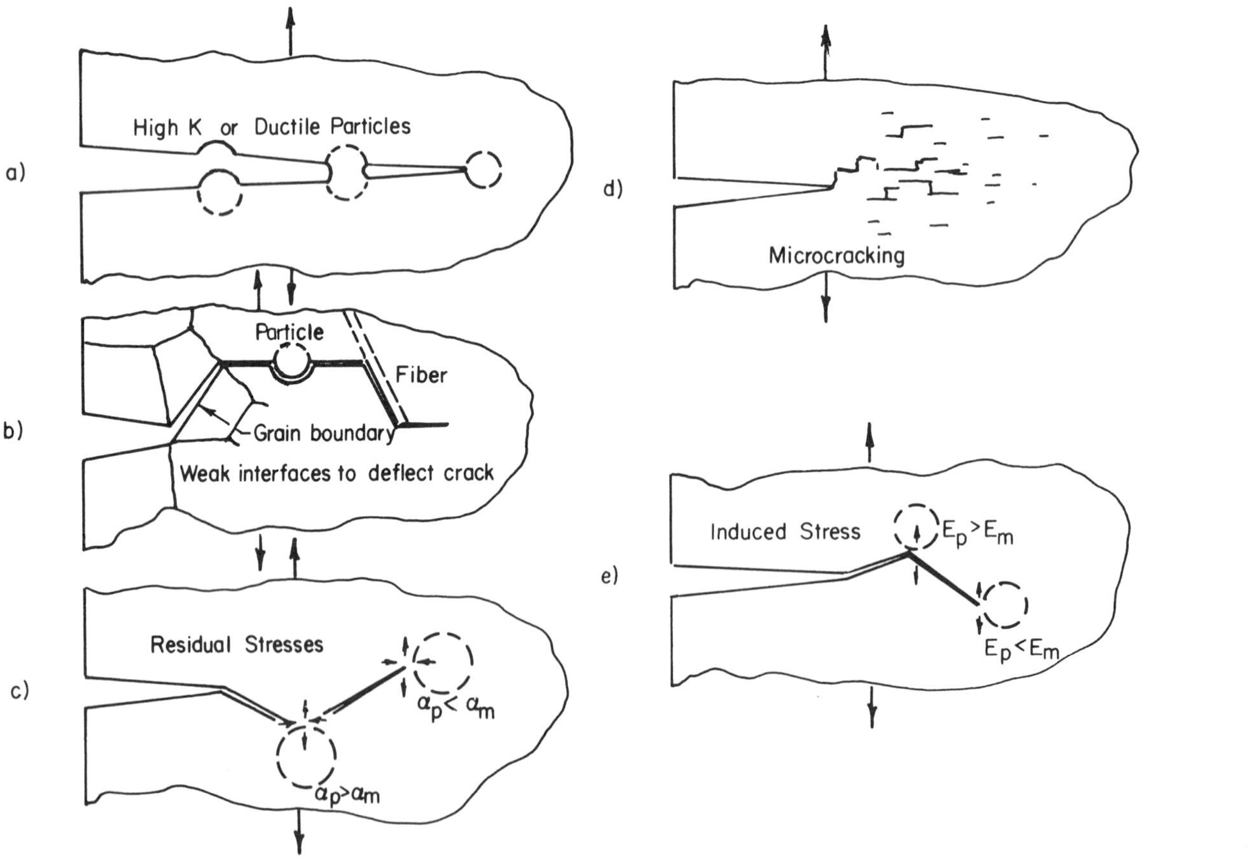

FIGURE 3 **Mechanisms for ensuring interaction between propagating crack and microstructure. (*a*) Tough second-phase particles, which hinder crack propagation. (*b*) Weak interfaces, which deflect crack. (*c*) Residual stresses at second phases, which affect crack propagation. (*d*) Microcracking, leading to crack branching and deflection. (*e*) Elastic mismatches between phases, leading to induced stresses and crack deflection.**

However, even here some of the effects may have detrimental results. In the case of high residual stresses, significant amounts of stored elastic strain energy are available to drive the crack, in addition to that supplied by the external loads. The contribution of the stored strain energy may *decrease* the apparent fracture energy, even below the true surface energy.

To summarize, various mechanisms are available to increase the fracture surface area and thus the apparent fracture surface energy in ceramic matrix composites. The most effective is the use of composites in which significant thermal expansion mismatches lead to large residual stress fields. Weak interfaces in certain composite materials (e.g., fiber and whisker) may also provide significant effects. The effects to be expected here are relatively small—increases of 1–2 times over the true surface energy—and may be masked by other toughening mechanisms with larger effects. However, as noted later, many of the same techniques that produce rough, high area fractures also lead to toughening by other means, and the net effect may be much greater than the modest effects from the area increase alone.

Energy Absorption by Irreversible Processes

Several techniques provide the opportunity for very large increases in fracture toughness by providing a mechanism for absorbing energy from the applied loads. This energy does not go into the creation of new crack surfaces, but is typically dissipated as heat in the bulk of the material. While the discrete processes involved may be reversible, the essential requirement is that the total process be irreversible. A mechanism that absorbs energy from the applied loads, stores it, and then returns it in a form that permits the energy to be used to drive the crack will not be beneficial. On the other hand, a process that converts the elastic strain energy associated with the applied loads or stresses to thermal energy will be beneficial because lattice vibrations or phonons will add to the total work required to propagate a fracture through the material.

Another feature common to this approach for fracture toughening is the concept of a "process zone." In all the toughening mechanisms discussed below, there is a region near the crack tip in which some process occurs, stealing energy from the advancing crack. The size of this zone, and the relative amount of energy absorption per unit volume within the zone, are critical parameters in the total effect on fracture toughness and may depend on: the number, density, and size of transformable particles; the number of microcrack nuclei; the degree of fiber bonding and the spacing of critical flaws on fibers (which govern the pullout lengths); and the number, density, and size of ductile particles in the material.

APPROACHES AND MECHANISMS. Some of the possible techniques for producing high fracture toughness via energy absorption in ceramic matrix composites listed below are shown schematically in Figure 4.

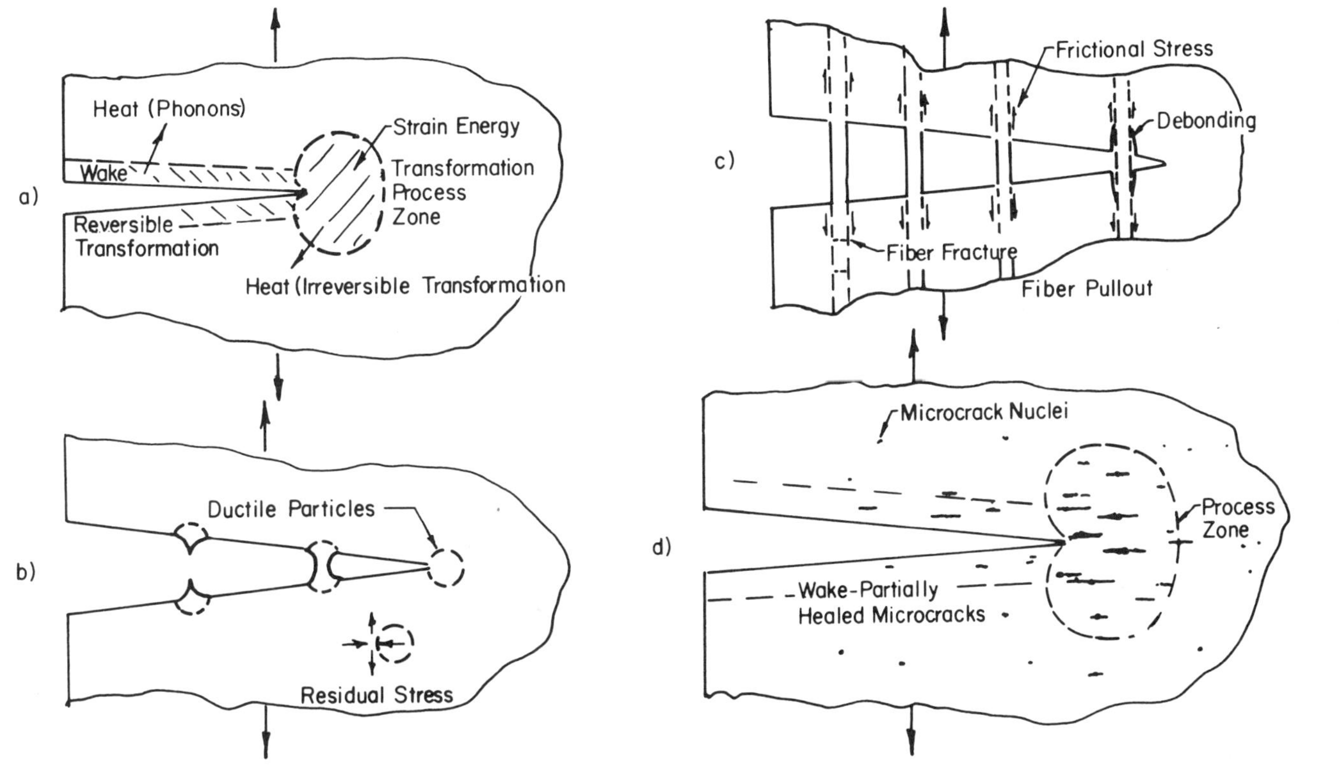

FIGURE 4 Processes leading to energy absorption and toughening in ceramic matrix composites. (*a*) Reversible or irreversible phase transformations. (*b*) Ductile failure in particulate phase. (*c*) Fiber pullout. (*d*) Stress-induced microcracking.

- Phase transformations—irreversible or reversible.
- Ductile flow in second phases.
- Whisker or fiber pullout against frictional stresses.
- Stress-induced microcracking at second phases.

One essential feature of all these mechanisms is the introduction of hysteresis into the stress–strain behavior of the material. Loading and unloading cycles, as occur on passage of a crack front, produce a net energy absorption per unit volume, given by the area in the stress–strain hysteresis loop (Fig. 5). Even in a process that itself is reversible, such as transformation and retransformation of an incoherent tetragonal zirconia particle in an alumina–zirconia composite, the net effect may be hysteretic. In the latter case, elastic strain energy is stored in and around the transformed zirconia particle by the advancing crack. When the crack has passed, and the stresses have decreased to sufficiently low levels, the particle retransforms, releasing the stored energy as elastic waves (phonons). Typically, however, these phonons are not available to aid in advancing the crack; most of them are simply dissipated as thermal energy in the bulk of the material.

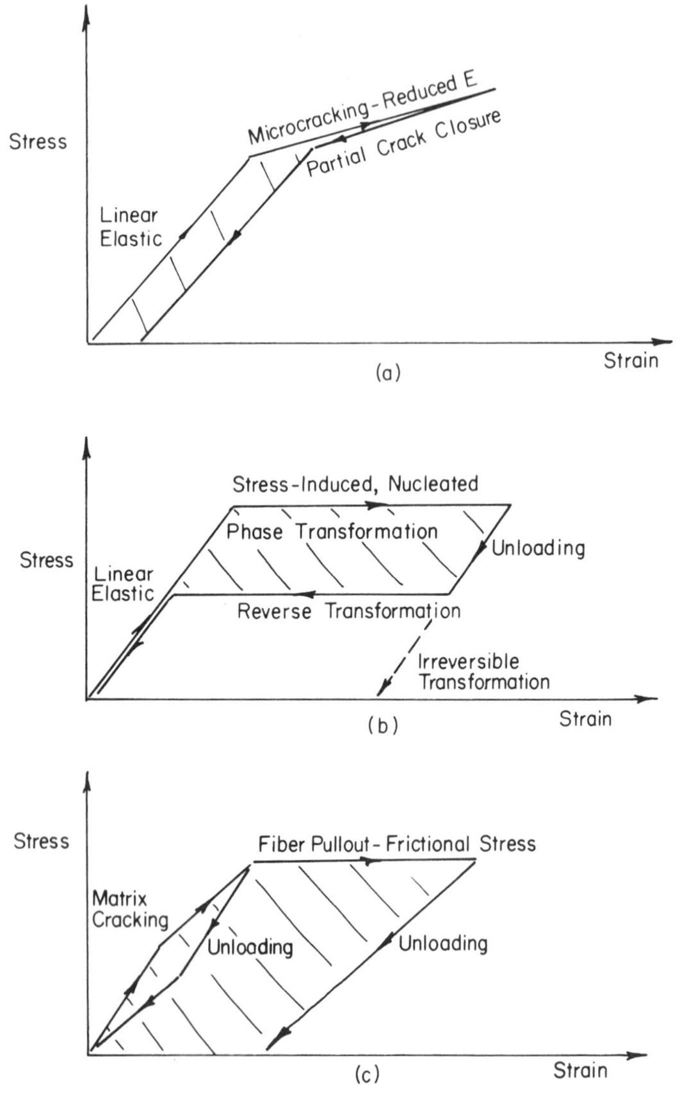

FIGURE 5 Hysteresis in stress–strain or load–deflection behavior leading to energy absorption during crack propagation.

The most successful use of the energy absorption technique to date has been with composites containing zirconia, using the martensitic monoclinic–tetragonal transformation. This class of transformation-toughened materials includes a variety of high modulus, high hardness ceramics containing dispersions of zirconia particles. If the zirconia is retained as the metastable tetragonal phase by the elastic constraint of the high modulus matrix, toughening by the phase transformation–energy absorption process is possible. Some of the materials, such as alumina–zirconia, exhibit combined hardness and toughness, properties that make them useful as cutting tools and possibly for other wear parts. Clearly, other materials can be toughened by the same technique, if chemical reactivity and processing difficulties can be managed. The toughness values obtained in these particulate composites are limited somewhat by the constraints on the volume fraction of zirconia that can be incorporated successfully and maintained in the tetragonal structure, but they may nearly double those of the matrix (e.g., 6–10 MPa · $m^{1/2}$ for alumina–zirconia vs. 4–5 for alumina, and 5–6 MPa · $m^{1/2}$ for mullite–zirconia vs. 2 for mullite itself).

Another class of these materials that might be construed to be a composite consists of large-grain cubic zirconia containing a high volume fraction of tetragonal zirconia precipitates (Fig. 6). These materials exhibit much higher values of toughness than the alumina–zirconias and similar toughness values (as high as 15–25 MPa · $m^{1/2}$), with significant *R*-curve behavior. The latter implies an increasing resistance to crack growth with increasing crack length, a very useful engineering property. These materials also exhibit surprisingly little scatter in strength (for ceramics), with variations of less than 1% in some large volumes of material. However, the strength values are typically limited to about 500–600 MPa by the large cubic grains (or associated defects), which act as the strength-limiting flaws.

While the transformation toughening effects in zirconia (and hafnia) have been the subject of the most investigation, various other methods offer significant performance gains in ceramic matrix composites without adversely affecting other important properties. In addition to the martensitic transformation in zirconia and hafnia, there may well be other phase transformations that will prove to be useful in toughening ceramics.

Another, and very effective technique for dissipating additional energy during fracture of a ceramic composite is through whisker or fiber pullout. Extracting fibers from a matrix in a composite, against frictional stresses, can absorb large amounts of energy, since the area involved (i.e., the fiber lateral area) is quite large, and the pullout distances can also be quite large. Sample stress–strain curves and the fracture surface of a ceramic matrix–ceramic fiber composite are shown in Figure 7. For example, for a composite with 50 v/o of 10 μm diameter,

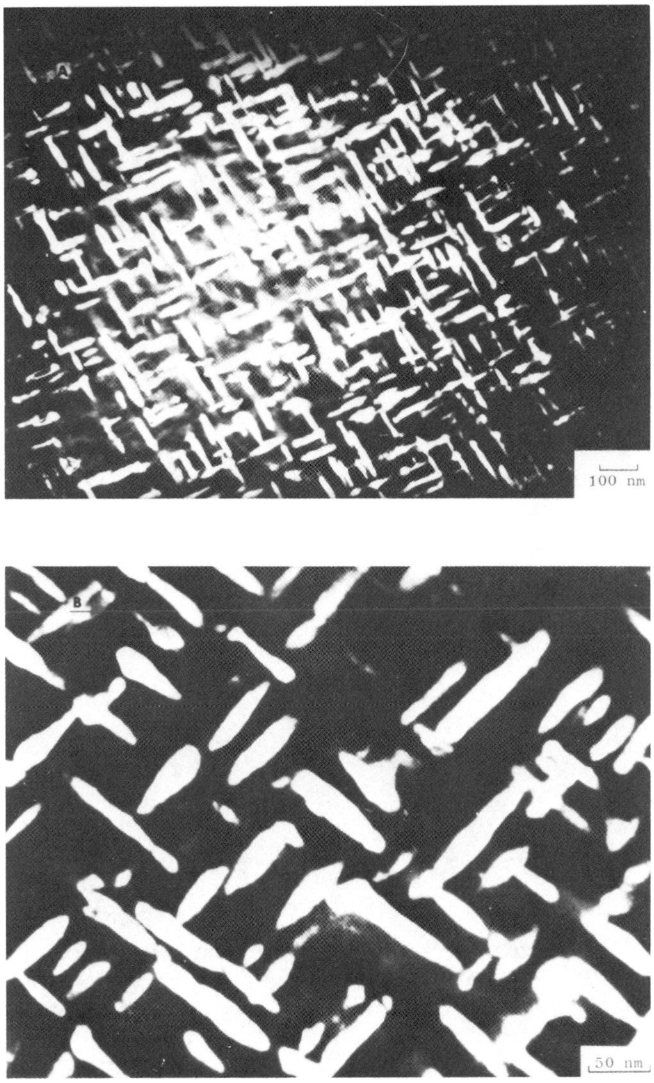

FIGURE 6 **Microstructure in Mg–partially stabilized zirconia, consisting of high volume fraction of metastable tetragonal precipitates (light regions) in cubic matrix, with high potential fracture toughness for optimal microstructure.**

continuous fibers, being fractured in a plane normal to the fibers, extracting the fibers to a length of 70 μm (the buckling length for a fiber stress of 2 GPa) against a frictional stress of 2 MPa (these numbers are typical of UTRC's Compglas) gives an additional work of fracture of 0.5×10^4 J/m^2. This is comparable to the values for the tougher metal alloys, but it represents the effects obtained for large cracks (the crack opening displacement must be on the order of 100 μm). Thus, this toughness would apply only in large components loaded to large strains; at smaller strains, or in smaller components, appropriately smaller values of fracture energy would pertain. This type of behavior, where resistance to crack propagation increases with increasing crack length, is called R-curve behavior and is more generally associated with tough metals rather than with brittle ceramics.

Microcracking, particularly the stress-induced form, is also thought to contribute to energy absorption during fracture of ceramic composites. Microcracking, typically resultant from residual stress fields from either thermal expansion mismatches or phase transformations (as with the monoclinic–tetragonal transition in zirconia), produces a hysteresis in the stress–strain curve, and thus a net absorption during the stress cycle associated with the passage of the fracture front. The opening of the microcracks, in a process zone around the crack tip, as a result of the combination of the residual stresses and the stress field associated with the crack, requires the contribution of energy from the applied loads. When the crack has propagated further, and the microcracks close, this energy is not recovered. This is analogous to the situation discussed with respect to reversible phase transformations. The possible fracture energy increases obtainable here are a function of the microcrack density (number/unit volume), the surface area of the microcracks, the size of the process zone, and contributions from the residual stress fields. In some materials, such as the alumina–zirconia composites, microcracking (rather than phase transformation) is thought to be the major contributor to fracture toughness increases, and values of K_{I_c} as high as 10 MPa $\cdot$ m$^{1/2}$ have been reported, corresponding to fracture energies of 125 J/m. These compare very favorably with values of approximately 4 and 25 MPa $\cdot$ m$^{1/2}$ for alumina alone.

The last technique discussed here is that of toughening via incorporation of ductile phases in a brittle matrix. This is a common technique for toughening brittle polymers, through block copolymerization with rubbery materials. It is also the major source of the toughness in metal-bonded carbide cutting tools, where the small fraction of the cobalt or other binder provides the bulk of the toughness for the material. In more general cases, attempts have been made to toughen other hard and brittle ceramics by incorporation of metal wires and metal particles. While some of these efforts have been moderately successful, they are typically limited by the high processing temperatures required for consolidation of the ceramic matrix, and the consequent problems with reactions between the ceramic and the metal.

When reactions are not a problem, or processing techniques permit consolidation or fabrication at lower temperatures, significant toughness increases have been obtained. One current example is the materials made by the Lanxide process, which in one case results in a continuous grain boundary phase of aluminum in an alumina body. This material exhibits roughly twice the fracture toughness of alumina, while maintaining most of the other desirable properties. In this particular instance, the alumina is formed by oxidation of the aluminum; if the process is not carried to completion, the residual aluminum remains at the grain boundaries and provides significant toughening through ductile failure during crack propagation.

Another example, also involving a nonreactive system, is that of alumina toughened by a dispersion of fine platinum particles, produced by decomposition of chloroplatinic acid added to the starting alumina powder. In

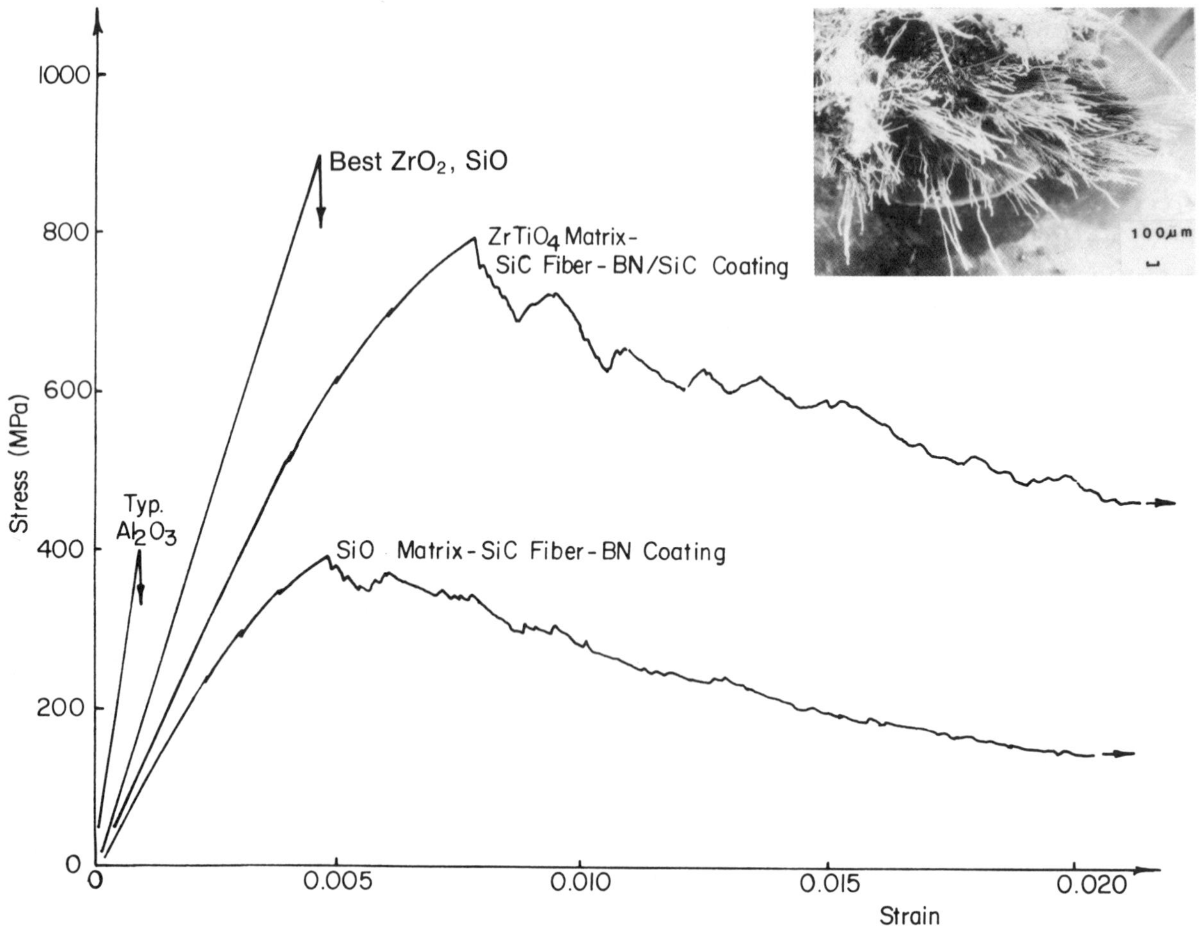

FIGURE 7 **Load–deflection curves in three-point bending for several conventional ceramics and several ceramic matrix–ceramic fiber composites; toughest composites exceed 25 MPa · $m^{1/2}$ for K_{I_c}. Inset photomicrograph shows fracture surface for SiC/BN-coated Nicalon fibers in $ZrTiO_4$ matrix with extensive fiber pullout associated with very high toughness.**

this case also, the platinum, which has a very low yield stress, and bonds to the alumina, fails in a ductile manner, and adds significantly to the fracture energy of the alumina, even for very small volume fractions of the platinum dispersion. The basic requirements, necessary for significant effects with the incorporation of ductile phases into a hard, brittle matrix, are for good bonding of the ductile material to the matrix, as achieved in the two cases above, a relatively high volume fraction of the ductile phase (this is normally limited by other considerations), and some mechanism to ensure crack interaction with the ductile phase (either residual stresses or a *continuous* ductile phase). The last is necessary to avoid the problems associated with the early work on cermets for gas turbine components, where brittle failure occurred by propagation of cracks through the continuous ceramic matrix phase, avoiding the ductile metal incorporated for toughening.

Change in Fracture Mode

One common feature of many of the toughening mechanisms discussed is a process that will cause the propagating crack to deviate from a planar mode. Since cracks in ceramics typically propagate in directions normal to the maximum tensile stress (i.e., in mode I), this deviation from planarity has some additional beneficial effects. Depending on the orientation of the out-of-plane portion of the crack front, the crack will typically be forced to propagate in a mixture of mode I and mode III or mode I and mode II (Fig. 1*c*). As the crack deviates from the original plane, the primary driving force (i.e., the K_I) decreases, and the crack must also be driven by the shear components of the stress field. In purely tensile loading, the net driving force for the crack would decrease as the crack deviated from its initial direction.

While the decrease in the crack driving force provides significant beneficial effects, the benefits of this mode change result also from the typically larger values of K_{II_c} and K_{III_c} relative to K_{I_c}. These imply not only a greater value of stress intensity required for crack initiation, but also, assuming linear fracture mechanics, a greater value of fracture energy for crack propagation, through the typical relationship $\gamma_f = K^2/2E$, where E is the Young's modulus of the material. This relationship, which clearly is valid only for materials with fracture energy independent of crack length and history, and for linearly elastic, brittle materials, would be expected to apply, in the general case, for small crack extensions. It is surely not valid for many materials incorporating the toughening mechanisms discussed above, where the fracture energy would clearly be a function of crack length, loading history, process zone size, stress state, etc.

Unfortunately, the effects here are closely interwoven with those of increased fracture surface area, and the individual contributions are difficult to separate, either analytically or experimentally. Some estimates have been made of the total effects, termed "crack deflection" or "crack–particle interaction" toughening, and these indicate that increases of 2–5 times in fracture toughness over that of the matrix are possible in the best cases.

APPROACHES AND MECHANISMS. Many of the approaches for inducing changes in fracture mode are common to the toughening mechanisms discussed previously and are not repeated. As noted earlier, the possible schemes include residual stresses, typically resulting from thermal expansion mismatches between matrix and second phases and thermal expansion anisotropy in the second phase; crack interaction in various ways with the components in composite or multiphase materials; and use of composite materials containing weak interfaces. The same cautions voiced earlier still apply. The residual stress fields can supply additional energy for crack propagation, reducing the work required from the external loads, and weak interfaces reduce the energy required for crack propagation. Presumably (or hopefully), the greatest toughening effects will occur at some intermediate values of residual stresses or interface weakness; otherwise no gains in fracture toughness will be realized.

Crack Shielding, Crack Bridging

Another mechanism for toughening hard materials, which is related to some of the effects discussed previously, is that of crack shielding. Its primary effect is to change the nature of the stress field in the vicinity of the crack tip, thereby reducing the driving force for crack propagation. This can be done by a process that operates in a zone around the crack tip, as in the process zones discussed previously with respect to energy absorption processes, or by a process that operates behind the crack tip. The latter, typically achieved by a process that tends to hold the crack surfaces together, is called crack bridging; here, one major beneficial effect is still the change in the crack tip stress field, and therefore a reduction in the driving force for crack propagation. In both cases we seek a mechanism to "shield" the crack from the applied stresses, hence the name "crack shielding."

APPROACHES AND MECHANISMS. Many of the toughening processes discussed before will also have beneficial effects relative to crack shielding or crack bridging, and thus greater total toughening effects. Phase transformations, especially those leading to dilatations (volume increases) or shears, will have significant effects on the stress fields adjacent to cracks, since the primary components of the near-tip stress field are biaxial or triaxial tension and shear. Thus, the martensitic transformation in zirconia and hafnia, which produces both large shear strains and dilatations, is particularly useful in this regard. Microcracking induced by combinations of residual stresses and applied stresses is also very effective, since there is a net dilatation associated with the opening of the microcracks.

Other effects that operate behind the crack tip and restrict the crack opening displacement will also have large effects. In the case of continuous fiber composites, where the fibers bridge the crack, *very* large effects can be obtained, though in practice, these are inextricably mixed with other toughening effects such as fiber pullout. The total increases in fracture energy for such fiber composites are large enough to produce in the composites fracture energies 10–100 times that of the unreinforced matrix material. For other types of material, such as those incorporating short fibers or whiskers, lesser though nevertheless significant effects would be expected.

Miscellany

Several techniques that have been proposed and evaluated for toughening ceramics have been omitted thus far because they do not fit neatly into any of the categories established. One of these, crack pinning, was originally conceived as an analogue to dispersion strengthening in metals; here a propagating crack is pinned at discrete locations by appropriate dispersed particles, which the crack cannot penetrate. This phenomenon could be achieved by, for example, dispersions of metal particles or fibers in a brittle matrix, or by dispersions of a relatively high toughness phase in a low toughness matrix. The most significant result of crack pinning is an increase in the length of the fracture front. In the case of dislocation lines in ductile materials, where there is an energy per unit length associated with the dislocation line, pinning by dispersed phases increases the length and therefore the total energy associated with the dislocation line. The localized pinning of cracks, by analogy, also should result in a net increase in the energy associated with the crack front. This would translate, on a macroscopic scale, to an effective increase in the stress or work necessary to propagate the crack.

While direct observation of such crack pinning has been made occasionally (e.g., by ultrasonic marking of fracture surfaces), there are many complications associated with the use of this technique for toughening ceramics. Effective toughening requires a small interparticle spacing to ensure interaction of the crack with a large number of particles; this in turn requires either a large volume fraction of dispersed phase, which will have adverse effects on other properties, or a smaller volume fraction of a dispersed phase of very small size. The latter is difficult to achieve because of dispersion and chemical reactivity problems. In addition to these difficulties, there must be some effective mechanism for ensuring that the crack front intersects the particles, rather than bypassing them. As noted before, this would probably require significant residual stresses, which would attract the propagating crack to the particles, in addition to strong bonding of the dispersed particles to the matrix; however, the use of high ductility particles would very likely relieve a major part of the residual stresses. Note that with all these restrictions satisfied, we would also have a system that would exhibit CPI toughening (increased fracture surface area and mode change) and toughening by energy absorption, and the contribution of the crack pinning process would probably be negligible.

Another technique, which has been little explored on a scientific basis until quite recently, is that of incorporating porosity to provide beneficial interactions with the crack front. Treating a porous material as a "composite" of ceramic and void spaces may be overextending this concept, but the same mathematics and mechanics apply with appropriate properties for the voids. The stress field at a pore, subjected to a macroscopic tensile stress or the stress field in front of a propagating crack, is such as to attract the crack to the equator of a spherical pore. The leading edge of the crack front at the pore then becomes the opposite side of the pore. For a pore or void of benign shape, with spherical or ellipsoidal geometry, this portion of the crack front is now effectively blunted and will not propagate until the adjacent, sharp portions of the crack have propagated well ahead of it. This localized pinning of the crack is countered, to some extent, by the reduction in new surface area which the propagating crack must create, but the net effect may be an increase in toughness.

A more subtle effect, which may be more significant than the modest effects of crack pinning, is that of the interrupted nature of the crack propagation, with large changes in speed, as indicated in Fig. 8. The fluctuations in crack speed must be accompanied by energy dissipation into the matrix (e.g., by phonons generated by the rapidly changing stress fields) and consequently will be associated with an increase in effective fracture energy.

However, efforts to improve the toughness of ceramics via incorporation of porosity have generally been hindered by the other detrimental effects of the porosity, and by the difficulty in introducing porosity in a controlled manner. If the pores introduced are random in size and spacing, the size and spacing distribution extremes will lead to large pores or to pore clusters, which will adversely affect the strength of the material. The largest pores or pore clusters will act as the strength-limiting flaws. Some limited success has been obtained in relatively low strength materials, such as refractories, where the normal flaw size is quite large. The pores in this case act primarily to arrest propagating cracks and prevent catastrophic failure of the refractory material.

A more recent development in this area, which has been applied successfully to a piezoelectric ceramic material, lead zirconate titanate, is the incorporation of an ordered array of anisotropic voids by a combination of

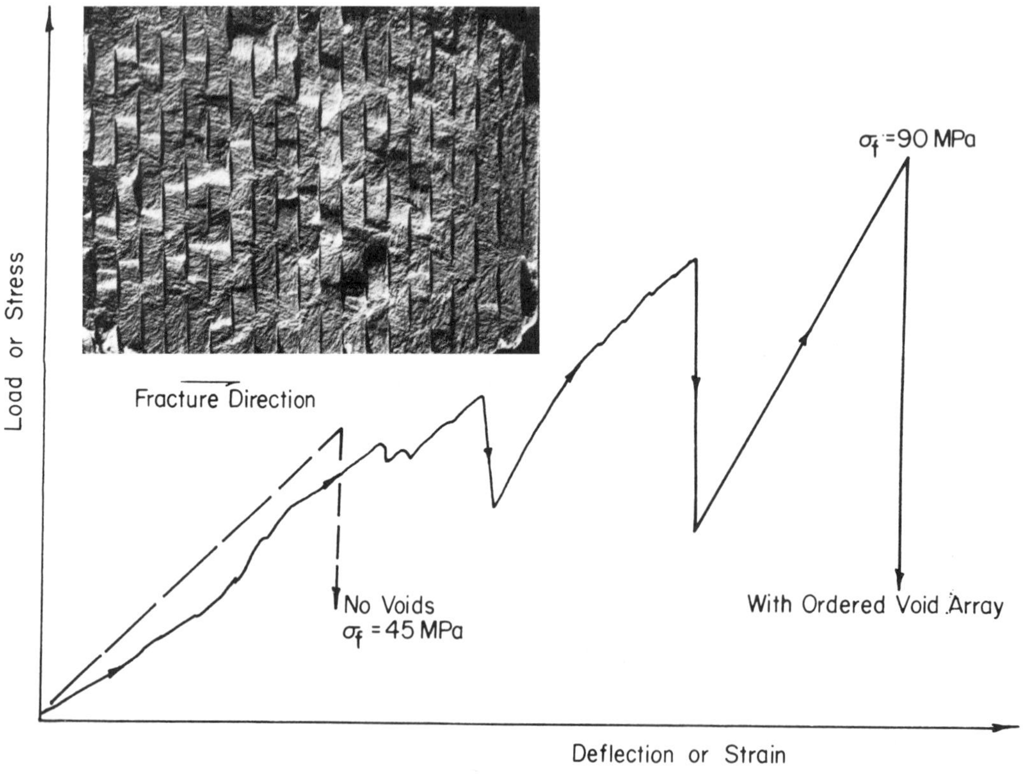

FIGURE 8 **Load–deflection curve for lead zirconate titanate material containing array of ordered voids and result for comparable material without voids, showing interrupted nature of crack propagation. Inset photomicrograph of fracture surface shows extensive crack–void interactions.**

tape casting, photolithography, and lamination. In this instance, the incorporation of appropriate arrays of ordered voids increases the apparent work of fracture by 1–2 orders of magnitude and the strength of the ceramic by about 100%. Fractographic analysis indicates a very complex fracture interaction with the void arrays, and 3D finite difference modeling of the material suggests that the void arrays significantly and beneficially affect the stress field at the surface cracks that commonly lead to failure. Thus the presence of ordered void arrays may lead to toughening by crack pinning, crack blunting, and crack shielding, the latter through the local changes in compliance associated with the voids.

Synergisms and Interactions

It is now evident that many of the composite toughening mechanisms and approaches are highly interactive. These interactions may well have beneficial effects, with appropriate design of composite microstructure, or they may instead have detrimental effects. An example of beneficial interactions is an alumina–zirconia material with phase compositions and phase sizes chosen to produce a combination of microcracking and phase transformation toughening (i.e., a duplex microstructure). In this case, process zones can be developed around each of the microcracks, and around the main crack, leading to a much greater extent of transformation toughening. In addition, the transformation toughening of the material around the microcracking regions should minimize the linkage of microcracks, thus minimizing one of the detrimental features of microcrack toughening. A similar result might be obtained with an appropriate combination of whisker toughening and phase transformation toughening or microcrack toughening, where it might be possible to take advantage of both toughening processes.

Another advantage of these multiple toughening schemes is the potential for high toughness over a wide temperature range. Typically, transformation toughening effects diminish with increasing temperature, and the toughening effects associated with many other schemes depend critically on the residual stress state. The residual stress state, as noted, depends on the difference between the operating temperature and the fabrication temperature of the material. Other toughening processes such as plasticity or viscous flow also have shown strong temperature dependence, becoming operative only at high temperatures. Clearly, it should be possible, with sufficient understanding of composite toughening mechanisms and composite processing, to produce materials with desired properties over a wide range of operating conditions. Such materials are critically required for a number of advanced applications such as the high performance gas turbine engine and structural materials for hypersonic aircraft, where the operating temperatures would range from ambient, or even cryogenic, to as high as 1000–1200°C, and high toughness over the entire operating range would be a necessity.

Obviously, one of the best ways to accomplish this would be through the effective use of multiple toughening mechanisms in ceramic composites, and through the use of the synergisms such approaches permit. A key element in this use of synergistic multiple toughening mechanisms is the precise control of processing and fabrication to permit precise microstructure and architecture control. The control of processing and fabrication is necessary to obtain the optimal microstructure or composite architecture *and* to control deleterious reactions between constituents. Such control, in addition to preventing or minimizing interdiffusion and reactions, also must ensure the existence of a controlled interfacial region or tailored interphases in the composite. There are many indications that effective control of interfacial regions is one of the key elements in the successful production and use of ceramic matrix composites.

Summary

It would be possible to carry this discussion to extreme lengths, as the subject is inherently complicated, becoming more so as the number of choices in composite systems increases and as the number of toughening mechanisms operative increases. One of the long-standing problems associated with the use of composites, and with ceramic composites in particular, is the unwillingness of investigators and users alike to accept the great complexity of these materials and their behavior. While complexity may be a great hindrance in terms of processing, characterization, analytical modeling, or simple understanding of the behavior of composites, it can also be a great advantage. The complexity of these materials and their behavior can be utilized in tailoring them to precise ends and in "designing" composite materials for particular demanding applications. This requires acceptance of the complexity and a willingness to use the tools necessary to deal with the consequent analytical and experimental difficulties.

Literature

To provide some other sources of information on ceramic composite approaches to toughening ceramics, a brief bibliography is attached, listing references in this article as well as items on many of the topics discussed.

David Lewis III

References and Bibliography

Fracture Mechanics

1. B. R. Lawn and T. R. Wilshaw, *Fracture of Brittle Solids*, Cambridge University Press, London, 1975, pp. 58–59.
2. D. Broek, *Elementary Engineering Fracture Mechanics*, Nijhoff, Boston, 1984, pp. 15–17.
3. S. W. Freiman, *Am. Ceram. Soc. Bull.*, 67(2), 392 (1988).

4. S. W. Freiman and E. R. Fuller, Jr., Eds., *Fracture Mechanics for Ceramics, Rocks and Concrete,* American Society for Testing and Materials STP 745, ASTM, Philadelphia, 1981.

Surveys on Toughening Ceramics

5. A. G. Evans and R. M. Cannon, *Acta. Metall., 34*(5), 761 (1986).
6. A. G. Evans, *Proceedings of the Third Intl. Conference on the Science of Hard Materials,* 1987.
7. F. F. Lange, "Fracture of Brittle Matrix, Particulate Composites," in L. J. Broutman, Ed., *Composite Materials,* Vol. 5, Academic Press, New York, 1974.
8. F. F. Lange, *J. Mater. Sci., 17,* 225 (1982).

Whisker Composites

9. G. C. Wei and P. F. Becher, *Am. Ceram. Soc. Bull., 64*(2), 298 (1985).
10. P. F. Becher and G. C. Wei, *J. Am. Ceram. Soc., 67,* C267 (1984).
11. T. N. Tiegs and P. F. Becher, "Whisker-Reinforced Ceramic Composites," in R. E. Tressler et al., Eds., *Tailoring Multiphase and Composite Ceramics,* Plenum Press, New York, 1986.
12. B. Budiansky et al., *J. Mech. Phys. Solids, 34*(2), 167 (1986).
13. D. B. Marshall and A. G. Evans, *J. Am. Ceram. Soc., 68*(5), 225 (1985).

Transformation-Toughened Ceramics

Items 5 and 8 above

14. R. P. Ingel and D. Lewis, "Physical, Microstructural and Thermomechanical Properties of ZrO_2 Single Crystals," in N. Claussen, A. Heuer, and M. Ruhle, Eds., *Advances in Ceramics,* Vol. 12, American Ceram. Society (1984).
15. R. P. Ingel and D. Lewis, *J. Am. Ceram. Soc., 65*(9), C108 (1982).
16. M. J. Readey et al., *J. Am. Ceram. Soc.,* (1987).
17. R. W. Steinbrech and A. H. Heuer, *Mater. Res. Soc. Symp. Proc., 600,* 469 (1986).
18. J. S. Moya and M. I. Osendi, *J. Mater. Sci., 19,* 2909 (1984).
19. P. F. Becher, *Acta Metall., 34*(10), 1885 (1986).

Microcrack Toughening

20. D. Lewis et al., *Ceram. Eng. Sci. Proc., 2*(7–8) (1981).
21. A. G. Evans and K. T. Faber, *J. Am. Ceram. Soc., 64*(7), 394 (1981).

Ceramic Matrix Composites, Carbon Fiber Reinforced

The development of tough, high-performance ceramics, capable of being economically fabricated into a variety of complex shapes, particularly with varying cross sections, has been a development goal for many years. During the 1970s and into the 1980s, the Energy Research Development Agency (ERDA), now the Department of Energy (DOE), recognized this need and funded major programs to improve the strength, fabricability, and toughness of monolithic ceramics. These programs have contributed greatly to understanding of monolithic ceramic microstructure, phase/interphase control, properties, process variables, performance, and design with brittle materials. However, obtaining meaningful increases in toughness continues to present an elusive goal. If ceramics are to be successfully utilized in the near term, a more thorough and deeper understanding of their raw materials, fabrication, process variables, and behavior is required. Use of available monolithic ceramic materials represents attempts to apply existing solutions via substitution rationales rather than attacking the limitations of these materials directly. Current ceramics, which are flaw sensitive and fail by brittle fracture mechanisms, will have to be replaced by tough, high-fracture energy ceramic material systems whose failure mode is more forgiving and predictable and which are able to meet currently planned operational goals.

Presently, the only known method for achieving practical high-performance ceramic material systems is to introduce cohesive elements or barriers to crack propagation into ceramic structures in the form of discrete particles or fibers; i.e. form a ceramic matrix composite. With such material systems an architectural approach for application design/material selection can be used. Identified design and operational requirements are matched with state-of-the-art materials, fabrication techniques, and processes using a product goal-oriented iteration of available technologies.

An example of a typical example of ceramic matrix composite (CMC) material systems is the development and production of operational automotive components. State-of-the-art automotive gas turbines presently require an additional 280°C above their normal steady-state operating temperature to allow for acceleration of the rotating components. The use of CMC for rotor and power train construction would reduce mass and inertia by at least 50%, thereby lowering the acceleration over-temperature requirement by at least 121°C, or providing improved throttle response while still providing for increased steady state temperature operation. This one-step improvement is equal to the turbine operating performance increases achieved through the past twenty-five years of metallurgical development.

Such achievements are within the current state of the art of the composite technologies involved. However, intrinsic material properties and limitations must be recognized and utilized within design and performance goal parameters. Raw materials, fabrication, and processing technologies have to be interfaced with the designer and equipment operator for materials/component/system development, with prototype demonstrations providing iterative feedback necessary to assure success.

Figure 1 is a composite representation showing strength vs. temperature for various materials. All material strengths drop off as temperatures increase. Since these rates of decrease are not the same, it provides designers with latitude in materials selection. The selection of materials for use at operating temperatures above

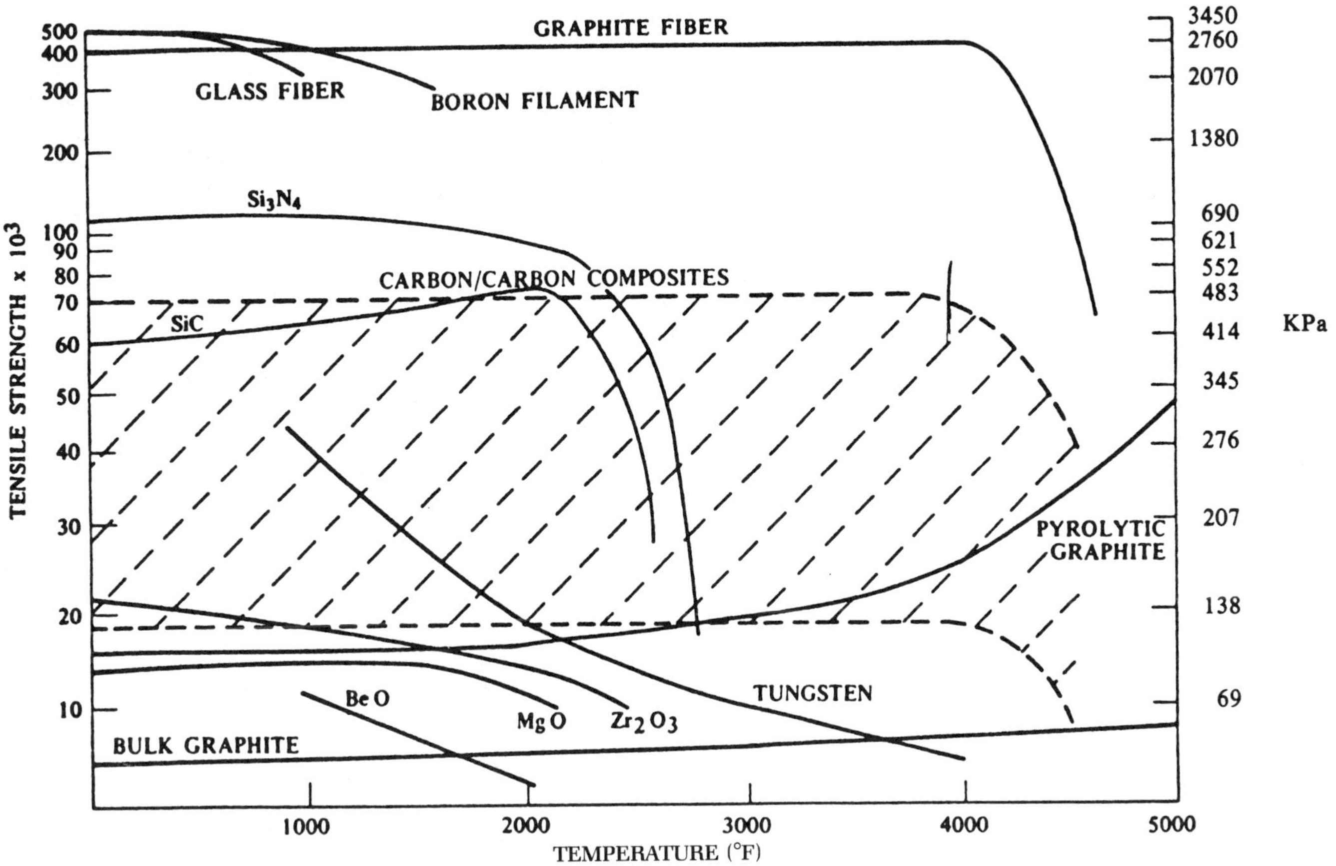

FIGURE 1 Strength vs. temperature for various materials. (From Ref. 9.)

1100°C is quite limited, and even at 1400°C, silicon nitride and silicon carbide are on the down slope of their strength curve. Conversely, carbon and its composite forms exhibit equal or increasing strength from room temperature up to 2200°C, thus providing a unique opportunity to meet high temperature structural needs.

Typical physical properties of some early carbon–carbon composites, which are brittle fiber, brittle matrix systems, are shown in Table 1. Some of the grades listed have been available for more than fifteen years. They are tough and somewhat forgiving materials. Data shows that although microcracking does occur at load levels below their ultimate failure loads, subsequent loading to equal or lesser stress levels will not propagate these microcracks. Fatigue life is almost infinite when these composites are loaded to stress levels of 55% of ultimate or less.

Composite Materials

Intrinsic flaw sensitivity and brittleness continue to impede the development and successful use of monolithic ceramic components. The large size and complex shapes of many high-temperature structures exceed the manufacturing limits of these materials. The development of fibrous reinforced carbon and ceramic matrix composite materials has progressed so that the brittle, catastrophic failure mechanism of monolithic ceramics has been replaced by a tough, microcracking matrix articulation and fiber pullout mechanism, which does not allow the composite to fail catastrophically.

Although the constituents of carbon and ceramic matrix composites are brittle, their fabrication, properties, and behavior are very similar to the more readily used fiber reinforced polymer (FRP) and metal matrix composite (MMC) material systems. Essentially, the load carrying and load transfer mechanism for each of these advanced composite material systems is identical—the fiber carries the load while the matrix serves to transfer the load from fiber to fiber.

Unlike the FRP and MMC composite systems, which have low modulus, high-strain capability matrices, the carbon and ceramic matrices have high moduli and very low-strain capabilities. Generally, the modulus of the matrix in these systems is equal to and sometimes greater than that of the reinforcing fiber. There is a common need for the matrix in composite materials to absorb strain and accommodate a uniformity of load transfer from fiber to fiber. To exhibit good toughness, the brittle and high modulus carbon and ceramic matrices must be made to appear to respond in a ductile manner.

In applying the theory of fracture mechanics to the thermal shock resistance of brittle materials, Rossi and

TABLE 1
Properties of Some Available Carbon/Carbon Composites

Material	Bond	Density g/cc	Fiber Vol%	Fiber Orien-tation	Tensile Strength MPa	Flexure Strength MPa	Inter-laminar Strength MPa
Ag Carb 101	Resin Char	1.45	66	flat	75.9	96.6	5.2
PTE	Pitch Char	1.57	> 60	flat	74.5	75.9	
Carbitex 530			66	flat	75.9	356	13.8
513	Resin Char	1.45	66	fil. wound	105–138		
515			66	unidir.	276	17.3	12.4
Pyrolarex	Resin Char	1.00		chopped	8.7	86.3	
Pyrolarex 350	Resin Char	1.20	> 60	flat	42.8	138	6.9
Pyrocarb 400	Char/CVD	1.30	> 60	flat	96.6		6.6
Haveg 513	Char/CVD					44.2	
41 G(L)	Char	1.45	> 60	flat	40	64.9	14.5
LTV-CG	Char/CVD	1.38	> 60	flat	64.2	82.8	15.2
RPG	CVD	1.75	5–7	felt	47.6		41.4
Pyro-Bond	CVD	1.34	50	fil. wound cir. wrap	828		
Pyco-Bond	CVD	1.65	34	flat		169	31.1
Pyco-Bond	CVD	1.45	44	fil. wound 75° wrap	to 391		

From Ref. 9.

Hasselman [1] describe the values of increasing the surface fracture energy of a ceramic through the use of reinforcing particles or fibers. Developments in graphite fiber reinforced carbon, silicon/silicon carbide, and glass matrix composite systems have demonstrated viability in achieving increases in surface fracture energy allowing economical production of high strength, intermediate to high temperature hardware tailored to meet specific applications.

Early efforts at Owens Illinois with glass reinforced glass and at United Technologies Research Center with tantalum reinforced silicon nitride and graphite/silicon carbide reinforced glass have shown that high-strength, impact resistant, tough composite components can be readily fabricated. Additional efforts at General Electric, Fansteel, Tylan, SEP, Amercom, and Refractory Composites have demonstrated that fibrous reinforced, net dimension ceramic composites can be produced free of detrimental porosity. General Electric and Fansteel have shown that fibrous reinforced Si/SiC composite strength approximates rule-of-mixture theory while producing materials that can be manufactured free of pore sensitivity (Table 2).

At UTRC, Brennan introduced tantalum fibers into a silicon nitride matrix, successfully increasing the Charpy impact strength from 0.68J to 4.8J energy required to cause cracking (Figure 2). Using graphite fibers in a glass matrix system, Prewo [2,3] demonstrated properties approximately equalling those of graphite/epoxy at room temperature, and even higher strengths at 550°C (Figure 3).

Although no chemical bonds exist in carbon–carbon systems, the brittleness in these "ceramic matrix" composite systems appears to be a function of the fiber matrix bond strength. In all cases, poor fiber matrix bonds produce tough composites, while good fiber to matrix bonds result in brittle, flaw sensitive materials.

Graphite fiber reinforced silicon nitride and silicon carbide matrix composite systems offer a unique potential for minimizing, if not overcoming, many of the limita-

TABLE 2
Types and Properties of Si/SiC

Designation	v/o SiC	Bond Strength MPa	Elastic Modulus MPa	Density (g/cm)
G. E. Reported Data[a]				
Type TH	80–85	484 (25°C)	393 (25°C)	3.30 (25°C)
Type THL	38–40	276 (25°C)	304 (25°C)	2.70 (25°C)
Type F	20–25	173 (25°C)	200 (25°C)	2.60 (25°C)

[a] Measured on 2.5 mm × 2.5 mm × 15.9 mm specimens tested in three-point bending.
[b] High Content (approx. 50%) staring carbon and silicon carbide from reaction.
From Ref. 9.

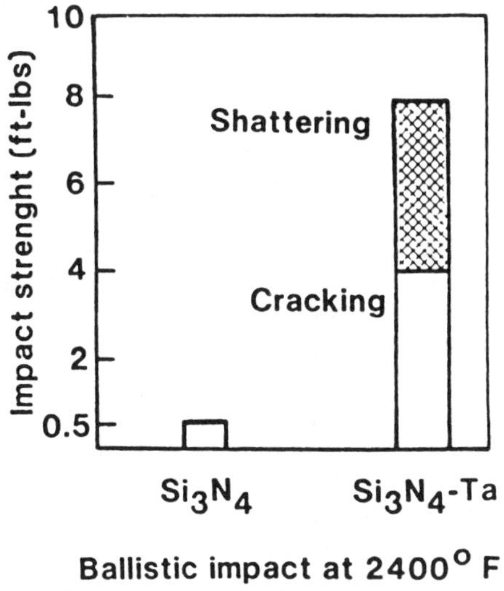

FIGURE 2 Impact properties of Si3N4 ceramics. (From Ref. 9.)

tions of monolithic ceramics. There are a myriad of approaches which involve fiber type and precursor as well as fiber volume, fabrication technique, thermal and CVI processing parameters, and the effect of the related variables. These are based upon existing experience and manufacturing capabilities in carbon–carbon composite technology presently used for commercial aircraft brakes and a wide variety of thermostructural aerospace applications.

Carbon–Carbon Composites

High-temperature aerospace component development for rocket nozzles, heat shields, nose tips, exit cones, etc., was paced by a need for components that were light weight and dimensionally stable at temperature, and had good thermal shock resistance and predictably uniform erosion behavior. The early components used combinations of monolithic graphite. The development of rayon based carbon and graphite fibers in the early 1960s provided a high-temperature replacement for glass fibers used in polymer matrix ablative systems. The subsequent development of carbon fiber reinforced carbon matrix (C–C) material systems further and very significantly extended the use temperature and structural properties.

The early C–C materials provided methods of manufacturing components that could not be produced from monolithic graphite. As fiber reproducibility and strength improved, they were coupled with better composite fabrication and processing techniques to produce

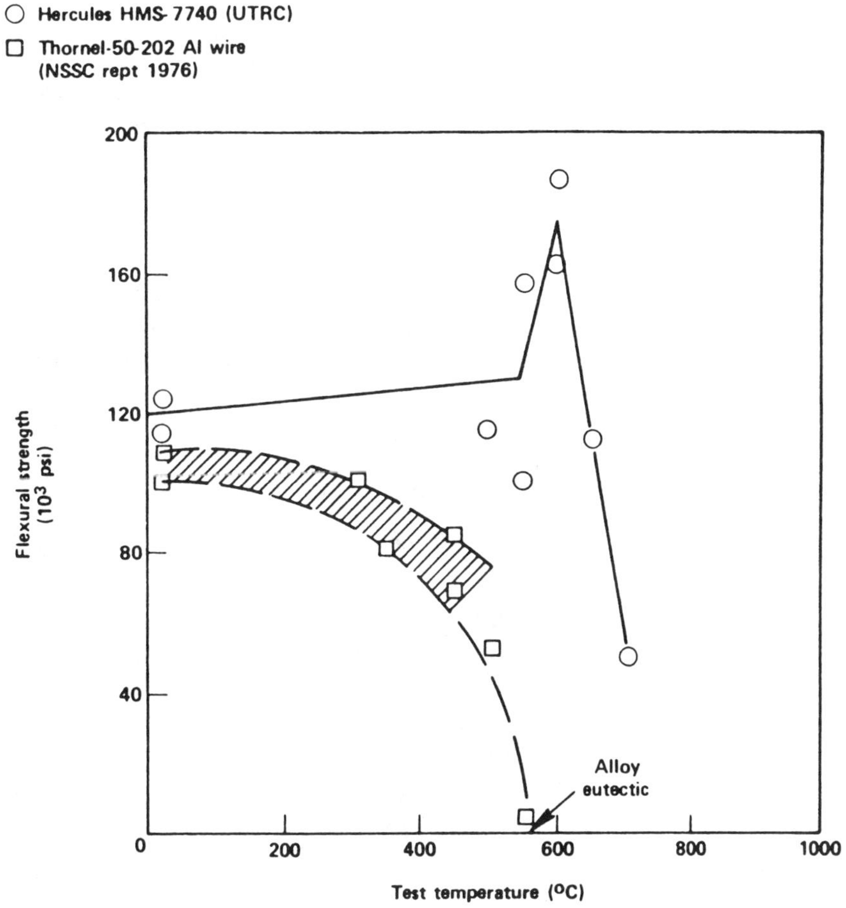

FIGURE 3 Graphite/glass composite strength vs. temperature.

FIGURE 4 **Carbon-carbon aircraft brakes (Photo courtesy of Bendix Wheels and Brakes Division, Allied-Signal Aerospace Company.)**

components that were capable of carrying some structural loads. By the mid to late 1960s, carbon–carbon structural components were being evaluated for a variety of structural applications from rocket engine and reentry components to aircraft brakes. Carbon–carbon composite brake systems are currently operational on Concorde, the space shuttle, B-1, F-14, F-16, F-18, 747, 757, 767, DC-10 and Airbus 300 (Figure 4).

Carbon Fiber Reinforcing

Because carbon fiber is strong, stiff, and lightweight, and has strength that increases with temperature, it is a logical choice for reinforcing ceramic matrix composite materials. Unfortunately, carbon readily combines with oxygen at temperatures in excess of 400°C, thus limiting its use unless oxidation resistant coatings or other appropriate form of oxidation protection are provided.

Since its introduction in the early 1960s, the properties and performance of carbon fibers have improved dramatically. Early rayon based fibers had low density (~ 1.5 g/cc), low strength (820–1050 MPa tensile) and low modulus (20–40 $\times 10^6$ KPa). These early fibers were characterized by a crenelated surface and internal porosity and a turbostratic crystallite orientation. Additionally, the properties, performance, reproducibility, and reliability varied widely from lot to lot and manufacturer to manufacturer. In the late 1960s Union Carbide introduced a heat stretched (stress-graphitized) grade of rayon based carbon fiber, Thornel 50, with a 345 GPa modulus of elasticity, 1965 MPa tensile strength and a 1.75 gm/cc density. This improvement was meaningful but expensive. It took 2.7 kg of rayon to make 0.5 kg of the unstretched carbon fiber.

During this time, substantial development effort was being expended in the area of alternate precursor materials, with polyacrylonitrile (PAN) and coal tar pitch showing the most promise. Because of their higher carbon yields, roughly 50 percent for PAN and 75 percent for pitch, they promised lower cost in production. Carbonized PAN exhibited higher strength and modulus than pitch, and rapidly replaced rayon as the precursor of choice. In addition, the PAN chemistry was easier to control and the carbonized fibers proved to be more uniform in cross-section, strength, modulus and density. Typical properties of carbonized PAN fibers ranged from 170–520 GPa in modulus, 210–280 MPa tensile strength, a fiber diameter of 8–10 μ and a density of 1.72–1.76gm/cc. A partial listing of PAN based carbon fibers and their properties available during the 1970s is shown in Table 3.

TABLE 3
Breaking Strength and Elongation for Yarn and Tow

		As-Received Ultimate Breaking		CVD Carbon Coated Ultimate Breaking	
Grade	Load, kg	Strength MPa	Elong. %	Load, kg	MPa
CY 2-1	2.2	69.0	1.7	3.1	97.3
CY 2-1	2.6	80.0	2.1	—	—
GSCY 2-5	3.9	58.7	1.5	7.8	94.5
KUREHA	3.6	111.1	2.7	3.8	116.6
GY 2-1	2.5	131.1	2.4	3.9	169.7
GY 5-1	5.4	80.7	1.6	11.1	133.9
GSGY 2-5	5.2	91.1	2.1	6.9	109.7
HMG 50	2.4	141.5	0.9	4.8	230.5
THORNEL 50	2.4	178.0	0.6	5.8	339.5
HTS	27.0	80.7	1.7	46.3	110.4

From Ref. 7.

The availability of light-weight carbon fibers that are stiffer and stronger than steel rapidly expanded their experimental use beyond the earlier ablative carbon phenolic, nonstructural carbon–carbon and structural carbon–carbon brake applications. When incorporated into epoxy matrices, in place of E or S glass fibers, the resulting composite properties were stiffer, stronger, and lighter weight. The toughness of the glass fiber reinforced system was also improved by the addition of carbon fibers. This dramatic improvement in fiber reinforced polymer (FRP) composite properties led to their expanded use and a substantial increase in carbon fiber production. Although high priced at $60–100/kg, the tough, light-weight, stiff, high-strength carbon FRP composites found a ready market in a variety of sporting goods applications (golf clubs, tennis racquets, bicycles, fishing rods, racing car bodies, and skis). Aerospace and industrial applications for this new class of high-performance advanced composite FRP materials also expanded rapidly (aircraft primary frame, control structures, automotive drive shafts, etc.).

Concurrent with this rapid expansion of PAN based carbon fiber manufacturing capacity and continuing carbon fiber improvement came a slower but equally meaningful improvement in carbon–carbon composite properties. Whereas, the earlier rayon based carbon fiber reinforced composites exhibited low strength and moduli (Table 1), the PAN based carbon–carbon systems demonstrated tensile strengths in excess of 21 MPa and elastic moduli of 69–104 GPa at densities of 1.66 to 1.76 gm/cc (Tables 4, 5).

The improvement in carbon–carbon composite properties was not nearly as rapid as that of the FRP systems. In general, this delay in carbon–carbon composite property improvement was due to the form in which the PAN fibers were produced. In order to achieve the high strength and modulus the PAN fibers must be oxidized under tension prior to being carbonized. This restraint, during oxidation, effectively results in a restructuring of the carbon of the polymer into a chain like backbone. When carbonized, a much greater degree of orientation of the carbon crystallites is achieved which, in turn, results in substantially improved fiber properties.

For economy and because these early PAN fibers were primarily developed for polymer reinforcement, they were processed in continuous tow bundles of 10,000–12,000 filaments. While this proved satisfactory for reinforcing low modulus epoxy type polymer systems, the coarseness of the preforms formed from the tows proved to be unsuitable for carbon–carbon composites. This is due to the difference in the matrix load transfer response mechanism in the two systems.

In the FRP system, the matrix has a much higher strain capability and much lower modulus than the carbon fiber. In fact, as the ratio of fiber modulus to matrix modulus increases, so does the strength and toughness of the composite. In the carbon matrix system, the carbon char matrix is brittle and has little strain capability. Also, it generally does not wet the fiber, only encapsulating it. The load transfer from fiber to fiber is achieved through a matrix microcracking or articulation and frictional load transfer mechanism. Thus, if the fibers are too widely spaced, the microcracking of the matrix will be excessive and not transfer the load and the composite will fail. The strength and performance of the composite system is, therefore, matrix limited.

As PAN based carbon fiber production increased, filament tows with finer fibers and smaller filament counts

TABLE 4
Flexural Property Data for Pyco-Bond® Pan Precursor Substrates

Material Concept	Final Density g/cc	Nominal Percent Fiber Volume	Flexural Strength MPa: E	Flexural Strength MPa: F	Modulus GPa	Load Span Support Span	Sample Orientation
MODMOR II TOW	1.78	37		248.4	66.6	1.00/2.00	B
0°/90° LAY-UP	1.78	37		155.9	27.1	.200/.600	
Celanese Tow	1.60	37		135.9	64.0	.200/.600	A
0° Lay-up							
HITCO HG-1900	1.73	36		324.3	70.7	1.00/2.00	
0°/90° Lay-up	1.76	36	329.1		52.9	1.00/2.00	
							B
	1.75	36		221.5	43.5	.200/.600	
	1.75	36	262.9		42.2	.200/.600	
HITCO HG-1900 0°	1.60	42		227.7	86.5	.200/.600	A
Lay-up	1.68	42	405.7		87.1	.200/.600	A
CELANESE GY-70	1.79	38	200.8		114.5	.200/.600	B
0°/90° Lay-up							

Note: Samples were tested by four-point loading.
A = parallel to 0° orientation
B = parallel to 0°/90° orientation
E = edgewise tested
F = flatwise tested
From Ref. 6.

TABLE 5
Flexural Strength for Pyco-Bond® PAN Precursor Fabric Substates

Material Concept	Sample Number	Final Density g/cc	Nominal Percent Fiber Volume	Flexural Strength MPa		Sample Orientation
				E	F	
HITCO HG-1900 0° Lay-up (unidirectional)	6–6	1.66	37	456.1	316.0	A
		1.66	37		125.6	B
	6–9	1.68	42	522.3	341.6	A
		1.69	42		166.3	B
HITCO HG-1900 0°/90° Lay-up	6–8	1.69	36	353.3	240.8	C
		1.72	36		251.2	C
HITCO HG-1900 0°/90° Lay-up	6–11	1.67	42	353.3	256.7	C
		1.70	42		278.1	C
Celanese GY-70 tape 0° Lay-up	6–12	1.71	37		289.8	A
	6–13	1.76	51		501.6	A
Celanese GY-70 tape 0°/90° Lay-up	6–14	1.80	38		155.3	C

Note: Samples were tested by three-point loading.
A = parallel to 0° orientation
B = perpendicular to 0° orientation
C = parallel to 0°/90° orientation
E = edgewise tested
F = flatwise tested
From Ref. 6.

became available. This reduction in tow filament count greatly reduced the tow-to-tow crossover discontinuities in woven fabric reinforced and filament wound substrates. When combined with improved carbon char, chemical vapor infiltration (CVI) and combination char-CVI matrix systems, dramatic improvements in carbon–carbon composite systems were achieved.

Ceramic Matrix Composites

During the past ten years, the earlier reported potential for achieving toughness and fabricability in ceramic matrix composite materials has been demonstrated to a level where component production is near term. Key to production is identifying the microstructural action and interaction of the constituents within the composite system(s) of interest. Microstructural control, macrostructural producibility, reproducibility, and reliability are all required. An important element is the use and integration of commercially available materials, fabrication and process technology coupled with available equipment design instrumentation and controls. Automation and in-process feedback systems insure process repeatability while reducing the overall labor input and cost.

The predominance of the open literature currently available for carbon and ceramic fiber reinforced ceramic matrix composites is from France. However, the baseline technology for ceramic matrix composite materials processing was acquired from the United States. Using chemical vapor infiltration (CVI) process technology from Super-Temp Company (STC) for producing aircraft brake materials, Societe Europeenne De Propulsion (SEP) proceeded to upgrade the original STC process by substituting silicon carbide for the carbon in the matrix. In 1977 SEP received a French patent (F. Christin [17] et al.) for producing carbon fiber reinforced ceramic matrix (silicon carbide) composites. With the exception of the UTRC glass matrix ceramic composites, this SEP work represents the first successful attempt to incorporate carbon fibers and carbon–carbon process technology in producing ceramic matrix composite materials.

SEP recognized the potential for ceramic materials and proceeded to support extensive investigations into the improvement of carbon fiber reinforced ceramic matrix composite materials and analog ceramic fiber reinforced CMCs. This work has been supported by the French government at both SEP and the University of Bordeaux. These efforts have been widely published internationally and currently represent a rather singular baseline for this class of materials.

It should not be assumed, however, that only limited domestic effort has been mounted in ceramic matrix composites during this same time frame as some excellent material development has been performed by Amercom, Refractory Composites and others domestically. Unfortunately, this domestic literature is, in general, restricted information controlled by the State Department Bureau of Munitions Control and not available for general release.

Ceramic Matrix Composite Component Development

Problem Definition

The development of carbon fiber reinforced ceramic matrix composite components is generally the same as developing FRP or MMC components in that a number of items must be defined: first, the component and the properties (physical, chemical, electrical, etc.) that are necessary in its operation; second, the operating environment (corrosive, erosive, high temperature, oxidizing, reducing, et al.), next, the geometry, overall size and complexity of the part and finally, the number of parts needed, the time frame and cost.

After these have been established the designer, materials specialist and applications engineer assess a matrix of possible material combinations, fabrication techniques and manufacturing process conditions that can be combined to produce an engineering prototype part. For composite materials systems, *many* materials, fabrication and process variables exist. Multiple variables are sometimes considered a problem, in that there are too many options open and not enough time or funds to examine them all. A directed product development effort minimizes the options and generally results in selecting a primary composite system with several alternative options to refine the system during optimization and manufacturing scale-up.

Design Requirements

Component design and composite constituent, fabrication, and process selection must be approached as a continuum rather than from conventional design handbook values. All composite materials are highly anisotropic and exhibit substantially different microstructural responses than the ductile isotropic metallurgical response most familiar to design engineers. Furthermore, the high modulus and brittle nature of the ceramic matrices require that it microcrack if an effective load transfer is to be made from fiber to fiber to produce a tough component. Figure 5 describes a typical iterative loop used to achieve viable carbon and ceramic matrix composite thermostructural components. At least one and more frequently two or more design/composite material iterations are necessary to achieve success in producing viable operational composite components.

Fabrication and Process Allowables

Control of fiber spacing and interfiber pore spacing are critical to producing both carbon and ceramic fiber reinforced carbon/ceramic matrix composites. In 1970 Bauer at Super-Temp [5], demonstrated the importance of this relationship in producing CVI carbon matrix composites and it is pertinent for ceramic matrices. In general, Bauer found that in carbon–carbon composites with fiber volumes below 35% the matrix would fail between fabric layers or fiber bundles and thus limit the maximum strength of the composite (Table 7). At 35% fiber volume and above the matrix cracking was impeded by the fiber reinforcing which, in turn, picked up the load and allowed the composite to exhibit substantially higher strengths than the lower fiber volume counterparts. Uniquely, once this load transfer mechanism was achieved, little improvement in strength was achieved by increasing the fiber volume loading. 2-D 8 H/S panels

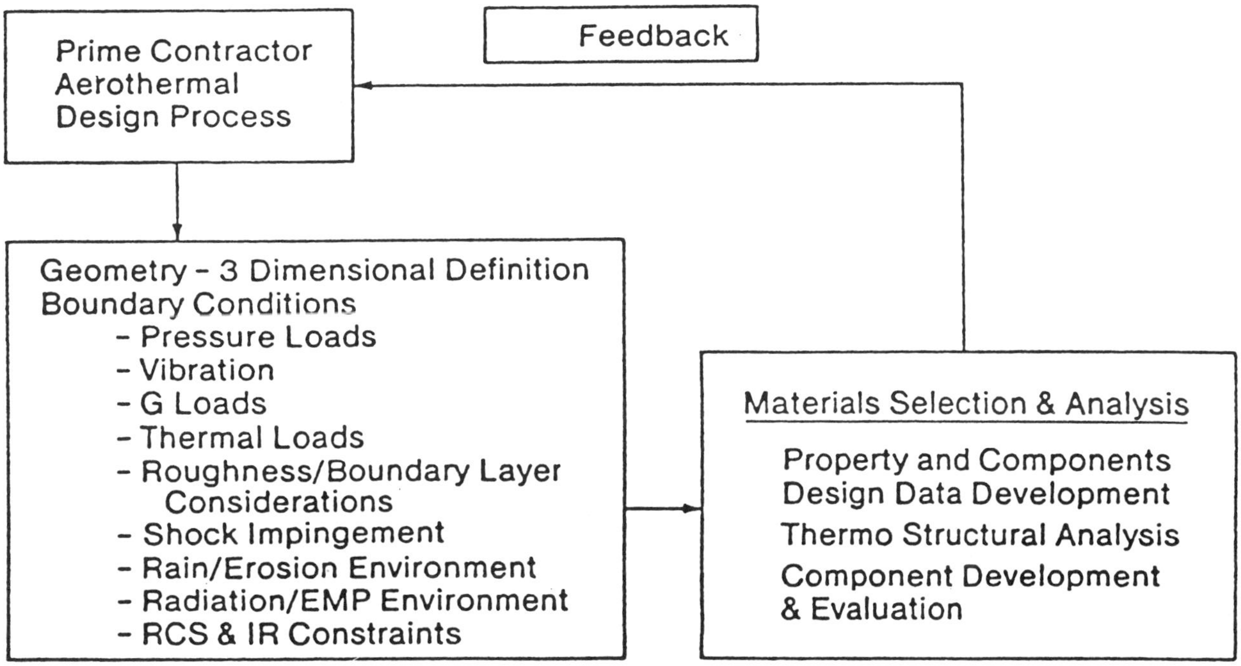

FIGURE 5 Thermostructural design process. Two and occasionally three iterative loops between the Customer and the Applications Engineering/Materials Development Organizations are required to create a component solution.

TABLE 6
Mechanical Property Data for CVD-Carbon Composites

Material Concept	Density Prior to Infiltration g/cc	Final Density g/cc	Fiber v/o	Fiber w/o	Flexural Strength MPa E	Flexural Strength MPa F	Interlaminar Shear Strength MPa	Impact Strength kJ/m of Notch
Mechanical Property Data for CVD-Carbon Infiltrated Tufted and Needled G-1550 8 H/S Graphite Cloth Substrates								
Hand tufted G-1550, 8H/S Graphite cloth, ¼″ Centers using 5-ply carbon cordage	.456	1.66	30.4	27.5	82.8	68.3	19.5	0.05
Machine tufted G-1550, 8 H/S Graphite cloth, 4.5 mm Centers using nylon served 2-ply CY2-1 carbon yarn	.4	1.69	26.7	23.7	124.9	109.0	22.6	0.03
	.5	1.63	33.3	30.7	133.2	116.6	22.8	0.03
Machine tufted G-1550, 8 H/S graphite cloth, 4.5 mm centers using 2-ply CY2-1 carbon yarn twisted into 6-ply cordage	.34	1.64	22.7	20.7	115.2	219.0	23.0	0.03
	.5	1.58	33.3	31.6	136.6	285.6	27.6	0.04
Needled G-1550, 8 H/S graphite cloth 500 needle penetrations/ 6.5 cm^2	.369	1.59	24.6	23.2	77.4	82.8	23.5	0.02
	.369	1.67	24.6	22.1	102.0	71.8	25.5	0.02
	.369	1.73	24.6	21.3	110.3	127.7	31.0	0.05
	.485	1.54	32.3	31.5	148.8	114.5	23.6	0.02
G-1550, 8 H/S Graphite cloth Lay-up	.504	1.63	33.6	30.9	168.6	128.8	27.6	0.04
	.678	1.53	45.2	44.3	156.8	130.9	26.8	0.04
	.717	1.64	47.8	43.7	156.6	123.5	—	—
	.813	1.56	54.2	52.1	142.5	134.6	—	—
	.842	1.49	56.1	56.5	149.4	125.4	28.2	0.05
	1.04	1.47	69.3	70.7	165.8	104.4	—	—
GSCC-8, 8 H/S Carbon cloth lay-up	.505	1.55	33.7	32.6	102.3	60.0	9.5	0.05
Square weave carbon cloth lay-up	.497	1.65	31.1	30.1	125.8	133.7	41.5	—

E = edgewise orientation
F = flatwise orientation
From Ref. 5.

TABLE 7
Constituent Materials Used to Make Glass Matrix Composites

Material	Density g/cm³	Elastic Modulus GPa	Tensile Strength MPa
Borosilicate Glass	2.23	63	—
Thornel 300	1.76	234	2930
Celion	1.76	234	2760
HM	1.80	350	2700
P-100	2.10	654	2070

From Ref. 2.

with 35 percent, 45 percent, and 69 percent volume fraction fiber loading were found to have essentially the same properties.

Uniform fiber spacing and an open, nonbottleneck pore structure are critical to producing high-strength, tough CVI carbon and ceramic matrix composites materials. As such, preform fabrication assumes an important role in designing, selecting constituent materials and processing desired components. In like manner, the fiber form; diameter, filament count, orientation and/or weave significantly influence the options open to the fabricator. Because the CVI matrix only deposits on the exposed surfaces of the fiber preform, long tunnel like (lenticular) pores are necessary to prevent a premature sealing of the internal or surface structure as the densification proceeds.

While pore structure is not quite so critical in polymer pyrolysis systems, variations in fiber diameter and fiber spacing have a considerable effect on the properties and performance of the composite materials and components produced. Large diameter fibers (20–150 micrometers) and high filament count fiber bundles tend to produce composite structures with large fiber crossover discontinuities. These crossover discontinuities result in matrix rich areas which, in turn, tend to respond to composite loading in much the same manner as the low fiber volume systems described by Bauer [5]. Conversely, fine fiber diameters (5–10 micrometers) coupled with low to intermediate filament count (1000–4000) fiber bundles provide a much greater uniformity in fiber spacing within the composite preform and a corresponding improvement in strength and property uniformity.

Critical Development and Design Issues

The importance of fiber diameter and fiber bundle filament count are of little use if a reinforcing fiber does not have thermochemical and thermomechanical stability. That is, a fiber must be able to be fabricated into a desired preform which is then thermally treated to a temperature somewhat in excess of its ultimate use temperature and densified by liquid, CVI or combination matrices without the fiber reacting with its process or (later) use environment. Additionally, the reinforcing fiber should have sufficient strength at the use temperature so that the composite can safely carry the required structural load.

In general, carbon fibers are unique in that they do not react with most corrosive environments, except oxygen, at high temperatures, and exhibit an increase in strength with temperature up to 2200°–2775°C, depending upon the precursor and/or heat treat temperature. Once processed, carbon fibers exhibit no changes in grain or crystallite size until they exceed their initial processing temperature. Thus, carbon fiber reinforced ceramic matrix composite materials can be used to very high operating temperatures without deteriorating. Many carbon fiber reinforced CMCs have shown strength increases of 25 percent from room temperature to 1400°C and higher. No materials other than carbon–carbon and carbon fiber reinforced ceramic matrix composites demonstrate this distinctive behavior.

Ceramic fibers, for example, have decreasing strength at higher temperatures (Figure 1). The difference between the different ceramic fibers relates to their composition, grain structure and softening point. Additionally, ceramic fibers tend to exhibit substantial grain growth when exposed to high temperatures for extended periods of time. This grain growth results in substantial strength reductions, as much as 95% of the fabricated room temperature tensile strength. Chemical reaction with the process and/or operational environment has also tended to limit the use of ceramic fibers for reinforcing ceramic matrix composites. Without carbon or other coatings both Nicalon, an organometallic polymer derived non-stoichiometric silicon carbide fiber, and Avco's SCS-6, a CVD silicon carbide fiber, have been found to deteriorate in strength during composite processing.

Other developmental issues that directly affect the design and use of carbon/ceramic fiber composite materials systems include:

1. *Fabrication & Processing.* Ceramic matrix composite materials are hard and costly to machine, and near net dimension fabrication and processing of components is necessary. Depending upon design requirements, a selection of fabrication techniques may be available, i.e. fabric lay-up, fabric molding, tape winding, filament winding, or combinations of these systems. More recently, three-dimensional braiding and weaving have been successfully used to obtain less anisotropic properties in the configured component. The finer fiber diameter carbon fibers have been the fiber of choice for fabrication ease, even with moduli of 552–690 GPa. With the exception of Nicalon, which has a 10–12 micrometer diameter, ceramic fibers have been found to be difficult to fabricate except for the most simple shapes.

2. *Load Transfer Efficiency.* Control of fiber-matrix interface chemistry and the fiber-matrix bond are critical to achieving a good ceramic matrix composite material. Any reaction between the matrix and the fiber is equivalent to notching the fiber. Fiber properties and interfacial fiber-matrix chemistry are critical to the composite being strong and tough. Because the ceramic matrix has an elastic modulus equal to or greater than that of the fiber, it is necessary for the matrix to microcrack to absorb strain in transferring a load from fiber to fiber. Energy is absorbed during the cracking of the matrix, more energy is absorbed as individual fibers become overloaded while additional energy is absorbed in fiber pull-out from the matrix (c.f., Figure 6). If there is a good bond between the fiber and the matrix, this energy absorption mechanism would not be able to function as any microcrack would propagate through the fiber without deflection.

Another reason for a weak or nonexistent bond between the fiber and the matrix involves the relative coefficient of thermal expansion (CTE) differences. If the fiber has a CTE equal to or less than that of the matrix then it

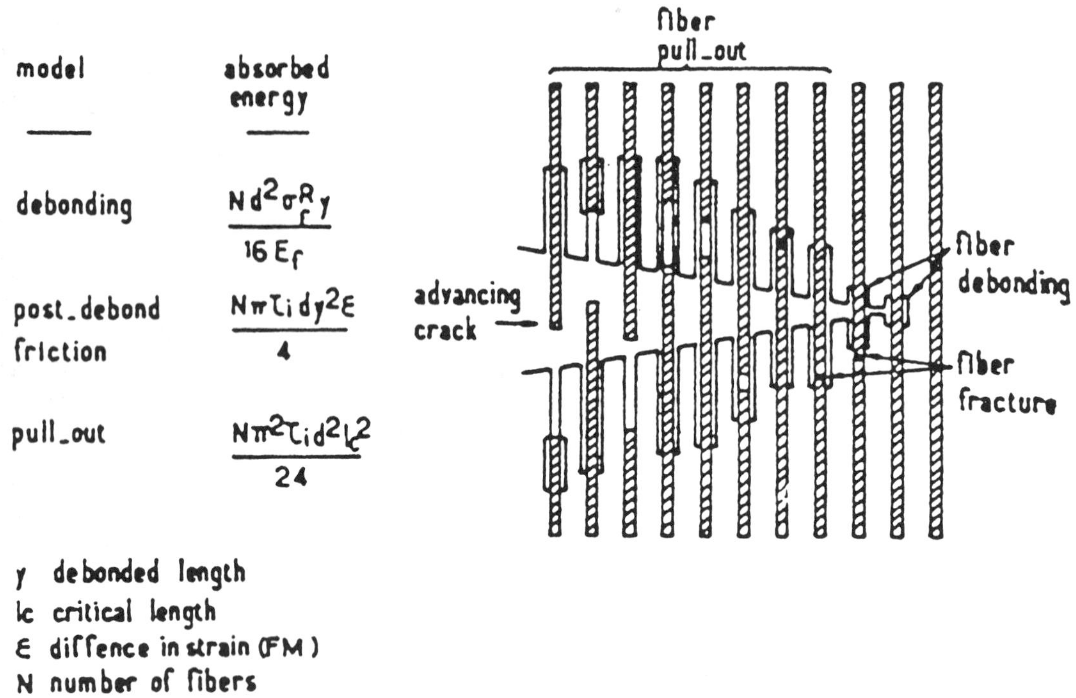

FIGURE 6 CMC energy absorbing mechanisms. (From Ref. 13.)

will generally be in compression when in use at operating temperature. If the fiber is not free to move independently of the matrix when it is under load, little resistance to crack propagation is available and the composite will lose low temperature toughness.

Carbon fibers typically have expansion coefficients somewhat less than most ceramic matrices. This difference in CTEs, which tends to put the interface in compression, is offset by the negligible bonding between the carbon fiber and the ceramic matrix, thus providing the limited bond desired for toughness, thereby increasing the energy absorbing capabilities of the composite (see Figure 7).

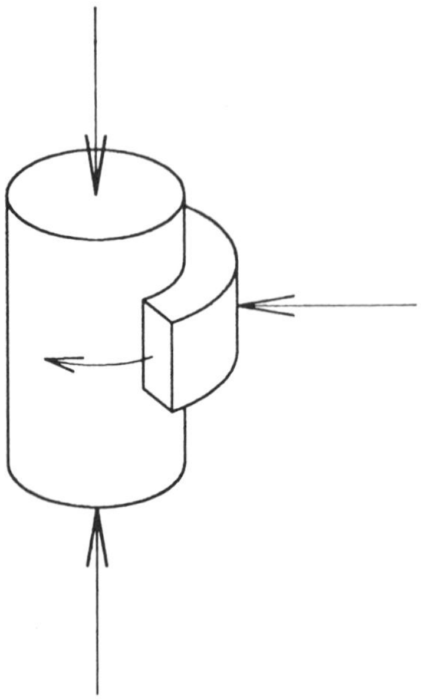

FIGURE 7 Fiber in radial longitudinal compression: interface in compression.

Design, Fabrication, and Process Integration

Composite materials incorporate a combination of constituent materials, phases, fabrication methods, and process variables. Each must be designed to meet specific operational needs as slight variations in any of these parameters produce significant changes in component producibility and performance. The issues of design, materials selection, fabrication and processing are application driven. Determination of what issues may be controlling in any given application are often difficult to determine. Figure 8 depicts typical design issues for structural aircraft components. Each design issue is compared against all of the other identified design concerns on a more important or less important basis. If the issue on the left is of more concern than that of each of the other issues listed across the top of the chart it is assigned a 1. If it is of less concern, it is assigned a 0. When compared with itself a + is shown. After all of the issues have been compared with each other and the totals added on the right a ranking of relative importance can be assigned for each.

These rankings of relative importance are then incorporated into the constituent materials, fabrication technique, and process variable selection, and used to opti-

	Sgth (Baseline)	Toughness	O_2 Resist	Fabric	Attach	Permeability	IR/Emissivity	RF/Dielectric	Pivot Tribology	Fatigue	Reproducibility	Reliability	Density	Stiffness	Total	Relative Order of Merit
Sgth (> Baseline)	-	O	O	O	O	O	I	I	O	O	O	O	I	I	4	11
Toughness	I	-	+	I	I	I	I	I	I	+	I	I	I	I	12	1
O_2 Resist	I	+	-	I	I	I	I	I	I	I	I	I	I	I	12	2
Fabric	I	O	O	-	O	O	I	I	O	O	I	I	I	I	7	6
Attach	I	O	O	I	-	I	I	I	+	O	I	O	I	I	9	5
Permeability	O	O	O	O	O	-	I	I	O	O	I	O	I	I	5	10
IR/Emissivity	O	O	O	O	O	O	-	+	O	O	O	O	O	II	2	12
RF/Dielectric	O	O	O	O	O	O	O	-	O	O	O	O	I	I	2	13
Pivot Tribology	O	O	O	O	+	I	I	I	-	O	O	O	II	6	7	8
Fatique	I	+	O	I	I	I	I	I	I	-	O	O	I	I	10	4
Reproducibility	O	O	O	+	O	I	I	I	O	O	-	+	I	I	6	7
Reliability	O	O	O	+	O	I	I	I	O	O	+	-	I	I	7	7
Density	O	O	O	O	O	O	O	O	O	O	O	O	-	O	0	14
Stiffness	O	O	O	O	O	O	O	O	O	O	O	O	O	-	0	15
Thermal Conduct.	I	O	O	+	I	I	I	+	O	+	+	I	I	I	11	3
Specific Heat	O	O	O	O	I	I	I	I	O	O	O	O	I	I	6	9

I Most Desirable O Least Desirable - Critical Issue

FIGURE 8 Critical design issues.

mize their efforts to meet identified component operational goals. A summary of the system elements that must be assessed and tuned during this process are shown in Figure 9.

CMC Physical Properties & Environmental Response

Glass Matrix Composites

The development of graphite fiber reinforced glass matrix composite material systems during the late 1970s provided the first demonstration of the potential of ceramic matrix composites. By reinforcing borosilicate glass with graphite fibers it was possible to demonstrate composite properties that compared favorably with resin matrix counterparts. Both high elastic modulus and strength were demonstrated at temperatures up to 600°C. Additionally, composite dimensional stability was superior to resin matrix systems due to the low thermal expansion behavior of the glass matrix.

Although a full range of graphite fibers and glass matrices were evaluated by Prewo [2] at UTRC only those constituents listed in Table 7 were considered for his later investigations. Table 8 summarizes both tensile and flexural test results for unidirectional and 0/90 cross ply materials. In both cases each composite contained a small amount of discontinuous graphite fiber scrim to help provide handleability to the unidirectional tape plys. In subsequent investigations, Prewo found that it was difficult to predict composite strength values. Table 9 presents actual and predicted strength values for both unidirectional and angle-ply composites. A very good prediction of strength was obtained when using the Tsai-Hill failure criteria and a value for t (an adjustable parameter used to provide a best fit of the unidirectional data) of 32 MPa.

In summary carbon fiber reinforced glass matrix composites have been shown to be high strength, high stiffness materials whose performance can be related to their constituent properties and the role of the fiber-matrix interface. Composite tensile strengths of nearly 1000 MPa (145 kSI) with an elastic modulus of 180 GPa (26×10^6 kSI) and an overall composite density of 2 g/cc make these composites important candidates for structural applications.

SYSTEM ELEMENTS

- COMPONENT DESIGN ITERATION
- SYSTEM & OPERATIONAL ANALYSES
- COMPOSITE CONSTITUENTS
  - o FIBER
  - o MATRIX
  - o COATING
- FIBER ORIENTATION & PREFORM FABRICATION
  - o 2-D
    - LAY-UP
    - RANDOM
    - WINDING
  - o MULTI-D
    - BRAID
    - ORTHOGONAL
    - < INTERLOCK
    - OTHER
  - o COMBINATIONS
  - o DENSIFICATION
    - POLYMER PYROLYSIS
    - ORGANOMETALLIC PYRO
    - CVI
    - SOL-GEL
    - COMBINATION
- EVALUATION
  - o MATERIAL
  - o COMPONENT
- PROCESS AND EQUIPMENT
  - o OPTIMIZATION
  - o AUTOMATION
- MANUFACTURING SCALE-UP

FIGURE 9 Composite materials system elements.

TABLE 8
Thornel 300 Reinforced Borosilicate (with scrim)

Unidirectional		0/90 Cross ply
E_{11}^{T*} = 168 GPa	E_{22}^{T} = 16.8 GPa	E_{11}^{T} = 72.1 GPa
σ_{11}^{T} = 580 MPa	σ_{22}^{T} = 29.0 MPa	σ_{11}^{T} = 246 MPa
$\varepsilon_f^{T_{11}}$ = 0.36%	$\varepsilon_f^{T_{22}}$ = 0.45%	$\varepsilon_f^{T_{11}}$ = 0.32%
ν_{12}^{T} = 0.21	ν_{21}^{T} = 0.012	ν_{12}^{T} = 0.04
$E_{11}^{F\dagger}$ = 124 GPa	E_{22}^{F} = 25.4 GPa	Vol% Fiber = 60
σ_{11}^{F} = 905 MPa	σ_{22}^{F} = 94 MPa	
Vol% Fiber = 54	ρ = 2.0 g/cm^3	
HM Reinforced Borosilicate (with scrim)		
E_{11}^{T} = 215 GPa	E_{22}^{T} = 10.2 GPa	E_{11}^{T} = 92.0 GPa
σ_{11}^{T} = 352 MPa	σ_{22}^{T} = 12.2 MPa	σ_{11}^{T} = 205 MPa
$\varepsilon_{f_{11}}^{T}$ = 0.17%	$\varepsilon_{f_{22}}^{T}$ = 0.26%	$\varepsilon_{f_{11}}^{T}$ = 0.23
ν_{12}^{T} = 0.22	ν_{21}^{T} = 0.022	ν_{12}^{T} = 0.023
E_{11}^{F} = 180 GPa		Vol% Fiber = 50
σ_{11}^{F} = 527 MPa		
Vol% Fiber = 67	ρ = 2.0 g/cm^3	

* superscript T denotes property measured in tension
† superscript F denotes property measured in three point flexure
From Ref. 2.

SEP—Carbon Fiber Reinforced Ceramic Matrix Composites

During the mid to late 1970s, SEP recognized the potential for overcoming the brittleness, crack sensitivity, and lack of thermal shock resistance of monolithic ceramics through the development of ceramic matrix composites derived from carbon–carbon by replacing part of the matrix with a ceramic matrix such as silicon carbide. Although very high bending strength (above 1000 MPa) was achieved on unidirectional material, the final goal was the development of two-dimensional composites with a reliable strength at both room temperature and high temperature. Subsequent investigations produced 2-D silicon carbide matrix analogs of carbon fabric reinforced C-C composites. This material, CERASEP, had much higher strengths than expected, especially at high temperature, above 1300°C (Fig. 10). Both carbon and ceramic (SiC) fiber reinforcements were evaluated, and the composites found to be very tough and exhibit very high fracture energies due to fiber pullout (Fig. 11). Figure 12 shows the fracture toughness for comparative four-point bending tests on notched specimens.

Figure 13 shows a typical stress-elongation curve for tensile specimens of C-SiC obtained at room temperature. Three zones can be observed on this curve. In the first zone, the behavior of the material is linear with modulus of 210 GPa up to an elongation of 0.035%. Then, the material becomes nonlinear in Zone 2, where there is believed to be a fracture of the matrix and a transfer of

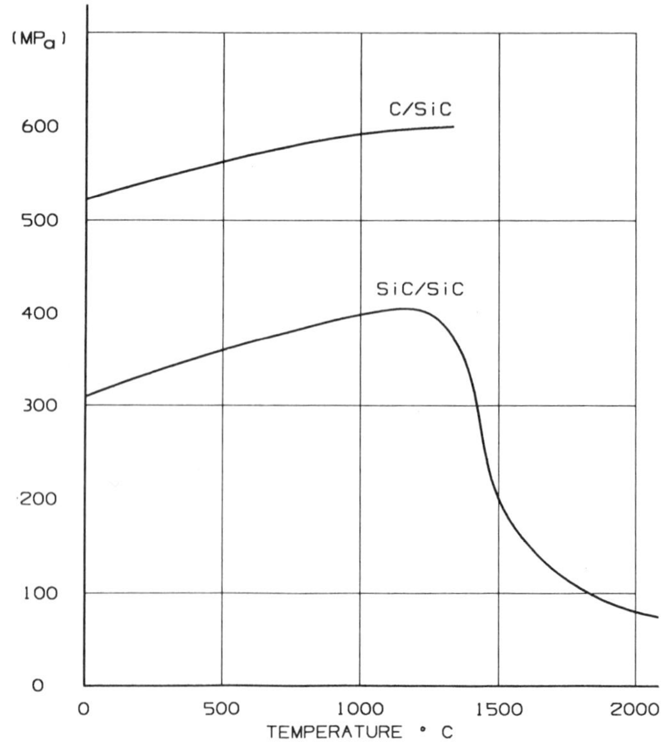

FIGURE 10 Bending strength of C and SiC/SiC CERASEP. (From Ref. 11.)

TABLE 9
Comparison of Actual and Predicted Composite Ultimate Tensile Strength

Composite	Experimental UTS (MPa)	Predicted Maximum-Stress UTS (MPa)[a]	Predicted Tsai-Hill UTS (MPa)[b]
0°	787	787	787
10°	171	145	176
30°	47	57	51
60°	23	23	22
[±10]	440	759	425
[±30]	187	106	109
[±45]	44	50	59
[±60]	24	28	26

[a] Using τ = 25 MPa.
[b] Using τ = 32 MPa.
From Ref. 2.

the load to the fibers. At about 0.2% elongation, there is a change of the direction of the curve (Zone 3) up to the ultimate strength around 190 GPa for an elongation of 0.6%. Although a complete explanation of this behavior has not been given, the "multiple fracture" theory which relates the differences between moduli and the elongation of the fibers and the matrix are taken into consideration.

Mechanical shock resistance using the IZOD method (ASTM 256) on a notched specimen show the CERASEP material to be ten times more resistant to mechanical shock than monolithic silicon carbide (Fig. 14). Water quench testing from temperatures at 200°–1000°C followed by residual four-point bend testing show that neither C-SiC or SiC-SiC CERASEP are greatly affected by thermal shock (Fig. 15).

In still other evaluations with ceramic matrices SEP was successful in demonstrating the versatility of the CVI ceramic matrix composite system. Figure 16 is the compressive stress and modulus for both rayon and PAN based carbon fibers in SiC and TiC matrices. Figure 17 shows shear strength and modulus for these materials. Variations in Youngs Modulus vs compacity in compression loading for SiC, TiC, B_4C & BN are shown in Figure 18. Figures 19 and 20 show the expansion vs temperature and the oxidation response of these materials.

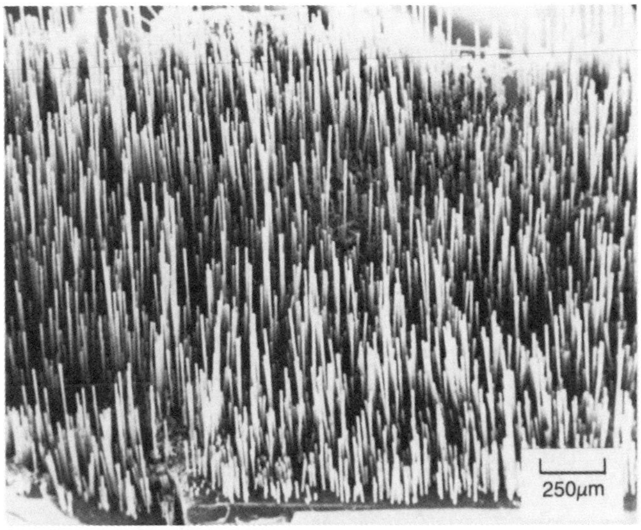

FIGURE 11 Fiber pull-out. (From Ref. 4.)

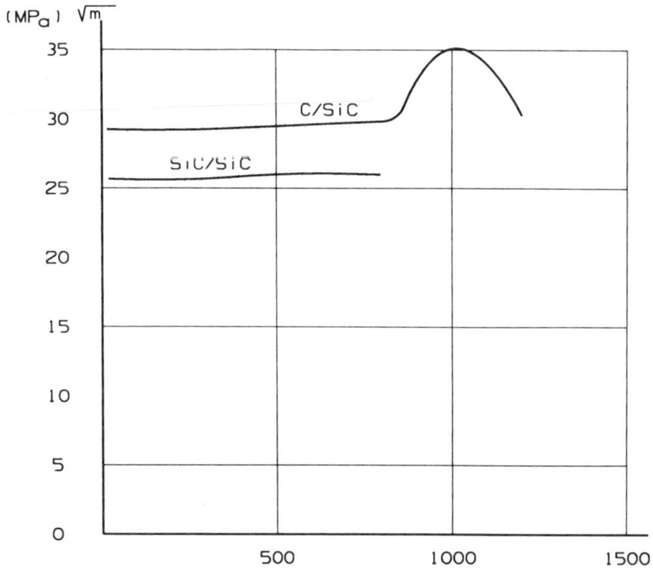

FIGURE 12 Toughness of C and SiC/SiC CERASEP.

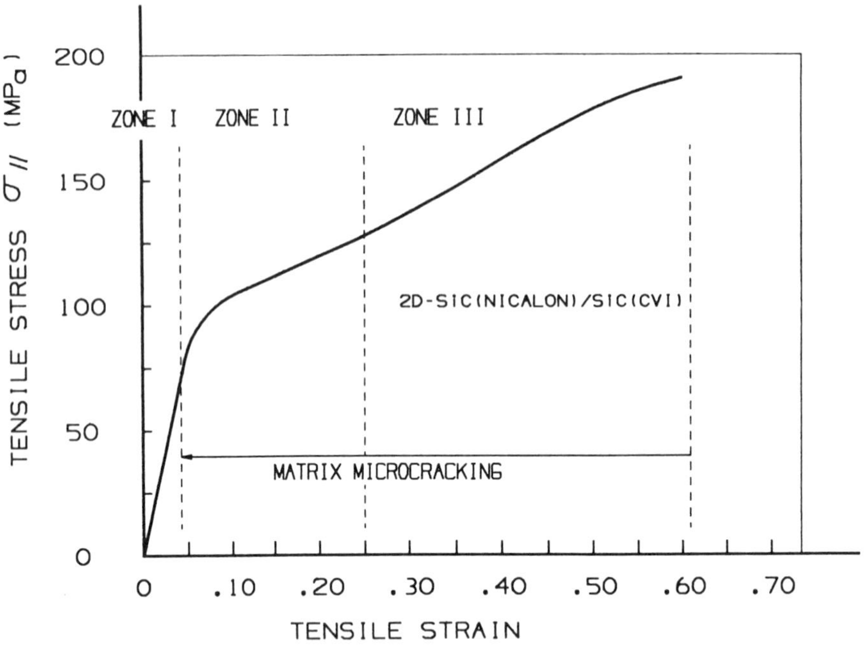

FIGURE 13 **Typical CERASEP tensile curve.** (From Ref. 11.)

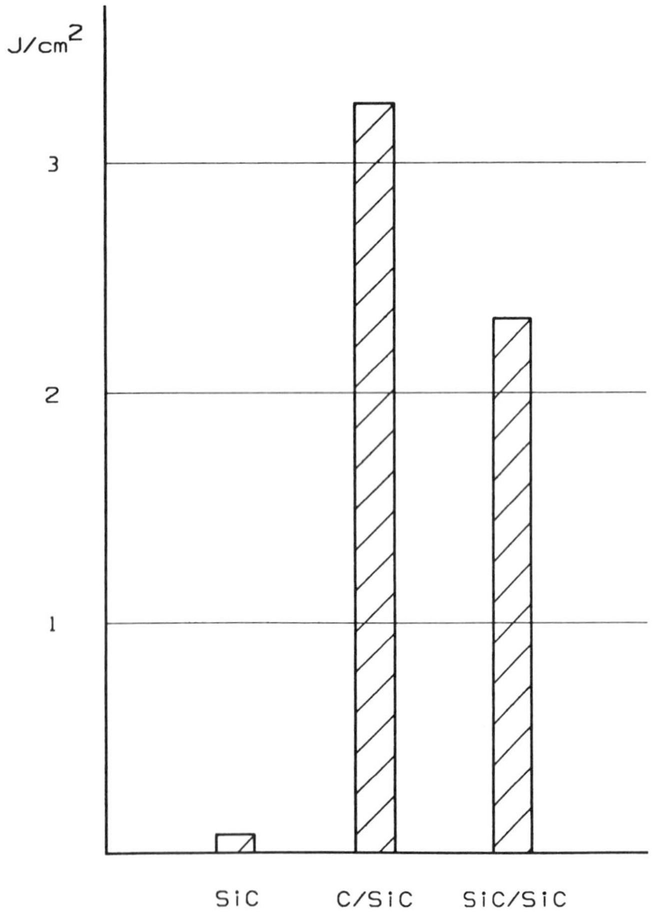

FIGURE 14 **IZOD impact resistance.** (From Ref. 11.)

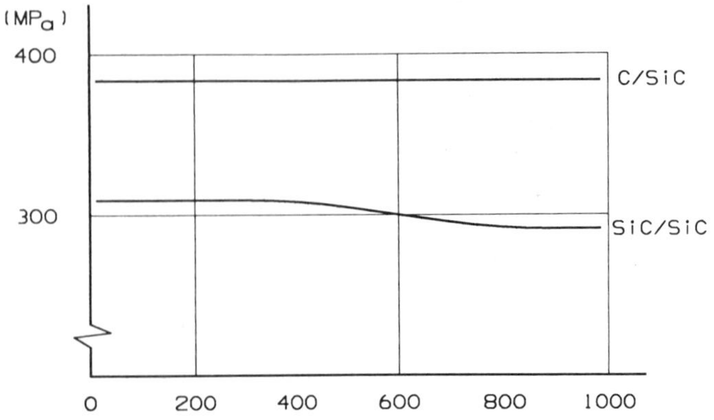

FIGURE 15 **CERASEP thermal shock resistance.** (From Ref. 11.)

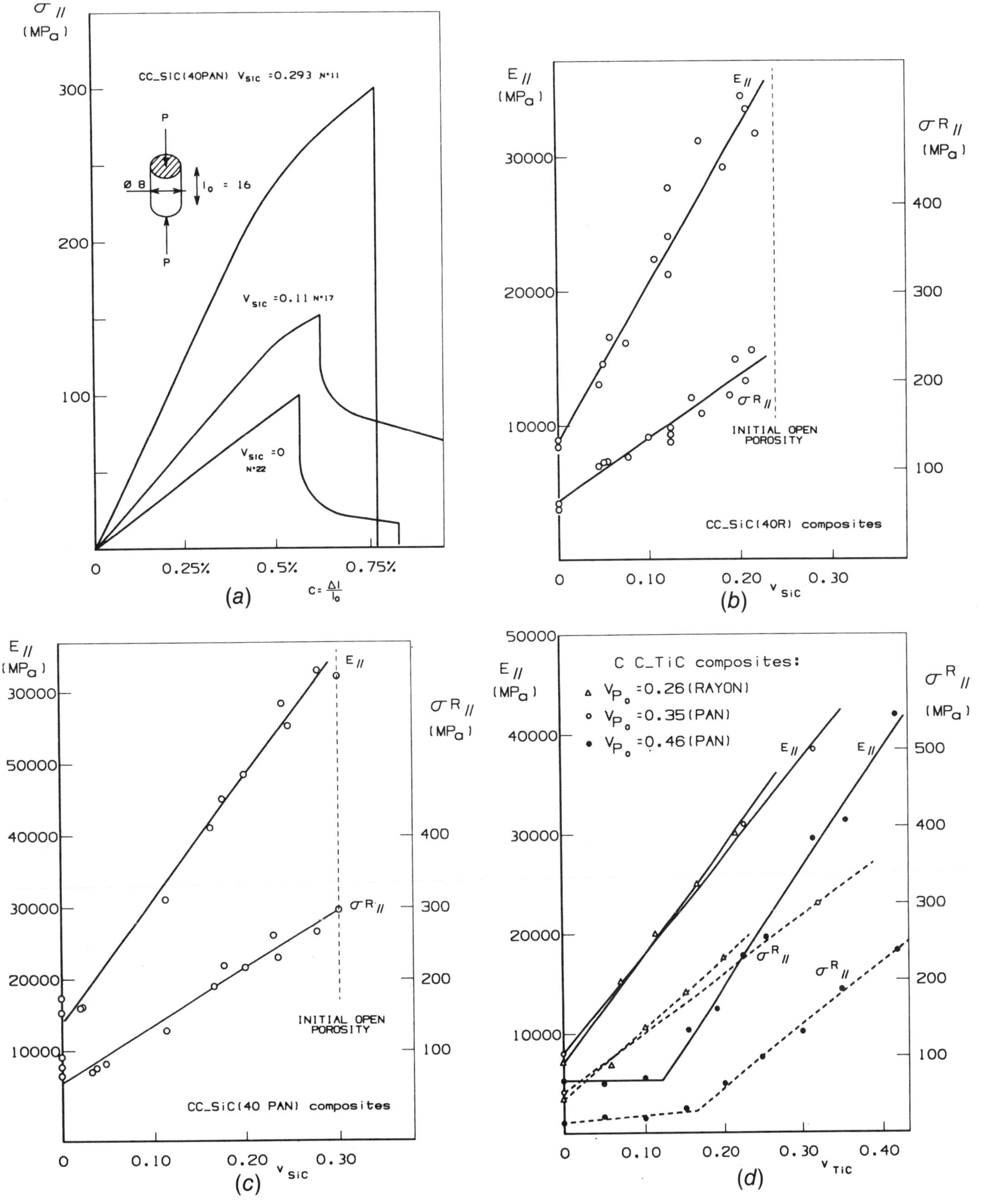

FIGURE 16 Compression tests on 2D-C-C/MC composites: (*a*) stress-strain curve (MC=SiC); (*b*) and (*c*) variation of $E_{\parallel}$ and $\sigma^{R}_{\parallel}$ vs. V_{SiC} (composites from rayon or PAN-based fibers); (*d*) variation of $E_{\parallel}$ and $\sigma_{\parallel}$ vs. V_{TiC}. (From Ref. 12.)

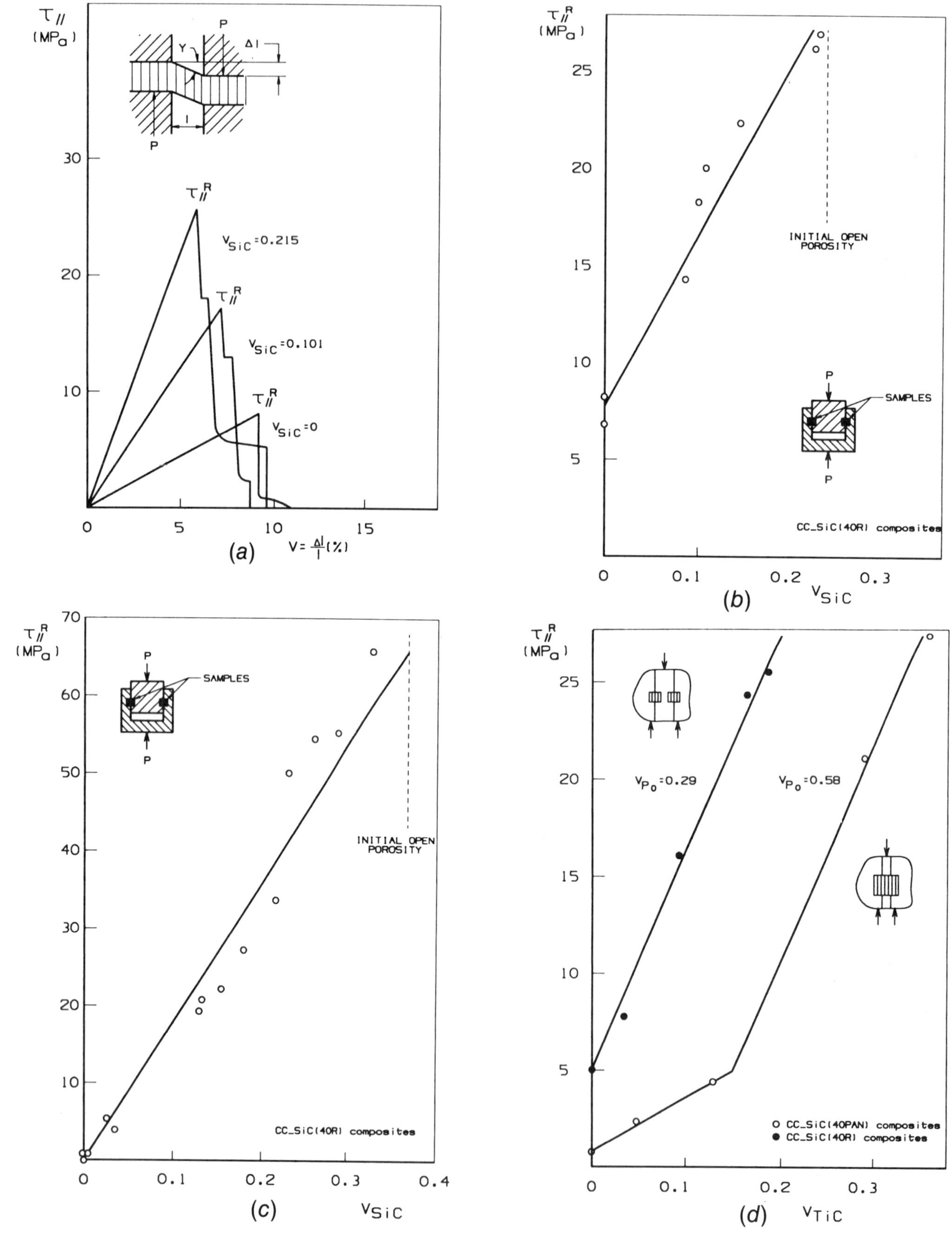

FIGURE 17 Shear tests on 2D-C-C/MC composites: (*a*) stress-strain curve (MC=SiC); (*b*) and (*c*) variation of $\tau^{R}_{\parallel}$ vs. V_{SiC} (composites from rayon or PAN-based fibers); (*d*) variation of $\tau^{R}_{\parallel}$ vs V_{TiC}. (From Ref. 12.)

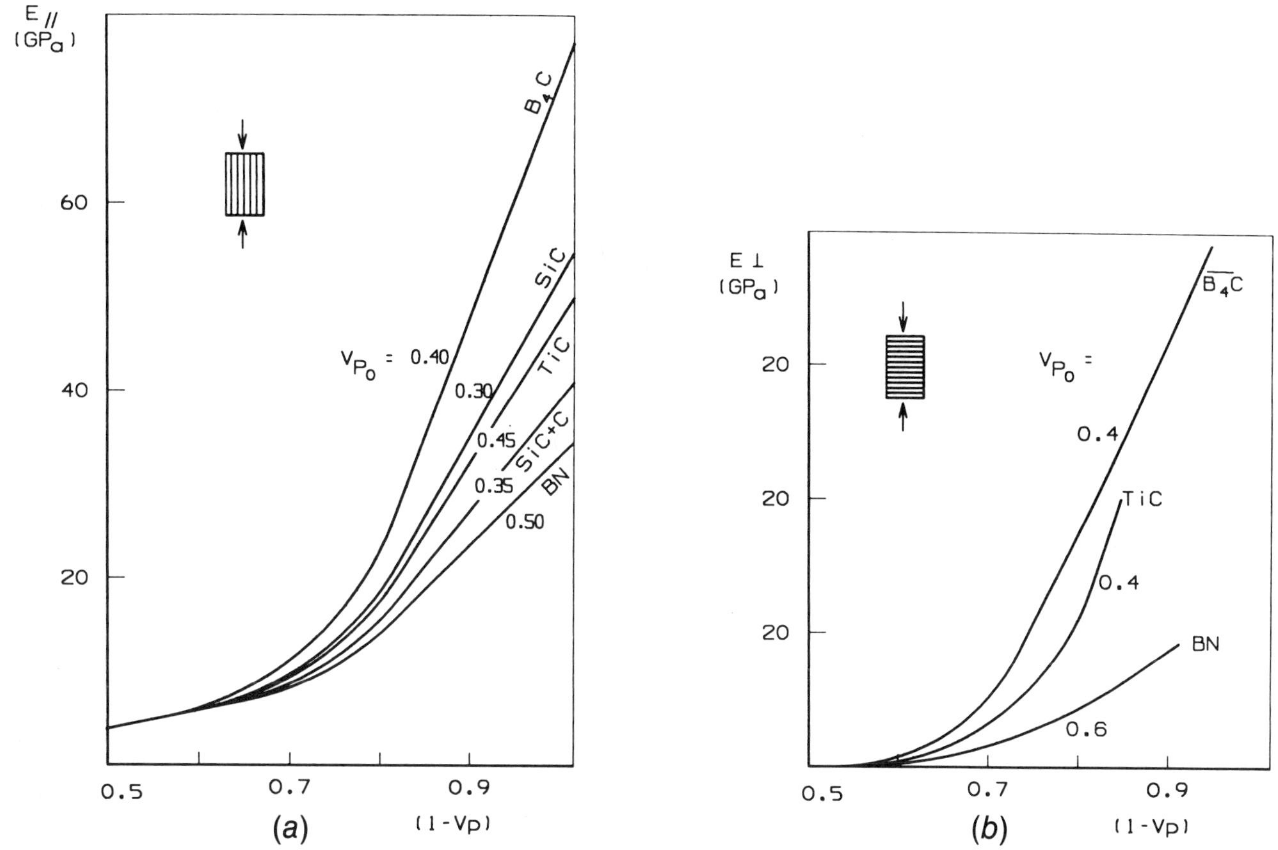

FIGURE 18 Variations of the Youngs modulus of 2D-C-C/ceramic composites vs compacity 1 − Vp under: (*a*) p-compression loading, (*b*) under o-compression loading. (From Ref. 13.)

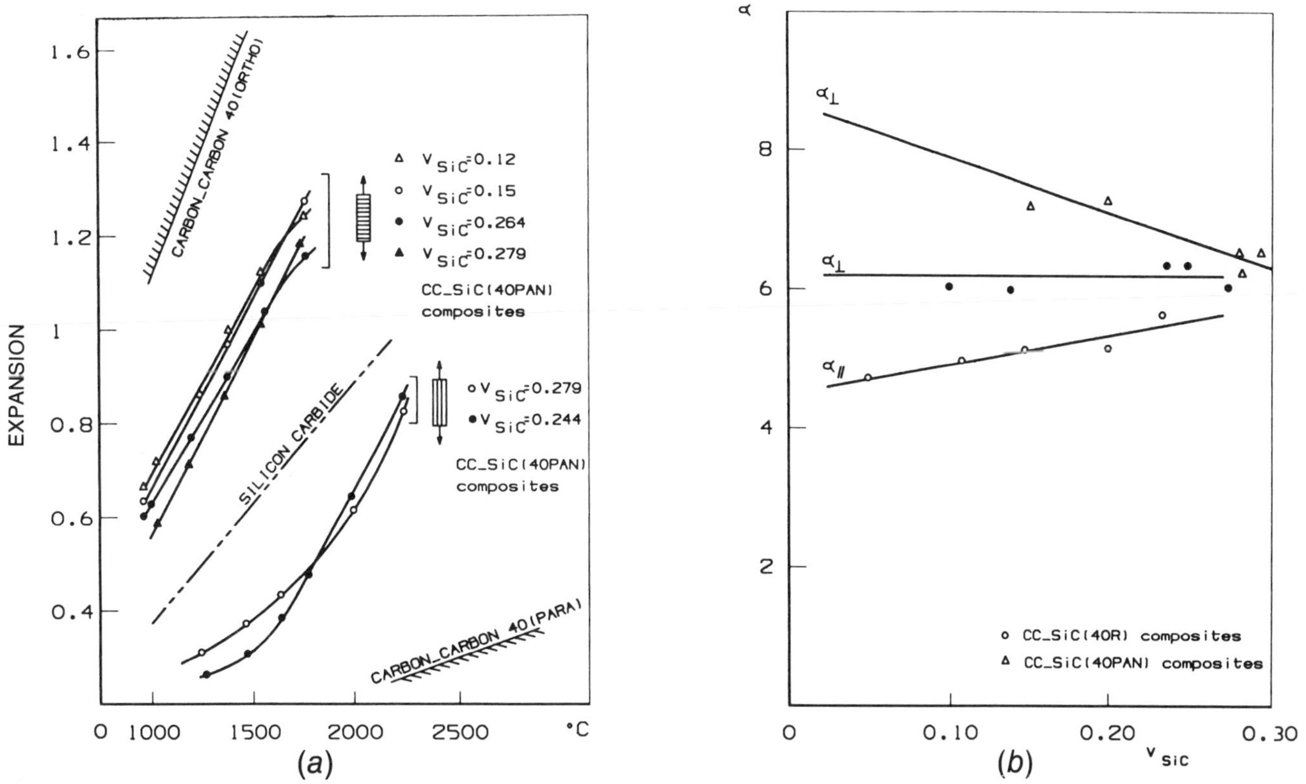

FIGURE 19 Thermal expansion of 2D-C-C/MC composites: (*a*) as a function of temperature (MC=SiC); (*b*) variation of the mean linear thermal expansion coefficients $\alpha_{\parallel}$ and $\alpha_{\perp}$ at 1030°C as a function of V_{SiC}. (From Ref. 12.)

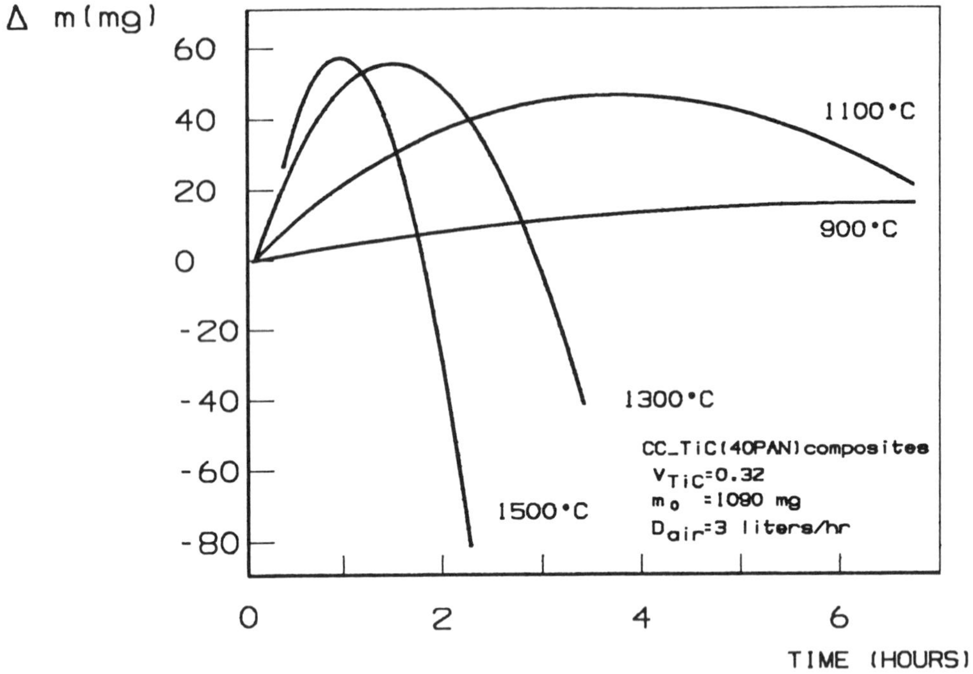

FIGURE 20 CERASEP oxidation behavior. (From Ref. 12.)

Conclusions

Carbon fiber reinforced glass, silicon carbide, boron carbide, boron nitride, and titanium carbide ceramic matrix composites have been shown to be high-strength, high-stiffness, and high-temperature materials whose performance can be related to their constituent properties and the role of fiber-matrix interface. Composite tensile strengths range from about 350 to nearly 1000 MPa. This combined with elastic moduli of 20 to 180 GPa and overall composite densities of 2.0–2.5 gm/cc make these composites important candidates for structural applications.

The retention of composite tensile strength to temperatures in excess of 1500°C is also notable in that, while significantly limited by carbon fiber oxidation, it provides significant potential advantage over traditional ceramics and metal matrix composites. The recent demonstration of internal oxidation inhibiting systems coupled with glass sealed oxidation resistant coatings further increase the potential for this class of materials.

In comparison with monolithic ceramics, fabrication reproducibility (Weibull analysis), toughness and shock resistance have been largely improved as a result of the introduction of the fiber network. To preserve fiber strength and achieve a high degree of toughness, however, it is necessary to have low fiber-matrix interfacial strength.

Because carbon fiber reinforced ceramic matrix composite materials have such a wide range of design material constituent selection, fabrication technique and processing variables, it is not expected that a few grades of material will be adequate to meet all identified component application needs. Rather, an iterative design-material system selection process that utilizes currently available materials, technology, equipment, instrumentation and controls to develop individual components on an application by application basis will provide the successful integration of this class of materials into a production reality.

Because of the anisotropic nature of these materials and the need to adjust design, material selection, fabrication technique and process variable selection for a specific application, an iterative component development process is necessary.

James W. Warren

References

1. R. C. Rossi, and D. P. H. Hasselman, "Application of Fracture-Mechanics Theory to the Thermal Shock Resistance of Brittle Materials," *Reviews on High Temperature Materials,* 1972.
2. K. M. Prewo, and E. J. Minford, *SAMPE Journal, 21*(2), 26 (1985).
3. V. C. Nardone, and K. M. Prewo, *Journal of Materials Science,* 23, 168 (1988).
4. K. M. Prewo, *Journal of Materials Science,* 23 2745 (1988).
5. D. W. Bauer, and W. V. Kotlensky, *Proceedings of the 23rd Pacific Coast Regional Meeting, American Ceramic Society,* 1970.
6. D. W. Bauer, T. S. Midolo, R. Harris and W. V. Kotlensky, *Proceedings of the 17th Natl. SAMPE Symposium,* 1972, IV-A-1, pp. 1–23.

7. W. V. Kotlensky, D. W. Bauer, J. W. Warren and W. H. Smith, *Proceedings of the Tenth Biennial Carbon Conference,* 1971.

8. J. W. Warren and C. D. Coulbert, "An Introduction to Carbon–Carbon Composite Materials," *Super-Temp Corporation,* 1971.

9. J. W. Warren, *Proceedings of the AIME Conference on Advanced Fibers and Composites for Elevated Temperatures,* 1979.

10. J. W. Warren, *Proceedings of the 9th Annual American Ceramic Society Conference on Composites and Advanced Ceramic Materials,* 1985.

11. M. Dauchier, G. Bernhart and C. Bonnett, *Proceedings of the 30th Natl. SAMPE Symposium,* 1985, pp. 1519–1525.

12. R. Naslain, J. Y. Rossignol and P. Hagenmuller, *Revue de Chimie minerale,* t 18, *5,* 544 (1981).

13. J. Y. Rossignol, J. M. Quenisset, H. Hannache, C. Mallet and F. Christin, *Journal of Materials Science, 22,* 3240 (1987).

14. R. Naslain, "Fibrous Ceramic-Ceramic Composite Materials Processing and Properties," *Journal De Physique,* supplement au n°2, *47,* (1986), C1–703.

15. R. Naslain and F. Langlais, *Proceedings of the University Conference on Ceramic Science.* 1986, pp. 145–164.

16. L. H. Heraud, F. Christin, R. Naslain and P. Hagenmuller, *Proceedings of the 8th International Conference on Chemical Vapor Deposition,* 1981, pp. 782–789.

17. F. Christin, et al, Title; Idem 71, Patent Numb. 77-26979, September 6, 1977; Republic of France, The National Institute of Patent Rights, Publication Numb. 2-401-888, to European Society of Propulsion.

Ceramic Matrix Composites; Ceramic Fiber Reinforced

Introduction

Ceramics have the capability of providing strength at high temperatures (to 2200°C) and considerable oxidation resistance. Advanced ceramics potentially might find application as engine components and in other areas of hypersonic vehicles subject to high temperatures. In advanced aircraft engines, higher thrust-to-weight ratios could be achieved via high specific strength ceramics and the higher operating temperatures they enable. In automotive engines, ceramics could achieve greater efficiency via higher operating temperatures, improved responsiveness because of their low density, and enhanced durability due to superior heat and wear resistance. They could provide lightweight superconductors with load-bearing capabilities for computers, magnets, and sensors, as well as materials for structural applications as heat exchangers, catalyst supports, turbochargers, and valve systems. However, the use of monolithic ceramics as structural materials has been limited by their tendency to fail catastrophically from small surface and/or internal flaws and by their sensitivity to thermal shock. Techniques for reducing flaw size and population have been employed, but, since the critical-sized flaws cannot be eliminated entirely, this approach is reaching its limitations.

The introduction of a second phase to produce a ceramic matrix composite (CMC) offers mechanisms for toughening the ceramic. The second phase may take the form of a second crystalline phase, as is the case in transformation-toughened zirconia, a particulate reinforcement, whiskers, or continuous fibers. Recent reviews of ceramic matrix composites have been presented by Schioler and Stiglich [1], Rice [2,3], Donald and McMillan [4], and Shetty [5].

The development of particulate and transformation-toughened composites is more mature than the fiber-reinforced area. Particulate composites are used commercially in the cutting tool industry. Transformation-toughened zirconia is the primary material of interest for the low heat rejection diesel engine. A variety of whisker-reinforced materials have been studied having whisker loadings of up to 40% [6–9]. The use of continuous fibers offers the opportunity for fabrication of composites with much higher fiber loadings ($\leq 65\%$), hence smaller interfiber distances for deflecting matrix cracks. For load-bearing applications in aircraft engines, such continuous fiber reinforced composites should offer higher strength and toughness than whisker-reinforced materials. This article focuses on continuous fiber reinforced CMCs.

Theory

Ceramics show relatively high strength and stability to high temperatures. Absence of stress-relieving mechanisms gives rise to brittle fracture, characterized by notch and flaw sensitivity and susceptibility to thermal shock. Incorporation of continuous fibers in a ceramic matrix provides several paths for energy dissipation and toughening not present in the monolithic material, including matrix microcracking, debonding at the fiber–matrix interface, crack bridging, pullout of fibers during crack extension, and crack deflection [10,11]. The analysis of matrix microcracking in a ceramic composite by Aveston, Cooper, and Kelly [12,13] suggests that the strain to failure of the matrix ε_{mu} is increased by the incorporation of fibers:

$$\varepsilon_{mu} = \left(\frac{12\tau_0\gamma_m E_f V_f}{E_c E_m^2 r_f V_f}\right)^{1/3} \tag{1}$$

where

τ_0 = fiber–matrix interfacial shear strength
γ_m = fracture surface energy of the matrix
E_f = modulus of the fiber
V_f = fiber volume fraction
E_c = composite modulus
E_m = matrix modulus
r_f = fiber radius
V_m = matrix volume fraction

The strain to failure of the composite therefore would be increased by a strong fiber–matrix interface, small diameter fibers, high fiber loading, and an enhanced matrix fracture energy.

The influence of fiber diameter, loading, and matrix flaw size is discussed in detail by Korczynskyj et al. [14], who propose an upper limit for fiber volume fraction V_f. As V_f increases, the composite strain at which a matrix crack will become unstable also should increase. The strain on fibers bridging the crack increases as well, and is greater than the overall composite strain. Therefore, the central bridging fiber can reach its ultimate failure strain and fracture at a point when the average composite strain is still less than the fiber failure strain. As V_f increases, a point can be reached at which the central bridging fiber attains its maximum fracture strain at the same time or before the matrix crack becomes unstable, resulting in catastrophic failure. Also, as the fiber volume fraction increases, the strain for propagation of existing matrix flaws increases, until eventually unstable matrix failure can result in fiber fracture. In addition, if V_f is too low, the fibers cannot support the load after matrix fracture occurs, again resulting in brittle failure.

A strong fiber–matrix interface is needed to redistribute the stress as the matrix microcracks, delaying further microcracking to higher stress levels. In contrast, fiber pullout is enhanced by a weak interface. If the interface is too strong, a matrix crack will propagate through the fibers, resulting in brittle fracture. Therefore some optimum interfacial strength must be determined.

For many systems (e.g., SiC/SiC or SiC/Si_3N_4), the moduli of the matrix and of the fibers are not very different, and therefore the modulus difference would not be expected to contribute significantly to strain enhancement or composite strengthening.

Chan and Patterson [15] have proposed that small interfiber spacings should limit the size of Griffith flaws, again suggesting the need for small diameter fibers.

Thermal expansion differences between the fiber and matrix should contribute substantially to the stress at the interface [4,5,11,16–19], hence the degree of toughening achieved. In the axial direction, the stress at the interface σ_a is given [16] as follows:

$$\sigma_a = (\alpha_m - \alpha_a)\,\Delta T E_m \tag{2}$$

where

α_m = average coefficient of expansion of the matrix
α_a = axial coefficient of expansion of the fiber
ΔT = temperature decrease below the temperature at which mechanisms for relieving stress can occur (for ceramics, typically approaching the processing temperature)
E_m = Young's modulus of the matrix at the final temperature

If the expansion of the fiber is greater than that of the matrix (σ_a is negative), the matrix will be in compression upon cooling from the fabrication temperature, and the fibers will act to resist crack opening in the matrix. However, a very large thermal expansion mismatch would be expected to result in debonding at the fiber–matrix interface, and possible matrix cracking in response to thermal stresses.

In the radial direction an analogous stress relationship is given by Sambell et al. [16] as follows:

$$\sigma_r = (\alpha_m - \alpha_r)\,\Delta T E_m \tag{3}$$

where α_r is the average radial coefficient of expansion of the fiber. If σ_r is positive, the matrix is in tension as a result of hoop stresses; if σ_r is negative, the fibers shrink away from the matrix and could act as holes, providing no toughening due to debonding, but still possibly acting as crack deflectors.

Moduli differences between fiber and matrix will contribute to the interfacial stress [11] as well. Geometry of the fiber array and fiber diameter also will influence the interaction of stresses and local composite modulus surrounding a given fiber. An interesting example of this is provided by Sambell et al. [17] in comparing matrix cracking in discontinuous and continuous carbon fiber reinforced borosilicate glass composites in which the continuous, higher fiber volume fraction material showed less matrix cracking than its chopped fiber counterpart.

Faber [11] and Faber and Evans [19] have analyzed crack deflection mechanisms at the interface. The length of the fiber is considered to be of importance when discussing crack bridging and fiber pullout. However, from a crack deflection standpoint the toughening that can be achieved cannot exceed twice the matrix toughness, and is independent of fiber length [11]. Cottrell [20] has shown that for discontinuous fibers, the work of fracture is maximized at some critical fiber length l_c and is increased with high fiber volume fractions and small fiber diameters. Shetty [5] points out that the same reasoning can be applied to continuous fiber composites if continuous fibers are assumed to have weak points along the length.

Although in theory the primary goal of introducing continuous fibers into a ceramic is to increase its toughness, fracture toughness in continuous fiber CMCs is a property that has proved to be difficult to quantify. The fracture strength σ_f is related to the crack length a by the Griffith [21] equation:

$$\sigma_f = \frac{K_{1_C}}{Ya^{1/2}} \tag{4}$$

where K_{1_C} is the fracture toughness (resistance to crack propagation) and Y is a material constant.

This relationship has been applied to the determination of fracture toughness in composites with discontinuous reinforcements (whisker or particulate). However, in the case of continuous fibers where the crack is deflected by the fibers, determination of crack length becomes difficult, and complete separation of the fracture surfaces does not occur [11]. Therefore toughness is only implied from the area under a stress–strain curve and the observation of fiber pullout, rather than being determined directly from Griffith theory.

Marshall and Evans [22] point out that classical mechanics can be applied only up until the occurrence of the first matrix crack. Frictional forces become important to fiber pullout and toughening once cracking has occurred.

It also has been argued by others [12,14,23] that the matrix strain is independent of crack length when crack growth is controlled primarily by fiber interaction (e.g., debonding, storage of strain energy by crack bridging, and frictional losses at the interface). Hence, in fiber-controlled fracture the composite is notch insensitive. Luh and Evans [24] have observed that, in Nicalon–lithium aluminosilicate (LAS) composites, changes occur with temperature at the fiber–matrix interface, such that at room temperature the interfacial bond is primarily frictional, and the material is notch insensitive, whereas at elevated temperatures reaction at the interface takes place, and the composite becomes brittle and notch sensitive.

Comparison with Resin Matrix Composites

The use of resin matrix composites in structural applications has grown to a mature technology and has led to a growing interest in extending composite usage to higher temperatures. The knowledge gained from resin matrix materials can serve as a basis for the development of structural ceramic composites. However, there are important differences between the two material systems in fabrication, microstructure, physical properties, and mechanical behavior, as summarized in Table 1.

Fiber modulus and strength in resin matrix composites typically are much greater than those of the matrix. The fiber, therefore, provides stiffness and load-bearing capability, while the matrix serves to distribute the load among the fibers. Hence a strong fiber–matrix interface is desirable to maximize stiffness and strength. Often, especially in the case of epoxy matrix materials, a coupling agent is used to increase the interfacial strength. The strain capability of the matrix usually is greater than that of the fiber; failure occurs by fracture of a fiber or bundle of fibers. The load is then redistributed among the remaining fibers until further fiber breakage occurs.

In contrast, in a ceramic matrix composite, the moduli of fiber and matrix may be similar. The fiber serves primarily to provide toughening mechanisms through crack deflection, crack bridging, debonding, crack blunting, and fiber pullout, rendering a relatively weak interface beneficial. The ceramic matrix is expected to have a higher population than its resin matrix counterpart of pores and process-induced flaws, as well as possible microcracks resulting from residual thermal stresses due to high processing temperatures. The strain capability of the fiber may exceed that of the matrix. If the interface is strongly bonded, matrix cracks will propagate through the fiber, resulting in brittle failure. However, if the interface is relatively weak, matrix cracks will be diverted at the interface, leading to debonding, the desired mode of failure to achieve toughening. Load transfer may involve large frictional contributions, as opposed to a strong fiber–matrix bond.

TABLE 1
Comparison of Ceramic Matrix and Resin Matrix Composites

Ceramic Matrix	Resin Matrix
$\sigma_f \sim \sigma_m$	$\sigma_f \sim 100\ \sigma_m$
E_f slightly $> E_m$	$E_f \gg E_m$
$\varepsilon_f \geq \varepsilon_m$	$\varepsilon_f < \varepsilon_m$
First crack matrix controlled; fibers increase strain capability by opposing displacement of matrix cracks	Fibers control strength and modulus; failure occurs by fiber fracture
Weak interface	Strong interface

In any composite system where there is a difference in coefficient of thermal expansion (CTE) between fiber and matrix, residual thermal stresses will be expected to develop both during composite fabrication and upon thermal cycling. For most fibers, expansion is anisotropic, differing along the fiber axis and in the radial direction. A pure β-SiC fiber might prove to be the exception.

When graphite fiber is used as the reinforcement, the CTE along the fiber axis is slightly negative. Residual stresses developed during fabrication place the fiber in compression and the matrix in tension. In a ceramic matrix composite it may be advantageous to choose a fiber with a higher CTE than that of the matrix, so that the matrix is in compression, providing resistance to matrix crack opening. However, if too large a fiber CTE is chosen, debonding at the interface will result. In any ceramic composite fabrication approach in which matrix shrinkage occurs during processing, as is the case for polymer and sol–gel derived matrices, the matrix will almost always be in tension. Since ceramic matrix composites can be expected to see high temperatures both in fabrication and in service, residual stresses will be much greater than in their resin matrix counterparts, possibly giving rise to microcracking on fabrication. In terms of designing with composites, this translates into a significant difference between resin and ceramic matrix materials. Whereas resin, glass, and probably reaction-bonded silicon nitride (RBSN) matrix composites can be designed as a cross-ply layup of unidirectional laminae, ceramic matrix composites in which matrix shrinkage occurs most likely will have to be fabricated from 2D cloth or 3D woven preforms to minimize the influence of shrinkage cracking.

Whereas in resin matrix composites matrix strain capability exceeds that of the fiber, in CMCs the reverse is the case. Decreased matrix strain capability relative to the fiber may decrease tensile strength, as the matrix cannot plastically deform to accommodate stress concentrations. Behavior under flexural loading might become more complex. Prewo [25], who has studied Nicalon–epoxy and Nicalon–LAS composites in both tension and flexure, notes that while the Nicalon–epoxy composites had the higher tensile strength, in flexure the LAS matrix material appeared to be stronger, as it "yielded" because of microcracking, shifting the neutral axis of the beam toward the compressive side of the test specimen. Thus some ceramic matrix composites may appear to have higher strength in flexure than their resin matrix counterparts as a result of both their higher matrix compressive strength and the pseudoductility imparted by microcracking. Conversely, a composite having low compressive and/or shear strength can produce strengths in flexure that are less than those in tension. This might be the case for cracked matrices or composites with matrices weak in shear or having a very low degree of interfacial bonding.

CMCs therefore are quite different from resin matrix composites with respect to material design considerations and from a mechanics standpoint. However, experience in fabrication of resin matrix systems could be applied to the newer ceramic composite materials.

Guidelines for Material Development

The discussion to this point has established the crucial nature of the interface to mechanical behavior, toughening mechanisms, and the nature of the composite fracture, that is, the determination of brittle versus gradual failure of the composite. The relationship between physical properties (e.g., relative moduli and thermal expansion coefficients of the fiber and matrix) have been reviewed. In addition, chemical compatibility effects must be considered.

At the high temperatures to which ceramic composites are exposed, both in processing and in service, chemical reaction between matrix and fibers is likely. Such reaction may result in a high degree of interfacial bonding, leading to brittle failure and/or a reduction in fiber strength as the result of reduced effective fiber diameter or the introduction of reaction products at the fiber surface, which can act as stress concentrations. Volatile decomposition products also might be formed, resulting in debonding as volatiles escape along the fiber interface.

Because the composite may still be load-bearing beyond the first matrix crack (indeed, matrix microcracking may occur upon processing), the interface could be exposed to the environment, hence must be resistant to environmental attack. Reactions with the environment could lead to embrittlement and/or degradation of fiber strength.

To control the strength of the interface for optimal mechanical behavior, minimize stresses due to thermal expansion mismatch, and blunt at the fiber surface stress concentrations that arise from matrix microcracking, as well as to avoid chemical reactions between matrix and fiber and fiber degradation from environmental exposure, it might be advantageous to coat the fiber. Several coating techniques are possible, including chemical vapor deposition (CVD), sol–gel techniques, and pyrolysis of polymer coatings.

One of the most successful demonstrations of the use of coatings to improve fiber strength and thermal and environmental stability is that of the Avco SCS SiC fiber, which utilizes a CVD carbon coating followed by a SiC overcoating.

This fiber has been used in reaction bonded silicon nitride matrix to produce a composite that is stronger than the RBSN monolithic and exhibits fiber-controlled failure [26–30]. Fracture occurs in the carbon-rich layer at the outer radius of the fiber, thus enhancing the material's toughness.

In some glass composite systems, additives may be introduced deliberately to form a reaction layer at the interface during processing. In the case of Nicalon–LAS composites [31] a carbon-rich zone is formed at the interface which is mechanically weak and serves to increase toughness. A study of the formation of this carbon interface from thermodynamic considerations has been presented by Spear, Benson, and Pantano [32,33].

The use of C, BN, and BN/SiC coatings on Nicalon in zirconium-based matrices has been demonstrated successfully by Bender et al. [34].

All the aforementioned CMC systems suffer from dependence on a readily oxidizable coating layer to impart toughness. Thus these systems are subject to rapid and severe environmental degradation under some exposure conditions.

Fibers

A variety of ceramic fibers are commercially available or under development. Mah et al. [35] provide a review of the state of fiber development as of 1987. Si and C, as well as Si-C-N-O and oxide fibers (primarily Al_2O_3) are available. Fiber properties are summarized in Table 2.

Only carbon fibers provide strength and long-term stability at temperatures above 1200°C, but these of course are limited by their susceptibility to oxidation, particularly if used in a ceramic matrix, which is likely to microcrack.

SiC fibers have been prepared as monofilaments by CVD [36,37] and as polymer-derived yarns. SiC is susceptible to oxidation; however, it does form a protective SiO_2 surface layer, with further fiber degradation controlled by diffusion of O_2 and CO [38,39].

The Avco SCS-6 fiber, produced by chemical vapor deposition of SiC onto a carbon core [40], and double-coated with carbon graded to silicon carbide, shows the greatest strength retention to high temperature but is limited to use in flat plate laminates as a result of its large (143 μm) diameter. All the other fibers listed show substantial decreases in strength at 1000–1200°C, as does the Avco CVD fiber without a coating or with a single coating. This strength decrease is *very* sensitive to the test environment. Above 1300°C, the Avco SCS fiber exhibits loss of CO, Al, Al_2O, Ca, and SiO, although the SiO loss is substantially less than that of Nicalon [41]. Degradation may be the result of calcium impurities. Creep also is observed at high temperatures. However, below 1400°C, creep strain is an order of magnitude less than that observed for Nicalon [42]. DiCarlo [42,43] has shown that thermal treatments at temperatures above 1400°C

TABLE 2
Properties of Ceramic Fibers From Data Supplied by Manufacturers: Gage Lengths and Test Techniques May Vary

Fiber	Composition (wt %)	Tensile Strength (MPa)	Tensile Modulus (GPa)	Diameter (μm)
Nonoxide				
Nicalon	59 Si, 31 C, 10 O	2520–3290	182–210	10–20
SCS-6	CVD SiC on C core	3920	406	143
Tyranno	Si, Ti, C, O	> 2970	> 200	8–10
Avco	Si, C	> 2800	280–315	6–10
MPDZ	47 Si, 30 C, 15 N, 8 O	1750–2100	175–210	10–15
HPZ	59 Si, 10 C, 28 N, 3 O	2100–2450	140–175	10
MPS	69 Si, 30 C, 1 O	1050–1400	175–210	10–15
Oxide				
Nextel 312	62 Al_2O_3, 14 B_2O_3, 24 SiO_2	1750	154	11
Nextel 440	70 Al_2O_3, 28 SiO_2, B_2O_3	2100	189	10–12
Nextel 480	70 Al_2O_3, 28 SiO_2, 2 B_2O_3	2275	224	10–12
FP	85 Al_2O_3, 15 SiO_2	> 1400	385	20
PRD-166	> 99 α-Al_2O_3	2100–2450	385	20
Sumitomo	Al_2O_3, 15–25 ZrO_2	1800–2600	210–250	9–17

Based on Ref. 35.

reduce creep strain, but at the expense of tensile strength [42,43]. This strength decrease might be attributable to loss of silicon [42].

The Nicalon fiber, a polycarbosilane-derived SiC, is manufactured by Nippon Carbon and marketed in the United States by Dow Corning Corporation. Cure behavior of the polycarbosilane, and the influence of curing conditions on the development of mechanical properties, have been discussed by Hasegawa, Okamura, Ichikawa, and coworkers [44–46]. Nicalon is available in several resistivity grades and with several different surface coatings. It is not a stoichiometric SiC, but contains nominally 10% O, leading to compositional and microstructural changes at elevated temperatures, including grain growth and creep [35,41,47–51].

Nicalon has been shown [47,52] to be comprised of amorphous or microcrystalline SiC, carbon microcrystals, and SiO_2. The presence of carbon crystallites suppresses growth of SiC crystals. Degradation is attributed to evolution of SiO and CO [41], accompanied by β-SiC grain growth [53]. It has been suggested by several of the workers cited that the use temperature of a polycarbosilane-derived fiber could be extended by elimination of oxygen in the precursor. In Nicalon the oxygen is introduced deliberately as part of the curing process [46]. Okamura et al. [54] have shown that γ-ray irradiation, rather than oxidation, might be used to cross-link polycarbosilane. However, when such irradiation is conducted in air, oxygen again is incorporated into the polymer. Irradiation in inert environments has not been reported.

Ube Industries' Tyranno fiber is produced from a titanium-doped (1.5–4.0 wt %) polycarbosilane using the same process as that used in the preparation of Nicalon. The manufacturer reports the fiber to be amorphous. Properties are similar to those of Nicalon [35,55], though loss of strength above 1000°C is somewhat less abrupt [56]. Levels of silicon, carbon, and oxygen also are similar to Nicalon, with somewhat higher oxygen levels (17% [56] vs. 10% for ceramic-grade Nicalon [57]) and more free carbon [56]. Above 1400°C the free carbon would be expected to react with Si—O bonds, evolving CO and accounting for the observed decrease in tensile strength [56].

An almost stoichiometric polymer-derived SiC fiber is under development by Dow Corning under U.S. government funding [57]. The MPS fiber is 69% Si, 30% C, 1% O by weight, and is 50–80% microcrystalline β-SiC. Fiber properties have not been reported in the open literature.

Also under joint development by Dow Corning and Hoechst–Celanese supported by U.S. government funding are two Si-C-N-O fibers: HPZ and MPDZ [57–59]. Both are amorphous. MPDZ forms a carbon-rich interface when used in oxide matrices; HPZ, which is lower in carbon content, does not. Both are processed from polymeric precursors and pyrolyzed in a nonoxidizing atmosphere [58]. Extensive fractography studies by Sawyer et al. [59] have shown that for a distribution of fiber strengths, low tensile strengths are the result of surface flaws and particulate defects within the fiber; strength can be increased by careful purification and filtering of the polymer precursor. The higher strength fibers showed internal granular defects attributed to local variations in stoichiometry. The analysis of this flaw–strength relationship probably can (and should) be applied to other glassy, polymer-derived fibers. There has been *significant* improvement in MPDZ and HPZ fiber strengths since publication of this paper, and HPZ now is being produced in quantities sufficient for composite fabrication on an experimental basis.

A number of oxide fibers ranging in diameter from 10 to 20 μm are available commercially. The use of oxide fibers is most likely limited to composites with oxide matrices, to avoid chemical reactions at the interface and interdiffusion between matrix and fiber. Even in oxide–oxide systems, fiber coatings may be required to provide an optimal interfacial bond.

Du Pont's Fiber FP is more than 99% small-grain polycrystalline α-Al_2O_3 and also is available with a SiO_2 coating. It has been used primarily in metal matrix composites. Above 1000°C Fiber FP exhibits grain growth and creep. PRD-166, also produced by du Pont, is an α-Al_2O_3 fiber having Y_2O_3 partially stabilized ZrO_2 as a second phase. The ZrO_2 addition inhibits grain growth and improves both room temperature strength and strength retention after exposure to elevated temperature [60].

Nextel 312, 440, and 480, manufactured by 3M Corporation, are sol–gel derived. Nextel 312 is a solid solution of aluminum borate and mullite. Nextel 440 and 480 are similar, but with less of the borate phase. Nextel 440 is comprised of an Al_2O_3 phase, with some mullite and amorphous SiO_2 present, while Nextel 480 is primarily mullite. The Nextel 480 fibers show loss of modulus above 900°C, and creep exceeding 10^{-4}/h above 1190°C [61].

Matrices and Fabrication Considerations

Suitability of a material as a composite matrix is governed by requirements for refractory behavior, compatibility with fibers, ability to wet the fiber and infiltrate a fiber tow, consolidation at temperatures low enough to avoid reactions with the fiber and/or fiber thermal degradation, and thermal expansion properties compatible with those of the fiber. Matrix and fiber candidate materials must be considered together; a "good" fiber for one matrix may perform poorly in another.

Possible matrix materials for CMCs include glasses, glass–ceramics, carbides, nitrides, oxides, and borides. Processing and properties of a number of ceramic matrix composite systems have been reviewed by Cornie et al. [62]. For optimum mechanical properties, it is desirable to have a composite in which the fibers are uniformly distributed and well infiltrated by the matrix and in which the matrix is uniform, dense, and free of large defects.

A number of glass and glass–ceramic systems, primarily lithium aluminosilicate (LAS), magnesium aluminosilicate (MAS), and barium magnesium aluminosilicate (BMAS) [63–66] and an oxynitride, Ba-Si-Al-O-N [67], have been studied. These are relatively easy to process because the glass undergoes viscous flow, enabling composites to be hot-pressed to full density at temperatures that may not degrade the fiber. Transfer molding and injection molding techniques also have been demonstrated in glass systems [66]. However, the use temperature then is limited by the softening point and creep behavior of the glass or glass–ceramic. Embrittlement of Nicalon–LAS composites tested in oxidizing environments at 900 and 1000°C has been observed as well [68].

Chemical vapor deposition (CVD) has proved to be a successful route for the fabrication of SiC matrix composites [69–76]. Stoichiometry of the matrix can be controlled and varied as needed to produce multilayered or graded compositions. Fabrication of complex shapes is possible. Deposition rates are slow, and closed pores can be formed. As a result, part thickness is limited, and surface machining often is employed to reopen the surface for further densification. SiC/SiC and C/SiC composites have been produced using this method by SEP (Société Européene de Propulsion, Bordeaux, France [73]), Amercom, Refractory Composites [74], Babcock and Wilcox [75], and others. Other matrices produced by CVD include BN, B_4C, TiB_2, Al_2O_3, ZrC, and TiC.

Reaction-bonded Si_3N_4 matrix composites reinforced with SiC fibers have been demonstrated successfully by Bhatt [26–30] at NASA-Lewis and by Corbin and coworkers [77] at Norton. Reaction bonding occurs without matrix shrinkage, resulting in a matrix that is not microcracked upon fabrication. Porosity levels of nominally 30% are present.

Reaction sintering of SiC (RSSC) also provides a possible route to matrix fabrication. However, reaction temperatures exceed those at which state-of-the-art SiC fibers, or graphite fibers, are chemically stable toward reaction with the matrix. Coatings of TiC, SiC, TiN, and pyrolytic carbon have been suggested as means of protecting the fiber [69]. However, recrystallization of the coating at the interface by the silicon melt could result in a brittle composite.

Hot pressing and sintering, traditionally used as processing techniques for monolithics, also require temperatures and pressures that result in fiber degradation.

Melt infiltration of ceramic matrices would provide greater flexibility in complex shape formation than does hot pressing and would reduce porosity. This approach has been demonstrated by Hillig [78,79] using CaF_2, $CaSiO_3$, $SrSiO_3$, and $SrAlSi_2O_8$ matrices. Dense composites were produced with little matrix shrinkage. However, the melting points of these matrices are quite high ($>$ 1500°C), producing chemical reactions with the SiC fibers. The approach might be possible with coated fibers, although, again, recrystallization of the coating would have to be considered.

The use of polymeric precursors to ceramic matrices provides low temperature composite processing to circumvent fiber degradation and the formation of reaction products at the interface. Complex shapes can be fabricated by established techniques used in processing resin matrix composites. Rheology can be controlled by tailoring the starting polymer; molecular structure and thermal processing history determine the stoichiometry of the ceramic that results upon polymer pyrolysis. A high degree of homogeneity can be achieved for single-phase matrices, and unique multiphase matrices can be formed.

However, during the conversion from polymer to ceramic the matrix can be expected to lose volatiles and shrink, resulting in porosity and microcracking. Some densification can be achieved by reimpregnation, in analogy to carbon–carbon processing. Despite shrinkages, matrix densities can be higher than those produced by CVI.

Many of the polymers studied as SiC precursors, including the polycarbosilane Nicalon precursor, do not produce stoichiometric SiC, and most incorporate some oxygen during processing, leading to thermodynamic instabilities of the pyrolyzed ceramic product at temperatures above 1200°C. Both char yield and final ceramic composition must be considered in the choice of a polymeric precursor.

The redistribution of methylchlorodisilanes to produce a methylpolysilane SiC precursor has been reported [80–83]. By modification of this reaction to form methyl- and phenylpolysilane copolymers, the Si/C ratio can be controlled to produce a ceramic with very close to SiC stoichiometry. Seyferth et al. [84,85] also have produced near-stoichiometric SiC by pyrolysis of mixtures of polymethylsilanes and organic or organosilicon polymers having unsaturated functional groups.

A number of polymeric precursors to Si_3N_4 have been investigated [57,58,86–95]. These tend to produce amorphous silicon oxycarbonitrides bonded in a random fashion [58]. The lack of stoichiometry leads to chemical reaction accompanied by loss of SiO, CO, and N above 1200°C, accompanied by crystallization. Some of the polysilazanes produce higher char yields than the majority of the polysilanes and polycarbosilanes, on the order of 80–85%. However, they are extremely moisture sensitive; thus composite fabrication would require handling under inert conditions.

Hatta et al. [96] have proposed an interesting mixed SiC/Si_3N_4 system fabricated by firing a mixture of silicon powder and polycarbosilane in a nitrogen atmosphere to form a porous structure of 5–20 wt % SiC in Si_3N_4. It is proposed that the porous texture would contribute to the thermal shock resistance of this material.

Sol–gel technology, which employs tetrafunctional alkoxides, has been used successfully for infiltration of oxide matrices, with subsequent densification by hot pressing. The use of gels provides more uniform infiltration than is offered by slurry techniques and minimizes the fiber damage that can result from slurry infiltration of particles. Densification may be conducted at lower temperatures than in particle infiltration. However, shrinkages tend to be large, and drying must be done carefully with the use of drying agents to minimize shrinkage cracking.

Condensation of trifunctional alkoxides has been used to form organosilsesquioxanes [97–102], which can be pyrolyzed to yield carbon-filled amorphous SiO_2, or, under different polymerization and pyrolysis conditions, SiC [99,100]. Composites can be fabricated at low temperatures and pressures compared with standard sol–gel techniques.

Jamet et al. [103] have used colloidal powders, including BN, Si_3N_4, and SiC, as fillers in thermosetting polymers to increase char yields on pyrolysis and to minimize matrix shrinkage and cracking. In some cases the char yield of the filled polymer determined experimentally exceeded that calculated on a rule-of-mixtures basis (char yield of the unfilled polymer plus weight of filler), probably the result of reduced surface area, which accompanies a reduction in microcracking on filler addition. Other work with filled polymers [104], however, has shown that depending on fiber loading, tow size, fiber shape, and diameter, the fibers can act as a sieve to filter the particulate, resulting in a layered material of polymer and filler and polymer plus fiber, with greatly different moduli between the layers. Particulate abrasion also may lead to diminished fiber strength.

Areas for Future Investigation

One of the greatest limitations to the structural use of CMCs at temperatures above 1200°C is the availability of a strong, thermodynamically stable, small-diameter ($< 20\ \mu m$) fiber suitable for forming complex shapes. A small-diameter, stoichiometric SiC fiber fabricated by CVD or polymeric routes would fill this gap. Another possibility is a microstructurally stable, creep-resistant oxide fiber. Alumina and mullite offer the greatest promise.

A second critical issue is the development of a theoretical understanding of requirements for the fiber–matrix interface, and reduction to practice by the identification of compatible coating materials and thicknesses necessary to protect the fiber from environmental attack, minimize residual thermal stresses, and produce composite-type, nonbrittle failure modes. The coating also is important to the wetting of the fiber by the matrix during some processing approaches. Surface modification of monofilaments has been studied, particularly for application in metal matrix composites [40]. Techniques for uniformly coating tows of small-diameter fibers also need to be established.

Methods of infiltrating and consolidating matrices under temperatures and environments that will not degrade the fiber or cause reaction at the interface are needed. The glass–ceramic approach is reaching the limitations of available matrices. Sialons offer some promise for surpassing state-of-the-art glass–ceramics. CVD has met with considerable success, but means for achieving high density and greater uniformity within practical and economical infiltration times must be identified.

The use of polymeric precursors with tailored molecular structure and molecular weight distributions may provide new materials, particularly for the formation of complex shapes, if an understanding of the pyrolysis process and its effect on ceramic stoichiometry can be established. Such materials would be expected to be analogues of carbon–carbon, having matrices that are microcracked during fabrication, but with inherently greater oxidation resistance than their carbon–carbon counterparts.

Ceramic approaches wherein densification does not result in volume change (e.g., reaction-bonded silicon nitride or silicon carbide) are promising, but residual matrix porosity is high. These approaches are most amenable to the use of monofilament fibers, and thus architecture and shape complexity are limited. Innovative approaches to the use of small diameter fibers in reaction-bonded matrices are needed.

Many other areas also remain to be addressed. Joining techniques for both ceramic to ceramic and ceramic to metal need to be developed, as do testing methodologies, life prediction techniques, and design approaches. Ceramic matrix composites are still in the very early stages of development as high temperature structural materials; however, many applications are open to such new materials.

Frances I. Hurwitz

References

1. L. J. Schioler and J. J. Stiglich, Jr., *Am. Ceram. Soc. Bull., 65,* 289 (1986).
2. R. W. Rice, P. F. Becher, S. W. Freiman, and W. J. McDonough, *Ceram. Eng. Sci. Proc., 1,* 424 (1980).
3. R. W. Rice, *J. Phys. Chem. Solids, 45,* 1033 (1984).
4. I. W. Donald and P. W. McMillan, *J. Mater. Sci., 11,* 949 (1976).
5. D. K. Shetty, "Ceramic Matrix Composites," *Current Awareness Bulletin 118,* Metals and Ceramics Information Center, Batelle Laboratories, Columbus, OH, 1982, pp. 1–4.
6. F. D. Gac, J. D. Petrovic, J. V. Milewski, and P. D. Shalek, *Ceram. Eng. Sci. Proc., 7,* 978 (1986).
7. J. V. Milewski, "Method of Producing Silicon Carbide Articles," U.S. Patent 4,513,030 (1985), to U.S. Department of Energy.
8. J. V. Milewski, *Adv. Ceram. Mater., 1,* 36 (1986).
9. P. D. Shalek, J. J. Petrovic, G. F. Hurley, and F. D. Gac, *Am. Ceram. Soc. Bull., 65,* 351 (1986).
10. R. W. Rice, *Ceram. Eng. Sci. Proc., 2,* 661 (1981).
11. K. T. Faber, *Ceram. Eng. Sci. Proc., 5,* 408 (1984).
12. J. Aveston, G. A. Cooper, and A. Kelly, "Single and Multiple Fracture," in *The Properties of Fiber Composites,* Conference Proceedings of the National Physics Laboratory, 1971, IPC Science and Technology Press, Guildford, Surrey, England, 1971, pp. 15–26.
13. J. Aveston and A. Kelly, *J. Mater. Sci., 8,* 352 (1973).
14. Y. Korczynskyj, S. J. Harris, and J. G. Morley, *J. Mater. Sci., 16,* 1533 (1981).
15. H. C. Chan and W. A. Patterson, *J. Mater. Sci., 7,* 856 (1972).
16. R. A. J. Sambell, D. H. Bowen, and D. C. Phillips, *J. Mater. Sci., 7,* 663 (1972).
17. R. A. J. Sambell, A. Briggs, D. C. Phillips, and D. H. Bowen, *J. Mater. Sci., 7,* 676 (1972).
18. M. Vedula, R. N. Pangborn, and R. A. Queeney, *Composites, 19,* 133 (1988).
19. K. T. Faber and A. G. Evans, *Acta Metall., 31,* 565 (1983).
20. A. H. Cottrell, *Proc. R. Soc. London, A282,* 2 (1964).
21. A. A. Griffith, *Phil. Trans. R. Soc., London, A221,* 163 (1921).
22. D. B. Marshall and A. G. Evans, *J. Am. Ceram. Soc., 68,* 225 (1985).
23. I. R. McColl and J. G. Morley, *J. Mater. Sci., 12,* 1165 (1977).
24. E. Y. Luh and A. G. Evans, *J. Am. Ceram. Soc., 70,* 466 (1987).
25. K. M. Prewo, *J. Mater. Sci., 17,* 3549 (1982).
26. R. T. Bhatt and R. E. Phillips, "Laminate Behavior for SiC Fiber-Reinforced Reaction-Bonded Silicon Nitride Matrix Composites," NASA TM 101350, 1988.
27. R. T. Bhatt, "Properties of Silicon Carbide Fiber Reinforced Silicon Nitride Composites," Army TR-88-C-027, 1988.
28. R. T. Bhatt and R. E. Phillips, "Thermal Effects on the Mechanical Properties of SiC Fiber Reinforced Reaction-Bonded Silicon Nitride Matrix Composites (SiC/RBSN)," Army TR-88-C-028, 1988.
29. R. T. Bhatt, "Effect of Processing Variables on the Mechanical Properties of SiC/RBSN Composites," NASA TM 88814, 1986.
30. R. T. Bhatt, "Mechanical Properties of SiC/RBSN Composites," in R. E. Tressler, G. L. Messing, C. G. Pantano, and R. E. Newnham, Eds., *Tailoring Multiphase and Composite Ceramics,* Plenum Press, New York, 1985, pp. 675–686.
31. J. J. Brennan, "Interfacial Characterization of Glass and Glass–Ceramic Matrix/Nicalon SiC Fiber Composites," in R. E. Tressler, G. L. Messing, C. G. Pantano, and R. E. Newnham, Eds., *Tailoring Multiphase and Composite Ceramics,* Plenum Press, New York, 1985, pp. 549–560.
32. K. E. Spear, P. M. Benson, and C. G. Pantano, "Thermomechanical Modeling for Interface Reactions in Carbon-Fiber Reinforced Glass Matrix Composites," in Z. A. Munir, D. Cubicciotti, and H. Tagawa, Eds., *High Temperature Materials Chemistry,* Vol. 4, Electrochemical Society, Pennington, NJ, 1988, pp. 345–354.
33. P. M. Benson, K. E. Spear, and C. G. Pantano, *Ceram. Eng. Sci. Proc., 9,* 673 (1988).
34. B. Bender, D. Shadwell, C. Bulik, L. Incorvati, and D. Lewis III, *Am. Ceram. Soc. Bull., 65,* 363 (1986).
35. T. Mah, M. G. Mendiratta, A. P. Katz, and K. S. Mazdiyasni, *Am. Ceram. Soc. Bull., 66,* 304 (1987).
36. E. Fitzer, D. Kehr, and M. Šahebkar, "CVD-SiC on Carbon Fibres," in R. C. Marshall, J. W. Faust, Jr., and C. E. Ryan, Eds., *Silicon Carbide—1973,* University of South Carolina Press, Columbia, 1974, pp. 33–41.
37. H. E. DeBolt, V. J. Krukonis, and F. E. Wawner, Jr., "High Strength, High Modulus SiC Filament via Chemical Vapour Deposition," in R. C. Marshall, J. W. Faust, Jr., and C. E. Ryan, Eds., *Silicon Carbide—1973,* University of South Carolina Press, Columbia, 1974, pp. 168–175.
38. C.-H. Andersson and R. Warren, *Composites, 15,* 16 (1984).
39. R. Warren and C.-H. Andersson, *Composites, 15,* 101 (1984).
40. F. E. Wawner, Jr., "Boron and Silicon Carbide/Carbon Fibers," in A. R. Bunsell, Ed., *Fibre Reinforcements for Composite Materials,* Elsevier, Amsterdam, 1988, pp. 371–425.
41. S. M. Johnson, R. D. Brittain, R. H. Lamoreaux, and D. J. Rowcliffe, *J. Am. Ceram. Soc., 71,* C132 (1988).
42. J. A. DiCarlo, *J. Mater. Sci., 21,* 217 (1986).
43. J. A. DiCarlo, "High Temperature Properties of CVD Silicon Carbide Fibers," presented at the International Conference of Whisker- and Fiber-Toughened Ceramics, 1988.
44. Y. Hasegawa and K. Okamura, *J. Mater. Sci., 18,* 3633 (1983).

45. K. Okamura, M. Sato, and Y. Hasegawa, *J. Mater. Sci. Lett.*, *2*, 769 (1983).

46. H. Ichikawa, H. Teranishi, and T. Ishikawa, *J. Mater. Sci. Lett.*, *6*, 420 (1987).

47. G. Simon and A. R. Bunsell, *J. Mater. Sci.*, *19*, 3649 (1984).

48. G. Simon and A. R. Bunsell, *J. Mater. Sci.*, *19*, 3658 (1984).

49. T. J. Clark, M. Jaffe, J. Rabe, and N. R. Langley, *Ceram. Eng. Sci. Proc.*, *7*, 901 (1986).

50. L. C. Sawyer, R. T. Chen, F. Haimbach IV, P. J. Harget, E. R. Prack, and M. Jaffe, *Ceram. Eng. Sci. Proc.*, *7*, 914 (1986).

51. T. J. Clark, E. R. Prack, M. I. Haider, and L. C. Sawyer, *Ceram. Eng. Sci. Proc.*, *8*, 717 (1987).

52. Y. Sasaki, Y. Nishina, M. Sato, and K. Okamura, *J. Mater. Sci.*, *22*, 443 (1987).

53. T. Mah, N. L. Hecht, D. E. McCullum, J. R. Hoenigman, H. M. Kim, A. P. Katz, and H. A. Lipsitt, *J. Mater. Sci.*, *19*, 1191 (1984).

54. K. Okamura, T. Matsuzawa, and Y. Hasegawa, *J. Mater. Sci. Lett.*, *4*, 55 (1985).

55. D. B. Fischbach, P. M. Lemoine, and G. V. Yen, *J. Mater. Sci.*, *23*, 987 (1988).

56. T. Yamamura, T. Ishikawa, M. Shibuya, T. Hisayuki, and K. Okamura, *J. Mater. Sci.*, *23*, 2589 (1988).

57. W. H. Atwell, W. E. Hauth, R. E. Jones, and R. M. Salinger, "Advanced Ceramics Based on Polymer Processing," AFWAL-TR-844061, under USAF Contract 33615-83-C-5006, June 1988.

58. J. Lipowitz, H. A. Freeman, R. T. Chen, and E. R. Prack, *Adv. Ceram. Mater.*, *2*, 121 (1987).

59. L. C. Sawyer, M. Jamieson, D. Brikowski, M. I. Haider, and R. T. Chen, *J. Am. Ceram. Soc.*, *70*, 798 (1987).

60. J. C. Romine, *Ceram. Eng. Sci. Proc.*, *8*, 755 (1987).

61. D. D. Johnson, A. R. Holtz, and M. P. Grether, *Ceram. Eng. Sci. Proc.*, *8*, 744 (1987).

62. J. A. Cornie, Y. M. Chiang, D. R. Uhlmann, A. Mortensen, and J. M. Collins, *Am. Ceram. Soc. Bull.*, *65*, 293 (1986).

63. K. M. Prewo and J. J. Brennan, *J. Mater. Sci.*, *15*, 463 (1980).

64. K. M. Prewo and J. J. Brennan, *J. Mater. Sci.*, *17*, 1201 (1982).

65. J. J. Brennan and K. M. Prewo, *J. Mater. Sci.*, *17*, 2371 (1982).

66. K. M. Prewo, J. J. Brennan, and G. K. Layden, *Am. Ceram. Soc. Bull.*, *65*, 305 (1986).

67. M. A. Herron and S. H. Risbud, *Am. Ceram. Soc. Bull.*, *65*, 342 (1986).

68. T. Mah, M. G. Mendiratta, A. P. Katz, R. Ruh, and K. S. Mazdiyasni, *J. Am. Ceram. Soc.*, *68*, C248 (1985).

69. E. Fitzer and R. Gadow, *Am. Ceram. Soc. Bull.*, *65*, 326 (1986).

70. A. J. Caputo and W. J. Lackey, "Fabrication of Fiber-Reinforced Ceramic Composites by Chemical Vapor Infiltration," ORNL/TM-9235, 1984.

71. A. J. Caputo and W. J. Lackey, *Ceram. Eng. Sci. Proc.*, *5*, 654 (1984).

72. A. J. Caputo, W. J. Lackey, and D. P. Stinton, *Ceram. Eng. Sci. Proc.*, *6*, 694 (1985).

73. P. J. Lamicq, G. A. Bernhart, M. M. Dauchier, and J. G. Mace, *Am. Ceram. Soc. Bull.*, *65*, 336 (1986).

74. J. W. Warren, *Ceram. Eng. Sci. Proc.*, *5*, 684 (1985).

75. H. H. Moeller, W. G. Long, A. J. Caputo, and R. A. Lowden, *SAMPE Qt.*, *17*, 1 (1986).

76. D. P. Stinton, A. J. Caputo, and R. A. Lowden, *Am. Ceram. Soc. Bull.*, *65*, 958 (1986).

77. N. D. Corbin, G. A. Rossetti, Jr., and S. D. Hartline, *Am. Ceram. Soc. Bull.*, *65*, 958 (1986).

78. W. B. Hillig, *Ceram. Eng. Sci. Proc.*, *6*, 674 (1985).

79. W. B. Hillig, *J. Am. Ceram. Soc.*, *71*, C96 (1988).

80. R. H. Baney and J. H. Gaul, "High Yield Silicon Carbide from Alkylated or Arylated Pre-Ceramic Polymers," U.S. Patent 4,298,559 (1981), to Dow Corning Corporation.

81. R. H. Baney and J. H. Gaul, "Method for Preparing Silicon Carbide," U.S. Patent 4,310,651 (1982), to Dow Corning Corporation.

82. R. H. Baney, J. H. Gaul, Jr., and Terrance K. Hilty, *Organometallics*, *2*, 859 (1983).

83. D. R. Bujalski, G. E. LeGrow, and T. F. Lim, *ACS Polym. Prepr.*, *2*, 396 (1987).

84. D. Seyferth, Y.-F. Yu, and G. E. Koppetsch, "Method for Forming New Preceramic Polymers Containing Silicon," U.S. Patent 4,719,273 (1988), to Massachusetts Institute of Technology.

85. D. Seyferth, T. G. Wood, and Y.-F. Yu, "Method for Forming New Preceramic Polymers for SiC and Si_3N_4/SiC Systems," U.S. Patent 4,645,807 (1987), to Massachusetts Institute of Technology.

86. J. H. Gaul, "Silazane Polymers from $(R_3Si)_2NH$ and Organochlorosilanes," U.S. Patent 4,312,970 (1982), to Dow Corning Corporation.

87. D. Seyferth, G. H. Wiseman, and C. Prud'homme, *J. Am. Ceram. Soc.*, *66*, C13 (1983).

88. D. Seyferth and G. H. Wiseman, *J. Am. Ceram. Soc.*, *67*, C132 (1984).

89. J. P. Cannady, "Silicon Nitride-Containing Ceramics," U.S. Patent 4,535,007 (1985), to Dow Corning Corporation.

90. L. A. Haluska, "Silicon Carbide Preceramic Vinyl-Containing Polymers," U.S. Patent 4,460,638 (1984), to Dow Corning Corporation.

91. Y. G. Blum, R. M. Laine, K. B. Schwartz, D. J. Rowcliffe, R. C. Benning, and D. B. Cotts, *Mater. Res. Soc. Symp. Proc.*, *73*, 389 (1986).

92. R. M. Laine and Y. Blum, "Method of Producing Polysilazanes," U.S. Patent 4,612,383 (1986), to SRI International.

93. Y. Blum and R. M. Laine, *Organometallics*, *5*, 2081 (1986).

94. A. W. Chow, R. D. Hamlin, Y. Blum, and R. M. Laine, *J. Polym. Sci. C, Polym. Lett.*, *26*, 103 (1988).

95. K. B. Schwartz, D. J. Rowcliffe, and Y. Blum, *Adv. Ceram. Mater.*, *3*, 320 (1988).

96. T. Hatta, H. Ueno, Y. Katsura, K. Fukuda, and H. Kubota, "Method of Making Flaky β-SiC," U.K. Patent Application GB 2,066,800A (1980), to Kurosaki Refractories Company Ltd., Japan.

97. F. K. Chi, and G. L. Stark, "Fiber Reinforced Glass Matrix Composites," U.S. Patent 4,460,639 (1984), to Dow Corning Corporation.

98. F. K. Chi and G. L. Stark, "Fiber Reinforced Glass Matrix Composites," U.S. Patent 4,460,640 (1984), to Dow Corning Corporation.

99. D. A. White, S. M. Oleff, R. D. Boyer, P. A. Budinger, and J. R. Fox, *Adv. Ceram. Mater.*, *2*, 45 (1987).
100. D. A. White, S. M. Oleff, and J. R. Fox, *Adv. Ceram. Mater.*, *2*, 53 (1987).
101. F. I. Hurwitz, L. H. Hyatt, J. Gorecki, and L. D'Amore, *Ceram. Eng. Sci. Proc.*, *8*, 732 (1987). Also NASA TM 89893, 1987.
102. F. I. Hurwitz, unpublished results, 1988.
103. J. Jamet, J. R. Spann, R. W. Rice, D. Lewis, and W. S. Coblenz, *Ceram. Eng. Sci. Proc.*, *5*, 677 (1985).
104. F. I. Hurwitz, *SAMPE J.*, *23*(2), 49 (1987). Also NASA TM 88812, 1986.

Ceramic Matrix Composites, Reaction Bonded

Lack of reliability has precluded ceramic materials from use in many important applications where their properties could be exploited to advantage. Although unacceptably wide variations have been observed in all properties, this issue is most vividly illustrated by structural components such as engine parts, where failure of even a small piece can ruin an entire engine. Flaws that cause premature failures almost always result from processing defects, service damage, or a combination of the two.

Materials research thrusts designed to achieve stronger, more reliable ceramics are illustrated in Figure 1, which predicts the tensile strength of brittle materials that are controlled by variable-size flaws having different fracture toughnesses. Materials scientists and processing engineers have strived to improve monolithic ceramic parts both by producing materials having increased K_{IC} values and by reducing flaw dimensions; they have accomplished important results. High K_{IC}'s have been achieved in both Si_3N_4 and ZrO_2 based materials; parts of many materials have been produced with small flaw sizes (down to 1–5 μm) by using highly perfect powders and clean processing techniques. This latter approach has been a focus of MIT's ceramics processing research programs [1].

Although K_{IC} values have been improved for ceramics used at low to moderate temperatures, there have been no solutions that apply for high temperature applications. The metastable phases responsible for the high toughness values in the ZrO_2 based materials are lost with single exposures to moderate temperatures ($\approx$ 600°C), and with repeated cycles to even relatively low temperatures. High fracture toughness values are achieved in Si_3N_4 by optimizing the α/β ratio in high density materials [2] that employ liquid phase sintering aids; typically these sintering aids limit the service temperature to a maximum of 1200°C. All the more refractory materials, such as SiC, have fracture toughness values (K_{IC} = 2.3–3.5 MPa $m^{1/2}$) that make them so sensitive to damage incurred during service that design stresses must be limited to levels far below the strengths of defect-free, as-processed samples.

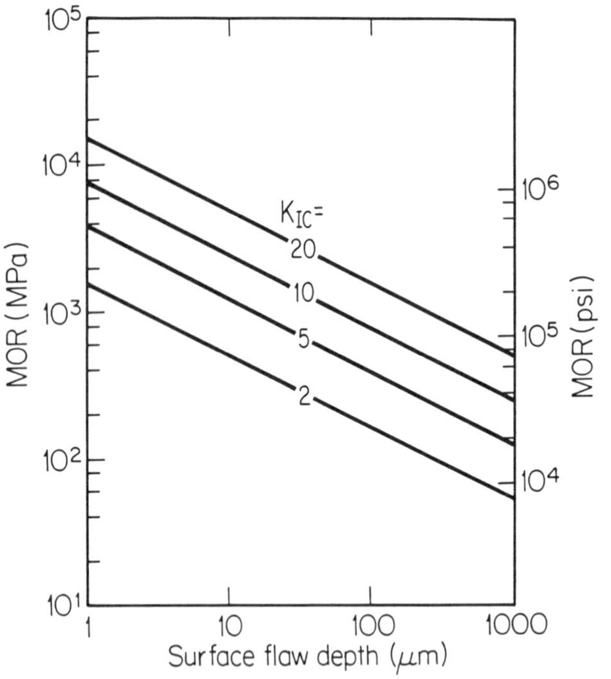

FIGURE 1 **Calculated tensile strengths of brittle materials having different fracture toughness K_{IC} values as a function of surface flaw depth.**

Composites represent an important opportunity for improving the toughness and reducing the flaw sensitivity of ceramic materials used at temperatures ranging from low to high values. Figure 2 is an idealized stress–strain curve [3] for a ceramic matrix composite; the toughening represented by the area under the stress–strain curve has in fact been observed in many cases. The toughening in these ceramic matrix composites (CMCs) results primarily from a combination of matrix microcracking and crack deflection for short-fiber and whisker reinforcements and a combination of fiber pullout and crack-bridging mechanisms for continuous or long-fiber reinforcements [4]. Although it is not entirely correct to discuss the toughening effects in terms of Figure 1 because K_{IC} has no precise meaning for highly oriented composites [5], the effects of increased area under the stress–strain curve [3], Figure 2, are analogous to an increased K_{IC}. For randomly oriented, short-fiber composites, fracture toughening effects have not been fully explored, but we would still expect K_{IC} to increase with crack length, analogous to highly oriented composites (*R*-curve behavior). Significantly increased fracture toughness values have been observed with short-fiber-reinforced crystalline and glass phase ceramic matrices [6,7]. Tough composite ceramics are tolerant of overstrains; they can be made completely insensitive to matrix flaw size, and they are resistant to impact, machining, and thermal shock damage. Properly designed composites should extend useful properties to very high temperatures.

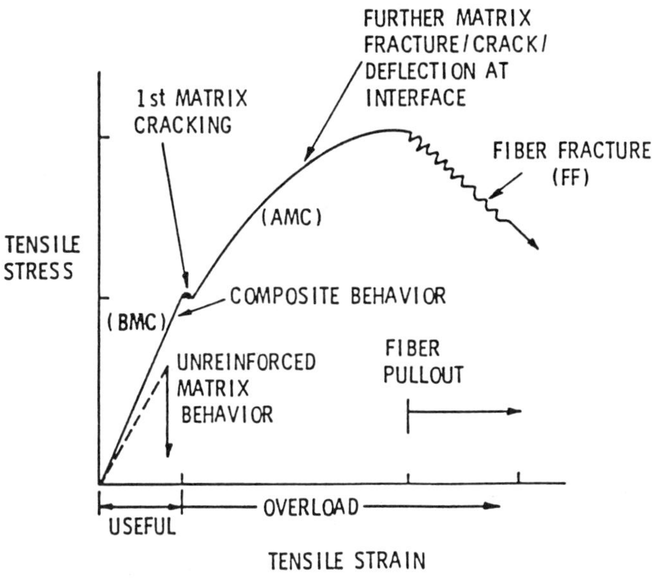

FIGURE 2 **Ideal stress–strain behavior of ceramic matrix composite, reinforced unidirectionally with long, weakly bonded fibers [3], showing before matrix cracking (BMC) and after matrix cracking (AMC) regions of the stress–strain curve.**

Ceramic composites based on crystalline matrices have not achieved the same level of improvement as those based on glass matrices [6]. Although these materials should exhibit superior mechanical and thermodynamically controlled properties, processing issues have made it very difficult to fabricate high density parts without resorting either to liquid phase sintering mechanisms, which degrade many important properties, or to combinations of high pressure and temperature, which damage the fibers.

Shrinkage of the matrix around the rigid, dimensionally stable fibers is a persistent, fundamental problem for crystalline matrix composites. Stresses that result can arrest the densification process and/or cause cracks to form in the matrix [8]. Stresses have been relieved by using a liquid phase sintering aid to permit rearrangement of powder particles, a feature that severely limits service temperature levels. With or without liquid phase sintering aids, many researchers have resorted to hot pressing or HIPing to achieve high densities. These pressure-induced densification processes are costly, cannot produce complex shapes, and do not provide precise dimensional control; also, the high pressures frequently damage the fibers. Reaction bonding circumvents these issues that have remained primary obstacles to successful application of crystalline matrix composites.

Reaction-Bonded Matrices

Reaction bonding—or reaction sintering or reaction forming, as the process is sometimes called—is an important option for fabricating ceramic matrix composites that is particularly applicable when net shape formation and high temperature capabilities are important. In the most highly developed reaction-bonded product, a green body made from Si powder is nitrided to form a reaction-bonded silicon nitride (RBSN) part. Reaction bonding of silicon carbide (RBSC) is also an established process, and other reaction-formed borides, carbides, nitrides, oxides, and silicides can be contemplated. Two important advantages of this type of process for composites are the absence of shrinkage during reaction bonding and the elimination of the need for liquid phase sintering aids. Consolidation in the net shape fabrication process occurs by filling the void space between Si particles with the Si_3N_4 reaction product.

Despite the potential advantages, to date only limited research with reaction-bonded ceramic matrix composites has been carried out by researchers at Georgia Tech, Los Alamos National Laboratory, MIT, NASA-Lewis, Norton, and Rutgers University. These studies have used RBSN matrices in combination with SiC fibers or whiskers as reinforcements because of the modulus differential and probable thermal expansion coefficient and thermodynamic compatibilities up to $\approx$ 1450°C. Low strengths and only moderate resistance to internal oxidation have been the principal sources of hesitancy about this approach for making CMCs. Fortunately there is important related research with SiC fiber reinforced hot pressed and sintered Si_3N_4 (HPSN and SSN) that defines the potential as well as difficulties that can be anticipated with the RBSN/SiC composites [7,9,10].

It is not possible to fully interpret and compare the results for these composite materials because the issues are inherently complex, not all the required properties were measured, and there were important differences between fiber architectures and sample preparation techniques for samples that were studied. The best HPSN matrix samples used short, randomly oriented, small-diameter, whisker-type fibers, while the most successful RBSN matrix samples employed large-diameter, unidirectional, continuous fibers made by the chemical vapor deposition (CVD) technique. This article illustrates the potential for CMCs of these types, cites results, discusses problems that have blocked achievement of maximum properties, and reports recent results that offer promise for the successful use of these materials.

Very significantly improved fracture toughness levels ($\leq$ 12.5 MPa $m^{1/2}$) were achieved using uncoated VLS SiC whiskers to reinforce HPSN [7] in one of the first investigations of the Si_3N_4/SiC CMC system. Demonstrating K_{IC} values that are approximately twice the maximum levels measured for highly optimized HPSN clearly indicates that this ceramic composite system has merit. More recent studies with materials of the same type have also demonstrated difficulties that are inherent to these high modulus, high temperature materials. Using rice-hull-derived whiskers in HPSN [9] gave maximum K_{IC} values ($\leq$ 6.5 MPa $m^{1/2}$) that are less than state-of-the-art levels [2] for HPSN ($\leq$ 8.9 MPa $m^{1/2}$). Using a different SiC whisker (unspecified process) in sintered Si_3N_4 (SSN), other authors [10] observed only degraded toughness and strength values. Creep rates for $T \geq$ 1200°C were unimproved or degraded with SiC whisker additions, effects that were attributed to liquid phase in the grain boundaries [11,12]. It is likely that the differences in results arise largely from the strengths of the reinforcement fibers that were used; however, other factors such as purity of the constituent materials and amount of sintering aid certainly played an important role. This first result should be viewed as a preliminary indication of the property levels that are accessible with RBSN/SiC composites if they are processed correctly; because these were only initial results for the composite, further improvements will almost certainly be achieved with optimizations of the fiber–matrix bonding, the fiber architecture, and processing details.

The RBSN/SiC system introduces several additional complications for making successful composites. The pores throughout the RBSN matrix must remain open and connected with the ambient atmosphere to permit complete nitridation of interior regions. One important consequence of the relatively high matrix porosity levels is that loads are largely carried by the SiC fibers, since the matrix's modulus of elasticity decreases proportionally with porosity. Consequently, it is critical that the fibers and their coatings not be degraded appreciably either by the nitriding process or by service exposures. The absence of matrix shrinkage makes it especially important to achieve defect-free green microstructures, since no void-filling rearrangements occur during consolidation. Typically, the low density levels will also result in poor matrix strength and uncertain resistance to oxygen penetration to the matrix–fiber interfacial region. While nitriding is undertaken at temperatures as much as 400°C below the hot pressing or sintering temperatures (typically 1650–1800°C) used for Si_3N_4, the times are long (up to 200 hours) and the atmospheres are chemically active [13]. The fibers and their coatings are vulnerable to interactions with the Si particles, at least during the initial stages of the nitriding process. Despite these additional problems, the initial results with RBSN/SiC CMCs are very promising, and they provide clear directions for future research.

One of the first reported, and most thorough, investigations of RBSN/SiC CMCs was undertaken at NASA-Lewis [14–17]. This work has focused on the use of the large-diameter ($\approx$ 144 μm) CVD SiC fibers produced by the Avco Company. These fibers were selected because of their high strength, thermodynamic stability, and low surface-to-volume ratio. Unidirectional and cross-ply composites were made by hand layup and pressing. Testing and characterizations have been done with unusual completeness. Using unidirectional as-received, SCS-6 type SiC fibers that had coatings $\approx$ 2–4 μm thick with a variable Si/C ratio on top of an amorphous carbon layer, Bhatt achieved very impressive improvements relative to the values he measured for monolithic RBSN made from the Si powders he used in his study. Ultimate flexural surface stress at failure (3 and 4 point loading) increased up to 9.5 times, the flexural stress at which the matrix first cracked increased up to 2.7 times, and the failures were not brittle even in tension. Thermal shocking by quenching from 1200°C into water had no effect on room temperature properties. Properties generally improved with increasing volume fraction of the fibers.

Microstructural analysis indicated that debonding during fracture occurred by splitting along the carbon-rich coating–matrix interface. Hence, the development of reduced nitriding schedules (as low as 1200°C, 40 hours) through the use of small, high purity Si powders was one of the important accomplishments of this research; Bhatt showed that properties degraded with longer time and higher temperature nitriding schedules and with thinner coatings on the fibers. The mechanisms responsible for the degradation were not completely identified, but increasing bond strengths between the fiber and matrix appear likely.

Cross-plied laminates also showed great promise. The strains at first matrix failure were unchanged from the unidirectional composites, indicating that the same level of matrix perfection had been achieved. Because the modulus was reduced, the composite strength at first matrix cracking was lower for the cross-plied case, as was the ultimate strength. Bhatt indicated that improved properties in cross-plied composites would come from improved matrix strength and fiber–matrix bonding; we would also suggest the use of smaller diameter fibers.

Simultaneously and independently, Norton Company investigated unidirectional RBSN/SiC composites based on similar CVD crystalline SiC fibers [18–20]. The results generally were the same as reported by Bhatt. The ultimate strength improvement ratios over monolithic values were not as great as Bhatt observed, largely because Corbin and his associates employed RBSN with better strengths. Work of fracture values for these composites improved 10–20 times relative to monolithic RBSN.

No transverse or shear data were reported for either RBSN study. It is likely that the transverse and shear strengths of the unidirectional RBSN composites also would be lower than those of the RBSN matrices. Weakly bonded fibers act as flaws to stress components aligned perpendicular to the fiber axes, thereby degrading transverse and shear strengths. For this reason and because of factors relating to formability, small diameters will likely be favored over large diameters for weakly bonded reinforcement fibers. The use of small-diameter SiC fibers will accentuate processing issues that must be resolved before these materials can be used to advantage in RBSN matrix composites.

Norton also has demonstrated the feasibility of using uncoated and coated polymer-derived SiC reinforcement fibers in an RBSN matrix. Key to this success was using nitriding schedules that had been reduced from 100–140 hours to 8 or 48 hours at 1350°C because each of the five fiber types studied still was at least partially amorphous. The means by which this reduction was accomplished was not revealed; however, both Norton and Danforth [21] indicated that nitriding kinetics are enhanced by the presence of the SiC fibers. Postulated mechanisms include enlarged channel dimensions into the interior of the matrix and the presence of accelerators such as Fe and SiO.

Using the uncoated Nicalon ceramic grade and Dow Corning MPDZ grade polymer-derived SiC fibers, Norton observed nonbrittle failures when the shorter (8 hour) nitriding cycle was used. All other combinations gave completely brittle failures at stress levels that were below strengths of monolithic RBSN. Any toughening should be viewed as highly encouraging, since these researchers had accessible only a very small range of process conditions to achieve complete nitridation without damaging the polymer-derived fibers. Subsequent studies with different coatings on Nicalon fibers showed that the thickest carbon coatings gave the best properties. All the coatings investigated (Al_2O_3, SiC, Si_3N_4, and C) reacted to some extent during nitriding, tending to bond the fibers to the matrix. The carbon coatings reacted with the unreacted Si, forming SiC. These coatings worked because they were thick enough to survive the nitriding schedule, and they formed a weak region at the location where Si was first detected in the reaction zone between the coating and matrix.

Three groups have studied short, randomly oriented, fiber-reinforced RBSN that should have the potential for matching the results achieved by Los Alamos with the HPSN/SiC(w) CMCs. Georgia Tech [22] investigated the use of several types of fiber in RBSN matrices, including polymer-derived (Nicalon) and rice-hull-derived whisker (Silar) SiC fibers. The nitriding schedule employed by Georgia Tech was so severe (150 hours with temperatures up to 1400°C) that all the fibers were degraded or completely converted to Si_3N_4. As a consequence, the mechanical properties were poor even by comparison to monolithic RBSN. Danforth [21] investigated SiC whisker reinforced RBSN composites made by colloidal pressing. His best materials achieved nearly complete reaction without damaging the SiC whiskers in parts with reasonably high matrix densities ($\leq$ 59%) and correct volume fractions (20–40%) of whiskers. Though strengths were 2.4–4.0 times higher than would be expected for the achieved matrix densities, they were not improved over high quality RBSN. Fracture toughness and work of fracture measurements were not made. Los Alamos [23,24] has studied RBSN/SiC composites that are based on the high purity, high strength SiC whiskers made by the VLS process at that laboratory. Using conventional nitriding schedules, the small-diameter whiskers were completely converted to Si_3N_4, precluding any hope of reinforcement. Gac found that the nitriding accelerators such as iron oxide actually enhance the attack, even though they reduce the severity of the nitriding schedule. In collaboration with Los Alamos, MIT incorporated the VLS SiC whiskers into very high purity Si powders and subjected the parts to the nitriding schedule in use at MIT at that time (1 hour, 1400°C). These preliminary results revealed that the combination of high purity and reduced nitriding conditions permitted complete nitridation without any evidence of attacking the uncoated fibers. The scanning electron micrograph of a fracture surface shown in Figure 3 reveals pullout lengths equal to many fiber diameters. This result eliminates any fundamental barrier for making these composites.

These results demonstrate the feasibility of RBSN. They also make clear several issues that must be resolved to make this composite practical. Most important, the nitriding temperature and time must be minimized to avoid damaging the fibers. Matrix strength must be improved to avoid premature matrix cracking, thereby exposing the reinforcements to damage from the ambient atmosphere. Similarly, the inherent oxidation resistance of the matrix must be improved.

FIGURE 3 **Scanning electron micrograph of fracture surface of SiC VLS whisker reinforced RBSN part made from laser-derived Si. Extensive pullout and an absence of fiber degradation are illustrated.**

Recent Findings for RBSN

MIT has undertaken an extensive ceramics processing research program specifically to reduce the incidence and size of defects in parts. One MIT study has synthesized nearly ideal Si powders from laser-heated silane (SiH_4) gas [25–27]. These high purity, spherical, uniform diameter, nonagglomerated powders can be made at commercially acceptable costs. Powder handling, dispersion, shaping, and drying techniques have also been developed that permit parts having random close packed structures to be made without introducing contaminants [28].

If the powders remain free of contaminants through all the fabrication steps, nitriding proceeds to completion at unusually low temperatures and in short times [28]. Our standard nitriding schedule employs a 1°C/min ramp to 1200°C followed by a 0.5°C/min ramp from 1200 to 1400°C and a 1 hour hold. The total exposure of 7 hours to nitriding temperatures ($T \geq 1200$°C) represents a major reduction from normal schedules. More recent results [29] show that this time period is substantially greater than is needed to fully nitride these powders. If as-synthesized purities are maintained until the firing stage, these Si powders will nitride to completion in even less severe combinations of time and temperature, such as 1 hour at 1150°C or 10 minutes at 1250°C. The use of a prenitriding nucleation step allows complete conversion in 1 hour at 1050°C. Nitriding times and temperatures have been reduced by using high purity, small-diameter powders and correct processing techniques.

Many of the RBSN properties have been measured already [28,29]. Results show that the deficiencies previously associated with RBSN as a CMC matrix largely have been corrected. A ball-on-ring biaxial strength test was used to determine the room temperature strengths of the nitrided samples in the as-processed condition and after exposures to 1000 and 1400°C air; hardness and fracture toughness were measured using a Vickers indenter. Oxidation resistance was measured for 1000 and 1400°C exposures to air for 1 and 50 hours.

Figure 4 shows average fracture stresses for MIT RBSN samples having a density of ≈ 65% (300 MPa) and ≈ 75% (531 MPa). The error bars correspond to the standard deviations in density (± 2–3%) and strength (± 55 MPa). Very recently, improved techniques have produced ≈ 75% dense RBSN samples with strengths up to 695 MPa. For comparison, Figure 4 also gives three different lines representing least-squares fits of exponential functions to RBSN, SSN, and HPSN strength data reported in the literature [13,31,32] for laboratory and commercial samples with all surface finishes. Silane-originating RBSN specimens that were diamond lapped to 4 μm exhibit average strengths that are 2.5–5.0 times the average reported values at both corresponding density levels. The highest strengths measured (450 MPa) for the 65% density samples were 60% higher than the highest average strength reported for an optimized 85% dense RBSN (290 MPa) [33] and 15% higher than any highest previously reported strength [34] for RBSN (389 MPa). The more recent 75% dense samples exhibited by far the highest strengths ever reported for RBSN. Exposure to 1000 and 1400°C air for 1 and 50 hours had no measurable effect on the room temperature strengths [30]. These room temperature strength levels are well into the range previously observed only with sintered or hot-pressed α-Si_3N_4; the absence of liquid phase sintering aids should yield the highest Si_3N_4 strengths at elevated temperatures.

Lapped specimens broke from processing flaws of two types. Fracture sources were usually either 5–15 μm diameter voids, or occasional lenticular cracks less than 50 μm deep, perpendicular to the stress axis. Application of the Griffin equation to the average strength levels yields flaw sizes of 4–10 μm.

Vickers hardness values were determined on polished surfaces using loads of 300 and 500 g. Observed values ranged from 3.5 to 11.0 GPa as density increased from 60 to 80%. The hardness of 77% dense RBSN (≈ 10 GPa) is higher than that of 85% dense, highly optimized commercial RBSN (≈ 8.3 GPa) [33]. Using the indentation technique, our dense specimens exhibited an average K_{IC} value of 2.8 MPa $m^{1/2}$, values that are higher than is characteristic of the optimized commercial RBSN (≈ 2.0 MPa $m^{1/2}$) [33] and approach those of sintered α-Si_3N_4.

Figure 5 summarizes the results of the oxidation studies along with the results reported by others for RBSN [35] and HPSN [36]. The SiH_4-originating RBSN exhibits up to 2 orders of magnitude lower weight gain than

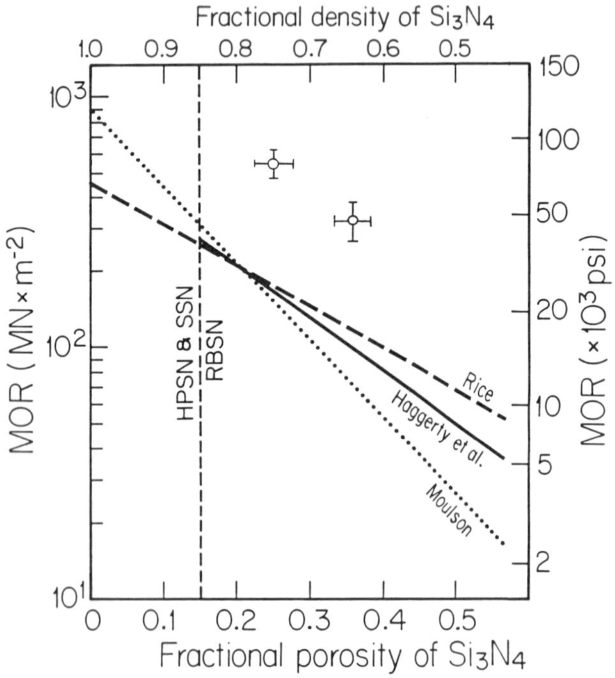

FIGURE 4 Room temperature strength of silicon nitride versus density.

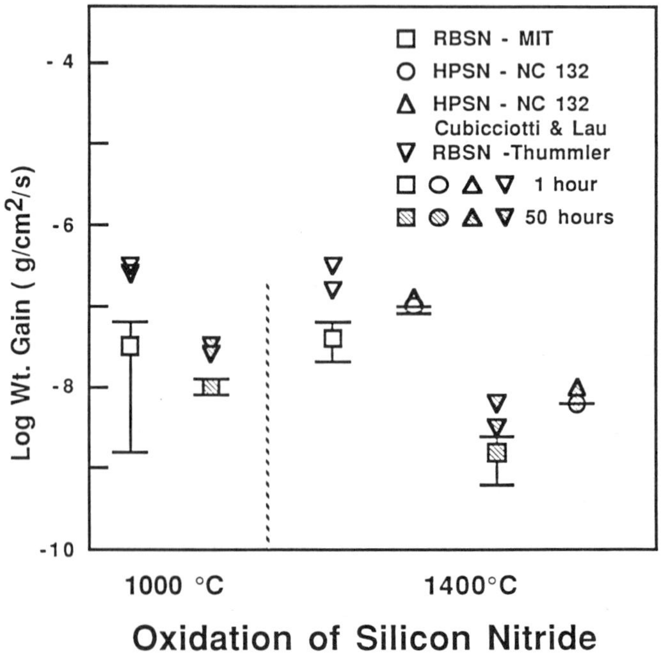

FIGURE 5 Oxidation weight gains for RBSN and HPSN samples at 1000 and 1400°C for 1 and 50 hours. (From Ref. 30.)

results for two more dense RBSN materials, and up to an order of magnitude better behavior than HPSN. Previous results with the SiH_4-originating RBSN shows that its oxidation resistance at 1250°C in flowing air is strongly dependent on the porosity level and pore size. As predicted by Porz and Thummler's [35] results at the same temperature, our low density specimens undergo significant internal oxidation before saturation is attained [28]. For densities exceeding 85% and Hg porosimetry pore radii less than 60 nm, Porz and Thummler showed that internal oxidation would be reduced to less than 1%, as a result of pore closure by the oxide product. The extent of our oxidation at 1250°C in air was less than 1% for specimens having densities as low as $\approx$ 75%. Although our results have not been analyzed mechanistically, it would appear that the same microstructural features that give high strengths in the SiH_4-originating RBSN also give it improved oxidation resistance relative to more dense RBSN. This superior performance at lower density levels very likely results from our samples having smaller pore diameters and more uniform microstructures.

These results indicate that the principal deficiencies exhibited by RBSN as a composite matrix have been resolved at least on an experimental basis for ceramic monoliths; nitriding kinetics, matrix strength, and oxidation resistance have all been improved to needed levels.

Conclusions and Summary

Reaction bonding offers an important opportunity for achieving the potential properties of crystalline ceramic matrix composites. The absence of densification during consolidation avoids the deleterious consequences of matrix shrinkage around dimensionally stable reinforcing media. Also, consolidation without shrinkage permits net shape fabrication processes. The results with HPSN/SiC composites show that the RBSN/SiC composite system has great potential if all the processing issues can be resolved. In addition to the other cited advantages, RBSN matrices should allow higher service temperatures than are feasible with SSN or HPSN matrices.

Reaction bonding introduces new processing difficulties. Although the processing temperatures are reduced, the combination of longer times, free Si, and frequently employed accelerators tends to degrade the fibers, whether coated or not. This potential degradation is critical, since fibers must be load bearing in the reduced-modulus, porous matrix. This degradation is particularly serious for the small-diameter fibers that must be employed to achieve adequate shear and transverse properties. The solution to these problems appears to be one of lessening the nitriding schedule through the route of using small-diameter, very high purity materials for the reinforcement and the Si powder. Appropriate high purity coatings could block adverse effects of lesser purity reinforcements while protecting them from attack. Once the Si has been converted to Si_3N_4, service temperature levels probably can be raised beyond those that degrade fibers during nitriding.

The procedures developed at MIT for making RBSN have already achieved the superior matrix properties needed for high performance composites and have reduced nitriding times and temperatures to values that do not cause either degradation of the Si whiskers or development of excessively strong matrix–fiber bonds. Strength ($\leq$ 695 MPa), fracture toughness ($\leq$ 2.8 MPa $m^{1/2}$), and hardness ($\leq$ 11 GPa) values essentially equal those of sintered and hot pressed α-Si_3N_4. Because pores are extremely small in diameter and are uniformly distributed, oxidation resistance of this RBSN is excellent. With their low specific gravities ($\approx$ 2.0), composites based on an RBSN matrix should exhibit exceptional specific properties for the entire temperature range of interest.

The RBSN/SiC composite is an important system based on the demonstrated properties of the constituent materials and the characteristics of the fabrication process. More important, it also serves as a model system for other potential reaction-bonded matrices. Using this processing strategy, we anticipate that the successes, considered to be routine with low temperature, reaction-bonded ceramic matrix composites like asbestos cement and reinforced concrete, will be fully accessible for the high temperature reaction-bonded CMCs.

John Haggerty

References

1. H. K. Bowen, *J. Mater. Sci. Eng., 44,* 1 (1980).
2. Y. Tajima, K. Urashima, W. Watanabe, and Y. Matsuo, *Proceedings of the First International Conference on Ceramic Powder Processing Science,* American Ceramic Society Meeting, 1987 (in press).
3. J. J. Mecholsky, *Am. Ceram. Soc. Bull., 65*(2), 315 (1986).
4. R. W. Rice, *Ceram. Eng. Sci. Proc., 2*(7–8), 661 (1981).
5. A. G. Evans, *Mater. Sci. Eng., 71,* 3 (1985).
6. K. M. Prewo, J. J. Brennan, and G. K. Layden, *Am. Ceram. Soc. Bull., 65*(2), 305 (1986).
7. P. D. Shalek, J. J. Petrovic, G. F. Hurley, and F. D. Gac, *Am. Ceram. Soc. Bull., 65*(2), 351–356 (1986).
8. A. G. Evans, "Inhomogeneous Sintering: Stresses, Distortion and Damage," in P. F. Becher, M. V. Swain, and S. Somiya, Eds., *Defect Properties and Processing of High-Technology Nonmetallic Materials, MRS Symposia Proceedings,* Vol. 60, Materials Research Society, Pittsburgh, 1986, pp. 63–78.
9. S. T. Buljan, J. G. Baldoni, and M. L. Huckabee, *Am. Ceram. Soc. Bull., 66*(2), 347 (1987).
10. R. Lundberg, L. Kahlman, R. Pompe, R. Carlsson, and R. Warren, *Am. Ceram. Soc. Bull., 66*(2), 330 (1987).
11. J. R. Porter, F. F. Lange, and A. H. Chokshi, in P. F. Becher, M. V. Swain, and S. Somiya, Eds., *MRS Symposia Proceedings,* Vol. 78, Materials Research Society, Pittsburgh, 1987, pp. 289–294.
12. R. D. Nixon, S. Chevacharoenkul, M. L. Huckabee, S. T. Buljan, and R. F. Davis, in P. F. Becher, M. V. Swain, and S.

Somiya, Eds., *MRS Symposia Proceedings*, Vol. 78, Materials Research Society, Pittsburgh, 1987, pp. 295–302.

13. A. J. Moulson, *J. Mater. Sci., 14*, 1017 (1979).
14. R. T. Bhatt, "Mechanical Properties of SiC Fiber-Reinforced Reaction-Bonded Si_3N_4 Composites," NASA Technical Report 85-C-14, July 1985.
15. R. T. Bhatt, "Effects of Fabrication Conditions on the Properties of SiC Fiber Reinforced Reaction-Bonded Silicon Nitride Matrix Composites (SiC/RBSN)," NASA Technical Report 86-C-2, January 1986.
16. R. T. Bhatt and R. E. Phillips, "Mechanical Behavior of SiC Fiber Reinforced Reaction-Bonded Silicon Nitride Matrix Composite Laminates," *J. Compos. Mater.* (to be published).
17. R. T. Bhatt and R. E. Phillips, "Thermal Effects on the Mechanical Properties of SiC Fiber Reinforced Reaction Bonded Silicon Nitride Matrix (SiC/RBSN) Composites," *J. Am. Ceram. Soc.* (to be published).
18. N. D. Corbin, G. A. Rossetti, and S. D. Hartline, *Ceram. Eng. Sci. Proc., 7*(7–8), 958 (1986).
19. N. D. Corbin, C. A. Willkens, and S. D. Hartline, *Proceedings of the NASA Conference on Metal Matrix, Carbon, and Ceramic Matrix Composites*, 1987 (in press).
20. N. D. Corbin, C. A. Willkens, and S. D. Hartline, *Proceedings of the NASA Conference on Metal Matrix, Carbon, and Ceramic Matrix Composites*, 1987 (in press).
21. F. Takao, W. R. Cannon, and S. C. Danforth, "Nonaqueous Colloidal Processing of Reaction-Bonded Silicon Nitride-SiC Whisker Composites," in G. L. Messing, K. S. Mazdiyasni, J. W. McCauley, and R. A. Haber, Eds., *Advances in Ceramics*, Vol. 21, *Ceramic Powder Science*, American Ceramic Society, Westerville, OH, 1987.
22. T. L. Starr and J. N. Harris, "Development of Advanced Fiber Reinforced Ceramics," Final Report for Oak Ridge National Laboratory, Subcontract 19X-43369C, U.S. Department of Energy Contract DE-AC05-840R21400, 1987.
23. F. D. Gac, G. F. Hurley, J. J. Petrovic, W. J. A. Parkinson, and P. D. Shalek, "Short Fiber Reinforced Structural Ceramics," Los Alamos National Laboratory, Los Alamos, NM, 1986.
24. F. D. Gac, "Investigation of VLS Silicon Carbide Whisker Reinforced Reaction Bonded Silicon Nitride," Ph.D. thesis, University of Washington.
25. W. R. Cannon, S. C. Danforth, J. H. Flint, J. S. Haggerty, and R. A. Marra, *J. Am. Ceram. Soc., 65*(7), 324 (1982).
26. W. R. Cannon, S. C. Danforth, J. H. Flint, J. S. Haggerty, and R. A. Marra, *J. Am. Ceram. Soc., 65*(7), 330 (1982).
27. J. H. Flint and J. S. Haggerty, *Proceedings of the First International Conference on Ceramic Powder Processing Science*, American Ceramic Society Meeting, 1987 (in press).
28. J. S. Haggerty, J. H. Flint, G. J. Garvey, J.-M. Lihrmann, and J. E. Ritter, "High Strength, Oxidation Resistant Reaction-Bonded Silicon Nitride from Laser-Synthesized Silicon Powder," in W. Bunk and H. Hauser, Eds., *Ceramic Materials and Components for Engines*, Verlag Deutsche Keramische Gesellschaft, 1986, pp. 147–154.
29. B. W. Sheldon and J. S. Haggerty, *Proceedings of the 12th Annual Conference on Composites and Advanced Ceramics*, Engineering Ceramics Division Meeting, American Ceramic Society, 1988 (in press).
30. J. S. Haggerty, A. Lightfoot, J. E. Ritter, S. V. Nair, and P. Gennari, *Proceedings of the 12th Annual Conference on Composites and Advanced Ceramics*, Engineering Ceramics Division Meeting, American Ceramics Society, 1988 (in press).
31. J. S. Haggerty, G. J. Garvey, J.-M. Lihrmann, and J. E. Ritter, "Processing and Properties of Reaction-Bonded Silicon Nitride Made from Laser-Synthesized Silicon Powders," in P. F. Becher, M. V. Swain, and S. Somiya, Eds., *Defect Properties and Processing of High-Technology Nonmetallic Materials*, Vol. 60, *MRS Symposia Proceedings*, Materials Research Society, Pittsburgh, 1986, pp. 51–62.
32. R. W. Rice, *J. Mater. Sci., 12*(7), L627 (1977).
33. S. C. Danforth and J. S. Haggerty, *J. Am. Ceram. Soc., 64*(4), C-58 (1983).
34. M. E. Washburn and H. R. Baumgarner, *Ceramics for High Performance Applications*, 1974, p. 479.
35. F. Porz and F. Thummler, *J. Mater. Sci., 19*, 1283 (1984).
36. D. Cubicciotti and K. H. Lau, *J. Am. Ceram. Soc., 61*(11–12), 512 (1978).

Ceramic Matrix Composites, SiC Whisker-Reinforced

Introduction

In contrast to metals, ceramics are brittle and are extremely sensitive to flaws in the materials. This property, which is a direct result of the generally low fracture toughness of ceramics, jeopardizes their reliability in high stress applications. Table 1 presents fracture toughness ranges for a variety of materials. From the Griffith relationship for fracture in brittle materials [1], we know that by increasing fracture toughness, the strength of ceramics is improved and the sensitivity to flaw sizes is reduced. Thus their reliability and lifetime can be increased.

Methods to improve the fracture toughness of ceramics include second-phase particulate additions [2,3], fiber reinforcement [4,5], and phase transformation toughening [6,7]. Recently, SiC whisker-reinforced ce-

TABLE 1
Typical Fracture Toughness Values for Various Materials

Material	Critical Fracture Toughness, $K1c$ ($MPa \cdot m^{1/2}$)
Glasses	< 1
Polycrystalline ceramics	3–5
Aluminum alloys	> 15
Steels	25–150

ramic composites have come into prominence for structural applications because of their potential for high strength and fracture toughness. Originally applied to the reinforcement of metal matrices, such as aluminum, the use of whisker reinforcement to ceramics was introduced in the alumina and mullite matrix systems [8,9].

Whiskers are usually defined as short, discontinuous, rod- or needle-shaped single-crystal fibers. While the size range can vary, the whiskers are normally 0.1–3.0 μm in diameter and 5–200 μm long. Because they are nearly single crystals, they typically have very high tensile strengths (up to 7 GPa) and elastic moduli (up to 550 GPa) [10]. To the naked eye, the whiskers appear like a normal ceramic powder. Under the scanning electron microscope, however, their rodlike morphology is readily apparent. Representative photomicrographs of current commercial SiC whiskers are shown in Figure 1.

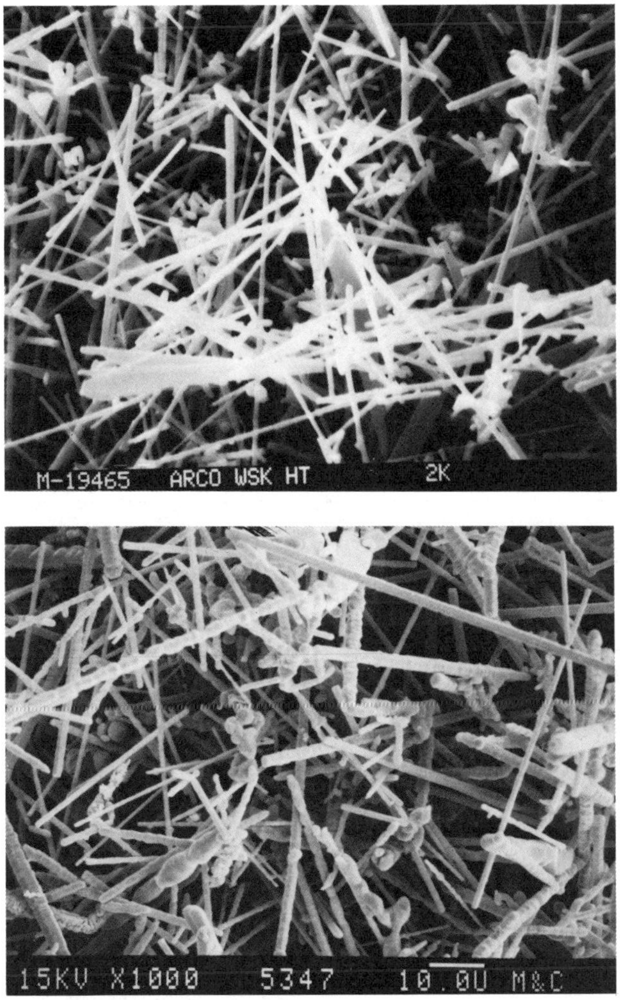

FIGURE 1 **Representative SiC whiskers examined with scanning electron microscope: (*Top photo*) from Advanced Composites Materials Corporation and (*Bottom photo*) from American Matrix.** (Photographs by Oak Ridge National Laboratory.)

Whisker Fabrication and Properties

Ceramic whiskers have been by-products of fabrication processes normally employed in the ceramic industry for many years. The first commercial source for SiC whiskers appeared in the early 1960s [11]. In the intervening years, interest in whiskers has increased, and many sources are presently available. Table 2 offers a noncomprehensive list of some current whisker sources. In addition, several other companies around the world are assessing products and may offer other whiskers, such as alumina, in the near future. At the present time, SiC whiskers have the most commercial value for use and thus have the greatest interest.

SiC whiskers have been grown by many methods and using numerous starting materials. Much of the early research and commercial production used vapor–liquid–solid reactions to produce small quantities of whiskers [12,13]. The cost of these whiskers, which tended to exceed \$2500/kg, was prohibitively high for their widespread application as reinforcing materials. Later production methods employing vapor–solid mechanisms used carbothermic reduction reactions of silica and carbon precursors [14–20]. Cheaper starting materials, such as rice hulls, were used, and large quantities of whiskers ($>$ 10,000 kg/year) could be produced at reasonable cost ($<$ \$400/kg) [11]. Thus, whiskers became viable as reinforcing materials for large-scale, high volume applications.

Present-day SiC whiskers vary in their physical and chemical characteristics from manufacturer to manufacturer and in some cases from batch to batch. Table 3 shows the variation in as-received physical characteristics from a number of current fabricators. Although differences in the physical dimensions are minor, it has been shown that these seemingly small differences can have major impacts in the performance of ceramic matrix composites [21]. In addition, during fabrication, the whiskers tend to get broken, and consequently the aspect ratio of the final composite is reduced.

The variations in bulk chemistry for selected SiC whiskers are given in Table 4. Again, minor variations are observed among the products of various manufacturers. The impurities present are attributable to the whisker starting materials and/or the catalysts used in the whisker growth processes. For example, the calcium associated with some of the whiskers has been linked with rice hulls, commonly used as a starting material [22]. Iron and cobalt, on the other hand, are listed as catalyst materials in some of the patents [15,16]. In any event, the bulk chemistry, as expected, is mainly a function of the processing route employed by the whisker fabricator.

The SiC whisker surface chemistry is also important in ceramic matrix composites because it influences the nature of the interface between the matrix and the whisker [23]. Current SiC whiskers can be classified generally into three surface chemistry categories based on results of X-ray photoelectron spectroscopy. The first type is rich in oxygen, and the surface appears as silica with strong silicon-to-oxygen bonding. The second classification would be best described as "clean" SiC with predomi-

TABLE 2
Commercial Sources of Ceramic Whiskers

Source	Type of Whiskers
Advanced Composite Materials Corporation[a] Greer, South Carolina	β-SiC
Alcan International, Ltd. Montreal, Quebec, Canada	β-SiC
American Matrix, Inc. Knoxville, Tennessee	β-SiC, α-SiC
Huber Corporation Borger, Texas	β-SiC
Nikkei Company Tokyo, Japan	β-SiC
Nippon Light Metals Company Tokyo, Japan	β-SiC
Shin-Etsu Chemical Company Tokyo, Japan	β-SiC
Tateho Chemical Company Ako-Shi, Japan	β-SiC, α-Si_3N_4
Tokai Carbon Company Sunto-Gun, Japan	β-SiC
Ube Industries, Ltd. Tokyo, Japan	α-Si_3N_4, β-Si_3N_4
Versar Manufacturing, Inc. Springfield, Virginia	β-SiC, α-Si_3N_4

[a] Formerly ARCO Chemical Company.

TABLE 3
Physical Characteristics of Selected Commercial SiC Whiskers

Manufacturer/Product	Diameter Range (μm)	Approximate Mean Diameter (μm)	Approximate Mean Aspect Ratio[a]
Advanced Composite Materials Corporation SC-9	0.5–0.8	0.65	30 : 1
Tateho SCW-1S	0.1–0.6	0.40	50 : 1
Tokai Carbon Tokamax	0.1–0.6	0.48	25 : 1
Huber 7117	0.1–0.7	0.54	
American Matrix Grade 1	0.3–5.0	1.5	
Versar VC1A	0.2–0.7	0.50	

[a] Aspect ratio is length to diameter.

TABLE 4
General Chemistry of SiC Whiskers[a]

	Manufacturer and Grade									
Element	ACMC, F-9	ACMC, SC-9	ACMC, SC-9	Tateho, SCW-1	Tateho, SCW-1S	Tokai Carbon, Tokamax	Versar, VCA-1	Huber, XPW-2	Huber, 7116	American Matrix, Ex[b]
Al	700	200	120	30	500	10	30	> 5000	30	200
B			< 1		1			1	< 1	> 1000
Ca	1000	100	150	300	500	20	300	250	5	600
Co			< 10	1	< 10	1000	< 1	< 10	< 10	< 1
Cr	1000	50	15	30	50	50	30	10	20	15
Fe	1000	100	60	50	300	50	50	200	300	25
K	200	20	10	30	25	3	30	5	< 1	< 1
Mg	300	300	90	30	120	3	50	5	5	200
Mn	1000	100	70	50	50	5	50	5	< 1	1
Na	30	30	10	5	15	50	5	10	1	5
Ni	70	30	< 10	10	< 10	10	10	< 10	< 10	20
Zr			< 10	10	10	3	10	< 1	< 10	10
O[c]	2.77	1.01	1.25		1.04			0.73	0.22	2.85

[a] Except as noted, analysis by spark source spectrometry; accuracy 50%.
[b] Ex = experimental.
[c] Values given in weight percent; analysis by neutron activation; accuracy 0.1.

nantly silicon-to-carbon bonding and a minor amount of silicon oxycarbide observed. The third major category has excess surface carbon, along with strong silicon-to-carbon bonding. In alumina–SiC composites, it was determined that the presence of excess surface carbon on the whiskers resulted in a weak interfacial bond and materials of high toughness and strength.

As with all materials that have been recently developed, the health effects and hazards of SiC whiskers are not well known. However, because of their similarity in size and shape to asbestos, they should be handled with extreme care, especially if there is a possibility of airborne contamination. The precautions that are sent with every batch of whiskers should be read carefully and observed.

Fabrication of Ceramic Matrix Composites

In contrast to other ceramic fiber composites [4,5], whisker composites can be fabricated using conventional powder processing techniques normally employed in current ceramic industry. This means that articles can be formed by standard procedures such as slip casting, dry-pressing, extrusion, and injection molding. Densification of the composite materials can also be carried out by conventional firing techniques. Early research used hot-pressing of powder mixtures to obtain dense materials for property studies [8,9,24]. At the present time, hot-pressing is still used commercially to fabricate blanks for alumina–SiC whisker cutting tools.

While hot-pressing of the alumina–SiC whisker composites produces dense materials with attractive mechanical properties, the ability to use pressureless sintering on the composites is desirable for some potential applications. Pressureless sintering is also necessary to provide complex and near-net-shape fabrication capability for these composites. Densification of alumina–SiC whisker composites by pressureless sintering was found to be severely inhibited because the whiskers interfere with particle rearrangement and shrinkage [25]. Thus as the whisker volume content increases, the ability to densify the composites decreases.

To promote sintering, a number of techniques were effective, such as lowering the whisker aspect ratio, adjusting the matrix particle size to optimize green density, and using a liquid phase sintering aid to encourage rearrangement. The same effects were also observed in silicon carbide, silicon nitride, and mullite matrix composite systems with SiC whisker additions [26–28]. For practical purposes, the whisker contents of pressureless sintered bodies generally must be below 20 vol %.

Dispersion of the whiskers and mixing them with the ceramic powders are the most critical processing steps in fabricating a composite with good properties. Ceramic composites with poor dispersion show whisker "clumping," which leads to low density regions that act as strength-limiting flaws in the materials. Good homogeneous whisker–ceramic powder mixtures have been obtained by a number of techniques, including high shear mixing [8,24,29], ultrasonic dispersion [30], and milling methods [24,31].

Composite Properties

In addition to the initial work in the alumina and mullite matrix systems previously mentioned, SiC whiskers have been used to reinforce other ceramic matrices such as silicon nitride [27,30–33], glass [34,35], magnesia–alumina spinel [36], cordierite [37], zirconia [38,39], alumina–zirconia [38–40], mullite–zirconia [39–41], and boron carbide [42]. Table 5 summarizes the effects on mechanical properties of SiC whisker additions to the various matrices. As shown, the addition of whiskers increases the fracture toughness of the ceramics in all cases, compared with the same monolithic materials. In many cases, improvements in the flexural strength were also observed. Significantly, these improvements are retained at elevated temperatures. Other physical properties of selected whisker-reinforced composites that have been determined to date are given in Table 6.

Research [21] into the behavior responsible for toughening in the composite materials shows that crack–whisker interaction results in crack bridging and deflection, the major toughening mechanisms. The effectiveness of these mechanisms is illustrated in Figure 2, which shows the rough fracture surface of a high toughness alumina/20 vol % SiC whisker composite. The whiskers are readily visible where debonding has taken place along the matrix–whisker interface. Whisker pull-out and crack pinning are believed to also contribute to the overall composite toughness. Analytical models and experimental evidence indicate that the toughening effect will increase with increases in whisker strength, volume fraction of whiskers, whisker radius, ratio of Young's modulus of the composite to that of the whisker, and ratio of the interfacial fracture energies of the matrix and the matrix–whisker interface. In addition to the increases in the fracture toughness and strength, other improvements in the performance aspects of whisker-reinforced ceramics have been observed. These are discussed individually below.

Slow Crack Growth

The resistance to slow crack growth was shown to increase markedly for an alumina/20 vol % SiC whisker composite versus unreinforced material [40]. This means that much higher stress intensities and stresses can be applied to these composites while still avoiding time-dependent failure. The improved ability to avoid time-dependent failures under stress was retained up to temperatures of 1100°C [45].

Thermal Shock

Thermal shock testing [44] of an alumina/20 vol % SiC whisker composite showed no decrease in flexural strength with temperature differences up to 900°C. Alumina, on the other hand, normally shows a significant decrease in flexural strength with temperature changes of exceeding 400°C. The improvement is a result of interaction between the SiC whiskers and microcracks in the matrix, which prevents coalescence of the cracks into critical flaws.

TABLE 5
Summary of Mechanical Properties of Whisker-Reinforced Ceramic Composites

Matrix	Whisker Content (vol %)	Fracture Toughness (MPa · $m^{1/2}$)	Flexural Strength (MPa)				Ref.
			25°C	1000°C	1200°C	1400°C	
Alumina	0	4–5	300–400				39,42
	10	6.1–7.1	450	320			39,42
	20	7.5–9.0	650–800	570–775	410–550		39,42
	30	8.5–10	700–800				39,42
Glass	0	1.0	77				34
	20	3.8–5.5	125–190				34
Mullite	0	2.2	200				9,39
	10	3.8	420				9,39
	20	4.7	425	410	395	245	27
Silicon nitride	0	4.6	770	585	490	200	26,30
	30	6.4	975	820	600	260	26,30
Spinel	0		320	300	275		35
	30		415	370	275		35
Cordierite	0	2.2	180	170			36
	20	3.7	260	245			36
Zirconia	0	6.8	1150	160			38
	20	10.2	600	340			38
	30	11.0	590	380			38
Mullite–zirconia	0	4.4					23,39
	20	5.1–7.0	440–580				23,29
Alumina–zirconia	0	6.2	1080				36
	20	7.8–13.2	700–750				23,36,39
Boron carbide	0	< 4					41
	20	5.3					41

High Temperature Creep

Creep measurements [46–49] demonstrated that additions of SiC whiskers to alumina, mullite, and silicon nitride significantly inhibit high temperature creep. For example, a 15 wt % loading of whiskers in an alumina matrix reduced the creep rate by up to 2 orders of magnitude at 1500°C [46]. The improvements in creep resistance in the mullite and silicon nitride matrix composites were approximately one order of magnitude [48,49].

Wear Resistance

Because of their high hardness, ceramics are used extensively for wear applications. Testing of ceramics under nonlubricating conditions showed that the wear coefficient for whisker-reinforced alumina was several orders of magnitude lower than for comparable monolithic ceramic materials [50,51]. Improvements in wear resistance also have been reported for SiC whisker-reinforced silicon nitride [52].

TABLE 6
Various Physical Properties of Selected Whisker-Reinforced Ceramic Composites

Material	Elastic Modulus (GPa)	Thermal Conductivity (W/m · K)	Thermal Expansion (10^{-6}/K)	Ref.
Al_2O_3	380	0.36	7.8–8.2	9
Al_2O_3/20 vol % SiC whisker	395	0.32	7.35	9
Glass	68			34
Glass/20 vol% SiC whisker	65–75			34
Si_3N_4	297	0.194	2.53	30
Si_3N_4/30 vol % SiC whisker	335	0.252	2.91	30
Mullite/20 vol % SiC whisker		0.007	5.60	9
Zirconia	205			36
Zirconia/20 vol % SiC whisker	232			36

FIGURE 2 **Fracture surface of alumina–SiC whisker composite showing extensive crack–whisker interaction.** (Photograph by Oak Ridge National Laboratory.)

Applications

Whisker-reinforced alumina is fabricated and marketed extensively as a cutting tool for high nickel alloys [53,54]. Cutting rates up to 10 times higher than conventional tools have been obtained [53]. Improvements in machining of cast iron and steel products also have been observed.

Because of the excellent wear resistance of whisker-reinforced ceramic composites, near-term applications include wear parts, such as pump seal rings, grit blast nozzles, and dies for metal extrusion and wire pulling. Future applications may use whisker-reinforced ceramic composites in advanced heat engines for items like valves and turbocharger rotors. Research on the joining of ceramics has shown that alumina–SiC whisker composite can be joined to itself and to metals using titanium-containing brazing filler metals [55].

Acknowledgments

This research was sponsored by the U.S. Department of Energy, Assistant Secretary for Conservation and Renewable Energy, Office of Transportation Systems, as part of the Ceramic Technology for Advanced Heat Engines Project of the Advanced Materials Development Program, under contract DE-AC05-840R21400 with Martin Marietta Energy Systems, Inc.

Terry Tiegs and Paul Becher

References

1. A. A. Griffith, *Phil. Trans., R. Soc. London, 221A,* 163 (1921).
2. R. W. Rice, *Ceram. Eng. Sci. Proc., 2,* (7–8) (1981).
3. F. F. Lange, *J. Am. Ceram. Soc., 56*(9), 445 (1973).
4. K. M. Prewo and J. J. Brennan, *J. Mater. Sci., 15,* 463 (1980).
5. G. A. Cooper, *J. Mater. Sci., 5,* 645 (1970).
6. R. Hannick and M. V. Swain, *J. Aust. Ceram. Soc., 18*(2), 53 (1983).
7. F. F. Lange, *J. Mater. Soc., 17,* 247 (1982).
8. P. F. Becher and G. C. Wei, *J. Am. Ceram. Soc., 67*(12), C267 (1984).
9. G. C. Wei and P. F. Becher, *Am. Ceram. Soc. Bull., 64*(2), 298 (1985).
10. J. J. Petrovic, J. V. Milewski, D. L. Rohr, and F. D. Gac, *J. Mater. Sci., 20,* 1167 (1985).
11. J. V. Milewski, "Whiskers," in J. V. Milewski and H. S. Katz, Eds., *Handbook of Reinforcements for Plastics,* Van Nostrand Reinhold, New York, 1987, p. 205.
12. I. Berman and C. E. Ryan, *J. Crystal Growth, 9,* 314 (1971).
13. G. A. Bootsma, W. F. Kippenberg, and G. Verspui, *J. Crystal Growth, 11,* 297 (1971).
14. J. G. Lee and I. B. Cutler, *Am. Ceram. Soc. Bull., 54*(2), 195 (1975).
15. I. B. Cutler, "Production of Silicon Carbide from Rice Hulls," U.S. Patent 3,754,076 (Aug. 21, 1973).
16. A. Yamamoto, "Process for Preparing Silicon Carbide Whiskers," U.S. Patent 4,500,504 (Feb. 19, 1985), to Tokai Carbon Company, Ltd.
17. M. Tanaka, T. Kawabe, and M. Kobune, "Method of Manufacturing Crystalline Silicon Carbide," U.S. Patent 4,504,453 (March 12, 1985), to Tateho Kagaku Kogyo Kabushiki Kaisha.
18. O. J. Horne, Jr., and D. E. Ramsey, "Production of SiC Whiskers," U.S. Patent 4,283,375 (Aug. 11, 1981), to Great Lakes Carbon Corporation.
19. T. Hayshi, F. Mochizuki, and M. Ozawa, "Process and Continuous Reaction Furnace for Production of β-type Silicon Carbide Whiskers," U.K. Patent Application GB 2162504 A (July 12, 1985), to Nippon Light Metal Company, Ltd.
20. R. L. Beatty and F. H. Wyman, "Continuous Silicon Carbide Whisker Production," U.S. Patent 4,637,924 (Jan. 20, 1987), to Atlantic Richfield Company.
21. P. F. Becher, C. H. Hsueh, P. Angelini, and T. N. Tiegs, "Toughening Behavior in Whisker Reinforced Ceramic Matrix Composites," *J. Am. Ceram. Soc.,* submitted.
22. N. K. Sharma, W. S. Williams, and A. Zangvil, *J. Am. Ceram. Soc., 67*(11), 715 (1984).
23. T. N. Tiegs, P. F. Becher, and L. A. Harris, "Interfaces in Alumina-SiC Whisker Composites," in *Ceramic Microstructures '86: Role of Interfaces,* Materials Science Research Series, Vol. 21, Plenum Press, New York, 1987, pp. 911–918.
24. T. N. Tiegs and P. F. Becher, "Whisker Reinforced Ceramic Composites," in *Tailoring Multiphase and Composite Ceramics,* Materials Science Research Series, Vol. 20, Plenum Press, New York, 1986, pp. 639–647.
25. T. N. Tiegs and P. F. Becher, *Am. Ceram. Soc. Bull., 66*(2), 339 (1987).
26. H. Abe and R. C. Bradt, *J. Am. Ceram. Soc., 58*(11), 525 (1975).
27. H. Yeh, J. Schienle, K. Karasek, and S. Bradley, "Development of Ceramic Matrix Composites for Application in the Ceramic Technology for Advanced Heat Engines," *Garrett Ceramic Components Division Report 87-60985,* 1988.
28. T. N. Tiegs, P. F. Becher, and P. Angelini, *Proceedings of the First International Conference on Mullite,* 1987.

29. J. Homeny, W. L. Vaughn, and M. K. Ferber, *Am. Ceram. Bull., 66*(2), 333 (1987).

30. R. Hayami, K. Ueno, I. Kondou, N. Tamari, and Y. Toibana, "Si_3N_4-SiC Whisker Composite Material," in *Tailoring Multiphase and Composite Ceramics,* Materials Science Research Series, Vol. 20, Plenum Press, New York, 1986, pp. 663–674.

31. S. T. Buljan, J. G. Baldoni, and M. L. Huckabee, *Am. Ceram. Bull., 66*(2), 347 (1987).

32. R. Lundberg, L. Kahlman, R. Pompe, and R. Carlsson, *Am. Ceram. Bull., 66*(2), 330 (1987).

33. P. D. Shalek, J. J. Petrovic, G. F. Hurley, and F. D. Gac, *Am. Ceram. Soc. Bull., 65*(2), 351 (1986).

34. K. P. Gadjaree and K. Chyung, *Am. Ceram. Soc. Bull., 65*(2), 370 (1986).

35. F. D. Gac, J. J. Petrovic, J. V. Milewski, and P. D. Shalek, *Ceram. Eng. Sci. Proc., 7*(7–8), 978 (1986).

36. P. C. Panda and E. R. Seydel, *Am. Ceram. Soc. Bull., 65*(2), 338 (1986).

37. N. Claussen and G. Petzow, "Whisker-Reinforced Zirconia-Toughened Ceramics," in *Tailoring Multiphase and Composite Ceramics,* Materials Science Research Series, Vol. 20, Plenum Press, New York, 1986, pp. 649–662.

38. N. Claussen, K. L. Weisskopf, and M. Ruhle, *J. Am. Ceram. Soc., 69*(3), 288 (1986).

39. N. Claussen, K. L. Weisskopf, and M. Ruhle, "Mechanical Properties of SiC-Whisker-Reinforced TZP," in *Fracture Mechanics of Ceramics, Vol. 7,* Plenum Press, New York, 1986, pp. 75–86.

40. P. F. Becher, T. N. Tiegs, J. C. Ogle, and W. H. Warwick, "Toughening of Ceramics by Whisker Reinforcement," in *Fracture Mechanics of Ceramics, Vol. 7,* Plenum Press, New York, 1986, pp. 61–73.

41. P. F. Becher and T. N. Tiegs, *J. Am. Ceram. Soc., 70*(9), 651 (1987).

42. G. C. Wei, "Silicon Carbide Whisker Reinforced Ceramic Composites and Method for Making Same," U.S. Patent 4,543,345 (Sept. 24, 1985), to U.S. Department of Energy.

43. T. N. Tiegs and P. F. Becher, *Proceedings of the 24th Automotive Technology Contractors' Coordinated Meeting,* 1986, Society of Automotive Engineers, 1987, Vol. P-197.

44. T. N. Tiegs and P. F. Becher, *Am. Ceram. Soc. Commun., 70*(5), C109 (1987).

45. P. F. Becher and T. N. Tiegs, *Adv. Ceram. Mater., 3*(2), 148 (1988).

46. J. R. Porter, F. F. Lange, and A. H. Chokshi, *Am. Ceram. Soc. Bull., 66*(2), 343 (1987).

47. J. R. Porter and A. H. Chokshi, "Creep Performance of Silicon Carbide Whisker-Reinforced Alumina," in *Ceramic Microstructures '86: Role of Interfaces,* Materials Science Research Series, Vol. 21, Plenum Press, New York, 1987, pp. 919–928.

48. R. D. Nixon et al., *Mater. Res. Soc. Symp. Proc., 78,* 295 (1987).

49. R. D. Nixon et al., *Proceedings of the First International Conference on Mullite,* 1987.

50. C. S. Yust, Oak Ridge National Laboratory, personal communication, 1988.

51. J. F. Rhodes, Advanced Composite Materials Corporation, personal communication, 1988.

52. K. F. Dufrane, W. A. Glaeser, and A. R. Rosenfield, "Studies of Dynamic Contact of Ceramics and Alloys for Advanced Heat Engines," Oak Ridge National Laboratory Rept. ORNL/Sub/84-00216/1, 1988.

53. K. H. Smith, *Mach. Tool Blue Book, 81*(1), 71 (1986).

54. Greenleaf Corporation, Sagertown, PA, company literature, 1986.

55. A. J. Moorhead, M. L. Santella, and H. M. Henson, "Brazing of SiC Whisker-Reinforced Ceramics," in *Proceedings of International Forum on Joining of Structural Ceramics,* in press.

Ceramic Matrix Composites CVI: Chemical Vapor Infiltration*

In the CVI process a matrix is chemically vapor deposited within a porous preform, which is often fibrous, to produce a composite material [1]. Thus, chemical vapor infiltration—CVI—is a specialized form of chemical vapor deposition (CVD). In the CVI process, deposition on surfaces located within the external surface of the preform is desired, whereas in CVD the goal is to deposit a coating onto the exterior surface of the substrate. Infiltration of fibrous preforms is emphasized here, since fiber-reinforced composites are of great technological interest. While it is possible to prepare ceramic and metallic composites by the CVI process, this description is limited to ceramic composites.

For a fibrous preform, the CVI process can be visualized as follows. The preform, which consists of a porous array of fibers, is placed in a high temperature furnace. Gaseous CVD reagents flowing into the furnace penetrate the pores of the preform and deposit as a solid coating onto the fiber surfaces. As this deposition process continues, the fibers get larger, and consequently the spaces between the fibers become smaller (Fig. 1). Eventually the fiber coatings interlock and the fibers are held together by the coating. This coating is the matrix, which along with the fibers constitutes the composite.

The large interest in the CVI process arises for several reasons. First, the process can be used to fabricate fiber-reinforced ceramic composites that have attractive mechanical properties, particularly enhanced fracture toughness. Importantly, compared with most other ceramic consolidation processes, the CVI process utilizes low stresses and temperatures. This minimizes damage to fibers, whiskers, particulates, or other reinforcing materials in the preform. A second advantage is that pre-

* Much of this article is taken, with permission, from Ref. 1.

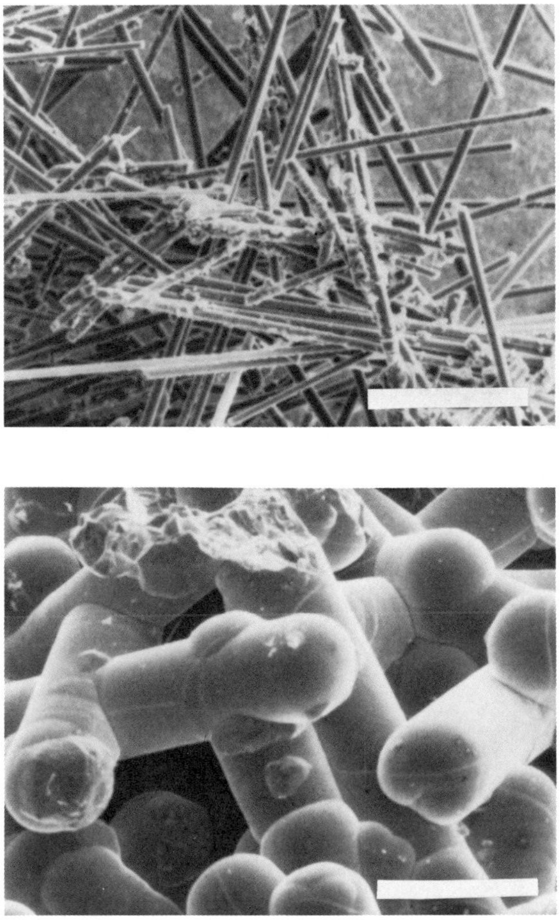

FIGURE 1 Scanning electron micrographs of SiC fiber preform after infiltration with Si_3N_4 showing (*Top photo*) minimal deposit on fibers and (*Bottom photo*) intermediate stage of infiltration, where appreciable deposit of Si_3N_4 on surface of fibers has caused reduction in porosity and interlocking of fibers. The precoated fiber diameter was about 12 μm in both instances (bar = 200 μm).

coating of the preform constituents (e.g., fibers), which might be required to tailor fiber–matrix bonding, can be performed as an initial step of the CVI process using different reagents but the same CVD equipment. A third advantage is that the CVI process can be used to further densify materials fabricated by other means. Internal pores or cracks can be filled as long as they are open to the surface to permit access by the gaseous reagents. The CVI process also permits the fabrication of components of irregular shape. A final advantage of CVI processing is that both purity and microstructure of the matrix are superior to what normally can be obtained with other fabrication methods. No sintering aids are required, and the ability to deposit a matrix having a fine-grained, controlled microstructure results in enhanced properties at low and high temperatures.

History of CVI

The first use of the CVI process was that of Bickerdike et al. in 1962 [2]. They increased the modulus and doubled the flexural and compressive strength of graphite by infiltrating it with carbon. Even before the experiments of Bickerdike et al., infiltration was used to make nuclear fuel elements by consolidating a bed of spherical fuel particles [3]. While this consolidation process differs from typical CVI processes, it could be referred to as CVI.

Apparently, the first use of the CVI process to prepare fiber-reinforced composites was the work of Jenkin [4], who infiltrated a fibrous alumina preform with chromium carbide. Williams [5] also did early work; he reinforced carbon felt by infiltrating it with carbon. During the next several years Taylor and Paluzelle [6], Pierson and Smatana [7], Withers [8], Kotlensky [9], and Theis [10] prepared, by the CVI process, fiber-reinforced composites containing carbon, Si_3N_4, SiC, or Al_2O_3 fibers in matrices of carbon, SiC, TiB_2, B_4C, or Al_2O_3. Filament-wound carbon fiber substrates were used as early as 1968 [6,10].

A good technical review of early CVI work is provided by Kotlensky [11], who discusses process types, densification of porous solids, and infiltration of fibrous preforms. A recent review of CVI technology is that of Lackey and Starr [1].

Chemical Vapor Deposition

Introduction to CVD

Since CVI is a specialized form of CVD, it is appropriate to describe, in generalities, the CVD process. Chemical vapor deposition can be defined as deposition of a solid onto a heated substrate surface as a result of chemical reactions in the gas phase. These reactions can, and frequently do, occur at or near the substrate surface. A system for chemical vapor deposition consists of three basic subsystems: reactant supply, reactor (furnace containing the substrate), and effluent gas handling equipment (Fig. 2). At room temperature, the reactants may be gases, liquids, or solids. Liquid and solid reactants are usually warmed sufficiently to raise their vapor pressure to a level at which carrier gas (Ar, He, H_2, etc.) flowing through a vessel containing the reactant carries its vapor into the reactor. Mass flow controllers provide accurate metering of gaseous reagents into the coating furnace. The reactors are usually heated either resistively (Figs. 3 and 4) or inductively. Control systems with feedback loops provide accurate control of furnace temperature and pressure. The substrate can be hung in the furnace (by a wire), it can lie on a platform, or it can be held by the gas injector. The effluent gas usually contains hydrogen, carrier gas, partially decomposed reactants, and HCl or HF, since frequently reactants are metal chlorides or fluorides. The acid vapor is removed by a solid or liquid caustic scrubber or cold trap.

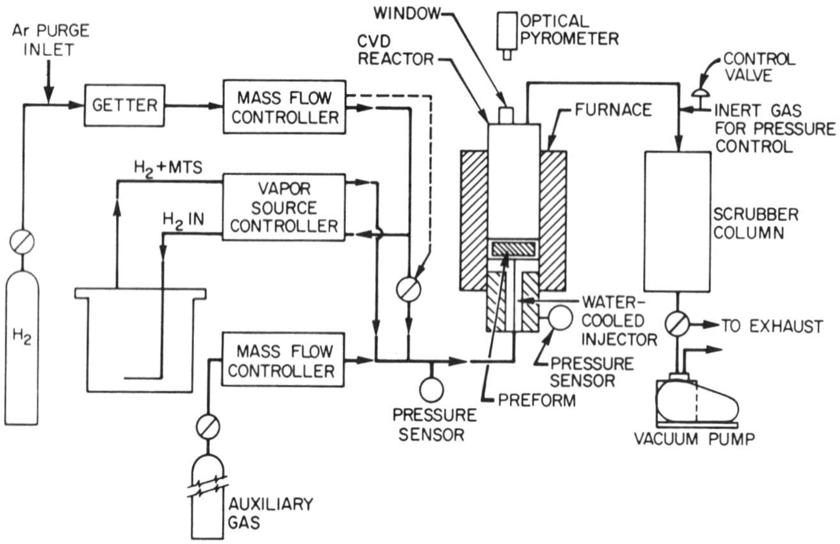

FIGURE 2 Equipment for CVD and CVI consists of three subsystems: reactant supply, furnace, and effluent gas handling.

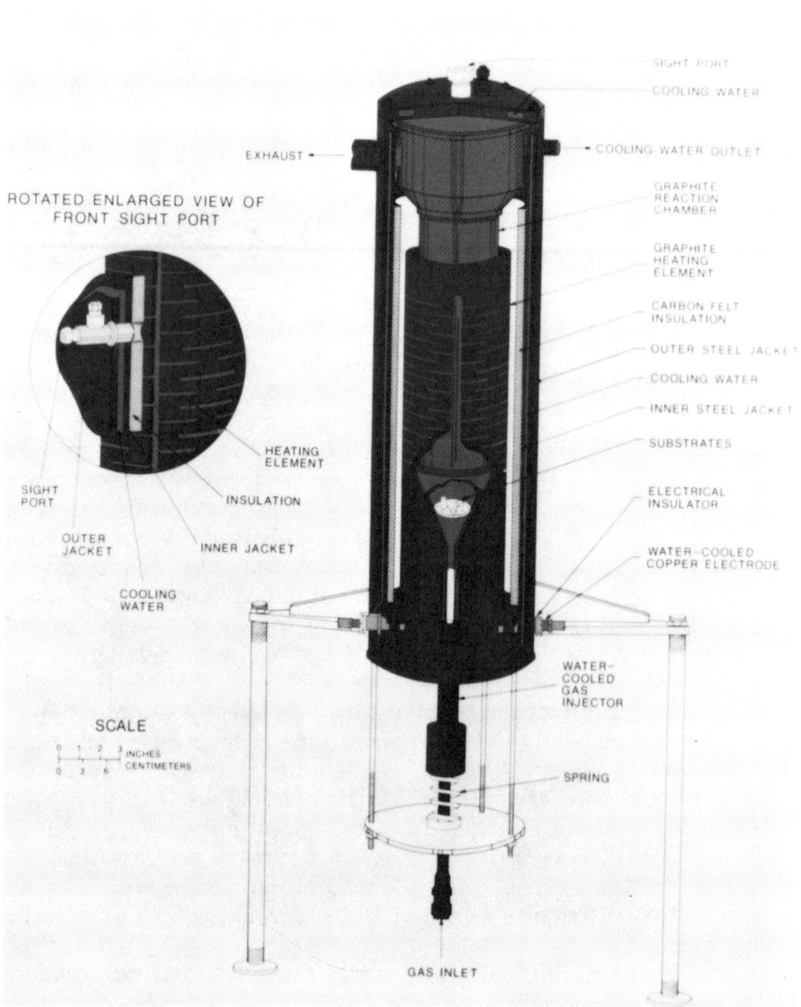

FIGURE 3 Laboratory scale chemical vapor deposition furnace.

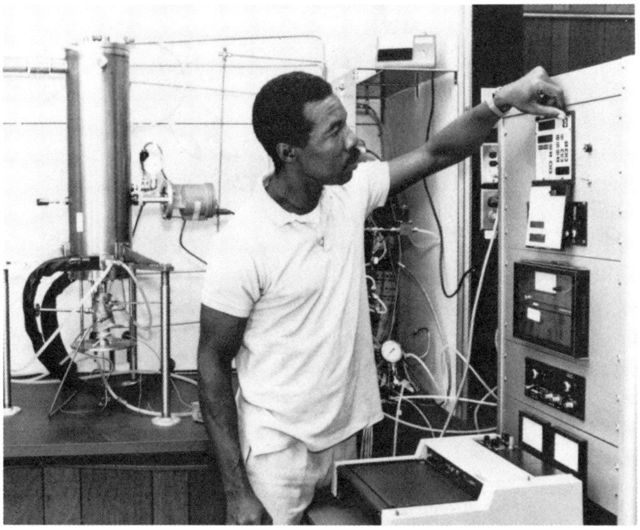

FIGURE 4 **Chemical vapor deposition system showing the reactor located in a fume hood behind the operator. The control panel contains instrumentation for automatic control of temperature, pressure, and reagent flow rates.**

Chemical Reaction Types

The deposition reaction can be of several types: reduction, thermal decomposition, or displacement. Reduction, the most common, can be accomplished with hydrogen or active metals such as sodium or calcium. Hydrogen reduction of metal chlorides or fluorides is by far the most common type of process. An example of such a reaction is:

$$TiCl_4(g) + 2BCl_3(g) + 5H_2(g) \rightarrow TiB_2(s) + 10HCl(g) \quad (1)$$

Thermal decomposition reactions, as opposed to hydrogen reduction, usually involve reactants that are metal carbonyls, hydrides, halides, or organometallics. An example of such a reaction involving a carbonyl is:

$$Ni(CO)_4(g) \rightarrow Ni(s) + 4CO(g) \quad (2)$$

An example of a displacement reaction is:

$$SiCl_4(g) + CH_4(g) \rightarrow SiC(s) + 4HCl(g) \quad (3)$$

A reaction of interest for CVI of a matrix of SiC is:

$$CH_3SiCl_3 \rightarrow SiC + 3HCl \quad (4)$$

Extra hydrogen over that stoichiometrically required is often added to the reactant gas stream. Since corrosive and explosive chemicals are involved, suitable safety and environmental procedures must be followed. Table 1 lists materials that can be deposited by CVD and the required reagents and deposition temperatures.

Kinetics

Frequently, CVD and CVI processes are described in terms of whether the process is limited by diffusion or reaction kinetics. Because of the importance of these concepts to selection of preferred process conditions, they are described next.

The mass transfer and kinetics of heterogeneous (surface) reactions may be conceptualized as occurring in several steps as follows.

1. Mass transfer of reactants from the main body of moving gas through a boundary layer to the deposition surface.
2. Adsorption of one or more reactants on the substrate.
3. Reactions on the substrate.
4. Desorption of gaseous reaction products.
5. Mass transfer of gaseous reaction products from the substrate to the main body of fluid.

The slowest of these steps controls the rate of the overall process. At low temperatures, reaction at the interface is normally slower than the diffusional steps, and therefore the deposition is said to be reaction or kinetically controlled. Increasing the temperature usually accelerates the interface reaction (exponential effect) more than it influences diffusion; thus at higher temperatures, the deposition process is normally limited by diffusion. It is common to plot the logarithm of deposition rate versus reciprocal temperature (an Arrhenius plot). Two straight lines are usually obtained: one for the kinetically controlled low temperature region and one of less slope for the high temperature region, where the rate of deposition is limited by diffusion.

Microstructure

Although the theory is not well developed, a great deal is known concerning how the microstructure of chemically vapor deposited coatings can be modified and controlled. Most CVD materials are crystalline, but use of extremely low temperatures can lead to amorphous deposits. Crystalline perfection and grain size normally increase with increasing deposition temperature. Retained volatiles (e.g., chlorine) usually diminish with increasing deposition temperature. Process variables can be controlled to influence nucleation and grain growth. High temperatures and low reagent concentrations are conducive to the growth of deposits having a single-crystal whisker morphology with a minimum of crystal imperfections. High temperatures and high reagent concentrations cause homogeneous nucleation of powders before the reagents reach the substrate. Such conditions result in soft, weak deposits. In general, intermediate deposition temperatures and reagent concentrations are preferred for CVD, since they result in deposits that contain small, equiaxed grains and are therefore hard and strong.

TABLE 1
Ceramic Materials Readily Deposited by CVD, with Typical Reagents and Deposition Temperatures

Coating	Reagents	Deposition Temperature (°C)
Carbides		
B_4C	$BCl_3—CH_4—H_2$	1200–1400
B_xC	$BCl_3—CCl_4—H_2$	1000–1500
	$B_2H_6—CH_4$	400[a]
Cr_7C_3	$CrCl_2—CH_4—H_2$	1000–1200
Cr_3C_2	$Cr(CO)_6—CH_4—H_2$	250–650
Cr_xCy	$Cr(AA)_3—H_2$[b]	600
HfC	$HfCl_4—CH_4—H_2$	900–1300
Mo_2C	$MoF_6—C_6H_6—H_2$	400–1000
	$Mo(CO)_6—H_2$	350–475
NbC	$NbCl_5—CCl_4—H_2$	1500–1900
SiC	$CH_3SiCl_3—H_2$	1000–1600
	$SiCl_4—CCl_4—H_2$	1300–1600
SiC	$(CH_3)_2SiCl_2—H_2$	1000–1600
	$SiCl_4—C_xH_y—H_2$	1000–1600
	$SiH_4—C_xH_y$	200–500[a]
TaC	$TaCl_5—CH_4—H_2$	1000–1200
TiC	$TiCl_4—CH_4—H_2$	900–1600
	$TiCl_4—CCl_4—H_2$	1300–1600
VC	$VCl_2—CH_4—H_2$	1050–1150
W_2C	$WF_6—C_6H_6—H_2$	400–1000
	$WCl_6—CH_4—H_2$	900–1150
	$W(CO)_6—H_2$	350–400
ZrC	$ZrCl_4—CH_4—H_2$	900–1300
	$ZrI_4—C_2H_6$	900–1400
Nitrides		
AlN	$AlCl_3—NH_3—H_2$	800–1200
	$Al(CH_3)_3—NH_3—H_2$	900–1200
	$AlBr_3—NH_3—H_2$	550–1200
	$AlBr_3—NH_3—H_2$	200–800[a]
BN	$BCl_3—NH_3—H_2$	900–2000
	$BF_3—NH_3—H_2$	1000–1300
	$B_3N_3H_6—Ar$	400–1000
	$B_2H_6—NH_3—H_2$	400–700[a]
HfN	$HfCl_4—NH_3—H_2$	900–1300
	$HfCl_4—N_2—H_2$	900–1300
NbN	$NbCl_5—N_2—H_2$	900–1300
Si_3N_4	$SiCl_4—NH_3—H_2$	1000–1550
	$SiF_4—NH_3—H_2$	1000–1550
	$SiH_4—NH_3—H_2$	700–1150
	$SiH_4—NH_3—H_2$	250–500[a]
	$SiH_xCly—NH_3—H_2$	1475
TiN	$TiCl_4—N_2—H_2$	800–1400
	$TiCl_4—N_2—H_2$	250–1000*
ZrN	$ZrCl_4—NH_3—H_2$	900–1300
	$ZrCl_4—N_2—H_2$	1100–1200
Oxides		
Al_2O_3	$AlCl_3—CO_2—H_2$	500–1100
	$Al(AA)_3—H_2$	340–500
	$Al(CH_3)_3—O_2$	300–500
	$Al[OCH(CH_3)_2]_3—O_2$	300–500
	$Al(OC_2H_5)_3—O_2$	300–500
BeO	$Be—HCl—Co_2—H_2$	1050–1550
	$Be_4O(CHO_2)_6$	400–1250
Cr_2O_3	$Cr(CO)_6—O_2$	400–600
Oxides (continued)		
HfO_2	$HfCl_4—CO_2—H_2$	800–1200
	$Hf(AA)—O_2—He$	450–700
	$Hf(TFAA)_4—O_2—He$[c]	400–750
SiO_2	$SiCl_4—O_2—H_2O—CO_2$	≤ 1000
	$SiH_4—CO_2—H_2$	200–600[a]
	$SiH_4—N_2O$	200–600[a]
Ta_2O_5	$TaCl_5—O_2—H_2—Ar$	600–1300
TiO_2	$TiCl_4—H_2O$	800–1000
	$TiCl_4—O_2$	25–1000[a]
	$Ti[OCH(CH_3)_2]_4—O_2$	25–700[a]
Y_2O_3	$YCl_3—CO_2—H_2$	−1200
	$Y(TMHD)—O_2—H_2O$[d]	400–750
ZrO_2	$ZrCl_4—CO_2—H_2$	800–1200
	$Ar(AA)—O_2—He$	450–700
	$Zr(TFAA)_4—O_2—He$	400–750
Borides		
MoB	$MoCl_5—BBr_3$	1400–1600
NbB_2	$NbCl_5—BCl_3—H_2$	950–1200
	$NbBr_5—BBr_3$	1500
Si_xBy	$SiCl_4—BCl_3—H_2$	800–1400
TaB_2	$TaBr_5—BBr_3$	1200–1600
ZrB_2	$ZrCl_4—BCl_3—H_2$	1000–1500
HfB_2	$HfCl_4—BCl_3—H_2$	1000–1600
Silicides		
$TiSi_2$	$TiCl_4—SiCl_4—H_2$	1400
Carbon		
C	CH_4	900–2250
	C_xH_y[e]	900–2250

[a] Plasma-assisted.
[b] AA = acetylacetonate [e.g. $Cr(C_5H_4O_2)_3$].
[c] TFAA = trifluoroacetylacetonate [e.g. $Hf(C_5H_4O_2F_3)_4$].
[d] TMHD = tetramethylheptanedionate [e.g. $Y(C_{11}H_{19}O_2)_3$].
[e] Numerous hydrocarbons have been used to deposit carbon [e.g., propane (C_3H_8), propylene (C_3H_6), acetylene (C_2H_2), and benzene (C_6H_6)].

CVI Processes

The different CVI processes may be classified into five types as shown in Figure 5. The bases for the classification are (a) whether the preform is uniformly heated (isothermal) or contains a thermal gradient and (b) the manner in which the gaseous reagents contact or flow through the preform. The different processes and their advantages and disadvantages, which to some extent depend on the type and number of parts being fabricated, are described below.

Type I: Isothermal

In the type I (isothermal) process, the preform or preforms are maintained at a uniform temperature and the reagents flow through the furnace (Fig. 5). There is no provision for forcing the reagents to flow through the preform. Reagents enter the preform by chemical diffusion. Similarly, gaseous reaction products must diffuse out of the preform; then, upon reaching the flowing gas stream, they are carried out of the furnace.

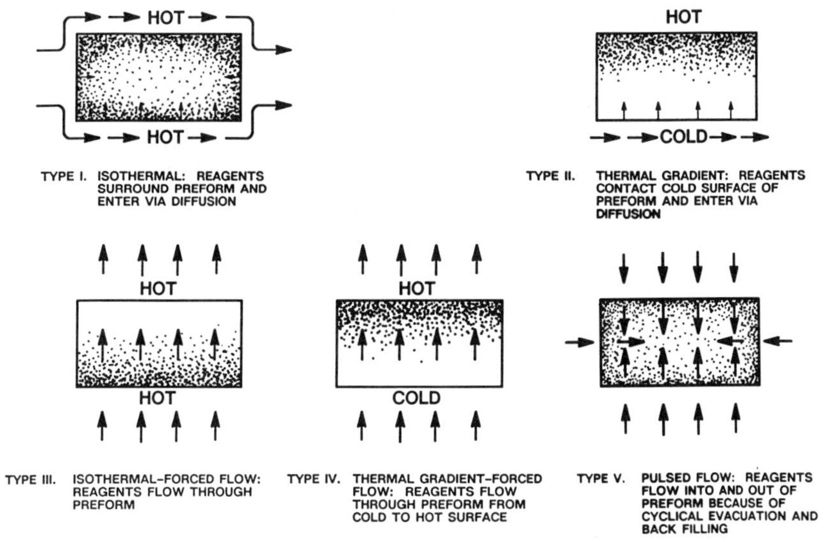

FIGURE 5 **CVI processes can be classified into five general types depending on whether the preform is uniformly heated and how the reagents contact the preform. The long arrows represent flowing reagent; the short arrows show reagent diffusion. The progression of densification is indicated by shading (heavier shading denotes more deposition).**

The deposition of coating (matrix) tends to occur preferentially near the outer surface of the preform (Fig. 5) for two reasons: (a) the concentration of reagent or reagents decreases on progressing into the preform and (b) the concentration of gaseous reaction products, which tend to retard deposition, increases within the preform. The process may be interrupted several times to permit removal, by machining, of a dense "skin" that tends to form near the exterior surface of the preform. If not removed, this relatively impermeable layer slows the rate of infiltration and also accentuates the natural tendency of materials prepared by the isothermal process to possess a density (i.e., porosity) gradient. That is, less deposition of matrix occurs with increasing depth within the preform.

Since diffusion is slow, the type I process tends to require long infiltration times; several weeks is not uncommon. Despite long infiltration times, this isothermal process is most frequently used commercially because it is readily applicable to a large number of parts per run, which can be of different sizes and shapes, including very irregular shapes of varying thickness. The isothermal process is used commercially on a large scale for fabricating carbon fiber–carbon matrix composites for high performance airplane brakes and aerospace applications, as well as for carbon or SiC fiber composites containing an SiC matrix. The process is also used commercially for further densifying carbon fiber–carbon matrix composites made by other methods.

Type II: Thermal Gradient

The type II (thermal gradient) process depicted in Figure 5 makes use of a steep thermal gradient through the thickness of the preform. The thermal gradient may be obtained by heating only a portion of the preform (e.g., the face of a disc or the outside of a tube). The part of the preform away from the source of heat is cooler because of the self-insulation provided by the porous preform. Even steeper temperature gradients can be obtained by deliberately cooling the cold portion of the preform. Heating is often accomplished by laying or wrapping a preform onto a graphite susceptor, which is heated inductively.

The supply of reagents is via a flowing gas stream over the cold surface of the preform. The reagents move through the preform by chemical diffusion from the cold to the hot surface. Because the rate of the deposition reaction usually increases exponentially with temperature, appreciable deposition does not occur until the diffusing reactants approach the vicinity of the hot surface. As deposition occurs at and near the hot surface, the amount of porosity present decreases, causing the thermal conductivity of the infiltrated portion of the preform to increase. This change in conductivity causes more of the preform to be heated to a temperature high enough to cause significant deposition to occur. Thus, in this manner, infiltration occurs progressively from the hot surface toward the cold surface. Long infiltration times (weeks) are required, since diffusion of reactants into the preform, and reaction products out, occurs by diffusion alone. During the course of an infiltration run, it is common to occasionally increase the temperature of the hot surface of the preform, to permit the attainment of sufficiently high temperatures further inside the preform. It is important that minimal deposition occur at the cold surface, since this is the area through which reagents enter the preform. At the conclusion of the infiltration run, the cold surface may be ground or otherwise machined to remove material that was not sufficiently infiltrated.

The type II process was used may years ago in the successful fabrication of carbon fiber–carbon matrix missile nose cones several feet long. Carbon fibers were wound around a graphite susceptor, which was inductively heated. Reagents flowed over the outer, cool surface of the preform. Infiltration times were on the order of 6 weeks. Additional details appear below (see Review of Composites Fabricated by CVI).

Type III: Isothermal–Forced Flow

A third type of CVI process relies on forced flow of reagents through a uniformly heated, essentially isothermal preform (Fig. 5). Deposition occurs throughout the preform. This type of process works well until some region of the preform reaches sufficient density (~90%) to become impermeable. This terminates the flow of reagent, even though other regions of the preform may be of much lower density. The tendency is for more deposition to occur near the preform surface through which reagents enter, because the reagent concentration is highest there. Further inside the preform, reagent concentration is lower; reagent depletion is a natural consequence of the previous upstream deposition. Alternatively, if an inner region of the preform is appreciably warmer or initially denser than the surface of the preform, this inner region may first become impermeable. Regardless of its location, formation of an impermeable region across the flow path prevents further downstream densification and the infiltration process is, for all practical purposes, terminated.

The isothermal–forced flow process has been used to consolidate fibrous or particulate materials by passing reagent gases through a tube furnace packed with the substrate. The process can also be used to infiltrate tubes or other shapes by forcing the reagents to flow through the wall of the preform.

Type IV: Thermal Gradient–Forced Flow

The type IV process simultaneously takes advantage of both a thermal gradient and a forced flow of reagents to enhance infiltration. As shown in Figure 5, reagents enter through a cold surface of the preform. Deposition occurs within regions of the preform that are hot enough to promote the deposition reaction. If a large thermal gradient is present—as is the case, for example, when a portion of the preform is heated while an opposing region is actively cooled—the majority of the deposition during the initial stage of the process occurs near the hot surface of the preform. As a greater portion of the preform becomes hot enough to promote deposition (as a result of changes in thermal conductivity or alteration of the heating and cooling), densification progresses from the hot face toward the cold face. Depending on the temperature of the cold face, densification may or may not occur in the coldest portion of the preform.

If process conditions are such that densification occurs predominantly in the vicinity of the hot surface, this region becomes impermeable. For an older thermal gradient–forced flow process [4], reagent gas flow, and therefore infiltration, ceased when a region of the preform became impermeable. However, an improved thermal gradient–forced flow process invented by Caputo and Lackey [23,24] provides alternative flow paths that permit continued forced flow of reagents through the preform even after some regions have densified to the point of impermeability.

Figure 6 illustrates schematically the principle of the improved process. In this example, a disc is being infiltrated. In addition to cooling the bottom of the disc, the sides are cooled. This causes deposition to initially occur preferentially at the top of the preform. As infiltration proceeds, a lens-shaped densified region (Fig. 6) progresses down through the preform. If sufficient cooling is provided, the bottom and lower portion of the sides of the preform remain uninfiltrated and provide paths for reagents to flow into and out of the preform. A sectional view of equipment designed for infiltrating a disc using the improved thermal gradient–forced flow process appears in Figure 7. The process is also applicable to the infiltration of hollow tubes and perhaps other shapes.

Process modeling performed by Starr [25,26] and experimental work at Oak Ridge National Laboratory [24,27–29] have shown that it is possible to select for the thermal gradient–forced flow process operating conditions that permit attaining essentially uniform deposition rates throughout the preform. Briefly, uniform infiltration is achieved by selecting process conditions (temperatures, temperature gradient, and reagent concentration, and flow rate) that balance the opposing tendencies for (a) greater deposition to occur in the hotter regions as a result of the influence of temperature on reaction kinetics and (b) greater deposition to occur in the cooler regions because of the higher reagent concentration.

The major advantage of the thermal gradient–forced flow process is greatly reduced infiltration time. Because of the forced flow of reagents, infiltration times are on the order of 10–24 hours even for parts several centimeters thick. Another advantage of this process is that a greater percentage of the reagent is utilized; that is, the

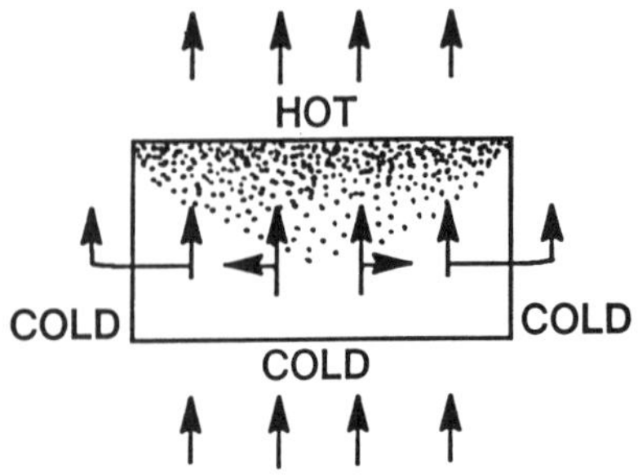

FIGURE 6 **Improved thermal gradient—forced flow CVI process. Reagents flow through the preform from the cold to hot surface initially, but later exit laterally when the hot surface becomes impermeable.**

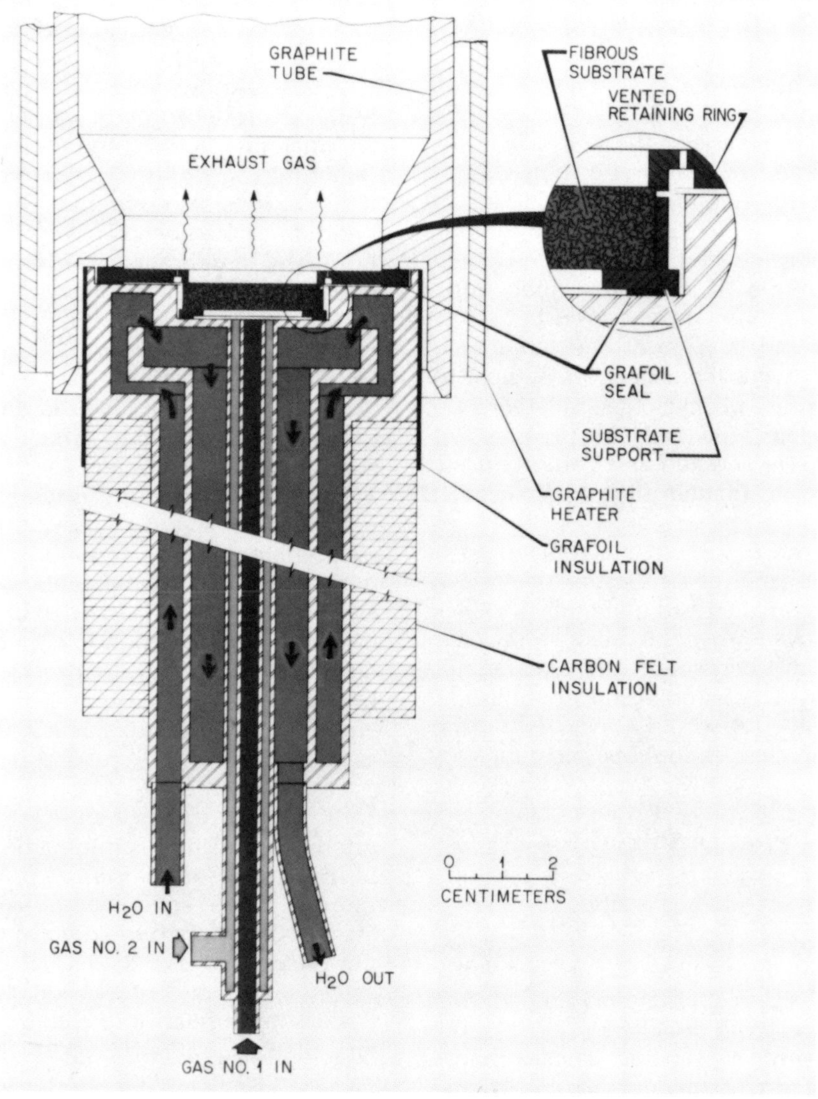

FIGURE 7 **Cross section of equipment for infiltrating a disc using the improved thermal gradient—forced flow CVI process.**

deposition efficiency is high. An additional advantage is the uniformity of infiltration (i.e., uniform composite density). The process has been demonstrated for discs and hollow tubes. A final advantage is that fabrication of preforms is simplified because preforms are held together by a refractory fixture such as graphite instead of requiring careful processing with binders. Fixtures, because they obstruct some diffusion paths, are detrimental to processes that rely on diffusion.

Type V: Pulsed Flow

A fifth and basically different type of CVI process relies on pulsed flow of reagents into and out of the preform. Pulsed flow is caused by cyclical evacuation of the reaction chamber and backfilling with reagents. The pulse CVI process has not been used extensively, but it is not new. Beatty and Kiplinger [30] used a vacuum–pressure pulsing process to infiltrate graphite with carbon. Weight increases up to 8% were achieved in graphite of original density of 1.86 g/cm^3 (18% porosity). Permeability to helium was greatly reduced. Pulse CVI has also been used to densify porous carbons and graphite with TiN [31] and SiC [32].

In the work of Beatty and Kiplinger [30], an inductively heated graphite substrate was cycled between 27 and 241 kPa butadiene for 2–30 hours at 750–950°C. The vacuum and pressure pulse periods used were 7.5–6.0 and 0.5–1 second, respectively. Weight increases of up to 8.0% were achieved, and helium permeabilities were reduced from 10^{-2} to less than 10^{-8} cm^2/s. The coating thickness on the walls of the larger pores was about 2 μm. The coating was thicker for the larger pores, since these pores had larger entrance dimensions and thus were not closed off early in the process, and also because the ratio of surface area to volume decreases with increasing pore size. Lower temperatures and longer periods at low pressure enhanced infiltration as opposed to overcoating. A shorter total processing time was possible for an intermediate temperature.

Sugiyama and Yamamoto [32] investigated the influence on deposition rate of varying the hold time (time system is pressurized per cycle) for the infiltration of porous carbon with SiC for pressures of 23 and 93 kPa. At the highest temperature (1400°C), limited infiltration occurred; the reagents reacted with the walls of the reactor and overcoated the sample. At the lowest temperature (1100°C), more uniform infiltration occurred but longer cycle times were required. Maximum deposition rates for a pressure of 23 kPa were about one-fourth those for 93 kPa. Preferred hold times per cycle depended on both temperature and pressure. Intermediate temperatures and short hold times were selected as a compromise to yield reasonable uniformity of infiltration and short total process time.

The practicality of the pulse process remains to be determined. The process may be applicable to final densification of materials fabricated and nominally densified by other means. A disadvantage of the pulse process may be increased equipment procurement and maintenance costs. If the efficiency with which reagents are utilized is low, then higher reagent costs would also be a disadvantage unless recycling of reagents could be accomplished.

Influence of CVI Process Variables

Key Variables

Optimization of key process variables such as temperature, temperature gradient, pressure, reagent concentration, and flow rate can reduce the time required for infiltration as well as improve the uniformity of infiltration from part to part and within a given part. These variables also influence the structure of the deposited matrix and therefore properties of the composite. For these reasons it is of interest to understand the influence on the process of the major variables. In this section we describe qualitatively the influence of important process variables. Analytical descriptions are available elsewhere [25,26,33,34].

Role of Kinetics

As briefly described earlier, the rate of coating deposition for a CVD or CVI process will be limited by either diffusion or reaction kinetics. The diffusion step may be visualized as transport of gaseous chemical reactant(s) (not necessarily the initial reagents) from the reagent gas stream into the preform for CVI processes not utilizing forced flow of reagents into the preform. For CVI processes utilizing forced flow of reagents, the diffusion step consists of transport of chemical species (reactants or products) through a stagnant gas boundary layer in contact with surfaces (not just external surfaces) of the preform. In either event, it is desirable for a CVI process to be controlled by surface reaction kinetics rather than diffusion, since reagents enter further into the preform before deposition. This obviously enhances infiltration (as opposed to overcoating of the exterior surfaces of the substrate).

It is well known that lower temperatures favor a kinetically controlled process as described previously. Thus, within limits, low temperatures are preferred because infiltration, as opposed to overcoating, is enhanced. Low temperatures also minimize thermal and chemical degradation of the fiber and minimize fiber–matrix interaction. Low temperature, however, invariably reduces the overall deposition rate.

Pressure Effect

Another process variable that has a strong influence on diffusional and kinetic processes is the total gas pressure. Diffusion is faster, as desired, at low pressures. Also, chemical reaction rates typically decrease with a reduction in reagent concentration or pressure. Both effects enhance infiltration, as opposed to overcoating of the preform. Another variable is the molecular weight of any diluent gas. Reactant diffusion is more rapid in a light gas; thus, use of a low molecular weight diluent is optimum. Frequently hydrogen is used: not only is it light, but its presence often is necessary for reaction with chemical constituents such as chlorine.

Reaction Type

Another factor that can influence infiltration as opposed to overcoating is the nature of the deposition reaction. Endothermic reactions with high activation energies are preferred, although freedom to exercise this choice may not often exist. For a thermal gradient process, a high activation energy enhances deposition at hot regions of the substrate relative to cooler regions. This is desirable for a thermal gradient process such as type II in Figure 5, which relies on diffusion to transport reagents into the preform.

Process Tradeoffs

Virtually all the process changes suggested above for enhancing uniform infiltration also unfortunately reduce the overall rate of matrix deposition. Lowering of temperature and reagent concentration and pressure particularly retard deposition. This accounts for the long infiltration times required for processes that rely on diffusion for transporting reagents into and reactants out of the preform. It is not as essential to use low temperatures and pressures in forced-flow CVI processes. This greater freedom in selecting process conditions with forced-flow processes is advantageous. It permits not only higher deposition rates but also increased flexibility in controlling the phase assemblage, microstructure, and properties of the deposited matrix. This is particularly important when depositing a dispersed-phase composite as the matrix [35]. Although this advantage of the forced-flow process is not generally appreciated, it is likely that future operations will be more ready to take advantage of it.

Preforms

Numerous types of preform are suitable for infiltration by the CVI process. The ease of infiltration is, of course, dependent on the permeability of the preform.

Constituents

Preforms may consist of continuous or chopped fibers, whiskers, spheres, flakes, or mixtures of these. The preform may also be a monolithic ceramic partially densified by conventional fabrication processes. However, common preforms are made from high strength ceramic fibers. Important characteristics of fiber preforms are fiber type, volume percent fiber, fiber diameter and length, and fiber orientation. The pore size distribution is important for any preform, since extremely large pores may not be fully infiltrated. Large voids in the final product are usually undesirable but are not as detrimental to the mechanical properties of a composite as they are for a nonreinforced ceramic.

Fabrication

Numerous preform fabrication methods are used. The preform may simply be a piece of felt. Stacking of multiple layers of tapes or cloth, with or without stitching to hold the layers together, is often employed, particularly in commercial fabrication of carbon–carbon composites. Fiber loadings of 70% can be achieved for cloth layup preforms, but 45% is more typical. Hybrid preforms containing two constituents, such as tapes and cloth or fibers and cloth or fibers and whiskers, have been prepared. Filament winding of uni- and multidirectional tubes and cones, and 2D and 3D weaves are also used. Braids or weaves are used to produce hollow tubes. Discs, tubes, cones, and other shapes have been prepared from chopped fibers by slurry molding, which is also called vacuum forming.

In the slurry molding process, chopped fibers or whiskers are first suspended in a liquid that may contain a binder. The slurry is then vacuum filtered to result in the formation of a disc or other shape consisting of about 15–25 vol % fibers. For example, Nicalon fibers and Silar and Tokamax whiskers (manufactured, respectively, by Nippon Carbon Company, Tokyo; Advanced Composite Materials Corporation, Greer, SC; and Tokai Carbon Company, Tokyo) have been molded into discs. The whisker preforms are relatively impermeable and resist the flow of reagents, compared with preforms containing larger diameter fibers.

Fiber Coatings

Purpose

To obtain optimum mechanical, chemical, and physical properties for a fiber-reinforced composite, it is often necessary to coat the fibers. This requirement is not unique to the CVI process. Fiber coatings may be used to control the strength of the bond between the fiber and matrix to permit debonding and/or fiber pullout during mechanical stressing of the composite (i.e., to enhance toughness). Coatings may also be required to improve fiber resistance to high temperature oxidation or to prevent chemical reactions between the fiber and matrix.

Choice of Coating

Ceramic coatings are often used because they are relatively inert and refractory. Carbon coatings are frequently used, but carbon contributes to oxidation of the composite during service. Another common coating material is boron nitride (BN). For many applications, BN also suffers from inadequate resistance to oxidation, although not as severely as carbon. Carbide, nitride, and boride coatings, such as TiC, TiN, and TiB_2, are also being developed. Oxide coatings promote excessively strong fiber–matrix bonds for oxide matrix composites. Oxide coatings may, however, prove useful for carbon–carbon or other composite compositions.

Coating Process

Fibers are typically coated using the CVD process because it permits uniform coverage of the fiber surface with a thin ($\sim$ 0.1 μm) coating, and it can be used to deposit numerous materials. Chopped fibers and continuous tows can be coated using established technology. Individual filaments of tows can be uniformly coated by selecting coating conditions that permit infiltration of the tow. Batch and continuous processes for coating of monofilaments and tows are possible. Coating of whiskers, which tend to be blown out of the coating furnace because of their small size, is more difficult. Recently, SiC whiskers were successfully coated with BN at the Georgia Institute of Technology.

Fiber coatings may be applied before or after preform fabrication. Fibers may be somewhat weakened and embrittled by the coating process, in which case weaving becomes more difficult. Organic coatings (sizing) applied during fiber manufacture are normally removed by solvents or oxidation before application of ceramic coatings.

Review of Composites Fabricated by CVI

Numerous composite types that have been fabricated by CVI are summarized in Table 2. A summary for carbon–carbon and SiC–SiC, which are two of the more technologically important composites, is presented here. Additional details for these two composites are available [1].

Carbon–Carbon Composites

Carbon fiber–carbon matrix composites are the most highly developed ceramic composites. Carbon–carbon is used for high temperature ($\sim$ 1000–2000°C) applications where high strength, stiffness, toughness, and thermal shock resistance are desired, as well as light weight. Car-

TABLE 2
Ceramic Composites Fabricated by Chemical Vapor Infiltration

Deposit	Preform	Ref.
Carbon	Carbon felt, foam, and fibers	5–7
Carbon	Filament-wound and woven carbon fibers	6, 10
Carbon	Monolithic carbon or graphite	2, 30
Carbon, SiC, or BN	Microspheres	3
SiC	SiC fibers	23, 24, 27–29, 36–46
SiC	Carbon fibers	8, 40, 42
SiC	Monolithic carbon, carbon–carbon, and SiC	36, 37
Chromium carbide	Al_2O_3 fibers	4
Si_3N_4	SiC fibers and monolithic SiC and Si_3N_4	36
B_4C	Si_3N_4 and carbon fibers	8, 41
SiC, TiB_2, Al_2O_3	Al_2O_3 fibers	8
TiB_2, B_4C, or TaC	Carbon fibers	47–49
BN	BN and SiO_2 fibers, carbon–carbon, and carbon felt	50–54
TiC	Monolithic carbon, carbon–carbon, and carbon fibers	31, 34, 51, 55
ZrC	Carbon fibers	56

bon–carbon has a strength-to-weight ratio superior to those of other high temperature materials. Protective coatings are frequently used to improve its poor resistance to oxidation. Carbon–carbon has been used for jet engine afterburner nozzles, nose cones, aircraft disc brakes, and leading edges such as on the space shuttle. Several gas turbine engine applications are under development. These include turbine rotors, combusters, exhaust nozzle flaps and seals, and augmenters.

The matrix of carbon–carbon composites is prepared commercially by two basic approaches: carbonization of a resin or pitch and deposition of carbon using the CVI process. Carbon–carbon made by the CVI process has better structural properties than material made by multiple impregnation–carbonization cycles and is thus preferred for critical applications. The remainder of this section on carbon–carbon composites is devoted solely to material made by the CVI process.

Carbon is readily deposited from a hydrocarbon such as methane (CH_4) and other hydrocarbons. A simplified overall deposition reaction is:

$$CH_4(g) \rightarrow C(s) + 2H_2(g) \quad (5)$$

Nitrogen, hydrogen, or inert diluent gases are used.

An excellent review of early carbon–carbon CVI work was prepared by Kotlensky [11]. Isothermal and thermal gradient CVI processes have been used. Commercially, large isothermal furnaces containing multiple parts predominate. Furnaces having a hot zone 1.8 m in diameter by 1.9 m high with a temperature capability of 1925°C have been used commercially for isothermal infiltration. Deposition temperature and pressure are in the range 950–1150°C and 0.1–20 kPa [11]. Typical values are 1100°C and 0.6–6 kPa. Aerospace components and brake discs are primarily made using the isothermal process. The thermal gradient process has been used to fabricate large nose cones.

Details of the infiltration process as well as the properties of the infiltrated composite depend largely on the type of fiber, volume percent fiber, and fiber orientation. Felt, tape, fabric layup, filament-wound shapes, and 2D and 3D woven structures have been successfully infiltrated. The carbon deposition rate for the isothermal process is typically 0.25 μm/h [11]. Infiltration is generally accomplished in multiple cycles, with machining between cycles to clean away surface coatings. Cycle times depend on a variety of factors such as preform thickness, pore size, and percent porosity, as well as on such process and equipment conditions as temperature, pressure, and furnace size. For carbon–carbon, optimum times are usually 60–120 hours per cycle [9].

A schematic drawing of a furnace for infiltrating a nose cone using the thermal gradient process appears in Figure 8. Heating is accomplished using an induction power supply. The graphite susceptor serves as a mandrel for the preform. A ceramic sleeve protects the water-cooled copper induction coil from the heat generated in the susceptor. The water-cooled steel jacket is usually insulated from the hot zone with carbon felt. Compared with the isothermal process, the thermal gradient process utilizes higher furnace pressures (generally 100 kPa or 1 atm), and the deposition rate is higher. For the thermal gradient process, carbon deposition rates at the hot face of a 1.3 cm thick carbon felt preform varied from 0.6 to 2.0 μm/h [11]. The deposition rate 1 cm away from the hot face was on the order of 0.1 μm/h. Higher flow rate enhanced deposition.

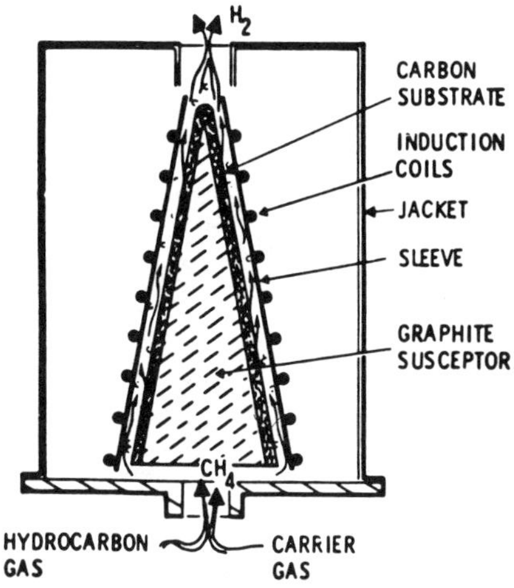

FIGURE 8 **Thermal gradient CVI process for fabrication of carbon–carbon nose cone. Carbon filament is wound onto graphite mandrel–susceptor, which is heated inductively.**

The structural features of carbon fiber preforms required for effective infiltration are small pore size, permeable structure, open porosity, and absence of bottleneck pores. For typical carbon cloth layup preforms with densities of about 1.2 g/cm^3, the average pore size between the fibers of tows is 1–4 μm and 10–30 μm between tows. There are many 100 μm pores and some pores as large as 400 μm. For low density felts the average pore size is 10–50 μm. Hundreds of hours is required to fill the larger pores. Infiltrated densities are about 1.9 g/cm^3 (i.e., about 84% of theoretical density).

The mechanical and physical properties of infiltrated carbons and fibrous preforms have been reviewed by Kotlensky [11]. Carbon–carbon retains its strength to about 2000°C. In general, composite strength increases with density and decreases with an increase in deposition temperature, the latter effect being caused by greater crystal size and preferred orientation of the matrix for higher deposition temperatures. Flexural strengths of up to 966 MPa have been observed for PAN-based unidirectional CVI composites. Ring tensile strengths of 414 MPa have been reported for filament-wound cylinders.

These values are much higher than typical values for commercially available 2D laminates (cloth layups). The flexural and tensile strengths for laminates are usually about 300 MPa with a Young's modulus and strain of about 110 GPa and 0.5–0.6%, respectively. Interlaminar shear strength depends on the test method and the thickness of the laminate. Typical values for 0.25, 0.64, and 0.89 cm thick samples are 9.3, 5.9, and 3.3 MPa respectively. Cross-ply tensile strengths vary from 4.1 to 5.5 MPa.

SiC–SiC Composites

Extensive efforts have been devoted to the development of SiC fiber–SiC matrix composites by the University of Karlsruhe, the University of Bordeaux, the Society of European Propulsion (SEP), Amercom, and Oak Ridge National Laboratory [23,24,27–29,36–46]. The interest in this system is due to the attractive high temperature properties of SiC, which include high strength and elastic modulus, light weight, low coefficient of thermal expansion, and high thermal conductivity, which contribute to good thermal shock resistance, resistance to wear and erosion, and chemical stability, including resistance to oxidation to about 1500°C. Fiber reinforcement is desired to increase fracture toughness. The major advantage of SiC–SiC over carbon–carbon is improved resistance to oxidation, erosion, and wear. The biggest disadvantages are thermal and chemical degradation of the SiC fiber.

Continuous SiC fibers are generally made from polycarbosilane polymer precursors (10–15 μm diameter) or by CVD (100–140 μm diameter) onto a carbon or tungsten filament. Very high strength whiskers (~ 0.6 μm diameter × 10–80 μm long) are often made from rice hulls. The polycarbosilane-derived fibers (Nicalon) are available commercially as continuous fibers, felt, and cloth fabrics. Nicalon cloth has been used extensively for reinforcement of SiC–SiC composites. The Nicalon fibers contain excess carbon and oxygen compared with SiC and lack the thermal stability desired in some applications. Depending on atmosphere and exposure time, they degrade at 1000–1200°C. The larger SiC monofilaments, produced by Avco Specialty Materials Division (Lowell, MA) and Sigma (Tübingen, Federal Republic of Germany), are stable to about 1400°C and have been used to prepare very high strength unidirectional composites. However their larger diameter is not as suitable for preparing ceramic composites possessing high toughness.

Methyltrichlorosilane (CH_3SiCl_3), which is abbreviated MTS, and hydrogen mixtures are normally used in the CVD and CVI of SiC. The deposit is typically crystalline β-SiC, which has a cubic crystal structure and a theoretical maximum density of 3.21 g/cm^3. Process conditions for CVI of SiC vary considerably. For isothermal infiltration of Nicalon 2D cloth preforms 4 × 5 × 50 mm in size and containing 40–50 vol % fiber, Fitzer and Gadow [42] recommend an infiltration temperature of 1050°C, a total pressure of 3 kPa (23 torr), and a time of 14 days.

Oak Ridge National Laboratory has extensively developed the thermal gradient–forced flow process for infiltration of SiC fiber (primarily Nicalon cloth) preforms with SiC [23,24,27–29,40,45,46]. A polished cross section of this material appears in Figure 9. The process has been demonstrated for discs up to 5 cm in diameter and 2.5 cm thick and, in a joint program with Babcock and Wilcox, for hollow tubes [44]. The experimental arrangement for discs was shown in Figure 7.

Two basic processes show promise. In one the gas flows upward through the preform; the process is terminated when the back-pressure reaches about 170 kPa (10 psig). The second process permits radial flow of the reagents when the hot surface of the preform seals as described previously. With either process, infiltration times (~ 20 hours) are much shorter than for the isothermal or thermal gradient processes. In some samples having av-

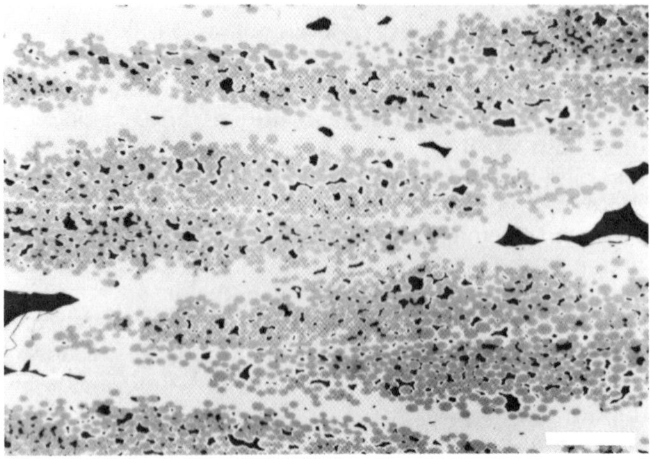

FIGURE 9 Polished section of multiple layers of Nicalon SiC cloth infiltrated with SiC. The fibers are gray, the SiC deposit is white, and pores appear black (bar = 200 μm).

FIGURE 11 Fracture surface of SiC–SiC cloth layup composite after four-point flexural testing showing multidirectional reinforcement (bar = 200 μm).

erage densities of about 86% of theoretical (i.e., 14% porosity), the variations in density throughout the disc were less than 1%.

While it is not clear what technique has been used, Amercom has made small and large tubes (~ 20 cm diameter × 2 m long). Both Nicalon SiC fiber and 3M's Nextel fiber have been used with a SiC matrix. The 3M Company, whose Nextel fibers are available in several compositions from the Al_2O_3-SiO_2-B_2O_3 system, also prepares a product called Siconex, which consists of Nextel fibers in a CVI SiC matrix. Tubes range in diameter up to 20 cm.

Flexural strengths, measured by four-point bending at room temperature, for the Oak Ridge National Laboratory 2D Nicalon cloth–SiC material ranged from 83 to 517 MPa for individual samples. Strength generally correlated with density and was typically 180 and 450 MPa for 65 and 90% of theoretical density, respectively. This material exhibits high strain tolerance and toughness in both flexural and tensile tests (Fig. 10), provided the extent of fiber–matrix bonding is controlled. In properly prepared material, extensive fiber pullout is apparent in fracture surfaces (Fig. 11). Fracture toughness values measured with the single-edge notch beam technique are in the range 7–10 MPa · $m^{1/2}$, depending on the orientation of the cloth. This is about a threefold improvement over nonreinforced SiC. The flexural strength at 1000°C in air is ~ 170 MPa.

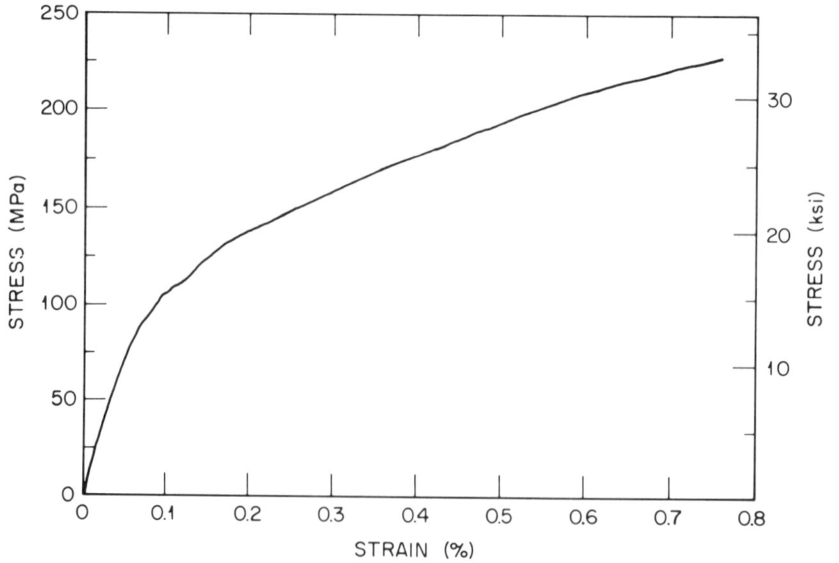

FIGURE 10 Stress–strain curve for SiC–SiC cloth layup composite tested under uniaxial tension.

The fiber-reinforced Nicalon SiC–SiC composite made by SEP is referred to as Cerasep. Room temperature flexural strength is reported to be 450–800 MPa for about 40 vol % fiber [38]. Most of the strength is retained at 1100°C, and about half is retained at 1400°C for short times in vacuum. The elastic modulus is 230 GPa.

Although the strength of SiC–SiC composites is comparable to that of monolithic SiC made by sintering or hot pressing, the CVI fiber-reinforced materials are much superior in toughness. For 2D Cerasep the fracture toughness is reported to be in excess of 25 MPa · $m^{1/2}$ for room temperature to 1000°C. The thermal shock resistance of the fiber-reinforced material is also much superior to that of monolithic SiC.

Acknowledgments

Ruth D. Thompson is gratefully acknowledged for typing the original manuscript and David P. Stinton, Oak Ridge National Laboratory, for contributing several visuals for use in the figures.

W. J. Lackey

References

1. W. J. Lackey and Thomas L. Starr, "Fabrication of Fiber-Reinforced Ceramic Composites by Chemical Vapor Infiltration, Their Structure and Properties," in K. S. Mazdiyasni, Ed., *Fiber Reinforced Ceramics*, Noyes Publications, Park Ridge, NJ, in press.
2. R. L. Bickerdike, A. R. G. Brown, G. Hughes, and H. Ranson, *Proceedings of the Fifth Conference on Carbon*, Vol. I, 1962, pp. 575–582.
3. L. L. Pepkowitz, B. L. Vondra, and F. S. Shipko, Paper 7 in USAEC Report TID-7622, Oct. 10–12, 1961.
4. William C. Jenkin, U.S. Patent 3,160,517 (Dec. 8, 1964), to Union Carbide Corporation.
5. R. M. Williams, *Proceedings of the Pyrolytic Graphite Symposium*, 1966.
6. A. J. Taylor and R. Paluzelle, "Pyrolytic Carbon Infiltration Studies," Y-DA-2123, Y-12 Plant, Oak Ridge National Laboratories, Oak Ridge, TN, Feb. 22, 1968. Also issued as Revision 1, Sept. 8, 1971.
7. H. O. Pierson and J. F. Smatana, *Proceedings of the Second International Conference on Chemical Vapor Deposition*, Electrochemical Society, 1970, pp. 487–505.
8. J. C. Withers, *Proceedings of the Second International Conference on Chemical Vapor Deposition*, Electrochemical Society, 1970, pp. 507–519.
9. W. V. Kotlensky, *Proceedings of the 16th National SAMPE Symposium*, 1971, pp 257–265.
10. J. D. Theis, Jr., *Proceedings of the Third International Conference on Chemical Vapor Deposition*, American Nuclear Society, 1972, pp. 561–573.
11. W. V. Kotlensky, "Deposition of Pyrolytic Carbon in Porous Solids," in P. L. Walker, Jr., and Peter A. Thrower, Eds., *Chemistry and Physics of Carbon*, Vol. 9, Dekker, New York, 1973, pp. 173–262.
12. A. C. Schaffhauser, Ed., *Proceedings of a Conference on Chemical Vapor Deposition of Refractory Metals, Alloys, and Compounds*, American Nuclear Society, 1967.
13. J. M. Blocher, Jr., and J. C. Withers, Eds., *Proceedings of the Second International Conference on Chemical Vapor Deposition*, Electrochemical Society, 1970.
14. F. A. Glaski, Ed., *Proceedings of the Third International Conference on Chemical Vapor Deposition*, American Nuclear Society, 1972.
15. G. F. Wakefield and J. M. Blocher, Jr., Eds., *Proceedings of the Fourth International Conference on Chemical Vapor Deposition*, Electrochemical Society, 1973.
16. J. M. Blocher, Jr., H. E. Hintermann, and L. H. Hall, Eds., *Proceedings of the Fifth International Conference on Chemical Vapor Deposition*, Electrochemical Society, 1975.
17. L. F. Donaghey, P. Rai-Choudhury, and R. N. Tauber, Eds., *Proceedings of the Sixth International Conference on Chemical Vapor Deposition*, Electrochemical Society, 1977.
18. T. O. Sedgwick and H. Lydtin, Eds., *Proceedings of the Seventh International Conference on Chemical Vapor Deposition*, Electrochemical Society, 1979.
19. J. M. Blocher, Jr., G. E. Vuillard, and G. Wahl, Eds., *Proceedings of the Eighth International Conference on Chemical Vapor Deposition*, Electrochemical Society, 1981.
20. H. O. Pierson, Ed., *Chemically Vapor Deposited Coatings*, American Ceramic Society, Westerville, OH, 1981.
21. McD. Robinson, G. W. Cullen, C. H. J. van den Brekel, J. M. Blocher, Jr., and P. Rai-Choudhury, Eds., *Proceedings of the Ninth International Conference on Chemical Vapor Deposition*, Electrochemical Society, 1984.
22. G. W. Cullen and J. M. Blocher, Jr., Eds., *Proceedings of the Tenth International Conference on Chemical Vapor Deposition*, Electrochemical Society, 1987.
23. W. J. Lackey and A. J. Caputo, U.S. Patent 4,580,523 (April 8, 1986), to U.S. Department of Energy.
24. A. J. Caputo, W. J. Lackey, and D. P. Stinton, *Ceram. Eng. Sci. Proc.*, 6(7–8), 694 (1985).
25. T. L. Starr, *Cer. Eng. Sci. Proc.*, 9(7–8), 803 (1988).
26. T. L. Starr, *Cer. Eng. Sci. Proc.*, 8(7–8), 951 (1987).
27. A. J. Caputo, R. A. Lowden, and D. P. Stinton, "Improvements in the Fabrication of Ceramic-Fiber–Ceramic-Matrix Composites by Chemical Vapor Infiltration," ORNL/TM-9651, June 1985.
28. David P. Stinton, A. J. Caputo, and Richard A. Lowden, *Am. Ceram. Soc. Bull.*, 65(2), 347 (1986).
29. R. A. Lowden, A. J. Caputo, D. P. Stinton, T. M. Besmann, and M. D. Morris, "Effects of Infiltration Conditions on the Properties of SiC/Nicalon Composites," ORNL/TM-10403, May 1987.
30. R. L. Beatty and D. V. Kiplinger, *Nuclear Appl. Technol.*, 8(6), 488 (1970).
31. Kohzo Sugiyama and Takashi Nakamura, *J. Mater. Sci. Lett.*, 6, 331 (1987).
32. Kohzo Sugiyama and Eiji Yamamoto, *Proceedings of the Tenth International Conference on Chemical Vapor Deposition*, Electrochemical Society, 1987, pp. 1041–1049.
33. E. Fitzer, D. Hegen, and H. Strohmeier, "Possibility of Gas Phase Impregnation with Silicon Carbide," in S. Somiya, Ed., *Advances of Ceramics*, KTK Scientific, Tokyo, 1985 [Lecture meeting, Tokyo Institute of Technology, Oct. 12–14, 1983].

34. J. Y. Rossignol, F. Langlais, and R. Naslain, *Proceedings of the Ninth International Conference on Chemical Vapor Deposition*, Electrochemical Society, 1984, pp. 596–614.

35. W. J. Lackey, Arlynn W. Smith, D. M. Dillard, and Douglas J. Twait, *Proceedings of the Tenth International Conference on Chemical Vapor Deposition*, Electrochemical Society, 1987, pp. 1008–1027.

36. E. Fitzer, D. Hegen, and H. Strohmeier, *Proceedings of the Seventh International Conference on Chemical Vapor Deposition*, Electrochemical Society, 1979, pp. 525–535.

37. F. Christin, R. Naslain, and C. Bernard, *Proceedings of the Seventh International Conference on Chemical Vapor Deposition*, Electrochemical Society, 1979, pp. 499–514.

38. L. H. Heraud, F. Christin, R. Naslain, and P. Hagenmuller, *Proceedings of the Eighth International Conference on Chemical Vapor Deposition*, Electrochemical Society, 1981, pp. 782–789.

39. R. E. Fisher, C. V. Burkland, and W. E. Bustamante, *Ceram. Eng. Sci. Proc.*, *6*(7–8), 806 (1985).

40. D. P. Stinton, A. J. Caputo, R. A. Lowden, and T. M. Besmann, *Ceram. Eng. Sci. Proc.*, *7*(7–8), 983 (1986).

41. Roger Naslain and Francis Langlais, "CVD-Processing of Ceramic–Ceramic Composite Materials," in Richard E. Tressler et al., Eds., *Tailoring Multiphase and Composite Ceramics, Materials Science Research*, Vol. 20, Plenum Press, New York, 1986, pp. 145–164.

42. Erich Fitzer and Ranier Gadow, *Am. Ceram. Soc. Bull.*, *65*(2), 326 (1986).

43. Pierre J. Lamicq, Gerald A. Bernhart, Martine M. Dauchier, and Jean G. Mace, *Am. Ceram. Soc. Bull.*, *65*(2), 336 (1986).

44. Helen H. Moeller, William G. Long, A. J. Caputo, and Richard A. Lowden, *Ceram. Eng. Sci. Proc.*, *8*(7–8), 977 (1987).

45. D. P. Stinton, *Proceedings of the Tenth International Conference on Chemical Vapor Deposition*, Electrochemical Society, 1987, pp. 1028–1040.

46. R. A. Lowden and D. P. Stinton, "The Influence of the Fiber–Matrix Bond on the Mechanical Behavior of Nicalon/SiC Composites," ORNL/TM-10667, December 1987.

47. L. R. Newkirk, R. E. Riley, F. A. Valencia, and T. C. Wallace, Sr., "Chemical Vapor Deposition Fabrication of Filament-Reinforced Composites for High Temperature Applications," in Hugh O. Pierson, Ed., *Chemical Vapor Deposited Coatings*, American Ceramic Society, Westerville, OH, 1981, pp. 82–101.

48. L. R. Newkirk, R. E. Riley, H. Sheinberg, F. A. Valencia, and T. C. Wallace, *Proceedings of the Seventh International Conference on Chemical Vapor Deposition*, Electrochemical Society, 1979, pp. 488–498.

49. L. R. Newkirk, R. E. Riley, H. Sheinberg, F. A. Valencia, and T. C. Wallace, *Proceedings of the Seventh International Conference on Chemical Vapor Deposition*, Electrochemical Society, 1979, pp. 515–524.

50. J. J. Gebhardt, *Proceedings of the Fourth International Conference on Chemical Vapor Deposition*, Electrochemical Society, 1973, pp. 460–472.

51. R. Naslain, H. Hannache, L. Heraud, J. Y. Rossignol, F. Christin, and C. Bernard, *Proceedings of the Fourth European Conference on Chemical Vapour Deposition*, Philips Centre for Manufacturing Technology, Eindhoven, The Netherlands, 1983, pp. 293–304.

52. H. Hannache, R. Naslain, and C. Bernard, *J. Less-Common Metals*, *95*, 221 (1983).

53. Hugh O. Pierson, *J. Compos. Mater.*, *9*(7), 228 (1975).

54. H. Hannache, R. Naslain, C. Bernard, and L. Heraud, *Proceedings of the Fourth European Conference on Chemical Vapour Deposition*, Philips Centre for Manufacturing Technology, Eindhoven, The Netherlands, 1983, pp. 305–312.

55. L. Aggour, E. Fitzer, and J. Schlichting, *Proceedings of the Fifth International Conference on Chemical Vapor Deposition*, Electrochemical Society, 1975, pp. 634–649.

56. A. R. Driesner, E. K. Storms, Paul Wagner, and T. C. Wallace, *Proceedings of the Fourth International Conference on Chemical Vapor Deposition*, Electrochemical Society, 1973, pp. 462–472.

Characterization

Characterization of materials is a very broad topic. The methods used to characterize composite materials are, for the most part, the same ones used to characterize homogeneous materials. Because of the inherent heterogeneity of composites and their frequent anisotropy, however, obtaining certain information about a composite may require the use of a greater number of methods, and more extensive characterization. This article offers an overview of what is involved in the characterization of composites. What is important to characterize? What techniques are used to do this? What are their current limitations? The details of subjects touched on here can be obtained from other articles in this encyclopedia, or from the references provided.

The material presented in this article is applicable to metal matrix and ceramic matrix composites as well as polymer matrix composites. More examples have been drawn from the latter class of composites, which reflects the more advanced state of development and more widespread use of polymeric composites.

Morphology

The morphology of an object is its structure, form, or makeup as it appears to an observer. We include in composite morphology all those commonly observable features, and all features that are part of the constitution of the composite, not a measure of its response to any externally applied field. Morphology can be studied on three levels: macroscopic, microscopic, and atomic/molecular. Macroscopic characterization encompasses visible observations such as thickness, shape, and color. Microscopy reveals finer details (individual fibers or particles, crystallites, phases, pores, etc.). The finest details concern the concentration and distribution of atoms, molecules, and chemical groups in space. From these one can determine such features as the state of cure, the nature of the fiber–matrix interface, and preferential orientation of crystals or molecules. Morphological characterization is very important for practical reasons such as quality control and process monitoring. It is also a critical tool in (a) identifying the source of problems with a given material, (b) determining the effectiveness of process variations or special treatments in generating a desired structural change, and (c) studying the fundamental dependencies of physical properties on morphological variables.

The dividing lines between macroscopic, microscopic, and atomic/molecular morphology are not distinct. We define as *macroscopic* characteristics that could be observed, at least in principle, without the aid of magnification. The limit of resolution of the naked eye is about 0.15 mm [1]. The macroscopic properties, aside from such things as the layup sequence, are usually examined by techniques not particular to composites, but applicable to any material. One of the most important properties of a composite material is its density, since composites are usually chosen over conventional materials because they can provide the required strength and stiffness at reduced weight. Generally it is the apparent density—which includes any void volume—that is desired. This can be determined from measurements of the weight and volume, or by flotation (ASTM D 1505) or buoyancy methods (ASTM D 792). The latter techniques are useful for small and/or odd-shaped specimens; however, they cannot be used if the composite imbibes the solvent by flow into any pores or by solvent diffusion into the composite materials themselves.

A macroscopic morphological property particularly important in applications on automotive exteriors is surface finish. A mirrorlike finish is inherently difficult to achieve with a system that contains fiber or filler. Often this difficulty is overcome by providing a resin-rich outer layer on the composite. Surface gloss is measured by reflectance (ASTM E 430 or D 523).

In many cases one cannot identify a given material as being a composite without examining it under a microscope. Here the gross morphological features of the multiphase system are revealed. Reflection techniques are generally the most useful. However, with careful cross-sectioning, transmission techniques can also be applied. Microscopy can be used to evaluate fiber and void content, and their distribution in a local region of the composite. This can be done by computer-aided analyses of video images from electron or optical microscopes [2]. The technique, called image analysis, can also be used to determine the grain size distribution in ceramics and metals, the aspect ratio of fillers, etc.

Determination of gross fiber content in a composite specimen generally requires weighing before and after removal of one phase. With organic matrices this usually can be done by matrix digestion in acid or base (ASTM D 3171). However, if microscopy reveals that there are few voids, an easier method of determining the fiber content is from the bulk density and the density of the phases (they must be, and usually are, significantly different). Similar calculations can be done on composites with a porous matrix if the porosity can be determined—for example, by mercury porosimetry.

The distribution of fibers in the composite is also important because it too will affect the physical properties. Matrix-rich layers between plies have been shown to toughen quasi-isotropic layups of graphite–epoxy [3]. On the other hand, fiber-bunching leads to stress concentrations, which can weaken a composite. These arise because the local value of the strain in the matrix increases dramatically as the space between fibers decreases (when the fibers are stiffer than the matrix) [4]. Consequently stress concentrations are generated in the matrix regions between closely spaced fibers. Thus an uneven fiber distribution can lead to a weaker composite. The fiber orientation in discontinuous fiber composites also will dramatically affect the local composite properties. Figure 1 illustrates how X-ray microradiography can be used to reveal the uneven fiber distribution generated by injection molding.

Radiography belongs to the class of characterization techniques known as nondestructive evaluation (NDE). Numerous nondestructive methods are used to evaluate composites, but the methods usually referred to as NDE techniques are those that probe a composite with pressure waves (sonic or ultrasonic), electromagnetic waves [X-rays, γ-rays, optical light (holography)], atomic particles (neutrons, protons), penetrants (dyes), or heat (thermal imaging). Ultrasonic probing is the most common: it is used to detect internal voids or cracks (both macroscopic and microscopic) in a laminate. The ultrasonic C-scan displayed in Figure 2 reveals the internal damage sustained in a composite panel that received the impact of a hemispherical tup. Under visual examination this panel appeared undamaged. NDE methods are used routinely for quality control and for field evaluation of structures.

At present, NDE assessments are crude and strictly empirical, but work is progressing in two areas: the development of techniques that can quantitatively measure the content, size, and position of flaws through the thickness of the composite, and the establishment of quantitative relationships for the degradative effects of flaws on mechanical properties [5]. An additional quantitative application of NDE techniques is the measurement of the resin content in prepregs, for which an ultrasonic instrument has recently become commercially available; it is accurate to within $\pm 1\%$.

The microscopic evaluation discussed so far has focused on characterizing the morphology of the composite system. These techniques can also be used to characterize the individual morphologies of the matrix and fiber in situ, which may be very different from their neat morphologies. For example, in two-phase matrices such as rubber-modified epoxies, thermoplastic blends, or interpenetrating networks, the type of morphology (particulate, lamellar, etc.), the size of the phases, and any variations of these adjacent to the fibers can be determined using optical microscopy (resolution ~200 nm), scanning electron microscopy (SEM: ~3 nm), or transmission electron microscopy (TEM: ~0.2 nm). With crystalline thermoplastics, crystallite size and percent crystallinity

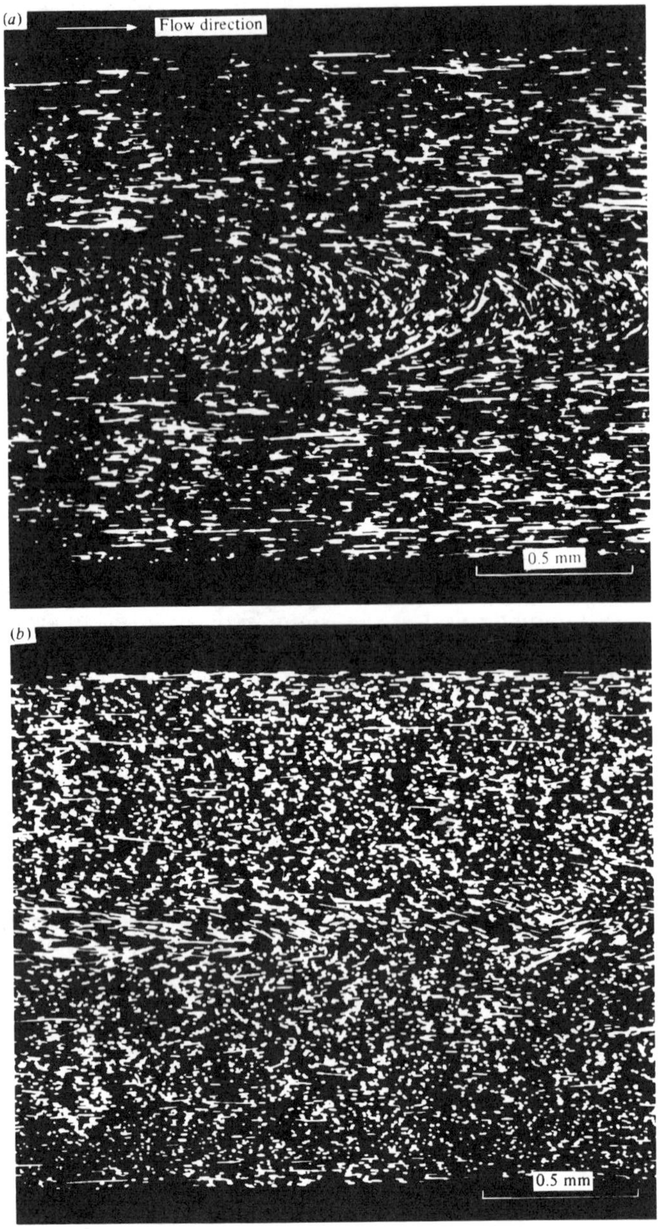

FIGURE 1 **Contact X-ray microradiographs of (*a*) longitudinal, and (*b*) transverse sections of injection-molded, glass-filled polypropylene containing 15 vol % fiber.** (Reproduced from M. J. Folkes and D. A. M. Russell, *Polymer, 21,* 1252 (1980), by permission of the publishers, Butterworth & Co. (Publishers) Ltd. ©.)

can be evaluated by X-ray diffraction. Spherulite size and nucleating sites are discernible using optical or scanning electron microscopy. Neat crystalline thermoplastics often have spherulites at least 10 μm across. In a carbon fiber composite with 60% by volume of fibers approximately 8 μm in diameter, the spacing between fibers is less than 2 μm; thus the presence of the fibers drastically alters the spherulitic morphology. Furthermore, spherulites are often nucleated at the surface of fibers (Fig. 3) and where fibers touch (Fig. 4). Another aspect of the

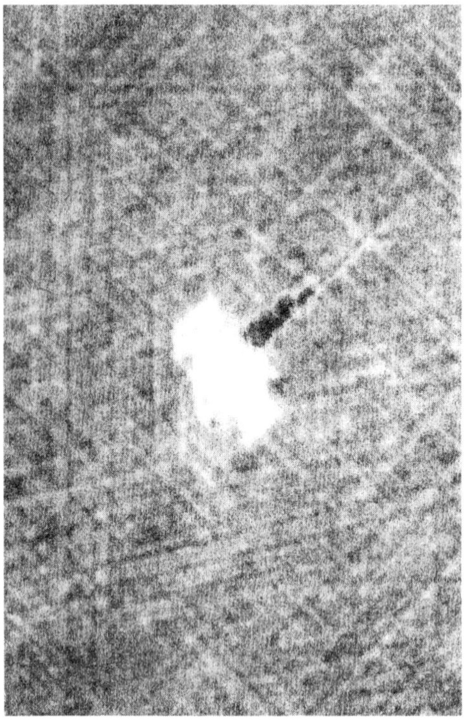

FIGURE 2 **Ultrasonic C-scan of an impacted composite panel revealing internal damage.**

FIGURE 3 **Photomicrograph taken using crossed polarizers, showing much more extensive nucleation of polyether ketone crystals at the surface of a carbon fiber than in the bulk polymer.** (Reproduced from A. J. Waddon, M. J. Hill, A. Keller, and D. J. Blundell, *J. Mater. Sci., 22,* 1773 (1987), by permission of the publishers, Chapman and Hall.)

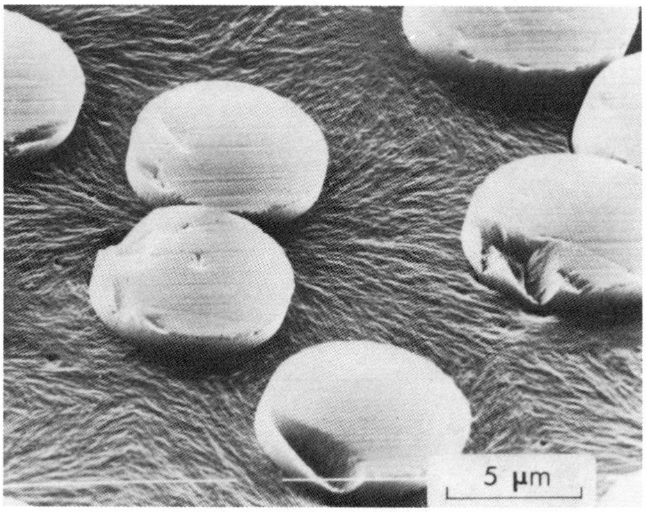

FIGURE 4 **Scanning electron micrograph of a polished and etched cross section, showing the spherulitic texture in a composite containing a crystalline polymeric matrix.** (Reproduced from R. A. Crick, D. C. Leach, P. J. Meakin, and D. R. Moore, *J. Mater. Sci.*, *22*, 2094 (1987), by permission of the publishers, Chapman and Hall.)

morphology of the in situ constituents that is different from their neat morphologies is the presence of molded-in stresses. These arise from any mismatch in the fiber and matrix coefficients of thermal expansion. The magnitude and orientation of the principal molded-in stresses can, in some cases, be determined by photoelasticity measurements using a polarizing microscope [6].

Often of more interest than the morphology of a composite as-prepared is its morphology along fracture surfaces. This information gives clues to the mechanism of failure in a given mechanical test. For example, if a fracture surface displays many fibers sticking out, with little or no resin on them, this is evidence of failure along the fiber–matrix interface. Other fracture surfaces require more experience to interpret correctly [7].

We next turn to the methods for characterizing the atomic/molecular morphology. These methods yield very detailed information about a composite and its constituents. Atoms, functional groups, molecules, and crystalline phases can be detected quantitatively, as well as qualitatively; furthermore, the distribution of such species can be spatially resolved in both two and three dimensions on a given sample. More often the researcher is limited by time and money rather than by the availability of adequate characterization techniques. Some of these are applicable to the final composite, others are useful for analyzing the constituents, and some can be applied to both.

First we mention some of the methods useful in characterizing organic resin matrices. Thermoset resins are typically characterized by manufacturers of composite parts as a quality control measure. An especially useful technique for this is gel permeation chromatography (GPC). The resin is dissolved out of the prepreg. This resin solution is injected into the chromatograph, which separates the constituents based on their molecular size, with the higher molecular weight species eluting first. The chromatogram serves as a fingerprint for a given commercial resin (Fig. 5). If the resin has advanced (partially cured) during storage, this is readily detected as an increase in higher molecular weight species. The identity of the various fractions is often known or can be determined by collection and analysis. Thus, the relative quantities of the constituents in the resin (curing agents, additives, etc.) can be determined. The same methods can be applied to the analysis of the sizings (coatings) applied to commercial fibers. Sizings are often thermoset resins, which also can advance during storage.

GPC is also used for its conventional application, which is to analyze the molecular weight distribution (MWD) of thermoplastic polymers. The MWD is an important quality control parameter because of its effects on the flow of the resin during composite consolidation and on the final properties of the resin. Although many of the crystalline thermoplastics used in composites are not soluble in the usual GPC carrier liquids, aggressive solvents or solvent mixtures, sometimes used at elevated temperatures, have been successfully employed. For an introduction to GPC the reader is referred to Tung and Moore [8].

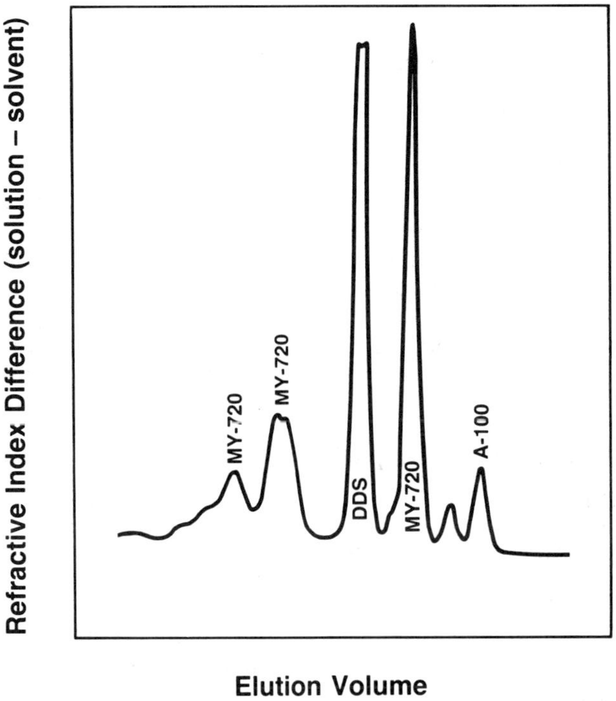

FIGURE 5 **GPC chromatogram of resin from a Fiberite 934 prepreg. Components of the epoxy (MY-720) are well separated from each other, and from the curing agent (DDS) and the modifier (A-100).** (Chromatogram courtesy of Fred N. Larsen, Bendix Kansas City Division of Allied-Signal.)

The cure of thermoset composites can be monitored in several ways. One potential method is dynamic mechanical analysis (DMA) using a flat plate geometry such as is available with a Rheometrics mechanical spectrometer. Here the modulus is monitored by applying an oscillatory shear to the resin between the plates at a fixed frequency. The increase in modulus with time at fixed temperature is an indication of the extent of the cure. This technique can follow cure through the liquid-to-solid transition. Cure can also be monitored by differential scanning calorimetry (DSC), provided the cure is sufficiently rapid and exothermic. Nuclear magnetic resonance (NMR) spectrometry could, in principle, also be used, with the additional advantage of elucidating the time-dependent chemistry of the cure cycle. We are not aware of this technique being applied to composites; however, it has been used successfully to monitor the cure of neat thermoset resins [9].

A recently developed technique for monitoring cure is by dynamic dielectric analysis [10]. In this technique the capacitance and conductance of the composite are recorded over a wide range of frequencies ($5–5 \times 10^6$ Hz) as cure proceeds. Sensors can be embedded in the actual part to be produced (they cannot be removed, however). From these measurements, one can determine the onset of resin flow, the time of maximum flow, and the extent of cure. In addition, when applied to thermoplastic composites, the following transition temperatures can be detected on heat-up and cool-down: glass transition temperature (T_g), melting point (T_m), and crystallization temperature (T_c).

Atomic/molecular characterization of neat fibers calls on many techniques that are also applicable to the composite. X-Ray diffraction can be used to determine the crystallinity, crystallite size, crystallite orientation, and crystalline phases present in a fiber. It can also determine the degree of graphitization in carbon fibers. X-Ray diffraction has been applied to composites as well; for example, it is capable of determining the percentage of crystallinity in carbon fiber reinforced thermoplastics despite the broad overlapping diffraction from the fibers [11]. This technique was also used [12] to show that the molecules of a liquid-crystalline polymer matrix oriented themselves parallel to the reinforcing fibers regardless of their orientation before impregnating the fibers.

Atomic/molecular characterization of fiber surfaces is frequently used to determine the effectiveness of surface modifications designed to improve fiber properties or to make the fiber compatible with the matrix. For instance, surface oxidation treatments of carbon fibers can be evaluated by techniques such as X-ray photoelectron spectroscopy (XPS or ESCA) [13,14], secondary-ion mass spectrometry (SIMS) [15], and ion-scattering spectroscopy (ISS) [15]. These techniques determine the atoms on the surface, and, with XPS, the chemical state of the atom can often also be discerned. In SIMS the surface is sputtered; consequently the elemental composition can be determined as a function of sputtering depth into the fiber (Fig. 6). Scanning tunneling microscopy is a new surface technique [16] that could be useful in analyzing conductive fibers. It is the only scanning technique that can produce surface images in which individual atoms are well resolved.

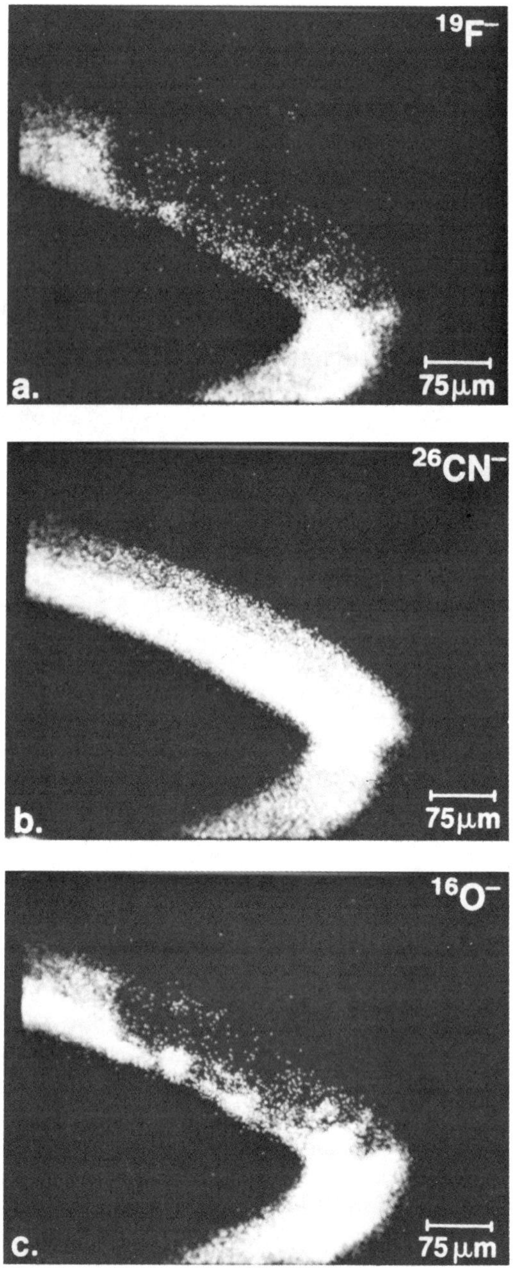

FIGURE 6 **SIMS images of a coated nylon fiber showing the surface distribution of the secondary ion generation of (*a*) $^{19}F^-$, (*b*) $^{26}CN^-$, and (*c*) $^{16}O^-$. The dark area in the center of the fluorine and oxygen images was eroded by about 100 nm by prolonged intense sputtering. These images show that the oxygenated and fluorinated species are both present only in the surface layer. (Photomicrograph courtesy of E. A. Leone of the Corporate Analytical Department of Allied-Signal.)**

One reason for modifying fiber surfaces is to promote wetting and adhesion of the fibers by the matrix, which can be a serious problem, especially in metal matrix systems [17]. The Wilhelmy technique can be used to measure the contact angle between a fiber and a liquid matrix, and thus the surface energy of the fiber [14], or to determine the surface tension of the matrix [18].

Composite cross sections can be analyzed to determine the effect of composite processing on the fiber, the matrix, and the interface. For example, functional group analysis can be performed on specific regions in an organic matrix composite using an infrared, Raman, or ultraviolet–visible microscope. Infrared microscopes can analyze areas larger than 10 μm^2. The microscope is used to fix the location on which a spectrum is to be taken. Higher resolution atomic maps can be generated by SEM or scanning transmission electron microscopy (STEM) using simultaneous X-ray fluorescence analysis, or by SIMS. These techniques can detect, for instance, reaction of a metal matrix with a reinforcing fiber leading to an interphase containing elements of each. Another use is the analysis of the distribution of the curing agent in an epoxy composite, since the curing agent usually contains nitrogen while the resin most often does not.

Mathematical Modeling

Mathematical modeling is the basis of mechanical characterization. Indeed, behind any quantitatively defined (numerically expressed) material property lies a mathematically processed physical idea. Simple ideas and simple mathematical treatments are used in the definition of properties such as density. The definition of density—namely, the mass of a unit volume of the material—is based on the assumption that the material has the same density throughout the volume. Density is an example of a scalar property: it has magnitude, but no direction in space.

More complex physical (specifically, thermodynamic and mechanical) and mathematical ideas are used in defining the mechanical properties of materials. For example, strength in material characterization has a special connotation. It is defined as the value of stress (a variable introduced in continuum mechanics) at which a specimen fails. Stress is a tensor, an abstract object representing a generalization of the concept of a vector. The strength of a molecular chain is expressed in units of force and is an example of a vector property: a physical property that has both magnitude and direction. Therefore, a test designed to determine a material property should be performed with a clear understanding of the extent and limitations of the model introduced. The concept of continuum is the most basic in the characterization of mechanical properties. In the paragraphs that follow we introduce the concept and some of the most important ideas derived from it.

Materials can be considered to consist of atoms, molecules, or their regular agglomerations—crystals. Typical engineering structures are much larger than individual crystals in a polycrystalline body or even the longest molecules in an amorphous polymeric material. If the dimensions of a structural part are at least about 2 orders of magnitude larger than the size of the relevant structural unit of the material from which it is made, the part can be characterized as if it were a continuum (i.e., structureless continuous matter). In a true continuum, the same properties are found irrespective of how small a volume element is tested. In the limiting case of infinitely small volumes, one can discuss the properties of points of material in the continuum. For an introduction to the concept of continuum, see Refs. 19 and 20.

The concept of continuum should be applied to composites with caution. Indeed, the elements defining the composite microstructure (the reinforcing inclusions) can be much larger than such elements (crystals) in conventional materials. Therefore, only when an engineering structure made of the composite is orders of magnitude larger than a characteristic dimension of the reinforcing element can the concept of continuum be applied to the material as a whole. An alternative is to apply the concept separately to the reinforcing inclusions and to the matrix, and study the material as an engineering structure having two or more phases. The theory developed by systematic application of such an approach is known as composite micromechanics. For an introduction see Refs. 4 and 21.

The concept of continuum is necessary to introduce the notions of stress and strain, important variables in material characterization. Elasticity, plasticity, creep, strength, and resistance to crack propagation (fracture toughness) of engineering materials, specifically composites, are characterized in terms of stresses and strains. Stress is a measure of the level of mechanical interaction in the point of the body between two planes created by an imaginary cut inside the body. Stress is defined as the force of such interaction per unit area. Strain is a measure of distortion (deformation) of the body caused by forces applied to the body. Linear strain is defined as the fractional change in distance between two points in a body. Shear strain is a measure of the angular distortions in a body.

Stresses and strains introduced in continuum mechanics are not scalars like density, or vectors like force; rather they are tensors: abstract objects representing a generalization of the vector concept. Stress, for example, is characterized by a set of three vectors, each corresponding to one of three mutually perpendicular directions, or nine scalar components, only six of which are independent. Continuum mechanics provides the rules used to describe the stress at a point—independent of the coordinate system—in terms of the stress invariants, which are quantities that are independent of the chosen coordinate system. Characterization of all mechanical properties should be conducted with a clear understanding of the tensorial nature of these and other variables of continuum mechanics. In some cases, the degree of complexity required to fully characterize a given property depends on whether the body in question is isotropic.

Isotropy is the concept according to which a body has the same properties irrespective of the directions in which they are measured. The absence of isotropy is called anisotropy. Different orders of physical symmetry introduce different degrees of anisotropy. Consider, for example, characterization of elastic properties. As is well known, the notion of elasticity is related to the ability of materials to recover from deformations imposed by external forces. Perfect elastic materials recover completely. All materials behave as linear elastics when the loads and respective distortions are small. In the case of linear elastic materials, the values of distortions are proportional to the applied forces. The property characterized is a material's rigidity or compliance. Correct and complete characterization of a material's rigidity requires the analyst to obtain a complete set of constants corresponding to a given level of material anisotropy. For an introduction see Ashton and Whitney [22].

Anisotropy is more typical for composites than for some traditional materials (metals, plastics). Composite materials reinforced by particles, platelets, or chaotically distributed short fibers can sometimes be nearly or quasi-isotropic. But even in the case of short fiber reinforcement, a certain degree of preferential orientation takes place, and as a result, the material becomes anisotropic. Much higher levels of anisotropy are obtained in high performance continuous-fiber composites.

In the case of anisotropic materials, not only is the stress state important, but also the orientation of the main stresses with respect to the directions of material anisotropy. Anisotropic composite materials should be characterized in all directions. For example, the elastic rigidity of a generally anisotropic material is characterized by 21 independent constants, versus only 2 constants for an isotropic material. Intermediate cases of materials with some level of symmetry are unidirectionally reinforced composites ("fiber symmetry") with 5 constants, and materials reinforced by fibers in three (or two) mutually perpendicular directions (orthorhombic symmetry) with 9 constants.

Like the elastic properties, other properties, such as viscoelasticity (related to the time-dependent reversible deformations), plasticity (related to time-independent irreversible deformations), and creep (related to time-dependent irreversible deformations), are also characterized in terms of relationships between stresses and strains. Complex phenomenological models—based on experimental observation of the phenomenon and speculations related to its physical nature—of viscoelasticity, plasticity, and creep have been developed, which constitute the theoretical bases for the respective characterizations. The reader will be referred to monographs devoted to each of these subjects. The objective of the models is to limit the number of experiments necessary. Each model introduces its own variables of characterization and defines the type and number of experiments required. Compared with the case of isotropic materials, much more complex phenomenological theories should be used for anisotropic composites.

The strength of materials is also characterized in terms of critical values of stresses and strains. Phenomenological theories of composite strength are based on ideas from thermodynamics, continuum mechanics, composite micromechanics, and the results of mechanical testing. The reader is referred to a recent review of the theories and an introduction to strength and other aspects of the characterization of composites [4]. Models of fracture mechanics, which characterize the resistance to crack propagation (i.e., fracture toughness) of materials, are a relatively recent development in the mechanics of materials. The literature on the fundamentals, and on specific applications of fracture mechanics as a means of characterizing the strength of different materials, including composites, is very extensive. For an introduction, we recommend two monographs [23,24]. Here we only underline the differences between the conventional approach to characterizing the strength of materials and the fracture mechanics approach. Fracture mechanics is actually the theory of sharp cracks; it is developed on the basis of continuum mechanics. Indeed, the stress distribution in solids depends on loading, the type of continuum (elastic, elastoplastic, etc.), and the configuration of the specimen. According to the laws of elastic continuum, the stresses in the areas surrounding a notch tip significantly exceed those in the areas remote from the tip. This phenomenon is known as stress concentration. The stress concentration is characterized by the ratio of the local stress to the stress in the absence of the notch. The sharper the notch, the higher the stress concentration.

For an infinitely sharp crack, the theoretical stress in the notch is infinite. Therefore, in the presence of an infinitely sharp crack, stress analysis predicts failure at the smallest loading. Experiments show that this is not the case. This failure of the conventional approach was resolved by fracture mechanics using two revolutionary ideas. One was taken from the theory of stress analysis of cracks in elastic bodies. It was found that in the case of absolutely sharp cracks the stress distribution around the tip is controlled by the local boundary conditions (the crack), but the stress intensity is controlled by the applied load and the boundary conditions of the specimen. The term in the expression for the stress distribution that is responsible for the stress intensity, the stress intensity factor, usually designated as K, was proposed as a parameter controlling crack propagation and, therefore, strength.

Another idea (historically the first) was to consider crack instability on the basis of the energy released and consumed in the process of crack propagation. In other words, it was suggested that for a crack to grow, the energy released in the elastic body during an increment of the crack growth should be at least as much as the energy consumed in creating the new crack surface. In this approach, a critical value of energy release rate (usually designated as G) becomes a new criterion of crack propagation. This concept (Griffith, 1921) was later expanded (G. Irwin, 1950) to incorporate the case of small plastic deformations in the crack vicinity. In such cases the energy release rate is calculated as in a perfect elastic body, but the energy necessary to propagate the crack includes that consumed in all inelastic processes in the tip vicinity: plastic deformations in metals, microcracking, breaks, debonding and pull-out of reinforcing fibers in composites, etc. For perfectly elastic solids, the stress intensity factor and the energy release rate are equivalent. These theoretical ideas constitute the basis for so-called linear fracture mechanics (LFM).

In some systems, including composites of certain types, the energy-dissipating zones are so big that the energy release rate cannot be calculated without considering distortions in the stress field created by plastic deformations. New ideas for characterization are used in this case. It was found that certain energy-related integrals calculated along contours drawn from one crack face to the other going around the crack tip in an elastoplastic continuum are independent of the chosen contour (or "path"). Therefore, the most convenient path can be chosen for the calculation. One of this class of integrals, the so-called *J*-integral [25], is equal to the energy release rate when the system is elastic. It is a popular practice to consider the *J*-integral to be a criterion in characterization of crack growth in elastoplastic systems when plasticity is not small.

Recently a more general model of crack propagation was introduced [26]. A crack in this model is considered to be a layer of transformed material. The physical justification for the model is that in many situations the crack stress field promotes irreversible changes (microcracks, plastic deformation) that surround the tip and, during the crack growth, remain as a layer adjacent to the crack. It was shown that *J*- and other (so-called *L*, *M*, etc.) path-independent integrals are the thermodynamic forces responsible for growth of elongated heterogeneities—"crack layers." The crack layer models efficiently characterize complex fracture situations [27,28]. These ideas constitute the basis for nonlinear fracture mechanics (NLFM).

A special class of mathematical models used for characterizing materials is based on statistics and the theory of probability. Statistical analysis of experimental data is routinely used in material characterization to quantify variability. This variability usually arises from experimental error and minor differences between test samples presumed to be identical. However, in some cases the properties themselves have a statistical nature. The strength of brittle materials is the most important example. In brittle materials the strength depends on the size, shape, and orientation of a single microdefect. Considering that the distribution of microdefects is random, different values of strength are obtained in different statistically identical specimens. A higher probability to meet a larger defect in a larger volume of material leads, according to this concept, to a lower value of strength for a larger specimen of the same material. A good example is the dramatically greater strength of glass fibers compared with glass objects of much larger dimensions (glass pane, glassware, etc.). As is well known, this concept led to the creation of fiberglass, the first and still one of the most important composite reinforcements.

Statistical models of failure do not give a single value for strength as opposed to other, deterministic models. Instead, the strength of a material is characterized by the probabilities of various levels of strength (i.e., by the distribution of expected strength values). These models generate special parameters of material characterization. For example, the popular Weibull theory calls for two independent parameters. The reader is referred to Ref. [29], a good study on such models. Statistical models are extremely important in the characterization of ceramics and naturally the ceramic matrix composites, and also the superstrong but brittle fibers such as carbon and glass, which are used as composite reinforcements. Besides the voids and other microdefects, which define the statistical nature of strength, other factors, such as the random distribution of reinforcing fibers or particles, lead to the use of statistical models in characterizing composites. Such models are being extensively developed for the characterization of the strength of short fiber reinforced composites and other materials. Promising models based on combining these statistical ideas with the ideas of fracture mechanics are being developed to characterize brittle materials [30].

Mechanical Characterization

Experimental Methods

Experimental methods for characterizing the mechanical properties of engineering materials can be generally classified as those based on measuring the response to (a) mechanical probing (traditional mechanical testing) and (b) nonmechanical probing, also known as nondestructive evaluation (NDE). Methods based on mechanical probing can also be subdivided into nondestructive and destructive techniques. If the load applied to a specimen does not exceed its elastic limit, no irreversible changes are introduced, and the characterization is essentially nondestructive. Nondestructive probing can, therefore, be used to obtain the elastic constants of a material or to perform an experimental stress analysis for micromechanical characterization. With composites containing elastomeric matrices, the viscoelastic properties can also be characterized without irreversible changes in the material. Destructive levels of mechanical probing are required to evaluate the strength, plasticity, fracture toughness, creep, and fatigue. We discuss next various aspects of mechanical probing methods, such as the loading rate, the type of loading (monotonous or repeated, cyclic), and the stress state. The implementation of the methods may differ for different types of composites (type of reinforcement, type of matrix).

LOADING RATE. Depending on the loading rate, a mechanical test method is classified as either quasistatic (or static) or dynamic. During static loading the loading rate is low enough to permit the role of acceleration of the material particles, and the consequent inertial forces, to be disregarded in the analysis. Many engineering structures are loaded slowly or subjected to a constant load in service, and that is why such a large number of methods have been developed for static characterization. Most of the common properties such as strength, rigidity, and plasticity are measured using static loading of test specimens.

The loading rate is an important parameter in the static characterization of time-dependent properties. The temperature at which the characterization is performed influences the degree to which the loading rate affects these material properties. Loading rate, for example, affects the response of viscoelastic materials (e.g., most plastics). In metals and ceramics, the effect of loading rate on creep properties is found at temperatures above 60% of the melting temperature.

There are two major types of dynamic characterization: modal analysis and impact testing. Modal or vibrational analysis is concerned with dynamic effects produced by moderate levels of periodic acceleration applied to engineering parts. Such periodic acceleration produces the dynamic response known as vibration. The patterns (modes), intensities, and frequencies of these deformations are studied by modal analysis. Impact is produced by high levels of acceleration. Several aspects of material behavior are characterized by impact testing (see below).

TYPE OF LOADING. Monotonous loading up to a certain level (a ramp) is the type of loading most often used. Periodic or repeated loading is used for characterization of fatigue properties (see FATIGUE) in the case of low acceleration (static loading) and for modal analysis in the case of medium levels of acceleration. Dynamic periodic loading is used to evaluate the ability of a material to dissipate energy (damping properties). From such tests, the complex dynamic modulus is obtained, which contains an elastic component and an energy-dissipating component. High damping materials display a large ratio of the latter to the former.

STRESS STATE. Mechanical properties of materials manifest themselves differently under different stress states. Before considering the stress states used in materials characterization, a discussion of the theoretical and technical limitations to obtaining various stress states is warranted.

First, there are an infinite number of possible stress states (represented by their three invariants and the orientation of the main stress axes with respect to the material axes of symmetry), which cannot all be explored by direct mechanical testing. Also, because of technical limitations, only a rather limited number of stress states can be achieved in practice. Material behavior under other stress states that cannot be arranged experimentally is characterized indirectly, by using mathematical theories of strength, plasticity, and creep.

Typical universal testing machines provide only unidirectional tension/compression loading, in which only one component of stress is present. A typical specimen is a rod (with a variety of cross sections: round, square, rectangular, etc.). Some universal testing machines can provide unidirectional tension/torque combined loading. Specimens for such tests can be round rods or thin-walled tubes. With these tests, a whole class of two-dimensional tensions and compressions can be generated, each with different orientations of the main stresses with respect to the material axes. Rare types of universal testing machine are provided, in addition, with a supply of moderate hydrostatic pressure. By applying the hydrostatic pressure to the center of thin-walled tubular specimens, it is possible to explore yet another class of two-dimensional stress states. Some universal testing machines can apply tension to flat specimens in two directions independently. There are also special fixtures that convert simple tension or compression into two-dimensional stress states.

High hydrostatic pressure is the primary means by which three-dimensional stress fields are created. Usually a high pressure chamber envelopes a solid cylindrical test specimen and also the clamps of the tension/compression or tension/torque tester. Of course, the three-dimensional states that can be created by this test represent a limited class of all possible three-dimensional stress states. Material behavior in other classes of three-dimensional stress, especially those close to three-dimensional tension, are studied using speculations on the mechanics/micromechanics of the materials and phenomenological theories.

Another important aspect in mechanical characterization is the homogeneity of the stress fields. By providing the same stresses throughout a specimen, one can significantly simplify the interpretation of the test results. In such cases, the specimen and every small volume [point (mathematically)] of the specimen is representative of the material at these stresses. Material specimens prepared for testing by a homogeneous stress often consist of two areas: the *gage* of the specimen, or the area where the required homogeneous stress is created, and the *clamping* area through which the load is transferred from the loading device (machine) to the specimen, and where the stress state is usually heterogeneous, complex, and unknown. A well-known example is the "dog bone" or "dumb bell" specimen shape used in tensile tests.

Simple loading schemes are often used because they are simple, even though they create heterogeneous stress fields. Theoretical stress analysis is used in such cases to relate the experimental results to material properties. Examples are the popular three- and four-point bending tests of material beams. These tests are used for indirect evaluation of the tensile or compressive strength. They also are used to characterize materials in bending (flexural properties). Other simple loading schemes include round rods loaded by torque, cylindrical springs and filament-wound rings loaded in tension, and plates loaded in bending. Various experimental techniques are discussed in the literature [31,32].

TYPE OF COMPOSITE. Approaches to the mechanical testing of composites may differ depending on the type of composite. First, the method of testing will generally depend on the distribution of the reinforcement. The methods used in testing a composite reinforced with continuous fibers may differ from those used in testing composites reinforced with regular or irregular particles or short fibers.

The choice of testing arrangement may also be influenced by the type of matrix. Polymeric matrices, for example, are forgiving with respect to small deformations, and, therefore, small imperfections in specimen geometry or in the testing arrangement do not normally affect results. Quite the opposite is true of ceramic matrices, where special arrangements should be made to compensate or minimize imprecisions in testing arrangements. The ability of a specimen to redistribute the applied stresses may also influence the choice of testing arrangements. Polymeric laminates, due to their strong anisotropy, are examples of such materials.

Composite Mechanical Properties

A property of a material can be defined as a certain aspect of its inner physical nature expressed in terms of its response to a given probing. Different methods of probing reveal different aspects of physical nature and also different facets of a given material property. We have discussed how mathematical modeling of physical ideas is used to establish the type and set of specific tests for the complete characterization of some properties, and how, in other cases, complete characterization is not possible due to technical limitations in testing, or insufficient understanding of the phenomenon. In the latter cases, the testing method remains an essential part of the property's definition. New ways of probing—new testing methods—reveal new facets of a property. These statements will be illustrated later on with examples of composite mechanical properties important in engineering practice.

The most important mechanical properties of engineering materials in general, and composites in particular, are as follows.

Elastic
Viscoelastic
Plastic
Creep
Strength
Impact
Fatigue
Resistance to crack propagation

Having mentioned these properties in classifying mechanical testing methods, we briefly discuss the importance of each one for engineering practice, the completeness with which each can be characterized, and the test methods used for characterization. A more complete description can be found elsewhere in the encyclopedia, or in the references provided.

ELASTICITY. This property is observed in all materials. Whenever possible, engineering structures are designed to function within elastic limits, to ensure that no irreversible changes in the material are introduced. The matrix of rigidity (or compliance) of a material is the complete set of elastic constants. A structure's ability to retain its shape under load—an ability so evidently important for engineering structures—depends on its matrix of rigidity. The theory of elasticity establishes sets of technically achievable tests that can be used to completely characterize a material's elastic properties. An introduction to elasticity can be obtained from texts [4,19,20,22]; see also two ASTM publications [33]: "Young's Modulus at Room Temperature" [ASTM E 111-61(78)] and "Poisson Ratio at Room Temperature" [ASTM E 132-61(79)].

VISCOELASTICITY. Viscoelasticity is observed only in high molecular weight materials: plastics and elastomers. When these materials are used as composite matrices, the composites are viscoelastic as well. Like elastic deformations, viscoelastic deformations are completely recoverable, but they are loading rate dependent. In contrast to elastic deformations, the recovery from a given deformation is not instantaneous. Viscoelastic deformations are reversible not only in terms of complete strain recovery, but also in terms of the internal structure of the material, which also completely recovers to its initial state. Characterization of viscoelastic properties is a significantly more complex task than the characterization of elastic properties. Each parameter of viscoelastic behavior is not a constant, but a function of loading rate and temperature. Some of these functions, and the variables defining the functions, can be determined from tensile tests conducted at various loading rates or by determining the strain history under constant load, or the relaxation of stress in a specimen extended a fixed length. These tests can be performed with temperature as a variable. For an introduction to the characterization of viscoelastic properties see Ferry [34].

CREEP. The property known as creep is observed in all materials; it is manifested as an irreversible strain produced by a constant load. Creep under normal circumstances is an undesirable property of engineering materials. Techniques for characterizing creep are similar to those used to characterize viscoelasticity; see, for example, two ASTM standards [33]: "Tensile, Compressive and Flex Creep and Creep Rupture" (ASTM D 2990-77) and "Stress Relaxation of Plastics" [ASTM D 2991-71 (78)]. However, the phenomenon is even more complex, and the functions relating strain rates to the stresses depend not only on temperature but also on such other parameters of the material's state as strains and work consumed in producing an irreversible strain. Considering that for complete characterization, such functions should be technically established for all stress states, the task is formidable. The theories (models) of creep allow one to reduce the task by speculating on the tensorial forms of such functions (phenomenological theories of creep). An excellent monograph on creep [35] will introduce the reader to the subject.

Additional difficulties in the case of composites stem from their anisotropy. A knowledge of the structure, however, often simplifies the task. For example, a 0/90 glass- or carbon-reinforced composite will not creep under tension in the directions of fiber orientation. Creep is an example of a property whose incomplete characterization we are forced to tolerate.

PLASTICITY. Various physical mechanisms in metals lead to plasticity: loading-rate-independent deformations that are irreversible. This independence of loading rate (time) is what, by definition, distinguishes plasticity from creep. The physical mechanism of plasticity is the temperature-independent movement of dislocations in metals. Plastic deformations can thus be observed in composites based on metal matrices. Irreversible, time-independent deformations can be observed in composites with other matrices, even though the matrices (plastic or ceramic) alone are not capable of plastic deformations. Apparent plasticity in such composites is caused by multitudes of internal discontinuities (microcracks in the matrix, fiber breaks, microcracks in the reinforcement–matrix interface) appearing at certain levels of loading. Additional displacements, exceeding the anticipated elastic displacements, which are also irreversible (not disappearing on unloading), are perceived as "plastic."

Since creation of the discontinuities absorbs energy, this apparent plasticity, like real plasticity, can toughen a composite consisting of low toughness (brittle nonplastic) components. Plasticity is desirable, because materials with plasticity resist catastrophic failure. While engineering structures are normally designed to avoid plastic deformations under normal loading conditions, if an occasional overloading takes place, and the yield stress limits are reached, plastic deformation occurs and redistributes the stress field. In this process, significant energy is absorbed. This energy-absorbing capability is particularly important when the overloading is brief, such as that caused by collision (impact). The geometrical distortion of the part caused by the plastic deformation can be a reasonable price to pay for avoiding the otherwise catastrophic failure.

Sometimes plastic deformations cannot be avoided even in the original design. The discs and blades in the hot section of gas turbines are good examples. In such parts, characterization of the plastic stress–strain relationship is necessary for design calculations of the history of stress–strain states to be expected in the parts. The results of such calculations are used to analyze the problem of low cycle fatigue. For metals (and metal matrix composites), a unified high temperature plasticity–creep characterization under cyclic loading can be performed [36,37]. Kachanov [38] offers an introduction to plasticity.

STRENGTH. One of the most multifaceted properties, strength is defined as a material's ability to withstand mechanical loading without failure. It usually is characterized simply by loading a material until it fails. It is important to note that different loading conditions can lead to failure by different mechanisms, and therefore the results of one test may give little or no information on the material behavior under another test. As already mentioned, by "loading conditions" we mean the type of applied stress state, the loading rate, and whether the loading is monotonous (ramp) or cyclic.

Static strength, dynamic strength (impact), and cyclic strength (fatigue) are discussed separately. The static strength of a composite is difficult to characterize completely because it may fail by a variety of mechanisms (e.g., breakage of the fibers, matrix, or interface, or a combination of these) depending on the stress state and the composite design. Since most composites are anisotropic, the strength depends on the direction of the applied stress components with respect to the material axes. The most useful tests for these anisotropic materials are designed such that the axes of the stress tensor coincide with the material's axes of elastic symmetry. Direct experimental characterization of static strength is limited to several basic evaluations. Tensile and compressive strength are two of the variables most commonly measured. The tests are performed in the uniaxial stress state.

Although apparently simple, each of these test types has its own difficulties, which are covered in detail by the ASTM (see individual standards cited below, which can be found in Ref. 33). In performing tension tests, the main difficulty is transferring the load from the grips to the gage section of the specimen such that the stress is homogeneous in the latter section. If a composite is quasi-isotropic, such as composites reinforced by particles or random short fibers, its tensile properties can be determined using "dog bone" specimens as are used for conventional materials ("Tensile Properties of Plastics," ANSI/ASTM D 638-80). For laminates, a different type of flat specimen (coupons) is used, with the ends for gripping reinforced by tabs ("Tensile Properties of Oriented Fiber Composites," ANSI/ASTM D 3039-76; "Tensile Properties of Fiber Reinforced Metal Matrix Composites," ASTM D 3552-77). For filament-wound composites, ring specimens are often used, with the tension created by internal pressure ["Fabrication of Ring Test Specimens for Glass Resin Composites," ASTM D 2291-76(80) and "Apparent Tensile Strength of Ring or Tubular Plastics and Reinforced Plastics," ASTM D 2290-76].

The reason for using specimen designs for laminates different from those used for random reinforced composites is the much lower shear rigidity of laminates. The shear modulus of random reinforced composites is high enough to redistribute stresses in a dog-bone specimen from the grip to the gage. In the case of laminates, the shear rigidity is too low, and the redistribution would not take place; consequently, the stress field in the gage would not be homogeneous. Special testing techniques and standards are used for characterizing the tensile strength of fibers because of their small cross section and often brittleness, as in the case of glass and carbon filaments ["Tensile Strength and Young's Modulus for High-Modulus Single-Filament Materials," ASTM D 3379-75; "Tensile Properties of Glass Fiber Strands, Yarn and Rovings used in Reinforced Plastics," ASTM D 2324-67(76); "Tensile Properties of Continuous Filament Carbon and Graphite Yarns, Strands, Rovings and Tows," ASTM D 4018-81].

Accompanying the evaluation of compressive strength is another problem, created by the usual practice of using composites as thin-walled bodies (plates and shells). Unidirectionally compressed thin specimens can fail for reasons unrelated to their true compressive strength. For example, thin specimens can fail by buckling (see below). The buckling force is related to the material's compliance and the specimen geometry, but not to its strength.

Several modes of failure related to the composite's true compressive strength can occur. Composites can fail by shearing of the reinforcing fibers, microbuckling, and separation of the plies, etc. To determine the true compressive strength, the specimen gage length should be sufficiently short. Still, the specimen geometry, and the way in which the compressive load is transferred to the specimen, may affect the result. Several standardized testing techniques and specimen geometries are commonly used (see, e.g., "Compressive Properties of Unidirectional or Cross-Ply Fiber–Resin Composites," ASTM D 3410-75). The test of choice may be determined by the availability of an appropriate specimen. For example, compressive strength of filament-wound composites is often obtained by compressing rings of the material cut from a filament-wound object.

The shear strength of composites can be characterized by several methods. In composites, this usually requires a separate test, whereas in conventional isotropic materials, shear strength often can be characterized from tensile/compressive measurements. Indeed, particularly in anisotropic composites, failure in shear is dominated by mechanisms little related to the mechanisms of tensile or compressive failure. As already emphasized, the directions in which the shear strength is measured are chosen on the basis of the directions of the reinforcing elements. Often there is not reinforcement against shear stresses, and the shear strength depends on the strength of the matrix and of the matrix–fiber interface. As a result, the shear strength is significantly lower than the tensile strength, which is dominated by the strength of the fibers.

Shear strength can be characterized by several methods, either by testing flat specimens (coupons) in tension, or by applying torsion to rods or tubes (see Ref. 33: "Interlaminar Shear Strength of Structural Reinforced Plastics at Elevated Temperatures," ANSI/ASTM D 3518-76; "In-Plane Shear Stress–Strain Response of Unidirectional Reinforced Plastics," ASTM D 3518-76; "In-Plane Shear Strength of Reinforced Plastics," ASTM D 3876-79; "In-Plane Shear Strength of Pultruded Glass-Reinforced Plastic Rod," ASTM D 3914-80). For a review of the methods, see Refs. 31 and 32.

Three- or four-point bending of beams is used to characterize the flexural strength of composites. Flexural strength, a practical characteristic of composite plates, shells, and beams subjected to bending, is related to either tensile, compressive, or interlaminar shear strength depending on the reinforcement, the matrix, their interface, and the test arrangement. Long beams fail in tension or compression (local buckling); short beams fail through interlaminar shear (see Ref. 33: "Flexural Properties of Plastics," ASTM D 790-71, "Apparent Horizontal Shear Strength of Reinforced Plastics by Short Beam Method," ANSI/ASTM D 2344-76). The tensile strength determined by extension of a specimen is, as a rule, lower than the tensile strength calculated from a flexural test. The discrepancy is due to the difference in the size of the specimen volume, which is tested in the two configurations, and the statistical nature of strength [29].

Characterization of composite strength in biaxial stress states is performed by using methods such as uniaxial loading along a direction other than one of the principal material axes, or by special arrangements using specimens such as a crossbeam sandwich, a flat plate, or a thin-walled tube. A description of the methods can be found in Refs. 31 and 32. Beyond the uniaxial, and the somewhat limited number of biaxial stress states, the designer must rely on phenomenological theories, which use the results of uniaxial and biaxial tests to predict the effects of triaxial loadings. For a review of phenomenological theories of the strength of composites, see Rowlands [39].

Another theoretical method for characterizing composite strength in two and three axial stress states, on the basis of the uniaxial tests, is the multiphase stress analysis known as composite micromechanics. This method uses the mechanical characteristics (modulus and strength) of the individual phases (fiber and matrix) and their interface to predict the moduli and the failure mechanisms of composites under a variety of stress situations. For more information on composite micromechanics, see the literature [4,21,40]. In general, experimental and theoretical stress analysis of composites is an important element of characterization. For an introduction, see Refs. 23, 32, and 41. For example, special experimental techniques and theoretical methods are used to characterize the interface, such as the single filament pullout test. Photoelastic experiments performed on model composite systems are an important means of experimental stress analysis [32].

IMPACT. By definition, impact is the striking of one body against another: a collision. The rapid change of the velocity of the bodies, as a result of the collision, produces extremely high accelerations, hence forces. These forces can lead to failure. Impact is more complex than static loadings, and the characterization of impact strength has even more limitations than the characterization of static strength. Impact properties are also more dependent on the particular testing arrangement than are static properties [42,43]. In general, most approaches to impact characterization are rather pragmatic; three such approaches are the falling weight, Izod, and Charpy tests.

In the falling weight test, composite plates are hit by a falling ball; the damage is visually observed and is correlated to the initial energy in the ball [see Ref. 33: "Impact Resistance of Rigid Plastic Sheeting or Parts by Means of a Tup (Falling Weight)," ASTM D 3029-78]. In the Izod and Charpy tests, composite beams, with or without notch, are struck by a swinging hammer, and the total energy absorbed in destruction of the beam is measured. The testing equipment and the specimen dimensions are standardized; therefore, different materials can be compared (see Ref. 33: "Impact Resistance of Plastics, and Electrical Insulating Materials," ASTM D 256-81). These methods provide useful information, but the results cannot be extrapolated to other testing conditions. Furthermore, they do not allow one to separate the energy absorbed in deforming the part from that consumed in fracturing it.

More advanced impact methods use an instrumented impactor, which records the values of force and displacement during the impact process. A regular Izod tester can be instrumented, but a special falling weight or a servohydraulic impactor is recommended instead. These devices permit a variety of loading schemes, including those with specimens similar to those used in tensile tests. One of the schemes is the split Hopkinson bar method. From the force–displacement diagram, the rigidity of the material (dynamic modulus), the maximum force, and the energy dissipated by the sample at different stages of the loading process are established. Several material properties are obtained from instrumented impact analysis: impact strength, damage tolerance, resistance to perforation, and dynamic response.

Impact failure is established by observation of evident destruction of a material sample or a structural component. Visible cracks, crazes, shattering, or excessive irreversible deformation, which make the component unfunctional or aesthetically unacceptable, are the signs of destruction. Instrumented impact gives the maximum value of dynamic force and the energy the component can absorb without destruction.

Damage tolerance is the ability of a component to experience an impact without significant structural changes and with a minimal effect on its designed load-bearing capabilities. A low energy, out-of-plane impact on the surface of a thin laminated composite plate, for example, can significantly reduce the in-plane compressive strength of the plate. Instrumented impact tests can yield information on how the damage tolerance is affected by other composite mechanical properties and the construction (layup) of the composite. Impact can reduce the fatigue life (see FATIGUE) of a component. Damage tolerance can be studied in terms of fatigue analysis.

Resistance to perforation is an important characteristic of load-bearing structures that also function as protection for personnel and equipment from potential perforating bodies or the surrounding media (gas, liquid). Examples of such structures are aerospace and marine hulls, and car bodies. Protection is the primary function of some structures, such as helmets and containers.

Yet another impact property is dynamic response. Two aspects of the dynamic response of a structure to impact can be studied using instrumented impact. First, at loads below those causing any damage, the force-to-deflection relationship (the dynamic modulus) and the damping characteristics of the component can be established. Second, during the damage process, the energy absorption capability through plastic deformation of the structure can be determined. The force–deflection diagram can provide clues to improving the energy-absorbing capabilities of a structure. Instrumented impact can also be used to study crack penetration (see FRACTURE) at high loading rate.

FATIGUE. Fatigue is the phenomenon of material failure under repeated loading at loads less than the static strength of the material. Fatigue characterization is used to establish the conditions under which a material can serve without failure in an engineering structure for a given number of loading cycles. Fatigue failure is caused by different micromechanisms depending on the material. However, the general feature of these micromechanisms is usually the accumulation with each loading cycle of irreversible deformations (in terms of continuum mechanics) or changes in microstructure (in physical terms). These, in turn, lead to the formation and growth of microdefects (microcracks or crazes) and, eventually, to macrocracks and mechanical failure. In some cases, repeated loading leads to the accumulation of heat, and failure through melting.

Depending on a material's application, samples tested for fatigue are either unnotched or with sharp notches (to provide a high level of stress concentration). In unnotched samples, failure initiates at voids or structural imperfections, which are always naturally present in materials. This test primarily characterizes the crack initiation phenomenon. If the loading conditions during testing are similar to the loading conditions the structural component will see in service, the method gives a direct evaluation of structural longevity. If not, phenomenological theories are applied to predict longevity. The method of testing samples with sharp notches relies on the concepts of fracture mechanics (originally developed for static loading). This test reveals information about crack propagation in the material.

Fatigue tests may be conducted in tension, in flex, or in torsion. A typical stress-controlled loading cycle is characterized by constant values of two parameters: either the maximum and the minimum stress during the cycle or the average of these two and the maximum deviation from the average (i.e., the stress amplitude), or the ratio of the minimum to the maximum stress and the amplitude. The shape of the cycle can be sinusoidal or any other type, and shape can affect results. Cycle frequency is also an important parameter.

Another type of cyclic loading program (called strain-controlled loading) keeps constant the maximum and the minimum values of strain. The raw data of fatigue characterization are the cycling stress–strain diagrams (hysteresis loops). The results of fatigue characterization are typically presented as the relationship between the stress amplitude and the number of cycles to failure. Such diagrams represent the basis of fatigue characterization of unnotched samples. [For details and an example, see Ref. 33: "Tensile–Tension Fatigue of Oriented Fiber, Resin Matrix Composites," ASTM D 3479-76; "Flexural Fatigue of Plastics by Constant Amplitude-of-Force," ASTM D 671-71(78)]. In notched samples, the results of characterization are presented as the relationship between the crack increment per cycle as a function of the stress intensity factor (the maximum and minimum values, or the average and the amplitude values); see, for example, Hertzberg and Manson [44]. For the definition of stress intensity factor, see FRACTURE.

Like many other mechanical properties, fatigue cannot be completely characterized because of the infinite variety of loading situations, many of which cannot be produced experimentally. However, theories of fatigue allow one to use a limited set of experimental data to predict results in other (untested) situations. Both deterministic and probabilistic theories are used. For basic studies and material design, mechanical fatigue analysis is especially effective in conjunction with macrofractography (using optical microscopy) and microfractography (using SEM): analysis of fracture surfaces and the damage zone surrounding the tip and the path of the crack. Other techniques include analysis of the decrease in static strength (determination of residual strength) and modulus of samples subjected to a number of loading cycles. Nondestructive evaluation techniques can also be useful in analyzing the defect formation process as cycling progresses.

Many variables influence fatigue behavior. Relevant variables for all classes of material include stress–strain history and current loading state, thermal history, temperature, environment, and specimen size. Composites introduce additional variables: reinforcement geometry (spheres, platelets, short fibers, continuous fibers), orientation, distribution, and volume fraction; adhesion of matrix and reinforcement; and the mechanical properties of these phases.

RESISTANCE TO CRACK PROPAGATION (FRACTURE TOUGHNESS). There are two aspects in the characterization of the resistance to crack propagation: resistance to sudden, catastrophic failure, and resistance to slow crack growth. In both cases, specimens with artificially introduced sharp notches, simulating initial cracks, are used. Resistance to sudden failure is characterized by the criteria already mentioned under FRACTURE MECHANICS. The specimens are loaded statically or by impact to the point of failure. Resistance to slow crack growth is measured by loading the specimens to lower levels of stress, where immediate failure does not take place, then by either holding the load constant or changing it periodically (fatigue testing). The first method is commonly used to study crack growth in aggressive environments. Resistance to slow crack growth is characterized by measuring crack velocities or crack increments per cycle, and the time or the number of cycles to failure as a function of the fracture mechanics parameters.

In using different criteria of fracture mechanics, one should be aware of the limitations of the models used to introduce them. First, almost all the models (except the "crack layer" model) consider perfectly sharp cracks. Second, some of the criteria (the critical value of the stress intensity factor K_c and the critical value of the energy release rate G_c) are theoretically valid only for linear systems (perfect elastic bodies). Other criteria (the path-independent integrals J, etc.) are valid for elastoplastic systems. In many elastoplastic materials the plastic deformations in front of the crack tip are often confined to a small zone. If this zone is sufficiently small (compared with the size of the crack and the specimen), the linear fracture mechanics (LFM) criteria can be used.

To comply with the theoretical restrictions of LFM and to provide uniform testing of materials by different laboratories, special experimental techniques have been developed. For example, crack sharpness is achieved by growing a natural crack from a notch by fatigue. Plastic deformations are restricted by the use of relatively large specimens. Another technique to restrict plastic deformations and provide crack sharpness is to introduce a so-called chevron notch (see, e.g., Ref. 45) into a specimen such as a rod, bar, or beam. This technique permits the use of smaller specimens and sets relatively low requirements for the sharpness of the notch. Some methods have been standardized (see, e.g., Ref. 33: "Plain–Strain Fracture Toughness of Metallic Materials," ASTM E 813-81). These tests are applicable to particulate and short fiber reinforced composites. Several other nonstandardized techniques are used to characterize isotropic and quasi-isotropic materials. These are based on single-edge- and double-edge-notch tensile specimens, single-edge notch beam, double-torsion beam, or the growing of a crack from an edge of a plate [23,24]. In the case of anisotropic composites (laminates, filament-wound, and pultruded materials), special techniques are used for two major reasons. First, these composite objects are usually thin; second, the resistance to crack propagation depends on the specimen orientation. To characterize the interlaminar resistance to crack propagation, the double-cantilever beam or the edge-notched beam methods are used. To characterize the intralaminar resistance, the double-torsion test is used [4,23].

Environmental Resistance

The complexity of composite materials makes it very difficult to predict the effects of environmental exposure and often even to simulate this exposure realistically. By "environmental exposure," we mean the in-service exposure of a composite to liquids, gases, radiation, and heat. The characterization of environmental resistance is complicated because (a) several types of environmental exposure may occur simultaneously, (b) their effects may be interdependent, and (c) these effects may be influenced by the mechanical history both during and before exposure. For example, the degree to which the properties of a composite specimen are degraded by contact with an organic solvent will depend on the temperature history during this exposure; the mechanical loading experienced also may play an important role. Nonetheless, there are a variety of simple tests useful for screening materials for a given application environment. We now discuss some of these tests, turning first to those useful in characterizing the effects of exposure to liquids.

Composites can be exposed to a variety of liquids, depending on their application. The most frequently encountered liquids are water, organic solvents, and aqueous acids and bases. Liquids can degrade the matrix, the fiber, or their interface through the processes of dissolution, absorption, or reaction. If possible, it is best to first screen fiber and matrix separately to determine which liquids affect each component. Commonly, a sample is immersed in a liquid at an appropriate temperature, and its weight gain, dimensional changes, and mechanical properties are monitored over time. Any change in modulus, glass transition temperature, or melting point as a result of exposure can be determined by dynamic mechanical analysis (DMA) [46]. This technique is applicable to various specimen configurations including fibers, films, and bars. Figure 7 displays DMA data taken on small bars (1.1 cm × 0.5 cm × 0.17 cm) of polypivalolactone (PPL). The elastic component of the modulus E' and the ratio of the loss component to the elastic component $E''/E' \equiv \tan \delta$ are shown as functions of temperature. The glass transition temperature T_g is evidenced by the peak in tan δ and occurs at about 7°C in the as-molded material. The PPL bar, which was immersed in the organic solvent methyl ethyl ketone (MEK) for 2 weeks, picked up 6.7% of this solvent, resulting in a drop in T_g to −20°C. The stiffness E' of this material at room temperature fell by about half after solvent exposure. Such a large drop in matrix stiffness would be expected to significantly lower the matrix-dependent mechanical properties of a composite, such as its compression strength and shear strength, and all its off-axis properties.

Changes in mechanical strength can occur without a significant change in modulus, and therefore it is also important to determine the degradation in strength upon exposure. Useful procedures along these lines are described in ASTM D 543 (for plastics) and ASTM C 581 (for glass fiber reinforced thermosets). If a chemical reaction is suspected, chemical analysis or molecular weight determination may be warranted. Even if a liquid does not degrade the fiber or matrix when in an unstressed state, it may degrade them, or their interface, in the presence of stress. For example, even nonsolvents can cause organic matrices to crack when stressed.

This phenomenon of stress cracking is not well understood and is difficult to predict [47]. It is believed to be caused by localized solvent absorption around flawed areas of the polymer, which act as stress concentrations. This absorption lowers the local strength of the polymer below the local level of stress, and thus a crack is initiated adjacent to the flaw. The crack generates a new stress concentration, and thereby the crack rapidly propagates across the sample. There are several simple methods of evaluating the stress-crack resistance of polymers [48]. Through this mechanism, residual stresses (molded-in stresses) and service stresses in a composite can lead to matrix cracking in the presence of a liquid, which, in an unstressed state, may not seriously degrade the matrix.

The problem is exacerbated if the liquid swells the matrix. In this case, swelling stresses are generated by the planar diffusion front and by the anisotropic diffusivity in composites. The diffusivity of water in the fiber direction of graphite–epoxy was determined by Blikstad [49] to be 10 times that perpendicular to the fibers. Part of this difference is attributed to rapid diffusion along the fiber–matrix interface, perhaps due to residual stresses—these are greatest at the interface. Thus diffusion and swelling can create a variety of inter- and intraply stresses, which may promote stress cracking in the matrix.

The combined conditions of stress and exposure to liquids can degrade the fibers as well as the matrix. For example, the strength of glass fibers is known to be reduced in the presence of certain acids while under stress [50]. Because of the possibility of stress-induced degradation, it is recommended that matrices be screened for stress cracking and that composite specimens also be tested for solvent resistance under stress. The latter can be accomplished by clamping composite pieces in a fixture where the strain is adjustable and known, and then immersing the fixture in the liquid of interest.

Ingress of a liquid can degrade a composite by more complicated mechanisms. One example is the blistering that can occur if a composite contains voids and leachable materials. Polyester–glass laminates, for instance, can collect water in voids. Leachable materials then tend to concentrate in the water-filled voids. Because of the

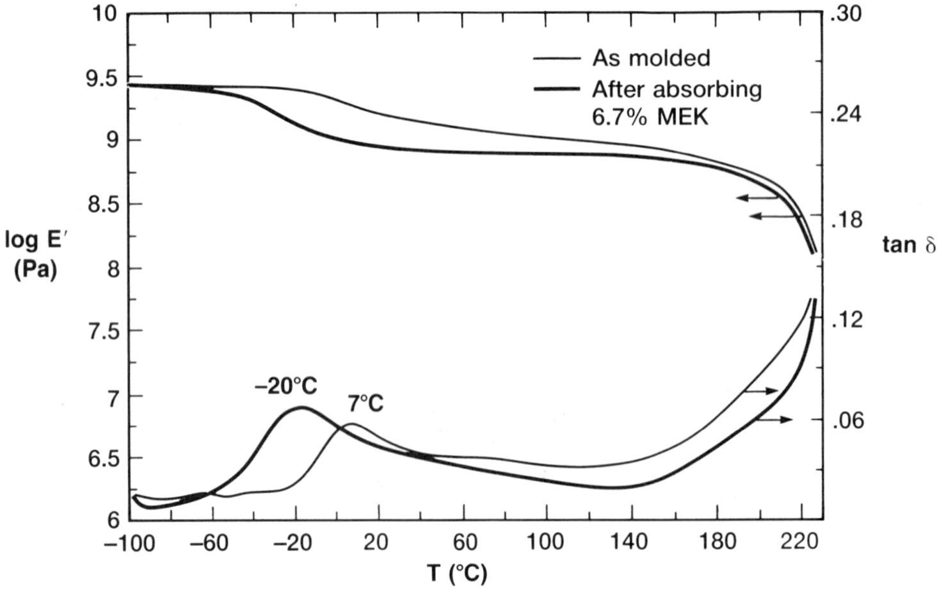

FIGURE 7 Dynamic mechanical analysis of polypivalolactone bars.

presence of these dissolved materials, osmosis occurs, with water continuing to diffuse through the composite into the voids, thereby raising the pressure in the voids well above atmospheric and generating blisters. Materials can also be leached from the fibers. The soluble oxides in glass fibers can be leached out by water, generating surface pits in the fibers and reducing their strength. These dissolved oxides can also lead to the development of osmotic pressure [4, p. 234]. Finally, the presence of a liquid may promote galvanic corrosion when both fiber and matrix are conductive; this occurs in carbon fiber reinforced aluminum when immersed in water [51].

$$\begin{aligned} &Al \rightarrow Al^{2+} + 2e^- \\ &\tfrac{1}{2}O_2 + H_2O + 2e^- \rightarrow 2OH^- \end{aligned} \qquad (1)$$

Electrons flow from the aluminum to the carbon, where they combine with water and dissolved oxygen to form hydroxyl anions. The aluminum matrix corrodes. Most of the unusual degradation processes described above can be detected by careful analysis of the morphology after environmental exposure.

The resistance to gases is characterized in a manner similar to that of liquids. The most common gases encountered are water and oxygen. Water vapor levels, of course, vary with season and location. For aircraft applications, it has been suggested that the worldwide worst environment that can be expected is 84% relative humidity [52]. *Accelerated testing* methods for composite coupons have been developed in which 200 flights are simulated per day of conditioning [53]. Thermal, load, and moisture cycling were done simultaneously. This particular study of a graphite–epoxy composite found no synergism between mechanical loading and hygrothermal degradation.

Empirical equations for predicting the effects of temperature and moisture uptake on polymer matrix properties have been proposed by Chamis [54]. The consequent effects on composite properties are then predicted through micromechanical formulas. This procedure assumes no degradation of the fiber or the interface. Even if these assumptions are valid, any such equations should be used with great caution: although useful in gauging the potential sensitivity of various composite properties to hygrothermal effects, they should not be used quantitatively for design purposes. Neither composite micromechanics nor the influence of hygrothermal exposure on polymer properties is sufficiently well understood at this time to provide reliable quantitative predictions.

Composites for space-based structures in low earth orbit are exposed to high levels of atomic oxygen, a reactive gas not found in significant levels at the earth's surface. At the proposed altitude for the space station, the atmosphere is 90% O, which because of the station's orbital velocity, would impact it at an energy of about 5 eV and at a flux of about 10^{15} atoms/(cm^2·s) [55]. Over its proposed 30-year lifetime, the station would receive a total flux of 10^{22}–10^{23} atoms/cm^2. The reactivity of various materials with oxygen atoms has been examined in experiments conducted on the space shuttle. As is to be expected, organic materials are generally quite reactive. Over the life of the space station, a graphite–epoxy part is predicted to lose 0.34 cm of thickness as a result of exposure to oxygen atoms [56]. Facilities to produce O beams of the required velocity are now being developed so that materials can be tested in the laboratory [55].

Space environments present other unusual conditions as well: ultraviolet radiation, high vacuum, electron radiation, and day/night thermal cycling. These are particularly hostile conditions for organic materials. Graphite–epoxy composites are candidate materials for the structural supports for antennae and reflectors to be placed into geosynchronous orbit for 20–30 years. However, the high level of electron radiation (about 10^{10} rads of 1 MeV electrons over 30 years) has been shown to chemically degrade epoxies, both in neat resin and composite forms, to the extent of decreasing T_g by about 60°C [57]. This degradation reduced the strength of composites at all temperatures, particularly the transverse (90°) tensile and compressive strengths of unidirectional laminates. Electron radiation has also been found to embrittle graphite–epoxy composites, causing microcracks to develop upon thermal cycling that altered the coefficient of thermal expansion to an unacceptable degree [58]. The cracks were detected using X-ray radiography by first soaking the sample in tetrabromoethane, which penetrates the cracks and is opaque to X-rays.

Thermal cycling alone can seriously degrade composites. Metal matrix materials, which see some of the largest amplitude thermal cycles in applications such as turbine blades, have exhibited a variety of effects. For example, chemical degradation can occur by diffusion of oxygen or an alloying element to the interface. Mismatch in the coefficients of thermal expansion of the reinforcement and the matrix can produce large stresses, which are concentrated at the interface. Thermal cycling can thereby weaken the interface or the fiber by fatigue. In addition, thermal stresses can lead to plastic flow along the interface. This phenomenon has been observed with W-Cu [59] and SiC-Al [60] composites, and manifests itself as changes in the dimensions of the specimen, with expansion of the composite occurring in the directions of fiber orientation.

The resistance of composites to heat has been discussed to some extent already. Retention of stiffness (modulus) with increasing temperature is readily determined by dynamic mechanical analysis (Fig. 7). The strength retention at elevated temperatures, and as a function of time at a fixed temperature, is evaluated by performing the mechanical test in one of a number of ovens especially designed for this purpose.

Resistance to the extreme temperatures of a fire can be characterized in many ways. To gauge whether a material helps promote an existing fire, there is an ASTM test (E 136) in which the sample is placed in a 750°C furnace through which air is flowing. Materials pass if they do not flame or heat up above the furnace air temperature. Other ASTM tests are available to evaluate the ignitability of materials (D 1929), the minimum oxygen concentration under which they will support flaming combustion (D 2863), and their extinguishing characteristics (D 3801), as well as numerous other fire-resistance-related properties.

A valuable test has been developed to determine the rate of heat and smoke release from materials (ASTM E 906). The specimen is placed in a chamber, and the surface that could be exposed to a fire in service is exposed to a radiant heat source. A constant flow of air passes through the chamber, but not directly over the sample. The rates of heat and smoke released into the passing air are recorded as a function of time. Ideally the test is repeated at several levels of incident radiation. The data have been input into a mathematical model of the progression of fires [61] as a tool for hazard assessment. Many composite materials are now being excluded from aircraft interiors on the basis of heat release rates that are above the 1986 Federal Aviation Administration standard.

The application in which composites see the highest temperatures is as ablative shields for spacecraft and missiles. These thermal protection devices usually function by several mechanisms. Like all materials, they absorb heat as their temperature is raised, because of their heat capacity, but they may also absorb heat by melting, vaporizing, chemically reacting, or fracturing. These shields can also provide thermal protection by radiating heat from their outside surface, and by thermally insulating the inside of the shield from the hot outer surface. The thermal protection mechanisms that are effective depend on the particular material and the ablative environment. For this reason, tests designed to gauge the capabilities of various materials must simulate the anticipated service conditions very closely.

The purpose of an ablative shield is to maintain the backface or inside temperature of a structure below a certain value. To do this, it ideally will consume a large amount of heat per unit mass ablated (the heat of ablation) and will have a low thermal diffusivity, $D_T \equiv k/\rho C_p$; that is, it will have a low thermal conductivity k, but a high specific heat C_p and density ρ. Methods for measuring the heat of ablation are described in ASTM E 458. The potential use of graphite–epoxy as both a structural material and an ablative shield, and a comparison of its capabilities relative to that of conventional ablative materials, can be found in an article by Sigur [62].

Other Properties

Electrical and Electromagnetic Properties

Composites are used in numerous applications in which their electrical and/or electromagnetic properties are important. They may serve as a structural support for an electronic device, as part of a device, or as an electromagnetic shield around a device. Methods by which some of the pertinent properties for these applications are measured are outlined below. The electrical and electromagnetic properties of composites, like their mechanical properties, are generally anisotropic due to nonrandom orientation of the reinforcement.

The most common electrical application of composites is as insulating structural members. For these applications a high electrical resistivity is, of course, critical. The volume resistivity ρ_v is the voltage drop V per unit thickness t for a given current i per unit area A:

$$\rho_v \equiv \frac{V/t}{i/A} \tag{2}$$

It is easily measured by various electrode configurations as described in ASTM D 257. In this test, the direct-current resistivity is measured after 60 seconds' exposure to 500 V; however, it should be recognized that ρ_v actually depends on time, voltage, humidity, and temperature. Electrons and ions may carry the current. The resistivity of insulating polymeric matrices is often very sensitive to ionic impurities. The amount of reinforcing material, its shape, and its orientation also influence the conductivity dramatically.

For some applications, such as circuit boards, the surface resistivity ρ_s may be as important as the volume resistivity:

$$\rho_s \equiv \frac{V/l}{i/w} \tag{3}$$

where l is the separation of the electrodes and w is their width. In such cases, the insulation resistance $R = V/I$ (where I is the total current), in the actual use configuration, may be the most relevant information. Procedures for determining R and ρ_s are outlined in ASTM D 257. As might be expected, ρ_s is strongly affected by any surface contamination.

The dielectric strength of a material sets its limitation as an insulator. At voltage gradients (volts per meter) greater than the dielectric strength, the insulator breaks down: its resistance drops abruptly and irreversibly. This strength should be measured at the frequency of interest. Measurement techniques are described in ASTM D 149. The strength generally decreases with increasing temperature and specimen thickness. Care must be taken that discharge in the surrounding medium does not precede discharge through the insulating material. Failure may be initiated by several mechanisms, including thermal breakdown, due to dielectric heating, and electromechanical breakdown, due to field-induced expansion of the dielectric material in the direction perpendicular to the applied field [63].

Dielectric breakdown may also occur along the surface of an insulator exposed to a large voltage gradient. Before breakdown, partially conductive paths are often formed on the surface near the electrodes; these paths are termed "tracks." Eventually the growth of these tracks leads to surface breakdown. Tracks are formed by localized surface arcing, which generates heat and subsequent thermal decomposition of the insulation. Generally, materials that decompose with little char tend not to form tracks, although they may erode. Surface contaminants such as dust and moisture will strongly affect a material's "tracking resistance." Characterization of a material's tracking properties can be achieved using ASTM methods D 2132, D 2303, and D 3638. Conductive tracks may also be formed by an arc discharge above the surface of the material. ASTM D 495 describes methods of measuring the arc resistance of materials.

The dielectric constant and dielectric loss of a composite material are critical properties for certain applications. The dielectric constant ε is the ratio of the capacitance (charge/voltage) of a capacitor (e.g., parallel-plate electrodes) in which the dielectric of interest is between the electrodes, compared with the capacitance with a vacuum between the electrodes. Measurement methods are described in ASTM D 150. The dielectric constant reflects the ability of the material to polarize in the presence of an alternating field. The polarization mechanisms depend on temperature and frequency. At material transition temperatures, new mechanisms of polarization may become operative, and therefore ε may change rather abruptly. Consequently, it is important to characterize ε over the range of frequencies, temperatures, and environmental conditions anticipated for a given application. The dielectric constant generally decreases with frequency at a given temperature and increases with temperature at a given frequency. For service as an insulator, ordinarily a low dielectric constant is desirable.

Dielectric loss occurs when the polarization of the dielectric does not keep in phase with the applied field. It results in an electrical power loss, which goes into heating the dielectric. The power loss is particularly objectionable at high frequencies since it is proportional to the frequency. The dielectric loss at a fixed frequency usually shows maxima at material transition temperatures. Outside transition regions, it can be generalized that materials of low dielectric constant tend to show lower losses. Procedures for measurement of the dielectric loss are described in ASTM D 150.

Detailed specifications, and techniques for determining compliance with these specifications, have been established for various grades of composite laminates for both electrical and nonelectrical applications (ASTM D 709). Specifications specifically for copper-clad laminates (circuit board precursors) have also been established (ASTM D 180).

Composite materials are sometimes chosen to house electrical devices because of their ability to contain electromagnetic waves emitted from the device, or, alternatively, to permit unimpeded passage of these waves out of and into the device housing. In the former application [electromagnetic interference (EMI) shielding], the composites selected consist usually of plastics filled or clad with a conductive material [48, p. 441]. Device housings from such materials are designed to reflect or absorb any electromagnetic waves generated by the device. The ASTM test method ES 7 has been developed for measuring shielding effectiveness.

For applications such as radomes (radar housings), it is desirable that the electromagnetic radar waves in the gigahertz range be minimally attenuated and distorted by the composite housing that protects the radar antenna from wind and the elements. Special equipment (e.g., waveguides) is required to determine the electrical properties at these high frequencies [64].

Thermal Properties

In designing parts from composite materials for application in environments where the temperature is variable, the primary concern of the designer may be the effect of temperature on mechanical properties. However, the designer may also need to know (a) how much heat energy is required to raise the temperature of the material to a certain level, (b) how fast it conducts heat, and (c) how much it expands when heated. To answer these questions the designer needs to know the following material properties, respectively: (a) specific heat, (b) thermal conductivity, and (c) coefficient of thermal expansion.

The specific heat C_p is the thermal energy required to raise the temperature of a unit mass of material one degree at constant pressure. Its units therefore are joules per gram per degree Celsius. Techniques for measuring specific heat are described in ASTM D 2766 and C 351. Specific heat is generally a rule-of-mixtures property; that is, it can be evaluated from the properties of the neat constituents (fiber and matrix) and their relative proportions in the composite. Since specific heats are expressed per unit mass, the composite specific heat is a weight fraction average of the constituent specific heats. The orientation of the phases relative to the direction of the heat flux is irrelevant.

The thermal conductivity, on the other hand, may be strongly affected by the orientation of the reinforcement with respect to the heat flux. This is not only because the conductivities of the fiber and matrix may be vastly different (e.g., glass fibers have 6 times the conductivity of typical epoxy resins), but also because many reinforcements have anisotropic conductivities (carbon fibers are 10 times more conductive axially than radially). The thermal conductivity k of a material is defined by Fourier's law for heat conduction:

$$q \equiv -k\frac{dT}{dx} \tag{4}$$

where q is the heat flux (J/(cm^2·s)), and dT/dx is the temperature gradient in the direction of the heat flux. Measurement of k involves generating a small temperature difference ΔT across the thickness of the material L and measuring the resulting steady-state heat flux. This flux must be purely conductive, it must be only in one direction, and it must be at steady state, so that $\Delta T/L \cong dT/dx$. A small temperature difference is used, primarily because k is a function of temperature. Testing details, and a discussion of the underlying assumptions in this measurement, can be found in ASTM C 177 and C 1045, respectively.

As with thermal conductivity, the coefficient of thermal expansion of a composite can be highly anisotropic. For this reason, the linear coefficients of thermal expansion α_L for a composite along its principal axes are more meaningful than the single coefficient of volume expansion α_V. The coefficient of linear thermal expansion is defined as the fractional change in specimen length per degree change in temperature (cm/(cm·°C)). Commercial instruments are available for measurement of α_L. These instruments follow ASTM techniques (E 831 and E 227) in using the movement of a push rod resting against the end of the sample to sense its thermal expansion. Composites can exhibit extremely low values of α_L, especially those with reinforcing fibers such as graphite and Kevlar, which actually contract axially when heated. Such composites are particularly useful in structures where it is critical to maintain constant dimensions despite temperature changes, such as supports for large reflectors and antennae in space. Measurement of extremely low values of α_L can be difficult. This difficulty can be overcome by testing longer samples over larger temperature changes, or by resorting to more elaborate optical techniques [58,65]. The thermal expansion coefficient, like k and C_p, is a function of temperature.

Wear Resistance

Addition of a rigid reinforcement to a softer matrix material is often effective in reducing its rate of wear in a given application. The extent of this beneficial effect depends on the loading of reinforcement and its orientation in the final part. The abrasion resistance of flat sheets of material can be characterized by ASTM D 1242. An abrasive grit, either bonded to a substrate or loose, is rubbed against the sample, and the loss in weight of the sample for a given number of abrasion cycles is determined. Results can be referenced to the weight loss from a zinc plate subjected to the same abrasive exposure.

The mar resistance of a composite can be characterized using ASTM D 673. In this test, silicon carbide grit is dropped onto the surface from a height of 63.5 cm and at an angle of 45°. The loss in surface gloss (ASTM D 523) is recorded for different amounts of abrasive dropped.

Acknowledgments

Charles W. Paul wishes to thank John E. Macur, Thomas J. Taylor, and Robert S. Kingsland for their reviews of the MORPHOLOGY, ENVIRONMENTAL RESISTANCE, and OTHER PROPERTIES sections, respectively.

Charles W. Paul and Igor Palley

See also CHEMICAL ANALYSIS AND CHARACTERIZATION OF POLYMER COMPOSITE MATERIALS; SURFACE ANALYSIS; TESTING

References

1. E. M. Chamot and C. W. Mason, *Handbook of Chemical Microscopy,* 3rd ed., vol. 1, Wiley, New York, 1958, p. 37.
2. A. C. Reimschuessel, J. E. Macur, and J. Marti, in J. P. Sibilia, Ed., *A Guide to Materials Characterization and Chemical Analysis,* VCH Publishers, New York, 1988, p. 160.
3. R. E. Evans, J. E. Masters, and J. L. Courter, in *Advanced Composites: The Latest Developments,* ASM International, Metals Park, OH, 1986, p. 249.
4. D. Hull, *An Introduction to Composite Materials,* Cambridge University Press, New York, 1981, p. 89.
5. J. H. Williams and S. S. Lee, *Mater. Eval., 43,* 561 (1985).
6. J. A. Nairn and P. Zoller, *J. Mater. Sci., 20,* 355 (1985).
7. D. Purslow, *Composites, 17*(4), 289 (1986).
8. L. H. Tung and J. C. Moore, in L. H. Tung, Ed., *Fractionation of Synthetic Polymers,* Dekker, New York, 1977, p. 545.
9. G. R. Hatfield and G. E. Maciel, *Macromolecules, 20*(3), 608 (1987).
10. D. Kranbuehl, S. Delos, M. Hoff, and L. Weller, in *Advanced Composites: The Latest Developments,* ASM International, Metals Park, OH, p. 169.
11. D. J. Blundell, J. M. Chalmers, M. W. Mackenzie, and W. F. Gaskin, *SAMPE Q., 16*(4), 22 (1985).
12. T. Chung, "Liquid Crystal Polyester-Carbon Fiber Composites," NASA-CR-172323, May 1984.
13. K. Waltersson, *Compos. Sci. Technol., 23,* 303 (1985).
14. G. E. Hammer and L. T. Drzal, *Appl. Surf. Sci., 4,* 340 (1980).
15. A. T. DiBenedetto and D. A. Scola, *J. Colloid Interface Sci., 64,* 480 (1978).
16. C. F. Quate, *Phys. Today, 39*(8), 26 (1986).
17. F. Delannay, L. Froyen, and A. Deruyttere, *J. Mater. Sci., 22,* 1 (1987).
18. H. W. Chang, R. P. Smith, S. K. Li, and A. W. Neumann, in H. Ishida and G. Kumar, Eds., *Molecular Characterization of Composite Interfaces,* Plenum Press, New York, 1985, p. 413.
19. W. Prager, *Introduction to Mechanics of Continua,* Ginn, New York, 1961.
20. S. Timoshenko and J. N. Goodier, *Theory of Elasticity,* 3rd ed., McGraw-Hill, New York, 1970.
21. G. S. Sihz and A. M. Scudra, Eds., *Failure Mechanics of Composites,* Elsevier, Amsterdam, 1985.
22. J. E. Ashton and J. M. Whitney, *Theory of Laminated Plates,* Technomic, Lancaster, PA, 1970.
23. B. R. Lawn and T. R. Wilshaw, *Fracture of Brittle Solids,* Cambridge Solid State Science Series, Cambridge University Press, New York, 1975.
24. R. W. Hertzberg, *Deformation and Fracture Mechanics of Engineering Materials,* 2nd ed., Wiley, New York, 1983.
25. J. Rice, in H. Liebowitz, Ed., *Fracture,* Vol. 2, Academic Press, New York, 1968, Ch. 3.
26. A. Chudnovsky, *Crack Layer Theory,* NASA-CR-174634, March 1984.
27. A. Chudnovsky, I. Palley, and E. Baer, *J. Mater. Sci., 16,* 35 (1981).
28. A. Chudnovsky, A. Moet, I. Palley, and E. Baer, in A. Hiltner, Ed., *Structure Properties Relationship in Polymer Solids,* Plenum Press, New York, 1983.
29. A. M. Freudental, in H. Liebowitz, Ed., *Fracture,* Vol. 2, Academic Press, New York, 1968.
30. A. Chudnovsky, *Int. J. Eng. Sci., 22,* 989 (1984).
31. Yu. M. Tarnopolski and T. Kincis, 1985, p. 215 in ref. 21.
32. I. M. Daniel, 1985, p. 277 in ref. 21.
33. *Annual Book of ASTM Standards,* American Society for Testing and Materials, Philadelphia, 1982.
34. J. D. Ferry, *Viscoelastic Properties of Polymers,* Wiley, New York, 1970.

35. L. M. Kachanov, *Theory of Creep*, Nauka, Moscow, 1973.
36. I. Palley, *Izvest. Akad. Nauk SSSR, Mech. Solids, 6,* 99 (1968).
37. I. Palley, *Eng. J. Mech. Solids* (*USSR*), *1,* 130 (1968).
38. L. M. Kachanov, *Foundations of the Theory of Plasticity,* North-Holland, Amsterdam, 1971.
39. R. E. Rowlands, 1985, p. 71 in ref. 21.
40. C. C. Chamis, *SAMPE Q., 15*(3), 14 (1984).
41. G. S. Holister and C. Thomas, *Fiber Reinforced Materials,* Elsevier, New York, 1966.
42. M. A. Macaulay, *Introduction to Impact Engineering,* Chapman & Hall, New York, 1987.
43. S. L. Kessler, G. C. Adams, S. B. Driscoll, and D. R. Ireland, Eds., *Instrumented Impact Testing of Plastics and Composite Materials,* ASTM STP 936, American Society for Testing and Materials, Philadelphia, 1986.
44. R. W. Hertzberg and J. A. Manson, *Fatigue of Engineering Plastics,* Academic Press, New York, 1980.
45. J. H. Underwood, S. W. Freiman, and F. I. Baratta, *Chevron-Notched Specimens: Testing and Stress Analysis,* ASTM STP 855, American Society for Testing and Materials, Philadelphia, 1984.
46. C. W. Paul, *J. Appl. Polym. Sci.,* 36 (3), 657 (1988).
47. Y. W. Mai, *J. Mater. Sci., 21,* 904 (1986).
48. *Modern Plastics Encyclopedia, 63* (10A), 419 (1986).
49. M. Blikstad, *J. Reinf. Plast. Comp., 5,* 9 (1986).
50. A. Akhtar and J. Y. Wong, *J. Compos. Technol. Res., 9*(3), 95 (1987).
51. M. G. Vassilaros, D. A. Davis, G. L. Steckel, J. P. Gudas, and D. W. Taylor, in J. E. Hack and M. F. Amateau, Eds., *Mechanical Behavior of Metal-Matrix Composites,* Metallurgical Society of AIME, New York, 1983, p. 335.
52. T. A. Collings, *Composites, 17*(1), 33 (1986).
53. E. Demuts and P. Shyprykevich, *Composites, 15*(1), 25 (1984).
54. C. C. Chamis, *39th Annual Conference, Reinforced Plastics/Composite Institute,* SPI, January 1984, Session 11-D, p. 1.
55. C. L. Johnson, R. G. Albridge, A. V. Barnes, R. K. Cole, D. J. Dean, R. F. Haglund, Jr., H. Pois, P. M. Savundararaj, N. H. Tolk, J. Ye, and A. F. Daech, *SAMPE Q., 18*(2), 35 (1987).
56. L. Leger, J. Visentine, and B. Santos-Mason, *SAMPE Q., 18*(2), 48 (1987).
57. C. T. Herakovich, D. J. Fox, and G. F. Sykes, *J. Eng. Mater. Technol., 110,* 146 (1988).
58. G. F. Sykes and D. E. Bowles, *SAMPE Q., 17*(4), 39 (1986).
59. S. Yoda, N. Kurihara, K. Wakashima, and S. Umehawa, *Metal. Trans., 9A,* 1229 (1978).
60. W. G. Patterson and M. Taya, *Proceedings ICCM-V,* Metallurgical Society of the American Institute of Mining, Metallurgical, and Petroleum Engineers, New York, 1985, p. 53.
61. E. E. Smith, *Fire Saf. J., 9,* 47 (1987).
62. W. A. Sigur, *SAMPE Q., 17*(2), 25 (1986).
63. M. E. Baird, *Electrical Properties of Polymeric Materials,* Plastics Institute, London, 1973, p. 35.
64. A. Ossin, E. A. Welsh, R. A. Mayor, and W. A. Seegers, "Millimeter Wavelength Radomes," AFML-TR-79-4076, July 1979.
65. S. F. Jacobs, J. N. Bradford, and J. W. Berthold III, *Appl. Opt., 9*(11), 2477 (1970).

Chemical Analysis and Characterization of Polymer Composite Materials

> Since the measuring device has been constructed by the observer . . . we have to remember that what we observe is not nature in itself, but nature exposed to our method of questioning.
>
> W. K. Heisenberg
> *Physics and Philosophy,* 1958

This article introduces and reviews contemporary methods for analyzing and characterizing the behavior of advanced polymer-based composite materials. Our objective is to provide a reasonably broad, yet sufficiently detailed perspective to enable the nonspecialist to ask the correct, or at least, the most pertinent questions of the analytical chemist. In few other fields of endeavor does the quality of the question or the level of the questioner's understanding affect the outcome as much as it does in the discipline of polymer characterization.

Virtually every chemical, physical, and mechanical property of a finished polymer matrix, advanced composite material depends strongly on the composition and structure of the starting or intermediate materials. Processibility, as well, of the intermediates to parts is critically dependent on certain physical and chemical attributes. For traditional structural materials such as iron and aluminum alloys, the structural and compositional features remain essentially constant through shipping and fabrication. This is not the case for a large number of matrix resins, molding compounds, adhesives, and prepregs. Such intermediate materials are derived largely from such thermoset resins as phenolics, polyesters, epoxies, bismaleimides, polyimides, and cross-linkable thermoplastics. These materials are manufactured intentionally as chemically reactive products whose structures and compositions change with time, temperature, and exposure to chemicals in the ambient environment (e.g., water). The changes continue until the intermediates are consolidated and polymerized (cured) to attain their final properties in composite form.

Thus, it should not be surprising that processibility into parts and attainment of requisite properties in the parts are critically dependent on the chemical state of the resin or the prepreg at the time of cure. The chemical state, in turn, depends on an enormous number of factors. These can be distilled simplisticly to three for a given material: the way the chemical formulation was assembled, the degree of advancement (partial cure) imparted during prepreg manufacture, and the degree of additional advancement developed in the prepreg prior to its use by the fabricator. Much the same can be said about thermoplastic-based materials, even though they are not chemically reactive at the time of fabrication. The processing characteristics and performance of advanced thermoplastics depend on the manufacturer's control of such properties as molecular weight and molecular weight distribution, content of solvents, plasticizers and impurities, and content of nucleating agents.

Chemical fingerprint analysis and in-depth characterization of polymer composite materials have assumed major importance in the development, specification, acquisition, and use of matrix resins and prepregs for advanced composites. There has been a remarkable proliferation of new materials during the past 20 years. Moreover, specification and use of these materials to replace metals as well as for initial designs have increased significantly. With these advances have emerged more stringent requirements for assuring the quality, consistency, and reliability of materials, as well as for more rationally developed part processing methods.

The increased and widespread use of sophisticated analytical instrumentation has revealed that in addition to the expected changes in materials with aging, there can be unacceptably large chemical differences from one batch of formulated resin to the next and from one lot of finished prepreg to the next. Such variability can create a host of problems ranging from prepreg tack level differences in the layup shop, which reduce manufacturing efficiency, to failure of a fabricated part to meet mechanical performance standards. The traditional measurements of gel time, resin flow, volatiles, resin content, etc. are no longer adequate, by themselves, for controlling product quality and for quality acceptance. These methods are still useful, but they are inexact and not sufficiently discriminatory for many materials.

The advent of relatively inexpensive computers and microprocessor-controlled analytical instrumentation has given rise to unparalleled ease in characterizing all forms of polymer composite materials. Complex analytical tasks now can be completed in minutes to several hours, often in the hands of staff technicians. Ten to fifteen years ago many of these analyses would have required days if they could have been done at all, even by chemical professionals holding advanced degrees.

State-of-the-art instrumentation and associated methods are yielding significant benefits. These include improved product quality and consistency, superior understanding of product chemistry and its relationship to fabrication processes, and improved product development for satisfying specific performance objectives. Use of these instruments poses risks as well. Owing to their great speed in processing data, coupled with the existence of user-friendly programmed analytical routines, modern analytical instruments tend to invite occasional mindless button pushing and poorly designed experiments by minimally trained personnel. Given the power these instruments possess, they do not interpret the data in relation to the need, the problem, or the question. The greater sophistication of chemical instrumentation increases, rather than decreases, the demand on the analyst for chemical and engineering knowledge and mathematical facility.

This article is limited to analysis and characterization* of completely formulated fiber-reinforced polymeric materials and the resulting composites. These materials, whether thermosetting or thermoplastic, are the direct precursors of advanced composites in final or fabricated form. By "advanced" is meant resin–fiber material combinations whose structural properties are comparable to or exceed those of traditional metallic materials, regardless of the application. Thus, the principal intermediates that are discussed to exemplify the various methods are materials in formulated resin and prepreg forms. Physical and chemical analysis of polymer raw materials and of fibers are covered elsewhere in this encyclopedia.

With respect to methods and instrumentation, this article discusses those that have demonstrated the greatest usefulness and acceptance for characterizing polymer composite materials. These methods fall into two categories: instrumental methods and the standard prepreg and solid laminate methods. The former include chromatography, spectroscopy, and thermal analysis (including dynamic mechanical analysis). The latter include such tests as resin content, gel time, and flow. Techniques and instruments for characterizing surfaces and microstructures (electron microscopy, photoelectron spectroscopy, X-ray diffraction, etc.) and for measuring optical, electrical, and mechanical properties are not discussed. Each of the chemical characterization methods is based, as much as possible, on studies in the open literature that provide explicit experimental details and interpretation of the results. The applications cited can be grouped into four categories of objectives or motivation for performing the analyses. These objectives are listed in Table 1 in increasing order of complexity and dimension.

As recently as 10 years ago, the vast majority of technical articles involving chemical characterization of resin systems, prepregs, and cured composites dealt with epoxy-based systems. Studies of epoxy-based materials still dominate the literature; however, a large body of information has also developed for other thermosetting systems including bismaleimides and polyimides and for a variety of amorphous and semicrystalline thermoplastics. This article strives to present balanced examples involving analysis and characterization of state-of-the-art materials.

No single test or single instrumental method is capable of providing all the typical kinds of characterization data desired for polymer composite materials. Only in the case of chemical fingerprint analysis for quality control and quality acceptance purposes does one method provide the required information. Yet, even for material-receiving purposes, many material specifications call for

* "Analysis" is used here to mean identification of a material (qualitative) and determination of its actual or relative quantity (quantitative). By "characterization" is meant description and measurement of aspects of behavior that are unique to or typical of a material, especially when molecular structures and/or compositions are not known.

TABLE 1
Objectives of Chemical Analysis and Characterization of Polymer Composite Materials

Quality Control/Quality Acceptance

Obtain chemical fingerprint to demonstrate product identity and/or lot-to-lot uniformity with respect to an identified or coded standard.

Aging Studies

Determine storage stability and potlife.
Correlate details of time, temperature, and chemical exposure with structure and composition.
Predict and explain property and processing variations.

Processing

Use selected chemical and physical properties or property changes to develop and optimize cure cycles and other fabrication processes.

Kinetics and Mechanisms

Correlate structure, processing, and properties.
Determine chemical reaction rates and pathways to predict reactivity, structural results, and better synthesis routes.
Prove structures, conformations, and compositions.

more than one kind of fingerprint and thus require variety in instrumentation. Well-designed, comprehensive studies of aging behavior, process development, and fundamental chemistry employ many methods and a systems approach to the partial or complete characterization. One such approach is outlined in Figure 1 with respect to the various analytical methods that are the subject of this article (Fig. 1*a*) and with respect to many of the kinds of analysis the methods yield (Fig. 1*b*). These diagrams are not intended to serve as methodological flowcharts, although they can. Rather, they illustrate relations between the methods and their typical applications for intermediate materials in as-received and in as-extracted states, and for finished composites. These outlines also show how one experiment can generate samples for subsequent analysis by other methods. For example, a mixture of gases generated during thermal analysis can be separated into its individual components by gas chromatography. These separate gases can then be identified by a spectroscopic method.

This article is concerned, almost entirely, with instrumental methods for analysis and characterization. These methods are discussed first, in terms of the basic principles involved, instrumentation and methods, and applications. Then the standard prepreg and laminate tests are presented briefly, chiefly as ASTM methodology.

Instrumental Methods

We begin by describing the three broad classes of instrumental methods shown in Figure 1*a*. The operating principles and constructions of the instruments for performing the various spectroscopic, chromatographic, and thermal analyses are quite different from one another. Yet, these instruments share common features with regard to their respective outputs. The output from any of these analytical instruments is simply an electronic signal from a detector or transducer. After appropriate electronic amplification, the signal is recorded on a chart recorder, or saved on a floppy or hard disc in the data acquisition system for subsequent plotting on a CRT or digital plotter. The recording (i.e., the results) generally takes the appearance of one or more positive or negative departures from an initial baseline. For most methods, these departures appear as collections of peaks and valleys, while for others, they appear as changes in the baseline that can be correlated to increases and decreases in the measured property. Chromatography, infrared spectroscopy, and differential scanning calorimetry are examples of the former; thermogravimetric analysis and thermomechanical analysis are examples of the latter. Peaks have reproducible heights, shapes, and patterns. They appear at times, temperatures, or frequencies that are characteristic of a particular material when the material is examined using a specific set of sample-related and instrument-related experimental conditions. While sampling methods and experimental conditions go beyond the scope of this article, it is absolutely vital to appreciate that they dictate the success or failure of any analysis. It is simply not possible to overestimate their importance.

In general, commercial materials show a complex response to the analytical testing done to characterize them. However, when analyzed using the same conditions, such materials always yield the same pattern. This pattern has a variety of names: a chromatogram for chromatography, a spectrum for spectroscopy, a thermogram for some forms of thermal analysis. Another name, recently made popular, is the chemical* fingerprint.

The chemical fingerprint can be considered to be the first level of chemical characterization for such applications as quality control, quality assurance, cure monitoring, and reaction mechanism surveys. The fingerprint, by itself, says nothing about the quantity of components present in the sample, nor about their identity, although an expert analyst often will be able to infer the possibilities. Nevertheless, the exact matching of a product chemical fingerprint with that for a standard is usually accepted as proof of chemical equivalence.

It is important to stress that instrumental analysis methods perform separations and detect energy excitations of molecules and of complex substances. In the absence of other information, these methods are not capable of identifying anything other than some change that occurred at some temperature, frequency, or time. Identification and quantitation are subsequent steps, performed using correlation charts, calibration stan-

* It is inaccurate to describe the pattern as chemical in origin, especially in thermal and rheological analyses. In most instances, the fingerprint reflects physical processes operating as a result of the chemical nature of the material. However, just as "chromatography" is a misnomer, "chemical fingerprint," a name in popular use, is graphic, hence useful.

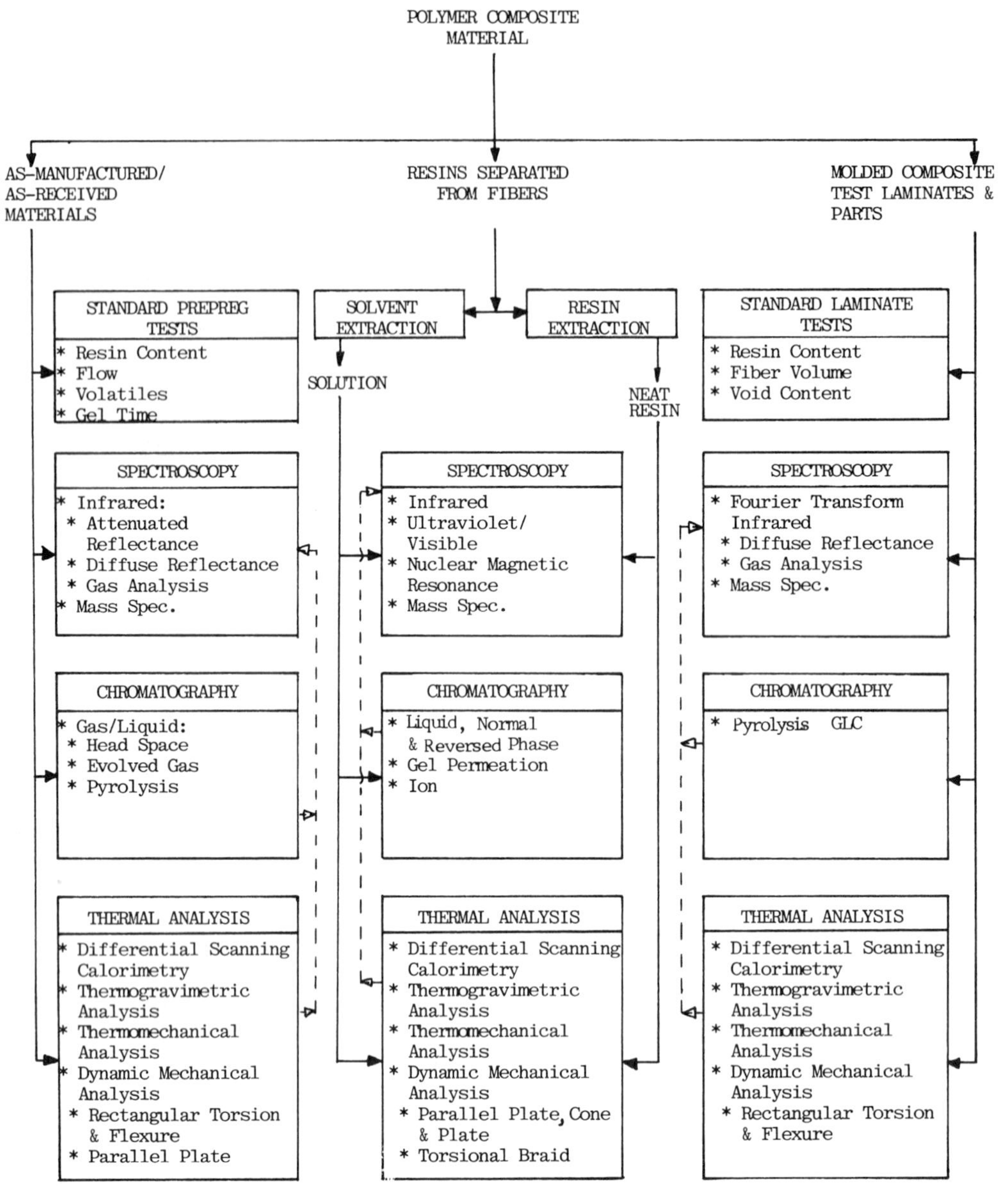

FIGURE 1a **Chemical analysis and characterization methods for polymer composite materials.**

dards, and comparison with the behaviors of previously characterized or reference materials.

Spectroscopy

Spectroscopic methods comprise the most powerful tools in the characterization arsenal. When applied at the most routine level, these methods can provide detailed chemical fingerprints and nearly unequivocal identifications of materials. At the other extreme, spectroscopic methods are used to determine the details of atomic and molecular architecture for elucidating structure, conformation, configuration, orientation, degree of branching, etc.

This discussion concentrates on infrared (IR) spectroscopy and nuclear magnetic resonance (NMR) spectroscopy, as these methods have the greatest utility for characterizing resins, prepregs, and formed composites. Methods involving ultraviolet (UV) and mass spectroscopy (MS) are discussed briefly. Their application to advanced composites thus far has been limited to detailed analyses of individual components, additives, metallic impurities, and molecular fragments produced by thermal or ionizing radiation.

GENERAL PRINCIPLES. Spectroscopic methods are based on some measurement of a reversible change in matter after exposure to electromagnetic radiation at a particular wavelength or range of wavelengths. Two events can occur when molecules are exposed to a source of energy. The first is absorption of the energy, which is accompanied by electronic, vibrational, or nu-

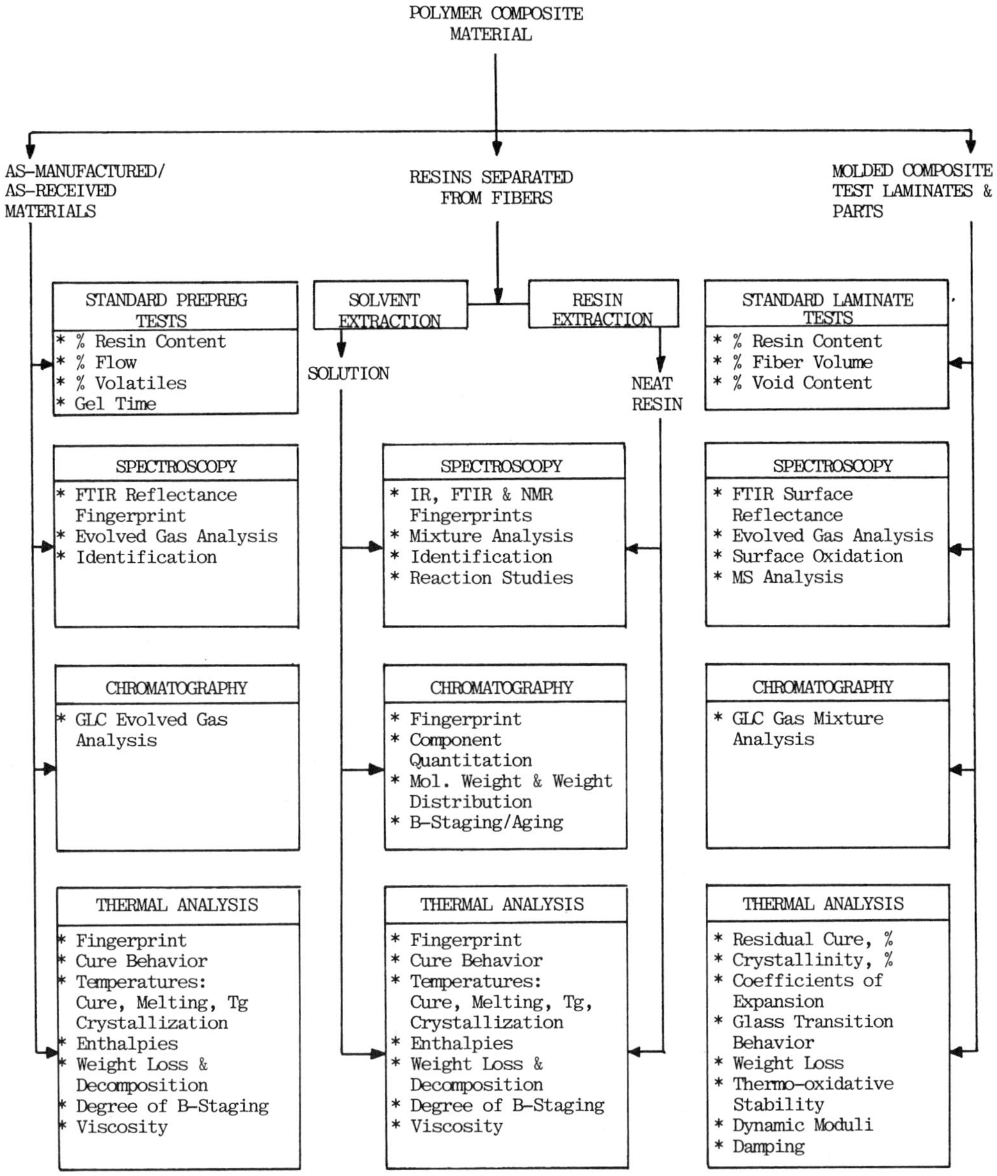

FIGURE 1b **Results obtained from chemical analysis and characterization methods.**

clear spin excitation, depending on the energy of the incident radiation. The second is the reverse process called emission, where the extra energy is dissipated, generally as heat or visible light. Absorption is measured as the difference between the magnitudes of the incident and transmitted radiation. Emission is measured directly. The two processes differ sufficiently in terms of production and measurement to permit their classification as absorption spectroscopy and emission spectroscopy. The former is the more useful in characterizing and studying microstructure.

An excellent way to understand absorption of electromagnetic radiation by molecules is to use an analogy for radio reception [1]. A simple radio is an oscillatory circuit with antenna, capacitor, and coil. When the antenna is exposed to a source of radiofrequency radiation (the signal), the radio receives or absorbs only the frequency that corresponds to the resonance frequency of the oscillatory circuit. All other frequencies from other transmitters will either pass through the radio, having no effect, or be reflected from it.

The different kinds of instrument for absorption spectroscopy are designed to measure the effects produced by molecular excitation processes of different kinds. However, while the mechanisms are different for the UV, IR, and NMR regions of the electromagnetic spectrum, whether molecules will or will not absorb energy is based on a fundamental proposition, namely: matter (as the radio) absorbs radiation selectively with respect to wavelength. Only wavelengths of incident radiation that correspond to the wavelengths required for exciting electronic, vibrational, or nuclear spin processes will be

TABLE 2
Characteristic Parameters and Excitations in Spectroscopy

Type of Spectroscopy	Wavelength, λ (nm)	Frequency, ν (Hz)	Energy, E (kj/mol)	Type of Excitation
UV–visible	$2 \times 10^2 - 8 \times 10^2$	1.5×10^{15}–0.4×10^{15}	63–628	Electron energy level
IR	$2 \times 10^2 - 1.6 \times 10^4$	1.5×10^{14}–2.0×10^{13}	1.3–126	Amplitude of chemical bond vibration
NMR	$10^8 - 10^9$	3.0×10^7–3.0×10^6	2.1×10^{-5}–2.1×10^{-6}	Change in orientation of nuclear spin

absorbed. For a particular kind of excitation, absorption of only one wavelength means that only one discrete or quantized amount of energy will be absorbed. Table 2 gives the characteristic ranges for wavelength, frequency, and the excitation process involved in UV, IR, and NMR spectroscopy. Because of the dependence of energy of electromagnetic radiation to its wavelength, as stated in Eq. (1), the levels of energy involved in the three absorption processes are different and therefore cause different types of excitation in matter:

$$E = h\nu = hc\bar{\nu} = \frac{hc}{\lambda} \qquad (1)$$

where

E = energy
λ = wavelength
h = Planck's constant, 6.62×10^{-34} J · s
c = the speed of light in a vacuum, 2.998×10^8 m/s^{-1}
ν = frequency
$\bar{\nu}$ = wavenumber

Infrared Spectroscopy

THEORY. The basis of IR spectroscopy is the increase in amplitude of natural stretching and bending vibrations of chemical bonds that are exposed to appropriate wavelengths. The atoms comprising molecules are in continuous motion. The atoms can be visualized as balls of differing masses connected to each other by springs having different stiffnesses to represent chemical bonds. The stiffness of a spring is given by its force constant, which is proportional to the force of attraction between the atoms. The naturally occurring or natural vibration of the spring (chemical bond) has a vibrational frequency that is proportional to both the force constant and the masses of the atoms it connects. Energy absorption in the infrared region will occur provided two conditions are satisfied:

1. The frequency of the incident radiation must be identical to that of the natural vibrational frequency of the chemical bond.
2. There must be a net change in dipole moment (distribution of electronic charge) to accompany bond stretching or bending.

When both conditions are met, the absorbed energy produces an increase in the amplitude of vibration. Symmetrical molecules such as O_2 and N_2 and symmetrical stretching of bonds between identical atoms such as the C—C group produce no absorptions in the IR range. The excitation is short-lived, and when the bond returns to its ground vibrational state, the absorbed energy is reradiated or emitted as heat.

The stronger the chemical bond between two atoms, the lower the absorption wavelength (the higher the energy) required for increasing its amplitude. Thus, for carbon–carbon bonds the triple bond absorbs between 2273 and 2000 cm^{-1}, the double bond between 1887 and 1493 cm^{-1}, and the single bond between 1299 and 800 cm^{-1}. The particular absorption wavelength depends on the atoms involved in the bond and other atoms in their immediate neighborhood. In IR spectroscopy it is of paramount importance that different functional groups of atoms absorb IR wavelengths that are characteristic, therefore, analytically diagnostic. For example, the O—H group for non-hydrogen-bonded alcohols and phenols absorbs at 3650–3600 cm^{-1}; the N—H group in primary and secondary amines absorbs at 3500–3100 cm^{-1} for stretching and at lower energy (1640–1550 cm^{-1}) for the easier bending vibrations. Stretching the C═O bond in anhydrides gives two characteristic absorptions, one at 1810 cm^{-1} and one at 1760 cm^{-1}. The C═O bond stretch for ketones occurs in the range 1725–1705 cm^{-1}, while in amides the absorption occurs at 1670–1640 cm^{-1}. Extensive charts and tables have been developed to correlate characteristic absorption frequencies with all the functional groups. One such chart is reproduced in Figure 2. Further detailed information can be found in texts and reference works on the subject [2–6].

The useful portion of the IR spectrum for direct analysis based on functional groups in organic materials is called the functional group region; it lies approximately in the 4000–1300 cm^{-1} range. The next region, 1300–900 cm^{-1}, is called the fingerprint region. Absorptions in this range originate chiefly from complex bending vibrations that are very difficult to assign to specific functional groups. Nevertheless, the patterns are quite distinctive and diagnostic for different molecules. The frequency range 900–650 cm^{-1}, called the aromatic region, contains out-of-plane bending absorptions for groups attached to benzene and other aromatic groups. This region is especially useful for determining the number, kind, and locations of substituents such as methyl or halogen groups on aromatic rings.

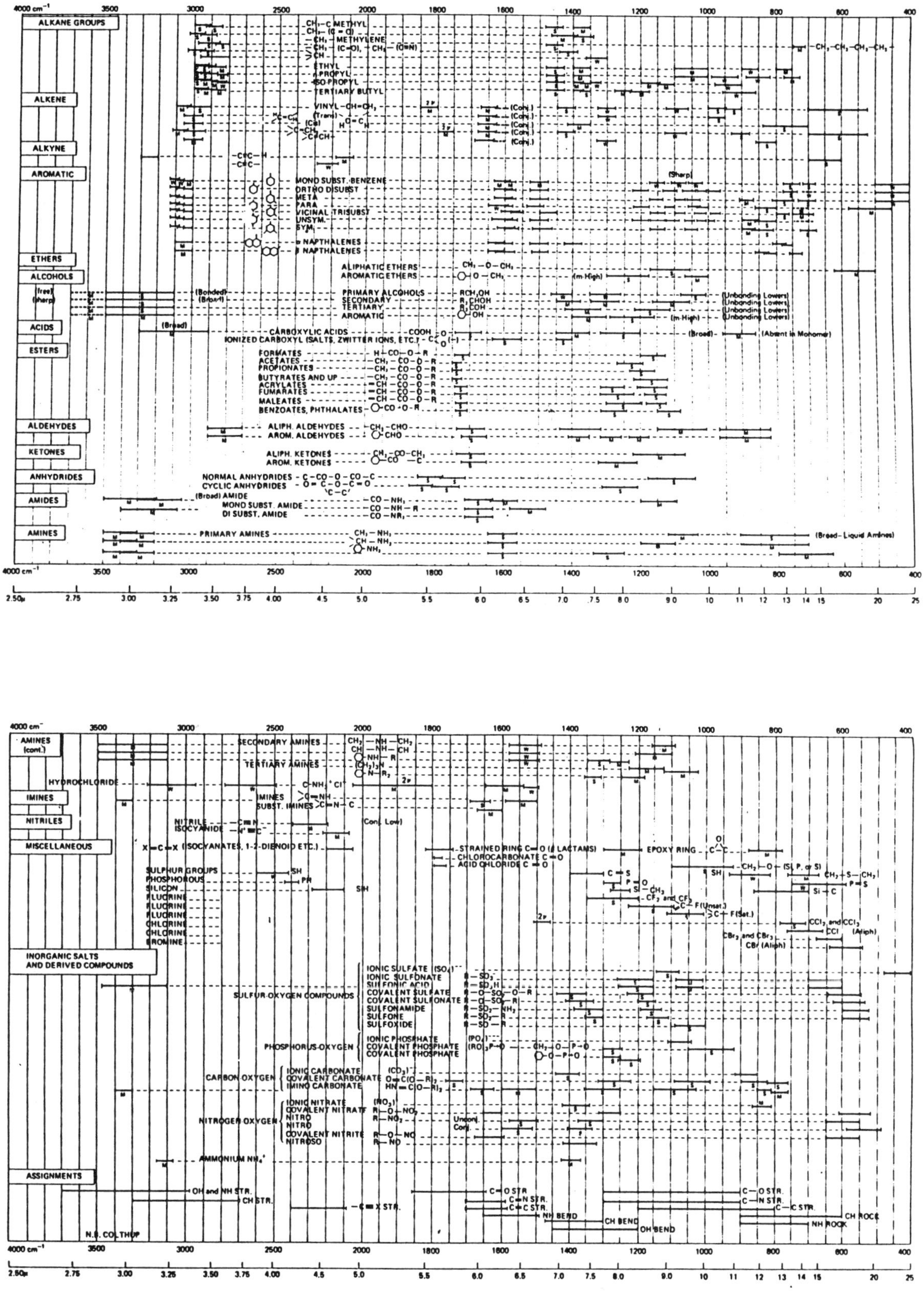

FIGURE 2 Characteristic infrared group frequencies. (N. B. Colthup, J. Optical Soc. Am., *40*, 397 (1950). Reprinted with the permission of American Cyanamid Company and the Optical Society of America. All Rights Reserved.)

An IR spectrum (Fig. 3) is obtained by measuring the amount of radiation transmitted by the sample in small frequency increments covering the frequency range of interest. The intensity of an absorption is proportional to the number of absorbing groups in the path of the radiation. The ordinate of the spectrum can be presented either as the percentage transmittance, which is the arithmetic ratio of transmitted energy I to the incident energy I_0, or

$$T = I/I_0 \tag{2}$$

or, alternatively as absorbance A, which is expressed as follows:

$$A = \log(1/T) = \log(I_0/I) \tag{3}$$

INSTRUMENTATION AND MEASUREMENT. There are now two kinds of spectrophotometer for acquiring infrared spectra. One is called "dispersive" because the instrument employs a prism or diffraction grating to disperse the wavelengths through a narrow slit to the detector. The other is called "Fourier transform IR" (FTIR) because the spectrum is generated by computer coding its optical transform produced using an interferometer and subsequently decoded using a fast Fourier transform. Practical FTIR spectroscopy is quite recent; it owes its development and current popularity to the availability of low cost computers. Details of the principles and mathematics of Fourier transforms are not relevant to this discussion. The interested reader is referred to Ref. 6.

Dispersive IR and FTIR spectrophotometer systems are illustrated schematically in Figure 4. The dispersive spectrometer is essentially a source of IR radiation plus an optical system of mirrors, light choppers, and filters for focusing and directing the IR beam through the sample; a rotating grating or prism system for separating or dispersing the wavelengths; a slit for minimizing the number of included wavelengths (e.g., 4 cm^{-1} resolution); and a detector. Typical radiation sources are molded ceramic rods that are heated resistively to glow at 1300–1500°C. The optomechanical system takes many forms, depending on instrument manufacturer. Double-beam systems are designed to compare the intensities of beams through the sample and a reference and to ratio the I and I_0 of each using an electronically controlled attenuator null balance circuitry.

The advent of small microprocessors has allowed simplification of rather complex optomechanical systems to the point where only a single beam is necessary. A reference spectrum can be stored and subtracted later from the sample spectrum. Detectors are generally of two types, thermal and photon. The thermal detectors are the thermistor, thermopile, and fluid expansion cell (Golay cell). Photon detectors are more sensitive than thermal detectors and are based on the movement of electrons and holes in semiconductor chips.

The FTIR spectrophotometer consists of a He-Ne laser, an infrared source, an interferometer, and a detector. The interferometer, which consists of a beam splitter, a stationary mirror, and a moving mirror, in conjunction with a modulated laser pulse, converts the polychromatic IR beam into a complex mixture of modulated frequencies. The sinusoidally modulated beam passes through the sample, but it is the radiation not absorbed by the sample that is focused onto the detector. The

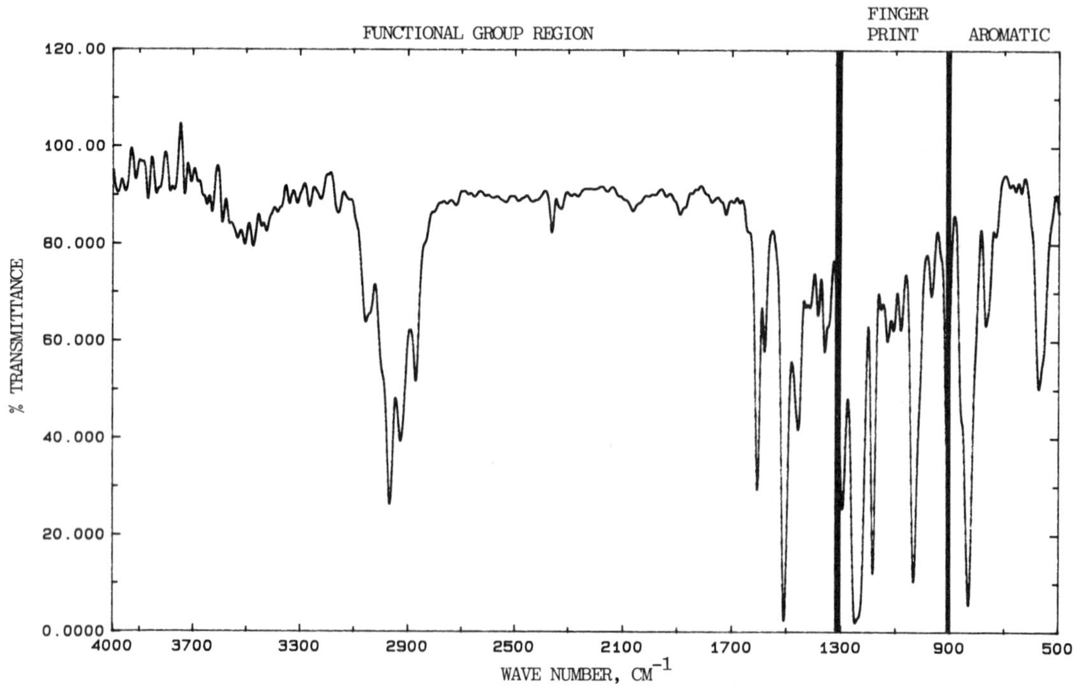

FIGURE 3 Infrared spectrum showing functional group, fingerprint, and aromatic regions.

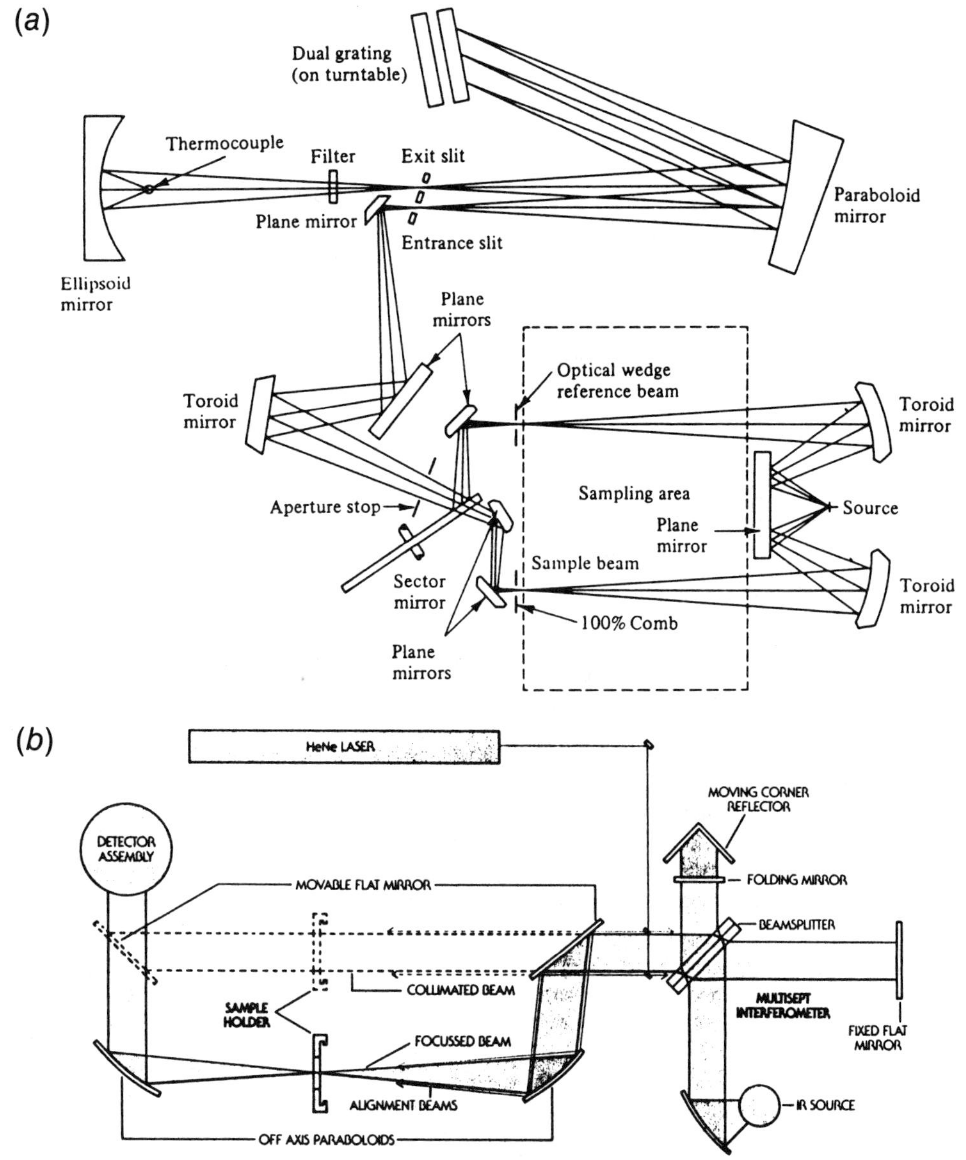

FIGURE 4 **Schematic diagrams of infrared spectrophotometers. (*a*) Dispersive.** (Courtesy of Perkin-Elmer Corp.) **(*b*) FTIR, Analect FX610.** (Courtesy of Laser Precision Analytical, Irvine, CA.)

detector energy is digitized synchronously with the positions of the movable mirror in the interferometer, and successive measurements (collected by a computer) yield an interferogram that encodes all the spectral information. An algorithm known as a fast Fourier transform is used to reconstruct the IR spectrum.

Dispersive and Fourier transform spectrometers acquire IR spectra quite differently. The dispersive instrument literally scans the intensities of the wavelengths presented to the detector from the prism (or grating) and the slit. Thus, the waveform drawn on the chart paper is a continuous recording of detector response versus wavelength obtained as wavelength increments defined by the slit width in front of the detector. In sharp contrast, the FTIR spectrometer "sees" all wavelengths and their associated intensities simultaneously. All the information is contained in one interferometric waveform or time-domain spectrum. The scanning is accomplished mathematically. The interferogram signal containing amplitude–mirror position coordinates is digitized and converted to a frequency-domain spectrum by the Fourier transformation shown in Eq. (4), which integrates over a positive range of movable mirror positions δ (0 to $+\infty$) to yield infinitely high resolution. In practice, the signal is measured over a limited δ causing the spectrum to have a finite resolution.

$$F(\bar{\nu}) = 2\int_0^{+\infty} I(\delta)A(\delta)\cos 2\pi\bar{\nu}\delta \cdot d\delta \qquad (4)$$

where $I(\delta)$ is the detected intensity at a point corresponding to the wavelength and instantaneous movable mirror position (the optical retardation n) and $A(\delta)$ is called the apodization function. Apodization is designed to suppress the oscillations on each side of the centerburst of the interferogram.

A typical transformation of an interferogram to a continuous IR spectrum is diagrammed in Figure 5. The interferogram shown at the top is obtained for the sample and is stored in a memory space in the computer.

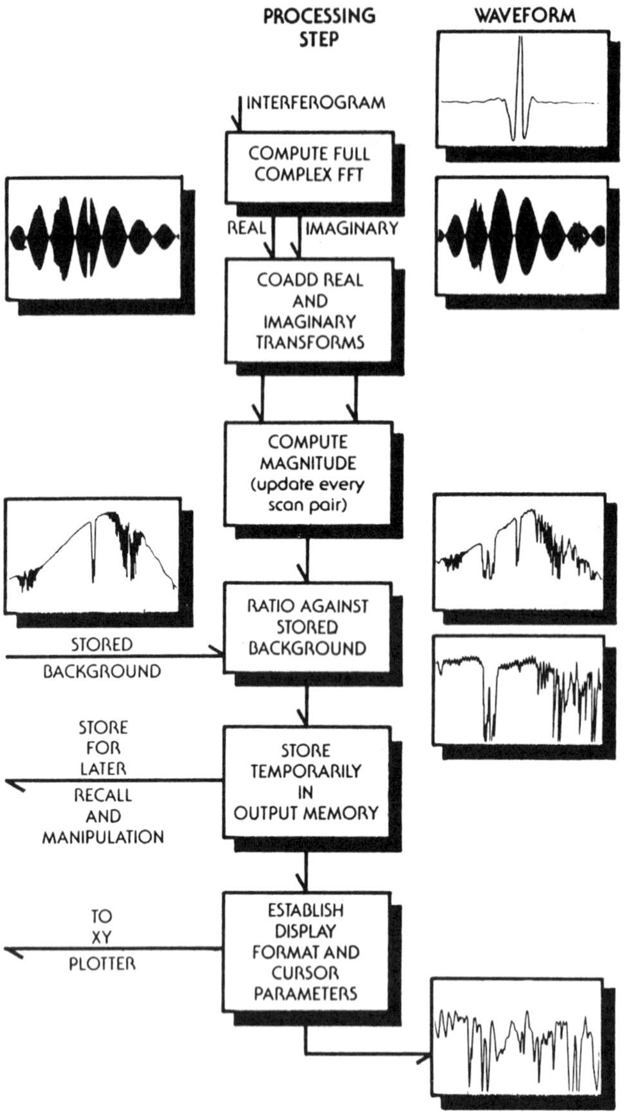

FIGURE 5 **Typical data processing steps in Analect's Multiprocessor Data Systems.** (Courtesy of Laser Precision Analytical, Irvine, CA.)

This waveform is subjected to fast Fourier transformation to obtain the IR spectrum in the Fourier plane. The spectrum is subsequently ratioed against a reference or background spectrum already stored in the memory to yield the spectrum in the real plane. This spectrum appears similar to the conventional infrared spectrum obtained from a dispersive instrument.

Since the complete process of generating an FTIR spectrum consumes only a fraction of a second, additional waveforms can be acquired in a reasonable amount of time. These scans can be averaged to maximize the signal-to-noise ratio. The background is typically a spectrum of carbon dioxide and water vapor present along the radiation path. Given a moderate resolution task of 32 scans, the entire process, from sample insertion to seeing the spectrum, takes about 2 minutes. A comparably resolved dispersion IR spectrum requires 15–25 minutes.

There has been a virtual revolution in the design of instrumentation and sampling accessories for IR spectroscopy. Even though FTIR methods have always been the choice for measuring high quality spectra, instruments were prohibitively expensive for common applications. This picture has reversed during the last decade owing to the reduced costs for microcomputers and other data acquisition equipment. Moderate quality FTIR spectrometers are now available in the \$25,000–\$50,000 range. Instruments that yield the highest quality spectra and have even greater sampling versatility are priced above \$75,000.

FTIR instruments provide a number of important advantages over dispersive IR instruments. Some of these advantages relate to the performance criteria for any type of spectroscopic measurement as well as to applications that are not otherwise feasible. Four particular advantages pertain to speed of analysis and signal-to-noise ratio (*S/N*). The condition called Fellgett's advantage exists because all wavelengths are detected simultaneously. An additional contribution to *S/N* is that more IR energy reaches the detector owing to the large aperture of the interferometer compared to energy-wasting slits. This sensitivity advantage is called Jacquinot's advantage. Laser measurement of the position of the movable mirror provides extraordinary wavenumber accuracy of 0.01 cm^{-1} over the mid-IR range. This is called Conne's advantage. Finally, an FTIR spectrometer has only one moving part, the movable mirror, which implies greater reliability and ruggedness. These advantages permit a widened scope of previously very difficult analyses, including monitoring changes in rapid chemical reactions, "on the fly" detection and identification of effluent peaks from gas and high performance liquid chromatography, sampling microscopic specimens, and more effective use of such methods as diffuse reflectance from solid surfaces, including those of cured or thermoplastic composite materials.

APPLICATIONS. One of the major strengths of infrared spectroscopy is the capability to examine and analyze materials as solids, liquids, and gases. Moreover, the solids may be particulate, thin films, and visually opaque surfaces. Depending on the material, IR spectra may be obtained by detecting transmitted or reflected energy. To obtain transparency to infrared radiation, sample cells must be made from special window materials. The various sampling techniques are described briefly in Table 3. Infrared spectroscopic applications can be divided broadly into the two major categories, quantitative analysis and qualitative analysis.

Quantitative analyses deal with mixture analysis and quantitation of the compositional and structural changes in a material detectable in qualitative analysis. Qualitative analysis emphasizes applications that involve fingerprinting, product identification, and monitoring of the structural and conformational changes that can occur during cure, crystallization, orientation, etc. Various applications related to both types of analysis for polymer composite materials are discussed briefly below.

TABLE 3
Sampling Methods for Infrared Spectroscopy

Method	Description
Solutions	Soluble materials can be dissolved in a suitable solvent and the solution placed in a cell having IR-transparent alkali halide (NaCl, KBr) windows. The sample space or path length is fixed to permit solvent compensation in double-beam instruments.
Liquid Films	Liquid materials such as uncured matrix resins can be analyzed as a capillary film by pressing a drop of fluid sample between two salt crystals. Liquids of very low viscosity may be examined as for Solutions (above).
Solid Dispersions	
Mulls	The solid is ground to $< 2\ \mu m$ particle size in a pure oil having the smallest number of strong IR absorbances. Two traditional media are Nujol (mineral oil) and Fluorolube. The interfering peaks can be removed by spectral subtraction.
KBr pellets	The solid material is mixed and ground with potassium bromide to give $\sim 1\%$ sample concentration. Application of sufficient pressure gives fused, IR-transparent discs. The usual sample size is a few milligrams. Samples as small as 0.1 mg have been analyzed.
Solids	
Solvent-cast films	Materials that can be dissolved may be cast directly onto a salt crystal and analyzed after the solvent has evaporated.
Pressed films	Thermoplastics may be heat-pressed from an end product (pellets, powder, etc.) to a sufficiently thin film. *Caution:* Thermal history may modify crystallinity-related effects.
Microtome sections	Effect is similar to pressed films, but does not destroy original thermal history.
Reaction-cast films	Thermoset resins may be polymerized and cross-linked directly on a salt crystal. Reactions can be studied dynamically with appropriate accessories for controlling the sample environment.
Opaque Surfaces	
Attenuated total reflectance (ATR)	Sample is placed into intimate contact with the flat surface of a specially ground and polished crystal having a high index of refraction. The high index causes the IR beam to traverse the crystal by multiple internal reflections. Reflection is not total at the crystal–sample interface for the wavelengths capable of absorption by the sample.
Diffuse reflectance (DR or drifts)	Sample holder is a special optical bench designed to collect and collimate reflected energy to the detector. IR energy not absorbed below the surface is scattered as a result of numerous collisions and is therefore diffuse rather than specular. Highly sensitive mercury–cadmium–telluride (MCT) detectors are required for this method because scattered radiation intensities are very low. Samples include KBr pellets, powders, prepreg, and solid composite materials.
Special or Hyphenated Techniques	
Microscopy-FTIR	Microscope attachments are available that allow direct analysis of particulate matter 10 μm in diameter and weighing less than 1 ng.
GC-FTIR	The sample is the effluent from a gas chromatograph column. The effluent is sent by heated transfer line to a gold-coated light pipe. The latter is the sample holder for IR analysis.
HPLC-FTIR	The sample is the effluent from an HPLC column. This technique is still developmental owing to the difficulty of disposing of the mobile phase that emerges with the separated solute. One method is to deposit the effluent onto a powder substrate, which moves on a conveyor as the solvent is evaporated. Analysis of the solute deposited on the substrate is accomplished by diffuse reflectance.

Quantitative Analysis

The goal of quantitative analysis is to determine how much material or how much of a particular property is represented by a given IR spectrum. It was stated earlier that the intensity of a particular absorption band is proportional to the number of absorbing groups in the path of the IR beam. The exact relationship is given by the Beer–Lambert law or, simply, Beer's law:

$$I = I_0 \exp(-abc) \tag{5}$$

where

I_0 = intensity of incident radiation
I = intensity of transmitted radiation
a = absorptivity of the functional group
b = sample cell path length
c = concentration (g/L), if a solution

Equation (3) for absorbance, defined earlier, and Eq. (5) can be combined to show the simple linear dependence of A on a, b, and c as follows:

$$\log(I_0/I) = A = abc \quad (6)$$

The simplest use of Beer's law is in the construction of a calibration curve based on measured absorbances for known concentrations of standard solutions. By holding constant, experimental conditions, including cell path length, Beer's law reduces to:

$$A = kc \quad (7)$$

where k is the slope of the calibration line. Since deviations from Beer's law can occur for instrumental, physical, and chemical reasons, extrapolation of the calibration line to higher than determined concentrations must be performed with great caution. A chemically caused deviation in concentrated solutions is molecular association through hydrogen bonding.

One deviation for solid materials that originates from molecular orientation and crystallinity is induced by physical properties: alignment and close, regular packing of long molecules cause interactions among functional groups, which can shift the maximum absorption frequencies for certain bands. In addition, close packing and orientation induce changes in molecular polarizability for the vibrational mode involved. Alignment of functional group dipoles causes the IR beam to become polarized, preventing some fraction of the transmitted beam from reaching the detector. This effect can be quite useful in the analysis of crystallinity and stress-induced orientation [5]. But, the potential to modify quantitative Beer's law measurements needs to be appreciated. For multicomponent systems, an excellent application of Beer's law is difference spectroscopy.

Difference spectroscopy or spectral subtraction is an especially useful technique for analyzing multicomponent mixtures [6,7]. For a mixture consisting of n components, which all contribute to the intensity of an absorption but do not interact with each other, the total absorption in a noninteracting solvent can be written as follows:

$$A = a_1bc_1 + a_2bc_2 + \cdots + a_nbc_n \quad (8)$$

It is necessary in manual calculations to compute scaling factors to compensate for differences in path length for the sample spectrum and for the reference spectra of the n components. Algorithms have been developed for computer-based subtraction routines that calculate scaling factors without reference to matched path lengths. This technique, especially for solids, is not without some potentially major problems [8].

Four major applications of difference spectroscopy [9] are illustrated in Figure 6. The first is identification of a component by spectral subtraction. The second case involves setting quality control or quality assurance tolerance limits to dictate the acceptance or rejection of a material. The third case involves scaled subtraction measurements to remove selected components, solvents, and impurities from the spectrum. The fourth application involves dynamic studies to monitor chemical and physical changes with time, temperature, radiation, stress level, etc.

Applications of Beer's law for single-component spectra may include the determination of concentration in a solvent, purity, degree of imidization for a polyimide oligomer, and degree of crystallinity in a thermoplastic. For reactive systems the purpose of analysis could be reaction kinetics; for multicomponent materials, typical objectives are to determine the number of components present and their relative or mass proportions. Since all these analyses require the identification of IR bands in the absorption spectra, these applications are discussed in more detail in the next section.

Qualitative Analysis

Qualitative analysis of polymeric composite materials generally has one of two principal objectives. The first involves analyzing a material having an expected identity or composition by comparing its IR spectrum (i.e., fingerprint) to a reference or standard spectrum for the material. This is a typical analysis for quality control and quality assurance purposes [10] to ensure that a formulated resin, prepreg, or other analyzate is what it is supposed to be and that there is no difference between it and previous lots of material. The sophisticated data handling capabilities of FTIR and certain dispersive instruments allow an analyst to compare electronically the sample and a stored reference spectrum on a CRT to determine the exactness of match. It is also possible to do these tasks and perform an automated subtraction of one spectrum from the other. An exact match will, in principle, produce a featureless line across the CRT. Any extra (or missing absorptions) or absorptions having larger or smaller intensities suggest one of several material discrepancies, including degree of advancement, undesired reaction product, too few or too many constituents, and degree of crystallinity.

The second objective is to identify unknown peaks associated with either unknown materials or structural changes in the material itself. This is, as might be expected, far more challenging to an analyst than comparing fingerprints for known materials. Significant analytical expertise is required. Most of the applications in this category use quantitative analysis to measure the identified change. Approaches for identifying pharmaceuticals, oils, plasticizers, and composite materials can be fairly similar. However, formulated resins, molding powders, and prepregs are based on multicomponent mixtures; thus their IR spectra can be quite complex. Therefore it is especially helpful to have additional characterization data from chromatographic instrumentation (e.g., HPLC and GPC) to know at least the minimum number of different kinds of component that might be present in the composite. Direct IR analysis of the separated and collected components would represent a very direct approach. Nevertheless, it is still possible to accomplish this task using spectroscopy alone. Algebraic methods known as factor analysis and least-squares regression analysis have been developed to allow the iden-

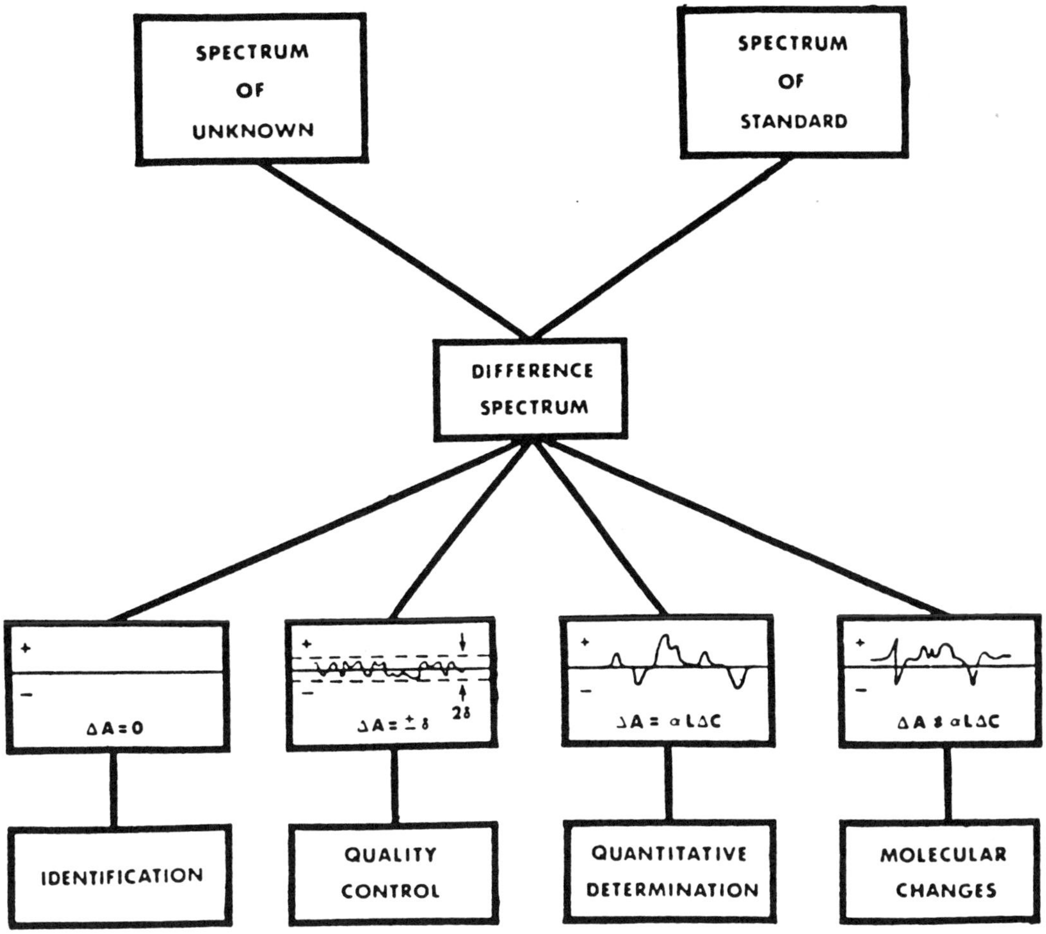

FIGURE 6 Applications of infrared difference spectroscopy. (Reprinted from J. L. Koenig, *Applied Spectroscopy, 29,* 293(1975), by courtesy of the Society of Applied Spectroscopy.)

tification of the number of components in a mixture as well as their concentrations.

The first step in analyzing a mono- or multicomponent IR spectrum is to identify the probable kinds of compound present based on analysis of the functional group frequencies in the 4000–1300 cm^{-1} range and of the patterns in the aromatic region of the spectrum. This analysis, by itself, may be sufficient to reveal that the partially cured or noncured material is a thermoset such as epoxy, phenolic, or polyimide, or a thermoplastic such as polyethersulfone (PES), polyphenylenesulfide (PPS), or polyether etherketone (PEEK). Spectra for these materials are reproduced in Figure 7.

The spectra pertaining to PMR-15 polyimide are reproduced in Figure 8. Spectra *a–c* are for the three pure constituents: monomethyl ester of norbornene anhydride (NE), dimethyl ester of benzophenone tetracarboxylic acid dimethyl ester (BTDE), and 4,4′-methylenedianiline (MDA). Spectrum *d* shows the mixture of the three before polymerization, and spectrum *e* illustrates the effect of curing to form the cured polyimide. Once the nature of the functional groups present has been ascertained, more rigorous analysis can be done by finding an exact or close match by comparison to the spectrum of a compound with known chemical structure (e.g., spectra on file in the laboratory or obtained from commercially available data bases and libraries of spectral collections).

Qualitative and quantitative methods have found a wide range of applications for advanced composites. Fiber coatings and composite surfaces have been analyzed frequently [11–13]. One of the more common applications of qualitative techniques is that involving chemical reaction characterization. One of the critical parameters for thermosetting resins and prepregs is shelf life. This makes outlife studies very important for optimum processing. Many studies have used IR to determine the effect of various environmental conditions on the outtime performance [14,15] of various thermosetting resins and prepregs. The properties of the cured laminates can also be significantly affected by the environmental factors, particularly moisture [16,17]. For reactive systems, the rate and the mechanism of change can be determined by monitoring the disappearance of the absorption bands related to the reactants and the appearance of bands to new reaction products [18–21]. Thermo-

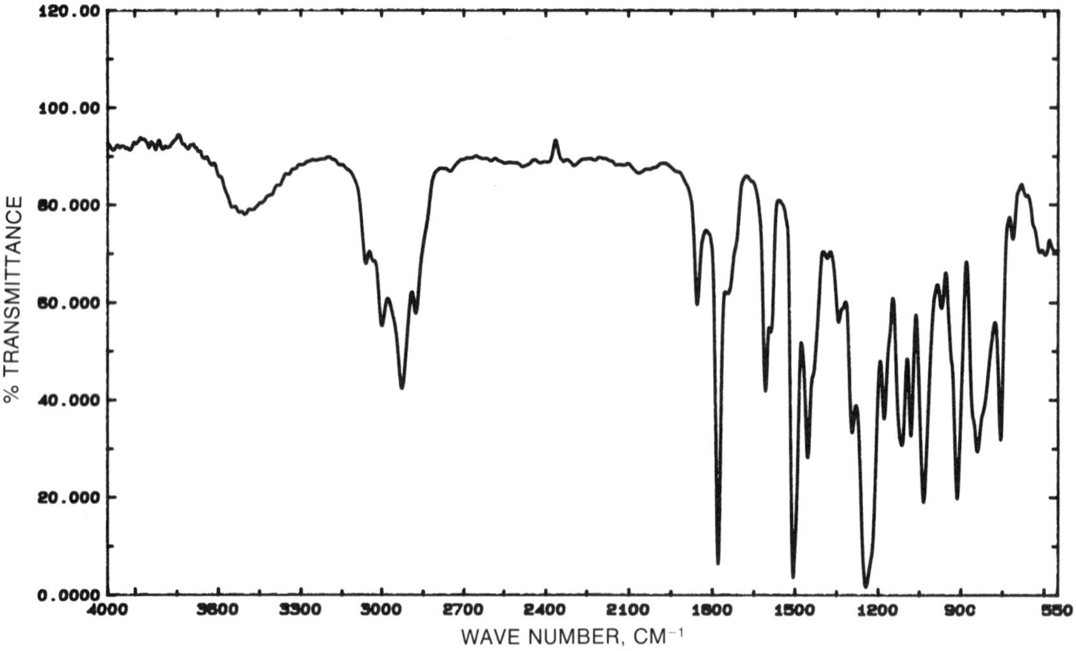

(a)

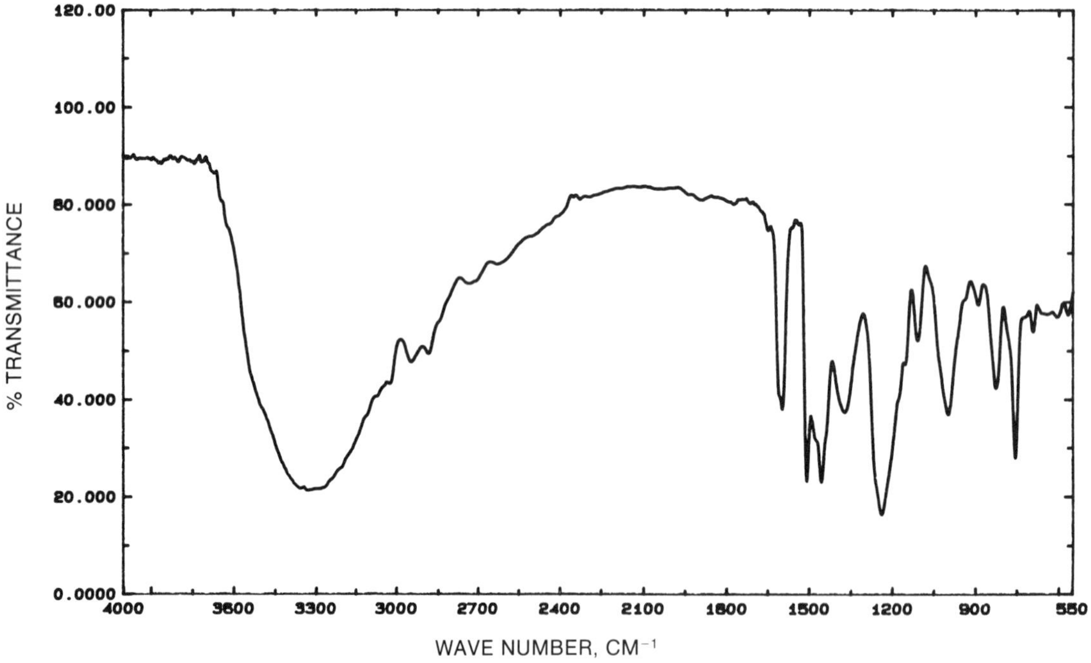

(b)

FIGURE 7 FTIR spectra obtained on an Analect FX6160: (*a*) E-293 epoxy, (*b*) CPH-2209 phenolic, (*c*) CPI-2249 polyimide, (*d*) I.C.I. Victrex 4100P polyethersulfone (PES), (*e*) Phillips Ryton R4 polyphenylenesulfide (PPS), and (*f*) I.C.I. polyether etherketone (PEEK). Plots show percent transmittance versus wavenumber for 200 scans, with Happ–Genzel apodization. Spectra *a*, *b*, and *c* were acquired from liquid films in transmission. Spectra *d*, *e*, and *f* were acquired from pressed films in reflectance. (Courtesy of Ferro Composites, Los Angeles.)

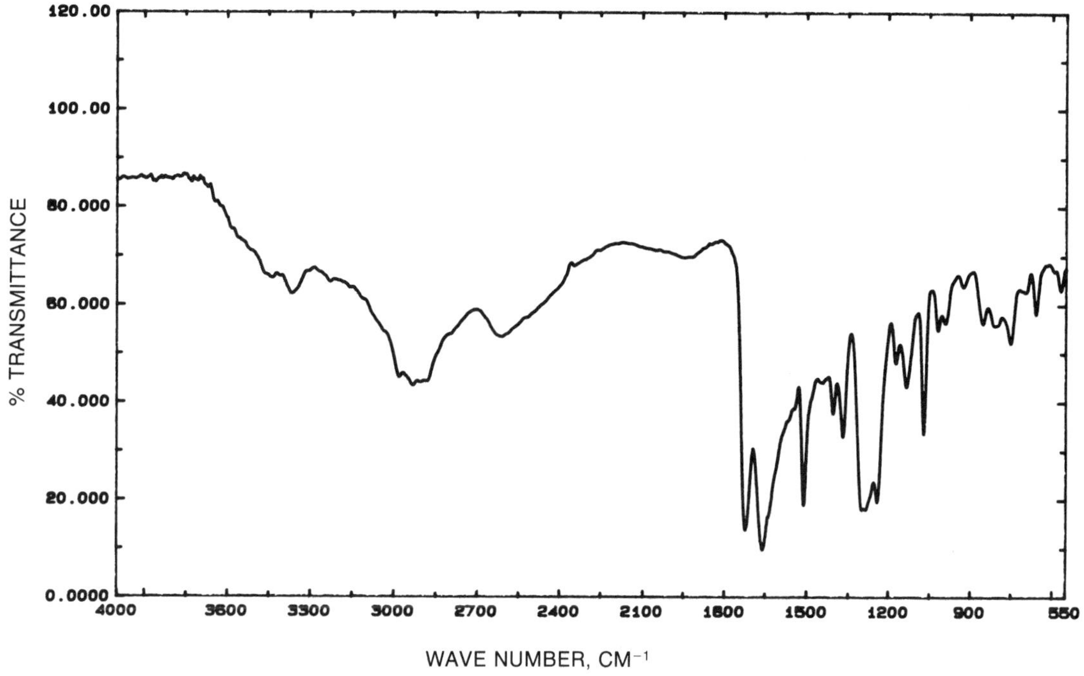

(*c*)

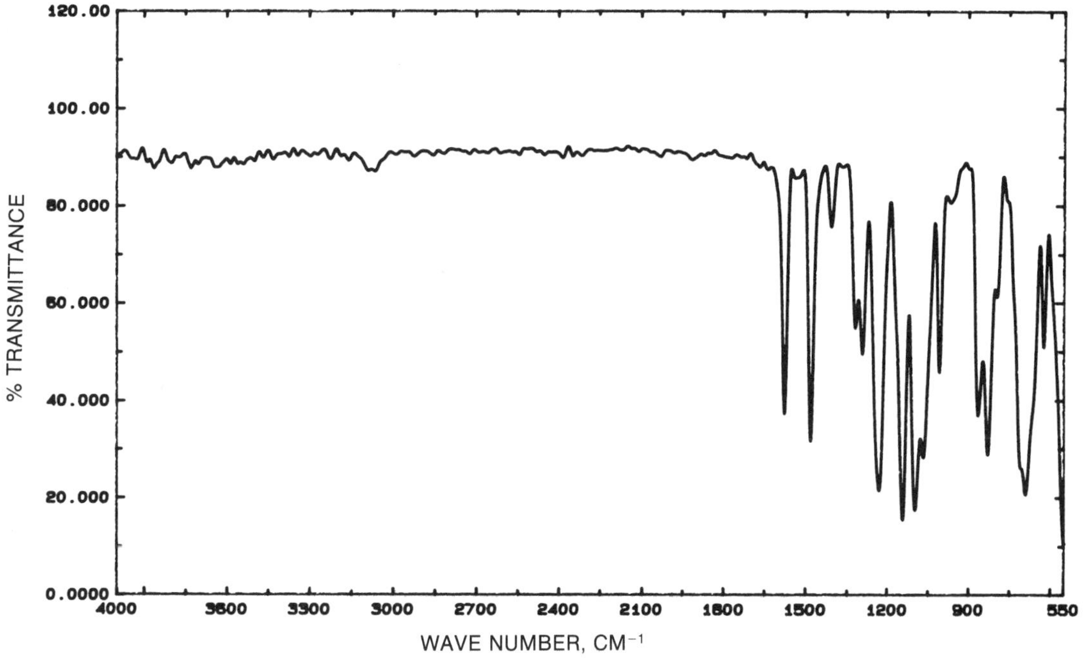

(*d*)

FIGURE 7 (continued)

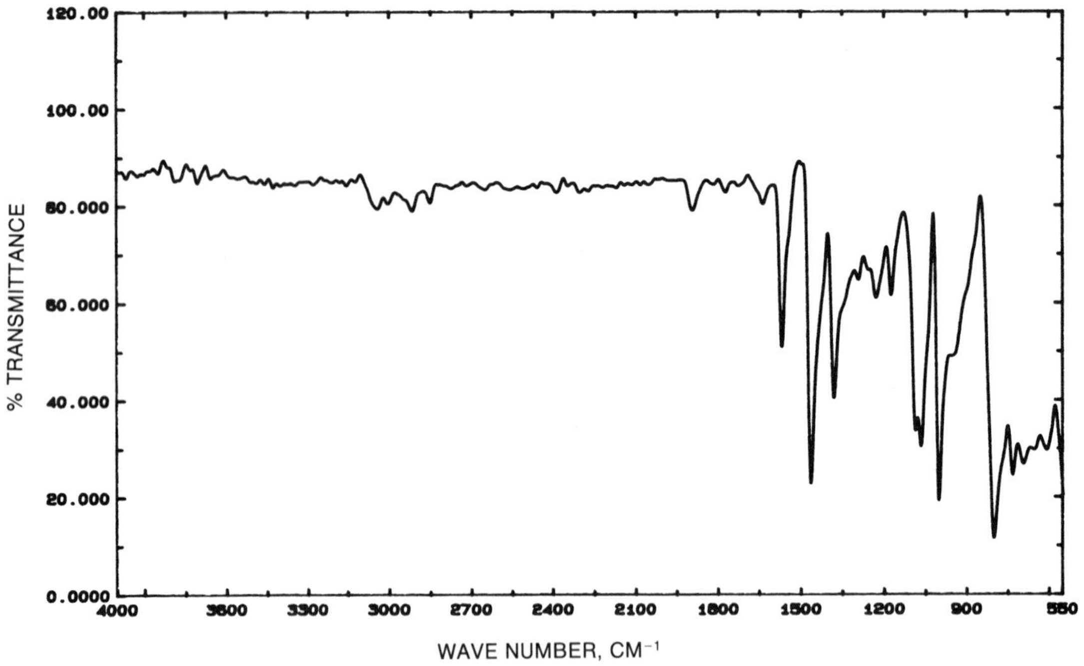

(*e*)

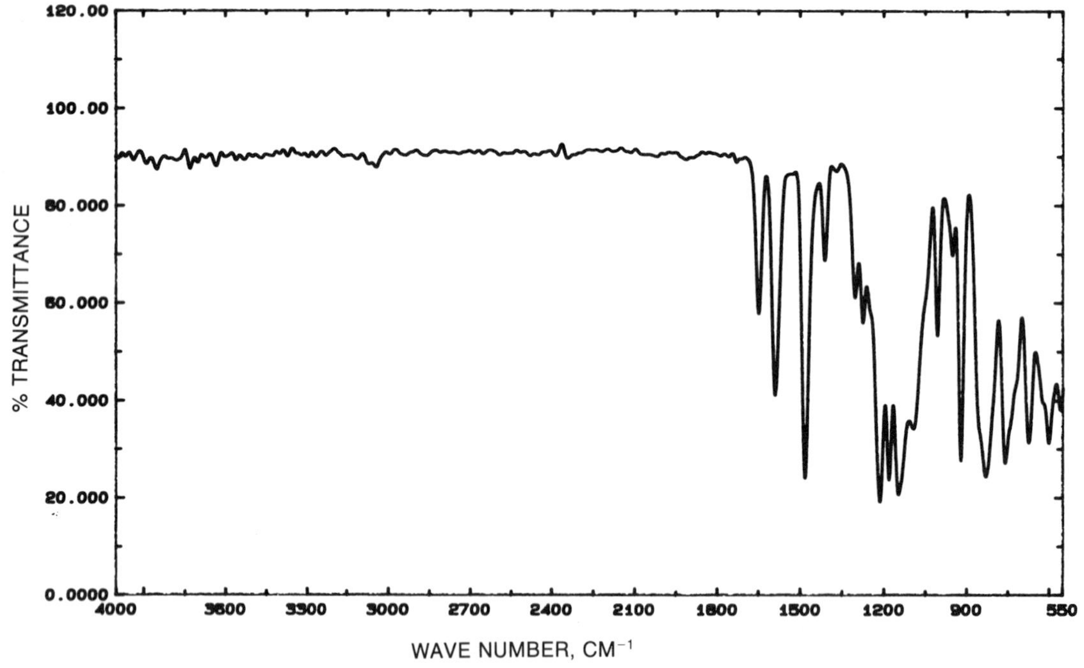

(*f*)

FIGURE 7 (continued)

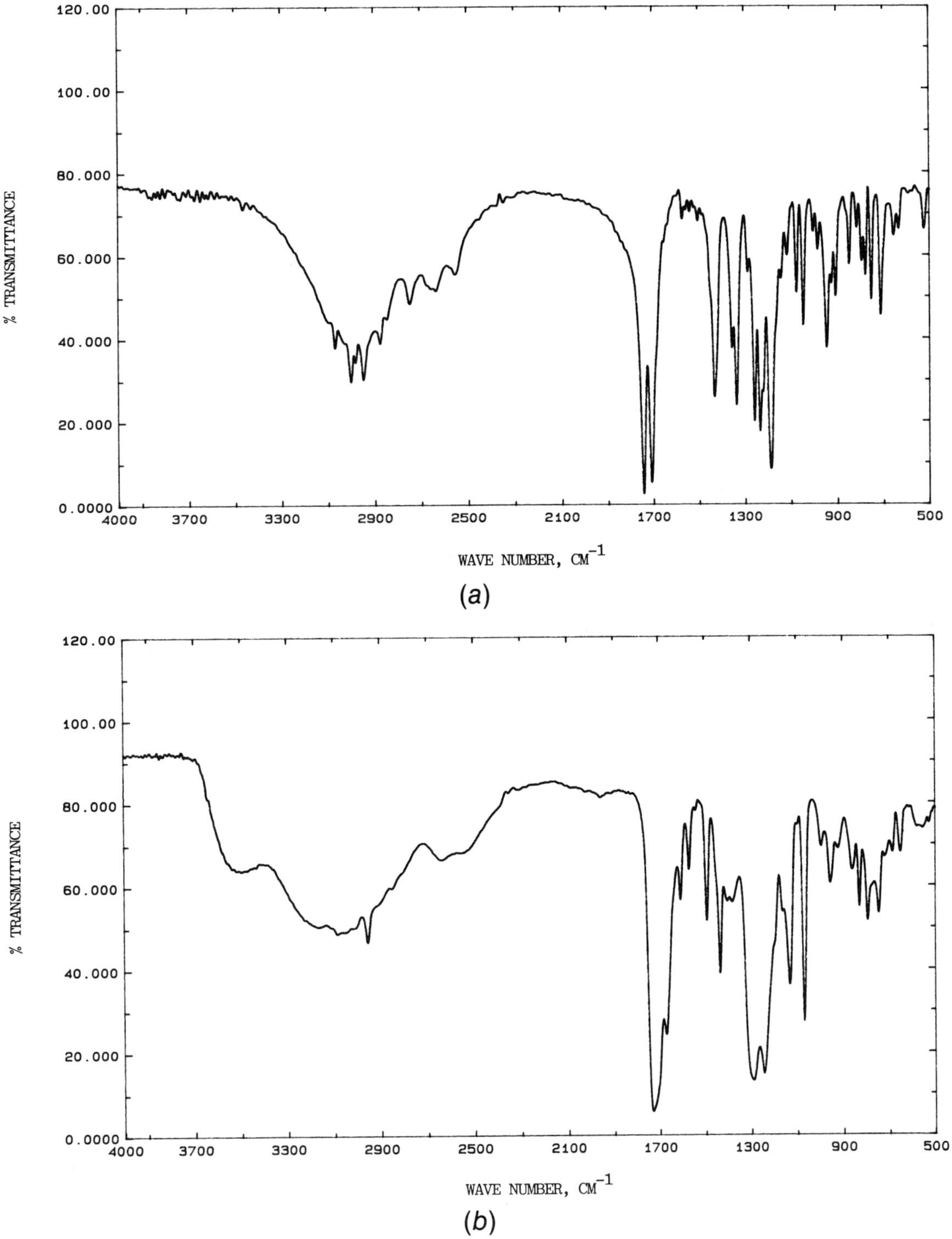

FIGURE 8 FTIR spectra obtained on an Analect FX6160: (*a*) nadic anhydride, monomethyl ester (NE), (*b*) benzophenone tetracarboxylic acid, dimethyl ester (BTDE), (*c*) 4,4′-methylenedianiline (MDA), (*d*) PMR-15, solution of *a*, *b*, and *c* in methanol, and (*e*) PMR-15, cured neat resin. Plots show percent transmittance versus wavenumber for 200 scans; KBr pellet Happ–Genzel apodization. (Courtesy of Ferro Composites, Los Angeles.)

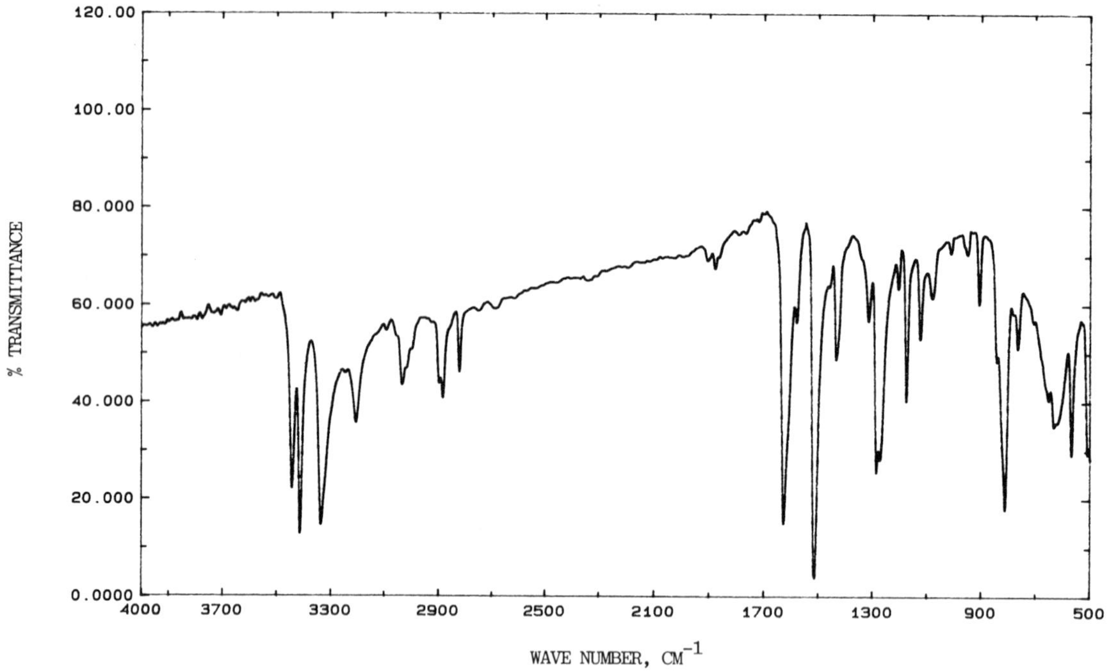

(*c*)

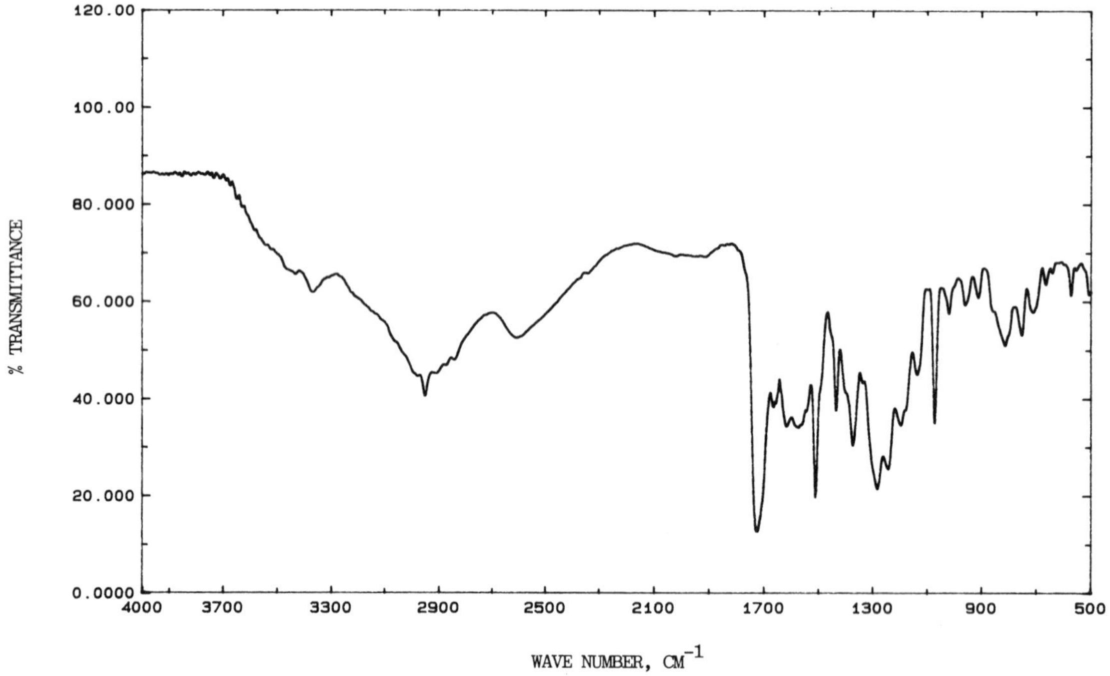

(*d*)

FIGURE 8 (continued)

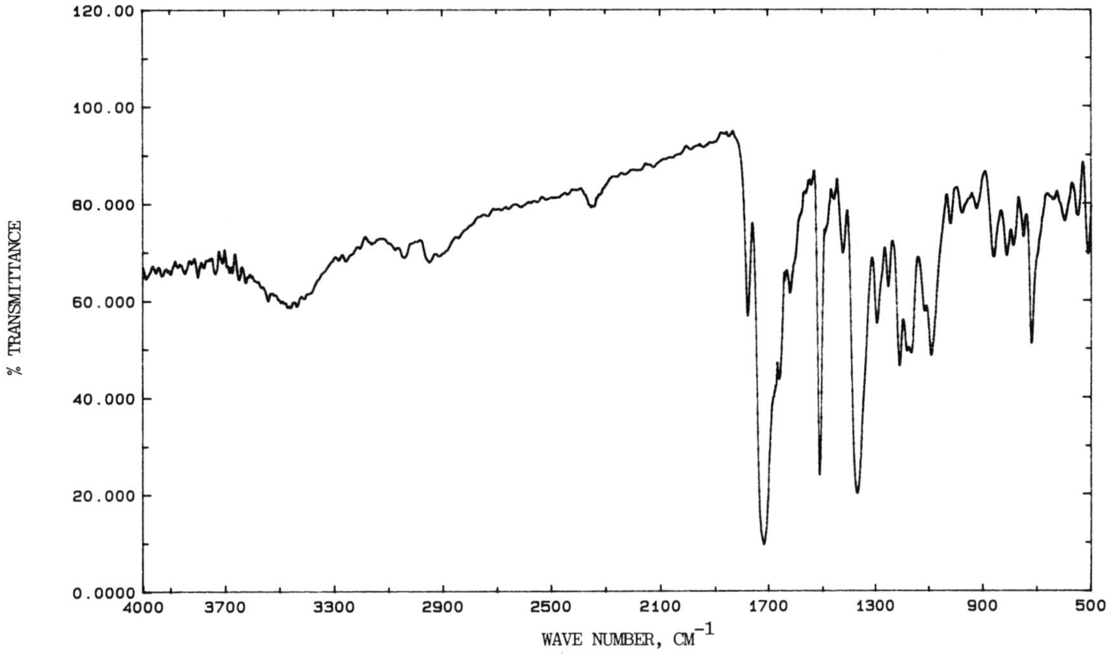

(*e*)

FIGURE 8 (continued)

oxidative stability of composite materials is of prime importance, particularly for resin systems designed for high temperature use. Many studies have employed IR methods to understand the nature of thermal or oxidative degradation in composite resin systems [22–25]. FTIR has also been used to obtain insight into the cross-link mechanism during curing [26–28].

The most often asked questions in composite processing pertain to the optimization of the cure cycle. FTIR has been used extensively along with differential scanning calorimetry and dynamic mechanical analysis to monitor or identify specific chemical reactions [29–34]. Figure 9 is an example of different IR absorbance characteristics for two lots of PMR-15 molding powders subjected to different thermal imidization histories [35]. IR absorption band assignments are shown for LARC 160 before and after imidization in Figure 10 [36].

FTIR characterization of surface reactions has contributed to the development of better interface bonding [12,13]. Many thermoplastic matrices used for composite materials are crystalline or have crystalline character. For such systems, the final properties are highly dependent on the degree of crystallinity, which in turn depends on the processing conditions used. Crystallization in semicrystalline materials has been shown to sharpen or split certain absorption bands that correspond to the collective vibrations of the functional groups present in the crystalline phase [4,5,37]. This property of the crystalline phase has been used to characterize the degree of crystallinity in such resin systems.

Nuclear Magnetic Resonance Spectroscopy

The objective of the nuclear magnetic resonance (NMR) method is to determine the number and arrangement of atoms in a molecule. Unlike IR spectroscopy, which measures absorption of energy by groups of atoms, the NMR technique measures the energy needed to excite the nucleus of individual atoms. The NMR approach owes its diagnostic power to the fact that the energy absorbed depends significantly on the location of the atom in the molecule with respect to the identity and number of neighboring atoms.

THEORY. NMR spectroscopy measures the absorption of radiofrequencies in the 100 MHz range by a molecule subjected to a strong magnetic field. Atoms that are capable of absorbing radiofrequency energy are those whose atomic numbers and mass numbers are not both even. Such atoms are ^{1}H, ^{13}C, ^{19}F, and ^{31}P. These are elements having nuclear spin quantum numbers that are odd multiples of $\frac{1}{2}$. That is, the axes of the nuclear spins of such atoms can be oriented in only one of two possible directions. The simple hydrogen atom, and the isotopes ^{13}C, ^{19}F, ^{31}P, etc., are said to have a spin-active nucleus. Nuclear spin generates a magnetic field, but in the absence of an external magnetic field, the spin axes are distributed randomly. In the presence of a strong imposed magnetic field H_0, the spin axes line up either parallel with the field or antiparallel (against) to it. Actually, the alignment is not perfect: the nuclear spin axes and the field direction are separated by a small, constant angle, with the nucleus resembling a spinning top that moves around a circular path having the field axis at the center of the circle (Fig. 11).

At ordinary temperatures the number of nuclei parallel with the field is marginally larger than the number having antiparallel spins. The energy of those nuclei with antiparallel spins is slightly higher (excited state) than the energy of those having parallel spins (ground state).

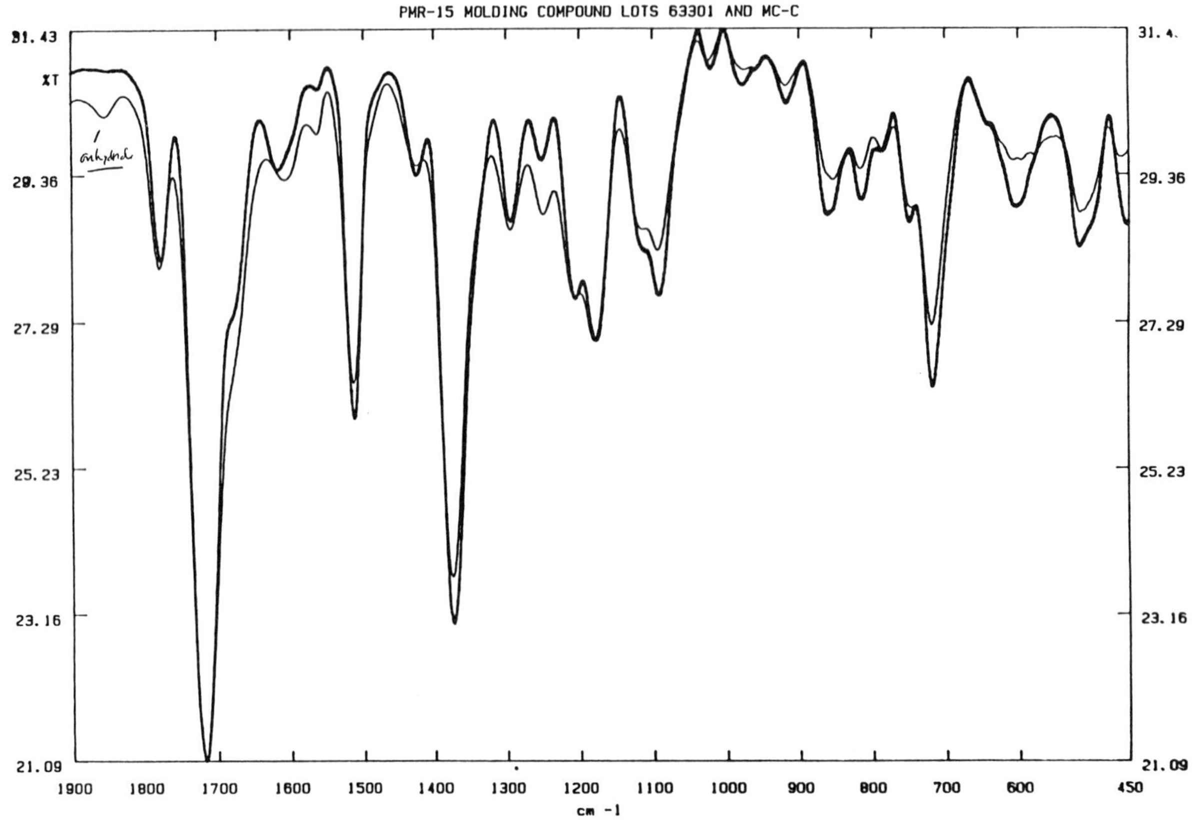

FIGURE 9 **FTIR spectra of two lots of PMR-15 molding powders having different thermal histories.** (From Ref. 35.)

This difference between the two energy states ΔE is proportional to the resonance frequency ν_0 such that

$$E = h\nu_0 \tag{9}$$

and

$$\nu_0 = -\frac{\alpha}{2\pi} - H_0 \tag{10}$$

where α is the magnetogyric ratio for a particular kind of nucleus and H_0 is the strength of the applied magnetic field.

Flipping nuclear spins from random to aligned orientations is far from a trivial task. A typical value for H_0 needed in NMR spectroscopy is 2.348 tesla [1 tesla (T) = 10^4 gauss]. Contrast this with the magnetic field strength of the earth at only about 10^{-4} T. Since the excitation lasts for a relatively long time, seconds, the relaxation time to the ground state can be measured easily. Relaxations in solids and viscous liquids occurs more quickly (broad absorption bands) than in low viscosity liquids (narrow line spectra).

One might think that the frequency at which protons or ^{13}C nuclei absorb radiofrequency radiation would be a constant for each particular kind of atom. This would be true provided the nuclei were highly separated from each other. When bound together in molecules, however, the different nuclei give rise to slightly different absorption frequencies, and these different absorptions produce the NMR spectrum. Slightly different absorption frequencies for a given kind of spin-active nucleus originate from shielding of the nucleus from the external magnetic field by the atom's electron cloud. These electrons create a magnetic field of their own under the influence of the external field, but in a direction opposed to the external field. The electron clouds around nuclei are polarized or distorted by chemical bonding, causing the nuclei to be shielded to different extents, depending on the immediate environment. Thus, for a particular value of H_0, chemically identical nuclei can resonate at different frequencies. Consider, for example, the terminal end of a long chain molecule such as linear polyethylene:

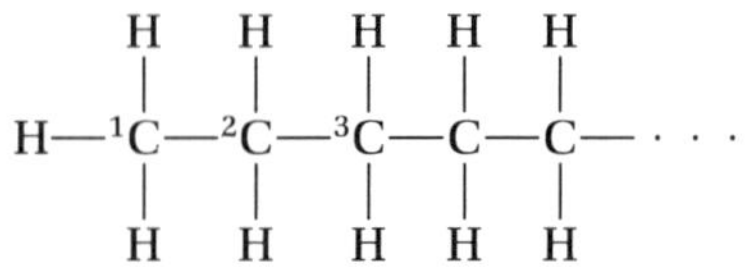

The protons attached to carbons 1, 2, and 3 appear to be equivalent, but in a strong magnetic field they are not. The electron clouds surrounding the carbon atoms, and particularly the cloud polarized in the direction of the polymer chain, shield the hydrogen nuclei from the full effect of the magnetic field.

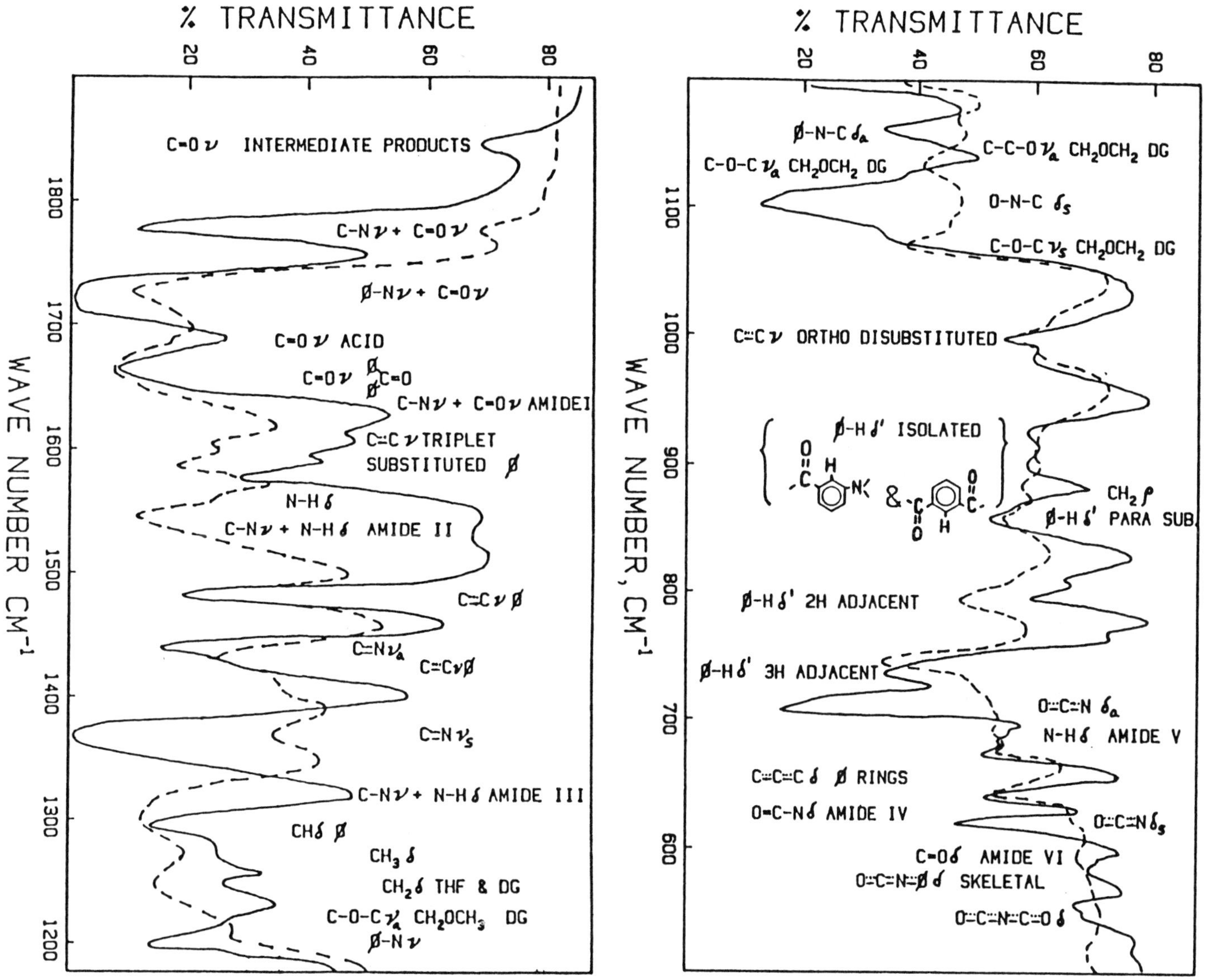

FIGURE 10 **Infrared spectra of LARC 160 before (dashed curves) and after (solid curves) imidization at 200°C.** (From Ref. 36, with permission.)

The extent of shielding, however, depends on neighboring atoms. Thus, the degree of shielding of the protons attached to carbon atom 1 (the methyl group) depends on shielding by both the electron cloud of carbon atom 1 and that of the methylene group, carbon atom 2. The protons attached to carbon atom 2 experience shielding from carbon atom 2 as well as the methyl group and the methylene group based on carbon atom 3. The analytical result is that the methyl and methylene groups in branched and in low molecular weight polyethylene can be distinguished because their protons resonate at different frequencies.

It is very difficult to assign an absolute value to a particular resonance frequency. It is easier experimentally to add to the sample a simple, noninteracting chemical compound having protons more highly shielded than those of the sample. This compound becomes the internal standard whose resonance frequency is the zero reference point for measuring the amount of frequency shift observed for various spin-active nuclei of the sample. The most universally accepted standard for ^{1}H NMR and ^{13}C NMR spectroscopy is tetramethylsilane [TMS: $(CH_3)_4Si$]. The distance on the frequency axis from the absorbance maximum for TMS to a particular sample resonance is called the chemical shift δ and is measured in parts per million (ppm) from TMS as defined in Eq. (11).

$$\delta_{\text{ppm}} = \frac{\nu(\text{sample}) - \nu(\text{TMS})}{\nu(\text{instrument})} \times 10^6 \qquad (11)$$

Typical ranges for chemical shifts for ^{1}H and ^{13}C NMR, respectively, are about 10 and 200 ppm.

Similar to IR functional group absorptions, the chemical shifts for protons and other spin-active nuclei occur in ranges that are specific to different chemical environments. Given strictly aliphatic molecules such as

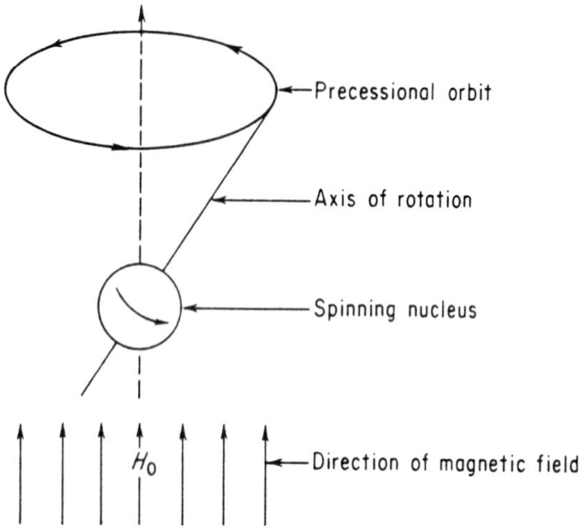

FIGURE 11 **Motion of a nucleus in a magnetic field.** (From John R. Dyer, Applications of Absorption Spectroscopy of Organic Compounds. © 1965, p. 59. Reprinted by permission of Prentice-Hall, Inc., Englewood Cliffs, NJ.)

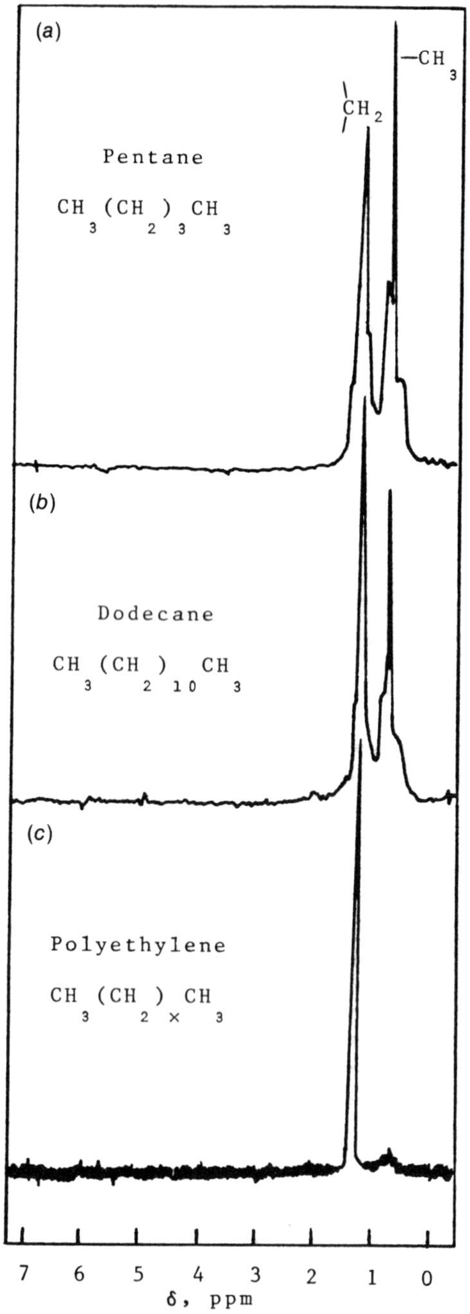

FIGURE 12 ^{1}**H NMR spectra for** (*a*) *n*-**pentane,** (*b*) *n*-**dodecane, and** (*c*) **polyethylene.** (Courtesy of Ferro Composites, Los Angeles.)

n-pentane and *n*-dodecane as extreme examples of ultralow molecular weight polyethylene and the polyethylene fragment cited above, one would observe the CH_3— resonance in the range 0.85–0.95 ppm, the —CH_2— resonance in the range 1.20–1.35 ppm, and the =CH resonance, if present, in the range 1.40–1.65 ppm downfield (at higher δ) from TMS. Proton NMR spectra for *n*-pentane, *n*-dodecane, and polyethylene are shown in Figure 12.

Another analytically useful result of the NMR technique pertains to the intensities of absorptions. Just as in the case of IR spectroscopy, the intensity of an absorption is proportional to the number of absorbing groups present. Each functional group, however, has its own absorptivity *a*. For ^{1}H NMR, since all protons absorb energy identically (identical absorptivity), the signal strength is exactly proportional to the number of protons bound to an atom. Using the "ultralow molecular weight" and the polyethylene molecule examples again, the ratio of absorbances (peak areas) for the CH_3 and CH_2 resonances for *n*-pentane and *n*-dodecane are 1:1 and 6:20, respectively. For polyethylene, the number of CH_2 units is effectively infinite compared with the number of terminal CH_3 units; therefore only one absorbance at 1.35 ppm is apparent. Higher resolution would show the small methyl resonance at 0.95 ppm.

Detailed empirical correlation charts giving chemical shifts for a wide variety of protons (and other spin-active nuclei) in different chemical environments can be found in NMR and instrumental analysis reference works [2,4].

INSTRUMENTATION AND MEASUREMENT. NMR spectroscopy is performed using the instrument shown schematically in Figure 13. The sample is dissolved in suitable solvent and placed in a tube between the poles of the magnet. The radiofrequency transmitter coils and receiver (amplifier) coils are arranged perpendicular to each other and around the sample tube. The frequency is varied continuously through the range of interest, and detected absorptions are displayed on a recorder. This mode of operation is called continuous wave (CW). In CW NMR work, the spectrum must be scanned slowly to avoid passing over narrow spectral lines. In reality, the CW instrument spends more time recording background than the desired signal. A typical ^{13}C analysis at 25 MHz requires about 83 minutes [4]. Modern Fourier transform spectrometers pulse the signal for times ranging from 1

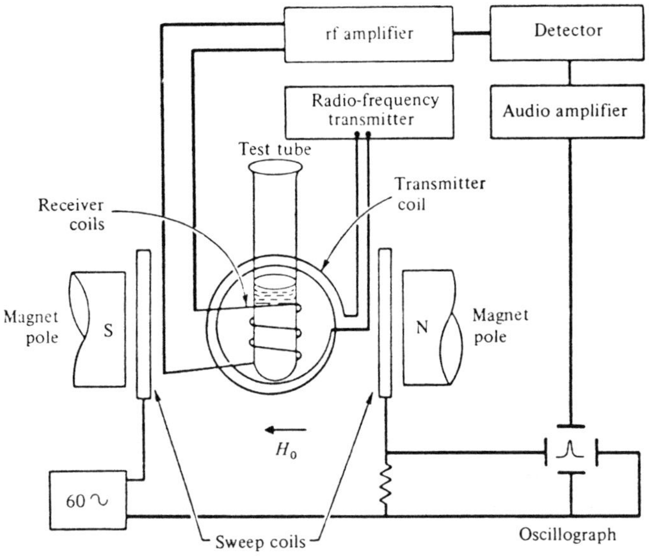

FIGURE 13 **Schematic diagram of an NMR spectrometer.** (Courtesy of Varian Associates, Palo Alto.)

to 1000 μs, permitting an entire spectrum to be acquired in less than a minute.

The most important element in an NMR spectrometer is the magnet. The magnet must be extraordinarily homogeneous to provide a constant field strength stable to one part in 10^{-8}–10^{-9} because detection of chemical shifts requires measurements at the parts per million level. To average out the effects of any field inhomogeneities, the sample tube is spun at several hundred rpm about the vertical axis.

APPLICATIONS. NMR spectroscopy has been used extensively for qualitative and quantitative analyses in organic chemistry. It is probably the most powerful technique available today for determining the details of molecular structure in polymeric materials. However, it cannot yet be viewed as an instrumental method for routine applications for formulated matrix materials. Moreover, NMR approaches cannot be used effectively to study highly cross-linked solids. The instrumentation is comparatively expensive, sample preparation can be complex, and NMR spectroscopists require advanced education and experience for interpreting spectra that can become extraordinarily complex.

A number of soluble polymers have been studied to elucidate the following kinds of information [37]:

- Monomer ratios and sequences in copolymers.
- Tacticity or configurational isomerism in homopolymers.
- Cis–trans isomerism.
- Isomerism in monomer linkages (e.g., head-to-tail).
- Chain branching and end groups.
- Progress of synthetic and degradative reactions.

Applications involving matrix materials for advanced composites are not numerous. Some of the more recent and noteworthy are as follows. Werstler applied ^{13}C NMR techniques to study the polymerization and condensation reactions on aging for phenol–formaldehyde [38] and resorcinol–formaldehyde resins [39]. Figure 14 illustrates the application of ^{13}C for quantitating the degree of aging for the phenol–formaldehyde resins. Spectra *a* and *b* refer to fresh resin and resin aged for 60 days at 33°C. Figure 14*c* gives the weight percents of phenol and formaldehyde remaining from zero to 14 days of aging as calculated from the ^{13}C NMR data. These studies include tabulations of chemical shift assignments for the various phenolic, bridgehead, and hemiacetal carbon atoms, and monitoring of different degrees of ortho- and parasubstitution during polymerization and subsequent aging.

Smith et al. [34,40] used ^{19}F NMR methods in conjunction with FTIR spectroscopy to determine the function of boron trifluoride monomethylamine (BF_3·MEA) in epoxy cure. This work showed that BF_3·MEA hydrolyzes rapidly at temperatures above 100°C to form fluoroboric acid, which was concluded to be the real catalyst in the system.

Poranski and Moniz [41] used ^{1}H and ^{13}C to determine the chemical composition of precursors for polyphthalocyanine polymers for adhesives and composites. They also found that ^{1}H NMR spectroscopy was especially useful for fingerprinting the intermediates for quality control purposes. The aging behavior of PMR-15 polyimide resin has been studied using ^{1}H NMR [42,43].

Lauver [42] was able to document and quantitate the formation of tri- and tetramethyl esters in solutions of the dimethylester of 3,3′,4,4′-benzophenone tetracarboxylic acid (BTDE) after 19 months of storage at room temperature. The NMR spectra of fresh and aged BTDE are compared in Figure 15. In another study involving PMR-15 and PMR-II resins subjected to ambient and subambient aging, Lauver et al. [43] showed that the diamines and diesters form salts that are responsible for phase separation in the resins and consequent diminished processibility of the prepregs. Figure 16 shows the NMR spectra for resins aged at various temperatures. In a very recent study, Clark and coworkers [44] used ^{13}C NMR spectroscopy to identify the various resonances in polyether etherketone (PEEK), as illustrated in Figure 17. The patterns for crystalline and amorphous PEEK are contrasted in Figure 18.

Ultraviolet Spectroscopy

Ultraviolet spectroscopy (UV) involves excitation of electrons from ground state energy levels to higher energy levels when molecules absorb electromagnetic radiation in the 185–400 nm range. A variety of electronic transitions are possible, but only a few can be detected in the ordinary UV range. Compounds whose atoms are bound by single covalent bonds require very high energies to promote outer bonding electrons; thus absorptions occur below 180–190 nm. Saturated linear hydrocarbons, for example, do not absorb in the ordinary UV region. They therefore make good solvents for absorbing compounds. Compounds that contain nonbonding electrons (oxygen, sulfur, halogens, etc.) and compounds that contain multiple bonds absorb strongly in the UV region.

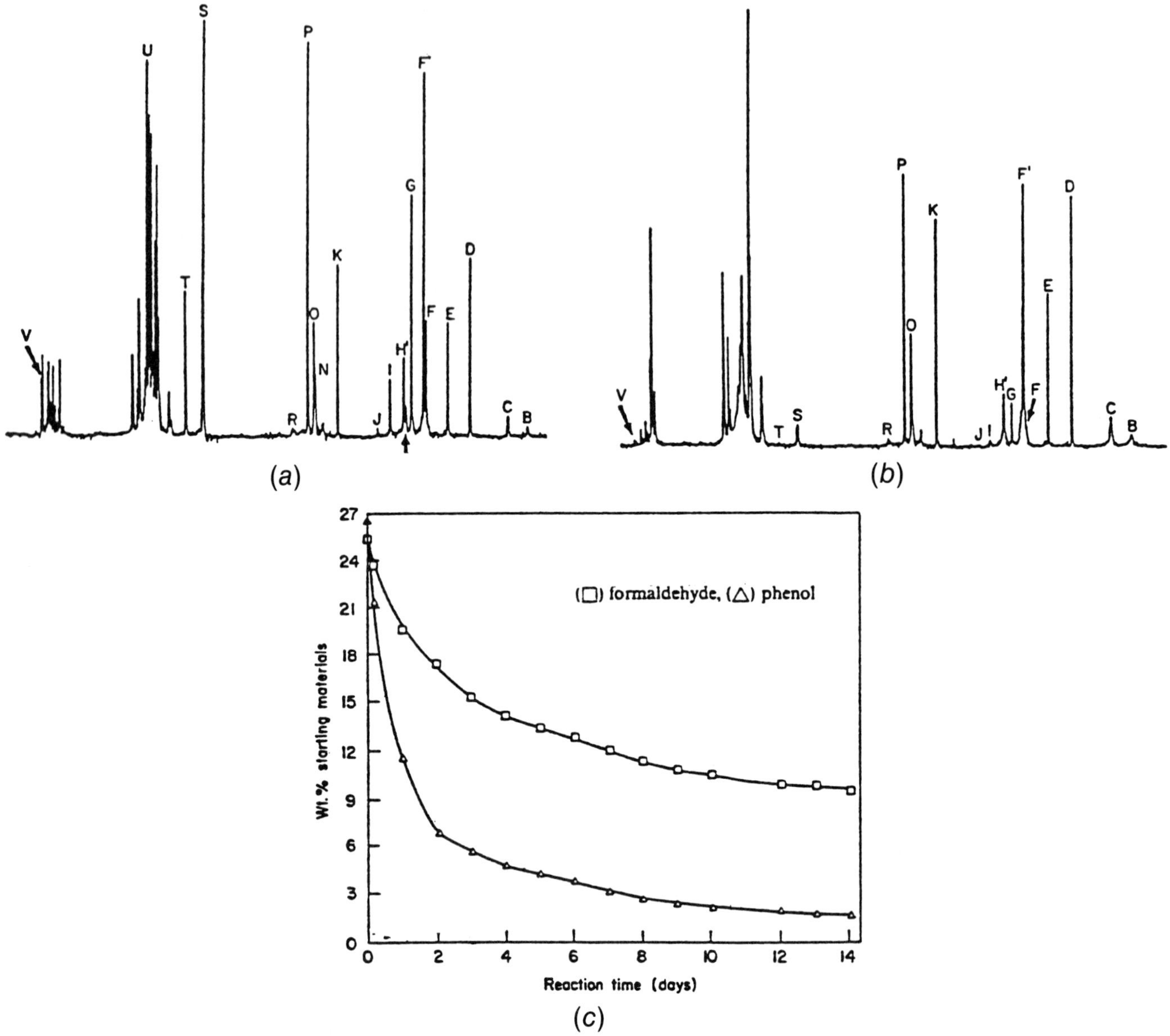

FIGURE 14 (*a*) ^{13}C NMR spectrum of a fresh phenol–formaldehyde resin. (*b*) ^{13}C NMR spectrum of same resin, aged for 60 days at 33°C. (*c*) Data from ^{13}C NMR spectra showing amounts of phenol and formaldehyde remaining unreacted at room temperature from 0–14 days. (From Ref. 38, with permission.)

The molecular groups responsible for absorption are called chromophores when maximum absorption occurs at a specific wavelength. Typical chromophores are double and triple bonds, aldehydes, ketones, acids, and amides. Multiple chromophores that are separated from each other by only a single bond (conjugated systems) have interacting electron clouds, and less energy is needed to promote the electrons to higher energy states. The result is that absorption maxima undergo a red shift; that is, the absorption peak shifts to longer wavelengths compared with the isolated chromophore. The red shifts can be so large in conjugated systems that absorptions occur beyond 400 nm into the visible spectrum (400–700 nm). Consequently, such compounds appear colored (e.g., dyes). Epoxies, phenolics, and polyimides all owe their colors to significant degrees of aromatic and other conjugated groups in the molecular backbones.

For molecules containing isolated chromophores, the effect on the intensity of absorption is additive. Beer's law can be used to make very sensitive and very accurate measurements of concentration in dilute solutions.

Unlike IR and NMR spectra, the UV spectrum for a molecule develops from its structure as a whole rather than from specific bonds or functional groups. A UV spectrum contains only one to several broad absorption bands, and unique fingerprints for quality control purposes are not possible [4]. One of the spectral characteristics that makes the technique useful analytically is that

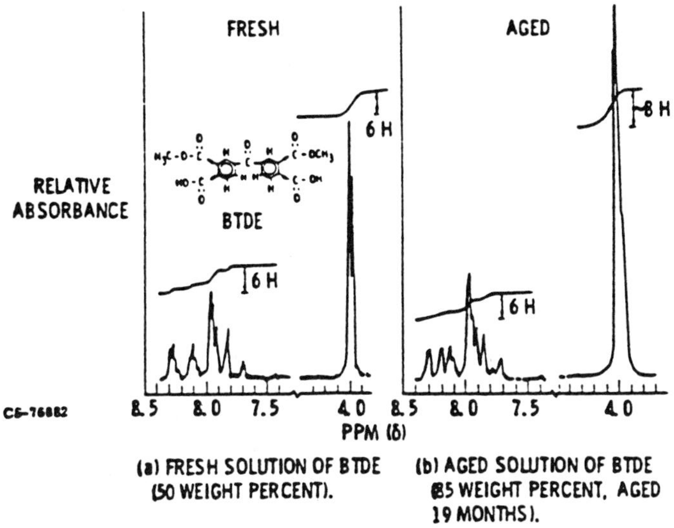

FIGURE 15 ^{1}H NMR spectra of (*a*) fresh BTDE and (*b*) BTDE aged 19 months. (From Ref. 42.)

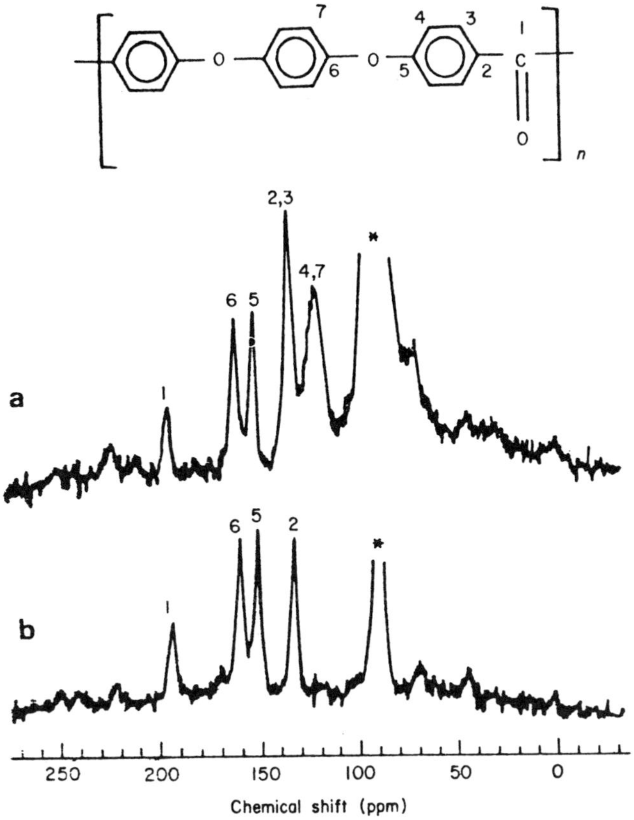

FIGURE 17 (*a*) Crystalline PEEK CP/MAS (cross-polarization, magic angle spinning) ^{13}C NMR spectrum, (*b*) dipolar dephased ^{13}C NMR spectrum. (From Ref. 44, with permission.)

molar absorptivities in the UV region can exceed those in the IR by greater than 10:1. UV spectroscopy is thus capable of analyzing very dilute solutions.

APPLICATIONS. Typical applications of UV spectroscopy involve molecular structure correlations based on substituent locations, stereochemistry, and aromatic character in small molecules. Applications to polymer molecules have been fairly limited so far, although one of the most important uses of the UV spectrometer for advanced composites is as a detector for HPLC and GPC instruments. Some work has been done using UV techniques to measure the degree of advancement and component purity in TGMDA/SU-8/DDS epoxy systems. Morse [45] used the digitized second derivative of the UV spectrum to quantitate all three components in the matrix resin using one analysis. Prior determinations of the concentrations of the three components required separate methods and took about 3 hours. The second-derivative UV method requires only 15 minutes and is capable of 10 ppm sensitivity. Plavin et al. [46] characterized advancement products and the degree of advancement for a similar epoxy–graphite prepreg system that was subjected to room temperature aging for 9, 16, 22, 30, and 39 days.

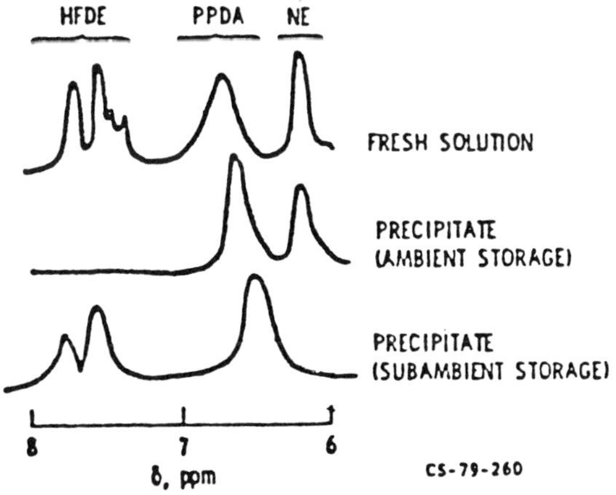

FIGURE 16 ^{1}H NMR spectra of PMR-II monomer solution and precipitates formed during storage [43].

Mass Spectrometry

Mass spectrometry (MS) is based on separating and identifying the molecular weights of ionic fragments of the original, unknown molecule. In effect, the MS analytical procedure becomes a conceptual synthesis of what the unknown material must have been before its destruction to produce the results obtained. This may sound like a rather indirect approach to analysis, but it is necessary because the principle involves separating and detecting ions according to differences in their mass-to-charge (m/z) ratios.

There are four key aspects to a MS analysis [47]:

1. Introduction of the unknown molecule in gaseous form.
2. Creation of (usually) positively charged fragments of the parent molecule.

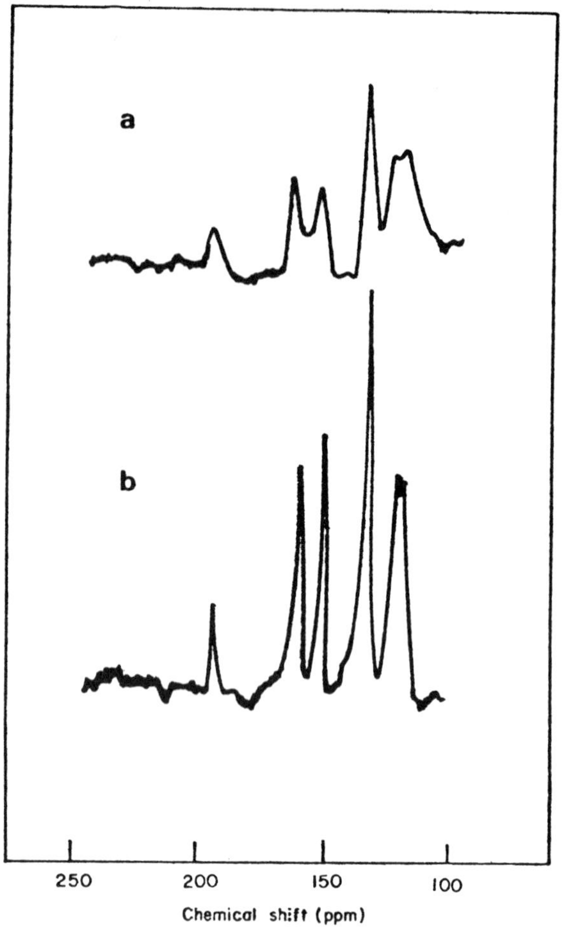

FIGURE 18 Solid state CP/MAS ^{13}C NMR spectra of (*a*) amorphous PEEK and (*b*) crystalline PEEK. (From Ref. 44, with permission.)

3. Separation of the ions according to their m/z ratios.
4. Detection of the separated ions.

Charged molecular fragments are produced in a mass spectrometer by one of two methods: direct electron bombardment of the unknown or indirect chemical ionization by ionic exchange mechanisms. Electron bombardment involves impacting the gaseous sample with an accelerated and collimated electron beam having an energy of 50–100 eV. These moderate energy electrons dislodge an electron from each molecule to form a positively charged molecular ion. Additional electron interactions with the ion cause the parent ion to break down into progressively smaller ionic fragments. In chemical ionization a reagent gas such as methane is ionized as above to form the CH_5^+ ion. Interaction of this ion and the sample causes protonation of the sample and restores the methane molecule to its original form. Since chemical ionization is not as harsh as direct electron impact, it produces fewer fragments. The positively charged fragments are accelerated into the mass analyzer (filter), which may be based on magnetic focusing, electronic field effects (quadrupole filtering), or velocity separation (time of flight). It is rare that a molecule would carry more than one charge under the conditions of MS. Thus, the mass analyzer ultimately separates according to the molecular weight of each fragment. A magnetic focusing mass spectrometer is illustrated schematically in Figure 19.

The fragmentation or cracking patterns produced depend on the ionization energy employed to generate them. For a particular set of experimental conditions, fragmentation patterns are generally reproducible and provide distinctive fingerprints for purposes of identification. A typical interpretation of a mass spectrum to elucidate structure requires a series of well-organized steps: determining the molecular weight as sketched above, determining the molecular formula based on an analysis of the isotope distribution pattern, and puzzling out the fragmentation pattern, largely from knowledge of chemical bond stabilities in various molecular environments.

APPLICATIONS. Detailed reports of mass spectroscopic analyses of polymer composite materials are still quite rare. The predominant use of MS is as a coupled method with gas chromatography. A number of successful applications for epoxy-based materials are reviewed by Mestan and Morris [48]. Stone, Ranganathan, and Zsolnay [49] have reported using pyrolysis–chemical ionization MS as a quality control tool for fingerprinting commercial RTV silicones. As part of their evaluation of thermal stability of thermoplastic matrices, Whitaker, Nease, and Yelton [50] reported using direct insertion probe MS to characterize PEEK (polyether etherketone), PES (polyethersulfone), and PEI (polyetherimide). In what is probably one of the earliest reports of successful coupling of HPLC and MS, Escott [51] confirmed the 1 : 1 and 2 : 1 reaction products formed between the monomethyl ester of nadic anhydride and 4,4′-methylenedianiline.

ASTM Methods for Spectroscopy

ASTM E 131	*Standard Definitions of Terms and Symbols Relating to Molecular Spectroscopy*
ASTM E 168	*Standard Recommended Practice for General Techniques of Infrared Quantitative Analysis*
ASTM E 204	*Standard Practices for Identification of Materials by Infrared Absorption Spectroscopy, Using the ASTM Coded Band and Chemical Classification Index*
ASTM E 334	*Standard Practices for General Techniques of Infrared Microanalysis*
ASTM E 573	*Standard Practices for Internal Reflection Spectroscopy*
ASTM E 386	*Standard Definitions of Terms, Symbols, Conventions and References Relating to High Resolution Nuclear Magnetic Resonance (NMR) Spectroscopy*

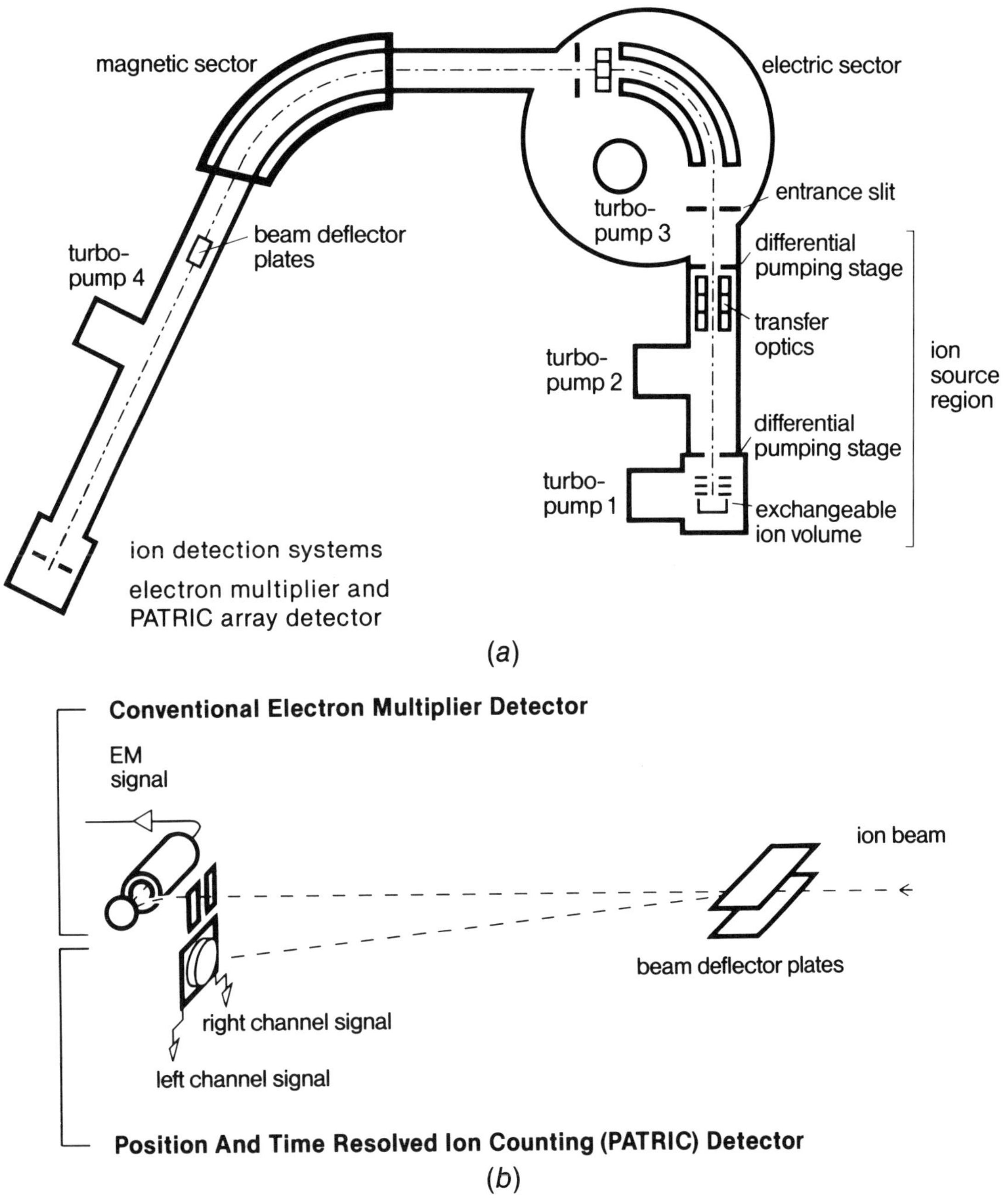

FIGURE 19 Schematic diagram of the Finnigan MAT 900 Magnetic Sector Mass Spectrometer showing the path taken by ions: (*a*) from the ion source region and through the magnetic sector; and (*b*) through the detectors. (Courtesy of Finnigan Corporation, San Jose, CA)

ASTM E 169 *Standard Practices for General Techniques of Ultraviolet–Visible Quantitative Analysis*

Chromatography

Chromatography is a collection of methods for separating and identifying the molecular components in a mixture or solution. The separation process is based on differing chemical properties of the components. The same and additional differences in chemical properties permit the subsequent determination of the relative or quantitative amounts of different chemical compounds in the material being analyzed.

The most widely applicable chromatographic methods for separating, fingerprinting, and quantitating the molecules in polymer composite materials are high performance liquid chromatography (HPLC), size exclusion chromatography or gel permeation chromatography

(SEC/GPC), and gas–liquid chromatography (GLC). Methods of additional kinds, for very specific applications, include column, ion-exchange, thin-layer, and paper chromatography; however, they have limited utility for characterizing polymeric materials and are not discussed.

The various kinds of chromatography differ significantly in terms of instrumentation and the mechanisms by which separations occur. Yet, the overall process and factors governing the chromatographic patterns or fingerprints obtained are essentially common to HPLC, GPC, and GLC.

At its simplest, the process of chromatography can be described as follows: A very small volume of dilute solution of the sample is injected into a moving stream of fluid (gas or liquid). This fluid or mobile phase carries the sample or eluent through a tubular column containing solid or liquid surfaces the eluent must contact. The solid or liquid material lining or packing the column is called the stationary phase. Individual molecular species separate from one another by a variety of mechanisms as the eluent passes through the column. Identical molecules group together and exit the column as effluent fractions. The fractions of like molecules move in separated "packets" through a detector, which is sensitive to some chemical property of the molecules. The detector response is continually recorded on a data acquisition system for further computations or to be used as a fingerprint.

General Principles

A simple chromatographic setup consists of sample injection port, column detector, and recorder (Fig. 20). While the microscopic details of separation in HPLC, GPC, and GLC differ fundamentally, the macroscopic treatment given in Figure 20 permits a clear description of the process. The sample consists of a dilute solution of A molecules and B molecules as a plug of A + B at some time after injection into the mobile phase. As the eluent moves through the stationary phase, such mechanisms as adsorption, selective filtration, and solute–solvent partitioning retard the forward velocity of B molecules more than A molecules. This retardation progresses by several hundred to thousands of idealized discrete separations, referred to as theoretical plates. The separate passage of A molecules and B molecules through the detector causes the distinctive peak response on the recorder. Differences in dispersion account for the widths of peaks on the chromatogram. A hypothetical chromatogram (Fig. 21) illustrates common measurements to facilitate the discussion that follows about the factors that influence a chromatographic separation.

Numerous factors influence the quality and repeatability of chromatographic separations. Those related to separation inside the column include flow rate, flow rate stability of the effluent, particle construction and size of the stationary phase, length of the column, and column diameter. Other factors that can influence resolution are the dead or void volume, the end fittings and tubular connections, the manner of sample injection, the sample

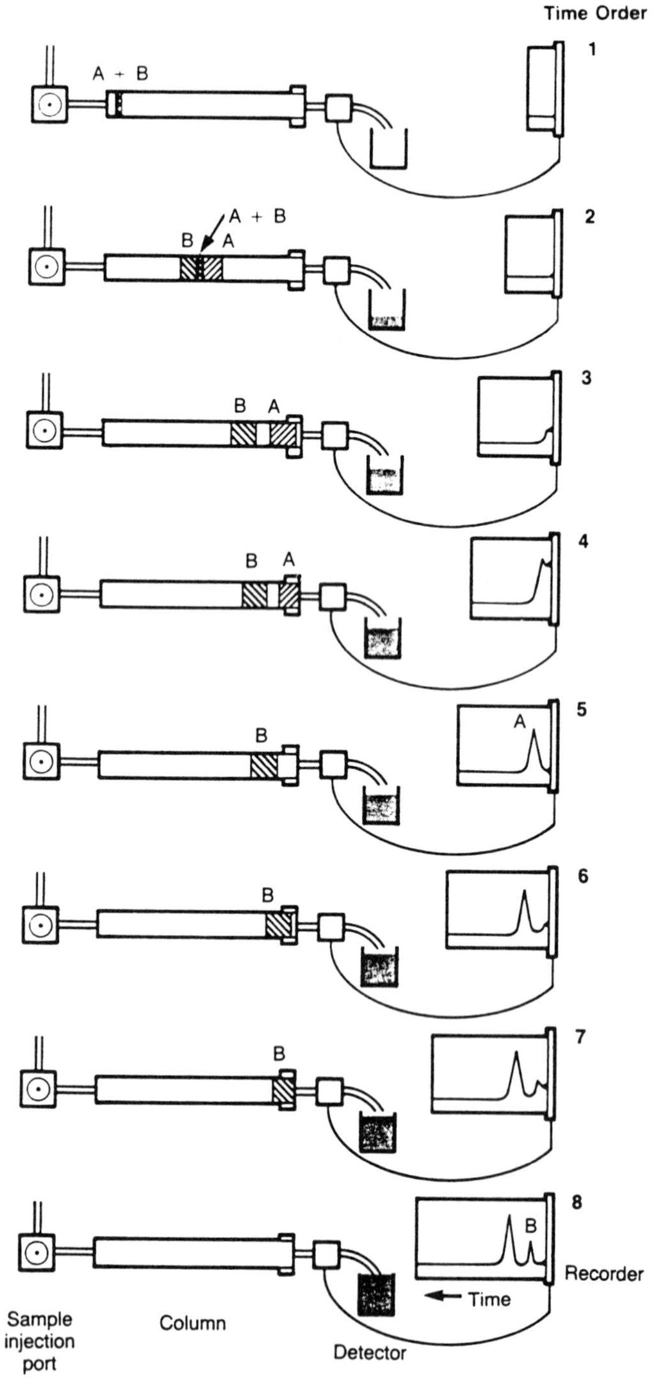

FIGURE **20** Simple chromatographic setup. (Reprinted with permission from W. R. Supina and R. S. Henley, *Chemistry, 37,* 15 (1964). Copyright 1964 American Chemical Society.)

size, the concentration of molecules in the sample, and numerous considerations involving the construction of the detector. The chromatographic separation can be understood in terms of three parameters: column capacity factor K', resolution R, and efficiency N. The capacity factors are governed by thermodynamic processes such as the mechanisms responsible for retention of mole-

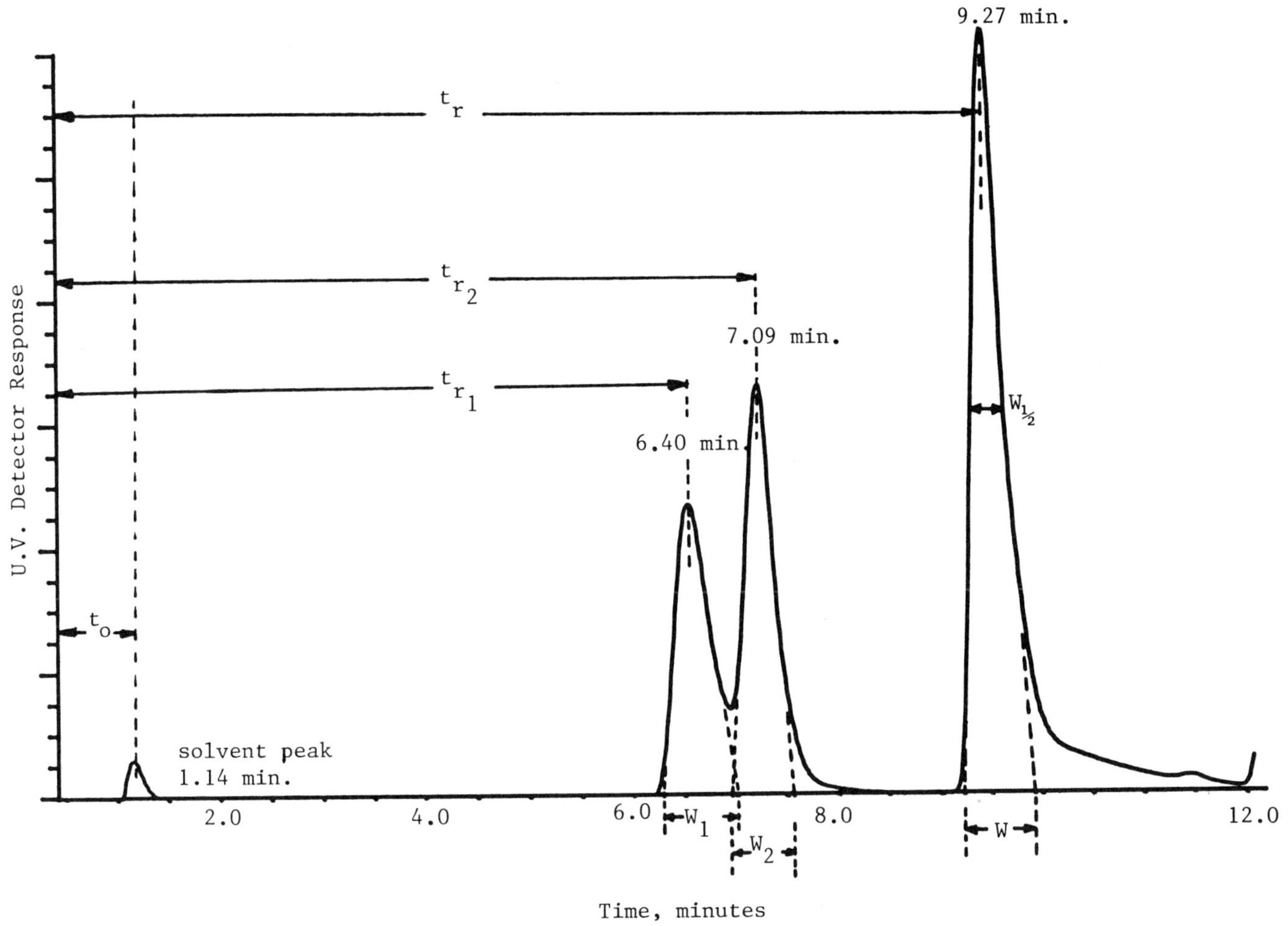

FIGURE 21 **Chromotogram for a phenl-formaldehyde resin showing conventional measurements.** (Courtesy of Ferro Composites, Los Angeles.)

cules on the stationary phase. Kinetic processes dictate the size of the volume element in which same-sized or monodisperse molecules are contained, hence influence the efficiency. Resolution, on the other hand, is determined by both kinetic and thermodynamic considerations. Considering the importance of these factors in chromatographic separations, it is appropriate to discuss each one briefly.

Column Capacity Factor (K')

The capacity factor expresses how the number of monodisperse molecules are distributed between the stationary phase N_s and the mobile phase N_m:

$$K' = \frac{N_s}{N_m} \quad (12a)$$

or

$$K' + 1 = \frac{N_s + N_m}{N_m} \quad (12b)$$

The column capacity factor affects the most fundamental measurement in a chromatographic separation, the retention time. Retention time t_r is the time required for the volume element containing monodisperse molecules to traverse the column. If there were nothing in the column to impede the forward velocity of the element, the time required would simply be the column length divided by the velocity of the element. This, of course, is not the case, since molecules carried by the mobile phase spend a finite time associated with the stationary phase. The retention time t_r is related to column capacity factor K' through t_0, the time needed for the mobile phase to traverse the column. At constant flow rate, the dependence can be expressed as follows:

$$t_r = t_0(1 + K') \quad (13)$$

Efficiency (N)

The efficiency of a column is evaluated in terms of the narrowness of the band of eluent (i.e., the width of the peak W). Small peak width generally reflects better or more efficient separations than large ones.

The efficiency is commonly expressed as the number of theoretical plates N or the height equivalent of a theoretical plate H; both variables are related simply through the column length L. For a Gaussian distribution, the following equalities hold:

$$N = \frac{L}{H} = 16\left(\frac{t_r}{W}\right)^2 = 5.52\left(\frac{t_r}{W_{1/2}}\right)^2 \quad (14)$$

where $W_{1/2}$ is the peak width at half-height. A more graphic way of appreciating the number N is to call it the number of individual extractions. The principal determinants of efficiency and the direction in which their magnitude influences it are tabulated below.

Variable	N Increases when
Molecule size	↓
Mobile phase viscosity	↓
Mobile phase flow rate	↓
Stationary phase size	↓
Pressure	↓
Column bore	↓
Temperature	↑

The resultant efficiency for a column used under a particular set of experimental conditions develops as a result of necessary practical tradeoffs among the variables mentioned above. Using HPLC as the example, reducing the stationary phase particle size allows one to reduce the column length to significantly lessen analysis time. There is a price to be paid for this benefit, however, because the primary pressure required to maintain the same volumetric flow rate increases with the inverse of the particle diameter; hence as pressure increases, efficiency decreases.

Resolution (*R*)

Separation of monodisperse bands of effluent implies resolution of the molecular species. This is the objective of chromatography. As a parameter, resolution is a measure of the completeness of separation of a pair of adjacent eluent bands or peaks. A pair of peaks is said to be completely resolved when the peak-to-peak distance divided by the average width of the peaks equals 1.5:

$$R = \frac{t_r(2) - t_r(1)}{(\frac{1}{2})(W(2) + W(1))} = 1.5 \quad (15)$$

Values of R smaller than 1.5 mean that the two peaks overlap, and quantitation becomes difficult. We can also define the separation factor $\alpha_{1,2}$ in terms of the retention times for peaks 1 and 2 and t_0.

$$\alpha_{1,2} = \frac{t_r(2) - t_r(1)}{t_r(1) - t_0} = K_2'/K_1' \quad (16)$$

Combining Eqs. (14)–(16) shows how resolution can be controlled by varying $\alpha_{1,2}$, N, and K':

$$R = \frac{1}{4}(\alpha - 1)\frac{K'}{1 + K'}N^{1/2} \quad (17)$$

where K' is the average column capacity for the two peaks.

To maximize the resolution, the three terms containing the separation factor, the capacity ratio, and the efficiency should be optimized separately [52]. For example, α varies as a function of the nature of the mobile and stationary phases, the capacity ratio varies with solvent strength, and the efficiency can be varied by changing certain variables (column length, mobile phase flow rate, temperature, etc.).

The sections that follow, dealing with theoretical aspects, equipment, and applications, concentrate chiefly on the HPLC and GPC techniques of liquid chromatography. Gas–liquid chromatography is not discussed with as much detail because it is not used directly for analyzing polymer molecules. Rather, the principal application of GLC is the detection and analysis of relatively low molecular weight compounds that may be present in a polymer resin formulation and prepregs (residual solvents, water, plasticizers, etc.), as well as degradation products that form on heating or pyrolyzing resins, prepregs, and composites. Very often, the GLC serves as an intermediate method for introducing separated volatile materials into equipment such as FTIR and MS instruments for identification.

Gel Permeation Chromatography

The objective of GPC is to determine characteristic molecular weight averages and the molecular weight distribution (MWD) of molecules present in a polymer matrix. Other physicochemical techniques are also used to determine molecular weights. These methods include light scattering, vapor pressure osmometry, and dilute solution viscosity. Each of these methods, however, gives information about certain average molecular weights and emphasizes only a portion of the molecular weight distribution. The power and utility of GPC is that it provides the various molecular weight averages ($\overline{M}_n$, $\overline{M}_w$, $\overline{M}_z$, and higher) and the MWD during one experiment relatively fast and reliably. A GPC chromatogram at the least sophisticated level provides a very graphic and visually useful chemical fingerprint that can reveal the presence of low molecular weight monomeric compounds, oligomers, and a myriad of resolved high molecular weight components or a broad envelope of the MWD.

The basis of GPC separation is a column containing particles whose internal porosity and packing create a series of tortuous channels as shown in Figure 22. The sizes of the channels and pores are controlled by the cross-link density in polymer gel particles. Many materials are used to form the gel particles, the most widely employed being a cross-linked styrene–divinylbenzene copolymer. These styrene-based copolymers are variously named, depending on the column manufacturer. Some examples are given in Table 4. Columns are typically rated by pore size and are offered in a range of increments ranging, for example, from 60 to 10^7 Å for

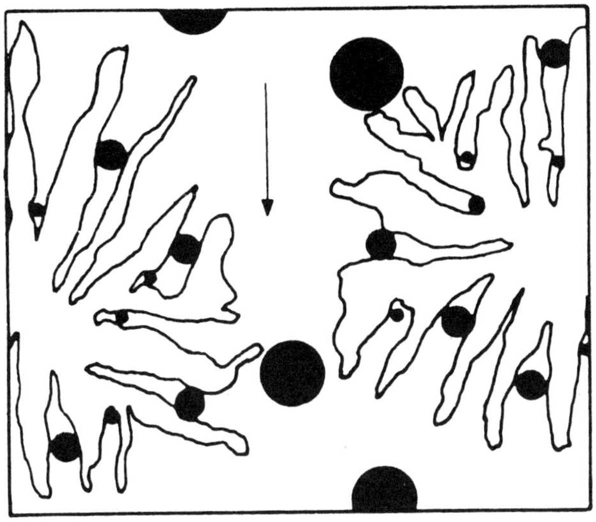

FIGURE 22 **Cross-sectional view of a porous particle, showing basis of GPC separation.** (Courtesy of Waters Chromatography Division, Millipore Corporation, Milford, MA.)

Waters Styragel columns. Pore size, by convention, is the size of a polystyrene molecule just large enough to be excluded from all the pores of a column. Thus, the column pore size represents the size exclusion limit. Columns of limited range are connected in series to overlap the expected molecular weight range of the sample.

Separation on a GPC column is illustrated schematically in Figure 23. As a highly dilute solution of molecules dissolved in the solvent or mobile phase is injected into the flowing mobile phase, large molecules find far fewer pores large enough to enter than are available for small molecules. Consequently, the larger molecules traverse the column more quickly than the small ones, entrained for longer periods in the pores. Figure 24 presents a hypothetical GPC chromatogram or fingerprint.

GPC does not provide molecular weight information directly. Rather, the column must be calibrated using several monodisperse standards whose molecular weights correspond to the expected range of separation for the column. Calibration with standards is necessary to relate the pore volume in the column to the effective or hydrodynamic volume occupied by a molecule having a particular molecular weight. This is not as simple as it may appear. Because of differences in molecular shape or chemical composition, two molecules occupying the identical hydrodynamic volume do not necessarily have identical molecular weights. For linear polymers, the computation of molecular weight and its distribution can be obtained using a universal calibration [52]. Molecular size is significantly affected by the branching characteristics of the molecule. Therefore GPC fractionation can differentiate between molecules with different degrees of branching. With proper calibration it is possible to estimate the degree of branching for branched or slightly cross-linked systems [52,53].

High Performance Liquid Chromatography

High performance liquid chromatography (HPLC) is now the most widely used analytical method for separating and analyzing mixtures. In fact, because of its extraordinary versatility for analyzing all kinds of mixtures that can be dissolved in some solvent, HPLC is the most pop-

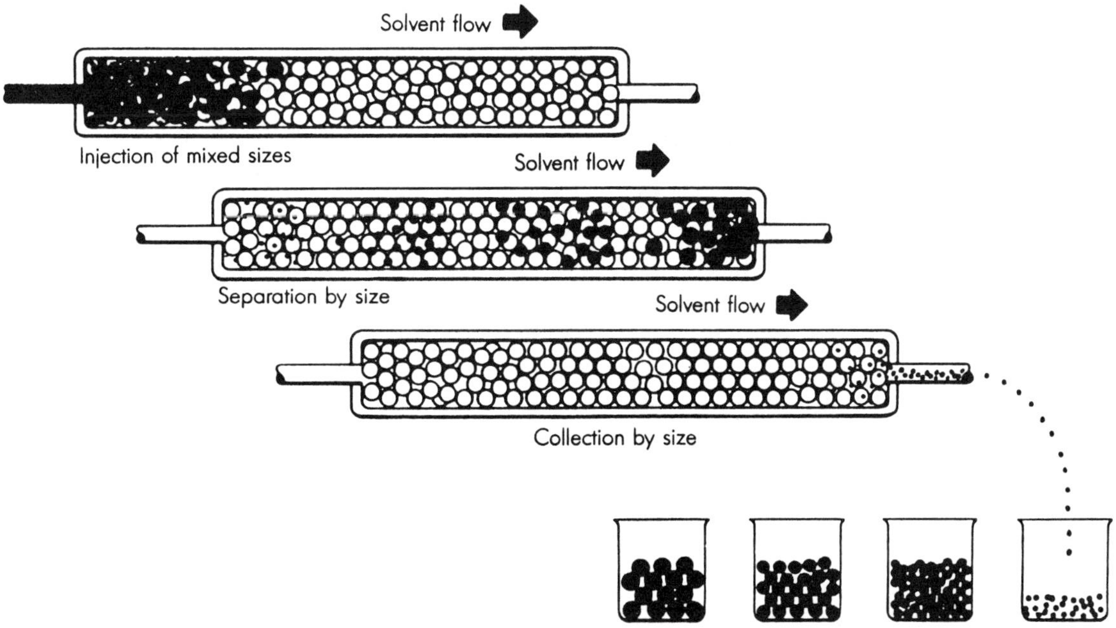

FIGURE 23 Illustration of size separation on a GPC column. (Courtesy of Waters Chromatography Division, Millipore Corporation, Milford, MA.)

TABLE 4
Packing Materials for Size Exclusion Chromatography

Name	Particle Size (μm)	Base Material[a]	Average Pore Size (Å)[b]	MW Exclusion Limit, or Range (PS)	Manufacturer
Organic-Soluble Samples					
μ Styragel	10	PS-DVB	100	50–1,500	Waters
Ultrastryagel	<10	PS-DVB	500	100–10,000	
			10^3	200–30,000	
			10^4	5,000–60,000	
			10^5	50,000–4,000,000	
			10^6	200,000–$>10^6$	
μ Spherogel	10	PS-DVB	50	<1,000	Beckman
			100	<4,000	
			500	$500–2 \times 10^4$	
			10^3	$1,000–4 \times 10^4$	
			10^4	$4 \times 10^3–4 \times 10^5$	
			10^5	$4 \times 10^4–4 \times 10^6$	
			10^6	$4 \times 10^5–4 \times 10^7$	
Water-Soluble Samples					
Zorbax-PSM-60	6	Silica gel	60	$0.01–4 \times 10^4$	DuPont
Zorbax-PSM-500			500	$0.1–5 \times 10^5$	
Zorbax-PSM-1,000			1,000	$0.03–2 \times 10^6$	
μ Bondagel	10	Silica gel	E-125	2,000–50,000	Waters
		(μPorasil)	E-500	5,000–500,000	
			E-1,000	50,000–2,000,000	
			E-High A	15,000–7,000,000	
			E-Linear	2,000–2,000,000	
μ Porasil GPC 60 A	10	Silica gel	60	1,000–10,000	Waters
Chromegapore	10	Silica gel	60	$<5 \times 10^4$	E. S. Industries
			100	$5–8 \times 10^4$	
			300	$1.5–3 \times 10^5$	
			500	$3–6 \times 10^5$	
			1,000	$6–1.4 \times 10^6$	
Spherogel PW	10	Polymer	1,000	5×10^2	Beckman
			2,000	5×10^3	
	13		3,000	6×10^4	
	13		4,000	5×10^5	
	17		5,000	8×10^5	
Supelcosil LC-Si	5	Silica	100		Supelco
			300		
			500		
SynChropak GPC	5,10	Silica gel	100	$10^3–10^5$	Syn Chrom
			300	$5 \times 10^3–8 \times 10^5$	
			500	$5 \times 10^3–4 \times 10^6$	
			1,000	$10^4–5 \times 10^7$	
			4,000		
Spherogel SW	10	Silica gel	2,000	9×10^3	Beckman
	10		3,000	4×10^4	
	13		4,000	2×10^5	
Bio-Sil TSK 125	10	Silica gel	125	500–20,000	Bio-Rad
Bio-Sil TSK 250	10	Silica gel	250	1,000–100,000	
Bio-Sil TSK 400	13	Silica gel	400	5,000–400,000	
Bio-gel TSK-10	10	HPR			Bio-Rad
TSK-20	10	HPR	50		
TSK-30	13	HPR	200	up to 100,000	
TSK-40	13	HPR	500	1,000–700,000	
TSK-50	17	HPR	1,000	10,000–2,000,000	
TSK-60	25	HPR		100,000–20,000,000	

[a] PS-DVB = polystyrene–divinylbenzene; HPR = hydroxylated polyether resin.
Reproduced with permission from M. T. Gilbert, *High Performance Liquid Chromatography,* Wright, an imprint of BUTTERWORTHS, Bristol, 1987 (Appendix, Table 4).
[b] 1 μm = 10,000 Å

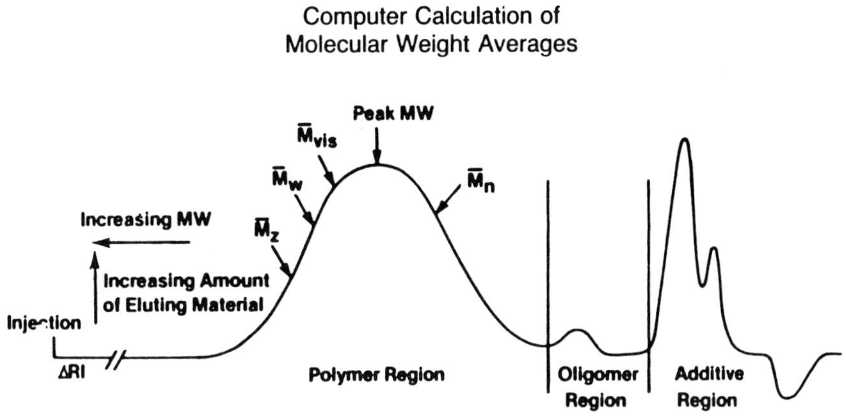

FIGURE 24 Hypothetical GPC chromatogram showing characteristic molecular weight distribution data. (Courtesy of Waters Chromatography Division, Millipore Corporation, Milford, MA.)

ular of the instrumental methods used in chemical laboratories.

The greatest utility of HPLC in the domain of polymer composite materials is for molecular weights less than 2000. Thus, it is ideally suited for characterizing complex thermoset resin formulations and B-staged prepregs that consist chiefly of monomeric and oligomeric materials. Virtually all the varieties of addition and condensation polymers can be analyzed. Examples of epoxy, phenolic, and polyimide chromatograms are given in Figure 25. On the other hand, fully polymerized thermoplastic resins generally exceed 2000 molecular weight and are more appropriately characterized by GPC.

Unlike GPC, which separates molecules according to their sizes, HPLC separates molecules according to their chemical identities. Subsequent objectives are to identify the various compounds present and to determine their relative or quantitative amounts in the mixture.

The basis for HPLC is that molecules in the mobile phase will be separated according to their respective chemical affinities for the mobile phase or for the solid support. Molecules that are attracted more strongly to the stationary phase will traverse the column more slowly than those either less strongly attracted or more highly associated with the mobile phase. Three principal kinds of stationary phase are used in HPLC. They are classified according to the type of chemical interaction that occurs between the solute and the stationary phase. These classifications are normal phase, reversed phase, and ion exchange.

In normal phase HPLC, the stationary phase is significantly more polar than the mobile phase and the separation is based on adsorption of the solutes. The most common stationary phases are hydrated amorphous silica (Fig. 26*a*) and alumina gels. Molecules having highly polar functional groups, such as cyano, amino, hyroxyl, and carbonyl, will be adsorbed more strongly and will be displaced by the mobile phase more slowly than less polar molecules. Thus, using a nonpolar mobile phase, the more polar the compound, the longer its retention time. The analyst, by judicious selection of solvent or solvent pairs on the basis of polarity, can exert considerable control over retention times and baseline resolution of the components in the mixture.

Reversed-phase HPLC utilizes solvents and solvent pairs having high polarity and stationary phases that are non-polar. Two kinds of solid support are used most commonly. The most popular is hydrated silica particles to which aliphatic hydrocarbons have been bonded (Fig. 26*b*). The typical hydrocarbon groups are methyl (C1), octyl (C8), or octadecyl (C18). These bonded hydrocarbon layers behave as a thin organic liquid, and the separation process can be described as a liquid–liquid extraction. This, however, is an oversimplification insofar as different mechanisms may dominate depending on the nature and composition of the mobile phase. Mechanisms that have been advanced include solvophobic interactions, solvophilic interactions with the stationary phase, interactions with accessible silanol groups, and secondary equilibria promoted by ionic additives in the mobile phase [54]. The second type of solid support is comprised of the styrene–divinylbenzene copolymer gels used for GPC.

Reversed-phase HPLC is more popular because of its greater versatility and simplicity. The adsorption–desorption processes tend to proceed faster than in normal phase HPLC; therefore, peaks exhibit less spreading and tailing. When normal and reversed-phase separations are performed using mobile phases of constant solvent power (eluotropic strength), the elution is said to be isocratic. The optimum separation cannot always be achieved this way, especially when molecules having a wide range of separation factors are present in the mixture. In such instances, separation, efficiency, and resolution may be vastly improved using gradient elution techniques. A gradient is created by mixing solvents having different eluting characteristics, beginning with the mobile phase composition that elutes most poorly and proceeding in linear, nonlinear, or stepwise fashion to a stronger eluting phase. For example, a gradient that

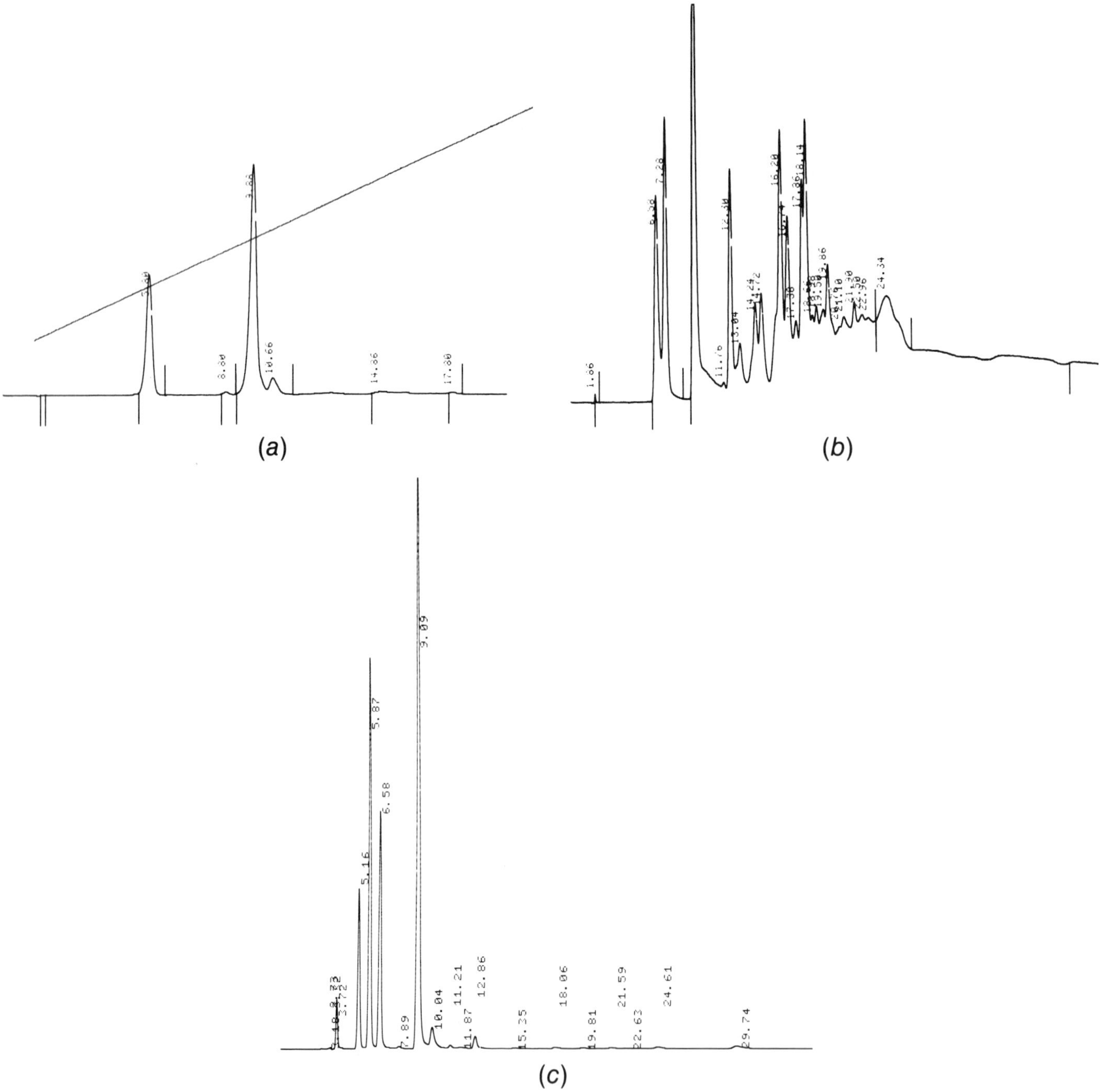

FIGURE 25 **HPLC chromatograms for prepreg resin extracts: (*a*) CE-9000 epoxy, (*b*) CPH-2209 phenolic, and (*c*) CPI-2237 (PMR-15) polyimide.** (Courtesy of Ferro Composites, Los Angeles.)

starts with its mobile phase composed of 80% acetonitrile and 20% water may end with 30% acetonitrile and 70% water.

Two other types of stationary phase are mentioned briefly because they have limited utility for polymeric matrices of the kinds encountered in composite materials. One is a version of reversed-phase HPLC in which the phase bonded to the solid support is a polar functional group such as thiol, diol, amino, nitro, or cyano. Separations are site specific and occur through hydrogen bonding or charge–transfer complexation. This form of HPLC is also called affinity chromatography. The second form is ion-exchange chromatography, using ion-exchange resin beads as the stationary phase. Ion-exchange chromatography finds its principal applications in inorganic analysis and analysis of biological materials.

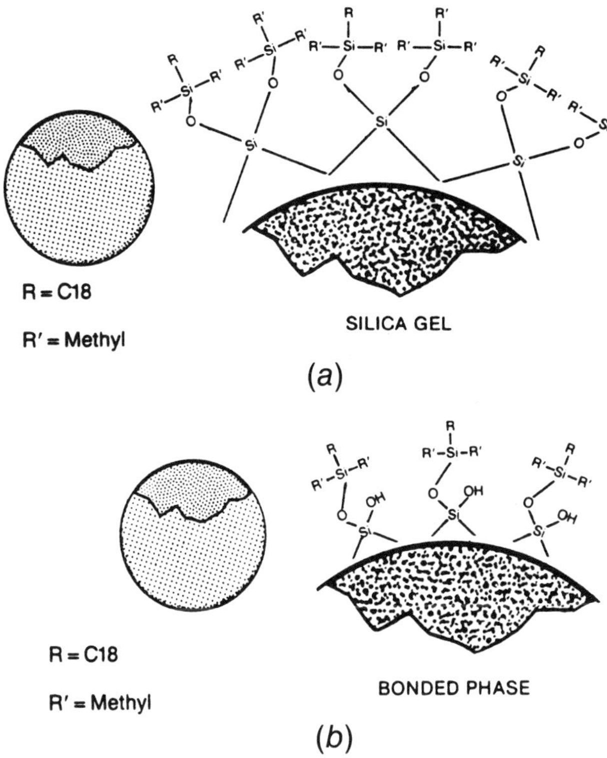

FIGURE 26 **HPLC solid supports used in (*a*) normal phase and (*b*) reverse phase chromatography.** (Courtesy of Ranin Instrument Company, Woburn, MA.)

Gas–Liquid Chromatography

GLC methods cannot handle most of the kinds of materials analyzed by HPLC and GPC, especially monomeric and polymeric materials having molecular weights that preclude their vaporization. Samples may be injected onto the GLC column as solutions or gas mixtures. The separation, however, must occur for solutes in their vapor state. One particular advantage of GLC is in the detection and analysis of relatively low molecular weight additives and contaminants in formulated resins and prepregs. Another is the separation of decomposition products formed from heating and oxidizing or pyrolyzing composite materials. The chemical or radiative destruction may be induced in another chamber or instrument downstream from the GLC or in a pyrolysis chamber attached directly to the injection port. Subsequent identification can be made using such instrumental methods as MS or FTIR. By far the most useful paired or hyphenated method is GLC-MS.

Just as for HPLC, the objective of gas chromatography is to separate chemical compounds in mixtures that may be tremendously complex. GLC is accomplished by injecting either a gaseous sample or a vaporizable solution into a stream of moving gas, the carrier gas or mobile phase. The vaporized sample is swept into the column, and components are separated by the same repetitive adsorption–desorption processes described earlier for HPLC. The effluent is then swept into one or more detectors for subsequent recording and data analysis.

The principal parameters for optimizing a GLC separation are similar to those of liquid chromatography: sample size, carrier gas flow rate, selection of separation factors based on the stationary phase or mixtures of stationary phases, thermal stability of the bound liquid phase, and column temperature. The latter factor is especially important because it is the analogue of the choice of mobile phase in HPLC. For mixtures whose components have a narrow range in boiling points, a single isothermal run can provide a well-resolved separation. Complex mixtures whose components have a wide range in boiling points cannot be separated well using one run, however. Two separate experiments can be run, one at moderate temperature to separate the moderate boiling point solutes and the other at some higher temperature to separate the higher boiling components. Such an analysis is normally performed far more satisfactorily using temperature programming.

Instrumentation and Measurement

HPLC AND GPC. The equipment needed to perform liquid chromatography is illustrated schematically in Figure 27. Modern chromatographic setups tend to be modular and to emphasize microprocessor control for each of the major modular components: the mobile phase delivery system, the injection system, the detector(s), and the data acquisition system. The modular approach to the construction of chromatographic equipment is especially suitable for performing routine analyses for a variety of reasons, including the following.

1. It is easier to isolate and troubleshoot operating problems.
2. Teaching use and techniques is straightforward.
3. A basic or lower cost system can be augmented by additional capabilities as needed.
4. The analyst can "mix and match" or interchange other manufactured components for specialty analyses or applications.

As shown in Figure 27, HPLC and GPC require a source of mobile phase, a way to mix and pump one or more solvents that comprise the mobile phase, an injection system capable of delivering highly repeatable and very small volumes of sample to the column, one or more columns (which may be enclosed in an oven), one or more detectors, and a way to display the chromatograms. Data acquisition and storage devices are by no means essential to the separation process. Yet their power for data reduction, report development, and storage of what could otherwise become huge quantities of paper make them nearly essential for high output, routine analyses.

MOBILE PHASE SUPPLY AND DELIVERY SYSTEM. The solvent supply is often diagrammed as a bottle, flask, or beaker; yet as a portion of the delivery system, it is

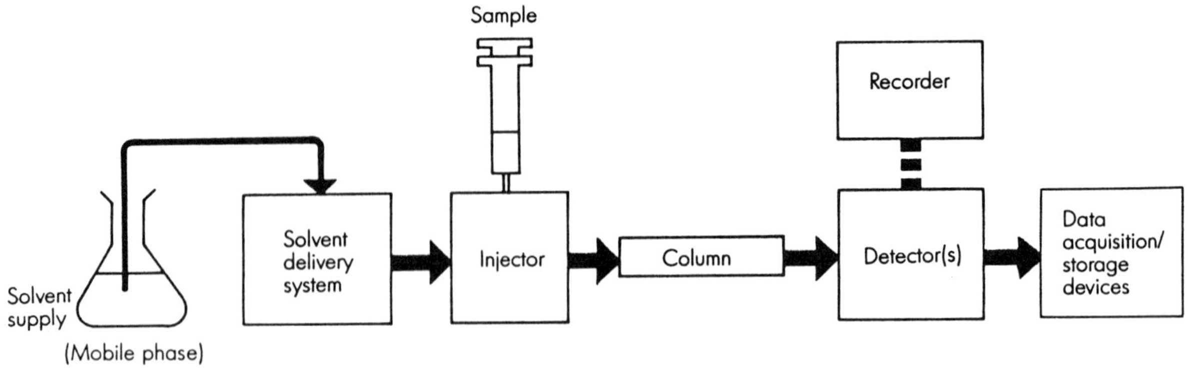

FIGURE 27 **Schematic diagram of a typical chromatography setup.** (Courtesy of Waters Chromatography Division, Millipore Corporation, Milford, MA.)

really more complex and important. The key function of the system ahead of the pump is to maintain the solvent at constant composition and free of particulate matter and dissolved gases. GPC and isocratic HPLC are carried out with a simple mobile phase. Gradient elution HPLC systems can be quite complex, requiring multiple solvent reservoirs and multiple pumps (or one pump and a variety of mixing valves and a mixing chamber for the mobile phase of constantly varying solvent strength).

The pumping and metering portion of the delivery system provides a repeatable supply of the mobile phase at a constant, low flow rate and as pulse-free as possible. The design and construction of pumps used in liquid chromatography must satisfy the following stringent requirements:

- Ruggedness and resistance to a broad range of solvents, including water.
- Operation at high pressures in the range 3.5–34.5 MPa (500–5000 psi).
- Pulse-free flow.
- Meter typically 0.5–3 mL/min at far better than 1% reproducibility.
- Easy access for maintenance of seals and valves.
- Small volume hold-up for gradient elution and rapid solvent changeover.

Metering the flow can be accomplished using constant volume and constant pressure pumps. The former, based on single or multiple reciprocating piston pumps, are the most widely used owing to lower cost and wide range of flow rates.

INJECTION SYSTEM. The injection system introduces the solution of monomer or polymer into the mobile phase. One key requirement is that it must deliver small volumes of sample reproducibly and with high precision. A second is that its presence must not disturb the continuous flow of mobile phase. A typical injector can normally deliver liquid volumes of 1.0–2000 μL. Special systems are available for delivering volumes as small as 0.06 μL to the larger volumes required for preparative-scale chromatography. The three principal types of injection system feature, respectively, syringe, manual sampling valves, and automated sampling valves.

THE COLUMN. The column is the most important component in the chromatograph, since this is where the separation occurs. Moreover, the column is the component over which the analyst has the greatest control and the most degrees of freedom in operation. HPLC columns are generally operated singly or in pairs. GPC of polymers having broad MWDs often requires several columns of different pore size ranges in series called a column bank. The low solubility of high molecular weight polymers and their attendant higher viscosity solutions require an oven for elevated temperature operation.

THE DETECTOR. The detector responds to some chemical or physical property of the separated fractions as they elute from the column. A typical detector will:

- Have a wide linear sensitivity range and very rapid response time.
- Be capable of detecting concentrations as small as 1 part of solute in 10 parts of mobile phase.
- Cause no remixing of components.
- Be insensitive to temperature changes.
- Allow the effluent to pass to other detectors with no destruction.
- Operate with low noise and very little signal drift.

There are two main types of detector: the bulk or physical property type and the selective chemical property type. The physical property type consists of differential refractometers of several varieties. It is the most widely employed detector for GPC, which is concerned with the number of molecular species of given molecular weight. The differential refractometer is often called a universal detector because all transparent substances refract light. Its extraordinary usefulness in GPC is based on the fact that above molecular weights of approximately 1000, solutions give constant refractive index at constant concentration. Alternatively, changes in refractive index are directly proportional to changes in concentration. Low-angle laser light-scattering photometry detectors are a relatively new development in the field. Applications include molecular weight, peak broadening calibration, determination of unperturbed chain dimensions, long chain branching frequency, and microgel content [53].

The selective chemical property type of detector is the most widely employed for HPLC. The chemical properties for which there are detectors include ultraviolet–visible (UV–visible) and infrared absorbance. The UV–visible absorbance detector is the most generally applicable for polymer matrix materials, although considerable interest is growing in the coupling of HPLC and FTIR instruments for analyses based on chemical functional groups. An ultraviolet detector is often used in conjunction with a refractometer during GPC to detect oligomeric materials as well as the monomeric compounds more commonly found in thermoplastic matrix resins. These materials include antioxidants, thermal stabilizers, and UV absorbers.

DATA RECORDING AND PROCESSING. The simplest representation of detector output is the chromatogram, drawn as a continuous trace on a chart recorder. The trace provides a permanent record of the analysis as well as a chemical fingerprint that can be compared visually with other chromatograms. In the absence of data processing equipment, all the quantitative aspects of the analysis can be performed from direct measurement of time and characteristic peak dimensions on the paper. It is an extremely tedious and time-consuming process for routine analyses, however.

The relatively low cost microprocessors developed during the past decade have vastly simplified data acquisition and processing and have improved reliability of the entire chromatographic process. Systems developed by such instrument manufacturers as Waters, Varian, and Perkin Elmer now provide almost trouble-free and extremely fast operation. A typical HPLC chromatogram and report (Fig. 28) contain a wealth of pertinent characterization data including retention times and integrated

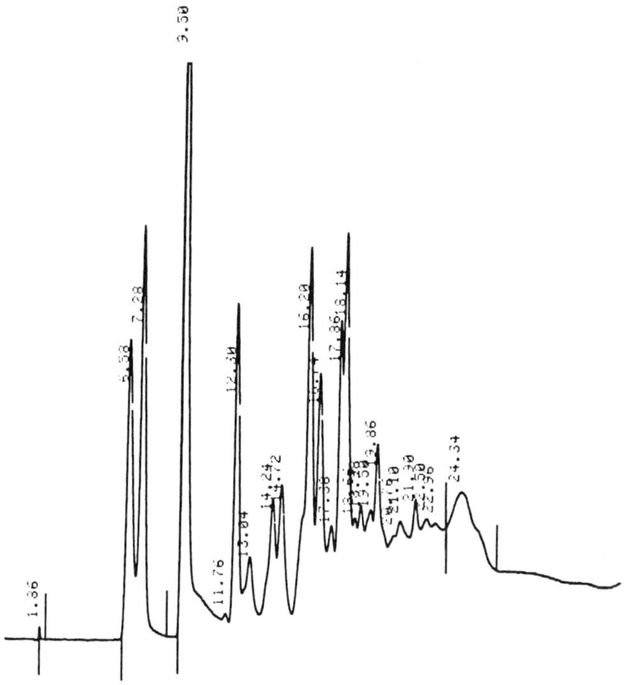


```
20          OFF  AUTO PARAMETER
21           12  PEAK WIDTH
22      250.000  NOISE REJECTION
23          200  AREA REJECTION
24        60.00  RUN STOP
JAN. 1A. 1988   12:45:44     CHART  0.50 CM/MIN     FLOW      2.00 ML/MIN
PRESSURE   2200.0            DETECTOR   280/100
                             RUN #4                     CALC #0
COLUMN                       SOLVENT                  OPR ID:       6

EXTERNAL STANDARD QUANTITATION
```

PEAK#	AMOUNT	RT	EXP RT	AREA		RF
	420.92200	1.86		420922	L	0.000000E0
	53516.60000	6.58		53516989	F	0.000000E0
	67690.50000	7.28		67690953	L	0.000000E0
	363676.00000	9.50		363680479	S	0.000000E0
	1457.31000	11.76		1457320	F	0.000000E0
	45406.00000	12.30		45406314	F	0.000000E0
	13806.70000	13.04		13806823	F	0.000000E0
	22908.80000	14.24		22908912	F	0.000000E0
	22710.70000	14.72		22710800	F	0.000000E0
	71268.70000	16.20		71269215	F	0.000000E0
	41333.80000	16.74		41334097	F	0.000000E0
	12879.70000	17.38		12879779	F	0.000000E0
	33567.60000	17.86		33567845	F	0.000000E0
	58235.60000	18.14		58236015	F	0.000000E0
	8723.71000	18.66		8723761	F	0.000000E0
	14019.40000	18.98		14019488	F	0.000000E0
	15736.10000	19.50		15736191	F	0.000000E0
	30033.70000	19.86		30033883	F	0.000000E0
	8975.86000	20.76		8975906	F	0.000000E0
	16857.20000	21.10		16857321	F	0.000000E0
	16204.40000	21.90		16204569	F	0.000000E0
	14813.70000	22.50		14813877	F	0.000000E0
	11881.80000	22.96		11881927	T	0.000000E0
	35774.00000	24.34		35774265	L	0.000000E0
TOTAL	981901.00000					

PEAK#	AMOUNT	RT	%
	420.92200	1.86	0.04
	53516.60000	6.58	5.45
	67690.50000	7.28	6.89
	363676.00000	9.50	37.03
	1457.31000	11.76	0.14
	45406.00000	12.30	4.62
	13806.70000	13.04	1.40
	22908.80000	14.24	2.33
	22710.70000	14.72	2.31
	71268.70000	16.20	7.25
	41333.80000	16.74	4.20
	12879.70000	17.38	1.31
	33567.60000	17.86	3.41
	58235.60000	18.14	5.93
	8723.71000	18.66	0.88
	14019.40000	18.98	1.42
	15736.10000	19.50	1.60
	30033.70000	19.86	3.05
	8975.86000	20.76	0.91
	16857.20000	21.10	1.71
	16204.40000	21.90	1.65
	14813.70000	22.50	1.50
	11881.80000	22.96	1.21
	35774.00000	24.34	3.64
TOTAL	981901.00000		99.99

FIGURE 28 HPLC and GPC reports. (Courtesy of Ferro Composites, Los Angeles.)

peak areas. With the appropriate software, detector response and calibration factors can be used to identify the components with their weight percent in the formulation. These programs additionally permit the automated comparison of several analyses to signal departures from expected or specification values.

GAS–LIQUID CHROMATOGRAPHY. The plumbing for GLC is much simpler than that required for the liquid chromatographic techniques. Mobile phase flow rates are adjusted and maintained using gas pressure regulators and flow meters.

There are two types of GLC column. One is an open, tubular or capillary column whose walls are lined with the liquid stationary phase. The other is a wider bore, packed column containing inert particles to which nonpolar or polar liquid phases have been bonded. Many of the stationary phases used for GLC are the same as used in HPLC columns. One of the main differences between GLC and HPLC columns is that the former are significantly longer, hence more efficient. Packed columns are typically several meters long, and capillary columns may be 150 meters long. Efficiency ratings per meter of column length of 3000–4000 theoretical plates are not uncommon. Thus, a capillary column can have in excess of 500,000 theoretical plates for separation.

The most popular detectors for gas chromatography include thermal conductivity (TC), flame ionization (FID), photometric ionization (PID), and electron capture (ECD).

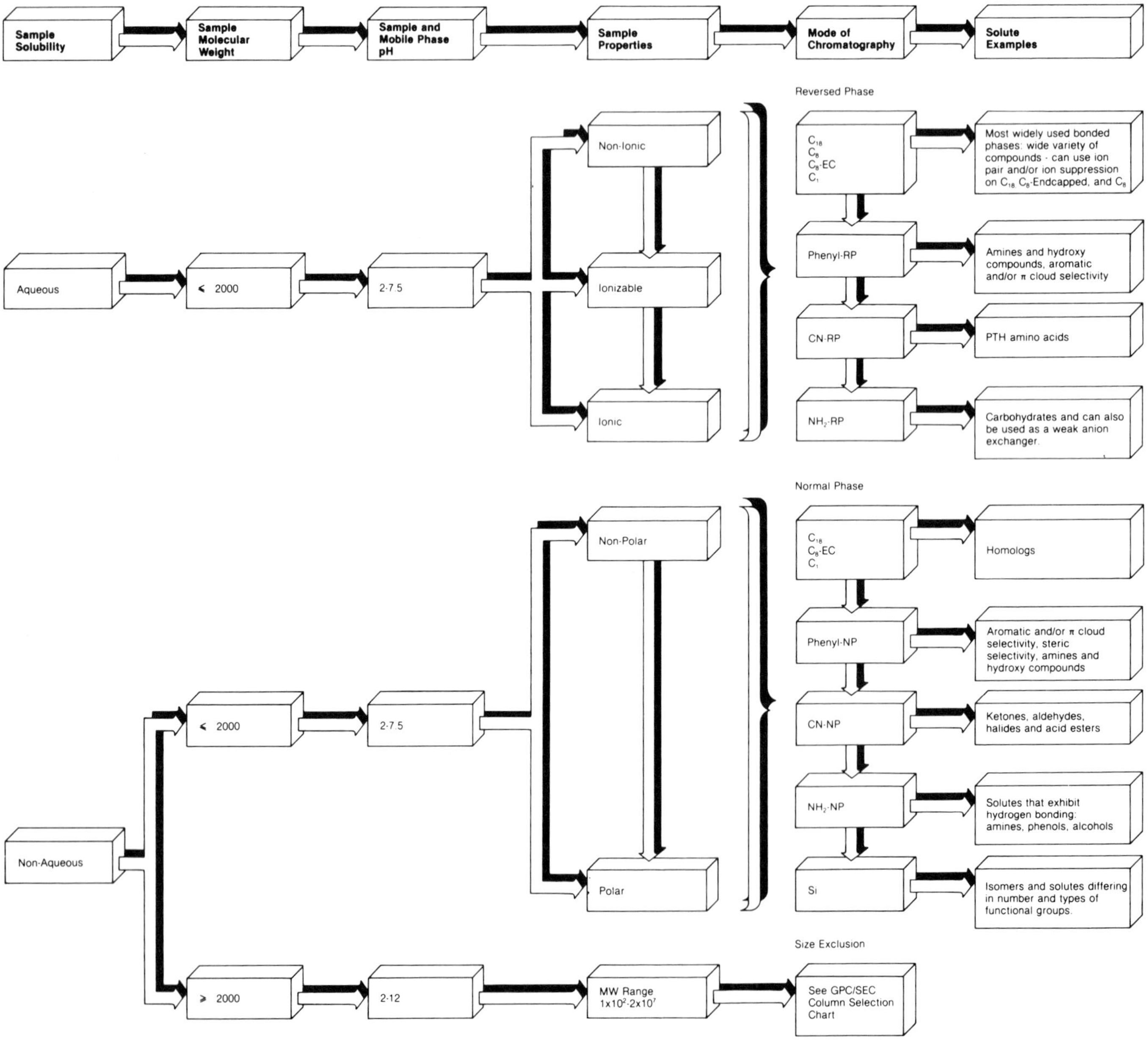

FIGURE 29 **Decision tree for selecting chromatography solid supports.** (Courtesy of Nicolet Instruments Corporation.)

APPLICATIONS. Chromatographic techniques are very powerful tools for mixture analysis. The applications range from quality control of a product to the characterization and development of new or existing products. Regardless of the type of application, the first task of liquid chromatography is to perform an optimum separation. This is achieved by using the correct combination of stationary and mobile phases. For polymers having molecular weight greater than 2000, GPC is preferred over other liquid chromatographic techniques. For more complex product mixtures with similar molecular sizes but different polar characteristics, reversed-phase or normal phase packings with suitable solvent system are preferred. However, if the sample is a simple mixture and has less than 2000 molecular weight but contains a wide distribution of molecular sizes, the preferable technique is still GPC. A decision tree for the selection of such methods is provided in Figure 29.

Quality control applications deal with the fingerprint of the raw materials, the resins, and the prepreg extracts. The first requirement is the solubility of mixture components in the mobile phase. Solubility plays the key role in selecting the mobile phase. Selection of the packing material, the flow rate, mobile phase composition, and the other operating parameters must be optimized to get the desired separation and detection. Optimization as it pertains to the detector response often means decreasing sensitivity in favor of detectability. Figure 30 gives an example based on an analysis of a TGMDA-novolac-DDS epoxy resin at one of three UV wavelengths [55]. The 230 nm wavelength is not as sensitive for the tetrafunctional epoxy and aromatic diamine constituents (TGMDA and DDS), but it does permit detection of the novolac at reasonable sensitivity. The main emphasis of a quality control analysis is detecting extra peaks due to undesirable components and deviations from the mixture composition. Numerous studies have demonstrated the feasibility of HPLC, GPC, or both techniques for quality assurance purposes [41,53,55–58].

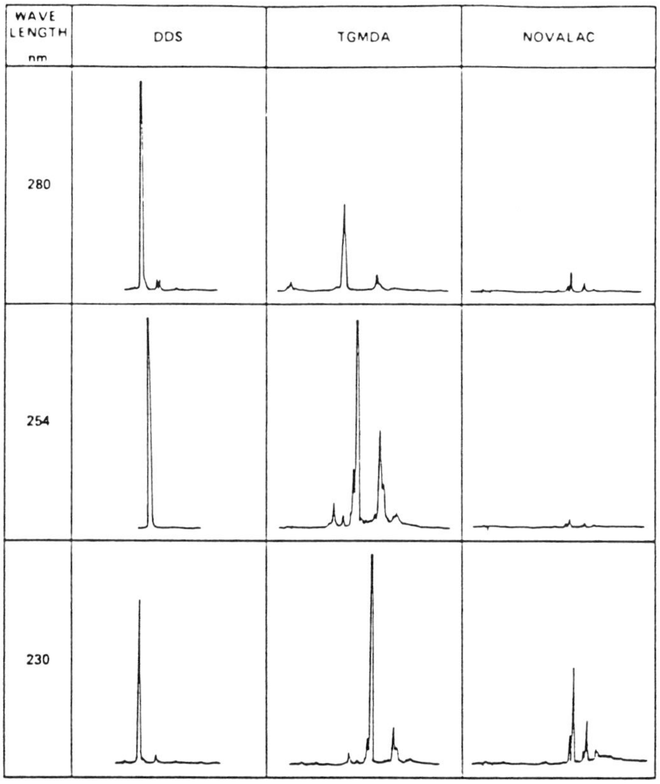

FIGURE 30 Chromatograms (HPLC) of TGMDA, DDS, and a novolac resin at 280, 254, and 230 nm UV wavelengths. (From Ref. 55, copyright ASTM. Reprinted with permission.)

An increasing number of prepreg and resin manufacturers now use chromatographic techniques to ensure the quality of their products. Indeed, almost all types of thermoset resin system can be analyzed by HPLC or GPC techniques. Figure 31 illustrates an HPLC analysis of "good" versus unacceptable prepregs. Operating conditions to achieve separation for epoxy-based systems are discussed in many references [58–61], including a comprehensive review article [48].

Chromatographic analyses of polyimide resins, in particular PMR-15 and LARC 160, have received considerable attention recently because of their increasing use as high temperature aerospace matrix materials [35,57,62–64].

Chromatographic techniques are finding extensive use in research and development applications. Some of the popular applications include understanding reaction kinetics [27], determining the extent of B-staging in prepreg materials [55,60], and aging studies on prepregs [14,58,60,62,64–70]. Figure 32 shows an example GPC study of 22°C aging of a TGMDA-DDS epoxy-based prepreg for 1 day and 40 days. Results of a similar analysis performed at −13, 22, and 49°C are shown in Figure 33.

Most of these applications address the issue of the change in the mixture composition due to the consumption of reactants or the formation of new products. Once the conditions for the separation of components within a mixture have been established, quantitation of a particular component is possible in absolute concentration (i.e., weight percent) or in terms of ratio with respect to the other components. Quantitative analyses of known constituents are performed by constructing Beer's law calibration plots of detector response (area units) versus known mass, at one or more wavelengths.

Figure 34*a* shows how the response factors for the constituents in PMR-15 vary with wavelength using UV diode array detector. Beer's law plots for 4,4′-methylenedianiline (MDA) are shown at five wavelengths in Figure 34*b* [71]. The feasibility of quantitation leads to quantitative assessment of kinetic parameters, information that can effectively be utilized for new product development and for optimizing the cure cycle for an existing product [56,70].

For thermoplastic matrices, molecular weight and molecular weight distribution are the most critical parameters for processing as well as for determining the final mechanical properties of the composite. Various molecular weight averages and their distribution can be measured using the GPC [72]. An example GPC analysis of PEEK conducted at 150°C is given in Figure 35.

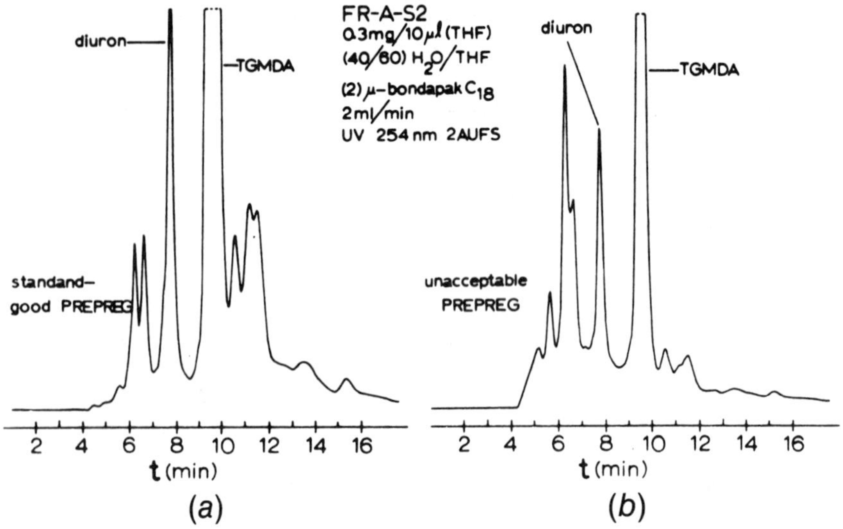

LC Analysis of FR-A-S2 Prepregs

sample	weight fraction with respect to component in standard prepreg	
	TGMDA	diuron
Standard-Good Prepreg	1	1
Acceptable Prepreg (stored 4 months at room temperature)	0.75	1.0
Acceptable Prepreg (stored 2 months in freezer)	0.61	1.0
Unacceptable Prepreg ("boardy" after 3 months storage in freezer)	0.36	0.67
Unacceptable Prepreg ("boardy" after 9 months storage in freezer)	0.32	0.67

FIGURE 31 HPLC chromatograms showing (*a*) good and (*b*) unacceptable epoxy prepregs. (Reprinted from ref. 58, p. 72, by courtesy of Marcel Dekker, Inc.)

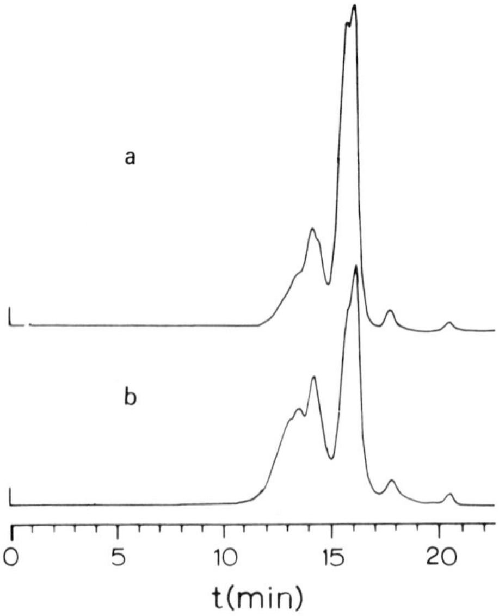

FIGURE 32 GPC chromatograms showing 22°C aging of a TGMDA-DDS epoxy-based prepreg for (*a*) 1 day and (*b*) 40 days. (From Ref. 65.)

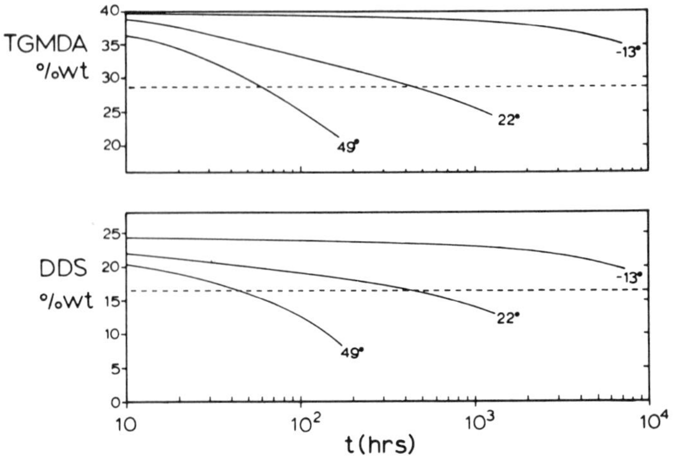

FIGURE 33 GPC analysis of TGMDA-DDS epoxy-based prepreg after aging at −13, 22, and 49°C. (From Ref. 65.)

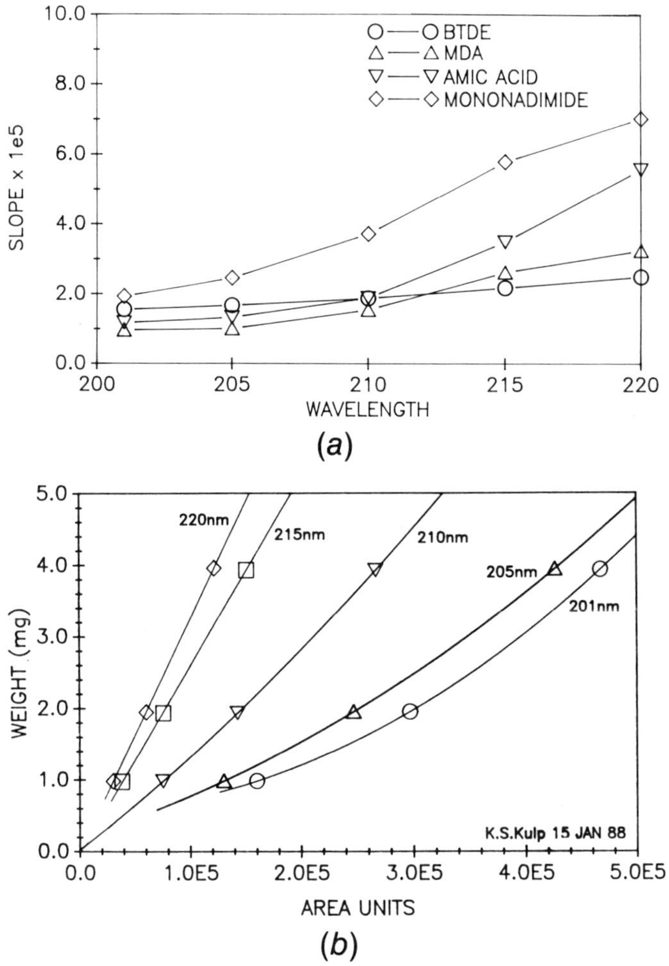

FIGURE 34 (*a*) Changes in the slope of the UV response curve as a function of wavelength for PMR-15 constituents and reaction products, (*b*) UV response curves at various wavelengths for MDA. (From Ref. 71.)

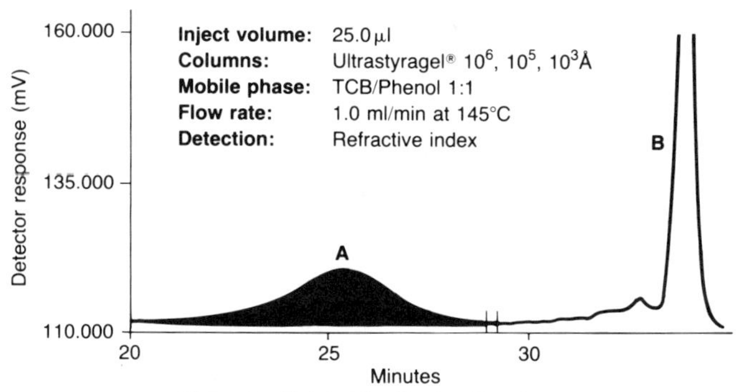

Molecular Weight Distribution Results

$\bar{M}w$	310791
$\bar{M}z$	707778
$\bar{M}v$	310791
$\bar{M}z + 1$	1195818
$\bar{M}n$	115093
$\bar{M}z/\bar{M}w$	2.277
Dispersity	2.770
Peak Molecular Weight	168420
$\bar{M}z + 1/\bar{M}w$	3.848

FIGURE 35 GPC chromatogram for PEEK. (From Ref. 72.)

ASTM Methods for Chromatography

ASTM E 682	*Standard Practice for Liquid Chromatography Terms and Relationships*
ASTM E 685	*Standard Practice for Testing Fixed Wavelength Photometric Detectors Used in Liquid Chromatography*
ASTM E 1151	*Standard Practice for Ion Chromatography Terms and Relationships*
ASTM D 3016	*Standard Practice for Use of Liquid Exclusion Chromatography Terms and Relationships.*
ASTM D 3593	*Standard Test Method for Molecular Weight Averages and Molecular Weight Distribution of Certain Polymers by Liquid Size-Exclusion Chromatography (Gel Permeation Chromatography—GPC) Using Universal Calibration*
ASTM E 260	*Standard Practice for Packed Column Gas Chromatography*
ASTM E 355	*Standard Practice for Gas Chromatography Terms and Relationships*
ASTM E 6697	*Standard Practice for Use of Electron-Capture Detectors Used in Gas Chromatography*

Thermal Analysis

Thermal analysis includes measurements of many different kinds, each designed to detect and measure changes in a material as it is heated or cooled in a precisely controlled fashion. Unlike chromatographic and spectroscopic methods, the various techniques comprising thermal analysis are not related by a common theoretical basis. Rather, they are related instrumentally in that changing the heat content of a material is intended to produce some chemical or physical change that can be converted to a measurable electrical signal. In effect, when a sample is subjected to programmed temperature heating, virtually any analytical method, including spectroscopy and chromatography, could be considered to be thermal analysis. Table 5 illustrates typical properties measured and gives the name of the thermal analysis technique used to initiate and measure the change. Certain characterization methods such as those of rheology have, with time, evolved a "free standing" body of theory, instrumentation, and applications. The measurement of mechanical properties as functions of frequency or time is not discussed in this article. This section covers rheological measurements involving temperature changes and termed "dynamic mechanical analysis."

Of the techniques listed in Table 5, four are the most generally useful for characterizing polymer composite materials. These are differential scanning calorimetry (DSC), thermogravimetric analysis (TGA), thermome-

TABLE 5
Thermal Analysis Methods

Physical Property Measured	Thermal Analysis Method
Mass	Thermogravimetric analysis (TGA)
	Evolved gas analysis (EGA)
Temperature	Differential thermal analysis (DTA)
Temperature and enthalpy	Differential scanning calorimetry (DSC)
Dimensions	Thermomechanical analysis (TMA); dilatometry
Mechanical (modulus and damping)	Dynamic mechanical analysis (DMA)
Acoustic	Thermosonimetry
Optical	Thermoptometry
Photochemical emission	Differential photocalorimetry (DPC)
Electrical	Thermoelectrometry
Magnetic	Thermomagnetometry

chanical analysis (TMA), and dynamic mechanical analysis (DMA). The reader requiring a comprehensive discussion of these and other methods of thermal analysis is encouraged to consult excellent monographs by Wendlandt [73] and Turi [74].

The DSC, TGA, TMA, and DMA approaches are of immense importance for characterizing polymer composite materials because the information obtained from these methods can be correlated directly with material processing and the performance of fabricated parts. Thermal analysis is useful for fingerprinting materials; but the results are not as distinctive and reliable as fingerprints from methods discussed previously. The chief virtue of thermal analysis methods is that they show how thermoset and thermoplastic resins, prepregs, and finished parts respond to temperature. More specifically, they measure and compare material properties that relate directly to resin and prepreg processibility, part fabrication, dimensional stability, service temperature limits, and durability in a variety of environments.

Thermal analysis measurements and their applications are given in Table 6. The listing is by no means exhaustive; but it is typical for polymer composites. The table also suggests that of the four methods, DSC and DMA are the more useful.

Contemporary instruments for thermal analysis are modular and microprocessor controlled. The instruments can be viewed as systems comprised of three subsystems (Fig. 36). One subsystem includes an extremely well-regulated temperature environment or sample cell, a transducer for converting temperature, mass, or linear dimension to an electrical signal, and a signal amplifier. The next subsystem is a temperature programmer for precisely heating, cooling, or maintaining the sample at constant temperature. The remaining subsystem consists of a dedicated computer for storing and analyzing the data and a variety of display devices (printers, plotters, LCD, CRT, etc.). Each of the four methods is discussed separately below with respect to theory of operation and instrumentation and methods. Selected applications are discussed for the group as a whole because many characterization studies utilize more than one method.

Differential Scanning Calorimetry

THEORY. Stated very simply, DSC measures heat flow into or out of a material. Instrumentally, the sample and an inert reference material (empty sample pan) are subjected to heating (or cooling) in a precisely tempera-

TABLE 6
Applications of Thermal Analysis Methods

Differential Scanning Calorimetry (DSC)
Fingerprint analysis for lot-to-lot uniformity
Characteristic temperatures for melting, crystallization, vaporization, cure, glass transition, and decomposition
Detection of metastable crystal forms
Detection of artificial nucleation
Construction of phase diagrams for composition analysis, purity, and screening new formulations
Heat (power) required to melt, crystallize, and cure
Degree of prepreg advancement and laminate cure
Evaluation of catalyst and catalyst concentration effects
Effects of pressure on the cure process
Kinetics and mechanism of cure and decomposition processes
Thermogravimetric Analysis (TGA)
Volatiles content
Thermal stability of materials
Kinetics and mechanism of reactions
Composition analysis for multicomponent systems
Thermomechanical Analysis (TMA)
Linear and cubical coefficients of expansion
Glass transition temperatures
Softening temperature
Evaluate creep
Gel time and resin flow
Residual strain in cured or solidified materials
Dynamic Mechanical Analysis (DMA)
Glass transition and other secondary transition temperatures
Determine damping
Determine storage and loss moduli
Distinguish copolymers and blends
Estimate concentrations of dispersed phases
Activation energy as a function of frequency
Viscosity versus temperature
Simulation of cure cycles

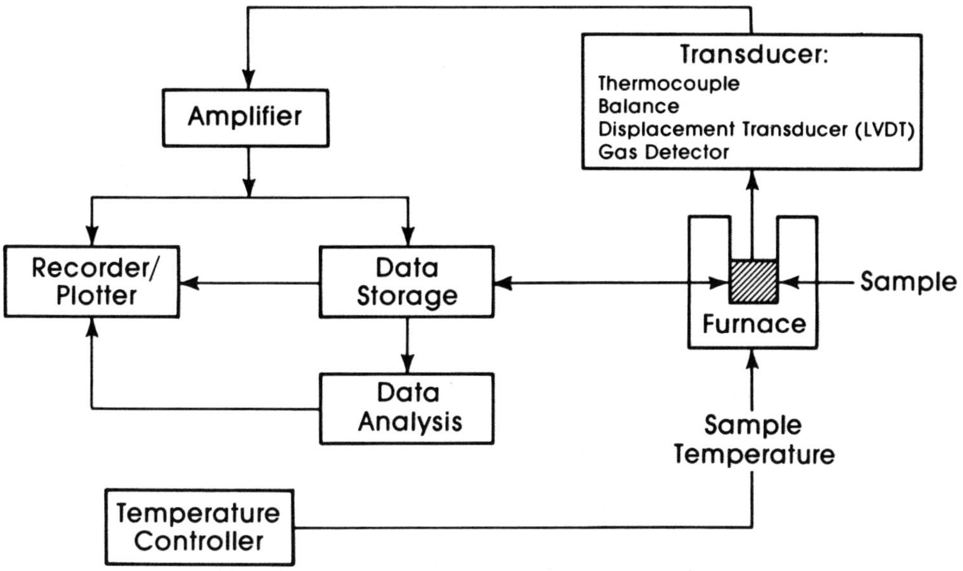

FIGURE 36 **Typical thermal analysis system.** (Courtesy of du Pont Instruments.)

ture-controlled chamber. As the sample undergoes a thermally activated change such as melting or polymerization, the instrument adds or subtracts sufficient energy to maintain the sample at the same temperature as the reference material. This difference required to maintain the electrical null balance is equivalent to the heat absorbed during endothermic transitions such as melting and reductive decomposition or evolved during exothermic processes such as crystallization, polymerization (cure), and oxidative decomposition.

The DSC method evolved from its predecessor differential thermal analysis (DTA) which, in turn, developed from the classical thermometric methods and calorimetry. In applying the classical methods familiar to undergraduate science students, phase changes are detected by noting the temperature at which the sample temperature exceeds or lags the bath or oven temperature. Low energy processes such as second-order transitions involving changes in heat capacity without phase change and transitions for milligram sample quantities are not easily detectable and measurable this way. Differential thermal analysis is based on electrically amplified temperature differences measured using opposed thermocouples in sample and reference materials located in the same temperature environment. This difference measurement may be used directly or subjected to electrical null balancing; it is capable of detecting transitions in microgram quantities of materials. A basic DTA thermocouple circuit is illustrated in Figure 37.

DTA can provide semiquantitative calorimetric data when standard reference materials are used for calibration. However, since DSC circuitry measures or proportions power directly, it has essentially displaced DTA instruments for studies at temperatures from subambient to about 750°C. DTA is still used at higher temperatures to investigate the thermal behavior of inorganic

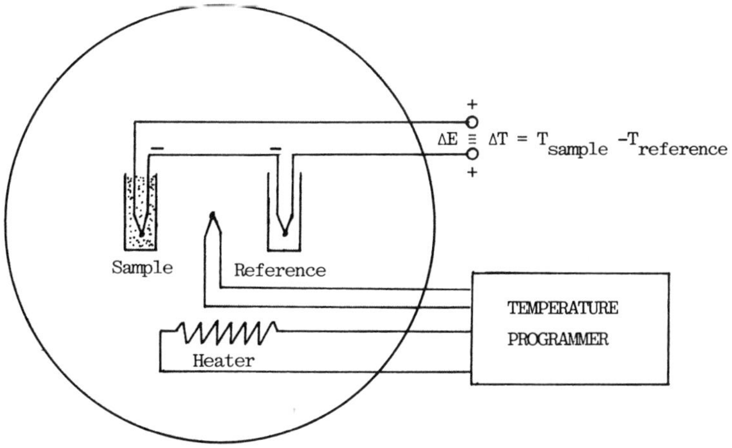

FIGURE 37 Schematic DTA thermocouple circuit.

salts, clays, and ceramics, however, because the equipment is constructed of more ragged materials.

Whereas DTA instruments measure the differential temperature between sample and reference, DSC instruments measure the differential heat flow (or heat flux, dq/dt) between the two. There are at least three different transducer designs in current instrumentation to accomplish this measurement. These are illustrated schematically in Figure 38. The original DSC (Perkin Elmer Corporation) transducer design (Fig. 38*a*) involves supplying heat independently to sample and reference, comparing their temperatures over small increments of time, and supplying power to the heaters, as needed, to maintain a zero temperature differential. This is called power compensation DSC. The two other transducer designs are called heat flux DSC and are found in the du Pont and Mettler instruments, both of which use one source of heat to heat the sample and the reference. In the du Pont cell (Fig. 38*b*), heat is transferred using a constantan disc as the base for sample and reference pans. At the base of each pan is a chromel–constantan thermocouple junction, and the different temperatures at each junction, coupled with an electronic calibration circuit, become the differential heat flow measurement. The sample and reference temperatures are also monitored separately using chromel–alumel junctions at the chromel wafer mounted on the underside of the constantan heat transfer disc. The Mettler instrument (Fig. 38*c*) employs a five-thermocouple ΔT thermopile to measure that flux directly. Despite these fundamentally different transducer designs, the devices are roughly comparable in terms of information developed and accuracies and precisions obtainable.

A hypothetical DSC curve or thermogram (Fig. 39) illustrates the appearance and origin of thermally activated processes customary for polymer composite materials. The material behavior shown consists of both thermoplastic and thermoset characteristics. The differential response (ordinate) is the difference in heat flow dq/dt between the sample and empty sample pan as a function of temperature.

Figure 39 indicates both the qualitative and the quantitative characteristics of DSC measurements. The overall thermogram can be used qualitatively as a fingerprint. It can also be a diagnostic tool to evaluate the nature of the material—that is, whether it is a thermoplastic, a cross-linkable thermoplastic, or a thermoset. In most instances, some independent knowledge of the material is

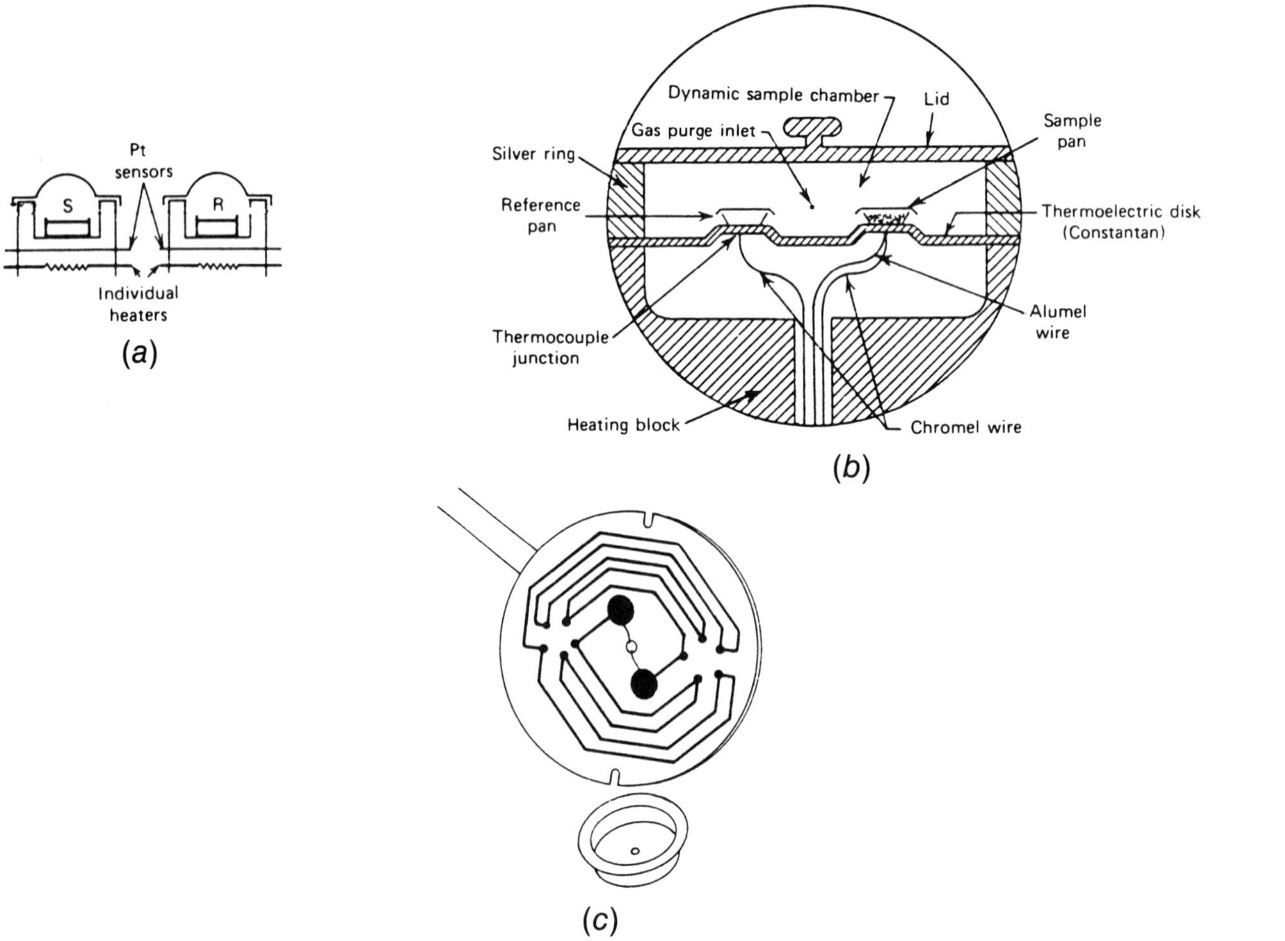

FIGURE 38 DSC transducer designs: (*a*) Perkin Elmer, (*b*) du Pont, and (*c*) Mettler. (Reprinted from D. D. Werstler, *Polymer, 27,* 752, 754, 755 (1986), by permission of the publishers, Butterworth & Co. (Publishers) Ltd. © 1986.)

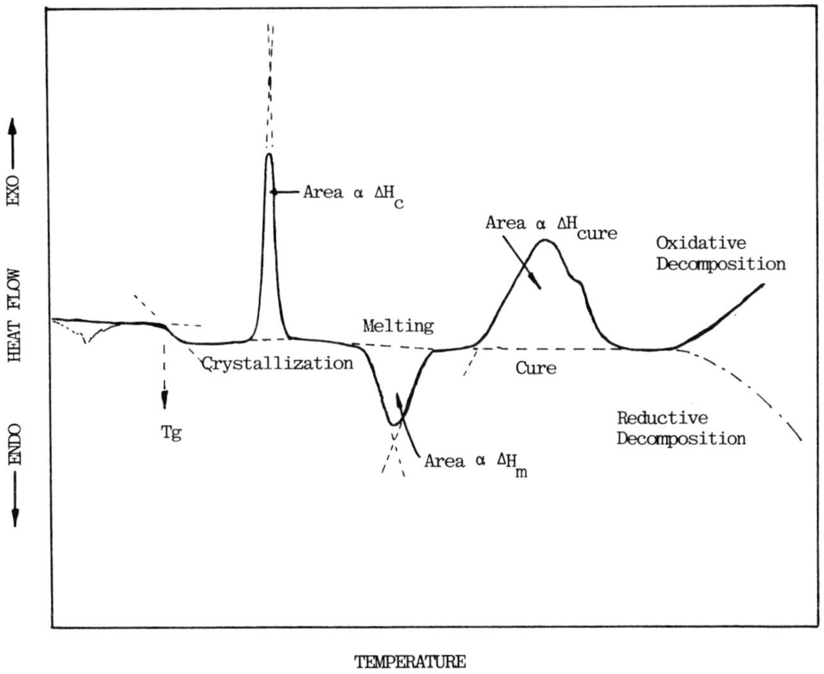

FIGURE 39 **Hypothetical DSC thermogram illustrating typical thermally activated processes in polymer composite materials.**

needed to analyze confidently the various thermal processes.

Some of the quantitative aspects of DSC are the temperatures at which the various thermally activated processes appear to initiate, attain their maximum rate or peak, and terminate, and the areas enclosed by the peaks. The heat or enthalpy ΔH or a particular transition is directly proportional to the area of the peak. In somewhat idealized form, we have:

$$\Delta H = \frac{KA}{m} \tag{18}$$

where m is the mass of material undergoing change and K is a constant that includes the heat flow calibration factor and geometric factors for the instrument. A mathematical description of the thermal processes responsible for peaks is a heat transfer problem and is not offered here. Note, rather, that in addition to sample characteristics, heat transfer considerations play a major role in governing the appearance of a DSC thermogram. To that extent, thermal analysis data must be viewed as relative to the instrument conditions employed. Another way to appreciate this is to realize that heating or cooling a material at analytically useful rates produces nonequilibrium or non-steady-state conditions. For example, the faster a sample is heated, the higher its apparent transition temperature. On the other hand, the more slowly it is heated, the lower the measurement sensitivity. An example of the heating rate effect for an epoxy prepreg is given in Figure 40.

The more important material variables and instrument variables are summarized in Table 7. As far as the instrument is concerned, relatively few of the variables are under the operator's control. Those that may be varied easily are the heating rate and the atmosphere in the sample chamber. Of the sample variables under the operator's control (including sample mass, packing density, type and amount of diluent, and particle size), the most important variable for typical polymer composite materials is the sample mass. The other variables pertain more appropriately to samples in powder or granular form.

Thermogravimetric Analysis

Thermogravimetric analysis is performed by continuously measuring the mass of a material as a function of temperature or time in an instrument called a thermobalance. The analytical result, or TGA curve, is a plot of the mass or the percentage of original mass remaining at temperature or time. TGA can be used to study any process that involves a mass change such as evaporation of water and solvents, condensation reactions, and decomposition processes. TGA is not sensitive to other thermally activated transitions during which the mass remains the same.

An idealized TGA curve for a thermoset prepreg appears in Figure 41. This is a very straightforward example showing the loss of a 2% solvent beginning at about 60°C followed by the pyrolysis of the resin beginning at approximately 315°C. Unlike transition temperatures and enthalpies measured by DSC, the weight losses that accompany devolatilization and decomposition processes in TGA do not occur at precisely defined transition temperatures. Rather, weight changes occur in ranges that are highly dependent on both the physical and chemical characteristics of the material as well as on instrumental

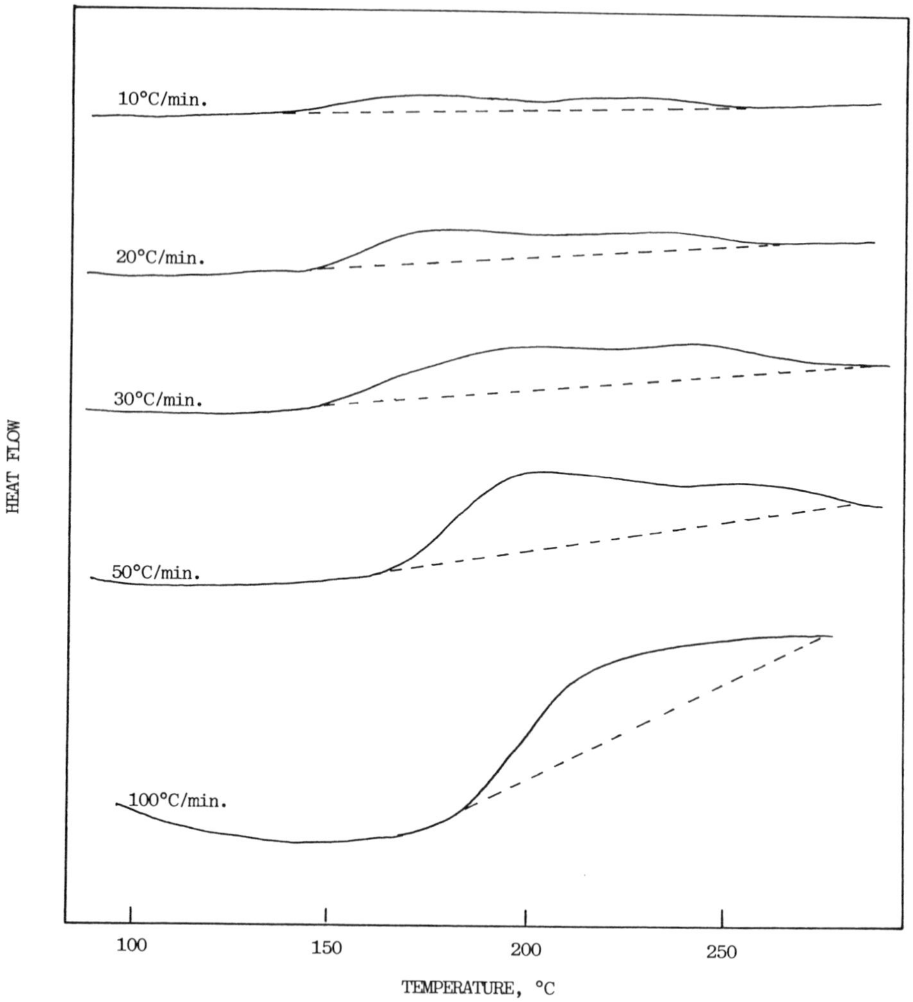

FIGURE 40 Effect of DSC heating rate on cure behavior for an epoxy prepreg (E-293/7781). (Courtesy of Ferro Composites, Los Angeles.)

TABLE 7
Instrument and Sample Variables Affecting Thermal Analysis Methods

Instrument-Related Variables
Sample chamber atmosphere
Sample chamber size and shape
Sample holder material
Sample holder geometry
Thermocouple type and junction size
Thermocouple location at the sample
Heating rate
Speed and response time of recording electronics
Sample-Related Variables
Particle size
Thermal conductivity
Heat capacity
Packing density
Swelling or shrinkage
Mass
Nature and mass of diluent
Degree of crystallinity or cure

and environmental factors, some of which are discussed below. Thus characteristic or onset temperatures that define some thermogravimetric event in a particular instrument have been called procedurally obtained temperatures [73].

TGA fingerprint and quantitative analyses of polymer composite materials are most commonly performed by subjecting the sample to temperature-programmed heating at 5–20°C per minute. Isothermal studies for materials of known composition are valuable for characterizing accelerated aging behavior in a variety of environments. Depending on the objective of the experiment, the heated sample can be bathed in an inert environment using such gases as nitrogen or argon, or it may be statically or dynamically contacted with an oxidizing or reducing gas.

Weight changes are measured using a highly sensitive null-type or deflection-type thermobalance capable of reproducing mass changes as small as $\pm 1.0\ \mu g$. One contemporary instrument is illustrated schematically in Figure 42 as an example of the null-type thermobalance. The sample is suspended from a quartz beam, which is

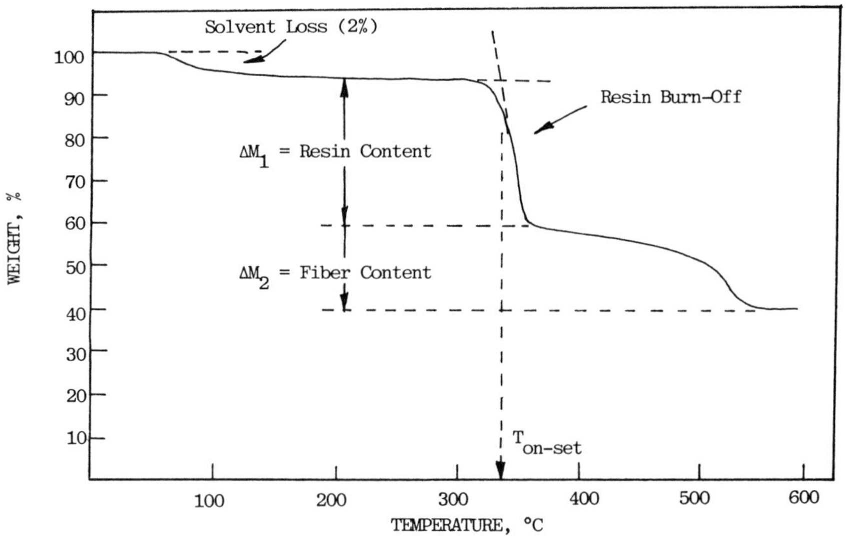

FIGURE 41 **Idealized TGA curve for a thermoset prepreg.**

fixed to an electromagnetic coil. The counterbalanced quartz beam is maintained in its null or zero position by regulating the current in the coil. At this point the single-pan laboratory balance is a good mechanical analogue of the thermobalance. As weight is lost from the sample on the left side of the beam, the right side deflects downward. This movement is detected by a pair of photodiodes that signal the supply of an increment of current or restoring force to the coil to return the beam to its null position.

As mentioned earlier, the TGA curve is governed by numerous material and instrumental factors. These factors affect the overall shape of the curve as well as reproducibility and accuracy of the results. Most of these factors were linked previously in Table 7 for DSC and, as in the case of DSC, the important factors over which the analyst can exert control are the heating rate, the sample mass, and the environment in the sample chamber. In general, the faster the heating rate, the higher will be the procedural transition or decomposition temperature,

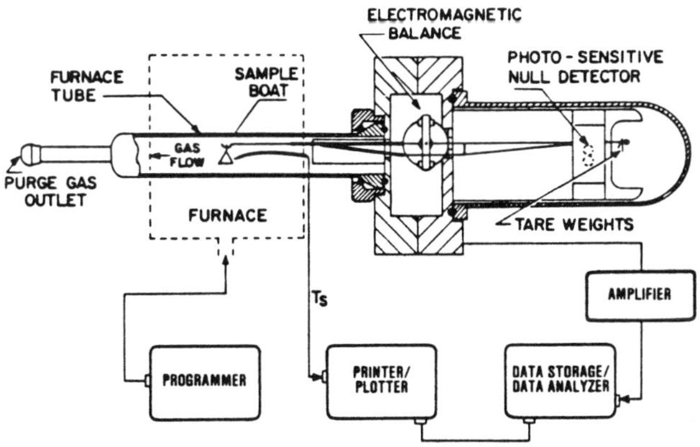

FIGURE 42 **Schematic diagram for a thermogravimetric analyzer.** (Courtesy of du Pont Instruments.)

and the lower will be the degree of apparent decomposition. Sample mass affects such issues as temperature change linearity deviations, thermal gradients, and degree of diffusion of product gases away from the sample.

Instrumental artifacts can be minimized by using the smallest sample consistent with experimental sensitivity. The sample environment is, of course, critical because it relates to the purpose of the analysis. An inert gas will emphasize transitions and reactions that are strictly thermally activated, while the use of an air or oxygen environment is designed to examine thermo-oxidative processes that may not occur otherwise. Other environment-related factors that can assume extraordinary importance and complexity are the gas flow rate and the partial pressures of product and reactant gases.

Thermomechanical Analysis

One of the simplest methods, thermomechanical analysis (TMA) measures changes in linear dimension, volume, or shape as a function of temperature. There are many chemical and physical reasons for all materials to undergo volumetric changes or expansion and contraction as they are heated. For example, solid–solid crystal transformations involving conversion from a metastable crystal form to a more stable crystal form are accompanied by reduction in volume (i.e., increase in density). In addition, there is the change in volume that accompanies relaxation processes such as reduction in fiber orientation and stress relief. Other reasons for simple dimension or shape changes include swelling, decomposition, warpage, and softening.

One of the most common behaviors consists of the expansion and contraction that occur directly with heating or cooling for most materials, and TMA is ideally suited for determining the linear coefficient of thermal expansion. For a polymeric material in particular, expansion increases markedly as the material is heated through its glass transition temperature T_g (Fig. 43). TMA

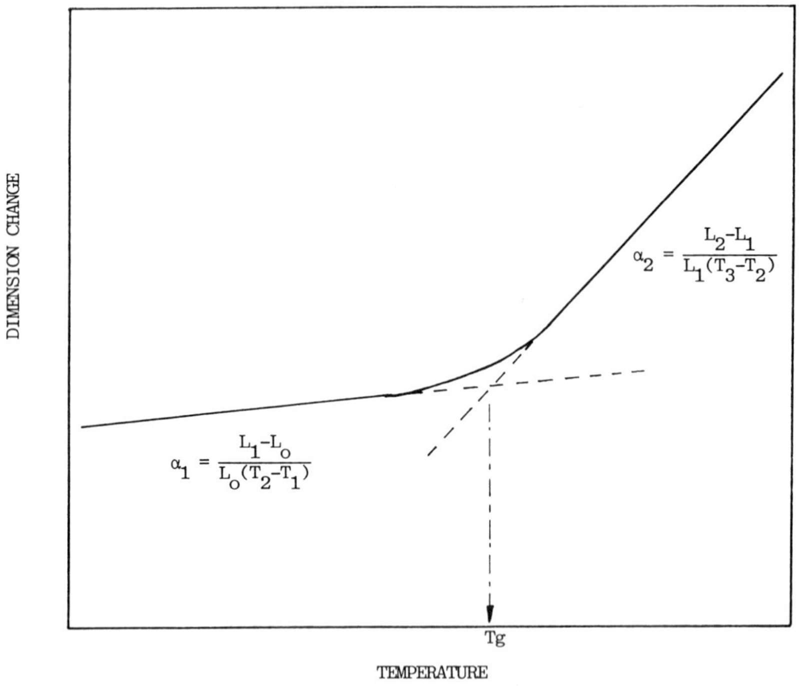

FIGURE 43 TMA curve.

is especially sensitive to the glass transition, and one experiment can provide the linear coefficient of expansion α directly from the slope of the expansion versus temperature trace as follows:

$$\frac{\Delta L}{L_0} = \alpha \Delta T \tag{19}$$

and the T_g from the intersection of extrapolations drawn from the expansion curves above and below the T_g. Such straightforward behavior will be observed for relatively homogeneous and chemically stable materials. Continuous fiber reinforced polymer composites will have different expansion coefficients, depending on which sample dimension is being measured. Moreover, thermosets having incomplete cure or containing residual strains can show sizable departures from linear expansion behavior.

INSTRUMENTATION AND METHODS. Commercial TMA instruments consist of a furnace for heating a specimen, a temperature-insensitive probe for transmitting specimen movement out of the furnace, and a sensing device for measuring small changes in specimen movement. The temperature-insensitive probe is a quartz rod having one of a variety of surface contours for specimen contact. These instruments operate typically from −170°C to as high as 1200°C, and quartz is dimensionally stable in this range. Positive and negative deviations of the probe's position on the specimen are sensed by a movable-core linear variable differential transformer (LVDT). As the specimen expands, contracts, or otherwise deforms, the core on the probe moves in the annular space of the LVDT. This relative movement produces a voltage change that is proportional to the linear displacement of the core. The signal is amplified and processed for reporting by x–y recorder or by a computer data reduction system. The temperature is monitored continuously by a thermocouple positioned as close as possible to the specimen. Schematic diagrams of the three most popular TMA instruments are given in Figure 44.

Different probe configurations allow the TMA to be used to produce and measure a variety of material responses to applied loadings of different kinds. The loadings possible include expansion, compression, penetration, and dilatometric (volumetric expansion). Additional accessories that are available permit such studies as stress relaxation, parallel-plate viscometry, and tensile elongation for fibers and films. A selection of the most commonly used probes is illustrated in Figure 45.

Dynamic Mechanical Analysis

In dynamic mechanical analysis, a rheological method is used to monitor and measure the ways in which a material responds to an applied, periodic stress as a function of temperature. "Rheology" is the name applied to studies of elastic, viscoelastic, and viscous deformation and flow behavior of materials. The word "viscoelastic" is generally used to describe the response of polymeric materials to applied strains. As strain is applied in a periodic or sinusoidal manner, the resulting deformation will also be sinusoidal. However, the applied and measured strains will be somewhat out of phase with each other. This is in contrast to the response of a perfectly elastic or Hookeian solid to an applied sinusoidal strain. Viscoelastic behavior gives rise to two dynamic properties: the dynamic storage modulus G' and the dynamic loss modulus G''. The former is related to the stiff-

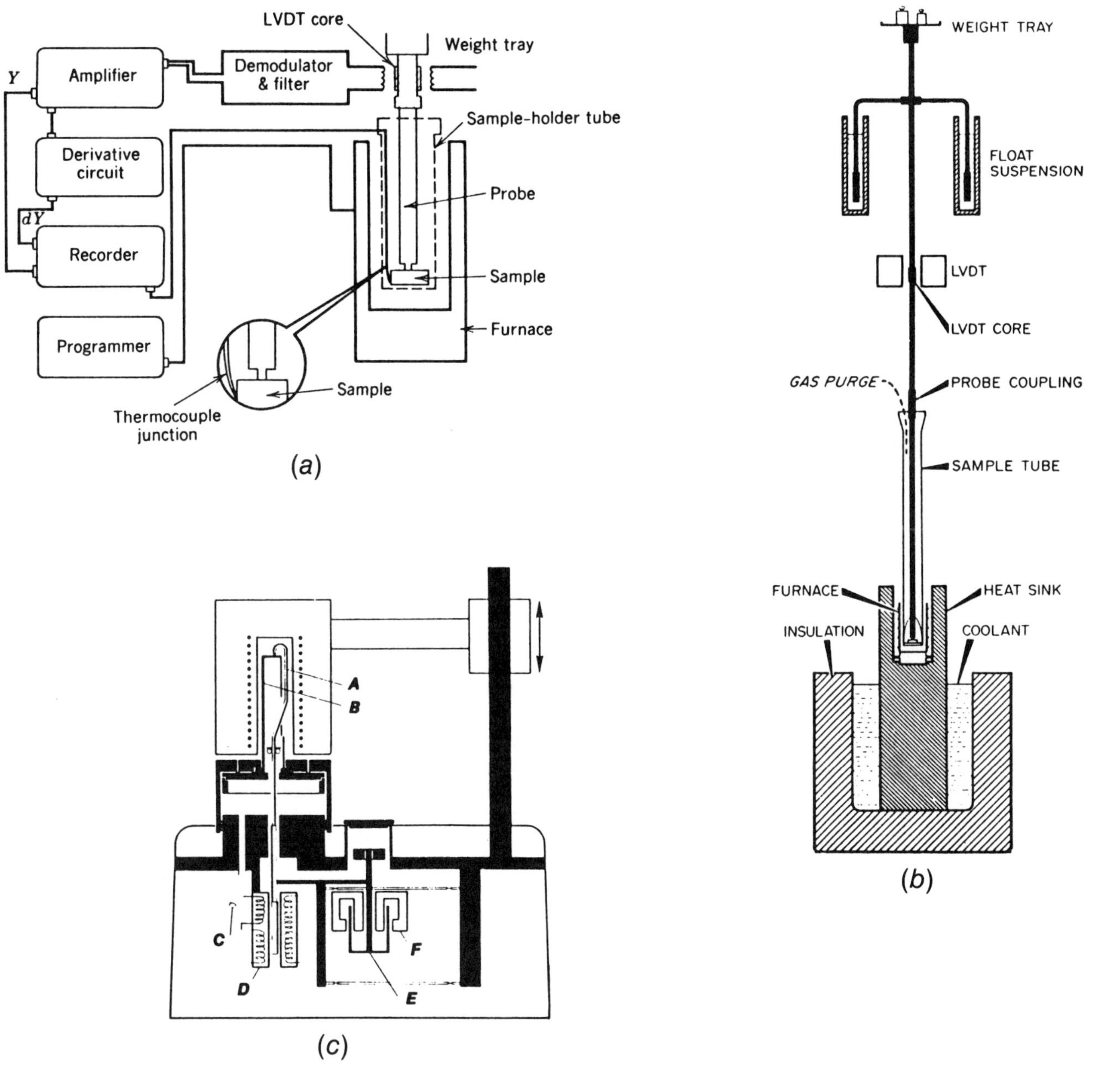

FIGURE 44 Schematic diagrams of three commercial TMA modules: (*a*) du Pont 943, (*b*) Mettler TMA40, and (*c*) Perkin Elmer TMS-2. (Reprinted from J. W. Clark, B. N. Jagannatham, and F. C. Jenning, *Polymer, 29*, 341 (1988), by permission of the publishers, Butterworth & Co. (Publishers) Ltd. © 1988.)

ness of the material, while the latter is a measure of its damping (i.e., energy-dissipating) characteristics. Measurements of the storage and loss moduli under the appropriate conditions for composite materials and their precursors can yield enormous insight about basic relationships between the microscopic architecture of the solid and its mechanical behavior. Moreover, these measurements can provide significant guidance for processing prepregs and related materials.

The use of rheological methods together with other thermal methods, spectroscopy, and chromatography represents a most comprehensive characterization scheme. Dynamic mechanical behavior is evaluated by measuring aspects of the elastic and the viscous components of deformation under shear stress. A wide variety of methods have been used to characterize the viscoelastic behavior of polymeric materials. These methods include creep and stress relaxation, free and forced vibration, and wave propagation (resonance). This discussion is concerned only with the free and forced vibration methods that have been used extensively for polymer composite materials.

THEORY. Polymeric materials can show the features of a glassy, brittle solid or a viscous liquid, depending on the temperature and the time scale of the measurement. Most polymers at room temperature behave as a combination of viscous liquids and elastic solids; hence they

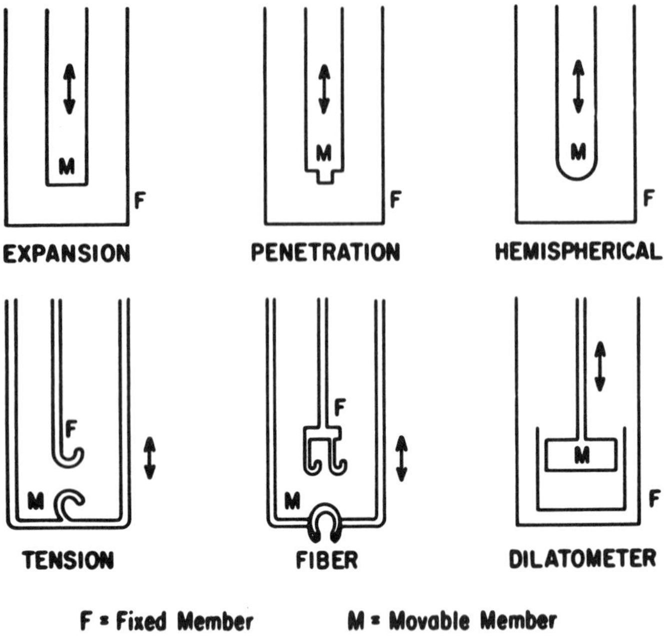

FIGURE 45 TMA probe configurations. (Courtesy du Pont Instruments.)

are called viscoelastic materials. Viscoelastic properties describe the time–temperature dependence of deformation as a combination of elastic solid and viscous fluid. Hooke's law defines the elastic part of deformation as follows:

$$(\tau_{xy})_{\text{elastic}} = G\varepsilon_{xy} \tag{20}$$

where $(\Gamma_{xy})_{\text{elastic}}$ is the shear stress, G is Young's modulus, and ε_{xy} is the strain.

Viscous flow under steady shear conditions is represented as follows:

$$(\tau_{xy})_{\text{viscous}} = \eta \frac{dV}{dy} \tag{21}$$

where η is the viscosity and the term dV/dy is the velocity gradient or shear rate.

Simple viscoelastic behavior can be described by combining Eqs. (20) and (21) to give:

$$\tau_{xy} = G\varepsilon_{xy} + \eta \frac{dV}{dy} \tag{22}$$

The assumption of viscoelasticity has proved to be very useful under the conditions of small deformation (i.e., linear viscoelasticity). A number of mechanical models represented by an elastic spring (Eq. 20) and a viscous dashpot (Eq. 21) in series (Maxwell element) or in parallel (Kelvin or Voigt element) or any combination of the two have been used to model linear viscoelastic behavior for polymer materials.

Most of the experimental work on linear viscoelasticity is performed using a single mode of deformation, usually a measurement of Young's modulus or the shear modulus. Stress can be applied to the sample in a number of ways. A measurement at constant stress is called creep testing, while one at constant strain is called stress relaxation. Alternatively, the specimen can be subjected to sinusoidal strain while stress is measured simultaneously as a function of time. For a linear viscoelastic material, stress and strain will both vary sinusoidally, but strain lags behind the stress (Fig. 46). The stress and strain are related to the angular frequency ω and phase angle θ as follows:

$$\varepsilon_{xy} = G_0 \sin(\omega t) \quad \text{and} \quad \tau_{xy} = G_0 \sin(\omega t + \theta) \tag{23}$$

Rearranging this set of equations yields a complex presentation of the shear stress in terms of dynamic moduli G' and G'', the storage modulus and loss modulus, respectively:

$$\tau_{xy} = G_0G' \sin(\omega t) + G_0G'' \cos(\omega t) \tag{24}$$

where $G' = (\tau_0/G_0)\cos(\theta)$ and $G'' = (\tau_0/G_0)\sin(\theta)$.

For purely elastic behavior, the stress and strain will be in phase at all times. However, for purely viscous behavior, the stress will be 90° out of phase relative to the applied strain. The ratio G''/G' is called tan δ, and describes the phase angle lag of the viscoelastic material. The quantities G', G'', and tan δ depend strongly on molecular structure and are thus affected by the strain, the strain rate, the temperature, and the rate of change of temperature.

Figure 47 describes the relationships between G', G'', and tan δ. The time–temperature superposition principle [75] describes the equivalence of low temperature testing with the high frequency testing and vice versa. Therefore, the shape of the curve obtained by increasing temperature is the same as that obtained by decreasing the frequency of testing.

Figure 48 helps to illustrate the following description. At low temperatures the polymer behaves like high modulus glass, the modulus being independent of the frequency. As the temperature increases, the polymer becomes viscoelastic and tan δ goes through a maximum. At this point the glassy polymer becomes leathery and

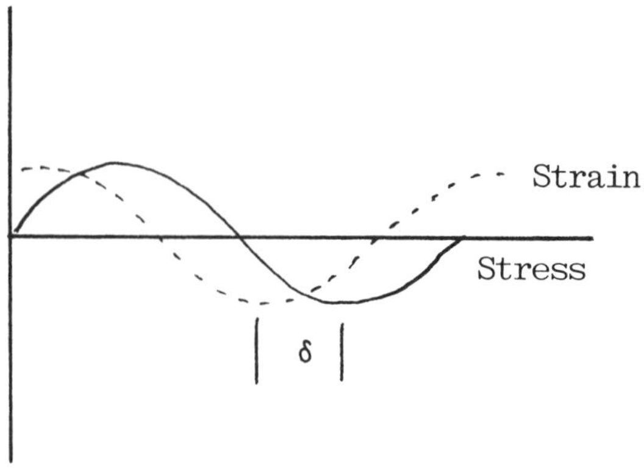

FIGURE 46 **Periodic stress–strain behavior for a viscoelastic material.**

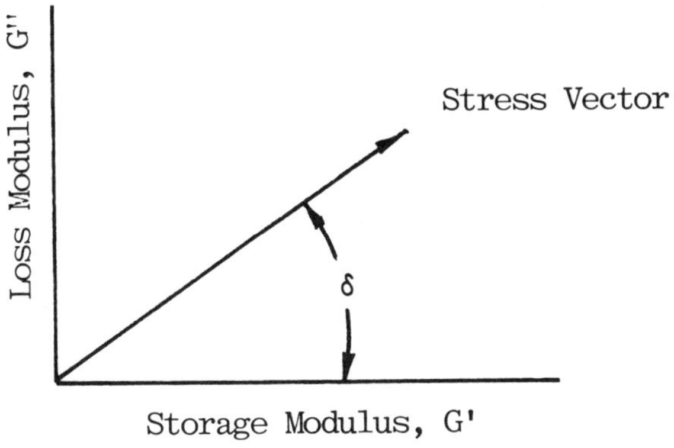

FIGURE 47 **Graph illustrating relationships for G', G'', and tan δ.**

the transition is defined as the glass transition. At still higher temperatures the polymer becomes rubberlike, exhibiting a low G' value, which is again independent of the frequency. Characteristic changes in G', G'', or tan δ values at the glass transition are used to measure the glass transition temperature.

INSTRUMENTATION AND METHODS. Dynamic mechanical analysis is generally performed in one of three modes: a torsional or flexural force is applied to a specimen which is then allowed to vibrate freely, caused to resonate at high frequency, or subjected to sustained vibration at relatively low frequency (0.1–100 Hz). The torsion pendulum shown in Figure 49 is an example of the oldest and simplest device based on free vibration. Such instruments subject a material in the form of cylindrical rod, rectangular beam, or braid-supported resin (torsional braid) to a frequency ranging between 0.01 and 50 Hz. One end of the specimen is fixed rigidly, while the other end supports an inertia member that oscillates

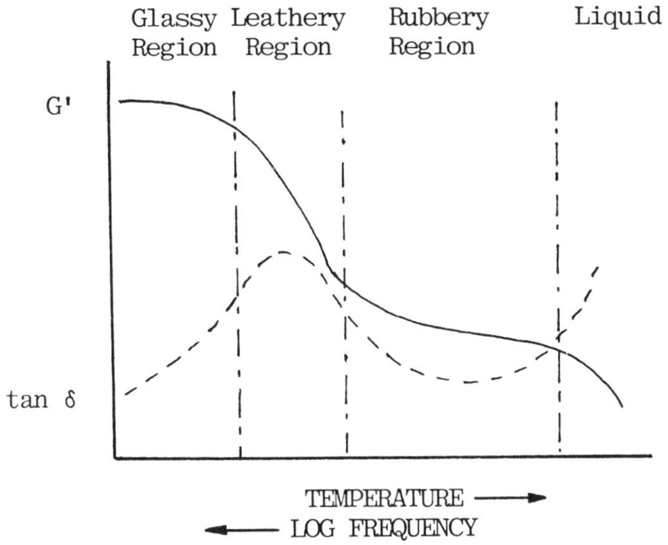

FIGURE 48 **Graph illustrating the dependence of G' and tan δ on temperature or frequency.**

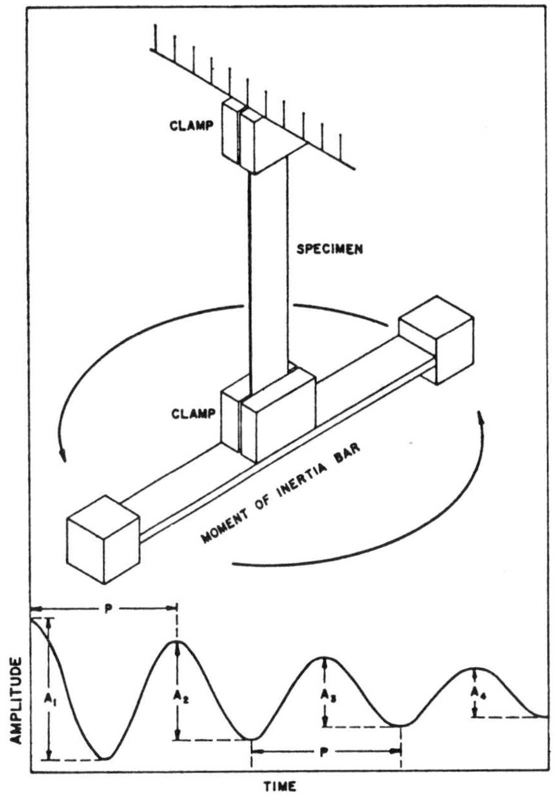

FIGURE 49 **Schematic diagram of a torsion pendulum.** (L. E. Nielsen, *Plastics Eng. J.* *16*, 525 (1960).)

freely. The specimen is set into motion and the amplitude of each successive vibration is measured with time (and temperature) as the amplitudes diminish because of damping. The natural logarithm of the ratio of two successive amplitudes is called the logarithmic decrement Δ:

$$\ln \frac{A_2}{A_1} = \Delta \tag{25}$$

and using this result with sample geometry terms and the frequency, the storage and loss moduli can be calculated using the following equations:

$$G' = \frac{2l}{\pi} r^{-4} M\omega^2 \quad \text{and} \quad G'' = \frac{2l}{\pi} r^{-4} M\omega^2 \frac{\Delta}{\pi} \tag{26}$$

where M is the moment of inertia of the system, r is the radius, and l is the length of the specimen.

From this information is obtained the phase angle lag of the specimen, as defined earlier:

$$\tan \delta = \frac{G''}{G'} \tag{27}$$

Resonance methods employ high frequencies whose wavelengths are smaller than the smallest specimen dimension. The storage and loss moduli are determined by measuring how the sample's vibration amplitude changes as a function of applied frequency. Vibrating reed or cantilever instruments are examples in this category.

Free damping and resonance methods, while simple and accurate, are generally not preferred for studying the viscoelasticity of advanced composites because the frequency changes as the temperature is varied. Forced oscillation methods, on the other hand, permit frequency and temperature to be varied independently.

A number of commercial instruments based on forced vibration are available. Three of the more commonly used are the Rheometrics Dynamic Spectrometer (RDS), the du Pont Dynamic Mechanical Analyzer (DMA), and the Polymer Laboratories Dynamic Mechanical Thermal Analyzer (DTMA). Figure 50 shows these instruments schematically to illustrate their modes of operation.

Specimens can range from viscous fluids to a variety of solid sample geometries, depending on available accessories. In general, the specimen is subjected to a torsional or flexural sinusoidal strain with a constant amplitude, and the resultant shear stress is measured simultaneously. The viscoelastic behavior is characterized by the phase angle lag and the relative amplitude of stress and strain waves.

The sample is held between two plates or two end clamps (Fig. 51) and enclosed in an environmental chamber that provides a heating and cooling capability [76]. A dc torque motor applies shear strain to the test sample. Strain, frequency, sample geometry, and temperature program conditions are automatically maintained by a microprocessor control unit. The resultant stress on the sample is measured by a precise transducer. Typical test frequencies range as low as 0.01 radian/s to as high as

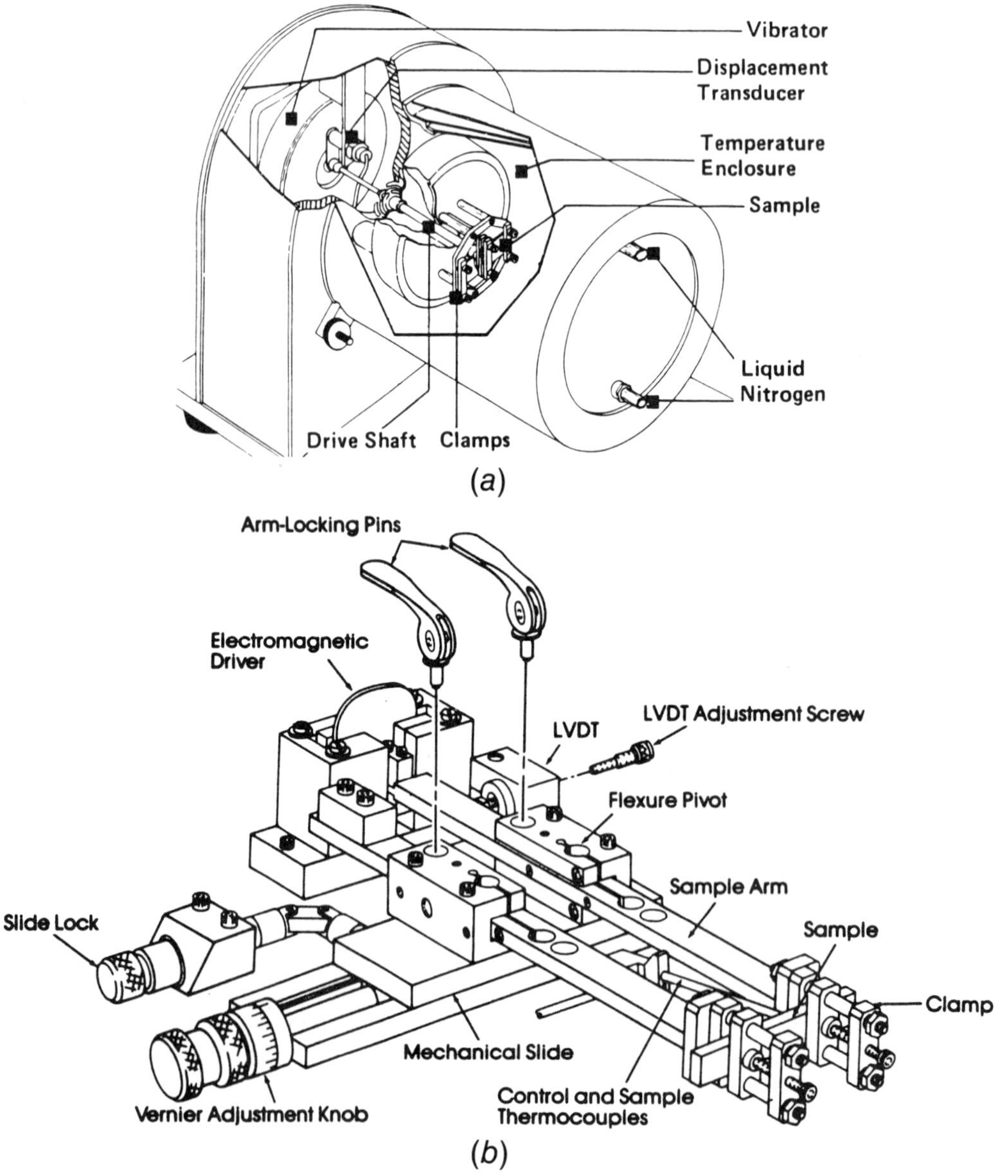

FIGURE 50 Schematic diagrams for commercial dynamic mechanical analyzers illustrating the specimen configuration: (*a*) Polymer Laboratories DMTA, (*b*) du Pont 983, and (*c*) Rheometrics RDS II.

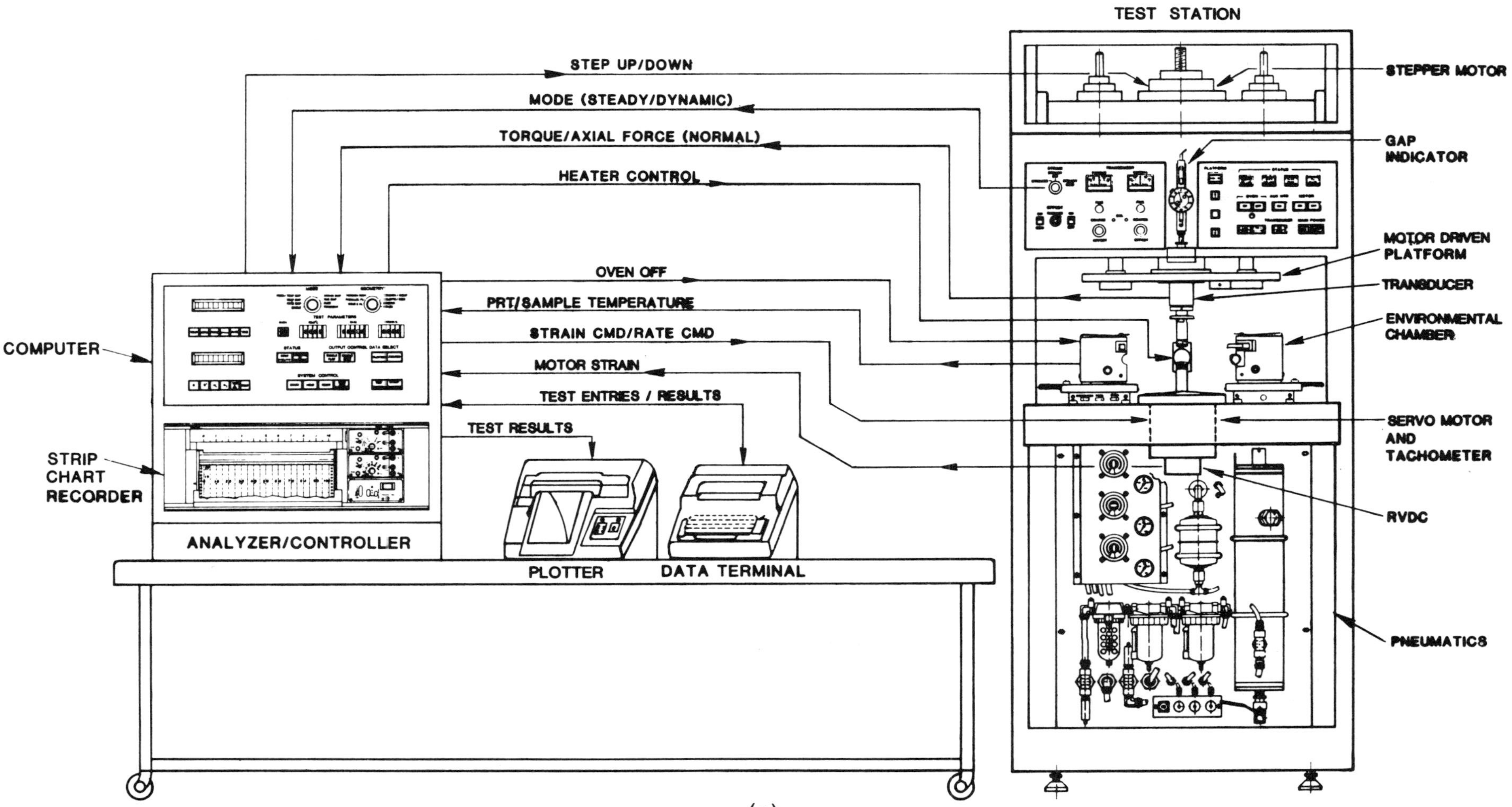

(c)

FIGURE 50 (continued)

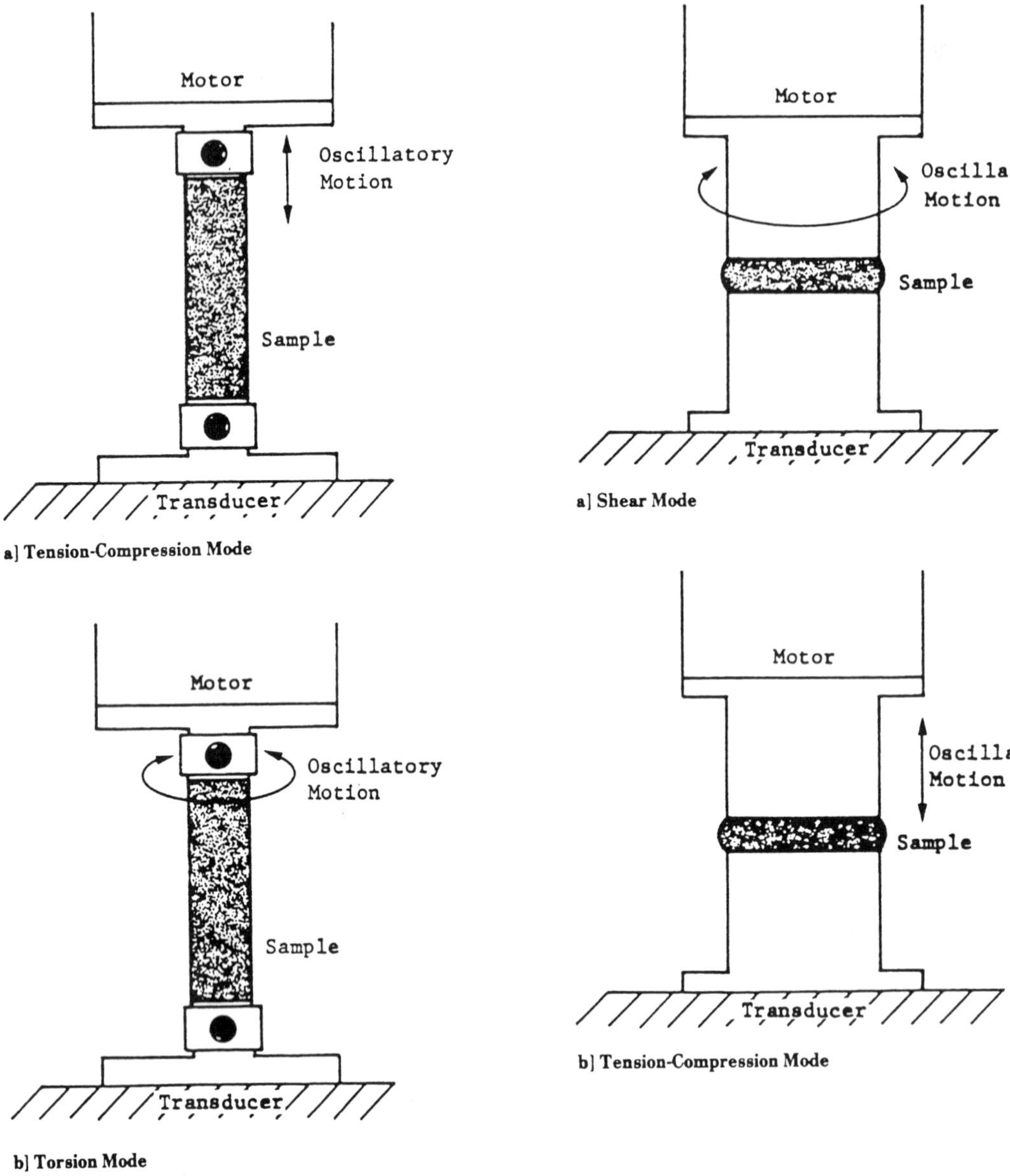

FIGURE 51 DMA test configurations for (*a*) rectangular torsion and (*b*) parallel plate. (From Ref. 76.)

500 radians/s. With state-of-the-art microprocessor systems, the complex calculations based on shear strain and phase lag are very easily performed in computer-based instruments, even for complex geometries (e.g., cone and plate and parallel plate).

APPLICATIONS. Thermal analysis methods are the most widely applicable for measuring and evaluating the physical properties of composite materials as functions of temperature. Exposure to various thermal environments is inevitable for composite materials, since elevated temperature processing is an integral step in their manufacturing as well as a criterion for end-use applications. As a result of heating or cooling, polymer composite materials undergo different thermal transitions, depending on whether they are based on thermoplastic or thermosetting resin systems. A thermosetting resin system will tend to cure rapidly after the cure onset temperature has been attained, and an amorphous thermoplastic will soften and flow above the glass transition temperature. A crystalline thermoplastic, on the other hand, will not begin to flow until the crystalline structure undergoes substantial melting.

The benefits of understanding these transitions are enormous for a wide variety of applications and users. For example, the glass transition temperature to a design engineer may define the maximum use temperature for a part, and to a process engineer it may represent the minimum temperature that would be needed to thermoform a composite based on an amorphous resin. Knowing the glass transition also has implications in the areas of materials selection for electrical and mechanical damping, fracture behavior, and analytical discrimination of blends and copolymers. In other words, applications of thermal analysis in the composites industry embrace manufacturing, processing, quality control, research and development, and structural design.

As outlined earlier in Table 6, the applications of thermal analysis provide a rapid means of studying cure kinetics for chemically reactive systems, crystallization kinetics for semicrystalline thermoplastic resins, glass transition behavior, thermal expansion, and thermal degradation and weight loss characteristics. Since the literature has become quite large, only a brief account of some of these applications and their associated thermally activated processes is attempted here.

Curing

Cure of thermosetting resins is characterized by the formation of a cross-linked network having infinite (essentially) molecular weight. Little subsequent processing is possible once the network has been formed. In general, the state of a thermosetting resin during the various stages of its processing can be classified into one of three categories:

Stage A: uncured resin.
Stage B: partially cured resin in intermediate forms, such as prepreg.
Stage C: completely cured material.

The uncured resin (stage A) is characterized by relatively low viscosity at room temperature to facilitate processing. For resin uses involving wet filament winding and resin injection, the base resin system and the curing agent are kept separate to avoid any reaction before the processing is started. The reactivity of the system is of critical importance once the two are combined. Resin reactivity and consistency can be characterized in terms of chemical composition, which can be fingerprinted by a suitable HPLC technique or an FTIR technique. DSC and DMA measurements have also been used in the quality control of the resin systems [77–80].

The resin present in most of the intermediate or prepreg forms is B-staged or partially cured. The properties of B-staged intermediate are strongly affected by composition and the degree of B-staging. Gross composition is characterized by resin content, volatiles content, moisture content, and fiber areal weight. All these properties are applicable to prepregs based on thermosetting as well as thermoplastic matrices. The degree of B-staging is commonly evaluated by such traditional methods as gel time, resin flow, tack, and drape of the prepreg. Much of this prepreg property data can provide quality control information, but the data are coarse and quite inadequate for developing or optimizing a cure cycle. Thermal and rheological measurements provide a wealth of pertinent information [81–84]. Dynamic mechanical analysis of cure rheology can provide a chemical fingerprint as well as the information needed to set autoclave or press cycles [85]. Chemical fingerprints thus obtained have been used to control resin quality as well as prepreg consistency for a variety of systems [64,86–90]. Cure characteristics and the processibility of the B-staged prepregs can also be determined for process optimization [91].

Figure 52 shows typical viscosity–time–temperature behaviors of two commercial thermoset resin systems based on cone-and-plate DMA [147]. As the resin gels, shear modulus and complex viscosity increase from a minimum. The dynamic viscosity is used as a reference parameter both for defining the appropriate time and/or temperature to apply pressure in an autoclave and for tailoring resin formulations. The onset temperature for gelation, the minimum viscosity, and the gel time as a function of temperature are a few of the parameters that can be obtained from dynamic viscosity experiments to characterize the processing of a thermoset resin system [84,92–94].

Cure characteristics of virtually all kinds of prepreg can be determined using similar analysis with parallel-plate geometry or a rectangular torsion setup [83,93,95–100]. DSC has been used extensively to characterize the extent of B-staging [92,101,102] as well as to optimize cure cycles for a variety of resin systems [56,103–107]. These optimization studies often include the simulating of cure cycles in the instrument and monitoring the residual cure exotherm.

Autoclave cure simulations are possible using a high pressure DSC cell or high-pressure DMA [108]. DSC characterization is finding increased use for fingerprinting and quality assurance applications [43]. For condensation reactions, TGA has proved to be useful for quantitating the mass losses during chemical changes [18]. Figure 53 illustrates a weight loss curve for a PMR-15/carbon prepreg as the condensation reaction evolves water and methanol. Thermal analysis techniques are also used to study the effect of moisture [109] and environmental aging [68] on the prepreg properties. Figure 54 shows how the complex viscosity versus temperature increases with 0, 7, 14, and 26 days of room temperature aging at 80% RH for an epoxy prepreg. This DMA work was done using a parallel-plate configuration. DSC curves for the same epoxy prepreg in as-received and aged conditions are shown in Figure 55.

In a finished composite product, the resin matrix should be in its fully cured state (C-staged). The importance of achieving complete conversion of the B-staged consolidated laminate into a completely cured product and knowing how to confirm the conversion cannot be overstated. The optimizing of cure cycles depends on reaction kinetics and limitations imposed by both mass and heat transfer processes. The chemistry of cure involves formation and linear growth of a molecular chain that branches and subsequently cross-links. As the polymerization proceeds, the molecular weight increases rapidly, and eventually several large chains link together to produce the beginnings of a network. The onset of this sudden, irreversible transformation from viscous liquid to elastic gel is called the gel point. Once beyond the gel point, the cross-link density increases rapidly to give a rigid network of essentially infinite molecular weight. The cross-link density determines, for the most part, the average molecular weight of molecular segments between cross-links, hence the potential mobility of seg-

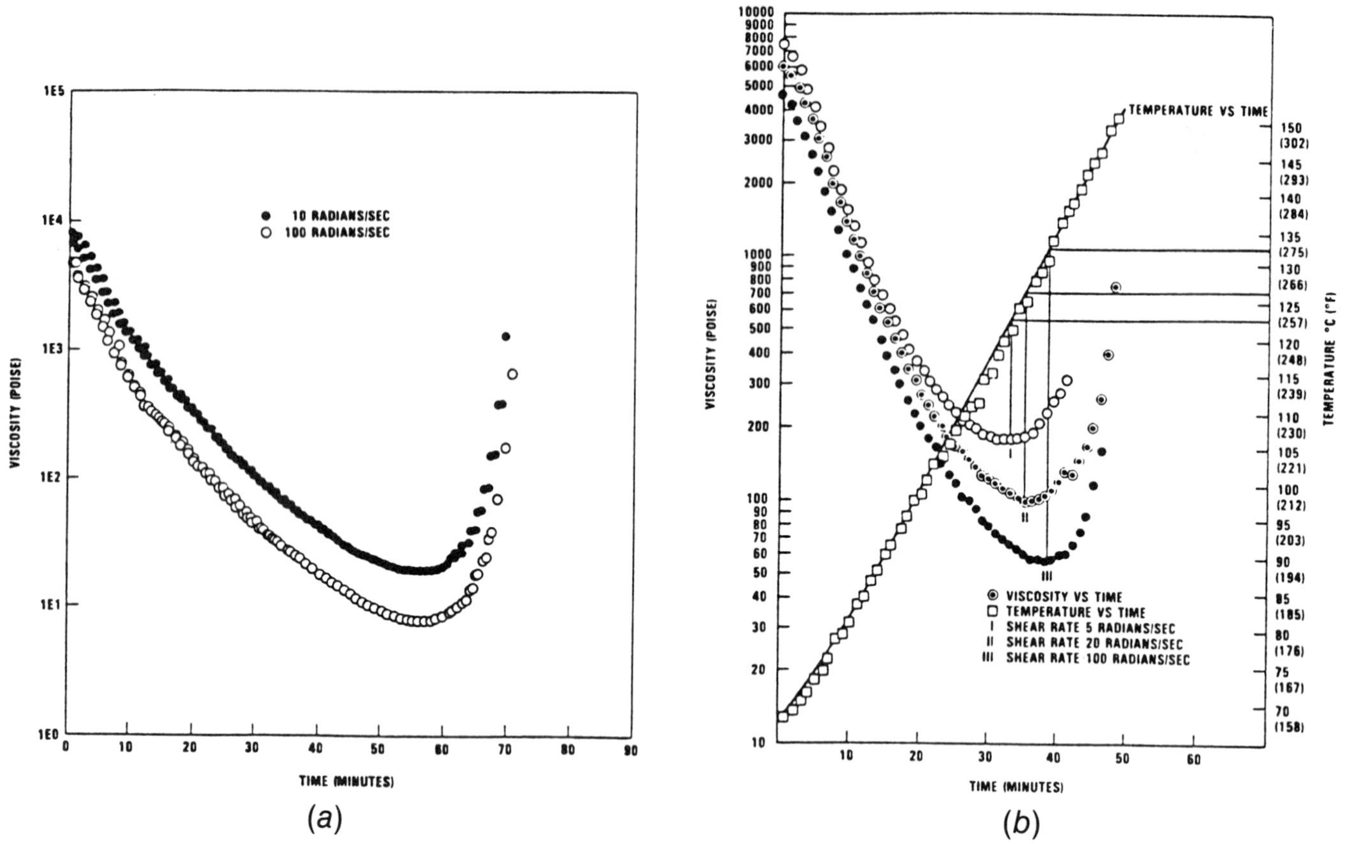

FIGURE 52 Viscosity–time–temperature behavior using a cone-and-plate configuration for (*a*) Hexcel HX 1504 and (*b*) Narmco (BASF Structural Materials) 5245C matrix resin systems.

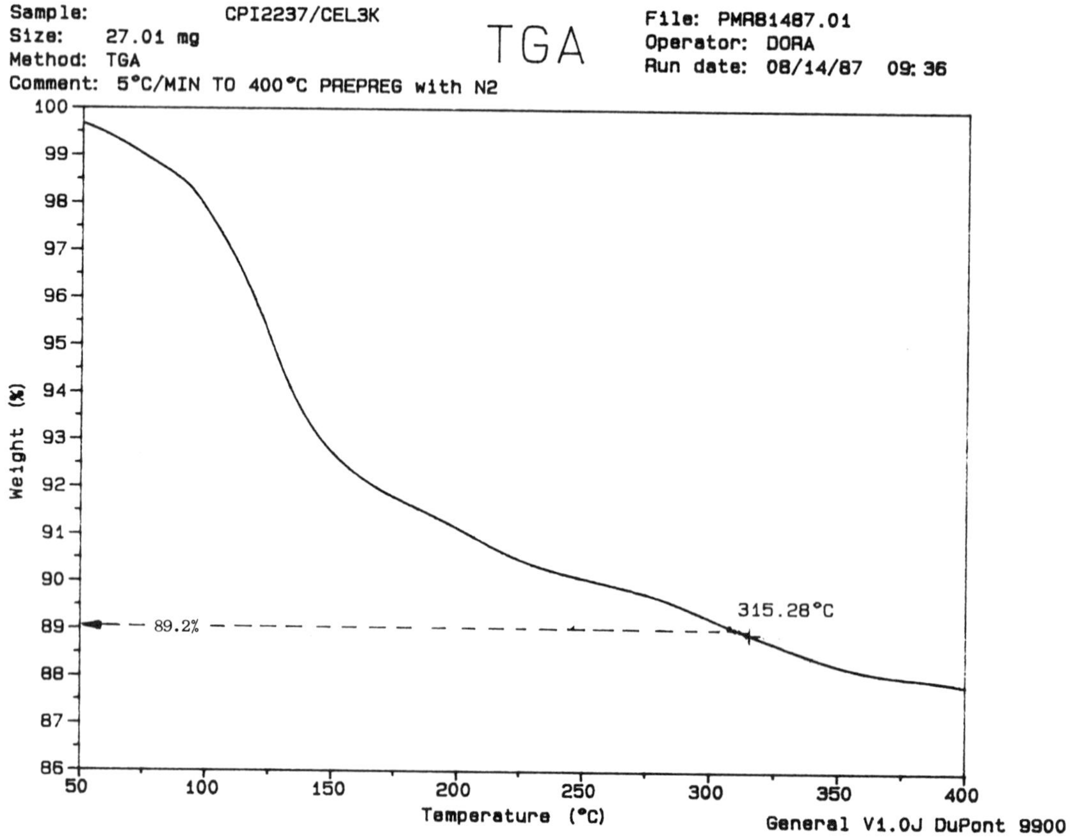

FIGURE 53 TGA weight loss behavior for a PMR-15/Celion, G30-500. Carbon fabric prepreg showing the effect of methanol and water devolatilization during imidization. (Courtesy of Ferro Composites, Los Angeles.)

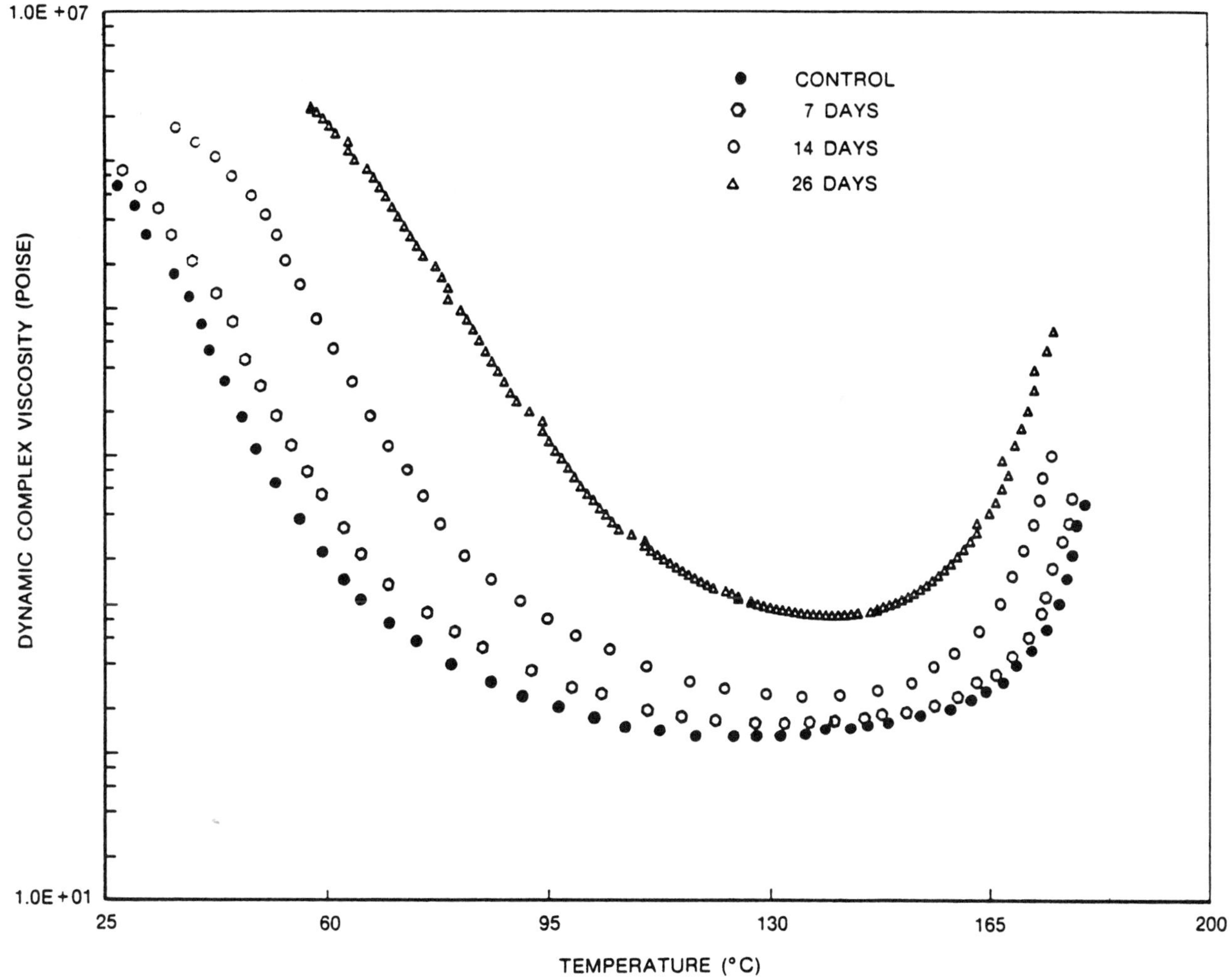

FIGURE 54 DMA parallel-plate viscosity versus temperature, illustrating increasing epoxy viscosity after 0, 7, 14, and 26 days of aging at room temperature and 80% relative humidity. (From Ref. 55.)

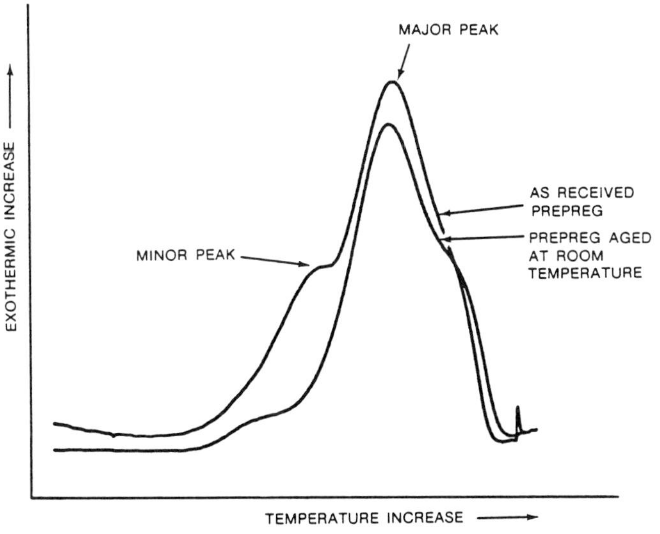

FIGURE 55 DSC cure behavior differences for the unaged and aged epoxy prepreg shown in Figure 54. (From Ref. 55.)

ments, which in turn governs the capability of the system to undergo volume changes. The smaller this capability, the higher the T_g and many of the ultimate physical and mechanical properties [21,110,111].

If the cure temperature is at the T_g of the system of growing chains, the resultant product is a glass, and the high viscosity causes a diminution in the amount of further reaction as a result of limits imposed by diffusion (i.e., a diffusion-controlled process). The onset of this transformation is known as the vitrification point. The handling, processing, and development of ultimate properties depend strongly on gelation and vitrification. Continuing the cure above the T_g increases segmental mobility and, therefore, promotes diffusion of molecular species and continuation of the cross-linking process.

Thermal analysis techniques have been developed and applied extensively as quality control methods to characterize the completeness of cure for polymer composite materials [112,113]. Incomplete cure can be observed as a residual exotherm in the DSC chromatogram. The ratio of the residual exotherm to the exotherm for uncured resin is one way to estimate the percentage of

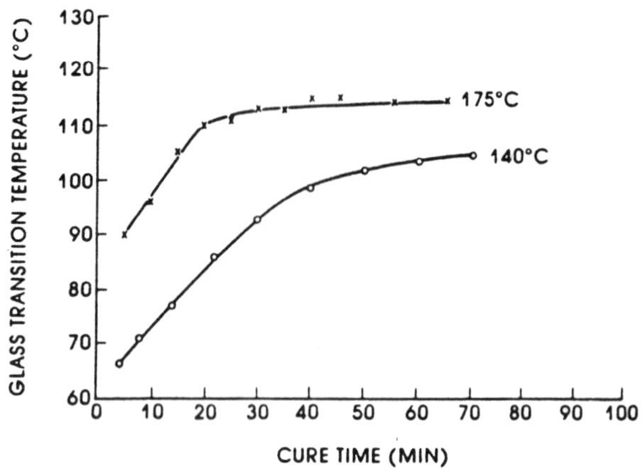

FIGURE 56 **DSC-measured glass transition temperature for a matrix resin cured at 140 and 175°C for the cure times indicated.** (Courtesy of du Pont Instruments.)

cure. Since T_g increases with the degree of cure, it too can be used for this purpose [82,112]. Figure 56 illustrates the effects of time and temperature of cure on the T_g as measured by DSC. Specialized techniques that have been developed to measure the degree of cure include dielectric analysis [114–116] and the viscosity-dependent fluorescent probe [117].

Reaction Kinetics

Kinetics calculations provide valuable guidance for product development, process development, and prediction of optimum cure cycles, as well as for problem solving. Other applications of kinetics include prediction of chemical and radiative decompositions, thermal stability, and thermo-oxidative processes. Mechanistically, polymerization reactions and decomposition processes can be modeled as either nth-order reactions or autocatalytic reactions. Equation (28) relates the rate of reactant conversion dc/dt at a constant temperature to some function of reactant concentration $f(c)$ through a rate constant k:

$$dc/dt = kf(c) \tag{28}$$

For chemical systems that follow nth-order kinetics, the rate of conversion is proportional to the concentration of the material that has yet to react, for example:

$$f(c) = (1 - c)^n \tag{29}$$

Autocatalytic reactions are characterized by attainment of a maximum reaction rate at about 30–40% of the conversion. These reactions can be described by the following relation:

$$f(c) = c^m (1 - c)^n \tag{30}$$

where m is also a reaction order.

The rate constant is temperature dependent and obeys the Arrhenius relation:

$$k = A \exp\left(\frac{-\Delta Ea}{RT}\right) \tag{31}$$

where A is the frequency factor, E is the activation energy, R is the gas constant, and T is the absolute temperature.

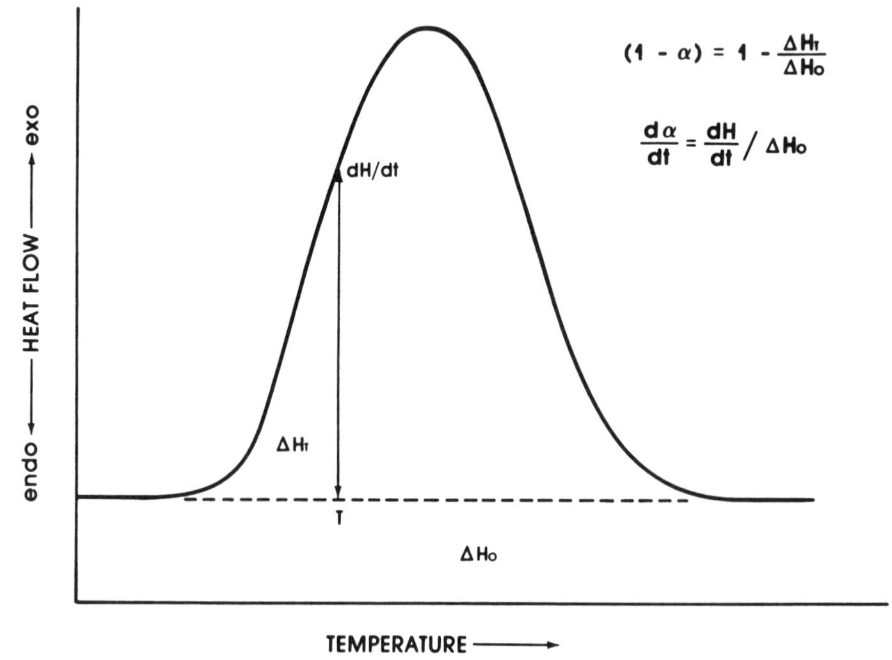

FIGURE 57 DSC cure exotherm measurements for applying the Borchardt–Daniels kinetics analysis. (Courtesy of du Pont Instruments.)

Differential scanning calorimetry has been used as the method of choice for characterizing cure kinetics because energies are determined directly. The Borchardt–Daniels calculation routine is the most rapid means of modeling a cure reaction [118]. DSC analysis of the cure exotherm provides the activation energy, the frequency factor, and the order of reaction from a single thermogram (Fig. 57). However, owing to certain simplifying assumptions, the Borchardt–Daniels model is not suitable for multiple reactions, autocatalyzed reactions, and degradative reactions. Isothermal methods [119–124] and methods requiring curing at multiple heating rates [121–125] give better accuracy and are suitable for autocatalytic reactions. Isothermal DSC has been used to examine the effect of reinforcements on the kinetics of neat resin cure [126]. Degradation and condensation reaction kinetics are studied more frequently using TGA [18,127,128]. The most widely used method is that of Flynn and Wall, which requires measuring the degradation profile at several heating rates between 1 and 20°C per minute [128].

Glass Transition

The glass transition is usually associated with the onset of segmental mobility in amorphous materials or in the amorphous phase of semicrystalline polymers. A number of bulk properties, including heat capacity, coefficient of thermal expansion, stiffness, hardness, and refractive index, undergo marked changes at the glass transition point T_g. This makes the glass transition a property of interest for scores of applications.

The most common way to determine T_g is by volume expansion using TMA [129], but a number of studies have been based on DSC and DMA [82,130]. Comparisons of various methods to determine T_g are made in Refs. 129, 131, and 132. The reinforcements in polymer composites often dilute the magnitude of change in such properties as specific heat and coefficient of thermal expansion, thus reducing the ability of DSC and TMA to measure them. Because the stiffness and damping properties change more dramatically than the heat capacity or the thermal expansion, DMA is a more sensitive method for measuring secondary transitions such as T_g.

A number of quality control techniques based on DMA have been developed. Figure 58 shows typical thermomechanical changes in a cured Kevlar–epoxy laminate in the glass transition region. The peak maximum in the loss modulus curve is often taken as the glass transition temperature. Figure 59 illustrates how G' and G'' change through the imidization and cross-linking reactions in a PMR-15/carbon cloth prepreg. Figure 60 shows how the

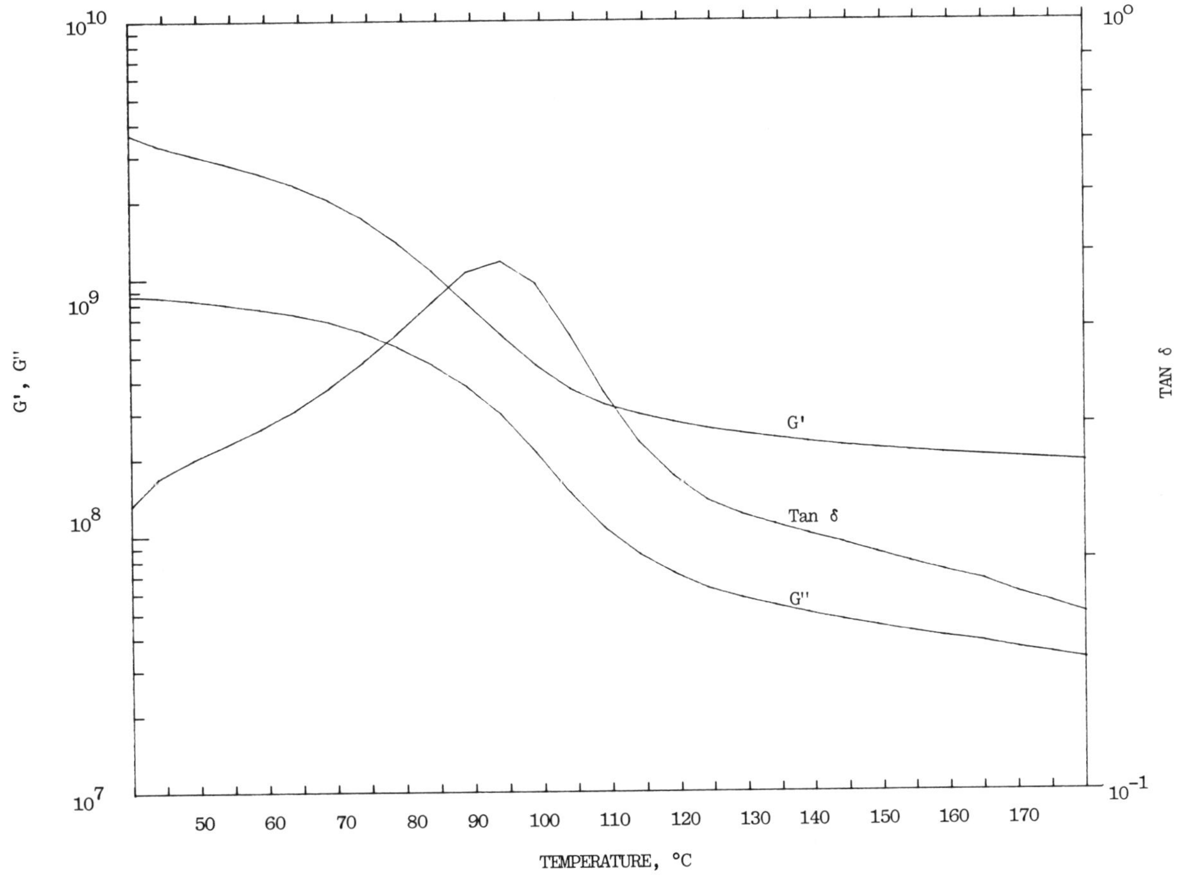

FIGURE 58 DMA analysis of an epoxy–Kevlar prepreg roving system showing how G', G'', and tan δ vary with temperature. (Courtesy of Ferro Composites, Los Angeles.)

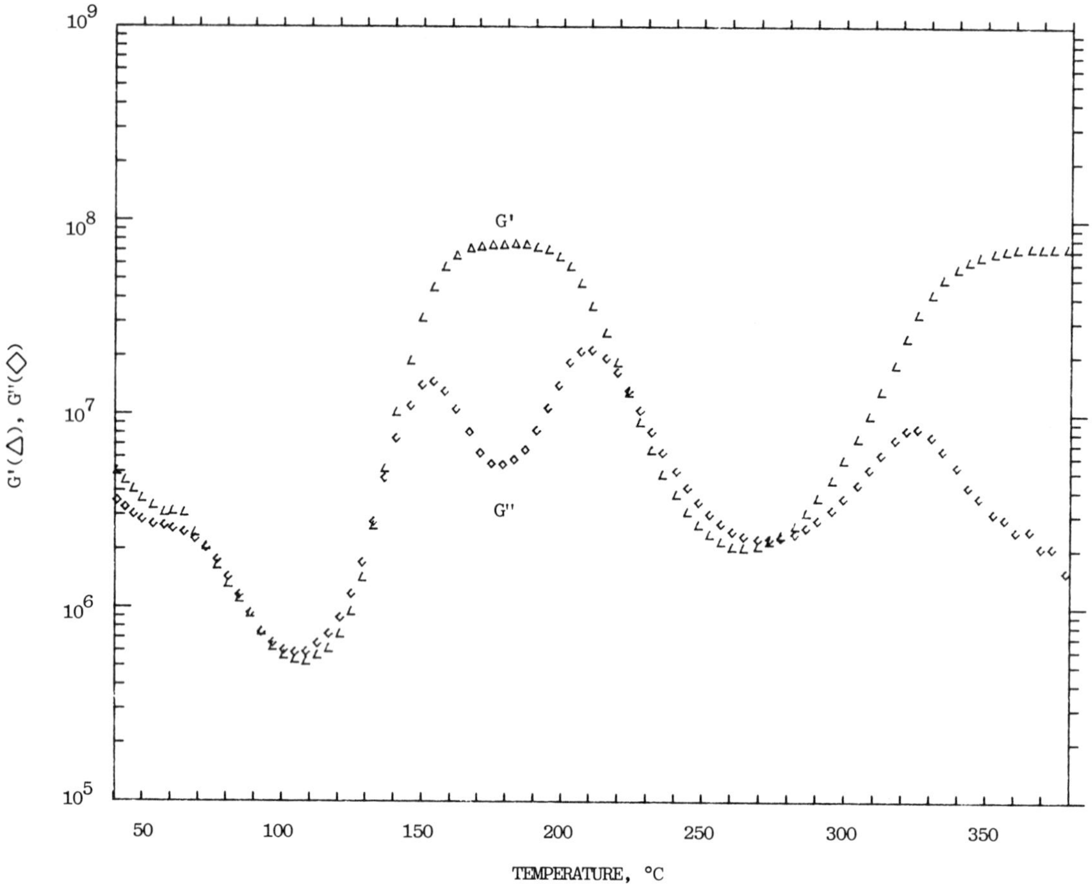

FIGURE 59 DMA analysis of a PMR-15/Amoco T300 woven prepreg system showing how G' and G'' vary with temperature. (Courtesy of Ferro Composites, Los Angeles.)

methanol content in PMR-15 prepreg affects the tan δ behavior [133]. Cured laminates can be tested for residual cure, which is exhibited by lower than required glass transition temperatures or shear modulus values [83,134].

The glass transition, being a kinetic phenomenon, is affected by the rate at which the sample is heated [131]. For thermosetting polymers, T_g values increase with the degree of cross-linking [82,112]. For thermoplastics (i.e., linear molecules), T_g increases with the number average molecular weight following Eq. (32):

$$T_g = T_{g\infty} - \frac{K}{\overline{M}n} \quad (32)$$

where $T_{g\infty}$ is the glass transition temperature of a polymer having infinite molecular weight, $\overline{M}n$ is the number average molecular weight, and K is a polymer-dependent constant.

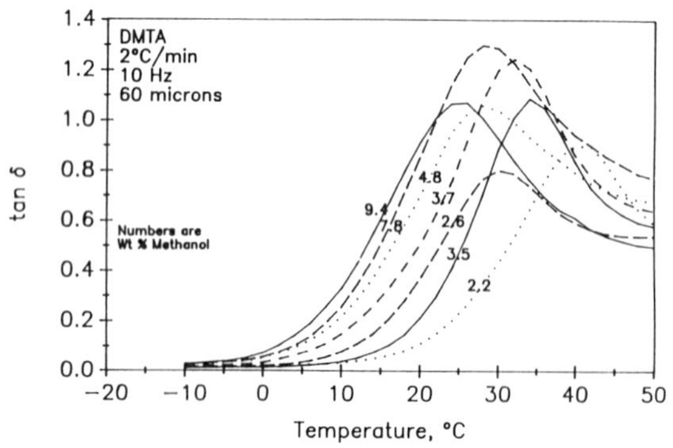

FIGURE 60 DMA analysis of PMR-15 prepregs showing variations in tan δ for prepregs having different methanol contents. (From Ref. 133.)

Crystallization and Melting

Semicrystalline thermoplastics exhibit one or more crystallization and melting transitions as well as the amorphous phase glass transition. In these materials, ordered regions consist of lamellar crystals that usually organize into larger shallow, pyramidlike structures called spherulites. The polymer chains are folded back upon themselves in the lamellar crystallites to produce parallel chains perpendicular to the face of the crystal.

Crystallizability and rigid linear molecular structure in polymers as polyether ether ketone (PEEK) are the origin of good solvent resistance and high use temperature properties. Figure 61 shows a DSC thermogram for commercially available amorphous PEEK film as it is

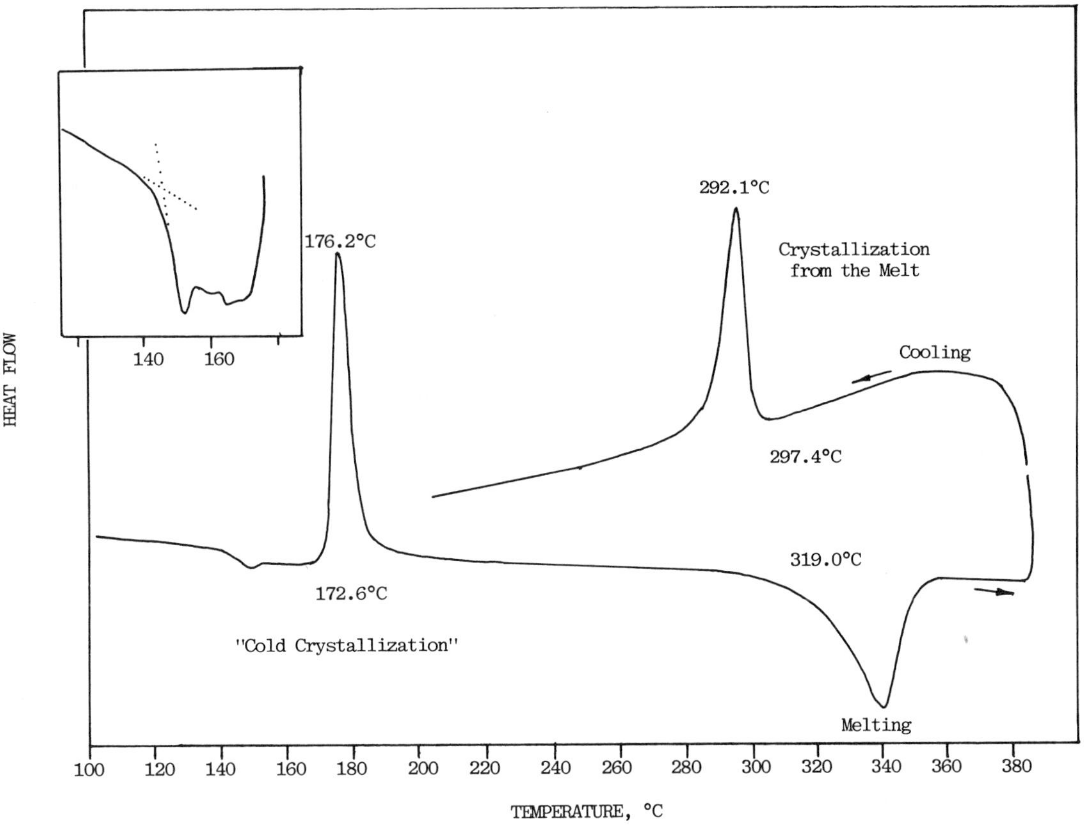

FIGURE 61 **DSC thermogram for commercially available PEEK film illustrating "cold" crystallization exotherm and melting endotherm on heating, followed by a recrystallization exotherm on cooling. *Inset:* amplified view of the glass transition.** (Courtesy of Ferro Composites, Los Angeles.)

heated at 20°C per minute. A sudden drop in the specific heat indicates the glass transition temperature around 145°C. The glass transition behavior is distorted by a small exothermic transition, which has been attributed in other polymeric glasses to enthalpy relaxation of the strained molecules [135]. Following the glass transition, rubbery PEEK shows a strong tendency to crystallize. This is shown by a sharp exotherm having its onset at 172.6°C. The crystals undergo reorganization as the temperature increases, and they eventually melt, as shown by a broad melting endotherm or peak at 340.7°C. The areas under the crystallization exotherm and the melting endotherm can be used to quantitate the degree of crystallinity. The degree of crystallinity has a profound effect on the bulk properties [136] as well as on the processibility of the resin [137,138]. A number of studies have attempted to explain the crystalline morphology of polymers [139–141]. DSC has been used to ascertain crystallization kinetics [142–145] and degree of crystallinity [127,146]. Specific heat and heat of fusion can also be calculated by using DSC.

ASTM Methods for Thermal Analysis

ASTM E 472 — *Standard Practice for Reporting Thermoanalytical Data*

ASTM E 473 — *Standard Definitions of Terms Relating to Thermal Analysis*

ASTM E 698 — *Standard Test Method for Arrhenius Kinetic Constants for Thermally Unstable Materials*

ASTM E 793 — *Standard Test Method for Heats of Fusion and Crystallization of Differential Scanning Calorimetry*

ASTM E 794 — *Standard Test Method for Melting and Crystallization Temperatures by Thermal Analysis*

ASTM E 831 — *Standard Test Method for Linear Thermal Expansion of Solid Materials by Thermomechanical Analysis*

ASTM E 928 — *Standard Test Method for Mole Percent Impurity by Differential Scanning Calorimetry*

ASTM D 2417 — *Standard Test Method for Heat of Fusion and Crystallization of Polymers by Thermal Analysis*

ASTM D 3418 — *Standard Test Method for Transition Temperatures of Polymers by Thermal Analysis*

ASTM D 4065 — *Standard Practice for Determining*

	and Reporting Dynamic Mechanical Properties of Plastics
ASTM D 4092	*Standard Definitions and Descriptions of Terms Relating to Dynamic Mechanical Measurements of Plastics*
ASTM D 4473	*Standard Practice for Measuring the Cure Behavior of Thermosetting Resins Using Dynamic Mechanical Procedures*

Standard Prepreg and Laminate Methods

The standard methods for determining prepreg properties involve the gel time and the percentages of resin content, volatile content, and flow. These are among the oldest of the methods used in the advanced composites industry for controlling and assuring prepreg quality and consistency. Virtually all government and industry material specifications for prepreg materials require conformity to at least these four properties. The specific values, of course, depend on the application, and the details for performing the tests vary depending on the kinds of resin and fiber involved. These tests are specified and described in detail in the ASTM Standard Test Methods cited at the end of this section. Some comments about sampling and relevance may be helpful before concluding this article.

It was stated in the Introduction that the standard prepreg methods are not sufficiently discriminatory to detect certain small differences in resin character that may lead to noticeable processing differences. In spite of this, the four methods named above are still the principal basis for regulating manufactured prepreg quality and for measuring conformity to product specification during receiving inspection in the fabricator's quality assurance laboratory.

The simple fact is, regardless of methodological flaws, the measurements correlate reasonably well with fabricators' optimized processes. Moreover, these methods require fairly large sample sizes, typically 51.6–103.2 cm^2, which are enormous compared with the sample sizes or weights taken for instrumental analysis. If the various instrumental methods produce chemical fingerprints, it can be said that the standard prepreg methods yield, by comparison, chemical footprints. This is not a facetious statement. Rather, it calls acute attention to the fact that the results of instrumental analysis are based, with few exceptions, on milligram-sized samples. Such samples are cut or punched from a roll of prepreg that may weigh 40 to several hundred pounds. In other words, DSC, HPLC, and FTIR samples are microscopic with respect to prepreg mass and surface area. Thus, given a roll of prepreg measuring, for example, 0.91 m wide and 45.7 m long and weighing 18,160 g, a 20 mg sample for DSC would represent one part per million on a weight basis and one part in five million on a surface area basis!

The samples for determining prepreg properties represent a significantly larger slice of the prepreg universe. One 10.16 cm^2 prepreg section represents about one part per 4000 by both weight and surface area, using the same arithmetic assumptions as above. The apparent odds of capturing the "correct" section of prepreg are about 250 times better using this sample size. This is not to imply that using the instrumental analysis methods is an exercise in futility, or that these approaches do not provide statistically meaningful data compared with using the standard prepreg methods. It is, however, to emphasize the importance of sampling and the importance of thinking through very carefully the functions that any analysis or characterization is to serve.

Standard Prepreg Methods

ASTM C 613	*Standard Test Method for Resin Content of Carbon and Graphite Prepregs by Solvent Extraction*
ASTM D 3529	*Standard Test Method for Resin Solids Content of Carbon Fiber–Epoxy Prepreg*
ASTM D 3530	*Standard Test Method for Volatiles Content of Carbon Fiber–Epoxy Prepreg*
ASTM D 3531	*Standard Test Method for Resin Flow of Carbon Fiber–Epoxy Prepreg*
ASTM D 3532	*Standard Test Method for Gel Time of Carbon Fiber–Epoxy Prepreg*

Standard Laminate Methods

ASTM D 3171	*Standard Test Method for Fiber Content of Resin-Matrix Composites by Matrix Digestion*
ASTM D 3355	*Standard Test Method for Fiber Content of Unidirectional Fiber–Resin Composites by Electrical Resistivity*

Melvin R. Kantz and Hemant K. Gupta

References

1. H. W. Siesler and K. Holland-Moritz, *Infrared and Raman Spectroscopy of Polymers*, Dekker, New York, 1980.
2. J. R. Dyer, *Applications of Absorption Spectroscopy of Organic Compounds*, Prentice-Hall, Englewood Cliffs, NJ, 1965.
3. P. R. Young and A. C. Chang, *Proceedings of 31st SAMPE Int. Symposium*, 1986, pp. 315–329.
4. H. H. Willard, L. L. Merritt, Jr., J. A. Dean, and F. A. Settle, Jr., *Instrumental Methods of Analysis*, 6th ed., Van Nostrand, New York, 1981.
5. R. Zbinden, *Infrared Spectroscopy of High Polymers*, Academic Press, New York, 1964.
6. P. R. Griffiths and J. A. DeHaseth, *Fourier Transform Infrared Spectroscopy*, Wiley, New York, 1986.
7. C.-C. Cheu and S. W. Ferla, *Proceedings of the 33rd SAMPE Int. Symposium*, 1988, pp. 12–25.
8. P. C. Painter and M. M. Coleman, *The Use of Absorbance*

Subtraction in FTIR Studies in Effective FTIR: A Workbook for Practical Applications. FTIR Spectroscopy in the Analytical Laboratory, available from Analect Instruments, Irvine, CA.

9. J. L. Koenig, *Appl. Spectrosc., 29,* 293 (1975).
10. S. P. Prince and M. C. Kage, *SAMPE Q., 14*(2), 25 (1983).
11. R. H. Pater and D. A. Scola, *Proceedings of the 11th SAMPE Natl. Technical Conference,* 1979, pp. 151–165.
12. S. R. Culler, H. Ishida, and J. L. Koenig, *Polym. Compos., 7*(4), 231 (1986).
13. S. R. Culler, H. Ishida, and J. L. Koenig, *J. Colloid Interface Sci., 109*(1), 1 (1986).
14. T. Vuong, *Proceedings of the 32nd SAMPE Int. Symposium,* 1989, pp. 839–852.
15. D. D. Schoon, pp. 433–443 in ref. 7.
16. M. K. Antoon and J. L. Koenig, *J. Polym. Sci. B, Polym. Phys., 19,* 198 (1981).
17. D. A. Scola, J. Vontell, and M. Felsen, *Polym. Compos., 8*(4), 244 (1987).
18. M. Navarre, in K. C. Mittal, Ed., *Polyimides: Synthesis, Characterization and Applications,* Vol. 1 Plenum Press, New York, 1983, pp. 429–442.
19. C. Di Giulio, M. Gautier, and B. Jasse, *J. Appl. Polym. Sci., 29*(5), 1771 (1984).
20. M. K. Antoon, "Fourier Transform Infrared Investigation of the Structure and Moisture Stability of the Epoxy Matrix in Glass-Reinforced Composites," available from University Microfilms International, Order 8013869, 1980, 251 pp.
21. G. A. Pasteur, H. E. Bair, and F. Vratny, "Curing of a Polyimide Resin," *Therm. Anal., Proc. Int. Conf., 2,* 1155 (1982).
22. S. C. Lun, B. J. Bulkin, and E. M. Pearce, *J. Polym. Sci. A, Polym. Chem., 17,* 3121 (1979).
23. G. A. George, R. E. Sacher, and J. F. Sprouse, *J. Appl. Polym. Sci., 21,* 2241 (1977).
24. R. A. Jewell and G. F. Sykes, in S. S. Labana, Ed., *Chemistry and Properties of Cross-Linked Polymers,* Academic Press, New York, 1977.
25. P. R. Young and A. C. Chang, pp. 538–550 in ref. 7.
26. M. K. Antoon and J. L. Koenig, *J. Polym. Sci. A, Polym. Chem., 19,* 549 (1981).
27. B. S. Rao, *J. Polym. Sci. C, Polym. Lett., 26,* 3 (1988).
28. P. H. Lindenmeyer and C. H. Sheppard, "Characterization of PMR-15 Polyimide Resin and Prepreg," NASA CR-168217, March 1984.
29. P. R. Young and A. C. Chang, *Proceedings of the 16th SAMPE Natl. Technical Conference,* 1984, pp. 136–147.
30. A. Wereta, Jr., and D. K. Hadad, in C. A. May, Ed., *Resins for Aerospace,* ACS Symposium Series 132, American Chemical Society, Washington, DC, 1980.
31. P. R. Young and A. C. Chang, *SAMPE J., 22*(2), 70 (1986).
32. G. Brown and C. A. Murray, *Plast. Technol.,* September 1965, pp. 1–3.
33. D. Garcia and T. T. Serafini, *J. Polym. Sci. B, Polym. Phys., 25,* 2275 (1987).
34. R. E. Smith, F. N. Larsen, and C. L. Long, *J. Appl. Polym. Sci., 29,* 3713 (1984).
35. R. Yungk and C. Watson, "PMR-15 Molding Compound Characterization," paper presented at High Temperature Workshop VIII, Riviera Beach, FL, Jan. 25–28, 1988; limited distribution through AFWAL/MLPJ, Weight-Patterson Air Force Base.
36. P. D. Frayer, in K. C. Mittal, Ed., *Polyimides: Synthesis, Characterization and Applications,* Vol. 1 Plenum Press, New York, 1983, pp. 273–294.
37. L. S. Bark and N. S. Allen, *Analysis of Polymer Systems,* Applied Science Publishers, Barking, England, 1982.
38. D. D. Werstler, *Polymer, 27,* 750 (1986).
39. D. D. Werstler, *Polymer, 27,* 757 (1986).
40. R. E. Smith, F. N. Larsen, and C. L. Long, *J. Appl. Polym. Sci., 29,* 3697 (1984).
41. C. F. Poranski, Jr., and W. B. Moniz, in C. A. May, Ed., *Resins for Aerospace,* ACS Symposium Series 132, American Chemical Society, Washington, DC, 1980.
42. R. W. Lauver, "Effect of Ester Impurities in PMR-Polyimide Resin," NASA Technical Memorandum TMX-73444, 1976.
43. R. W. Lauver, W. B. Alston, and R. D. Vanucci, "Stability of PMR-Polyimide Monomer Solutions," NASA Technical Memorandum 79063, 1979.
44. J. N. Clark, R. N. Jagannathan, and F. G. Herring, *Polymer, 29,* 341 (1988).
45. G. A. Morse, pp. 721–732 in ref. 29.
46. I. Plavin, H. Rosenthal, and U. Arnon, *Proceedings of the 16th SAMPE Natl. Technical Conference,* 1984, pp. 696–704.
47. W. R. Supina and R. S. Henley, *Chemistry, 37,* 12 (1964).
48. S. A. Mestan and C. E. M. Morris, *Rev. Macromol. Chem. Phys., C24*(1), 117 (1984).
49. F. N. Stone, B. N. Ranganathan, and A. Zsolnay, *SAMPE Q., 16*(2), 7 (1985).
50. R. B. Whitaker, A. B. Nease, and R. O. Yelton, "Characterization of Polyetheretherketone and Other Engineering Thermoplastics," Mound Laboratory Report MLM-3141 (DE84-010668), April 13, 1984.
51. R. E. A. Escott, *Proceedings of the 19th SAMPE Int. Technical Conference,* 1987, pp. 398–409.
52. L. R. Snyder and J. J. Kirkland, *Introduction to Modern Liquid Chromatography,* 2nd ed., Wiley, New York, 1979.
53. A. E. Hamielec and H. Meyer, *Developments in Polymer Characterization,* Vol. 5, Elsevier Applied Science Publishers, London and New York, 1986.
54. M. T. Gilbert, *High Performance Liquid Chromatography,* Wright-IOP, Bristol, 1987.
55. T. A. Sewell, in C. E. Browning, Ed., *Composite Materials: Quality Assurance and Processing,* ASTM STP 797, American Society for Testing and Materials, Philadelphia, 1983, pp. 3–14.
56. D. A. Carey, "Screening Testing for Composite Prepregs," Contract DE-AC04-76-DP00613, Final Report January 1987, Allied Bendix Aerospace, BDX-613-S642.
57. C. H. Sheppard, J. T. Hoggatt, and W. A. Symonds, "Quality Control Developments for Graphite/PMR-15 Polyimide Composites Materials," NASA CR-159182 (Final Report), June 1979.
58. G. L. Hagnauer and I. Setton, *J. Liquid Chromatogr., 1*(1), 55 (1978).
59. F. N. Larsen and D. A. Spieker, "Characterization of DGEBA Epoxy Resins," Bendix Kansas City Division Report DBX-613-3644, April 1987.

60. H. K. Gupta and M. R. Kantz, in *Proceedings of the American Society for Composites, Second Technical Conference*, Sept. 23–25, 1987, Technomic, Lancaster, PA.

61. K. C. Cole, D. Noel, and J.-J. Hechler, *Proceedings of the 30th SAMPE Int. Symposium*, 1985, pp. 624–637.

62. P. R. Young and G. F. Sykes, in C. A. May, Ed., *Resins for Aerospace*, ACS Symposium Series 132, American Chemical Society, Washington, DC, 1980, pp. 479–490.

63. G. D. Roberts and R. W. Lauver, *J. Appl. Polym. Sci., 33*, 2893 (1987).

64. G. D. Roberts and R. D. Vanucci, *SAMPE J., 22*(2), 24 (1986).

65. G. L. Hagnauer, J. M. Murry, and B. M. Bowse, "HPLC Monitoring of Graphite–Epoxy Prepreg Aging," AMMRC TR 79-33, May 1979.

66. C. L. Hammermesh and P. J. Dynes, in T. T. Serafini, Ed., *High Temperature Matrix Composites*, Noyes Data Corporation, Park Ridge, NJ, 1987.

67. D. Noel, K. C. Cole, and J.-J. Hechler, *J. Appl. Polym. Sci., 32*, 3097 (1986).

68. D. J. Crabtree, in C. A. May, Ed., *Resins for Aerospace*, ACS Symposium Series 132, American Chemical Society, Washington, DC, 1980.

69. R. Hinrichs and J. Thuen, *SAMPE J., 15*(6), 12 (1979).

70. S. A. Johnson and N. K. Roberts, pp. 868–879 in ref. 14.

71. G. Mayorga, paper presented at High Temperature Workshop VIII, Riviera Beach, FL, Jan. 25–28, 1988; limited distribution through AFWAL/MLPJ, Wright-Patterson Air Force Base.

72. J. K. Del Rios, *J. Anal. Purif.*, March 1986, pp. 71–72.

73. W. W. Wendlandt, *Thermal Analysis*, 3rd ed., Wiley, New York, 1986.

74. E. A. Turi, Ed., *Thermal Characterization of Polymeric Materials*, Academic Press, New York, 1981.

75. I. M. Ward, *Mechanical Properties of Solid Polymers*, 2nd ed., Wiley, New York, 1979.

76. L. D. Lauer, *SAMPE Q., 15*(1), 31 (1983).

77. A. Apicella et al., *J. Therm. Anal., 30*(6), 1349 (1985).

78. P. S. Patel, P. P. Shah, and S. R. Patel, *Polym. Eng. Sci., 26*(17), 1186 (1986).

79. C. M. Tung and P. J. Dynes, in C. E. Browning, Ed., *Composite Materials: Quality Assurance and Processing*, ASTM STP 797, American Society for Testing and Materials, Philadelphia, 1983.

80. D. J. Crabtree, *Proceedings of the 12th SAMPE Natl. Technical Conference*, 1980, pp. 614–636.

81. C. D. Han and L. Kwok-Wai, *J. Appl. Polym. Sci., 29*(5), 1879 (1984).

82. P. Burroughs and J. N. Leckenby, *Compos. Struct., Proc. Int. Conf., 1*, 438 (1981).

83. P. S. Gill and R. L. Blaine, *Therm. Anal., Proc. Int. Conf., 2*, 994 (1982).

84. T. H. Grentzer, R. M. Holsworth, and T. Provder, *ACS Polym. Prepr., 22*(1), 254 (1981).

85. C. G. Ford, G. Anderson, F. W. Kraushaar, and P. L. Riley, *Proceedings of the 31st SAMPE Int. Symposium*, 1986, pp. 1300–1312.

86. D. H. Bowen, R. Davidson, R. J. Lee, and M. Aguirre, "Dynamic Mechanical Analysis as an Aid to the Characterization of Composite Materials," European Space Agency, Special Publication ESA SP-243, Compos. Des. Space Appl., 39–46, 1986.

87. K. E. Reed, *Polym. Compos., 1*(1), 44 (1980).

88. R. L. Hassel, Industrial Research & Development *20*(10), 160, (1978).

89. P. S. Gill, Industrial Research & Development, March, 104 (1983).

90. S. W. Street, pp. 822–838 in ref. 14.

91. J. F. Carpenter and R. J. Juergens, pp. 168–175 in ref. 85.

92. S. T. Bhi et al., pp. 1114–1118 in ref. 14.

93. M. R. Dusi, R. M. Galeos, and M. G. Maximovich, *J. Appl. Polym. Sci., 30*(5), 1847 (1985).

94. P. J. Dynes, T. T. Liao, C. L. Hammermesh, and E. F. Witucki, pp. 311–326 in ref. 15.

95. F. A. Myers, *Proceedings of the Second International Technical Conference on Polyimides*, 1985.

96. C. G. Ford and F. W. Kraushaar, pp. 609–617 in ref. 51.

97. J. K. Gillham, in C. A. May, Ed., *Resins for Aerospace*, ACS Symposium Series 132, American Chemical Society, Washington, DC, 1980.

98. M. G. Maximovich and R. M. Galeos, in T. T. Serafini, Ed., *High Temperature Polymer Matrix Composites*, Noyes Data Corporation, Park Ridge, NJ, 1987.

99. W. McCormack and F. A. Myers, "Cure Rheology and Chemistry of PMR-15 Prepreg During Imidization," High Temperature Workshop VI, Cocoa Beach, FL, March 3–7, 1986, AFWAL/MLSE, Wright-Patterson Air Force Base.

100. M. G. Maximovich and R. M. Galeos, *Proceedings of the 28th SAMPE Natl. Symposium*, 1983, pp. 568–580.

101. P. Marek et al., *Plasty Kauc., 22*(3), 87 (1985).

102. B. K. Appelt and P. J. Cook, *Anal. Calorim., 5*, 57 (1984).

103. J. M. Barton, *Br. Polym. J., 118*(1), 44 (1986).

104. J. A. Koutsky and R. E. Ebewele, in S. S. Labana, Ed., *Chemistry and Properties of Cross-Linked Polymers*, Academic Press, New York, 1977.

105. R. J. Morgan, C. M. Walkup, and T. H. Hoheisel, *J. Compos. Technol. Res., 7*(1), 17 (1985).

106. B. N. Ranganathan, *SAMPE Q., 15*(2), 38 (1984).

107. J. T. Gotro, B. K. Appelt, and K. I. Papathomas, *Polym. Compos., 8*(1), 39 (1987).

108. C. Gramelt from *American Laboratory*, January, 102(1984).

109. J. M. Barton and D. C. L. Greenfield, *Br. Polym. J., 18*(1), 51 (1986).

110. E. M. Woo, J. C. Seferis, P. R. Lopez, and L. B. Ilcewicz, pp. 1073–1083 in ref. 14.

111. C. M. Tung, ACS Polym. Prepr., *28*(1), 7 (1987).

112. J. Galy, J. P. Pascault, and M. F. Grenier-Loustalot, "The Effect of the Prepolymer Structure on the Cure Kinetics of Epoxy–Amine Networks, Cross-Linked Epoxies," 1987, pp. 169–177.

113. R. L. Hassel, *Coal Mining & Processing, 13*(5), 74(1976).

114. D. Frisch and R. Ciccarone, *Circuits Manufacturing, 17*(7), 54(1977).

115. S. A. Bidstrup, N. F. Sheppard, and S. D. Senturia, "Dielectric Analysis of the Cure of Thermosetting Epoxy/Amine

Systems," Report TR-8, Order AD-A177569/1/GAR, 15 pp. NTIS: Government Report Announcement Index (U.S.) 1987, 87(12), Abstract 724,838, 1986.

116. L. Leonard, *Adv. Compos.*, January/February 1987, pp. 49–54.

117. R. L. Levy and D. P. Ames, *Polym. Sci. Technol.*, *29*, 245 (1984).

118. L. Elegant et al., *J. Therm. Anal.*, *31*(6), 1351 (1986).

119. M. N. Khalaf, A. S. Matooq, and G. A. Adam, *Thermochim. Acta*, *111*, 115 (1987).

120. M. R. Keenan, "Autocatalytic Cure Kinetics from Isothermal DSC Measurements," Report SAND-86-1378C, Order DE86012280, 12 pp. NTIS: Energy Research Abstract 1986, 11(22), Abstract 51441, 1986.

121. M. R. Keenan, *J. Appl. Polym. Sci.*, *33*, 1725 (1987).

122. T. Kamon and K. Saito, *Kobunshi Ronbunshu*, *41*(5), 293 (1984).

123. Y. J. Huang and L. J. Lee, *Polym. Mater. Sci. Eng.*, *49*, 638 (1983).

124. W. J. Sichina, "Autocatalyzed Epoxy Cure Prediction Using Isothermal DSC Kinetics, Thermal Analysis," Application Brief, du Pont TA-93, 1986.

125. M. Cizmeciogly and A. Gupta, *SAMPE Q.*, *13*(3), 16 (1982).

126. J. Mijovic, *J. Appl. Polym. Sci.*, *31*(5), 1177 (1986).

127. W. E. Collins, "Characterization of Thermoplastic Polymers," presented at the Education Seminar, Pittsburgh Conference on Analytical Chemistry and Applied Spectroscopy, Cleveland, March 1–4, 1970.

128. J. H. Flynn, in H. H. G. Jellinek, Ed., *Aspects of Degradation and Stabilization of Polymers*, Elsevier, Amsterdam, 1978.

129. T. H. Brayden and C. C. Hays, *Proceedings of the 19th SAMPE Int. Tech. Conf.*, 1987, pp. 598–608.

130. M. Cortazar et al., *Polym. Bull.* (*Berlin*), *18*(2), 149 (1987).

131. R. Reisen, *Thermochim. Acta*, *93*, 213 (1985).

132. E. Kaisersberger, *Thermochim. Acta*, *93*, 291 (1985).

133. G. Mayorga, "The Effect of Methanol Content on the Viscoelastic Properties of PMR-15 Polyimide Prepreg," paper presented at High Temperature Workshop VIII, Rivera Beach, FL, Jan. 25–28, 1988; limited distribution through AFWAL/MLPJ, Wright-Patterson Air Force Base.

134. R. L. Hassel and R. L. Blaine, *Proceedings of the 24th SAMPE Natl. Symposium*, 1979, pp. 342–350.

135. S. Kumar and R. Salovey, *J. Appl. Polym. Sci.*, *30*, 2315 (1985).

136. L. L. Zhu, G. Wegner, and U. Bandara, *Makromol. Chem.*, *182*(12), 3639 (1981).

137. G. M. K. Ostberg and J. C. Seferis, *J. Appl. Polym. Sci.*, *33*(1), 29 (1987).

138. J. C. Seferis, *Polym. Compos.*, *7*(3), 158 (1986).

139. C. M. Tung and P. J. Dynes, *J. Appl. Polym. Sci.*, *33*(2), 505 (1987).

140. Y. Lee and R. S. Porter, *Polym. Eng. Sci.*, *26*(9), 633 (1986).

141. D. J. Blundell and B. N. Osborn, *SAMPE Q.*, *17*(1), 1 (1985).

142. B. Chabert, J. Chauchard, and J. Cinquin, *Makromol. Chem., Macromol. Symp.*, *9*, 99 (1987).

143. C. N. Velisaris and J. C. Seferis, *Polym. Eng. Sci.*, *26*(22), 1574 (1986).

144. J. F. Carpenter, *SAMPE J.*, *24*(1), 36 (1988).

145. J. D. Muzzy, D. M. Abouelnasr, and A. O. Kays, pp. 1319–1330 in ref. 7.

146. C.-L. Ong, W.-L. Chung, M.-F. Sheu, and J.-H. Lin, pp. 1307–1318 in ref. 7.

147. R. H. Boschan, Y. A. Tajima, G. A. Forseberg, G. H. Hull, and J. Harper-Tervet, in N. J. Johnson, Ed., *Toughened Composites*, ASTM STP 937, American Society for Testing and Materials, Philadelphia, 1987, pp. 437–452.

Circuit Board Laminates

Circuit board laminates are unique composite materials by virtue of their special construction and their special applications. The materials selected for use in printed circuit board laminates must meet the requirements of extremely varied environmental conditions, functional restrictions, and design parameters. The wide range of applications for printed circuits dictates that designers be flexible in selecting printed circuit materials and in stipulating the manner in which they are used in various products. The classification of printed circuit materials tends to be based on the construction of the printed circuit board as well as on the properties of the materials used. For example, the Institute of Printed Circuits (IPC) tends to designate printed circuits as either flexible or rigid. Further distinctions are drawn, in IPC literature, between single-sided, double-sided, and multilayer circuits. These designations have to do with the placement of the metal (usually copper) conductors on the insulating base material. Finally, the IPC recognizes variations in the composition and construction of the base (insulating) materials. These variations usually involve the heat resistance, flame retardance, or mechanical stability of the base material.

Printed circuit designers are normally responsible for selecting the construction and composition of laminates used in their products. Electronics assemblies, in the recent past, did not require the designer to give a great deal of thought to the selection of laminates. Cost and heat resistance were the primary considerations at that time. The complexity of circuit board design is changing very rapidly now, however, primarily because of the extremely high packaging densities in modern products, and the use of surface mount components. High packaging densities call for more conductor layers and finer lines. Surface mount components call for closer control of the thermal coefficient of expansion of base materials and of the surface geometry of the circuit board.

In spite of these complexities, circuit board laminates tend to fulfill basic functions in the construction of electrical and electronic products. While the circuit board designer must be mindful of subtle requirements, basic requirements must not be forgotten. These basic requirements are very important to laminate construction and composition and, therefore, to the designer's selection of a laminate material. Among these requirements are:

1. The use environment (ambient conditions)
2. Life expectancy
3. Operating parameters (voltage, component heat, etc.)
4. Mechanical load (component weight and distribution)
5. Functional requirements (signal frequency, wave shape, etc.)
6. Cost restrictions

Printed circuit boards serve two basic functions: to interconnect electrical components and to provide mounting support for these components. Sometimes they are also used for other functions—for example, as part of the structure of electronic cabinets or as part of a mechanical assembly. In earlier designs, no attempt was made to integrate functions of printed circuits because of their structural and material limitations. With improved materials and manufacturing methods, it is possible to integrate mechanical and electrical functions in printed circuit functions into the printed circuit design while, at the same time, providing the designer a variety of options for mounting and interconnecting electronic components.

While these options are intriguing to the designer, they also present a significant number of challenges that did not always exist in the early days of printed circuit design. For instance, the circuit board designer must understand how the product will be fabricated and assembled as well as how it will be used. In printed circuit fabrication, as in few other processes, the materials used for processing sometimes influence the final product's characteristics as much as its structural materials. For this reason, it is common practice for printed circuit designers (or their support associates) to specify manufacturing processes as well as construction materials. In fact, in the case of printed circuits requiring controlled impedance on part or all of the signal circuits, the designer must have process control to ensure acceptable products.

Printed circuit geometry has a relatively simple basis but may evolve into complex issues. The utilitarian function of printed circuits has already been stated, and from this aspect, we are considering a simple product (i.e., supporting base material and conductors). The circuit board design dictates what materials will be selected for its construction, but these materials are either insulating base material sheets or metal conductors. In the case of single-sided or double-sided conductors, the construction material is almost always a simple copper-clad laminate. This laminate consists of a sheet of insulating material with copper foil bonded to one or both of its surfaces. Figure 1 illustrates this construction principle. In single- and double-sided construction (i.e., conductors on one or both sides of the insulating base) the copper-clad laminate is relatively thick (18.54 cm). For multilayer core materials, the copper-clad laminate can be as thin as 0.24 mm, with bonding layers as thin as 0.015 cm. The thickness of the copper foil (see below) may be selected to suit a wide variety of design requirements.

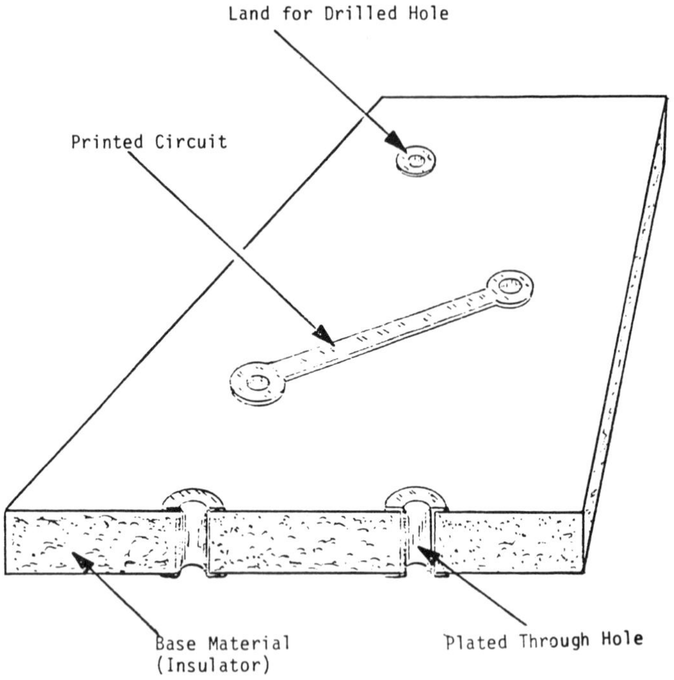

FIGURE 1 **Printed circuit board features.**

Printed circuit designers expect the laminates they specify to have high quality metallic foils (usually copper) bonded to high quality base (substrate) materials (usually glass/epoxy). The metal foil requires high purity, uniformity, and an absence of defects. The base material must be strong, an excellent insulator, and of uniform quality. Some base materials are rigid and some are flexible. Flame retardancy, heat resistance, and machinability are characteristics selectable by the designer, depending on the material chosen.

Laminate Construction

A wide variety of circuit board laminates are available to the designer, and these are selected on the basis of electrical and mechanical requirements. Table 1 lists the more common types of circuit board laminate along with some of their more important characteristics. Most of these materials are quite stable over extended periods of normal exposure to ambient conditions. Table 1 serves only as a general statement of the characteristics of circuit board laminates.

The more modern approach to the design of laminates for highly sensitive requirements is to stipulate specific circuit function parameters (e.g., impedance), then to tailor the laminate composition and geometry to meet these needs. In many cases, standard laminate construction will suffice, but not always. It is not possible to satisfy every need with a single grade of glass–epoxy, a single grade of phenolic, or a single grade of

TABLE 1
Common Circuit Laminate Materials

Name/ Grade	Rockwell Hardness, M	Specific Gravity	Flexural Strength (MPa)	Water Absorbency (%)	Dissipation Strength (× 1 MHz)	Tensile Strength (MPa)	Compound Strength (MPa)	Dielectric Breakdown (kV)	Dielectric Construction (× 1 MHz)
X	110	1.36	162	6.00		124	186	40	
XP	95	1.33	83	3.60		72	172	40	
XPC	75	1.34	62	5.50		66	152	40	
XX	105	1.32	100	2.00	0.045	100	234	40	5.5
XXP	100	1.32	90	1.80	0.040	67	172	60	5.0
XXX	110	1.30	94	1.40	0.038	93	221	50	5.3
XXXP	105	1.31	78	1.00	0.038	76	172	60	4.6
XXXPC	105	1.31	78	0.75	0.038	76	172	60	4.6
G-3	100	1.65	310	2.70		159	345		
G-5	120	1.90	310	2.70	0.020	255	483	23	7.8
G-7	100	1.68	310	2.70	0.003	159	310	32	4.2
G-9		1.90	345	0.80	0.017	276	448	60	7.2
G-10	110	1.75	379	0.25	0.025	241	483	45	5.2
G-11	110	1.75	379	0.25	0.025	241	483	45	5.2
FR-1			90	3.60				40	
FR-2	105	1.30	78	0.75	0.038	85	172	60	4.6
FR-3		1.45	124	0.65	0.035	83	193	60	4.6
FR-4	110	1.75	379	0.25	0.025	241	483	45	5.2
CEM-1			214	0.30	0.035			45	5.0
CEM-3			248	0.25	0.025			45	5.2

high temperature material. Most printed circuit insulating materials are thermoset compounds, taking advantage of a property that enhances their usefulness during soldering and high temperature applications. Glass–epoxy construction is the most common for circuit board use. The versatility of glass–epoxy, together with its excellent electrical and mechanical characteristics, makes this material highly useful for circuit board insulation (note the several grades listed in Table 1).

Figures 2 and 3 show how woven glass cloth impregnated with semicured epoxy resin is laminated between layers of copper foil to form the sheet material from which common printed circuit boards are made. In the printed circuit industry, these sheets are referred to as copper-clad laminates. When these materials are laminated under heat and pressure, the epoxy resin polymerizes to provide a very high quality base material for the copper foil. This laminate structure, wherein a polymer is used in conjunction with a structural material (glass fibers), is common in printed circuit construction. More is said later about specific requirements for the glass cloth and how it is woven into the specific construction required for this application.

Table 2 lists the more common thicknesses of copper foil in general use for printed circuits. Note that these foils are listed according to their weight per square foot

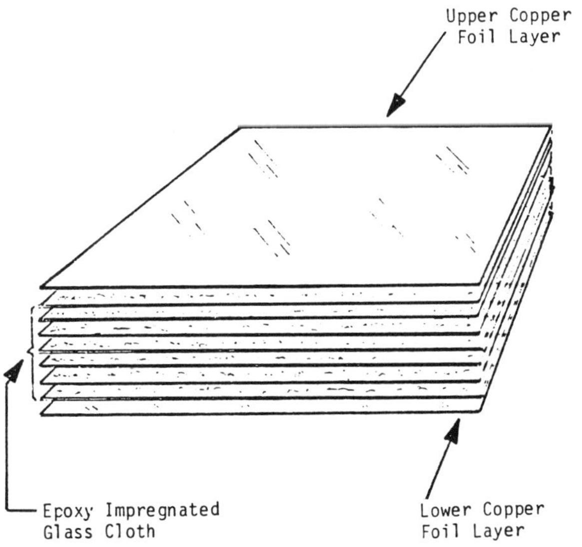

FIGURE 2 Basic circuit laminate construction materials.

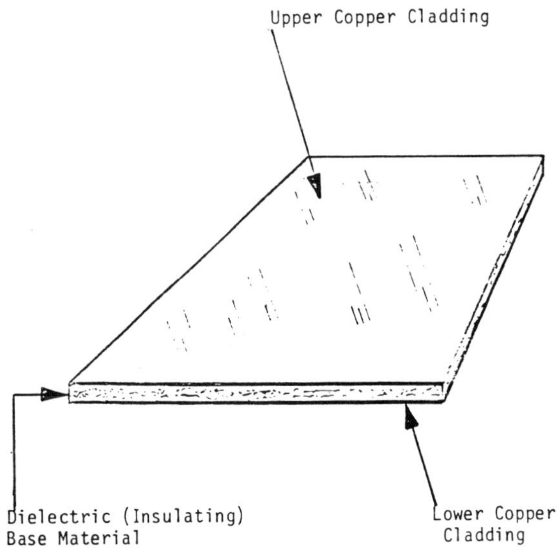

FIGURE 3 Circuit board laminate construction.

TABLE 2
Copper Thickness for Circuit Laminate

Copper (g/m^2)	Thickness mm
305	0.0356
610	0.0711
915	0.1068
1,221	0.1424
1,525	0.1780
1,831	0.2136
2,136	0.2492
2,441	0.2848
2,746	0.3204

(the common method for specifying copper foil) along with the conversion of this weight into actual thickness. The user is primarily concerned with the foil thickness, its purity, and how well it adheres to the base material. As shown later, there are other considerations of interest to the laminate manufacturer.

Table 3 lists the common types of base material together with the recommended usage for each type. These materials may be compared with the NEMA grades listed in Table 1. Special applications such as surface mount assemblies and products used in harsh environments sometimes dictate the need for special base materials. Sometimes the properties of base materials can be enhanced by mechanical design considerations and the use of special coatings for the printed circuit board.

As Tables 1 and 3 indicate, there is wide variation in the properties of printed circuit insulating (base) materials in common use. Phenolics are entirely satisfactory for many consumer products and have a distinct advantage in their compatibility with piercing and blanking as a fabrication method. If compliance with fire codes and safety regulations is mandatory (or if electrical insulation requirements are stringent), one of the grades of glass–epoxy is more appropriate. Polyimide is for higher temperature applications. It is acceptable to use almost any insulator as a base for printed circuits (e.g., glass or ceramics), but the possibilities are too extensive to consider in this context.

TABLE 3
Common Circuit Laminates

Type	Use
Phenolic	Pierced and blank construction for inexpensive printed circuits
Glass–epoxy composite	General printed circuit use
Glass–epoxy sheet	High reliability printed circuit use, including multilayers
Glass–epoxy FR	High reliability flame retardant
Polyamide	High temperature printed circuits, including flexible

The overall laminate thickness—copper plus the base material—can be selected as desired, but the common thicknesses (0.157, 0.236, 0.317 cm, etc.) are less expensive than special thicknesses. Thickness tolerances for standard production laminates tend to be ±10% of the laminate thickness. Precision laminates require thickness tolerances that are as close as ±5% of the specific thickness. Tighter thickness tolerances will naturally result in increased costs, although materials specified at ±5% thickness will normally be selected from standard laminate production runs. Laminate thickness measurements are normally made electronically, allowing the thickness gradient of a production sheet to be quickly characterized. Laminates can be supplied in sheets of 1.2 m × 1.2 m or cut to smaller size as needed.

Laminate Materials

When glass cloth is used for structural integrity in printed circuit laminates, special treatment is required. To begin, a special grade of glass is used for the fibers in the woven cloth (Table 4). In its natural state, glass cloth is not suitable for use in printed circuit manufacture. This is primarily because epoxy will not readily bond to the surface of the glass. The glass tends to move with respect to the epoxy because of the difference in thermal coefficient of expansion between the two materials, and to restrict this movement, it is important to maintain a rigid bond between the epoxy and the glass. This is critical during soldering and other high temperature operations. The required bond strength is facilitated by treating the glass with silane coupling agents. Characteristics of these coupling agents are shown in Table 5.

As mentioned previously, the glass used in circuit laminates is made from a special grade of glass (E-glass), formulated to give the best service in this application. Some of the characteristics of E-glass are listed in Table 6. In addition to using this special grade of glass, circuit board laminate designers must be careful to select the correct weave for the cloth and to specify the way that the cloth is oriented in the laminating press. The precleaning and epoxy impregnation of the cloth (after silane treating) must follow proper alignment of the glass cloth, wherein warp and woof are oriented correctly from layer to layer of the laminate. From Figure 2 it can be seen that, depending on the thickness and design of the laminate, many layers of glass cloth may be required. In a typical thick laminate 0.157 cm thick, it is common practice to use 7–9 layers of glass cloth.

The number of layers of glass cloth used in a particular circuit board laminate depends on several factors associated with the laminate's end use. At the same time, there are several options available in the construction of the glass cloth itself. The cloth is usually designated as 0.025 mm, 0.050 mm, or 0.075 mm thick. To achieve these thicknesses, the size of the glass fiber yarns is adjusted. A square weave is preferred for glass cloth used in printed circuits. The woven cloth carries an ASTM or textile industry designation that indicates its general construc-

TABLE 4
Classification of Glass Fibers

Symbol	Example
Typical Designation: CS D 100$\frac{1}{2}$	
C = glass composition	C = chemical
S = structural	
E = electrical	
S = type of yarn	S = stable filaments
	C = continuous fibers
D = filament diameter (mm)	D = 0.005
	DE = 0.006
	E = 0.007
	G = 0.009
100 = strand weight in 0.9 m/16 × 100	
$\frac{1}{2}$ = yarn construction	Numerator = number of strands twisted together
	Denominator = number of twisted strands plied together

TABLE 5
Silane Coupling Agents

Name	Formula[a] Inorganic	Organic
Cationic vinyl benzyl	$(CH_3O)_3$—Si—R—	$CHCH_2 \cdot HCl$
Epoxide	$(CH_3O)_3$—Si—R—CH$\overset{O}{\triangle}$CH$_2$	
Amino	$(CH_3O)_3$—Si—R—NH_2	

[a] R = Resin (Epoxy)

TABLE 6
E-Glass Composition and Properties

Compound	Proportion (wt %)
Silicon dioxide (SiO_2)	54
Calcium oxide (CaO)	20
Aluminum oxide (Al_2O_3)	14
Boron oxide (B_2O_3)	7
Magnesium oxide (MgO)	3
Sodium and potassium oxide (Na_2O and K_2O)	1
Iron (Fe)	1
Titanium oxide (TiO_2)	0.8
Ferric oxide (Fe_2O_3)	0.4
Property	**Value**
Moisture absorption	3%
Volume resistivity	10 MΩ/cm
Specific gravity	2.54
Elongation at separation	5.0%
Modulus of elasticity in tension	10.5×10^6
Coefficient of thermal expansion	5.04 mm/mm K $\times 10^{-6}$
Tensile strength	1723 MPa
Power factor at 1 MHz	0.0002

tion features. These designations are listed in IPC-L-110A, together with their associated information.

It should be noted that glass mats are sometimes impregnated with epoxy to produce base materials for printed circuit laminates. These laminates, referred to as "composites," are normally less expensive than glass cloth laminates, while retaining most of the necessary properties. Combinations of material are also used to produce laminates with unique properties. The excellent adhesive qualities of epoxy make it useful in fabricating laminates in combination with other insulating sheets and with metal cores (for heat transfer and shielding).

Various types of metal foil may be used for the conductor layers on printed circuit panels. However, copper is by far the most useful and most versatile. When used in this fashion, the copper foil is usually first electroplated on a rotating steel drum. To either cold-form or hot-form copper into large sheets suitable for use in printed circuit lamination would be very difficult. Modern proprietary techniques of electroforming copper sheets simplify production and provide better controls over the metal purity as well as its mechanical uniformity and grain structure. After the copper foil has been electroformed to the correct thickness, it is coated with oxide to enhance adhesion of the copper to the epoxy (primarily through the formation of oxide crystals on its surface). Then the foil is removed from the drum, and its shiny, smooth inner surface becomes the laminate surface copper, while its copper oxide surface is bonded to the insulating base material. In actual practice, copper foil is electroformed and oxidized in a continuous process.

The circuit board laminate structures shown in Figures 1–3 are typical of the more common types used in the electronics industry; however, there are many other options in laminate materials and construction available for circuit board use. A designer may be influenced to select something other than glass–epoxy laminate for various reasons. For instance, other materials may offer one or more of the following advantages:

Lower cost
Easier machinability
Higher heat resistance
Specific chemical resistance
Better availability

Some of the reasons for selecting specific materials are listed in Table 3.

Many studies have been made on the properties of the materials listed in Table 3, along with their uses. The data generated by these studies have generally delineated the usefulness of a particular material for a given application. As a result of years of study and data collection, most laminate designers have a clear basis for selecting the proper materials to suit each application. Furthermore, criteria have been developed for qualifying materials electrically, mechanically, and chemically. These criteria are available in the various publications of the IPC (Table 7). In addition to IPC specifications, various trade organizations (e.g., NEMA) and government procurement organizations (Mil Specs) contain extensive functional data regarding circuit board laminate materials.

TABLE 7
Mechanical and Electrical Properties of G-10 Laminates

Properties	Laminates	
	G-10	G-10 FR
Mechanical Properties		
Glass transition temperature, °C	110	125
Dimensional change (temp. shock)	0.0004	0.00035
Flexural strength, MPa		
Length	572	607
Cross	524	552
Flexural modulus, GPa		
Length	18.6	19.3
Cross	16.5	17.2
Shear strength, MPa	145	148
Coefficient of thermal expansion, mm/mmK		
x Axis	3.2×10^{-5}	1.8×10^{-5}
y Axis	2.5×10^{-5}	2.3×10^{-5}
z Axis	11.7×10^{-5}	9.9×10^{-5}
Electrical Properties		
Dielectric constant at 1 MHz	4.8	4.8
Dissipation factor at 1 MHz	0.022	0.02
Volume resistivity, MΩ/cm		
Humidity	1.5×10^{8}	2.0×10^{8}
At 125°C	1.0×10^{6}	5.0×10^{6}
Surface resistivity, MΩ		
Humidity	15.0×10^{6}	5.0×10^{6}
At 125°C	10.0×10^{6}	50.0×10^{6}
Arc resistance, s	120	100
Dielectric breakdown, kV	55	55
Electric strength, V/mil (kV/mm)[b]	27.6	29.5
Capacitance drift, %	4	4

Flexible Circuits

Many printed circuit applications require that the finished product be flexible and capable of being bent or routed through restricted areas. When the printed circuit application has this requirement, there is an implied need for the base material to be flexible and for the copper foil to be ductile. Along with the flexibility, there may be an implied need for the completed circuit to be thin. The normal materials used for rigid printed circuits are not suitable for flexible circuits. No matter how thin the base material, neither phenolics nor glass–epoxy can be suitably formed to meet flexible circuit requirements.

A variety of materials have been specified for flexible circuits. Each of these has performed satisfactorily as an insulating substrate and should be considered when candidates for this application are evaluated. These materials include:

Polyvinyl chloride
Polyvinyl fluoride
Polyethylene terephthalate
Polychlorofluoroethylene
Polyethylene propylene, fluorinated
Polytetrafluoroethylene
Polyamide
Polyamide (aromatic)
Polyamide-imide
Epoxy (limited flexibility)

These materials may be modified in the following ways to enhance their usefulness:

Nonreinforced
Reinforced with woven filaments
Reinforced with nonwoven filaments

Reinforcements are not limited to glass fibers but may be constituted of any of the following material:

Glass
Polyester
Polyamide
Cellulose

In general, flexible printed circuit base materials are thinner than rigid materials. For this reason, they are not normally used for component mounting. The difficulty of soldering discrete components to flexible circuits, together with the problems associated with substrate movement, essentially precludes the use of flexible circuits for component mounting. On the other hand, flexible circuits provide an excellent means for permanently connecting two rigid circuit boards, which cannot occupy the same plane in the completed product. Therefore, it is common to see rigid and flexible printed circuits combined into a single assembly. Flexible circuits offer several advantages over flat cables and hard wiring, primarily in the areas of design flexibility, reliability, and cost reduction.

Flexible circuits usually require special materials to cover their surface conductors and thus provide good electrical insulation. Rigid circuit boards usually call for solder mask, which protects the circuits during assembly as well as throughout the product life. Solder mask is not appropriate for most flexible circuits. Instead, various types of plastic sheets may be bonded by means of adhesive to their surface (or liquid coatings applied). Some conformal coatings and insulating sheets are flexible when cured and are therefore useful as coatings for flexible circuits, provided they can be successfully bonded. Sheet materials either must be machined (cut) to the proper configuration for covering the circuit or they must be capable of being photographically formed to the correct image. Liquid coatings are usually screened onto the surface of the circuit in the correct pattern. Fundamentally, flexible circuits must be capable of being bent or twisted without experiencing significant degradation or malfunctioning. Implicit in this requirement is the ability to terminate the flexible circuit to a rigid panel or connector and to subject it to other manufacturing steps as necessary.

Multilayer Circuits

The concept of multilayer circuits calls for the rigid circuits to be bonded into a laminated assembly using adhesive sheets of material of the same general construction as the cores of the rigid cases. This concept is shown in Figure 4. After lamination, this assembly is processed in the same manner as a normal double-sided circuit board. From a material standpoint, there is little difference. Since the printed circuit industry tends to regard the product in terms of design and fabrication requirements, multilayer circuit boards are treated as a separate topic. To meet design restrictions, multilayer circuits are almost always constructed of glass–epoxy materials. They sometimes incorporate metal or graphite cores to satisfy thermal conductivity or mechanical stability requirements. The metal or graphite may be bonded inside the circuit layers or it may constitute a surface layer, as needed.

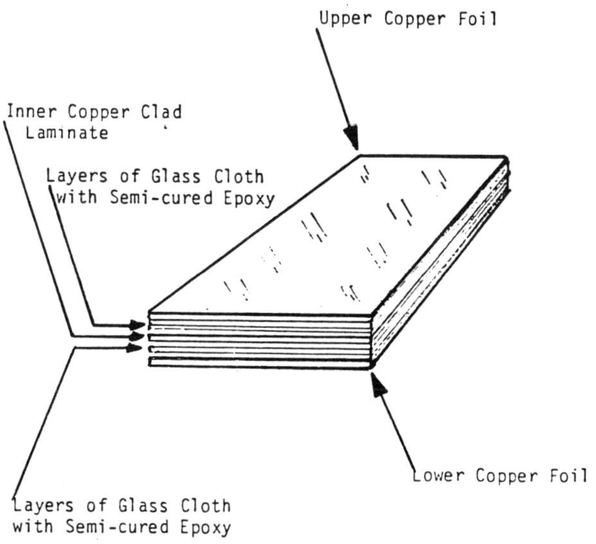

FIGURE 4 **Multilayer circuit laminate material.**

Multilayer circuit boards may be very small (2.54 × 2.54 cm) or very large (0.92 × 0.92 m). They may have as few as 4 conductive layers or as many as 30. The tendency to combine as much circuitry as possible into a single circuit board seems to be gaining favor with designers. Given that most designs are computer-generated and that many manufacturing processes are also computer-controlled, the complexity of circuit board design (as in the case of dense multilayer circuit boards) is not a serious problem. It does force printed circuit manufacturers to maintain tight controls over their processes to enhance yields, however.

The high frequencies and strict control of electrical signal characteristics that are required by computers, telecommunications, and other electronics users have caused more attention to be paid to printed circuit geometry. For low frequency and power circuits, there is little impetus for concern of the static or dynamic impedance of printed wiring. With high speed digital circuits, the geometry of a printed circuit board can mean the difference between success and failure of an electronic design. Many multilayer circuit boards require that the impedance of certain printed circuits be controlled very closely. Controlling impedance simply means controlling the geometry of the circuit with respect to adjacent conductors, together with controls of dielectric material associated with these conductors. This aspect of multilayer manufacture is relatively new with some firms and fairly settled with others. The most common controlled impedance circuit is a stripline (buried conductor), but microstrips (surface conductors) are also used. In either case, the objective is to match the impedance of the conductor with the impedance of an electronic component or the impedance of a transmission cable, to prevent the distortion of transmitted signals by distributed capacitance or inductance in the printed circuit. When the printed circuit manufacturer must control the impedance of circuits, imaging, etching, and dielectric thickness control become very important.

Surface Mount Printed Circuits

Surface mount technology (SMT) printed circuits are sometimes addressed as if they were entirely new. The only unique feature of SMT circuit boards is in their assembly: the components are soldered to their surface pads instead of having leads soldered in plated holes. Since this method of mounting attaches the component body rigidly to the circuit board (instead of indirectly through wire leads), there must be a relatively close match between the component thermal coefficient of expansion T_{CE} and the printed circuit T_{CE}. Beyond this requirement, there is little difference between an SMT circuit board and other dense multilayer boards.

Assembly methods used to pick and place leadless components on solder-paste-filled surface pads dictate that a reasonable amount of control be exercised in designing and fabricating SMT circuit boards. The x, y, and z coordinates of component mounting pads are slightly more critical in SMT applications than for through-hole components. Even though SMT components tend to weigh less than their leaded counterparts, a thicker circuit board may be used to enhance z-axis stability.

Weldable Circuits

The vast majority of printed circuits are copper foil with some type of coating (solder mask or electroplated metal); some require that an alternate material be used. The copper circuit with tin-lead (solder) coating is ideal for soldering and for good electrical conductivity. However, when there is a need to weld component leads to the printed circuit, neither copper nor tin-lead is appropriate. In such cases, either the etched foil must be a weldable material or the copper foil must be electroplated with the weldable metal (e.g., nickel).

Summary

Circuit board laminates are diverse and tend to be highly process-related. They are therefore difficult to define concisely. In comparison with other composites, their use and composition is changing rapidly. Furthermore, since they almost always work behind the scenes, they do not have high visibility in consumer products. Their reliability often belies their importance in the function in an electronic system. This low-key participation in electronic products ceases the moment that circuit boards are taken for granted and mistakes are made. When a high complexity circuit board fails, its materials, its design, and its manufacturing processes become very important.

Fred W. Kear

Coatings

See Protective Coatings; Thermal Spray Coatings

Cold Flow

See Creep

Cold Press Molding

See Molding

Colorants for Composites*

Color, simply stated, is light energy, reflected from or passing through an object, and viewed by a human observer or instrument detector. Colorants are the chemicals that may be added to, or naturally occur in objects that modify the light energy producing the object's color. Colorants contain certain groups of atoms that selectively absorb visible light and reflect the complementary color. These chemical groups are called chromophores. Other chemical groups called auxochromes may be present in a colorant molecule to modify the color and provide undertones.

There are two major classifications of colorants used for the coloring of composites. They are dyes and pigments. Dyes are soluble in the medium used and only produce color by light absorption. Pigments are essentially insoluble in the plastic medium and must be mixed in a fine dispersion to provide uniform coloration.

Pigments can be further divided by chemistry into organic and inorganic types. As with dyes, pigments produce effects by the mechanism of light absorption, but since many pigments (inorganic) have refractive indices greater than the plastics in which they are dispersed, pigments produce color effect by light scattering as well. Dyes, due to their fine particle size, tend to be transparent. Pigments have varied opacifying power.

Although dyes are gaining increased acceptance, especially among the transparent plastic users (e.g., polycarbonates), their general industrial use has been limited. However, it must be stressed that if the plastic application requirements include bright shades, transparency, and good tintorial strength, dyes should be investigated for the application. Because of the limited use of dyes in plastics, the remainder of this discussion is directed to pigment properties and types.

Pigment Properties

Pigments are selected partly for their hue and shade, and depending on the quantity present in the plastic formulation, they can produce colors ranging from pale pastels to deep shades, with a myriad of midtones. However, color cannot be the only determining factor in pigment

* Reprinted with minor changes from 43rd Annual Conference, Composites Institute, The Society of the Plastics Industry, Inc. February 1–5, 1988.

selection. Additional considerations should include, but not be limited to, dispersibility, heat stability, light fastness, chemical resistance, opacity, and toxicity restrictions. The following discussion of pigment properties is designed as a generic overview for thermoset and thermoplastic applications.

Dispersibility

Achieving the optimum pigment particle size and determining the proper dispersion process can be one of the most difficult problems a pigment user must solve. Pigment particle size and shape are critical to the performance and acceptance of a colored resin system. The dispersing process requires the dry powder to be blended into a resin or binder vehicle to "wet out" the particle and produce a homogeneous mix.

The "wetting" process involves replacing dry pigment surface areas with the resin or binding vehicle. The pigment's ease of "wetting" is determined by the general particle surface, size, and shape and the character of the vehicle. Often a surfactant will be used to enhance the wetting process.

Pigment particles can be separated into three distinct groups: primary particles, aggregates, and agglomerates. Generally, the term "particle size" describes a single discrete particle, identifiable by microscopic techniques. These particles are formed naturally or during synthesis and represent the fundamental building block in the particle chain. Aggregates are tightly held primary particles, oftentimes called aggregate crystals. Agglomerates are loosely held aggregates and primary particles.

The colorant user works to reduce the agglomerates and break the aggregates to develop the size dictated by pigment strength and shade, system opacity, weathering, and rheological and economic requirements.

Heat Stability

Some pigments darken or decompose at excessive temperature. Pigment heat stability is determined by subjecting the pigment and resin matrix to specified temperatures for prescribed time periods, then comparing the results to standards. The temperature profiles may range from 120 to 425°C, depending on the polymer and molding requirements. In general, inorganic pigments demonstrate greater heat stability than organic pigments, which degradate or sublime when exposed to high temperature for extended periods of time.

Light Fastness

Light, especially daylight, has a destructive effect on pigments used to color composites. Depending on the colorant chemistry, pigments can lose color intensity, darken, or in some instances decompose. As a result, light fastness requirements vary according to end-use application. For example, a child's toy designed for interior use may need minimal light fastness qualities, while an automotive coating may have stringent light fastness requirements.

The light fastness of the pigment itself cannot be evaluated without due consideration to the resin system. No doubt the physical and chemical properties and concentration of the colorant play a major role in the weatherability of a colored plastic. However, the resin matrix chemistry is also susceptible to light and may degrade, producing color shift or loss.

Exterior exposure tests are conducted in Florida, Arizona, and southern California to determine the effects of daylight on colorants and their resin system. A standard outdoor durability test may include a protected control and four or five like panels used for timed sequence evaluations (3, 6, 9, 12, and 24 months). The light fast and weathering results are compared to the control for color permanency.

An alternate method of determining relative light fastness is the use of xenon or carbon arc fadeometers or weatherometers to accelerate the color loss. The degree of loss is usually a subjective comparison to previously acceptable standards and conditions.

Chemical Resistance

Chemical resistance to color loss upon exposure to solvents, acids, alkalies, etc., is a chemical property of the pigment. However, as in the case of light fastness, the chemical resistivity is subject to the durability of the polymer system.

As a rule, neutral organic pigments are resistant to most chemicals. Inorganics and organic metal salts demonstrate varying degrees of chemical sensitivity. To determine the effect of the chemical on the colored plastic, tests are conducted by exposing the plastic to the chemical in question under specified conditions.

Opacity

"Opacity" and "hiding power" are terms usually used to describe the ability of a pigment to scatter light and hide the resin matrix. Pigments can be described as opaque, semiopaque, or transparent. The property of a pigment that determines its opacity is the refractive index. The greater the difference in refractive indices between the pigment and the matrix vehicle, the greater the hiding power. The scatter of light and resulting opacity can be controlled, to some degree, during pigment synthesis by particle size. As a rule, the larger the pigment—more scattering, greater hide; the smaller the pigment—greater transparency.

Toxicity Restrictions

Health and environmental issues are of major concern to the plastic compounder. With tightening governmental requirements, the compounder is turning to the dispersion house and/or raw material supplier for help. With the institution of "right to know" regulatory compliance, November 1985 and May 1986, the colorant user is provided with detailed information (by way of container labels and Material Safety Data Sheets) concerning the physical and health hazard of the pigment. Some pigments (such as lead chromates or cadmiums) do have human health restrictions, while other pigments (like titanium dioxide and iron oxides) provide only nuisance dust hazards.

For plastic applications requiring regulatory status, the formulator and colorant supplier must work together to select pigments that provide good color and system compatibility without jeopardizing the compliance status of the molded product.

The molder is responsible for complying with regulations promulgated in the field. For food contact applications, the molder may be required to have extraction tests run at an independent laboratory. The extraction test will measure the nonmigratory integrity of the molded compound including the quantity of colorants leached from the molding. The colorant supplier can assist the molder in this regard, but only the molder has overall control of the finished part; therefore, the molder is responsible for the regulatory compliance of the finished part.

Pigment Classifications

Pigments are the primary components used to color composite parts. Today's colorant user can, for the most part, choose the form (dry powder, pellet concentrate, liquid color, etc.) and concentration in which the colorant material is supplied. For the purpose of discussion, pigments hereafter are described, generally, as dry powders used to impart color.

As mentioned earlier, pigments are divided into two chemical categories, inorganic pigments and organic pigments. Inorganic pigments are metallic compounds and occur naturally or as the result of chemical synthesis. Organic pigments, by definition, are compounds containing carbon. Some organic molecules may contain metals or metal salts; however, the inorganic composition is considered to be secondary (responsible more for undertones than primary hue color). The following pigment review has been divided according to chemical constitution or chemical family.

Inorganic Pigment Families

Inorganic pigments are larger molecules and have greater hiding power than organic pigments; however, compared on a weight basis, they have less tinting strength than organic pigments of similar color. The most commonly used inorganic pigments are as follows.

IRON OXIDE PIGMENTS. Iron oxides can occur naturally or are prepared through chemical synthesis. Iron oxides come in four dull shades: yellow, black, red, and tan (a mixture of the first three). Most yellow and black shades are stable to 175°C. Temperatures exceeding 175°C will cause the pigment to lose chemically bound water and change to the red shade. The red shade is stable to temperatures in excess of 982°C. Iron oxides are reasonably inexpensive and easily dispersed. They have a relatively high degree of opacity, although special transparent grades exist.

CHROMIUM OXIDE PIGMENTS. Chromium oxide is a dull olive green pigment, synthesized by heating and reducing hexavalent chromium compounds. Chromium oxide green is a fairly inert pigment with good heat, light, and chemical stability, but unlike other inorganic oxides, it has poor hiding and tint strength. Also, chromium oxides can be difficult to deagglomerate. The hydrate form, containing 20–25% water, is a brighter shade of green than the normal chromium oxide, but it can produce rheology problems with some plastic systems.

LEAD CHROMATE PIGMENTS. Lead chromate pigments come in a variety of yellow and orange shades. The color results from the presence of lead chromate molecules (orange), while the shade is controlled by the coprecipitation of lead sulfate (yellower) or lead molybdate (redder). As a family, lead chromates are inexpensive and provide good tint strength and opacity. Lead chromates can darken upon exposure to light and heat (> 200°C) and in the presence of sulfides. They also demonstrate sensitivity to acids and bases. Many lead chromates are treated by the manufacturer to improve chemical resistance, heat stability and light fastness. A disadvantage is the human toxicity and environmental leaching of lead and chromium. Lead chromates are under surveillance as a potential human carcinogen.

IRON BLUE PIGMENTS. Iron blue is ferric ammonium ferrocyanide and is known by numerous names, such as Milori blue, Prussian blue, and bronze blue. These pigments are inexpensive, providing dull shades with good tint strength, moderate light fastness, and fair heat stability. Iron blues are susceptible to alkalies and can react with thermoset resins to initiate free-radical polymerization and premature gelation.

CHROME GREEN PIGMENTS. Chrome green pigments are coprecipitates of lead chromate yellow and iron blue or physical mixtures of the two. The resulting color varies from shades of greenish yellow, to medium shades of olive green, depending on the quantity of blue and shade of yellow chromate.

Chrome greens are fairly inexpensive and have good hiding power. They can be reactive in thermosets, and they initiate premature gelation. Their level of toxicity varies according to the quantity of lead chromate used to provide shades.

ULTRAMARINE BLUE PIGMENTS. Ultramarine blue is a complex compound comprised of a combination of sodium, aluminum, silicon, and sulfur compounds, calcined at temperatures in excess of 800°C. Ultramarine blues are bright and intense; however, their tinting strength is 0.1–0.05 that of an equal quantity of organic blues. Ultramarine blues are reasonably inexpensive. They have good light stability and chemical resistance, except to strong acids.

CADMIUM PIGMENTS. Cadmiums are moderately priced pigments, commercially available in a variety of yellow, orange, and red shades. Cadmium pigments are actually classified according to their sulfide, sulfoselenide, or mercury complexes. Pure cadmium sulfide is an orange shade of yellow, but when zinc sulfide is added, the shade shifts to greenish yellow. On the other hand, when selenium is added to the molecule complex (cadmium sulfoselenides), the shade shifts to red and maroon. Cadmium mercury sulfides (called mercadmiums) are used to vary the shade from reddish orange to bluish red. Cadmiums are available as a pure pigment or in mixtures of barium sulfate (~ 60%) called lithopones.

As a family, cadmiums are sensitive to acids. They exhibit brilliant shades, have adequate heat stability, and good light fastness in mass tone and acceptable fastness in tints with titanium dioxide. Light fastness of the yellow shade tints is suspect, and they should not be used for exterior exposure. Cadmium, selenium, and mercury are toxic, and the user should exercise restraint when formulating with cadmium pigments.

TITANIUM DIOXIDE (AND OTHER WHITES). Titanium dioxide (TiO_2) is the white pigment of choice in most composite applications. Titanium dioxide is inexpensive and provides excellent brightness, hiding power, and heat stability. The high refractive index is responsible for the scattering of light and good opacity.

Titanium dioxide is available in two crystal forms: anatase (bluer shade) and rutile (yellower). Rutile is the preferred form for most composites, although anatase is used for applications requiring bluer undertones. Additionally, anatase is more reactive than rutile and is not recommended for exterior exposures.

Rutile and anatase titanium dioxide are not the only white pigments used in plastics. Titanium dioxide tends to be abrasive and in applications where abrasiveness is of concern, titanium dioxide must be substituted with other white pigments such as zinc sulfide. Zinc sulfide is not as great an opacifier as titanium dioxide. Other white pigments include zinc oxide, magnesium oxide, and barium sulfate.

MIXED METAL OXIDE PIGMENTS. Mixed metal oxides are pigments that have metal ions substituted in the crystal structure of titanium dioxide formed at extreme high temperatures. These metals include nickel, aluminum, antimony, cobalt, chromium, iron, and copper.

These pigments have excellent weathering properties and good chemical resistance but they have low tinting strength and poor hiding power when compared to titanium dioxide.

EXTENDER PIGMENTS. Extender pigments are used as partial replacement for more expensive colorants. Extender pigments are white powders with refractive indices similar to the composite resin materials. As a result, extender pigments become more or less transparent when surrounded ("wet out") by the resinous material.

The extender pigment mechanism is simple. The primary pigment contributes color and opacity, and the extender pigment provides incidental pigmentation and formulation properties required for the resin matrix and desired molded part. The most common extender pigment is calcium carbonate. Other examples include hydrated magnesium silicate (talc), hydrated aluminum, silica, mica and others.

Organic Pigment Families

As described earlier, organic pigments are composed of carbon and contain complex structures, including aromatic rings. Organic pigments offer brilliant shades, excellent tint strength, but varied resistance to heat and light. As a rule, organic pigments are of smaller particle size and have poorer opacifying properties than their inorganic pigment counterpart. The major types of organic pigments are as follows.

AZO PIGMENTS. An azo molecule contains one or more azo linkages —N=N—, which are strong absorbers of light energy. Azo pigment colors range from greenish yellow to bluish red. Azo pigments can be further divided into three broad categories: monoazo, polyazo, and metallized azo pigments. Monoazo pigments contain one azo group and, as a family, have poor light fastness and heat stability. They also tend to bleed. Bleeding can be described as the diffusion or migration of color from one surface to a surrounding surface as a result of the colorant's solubility. Pigment manufacturers offer special grades which contain chemical modifications to the basic molecule, or add special "aftertreatments" to improve the pigment's capabilities. Examples of monoazo pigments are Hansa Yellow, Para Red, and Toluidine Red.

Polyazo pigments contain more than one azo group and, as a result, have better heat and lightness capabilities, improved bleed resistance, and brighter shades than the monoazo pigments. Examples of polyazo pigments include diarylide yellows and oranges and parazolone oranges and reds.

Metallized azo pigments, primarily red shades, contain at least one azo group and an anionic group that, when ionized, will react with a positively charged metal to produce a salt. Metals commonly precipitated into azo pigments include calcium, barium, and manganese. The metal improves the pigment's properties, particularly bleeding, migration, and light fastness. However, the metal ions make the pigments less stable to acid and bases. When compared with nonazo organic pigments, metallized azo pigments have only fair heat stability and marginal light fastness in mass tone and poor light fastness in tint. Examples of metallized azo reds are Lithol Rubine, 2B Red, Bon Red, and Red Lake C.

CARBON BLACKS. A multitude of carbon blacks, such as channel blacks, furnace blacks, bone black, and lamp black, are available to colorant users for plastic applications. As the name implies, carbon blacks are black, but the degree of jetness, shade of undertone, and tint strength are functions of particle size and surface area. Carbon blacks are inexpensive and are marketed in two forms: dry powder and compressed pellet; the latter is less dusty and easier to handle. Carbon blacks have good heat and light stability.

PHTHALOCYANINE BLUES. The primary organic blue used in the plastic industry is copper phthalocyanine blue. Phthalocyanine pigments are of two forms, the red shade (α form) and the greener shade (β form). Phthalocyanine blue is a very strong and intense pigment, and incomplete dispersion results in specks and streaks.

In thermosets, these pigments tend to flocculate. To counteract this phenomenon, flocculation resistant grades have been introduced with varying degrees of success.

Phthalocyanines are relatively low in cost and have excellent properties, including: light fastness, heat stability, chemical resistance, and exceptional hiding power. Additionally, they provide bright shades and at low loading levels, satisfactory transparency. Phthalocyanine blues are difficult to disperse in some plastic systems.

PHTHALOCYANINE GREEN. Chemically, phthalocyanine green is very similar to its blue counterpart. The primary difference, however, is not the crystal structure but the degree of chlorination (blue shade) and/or mixed halogenation from partial replacement of chlorine by bromine to complete bromination (yellow shade) about the periphery of the molecule. As with phthalocyanine blue, the green pigment is reasonably priced, has good tint strength, and is very stable to heat and light.

QUINACRIDONE PIGMENTS. Quinacridone pigments come in a variety of reddish shades, ranging from orange-red to violet, depending on the crystalline structure. Quinacridones have heat stability to 400°C. They weather well and have good tint strength and very good resistance to chemicals; however, they are expensive.

DIOXAZINE PIGMENTS. Dioxazine pigments are violet and have excellent tint strength and light fastness qualities, even in tints. Because of their price, however, dioxazine pigments are used sparingly. Their heat stability is limited, and the molecule will decompose at elevated temperatures for prolonged periods. The most common dioxazine pigment is carbazole violet.

ISOINDOLINONE PIGMENTS. Isoindolinone pigments are premium pigments available in yellow, orange, and red shades. These pigments are light stable and have good chemical resistance. Their heat stability is good to 300°C. They are very expensive and, as a result, have been limited in their applications.

VAT PIGMENTS. Vat pigments are complex molecules, and many of them include anthraquinone derivatives. They provide a variety of intense hues at a wide range of cost. Generally speaking, vat pigments have excellent tint strength and transparency, but the heat and light stability should be tested in the laboratory before use.

SPECIALTY APPLICATION PIGMENTS. Specialty pigments, such as fluorescents, phosphorescents, and metallic pigments, are selected not for their color, but for their effect on the appearance of the finished part.

Fluorescent pigments are organic dyes affixed to an inorganic structure which is insoluble in the resin. Fluorescent pigments will absorb visible light and nonvisible energy at shorter wavelengths and reemit the light energy at longer wavelengths. Fluorescence is dependent on the light source used for viewing. If insufficient energy is present, the pigment will not fluoresce; on the other hand, if the light energy is too intense, the effect is exaggerated. When the light source is turned off, the fluorescence ceases.

Phosphorescent pigments "glow in the dark," and are inorganic materials, which absorb light energy through the activation of a light source, then continue to emit the energy after the light source is turned off. Phosphorescence is dependent on the ability of the pigment to absorb the energy and emit it as light.

Metallic pigments are available in various shapes, sizes, and forms for a multitude of applications. Aluminum powders or flakes are available for silver-type effects, while copper and copper alloys are available for gold/bronze effects. These pigments provide sparkle to the plastic part and are manufactured with a partial colored coating, or they can be incorporated into a colored resin system to provide color and glitter. As with other specialty pigments, metallics can be difficult to process.

Summary and Conclusion

The color on the finished plastic part is primarily the function of the pigment. However, the colorant user must consider the color effects contributed by the resin system, the matrix formula, and process temperatures.

Also, the user must consider chemical resistance as well as end-use requirements for heat stability, light fastness, and toxicity before selecting the colorant.

The decision can be confusing and many times overwhelming. However, the colorant supplier can help. When the user plans a new design or institutes a color change in the conventional product line, the colorant supplier should be consulted so that the selection process will flow smoothly and the end product results will be more desirable.

Selby M. Brannon

Bibliography

"Colorants," in *Plastics Compounding for Resin Producers, Formulators, and Compounders,* 1987/88 Redbook, *10*(4), 52 (1987).

Dry Color Manufacturers' Association, *Am. Ink Maker,* *62*(6), 15 (1982).

Kurtz, W. *Am. Ink Maker,* *65*(6), 21 (1987).

Lele, S. S. *Plast. Compd.,* *9*(3), 14 (1986).

Lewis, P. A. in *Am. Paint Coatings J.,* March–December, 1984.

Lewis, P. A., ed., *Pigment Handbook,* Vol. 1, 2nd ed., John Wiley & Sons, New York, 1988.

Reeve, T. B. *Plast. Compd.,* *9*(3), 48 (1986).

Simms, J. A. *J. Coatings Technol.,* *59*(748), 45 (1987).

Smith, H. M. *Am. Ink Maker,* *64*(6), 19 (1986).

Sonn, G. *Am. Ink Maker,* *65*(6), 54 (1987).

Speelman, D., and G. Wildeman, *Plast. Compd.*, 7(13), 30 (1984).
Stemple, T. W. *Plast. Compd.*, 9(3), 32 (1986).
Surface Coatings, Raw Materials and Their Usage, Vol. 1. 2nd ed., Chapman & Hall, New York, 1983.
Webber, T. G., ed. *Coloring of Plastics*, Wiley-Interscience, New York, 1979.
Webber, T. G. *Plast. Compd.*, 3(4), 37 (1980).
Wigotsky, V., et al. *Plast. Eng.*, 42(10), 21 (1986).

Compaction

See Shock Compaction Preparation

Composite Materials, Ablative*

The word "ablation" is used to describe the erosion and disintegration of meteors resulting from the intense heat generated by passage through the atmosphere at high velocities. During development of the first operational ballistic missiles at Peenemunde, Germany, aerodynamic heating of the missile skin caused many to disintegrate during test; even with redesign, about 650 of the 3550 operational V-2s launched failed to reach their targets because of such air disintegration [1]. The development and study of materials which protect the payload and structure of missiles, earth satellites, or space probes from damage during ablation is a relatively new area of chemical technology. The same principles of thermal protection also apply to other applications, such as the protection of rocket nozzles from attrition by propellant gases and protection from laser beams.

The Ablation Environment

The selection of materials to resist ablation is highly dependent on the intensity and duration of heating, which can be much more severe than the hot environments encountered in other technologies. Figure 1 illustrates schematically some typical reentry velocities and heating rates at different altitudes, representative of different types of reentry bodies [2].

Near the leading point, or stagnation point, of a hypervelocity vehicle, gas molecules are abruptly accelerated to the velocity of the vehicle. This translational energy appears as a temperature rise, although dissociation and reactions of the gas molecules also occur; the specific energy content of the gas is approximately proportional to the square of the velocity. The gross heating rate depends not only on this energy but on the local pressure, which varies with velocity, altitude (gas density), and position on the vehicle, decreasing from stagnation point to trailing edge. The heat transfer rate from gas boundary layer to surface also depends on whether the stream is in laminar flow or is turbulent; in the latter case, heating is more severe. The duration of intense heat is longer if the flight path has a low angle with the horizontal, although the rate of heating is less because the body decelerates at high altitudes where pressure is low. Heating rate also depends on the vehicle design, since a light blunt body decelerates at a higher altitude than a heavy pointed body having a higher ratio of weight to drag.

In Figure 1, the numbered curves are lines of equal heating rate calculated for laminar heating of a blunt body, as described in the legend, at the different combinations of altitude and velocity. The lettered curves show velocity–altitude trajectories for typical reentry conditions. Note that the most severe condition B develops temperatures in excess of 4000°K for radiation equilibrium of the surface, which limits the temperature of the material.

Simulation of the more severe environments for laboratory tests has been the subject of considerable development. Various chemical flames, plasma arcs, solar furnaces, lasers, and electric arc air heaters have been used as heat sources, often in combination with wind tunnels. Feeding devices have often been incorporated to maintain samples in a controlled or constant heat flux while they recede. In some tests, samples have been fired down a ballistics range and profiles photographed at successive stations while the surface recedes. In almost all cases, some aspects of the real environment are lacking, however, so that full-scale tests are eventually necessary to confirm the selection. A reliable prediction of performance under conditions that cannot be adequately simulated usually requires study of the mechanism involved in ablation.

Because of the diversity in environments, many types of materials have been utilized where ablation is involved. During reentry, a protective heat shield must perform as an aerodynamic body, a predictably decomposing ablative material, a structural component, and an insulator [3,4]. In missile nose tips, where shape change may affect the trajectory, a material such as graphite may be selected primarily to minimize recession under very high heat fluxes [5,6]. If the environment involves erosion by snow or ice particles, a refractory metal may be considered [7]. On the other hand, ability to insulate or protect the payload from the heat generated, with minimum weight, is ordinarily the primary consideration, and a variety of ablative plastics and fiber-reinforced plastic composites have been developed [8–10]. This article reviews the principles involved in material selection for the most common areas of application.

Metals and Ceramics as Ablative Materials

If one surface of a large slab of solid metal is exposed to a very hot stream of inert gas, its surface temperature rises at a rate proportional to the difference between the rate at which heat reaches the surface and the rate of heat

* Reprinted from Kirk-Othmer, *Encyclopedia of Chemical Technology*, Third Edition, © 1978, reprinted by permission of John Wiley & Sons, Inc.

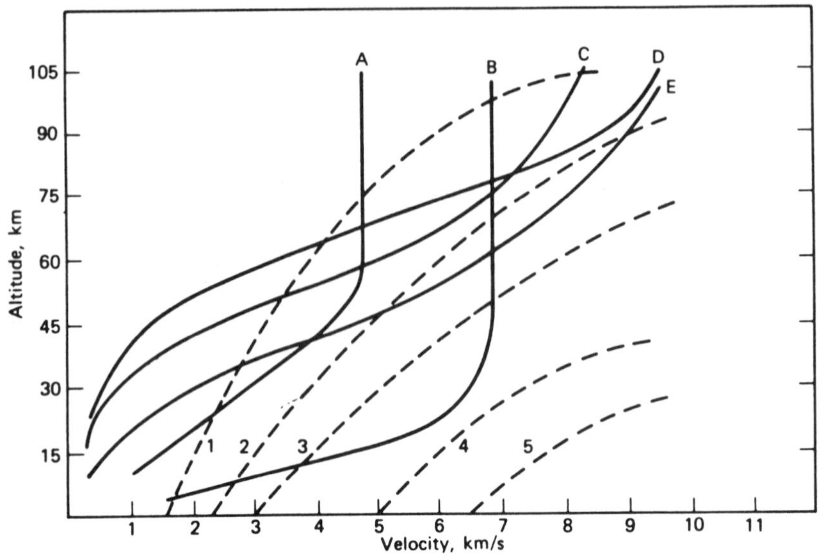

FIGURE 1 Hypervelocity flight environments [2]. Curves: A, intermediate range vehicle; B, intercontinental range vehicle; C, manned orbital vehicle; D, satellite; E, space probe. The dashed lines represent calculated heating rates for laminar flow about a body with a nose radius of 15.24 cm, and corresponding radiation equilibrium temperatures (calculated for an emissivity of 0.9) for each numbered curve are:

Curve	Heating Rate (W/cm^2)	Radiation Equilibrium Temperature (K)
1	57	1830
2	230	2590
3	570	3260
4	2300	4590
5	5700	5810

conduction into the material. Eventually, surface melting will occur, and hot liquid may drip or be blown off the surface. After this, the face of the solid will recede at a steady rate, which depends on the latent heat of melting the metal, its heat capacity, and its thermal conductivity.

The initial, transient phase of this process is the basis for protection against heat pulses of relatively short duration by using a heat sink. Beryllium, copper, tungsten, molybdenum, and even steel have functioned in this way. Refractory carbides, borides, nitrides, and oxides, beryllium oxide in particular, have also been considered because of high melting point, high thermal conductivity, and high specific heat. In some designs, fluid is forced through the pores of such a material to provide transpiration cooling.

Tungsten and molybdenum have been evaluated as nose tip materials, and recession rates of samples ablated in both flight and laboratory tests have been correlated by analysis of the mechanisms involved [7]. The formation of volatile oxide species is significant with these materials, and it could be assumed that the melt layers on the surface, either of oxide at low temperatures or of the metal in the steady state region, were infinitesimally thin.

The Ablation of Graphite and Other Carbons

The element carbon, due to its high heat capacity per unit weight, high energy of vaporization, and high temperature and pressure required for melting, has the highest "heat of ablation" (energy absorbed per mass lost) of any material, provided mechanical removal of particulates does not occur. Factors involved in predicting recession rates, which usually must be minimized, are reviewed.

The melting temperature and pressure of the solid–liquid–vapor triple point is important because reradiation of energy absorbed from the gas stream is limited by the surface temperature. Several studies have shown that wetting droplets of liquid form only if the pressure of the surrounding gas exceeds 10 MPa; solidified liquid is identified by graphite crystals having basal planes that grow in the direction of solidification [11]. The most recent determination, by laser heating of pyrolytic graphite, indicated a triple point at 4130 ± 30°K and 12 ± 1 MPa, with little change in temperature up to 21.5 MPa [12].

However, other studies of laser-heated spinning pyrolytic graphite rods [13–15] have shown evidence of liquid formation at a lower temperature, 3780 ± 30K, at gas pressures as low as 20 kPa; the phenomena were associated with the formation of a variety of nongraphitic carbon structures, primarily linear polymorphs, above about 2600K with vapor pressures different from that of graphite. Similar studies on industrial graphites [15] indicated that liquid formation at grain boundaries during ablation may contribute to the ejection of particulate material observed in many types of graphite heated to vaporization temperatures [15–17] and postulated as an important mechanism in carbon arcs [18]. Spontaneous emission of particulates may also result from the expansion of gas trapped in closed pores during manufacture of graphite [19,20].

Such mechanisms of ablation in materials composed of pure carbon have introduced uncertainties into calculations based on much of the vaporization data in the literature [21]. Although modified data for the free energy functions and vapor pressure of the molecular products of vaporization (primarily C_3) have appeared recently [22–25], the JANAF tables [26] have been used in most studies and have permitted good agreement with vaporization rates in arcs and laser beams [16,27]. Calculated rates have also generally matched tests intended to simulate reentry conditions, where chemical reaction with dissociated species in the boundary layer of gas is important [28–30]. The calculated heating of carbon–carbon heat shields during flight has provided good agreement with temperatures measured by thermocouples on the backface [30]. However, these comparisons have not covered some conditions of heat flux and pressure that may be encountered under more severe reentry trajectories.

Surface roughening of graphite affects the surface heating, if flow changes from laminar to turbulent conditions, and thus can increase the recession rate. Preferential vaporization in subsurface porosity is a mechanism for such roughening [31], and the proportion of large pores in the material affects ablation rate [32]. In fiber-reinforced composites, spalling of bundles from the surface or large porosity can affect ablation [33], but performance is otherwise similar to graphite [32]. Bonding of the constituents can be important, and a carbon–carbon composite processed at low temperatures has shown resistance to laser beam penetration superior to industrial graphites [34]. Pyrolytic graphite, which is free of open porosity, is superior to the other types in resistance to laser penetration [34].

Ablation in environments like the combustion gases in rocket nozzle throats can be predicted, as in reentry, by analysis of the concentration of reacting species available by diffusion through a boundary layer of reaction products [35]. The relative temperatures of surface and gas are important, because the proportion of reactive, dissociated species among the molecules that reach the surface is affected [36]. For example, reaction rates of pyrolytic graphite plates heated in a cold gas stream *decreased* with increasing temperature in the range 1800–2500°K for CO_2 at 110 kPa and in the range 2400–2700K for steam at 130 kPa [37]. Rotating pyrolytic graphite discs, inductively heated in oxygen–argon mixtures, showed more rapid reaction on cut surfaces (edges of graphite basal planes exposed) than on as-deposited surfaces, which had fewer reacting sites [38]. In contrast, tests by one of the authors (Stover) demonstrate that pyrolytic graphite heated in hydrogen–oxygen flames, where reactive species are present in the gas initially, instead of being created on the surface by dissociation of molecules, shows no decrease with increasing temperature, and no effect of AB (edge) versus C (basal plane) orientations (Fig. 2). The tests, in which penetration rates of slabs heated in opposed flames were measured, showed a higher rate for a porous industrial graphite (ATJ) susceptible to loss of particles, and a lower rate for tungsten, where more oxygen is required to remove an atom [W reacting to WO_3]. Rate also varied with flame composition at the same surface temperature. Thus, proper simulation of the environment is essential in tests of graphite ablation below temperatures where vaporization is predominant.

Ablative Plastics and Organic Insulations

Gas evolution from an ablative material is important in reentry because it thickens the boundary layer and interferes with the convective transfer of heat to the surface. Such blocking action can reduce the net heating of the solid by more than 50% in many practical situations.

This principle is utilized with thermoplastic ablative materials such as polytetrafluoroethylene or Teflon, which decomposes above about 500°C to form the volatile monomer without leaving a solid residue. This is useful where uniform, clean removal of the ablative material is desired. A low weight heat shield for a Venus entry probe has been designed from Teflon with a reflective coating on the back face for further insulation [39].

In most cases, however, formation of a porous solid at the surface is desired. Elastomeric shield materials (ESM) were developed for applications where heat flux and aerodynamic shear forces are relatively low and thermal insulation is a primary requirement. For example, General Electric Company's RTV 560 is a foamed silicone elastomer loaded with silicon dioxide and iron oxide (Fe_2O_3) particles, which decomposes to a similar foam of SiO_2, SiC, and $FeSiO_3$ [40]. Elastomers may contain hollow microspheres of silicon dioxide or phenolic resin to control void size and are often contained in fiber-reinforced phenolic honeycombs [41]. As shown in Figure 3, silicone resins are relatively resistant to decomposition, and the silicon dioxide forms a viscous liquid when molten.

A low cost, thermally insulating ablative material, which can be used in mild environments, is cork; the porous carbon char permits high surface temperatures, and the gaseous products provide transpiration cooling of the outer layers.

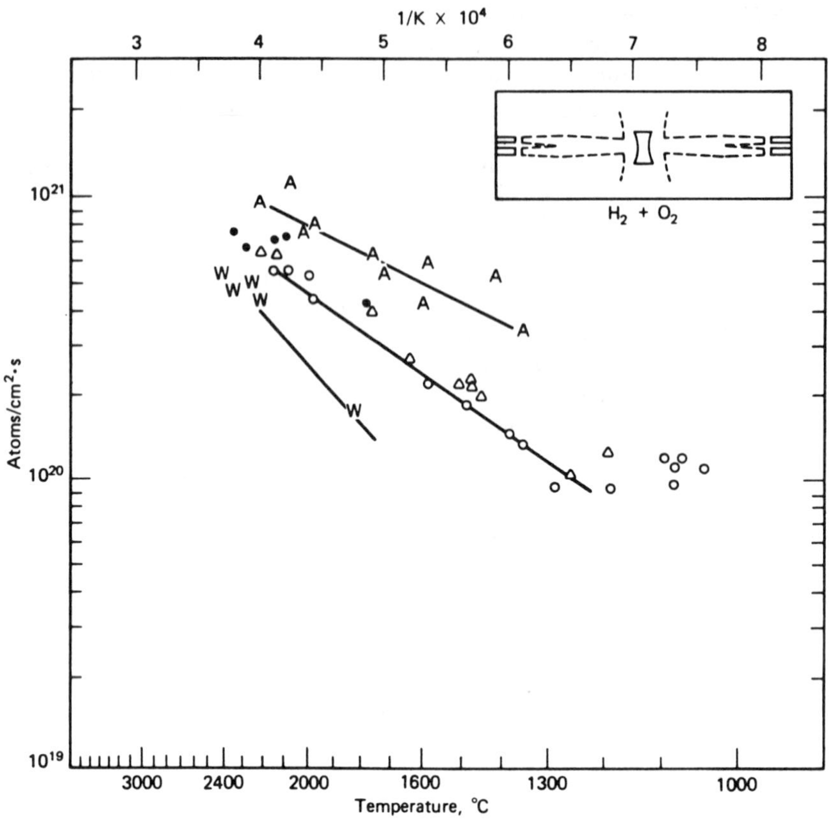

FIGURE 2 Rate of penetration of plates by flames from opposing hydrogen–oxygen welding torches, multiplied by bulk density and atomic weight, and plotted versus surface temperature at the point of impingement, corrected for emissivity; rates for pyrolytic graphite (PG) slabs cut from thick plate (A direction), ●, are compared with plates with the deposition surface exposed (C direction) without boron, ○, and with the addition of 0.6 wt % boron, △, which improves resistance to oxidation in air; A, ATJ graphite (Union Carbide Corp.); and W, tungsten data under the same conditions.

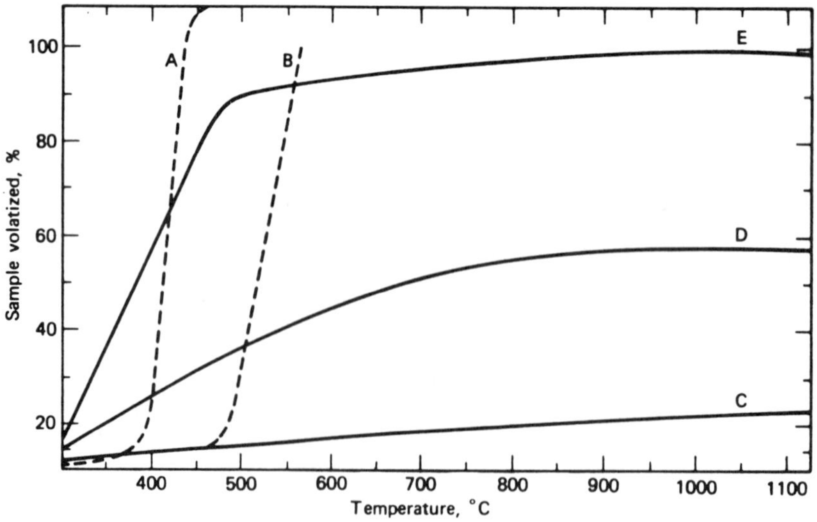

FIGURE 3 Decomposition of polymers as a function of temperature during heating [42]. Curves: A, polymethylene; B, polytetrafluoroethylene; C, silicone; D, phenolic; E, epoxy.

Ablative Plastic Composites

Among the common plastics, the phenolic resins give the highest yield of carbon during pyrolysis, and they have been widely used as surface charring ablative materials. Since a char is relatively weak, and is removed mechanically by high shear forces associated with reentry, fibers of carbon, silicon dioxide, refractory oxides, mineral asbestos, or even glass have been added to assist the char retention.

When fiber-reinforced plastic ablators are exposed to ablative environments, they first act as heat sinks. As heating progresses, the outer layer of polymer may become viscous and then begins to degrade, producing a foaming carbonaceous mass and ultimately a porous carbon char. The char is a thermal insulation; the interior is cooled by volatile material percolating through it from the decomposing polymer. The process is schematically illustrated in Figure 4. During percolation, the volatile materials are heated to very high temperatures with decomposition to low molecular weight species, which are injected into the boundary layer of gases. This mass injection creates a blocking action, which reduces the heat transfer to the material. Thus, a char-forming resin acts as a self-regulating ablation radiator, providing thermal protection through transpiration cooling and insulation. The high efficiency, in terms of heat absorbed per weight of material lost of many such materials, is about 40 times that of a copper heat sink.

As the layer of char thickens, the heat reaching the moving interface, in Figure 4, is reduced by the insulating effect of the porous char combined with the cooling effect of the diffusing gases. The rate of decomposition and gas formation is also reduced. Ablation of the char occurs by sublimation of carbon and oxidation in the boundary layer with concomitant vaporization. The rate of ablation slows as the rate of char surface recession equals or exceeds the recession rate of the interface between the char and the virgin polymer.

An example of the response of a carbon fiber reinforced phenolic composite to heat fluxes typical of reentry environments is shown by the test data in Figure 5 [44]. A slab of the composite was exposed to arc-heated air in a graphite channel in such a way that the surface was first exposed to a low heat flux, 495 kW/m^2, for 230 seconds and then to 4950 kW/m^2 for about 100 seconds. The temperatures recorded by thermocouples at various depths from the surface are shown; the deepest thermocouple, 10.67 mm, indicated that decomposition was just beginning at the end of test. Separately measured surface temperatures were approximately 1200K before the increase in heat flux and 2000K afterward.

The resinous component of a typical reinforced plastic ablator comprises about 35% of the total weight. Since this part provides the essential gas for transpiration cooling of the char, the decomposition products are important. In phenolic resins, water is the principal product up to about 350°C; depolymerization then occurs, with release of phenol and larger species; however, above 530°C, lower molecular weight products are evolved, with H_2 as the primary product (in the absence of oxygen) at 700°C and above [45]. Several phenolic com-

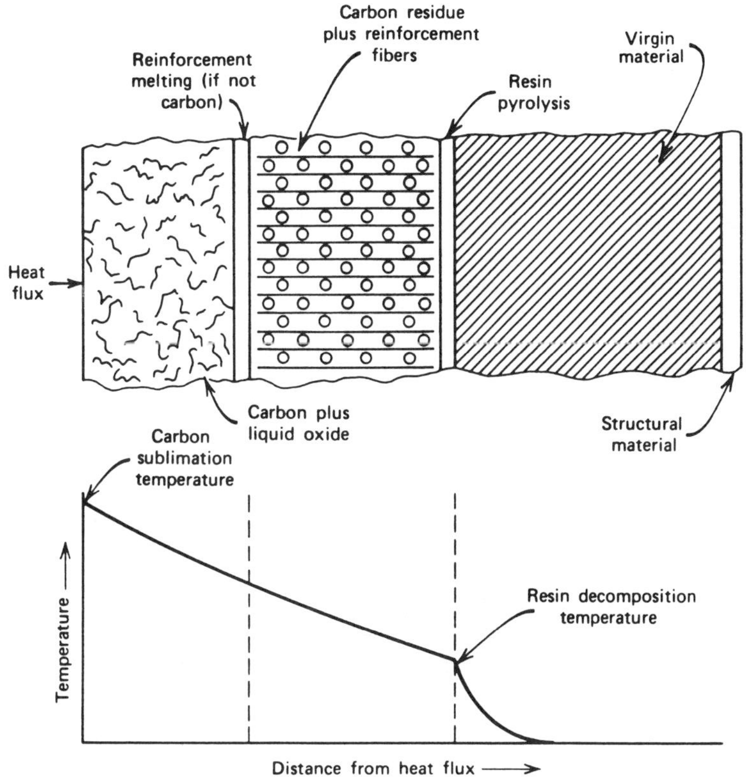

FIGURE 4 Schematic diagram of the ablation process. (From Ref. 43.)

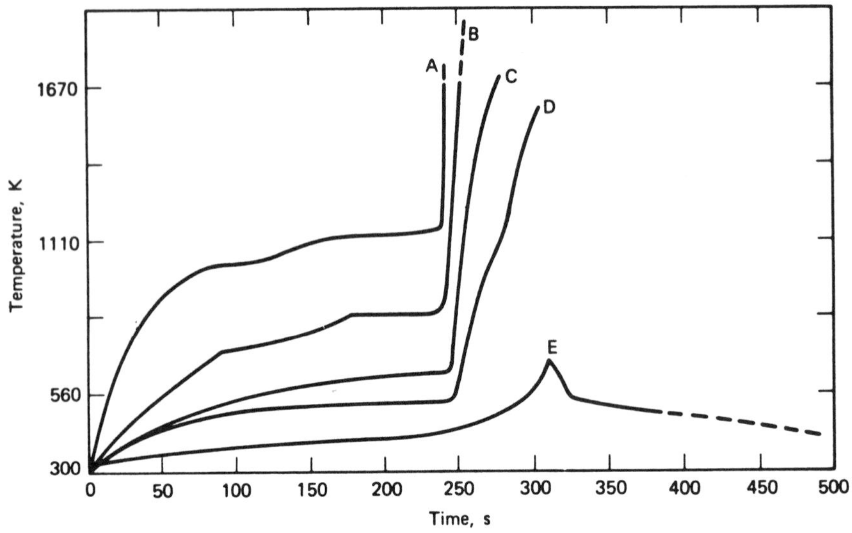

FIGURE 5 **Subsurface temperature response of carbon–phenolic composite during ablation test. Peaks representing thermocouple depth (mm): A, 2.26; B, 4.07; C, 6.20; D, 7.82; E, 10.67.**

pounds have the combination of high carbon content to provide high char yield, low oxygen to minimize CO and CO_2 formation, and many carbon–hydrogen bonds, which provide the evolution of H_2 and CH_4 for transpiration cooling [46]. The plastic must also have good forming characteristics, good adherence to the fibers, and strength in the composite, and it must form a char that adheres well to the fiber reinforcements. During ablation, some pyrolytic deposition of carbon from the CH_4 or other gases in the outer portions of the char can contribute to the strength and resistance to mechanical removal from the flow field. Addition of particulate carbon or other additives to the resin can also contribute to strength of the char. Very good char retention has been obtained with carbon phenolic heat shield materials [47].

Other resins besides the phenolics have even higher char yields and higher decomposition temperatures, hence have been evaluated for ablative composites. A polyimide resin, for example, was found to show very little degradation at 400°C for 2 hours, and rapid degradation at 600°C, with evolution of low molecular weight gases [44]. Other resins considered have been polybenzimidazole (PBI), poly(*p*-diphenylene oxide) (DPO), polyphenylenes, *p*-phenylphenols, and substituted heterocyclic polymers. Composite processing is a limitation in some cases.

When silica fiber reinforcements are used, as in less severe reentry trajectories, the mechanisms of ablation are somewhat more complex, as indicated in Figure 4. For example, the Apollo heat shield material [48] was a low density ablator consisting of an epoxy–novolac resin containing phenolic microballoons and a silica fiber reinforcement in the cavities of a fiberglass-reinforced phenolic honeycomb. Upon heating to about 1600K, agglomerated silica fibers first appeared on the surface. Silica globules then partially covered the surface, decreasing as silicon monoxide vapor formed upon further heating to about 2100K. Above 2100K, silica melt was no longer found on the surface, and the material suffered mechanical removal of particles at pressures above 101 kPa.

Thermal–Structural Considerations in Application of Ablators

The mechanical and thermal properties of the ablative materials can limit their application, due to structural requirements or stresses resulting from the thermal gradients. The plastics and plastic composites are relatively simple to attach to structures because the back face remains at low temperatures. The fiber reinforcement must provide good mechanical properties, however. Table 1 lists typical properties in the fiber directions of cloth-reinforced phenolic plates [49].

TABLE 1
Typical Properties of Ablative Composites Used for Heat Shields

Property at Room Temperature	Carbon Phenolic	Silica Phenolic
Density, g/cm^3	1.4–1.5	1.7
Tensile strength, MPa	138	103
Elastic modulus, GPa	19	17
Compressive strength, MPa	359	214

From Ref. 49.

Laminated composites of this type have usually been made as heat shields by tape-wrapping cloth impregnated with the resin, then molding and curing. For example, an 8-harness satin cloth may be layed up with fibers at 45° to the wrapping direction (bias orientation) and then layed up on a cylinder or frustum so that the plane of the tape makes an angle (of 20°, e.g.) to the tangent plane. This arrangement provides resistance to delamination, which may occur with a simple cylindrical (scroll) wrap, and allows the fibers to make a low angle to the ablating surface, thus minimizing thermal conduction.

As previously discussed, pyrolytic graphite has shown the best resistance to surface recession under very high heat flux conditions. This material is made as flat plates up to 3 cm thick or monolithic shapes by chemical vapor deposition (CVD) of carbon from methane or other carbonaceous gases onto industrial graphite substrates, usually at 2100–2600K [50]. Unlike "aerospace grades" of industrial graphites, which are processed from granular material and pitch and have densities in the range 1.7–1.9 g/cm^3, pyrolytic graphite is impervious to gas and has densities of 2.1–2.2 g/cm^3 as deposited; upon "graphitizing" at 2500°C or above, densities approaching the graphite single crystal, 2.26 g/cm^3, are obtained.

However, the high degree of preferred orientation of graphite basal planes parallel to the substrate results in a severe anisotropy in mechanical and thermal properties. These can vary widely, depending on the microstructure resulting from different deposition conditions [51], but tensile strength is typically 100–200 MPa in the plane of deposit (AB orientation) and 5–10 MPa through the thickness (C), compared with 30–60 MPa for aerospace graphites. Thermal conductivity is only about 1% as high through the thickness as in the plane of deposit. The most critical property affecting thermal stresses, however, is the thermal expansion, compared in Table 2 [52,53].

The high thermal expansion anisotropy of pyrolytic graphite means that any curvature in the deposit will either distort (if free) or develop bending stresses (if restrained) during heating or cooling. Such stresses place a limit on the thickness-to-radius ratio that can be deposited without cracking during cooling after deposition (typically 0.1 thickness-to-radius). Further distortion will also occur if the deposit graphitizes at higher temperatures, since not only does the anisotropy increase, but the deposit shrinks through the thickness (usually 10% or more) and expands radially (typically 3–8%, depending on initial microstructure). This has limited the use of pyrolytic graphite coatings on rocket nozzles [53]. Designs with minimum recession throats have often utilized stacks of plates with the C direction (thickness) parallel to the nozzle axis. Although the anisotropy can be greatly reduced by codeposition of 14 vol % SiC [54], such "alloying" generally results in an increase in vaporization rate.

The thickness limitations inherent to pyrolytic graphite are absent in industrial graphites; both ATJ-S (Union Carbide) and AXF-5Q (Poco Graphite Inc.), in Table 2, have been used in reentry protection systems [5,6]. However, the biaxial stresses developed in heated rocket nozzles or nose tips cause catastrophic cracking in all particulate-processed graphites, if the tensile strain at failure is exceeded. Some special grades have been made with improvements in this property, combined with adequately low thermal expansion coefficients and elastic moduli, by modifications and quality control of the constituents and processing, together with nondestructive evaluation to eliminate crack initiators in the material [55].

Carbon–Carbon Composites

The thermal stress limitations of industrial graphites have been overcome without significant degradation in ablation performance, although at significantly higher cost, by incorporating 40–50 vol % of high strength graphite fibers as a three-dimensional reinforcement [56–59]. Woven preforms, usually having 5–30 cm dimensions with center-to-center bundle spacings of 0.3–3 mm [60], have been made with several types of geometry; examples are shown in Figure 6 [61]. The three-directional packing has been made as cylinders or frusta with polar coordinate geometry, axial, radial, and circumferential reinforcements [62]. It is also made, in principle, by laminating woven cloth and injecting rods or yarn perpendicular to the "pierced fabric" with 2.5 mm or less between centers [63].

In all cases, the fiber assemblies are impregnated with resin or pitch and carbonized in several cycles to reduce porosity to levels where acceptable ablation performance is obtained. Usually, graphitizing temperatures (normally 2700–2800°C) are employed after all or most of the cycles. In some cases, gaseous infiltration and deposition of pyrolytic carbon is employed. The four-directional and seven-directional geometries (Fig. 6) permit better processing, in this way, because of the porosity continuity [61].

TABLE 2
Typical Thermal Expansion of Graphite from 300 to 2500K

Orientation	Pyrolytic Graphite (%)	Union Carbide ATJ-S (%)	Poco AXF-5Q (%)
Radial or AB direction	0.5	0.9	1.9
Axial or C direction	6	1.2	1.9

The properties of the three-directional composites on a macroscopic scale are determined primarily by small unidirectional composites, which fill the reinforcement prism positions in Figure 6. About 60–65% of the volume of the fiber bundles is filled by filaments of the yarn or tow, with graphite or carbon obtained from repeated impregnation–graphitization cycles comprising the rest of the material. The fibers, prepared from rayon, acrylic, or pitch precursors, are usually processed to high modulus of 300–500 GPa and strength (1–3 GPa) and have a high preferred orientation of the graphite layers. This results in a low thermal expansion coefficient in the bundle filament direction (about 0.5% from 300 to 3000K) and a high expansion coefficient transverse to the bundle (about 3% from 300 to 3000K). When the material is heated to about 3000K at the conclusion of processing, the transverse bundle contraction upon cooling is resisted by bundles in opposing directions, so that fissures tend to open around or within the bundles because of cracking in tension across the layers. Upon reheating, these spaces accommodate the high transverse expansions, so that the composite as a whole has a low coefficient of expansion, similar to that of the high modulus fibers. The fissures are probably the primary reason for resistance to crack propagation, which prevents thermal stress failure even though tensile strain to failure (0.1–0.4%) is less than that of the brittle aerospace graphites (0.3–0.7% at low temperatures). Strength and modulus are higher than the graphites and depend on the proportion of fibers reinforcing the direction of test; 100–300 MPa tensile strengths are found in the three-directional carbon–carbon composites [56–59].

Most published literature on carbon–carbon composites is concerned with two-dimensional reinforcements made by laminating cloth or unidirectional tape oriented in different directions, or by filament winding, and with relatively isotropic, low filament density composites made by chemical vapor deposition/infiltrating of carbonized felt that has been formed to shape [64–66]. In the former case, a resin matrix is often prepared initially,

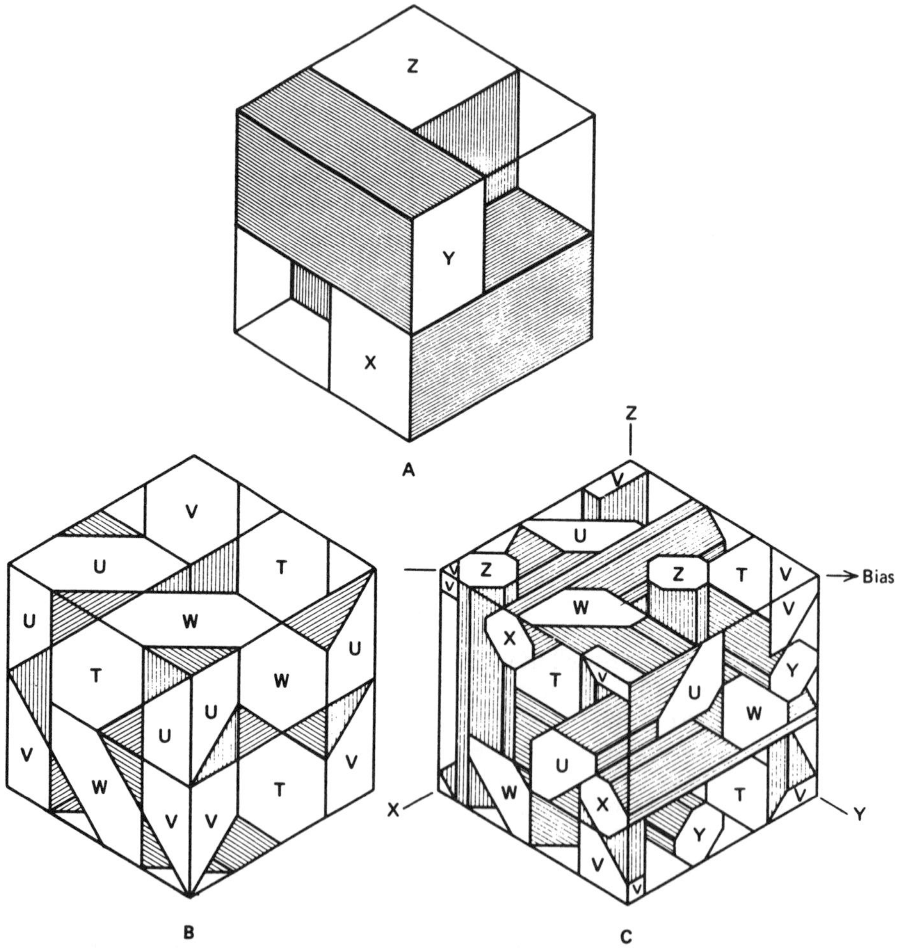

FIGURE 6 Three-dimensional packing geometries used in carbon–carbon composites; note the shape of the continuous prism positions filled by fibers [61]. (*A*) Three-directional geometry, parallel to cube sides (space filled ≥ 75%). (*B*) Four-directional geometry, parallel to cube diagonals (space filled = 75%). (*C*) Seven-directional geometry [three- and four-directional combination (space filled = 60%)].

and the carbon fiber–resin composite is precharred or carbonized. This processing is often followed by CVD infiltration to increase density near the surface and interlaminar shear strength [67,68]. Such materials have been evaluated for rocket nozzles. With additions of silicon carbide to improve low temperature oxidation resistance, they have been utilized as the leading edge thermal protection for the space shuttle [69]. In the CVD felt composites, the microstructure of the deposit may be varied by changing deposition conditions to control the properties after graphitization [70]. Both CVD felt and CVD/infiltrated filament-wound frusta have been flight tested successfully [71]; the properties can vary considerably, depending on the filament winding orientation or degree of needling the felt (which pushes some filaments perpendicular to the layers); but both materials have a fracture toughness superior to graphite [72].

One example of a successful "2D" composite was the protective container around the radioactive fuel capsule used to power the Apollo lunar geophysical stations. High temperatures generated by the fuel during storage prevented use of a carbon–phenolic ablative material. The composite was fabricated from phenolic resin impregnated cloth, made from low modulus carbon fibers, layed in a rosette pattern 6–9° to the tangent of the cylinder (Fig. 7). After molding and carbonizing, it was heated to about 2500K and then CVD infiltrated at about 1300°K to densify the outer layers to about 1.5 g/cm^3; the center remained at about 1.3 g/cm^3. The lower elastic modulus associated with the lower central density contributed to a reduction of thermal stresses. Although relatively low in tensile strength ($\sim$ 90 MPa axial, $\sim$ 30 MPa circumferential, and $\sim$ 6 MPa through the thickness), this material had much better resistance to crack propagation than graphite, thus providing protection against fuel dispersal upon impact as well as surviving the ablation and thermal stress environment during reentry [73]. Such reentry occurred during the aborted Apollo 13 mission.

Composite Dielectric Ablative Materials

As in the case of carbon heat shields, the antenna window portion of a reentry thermal shield has been subject to development to improve mechanical properties by incorporating fiber reinforcements. In these cases, either high purity silicon dioxide or boron nitride (BN) are used to avoid forming a conducting char.

Table 3 summarizes the pertinent properties of both composite grades and typical ceramic materials that have been considered. Fused silica has been used in optically transparent and porous slip-cast grades. Hot-pressed boron nitride has higher resistance to ablation but the disadvantage of high thermal conductivity; the lower interlayer shear strength of pyrolytic (CVD) BN contributes to more rapid ablation, under high shear environments.

The fiber-reinforced composites in Table 3 were developed for improved mechanical toughness. Significant improvements in tensile strain to failure and relative impact resistance have been obtained, as shown in the table. The Aeronutronic-Ford AS-3DX utilizes a three-directional orthogonal arrangement of silica fibers, densified by repeated impregnation with an aqueous suspension of colloidal silica and heating to 900K, until about 1.7 g/cm^3 bulk density is obtained [75]. A four-directional composite, having the geometry in Figure 6, utilizes the same high purity fibers and a similar densification process [76]; it is known as General Electric ADL-4D6. A three-directional BN composite, Aeronutronic-Ford BN-3DX [77] is based on high purity Carborundum

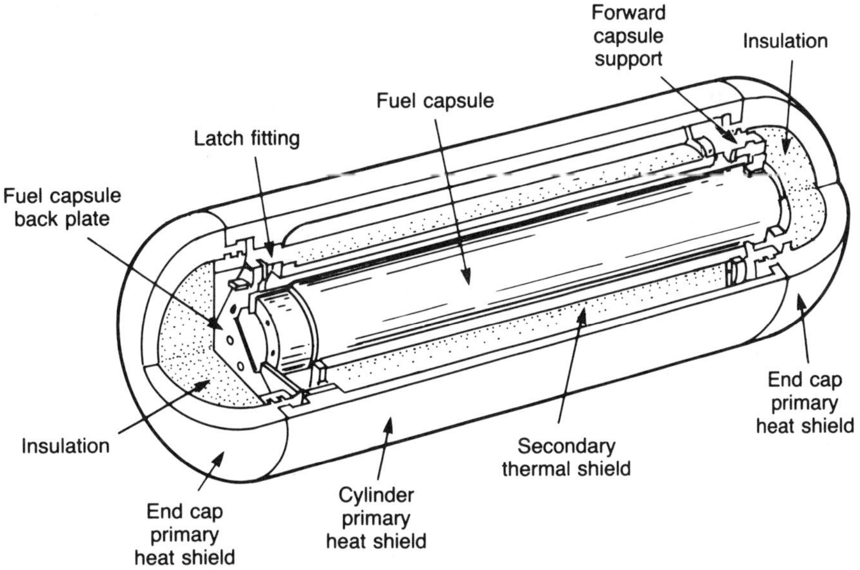

FIGURE 7 Design of the carbon–carbon reentry primary heat shield used for protection of the radioactive fuel source carried on the Apollo lunar excursion modules.

TABLE 3
Properties of Ablative Materials for Antenna Windows at About 300K

Material Description[a]	Bulk Density (g/cm^3)	Thermal Conductivity [$MW/(m \cdot K)$][b]	Dielectric Constant	Loss Tangent, $\times 10^{-3}$	Tensile Properties[c] Modulus (GPa)	Strength (kPa)	Strain (%)	Relative Impact Strength[d]	Heat of Ablation (MJ/kg)
Fused SiO_2, transparent (Corning 7940)	2.2	1.4	3.8	1.6	69	47		0.5–1.5	7–12
Slip-cast SiO_2 (Corning 7941)	1.9–2.1	0.9	3.3	0.2	58	37–62		0.5–1.5	7–12
Three-directional SiO_2–SiO_2 (A-F AS-3DX)	1.7	0.7	2.9	8.0	28	30 (35 flexure)	0.2	4.0–6.0	10–26
Four-directional SiO_2–SiO_2 (GE ADL-4D6)	1.6	0.7	2.8–3.1	6.0	(14 flexure)	(35 flexure)	1	4.0–8.0	14–26
Four-directional SiO_2–silicone (GE ADL-10)	1.8	0.6	3.0–3.3	2–5	7–14	70–140	1–2	5.0	
Boron nitride									
Hot pressed (UCC HBC)	1.9–2.05	62 AB 44 C direction	4	0.26	41–76	41		0.9 max	26
Pyrolytic (UCC Boralloy)	2.1	62 AB 1.6 C direction	5.1 AB 3.4 C	0.15 0.10 max	21 AB	83 AB 41 C	0.4 AB		HBC-AB Lower-C
0.4 BN–0.6 SiO_2 (Carborundum M)	2.12	26 AB 10 C direction	3.7 AB 4.2 C	0.5 1.0	106	97			
Three-directional BN–BN (A-F BN-3DX)	1.5–1.6	8.7	2.9–3.2	3.0	15	25 (58 flexure)	0.2	3.0	26–56

[a] Manufacturers: AF, Ford Aerospace and Communications; UCC, Union Carbide Corporation; GE, General Electric Company, Re-entry & Environmental Systems Division.
[b] Measured at 340K.
[c] Relative spall threshold in plate-slap tests.
From Ref. 74.

Company boron nitride fibers and the formation of boron nitride from boron oxide and ammonia in the pores. A reinforced four-directional, SiO_2 containing a silicone–resin matrix [74] has also been tested but is limited to 1800K surface temperature to avoid forming an electrically conducting char.

E. R. Stover
P. W. Juneau, Jr.
J. P. Brazel

[*See also* CARBON-CARBON COMPOSITES; TRIBOLOGICAL APPLICATIONS.]

References

1. D. K. Huzel, *Pennemunde to Canaveral,* Prentice-Hall, Englewood Cliffs, NJ, 1962.
2. D. L. Schmidt, "Ablative Plastics for Re-entry Thermal Protection," WADD TR 60-862, USAF ASD, Wright-Patterson AFB, Ohio, August 1961.
3. H. J. Allen, *J. Aeronaut. Sci., 25,* 217 (1958).
4. M. C. Adams, *Jet Propul., 29,* 625 (1959).
5. A. M. Morrison, *J. Spacecr. Rockets, 12,* 633 (1975).
6. P. J. Schneider et al., *J. Spacecr. Rockets, 10,* 592 (1973).
7. H. L. Moody et al., *J. Spacecr. Rockets, 13,* 746 (1976).
8. D. L. Schmidt, in D. V. Rosato and R. T. Schwartz, Eds., *Environmental Effects on Polymeric Materials,* Wiley-Interscience, New York, 1968, pp. 487–587.
9. D. L. Schmidt, in G. F. D'Alelio and J. A. Parker, Eds., *Ablative Plastics,* Dekker, New York, 1971, pp. 1–39; *J. Macromol. Sci. Chem., 3,* 327 (1969).
10. M. L. Minges in G. F. D'Alelio and J. A. Parker, Eds., *Ablative Plastics,* Dekker, New York, 1971, pp. 287–313.
11. N. S. Diaconis, E. R. Stover, J. Hook, and G. J. Catalano, "Graphite Melting Behavior," AFML-TR-71-119, ASD, Wright-Patterson AFB, Ohio, July 1971.
12. N. A. Gokcen et al., *High Temp. Sci., 8,* 81 (1976).
13. L. S. Nelson, A. G. Whittaker, and B. Tooper, *High Temp. Sci., 4,* 445 (1972).
14. A. G. Whittaker and P. L. Kintner, *Proceedings of the 12th Biennial Conference on Carbon, Extended Abstracts,* University of Pittsburgh, 1975, pp. 45–47.
15. A. G. Whittaker and P. L. Kintner, *Carbon, 14,* 257 (1977).
16. J. H. Lundell and R. R. Dickey, *AIAA J., 11,* 216 (1973).
17. J. H. Lundell and R. R. Dickey, *AIAA J., 13,* 1079 (1975).
18. J. Abrahamson, *Carbon, 12,* 111 (1974).
19. M. L. Lieberman, *Carbon, 9,* 345 (1971).
20. R. T. Meyer and A. W. Lynch, *High Temp. Sci., 4,* 283 (1972).
21. H. B. Palmer and M. Shelef, "Vaporization of Carbon," in P. L. Walker, Jr., Ed., *Chemistry and Physics of Carbon,* Vol. 4, Dekker, New York, 1968, pp. 85–135.
22. H. R. Leider, O. H. Krikorian, and D. A. Young, *Carbon, 11,* 555 (1973).
23. W. E. Pearson and W. C. Davy, *AIAA J., 11,* 1207 (1973).
24. R. T. Meyer and A. W. Lynch, *High Temp. Sci., 5,* 195 (1973).
25. P. D. Zavitsanos and G. A. Carlson, *J. Chem. Phys., 59,* 2966 (1973).
26. *NSRDS-NBS-37,* 2nd ed., National Bureau of Standards, Washington, DC, 1971.
27. J. H. Lundell and R. R. Dickey, *Proceedings of the 14th Aerospace Sciences Meeting,* AIAA, 1976, Paper 76-166.
28. S. M. Scala and L. M. Gilbert, *AIAA J., 3,* 1625, 2124 (1965).
29. J. A. Segletes, *J. Spacecr. Rockets, 12,* 251 (1975).
30. K. E. Putz and E. P. Bartlett, *J. Spacecr. Rockets, 10,* 15 (1973).
31. K. M. Kratsch et al., AFML-TR-70-307, Vol. IV, ASD, Wright-Patterson AFB, Ohio, May 1973.
32. I. Auerbach et al., *J. Spacecr. Rockets, 14,* 19 (1977).
33. E. Frye, *Nucl. Technol., 12,* 93 (1971).
34. J. H. Lundell, R. R. Dickey, and J. T. Howe, *Proceedings of the ASME Conference on Environmental Systems,* 1975.

35. J. W. Schaefer, H. Tong, and R. J. Bedard, *J. Spacecr. Rockets, 12,* 552 (1975).

36. L. M. Gilbert, *J. Spacecr. Rockets, 12,* 184 (1975).

37. J. C. Lewis, I. J. Floyd, and F. C. Cowlard, *Proceedings of the Third Conference on Industrial Carbons and Graphite (1970),* Society of Chemical Industry, 1971, pp. 282–296.

38. T. R. Acharya and D. R. Olander, *Carbon, 11,* 7 (1973).

39. D. L. Petersen and W. E. Nicolet, *J. Spacecr. Rockets, 11,* 382 (1974).

40. T. F. McKeon, "Ablative Degradation of a Silicone Foam," in G. F. D'Alelio and J. A. Parker, Eds., *Ablative Plastics,* Dekker, New York, 1971, pp. 259–286.

41. S. S. Tompkins and W. P. Kabana, "Effects of Material Composition on the Ablation Performance of Low-Density Elastomeric Ablators," NASA TND-7246, NASA Langley Research Center, Hampton, VA, 1973; *Chem. Abstr., 79,* 32555G.

42. P. W. Juneau, Jr., presentation at the Third Annual Polymer Conference Series, Program VII, 1972, University of Detroit, Detroit, Mich.

43. P. W. Juneau, Jr., *Proceedings of the SPE ANTEC, 12,* 1966, Paper XVII–3.

44. R. W. Farmer, "Extended Heating Ablation of Carbon Phenolic and Silica Phenolic," AFML-TR-74-75, ASD, Wright-Patterson AFB, Ohio, September 1974.

45. G. F. Sykes, NASA TN D-3810, National Aeronautics and Space Administration, February 1967.

46. R. P. Rastigi and D. Deepak, *AIAA J., 12,* 1146 (1974).

47. J. W. Metzer, *Proceedings of the AIAA–AIME–SAE 13th Structures, Structural Dynamics, and Materials Conference,* 1972.

48. E. P. Bartlett and L. W. Anderson, *J. Spacecr. Rockets, 8,* 463 (1971).

49. *Ablative Materials Handbook,* U.S. Polymeric, Inc., Santa Ana, CA, 1964.

50. W. H. Smith and D. H. Leeds, "Pyrolytic Graphite," in *Modern Materials,* Vol. 7, Academic Press, New York, 1970, pp. 139–221.

51. J. C. Bokros, "Deposition, Structure and Properties of Pyrolytic Carbon," in P. L. Walker, Jr., Ed., *Chemistry and Physics of Carbon,* Vol. 5, Dekker, New York, 1969, pp. 1–118.

52. C. D. Pears, *Proceeding of the Conference on Continuum Aspects of Graphite Design, 1970,* CONF-701105, NTIS, Springfield, VA, 1972, pp. 115–136.

53. J. G. Baetz, "Characterization of Advanced Solid Rocket Nozzle Materials," SAMSO-TR-75-301, Air Force Rocket Propulsion Laboratories, Edwards AFB, CA, December 1975.

54. R. H. Singleton, E. L. Olcott, and C. Pears, *Proceedings of the 11th Biennial Conference on Carbon,* CONF-730601, NTIS, 1973, p. 192.

55. C. A. Pratt, Jr., "Petroleum Derived Carbons," in M. L. Deviney and T. M. O'Grady, Eds., ACS Symposium Series 21, American Chemical Society, Washington, DC, 1976, pp. 203–211.

56. A. R. Taverna and L. E. McAllister, in J. Buckley, Ed., *Advanced Materials, Composites and Carbon, 1971 Symposium,* American Ceramic Society, Columbus, OH, 1972, pp. 203–211.

57. K. M. Kratsch, J. C. Schutzler, and D. A. Eitman, *Proceedings of the AIAA–ASME–SAE 13th Structural Dynamics and Materials Conference,* 1972, Paper 72365.

58. E. R. Stover et al., *Proceedings of the 11th Biennial Conference on Carbon,* CONF-730601, NTIS, 1973, pp. 277, 335–336.

59. J. L. Perry and D. F. Adams, *Carbon, 14,* 61 (1976).

60. Fiber Materials, Inc., Biddeford, ME, product data, January 1975.

61. J. J. Gebhardt et al., in M. L. Deviney and T. M. O'Grady, Eds., ACS Symposium Series 21, American Chemical Society, Washington, DC, 1976, pp. 212–227.

62. R. W. King, U.S. Patent 3,904,464 (Sept. 9, 1975).

63. P. G. Rolincik, in *11th Biennial Conference on Carbon, 1973,* CONF-730601, NTIS, Springfield, VA, pp. 343–344.

64. D. L. Schmidt, *SAMPE J.,* 8 (3) 9 (1972).

65. H. M. Stoller et al., in J. Weeton and E. Scala, Eds., *Composites: State of the Art,* Metropolitan Society AIME, New York, 1974, p. 69.

66. W. V. Kotlensky, in P. L. Walker, Jr. and P. A. Thrower, Eds., *Chemistry and Physics of Carbon,* Vol. 9, Dekker, New York, 1973, pp. 173–261.

67. H. O. Davis, *J. Spacecr. Rockets, 13,* 456 (1976).

68. L. Boyne, J. Hill, and K. Turner, in J. M. Blocher et al., Eds., *Proceedings of the Fifth International Conference on Chemical Vapor Deposition,* Electrochemical Society, Princeton, NJ, 1975, pp. 577–588.

69. J. M. Williams and R. J. Imprescia, *J. Spacecr. Rockets, 12,* 151 (1975).

70. H. O. Pierson and D. A. Northrup, *J. Compos. Mater., 9,* 118 (1975).

71. H. M. Stoller et al., "Properties of Flight Tested CVD-Felt and CVD-FW Composites," SC-DC 71-4046, Sandia Laboratories, Albuquerque, NM, 1971; *Summary of Papers, Tenth Biennial Conference on Carbon, 1971,* pp. 90–91.

72. T. R. Guess and W. P. Hoover, *J. Compos. Mater., 7,* 12 (1973).

73. R. Gavert, *SAMPE Q., 1*(1), 56 (1969).

74. J. P. Brazel, AMMRC CR-76-4, U.S. Army Materials and Mechanics Research Center, Watertown, MA, February 1976.

75. T. M. Place, *Proceedings of the 12th Symposium on Electromagnetic Windows,* Georgia Institute of Technology, 1974, pp. 47–51.

76. J. P. Brazel and R. Fenton, *Proceedings of the 13th Symposium on Electromagnetic Windows,* Georgia Institute of Technology, 1976, pp. 9–16.

77. T. M. Place, *Proceedings of the 13th Symposium on Electromagnetic Windows,* Georgia Institute of Technology, Atlanta, 1976, pp. 17–22.

Composite Micromechanics

While overall properties of a material must be known for the design of any component, an additional dimension of composite material behavior is the unique problem of the complex stress state present on the microscopic scale. When a fiber of high strength and modulus is embedded in a matrix material of relatively low strength and modulus, local stress concentrations are induced that are significant to the overall behavior of the composite. The mathematical characterization of this situation is known as a micromechanics analysis, while a larger scale analysis (i.e., lamination theory) is known as macromechanics.

Macromechanics, which is concerned with the overall properties of a composite laminate, can be used to determine the contribution of each ply to the overall response. In contrast, micromechanics attempts to predict such phenomena as matrix yielding, crack initiation, and unidirectional composite ply behavior, using an analysis concerned with the local scale. This is important for understanding particular failure initiation modes and predicting properties of unidirectional composites.

Early micromechanics analyses relied on simple models and gross approximations to physical reality. These are often termed phenomenological approaches. For estimating certain composite material properties, these approximate methods can be very useful, and thus they are still frequently used in engineering design. However, most current research effort is being focused on the development of physically based micromechanics analyses. Currently, these are numerical analyses, typically employing a finite element numerical scheme. They use the material properties of the fiber and matrix to predict the microstress distributions in the composite, usually assuming a state of generalized plane strain. That is, they are two-dimensional analyses. These rigorous analyses can also be performed using a full three-dimensional finite element analysis, although much more computer time and memory are necessary.

While earlier two-dimensional analyses (viz., generalized plane strain analyses) allowed only a normal stress in the third (fiber) direction, some present analyses permit a longitudinal shear loading capability also. As Miller and Adams [1] predicted in 1977, "more attempts will be forthcoming to marry micromechanical analyses . . . with various lamination analyses," and this is one of the primary reasons for the addition of a longitudinal shear loading capability. Results of a laminate point stress analysis can then be analyzed further by using such a micromechanics formulation. The laminate analysis will reveal the stress state in each ply, from which the microanalysis can predict local inelastic behavior and reveal the complex stress state within the matrix material. This type of analysis can be considered to be an important first step in bridging the gap between micromechanics and macromechanics.

An indication of the progression of micromechanics analyses available, from simple to complex (i.e., from approximate to rigorous), is given in Figure 1. Each of the levels of analysis shown is be discussed in the sections that follow.

Simple Model Analyses

Definition of a Simple Model

For our purposes, a "simple model" will be defined as one that either utilizes gross approximations to physical reality or restricts the applicability of the analysis to specific, relatively simple, physical configurations. The former type implies an applicability to general problems but offers no indication of the accuracy achieved. Thus, unless a rigorous analysis is also available for comparison of results, this type of simple model provides, at most, a general, qualitative indication of material behavior. Quantitative results are available, but are of little significance.

On the other hand, the latter type of simple model can provide reliable quantitative results. Unfortunately, the range of practical application to physically realistic problems is very limited.

Examples of both types of simple model analysis are presented here. The analyses included are considered to

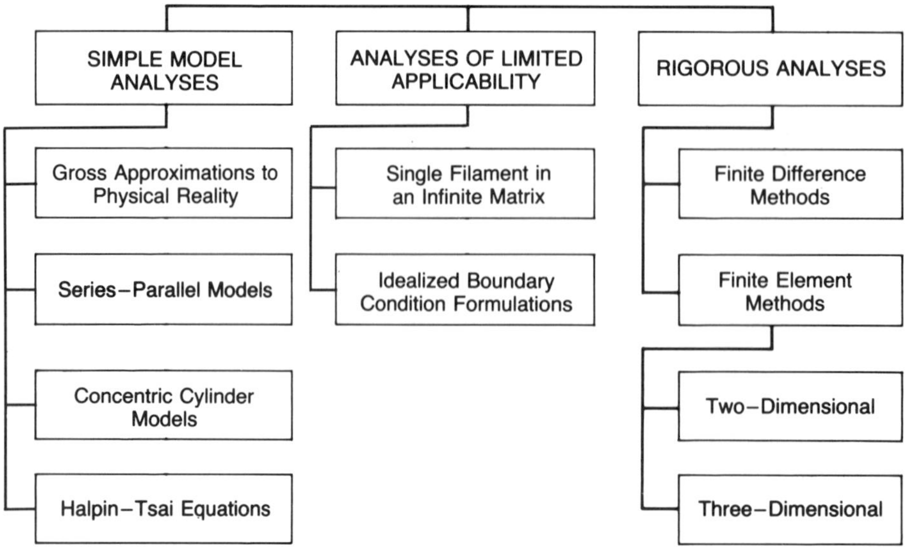

FIGURE 1 Levels of micromechanical analysis currently available.

be representative of those most frequently referred to in the literature and/or of the greatest significance.

Simple model analyses of the elastoplastic behavior of composite materials have been developed for the same reasons found in any other area of investigation—the physical phenomenon was well recognized long before the basic mathematical descriptions and computational schemes required to develop a rigorous method of analysis were available. During the mid-1960s, when rigorous elastic analyses were being published, the incremental loading procedures and large capacity digital computers required to extend these analyses to elastoplastic material behavior were not yet well established. Thus, many investigators reverted to the prior simple model elastic analyses, extending them to consider elastoplastic material behavior also. Since rigorous elastic analyses were already available, it was well known that such efforts were, at best, an interim step, and of only limited quantitative value. Nevertheless, they can often provide some physical insight to actual material behavior.

In addition to the two basic types of simple model analysis defined above, which are discussed in greater detail in subsequent paragraphs, another approach is available for predicting the stress–strain response of a composite material in terms of the properties of its constituents. This is essentially a semiempirical method. Various equation forms (particularly power laws) containing undefined constants are fitted to known experimental results to obtain empirical relationships. Such formulas typically relate stress to strain in terms of physical parameters such as the filament volume content and the material properties of the constituents. These stresses and strains might be selected local values, but most often they are average or composite values. In either case, such analyses are not truly micromechanical analyses because they do not provide a description of the local stress and strain gradients within the composite. Their utility is primarily as a simple-to-evaluate engineering estimate of the basic composite stress–strain response, which is information required as input to most macromechanical or laminate analyses.

Bert [2] presented one such semiempirical analysis. His work is cited here particularly because it contains an extensive list of references to similar works of this type, providing the interested reader with an excellent summary and source of information. Like the authors of all such studies, Bert chose to make several very severe assumptions to keep the working formulas relatively simple. Thus, it is not surprising that the correlation with experimental data other than those used to evaluate the empirical constants is frequently not very good. However, until the rigorous analyses discussed below are adequately developed and can be readily used for routine engineering computations, results such as those presented by Bert serve a useful function.

Gross Approximations to Physical Reality

SERIES–PARALLEL MODELS. Since composite materials consist of distinct constituent material phases bonded together, the concept of a model consisting of elements connected together in series and parallel arrays has long been popular. Perhaps the first attempt to apply a series–parallel model to elastoplastic material behavior was made by Shaffer [3]. In an extension of his earlier elastic analysis [4], Shaffer assumed linearly elastic–perfectly plastic material behavior in analyzing transverse normal loading. A typical repeating unit of the composite cross section is isolated, as indicated by the dashed lines in Figure 2. This unit is then divided into elemental strips of differential width and constant length p as indicated by the dotted lines in Figure 2. For a

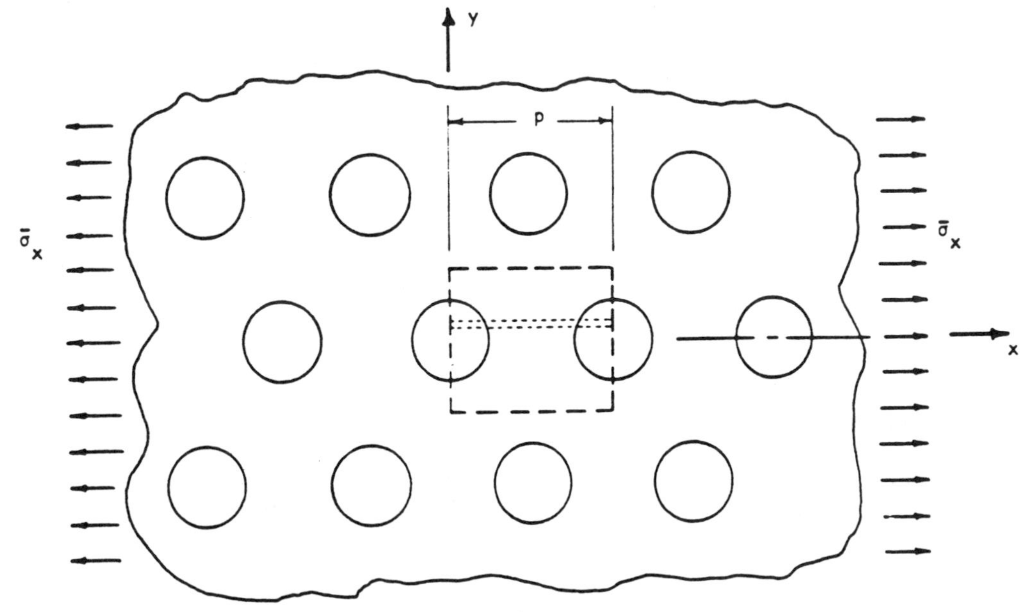

FIGURE 2 **Composite cross section indicating a typical repeated unit (dashed lines) and an elemental strip (dotted lines).**

filament volume less than 68%, some elemental strips of matrix material will exist that contain no filament material, as is obvious in Figure 2. In this case the model consists of elemental strips of matrix material arranged in parallel with strips consisting of a series array of matrix and filament regions. For filament volume contents greater than 68%, the continuous strips of matrix material vanish and the model reduces to a parallel array of series elements, the geometry of which becomes more complicated.

Several gross approximations are made in this analysis. Two of the most severe are the complete neglect of interactions between the surfaces of adjacent elemental strips and the assumption of uniformity of the stresses and strains within each elemental strip of length p. The first assumption reduces the problem to a one-dimensional stress analysis; that is, σ_x is the only nonzero stress component. This is not physically realistic and reduces the yield criterion to the trivial case, whereby yielding is predicted to occur when σ_x attains the yield value as determined from a simple uniaxial test (i.e., when $\sigma_x = \sigma_{yp}$). In actual composites, significant biaxial and triaxial stress states occur. The first approximation combined with the second also results in the prediction that the stress in any elemental strip (all of which are of length p, the filament center-to-center spacing) is constant. Thus, even for elemental strips that extend across both filament and matrix material regions, the stress does not vary within the element. That is, the model does not predict gradients in the x direction.

In spite of these gross approximations, certain numerical results obtained from the original elastic analysis agreed quite well with the more rigorous analyses of other investigators (see Ref. 4). It is possible that similar correlations may be achieved when results of rigorous elastoplastic analyses become more readily available. The problem is that before the simple model can be used with any degree of confidence, a rigorous analysis (or very extensive experimental data) must be available for comparison and verification purposes. Hence, the principal value of a simple model is, and will continue to be, as an easily used engineering tool, to be utilized only within the range of parametric variations for which it has been verified. Thus, it is in a sense comparable in its applicability to the semiempirical models such as those discussed by Bert [2], even though it is not in itself a semiempirical model.

CONCENTRIC CYLINDER MODELS. As for the series–parallel models, the concentric cylinder models represent a gross approximation to actual composite behavior and yet are very frequently used. This is because they permit relatively simple analyses and lead to closed-form solutions. A typical model consists of a single filament, usually of circular cross section, embedded in a concentric cylinder of matrix material. The gross approximation is associated with the assumed boundary conditions (explicit or implicit) on the surface of the outer cylinder. In an actual composite, adjacent filaments interact, resulting in a complex stress distribution in the intervening matrix. In the model, a simple, circumferentially and axially uniform stress or displacement condition is assumed.

One such approach is to use limit analysis methods. Two distinct sets of boundary conditions are solved for: one can be proved to give a lower bound on the desired result, and the other an upper bound. The challenge is to establish boundary conditions that will result in upper and lower bounds that are reasonably close together. This of course is not always easy to achieve.

One of the earliest applications of this methodology was by Dow and Rosen [5], who assumed elastic–perfectly plastic matrix material behavior and elastic filaments. A more refined presentation was later published by Shu and Rosen [6], although they assumed rigid rather than elastic filaments. Because the assumed boundary conditions represent average rather than local values, such analyses are typically used to estimate gross or composite properties rather than local stresses and strains. When so used, they cannot truly be considered to be micromechanical analyses, since they do not predict local behavior.

Another analysis using a concentric cylinder model is that of Ebert and Gadd [7]. They performed a reasonably straightforward continuum analysis for two concentric circular cylinders of infinite length subjected to an axial load. Elastic–perfectly plastic material behavior was assumed, the investigators' particular interest being the magnitude of the transverse stresses induced. A minor extension of the original analytical development was later presented, along with experimental results, by Ebert et al. [8].

The gross approximation that makes this analysis of little more than academic interest is that the surface of the outer cylinder is assumed to be stress free; that is, interactions with adjacent filaments that would be present in a real composite are not modeled. Thus, the analysis is a reasonable approximation only for composites of very low (< 10%) filament volume, for which the filaments are widely spaced.

Hill [9] also considered the concentric cylinder model for an assumed plane strain condition and uniform lateral pressure loading. Hill's general methodology is of interest even though he emphasized the estimation of composite stress–strain response rather than local stress distributions.

HALPIN–TSAI EQUATIONS. Perhaps recognizing the limitations of the various simple models then in existence, Halpin and Tsai in the late 1960s proposed an empirical approach for use in estimating the unidirectional composite stiffness properties, using equations of simple form containing fit parameters [10–12]. These fit parameters could be evaluated from available experimental data, or fit to the results of the few rigorous solutions becoming available at that time. The equations for the fiber-dominated stiffness properties (i.e., axial modulus E_z and major Poisson's ratio ν_{zx}) were simple rule-of-mixtures forms, namely:

$$E_z = E_f V_f + E_m V_m \tag{1}$$

and

$$\nu_{zx} = \nu_f V_f + \nu_m V_m \tag{2}$$

To estimate the matrix-dominated stiffness properties (i.e., transverse modulus E_x, in-plane shear modulus G_{zx}, an expression of the following form was used:

$$\frac{\bar{p}}{p_m} = \frac{(1 + \zeta\eta V_f)}{(1 - \eta V_f)} \tag{3}$$

where the constant r is defined as:

$$\eta = \frac{p_f/p_m - 1}{p_f/p_m + \zeta} \tag{4}$$

In Eqs. (3) and (4), the quantities $\bar{p}$, p_f, p_m, V_f, and ζ are identified as follows:

$\bar{p}$ = composite stiffnesses E_x, or G_{zx}
p_f = corresponding fiber stiffnesses E_f, G_f, or ν_f
p_m = corresponding matrix stiffnesses E_m, G_m, or ν_m
V_f = fiber volume
ζ = an empirical fit parameter

The problem is to determine ζ. However, because of the ease of using the Halpin–Tsai equations, they are still used extensively, even though exact solutions are now becoming more widely available.

Analyses of Limited Applicability

SINGLE FILAMENT IN AN INFINITE MATRIX. The gross approximation to real composites of assuming a single filament in an infinite matrix is obvious. Actually, the model of Ebert and Gadd [7] discussed above approaches this case, since the diameter of the other cylinder is increased relative to that of the inner one.

The single-filament approximation is often assumed primarily because it permits the use of any one of several well-defined analytical methodologies. An example is the analysis presented by Mendelson [13], who assumed a circular filament, equal biaxial tension at infinity, and a total deformation theory. Another presentation has been given by Sendeckyj [14].

In addition to analytical methods, numerical techniques have been used. For example, Owen et al. [15] used a concentric cylinder, finite element formulation to study the stress distributions in an elastic–perfectly plastic matrix in the gap region between two collinear filaments subjected to an axial loading. The matrix region is actually of finite extent, but it can be assumed to be large relative to the filament diameter. Thus, this model approximates a single (in this case discontinuous) filament in an infinite matrix adequately well for most purposes.

IDEALIZED BOUNDARY CONDITION FORMULATIONS. A major problem in correctly modeling composite material behavior is adequately representing the complex boundary conditions that are introduced when a local material region is isolated for analysis. For example, a single filament and the matrix material immediately surrounding it might be isolated from the composite. Stress and/or displacement conditions must then be established at the isolating surfaces, which correctly represent the actual physical conditions. However, the stress and displacement distributions throughout the composite are very complicated because of the multiple interactions between filaments. Thus, the stresses and displacements along the isolating cutting planes, which become the boundary conditions of the analysis, are generally unknown distributions.

Necessary boundary conditions can often be established, however, either by idealizing the material behavior, as discussed in this section, or by limiting the filament packing geometry to special configurations, as discussed below (see ELASTOPLASTIC BEHAVIOR). Actually, the concentric cylinder models described previously, when they are used in an attempt to represent a filament array, also utilize idealized (uniform) boundary conditions.

An example of an idealized boundary condition formulation is the analysis by Owen et al. [15], which is an extension, presented in the same paper, of these authors' previously cited work. Elastic–perfectly plastic material behavior was assumed in modeling the axial loading of a composite containing a discontinuous filament adjacent to continuous filaments (Fig. 3). A finite element numerical technique was used, the region indicated in Figure 3*b* being divided into a network of constant strain triangular elements. The particularly limiting boundary condition idealization is the assumption of plane strain. That is, the material is assumed to be uniform and of infinite extent in the x direction. Thus, the model is actually that of a discontinuous flat plate sandwiched between two continuous plates. This is not a very satisfactory approximation to the actual geometry of a filament-reinforced composite. A three-dimensional analysis would be required to model correctly the actual composite in this case.

Another example of an idealized boundary condition formulation is that of Huang [16], although it is not a complete micromechanics analysis in that the author attempted to predict the composite stress–strain response only, not local stress and strain distributions. Huang incorporated two very limiting boundary condition idealizations in analyzing the transverse normal loading problem. He assumed the filaments to be rigid, so that a simple boundary condition (viz., zero normal displacement and strain at the filament–matrix interface) is obtained. The second idealization is that the filaments are uniaxially oriented but randomly distributed within the matrix, so that macroscopically the composite is transversely isotropic. Huang then utilized the so-called self-consistent approach, the basic assumption of which is that an isolated filament can be regarded as being embedded in an "equivalent" homogeneous medium whose properties are the unknowns to be determined. This procedure averages out the effect of interactions between individual filaments, leading to average results rather than local values. This becomes a particularly limiting assumption for high filament volumes,

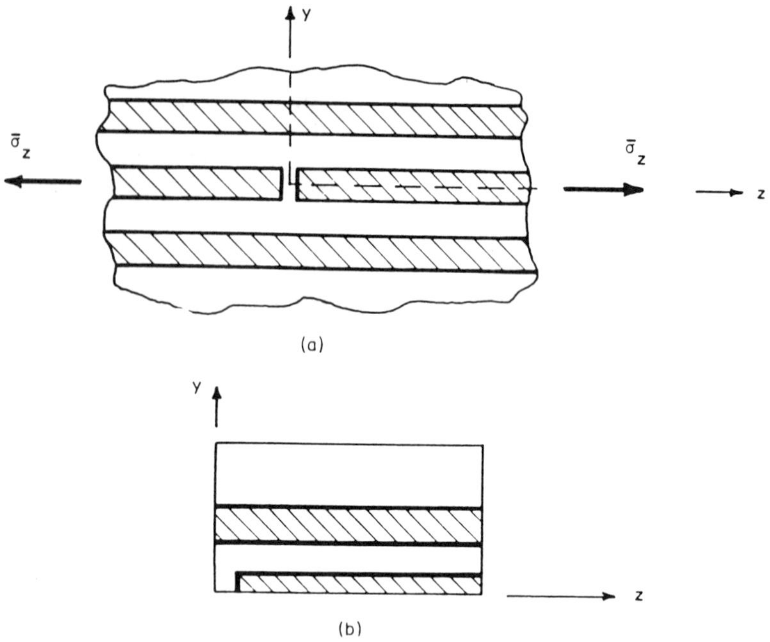

FIGURE 3 **Discontinuous filament adjacent to continuous filaments: (*a*) general arrangement and (*b*) local region analyzed.**

when filaments are closely spaced, hence interacting strongly.

Rigorous Analyses

A "rigorous analysis" is defined here as one that satisfies the mathematical laws of analytical mechanics and at the same time accurately models the known physical constraints. These physical constraints are associated with the material characteristics of the individual constituents and their geometric arrangement within the composite.

Simple model analyses were defined and discussed above. In all cases, the true representation of actual physical behavior was severely compromised to simplify the analysis. Within the past several years, somewhat more rigorous representations have begun to be developed. Because of the complexities contained in these formulations, numerical solution techniques are required, typically involving the related development of digital computer programs. These programs tend to become very large as the accuracy of the model representation is increased, requiring the use of large capacity computers and large amounts of computer running time. Thus, the programs are time-consuming to develop and can be costly to use.

Nevertheless, at the present time the computer route appears to be the only way of accurately representing the actual elastoplastic behavior of composite materials. It should be noted that the same progression of analysis methods, from simple models to rigorous analyses, was followed in past years in developing analyses of the linearly elastic behavior of composite materials. It is anticipated that the same will be true of the rigorous elastoplastic analyses. Improvements in this area will continue to appear, and such analyses will represent the best approach available for many years to come.

The basic mathematical laws governing elastoplastic material behavior are presented next, followed by a discussion of their practical application to rigorous analyses.

Elastoplastic Behavior

General working theories governing elastoplastic material behavior are a relatively recent development. Thus, specific applications to composite material micromechanical behavior are as yet somewhat limited. Nevertheless, it has been recognized that because of the high stress and strain concentrations that exist in local regions between filaments, the elastic limit of the matrix material may be exceeded long before the gross behavior of the composite exhibits an obvious degree of nonlinearity. This localized elastoplastic behavior has a significant influence on the redistribution of stresses within the composite, hence the initiation of composite failure. The importance of this behavior in engineering applications is reviewed, and a discussion of possible approaches to the problem is given. The elastoplastic constitutive relations are summarized and the several commonly utilized yield criteria compared. Existing applications of these methods of analysis are categorized as being either simple models or rigorous analyses. They are discussed in terms of their range of applicability, accuracy of representation, and assumptions incorporated. Particular attention is given to the rigorous analyses, this being the expected direction of future investigations.

Definition of Elastoplastic Behavior

Most studies of the micromechanical behavior of composite materials have assumed, either explicitly or implicitly, that the individual constituent materials remain linearly elastic. However, many materials, and particularly those typically selected for use as matrices in composite systems, do not remain linearly elastic when subjected to an applied loading.

For some materials this nonlinear response may be due, at least in part, to viscoelastic (i.e., time-dependent) effects as discussed in Refs. 17–20. On the other hand, most materials exhibit a nonlinear, time-independent, deformational response when the applied loading exceeds a limiting value (viz., the yield point of the material). This latter type of nonlinear response, exhibited beyond the elastic range, is termed plasticity. Hence the terminology "elastoplastic behavior" is commonly used to denote the full range of loading.

Typical shear stress–strain curves for three commonly used polymer matrix materials are shown in Figure 4. Nonlinear behavior is indicated. It is obvious that an elastic analysis is applicable only in the initial portion of the stress–strain response.

Significance in Composite Applications

The reinforcing filaments in a composite material are typically selected on the basis of their high strength and stiffness characteristics. A primary role of the matrix material is to bind the filaments together into a solid mass,

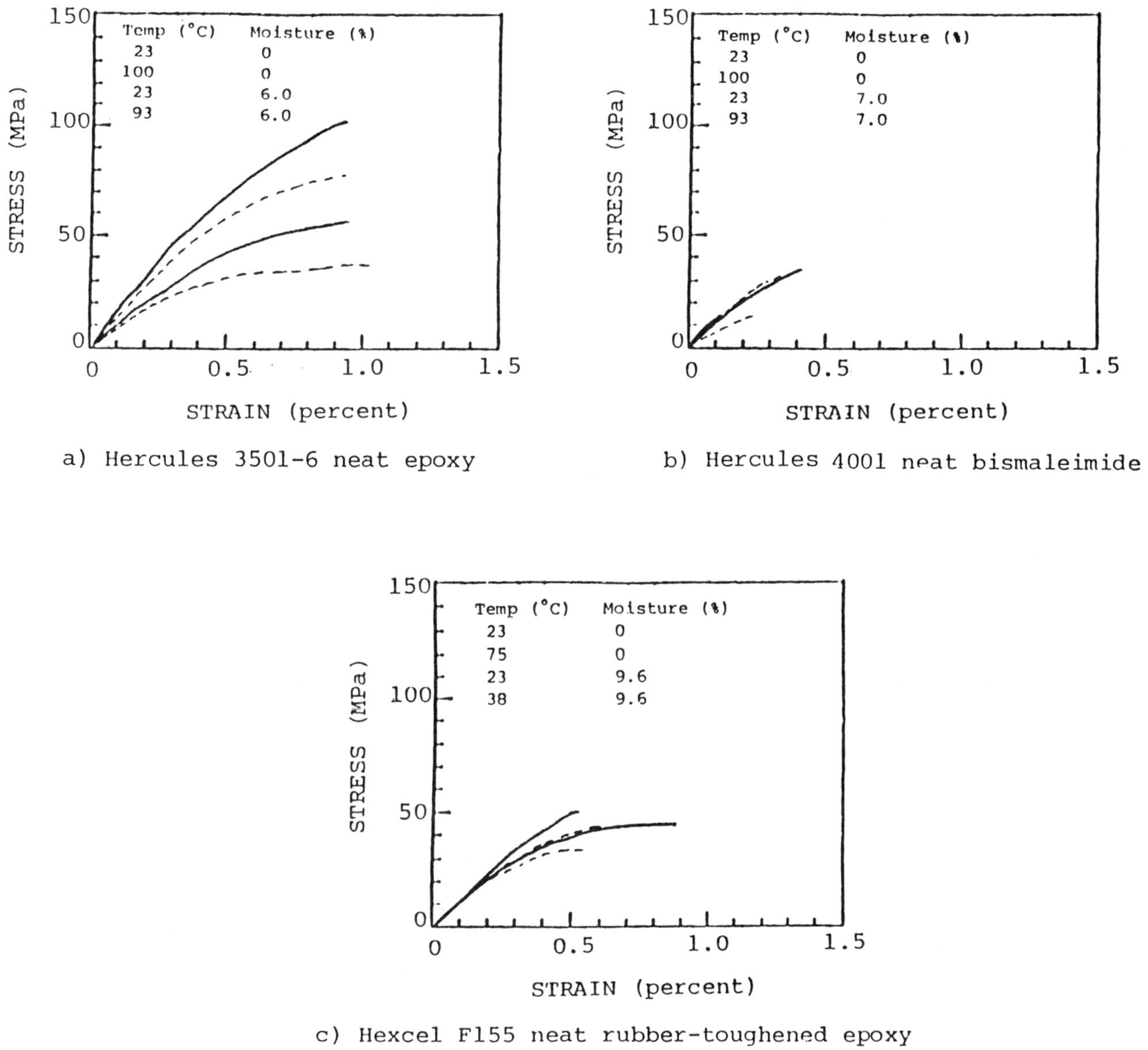

FIGURE 4 **Neat resin matrix shear stress–strain curves at various combinations of temperature and moisture: (*a*) Hercules 3501-6 neat epoxy, (*b*) Hercules 4001 neat bismaleimide, and (*c*) Hexcel F155 neat rubber-toughened epoxy.**

while permitting ease of fabrication into a desired configuration. Implicit in this function of the matrix is its ability to transfer loads from filament to filament. Typically the fabricability requirement results in the selection of a matrix material having a low stiffness (relative to the filament reinforcement) and correspondingly a low strength. The resulting mismatch of stiffness properties between the filaments and the surrounding matrix material leads to the development of high local strain concentrations in the matrix when loads are applied. Thus, even when the average applied stress on the composite material is relatively low, local stresses and strains within the composite material may already have exceeded the elastic limit values. For this reason, the consideration of elastoplastic micromechanical behavior is important even for composite materials subjected to applied stresses that are intended to remain below the apparent elastic limit of the composite. For structures designed to an ultimate strength criterion, the need for consideration of elastoplastic material behavior is even more apparent.

To accommodate high local strains without inducing local failures, a matrix is normally selected that has a high strain to failure. This permits local plastic flow to occur in regions of high strain concentration, causing a redistribution of stresses (stress relief). This ability to accommodate large local strains is also important in regions of load path discontinuity, such as around the ends of noncontinuous reinforcements (e.g., chopped fibers, whiskers, and particulates).

Approaches to the Problem

An analytical study of the problem of elastoplastic behavior of composite materials can be approached from any one of a number of different levels of modeling accuracy. Historically, as is the case for most analyses, the initial analytical investigations of elastoplastic material behavior relied on very simple models. These models were known to be very crude representations of the actual material system. However, they were relatively easy to formulate, while providing at least qualitative indications of theretofore unknown behavior. Subsequent analyses have used somewhat more accurate models, having a broader range of applicability and providing numerical results of a greater degree of reliability.

The more rigorous analyses are discussed below (see METHODS OF IMPLEMENTING ELASTOPLASTIC THEORY). First, however, it is informative to consider the basic principles of the general theory of plasticity, to indicate the characteristics and basic assumptions that distinguish elastoplastic behavior from elastic and viscoelastic material behavior.

Elastoplastic Constitutive Relations

The General Concepts of Plasticity

When a material is loaded beyond its elastic limit, the state of stress at any time is dependent on the loading history. This is unlike elastic behavior, for which the instantaneous state of stress is dependent only on the current state of strain. Thus, it is necessary to first establish a criterion for yielding—that is, to define the limits of applicability of elastic (Hookeian) behavior—and then to establish a suitable stress–strain relationship for material behavior beyond the elastic limit.

Since, up to the elastic limit, strains can be expressed directly in terms of stresses, it is obvious that a yield criterion can be expressed in terms of stresses only. Likewise, if the material is assumed to be isotropic, the functional relationship between stress components that expresses this criterion must be independent of the coordinate system selected; that is, the functional relationship will be an invariant of the stress tensor. Thus, the yield criterion is governed by the various laws of elastic behavior.

Once the material has been stressed beyond its elastic limit, behavior during unloading and subsequent reloading is not necessarily governed by the theory of elasticity, since the strain field is no longer uniquely determined by the state of stress. However, experiments verify that for most materials it is a reasonable approximation to assume that unloading and subsequent reloading to the previous state of stress do occur elastically and that hysteresis effects are negligible.

Experiments also verify that a hydrostatic pressure does not produce appreciable plastic deformation. Thus, it can be assumed that the plastic component of strain is a function only of the deviatoric (shear, component of the stress tensor, being independent of the dilatational component.

These then are the basic distinguishing characteristics and assumptions of the general theory of plasticity. For a detailed presentation, the interested reader is referred to a text such as that by Hill [21] or Prager and Hodge [22].

Comparisons of Various Yield Criteria

Various yield criteria have been proposed through the years, but most of them violate one or more of the basic experimental observations mentioned previously. In particular, most predict that the dilatational component of the stress tensor influences yielding and plastic flow. Two theories that do not are those attributed to Tresca and von Mises. Both are in common use, as much because of their relative simplicity as their demonstrated accuracy.

The Tresca yield criterion is the more complicated of the two. It assumes that yielding occurs when the maximum shear stress reaches a given value. The complication of this theory arises because, for a given problem, it is not known beforehand which of the three principal shear stresses will be the maximum. Thus it becomes necessary to use the more general form of the yield criterion, that is:

$$[(s_2 - s_3)^2 - 4k^2][(s_3 - s_1)^2 - 4k^2][(s_1 - s_2)^2 - 4k^2] = 0 \quad (5)$$

where $s_1 = \sigma_1 - \frac{1}{3}(\sigma_1 + \sigma_2 + \sigma_3)$, etc., represent the principal components of the deviatoric stress tensor; σ_1,

σ_2, σ_3 are the principal components of stress; and $2k$ is the yield value of the maximum shear stress. Expressed in terms of invariants of the deviatoric stress tensor, Eq. (5) becomes:

$$4J_2^3 - 27J_3^2 - 36k^2J_2^2 + 96k^4J_2 - 64k^6 = 0 \tag{6}$$

The Tresca yield criterion is therefore somewhat difficult to implement in either of the forms represented by Eq. (5) or (6).

The von Mises yield criterion is considerably simpler; it is expressed as follows:

$$J_2 - k^2 = 0 \tag{7}$$

where k is the yield limit of a simple shear test. In fact, von Mises originally proposed Eq. (7) as a simple approximation to the Tresca yield criterion, which was in common use at the time. The two criteria differ by only a relatively small amount (see, e.g., Ref. 21, p. 21). However, subsequent experimental work demonstrated that the von Mises yield criterion is actually in closer agreement than the Tresca criterion with experimentally observed material behavior.

Most current practical applications of plasticity theory, including those related to composite materials, use the von Mises yield criterion or one of its variations because this model is both simple to use and correlates well with experimental observations of material behavior. Nevertheless, it is of interest to attempt to attach some physical significance to it. For example, Eq. (7) can be rewritten in terms of principal components of the deviatoric stress tensor as follows:

$$s_1^2 + s_2^2 + s_3^2 = 2k^2 \tag{8}$$

As Nadai [23] first proposed, Eq. (8) implies that yielding begins when the shear stress acting on the octahedral planes reaches a limiting value; the octahedral shear stress has the value $(2k/3)^{1/2}$ at the yield point. A more detailed discussion of this and other physical interpretations of Eq. (7) is given by Prager and Hodge [22, p. 26]. This discussion of physical significance is introduced here primarily because an octahedral shear stress criterion is used later in deriving the required flow rule.

Yield criteria other than those of Tresca and von Mises have also been established, some of which provide even better agreement with experimental data. However, they are generally too complex for practical applications to most plasticity problems.

The Prandtl–Reuss Flow Rule

Having established a yield criterion to indicate the onset of plastic flow, it is necessary to establish appropriate stress–strain relationships which govern plastic flow. The Prandtl–Reuss flow rule is most commonly used; its key assumption is that the rate of change of the plastic strain is at any instant proportional to the deviatoric stress, that is:

$$\dot{\varepsilon}_{ij}^{(p)} = \dot{\lambda} s_{ij} \tag{9}$$

where $\dot{\lambda}$ is a positive scalar function, in general a function of both time and the spatial coordinates, and

$$s_{ij} = \sigma_{ij} - \tfrac{1}{3}\sigma_{kk}\delta_{ij} \tag{10}$$

is the deviatoric component of the stress tensor σ_{ij}, where δ_{ij} is the Kronecker delta. Latin indices have the range 1, 2, 3 and Greek 1, 2. The usual conventions of indicial notation apply.

To implement Eq. (9), the rate of change of the total strain can be separated into an elastic (recoverable) and a plastic (irrecoverable) component, that is:

$$\dot{\varepsilon}_{ij} = \dot{\varepsilon}_{ij}^{(e)} + \dot{\varepsilon}_{ij}^{(p)} \tag{11}$$

The elastic component follows the same generalized Hooke's law as for completely elastic material behavior, that is:

$$\dot{\varepsilon}_{ij}^{(e)} = \frac{1 - 2\nu}{3E}\dot{\sigma}_{kk}\delta_{ij} + \frac{1 + \nu}{E}\dot{s}_{ij} \tag{12}$$

where E and ν are the material Young's modulus and Poisson's ratio, respectively, and $\dot{s}_{ij}$ bears the same relationship to $\dot{\sigma}_{ij}$ as s_{ij} to σ_{ij}, which is indicated in Eq. (10).

For Eq. (9) to be useful, it is necessary to define $\dot{\lambda}$ in terms of known quantities. One method is as follows [24]:

Equation (9) is multiplied by itself and solved for $\dot{\lambda}$ to give:

$$\dot{\lambda} = \frac{(\dot{\varepsilon}_{ij}^{(p)}\,\dot{\varepsilon}_{ij}^{(p)})^{1/2}}{(s_{kl}s_{kl})^{1/2}} \tag{13}$$

Defining the octahedral plastic shear strain rate as

$$\dot{\varepsilon}_0^{(p)} = (\tfrac{1}{3}\dot{\varepsilon}_{ij}^{(p)}\,\dot{\varepsilon}_{ij}^{(p)})^{1/2} \tag{14}$$

and the octahedral shear stress as

$$\tau_0 = (\tfrac{1}{3}s_{ij}s_{ij})^{1/2} \tag{15}$$

Eq. (13) can be rewritten as follows:

$$\dot{\lambda} = \frac{\dot{\varepsilon}_0^{(p)}}{\tau_0} \tag{16}$$

thus defining the positive scalar function $\dot{\lambda}$ in Eq. (9), which becomes:

$$\dot{\varepsilon}_{ij}^{(p)} = \left(\frac{\dot{\varepsilon}_0^{(p)}}{\tau_0}\right) s_{ij} \tag{17}$$

It remains to establish a relationship between $\dot{\varepsilon}_0^{(p)}$ and τ_0. This can be done by applying Eq. (17) to a problem for which both $\dot{\varepsilon}_0^{(p)}$ and τ_0 are known. For example, the results of a simple uniaxial test can be employed, for which we can write:

$$\sigma_{22} = \sigma_{33} = \sigma_{12} = \sigma_{13} = \sigma_{23} = 0 \qquad (\sigma_{11} \neq 0) \quad (18)$$

and

$$\dot{\varepsilon}_{22}^{(p)} = \dot{\varepsilon}_{33}^{(p)} = -\tfrac{1}{2}\dot{\varepsilon}_{11}^{(p)}$$
$$\dot{\varepsilon}_{11}^{(p)} \neq 0 \quad (19)$$
$$\dot{\varepsilon}_{12}^{(p)} = \dot{\varepsilon}_{13}^{(p)} = \dot{\varepsilon}_{23}^{(p)} = 0$$

Substituting Eqs. (19) into Eq. (14) and solving for $\dot{\varepsilon}_{11}^{(p)}$ gives:

$$\dot{\varepsilon}_{11}^{(p)} = 2\dot{\varepsilon}_{0}^{(p)} \quad (20)$$

Since σ_{11} and $\dot{\varepsilon}_{11}^{(p)}$ are uniform throughout the test specimen, they are functions of t only, and Eq. (20) can be rewritten as follows:

$$\dot{\varepsilon}_{11}^{(p)} = \frac{2d\varepsilon_0^{(p)}}{dt} = \frac{2d\varepsilon_0^{(p)}}{d\tau_0 d\tau_0/dt} = \frac{2/\dot{\tau}_0}{2M_T} \quad (21)$$

where, by definition,

$$2M_T = \frac{d\tau_0}{d\varepsilon_0^{(p)}}$$

is the tangent modulus of the octahedral shear stress–octahedral plastic shear strain curve for the elastoplastic material, and

$$\dot{\tau}_0 = \frac{s_{ij}\dot{s}_{ij}}{3\tau_0} \quad (22)$$

is obtained by differentiating Eq. (15). Substituting the rate of plastic strain values defined by Eqs. (19) and (21) into Eq. (14), we obtain:

$$\dot{\varepsilon}_{0}^{(p)} = \frac{\dot{\tau}_0}{2M_T} \quad (23)$$

Substituting Eq. (23) into Eq. (17) gives:

$$\dot{\varepsilon}_{0}^{(p)} = \frac{\dot{\tau}_0 s_{ij}}{2_{\tau_0} M_T} \quad (24)$$

This is the form of the Prandtl–Reuss flow rule desired.

Substituting Eqs. (12) and (24) into Eq. (11), the constitutive relations can be expressed as follows:

$$\dot{\varepsilon}_{ij} = \frac{1-2\nu}{3E}\dot{\sigma}_{kk}\delta_{ij} + \frac{1+\nu}{E}\dot{s}_{ij} + \frac{\dot{\tau}_0 s_{ij}}{2\tau_0 M_T} \quad (25)$$

Substituting Eqs. (10) and (22) into Eq. (25) and rearranging terms gives:

$$\dot{\varepsilon}_{ij} = \frac{1+\nu}{E}\dot{\sigma}_{ij} - \frac{\nu}{E}\dot{\sigma}_{kk}\delta_{ij} + \frac{s_{ij}s_{kl}\dot{\sigma}_{kl}}{6\tau_0 2M_T} \quad (26)$$

In obtaining the last term, it has been recognized that no plastic work is done by the hydrostatic component of the applied stress, that is:

$$s_{kl}\dot{s}_{kl} = s_{kl}(\dot{\sigma}_{kl} - \tfrac{1}{3}\dot{\sigma}_{mm}\delta_{kl}) = s_{kl}\dot{\sigma}_{kl}$$

since

$$s_{kl}\dot{\sigma}_{mm} = 0$$

Equation (26) is the general constitutive relation governing elastoplastic material behavior, relating the total strain and stress rates at any instant in terms of the octahedral shear stress τ_0.

Often this general constitutive relation is alternately expressed in terms of a quantity $\bar{\sigma}$ (which is variously known as the "effective stress" or "equivalent stress") rather than in terms of the octahedral shear stress τ_0 defined in Eq. (15). This effective stress is defined as follows:

$$\bar{\sigma} = (\tfrac{3}{2}s_{ij}s_{ij})^{1/2} \quad (27)$$

and thus differs from τ_0 only in the magnitude of the constant coefficient.

Various other expressions of an elastoplastic constitutive relation are available, which incorporate a yield criterion other than that of von Mises (e.g., that of Tresca) and/or a flow rule other than that of Prandtl–Reuss. Few of them are actually being used at present, however, primarily because of the complexities they introduce into the general analysis.

Perfectly Plastic Material Behavior

The general theory of plasticity just discussed implies the consideration of a material stress–strain response of a general nature. That is, beyond the elastic limit, both an elastic and a plastic component of strain increment are present, as indicated in Eq. (11). Since by definition the elastic component of the strain increment is associated with a proportional increment of stress (Eq. 12), the resulting total stress–strain behavior (as represented, e.g., by Eq. 26) will exhibit a variation of stress with increasing strain beyond the elastic limit (as indicated in Fig. 4). This is frequently referred to as strain-hardening behavior.

Because of the complexity associated with the introduction of a general flow rule, such as represented by Eq. (9), a simpler idealization was introduced early in the development of the plasticity theory, namely, perfectly plastic material behavior. The simplification is that beyond the elastic limit the elastic component of the strain is assumed to be zero. Thus, the stress–strain curve as represented by a simple uniaxial loading is bilinear, with the second segment of the curve being parallel to the strain axis, as indicated in Figure 5. This eliminates the need for a flow rule: at strain levels below the elastic limit or yield strain ε_{yp}, the stress is directly proportional to the strain; at strain levels greater than ε_{yp}, the stress remains constant.

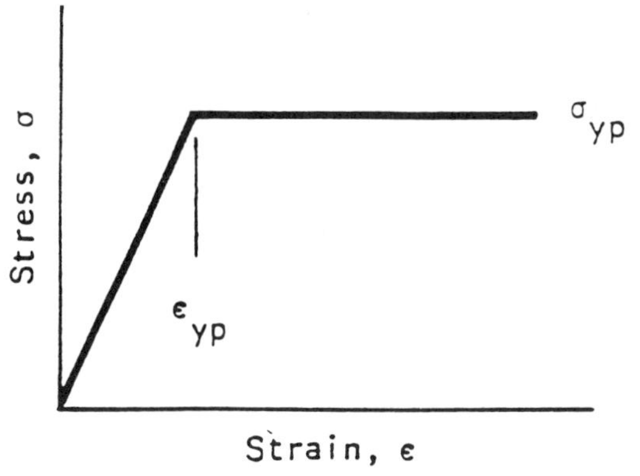

FIGURE 5 Elastic–perfectly plastic material behavior.

The stress–strain behavior of most engineering materials, including those of interest for use in composites (see, e.g., Fig. 4), is not accurately represented by the simple bilinear stress–strain relation of Figure 5. This is particularly true when the material is uniformly stressed, so that all regions reach the yield value simultaneously. The idealization would predict unrestricted plastic flow (i.e., unlimited strain with no increase in stress). However, when a loading creates stress gradients within the material, the most highly stressed regions will yield first. However, this plastic flow is restricted, since the surrounding material has not yet yielded. This is termed "contained plastic flow" and is typical of composites, since high stress gradients usually exist as a result of the difference in stiffness properties of the constituents. Thus, if the actual stress–strain curves of the constituents (as represented by Fig. 4) are crudely approximated by a bilinear curve of the form of Figure 5, an elastic–perfectly plastic analysis can be performed.

Perfectly plastic material behavior was introduced here because such an assumption frequently has been used in analyses of composite material behavior, particularly in conjunction with the simple model analyses.

Methods of Implementing Elastoplasticity Theory

Elastoplastic constitutive relations incorporating the Prandtl–Reuss flow rule were presented above, the results being represented by Eq. (26). Since the deviatoric components of stress s_{ij} and the octahedral shear stress τ_0 appear as nonlinear terms that change in value as the stress and strain rates $\dot{\sigma}_{ij}$ and $\dot{\varepsilon}_{ij}$ vary, Eq. (26) is nonlinear. To avoid the mathematical complexity of solving a system of nonlinear equations, the loading can be applied to the composite material in small increments. The nonlinear terms involving s_{ij} and τ_0, as well as the linear term M_{T}, can then be evaluated at the beginning of each small load increment and assumed to remain constant during the load increment. Thus, by utilizing an incremental loading procedure, the nonlinear problem is reduced to a series of linear problems. While this does represent an approximation to the basic mathematical laws, its extent can be governed by controlling the size of the load increments permitted.

Two basic methods have evolved for incorporating this incremental loading procedure. They are generally referred to as the "method of initial strains" and the "tangent modulus method."

Method of Initial Strains

The method of initial strains (e.g., see Ref. 25) is the older and perhaps more commonly used of the two methods available for analyzing incremental loading problems. The elastic and plastic components of the total strain increment as given by Eq. (26) are kept separate when constructing the set of matrix equations, so that the stiffness matrix includes only the elastic components of the strain increment; that is, it contains only the elastic material constants E and ν. Since these do not change from one load increment to the next, this stiffness matrix need be inverted only once. The increments of plastic strain, represented by the last term of Eq. (26), are treated as unknown constants.

An iterative solution technique is utilized within each load increment. The plastic strain increment at each material point is first estimated and treated as an initial condition. A linearly elastic analysis is performed, subject to this initial condition, to obtain the elastic strain increment field. The total strain increments (the sum of the initial plastic strain increments and the computed elastic strain increments) are then substituted into the inverse of Eq. (26) to obtain the total stress increments $\dot{\sigma}_{ij}$. A revised estimate of the plastic strain increment field is then obtained by using those values of $\dot{\sigma}_{ij}$ to evaluate the plastic strain increment components (i.e., the last term of Eq. 26). These new values of $\dot{\varepsilon}_{ij}^{(p)}$ then become the initial conditions for the next iteration cycle. This iterative procedure is repeated until some prescribed degree of convergence is attained.

There are actually two options available for determining the revised estimate of $\dot{\varepsilon}_{ij}^{(p)}$ within each iteration, as discussed by Lansing et al. [26]. The computationally quicker "constant stress method" has been found to be subject to "sudden and catastrophic divergence," the onset of which is associated with the size of the load increment assumed. The "constant strain method," although converging much less rapidly, is much more stable and is therefore generally used.

An early use of the method of initial strains was that by Padlog et al. [27]. However, one of the first applications specifically to composite materials was made by Foye and Baker [28]. Foye actually outlined the method of application in unpublished reports as early as 1966.

Tangent Modulus Method

The tangent modulus method (e.g., see Ref. 29) permits the use of the basic procedures developed earlier for linearly elastic analyses. The elastoplastic constitutive relation represented by Eq. (26) is utilized in place of the generalized Hooke's law (Eq. 12), the complete loading history being obtained as the sum of individual linear (but not elastic) solutions. The quantities s_{ij}, τ_0, and M_T in Eq. (26) are reevaluated at the beginning of each load increment and treated as constants, resulting in a linear relation between the variables $\dot{\varepsilon}_{ij}$ and $\dot{\sigma}_{ij}$, the total strain and stress components at each point within the material, respectively. Thus, a direct solution can be obtained for each load increment, with no iteration being required.

As described previously, numerical rather than closed-form analytical solutions are necessary for all the rigorous analyses to be discussed here. This usually requires that a set of linear algebraic equations be solved to obtain values of $\dot{\sigma}_{ij}$ or $\dot{\varepsilon}_{ij}$ at each material point. In terms of matrix algebra, a stiffness matrix must be constructed and inverted. The tangent modulus method requires that this stiffness matrix be constructed and inverted for each applied load increment, since new values of s_{ij}, τ_0, and M_T must be introduced each time.

Comparisons and Relative Merits

Marcal [30], in making a comparative study, presented a derivation of the equations governing the method of initial strains directly from those of the tangent modulus method. He demonstrated that the method of initial strains does not converge and therefore is not applicable to the special case of an elastic–perfectly plastic material. That convergence becomes very slow as elastic–perfectly plastic material behavior is approached (i.e., for materials exhibiting a very small change in stress with increasing strain for strains beyond the elastic limit) was confirmed by Foye and Baker [28]. On the other hand, Adams [31,32] found that the tangent modulus method provides good results in this case.

Marcal [30] indicated that for one specific numerical example he considered, the computing time was somewhat less when using the tangent modulus method. However, such a comparison is undoubtedly a function of the problem geometry and the assumed material stress–strain curve. A valid comparison for composite material micromechanical analysis applications would match the tangent modulus method as utilized by Adams [31,32] against the method of initial strains as utilized by Foye and Baker [28]. Published reports, however, give insufficient information concerning accuracy achieved and computing time requirements to make such a comparison possible.

Formulation of the Boundary Value Problem

Having selected a method of implementing the general theory governing elastoplastic material behavior, whether it be the tangent modulus method or the method of initial strains, it remains to formulate a boundary value problem that is representative of the physical geometry to be modeled by the analysis. A representative local material region must first be isolated from the bulk composite. Advantage is usually taken of some assumed filament packing symmetry to establish displacement and shear stress boundary conditions. That such conditions necessarily exist can be argued from simple physical considerations. The assumed symmetry is usually with respect to the filament cross-sectional shape and the packing geometry. For example, it is often assumed that the filament shapes are symmetric about orthogonal axes, which are also both axes of filament packing symmetry and principal load axes. This is not as restrictive as it may first appear. For example, the rectangular and diamond packing arrays shown in Figures 6*b* and 6*c* both satisfy these conditions, symmetry being maintained about both the x and y axes. The filaments need not be circular in cross section; any shape that is symmetrical about both the x and y axes is admissible. A complete discussion of boundary conditions associated with Figure 6 can be found in the works of Adams et al. [33] and Foye [34].

Ideally, a full three-dimensional analysis of the boundary value problem would be preferred. This would permit the consideration of stress and strain gradients in the direction of filament reinforcement (the z direction indicated in Fig. 6*a*) as well as in the transverse x–y plane. For example, this would permit the analysis of the behavior of discontinuous reinforcements such as whiskers and chopped fibers (e.g., the transfer of stresses at the fiber ends and into adjacent fibers) and particulate-reinforced composites. It would also facilitate the modeling of voids, local debonding, and similar manufacturing defects. The analytical formulation of a truly three-dimensional model is straightforward. However, the complexity of the boundary value problem makes a numerical solution technique mandatory. Large amounts of digital computer storage are required, and the computer running times are relatively long. Most versions of the present generation of digital computers do not have sufficient storage capacity and computational speed to make the consideration of practical three-dimensional analyses feasible. However, just a few years ago the same disadvantages were associated with two-dimensional analyses—it was difficult to overcome storage limitations, and computer running times required were prohibitive in many cases.

In the future, solutions of full three-dimensional problems will undoubtedly be routinely performed. At present, however, it is often necessary to make certain idealizations to render a problem amenable to practical solution. One approach is to reduce the problem to a two-dimensional formulation. For example, for continuous filament reinforced composites, it is reasonable to assume that stress and strain gradients in the axial direction (the z direction in Fig. 6*a*) are negligible relative to those in the transverse plane. This is the classical plane stress or plane strain condition. Because displacement-type boundary conditions are typically encountered in composite material micromechanics problems, it is often convenient to formulate the numerical analysis in terms of displacement components.

When the analyst has defined the material region to be analyzed, the loading conditions, the boundary conditions, and the governing constitutive relations, there

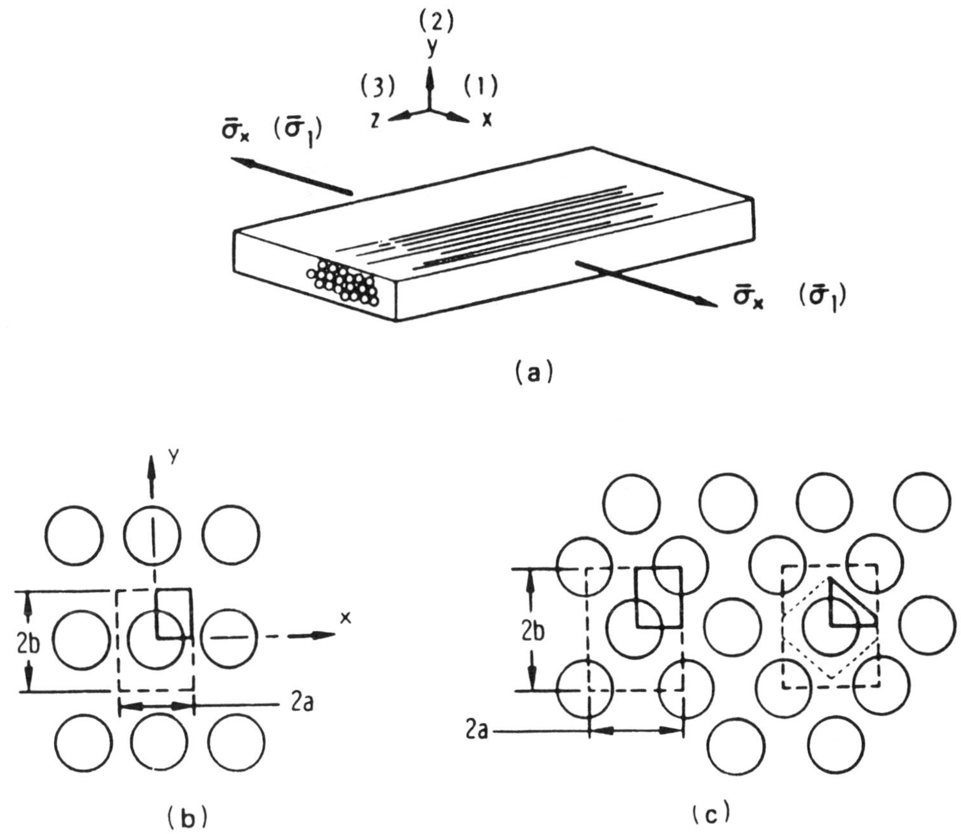

FIGURE 6 Filament packing geometries: (*a*) transverse normal loading of a unidirectionally reinforced composite, (*b*) rectangular array, and (*c*) diamond array.

remains the task of selecting a specific solution technique.

Solution Techniques

As discussed previously, rigorous analyses typically require a numerical solution technique of some type. Early works, not necessarily related directly to composite materials analyses, utilized finite difference methods extensively, while current emphasis is clearly toward the use of finite element techniques. In addition, there have been occasional attempts to adopt stress function approaches directly.

Finite Difference Methods

Finite difference methods were first applied to plane elastoplastic (more specifically, elastic–perfectly plastic) problems by Allen and Southwell [35], at which time hand computations using relaxation methods were being used to obtain numerical results. Relaxation methods were described by Southwell [36] and in more detail later by Allen [37]. Spencer [38] extended the work of Allen and Southwell [35] to the treatment of multiply connected plane bodies, a key step toward subsequent applications to composite material problems. Spencer's analysis was also restricted to the consideration of elastic–perfectly plastic material behavior, no unloading, and a statically determined problem (this latter condition being associated with the assumption of elastic–perfectly plastic material behavior).

Havner [39] later presented an alternate approach that is applicable to general elastoplastic material behavior. He obtained a finite difference discretization through a minimization of the total potential energy function, which leads to a positive definite symmetric matrix. He did not present numerical results, however, and he did not consider the two-phase material problem, which is representative of a composite system.

In fact, no applications of the finite difference methods specifically to elastoplastic analyses of composite material problems have been published. Adams et al. [33], who utilized the finite difference method in developing elastic analyses of composite material behavior, subsequently switched to a finite element technique when considering elastoplastic material behavior [31,32].

The principal reason for the lack of applications of finite difference methods to elastoplastic material behavior of composites is not associated with any unique mechanical properties of the constituents. Rather, it is a geometry problem, the same difficulty that is encountered in any application of the finite difference method to problems involving curved boundaries, namely, constraining the finite difference node points to lie on the curved boundaries. Even the use of irregular grid spacings does not completely eliminate this problem, as discussed by Adams et al. [33]. The use of triangular finite element grids completely avoids this major problem, however, making the finite element method a much more versatile approach.

Finite Element Techniques

A significant number of applications of the finite element method to elastoplastic material behavior have been published during the past few years. In fact, much of the work relating to the development of practical methods of implementing plasticity theory, in particular the method of initial strains and the tangent modulus method discussed previously, has been conducted using finite element analyses. Several of these applications are directly associated with composite materials investigations, while others could be adapted readily to this purpose.

Finite element analyses had their beginning in the development by Turner et al. [40] of the "method of direct stiffness calculation." This in turn was an extension of the computation of influence coefficients by Levy [41]. In the direct stiffness method, the displacements of specific points within the body are considered to be the unknowns. An infinite number of systems of mutually compatible deformations are possible; the correct displacement field is that for which the governing equations of equilibrium are satisfied. Gallagher et al. [42] first applied the direct stiffness method to elastoplastic material behavior, using the method of initial strains. For a chronology of subsequent applications of the direct stiffness method, using both the method of initial strains and the tangent modulus method, see, for example, the work of Marcal [43]. A number of these studies were of stress distributions around holes, notches, and similar discontinuities in flat plates subject to in-plane loadings. By assuming the discontinuity to be an inclusion (e.g., a filament), rather than a cutoff of some type, a direct extension to composite materials is achieved. The finite element method, in its current state of development, is described in detail in many texts.

The first rigorous application to elastoplastic composite material behavior of the finite element method using the method of initial strains was by Foye [34]; subsequently Foye and Baker [28] provided greater detail. The first rigorous application using the tangent modulus method was by Adams [31,32], who later offered greater detail [44].

METHOD OF INITIAL STRAINS. The basic procedure outlined above for implementing the method of initial strains is directly applicable to the finite element technique. Details of the computational procedure and an actual computer program are given by Foye and Baker [28]. Constant strain triangular elements were used, the analysis being a direct extension of a prior elastic analysis [34]. Generalized plane strain was assumed, the analysis being readily modifiable to handle the assumption of plane stress also, if desired. The generalized plane strain condition permits combined axial and transverse normal loadings. In addition, longitudinal shear loading is included. This is readily achieved, since the generalized plane strain solution involves only the in-plane (x,y) components of the node point displacements, while the shear loading involves only axial (z) components. Thus, a complete combined loading problem can be analyzed.

TANGENT MODULUS METHOD. The basic procedure outlined above for implementing the tangent modulus method is directly applicable to the finite element technique. For a detailed presentation of the general method of analysis and a description of the required computational procedure, see Ref. 44.

Typical Solution Method

A two-dimensional finite element analysis and related computer program (WYO2D) in current use at the University of Wyoming has evolved over a number of years [1,17–20,31,32,44–53]. It is a generalized plane strain formulation, utilizing constant strain triangular elements. Its special features include the ability to model fiber anistropy, matrix stress–strain nonlinearity, temperature- and moisture-dependent constituent material properties, and incremental temperature, moisture, and mechanical loadings. These loadings can be applied in any combination. The ability to model longitudinal shear loadings in combination with normal stress loadings was originally developed by Crane and Adams [50–52]. Local failures in the form of microcracks are predicted, based on a failure criterion selected from a menu by the user. These microfailures are then propagated in subsequent loading increments, using a failed element scheme originally published in Ref. 54.

Propagation of a crack completely across the model defines failure of the composite and its ultimate strength. Since the increments of composite strain corresponding to each increment of applied loading are obtained as part of the incremental solution technique, the complete composite stress–strain curve to failure is obtained. In addition, properties of the unidirectional composite such as Poisson's ratios and axial and transverse coefficients of thermal expansion and moisture expansion are available. That is, the properties of the unidirectional composite can be fully predicted.

By using a finite element mesh that is very fine at the fiber–matrix interface, interface degradation of the presence of an interphase zone can be modeled. The use of simple constant strain triangular elements permits fine meshes while keeping computer running time reasonable. The computer program also incorporates reduced integration and a frontal solution technique to enhance computational efficiency [55].

Typical Micromechanics Procedures

In practical analyses, a square packing array of circular fibers is often assumed because this simple packing geometry is known to give satisfactory results [47]. By making this assumption, symmetry conditions [45] permit the analysis of only one quadrant of the circular fiber cross section and surrounding matrix, as shown in Figure 7. The mesh refinement at the fiber–matrix interface is evident.

Constituent Material Properties

Constituent material properties can sometimes be obtained from the published literature (e.g., see Refs. 56–61). Transverse fiber properties are very difficult to measure directly; therefore, it is usually necessary to estimate these values based on experience.

For use in the micromechanics analysis, it is necessary to express these data in a form that can be input to the associated computer program. The experimental data could be input point by point (i.e., by expressing each stress–strain curve as a series of point values). This would be tedious. Also, since matrix tests are performed at discrete temperature and moisture conditions and the analysis is capable of predicting composite response for any combination of conditions, it is necessary to interpolate (and sometimes even to extrapolate) the available experimental data to other hygrothermal conditions. Thus, it is convenient to fit the temperature- and moisture-dependent data to an equation. This permits easy interpolation, and at the same time provides a means, by linear regression, of averaging the typically five individual stress–strain curves at each test condition together. This data management can be performed, for example, by using a three-parameter experimental equation of the form first suggested by Richard and Blacklock [62].

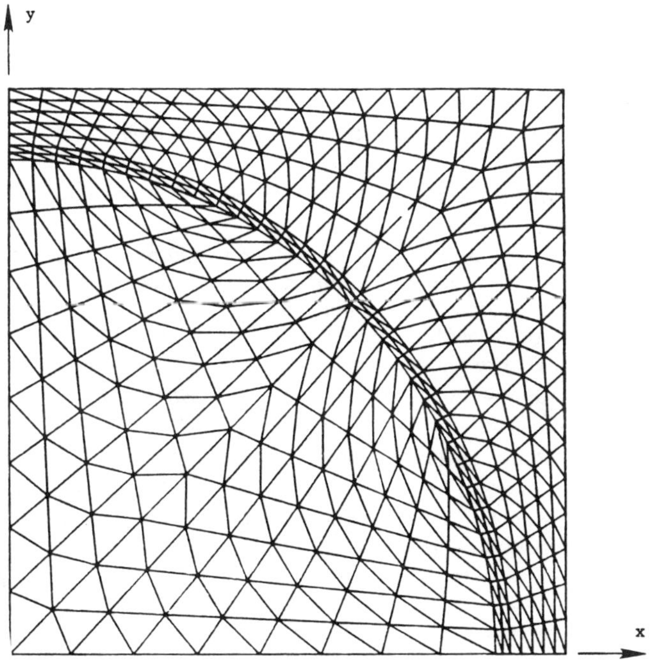

FIGURE 7 Typical finite element model of one quadrant of the repeating unit cell of a unidirectional composite material in a square fiber packing array.

Discussion

The WYO2D finite element analysis is capable of modeling any loading of a unidirectional composite. However, such finite element micromechanics analyses and related computer programs have been developed relatively recently and therefore are not yet being used extensively. However, results available, and their correlations with actual experimental data, demonstrate the great potential of such tools. Not only can these analyses predict actual unidirectional composite response, they can be used to estimate the influence of potential variables. In fact, it is this capability that is of particular significance. If the actual testing has already been done, an analysis to predict these measured properties is superfluous. However, testing is both extremely time-consuming and expensive; moreover, if it does not lead to useful results, it is wasteful of resources. The micromechanics analysis can be used to screen candidate material combinations under various loading and environmental conditions, including those that would be difficult to achieve in the laboratory. Only the most promising combinations need be actually fabricated and tested.

In summary, the basic methodology required to perform meaningful micromechanical analyses of composite material behavior is now available. Simple model analyses have already been developed to about the fullest extent practicable. They cannot be expected to provide the accurate, detailed information required in future composite materials developments. On the other hand, the rigorous micromechanical analyses, which can provide this information, have not been adequately exercised, although the basic methodology is now in hand and the required computational schemes and computer capacity are available.

There are several reasons for this lack of widespread use of the rigorous methods up to the present time. For example, the rigorous methods have not been available for very long. They require the use of large-scale, complex computer programs, and acquiring familiarity with the software requires some effort even when there is thorough documentation, which is not always the case. Most of these computer programs require a relatively large computer facility and significant amounts of computer time. Thus, they can be expensive to utilize extensively. The significant effort required to develop such a computer program has limited the number of investigators active in this area of research. On the other hand, the most difficult part of the basic development work has now been completed, and it has been demonstrated that these rigorous analyses can be successfully utilized.

Donald F. Adams

References

1. A. K. Miller and D. F. Adams, "Micromechanical Aspects of the Environmental Behavior of Composite Materials," Report UWME-DR-701-111-1, Department of Mechanical Engineering, University of Wyoming, January 1977.
2. C. W. Bert, "Plasticity and Creep Analysis of Filmentary

Metal-Matrix Composites," Sandia Corporation Report SC-DR-720055, 1972.

3. B. W. Shaffer, *AIAA J., 6,* 2316 (1968).
4. B. W. Shaffer, *AIAA J., 2,* 348 (1964).
5. N. F. Dow and B. W. Rosen, "Evaluation of Filament-Reinforced Composites for Aerospace Structural Applications," NASA Contractor Report CR-207, 1965.
6. L. S. Shu and B. W. Rosen, *J. Compos. Mater., 1,* 366 (1967).
7. L. J. Ebert and J. D. Gadd, in *Fiber Composite Materials,* American Society for Metals, Metals Park, OH, 1965, Ch. 5.
8. L. J. Ebert, S. S. Hecker, and C. H. Hamilton, *J. Compos. Mater., 2,* 458 (1968).
9. R. Hill, *J. Mech. Phys. Solids, 12,* 199 (1964).
10. J. C. Halpin and S. W. Tsai, "Effects of Environmental Factors on Composite Materials," AFML-TR-67-423, June 1979.
11. R. M. Jones, *Mechanics of Composite Materials,* Hemisphere Publishing, Philadelphia, 1975.
12. B. D. Agarwal and L. J. Broutman, *Analysis and Performance of Fiber Composites,* Wiley, New York, 1980.
13. A. Mendelson, "Elastoplastic Analysis of a Circular Cylindrical Inclusion in a Uniformly Stressed Infinite Homogeneous Matrix," NASA TN D-4350, 1968.
14. G. P. Sendeckyj, "Integrated Research on Carbon Composite Materials," AFML-TR-66-310, Part III, 1969, p. 259.
15. D. R. J. Owen, J. Holbeche, and O. C. Zienkiewicz, *Fibre Sci. Technol., 1,* 185 (1969).
16. W. Huang, *J. Compos. Mater., 5,* 320 (1971).
17. B. G. Schaffer and D. F. Adams, "Nonlinear Viscoelastic Behavior of a Composite Material Using a Finite Element Micromechanical Analysis," Report UWME-DR-001-101-1, Department of Mechanical Engineering, University of Wyoming, June 1980.
18. B. G. Schaffer and D. F. Adams, *J. Appl. Mech., 48*(4), 859 (1981).
19. R. Mohan and D. F. Adams, "Nonlinear Viscoelastic Characterization of Unidirectional Composites in Hygrothermal Environments," Report UWME-DR-301-107-1, Department of Mechanical Engineering, University of Wyoming, December 1983.
20. R. Mohan and D. F. Adams, *Exp. Mech., 25*(3), 262 (1985).
21. R. Hill, *The Mathematical Theory of Plasticity,* Oxford University Press, London and New York, 1950.
22. W. Prager and P. G. Hodge, Jr., *Theory of Perfectly Plastic Solids,* Wiley, New York, 1951.
23. A. Nadai, *J. Appl. Phys., 8,* 205 (1937).
24. J. L. Swedlow and W. H. Yang, "Stiffness Analysis of Elasto-Plastic Plates," California Institute of Technology Report GALCIT SM 65-10, 1966.
25. A. Mendelson and S. S. Manson, "Practical Solution of Plastic Deformation Problems in the Elastic-Plastic Range," NASA TR R28, 1959.
26. W. Lansing, W. R. Jenson, and W. Falby, in J. S. Przemieniecki et al., Eds., *Matrix Methods in Structural Mechanics,* AFFDL-TR066-80, 605, 1966.
27. J. Padlog, R. D. Huff, and F. G. Holloway, "Unelastic Behavior of Structures Subjected to Cyclic, Thermal and Mechanical Stressing Conditions," WADD TR-60-271, 1960.
28. R. L. Foye and D. J. Baker, "Design/Analysis Methods for Advanced Composite Structures," AFML-TR-70-299: Vol I, *Analysis;* Vol. II, *Computer Programs,* 1971.
29. P. V. Marcal and C. E. Turner, *J. Mech. Eng. Sci., 5,* 3 (1963).
30. P. V. Marcal, *AIAA J., 6,* 157 (1968).
31. D. F. Adams, *J. Compos. Mater., 4,* 310 (1970).
32. D. F. Adams, "Inelastic Analysis of a Unidirectional Composite Subjected to Transverse Normal Loading," Rand Corporation Report RM-6245-PR, 1970.
33. D. F. Adams, D. R. Doner, and R. L. Thomas, "Mechanical Behavior of Fiber-Reinforced Composite Materials," Report AFML-TR-67-96, 1967.
34. R. L. Foye, "Advanced Design Concepts for Advanced Composite Airframes," Report AFML-TR-68-91, *1,* 1968.
35. D. N. de G. Allen and R. V. Southwell, *Phil. Trans. R. Soc. London, A242,* 379 (1950).
36. R. V. Southwell, *Relaxation Methods in Engineering Science,* Oxford University Press, London and New York, 1940.
37. D. N. de G. Allen, *Relaxation Methods,* Oxford University Press, London and New York, 1954.
38. A. J. M. Spencer, *Appl. Sci. Res., A12,* 391 (1962).
39. K. S. Havner, *Q. Appl. Math., 23,* 323 (1966).
40. M. J. Turner, R. W. Clough, H. C. Martin, and L. J. Topp, *J. Aerosp. Sci., 23,* 805 (1956).
41. S. Levy, *J. Aerosp. Sci., 20,* 449 (1953).
42. R. H. Gallagher, J. Padlog, and P. P. Gijlaard, *Am. Rocket Soc. J., 32,* 700 (1962).
43. P. V. Marcal, "Finite Element Analysis with Material Nonlinearities—Theory and Practice," Brown University, 1969.
44. D. F. Adams, "An Elastoplastic Analysis of Crack Propagation in a Unidirectional Composite," Rand Corporation Report R-1070-PR, 1973.
45. D. F. Adams and D. R. Doner, *J. Compos. Mater., 1*(2), 152 (1967).
46. D. F. Adams and D. R. Doner, *J. Compos. Mater., 1*(1), 152 (1967).
47. D. F. Adams and S. W. Tsai, *J. Compos. Mater., 3*(3), 4 (1969).
48. D. F. Adams and A. K. Miller, *J. Compos. Mater., 11*(3), 285 (1977).
49. A. K. Miller and D. F. Adams, *Fibre Sci. Technol., 13*(2), 135 (1980).
50. D. A. Crane and D. F. Adams, "Finite Element Micromechanical Analysis of a Unidirectional Composite Including Longitudinal Shear Loading," Report UWME-DR-101-101-1, Department of Mechanical Engineering, University of Wyoming, February 1981.
51. D. F. Adams and D. A. Crane, *Comput. Struct., 18*(6), 1153 (1984).
52. D. F. Adams and D. A. Crane, *Composites, 15*(3), 181 (1984).
53. D. F. Adams and D. P. Murphy, "Analysis of Crack Propagation as an Energy Absorption Mechanism in Metal Matrix Composites," Report UWME-DR-101-102-1, Department of Mechanical Engineering, University of Wyoming, February 1981.
54. D. F. Adams, *J. Compos. Mater., 8*(1), 38 (1974).
55. R. F. Cilensek, "WYO2D Finite Element Analysis Computer Program User Instructions and Documentation," Composite Materials Research Group, University of Wyoming, Laramie, 1984.
56. D. F. Adams, R. S. Zimmerman, and E. M. Odom, "Polymer

Matrix and Graphite Fiber Interface Study," Report UWME-DR-501-102-1, Department of Mechanical Engineering, University of Wyoming, June 1985.

57. "Magnamite Graphite Fiber Type AS4," Hercules Product Data Sheet 847-3, Hercules, Inc., Magna, UT, 1981.

58. D. S. Cairns and D. F. Adams, *J. Reinf. Plast. Compos.*, 2(4), 239 (1983).

59. R. S. Zimmerman and D. F. Adams, "Mechanical Properties Testing of Candidate Polymer Matrix Materials for Use in High Performance Composites," Report UWME-DR-401-104-1, Department of Mechanical Engineering, University of Wyoming, August 1984.

60. R. S. Zimmerman, D. F. Adams, and D. E. Walrath, "Investigation of the Relations Between Neat Resin and Advanced Composite Mechanical Properties," Report UWME-DR-301-101-1: Vol. I, *Results;* Vol. II, *Appendices*, Department of Mechanical Engineering, University of Wyoming, May 1983.

61. D. F. Adams, R. L. Ramkumar, and D. E. Walrath, "Analysis of Porous Laminates in the Presence of Ply Drop-Offs and Fastener Holes," Northrop Technical Report NOR 84-113, Naval Air Systems Command Contract N00019-82-C-0063, May 1984.

62. R. M. Richard and J. M. Blacklock, *AIAA J.*, 7(3), 432 (1969).

Composite Structures

See Marine Structures; Spacecraft Applications; Spacecraft Structural Development; Structural Aircraft, Noncommercial and Homebuilt

Concrete Composites, Fiber Reinforced

Definition

Fiber-reinforced concrete is concrete made of hydraulic cements with aggregates of various sizes and amounts and reinforced with discontinuous and discrete fibers. Fibers suitable for reinforcing concrete have been produced from steel, glass, and organic polymers. Naturally occurring asbestos fibers and vegetable fibers such as sisal and jute are also used for reinforcement. Fibers are available in various sizes and shapes. Generally, the length and diameter of the fibers used for concrete do not exceed, respectively, 50 and 0.5 mm.

Historical Perspective

Fibers have been used to reinforce brittle materials since ancient times; straw was used to reinforce sunbaked bricks, and horse hair was used to reinforce masonry mortar and plaster. More recently, large-scale commercial use of asbestos fibers in a cement paste matrix began with the invention of the Hatscheck process in 1900. Asbestos cement construction products are widely used throughout the world today. However, partly because of the health hazards associated with asbestos fibers, different types of fiber have been developed in the past 20–30 years.

The use of glass fibers in concrete was first attempted in the Soviet Union in the late 1950s [1]. It was quickly established that ordinary glass fibers, including the borosilicate E-glass fibers, are prone to attack by the alkaline environment of cement paste. Considerable development work was directed toward producing a form of alkali-resistant glass fibers containing zirconium by Majumdar and his colleagues at the Building Research Establishment in England [2]. This work has stimulated considerable commercial activity in glass-fiber reinforced concrete building products.

Experimental trials and patents involving the use of discontinuous steel reinforcing elements (e.g., nails, wire segments, and metal chips) to improve the properties of concrete date from 1910. In recent times, the first major attempt to evaluate the potential of steel fibers as a reinforcement for concrete was the investigation by Romualdi and Batson [3] in the United States in the early 1960s. Since then, a substantial amount of research, development, experimentation, and industrial application of steel fiber reinforced concrete have occurred.

Initial attempts at using synthetic fibers (nylon, polypropylene) were not as successful as those using glass or steel fibers [4,5]. However, better understanding of the concepts behind fiber reinforcement, new methods of fabrication, and new types of organic fiber have led researchers to conclude that both natural and synthetic fibers can successfully reinforce concrete [6,7].

Considerable research, development, and applications are still taking place all over the world. The enthusiasm and interest shown in the new fiber-reinforced concrete construction materials are evidenced from the numerous research papers and several international symposia and the state-of-the-art reports issued by professional societies. The American Concrete Institute's technical committee on fiber-reinforced concrete wrote the initial report in 1973 [8] and has recently revised the state-of-the-art report. RILEM's committee on fiber-reinforced cement composites has also published a report [9]. A recommended practice for glass-fiber reinforced panels has been published by the Prestressed Concrete Institute [10]. Two recent symposium proceedings provide a good summary of the current development [11,12].

Fiber-Reinforced Concrete and Reinforced Concrete

Plain, unreinforced concrete has a low tensile strength and a low tensile strain before fracture. These shortcomings are traditionally overcome by incorporating reinforcements. Reinforcing bars are continuous and are carefully placed in the structures to optimize their performance. Fibers are discontinuous and generally are randomly distributed throughout the concrete matrix. As a result, the reinforcing performance of steel fibers, for example, is inferior to that of steel reinforcing rods. In addition, fibers are likely to be considerably more expensive than conventional reinforced concrete. However, because of the flexibility in methods of fabrication, fiber-reinforced concrete can be an economic and useful construction material. For example, thin (10–20 mm) precast fiber-reinforced concrete panels appear to be economically viable in the United States and in Britain for curtain walls of buildings and other applications involving thin sheets. In this and other uses of fiber-reinforced concrete, fibers are not replacing conventional reinforced concrete; rather, they are making plain concrete more versatile, competitive, and useful.

Basic Concepts and Mechanical Properties

Glass Fiber Reinforced Polymers

The current interest in microreinforced composite materials stems largely from research by Griffith on the strength of glass fibers. In a paper published in 1920, he reported that the strength of glass fibers increased manyfold as their diameter decreased. Although these small-diameter glass fibers are very strong in tension, they are brittle and notch-sensitive, and they cannot be used by themselves in larger lengths. When glass fibers are embedded in a polymeric matrix such as polyester resin, it is possible to take advantage of the high tensile strength of glass and at the same time provide a notch-insensitive composite. The highly successful composites are sold as "Fiberglas." The volume fraction of glass in such composites is up to 40% of the total volume. The relative stress–strain curves in tension for glass fibers and an epoxy matrix is shown in Figure 1*a*. The incorporation of glass fibers increases both the stiffness and the strength of the composite. Since the failure strain of the matrix is higher than that of the glass fibers, it is possible to realize the full potential strength of fibers.

Strong Fibers and Brittle Matrices

The relative strain curves in tension for a strong fiber (steel, glass, asbestos, polypropylene, sisal, Kevlar, etc.) and a brittle matrix (portland cement, gypsum, etc.) are shown in Figure 1*b*. The strain at fracture for the brittle matrix is considerably smaller ($< \frac{1}{50}$) than that for the fiber. As a result, when a fiber-reinforced brittle matrix composite is loaded, the matrix will crack long before the fiber can be fractured. Once the matrix has cracked, the composite fails in one of the following ways.

1. The composite fractures immediately after the matrix cracking. This type of behavior is shown in Figure 2*a*.

2. Although the maximum load on the composite is essentially the same as that of the matrix alone, the composite continues to carry decreasing load after the peak, as shown in Figure 2*b*. The postpeak resistance is primarily provided by the pulling out of fibers from the cracked surfaces. Although no significant increase in the tensile strength of the composite is observed, a considerable increase in the toughness of the composite (area under the complete stress–strain curve) can be obtained.

3. Even after the cracking of the matrix, the composite continues to carry increasing tensile stresses; the peak stress and the peak strain of the composite are greater than those of the matrix alone; and during the inelastic range (between the first cracking and the peak) multiple cracking of the matrix, fiber debonding and slip may occur (Fig. 2*c*). It is clear that this mode of failure results in the optimum performance of both the matrix and the fibers. Various aspects of this type of behavior are discussed below (for composites subjected to uniaxial tension). Figure 3 shows some experimentally observed uniaxial tensile stress–strain responses of portland cement composites reinforced with steel, glass, and polypropane fibers [13–15]. The three stages of the behavior—linearly elastic, nonlinear prepeak response, and postpeak response—can be identified from these curves.

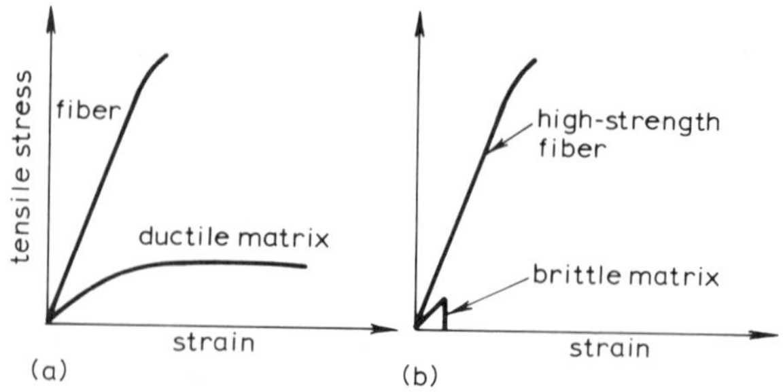

FIGURE 1 Tensile stress–strain curves for fibers with ductile and brittle matrices.

Elastic Range

Until the initial cracking of the matrix, it is reasonable to assume that both the fibers and the matrix behave elasti-

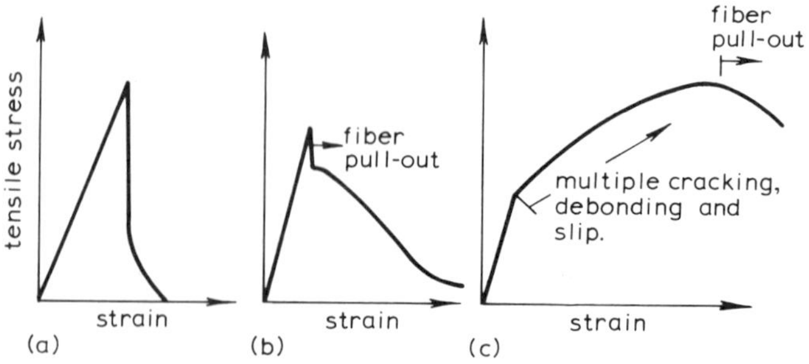

FIGURE 2 Three possible composite stress–strain curves for fiber-reinforced brittle matrices.

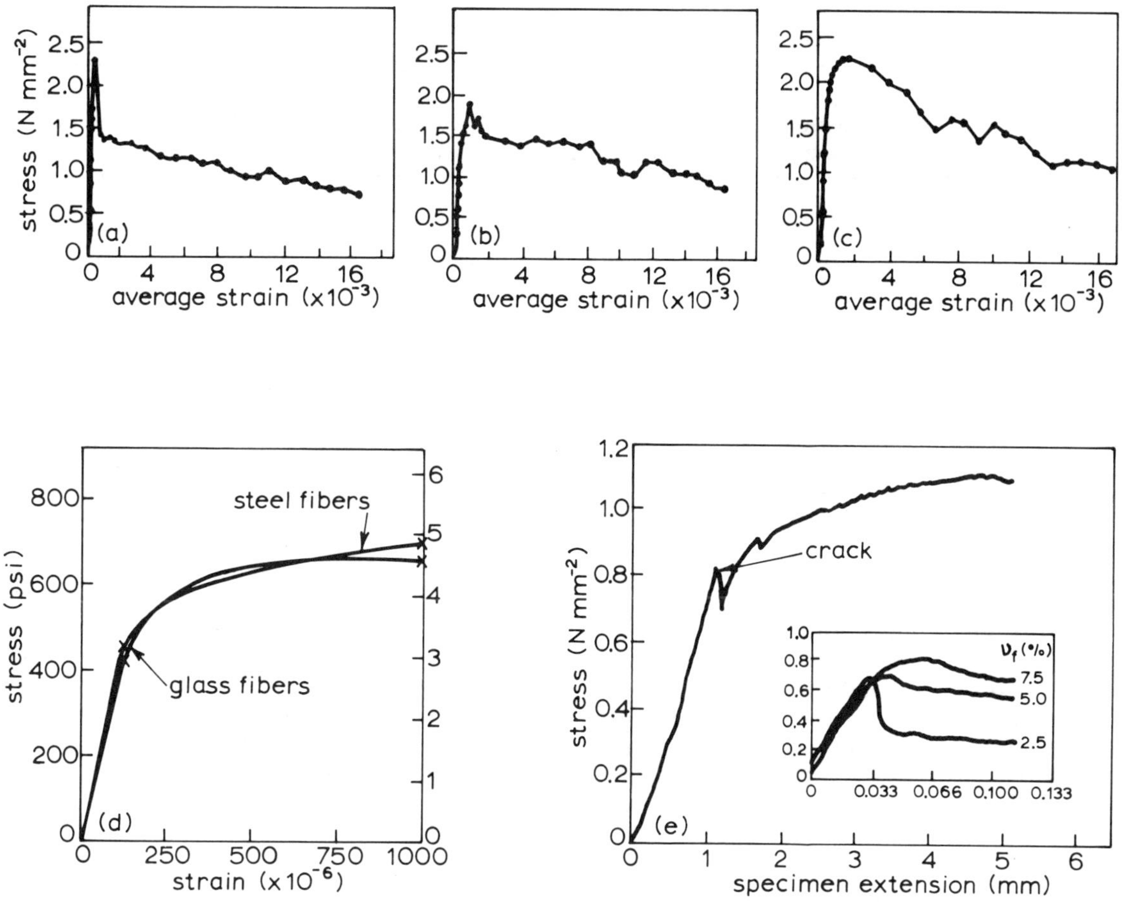

FIGURE 3 **Experimentally observed tensile behavior of fiber-reinforced concrete composites made using (*a–c*) steel fibers of different types (from Ref. 13), (*d*) smooth steel and glass fibers (from Ref. 14), (*e*) monofilament polypropylene fibers [inset shows response of fibrillated polypropylene fibers] (from Ref. 15). (*a*) Smooth, straight steel fibers with $V_f = 1.23\%$, $l =$ 25 mm, $2a = 0.38$ mm. (*b*) Hooked ends, $V_f = 0.90\%$, $l = 30$ mm, $2a = 0.40$ mm. (*c*) Enlarged ends, $V_f = 1.40\%$, $l = 50$ mm, $2a = 0.75$ mm. (*d*) Smooth, straight steel fibers, $V_f = 2\%$, $l = 25$ mm, $2a = 0.43$ mm, glass-fiber (bundles) $V_f = 2\%$, $l = 13$ mm; the first crack and the maximum loads are also indicated. (*e*) Polypropylene monofilaments, $V_f = 5\%$, $l = 26$ mm, $d = 130$ μm; inset: fibrillated fibers, $l = 6$ mm, average thickness 72 μm.**

cally and that there is no slippage between the fibers and the matrix. In that case, the stress in the composite just before the first cracking in the matrix σ_{cc} and the corresponding strain in the matrix ε_{mu} are related as follows:

$$\sigma_{cc} = E_m(1 - V_f)\varepsilon_{mu} + E_f^* V_f \varepsilon_{mu} \tag{1}$$

where E_m is the modulus of elasticity of the matrix, V_f the volume fraction of fibers, and E_f^* the equivalent modulus of elasticity of the fibers in the direction of tensile loading. Only if the fibers are sufficiently long and are aligned in the direction of loading does $E_f^* = E_f$, the modulus of elasticity of the fibers. Otherwise, E_f^* can be as little as one-third of the actual modulus. The properties of various fibers (including their elastic moduli) used for reinforcing concrete are shown in Table 1. Typical properties of fibers that are used in concrete construction are shown in Table 2. Combining the information given in Tables 1 and 2 and Eq. (1), it can be seen that, in general, the composite stress at first cracking will not be substantially different from the tensile strength of the matrix. The experimental evidence (Figs. 4 and 5) confirms that the presence of fibers does not significantly alter the properties of the unreinforced matrix until the first cracking [16]. Note that the determination of the stress at first cracking is dependent on the definition of first cracking and the methods of measurement (Fig. 4).

TABLE 1
Some Typical Fiber Properties

Fiber	Diameter (μm)	Density (× 10³ kg/m³)	Young's Modulus (GPa)	Tensile Strength (GPa)	Elongation at Break (%)
Asbestos					
Chrysotile	0.02–20	2.55	164	3.1	2–3
Crocidolite	0.1–20	3.37	196	3.5	2–3
Polypropylene	20–200	0.09	5–10	0.5	10–20
Nylon (high tenacity)	> 4	1.14	4	0.9	~ 15
Kevlar					
PRD 49	~ 10	1.45	133	2.9	2.6
PRD 29	12	1.44	69	2.9	4.0
Sisal	10–50	1.5	5–10	0.8	~ 3
Glass	9–15	~ 2.6	~ 80	2–4	2–3.5
Steel	5–500	7.8	200	1–3	3–4

Inelastic Range

The composite will carry increasing loads after the first cracking of the matrix only if the pullout resistance of the fibers at the first crack is greater than the load at first cracking. The pullout resistance of the fibers will depend on the average bond strength τ between the fibers and the matrix, the number of fibers crossing the crack, and the length and diameter of the fibers. Thus it is not surprising that the product of the quantity V_f and the aspect ratio l/d (where l is the length and d the diameter of the fibers) has been found to greatly influence the postcracking behavior of fiber-reinforced concrete (Fig. 5). If either the bond strength τ or the quantity $V_f l/d$ is too low, the behavior shown in Figure 2*b* is observed.

Postpeak Range

The prepeak inelastic response of fiber-reinforced concrete results from matrix cracking, fiber debonding, and fiber–matrix slip. This stage is terminated when the peak of the stress–strain curve (often termed the composite strength of the composite failure stress) is reached. The postpeak response is characterized primarily by the widening of a single crack. To better design fiber-reinforced composite, it is essential to understand prepeak and the postpeak nonlinear behavior. Some aspects of the postpeak nonlinear response can be understood by examining results of tension tests conducted by the author [17,18] on double-edged notched specimens of steel fiber (1 inch long, smooth fibers with an aspect ratio of 100) reinforced concrete (Fig. 6). Closed-loop, stable, direct tension tests on the unreinforced matrix and three different steel fiber reinforced concrete (SFRC) composites (V_f = 0.5, 1.0, and 1.5%) were conducted with a monotonically increasing notch mouth opening displacement at a rate corresponding to one microstrain per second. The following observations are pertinent to the subject of this article.

TABLE 2
Typical Properties of Fiber-Reinforced Concrete

Fiber Type	Length (mm)	Diameter (μm)	Volume (%)	Matrix Proportions by Weight	Flexural Strength (MPa)
Asbestos	5–10	0.02–20	8–16	Only portland cement, no coarse aggregates	15–25[a]
Glass	20–50	500–500[b]	4–6	Sand/cement ~ 0.5 Water/cement ~ 0.3	20–30[c]
Steel	12–25	~ 500	1–2	Sand/cement ~ 2.0 Water/cement ~ 0.45 admixtures	8–12[d]
Polypropylene	12–25[e]	20–200	2–4	Sand/cement ~ 1.0 Water/cement ~ 0.35 admixtures	6–15[f]

[a] May vary with the direction of testing.
[b] Dimensions of fiber bundle made up of about 200 filaments.
[c] Properties of about 10 mm thick precast sheets.
[d] Conventionally mixing procedure.
[e] Continuous sheets are also used.
[f] Depends on method of manufacturing.

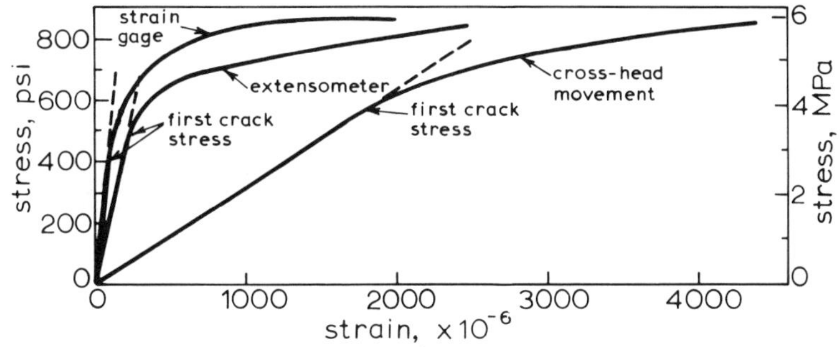

FIGURE 4 **Stress–strain curves in tension for a steel fiber reinforced specimen: $V_f = 3\%$, $l = 25.4$ mm (lin), $d = 0.25$ mm (0.01 in.).**

TENSILE STRESS–DISPLACEMENT RESPONSE. Plain mortar matrix specimens exhibit linear elastic behavior up to about 50% of their tensile strengths. The onset of inelastic behavior prior to the peak load suggests the initiation of the microcracking process, although no cracks are detected prior to the peak loads (optically, at a 100% magnification). Almost immediately after peak loads, deformations become localized (nonunique local strains) with the eventual widening of a single crack. The unreinforced matrix exhibits significant traction capacity at displacements several times that observed at the peak load (Fig. 6). More detailed information on the fracture characteristics of the unreinforced mixes can be found in Ref. 18. SFRC behavior is linearly elastic up to about 80% of the matrix tensile strength. Nonlinear deformations take place beyond this point. Composite peak stress and corresponding displacements (or strains) are larger than the corresponding values for the unreinforced matrix mix ($\leq$ 25–30% for the 1.5% SFRC mix). After composite peak stress, the load-carrying capacity abruptly drops to a stress level termed by some investigators the postcracking strength. One single crack becomes visible at the critical section. Upon further increases in displacement, the load-carrying capacity gradually drops (in a linear fashion) with increases in the displacement. A frictional type of pullout resistance (due to interfacial bond slip) is perhaps responsible for the linear σ–δ relation with a negative tangent modulus.

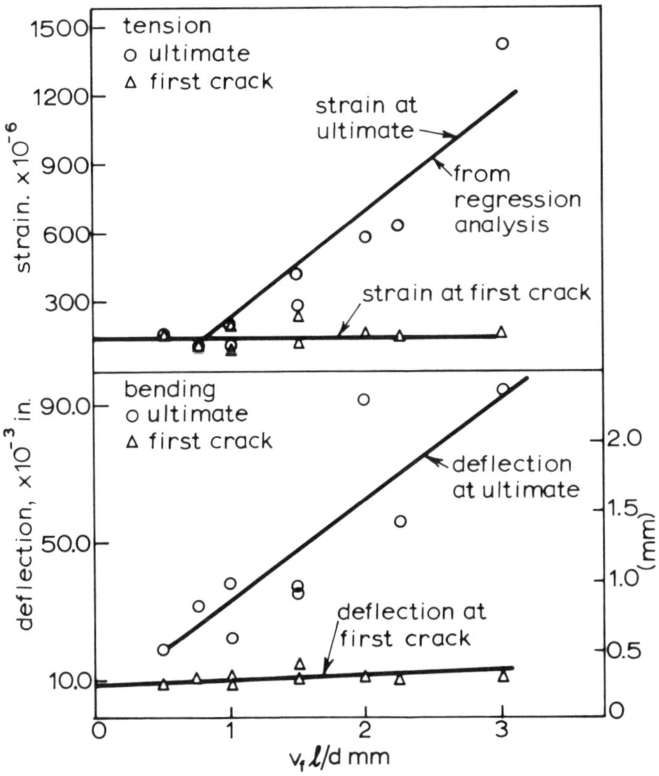

FIGURE 5 **Strains and deflections for steel fiber reinforced specimens.**

LOCALIZATION OF DEFORMATION. Based on tests on notched rectangular prismatic SFRC specimens, Shah et al. [13] have reported that fracture of such composites (using short, random steel fibers) is due to the formation of a single major crack near the peak stress level. Strains at various locations on the specimen were monitored during a monotonic test conducted on an SFRC specimen ($V_f = 1\%$). Results from this test are illustrated in Figure 7. The behavior of the fracture zone and the zone outside the fracture zone were studied to enable a better understanding of the fracture process in SFRC. Although crack widths could be measured only after peak load levels, the strain reading at location 6 (Fig. 7) was significantly lower from those at locations 1–5 at the peak stress, suggesting the formation of a major crack at the critical section. Rapid increases in strain values at this section and unloading of gages 6 and 7 was observed, similar to the softening behavior of the plain matrix [18]. However, the unloading of gages 6 and 7 was not as significant as that observed for the plain matrix [18]. This is probably because the composite, unlike the plain matrix, carries substantial loads even beyond its peak stress. Additionally, it is believed that microcracks that form in the zones before formation of the major crack are restrained from closing up because of the presence of fibers, unlike in the unreinforced matrix. This is evidenced in the size of permanent deformations (strains) at location 6 and 7—they are larger than that observed for the plain matrix. Final fracture still occurs because of the widening of a single major crack, which has a more tortuous profile as a result of the presence of fibers. Confir-

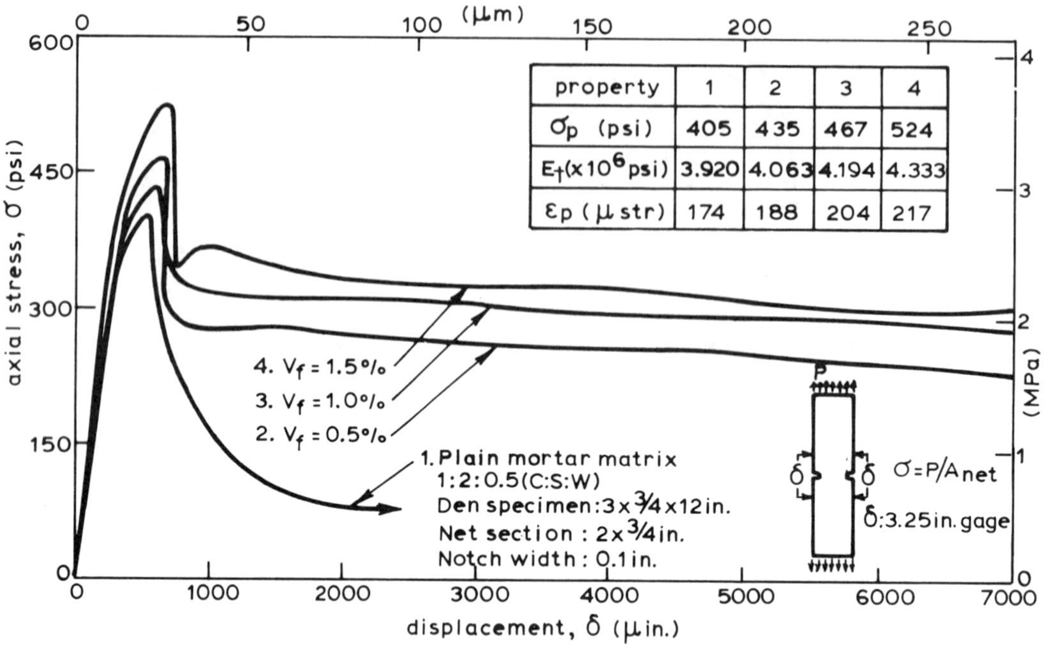

FIGURE 6 Typical results of σ–δ curves obtained from direct tension tests on plain mortar matrix and SFRC mixes.

mation of this behavior was also obtained from optical crack width measurements.

OPTICAL CRACK WIDTH MEASUREMENTS. Microscopic measurements indicate that (a) a single crack of measurable width becomes visible across the critical section almost immediately after peak load, and (b) the postcracking resistance observed in SFRC is due to the widening of a single crack. Average crack widths computed from displacement measurements ω_δ (defined in the inset of Fig. 8), matched optically measured crack widths ω_0 (Fig. 8). The line connecting all the points (ω_δ, ω_0) is farther away from the 45° line for SFRC than for the plain matrix, again confirming larger permanent deformations in the noncritical zones for SFRC specimens. It has been shown [18] that the softening behavior of the unreinforced matrix can be uniquely represented by an exponential stress–crack width relation.

RESIDUAL DEFORMATION. Unloading–reloading tests conducted in the postcracking regime showed little degradation in the apparent material stiffness compared

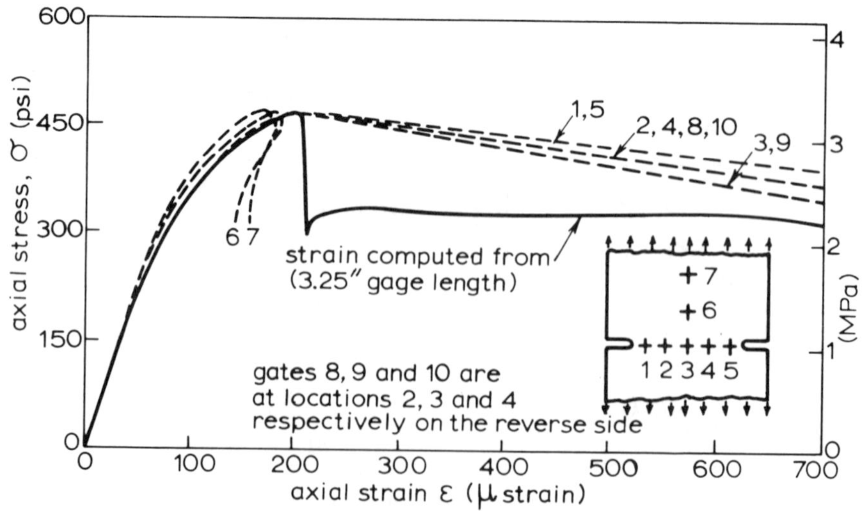

FIGURE 7 Results from a test that monitored local strains for an SFRC specimen. Net axial stress at the critical section is plotted versus strain computed from δ and is compared with the local strain behavior. Dashed lines are used for the local strain behavior to indicate that these were measured only at discrete points.

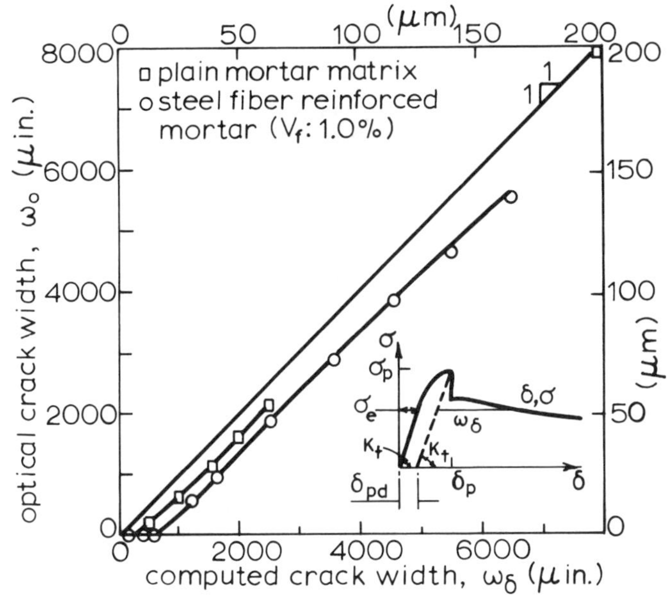

FIGURE 8 **A comparison of optical crack width with computed crack width. Inset illustrates method used to compute crack width from displacement.**

with that observed for plain matrix, even at large unloading displacements (Fig. 9). This results from interfacial failure of the fibers and the irreversible nature of fiber slip across the crack. Therefore, and because of the frictional restraint to crack closing provided by the fibers, residual deformations are very large.

Prediction of Failure Stress

Experimentally observed average tensile stress–strain response of fiber-reinforced concrete shows that the peak stress (i.e., the failure stress) is preceded by transverse matrix cracking, fiber debonding, and fiber–matrix slip. Empirical relationships have been prepared to predict composite failure stress in terms of volume fraction of fibers, aspect ratio of fibers, and the average bond strength of the fiber–matrix interface. The equations proposed are of the following type:

$$\sigma_{cu} = \alpha\sigma_{mu}V_m + \beta\tau V_f l/d \tag{2}$$

where σ_{cu} is composite failure stress, σ_{mu} is matrix tensile strength, V_m and V_f are volume fraction of matrix and fiber, l/d is the fiber aspect ratio, τ is the average bond strength of the interface, and α and β are empirical constants determined from the test results. The nonzero value of α represents the strain-softening contribution of matrix as indicated in Fig. 6. The empirical constant β accounts for the effects of spatial distribution of fibers and other simplifying assumptions such as those involved in defining average bond strength from a single fiber pullout test. The approach outlined above can be valid only for a particular type of test and a type of composite. The author has been developing a more general model to relate the interface–bond properties to the failure stress of the composite.

Fracture Mechanics Approach

To evaluate the results of single-fiber pullout tests and to determine the failure load on composites, an approach based on fracture mechanics is being studied. No assumption regarding the distribution of shear stresses at the interface is explicitly made. In the analytical model, the zone of debonding is treated as an interfacial crack. Failure modes such as fiber yielding, fiber fracture, and further transverse matrix cracking are disregarded. It is assumed that a transverse crack has already been

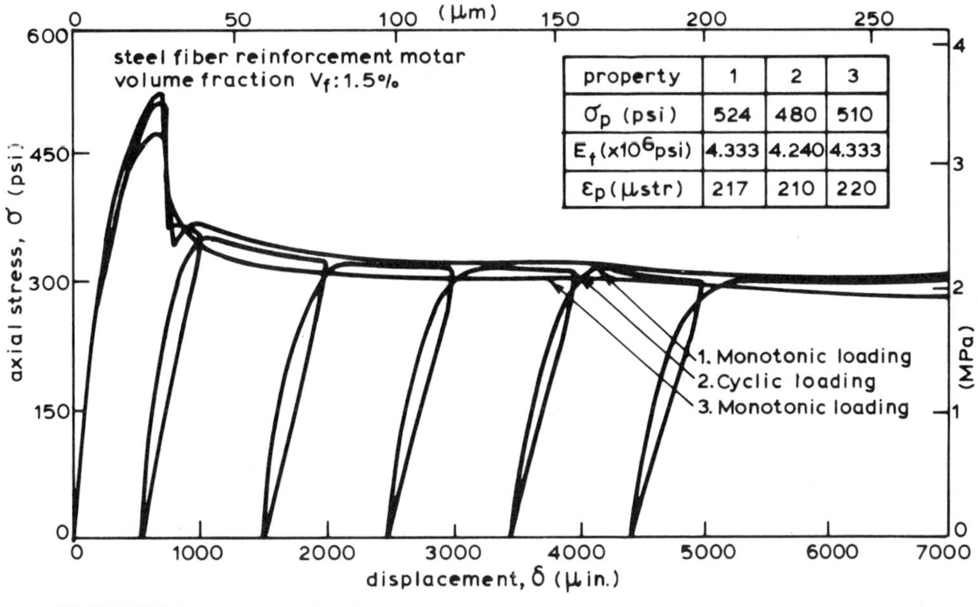

FIGURE 9 **Stress–displacement curves for SFRC specimens (V_f = 1.5%) subjected to monotonic and cyclic loading.**

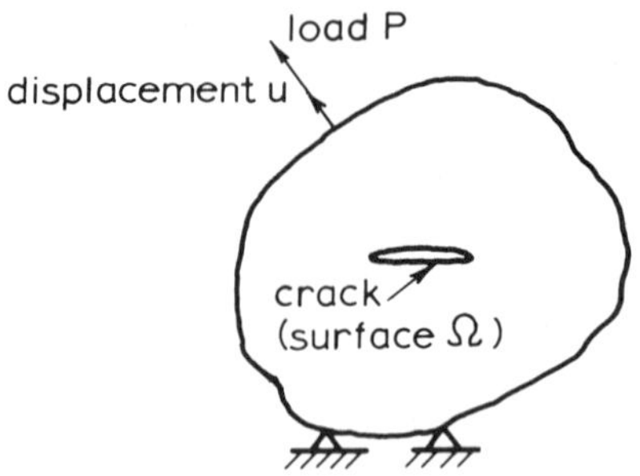

FIGURE 10 **Definition of some parameters used in the fiber pullout problem.**

formed and that the debonding starts at the transversely crack matrix surfaces.

Consider a linear elastic body with a crack (Fig. 10). The surface of the crack Ω is determined by a single parameter δ, such that

$$\Omega = \Omega(\delta) \tag{3}$$

The displacements are described on a part of the boundary of the body and the body is loaded with a single force P. The displacement of the boundary at the force P in the same direction as the force is called u and is governed by

$$u = CP \tag{4}$$

where C is the compliance of the body. The compliance is, among other parameters, a function of the crack size:

$$C = C(\delta, \ldots) \tag{5}$$

It can be shown that the total energy released (work done by the force minus the change in strain energy of the body) when the crack grows an infinitesimal amount $\partial\delta$ is given by

$$\frac{1}{2}\frac{\partial C}{\partial \delta}p^2 d\delta = \text{energy released} \tag{6}$$

A simple Griffith-type criterion for the crack growth load P_{cr} can then be written:

$$\frac{1}{2}\frac{\partial C}{\partial \delta}(P_{cr})^2 = \frac{d\Omega}{d\delta}\gamma_i \tag{7}$$

where γ_i is the interfacial specific work of fracture. This criterion can be used to determine γ_i from the single-fiber pullout test and to determine the failure load of FRC composites.

Single-Fiber Pullout Test

In the classical fiber pullout test (Fig. 11), the debonding zone b is considered to be stress-free, at least compared with the bonded part of the interface. This means that the debonded zone can be characterized as an interfacial crack with a crack length b ($b < l$) and a total surface

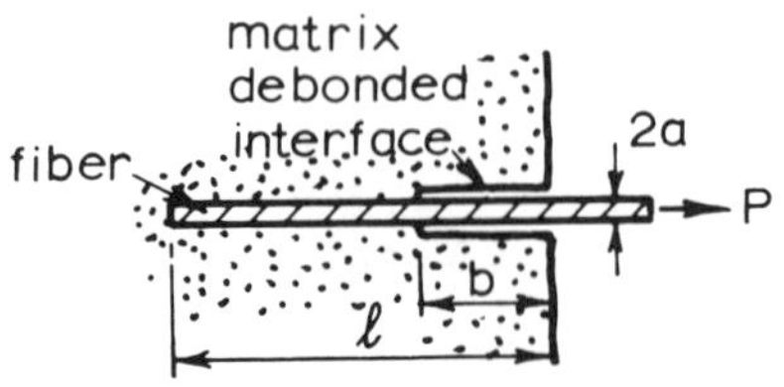

FIGURE 11 **The classical fiber pullout problem.**

of $2(2\pi ab)$. The fiber is assumed to be circular with diameter $2a$ and length l and linearly elastic with Young's modulus E_f and Poisson's ratio ν_f. The matrix is also considered to be linearly elastic with the corresponding properties E_m and ν_m.

The maximum pullout will be reached when the interfacial crack growth criterion (Eq. 7) is satisfied. That is, for the pullout problem, using b as a crack length, Eq. (7) becomes

$$\frac{1}{2}\frac{\delta C}{\delta b}(P_{cr})^2 = 2\pi a 2\gamma_i \tag{8}$$

where γ_i is the specific interfacial work of fracture.

No exact solution is available for the $C = C(b, \ldots)$. However, partly based on approximate analysis using an asymmetric analysis as well as on finite element analysis, Stang and Shah [19] have shown that for steel fibers in a matrix based on portland cement, the following equation can be used to determine the maximum pullout load:

$$\frac{1}{2}\frac{P_{cr}^2}{E_f \pi a^2} = 2\pi a 2\gamma_i \tag{9}$$

For simplicity of presentation, the effect of friction (after debonding) is ignored, although it was considered in the original analysis.

Using Equation (9), one can determine γ_i, the specific interfacial work of fracture from a single-fiber pullout test. It is suggested that γ_i is a more accurate measure of bond strength than usually reported average bond strength (τ).

Uniaxial Tensile Specimen

To predict the tensile strength of fiber-reinforced, cement-based composites whose ultimate strength is dictated by the strength of the relatively weak fiber–matrix interface, the analyses described for the single-fiber pullout case have been applied. For the specimen considered (Fig. 12), it is assumed that (a) one, macroscopic, transverse matrix crack is formed, (b) interfacial cracks have initiated, and (c) the fibers are continuous (i.e., sufficiently long) and are aligned in the direction of the applied tensile stress. Only the case of steel fibers is presented, and friction is ignored.

The length of the specimen is $2l$ and the length of the uncracked (bonded) region is $2(l - b)$. Let the cross section of the specimen be A and the volume fraction of

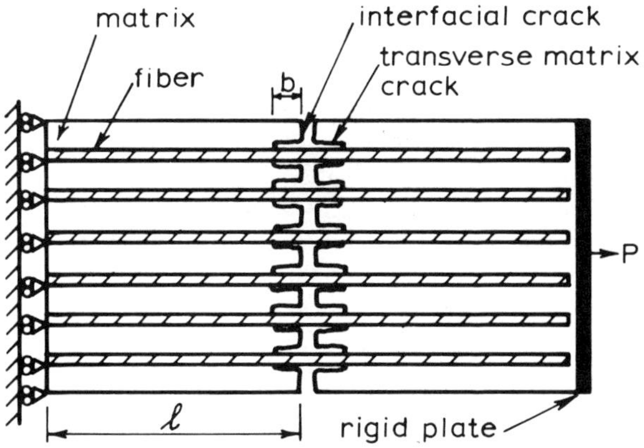

FIGURE 12 **Idealized tensile specimen considered for the composite model.**

fibers V_f. The compliance of the specimen shown in Figure 11 can be expressed as

$$C = 2\left[\frac{b}{V_f E_f A} + (l - b)\frac{1}{E_c A} + H\right] \quad (10)$$

where the first term describes the compliance of the debonded zone and the second term describes the compliance of the bonded region, with E_c defined as the Young's modulus of the composite material; by the rule of mixtures, we have

$$E_c = E_f V_f + (1 - V_f)E_m$$

The third term H describes the compliance of the zone where the interfacial cracks end, and for simplicity it is assumed to be independent of b. The total interfacial crack surface Ω can be written as

$$\Omega = 4bAV_f \frac{2}{a} \quad (11)$$

Using once more the criterion proposed in Eq. (7), the following relationship for the interfacial crack growth and the tensile stress (σ_{cr}) is obtained:

$$\frac{1}{2}2A\left(\frac{1}{V_f E_f} - \frac{1}{E_c}\right)\sigma_{cr}^2 = 4AV_f \frac{2}{a}\gamma_i$$

or (12)

$$\sigma_{cr} = 2V_f\left(\frac{E_c E_f}{E_c - V_f E_f}\frac{2\gamma_i}{a}\right)^{1/2}$$

Note that if σ_{cr} is less than the tensile strength σ_u of the unreinforced matrix (which is approximately the same as the limit of proportionality for the commonly employed FRC composites), then the stress–strain curve labeled (*a*) in Figure 13 is observed for the composites. On the other hand, if σ_{cr} calculated from Eq. (12) is greater than σ_u, the stress–strain curve labeled (*b*) in

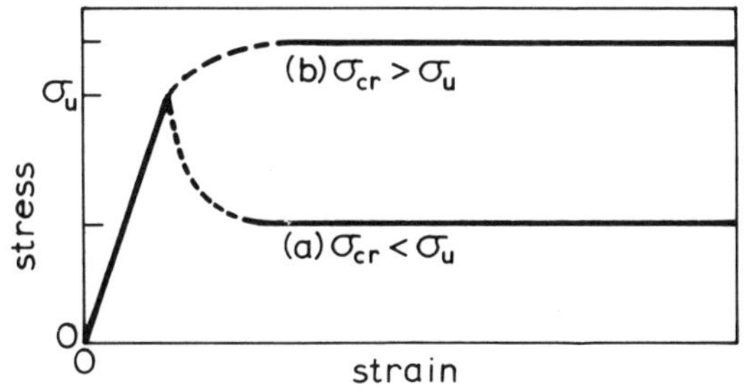

FIGURE 13 **Possible model-predicted solutions for composite tensile stress–strain behavior.**

Figure 13 is predicted from the proposed analysis. Curves of both these types have been reported for FRC composites (Fig. 3).

Some Comparisons with Experimental Results

To examine the applicability of the analysis proposed, the prediction of the tensile strength of the composite was compared with the observed data for steel fiber reinforced concrete.

Naaman and Shah [20] have conducted pullout tests on straight and smooth steel fibers. The average peak pullout loads P_{cr} for fibers with embedment length about 13 mm and diameters 0.4, 0.25, and 0.15 mm were observed to be 42, 26.5, and 6.20 N. For these values of P_{cr}, using Eq. (9), the values of γ_i = 13.3, 21.7, and 5.5 J/m^2 are obtained. Using these values of γ_i and for $V_f = 0.01$, values of σ_{cr} = 3.51, 5.67, and 3.68 MPa are obtained by using Eq. (12). These values of tensile strength for cement-based matrix reinforced with aligned fibers appear to be reasonable. Also, note that the value of γ_i = 6–22 J/m^2 for the steel fiber–matrix interface seems to be reasonable when compared with the values of 20 and 60 J/m^2 observed for unreinforced matrix in mode I crack propagation [21].

Equation (12) points toward a simple, approximate, yet radical method of determining the failure stress for fiber-reinforced cement composites. The value of γ_i can be determined from a single-fiber pullout test. It has been shown by the author that when the effect of friction is included, a single value of γ_i is obtained regardless of the embedment length of the fibers. However, formulations based on average bond strength are dependent on the embedment length [17].

Impact Loading

Adding fibers greatly improves the fracture and impact resistance of concrete, although few quantitative studies on the dynamic behavior of FRC are available. Committee 544 of the American Concrete Institute recommends the use of a "repeated impact" drop-weight test for the qualitative estimation of the impact resistance of such composites. Although useful for ascertaining the relative merits of the different mixes, the mechanical properties essential for design purposes cannot be obtained.

Results also depend on the specimen geometry, loading configuration, and the prescribed failure criterion. Consequently, it would be more useful to prescribe impact tests for FRC that (a) yield mechanical properties useful for design, such as the modulus of rupture (MOR) or the tensile strength, (b) are quantitative, and (c) provide reproducible results that are relatively independent of the test setup. The author has developed an instrumented impact test that meets all these requirements [22].

Instrumented Charpy Testing System

A conventional Charpy impact machine was modified and instrumented to facilitate tests on FRC specimens at different impact velocities (Fig. 14). Among the primary modifications were (a) instrumentation of the striker and the supporting anvil, (b) seating arrangement to accommodate large specimens [76 × 25 × 229 mm in lieu of 10 × 10 × 50 mm used for metals], (c) a fixture to enable tests at different impact velocities [up to 3 m/s], and (d) a device to monitor continuously the midpoint deflection of the beam.

During impact, the load experienced by the beam can be evaluated from that recorded by the instrumented tup (striker attached to the pendulum) or from that recorded by the two anvils (beam supports). However, inertial oscillations may render these loads significantly different at a given time. To obtain a valid flexural test during impact, these inertial effects must be reduced. Problems encountered during impact testing that are due to inertial oscillations for a Charpy setup and their solutions are detailed in Ref. 22. Using a two-degree-of-freedom lumped-mass system, originally proposed in Ref. 23, simple and approximate expressions were derived to predict the maximum difference between the recorded tup and anvil loads. This difference, and thus the adverse inertial effects, can be minimized by reducing the velocity of impact and increasing the ratio of the beam stiffness to the effective stiffness of the contact zone. The testing procedure used in this study was designed based on the guidelines outlined in Ref. 22. Simultaneous measurements of the tup and anvil loads ensured that the effects of inertia were indeed insignificant.

Typical results obtained from an impact test on a mortar and on FRC specimens are shown in Figure 15. The values of the load recorded from the instrumented tup, the sum of the two anvil loads, load–point deflection, and the measured strain values are plotted with respect to time. The results are for specimens impacted at a velocity of 2.45 m/s. The peak load was reached within about one millisecond. The loads measured by tup and anvil were comparable. Inertial oscillations were present, as expected, but as was designed, their amplitudes around mean values were not significant.

The strain–time plot for the notched specimens was initially linear, but just before the peak load, crack propagation caused a very rapid increase in strain. The linear part of the strain–time plot of the unnotched specimens was used for calculating the strain rate and the modulus of elasticity. For unnotched specimens, since the strain was measured away from the critical section, the strain reduces beyond the peak load because of loss of load-carrying capacity on cracking. For both sets of specimens, deflections continue to increase, at a higher rate, beyond the peak load.

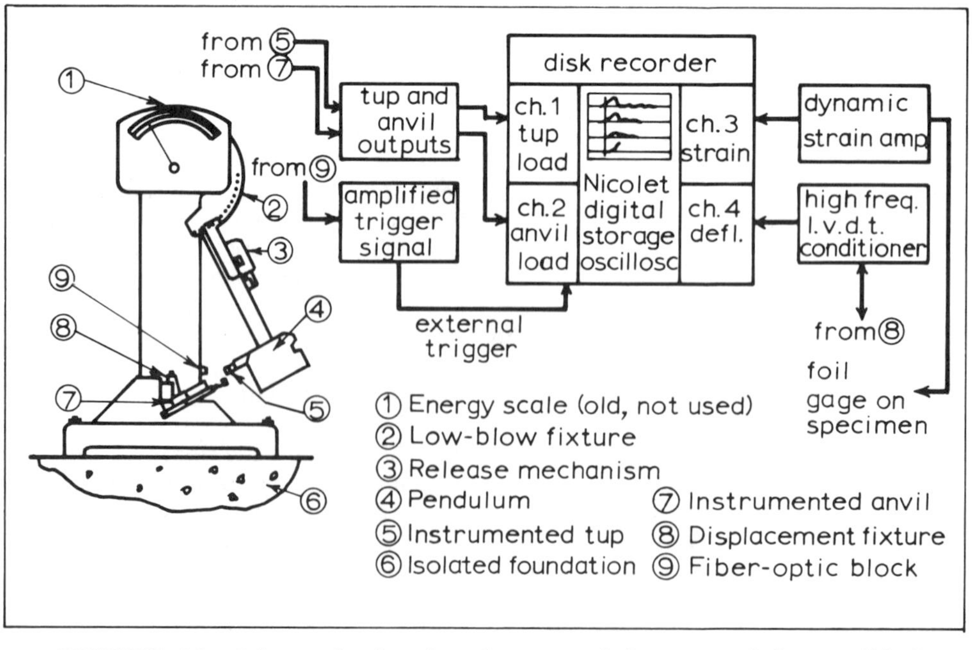

FIGURE 14 Schematic showing the general features of the modified instrumented Charpy test.

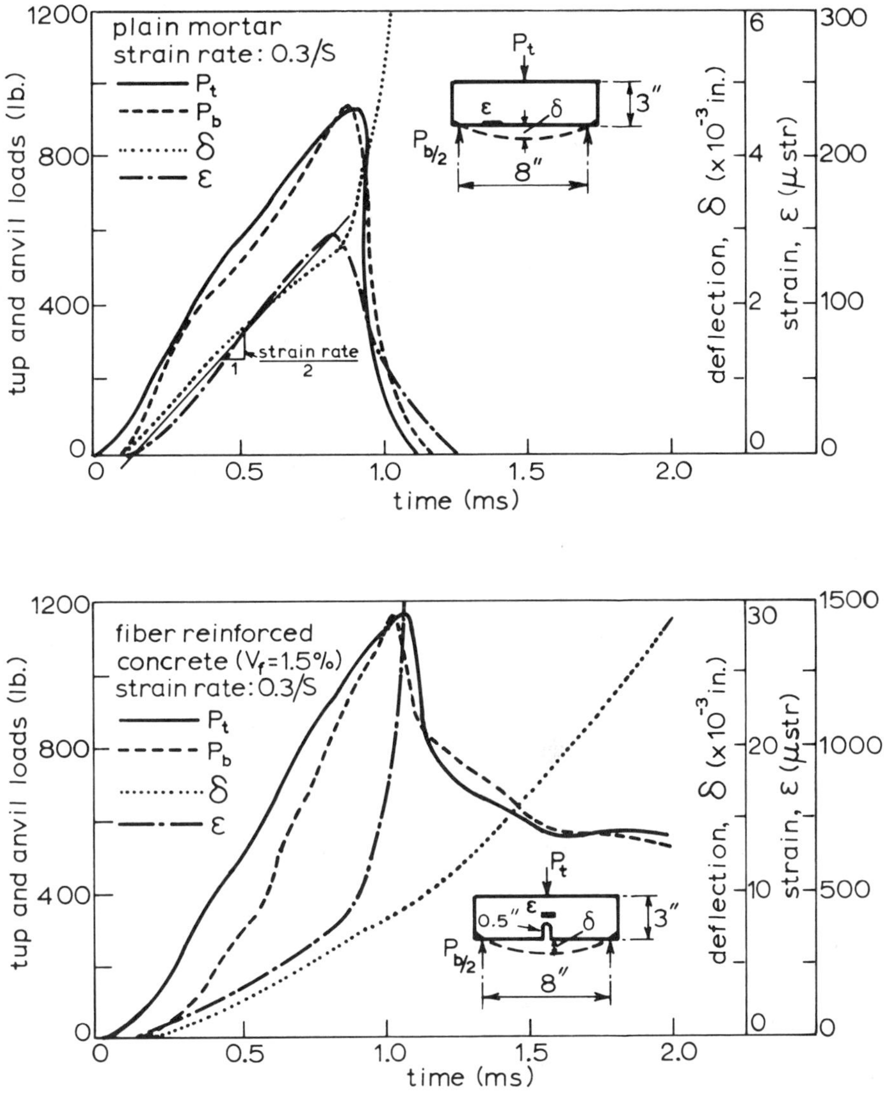

FIGURE 15 **Typical histories of tup load, anvil load, midpoint deflection, and beam strain recorded during an impact event at a strain rate of 0.3/s. Strain rate is computed as twice the slope of the quarter-point strain–time response. (*top*) Unnotched plain mortar beam and (*bottom*) notched FRC beam (V_f = 1.5%, l/d = 63).**

Effects of Strain Rate

Load–deflection curves for steel fiber reinforced concrete specimens subjected to five different strain rates are shown in Figure 16. Steel fibers having a diameter of 0.4 mm were smooth, straight, and 25 mm long. The relative values of peak loads (average of four specimens each) at various strain rates (normalized by the corresponding values at the slowest rate) are shown in Figure 17 for specimens of unreinforced mortar as well as those reinforced with different amounts of fibers. Figure 17 shows that the effect of strain rate is higher for FRC specimens and even more so the higher volume fraction of fibers. For example, the modulus of rupture for FRC specimens made with 1.5% at the highest strain rate is 15.43 MPa, compared with 7.28 MPa at the lowest strain rate. For mortar specimens, the values at the highest and lowest rates were 8.55 and 5.15 MPa, respectively. The higher strain rate sensitivity of FRC specimens is probably due to additional cracking (both transverse matrix cracking and interfacial cracking on debonding) generally associated with FRC specimens and to the fact that the strain rate sensitivity in cement-based composites is related to crack growth.

The strain rate sensitivity of FRC also increases with an increasing aspect ratio, as shown in Figure 18. The data are for specimens reinforced with 2% of steel fibers (the same type used in the present study) and made with fibers of three different aspect ratios. Figure 18 plots the nondimensionalized strength values versus strain rate.

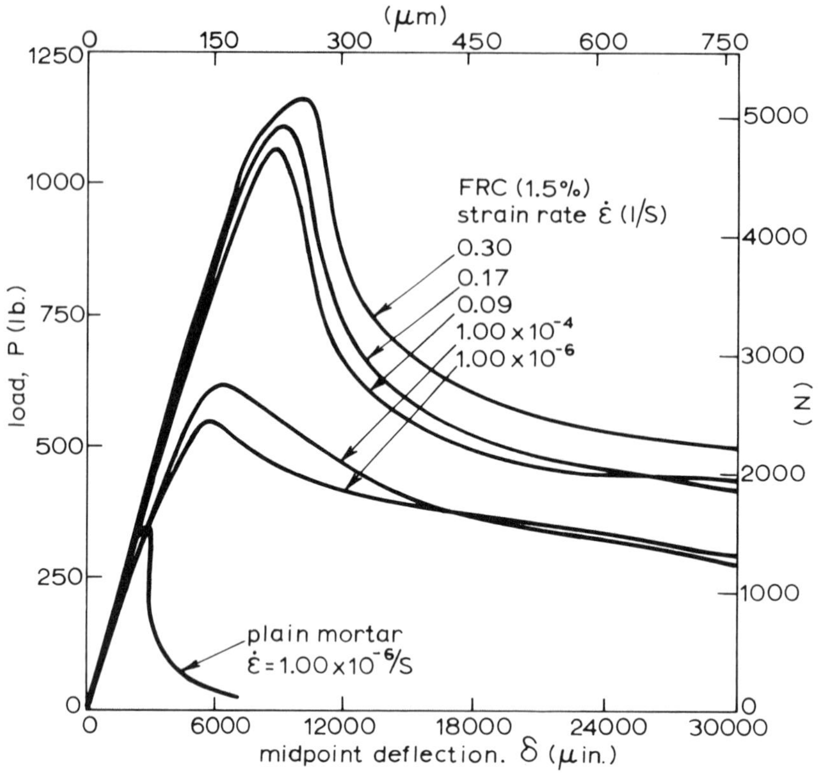

FIGURE 16 **Typical load–deflection response of notched beams [notch depth = 0.5 in. (13 mm), width = 0.1 in. (2.5 mm)] at the different rates of loading (V_f = 1.5%, l/d = 63).**

Comparison of Relative Performance of SFRC

The relative improvements in impact resistance and in static fracture energy measured by several methods are shown in Table 3. A drop-weight test using a 6 in. (152 mm) diameter, 2.5 in. (64 mm) high cylinder has been proposed by Schrader [24]. The number of blows required to produce first crack and to induce a fixed amount of diametrical expansion are used as criteria for quantifying impact resistance. The area under the load–deflection curve up to a fixed value of deflection for FRC specimens obtained by testing beams under static loading rates has been suggested as a measure of fracture toughness by Johnston [25] and Henager [26]. This area for FRC specimens, when compared with the area up to the peak load of unreinforced matrix (which is some-

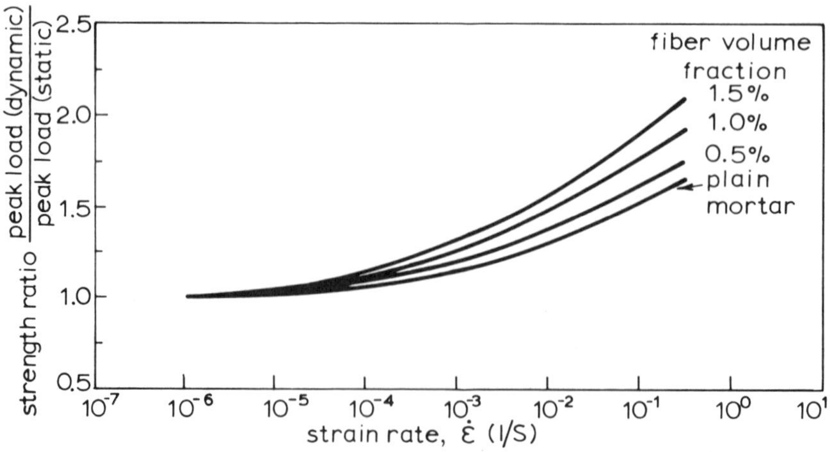

FIGURE 17 **Effect of strain rate on the flexural strength of unreinforced and fiber-reinforced mortar specimens. The influence of fiber content (l/d = 63) is highlighted.**

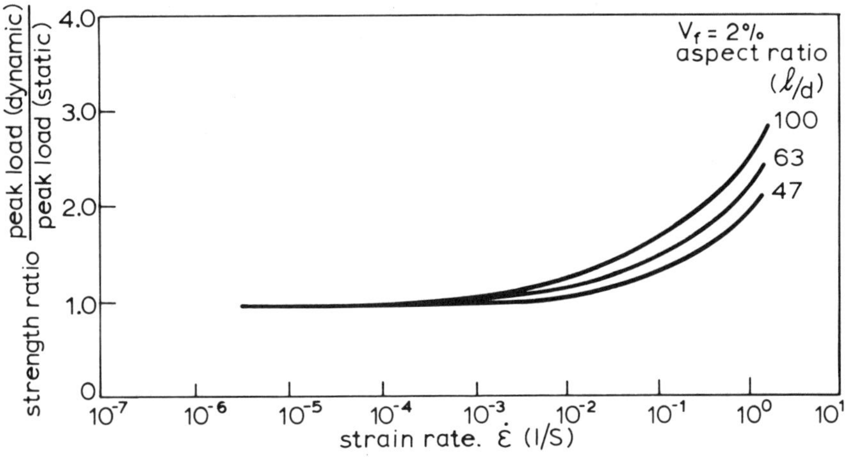

FIGURE 18 **Effect of strain rate on the flexural strength of unreinforced and fiber-reinforced mortar specimens. The influence of fiber aspect ratio (V_f = 2%) is highlighted.**

times taken as equal to that up to the first cracking load for FRC specimens), is termed the toughness index and is taken as an indication of the relative performance of FRC specimens. Table 3 reports relative performance obtained using a drop-weight method, the standard Charpy test, and the static toughness index, as well as that observed from the instrumented impact tests. These results are for straight, smooth steel fibers. The corresponding amounts of fibers, given in Table 3, were within a comparable range.

The values of the toughness index are close to the relative performance as measured accurately by the instrumented impact test. This is not surprising, since both methods use the area under the load–deflection curve as a criterion for evaluating the performance of FRC. The other methods generally underestimate the relative performance of FRC.

Theoretically, the area under the complete load–deflection curve should correspond to the energy required to fracture the critical cross section (under the load point in the three-point test). However, there are some experimental difficulties in accurately obtaining this value.

1. The postpeak load–deflection response is sensitive to the relative stiffness of the testing machine. Testing in a closed-loop mode as performed in this investigation can reduce the parasitic testing system interaction.

2. If cracks and nonlinear deformations occur in regions other than the critical section, the area under the load–deflection curve will overestimate the true fracture energy of the material [21].

3. Unless the deflections are recorded in such a manner that the local deformations occurring under the load point and supports are eliminated, the measured energy (area under the load–deflection curve) may overestimate the true fracture energy. This is especially critical for unreinforced specimens or for reinforced specimens up to the first-cracking load.

A similar observation was made in the present study. The midpoint deflection was measured as a relative displacement between a fixed support and the center of the beam. The average modulus of elasticity calculated from the linear part of the load–deflection curve (after making corrections due to shear deformations) for mortar specimens tested at the slowest rate was 17.1 Gpa. For the same set of specimens, the modulus calculated from the strain measurements (quarter-point: see Fig. 15) was 24.4 GPa. This value was identical to that observed from uniaxial compression and uniaxial tension tests.

All these points should be considered when using the area under the load–deflection curve to evaluate the relative performance of FRC.

Long-Term Durability of Glass Fiber Reinforced Concrete

Glass fiber reinforced concrete (GFRC) panels are increasingly being used as curtain walls in the United States. The volume of GFRC panels sold in the United States exceeded $70 million per year in 1985. These panels are about 12 mm thick and can be as large as 10 meters long and a few meters wide. They are precast by spraying chopped glass fibers and cement slurry onto a mold of desired shape. Typical glass fibers used are about 37 mm long with a volume fraction of 5%.

The glass fibers used are alkali-resistant (AR) glass fibers. However, it is known that GFRC panels made with AR glass fibers and portland cement can show a strength retrogression if subjected to prolonged exposure to a wet and hot environment.

This strength deterioration is demonstrated in Figure 19, which plots the results of flexural tests of GFRC panels subjected to an accelerated aging environment (50°C, under lime-saturated water) [27]. These results are from a continuing study at Northwestern University being conducted in collaboration with the Prestressed Con-

TABLE 3
Relative Increase in Energy Absorption Capacity Due to the Incorporation of Fibers Measured by Conventional Tests and Instrumented Impact Tests

	First Crack		Ultimate Failure	
	Number of Blows	Relative Increase	Number of Blows	Relative Increase
Ramakrishnan et al.[a,b]				
Mortar 1:3.66:0:0.46	10	1.0	13	1.0
SFRC $\nu_f = 1\%$ $l/d = 70$	48	4.8	155	11.9

	Charpy Impact Resistance	
	kJ/m²	Relative impact
Krenchel[c,d]		
Mortar 1:2:0:0.45	2.9	1.0
SFRC $\nu_f = 2\%$ $l/d = 100$	20.0	6.9
Static flexural toughness	Energy absorbed	Toughness index
Henegar[e,f]		
SFRC $\nu_f = 1.5\%$ $l/d = 61$	—	30.4
Ramakrishnan et al.[a]		
SFRC $\nu_f = 1\%$ $l/d = 70$	—	16.0

	Static Flexural Toughness		Instrumented Impact Toughness	
	kJ/m²	Relative Increase	kJ/m²	Relative Increase
Gopalaratnam et al.[h]				
Mortar 1:2:0:0.5	0.075	1.0	0.102	1.0
SFRC[j] $\nu_f = 1\%$ $l/d = 63$	1.711	22.8	3.013	29.5
Suaris and Shah[k]				
Mortar 1:2:0:0.5	0.75	1.0	0.177	1.0
SFRC[a,l] $\nu_f = 1\%$ $l/d = 100$	2.773	37.0	5.158	29.1

[a] Ref. 29.
[b] ACI recommended drop-weight test in a compressive configuration on 152 × 38 mm (1.5 cm diameter, 4 cm high).
[c] Ref. 30.
[d] Conventional Charpy test on flexural specimens, cross section 40 × 40 mm, span 140 mm, impact velocity 3.5–4 m/s, pendulum energy 50 J.
[e] Ref. 26.
[f] Third-point loading, cross section 102 × 102 mm, span 305 mm, loading rate 0.5–1 mm/min, matrix composition 1:2.4:0:0.47, $\sigma_{MOR} = 6.13$ MPa, ACI toughness index up to $\delta = 1.9$ mm.
[g] Third-point loading, cross section 76 × 76 mm, span 305 mm, matrix composition 1:3.66:0:0.46, ACI toughness index up to $\delta = 2.5$ mm.
[h] Ref. 31.
[i] Three-point flexural tests at static strain rate of 1×10^{-6}/s and impact strain rate of 0.3/s.
[j] Toughness computed up to 0.25 cm deflection.
[k] Ref. 23.
[l] Three-point flexural tests at static strain rate of 0.67×10^{-6}/s and impact strain rate of 0.27/s.

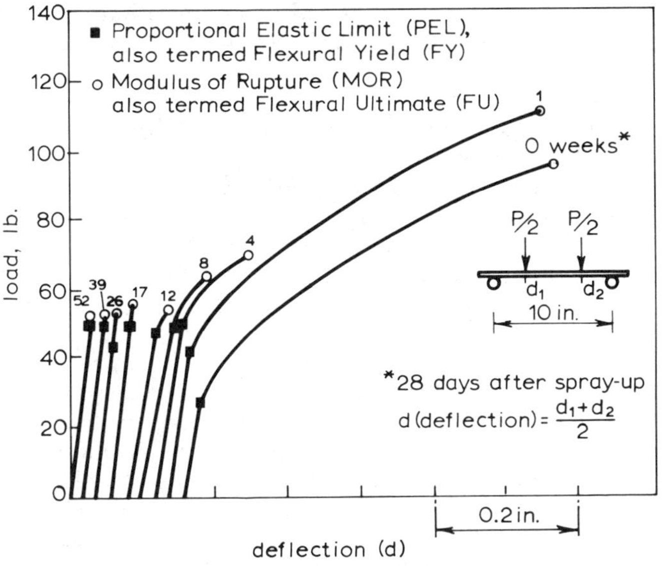

FIGURE 19 **Load–deflection plots of GFRC panels tested in flexure after accelerated aging from 0 to 52 weeks.**

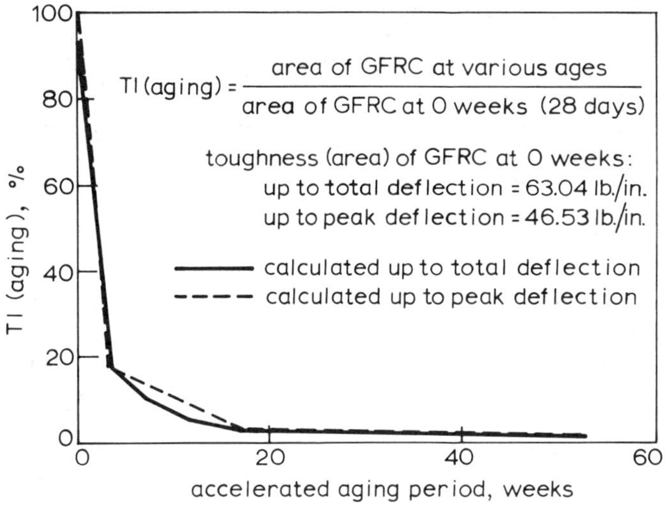

FIGURE 21 **Changes in the proposed toughness index with aging.**

crete Institute and supported by the National Science Foundation to examine durability of GFRC. It can be seen from Figure 19 that after about a year of accelerated aging, the modulus of rupture of GFRC panels reduces to that of a corresponding unreinforced matrix.

Even more dramatic than the reduction in strength is the decrease in toughness with aging. This is shown in Figure 20. Toughness was defined as the area under the complete load–deflection curve. A toughness index TI(aging) has been proposed by Shah et al. [27] to characterize the ductility–durability of GFRC (Fig. 21). It can be seen from Figure 21 that TI(aging) reduced from 100 to about 1 after prolonged exposure to an adverse climate.

To reduce this embrittlement of GFRC panels, several approaches are being studied. These include addition of polymer latex, use of a new type of lime-free cement, and application of coatings to AR glass fibers [28].

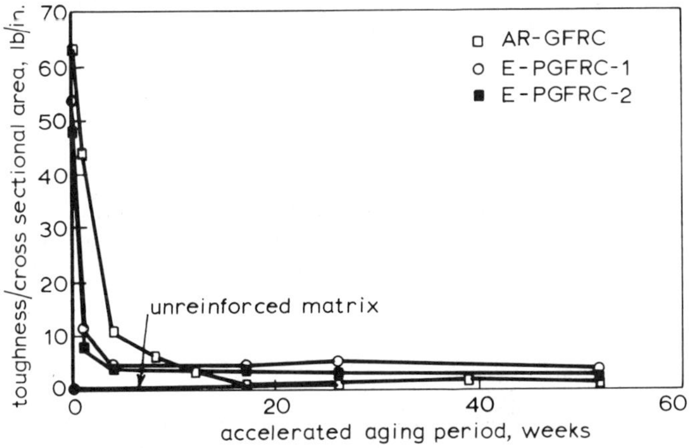

FIGURE 20 **Changes in toughness of GFRC panels due to aging.**

Applications

Asbestos Fibers

The naturally available mineral fibers, asbestos, have been successfully combined with portland cement paste to form the product called asbestos cement. Asbestos cement has been the most widely used fiber-reinforced concrete composite. The world consumption of asbestos fibers for making building products such as sheets, shingles, pipes, tiles, and corrugated roofing elements is about 3 million tons/year. The reasons for the wide use of asbestos fibers are many. They are naturally available and as a result are relatively inexpensive. They have desirable properties of thermal, mechanical, and chemical resistance. Moreover, the methods of manufacturing asbestos cement products, such as the Hatschek or Magnani process, are well developed and suited for precast factory products.

Asbestos cement products contain about 8–16% by volume of asbestos fibers. The flexural strength of asbestos cement board is roughly 2–4 times that of the unreinforced matrix (Table 2). However, since the fibers are relatively short (10 mm), the composite has a brittle behavior (Fig. 2*a*) and a low impact strength).

Only a few countries are major producers of asbestos fibers, and in these countries the better grades are being depleted. Because of this depletion and a lung cancer called asbestosis (associated with asbestos), a great demand exists for a suitable substitute.

Glass Fibers

Glass fibers are used in large quantities with a resin matrix in the production of Fiberglas composites. The common glass fibers, called E-glass, are believed to be chemically attacked by the highly alkaline environment of cement paste, and as a result the glass–fiber reinforcement cement composites lose their strength with time.

The strength reduction depends on the chemical nature of the glass and the cement paste as well as on the environment. The rate of strength reduction can be reduced by using the so-called alkali-resistant glass fibers, by using an organic coating over the fibers, or by reducing the relative humidity of the environment. Considerable research is being done to establish the long-term properties of GFRC products [27,28].

Glass fibers are made from 200–400 individual filaments, 5–15 μm in diameter, which are lightly bonded to make up a strand. These strands can be chopped into various lengths or combined to make cloth, matt, or tape. Commercial applications of glass fibers have involved the addition of fibers either by the conventional mixing procedure or by the spray-up process. When glass fibers (or any other fibers) are incorporated using the fabrication techniques that are standard for normal concrete, it is not possible to mix more than about 2% by volume of fibers of up to about 25 mm long. With this amount and length of fibers, the flexural strength is at most about double, and the ductility, toughness, or impact resistance is increased about tenfold.

With the spray-up process, it is possible to incorporate higher volume (6%) and length (50 mm) of fibers with a low water/cement ratio matrix (0.4), resulting in a superior composite (Table 2). In this process, both the chopped glass fibers and the cement slurry are sprayed with pressurized air onto a surface. If the process is done in a factory, the excess water can be removed by applying suction.

Applications of glass fiber reinforced concrete (mortar) include a variety of precast products, curtain wall, and surface bonded masonry construction.

Steel Fibers

Steel wire is produced by a series of hot- and cold-working methods. Round steel fibers are produced by cutting or chopping the wire, typically having diameters in the 0.25–0.76 mm range. Flat sheet fibers having typical cross sections 0.15–0.41 mm thick and 0.25–0.90 mm wide are produced by shearing sheets. Fibers also have been produced from hot-melt extract. Deformed fibers, which are loosely bonded with a water-soluble glue in the form of a bundle, are also available. Since individual fibers often cluster together, their uniform distribution in the matrix is often difficult. This may be avoided by adding fiber bundles, which separate during the mixing process. Since the failure of fiber-reinforced concrete is by fibers pulling out, the use of deformed fibers increases the pullout strength and, consequently, the mechanical properties of the composite.

Normal mix proportions of steel fiber reinforced concrete consist of about 1–2% by volume of fibers up to 25 mm long (aspect ratio of 75–100) and a higher than usual cement content (Table 2). The larger volume of long fibers would give higher mechanical performance if the fibers could be uniformly distributed. But problems of workability and uniform distribution increase with increasing volume and length of fibers. The improvements with steel fibers include about a two to threefold increase in flexural strength and, more important, up to a twentyfold increase in crack resistance or toughness. The corrosion of steel fibers has not been observed to be a problem, perhaps because of the high cement content and the large number of fibers. However, the long-term aspects of the corrosion of fibers at a cracked surface have not been fully investigated [11].

It is possible to incorporate a large volume of steel fibers by a process known as SIFCON [11]. It is also possible to cut and spray very long lengths of steel fibers [11].

Organic Fibers

Organic fibers include synthetic fibers such as polypropylene and natural fibers such as sisal. Organic fibers are lighter in weight and may be more chemically inert than either steel or glass fibers. They also may be cheaper, especially if they are natural. However, since these fibers have poorer bond and a lower modulus of elasticity than mineral or metallic fibers, they do not show similar improvements when used in similar proportions. However, our recent work [6,7] has shown that it is possible to obtain multiple cracking and stress–strain curves of the type shown in Figure 2*c* with organic fibers, provided a large enough volume fraction (3%) is used. The problems of mixing and uniform dispersion were solved by adding a superplasticizer. Although additional research is needed, it appears that organic fibers can be substituted for mineral or metallic fibers in many applications.

Acknowledgment

This paper was prepared while the author was a visiting professor at the University of Sydney. The author appreciates the support provided by Prof. Karihaloo at the University of Sydney, Department of Civil and Mining Engineering. The author appreciates helpful comments from Barzin Mobasher, Research Assistant, Northwestern University.

Surendra P. Shah

References

1. K. L. Biryukovich and D. L. Yu, "Glass Fiber Reinforced Cement," translated by G. L. Cairns (CERA Translation, 12), Civil Engineering Research Association, London, 1965, 41 pp.
2. A. J. Majumdar, "Properties of Fiber Cement Composites," *Proceedings of the RILEM Symposium*, London, 1975, Construction Press, Lancaster, 1976, pp. 279–314.
3. J. P. Romualdi and G. B. Batson, *J. Eng., Mech. Div. ASCE, 89* (EM3), 147 (1983).
4. G. E. Monfore, *J. Res. Dev. Lab. Portland Cement Assoc., 10*(3), 36 (1968).
5. S. Goldfein, "Plastic Fibrous Reinforcement for Portland Cement Concrete," U.S. Army Corps of Engineers Ohio River Division Laboratory, Cincinnati, Technical Report 1757-TR, 56 pp.

6. H. Krenchel and S. Shah, "Applications of Polypropylene Fibers in Scandinavia," Concrete International, March 1985.
7. A. Naaman, S. Shah, and J. Throne, "Some Developments in Polypropylene Fibers in Concrete," American Concrete Institute SP-81, pp. 375–396.
8. ACI Committee 544, "State-of-the-Art Report on Fiber Reinforced Concrete," 544.1 R-82 reapproved 1986.
9. RILEM Technical Committee 19-FRC, *Mater. Struct., Test Res., 10*(56), 103 (1977).
10. PCI Committee on Glass Fiber Reinforced Concrete Panels, "Recommended Practice for Glass Fiber Reinforced Concrete Panels," Prestressed Concrete Institute, Chicago, 1987.
11. S. Shah and A. Skarendahl, Eds., *Steel Fiber Reinforced Concrete,* Elsevier Applied Science Publishers, New York, 1986.
12. G. Batson and S. Shah, Eds., "Properties and Application of Fiber Reinforced Concrete," American Concrete Institute Special Publication SP 105, 1988.
13. S. P. Shah, P. Stroeven, D. Dalhuisen, and P. Van Stekelenburg, in *Proceedings of the RILEM Symposium on Testing and Test Methods of Fiber Cement Composites,* Sheffield, September 1978, Construction Press, Lancaster, 1978, p. 399.
14. S. P. Shah and A. E. Naaman, *J. Am. Concrete Inst., 73,* 50 (1976).
15. R. Baggot, *Int. J. Cement Spec. Compos. Lightweight Concrete, 5,* 105 (1983).
16. S. P. Shah and B. V. Rangan, *J. Am. Concrete Inst., Proc., 68*(2), 126 (1971).
17. V. S. Gopalaratnam and S. P. Shah, "Tensile Fracture of Steel Fiber Reinforced Concrete," *ASCE J. Eng., Mech. Div.,* May 1987.
18. V. S. Gopalarathnam and S. P. Shah, *Am. Concrete Inst. J., 82*(10), 310 (1985).
19. H. Stang and S. P. Shah, *J. Mater. Sci., 21,* 953 (1986).
20. A. E. Naaman and S. P. Shah, *ASCE J. Struct. Div., 102*(ST8), 1537 (1976).
21. Y. S. Jenq and S. P. Shah, *Eng. Fract. Mech., 21*(5), 1055 (1985).
22. V. S. Gopalaratnam, S. P. Shah, and R. John, "A Modified Instrumented Charpy Test for Cement Based Composites," *Exp. Mech.,* June 1984, pp. 102–111.
23. W. Suaris and S. P. Shah, *Cement Concrete Aggregates, 3*(2), 77 (1981).
24. Ernest K. Schrader, *ACI Proc., 78*(2), 141 (1981).
25. C. D. Johnston, *Cement Concrete Aggregates, 4*(2), 53 (1982).
26. C. H. Henager, "A Toughness Index for Fiber Concrete," *Proceedings of the RILEM Conference on Testing and Test Methods of Fiber Cement Composites,* Construction Press, Lancaster, 1978, pp. 79–86.
27. S. P. Shah, J. I. Daniel, and D. Ludirdja, "Toughness of Glass Fiber Reinforced Concrete Panels Subjected to Accelerated Aging," *J. Prestressed Concrete Inst.,* September–October 1987, pp. 82–99.
28. *Proceedings of the Durability of Glass Fiber Reinforced Concrete Symposium,* Sidney Diamond, Ed., Prestressed Concrete Institute, Chicago, 1985.
29. V. Ramakrishnan, T. Brandshaug, W. V. Coyle, and K. Schrader, *ACI Journal, Proceedings* V. 77(3), 135 (1980).
30. Herbert Krenchel, "Fiber Reinforced Brittle Matrix Materials," Fiber Reinforced Concrete, SP-44, American Concrete Institute, Detroit, 1974, pp. 45–77.
31. V. S. Gopalaratnam and S. P. Shah, *Am. Conc. Ins. J., 83*(14), 117 (1986).

Conductive Composites

When an electrically conducting dispersed phase is added in sufficient quantity to a polymeric resin, a conductive composite is formed. The unique properties of such composites make them technologically superior to or more cost effective than alternative materials in a variety of applications. The inevitable link between conductive composites and the electronics industry has provided much of the impetus for their rapid development in the 1980s.

Conductive composites, broadly defined, include any composites having significant electrical or thermal conductivity. (Defined simply, electrical conductivity is a measure of how easily electrical charge can move through a material under the influence of an applied electric field.) However, in practice, the term "conductive composites" is usually reserved for materials for which the conductivity not only is significant but is a primary characteristic of the composite and is specifically exploited in the end use. Following this practice, this article will not discuss composites such as graphite fiber reinforced epoxy resin composites used for structural reinforcement without regard for their electrical properties. Also, this discussion excludes conductive composite structures that are not composed of discrete particulate additives in a polymeric matrix. For example, expanded metal sheets or foils embedded in a polymer are not considered.

Electrical Conduction in Composites

Conduction Mechanism

Electrical conduction occurs in two ways in conductive composites. In the first and more efficient, electronic conduction proceeds through a network of touching filler particles. In metallic fillers, conduction proceeds through the conduction bands established by the continuum of metal atoms, for example. Direct interparticle contact provides the maximum conductivity for a given filler–matrix combination. Another means of conduction is required when the conductive particles are separated from each other by gaps of poorly conducting or nonconducting material. Electrical conduction must then proceed across these gaps, which represent electrical potential barriers to conduction. When the gaps are narrow enough, electrons can cross them, although with difficulty, by tunneling or hopping. Tunneling is a process by which electrons on one side of the barrier jump to the other side without an energy input. Hopping is similar, but the electron energy must be raised before the barrier can be crossed; that is, the process requires an activation energy.

The probability of tunneling drops rapidly with increasing gap width, in an inverse exponential relationship. When the gap is wide, the composite behaves like an electrical insulator. As a result, a network of conductive particles with either direct interparticle contact or very narrow interparticle gaps conducts electricity much more efficiently than a distribution with wide gaps of insulating resin between the particles. The factors that influence the formation of these networks comprise the main area of fundamental research in conductive composites. They are reviewed in the remainder of this section. More thorough treatments of the subject of electrical conduction in composites are given in Refs. 1 and 2.

Effect of Composite Properties on Conduction

The electrical conductivity of a conductive composite depends on intrinsic properties of the filler material as well as matrix–filler interactions and processing conditions. Other factors being equal, composites made with metallic fillers, which have a high volume conductivity, will be more conductive than those made with nonmetallic fillers such as carbon black, once enough filler has been added to form a conductive network. Similarly, highly conductive metals such as silver will be more effective fillers than less conductive metals such as stainless steel.

The relationship between filler concentration and composite conductivity has been studied in detail. ("Electrical conductivity" and "resistivity" are reciprocal quantities used interchangeably in the remainder of this article.) For randomly dispersed conductive particles in a nonconductive matrix, the relationship between resistivity and filler concentration (as a volume fraction or percentage) is described by percolation theory and takes the form shown in Figure 1. Up to a critical filler concentration, the resistivity of the composite is hardly reduced by the filler. In the region of the critical concentration, resistivity drops sharply. Further increases in filler concentration continue to reduce the resistivity of the composite, but at a much lower rate. This behavior can be explained as follows. At low filler concentration, too little filler is present to form a conductive network; that is, very few of the filler particles are nearly or actually in contact with other conductive particles. As a result, the conductive path includes many large gaps between filler particles, where conduction across the highly resistive matrix resin is necessary. The composite resistivity is dominated by the insulating matrix material. At a critical concentration, dependent on filler morphology and other factors as discussed below, the number of interparticle contacts increases sharply over a narrow range of filler concentration. The conductive path now consists of a network of filler particles that either are touching or are separated by very small gaps, and the resistivity falls sharply as a result. Once an extensive conductive network has formed, further increases in filler concentration increase the average cross section of the conductive network by increasing the number of parallel pathways, and the volume resistivity gradually falls accordingly. The efficiency of continuous network formation for a particulate material is inversely related to how densely it packs. Those that pack densely are inefficient network formers, while those that pack loosely are efficient network formers.

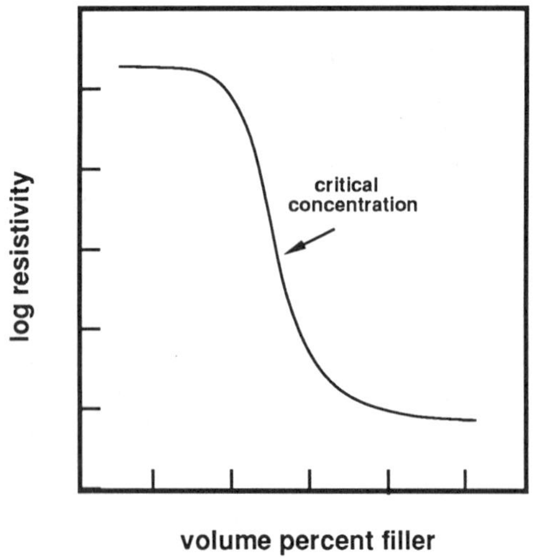

FIGURE 1 Relationship between composite resistivity and the loading of conductive filler particles.

The critical concentration of a given filler depends strongly on its aspect ratio, defined as the ratio of the long dimension to the short dimension of the filler particles. For example, the aspect ratio of a cylindrical rod is equal to its length divided by its diameter. In practice, the aspect ratio can vary from 1, for spherical particles, to several hundred or more for fibers. The shape of the resistivity versus filler concentration relationship (Fig. 1) remains about the same as the aspect ratio changes. However, network formation begins at lower filler concentrations as the aspect ratio increases, so that the curve in Figure 1 is shifted to the left. The effect of aspect ratio on network formation can be seen in Figure 2, which shows three computer simulations of two-dimensional composites. In each of the three frames, the filler particles occupy 20 area percent and have the same randomly assigned positions and orientations, but their aspect ratio varies from frame to frame. At an aspect ratio of 1, network formation is very short order; most particles are not in contact with any other filler particles. Some additional interparticle contacts are made at an aspect ratio of 4. When the aspect ratio is increased to 64, the particles are joined into an effective network that contains multiple conduction paths.

Network formation is generally most efficient when the filler particles are all the same size. Many fillers are not homogeneous, however, but contain a distribution of filler particle sizes and shapes. Brittle fibers such as graphite will break during compounding and molding, affording a distribution of fiber lengths in the finished

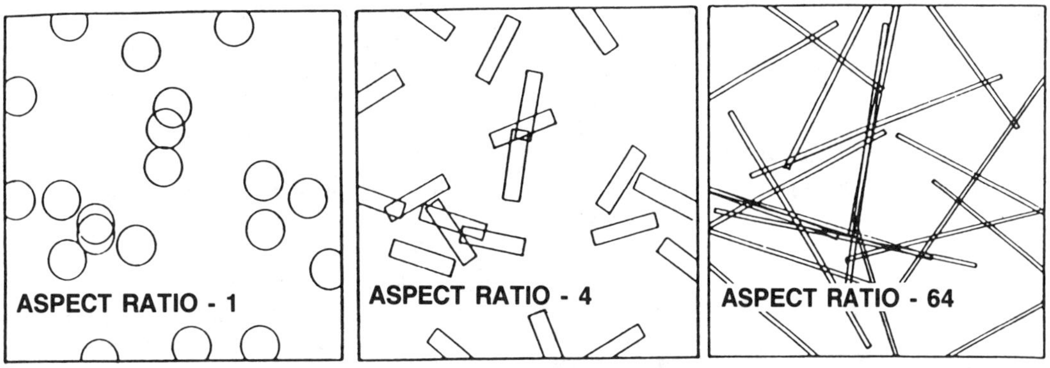

FIGURE 2 **The effect of filler particle aspect ratio on network formation. Each frame contains 20 area percent filler. Filler particles have the same randomly assigned positions and orientations.** (From Ref. 3.)

composite. Blending two or more types of filler particles with different morphologies may be advantageous in reducing materials costs or controlling electrical and physical properties simultaneously. Network formation in such systems is more complicated than in those with a single particle geometry.

The tendency of nonhomogeneous distributions of filler particles to be less efficient network formers (i.e., to pack more densely) is illustrated most easily by a bimodal distribution of spherical fillers. As small spheres are added to an array of larger spheres, they fill in the voids between the larger spheres, thus increasing packing density. Beginning from the other extreme, replacing a volume of small spheres and their voids with a single large sphere also increases packing density. At some intermediate bimodal particle distribution, when the small spheres are packed to maximum density within the voids between the large spheres, maximum packing for the bimodal distribution occurs. The network formation efficiencies of mixtures of two different sizes of spherical particles, and of mixtures of spheres and cylindrical rods, have been studied [4]. Packing density was shown to depend on the relative amounts of the two types of filler as well as geometrical factors. The latter included the ratio of diameters in the case of bimodal spherical distributions, and the aspect ratio of the cylindrical rods in the case of sphere–rod mixtures.

The tendency of particles to form clusters or aggregates can affect the critical filler concentration. When particles cluster together into low aspect ratio aggregates, network formation is hindered. As a result, the critical concentration increases; that is, the curve in Figure 1 is shifted to the right. This effect of clustering has been substantiated by transmission electron microscopy of carbon-filled polycarbonate composites [5]. The sharpness of the transition from insulating to conductive behavior shown in Figure 1 may also be affected by aggregation. This transition, quite sharp with many metallic fillers, is spread over the concentration range of 15–35% with carbon black in polymer matrices [6].

In contrast, nonrandom "segregated" distributions of filler particles can be more efficient than strictly random distributions in forming conductive networks. In a segregated distribution, the conductive particles are restricted to certain sites, promoting formation of reticulated filler structures [7]. Such distributions can be obtained when large polymer particles and small filler particles are combined; the smaller filler particles locate at interstices and contact points around the polymer particles. This can be seen in Figure 3, a schematic cross section through a three-dimensional reticulated network of a fine dispersed phase (black) in a matrix formed from larger polymer particles. Clearly, the polymer particles must retain their integrity to a high degree to prevent randomizing of the filler. As a result, conductive composites with segregated distributions are generally obtained by compression molding, a low shear process. Random distributions are typical of injection molding and extrusion, in which polymer particles are subjected to high shear stresses. (Compression and injection molding and extrusion are discussed in more detail below; see PROCESSING.) Segregated filler distributions can, however, also form in injection molding when the matrix polymer is semicrystalline and the filler particles are very small. The filler particles

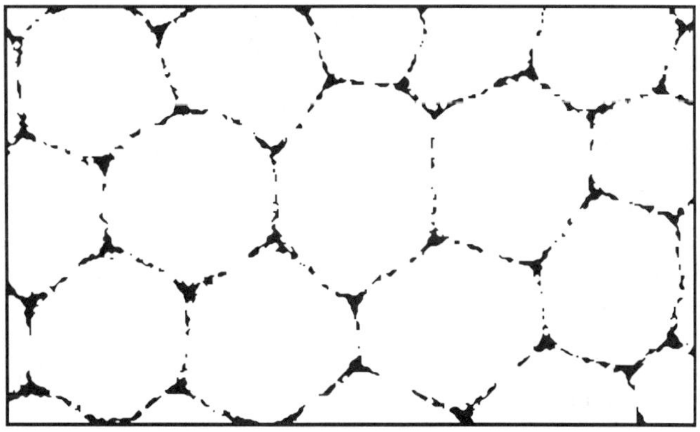

FIGURE 3 **Sketch of a cross section through a segregated distribution, showing formation of a reticulated network of conductive filler material (black). (A continuous three-dimensional network is not necessarily continuous in cross section.)**

are pushed into the interspherulitic boundaries to form a network during recrystallization [8]. The influence of segregation on the electrical properties of the composite can be dramatic. For example, the critical concentration of filler in a random distribution is about 35–37% when the diameter of the nonconductive particle is twice that of the filler particle. The critical concentration drops to 6% in a segregated distribution when the filler particles are one-thirtieth the size of the nonconductive particles [7].

Even when enough filler has been added to the composite to create a conductive network, the filler particles are not necessarily touching each other, so that numerous small gaps may have to be crossed by tunneling or hopping. The electrical conduction between one filler particle and the next can thus be influenced by the degree of wetting of the filler by the polymer. Highly wetted fillers will be coated by a thin layer of polymer, preventing direct contact between adjacent filler particles. For metallic filler particles, the nature and thickness of the surface oxide must also be taken into account. Because some metal oxides are much less conductive than the corresponding metals, they can greatly reduce the conductivity of the continuous network. Surface oxide formation can be reduced by minimizing exposure to heat and moisture during production of the filler and the composite. The presence of adsorbed species on the surface of carbon black can also impede electrical conduction.

Conductive Composite Materials

A conductive composite can be made with any combination of matrix polymer and conductive particulate filler, provided enough filler is used to form an electrically conductive network. From a practical standpoint, however, a number of materials criteria must be satisfied to make a useful composite. Incompatibilities between matrix and filler materials can prevent a well-dispersed, stable conductive filler network from being formed. For example, the filler material may interfere with resin-curing mechanisms, as observed with aluminum powder in polyester resin. Also, the cured matrix resin and the filler should be chemically inert toward each other, to ensure that the physical and electrical properties of the composite do not degrade with time. For example, the polymer matrix should not oxidize the surface of metal fillers. In practice, some matrix–filler combinations work much better or are more cost effective than others, and have been commercialized. This section covers several of the more important (and some unusual) filler and matrix materials, other additives, and some novel approaches to making conductive composites.

Filler Materials

The selection criteria for filler materials depend on the application. In commercial electromagnetic interference (EMI) shielding, discussed in more detail later in this chapter, cost effectiveness is the dominant factor. In military and other critical applications, meeting demanding technical specifications may be more important than cost. In aerospace structural applications, physical as well as electrical properties are emphasized. As a result, different fillers predominate in different applications.

METAL PARTICLES. The most electrically and thermally conductive class of fillers consists of metal particles. Metals of primary importance are aluminum, copper, nickel, iron, stainless steel, and silver [1]. Silver is the most expensive of the group, but it is also the most conductive. Iron, copper, and aluminum are relatively inexpensive but all oxidize readily. The oxide layer on the particle surface, as discussed earlier, can reduce effective conductivity significantly. Nickel and stainless steel, while resistant to heavy oxidation, are expensive compared with copper, iron, or aluminum. Stainless steel is also considerably less conductive than the pure metals. Low loadings of metal particles are desirable because the metals are denser than the polymer matrix resins, thus increasing the weight of composite parts. Also, most metal fillers provide no structural reinforcement.

Aluminum flakes. Rapidly solidified aluminum flakes have been made with dimensions of about 1 mm × 1.2 mm × 0.03 mm [9]. The resulting aspect ratio of about 40, compared with an aspect ratio of only about 3 for aluminum powder, is high for a nonfiber filler. The flakes are ductile and align in the direction of resin flow during molding. The high aspect ratio aluminum flake is usually compounded with matrix resin and extruded. Alternatively, a dry blend of powdered resin and Al flake can be used for injection molding [10]. The relatively low cost of aluminum flakes is an advantage, but high loadings are required; the critical concentration is 18–22% by volume [9], or about 40% by weight for 40 dB shielding. This high loading can lead to several problems [11]. It reduces impact strength, which is vital for large business machine housings, thus limiting penetration in this market. In addition, because of flow problems that can occur in processing, molding requires special care. High concentrations of aluminum flake can produce a mottled appearance. Thus, molded articles may have to be painted or coinjection molded as a core between two layers of unfilled material; both alternatives add to cost.

Stainless steel fibers. Fibers as fine as 4 μm in diameter have been drawn [9]. The continuous fiber is chopped to produce high aspect ratio stainless steel filaments that are strong and ductile enough to resist length reduction during compounding and molding. Stainless steel fiber is generally considered to be the most effective filler commonly used for EMI shielding. The major advantage of stainless steel fiber, arising from its metallic conductivity and high aspect ratio, is high EMI attenuation at low loadings compared with many other fillers. Five to ten percent fiber by weight (only a percent or two by volume) is all that is usually required for most shielding applications. Figure 4 plots the relationship between the volume resistivity and the filler content of conductive

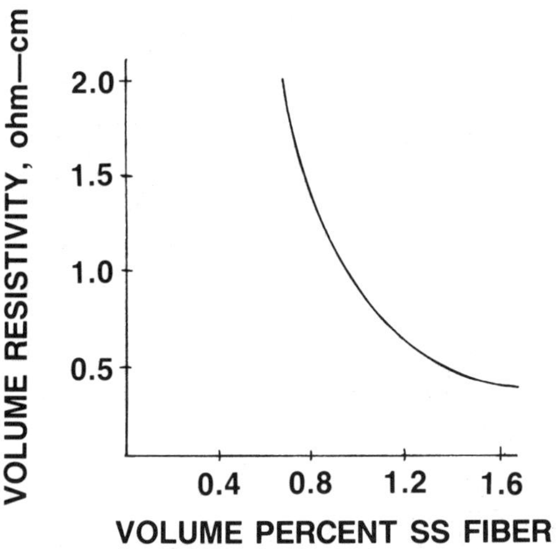

FIGURE 4 **Volume resistivity of stainless steel filled polycarbonate composite versus volume loading of stainless steel fiber.** (From Ref. 12.)

composites containing polycarbonate resin and stainless steel fiber.

Because low filler loadings are used, the fiber itself provides little reinforcement; however, the resin's mechanical properties are retained to a high degree. Composites containing stainless steel fibers at low loadings can be pigmented easily [13]. Because of their low stiffness, the fibers tend to entangle during processing. This is a potential cause of both compounding and molding problems, and can cause erratic and unexpectedly low shielding. The fibers are also relatively expensive.

METAL-COATED INSULATORS. In this class of fillers, the insulator particle serves as a carrier for the metal coating. As a result, the metal is put into a form with a higher effective aspect ratio. The insulating carrier generally has a lower specific gravity than the metal, an advantage of this type of filler when weight savings are important. In some cases, the carrier particle is also a reinforcing material such as glass fiber, so that structural property degradation may be minimized; some structural properties may actually be improved. Metal coatings may be applied to insulators by a number of methods, including vacuum deposition and electroless deposition.

Metal-coated mica particles. Mica is an inexpensive material with a modest aspect ratio, although values as high as 20–50 have been claimed [3]. However, the metal coating considered alone can have a very high aspect ratio, and thus the metal, as in most metal-coated nonconducting particles, is used very effectively. Metal-coated mica particles must be used at high loadings (e.g., ~ 30–40%) to reach the critical concentration. As with other conductive fillers requiring high loadings, loss of mechanical properties is a concern. Mica flakes coated with 43% by volume nickel performed almost identically to aluminum flakes in providing EMI shielding [3].

Metal-coated glass beads and metal-coated hollow glass microspheres. These low aspect ratio fillers must be used at high loadings. Critical concentrations measured for three types of silver-coated glass spheres in epoxide resin were in the range of 30–35% [6]. As with other low aspect ratio fillers, overall loading of the conductive filler can be reduced by blending in a second filler with a higher aspect ratio, such as stainless steel fiber. Hollow glass spheres up to 105, μm in diameter have been coated with between 0.02 and 0.1 μm of silver, equivalent to silver contents of 4–12% [1]. The hollow microspheres may be too fragile for injection molding [11]. This is particularly true in the case of thin silver coatings, which are brittle and can be stripped off during mixing.

Metal-coated glass fibers. A variety of methods, including vacuum deposition and electroless plating, can be used to metalize glass fibers. The primary selection criteria for the coating method are coating quality (coverage, uniformity), metal–glass adhesion, and economics [1]. Glass fibers are generally larger in diameter than graphite fibers. As a result, metal-coated glass should be a somewhat less effective network former than metal-coated graphite, assuming comparable fiber particle length and metal loading [14].

Metal-coated organic fibers. Organic fibers can be used as inexpensive carriers for metal in high aspect ratio form. Metals can be deposited on organic fibers by electroless deposition, or, when the fibers have been treated to provide surface conductivity, by electroplating [15]. Because of the mismatch in physical properties between the fiber and the metal coating, coating integrity may limit applications to those in which subsequent processing does not involve high shear forces. Because most organic fibers melt or degrade at high temperatures, low temperature processing may be necessary.

GRAPHITIC AND CARBON-BASED MATERIALS. These materials span a broad range of prices and particle morphologies. The carbon-based fillers are produced by controlled pyrolysis of carbonaceous materials, primarily hydrocarbon fuels in the case of carbon blacks and organic polymer fibers in the case of graphite fibers. The fiber materials are available in a variety of product forms, including continuous filament, chopped fiber, and nonwoven mats of varying weights.

Carbon black. Many types of carbon black with different properties have been developed. Some blacks with a small particle size form long, fibrous aggregates, which promote formation of conductive networks. Others, composed of highly porous particles, are even better network formers. Composites made with these porous blacks may be orders of magnitude more conductive than composites made with other blacks [16]. Without complementary fillers, however, high loadings of carbon black must be used because of the material's modest conductivity and low aspect ratio. Because the highly loaded molded composites are black, they must be painted when decorative surfaces are needed on the finished parts [11]. The fine, messy powder presents handling difficulties in manufacturing. Carbon black, because of these drawbacks, is not popular as a conductive filler for many EMI shielding applications. However, it has been used as the primary filler additive in making directly platable plastic parts, described in more detail later. Because of its long use as a reinforcement in vehicle tires, carbon black has been studied considerably more than most other filler materials [2].

Graphite fibers. Because it has a higher aspect ratio than carbon blacks, graphite fiber is a better choice than carbon black when highly efficient EMI shielding is required. In addition, graphite fiber has excellent mechanical properties and provides reinforcement when chopped and dispersed in a composite. However, the shielding efficiency is still fairly low because of the material's low electrical conductivity. Compounding must be done properly to minimize fiber breakage; otherwise, the high aspect ratio of the fiber is reduced. Although some colorization is possible with pigment, molded parts are still black, and painting may be necessary [11].

Graphite fibers can be modified by treating them with various chemicals that enter the graphite structure between the graphite planes in a process called intercalation. By intercalating pitch-based graphite fibers with bromine, for example, electrical conductivity can be increased by a factor of 5, giving the fibers about the same conductivity as stainless steel [17].

Nickel-coated graphite fibers. Electroplating a thin nickel coating on graphite fiber greatly increases its electrical conductivity. The calculated and measured electrical resistance per unit length of continuous nickel-coated graphite fiber is plotted against the weight percent nickel in Figure 5. The calculated curve assumes that the graphite fiber and the nickel coating are two concentric cylindrical resistors in parallel; the experimental values contain some contact resistance and indicate some deviation from the idealized model. Depending on the thickness of the nickel coating, the electrical conductivity of the coated fiber can be on the order of 30–50 times higher than the conductivity of uncoated graphite. This, in turn, makes nickel-coated graphite significantly more effective than uncoated graphite fibers as a filler for EMI shielding. For example, in one study [19], 25% nickel-coated graphite in polycarbonate resin produced 40–50 dB shielding, compared with 40 dB for 40% uncoated graphite. Because less nickel-coated graphite is necessary to provide shielding, the degree of reinforcement is generally lower than when uncoated graphite fiber is used. A new form of NCG has been developed that can be added directly at the molding machine, without prior compounding. This reduces fiber breakage and also produces manufacturing cost savings [11]. The product consists of pellets of quarter-inch chopped nickel-coated graphite fiber, held together with a binder. During injection molding, the pellets separate into individual fibers that disperse through the resin [20].

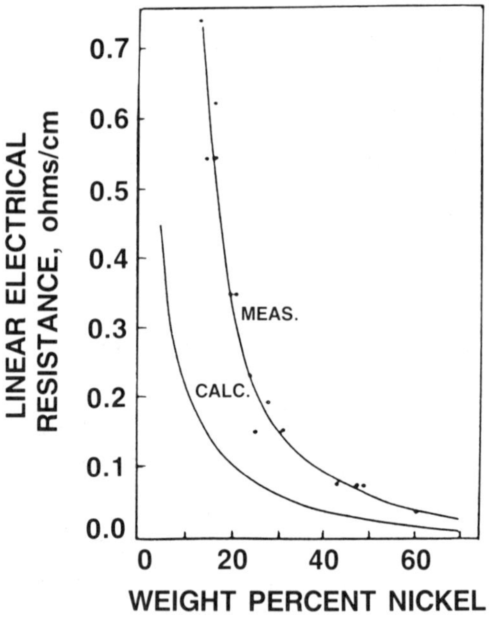

FIGURE 5 **Electrical resistance per unit length of a 3000 filament tow of nickel-coated graphite fiber versus the weight percentage of nickel in the coated fiber. Both calculated and measured resistances are shown.** (From Ref. 18.)

LOW MELTING POINT METAL ALLOYS. The alloys melt during the molding process to produce a conductive network in the polymer [11].

METAL-COATED POLYMER PARTICLES. Not strictly a filler material, metal-coated polymer particles form both the matrix and conductive network. During compression molding, the metal shells crack and molten polymer squeezes out to form a continuous matrix. The metal shell of each particle comes into contact with those of its nearest neighbors, forming a continuous conductive network similar to the reticulated structures arising from segregated networks. In one example [21], copper, nickel, and copper–nickel coatings were applied by electroless deposition to polystyrene particles about 1 mm in diameter. The critical thickness of the nickel coatings was about 0.7 μm, equivalent to a critical filler concentration on the order of only 1% by volume.

Polymers

The choice of matrix resin depends primarily on the physical, chemical, and thermal properties required in the application, as well as on economics. For example [9], business machine housings require rigidity and strength; a demanding application such as under-the-hood automotive electronic components requires high resistance to thermal distortion and chemical attack coupled with good physical properties.

Polymer matrix material choices range from styrenic resins and polyolefins to high grade engineering resins.

Both thermoplastic and thermoset resins can be used. Acrylonitrile–butadiene–styrene (ABS) and polycarbonate are widely used for business machine and electronic equipment housings, applications in which impact strength is important. Epoxy resins are commonly used in aerospace structural composites. Polypropylene copolymer has been used to make directly platable plastic, because of the cost sensitivity of the end use as well as physical properties. Conductive composites have also been made with polyurethane, nylon, polysulfone, polyphenylene sulfide, and other common polymers.

Coupling Agents

Coupling agents may be added to conductive composites to improve the compatibility of the resin and filler. Common choices include silanes, organotitanates, and chemicals with reactive organic functional groups, such as fatty acid chlorides [22]. The advantages of using coupling agents can include reduced melt viscosity and improved dispersion of the conductive particles, leading to better electrical conductivity at a given filler loading.

The mechanism of coupling agent action is not always well understood and probably varies from one type of coupling agent to another. When inorganic filler materials are added to organic resins, coupling agents such as titanates are claimed to replace water of hydration at the inorganic interface with a monomolecular layer of organofunctional titanate [23]. As a result, air is more readily displaced in agglomerates of the inorganic phase, promoting formation of a good conductive network. The replacement of poorly conducting surface layers on the filler particles, together with deagglomeration, can reduce both dc and high frequency ac electrical resistance.

Conductive Composites Made with Conducting Polymers

A material similar in essential respects to conductive composites can be made by intimately mixing nonconducting and conducting polymer phases in an interpenetrating polymer network (IPN). The development of IPNs with electrical conductivity has begun to blur the distinction between conducting composites and intrinsically conducting polymers. In the future, conductive composites and interpenetrating networks may become more similar in their properties and in their end uses.

Interpenetrating networks are combinations of two polymers in intimate contact on a molecular level [24]. At least one of the polymers is synthesized or cross-linked in the immediate presence of the other [25]; thus the polymers interlock with domains smaller than are found in a true blend or composite. The two polymers are distributed so that each forms a continuous network through the system. IPNs were developed in an attempt to combine high electrical conductivity with good mechanical properties [26]. The electrical conductivity of single-phase ionically conducting polymers can be improved by lowering the molecular weight. However, this also degrades the mechanical properties of the polymer. Conversely, cross-linking can be used to improve mechanical properties, but it typically results in a conductivity decrease. In IPNs, each of the two continuous phases provides one of the desired properties.

In one example of an interpenetrating network, epoxy resin was used as the nonconducting phase, while a polyethylene oxide–lithium perchlorate complex was used as the conducting phase [26]. The electrical conductivity of IPNs containing up to about 30% epoxy was roughly equal to that of the pure conducting phase, as shown in Figure 6. Similarly, the modulus of the IPN containing 30% epoxy approached that of the pure epoxy, as the figure also shows.

Electropolymerizating pyrrole in a host membrane of polytetrafluoroethylene impregnated with perfluorosulfonate ionomer [27] produced a membrane composite with nearly the same electronic conductivity as polypyrrole and the mechanical properties of the host membrane. Because the growth of the conductive material within the host membrane can be controlled, membranes can be made conductive in one surface region only, at both surfaces independently, or throughout. Mechanical properties can be varied in some conductive composites of this type. For example, polypyrrole–polyvinyl chloride composites can be altered by adding plasticizers to the PVC, permitting the mechanical properties to be varied over a broad range [28].

Processing

Molding methods and equipment for producing conductive composites are similar to those used with un-

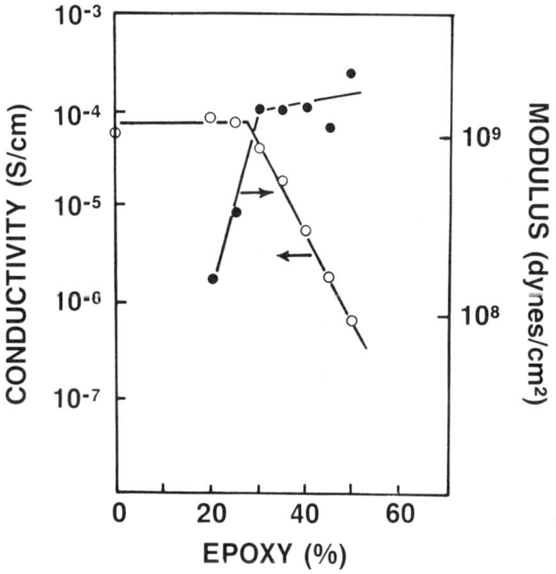

FIGURE 6 **The electrical conductivity of epoxy–polyethylene oxide–lithium perchlorate interpenetrating network as a function of epoxy content [26]. The straight lines are the least-squares fits for data above and below 30 w % epoxy. [Reproduced by permission of the publishers, Butterworth and Co. (Publishers) Ltd.]**

filled polymers and filled, nonconductive composites. Conductive composites are usually made by injection molding. In this process, a thermoplastic material is heated until it flows under pressure and is then injected through a gate into a mold. After cooling, the hardened plastic part is ejected. Compression molding and extrusion have also been used to make conductive composites. In compression molding, a thermosetting plastic is heated to soften it enough to assume the mold shape. It is held under pressure until polymerized or crosslinked. Extruded conductive composites are made by forcing a viscous melt under pressure through a shaping die in a continuous stream.

Operating parameter adjustments are often necessary to maximize the effectiveness of the conductive filler. For example, the processing conditions must be made less aggressive for some of the higher aspect ratio filler materials to prevent a decrease in aspect ratio. Graphite and metallized glass fibers, in particular, will break under high shear conditions. Stainless steel fibers, however, are strong and much more ductile and are thus less susceptible to breaking. Aluminum flakes can fold or tear. At the same time, mixing must be thorough enough to achieve uniform dispersion of the filler throughout the part. The barrel temperature, injection speed, and back pressure are all likely to affect shear during injection molding; the effects of these factors have been discussed recently [29]. Processing conditions for the reaction injection molding (RIM) of conductive composites for EMI shielding applications have also been described [30]. Mold designs should permit uniform dispersion of the filler and should prevent formation of loose clumps ("bird's nesting") of high aspect ratio fillers. Wider gates and flow channels may be required.

The influence of processing parameters on the properties of conductive composites was demonstrated in a recent study of carbon black in low density polyethylene, polymethyl methacrylate, and PVC–vinyl acetate copolymer matrices [31]. Using compression molded sheets 0.5 mm thick as test specimens, the critical filler concentration was shown to increase with the melt viscosity of the matrix material. At a given melt viscosity, the critical concentration increased with increasing surface tension of the polymer.

Some processing-related phenomena are unique to conductive composites. In molding processes, the polymer flow orients high aspect ratio fillers parallel to the plane of the part. As a result, the electrical conductivity in that direction can be as much as 10 times as high as the conductivity normal to the plane [9].

Environmental Degradation

The performance of conductive composites has been studied most extensively in the as-fabricated condition. However, the effects of environmental conditions on the electrical and physical properties of the composite are important in determining suitability for a given end use. Degradation of properties can occur by exposure to chemicals, moisture, heat, light, and mechanical stresses. Unfortunately, there are few published studies on the effects of these environmental factors.

Conductive composites can be exposed to chemicals as cleaning solutions, spills, or as liquids or vapors in a manufacturing environment. In one study [32], the effects of a number of chemical environments on bulk electrical conductivity of conductive composites were found to depend on the type of filler material. Conductive composites were exposed to detergent–water solutions at 25 and 80°C, to isopropyl alcohol, and to seven chemical plant environments. Under these conditions, composites containing either carbon black or carbon fibers were very resistant to degradation. When metal fillers were used, the conductivity loss depended on the type of polymer matrix. Conductive composites made with polymers with low glass transition or melting temperatures lost most of their bulk conductivity; thermosetting polymers and those with high glass transition temperatures performed somewhat better.

When conductive composites are exposed to thermal stresses, differential expansion between the filler and the matrix can disrupt the conductive network of filler particles. Particles in contact may be separated by a small distance, or an existing small separation may grow larger. In both cases, electronic conduction becomes more difficult, as discussed earlier. A study of thermal cycling effects on the EMI shielding performance of several matrix–filler combinations [33] produced results essentially similar to those of the chemical exposure study cited above. Composites with carbon-based fillers were relatively unaffected by thermal cycling between −20 and 80°C. Composites with metal fillers lost most of their conductivity if polymers with low melting or glass transition temperatures were used, and were degraded somewhat less if thermosetting or high glass transition temperature polymers were used.

In aerospace and other structural applications of conductive composites, the question of materials compatibility must be considered. Composite parts will sometimes be in contact with metallic parts. When two such dissimilar, electrically conductive materials are exposed to hostile environments such as marine atmospheres or high temperature and humidity, galvanic cells can be set up in which one of the materials will corrode. The tendency for this to occur is determined by the open-circuit potential difference established between the materials in the test environment, which represents the driving force for galvanic corrosion. A study of the potentials of a variety of graphite, nickel-coated graphite, and stainless steel reinforced composites showed that the driving force for corrosion of aluminum would be substantial in contact with any of the composites [34]. As a result, corrosion prevention measures would be needed in critical applications.

Uses of Conductive Composites

The survey that follows, which is not intended to be exhaustive, illustrates the wide range of uses for conductive composites that have been or are being explored.

Electromagnetic Interference Shielding

The primary applications for conductive composites involve EMI shielding. Electromagnetic interference can be propagated by both conduction and radiation. Shielded enclosures, including those made of conductive composites, are primarily designed to protect against the latter. Radiated EMI is the unintentional and undesirable broadcast of electromagnetic signals [35], which can produce spurious currents and voltages when received. The sources of EMI are numerous, many of them arising from electronic circuitry. High speed switching in microcircuitry may produce EMI in the form of short electromagnetic pulses. Sources include common devices such as radios and calculators, cash registers, computers, and video games. Electric motors and devices that contain them, such as power tools, are also EMI generators. Fluorescent lighting, power lines, and lightning can also produce unwanted electromagnetic signals.

The largest market in the EMI shielding area is housings for business machines. In addition, conductive composites can be used to house individual electronic components, which are either particularly noisy or particularly sensitive to outside noise. Other potential EMI-related applications include classified military and aerospace equipment, avionics instrument enclosures, electric meter housings, and printer machine bases. As a shielding for electrical wiring, conductive composites have been proposed as a replacement for woven metal wire [36]. Conductive composites made with elastomeric polymers can be used to make shielding gaskets.

In the EMI shielding market, conductive composites compete with a number of other technologies based on conductive surface coatings. These methods include vacuum metallizing, sputter coating, zinc arc spray–flame spray, nickel-based paints and coatings, and electroless metal coatings. Because the coatings are usually applied to the interior surfaces of plastic parts, they have no adverse effect on the appearance of finished parts. However, the coatings have disadvantages [3]. The coating process adds the costs of an extra manufacturing step, as well as the outlay for capital equipment. Coating the recesses of a complex part can be difficult. Also, the strength, adhesion, and corrosion resistance of the coating are critical, because any loss of coating during the lifetime of the part will cause an EMI leak.

NEED FOR EMI SHIELDING. Emissions of electromagnetic radiation from electronic devices occur over a broad frequency range. Emissions between 30 MHz and 1 GHz can interfere with broadcast signals; EMI in this range is covered by Federal Communications Commissions regulations, as discussed in more detail elsewhere [1]. EMI can interrupt communications, erase information stored electronically in computers, and cause equipment malfunctions.

The need for effective EMI shielding methods has resulted from the switch from metal to plastic device enclosures and from stricter regulations on electromagnetic emissions. The switch to plastic housings has been driven by the low cost, light weight, aesthetic advantages, and design flexibility they provided [14]. These were all important factors as electronic equipment became widely available for general office use and personal computing.

Shielded enclosures fulfill either or both of two main functions, depending on the specific application. Frequently, the purpose of shielding is to contain the device emissions inside the enclosure. In other instances, the primary goal is to prevent external radiation from penetrating and damaging the internal device. An example of the latter is shielding of avionics boxes and other navigation equipment.

EMI SHIELDING EFFECTIVENESS. An EMI shield is a barrier made of conductive material that attenuates radiated electromagnetic energy. Shielding effectiveness (SE) is a number that quantifies the attenuating ability of a particular material in a particular configuration. It is defined as the logarithm of the ratio of incident power P_{in} to transmitted power P_{tr} and expressed in decibels (dB) according to:

$$SE = 10 \log (P_{in}/P_{tr})$$

Thus, for example, a shielding effectiveness of 10 dB corresponds to attenuation of 90% of the incident power, while a shielding effectiveness of 30 dB corresponds to attenuation of 99.9% of the incident power. This relationship is plotted in Figure 7.

Shielding is the sum of three mechanisms for interrupting the transmission of electromagnetic radiation: reflection, absorption, and multiple internal reflection. In conductive composites, reflection is by far the dominant mechanism, accounting for as much as 85–90% of the shielding [10]. The measurement of shielding effec-

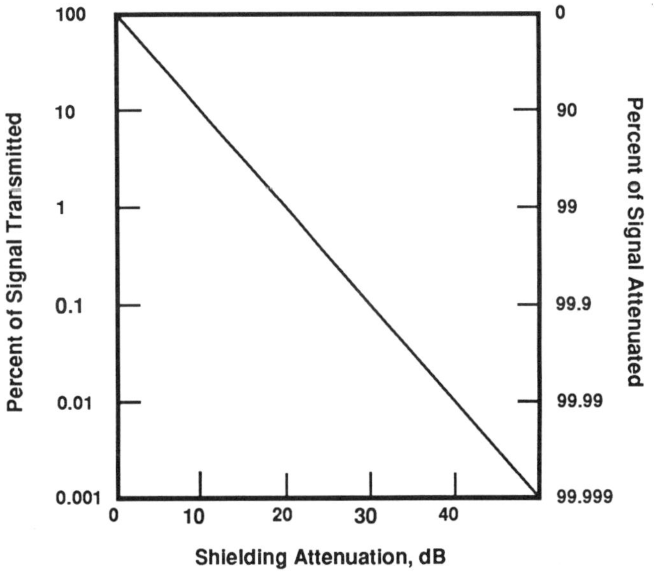

FIGURE 7 The relationship between the percentages of signal transmission/attenuation and the shielding (dB).

tiveness, outside the scope of this article, is discussed in detail elsewhere [1,37].

Electrical conductivity is a prerequisite for shielding. Electrical insulators are transparent to electromagnetic radiation and thus do not provide shielding. Generally, a material's shielding effectiveness is improved by increasing its electrical conductivity. However, the relationship is not a simple one in all cases, particularly given the nonhomogeneous character of a conductive composite on a microscopic scale. The reflection and absorption coefficients of the filler, and the size, shape, and distribution of the voids in the conductive filler network, will have an influence on the shielding effectiveness. In most cases, however, the shielding effectiveness plotted against the volume resistivity of the material falls within a fairly narrow envelope, as illustrated by Figure 8. As the figure shows, a bulk resistivity on the order of 1 Ω-cm provides about 40 dB of shielding; this would suffice for a large share of present commercial requirements. In recent years, work has been directed toward developing highly conductive plastics with resistivities in the range from 10^{-3}–10^{2} Ω-cm [14].

The design of the part will also affect the degree of shielding it provides in two principal ways. First, if the part contains louvers or other breaks in the enclosure, electromagnetic radiation can escape through them. Second, if the part design or processing conditions prevent the proper distribution of filler material in the conductive composite, areas of undesirably high radiation leakage may be formed. As an example, Figure 9 represents a rectangular part in which a network of fibrous filler is continuous from one end of the part to the other through the denser distribution around the part edges, while a poorly shielded region remains in the center. In this case, measuring the electrical resistance along the length of the part would lead to the erroneous conclusion that the composite had good shielding characteristics.

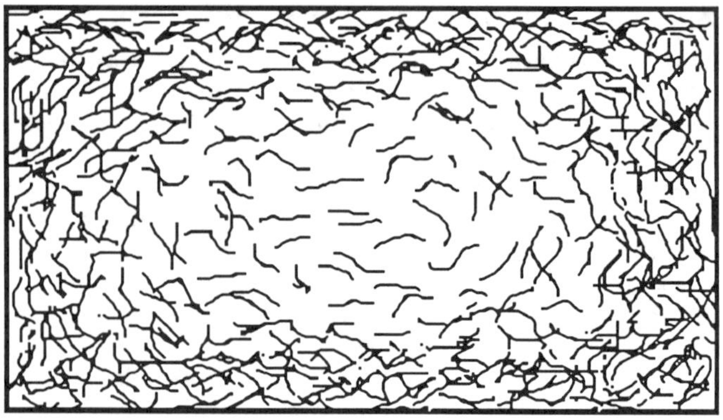

FIGURE 9 **A distribution of conductive fibers in a composite that will show high electrical conductivity from end to end but poor shielding.** (Adapted from Ref. 1 and published by courtesy of Marcel Dekker, Inc.)

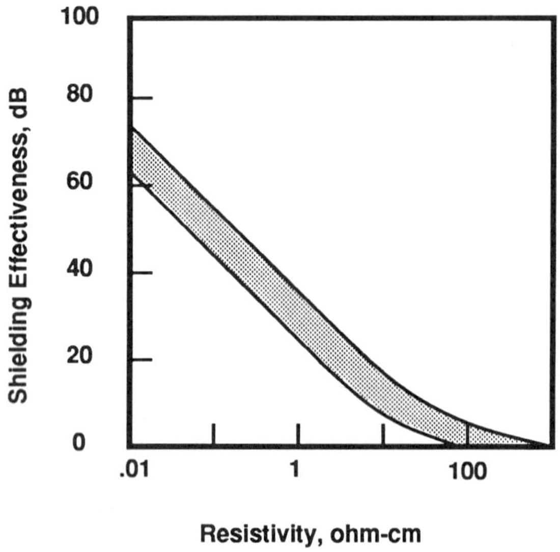

FIGURE 8 The rough relationships between shielding (dB) and volume resistivity characteristic of many materials. (From Ref. 3.)

ECONOMICS OF CONDUCTIVE COMPOSITES FOR EMI SHIELDING. The commercial EMI shielding market is highly cost sensitive. The various conductive fillers compete with each other and with other shielding methods, including conductive surface coatings, primarily on the basis of cost per unit area for the desired level of shielding. Because the field of conductive composites is developing rapidly, and producers of conductive filler materials continue to develop new product forms and to introduce cost savings in their manufacturing, the relative cost effectiveness of various shielding treatments can change markedly and quickly. Published data, even if very recent, must be used with caution. The set of estimates of the cost per square foot to shield polycarbonate with conductive fillers or coatings given in Table 1 indicates that the more economical conductive composites are competitive with conductive coatings. Note, however, that because of recent price adjustments, costs for stainless steel and graphite fibers may be high by about 10%. Furthermore, the nickel-coated graphite fiber cost given in Table 1 is for chopped fiber compounded with resin and subsequently molded. A newer product form, the recently developed pelletized nickel-coated graphite fiber for dry blending at the molding machine, discussed earlier, is claimed to lower shielding costs. The two lowest cost conductive composite alternatives in Table 1 illustrate that considerably different approaches may be taken to reach the same result. Stainless steel fiber is a low cost alternative because it is highly conductive and has a low aspect ratio. It is cost effective despite its high material cost because it can be used in small quantities. Aluminum flake, also highly conductive, has a low aspect ratio and must thus be used in large amounts, but the material itself is inexpensive.

A second cost summary (Table 2), included a greater variety of conductive fillers. While Tables 1 and 2 report cost in different ways, the relative costs of the four fillers

TABLE 1
Cost to Shield One Square Foot of Polycarbonate with Conductive Additives or Coatings[a]

Method	Shielding Level (dB)[b]	Cost ($/ft²)
Conductive fillers		
30% Carbon fiber	30	6.23
15% Nickel-coated carbon fiber	40	6.00
5% Stainless steel fiber	40	3.30
40% Aluminum flake	40	2.80
Coatings[c]		
Nickel paint	50	2.77
Zinc spray	70	3.97
Electroless nickel	90	2.67

[a] Thickness assumed to be 0.10 in.
[b] Shielding effectiveness values given as average over range 30–1000 MHz.
[c] From *Plastics Design Forum*, November–December 1985, p. 43.
From Ref. 37.

they have in common can be compared. In both tables, the order of costs is the same. Aluminum flake is least expensive, followed by stainless steel fiber, nickel-coated graphite fiber, and graphite fiber in that order. While the shielding cost of nickel-coating graphite fiber is only marginally lower than that of graphite fiber in Table 1, it is considerably more cost effective in Table 2, where the assumed loading has been reduced from 15% to 10%. Among the carbon blacks, filler costs are very low.

Material costs alone, however, do not take into account several factors that can influence the overall economics. These include effects on the manufacturing process (e.g., injection molding rate), alterations of physical properties of the composite, and the need for postprocessing such as painting.

Static Dissipation

A related and less demanding conductive composite application is for electrostatic dissipation (ESD), that is, to bleed off charge continuously as a means of preventing harmful arcing discharges. The requirements for ESD are similar to those for EMI, but the level of conductivity required is lower: resistivities of 10^2–10^6 Ω-cm are adequate to bleed off static charges rapidly. ESD protection is essential in manufacture and shipping of electronic equipment, as well as in its use. Applications include circuit board carriers for use in subassembly operations. Films of polymer filled with conductive particles are used to make gloves for handling electronic components, as well as packaging for protection of electronic equipment during shipping. Recently, an electrically conductive, carbon-impregnated, cross-linked polyethylene foam was developed for physical and ESD protection of integrated circuits [39]. Gears, spindles, guides, and parts of sensitive electrical equipment make use of composites with modest electrical conductivity [9].

ESD materials are needed to guard against ignition of flammable substances and in handling explosives. Such hazardous substances include not only hydrocarbon fuels, but high surface area powders and dusts. Antistatic protection is also required when relative motion between dissimilar materials occurs [7]. Parts such as conveyor belts, weaving machine arms, and airplane tires must be protected.

Directly Platable Plastics

Directly platable plastics are conductive composites that are electrically conductive enough to be electroplated

TABLE 2
Conductive Filler Cost to Achieve 40 dB Shielding or 1 Ω/□ Resistivity

Filler	Raw Material Cost ($/lb)	Typical Loading (%)[a]	Additive Cost ($/lb of compound)
Aluminum			
Fibers	3.00–6.00	30	0.90–1.80
Flakes	2.00–2.87	40	0.80–1.15
Powder	1.10	30	0.33
Coated glass fiber	3.00–6.00	30	0.90–1.80
Carbon			
Carbon black	0.32	35	0.11
Furnace black	0.55	30–40	0.17–0.22
Graphite mat	2.40	35	0.84
Ketjenblack EC[b]	4.15	5–15	0.21–0.62
PAN fiber	20.00	30	6.00
Copper powder	2.50–3.50	35	0.88–1.23
Metallized glass fiber	2.55	30	0.77
Nickel-coated graphite fiber	26.00–30.00	10	2.60–3.00
Stainless steel fiber	36.00–45.00	5	1.80–2.25

[a] To achieve approximately 40 dB attenuation or 1 ohm/□ resistivity.
[b] Trade name of Akzo Chemie America, Noury Chemicals Division.
From Ref. 11 (per Business Communications Company).

directly with a metal coating. As a result, prior metallizing or other surface conductivity treatments are unnecessary, leading to cost savings. Because many of the potential applications of platable plastics are very cost sensitive, the fillers are commonly carbon black and other inexpensive materials. More expensive additives may be used in smaller quantities to improve various properties such as adhesion of the plated metal. One early entry into the field of directly platable plastics was a carbon-filled polypropylene copolymer containing proprietary additives to produce a strong metal-to-composite bond [40].

For direct plating to occur, the plastic itself must be electrically conductive enough to carry the current during the initial stages of metal deposition. Once plating has been initiated (with a preliminary low current "strike" if necessary) and coverage of the plastic surface with metal is complete, the growing electrodeposited metal layer becomes the primary current carrier. As a result, only modest electrical conductivity is required for the plastic: while a volume electrical resistivity of less than 10 Ω-cm is desirable, materials with up to about 1000 Ω-cm resistivity can be plated [41].

Directly platable plastics have potential applications in small decorative parts such as cosmetic packages, bottle caps, toys, and ornamental trim of various kinds. The automotive industry represents a large potential market for metal-coated plastic parts for both interior and exterior use. The major advantage of directly platable plastics over conventional plastics such as ABS is economic: they do not have to go through expensive preplate procedures to render their surfaces electrically conductive. Other advantages have also been claimed. For example, the carbon black filler significantly increases the thermal conductivity of the material as well as its electrical conductivity. As a result, molding cycle times for one directly platable polypropylene-based formulation could be reduced by as much as 15–20% relative to unfilled polypropylene [40], leading to additional cost savings at an earlier stage in the part manufacturing process.

General Motors evaluated directly platable plastic for exterior automotive parts during the late 1970s. GM carried out an extensive field and laboratory test program of parts such as grilles, hubcaps, and headlight bezels [42]. Although the addition of conductive filler was claimed to reduce mold shrinkage compared to unfilled polypropylene, GM found high mold shrinkage. As a result, early molded parts suffered from cosmetic defects such as warpage and sink marks. However, hundreds of parts were eventually molded and plated for field evaluations on taxi fleets and at the GM Proving Grounds.

Despite optimism at GM and elsewhere in the late 1970s that directly platable plastics would have a major impact on the chromium plating industry in the 1980s, such materials have failed to achieve their commercial potential. The technology has been unsuccessful in competing with other methods, notably vacuum metallizing, of coating plastics with metals. However, because of the size of the market, this application for conductive composites continues to have promise. Success in the automotive industry will require a material that molds without defects, has a smooth surface free from irregularities, accepts electrodeposited metal readily and uniformly, and forms a strong metal-to-composite bond. For external use on automobiles, the plated parts must survive rigorous environmental degradation tests (e.g., thermal cycling and salt spray corrosion tests) designed to simulate end-use conditions.

In a related development with applications in patterned electronic circuits and elsewhere [43], a polymer resin containing epoxy, curing agent, and wetting agent was loaded with fine metal powder, including some zinc, up to eight times the weight of resin. The loaded resin, in solvent, was screen printed onto substrates and cured. The resulting conductive surface was then plated with an overlayer of copper by a chemical displacement process, in which dissolving zinc was replaced at the conductive composite surface by a layer of copper. The copper grew laterally to form a continuous film with a surface resistivity comparable to that of fired noble metals (i.e., on the order of $3–10 \times 10^{-3}$ Ω/□). The replacement plating process possible with the conductive composite was claimed to be considerably faster than conventional surface metalizing by electroless plating.

Aerospace and Structural Conductive Composites

Fiber-reinforced composites are becoming more widely used in making aircraft components, including structural members. Structural assemblies for aerospace use include body and wing sections, where the composites are replacing metallic parts. The use of graphite fiber as a reinforcement is attractive because of the low density, high tensile strength, and high modulus of the material. Glass, aramid, and other fibers may also be used. Glass and aramid fibers are electrical insulators, while graphite fiber is a modest conductor. However, even the graphite–epoxy composites are much poorer electrical conductors than the metal parts they replace. As a result, while aluminum skins and airframes provide sufficient conductivity to dissipate lightning strikes, the composites are inadequate for withstanding the high currents encountered (200,000 A for a zone I lightning strike).

Lightning strikes on composite parts may damage them catastrophically, or produce arcing. If arcing occurs near sections of the aircraft containing fuel, the fuel–air mixture in the partially filled fuel tank can be ignited. Thus, composite parts require the incorporation of some form of conductive material as either a coating or an internal addition. Woven screens of metal wire, for example, may be used in an outer layer of the composite. Recently, the use of nickel-coated graphite fiber as a conductive element was described [44]. Incorporating the fiber into the outer one or two plies of a conventional graphite–epoxy composite, it was possible to make composites that withstood zone I lightning strikes with a minimum weight penalty. Damage was limited to a small area on the composite surface. The physical properties of the conductive composites followed rule of mixtures behavior—that is, they were dependent on the relative volumes of the resin and fiber phases.

TABLE 3
Tensile Properties[a]

Material System	At Room Temperature Stress (MPa)	At Room Temperature Modulus (GPa)	At Room Temperature Strain (%)	At 104°C Stress (MPa)	At 104°C Modulus (GPa)	At 104°C Strain (%)
PEEK/40NiGr	156.5	15.2	1.12	113.8	15.9	1.01
PEEK/30NiGr/20NiFlk	135.1	16.8	0.88	104.1	16.8	0.81
PEEK/25NiGr/10SS	135.1	13.9	1.22	92.4	13.6	1.07
PPS/40NiGr	133.1	19.1	0.64	75.2	—[b]	1.19
PPS/30NiGr/20NiFlk	132.4	16.3	0.87	68.9	11.0	1.11
PPS/40Gr	125.5	24.1	0.62	84.9	—[b]	1.17
PPS/40NiGr/15NiFlk	125.5	20.1	0.98	71.7	11.9	1.15
PPS/25NiGr/10SS	77.9	10.8	0.76	43.4	7.1	1.47
PPS/30NiGr/10SS	69.6	11.9	0.61	38.6	8.5	1.09
LCP300/30Gr	68.9	17.7	0.51	49.6	12.6	0.72
PPS/25NiGr/15SS	64.8	10.2	0.72	37.9	6.3	1.51
PPS/30NiGr/15SS	60.7	11.9	0.51	38.6	7.4	1.26

[a] Filler loadings are reported as weight percentages.
[b] No linear portion of stress–strain curve.

Two recent studies reported extensive physical and electrical characterizations of conductive composites. One study [44] was confined to the properties of composites containing nickel-coated graphite fiber. The second [45] discussed injection-molded composites containing graphite (Gr), nickel-coated graphite (NiGr), and stainless steel (SS) fibers, as well as nickel flakes (NiFlk), both individually and in various combinations. Matrix resins included polyphenylene sulfide (PPS), polyether ether ketone (PEEK), and liquid crystal polymers (LCP). Tensile, flexural, and drop-weight impact tests were carried out. Results of the tensile properties studies are summarized in Table 3. Materials are listed in order of decreasing room temperature tensile strength. Table 4 shows impact and flexural property measurements in order of decreasing impact strength. Water absorption was less than 1% for all the composites except 40% nickel-coated graphite/15% nickel flake and 30% nickel-coated graphite/10% stainless steel in PPS. Composite densities and electrical resistivities were also measured. Densities ranged from 1.47 g/cm^3 for LCP300/30NiGr to 1.98 g/cm^3 for PPS/30NiGr/20NiFlk. PPS composites containing high loadings of nickel-coated graphite fiber plus stainless steel fiber provided at least 80 dB of shielding in the frequency range 80–300 MHz and had electrical resistivities as low as 0.60 Ω-cm.

Other Applications

As traditional metal joining methods such as welding have been replaced by more economical adhesive bonding, conductive adhesives have become more widely

TABLE 4
Impact and Flexural Properties[a]

Material System	Energy Absorbed (J)	At Room Temperature Stress (MPa)	At Room Temperature Modulus (GPa)	At 104°C Stress (MPa)	At 104°C Modulus (GPa)
PEEK/25NiGr/10SS	9.10	238.6	13.86	168.2	11.45
PEEK/40NiGr	8.69	226.1	14.82	199.9	13.03
PPS/30NiGr/20NiFlk	8.08	229.6	19.17	116.5	9.72
PEEK/30NiGr/20NiFlk	7.31	215.8	15.31	183.4	11.58
LCP300/30Gr	6.97	145.5	13.31	86.9	7.17
PPS/30NiGr/15SS	4.70	155.1	16.75	75.8	7.86
PPS/25NiGr/15SS	4.41	137.9	15.38	77.9	5.38
PPS/25NiGr/10SS	4.28	158.6	14.34	84.1	5.65
PPS/40NiGr/15NiFlk	3.92	180.6	19.58	107.6	8.20
PPS/40NiGr	3.89	217.2	23.72	137.9	12.82
PPS/30NiGr/10SS	3.80	153.1	16.34	74.5	6.27
PPS/40Gr	3.67	191.0	17.72	160.6	16.34

[a] Filler loadings are reported as weight percentages. From Ref. 45.

used. The cured adhesive produces an electrically conductive bond between conductive parts. Conductive adhesives are useful in applications such as joining auto body panels. Auto bodies are commonly painted by electrophoretic deposition processes, which require that all areas of the receiving surface be electrically conductive and in electrical contact. Conductive adhesives allow the paint to bridge the adhesive zone between metal parts to produce a continuous film. Conductive adhesives are also useful in joining heat-sensitive materials that would be adversely affected by the heat of welding or brazing. Conductive caulks are used to join larger gaps when electrical or thermal continuity is needed, or when shielding is required.

"Hyperconductive" filled polymers are useful in electronic applications as adhesives and circuit elements, particularly as conductors in polymer-based electronic circuits [46]. These materials consist of silver powder incorporated into a variety of polymers, including epoxy, silicone, polyurethane, and polyimide. Their utility arises from their very low electrical resistivities (on the order of 1×10^{-5} Ω-cm) and high thermal conductivities (3–5 times higher than conventional silver-filled epoxies).

Metal-filled polymers have been studied for use as switchgear for power engineering, where contact separation under electrical load is accompanied by an electrical arc [47]. Metal-filled polymer electrodes are able to extinguish arcs quickly, because gaseous decomposition products from the metal-filled polymer produce an axial gas blasting action on the arc column to quench it. Because the metal-filled polymer electrodes deteriorate during high current flow, they must be moved into place by a sliding contact mechanism immediately before switching. In a related application, the positive temperature coefficient of resistance of conductive composites is used to provide overcurrent protection [48]. As a composite with a crystalline polymer is heated by the passage of electrical current, the volume expansion in the temperature region of the polymer melting point disrupts the conductive filler network, causing the resistance of the composite to increase greatly. The composite thus shows two distinct, current-dependent behaviors. At low current, it conducts electricity efficiently, but it becomes a current-limiting device as it is heated by higher currents.

Architectural shielding is among the more novel uses of conductive composites [49]. This application ranges from eliminating the effects on television reception of nearby radio transmitters to preventing external monitoring of communications in "secure" areas. Nonwoven mats of high aspect ratio (up to 4000) fibrous conductive filler are laid up, using either wet or dry processes. These are then laminated with polymer films to produce a conductive composite, which can be used to replace conventional wall coverings.

All the applications above take advantage of the electrical conductivity of the composites. However, the addition of electrically conductive fillers usually produces a marked increase in thermal conductivity as well. In some applications, the thermal conductivity is used to counter heat buildup and thermal distortion (i.e., heat sink applications). An advantage of this approach is that heat sinks can be molded as an integral part of an equipment housing, rather than being added as an extra piece of hardware. In other applications, the conductive composite is used as a heating element, in heated tools for blow molding, vacuum forming, sheet molding, and mat molding [9], for curing parts, and in wire heating elements. Conductive composites also can be used to make self-heating pipes and ducts useful for preventing liquid freeze-up and for transport of viscous liquids [7]. The thermal conductivity of the composite can be controlled over much of the range defined by the limits set by the pure filler and matrix materials.

Market Size

The market for conductive composites and its expected growth vary from one forecast to another. In addition, published numbers are sometimes defined in such a way that the desired information must be inferred. However, there is general agreement that the market, dominated by electronics applications, is both substantial and increasing rapidly. For example, one study [50] estimated that the 1985 market for protecting plastic enclosures for electronic equipment against electromagnetic interference and electrostatic discharge was more than $100 million. A 16% growth rate was predicted, with the market exceeding $200 million by 1990. Conductive composites, as mentioned above, compete against alternatives such as metal paint and flame-sprayed metal coatings for this market.

A 1984 study [51] predicted that consumption of conductive plastics would increase at 58% per year, from 10.3 million pounds in 1983 to more than 100 million pounds in 1988. The predicted usages included a contribution from inherently conductive plastics in addition to conductive composites. In rough agreement with this, other studies predicted a market of 50 million to 60 million pounds per year for conductive composites by 1987 [9].

Initial market penetration and growth were slower than expected, however, and conductive composites have not met earlier expectations in the electronics industry. In 1985, one estimate for total annual sales of conductive composites was no more than 250,000 pounds, of which less than 100,000 pounds was for thermoplastic compounds [11]. As a result, some producers have set aside plans to pursue the business machine housing market and have focused or specialty or "niche" markets, in which higher cost, more sophisticated approaches to EMI shielding and ESD can be justified.

Dale E. Hall

[*See also* CONDUCTIVE COMPOSITES.]

References

1. D. M. Bigg, in S. K. Bhattacharya, Ed., *Metal-Filled Polymers: Properties and Applications*, Dekker, New York, 1986.
2. E. K. Sichel, Ed., *Carbon Black–Polymer Composites: The*

Physics of Electrically Conducting Composites, Dekker, New York, 1982.

3. M. T. Kortschot and R. T. Woodhams, *Polym. Compos., 6*(4), 296 (1985).
4. J. V. Milewski, *Ind. Eng. Chem. Prod. Res. Dev., 17*(4), 363 (1978).
5. T. A. Ezquerra, J. Martinez-Salazar, and F. Balta Calleja, *J. Mater. Sci. Lett., 5,* 1065 (1986).
6. J.-P. Reboul, in S. K. Bhattacharya, Ed., *Metal-Filled Polymers,* Dekker, New York, 1986.
7. R. P. Kusy, in S. K. Bhattacharya, Ed., *Metal-Filled Polymers,* Dekker, New York, 1986.
8. D. M. Bigg, *Polym. Compos., 4*(1), 40 (1983).
9. R. A. Crossman, *Polym. Eng. Sci., 25*(8), 507 (1985).
10. A. L. Holbrook, *Int. J. Powder Metall., 22*(1), 39 (1986).
11. E. Kiesche, *Plast. Technol.,* November 1985, pp. 77–81.
12. S. R. Gerteisen, Northcon/85, Oct. 22–24, 1985.
13. K. Nangrani and S. Gerteisen, *Res. Dev.,* July 1985, pp. 60–63.
14. D. E. Davenport, *Polym. News, 8,* 134 (1982).
15. D. E. Hall and H. Ando, *Proceedings of the 31st Int. SAMPE Symposium,* 1986, pp. 1340–1349.
16. D. M. Bigg, *J. Rheol., 28*(5), 501–516 (1984).
17. D. A. Jaworske, J. R. Gaier, C. C. Hung, and B. A. Banks, *SAMPE Q., 18*(1), 9 (1986).
18. D. E. Hall, in D. R. Turner and L. T. Romankiw, Eds., *Symposium on Electrodeposition Theory, Technology, and Practice,* Fall 1986 Meeting of the Electrochemical Society, San Diego, CA, Pro. Vol. 87-17, The Electrochemical Society, 1986, p. 684.
19. J. M. Crosby and J. E. Travis, *Rubber World, 193*(2), 30 (1985).
20. M. V. Murthy, B. A. Luxon, R. S. Carey, and J. D. Kuc, *Proceedings of the Fourteenth Annual Structural Foam Conference and Parts Competition,* Society of the Plastics Industry, 1986.
21. M. Narkis, J. Yacubowicz, A. Vaxman, and A. Marmur, *Polym. Eng. Sci., 26*(2), 139 (1986).
22. P. Datta and R. N. Friel, *Polym. Prepr., 24,* 193 (1983).
23. S. Monte and G. Sugerman, *Rubber World, 190*(1), 20 (1984).
24. M. E. Galvin and G. E. Wnek, *Polym. Commun., 23,* 795 (1982).
25. A. J. Klein, *Adv. Mater. Processes, 2*(7), 25 (1986).
26. C. K. Chiang, B. J. Bauer, R. M. Briber, and G. T. Davis, *Polym. Commun., 28,* 34 (1987).
27. R. M. Penner and C. R. Martin, *J. Electrochem. Soc., 133,* 310 (1986).
28. M. DePaoli, R. Waltman, A. Diaz, and J. Bargon, *J. Polym. Sci., Polym. Chem. Ed., 23,* 1687 (1985).
29. S. D. Gerbig, *Eval. Eng., 24*(5), 84 (1985).
30. T. C. Wilkinson, S. F. Chappell, and W. L. Kelly, *Polyurethane: New Paths, Progress in Marketing and Technology, Proceedings of the 6th SPI Int. Technical and Marketing Conference,* 1983, pp. 125–127.
31. M. Sumita, H. Abe, H. Kayaki, and K. Miyasaka, *J. Macromol. Sci.-Phys., B25*(1&2), 171 (1986).
32. D. M. Bigg, *Polym. Compos., 8*(1), 1 (1987).
33. D. M. Bigg, *Polym. Compos.,* 7(2), 69 (1986).
34. L. J. Buckley, I. Schaffer, and R. E. Trabocco, *Proceedings of the 29th Annual SAMPE Symposium,* 1984, pp. 1127–1140.
35. B. A. Luxon, *Abstracts of Chicago Section Technical Conference,* Society of Plastics Engineers, June 19–20, 1984.
36. R. M. Simon, *Ind. Res. Dev., 24*(6), 104 (1982).
37. D. M. Bigg, W. Mirick, and D. E. Stutz, *Polym. Test., 5,* 169 (1985).
38. S. R. Gerteisen, personal communication, January 1988.
39. *Plast. Eng.,* November 1986, p. 62.
40. N. Fountas, *Plast. World,* June 1978, pp. 66–67.
41. D. Luch, U.S. Patent 4,009,093 (Feb. 22, 1977), to The International Nickel Company.
42. D. M. Lindsey, *Prod. Finish.,* July 1979, pp. 34–43.
43. A. Auerbach, C. W. Eichelberger, and R. J. Wojnarowski, *J. Electrochem. Soc., 130,* 1111 (1983).
44. R. E. Evans, D. E. Hall, and B. A. Luxon, *SAMPE Q., 17*(4), July 1986.
45. J. J. Reilly, S. J. Thoman, and W. W. Lin, *Proceedings of the 18th Int. SAMPE Technical Conference,* 1986, pp. 22–31.
46. *Mater. Processing Rep., 2*(6), 1 (1987).
47. B. Gengenbach, U. Mayer, R. Michal, and J. Radbruch, *IEEE Trans. Compon. Hybrids Manuf. Technol., CHMT-8*(1), 52 (1985).
48. F. A. Doljack, *IEEE Trans. Compon. Hybrids Manuf. Technol. CHMT-4*(4), 372 (1981).
49. D. Pendzimas and Z. Mate, "Architectural Shielding with Advanced Nonwoven Structures," Index 87 Congress, Brussels, 1987.
50. *Journal of Commerce,* Aug. 13, 1985, p. 21B.
51. *Journal of Commerce,* Oct. 15, 1984, p. 22B.

Contact Molding

See Molding

Continuum Damage Mechanics

Fiber-reinforced composite materials display distributed damage when they are subjected to thermomechanical loads. Consequently, their properties (e.g., stiffness, electric conductivity, damping coefficient) change permanently, and the subsequent response to a given loading is determined by the new properties. A discipline of mechanics that addresses the problems of describing the state of damage and its evolution, and the relationships between materials response characteristics (properties) and damage has come to be known as (continuum) damage mechanics.

Basically two approaches are taken to deal with the problems: one is micromechanical, the other phenomenological. In the first approach, typical details of damage (e.g., a crack or a void) are analyzed to obtain local stress and strain fields, from which the overall response characteristics are estimated by suitable averaging schemes. This approach can treat relatively simple situations and, considering the observed complexities of damage in composites, does not appear to be of much practical use. In the phenomenological approach, on the other hand, general relationships between materials response characteristics and the (internal) state of damage are formulated. The relationships involve certain coefficients that relate the state variables to the materials response characteristics. These coefficients, described as phenomenological constants, are then evaluated experimentally by activating the physical phenomena of interest (e.g., elastic deformation, yielding, creep) and measuring the response characteristics for selected material states. The generality of the principles underlying the relationships assures the validity of the coefficients to all material states within the considered phenomena, and the approach has potentially wide applicability. This article focuses on the latter approach to damage mechanics.

Mechanisms of Damage

Damage in composite materials has been studied for more than two decades. The basic mechanisms operating under mechanical loads in thermosetting polymeric matrix composites are fairly well understood. These mechanisms are described briefly, with separate headings for short-fiber composites and for long-fiber composites.

SHORT-FIBER COMPOSITES. The early studies in this area, conducted in the 1960s, were primarily concerned with damage under static and cyclic loading of chopped strand mats of E-glass in a polyester resin. The first mechanism of damage to initiate was debonding of fibers lying normal to the applied tensile stress. Increased stress or increased cycling of a given stress amplitude led to debonding of fibers of orientations deviating increasingly from the normal to the stress axis. The next mechanism to initiate was microcracking of resin in the resin-rich areas in the vicinity of the fiber debonding zones. Figure 1 illustrates the fatigue life spent in initiating fiber debonding, matrix cracking, and total separation of specimens of a chopped strand mat reinforced polyester.

More recent studies have attempted to quantify the distribution of crack length and crack orientation in short fiber composites under mechanical loads. The cracks observed were the matrix cracks, the fiber–matrix interface cracks, and the fiber end cracks. The distributions of the length and orientation of all these cracks taken collectively displayed certain characteristic features. For example, the crack size and the crack number density were found to be functions of the applied stress amplitude; high amplitudes give numerous cracks of small size, while low amplitudes lead to fewer cracks of larger size. In addition, most cracks form within 30° off the normal to the loading direction.

LONG-FIBER COMPOSITES. In unidirectional composites loaded along fibers in tension, the fibers break at their weak points. The fiber failures occur at isolated locations spread randomly in the volume of a composite. Increased stress causes an increase in the density of

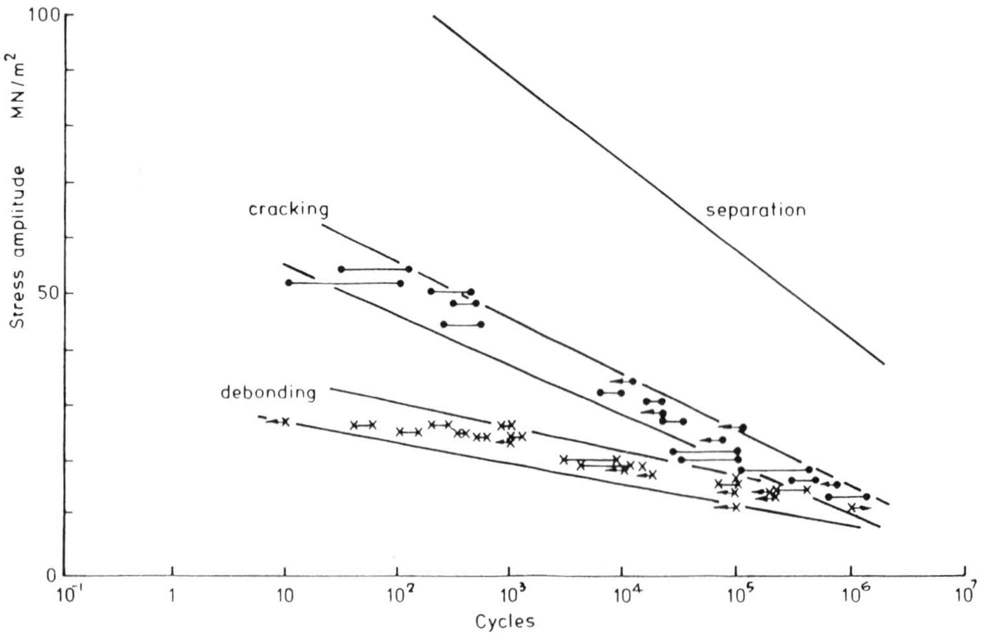

FIGURE 1 Fatigue life expended in initiating debonding, matrix cracking, and final failure (separation) of chopped strand mats reinforced polyester. (From Ref. 1.)

fiber breaks and, at the same time, fiber breaks tend to lump together in some areas. When the number of adjacent fiber failures reaches a critical value, which is typically 5–7, a large crack initiates from the failed fiber zone and spreads laterally. The crack also causes some interfacial failure, depending on the interfacial strength, and leads to separation of the specimen.

In tensile fatigue of unidirectional composites loaded along fibers, the damage mechanism operating under a low stress is matrix microcracking normal to the fiber axis. The cracks are initially arrested by fibers and, if the stress is below a certain limit, there is no further progress in crack growth. However, higher stress may result in fiber failures, which may then assist matrix crack growth. A growing matrix crack cuts through some fibers, is bridged by others and, at later stage, causes interfacial cracking. At still higher applied stress, extensive fiber failures may cause failure governed by a statistical process that does not have an underlying progressive mechanism. Figure 2 shows the basic damage mechanisms that operate either separately or with mutual interaction in different stages of the fatigue process.

Mechanisms of damage in laminated composites have been studied extensively since the early 1970s. Despite the complexities of details, some general features of the mechanisms, valid for a large class of composites, can be described. Figure 3 offers a good summarizing picture. The damage process is seen as having certain stages. The first stage is dominated by matrix cracking along fibers in plies inclined to the loading direction. The matrix cracks develop as arrays of parallel cracks whose average spacing reduces with loading and approaches a saturation state, unless failure of the specimen occurs by another cause. The saturation state, where all off-axis plies have attained their minimum spacing, has been found to be a characteristic property of the laminate independent of the actual loading history. This state has been called a characteristic damage state (CDS) [2]. In the stage that follows CDS the intralaminar cracks divert locally into the interlaminar regions common to intralaminar cracks of adjacent plies. This causes local delamination, which grows with further loading and on merging with neighboring delaminations causes large-scale delamination. The final stage of damage usually is dominated by unstable growth of locally failed regions and has extensive fiber failure activity.

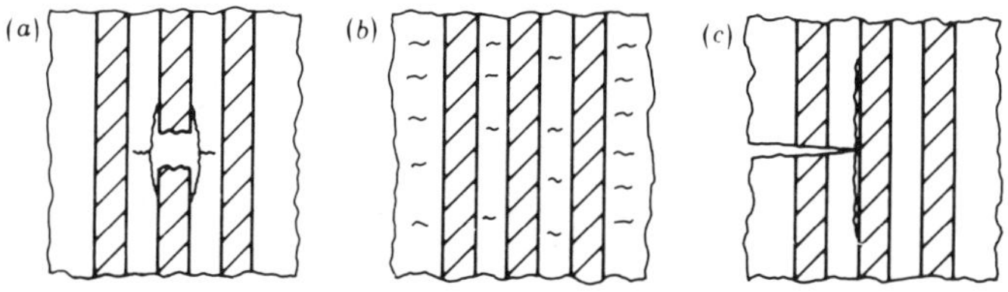

FIGURE 2 **Damage mechanisms in unidirectionally reinforced fiber composites loaded cyclically in tension parallel to fibers. (*a*) Fiber breakage, interfacial debonding. (*b*) Matrix cracking. (*c*) Interfacial shear failure.**

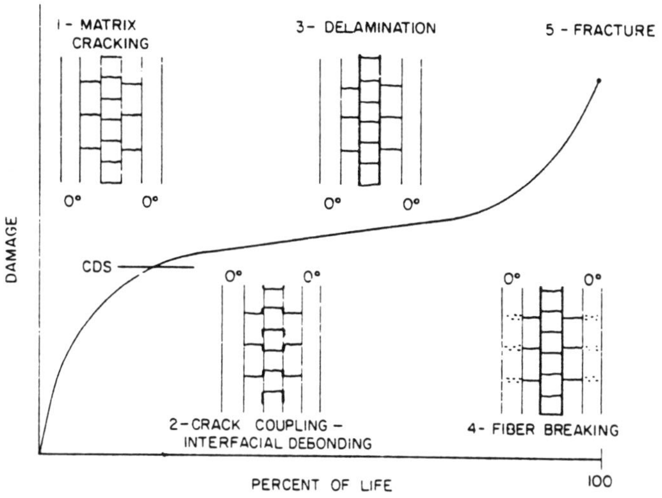

FIGURE 3 **Development of damage in composite laminates.** (From Ref. 2.)

Characterization of Damage

There are basically two approaches for characterizing damage. One, which may be called a direct or geometrical characterization, is based on the size and orientation features of cracks, voids, debonds, etc. The other approach is to use change of a certain property (e.g., stiffness, damping, electric conductivity) to construct a measure of damage. The first approach should be preferred, whenever possible, on the grounds that it does not a priori assume relationships between damage and property degradation. Indeed, the primary objective of characterizing damage is determination of such relationships. We discuss the geometric approach to damage characterization.

GEOMETRICAL CHARACTERIZATION: DEFINITIONS. Two definitions are given here to facilitate a particular characterization to be discussed later.

A *damage entity* is defined as an individual microstructural change brought about in a material by an internal dissipative mechanism. Once such microstructural change has occurred, it is irreversible. Examples of damage entities in metals are slipped surfaces caused by intragranular slip and voids caused by grain boundary sliding and diffusion. In thermosetting polymeric composites, typical damage entities are matrix cracks, fiber–matrix interfacial debonds, and fiber breaks.

A *damage mode* is a collection of damage entities having the same (or similar) geometrical features. More precisely, a set of damage entities whose formation and growth is governed by the same mechanism may be grouped together to be treated collectively as a damage mode.

A TENSORIAL CHARACTERIZATION. At first we need some way to characterize (i.e., to quantify the existence of) a damage entity. Physically, a damage entity can be considered to consist of a surface bounding a microvolume in a given material. The microvolume may be empty (i.e., a void) or filled with air, liquid, or a solid material. We may account for the generally irregular surface, to some extent, by assuming it to be an ellipsoid with its characteristic planes oriented at some angles to a reference frame of axes. A vector-valued function will be needed to express the size of an ellipsoid and its orientation in space. In the special case of a set of ellipsoids oriented randomly, an average characterization will reduce to a scalar-valued function. However, a conceptually different and perhaps more rational characterization may be achieved if the *influence* of a damage entity is included along with its geometry (i.e., size and orientation) in the characterization.

We rationalize this by pointing out that a damage entity makes its existence felt only by influencing the surrounding medium in some way. Only viewed through that influence can the damage entity be seen as existing. For example, cracks of certain orientation in a solid may not affect some deformational properties (e.g., stiffness), but they may change electric conductivity or other properties. A quantification of the influence of a damage entity will thus depend on what materials response characteristics are to be described. For mechanical response measured in terms of stress and strain, the influence of a damage entity should be viewed in terms of the local stress and strain fields. Or more precisely, the disturbance caused in the local stress and strain fields by embedding a damage entity can be quantified in some way to provide measure of the influence sought. Such quantification is the concern of a micromechanics approach, which then uses it in an averaging scheme to calculate the altered overall properties of a damaged solid.

In the phenomenological approach, however, exact quantification is not necessary; only an appropriate form of the influence is required. The phenomenological coefficients associated with the damage entities, to be determined experimentally, then determine the extent to which the influence has been translated into the overall properties. The choice of the form of the influence must be made judiciously, accounting also for the feasibility of determining the resulting phenomenological coefficients.

Consider now a generic point P in a damaged composite solid with reference to a fixed Cartesian coordinate axes x_i, $i = 1, 2,$ and 3 (see Fig. 4). Consider further an element of volume V about the point P containing a statistically representative set of damage entities in the neighborhood of point P. Focus attention on a single damage entity of surface S. To a point on the surface assign two vectors with components a_i and n_i in the coordinate system. The vector (with components) a_i represents the influence of the considered point, and n_i is the outward normal to the surface at the point. By taking a vectorial quantity for the influence we intend to incorporate both the severity and the directionality of the influence. Thus, if the deformational response of a solid is of interest, the influence vector a_i may be seen as being the displacement vector at the considered point. Having two vectors a_i and n_i at each point on the surface of a damage entity, then, we may define a dyad (i.e., a second-order tensor) given by

$$d_{ij} = \int_S a_i n_j \, dS \tag{1}$$

and call it a *damage entity tensor*.

The state of damage at the generic point P may be represented, in an average sense, by the set of *damage mode tensors* $D_{ij}^{(\alpha)}$, where $\alpha = 1, 2, \ldots, n$ designates a damage mode, given by

$$D_{ij}^{(\alpha)} = \frac{1}{V} \sum (d_{ij}^{(\alpha)})_k \tag{2}$$

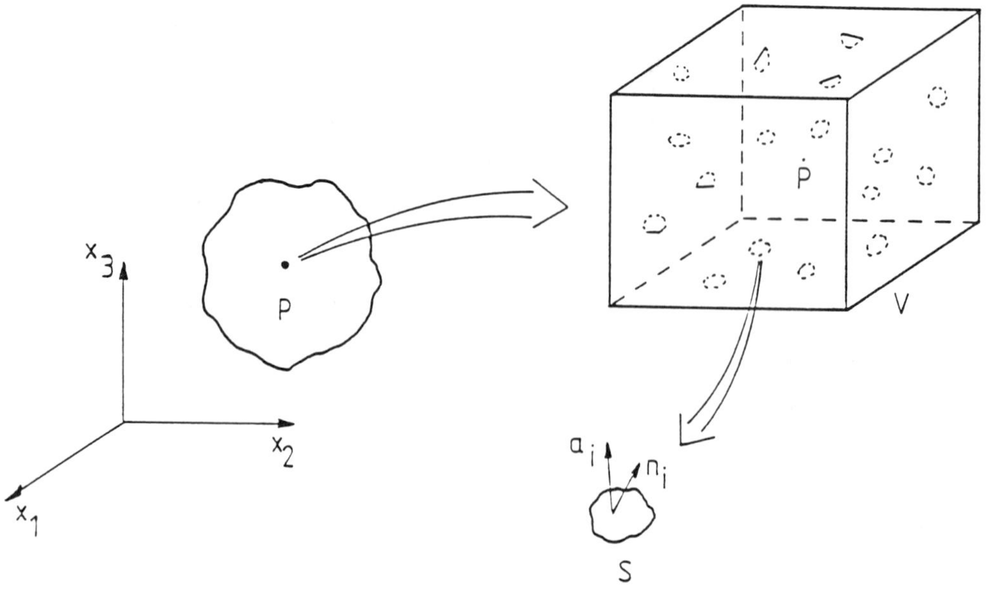

FIGURE 4 A tensorial characterization of damage.

where the summation on the right-hand side of the equation is carried over k damage entities, all belonging to a damage mode designated by α.

Mechanical Response of Composites with Damage

The phenomenological approach of damage mechanics postulates that the response of a material participating in certain phenomena—physical, chemical, or both—can be described by a set of response functions. For materials bearing purely mechanical loads, the assumed response functions are the stress and the Helmholtz free energy. These functions are assumed to depend on the deformational state of material given by the deformation gradient or, for small deformation gradients, by the strain. If the microstructural constitution of a material undergoes changes during mechanical loading, an additional variable or set of variables called the internal state variables is needed to represent this altered state. The damage entities in a composite material usually are of such dimensions that they may be regarded as microstructural entities, and the set of damage mode tensors $D_{ij}^{(\alpha)}$ defined in Eq. (2) may be taken as a set of internal state variables.

From thermodynamic considerations it may be shown that stress is derivable from the Helmholtz free energy. Thus either of the two response functions can be taken to describe the mechanical response. It is preferable to use energy since it is a scalar quantity. Thus, we have

$$\psi = \psi(e_{ij}, D_{ij}^{(\alpha)}) \tag{3}$$

and

$$\sigma_{ij} = \rho \frac{\partial \psi}{\partial e_{ij}} \tag{4}$$

where ψ is the Helmholtz free energy per unit mass, e_{ij} is the strain tensor, σ_{ij} is the stress tensor, and ρ is the mass density.

The function ψ may now be expanded in terms of its variables as a polynomial up to the number of terms desired. The coefficients associated with the terms would be the phenomenological constants, which take specific values for a given material. The number of such constants can become very high in a general case, thereby rendering their experimental determination impractical. It is, however, possible in most situations to reduce this problem by resorting to certain restrictions that may be placed on the function ψ.

The most important of these restrictions is given by the initial symmetry, if any, of the material. Thus, if the material is initially isotropic, which is typically the case for short-fiber composites, the polynomial for ψ can be expressed in terms of the invariants of e_{ij} and $D_{ij}^{(\alpha)}$. The number of resulting phenomenological constants is then drastically reduced. Another example is that of laminated composites, which are often constructed to have orthotropic symmetry. Here, too, the number of phenomenological constants become manageable in comparison to the case of general anisotropy. Another restriction to ψ can be placed by establishing certain initial conditions that must be satisfied. For instance, the material can be required to be stress-free and undamaged in the initial state.

The author has developed a phenomenological theory of mechanical response of composites with damage

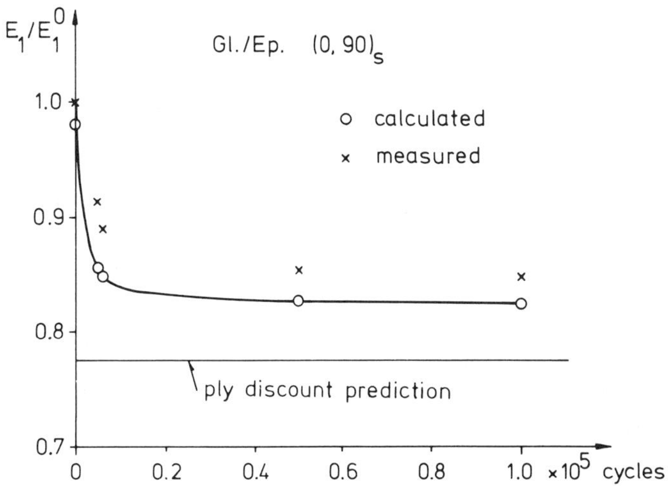

FIGURE 5 **Variation of the longitudinal Young's modulus E_1 normalized by its initial value E_1^0 of a glass–epoxy $(0,90)_s$ laminate under tensile fatigue. The calculated values are by Talreja [5] and the bound to modulus change is given by an engineering scheme called ply discount method.**

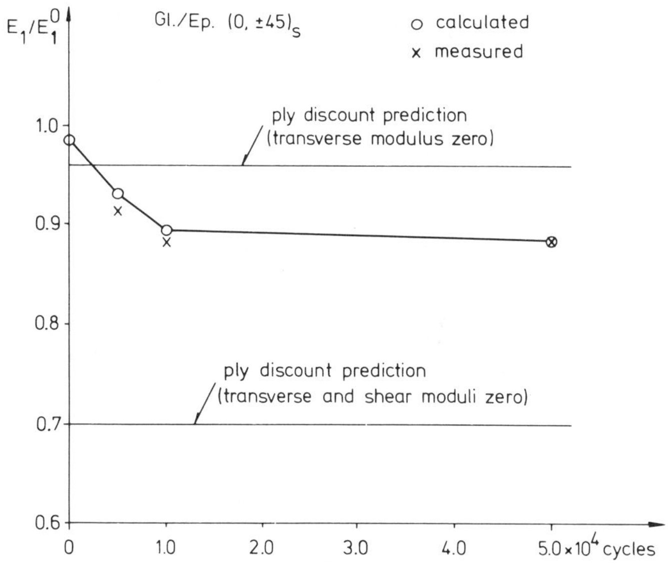

FIGURE 6 **Variation of the longitudinal Young's modulus of a glass–epoxy $(0, \pm 45)_s$ laminate under tensile fatigue. The ply discount method gives two different and inaccurate predictions, depending on the assumptions made regarding loss of properties of a cracked ply.**

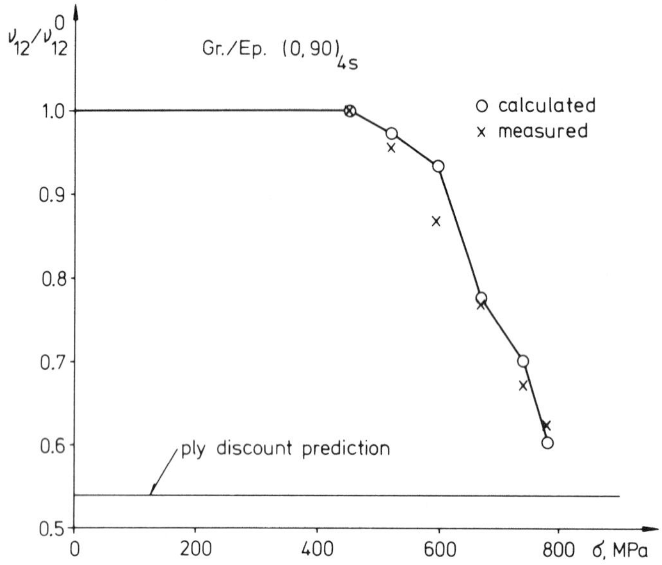

FIGURE 7 **Variation of the Poisson ratio ν of a graphite–epoxy $(0{,}90)_{4s}$ laminate under monotonically increasing tensile stress.**

along the lines described above [3–6]. A useful result of the theory is the description of the changes in the stiffness properties caused by damage. The evaluation of the phenomenological constants from experimental data has been illustrated for laminates of glass–epoxy and graphite–epoxy, and predictions of stiffness changes with damage in these materials have been found to compare favorably with experimental results. Typical examples of these comparisons are shown in Figures 5–7, which also illustrate the development of damage in these composites.

R. Talreja

References

1. T. R. Smith and M. J. Owen, *Mod. Plast., 46*(5), 128 (1969).
2. K. L. Reifsnider, E. G. Henneke, W. W. Stinchcomb, and J. C. Duke, *Mechanics of Composite Materials—Recent Advances,* Pergamon Press, Elmsford, NY, 1983, pp. 399, 420.
3. R. Talreja, *Proc. R. Soc. London, A399,* 195 (1985).
4. R. Talreja, *Advances in Fracture Research,* Vol. 4, Pergamon Press, Elmsford, NY, 1984, pp. 3013, 3020.
5. R. Talreja, *J. Compos. Mater., 19,* 355 (1985).
6. R. Talreja, *Eng. Fract. Mech., 25*(5–6), 751 (1986).

Copper Matrix Composites

See Metal Matrix Composites

Core Composite and Sandwich Structures

The typical arrangement of the materials used in core composite and sandwich structures is shown in Figure 1. There is no required relationship between the thickness of the facings and the thickness of the core. However, in most practical cases, the core is at least three or four times the thickness of the facings, and the more efficient sandwich structures are normally observed to have cores an order of magnitude thicker than the facings. Nevertheless, a number of recent designs of advanced composites illustrate that significant advantages are realized when the core is only two or three times the facing thickness. This is particularly true in the case of carbon fiber facings, where the bending stiffness of the laminate often needs only a modest increase to achieve design objectives.

The relative amount of the increase in both stiffness and strength offered by the use of a composite core is easily illustrated by variations in the simple sandwich shown in Figure 2. For the purpose of illustration, a composite facing is shown as composed of two members, each forming half the total thickness. The increase in bending stiffness and strength that can be achieved by the addition of a composite core placed in between the two half-thicknesses of the original laminate can be quite large, considering the small increase in weight required.

Although the values in the example were originally calculated by Hexcel for aluminum facings, with a highly efficient honeycomb core material and adhesive system, quite similar results are obtained in nonaircraft struc-

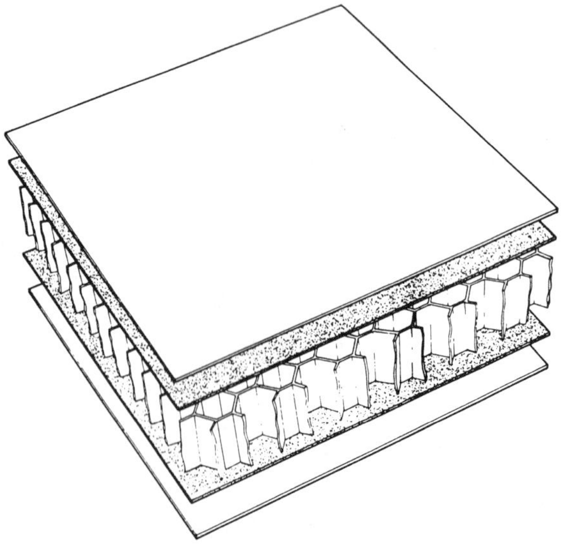

FIGURE 1 **A sandwich structure must have thin, dense, high strength facings, a low density core, which is usually thick and relatively weak, and a rigid and strong attachment between them. If any one of these key requirements is missing, the structure is not a sandwich.**

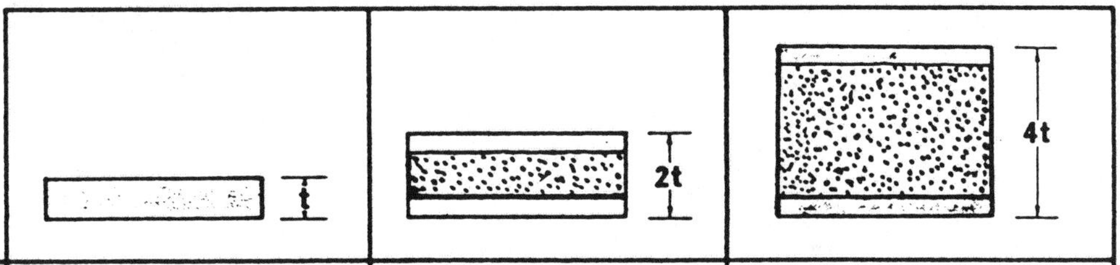

Relative Stiffness (D)	1	7.0	37.0
Relative Strength	1	3.5	9.2
Relative Weight	1	1.03	1.06

FIGURE 2 The overriding reason for choosing a sandwich structure over a rib-stiffened or other simpler structure is nearly always the added stiffness and strength that the sandwich offers. The chart displays the remarkable improvement that a sandwich offers when more stiffness or strength is needed in a beam, a column, or a plate.

tures when one of the low cost foams or wood materials is used to make the core. In many cases, a sandwich design approach is the most efficient in either stiffness or strength for large surfaces expected to carry bending, in-plane shear, or in-plane compressive loads. In normal design practice, determination of the actual thickness and construction details that will yield the optimum design becomes a rather complex problem. Figures 3 and 4 show the results of such an optimization study. This study was performed to establish relative efficiency of several design approaches for cylinders under external pressure. Additional detail is provided in Ref. 1.

The use of core materials to make sandwich structures is one of the earliest examples of composite structures to reach broad usage. Radomes on military aircraft were built in this manner during and after World War II, and by the late 1940s were accepted as the norm by both military and commercial aircraft users. By the mid-1950s, virtually all commercial and military aircraft employed radomes having core materials of glass fiber honeycomb, urethane foam, or rectangular glass fiber tubes.

The glass fiber cores consist of composite tubes of rectangular cross section, laid parallel to each other, with the axes of the tubes also parallel to the plane of the facings. They are also called "lost wax cores" after the method by which they are produced. This arrangement permits circulation of warm air through the sandwich core, so that anti-icing capability may be built into the radome structure.

With the development of resin systems having both higher strength and higher toughness, as well as strength at much higher temperatures, the composite sandwiches worked their way into a few nonradome structures in the 1960s. The development of S-glass fibers and Kevlar and carbon fiber materials combined to make these structures even more attractive in the 1970s. As a

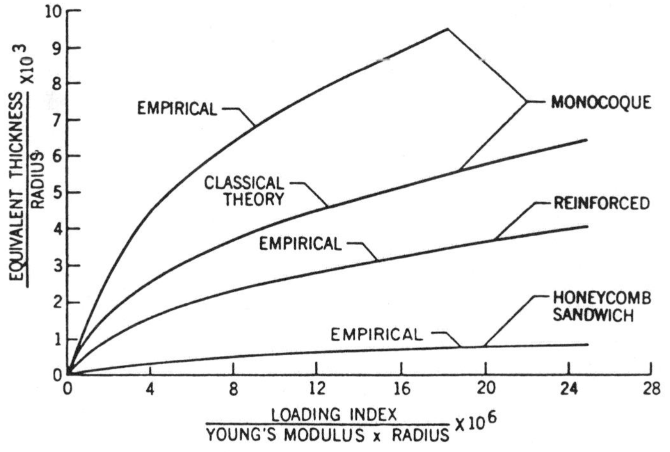

FIGURE 3 Strength-to-weight comparison of deep spherical caps under external buckling pressure.

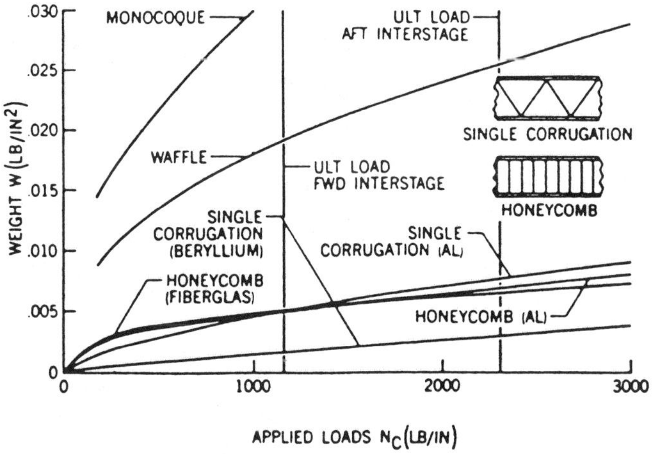

FIGURE 4 Comparison of minimum weight designs for various structural systems and materials in an axially loaded cylinder.

result, composite structures having many different composite core materials now see wide use in nearly all military and commercial aircraft.

The advances in these areas created a body of technology and manufacturing experience that made composite sandwich structures economically feasible in other product areas, such as boat hulls, military vehicles, small aircraft (particularly amateur-built aircraft), truck bodies, and large tanks, as well as many other very ordinary applications.

To form a true "structural sandwich," both facings and core must be present, in addition to a strong and rigid attachment between them. This attachment is so important that sandwich-type structures made with rubber cement or other soft adhesives do not qualify as structural sandwiches, even if the core and facings are properly chosen. Such "rubbery" structures do have a place in the scheme of things, and some very good baggage compartment liners are made this way. However, these structures do not follow the usual formulas for sandwich loads and deflections.

Figure 2 points out the importance of making the core as thick as practicable, and in most sandwiches the core is much thicker than the facings. This relatively thick core means that small differences in core density are important in the overall weight of the sandwich. As a result, it is quite common to compare core materials to each other by quoting the strength at a given density. The strength most often quoted is the compressive strength, as it is the easiest to visualize and to test. The engineers who design with these materials, however, usually also look at the shear strength and the shear modulus.

In the sandwich structures field, it has been found that a great many materials make excellent sandwich cores. Following are descriptions of some of the currently used cores, along with a brief discussion of their differences.

Types of Core

Woods as Composite Cores

Some of the most commonly used and lowest cost cores are light woods. The most popular is balsa, but substantial use has also been made of spruce, mahogany, redwood, pine, fir, and many others.

The reason for the popularity of balsa is its good strength at densities as low as 128 kg/m^3. When oriented in the sandwich so that the grain direction is perpendicular to the sandwich facings, the compressive strength of balsa, when measured at equal density, is higher than nearly all other cores, including some of the high performance honeycombs.

Balsa, as well as most other woods, suffers from having a shear strength that is rather low compared with its outstanding compressive strength, as well as from the usual wood problems of sensitivity to fungus, rot, and moisture absorption. Because the material has excellent distribution and reasonable cost, it has been used in many commercial, industrial, and marine applications—many more than all the other wood cores combined.

Expanded Polymer Foams

The foamed plastic composite core materials are widely used in surfboards, wind surfers, and boat hulls, as well as in commercial and amateur-built aircraft, primarily because they are both reasonable in cost and easy to work with. Each of these material families has a unique set of mechanical properties, physical properties, and working and handling characteristics, as well as its own cost structure. It is nonsense to think that one of them will be just like another.

In addition to the differences from one chemical type to another, there are major differences between the low density and high density members of the same family. A 32 kg/m^3 polyurethane will turn out to be vastly different from a 320 kg/m^3 piece having exactly the same chemistry and appearance. The resistance to gasoline and solvents, of course, will be about the same, since both materials are made from the same polymer.

A few of the more commonly used foam materials are listed in sections that follow, along with some of the characteristics which distinguish them.

POLYSTYRENE FOAM. This is the material used in the wings and control surfaces of the VariEze, Long-EZ, Dragonfly, and the many other homebuilt airplanes sparked by the innovative designs and manufacturing methods of Burt Rutan. It even appeared in the last few feet of the wings of *Voyager*, the famous homebuilt airplane that flew around the world on a single tank of gasoline.

Polystyrene foam is used because its strength properties are well matched to the needs of these particular structures and because it can be hot-wire contoured without, like the urethanes, generating poisonous gases. It has the strong disadvantage of being soluble in styrene monomer, which is a major part of polyester and vinyl ester resins, and it is softened badly by exposure to gasoline and many other solvents.

The most well known of these materials is styrofoam, produced as a continuous slab by the Dow Chemical Company. It is quite uniform and reliable in a density of 32 kg/m^3. Be careful not to inadvertently use any "bead foam" material, as found in the manufacture of picnic cooler chests, disposable cups, and many other low cost items made by exposing styrene beads, or granules, to live steam while they are contained in a closed mold. This procedure causes the beads to expand and bond to each other. However, the bonds may vary greatly in strength, and as a consequence, the material is usually not suitable for use as a sandwich core material.

In addition to an unpredictable strength level, the pattern of the expanded styrene beads usually leaves an interconnecting web of tiny spaces, which makes the finished material permeable to most liquids and gases after the skin next to the mold is cut away. Definitely *NOT* the material to use in a critical composite structure.

POLYURETHANE FOAM. This family of materials includes both rigid and flexible versions, a great many choices of density, and several chemically different types. Care must be taken in selection of these materials, since many of the urethane foam producers are principally in the cushion, mattress, and padding business and do not understand the unique requirements of the sandwich core type of application.

Typical problems of the off-breed urethanes are excessive brittleness, serious density variations within a single piece, internal cracks, and soft areas. Fortunately, there are some good suppliers who specialize in composite core supply, and their materials rarely have such defects.

Two materials in this latter group are Last-A-Foam, made by General Plastics, and Clark Foam, made by Clark Foam Products. Last-A-Foam is light brown; it is slightly more brittle than the Clark product and possesses better elevated temperature properties. Clark Foam is a brilliant white, and quite tough.

Both companies make only structural foams. Clark is an ester-based material, while Last-A-Foam and most other urethane foams are based on polyether. Densities from about 40–320 kg/m^3 or higher are offered, and both suppliers make some excellent literature available to users of the materials. Clark Foam will also have a slightly smaller and more uniform cell size, and is noticeably tougher than nearly any of the other rigid urethane foams available.

A few words of caution are in order regarding all the urethane foam materials.

First, they are usually very flammable and will produce extremely poisonous gases when burning. For this reason it is quite inadvisable to use hot-wire contouring for generating shapes from the material. In addition, these materials should never be incinerated or burned with other trash, even in an open area. The poisonous gases given off are simply too dangerous.

The other caution regards the practice of foaming in place to provide contoured cores. Many boat or plastic supply shops offer two-part, foam-in-place urethane liquids for use in providing flotation in boats. The process seems so simple and obvious that many unwary users (including some of the largest of the big airframe makers) have tried to make the process work to produce a structural core for a sandwich having a complex shape.

The success rate of these many attempts has been almost zero. The problem lies in the extreme complexity of urethane chemistry, which depends not just on the two components mixed in the bucket, but also on humidity, barometric pressure, exact temperature of each of the components, exact temperature of both bucket and mold, and many other difficult-to-control factors.

This wide combination of variables makes it rather unlikely that a first-time, pour-in-place operation will succeed. It also emphasizes that in the composite structures field, much of the success of the final structure invariably depends on the exact outcome of a chemical reaction. This step must be carefully and reliable controlled. Where the situation makes the necessary care and reliability difficult or unlikely, the chemical reaction should be handled by the supplier in his factory, to permit his resident chemist to use the in-house laboratory facilities to make sure that the chemical reaction gives the mechanical properties expected in the final product.

One additional caution regarding the urethane materials should be noted. Some suppliers offer very low densities, and the mechanical property information supplied may include ranges down to 16 or 32 kg/m^3. The problem is that most urethane foam materials become extremely fragile and somewhat unreliable at densities under 60 kg/m^3. It is standard practice, therefore, to use 60 kg/m^3 density as the minimum for acceptable structural performance.

POLYVINYL CHLORIDE FOAMS. PVC foams are based on the same chemical family as the familiar garbage bags, plastic pipe, and plastic films in common use, but are produced quite differently. The manufacturing process is more difficult than those used to produce most other foams, and the materials are consequently higher priced. Several are available in this country, but all are made overseas and imported into the United States.

The most commonly used cores of these foams are Termanto, made in Italy, Klegecell, made in France, and Divinycell, made in Sweden. A fourth material, Airex, made in Switzerland, is more suited to heavier boat hulls than to aircraft or industrial structures. It is also much more expensive than the rigid versions.

Klegecell, Divinycell, and Termanto are fully cross-linked PVC foams and make very good structural cores. Their cell size is somewhat larger than most other types of foam core, and they are not quite so resistant to the wide range of solvents as the urethanes. Densities down to as low as 40 kg/m^3 are available, and properties at the low end of the range are quite good. Surprisingly, strengths of foams at densities under 96 kg/m^3 are somewhat better than for the urethanes, while at densities above 96 kg/m^3 the better urethanes have higher strength than the PVC materials. This crossover in strength properties is usually not seen in other cores.

One problem encountered with the PVC cores (and sometimes other foams too) has been the infrequent occurrence of an undercured area in the material. This shows up after layup of a sandwich as a soft area, having been attacked by the laminating resin system. Such instances, while very perplexing at the time, are rather rare.

Table 1 lists mechanical properties of a representative PVC core, Divinycell.

POLYMETHACRYLIMIDE FOAM. Don't be put off by the impossible name—this is a very high performance material, known by the trade name of Rohacell and available in densities down to a feather weight 30.4 kg/m^3. It is extremely fine in cell size and very uniform.

The mechanical properties of Rohacell are well above those of either the urethanes or the PVC foams at all densities, and it is a brilliant white color, much like Clark Foam. The materials are resistant to nearly all solvents and chemicals, and are good structures up to well above 121°C. They also have the rather unusual property of expanding during cure of the facings, if the cure temperature is raised to 177°C or higher. This allows a perfect fit between facings and core by simply making an approximate shape for the core and curing the entire sandwich in a closed cavity mold, using a cure temperature in excess of 177°C.

This material has no closely equivalent competitor, but it is quite generally available from Rohm Tech, a

TABLE 1
Typical Mechanical Properties Developed by PVC "Divinycell" Foam Core

			Strength or Modulus at 20°C (MPa)							
Quality[a,b]	Applications	Density (kg/m^3)	Compressive Strength, ASTM D 1621	Compressive Modulus, ASTM D 1621	Tensile Strength, ASTM D 1623	Shear Strength, ASTM C 273	Shear Modulus, ASTM C 273	Flexural Strength, ASTM D 790	Flexural Modulus, ASTM D 790	Water Absorption[c] at + 20°C, ASTM D 28425
H 30	Structural insulation core for lightweight sandwich construction in reefers, mobile homes, shelters, buildings, etc.	36	0.36	30	0.55	0.25	8	0.45	12	0.14
H 45	Lightweight structural core for boats, reefers, doors, containers, shelters, etc.	48	0.55	40	1.0	0.45	15	0.8	22	0.097
H 60	Lightweight structural core for boats, reefers, doors, containers when higher physicals than H 45 are required.	60	0.75	50	1.3	0.60	22	1.3	30	0.072
H 80	Structural core for high loadings when thin skin surfaces are used such as trailer, mobile home, and railroad boxcar floors.	80	1.2	80	1.8	0.90	30	2.0	40	0.048
H 100	Structural core for high loadings when thin skin surfaces are used and higher physical properties than H 80 are required.	100	1.8	120	2.5	1.5	45	3.2	70	0.036
H 130	High density core used in larger boats requiring superior physical properties such as minesweepers up to 200 ft long.	130	2.6	160	3.0	1.9	55	3.5	80	0.029
H 200	High density core used for buoyancy at great depths and in the construction of cargo pallets, snow skis, large ships, etc.	200	4.0	230	4.4	3.0	80	6.0	120	< 0.02
H 250	Used in same applications as H 200 when higher physical properties are required.	250	5.5	320	6.0	4.0	100	8.0	150	< 0.02
H 300	Used for buoyancy applications at extreme depths down to 3000 ft.	300	9.0	500	9.0	150	12.0	200	< 0.02	
HT 50	Especially formulated for the aircraft	50	0.7	70	1.1	0.5	17	0.8	22	0.12
HT 70	industry, where higher temperature	70	1.1	90	1.8	0.8	25	1.6	35	0.09
HT 90	and higher resistance properties are	90	1.6	120	2.5	1.2	35	2.5	50	0.07
HT 110	required per given density.	110	2.1	160	3.0	1.6	50	3.0	70	0.05

[a] Data and recommendations for Divinycell are derived from tests and experience but are given without liability to the company.
[b] Additional quality data:
Operating temperature range
H qualities − 200 to + 70°C
HT qualities − 200 to + 120°C
Maximum continuous operating temperature for all qualities, + 160°F (+ 70°C)
Minimum heat-forming temperature
H qualities + 100°C
HT qualities + 140°C
Coefficient of linear expansion, ASTM D 696, approx. 35 × 10^{-} K
Thermal conductivity, ASTM C 518-76 for H 30–H 80 and HT 50 and HT 70, 0.027 W/mK
[c] kg/m^2 of cut surface under 3 m head for 48 hours.
[d] Strength too high to be measured by ASTM C 273 (bond failure).
Courtesy of Diab-Barracuda Corp.

subsidiary of Rohm Chemische of Darmstadt, Germany. This material costs somewhat more than most other foams, and as a result, it tends to be used in the more weight-sensitive or critical applications. Of all the foam cores, it is the only one that is closely competitive, on a strength-to-weight basis, with the Nomex honeycombs in many applications in commercial aircraft. Table 2 lists mechanical properties of several Rohacell materials.

POLYIMIDE FOAM. The polyimide resins are similar in appearance to the phenolics but have much higher temperature resistance. A foam based on this resin system, Solimide, has been developed and is now marketed by the Imi-Tech Corporation. The material will not burn, even after prolonged exposure to temperatures of 816°C. It is available in a density of 96 kg/m^3 and is beginning to see use in firewalls and flame barrier structures. A flexible version of this material is also used in the manufacture of fire-resistant flexible insulation.

Syntactic Foam Cores

The word "syntactic" simply indicates that a foam is made by mixing microballoons, ceramic spheres, or other lightweight aggregate, with a resin system. This provides a material whose density before cure is the same as the density after cure. This is different from most other foams, which are expanded during cure to make the final density much lower than the density of the mixed resin system just before the expansion, or "blowing," which takes place as cure proceeds.

These materials may be made by mixing Q-cells, microspheres, or ground-up low density material of almost any nature with the resin being used. They also may be purchased already mixed with either a one-part or two-part resin system. Many versions are commercially available, but they are so simple to make that most users simply mix them as needed. The usable strength of these foams will depend on the resin system employed, the particular type of low density additive used, and the ratio of resin to filler in the mix. Since the strongest of the range may be several times as strong as the weakest you might use, it is a good idea to stick with the exact mix specified if you are building to somebody else's plans, or, if you are striking out on your own, to do some testing.

A word of caution: Some of the bubbles are provided with a surface finish, just as in fibers, and this finish may be compatible with only one resin system. Be sure that the material you are using is intended for use with the resin system with which you are mixing it. Using the wrong resin system will result in *very* low strength of the final product. Even worse, this low strength may not appear until the finished structure has aged.

The Honeycomb Cores

There are about as many members in the honeycomb family as there are in the foam family of materials, and just as one might expect, the differences between them are rather wide. Most honeycombs appear to be composed of a nested array of hexagonal cells, made from any one of the many standard papers, plastic films, woven fabrics, or metal foils. Many nonhexagonal honeycombs have been used commercially, but the hexagonal cores exhibit both higher strength and a better balance of properties than most other configurations.

The two usual methods of manufacture (Fig. 5) allow the finished core material to be furnished in large pieces, usually 122 × 244 cm, cut to the exact thickness specified by the user. Some of the various configurations of honeycomb core materials are shown in Figure 6.

CELL SHAPE. The primary variables in honeycomb are the material of manufacture and the density of the core. A third important variable is cell configuration. Variations in cell shape arise from problems in manufacture, producing unwanted local or general changes in mechanical properties, or they may be introduced intentionally, offering some specific advantage over the original configuration.

Some of these variations are slight enough to go unnoticed by the casual observer, but the effect on one or more of the mechanical properties of an errant cell shape can be all out of proportion to the change in appearance.

UNWANTED VARIATIONS. The hexagonal honeycombs are produced either by expanding or by the corrugation process, as shown in Figure 5. Each of these manufacturing methods holds its own family of errors in cell shape, and they must be considered separately. The basic cell in both materials can be described as in Figure 7. A true hexagon would have sides *a, b, c,* and *d* all of equal length. Such is rarely the case in production core

TABLE 2
Typical Mechanical Properties of Rohacell IG and Rohacell WF[a]

Property	Units	Rohacell Industrial Grades[b]						Rohacell WF Grades					ASTM Test Method
		31 IG	51 IG	71 IG	110 IG	P170	P190	51 WF	71 WF	110 WF	200 WF	300 WF	
Density	Kg/M^3	30.4	49.6	70.4	110.4	169.6	190.4	49.6	70.4	110.4	200	299.2	D 1622
Tensile Strength	MPa	0.98	1.86	2.74	3.43	7.38	8.34	1.57	2.15	3.62	5.87	10.28	D 638
Compressive Strength	MPa	0.39	0.88	1.47	2.94	6.37 (2.74)	7.65 (3.13)	0.88	1.47	3.52	8.21	15.67	D 1621
Flexural Strength	MPa	0.79	1.57	2.45	4.41	10.27 (9.56)	12.27 (11.78)	1.57	2.84	5.09	11.75	19.58	D 790
Shear Strength	MPa	0.39	0.79	1.28	2.35	4.41 (2.9)	5.39 (2.9)	0.79	1.28	2.35	4.90	7.83	C 273
Modulus of Elasticity	MPa	35.30	68.61	90.32	156.52	313.72	372.33	73.43	102.80	176.24	234.43	365.44	C 638
Shear Modulus	MPa	13.72	20.62	29.44	56.88	117.22	181.34	19.58	32.31	56.79	0.00	0.00	D 2236
Shear modulus	MPa	12.76	18.62	28.41	49.02	86.19	97.91	18.60	28.39	48.95	144.80	293.73	C 273
Elongation at Break	%	3.5	4.0	4.5	4.5	5.0	6.0	3.0	3.0	3.0	3.5		D 638

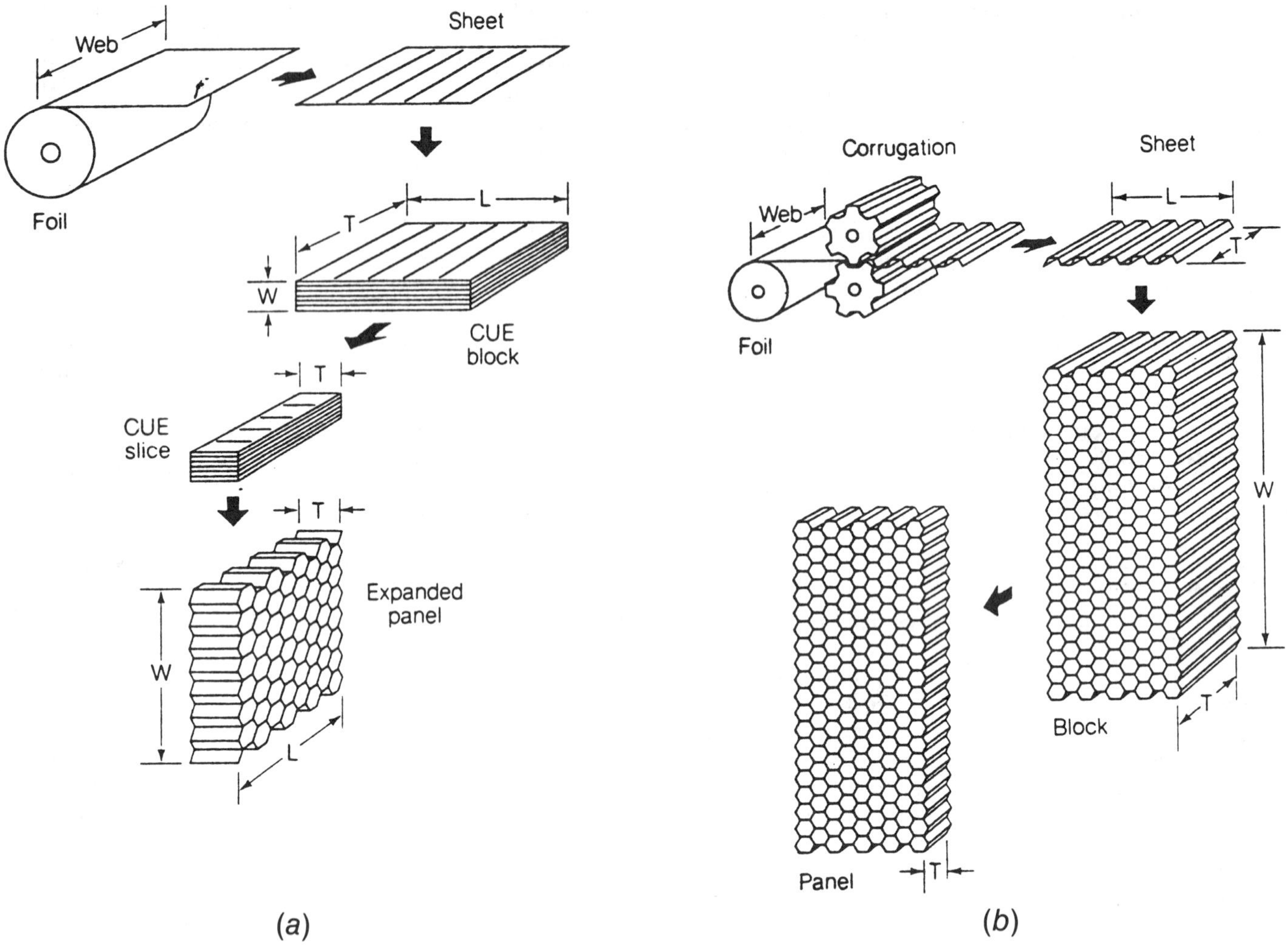

FIGURE 5 Two typical methods of making honeycomb: (*a*) expansion process and (*b*) corrugation process.

materials. The usual cell shape is designed such that *a* and *b* are of unequal length; typically, b/a is about 0.7. This design ratio enables a better balance of strong direction (L direction) strength and modulus to strength and modulus of the weak (W) direction.

The b/a ratio is subject to some unwanted changes because the adhesive that holds the ribbons together at the node line or double-wall portion of the cell (the *b* and *d* sides), is subject to a certain amount of peelback at the time the core is expanded. The amount of peelback may be nearly zero for a 16 or 32 kg/m^3 core, but approaching, perhaps, 20% of the intended *b* dimension for an 128–160 kg/m^3 core. Thus, a separate set of tooling is often used to produce a high density material. Of the three, only the second, or open offset ribbon, is usually of enough consequence to merit serious examination. This aberration has little effect on the density, but it can result in sharply lower strong direction properties. Figure 8 indicates that such problems in the production of honeycomb cores can lead to unusual cell configurations.

KRAFT PAPER HONEYCOMBS. Kraft paper is the only member of the honeycomb family cheap enough to be an attractive candidate for a cost-sensitive project, and it is available from a number of manufacturers. The problem is that all these products, even the rather expensive resin-impregnated paper cores, are grossly affected by the presence of water vapor. At close to 100% humidity of the surrounding air, they lose about 75% of their dry strength (Fig. 9). Since this is considered to be an additional problem beyond the fungus, mildew, and rot problems, the usual decision is to avoid use of these materials in any aircraft structure, or in any structure of a critical nature.

One of the paper honeycombs, Hexcel's WR-II, is used extensively in military shelters and other such structures where the tradeoff of some extra weight for a little less cost is attractive.

GLASS FIBER–PLASTIC HONEYCOMBS. Glass fiber–plastic honeycomb materials have been in use for more than 40 years as cores for radomes and antenna windows, but high cost has generally restricted their use, even in commercial and military aircraft. Many variations exist, and some have remarkable electrical and structural properties. However, recent interest in "low observabil-

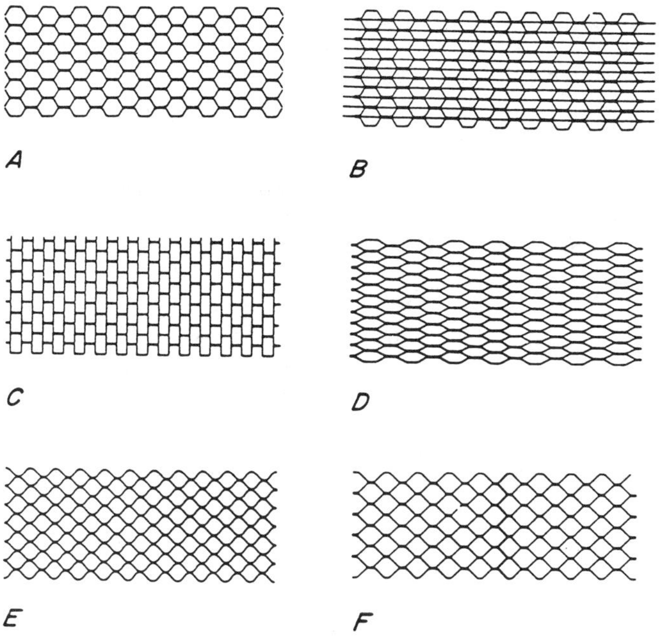

FIGURE 6 The most commonly used cell shapes in current honeycomb products. (*A*) Near-hexagonal shape seen in all honeycomb materials except stainless steel; it is not usually a true hexagon, but tends slightly toward (*E*) or (*F*). (*B*) Reinforced core, produced only by the corrugation method. The straight ribbons may be the same or heavier than the corrugated ribbons. (*C*) Overexpanded core, produced by expanding the honeycomb in (*A*) farther than the amount needed to produce a hexagonal cell shape. (*D*) Cell shape that is possible because all expanded cores can be somewhat underexpanded. (*E*) Usual appearance of fiber glass core, particularly in the higher densities. In the lower densities, fiber glass core usually looks more like (*A*). (*F*) Typical structure of welded stainless steel, titanium, or refractory alloy foil cores, usually the *b* side is very short.

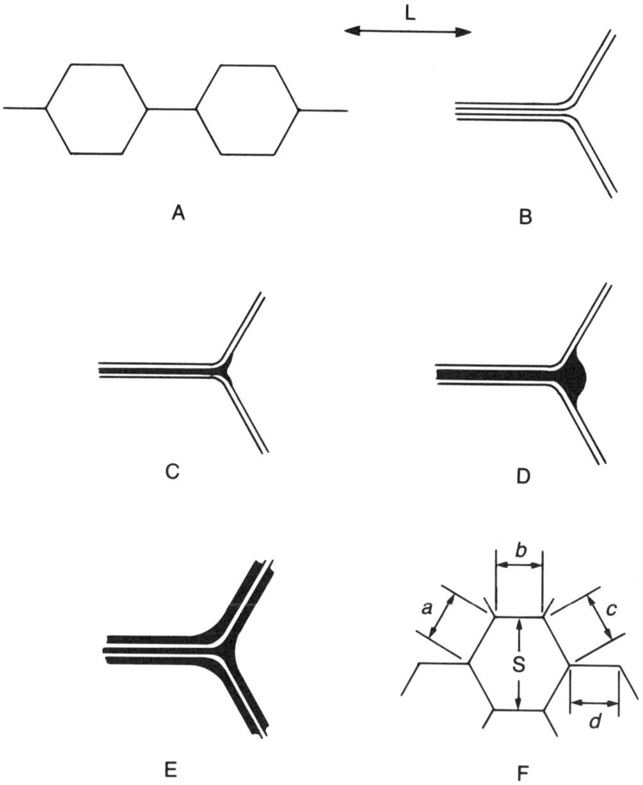

FIGURE 7 The hexagonal cell shape, usually is a modification of the true hexagon (*A*), with the dimensions described as in (*F*). The bonded joint on the *b* side may look like the version shown in (*B*) or (*C*) if the core is produced by the expansion method. If it is produced by the corrugating method, (*D*) is much more common. The difference in weight of the adhesive in the node joint is quite large, with a factor of 50-200 quite common.

ity" or "stealth" aircraft by the military here and overseas has given a sharp impetus to these materials. As a consequence, bias weave glass fabric honeycombs, in which the 45° orientation of the yarn increases the shear modulus performance, are now available from Hexcel (HFT cores) and from Ciba-Geigy (HTX cores). Both firms also make special low observability cores, but all details are classified at this time.

Mechanical properties of several of these families of cores are shown in Table 3.

ARAMID PAPER HONEYCOMBS. Aramid paper honeycombs are based on Nomex paper made by du Pont. This very unusual material makes a remarkably tough and damage-resistant sandwich, even when the core is at less than 32 kg/m^3, and the facings are a single ply of glass cloth only 0.25 mm thick. That particular combination makes up a large part of the cabin interior structure in nearly all large commercial aircraft. Even in the floor panels, the core density is only 64–96 kg/m^3, except in the aisles of the larger planes, where densities up to 144 kg/m^3 are sometimes used to resist the wheel loads imposed by food service carts.

Even with their relatively high cost, the aramid paper honeycombs offer an unusual performance, which has led to very large volume usage. This large usage, however, is confined almost entirely to aircraft structures. Mechanical properties of several of these materials are shown in Table 3.

CARBON FIBER AND KEVLAR HONEYCOMBS. Up until a very few years ago, carbon fiber materials and Kevlar were seen only in laboratories, and even then, only as curiosities. The same renewed attention that has revived the glass fiber honeycombs has now also led to wide interest in both aramid fiber and carbon fiber mate-

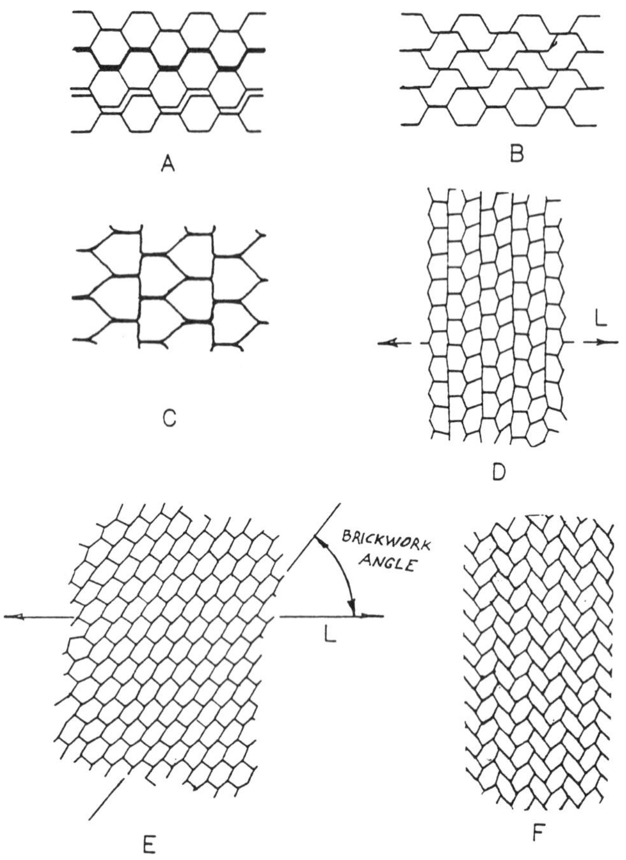

FIGURE 8 **Unusual cell configurations. (*A*) In the corrugation process, a corrugated ribbon may slip sideways, falling entirely or partially into the ribbon below, as they are being stacked up after having the adhesive applied to the node area, as shown. (*B*) If the corrugated ribbon slides only slightly to the side, but does not actually fall into the ribbon below, the defect may look like this. (*C*) An offset adhesive line may cause this pattern of nested trapezia. When the area involved is more than a few square inches of the classical form shown, the structural properties may be adversely affected to the extent that the material will fail to meet the shear strength or shear stiffness minima for the core type. (*D*) Print taken from an actual sample of core exhibiting the problem shown in (*C*). (*E*) "Brickwork": if the brickwork angle is less than about 60°, the shear properties may be in question. (*F*) The "herringbone" pattern, which is not as serious as the brickwork or offset problems. In extreme cases, however, it also may lead to below-minimum properties.**

rials. A rather high cost makes their use feasible only in applications that urgently need the unique performance characteristics they offer.

In the case of carbon fiber cores, the dominant feature sought is usually the near-zero coefficient of thermal expansion. Where space structures of exceptional stability and low mass are needed, this choice offers one of the few feasible solutions to the problem.

Kevlar honeycombs have features similar to those found in carbon fiber cores, but the resulting structure is electrically transparent.

In the case of both carbon fiber and Kevlar materials, the resin system employed may be modified to achieve specific absorption capabilities to enhance low observability objectives.

STAINLESS STEEL, TITANIUM, AND SUPERALLOY HONEYCOMBS. Honeycombs of stainless steel, titanium, and superalloys share some of the high cost and unusual features noted for the other materials described, but they are usually employed only where very high temperature resistance is required, along with very low weight and/or high resistance to sonic fatigue. Common applications are seen in the hot areas near or in jet engine exhaust, inside engines, and in other such areas. The materials are all very expensive compared with most other honeycomb sandwich materials and are used only when nothing else will fill the needs. These cores are almost never used with composite facings; rather, they tend to have metal facings of alloys the same as or similar to the alloy used in the core. Facings may be attached with brazing alloy, or they may be resistance welded to a flange on the cell edge.

Strength Comparisons of Core Materials

Comparing the strength of various cores is a little bit dangerous because the results depend not only on the test method but also on what has been done to the core during fabrication of the sandwich. If the core is one that is softened by polystyrene, and a polyester or vinyl ester resin system is used in constructing the sandwich, the final product will be much weaker than the core producers' literature indicates should be the case. Nearly all the core materials have some such Achilles' heel, and one must be sure to fully understand all the many relevant facts concerning the materials to be used before starting the work.

Assuming that after proper research, these facts are understood and problem areas have been avoided, Figures 10–12 may be consulted for the relative strength, in graph form, of some of the core materials mentioned previously. The strength can clearly be seen to increase with higher density. This is true for all cores, not just for the properties shown, but for values of most other properties as well.

Cost Comparisons of Core Materials

The various core materials seem to differ in cost as greatly as in strength, and placing any reliance on a summary of costs can be a risky business. The real cost is what you must actually pay for the material when and

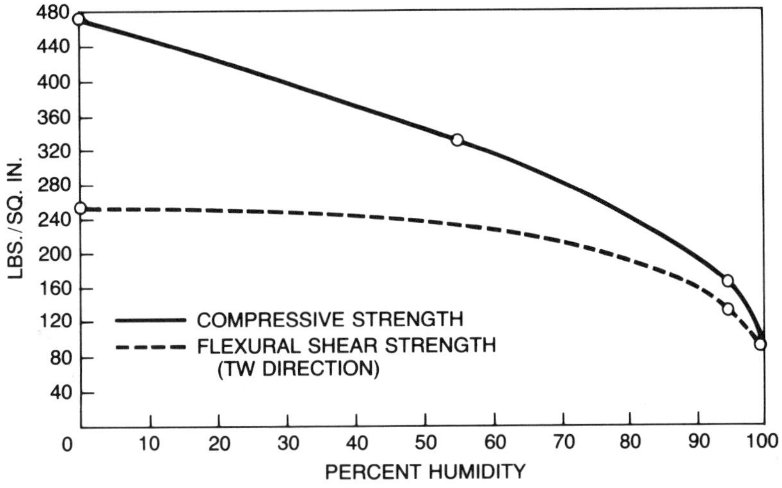

FIGURE 9 Variations in compressive and shear strengths of aircomb style 125-35 (125lb Kraft paper, 35% resin content, 7/16 cell, 3.7 pcf density) Type 20 with changes in humidity. (Graph courtesy of Douglas Aircomb.)

TABLE 3
Comparison of Properties of Honeycomb Core Materials

HRP Glass Reinforced Phenolic Honeycomb

Hexcel Honeycomb Designation	Compressive					Plate Shear					
	Bare		Stabilized			"L" Direction			"W" Direction		
	Strength MPa		Strength MPa		Modulus MPa	Strength MPa		Modulus MPa	Strength MPa		Modulus MPa
Material-Cell-Density	typ	min	typ	min	typ	typ	min	typ	typ	min	typ
HRP-3/16-4.0	3.45	2.41	3.86	3.31	0.39	2.00	1.45	0.08	1.10	0.90	0.03
HRP-3/16-5.5	5.52	3.93	6.00	5.17	0.65	3.00	2.41	0.13	1.65	1.38	0.06
HRP-1/4-3.5	2.59	1.93	3.14	2.76	0.32	1.59	1.17	0.08	0.90	0.69	0.02
HRP-3/8-3.2	2.21	1.79	3.03	2.41	0.26	1.38	1.10	0.08	0.72	0.59	0.02

HFT Glass Reinforced Phenolic Honeycomb

Hexcel Honeycomb Designation	Compressive			Plate Shear					
	Stabilized			"L" Direction			"W" Direction		
	Strength MPa		Modulus MPa	Strength MPa		Modulus MPa	Strength MPa		Modulus MPa
Material-Cell-Density	typ	min	typ	typ	min	typ	typ	min	typ
HFT-1/8-5.5	6.14	4.41	476	3.17	2.48	276	1.59	1.24	97
HFT-3/16-4.0	4.07	2.96	303	1.90	1.45	207	0.97	0.79	97

HRH-327 Glass Reinforced Polyimide Honeycomb

HRH 327-3/16-4.5	3.59	2.76	400	2.21	1.52	228	1.03	0.76	76
HRH 327-3/16-8.0	8.34	6.89	869	4.83	3.38	427	2.90	2.07	152

TABLE 3 (*continued*)

WR II Shelter Core

Hexcel Honeycomb Designation	Compressive			Plate Shear			
	Bare	Stabilized		"L" Direction		"W" Direction	
	Strength MPa	Strength MPa	Modulus MPa	Strength MPa	Modulus MPa	Strength MPa	Modulus MPa
Material-Cell-Density	typ	typ	typ	typ	typ	typ	typ
WR II-3/8-2.5	1.79	2.34	228	1.17	90	0.69	48
WR II-3/8-3.8	3.55	3.93	379	1.62	131	1.21	62

5052 Alloy Hexagonal Aluminum Honeycomb

Hexcel Honeycomb Designation	Nominal Density	Compressive					Crush Strength	Plate Shear					
		Bare		Stabilized				"L" Direction			"W" Direction		
	Kg/M³	Strength MPa		Strength MPa		Modulus MPa	MPa	Strength MPa		Modulus MPa	Strength MPa		Modulus MPa
Material-Cell-Guage		typ	min	typ	min	typ		typ	min	typ	typ	min	typ
1/8-5052-.0007	49.6	1.86	1.38	2.00	1.48	517	0.90	1.45	1.07	310	0.90	0.62	152
1/8-5052-.001	72.0	3.59	2.59	3.76	2.79	1034	1.79	2.34	1.96	483	1.52	1.16	214
1/8-5052-.0015	97.6	6.00	4.48	6.27	4.69	1655	3.10	3.48	3.14	676	2.21	1.88	283
1/8-5052-.002	129.6	9.65	6.89	10.14	7.58	2413	5.17	5.00	4.62	931	3.14	2.76	372
3/16-5052-.001	49.6	1.86	1.38	2.00	1.48	517	0.90	1.45	1.07	310	0.90	0.62	152
3/16-5052-.0015	70.4	3.45	2.48	3.62	2.65	1000	1.72	2.28	1.93	469	1.48	1.10	207
3/16-5052-.002	91.2	5.31	3.86	5.58	4.14	1517	2.69	3.17	2.83	621	2.07	1.68	265
1/4-5052-.001	36.8	1.14	0.83	1.21	0.90	310	0.52	0.97	0.69	221	0.59	0.39	112
1/4-5052-.0015	54.4	2.21	1.65	2.34	1.72	621	1.03	1.62	1.24	345	1.03	0.72	165
1/4-5052-.002	68.8	3.31	2.41	3.48	2.55	965	1.59	2.21	1.83	455	1.45	1.07	205
1/4-5052-.004	126.4	9.38	6.69	9.79	7.24	2344	5.00	4.83	4.48	896	3.03	2.69	364
3/8-5052-.0007	16.0	0.21	0.14	0.31	0.14	69	1.76	0.31	0.22	83	0.21	0.14	48
3/8-5052-.007	48.0	2.07	1.31	2.17	1.38	483	0.83	1.38	1.00	296	0.86	0.59	146

5056 Hexagonal Aluminum Honeycomb

1/8-5056-.0007	49.6	2.34	1.72	2.48	1.79	669	1.17	1.72	1.38	310	1.07	0.76	138
1/8-5056-.001	72.0	4.34	3.27	4.62	3.45	1276	2.21	2.93	2.41	483	1.76	1.41	193
3/16-5056-.001	49.6	2.34	1.72	2.48	1.79	669	1.17	1.76	1.38	310	1.07	0.76	138
1/4-5056-.001	36.8	1.41	1.00	1.45	1.07	400	0.69	1.17	0.90	221	0.72	0.43	103

Aramid Fiber Reinforced Honeycomb

Hexcel Honeycomb Designation	Compressive					Plate Shear					
	Bare		Stabilized			"L" Direction			"W" Direction		
Material-Cell-Density Gage	Strength MPa		Strength MPa		Modulus MPa	Strength MPa		Modulus MPa	Strength MPa		Modulus MPa
Hexagonal	typ	min	typ	min	typ	typ	min	typ	typ	min	typ
HRH10-1/8-1.8(1.5)	0.76	0.48	0.90	0.59	—	0.62	0.45	26	0.34	0.25	14
HRH10-1/8-6.0(3)	7.41	5.52	7.76	5.69	414	2.55	1.79	90	1.38	0.93	41
HRH10-3/16-3.0(2)	2.07	1.45	2.55	1.86	138	1.10	0.90	40	0.62	0.48	24
HRH10-3/8-2.0(2)	1.03	0.55	1.17	0.72	76	0.76	0.50	29	0.38	0.25	15
HRH10/0X-3/16-3.0(2)	2.52	1.76	2.76	1.86	117	0.86	0.65	21	0.97	0.72	41
FLEX-CORE	0.00	0.00	0.00	0.00	0	0.00	0.00	0	0.00	0.00	0
HRH10/F35-2.5(3)	1.38	0.97	1.62	1.21	83	0.90	0.72	28	0.45	0.28	13
HRH10/F50-3.5(3)	2.45	1.69	3.00	2.34	165	1.31	0.86	39	0.55	0.38	28

From Refs. 4 and 5.

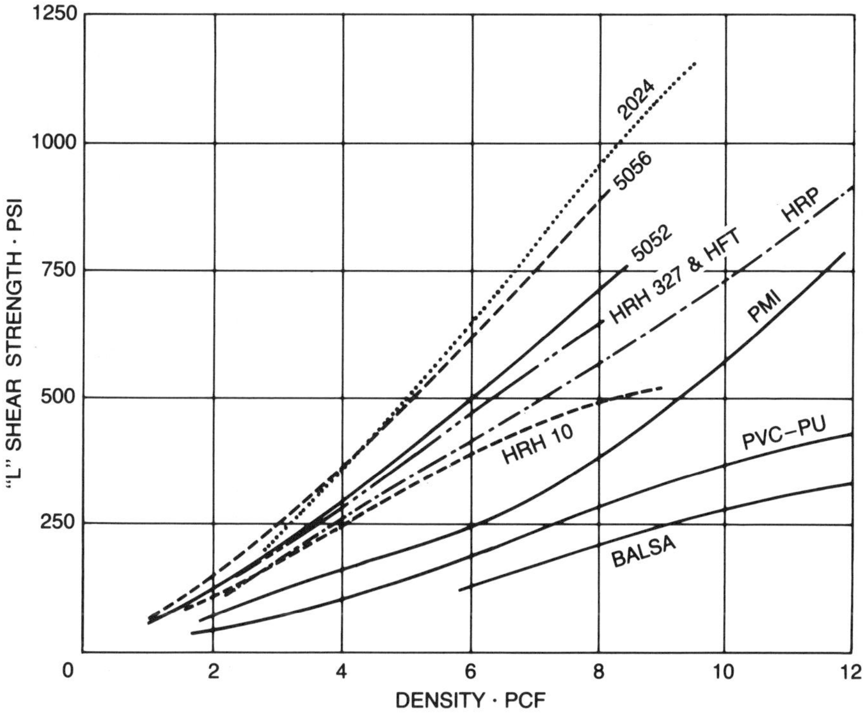

FIGURE 10 Manufacturers' published data on typical strong direction shear strength for Rohacell foam core (curve R), typical polyvinyl chloride and polyurethane foam cores (curve PVC-PU), and balsa wood core (curve B).

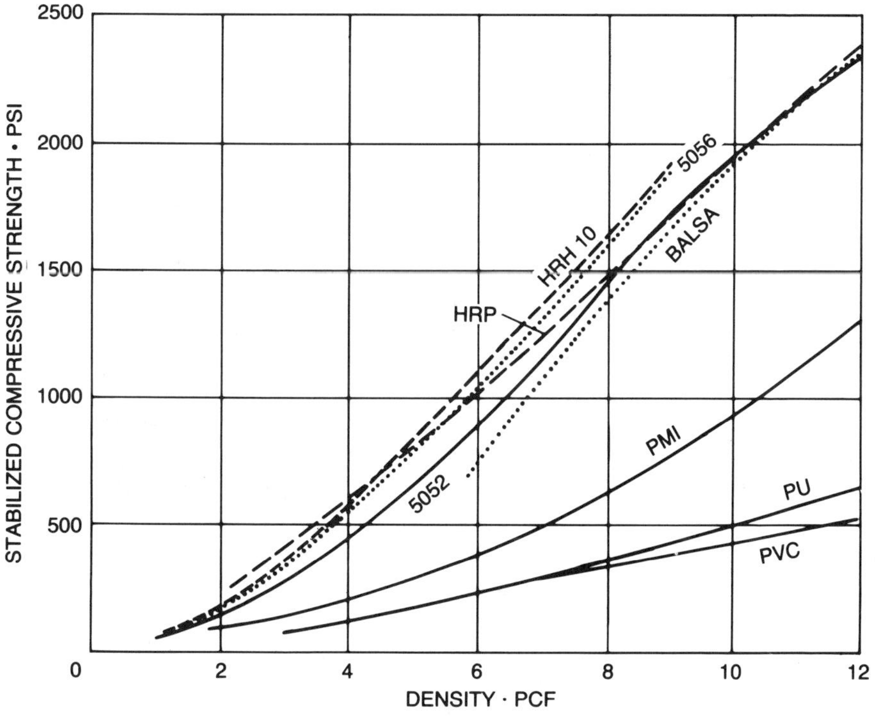

FIGURE 11 Manufacturers' published data on typical stabilized compression strength for materials represented in Figure 10.

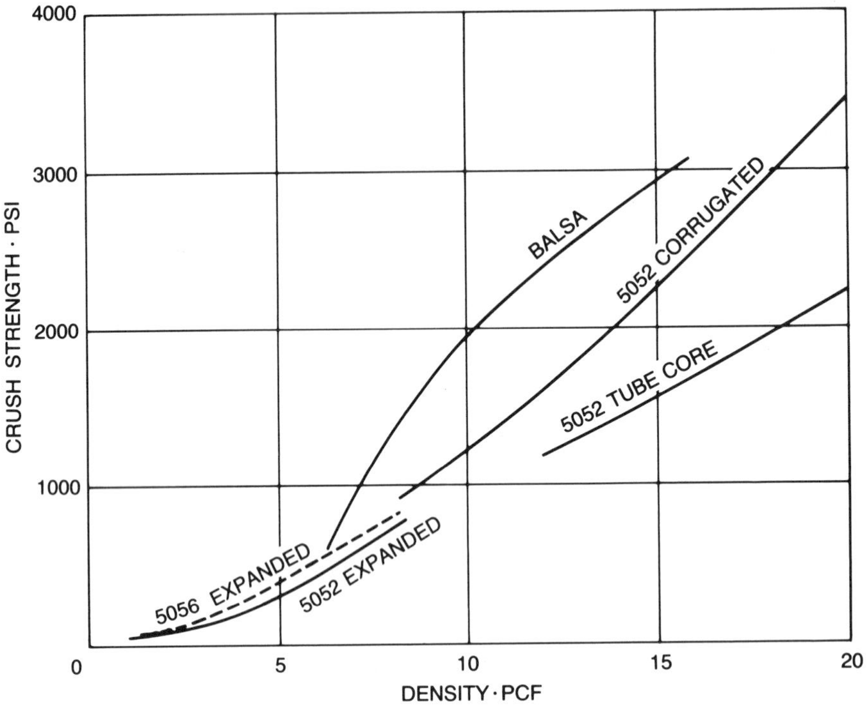

FIGURE 12 Manufacturers' published data on typical crush strength of balsa (curve B) and other labeled core materials. Note that foam cores are not generally suited for crush applications.

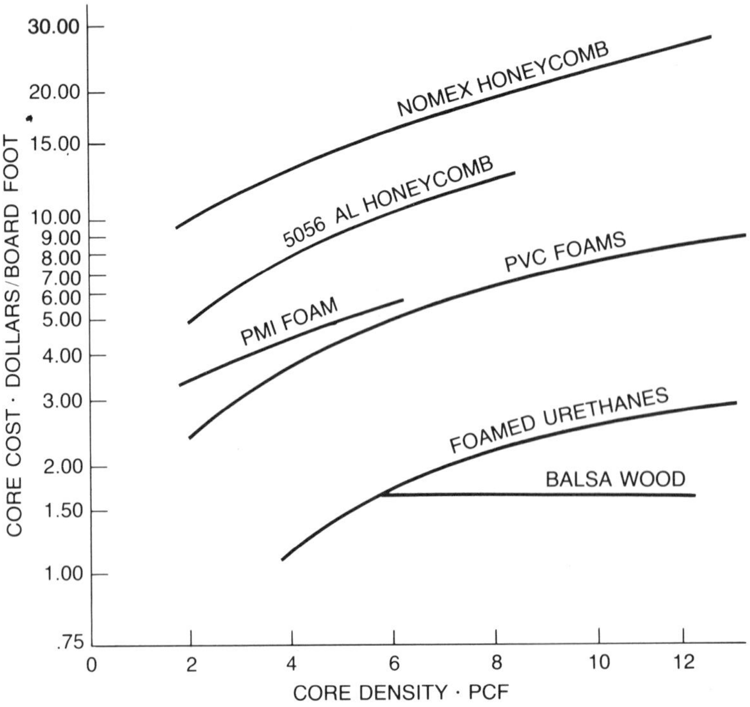

FIGURE 13 Relative core cost (note nonlinear scale).

where it is purchased. Several factors remain constant, however.

1. Polymer cores, such as the various foams, vary in cost directly with the density. In other words, you are really buying the material by the pound.
2. The cost of metal honeycombs reflects the cost of the material from which they are made.
3. The cost of glass fiber honeycombs is primarily affected by how many dip coats of resin were applied to the expanded, "prime cure" honeycomb.
4. The cost of balsa core varies with the range in which the density is allowed to vary but does not vary with density. That is, a very tight tolerance on density will result in a higher cost.

Figure 13 presents an approximation of the cost of several core materials.

Facing Materials

Facings of sandwich structures may be made of any material that is reasonably strong and reasonably thin. Examples exist in which materials as diverse as titanium alloy or birch veneer have been used quite successfully as facings. In the area of composites, all the glass fibers, carbon fibers, Kevlar fibers, and many others have seen, and continue to see, very satisfactory service. Several factors are quite significant in the consideration of what material and orientation will do a specific job most effectively. Some of the more important are discussed in the sections that follow.

Wood Veneer Facing Material

Wood veneer is surprisingly effective, particularly when combined with a light woven glass fabric. Combinations like this are often used where resistance to impact is important and deterioration of the wood is not a problem.

Glass Fiber

Facings based on glass fiber are among the most cost-effective answers, and nearly the entire small boat industry is based on this choice. The particular way in which the fibers are arranged, however, can exert a major influence on the strength and stiffness of the part, as shown in Figure 14. This indicates the typical difference in strength developed as the fibers are changed from the small, randomly oriented fibers in SMC (sheet molding compound), through continuous strand mat, woven roving, and fine woven fabrics.

If the fibers are to be woven into fabric, which not only makes them much easier to handle but also improves strength, the exact choice of the weave style is also important. Figure 15 shows the trend in strength as we move from the coarsest woven roving up to the good structural styles, such as 7781.

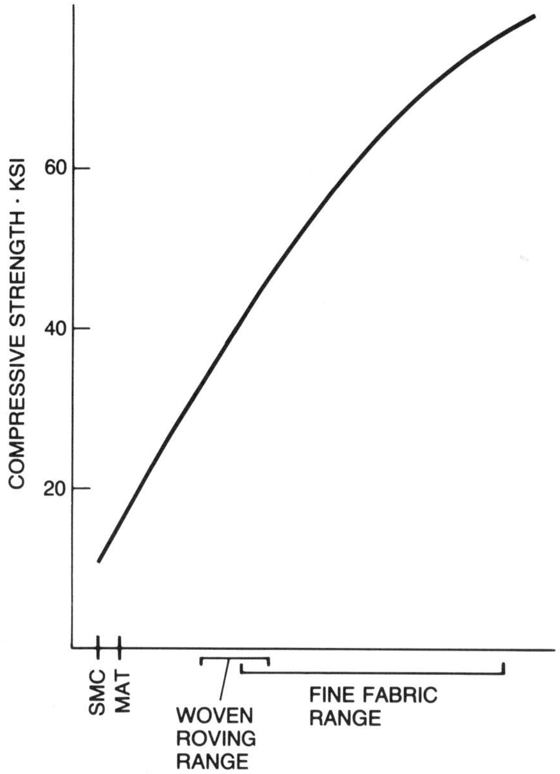

FIGURE 14 **The length, direction, and degree of straightness all have a direct effect on the level of strength that can be developed by a fiber-resin composite structure. The general levels expected of the various fiber arrangements are shown. Exact values cannot be taken from this curve, since many other factors also affect the actual strength measured in the final structure.** (From *Handbook of Composites.*)

Resin content of the laminate also affects strength, as shown in Figure 16. It should be noted that the minimum resin content that can be achieved depends on a great many factors, including both the weave style and the manufacturing method. The curve merely indicates a general trend.

In addition to the foregoing trends, the resin choice itself makes a significant contribution to the part strength, as shown in Figure 17.

One final observation should be made in regard to the effect of the reliability of the manufacturing method, considered in its broadest sense, on the strength and performance to be expected from the finished product. Most real manufacturing processes are made up of a great many individual steps and procedures, each of which will influence the final part in some respect. In the end, one part will exhibit a strength level different from that of another part made of exactly the same materials and by the same process. As more parts are made and measured, we can soon determine with fair accuracy what the entire range in these values will turn out to be. If we then plot the values on a chart, using the strength

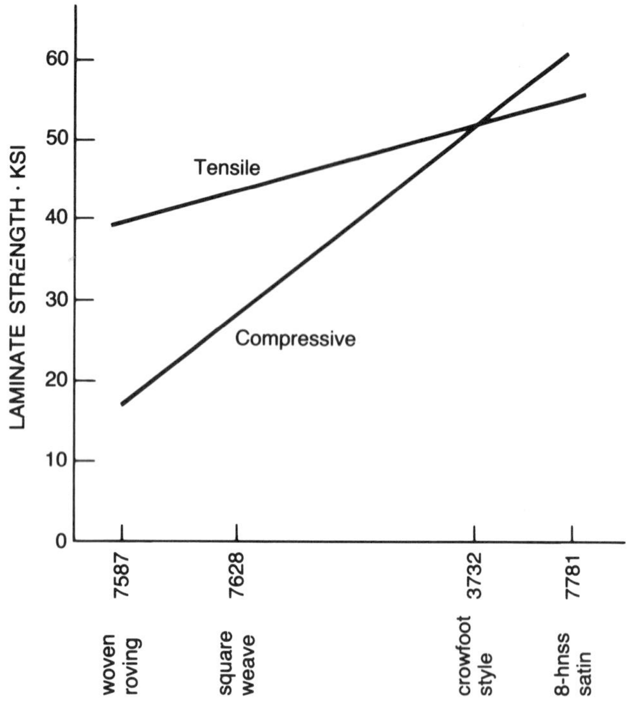

FIGURE 15 **When using woven fabrics as a composite reinforcing system, the weave style will have a direct effect on the final developed strength. The curve shows compressive strength higher than tensile strength for some styles, but this is rarely seen, except at very low resin content.**

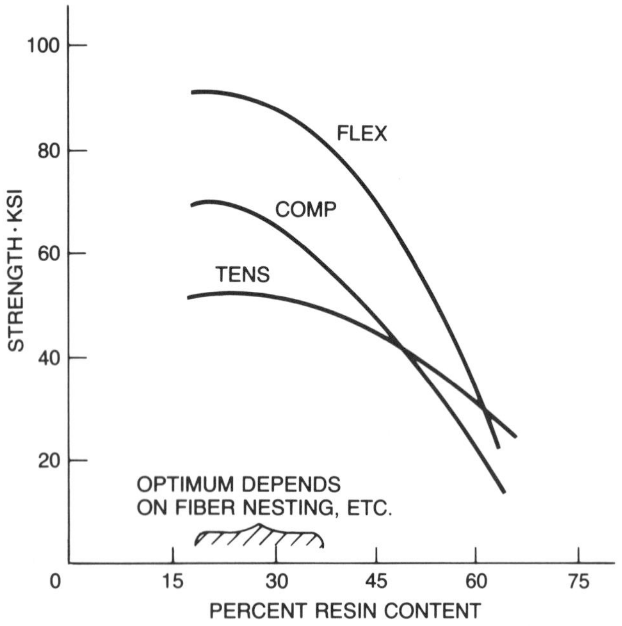

FIGURE 16 **Effect of resin content on laminate strength (cross-plied E-glass).** (From *Handbook of Composites.*)

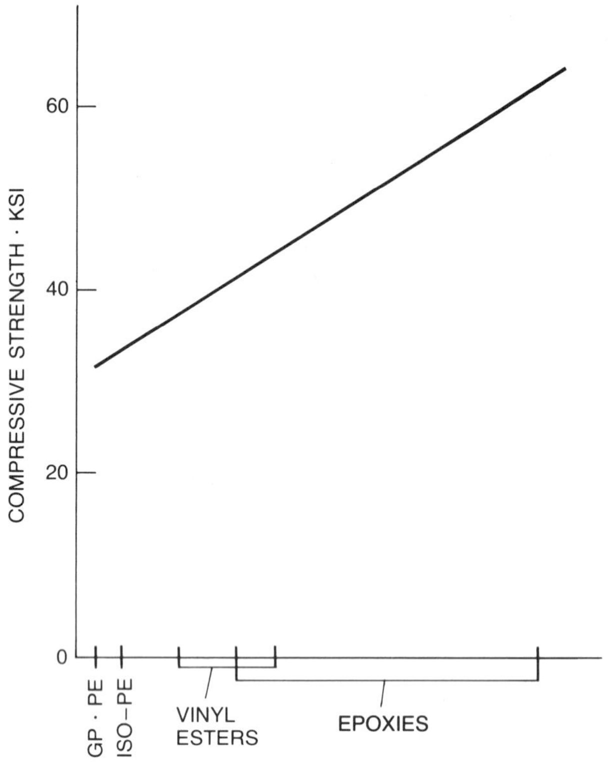

FIGURE 17 **Effect of Resin system on laminate strength.** (From *Handbook of Composites.*)

value for the baseline, and the height of the plot as the number of specimens having a value near this range, we can obtain a chart looking somewhat like Figure 18.

Figure 18 was constructed by using an imaginary part with a production run of 400 pieces. Curve A plots a tightly controlled process in which 300, or 75% of the parts, had a strength of at least 91% of the average strength. Curve B shows the result of a process that was not nearly as well controlled, which resulted in a much wider spread of values. In this case, the weakest of the 75% group would be only at 47% of the average value. This sort of problem has led to the use of statistical analysis and control by nearly all producers of composite parts in the military and aerospace fields, as well as its adoption by any producer who has a serious interest in tracking the performance of the items he makes.

Kevlar Fiber

A common choice of material among composite facings, especially where a high degree of toughness is desired, is Kevlar, an aramid fiber developed and patented by du Pont. The material is available in the usual variety of yarn or tow weights, but it comes in three distinctly different materials. Kevlar 29 is used where the highest degree of toughness is required. It has about the same strength as the usual structural fiber, Kevlar 49, but only half the modulus. Where the highest modulus or strength is desired, the newer Kevlar 149 is used.

In addition to their remarkable resistance to brittle failure, these aramid materials are similar to carbon fiber

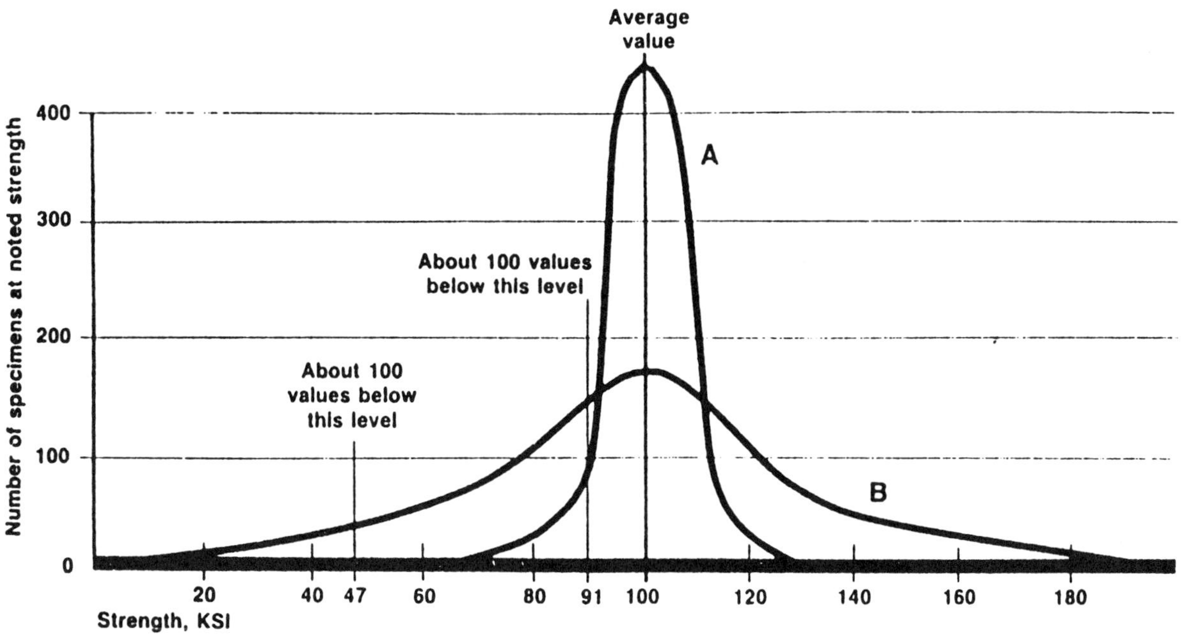

FIGURE 18 Normal distribution curves for two typical structural materials (about 2000 specimens tested for each material).

facings in that they can be tailored to show either a zero or a slightly negative coefficient of thermal expansion. Care must be taken in choosing a resin system for use with the Kevlar materials, since some resins bond poorly to the surface of the fibers, resulting in lower than expected strength of the laminate.

Another caution is necessary because all the Kevlars are very difficult to cut, either in raw fabric form or as finished laminate. Proper planning is required to have the appropriate cutting tools and equipment at hand when that step is to be taken. This difficulty in cutting problem extends to sanding or otherwise dressing the finished molded surface. Thus many designers include on the outer surface a layer of glass fabric, which can be smoothed or abraded for finishing.

A final concern regarding Kevlar involves the tendency of the fibers to absorb moisture within the molecular structure. In a few days of storage in a normal warehouse, enough moisture will be absorbed to risk having layup and oven curing result in very bad parts. The quantity of water given off can result in surface pinholes, or inhibit the cure of some moisture-sensitive resin formulations. It is normally recommended, therefore, that the dry fiber or cloth be stored in a dry environment. Since the materials are not adversely affected by moderate heat, the usual solution is to simply raise the temperature in a closed room, rather than incur the expense of dehumidifying equipment.

Carbon Fiber

A large number of manufacturers currently make and sell a bewildering profusion of the new carbon fiber facing materials. Details of the materials are covered elsewhere in this encyclopedia, but a few comments should be made concerning their use as sandwich facings. Nearly all the products offered are characterized by a rather high modulus (vs. glass or Kevlar), and this modulus is so high that the materials are in a class by themselves among structural materials.

Most combinations of carbon fiber and structural resin systems yield parts in which the failure, when they finally break, is catastrophic. This behavior is not really universal, but it is common enough that unless another fracture mode has been well established, it should be expected.

Some resin systems are not well suited to the strain characteristics of the fiber at hand and will yield very low strength, particularly compressive strength, in finished laminates. This problem is very common in open layups with the common polyester, vinyl ester, and epoxy room temperature curing resins. In this sort of application, therefore, the carbon fiber should be thought of as a device to add stiffness, rather than strength. This low strength problem is alleviated somewhat by keeping the fiber volume as high as practicable, but it is usually found that strength remains low no matter what steps are taken. The real solution consists of changing resin systems to achieve better compatibility, and using an autoclave cure, if possible.

Table 4 lists mechanical properties developed by some of the facing systems just discussed. Antrim Associates, a structural design group specializing in composite structural design for small boats and other structures, uses these values as a starting point for structures based on open molds and hand layup technology. All the values shown are felt to be close to those that will actually be developed in practice, but experience shows that a large amount of variation in the strength of the parts made is to be expected from one shop to another. Struc-

TABLE 4
Typical Composite Facing Properties: Hand Layup Using Vinyl Ester or Epoxy Resin

Material Designation	Weight gm/M²	Fiber Volume	Ply Thickness mm	Ply Weight Kg/M²	Poisson's Ratio	Compressive Properties Sc-0	Ec-0	Sc-90	Ec-90	Tensile Properties St-0	Et-0	St-90	Et-90
Orcon Unidirectional Materials						MPa	KMPa	MPa	KMPa	MPa	KMPa	MPa	KMPa
O-750	322	0.350	0.353	0.571	0.300	613.2	22.7	141.2	5.9	592.5	31.0	17.0	5.6
S-300	163	0.350	0.188	0.294	0.300	427.2	37.9	124.0	6.1	881.9	37.9	34.5	9.0
S-500	273	0.350	0.315	0.494	0.300	427.2	37.9	124.0	6.1	881.9	37.9	34.5	9.0
G-450	161	0.350	0.264	0.346	0.250	668.3	67.5	68.9	6.2	930.2	86.1	55.1	10.3
G-900	322	0.350	0.531	0.694	0.250	668.3	67.5	68.9	6.2	930.2	86.1	55.1	10.3
G-650	220	0.350	0.361	0.474	0.250	789.6	79.8	68.9	6.2	1099.0	101.6	55.1	10.3
KS-400	215	0.350	0.373	0.477	0.350	234.3	28.9	68.9	6.2	668.3	40.0	15.2	4.1
S-1000	501	0.350	0.577	0.905	0.300	427.2	37.9	124.0	6.1	881.9	37.9	34.5	9.0
Commercial E-Glass Boat Materials													
Veil mat	93	0.350	0.277	0.352	0.300	140.6	7.3	140.6	7.3	72.3	6.0	72.3	6.0
0.5 oz. mat	153	0.350	0.452	0.576	0.300	140.6	7.3	140.6	7.3	72.3	6.0	72.3	6.0
0.75 oz. mat	229	0.350	0.678	0.866	0.300	140.6	7.3	140.6	7.3	72.3	6.0	72.3	6.0
1 oz. mat	305	0.350	0.904	1.155	0.300	140.6	7.3	140.6	7.3	72.3	6.0	72.3	6.0
1.5 oz. mat	458	0.350	1.356	1.733	0.300	140.6	7.3	140.6	7.3	72.3	6.0	72.3	6.0
2 oz. mat	611	0.350	1.808	2.310	0.300	140.6	7.3	140.6	7.3	72.3	6.0	72.3	6.0
6 oz. fabric	203	0.350	0.267	0.408	0.107	168.5	14.1	168.5	14.1	189.7	12.5	176.2	9.9
8 oz. fabric	271	0.350	0.356	0.543	0.107	168.5	14.1	168.5	14.1	189.7	12.5	176.2	9.9
10 oz. fabric	339	0.350	0.447	0.680	0.107	168.5	14.1	168.5	14.1	189.7	12.5	176.2	9.9
12 oz. fabric	407	0.350	0.536	0.815	0.107	168.5	14.1	168.5	14.1	189.7	12.5	176.2	9.9
18 oz. roving	610	0.350	0.803	1.223	0.107	181.2	15.2	181.2	15.2	203.9	13.5	189.5	10.6
24 oz. roving	814	0.350	1.069	1.632	0.107	181.2	15.2	181.2	15.2	203.9	13.5	189.5	10.6
Commercial Polyester Filler Putty													
Bondo	1304	0.350	2.118	1.304	0.300	10.0	1.0	10.0	1.0	10.0	1.0	10.0	1.0
Kevlar Fabrics													
Style 243	227	0.350	0.538	0.638	0.300	163.3	24.1	84.7	7.6	489.2	33.2	77.2	4.7
Style 285	169	0.350	0.396	0.470	0.200	124.0	15.8	124.0	15.8	337.6	18.9	337.6	18.9
Knit-Stabilized Nonwovens													
10 oz. 0-90	339	0.350	0.447	0.680	0.107	188.1	15.8	188.1	15.8	213.6	13.8	213.6	13.8
12 oz. 0-90	407	0.350	0.536	0.815	0.107	188.1	15.8	188.1	15.8	213.6	13.8	213.6	13.8
18 oz. 0-90	611	0.350	0.803	1.223	0.107	188.1	15.8	188.1	15.8	213.6	13.8	213.6	13.8
10 oz. Uni	339	0.350	0.447	0.680	0.300	420.3	18.6	117.1	4.8	602.9	25.5	17.0	4.8
12 oz. Uni	407	0.350	0.536	0.815	0.300	420.3	18.6	117.1	4.8	602.9	25.5	17.0	4.8
26 oz. Uni	899	0.350	1.189	1.805	0.300	303.8	19.3	110.2	11.0	507.8	24.2	17.2	4.8
12 oz. +45-45	411	0.350	0.538	0.821	0.523	219.8	8.7	219.8	8.7	35.8	8.1	35.8	8.1
17 oz. +45-45	577	0.350	0.759	1.157	0.523	219.8	8.7	219.8	8.7	35.8	8.1	35.8	8.1
24 oz. +45-45	815	0.350	1.069	1.632	0.523	219.8	8.7	219.8	8.7	35.8	8.1	35.8	8.1
Gel Coats													
Vinyl ester	329	0.350	0.257	0.329	0.300	41.3	2.1	41.3	2.1	34.5	1.7	34.5	1.7
Polyester	488	0.350	0.386	0.489	0.300	34.5	2.1	34.5	2.1	27.6	1.7	27.6	1.7
Marine-Grade Lumber													
Mahogany	715	0.350	1.588	0.715	0.300	45.7	10.4	5.6	0.7	79.9	10.4	5.2	0.7
W R cedar	508	0.350	1.588	0.508	0.300	31.4	7.6	3.2	0.5	51.7	7.6	1.5	0.5
Spruce	699	0.350	1.588	0.699	0.300	37.7	9.2	3.2	0.6	67.5	9.2	2.5	0.6
Birch	1513	0.350	1.588	1.513	0.300	56.3	13.8	6.7	0.9	114.4	13.8	6.3	0.9

Courtesy of Antrim Associates.

tural tests should be conducted on any part whose performance is critical to the success of the project.

Metal Facings and Inserts in Composite Layups

Common practice allows the use of metal facings or buried inserts in many composite structures. When the surface treatment of the metal is carefully chosen, so that no degradation at the interface of the metal and resin occurs with time and exposure to moisture, very satisfactory results can be expected. If this step is neglected or improperly accomplished, however, the attachment between the metal and the rest of the composite structure soon degrades to the point of delamination. The presence of moisture and even moderate heat hastens this process, such that improper surface preparation can be identified as a certain cause of a very short structural lifetime in a large percentage of the parts produced under such ill-advised conditions.

The most satisfactory metal for producing reliable and long-lived parts in or on a composite structure is aluminum. Most of the common alloys can be treated and primed so that their attachment can be regarded as permanent. The most reliable process is the phosphoric acid anodizing and priming system covered by Boeing Specification BAC 5555, or the Society of Automotive Engineers Specification ARP-1524.

A very simple test has now been almost universally adopted to monitor the effectiveness of the surface preparation and bonding of the aluminum. This is ASTM Standard Method 3762, shown in Figure 19.

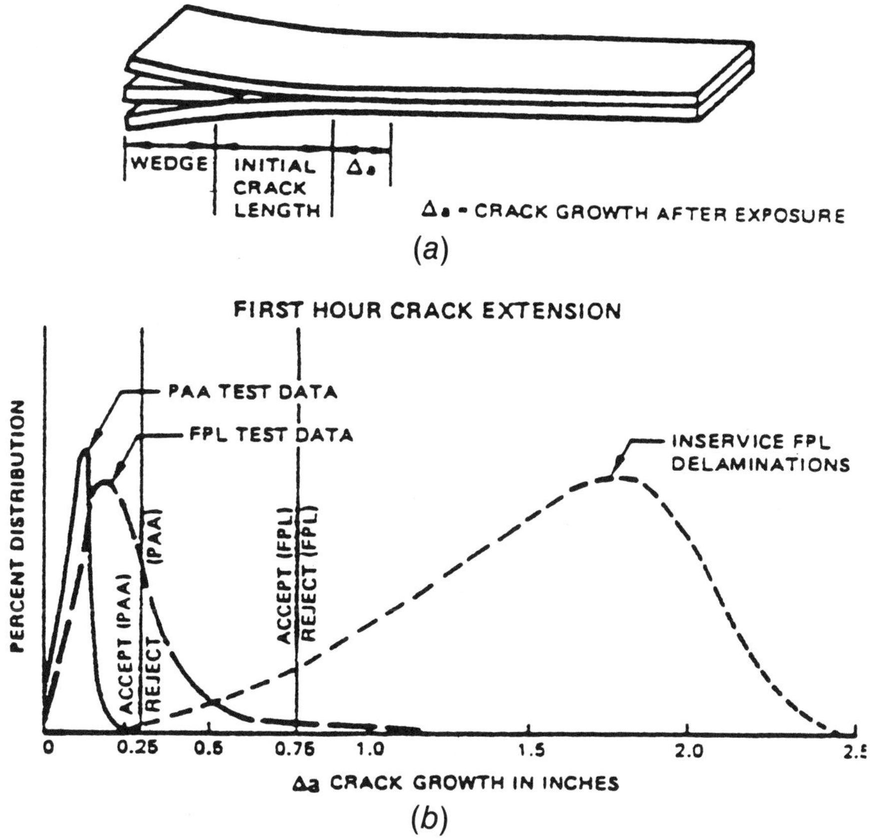

FIGURE 19 Durability testing. (*a*) Wedge test coupon typical of the tens of thousands tested by Boeing in the development and maintenance of the PAA process. Developed in 1970, it closely correlates laboratory tests to actual in-service failures. (*b*) Curves comparing wedge test data derived from laboratory coupons as well as coupons taken from in-service delaminated panels. Note: The improved bond durability provided by PAA allows far more stringent acceptability levels than can be obtained using the FPL etch process. (Diagrams courtesy of Boeing Aircraft Company.)

Resins and Adhesives

The third vital element in any structural sandwich is the adhesive attachment holding the facings firmly in place on the surface of the core material. In the case of composite structures, this attachment may be formed using any one of the many excellent structural adhesives, or it may be formed by the same resin used as the matrix resin in the facing laminate. Frequently the resin used for the facing laminate, chosen for its performance as a laminating resin, is not a very good adhesive. The problem is solved either by precuring the facings and later bonding them to the core with a suitable adhesive, or by prime coating the core with a different, but compatible, resin, which will improve the core bond to an acceptable level of performance.

The particular adhesive or resin best suited for a specific application is the subject of much disagreement among qualified process engineers, but some general comments are appropriate.

Room Temperature Curing Systems

POLYESTERS. A number of excellent laminating resin systems are available that both wet out well and reach full cure in a reasonable time at room temperature. The most common is the polyester group, used widely in the manufacture of boats, shower stalls and bath tubs, underground gasoline storage tanks, industrial tanks, and many other similar applications. Several new versions, made by Ashland, Reichold, and others, offer improved performance at a nominal price premium. Most of these resins are very effective in glass laminates but have a poor bond to fibers such as Kevlar or carbon. They have almost no strength at all in attaching metals. They are also widely used for making boat molds or other large tools where cost is a factor. However, they shrink substantially upon curing (5% or more), and still more as they age in storage or use, and thus are a poor choice for an application that can support the higher cost of a good tooling resin.

VINYL ESTERS. These systems represent a large improvement over the various polyesters, the most impressive being their higher strength. In most applications the vinyl esters approach the epoxies in the level of strength actually delivered by the finished part. This impressive performance is as much a product of the easier working characteristics of the vinyl esters as it is their basic properties. Their excellent wetting and low viscosity make it easier to demonstrate a level of craftsmanship higher than is possible with the more viscous and hard-to-use epoxies usually seen. Also, the vinyl esters represent for production workers a lower level of toxic exposure than even the best of the epoxies. An additional advantage, at least in the use of the Dow Derakane series of products, is the availability of an additive that delays the start of the cure. This additive, 2,4-pentanedione, permits a layup to be continued over many hours without the resin starting to set up. The peculiar nature of 2,4-pentanedione is such that the cure is first delayed by a factor of 10 or more, and actually gels much faster when the cure finally starts!

One final advantage of the vinyl esters consists of their remarkable adhesive properties. In many cases, the adhesive strength of secondary bonds between composites and metals made using Dow Derakane 411-45 has been measured at much higher values than were obtained using either epoxy laminating resin or epoxy paste adhesive. If the secondary bond is to a fully cured polyester laminate, there is simply no equal to the performance of the vinyl esters.

EPOXY. The epoxy systems include many hundreds of combinations of the numerous basic resins available, combined with the even greater number of hardeners available. When one considers that many of the systems also employ additives to alter the viscosity, thixotropy, toughness, elevated temperature performance, solvent resistance, or any of hundreds of other properties, the number of different systems that are waiting to be developed staggers the imagination.

Two families of the room temperature curing epoxies merit separate comment. The first group has many members, each with a different claim to unique attributes, but all with a common problem. That problem is that they cannot withstand any substantial load at even modestly elevated temperatures. The usual limit is a maximum of 66–71°C. Even at levels well below this, low loads will cause substantial creep deformation in the structure. This applies equally to both the adhesives and to the laminating resins.

The second group is an unusual family of materials that can be cured at room temperature, and then, after a suitable postcure without fixtures, can be used at temperatures of 177°C or higher! Many suppliers furnish various candidate materials in each of these families.

TOXICITY AND HEALTH CONSIDERATIONS. All the epoxy wet-resin and adhesive systems use materials having varying degrees of toxicity. At best, some 50% of all who work with them will become sensitized to the materials and will develop serious skin rash or other health problems. Some of the most serious offenders are the very best tooling resins that have been developed, and the resin producers have worked out many methods of coping with these problems, so that these resins can be effectively employed in the manufacture of high quality tooling. Even with these efforts, the trend has recently been toward the use of prepreg, in place of wet-resin systems, for the manufacture of tooling.

Even with all the problem areas cited above (and many more not covered), the advantages of the epoxy systems far outweigh the disadvantages. These materials offer the broadest range of performance parameters of any resin family yet developed, and it is quite likely that they will be with us for another generation, at the very minimum. Despite the attention given the development of newer polymers, some of which offer great promise in the areas of higher temperature performance or more rapid manufacturing processes, development continues on that workhorse of the structural polymers, the epoxies.

Elevated-Temperature Curing Systems

POLYESTERS. This elderly member of the family is still used as a prepreg in a few low cost commercial products. Virtually no new polyester products have appeared in recent years, however, and the product will one day disappear. In wet-resin form, it still is a viable competitor for low cost structures in which the short cure time offered by the elevated temperature is attractive.

PHENOLICS. Even older than the polyesters, phenolic materials are beginning to find new uses in aircraft interiors, where low smoke emission, low toxic gas emission, low burn rate, low fuel contribution, and low price combine to make them outstanding candidates. Phenolic is a poor adhesive in almost any form, so most applications are as precured facings for cabin interior structures.

Adhesive versions of this polymer are still found in the form of nitrile-rubber-modified, supported films used for metal to metal bonds.

EPOXIES. The many formulations of the epoxy family comprise the foremost group in both laminating resins and adhesives. No other material has seen such broad use as a core-to-face bond for honeycomb and other core materials. In the area of structural laminates for facings, this resin dominates all others. Again, publicity regarding the newer polymers as laminating resins hides the fact that the epoxies are used for more applications than perhaps all other resins combined.

THE THERMOPLASTIC RESINS. A new class of polymers, the thermoplastic resins, offer the possibility of "instant cure," replacing the present long times in an autoclave with a simple heat-up/cool-down cycle. Such a simplified system seems to be possible, since these resins are already fully cured, but turn liquid at high temperature and turn solid again when the temperature is lowered. Much development work is currently in progress, but not many production examples exist as yet. Most resins now being tested are not good adhesives in the normally used sense, but they seem to develop good adhesion to the reinforcing fibers when melted in contact with them. Little application has so far been suggested in connection with core composites or sandwich structures.

Andrew C. Marshall

References

1. L. H. Abrams and M. J. Lows, "Shell Instability Problems as Related to Design," NASA TN D-1510, date unknown.
2. A. C. Marshall, in George Lubin, Ed., *Handbook of Composites*, Van Nostrand Reinhold, New York, 1982, pp. 557–601.
3. A. C. Marshall, *Composite Basics*, Marshall Consulting Publisher, Walnut Creek, CA, 1985.
4. "TSB 120, Mechanical Properties of Hexcel Honeycomb Materials," Hexcel Corporation, Dublin, CA, 1987.
5. "TSB 124, The Basics on Bonded Sandwich Construction," Hexcel Corporation, Dublin, CA, 1987.
6. "Rohacell PMI Rigid Foam," Cyro Industries, Orange, CT.
7. A. C. Marshall, *Cores*, a section of the course material accompanying the seminar "Practical Sandwich Structures," 1978–1985.
8. "Preparing the Surface for Bonding," Bulletin G1-600, Hysol Division, Dexter Corporation, Pittsburg, CA, 1988. A practical description of the steps to be taken in making durable bonds to many types of materials.
9. Society of Automotive Engineers, ARP-1524, A process specification covering the phosphoric acid anodizing and adhesive priming of aluminum alloys for inclusion in a composite assembly.
10. *Structural Sandwich Composites*, MIL-HDBK-23, Government Printing Office, Washington, DC. A complete manual for designing sandwich structures for approval of any federal agency.
11. *Nomex Honeycomb Specification*, MIL-C-81986. Government Printing Office, Washington, DC.
12. *Glass Fabric Honeycomb Specification*, MIL-C-8073, Government Printing Office, Washington, DC.
13. *Aluminum Honeycomb Specification*, MIL-C-7438, Government Printing Office, Washington, DC.
14. *Sandwich Test Methods*, MIL-STD-401, Government Printing Office, Washington, DC.

Corrosion

See Polymer Composites, Corrosion; Metal Matrix Composites, Corrosion

Coupling Agents

A coupling agent is primarily a material that provides a water-resistant bond between an organic polymer and an inorganic substrate [1]. Modification of the interface may produce other changes in composites that are very desirable. Surface treatment of a mineral may improve wetout with a polymer, as well as dispersion of fillers in liquid polymers. The treatment protects the mineral against cleavage and may even heal flaws in the surface. The polymer layer may be strengthened by interface modification that promotes alignment of polymer segments and interdiffuses with the polymer, possibly forming interpenetrating polymer networks (IPNs). Surface modification may overcome inhibition of polymer cure by mineral surfaces. A silane treatment on a metal oxide surface (e.g., alumina) stabilizes the oxide against recrystallization into a weak oxide form. All these factors may be important in determining total practical adhesion, but the test of a true coupling agent is whether it promotes retention of adhesion under wet conditions.

Although bonding of organic polymers to inorganic surfaces has long been a familiar operation (e.g., protective coatings on metals), a major need for new bonding techniques arose in 1940 when glass fibers were first used as reinforcement in organic resins. The specific strength-to-weight ratios of early glass–resin composites were higher than those of aluminum or steel, but the new materials lost much of their strength during prolonged exposure to moisture. This loss in strength was attributed to the debonding of resin from hydrophilic glass by the intrusion of water. All fiberglass used for polymer reinforcement now has an appropriate silane coupling agent applied to the surface by the manufacturer. Although silane coupling agents are known to improve the adhesion of coatings and adhesives to metals and ceramics, commercial applications in these areas are not as universal as on fiberglass reinforcements.

Discussion

Silane coupling agents are ambifunctional molecules of general structure Y—A—$Si(OCH_3)_3$, where Y is an organofunctional group bonded to silicon through divalent organic radical A by stable, covalent bonds. For many composites a silane is chosen in which Y has maximum reactivity with the matrix resin. Commercial silanes are available for bonding virtually any polymer to mineral surfaces (Table 1). Dow Corning, Union Carbide, Petrarch Systems Division of Dynamit Nobel, and SCM are the major suppliers in this country.

Silane coupling agents provide chemical bonds between organic polymers and mineral surfaces. Methoxysilanes may react directly with hydroxyl groups on a mineral surface to form oxane bonds. The silane may also be prehydrolyzed to form silanetriols, which condense with surface hydroxyl groups and with each other to form an organosilicone-bonded surface. The fiberglass

TABLE 1
Representative Commercial Silane Coupling Agents

Commercial Designation (Dow Corning)	Organofunctional Group
Q9-6300	Vinyl
Z-6076	Chloropropyl
Z-6040	Epoxy
Z-6030	Methacrylate
A-1100 (UCC)	Primary amine
Z-6020	Ethylenediamine
Z-6062	Mercapto
Z-6032	Cationic styryl
X1-6100	Mixed phenylsilane [$O(CH_3)_3$]/Z-6020
Q1-6106	Z-6040-Modified melamine resin
S-3076S (Hercules)	Azide
X1-6145	Cross-linker

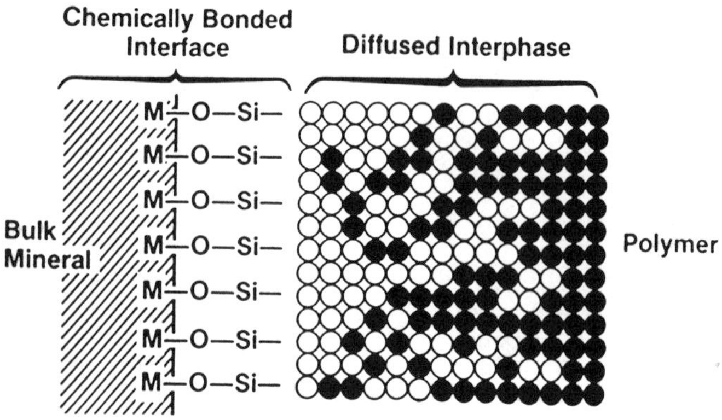

FIGURE 2 **Bonding through silanes by interdiffusion: open circles indicate regions of coupling agent; solid circles indicate regions of polymer.**

industry invariably applies silanes from aqueous solution to deposit about 0.5% by weight of siloxane on the fibers, as shown in Figure 1.

Coupling agents may also be added to resins to obtain unprimed adhesion to mineral surfaces. Silanes may be formulated into primers by partial hydrolysis or by modification with polymer precursors. Primers on massive mineral surfaces are related to coupling agents applied to fiberglass or particulate minerals but are often applied in much thicker layers.

Commercial applications of silane coupling agents are not as simple as shown in Figure 1. From practical tests it has been determined that good adhesion of polymers to minerals often requires much more than a monolayer of coupling agent. It is common practice to put up to a hundred monolayer equivalents, or more, of silanes on the surface. Furthermore, it has been observed that silane primers are very effective in bonding certain polymers to minerals where no chemical reaction between the organofunctional group Y and the polymer is observed [2].

Bonding through silane coupling agents by other than simple chemical reactivity is best explained by interdiffusion and IPN formation in the interphase region [3]. Interdiffusion and cross-linking of silane adhesion promoters was recently identified as the mechanism of adhesion of polyvinyl chloride plastisols to silane-treated mineral surfaces [4] (Fig. 2).

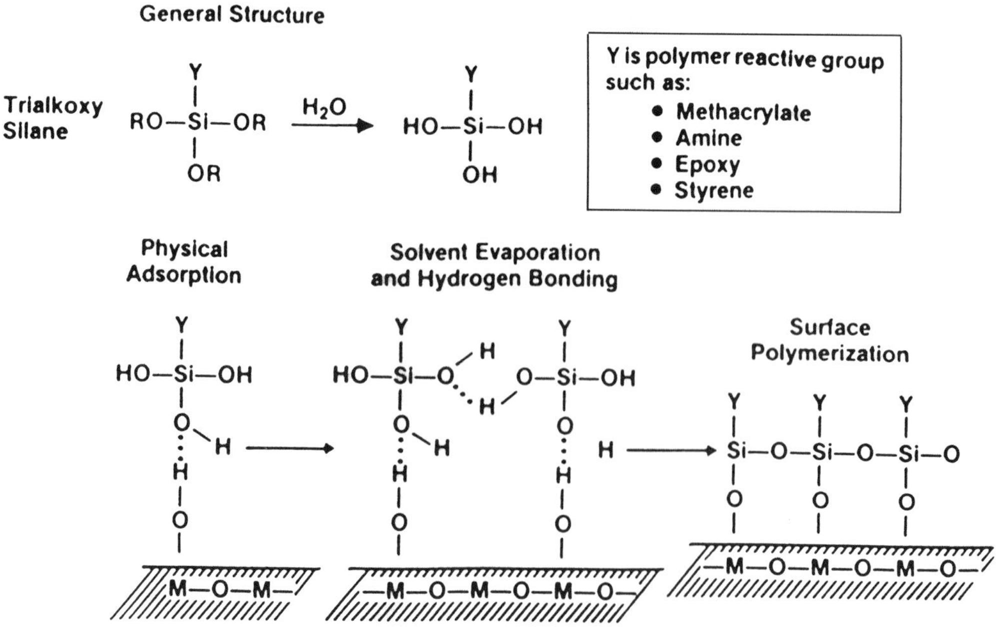

FIGURE 1 Monolayer model for bonding in mineral–polymer composites.

It has been determined that silane coupling agents form covalent oxane bonds with silica and glass reinforcements [5]. These bonds are hydrolyzed by water, but they re-form rather easily. The result is that adhesion lost under humid conditions is largely regained when the composite dries. It was recently demonstrated [6] that hydrolysis and re-formation of siloxane bonds of silane coupling agents are in true equilibrium. The rates of both hydrolysis and condensation are catalyzed by acids and bases.

Equilibrium constants for the hydrolysis of silanols and alcohols on a silica surface are shown in Table 2. Constants for two silanes are derived from Pohl's study with model compounds [6], while the others are estimated from the trend of variation with structure. It is estimated that bonding to silica in the presence of moisture is 10^3 times more favorable for a standard silane coupling agent than for a hydroxyfunctional organic compound.

It is believed that similar equilibrium mechanisms apply for bonding silane coupling agents to metal oxide surfaces, although equilibrium constants with various metals are not known.

Equilibrium conditions across the interface are best maintained if the interphase region is highly cross-linked or very rigid, as in high temperature resins. Silane coupling agents also improve the bonds of vulcanized elastomers and of room temperature curing polymers to glass and metals, but the bonds have limited resistance to water. In bonding Pyrex blocks with an aliphatic amine-cured epoxy resin, Andrews et al. [7] showed that silane could delay the time of failure in water by more than a thousand-fold, but all bonds ultimately failed (Fig. 3).

It would seem that a trifunctional coupling agent (silanetriol) with added organic functionality would supply adequate cross-linking at the interface even with elastomers and room temperature curing polymers, but triols with large organic groups on silicon have a strong tendency to form cyclic siloxanes rather than to cross-link. Primers based on Z-6030, Z-6040, Z-6020, etc., dry at room temperature to soluble, fusible oligomers. When baked at 175–200°C they become highly cross-linked through siloxane rearrangement.

TABLE 2
Equilibrium Hydrolysis of Coupling Agents on Silica:

$$\equiv\text{Si—O—SiR}_n(\text{OH})_m + \text{H}_2\text{O} \rightleftharpoons \equiv\text{Si—OH} + \text{HOSiR}_n(\text{OH})_m \qquad n + m = 3$$

Coupling Agent on Silica	Equilibrium Constant, K_{eq}
Si—O—Si(OH)$_3$	10^{-5} (estimate)
Si—O—Si(OH)$_2$R	10^{-4} (estimate)
Si—O—Si(OH)R$_2$	2.75×10^{-3}
Si—O—SiR$_3$	5×10^{-2}
Si—O—R	10^{-1} (estimate)

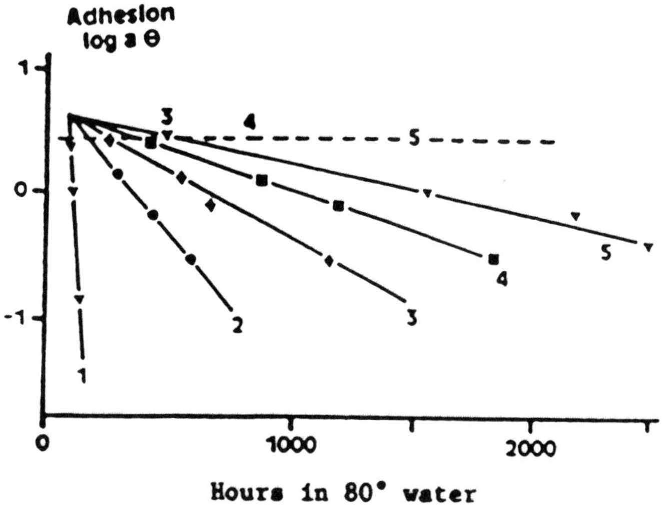

FIGURE 3 **Pyrex blocks bonded with epoxy adhesive containing silanes of Table 1. Curves: 1, no silane; 2, 0.05% Z-6040; 3, 0.05% Z-6020; 4, 0.10% Z-6040; 5, 1.0% Z-6011. The broken line indicates time when corresponding specimens showed transition from cohesive to adhesive failure [7]; *a* = plasticization factor for epoxy saturated with water, θ = adhesive energy (J/m²).**

It has recently been disclosed that performance of any silane adhesion promoter can be improved by adding to it a cross-linking silane [8]. The silane X1-6145 (Silane L), bistrimethoxysilylethane [$(\text{MeO})_3\text{SiCH}_2\text{CH}_2\text{Si(OMe)}_3$], was considered to be the most likely to succeed as a commercial cross-linker and was tested with Z-6020 and with Z-6040 in primers for room temperature curing epoxies (Table 3). With this cross-linker, it was possible to formu-

TABLE 3
Room Temperature Cured Epoxy Films on Glass

Silanes of Table 1 Primer on Glass[b]	Time (h) in 70°C Water to Lose Adhesion[a] for:		
	Polyamide	DMP-30	DEH-24
None	< 1	< 1	< 1
Z-6020 (F)	1	1	1
9/1 Z-6020-L	2	3	1
8/2 Z-6020-L	2	3	1
6/4 Z-6020-L	3	3	2
1/1 Z-6020-L	4	4	8
1/9 Z-6020-L	c	c	c
L alone	c	6	c
Z-6040	c	8	c
9/1 Z-6040-L	c	36	c
8/2 Z-6040-L	c	c	c

[a] c = No failure after one week in 70°C water.
[b] L = $(\text{CH}_3\text{O})_3\text{SiCH}_2\text{CH}_2\text{Si(OCH}_3)_3$.

late primers that provided bonds that resisted 70°C water for more than a week. Primer Z-6040 required less cross-linker than did Z-6020 to provide water-resistant bonds to glass. The cross-linker itself was a creditable adhesion promoter in these systems, although it is not effective alone with some polymers.

From Table 2, it is seen that orthosilicates have the most favorable equilibrium constant for retention of adhesion in water. Numerous primers based on orthosilicates, as well as orthoesters (alkoxy derivatives), of titanium, zirconium, and aluminum have been recommended for water-resistant bonds of polymers to metals and glass [9]. The new oxide deposited on a surface by hydrolysis and condensation of orthoesters often is more reactive with resins than the native oxide on a metal. In many cases, however, it is necessary to apply the resin while the primer is still soluble and fusible, to develop an interpenetrating polymer network in the interphase region. This allows a narrow "window" of primer open time that is dependent on temperature and humidity. When priming conditions for orthoesters are optimized, these primers provide an interpenetrating oxide structure of essentially unlimited heat stability, even compared with high temperature organic resins.

Screening Procedures

It is not possible to predict which silane is best for bonding any particular resin, but a rapid screening test on glass microscope slides is helpful in making a good choice.

Commercial microscope slides provide convenient clean surfaces for observing adhesion of resins to glass. The slides may be dipped in dilute (e.g., 0.5%) aqueous solutions of silane or wiped with a 10–25% solution of prehydrolyzed silane in an alcohol. Thermoplastic resins are then fused under light pressure against the treated slides. Thermosetting resins are applied as liquids and cured against the treated slides. The films on glass are observed for initial adhesion and then soaked in water at room temperature or at elevated temperature to determine resistance of the bond to intrusion by water.

Adhesion of flexible polymers may be measured directly as peel strength. Rigid polymers cannot be peeled. Attempts to lift the film will indicate a qualitative degree of adhesion up to cohesive failure in the polymer. When thick rigid films with good adhesion to glass are immersed in water, the film may suddenly pop off the slide but carry a complete layer of glass with it. Adhesion provided by the silane may be rated as the length of time a film may be soaked in water, assuming there was good dry adhesion, before there is loss of adhesion. In general, there is very good correlation between the adhesion observed in microscope slide tests and the performance a silane will provide in mineral-filled elastomers or resins.

Results of such screening tests with a number of thermosetting and thermoplastic polymers are shown in Tables 4 and 5. Primers were rated according to the relative length of time the polymer retained adhesion in water. Ratings were:

− *Not recommended,* although performance may have been considerably better than a control.
\+ *Good,* significantly better than the control—as much as a hundred-fold increase in time to failure.
++ *Very good,* up to a thousand-fold improvement over the control.
+++ *Outstanding,* best of the series; when several primers showed no failure, all were rated ++.

Several commercial epoxy formulations [10] and many other polymers [11] have been evaluated in a similar manner.

TABLE 4
Primers for Thermosets

	Resin[a]						
Silane Primer	Polyester Bz_2O_2	Polyester MEK per	Epoxy DETA	Epoxy Amide	Epoxy Anhydride	Epoxy *m*-PDA	Epoxy Dicyanate
Cure (°C):	130	70	RT	RT	150	150	175
Z-6076	−	−	++	+	+	+++	++
Z-6040	+	−	+++	++	+	++	++
Z-6030	++	++	++	−	+	++	++
A-1100	−	−	−	−	−	−	+
Z-6020	−	+	−	+	−	+	++
Z-6062	+	+	++	+	−	+	++
Z-6032	+	+++	+	+	+	++	+
X1-6100	−	−	++	−	+	++	++
Q1-6106	+	+	++	+++	++	−	+
$Al[Os(C_4H_9)_3]$			+			−	+
Limits of H_2O test	−48 hour boil		7 day 70°C		(7 day boil)		

[a] B_2O_2 = benzoyl peroxide
MEK = methyl ethyl ketone peroxide
DETA = diethylenetriamine
m-PDA = miphenylone diamine.

TABLE 5
Primers for Thermoplastics

Primer on Glass		Polymers[a]					
	PVC	Urethane	HDPE	HDPP	Nylon	Polycarbonate	PEEK
Fuse (°C):	180	175	240	250	275	275	400
Q9-6300	−	−	+	−	−	−	−
Z-6076	−	+	+	+	+	+	−
Z-6040	−	++	−	−	+	+	−
Z-6030	−	−	+	−	+	+	−
A-1100	+	−	−	−	+	+	−
Z-6020	+	−	−	−	+	+	−
Z-6062	−	++	−	−	+	−	−
Z-6032	++	++	++	++	++	+	+
X1-6100	++	++	−	−	++	++	+
Q1-6106	+	+++	−	−	++	++	++
S-3076S	−	−	+	++	−		

[a] PVC = polyvinyl chloride plastisol; HDPE = high density polyethylene; HDPP = high density polypropylene; PEEK = polyether ether ketone.

Conclusion

It is encouraging to observe that proper formulation of silane adhesion promoters may provide very water-resistant bonds between polymers and glass or metals. A combination of cross-linkable, hydrophobic siloxanes at the interface with interpenetration and cross-linking in the polymer provides a well-bonded interphase region that accomplishes on a molecular scale what mechanical interlocking to anodized aluminum does on a microscale. An optimized silane primer for a given polymer will be effective on all siliceous and metallic surfaces. A clean surface such as can be obtained by scrubbing with Scotchbrite or with kitchen cleanser is all the surface preparation required.

Silane coupling agents are moderately costly materials, selling for \$16–\$33/kg, but the proportion of silane in most composites is so small that it adds little to the final cost. The total market for silane coupling agents in the United States for 1986 was about 1.36 mg., of which the fiberglass industry consumed more than 50%. Japan and Western Europe are other major markets for coupling agents and are supplied largely by local manufacturers. Figures for Eastern Europe are not available, but industries there buy some silanes from the West. China does not have silane manufacturing and the present demand for imported silanes is small, but is expected to grow rapidly in the next 10 years.

Edwin P. Plueddemann

References

1. E. P. Plueddemann, *Silane Coupling Agents,* Plenum Press, New York, 1982, preface.
2. E. P. Plueddemann, *Silane Coupling Agents,* Plenum Press, New York, 1982, p. 131.
3. E. P. Plueddemann and G. L. Stark, 35th Annual Conference of the Composites Institute, SPI, 1980, 20-B.
4. M. K. Chaudhury and E. P. Plueddemann, *J. Adhes. Sci. Technol., 1,* 30 (1987).
5. H. Ishida, *Molecular Characterization of Composite Interfaces,* Plenum Press, New York, 1985, Ch. 2.
6. E. R. Pohl and F. D. Osterholtz, in D. E. Leyden, Ed., *Silanes, Surfaces, and Interfaces,* Gordon & Breach, Amsterdam, 1986, pp. 481–500.
7. E. Andrews, H. Sheng, H. Majid, and C. Vlackos, *Proc. Royal Soc. London* A *381* 345(1982).
8. E. Plueddemann and P. Pape, *Proceedings of the 42nd Annual Conference of the Composites Institute,* SPI, 1987, 21-E.
9. R. Pike and F. Lamm, *Proc. ACS Div. Polym. Mater. Sci. Eng., 56,* 199 (1987).
10. E. P. Plueddemann, *J. Adhesion Sci. Technol.* 2(3) 179(1988); *Symposium on Structural Adhesive Bonding,* 1987, pp. 168–176.
11. E. P. Plueddemann, *Proceedings of the American Society for Composites, First Technical Conference,* 1986, pp. 264–279.

Cracking

See Fractography; Fracture, Interlaminar; Fracture Mechanics; Laminates, Matrix Cracking

Creep

Overview

Creep is a special case of dimensional instability. It is generally described as a time-dependent strain caused by a constant applied stress at constant temperature. Variable applied stresses also qualify, such as creep during fatigue conditions. Temporal instability is a special case of creep, being defined as permanent deformation with time with zero applied stress (but with the expected influence of residual stresses and gravity). The term "creepocity" has been used to denote the percentage increase in strain during some time period [1]. We may distinguish between normal creep and microcreep, with the former implying large deformation rates up to 1%/min, with damage accumulation in the material eventually leading to stress rupture. Microcreep is creep that occurs at stress levels near the microyield strength (MYS). The MYS is defined as that stress level which, after repeated short-term and increasing stress level loadings, results in a permanent strain of 10^{-6}.

Interest

Creep is of concern in a variety of scientific and engineering endeavors. The use of adhesives frequently raises questions of creep resistance [2]. Creep causes a loss of compressive stresses in fasteners, resulting in a more easily deformed structure later. It has caused misalignment of telescope and optics systems, resulting in changes of focus, loss of figure control, image aberrations, and a continued need for adjustments. Creep alters the effectiveness of metering and pointing assemblies in space systems. Composite creep can also be a problem to the maintenance of the surface smoothness of optical components [3]. Loss of integrity of press fit assemblies can occur. Creep is partly responsible for dimensional hysteresis during thermal cycling of composites and is a source of calibration drifts and gradual misalignment of precision instruments such as gyroscopes, accelerometers, and laser housings. [4–6].

Creep of high temperature components such as turbine blades is a major concern for jet engine reliability [7,8]. Inertial guidance systems and lightweight optics are prime candidates for creep-resistant, low cost composites such as silicon carbide fibers in aluminum [9,10]. Composite pressure vessels are subject to hoop strain changes as a result of creep [11,12]. Rocket propellants show shifts in stress-free temperatures as a result of relaxation processes [13]. General creep resistance of composites is of concern in automobiles [14] and in aircraft [15,16].

History

Creep and recovery studies have been reported in the technical literature since at least the early 1920s. Creep as such was associated with internal stresses and "overstrain" [17], while recovery of size and shape was regarded as "subpermanent elasticity or reactivity" after stress was removed [18]. After wider recognition of the industrial importance of creep in furnace construction, the ASME and ASTM in 1924 established a standing research committee to study the effects of temperature on metals, and their work on creep was summarized in 1938. The ASTM set up a recommended practice for creep testing (E 22-41), which has been replaced with (E 139-83) 03.01 (1985) *Recommended Practice for Conducting Creep, Creep Rupture and Stress Rupture of Metallic Materials. Test Methods for Tensile, Compressive, Flexural Creep and Creep Rupture of Plastics* is the title of D 2990-77 (1982) 08.02, formerly D 674. Recommendations for related behavior—stress relaxation and compression of plastics, and notched-bar rupture testing—are covered in (D 2991) 08.02, (D 3410) 15.03, and (E 292-83) 03.01, respectively. The use of thermocouples in creep and stress rupture testing to 1000°C in air is covered in (E 633-87) 14.01.

Speakman, in 1926 [19], was one of the first to study the slow deformation that occurs when a constant stress is applied to a fiber. Early work on creep of fibers is summarized in Ref. 20, while some early studies of the creep of composites are reviewed in Refs. 2 and 21–23.

Discussion

General Creep Behavior

Figure 1 shows typical response of metals and some nonmetals to a constant applied load. The resulting creep curve consists of four stages: (*a*) an elastic, or fully recoverable deformation, (*b*) a primary region or transient creep, or stage I, where the creep rate continuously decreases, (*c*) a secondary stage or steady-state creep, where the creep rate is constant, and (*d*) a tertiary or stage III creep, where the creep rate increases rapidly and fracture or stress rupture results. Step *a*, although not actually creep, is unavoidable and will contribute to the allowable strain in any engineering situation. The behavior illustrated in Figure 1 is also called high temperature creep, because for most metals it occurs above about 40% of the melting point. At lower temperatures, primary creep may extend over long time periods. The mechanisms of creep, in terms of motions of atoms, dislocations, grain boundaries, etc., may be different in each stage.

Creep of Constituents

Much effort has gone into development of predictive models for creep phenomena, and a major input here is knowledge of the constituent behavior. Purely elastic fibers are represented by graphite, boron, glass, and ceramics at ambient or low temperatures. Creeping fibers, such as metals at high temperatures or organic fibers in moist atmospheres, are less common. Metallic, polymeric, and many glass matrices may be elastic at low loads and/or temperatures, but they rapidly become viscous, viscoelastic, or plastic under common service conditions. The situation is complicated by the increased use of weaves, felts, pseudoisotropic fabrics, braids, and cloths as reinforcements. Hybrid composites, in which two or more fiber materials are used, also make creep predictions increasingly complex.

The basic mechanical properties of the constituents, such as Young's moduli in the principal directions, the Poisson ratios, and shear moduli must be known for any analysis. The proportional limits and microyield strengths are helpful to know. Additional factors that

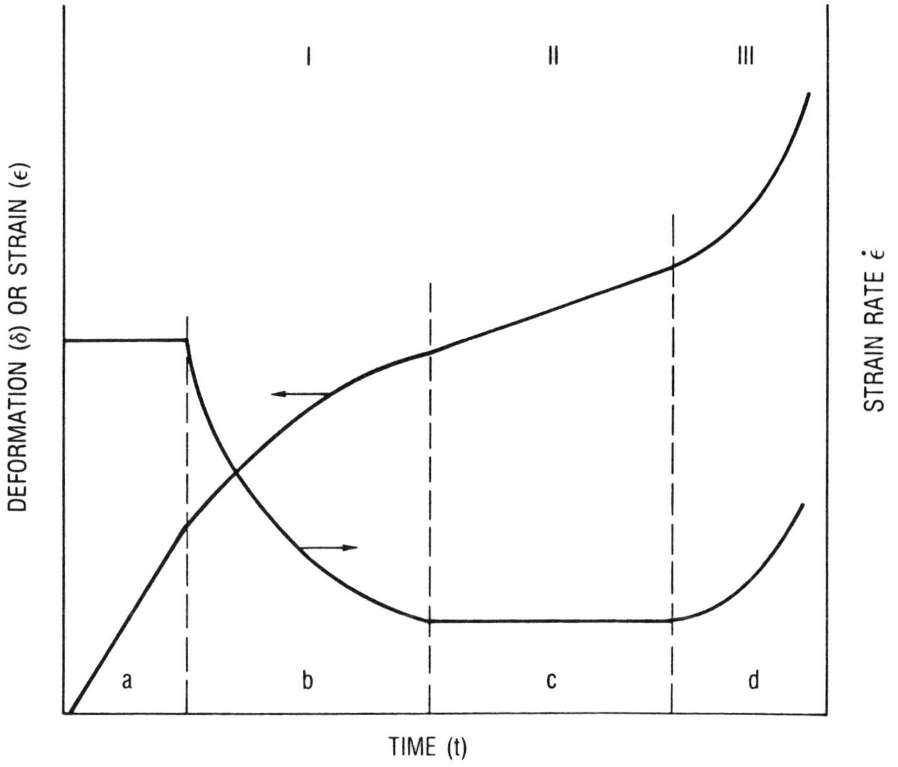

FIGURE 1 **Typical strain response of a material with time to a constant load.**

may influence the creep behavior include coatings, such as sizing for glass fibers, and fiber–matrix interactions.

Typical creep data for reinforcements are available in the literature as follows: boron fibers [24–26], carbon or graphite fibers [27–29], glass fibers [30–32], Kevlar fibers [32–35], for SiC fibers [36], and tungsten wires [37]. Creep of metal matrices also is given for aluminum [36,38–40], for silver [41,42], for copper [38,42], and for nickel [42]. Polymeric matrix creep data are given for epoxies in Refs. 24, 25, 40, and 43–45, while creep data for ceramic matrices are given in Refs. 40 and 46. For a more complete list, see Ref. 47.

Causes of Creep

The internal geometry of a composite is the major factor influencing its creep behavior. For unidirectional reinforcement with continuous fibers, the creep rate is a minimum in the fiber direction and a maximum at an angle to the fibers. A cross-ply laminate requires a higher stress to achieve the same minimum creep rate than a unidirectional sample loaded off-axis at the same angle (especially when compared at higher temperatures). However, the cross-plied material has a larger primary stage creep. Plastic or viscous flow of the matrix may lead to rotation of the fibers into the load direction with time. Hence, a changing geometry may lead to irrecoverable creep strain (e.g., in boron–aluminum [48] or graphite–epoxy [49]).

With discontinuous fiber reinforcements, the aspect ratio (the ratio of length to diameter l/d), determines the ability of the matrix to transmit loads via shear stresses to the central portion of the fiber. This may be handled analytically by postulating a supplementary creep rate dependent on the volume fraction V_f and l/d. Fiber and platelet shapes, even with identical volume fractions, will exhibit different creep strengths (or stress rupture lifetimes [50]). Anisotropic creep behavior is normal, since it is difficult to fabricate a truly random discontinuous fiber orientation, especially in three dimensions.

"In situ" composites are produced in a single-stage process by unidirectional solidification of alloys near the eutectic composition. They suffer from limited stability of the microstructure at high temperatures, and an increase in the fiber diameter or lamellar thickness reduces the creep strength. Creep in the fiber direction may also occur in composites with continuous reinforcements as a result of fiber fracture and straightening of crimp [51]. Matrices may change their microstructure as a result of composite fabrication—for example, thermoplastics consist of crystalline as well as amorphous phases; segregation of the former at the fiber interface alters the creep behavior [52]. Physical aging of resin systems has also been shown to influence creep [53]. Preconditioning through thermal or stress cycling is commonly done to "stabilize" defects. The use of progressively increasing (load–unload cycle) periods, instead of constant periods, reduces the time required to "condition" the composite to stable behavior [54]. Other effects of microstructure are discussed in Refs. 8, 39, 55, and 56.

Residual stresses are normally present in a composite

because of the differences in coefficients of expansion of the constituents. At the same time, creep during cooldown and storage at ambient temperatures can reduce the level of residual stresses. Computer simulation of creep/relaxation data without considering residual stresses may lead to overestimation of creep rates [57]. Relief of residual stresses through creep may be confused with other processes that lead to temporal instability. These include postcuring of polymer, thermal aging [58], physical and/or chemical aging [57,59], phase transformations, and such metallurgical processes as annealing, settling due to gravity, and slow microcracking as a residual stress relief mechanism. Moisture absorption is an important modifier of creep behavior in polymer matrix composites, the general effect being to increase creep rates in matrix-dominated directions. All these effects may cause dimensional instabilities on the same order of magnitude as microcreep from applied stresses. A variety of tests have been used to differentiate these effects, including weight change measurements, metallography, and surface roughness tests [3].

Theory and Mechanisms

Elastic Fibers in Plastic Matrices

Let us first consider creep parallel to continuous, stiff fibers in a plastic matrix (such as a metal) in response to a single tensile load in the fiber direction. Load equilibrium requires that the stress needed to maintain creep satisfy the relation:

$$\sigma_c = \sigma_f V_f + \sigma_m V_m \tag{1}$$

where σ_f and σ_m are functions of the strain rate $\dot{\varepsilon}$ and V represents the volume fraction of the fibers (f) or matrix (m). Strain compatibility requires that:

$$\dot{\varepsilon}_f = \dot{\varepsilon}_m = \dot{\varepsilon}_c \tag{2}$$

It has been found that most materials exhibit a power law relationship between strain rate and applied stress [22]. The basis for this is the diffusion-controlled climb of dislocations and the accompanying formation of dislocation cells [42]. The generalized expression then is:

$$\dot{\varepsilon} = A(\sigma - \sigma_0)^n \exp\left(\frac{-Q}{RT}\right) \tag{3}$$

Alternate forms of this equation include:

$$\dot{\varepsilon} = \varepsilon_0 \left(\frac{\sigma}{\sigma_0}\right)^n \tag{4}$$

$$\dot{\gamma} = \gamma_0 \left(\frac{\tau}{\tau_0}\right)^n \tag{5}$$

where γ and τ denote shear strain and stress, respectively. The friction or backstress σ_0 [60] is initially related to the matrix flow stress and to matrix creep mechanisms such as dislocation climb. The stress–temperature deformation mechanism maps of Ashby [61] are useful guides to metallic matrix creep. For example, another common mechanism is dislocation glide aided by vacancy diffusion. Here, models by Orowon and Bailey describe creep as a competition between work-hardening and softening processes (see Ref. 62). The effect of heat treatment is to change the matrix strength, stiffness, and possibly predominant creep mechanism [50,63]. In a composite, however, the fibers interfere and additional mechanisms, such as Orowon bowing between fibers, must also be considered. The other major creep laws are the exponential law:

$$\dot{\varepsilon} = \dot{\varepsilon}_0 \exp\left[\frac{-Q}{RT}\left(1 - \frac{\sigma}{\sigma_0}\right)\right] \tag{6}$$

and the logarithmic law [64]:

$$\dot{\varepsilon} = \frac{A\sigma^n}{t} \tag{7}$$

or alternatively:

$$\in(t) = \varepsilon_0(\sigma) + B(\sigma) \log t \tag{8}$$

The assumption that the creep laws for the constituents do not change once they are in a composite is reasonable if the reinforcing phase has fairly large dimensions. For example, the diameter of the fiber should exceed 10 μm [36]. If we substitute Eq. (3) into Eq. (1) and note Eq. (2), we get:

$$\sigma_c = V_f \left[\frac{\dot{\varepsilon}_c}{A_f \exp(-Q_f/RT)}\right]^{1/n} + V_m \left[\frac{\dot{\varepsilon}_c}{A_m \exp(-Q_m/RT)}\right]^{1/m} + \sigma_{0_f} V_f + \sigma_{0_m} V_m \tag{9}$$

where n and m, respectively, are the stress sensitivity exponents for the fibers and the matrix, which are unlikely to be identical. Thus there is no unique stress sensitivity for the composite, only for the constituents.

If the fibers deform elastically, they will carry increasing amounts of the applied stress, as expected from Hooke's law. The lower stress on the matrix will gradually decrease. When this strain rate becomes zero, all the stress is carried by the fibers, and we have:

$$\dot{\varepsilon}_c = \frac{\dot{\sigma}_c}{E_f V_f} \tag{10}$$

This critical strain rate is approached asymptotically with time as a result of decreasing matrix creep rate; strictly speaking, thus, there is no second-stage creep. A plane–stress continuum model was employed to account for multidimensional loading effects on graphite–aluminum, taking the thermal residual stresses and microstructural inhomogeneities into account [39]. At high temperatures, especially with high concurrent stresses, fiber–matrix interactions lead to reductions in creep strength and creep rupture life.

Viscoelastic Matrices

Most polymeric matrices exhibit a combination of viscous flow (e.g., Newtonian fluid, where the strain rate is proportional to the applied stress) and elastic behavior (where stress is proportional to strain). Creep and creep recovery tests provide the most direct means of obtaining the thermoviscoelastic properties of a material. (Other methods include stress relaxation and constant rate straining or stressing.) General viscoelastic behavior is in Article 253. The relation to plastic flow, where strain is proportional to the stress to some power, is discussed in Ref. 65.

When the ratio of stress to strain in creep is a function of time only, not of stress or strain magnitude, we may usually speak of linear (vs. nonlinear) viscoelastic behavior. Linear behavior is found at low stress and strain levels, with a low moisture content in the polymeric matrix, and with reinforcement-dominated directions. Nonlinear behavior is common at high temperatures, stresses, radiation dosages, and moisture contents, and in directions at an angle to the fibers [25,66,67]. It is caused by changes in the defect nature of the matrix or fiber–matrix interface, and it appears when there is viscous flow in the matrix. Tensile tests help to define the boundaries between linear and nonlinear behavior [45]. More precise measurements of creep suggest that linear behavior may be quite rare [57]. Even a uniaxial carbon fiber reinforced plastic may be a nonlinear material [45].

Figure 2 illustrates the major strain parameters that need to be specified for the prediction of creep and recovery behavior. Additional terms are needed to describe nonlinear behavior. Most models for the prediction of laminate creep, however, are based on linear viscoelasticity theory coupled with classical lamination theory. A common approach involves the Boltzmann (time–temperature) superposition principle (TTSP), or modifications such as the time–stress and time–temperature–stress superpositions (TSSP and TTSSP). It is generally assumed that laminate plate theory, developed for static loads and strains, can be applied during creep and recovery conditions. For a stress history given by:

$$\sigma = \sigma_0[H(t) - H(t - t_1)] \tag{11}$$

where H is the Heaviside function, the time-dependent strain of a ply with off-axis fibers is given by:

$$\varepsilon(t) = g_0 S_0 \sigma(t) + g_1 \int_0^t \Delta S(\psi - \psi') \frac{dg_2\sigma}{d\tau}\, d\tau \tag{12}$$

where S_0 and $\Delta S(\psi - \psi')$ are the initial and transient components of the creep compliance, and g_0, g_1, g_2, and "a" (see Eq. 13) are the stress-dependent material functions. A power law [68] describes the relation between creep strain and time. Application of the theory, including methods to derive the nonlinear material functions, are given in Refs. 55, 69, and 70. If the nonlinear parameters are combined, so that

$$m = \frac{g_1 g_2 C\sigma}{a_\sigma^n} \tag{13}$$

the compliance $S(t)$ may be written as follows:

$$S(t) = \frac{\varepsilon_0 + mt^n}{\sigma} \tag{14}$$

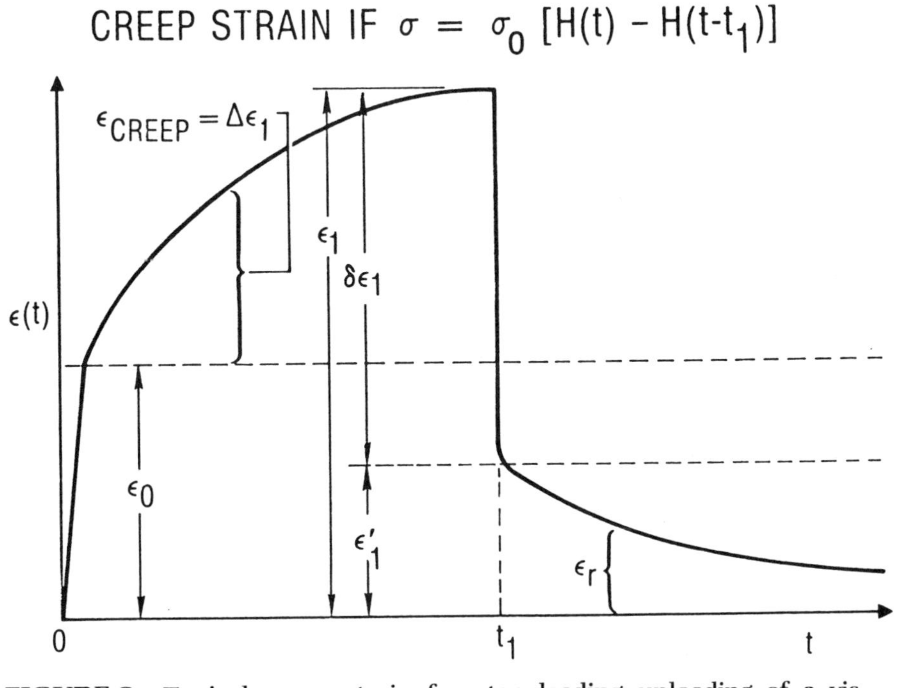

FIGURE 2 Typical creep strain for step loading–unloading of a viscoelastic matrix composite.

These terms are evaluated by short-term creep data to fit into a general laminate equation [71]. The possibility of predicting the five mutually independent components of the viscoelastic compliance tensor under conditions of creep in monotropic unidirectional composites has been explored [72]. Principal contributors in this area have been Brinson [69,73,74], Crossman [57,75], Schapery [55], and Weitsman [2,76]. Hashin [77] has extended the theory to any temperature variation and any thermorheological matrix.

Creeping Fibers

Many "in situ" composites, high temperature systems such as nickel-based matrices reinforced with refractory metal wires, and organic fiber composites show fiber as well as matrix creep. The volume fraction of the reinforcement in this case is determined by the relevant phase equilibria [8]. The determination of creep exponents and activation energies requires attention to the fraction of the applied stress which is available to each constituent. At high stress levels, where fibers may be expected to creep, the exponential law of Eq. (5) may be applicable. Combination of Eqs. (1), (2), and (6) leads to:

$$\sigma_c = \left[\frac{V_f}{\alpha_f} + \frac{V_m}{\alpha_m}\right] \ln \dot{\varepsilon}_m - \left[\frac{A_f V_f}{\alpha_f} + \frac{A_m V_m}{\alpha_m}\right] \tag{15}$$

where

$$\alpha = \frac{Q_0}{RT}\left[\frac{1}{\sigma_0} - \frac{1}{\sigma}\right] \tag{16}$$

and

$$A = \ln \dot{\varepsilon}_0 \tag{17}$$

The transition from rigid (noncreeping) to creeping fibers depends on the volume fraction V_f, the fiber aspect ratio l/d, and the applied stress and temperature. Figure 3 shows that a plot of log $\dot{\varepsilon}$ versus σ helps to determine whether the fiber is creeping [36]. With high temperature systems, it is difficult to judge whether fibers are in fact creeping because changes in fiber microstructure are possible during composite fabrication or during the creep process.

Organic fiber creep introduces viscoelastic behavior. Kevlar fibers follow a logarithmic creep law [44]:

$$\varepsilon_f(t) = \varepsilon_f(\sigma) + A_f \sigma_f^n \log t \tag{18}$$

By combining this expression with similar ones for an epoxy matrix, it has been shown that stress redistribution can occur, at least until the terms A are equal in both phases [44,64]. The viscoelastic Schapery model has been applied to Kevlar 49/Hercules 3501-6 resin for a uniaxial stress state [78]. Increasing stress and temperature levels were found to require increasing use of nonlinear parameters in the theory. The creep rates of Kevlar–epoxy are similar to those for glass–epoxy and are lower than those with other organic fibers such as nylon or Dacron polyester [79].

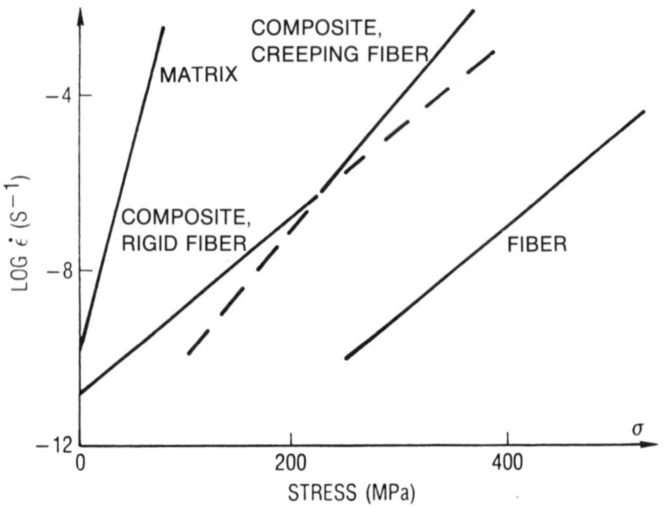

FIGURE 3 **Transition between rigid (noncreeping) and creeping fiber behavior; schematic curves for fiber volume fraction 0.4 and fiber aspect ratio of 60.** (From Ref. 36, Fig. 4.)

Discontinuous Fibers

Transitions in the creep rates are to be expected with discontinuous reinforcements, depending on the aspect ratio, volume fraction, and temperature. The full stress of the applied load is carried only by the central portion of the fiber because of matrix shear near the fiber ends. The creep strength is reduced in two ways: not all of each fiber works to resist the creep load, and differential straining between the constituents is now possible. A basic model for short aligned fibers loaded in the fiber direction suggests the relation:

$$\dot{\varepsilon}_c K_c = \dot{\varepsilon}_m = f(\sigma_m) \tag{19}$$

This means that the composite creep rate equals that of the matrix except for the factor K_c, and the functional stress dependence $f(\sigma_m)$ of the matrix is preserved [36].

$$K_c = [4(d/l_f) + \tfrac{1}{2}]^{-1}[1 - V_f^{1/3}]^{-1} \tag{20}$$

The term $\frac{1}{2}$ allows for a possible constrained tensile deformation of the matrix. Expressions for the fiber-reinforced composite stress during creep must be scaled relative to that of the pure matrix. The assumption that the creep laws of the constituents may be transferred to expressions describing the composite behavior may not always hold. For example, a pure metal may deform by diffusional creep, but with dispersion hardening, the mechanism may change to climb of dislocations at particles, leading to a different creep law. Composite forming, such as hot rolling or extrusion, also results in dislocation densities different from those characteristic of normal alloy processing, and this may alter the creep behavior. Changes in reinforcing particle volume fraction or shape [80] and matrix flow at the interface [41] or around fiber ends may also influence creep behavior [81,82].

Creep of SiC whiskers in aluminum is lower than in pure aluminum and many Al alloys (Fig. 4). The rate-controlling mechanism is the glide of dislocations [9]. However, the creep rate may not be significantly less than in Al alloys at ambient or at low temperatures. The observation that the stress sensitivity of the creep rate ($d \log \dot{\varepsilon}/ds$) is larger for the composite than for the matrix may be interpreted as an indication of possibly complex interactions between the fibers and the precipitation process in the alloy matrix [4].

Ceramic Matrices

There are few data on the creep of ceramic matrix composites, probably because their fracture and thermal shock resistance is of greater concern. Adhesion between the fiber and the matrix is a major problem [83], and indeed, lack of it is desirable for fracture toughness. The creep of optical ceramics and the temporal instability of mirror substrate ceramics and composites have been examined [84].

In general, the minimum creep rate is proportional to stress raised to some power "n," where the value of "n" is characteristic of the deformation mechanism. With viscous flow or diffusion processes, n is only slightly above unity. With dislocation movements, n tends to be closer to 3–4. Values of $n = 1$–2 were found via torsional creep studies for glass–ceramics, such as partially crystalline lithia–alumina–silica systems [46]. Here heat treatment is critical, for it determines the ratio of crystalline to amorphous or glassy phases. Since voids also increase the likelihood of linear viscous flow, the rate of deformation in second-stage creep is directly proportional to the applied stress. Viscous flow of glassy phases at grain boundaries leads to time-dependent changes in the stiffness of the material. At high temperatures and/or high stress levels, the strain takes long times to come to an equilibrium value. In composites with brittle matrices such as SiC, crack formation and propagation are also common if the strain-to-failure of the pure matrix is exceeded.

Steel wires provide reinforcement in cement-based composites mainly through the friction at the steel–cement interface; thus shrinkage of the matrix during fabrication is a major factor in determining the creep rates [85]. In SiC reinforced glass–lithia–alumina–silica composites, the SiC fiber creep dominates at lower temperatures, while at higher temperatures glassy phases determine the creep rate [86]. There may also be an extra phase in the matrix of carbon–carbon composites as a result of the mode of fabrication, which causes a higher creep rate [87]. Composites made by impregnating single bundles of fibers with pitch and carbonizing the matrix show easily measurable creep rates above 2200°C [27,29]. These rates are 10–20 times lower than those found in dense polycrystalline graphite at similar stresses. Although the fibers provide most of the creep resistance, stress transference does occur between filaments and modifies the creep behavior of the composite.

Creep of NaCl containing a dispersed phase of alumina occurs above a threshold stress, where the creep rate varies linearly with applied stress [88]. This threshold stress decreases with temperature and is proportional to the volume fraction of the dispersoid.

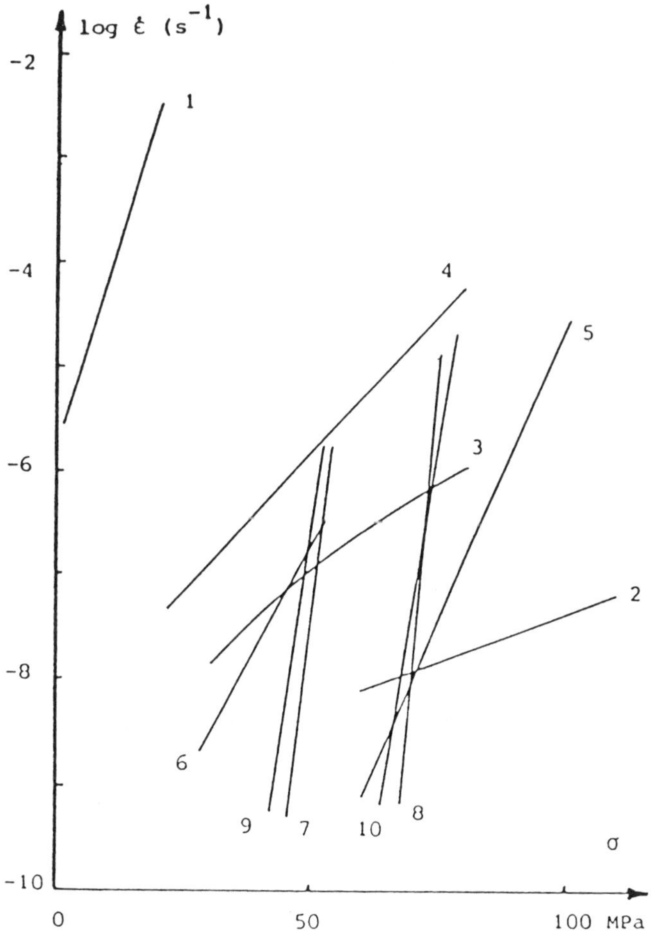

FIGURE 4 **Creep rates in stage II for Al, Al 2124 composites and Al 6061 composites.** (From Ref. 4, Fig. 4.)

Consequences of Creep

Recovery and Relaxation

Creep recovery is a dimensional response to a reduction in an applied creep stress. This tends to diminish with time, and it depends on both the former applied stress level and the duration of stress application. Creep recovery occurs much more slowly with viscoelastic than with plastic matrix composites. Related behavior includes stress relaxation when the material is subjected to a constant strain. Residual stress relaxation may also have a minor influence on creep behavior, especially when large temperature excursions immediately precede the occurrence of creep.

In principle, most of the strain accumulated during creep of a composite with elastic fibers in a plastic matrix should be recoverable. The same is not true if the fibers creep as well. The differing creep rates in the latter case will result in residual stresses upon unloading, and a partial anelastic contraction may be expected. Temporary reduction in applied stress may result in a zero creep strain even if the load does not equal zero; this is because of partial recovery. Indeed, internal stress relaxation may serve to increase the creep strain above that due to the applied load alone [51]. Postcuring reduces residual stresses, hence stress relaxation (see, e.g., Refs. 76 and 89). Nonlinear behavior in viscoelastic matrices may also promote stress redistribution during relaxation [82]. The importance of recovery data for adhesive use is outlined in Refs. 2 and 6. Typical strain recovery data for glass and Kevlar–epoxy composites are given in Ref. 79.

Constituent–Composite Failure

Constituent damage may be present as a result of "preconditioning," an often poorly characterized procedure based on thermal–stress cycling. Damage mechanisms contribute to the second-stage creep and ultimately to stress rupture. Much nonlinear behavior may be attributed to internal crack growth within the matrix or at the fiber–matrix interface. Secondary effects of microcracks include a reduction in matrix stiffness and increased permeability for water into the composite. The former will influence loading rates and cyclic effects, such as creep during fatigue. Irreversible creep effects (e.g., from hydrogen bond displacement) occur if moisture contents and stress levels exceed certain limits in polymeric matrix composites [90].

Other damage mechanisms include microstructural changes [63], fiber necking, cumulative fiber fracture [8], matrix microcracking [91], stress corrosion [51], and recrystallization of the fibers. The power law relationships for second-stage creep do not predict the form of tertiary creep nor the time for rupture. Typical data for the stress rupture behavior of various composites are available in the literature composites of tungsten–superalloys [92], graphite–aluminum [93], graphite–epoxy [49,94], glass–epoxy [49,51,56], glass–thermoplastics [94], and Kevlar [33,95].

Composite Creep Data

Some typical creep data are given as follows for common composite fiber–matrix systems (a more complete list appears in Ref. 47): boron–aluminum [24,39,48,72], carbon–carbon [27,29], glass–epoxy [1,11,25,54,55,56,79], glass–polyester [1,14,51,54,71,96], graphite–aluminum [39,93], graphite–epoxy [12,16,49,70,72,90,94,108], graphite–magnesium [97], Kevlar–epoxy [12,44,79,95,98], SiC–aluminum [4,5,9,36], tungsten–superalloys [8,22,37,92], and wood [21,98]. Figure 5 gives compressive creep data for 6 in. long four-ply graphite–epoxy tubes with pitch 75 fibers and different laminate orientations [99].

Measurement of Creep

Application of Stress

Creep of composites is a relatively difficult property to measure. The anisotropic nature of the materials assures

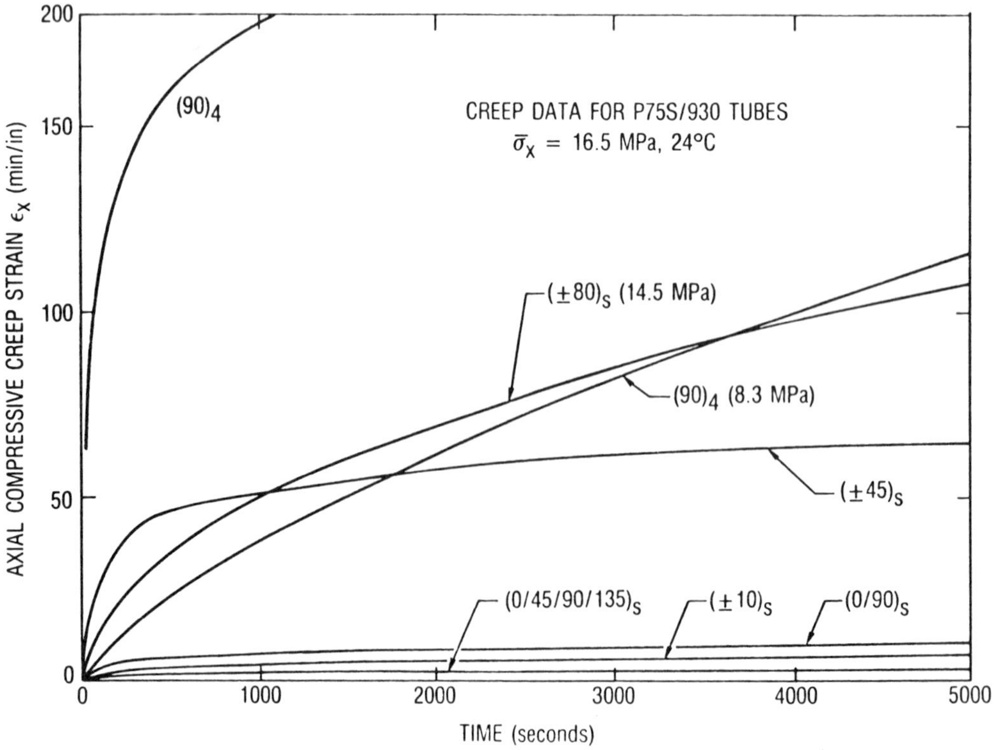

FIGURE 5 **Short-term compressive creep curves for 6 in. tubes of four-ply layups of pitch 75S/934 resin, loaded at 16.5 MPa.** (From Ref. 99.)

a wide range of creep strains. Temperature, load, and moisture content must be held constant over long periods of time. The strains are small, and it is difficult to determine the boundary between the initial elastic loading and the first-stage creep region.

Problems specific to the type of loading include tension [30,100,101], compression [5,60,70,87,98,102,103], torsion [43,46], flexure [104], bending [21,46,86,98], and shear [6,25,72]. Grip shearing is discussed in Ref. 22. Stress relaxation after creep may be measured by clamping the sample over a fixed mandrel for a prolonged time and measuring the deflection of the free end with a micrometer after clamping has been discontinued.

Multiaxial load studies are discussed in Refs. 7 and 55. Methods for specialized shapes, such as yarns, are reviewed in Ref. 32. Creep testing of cylinders or tubes is presented in Refs. 12 and 72. Problems due to eccentric and/or off-axis loading are outlined in Refs. 8, 56, and 100. Constant stress instead of constant load is discussed in Ref. 7. One consequence of constant stress is the prolongation of various stages of creep, and since the rupture mechanisms tend to be the same, the use of constant stress rather than constant load needs to be well justified. Variable load measurements are described in Ref. 96.

Strain Measurements

The strain-measuring system must itself be constant in behavior over long time periods; thus a constant temperature and humidity room is recommended. A frequent auxiliary requirement is the periodic determination of net creep strains after arbitrary recovery periods. Thus methods to "recall" creep strains must be built into the recording system. Because of the prolonged times involved, expensive instruments may be tied up unless multiple loading stations can be arranged, together with scanning capabilities of the strain detector.

Strain transducers may be divided into broad categories such as electrical and interferometric and noninterferometric electro-optical devices [105]. In the first category, the electrical resistance strain gage is commonly used for strain resolution down to about 10^{-7}. Balanced bridge circuits with compensating dummy (unloaded) samples in alternate branches and automatic temperature compensation are required. Without postcuring, however, there will be long-term stability problems. Even the most stable of gages—such as Micro-Measurements type WK [106]—may show a drift of more than 10 microstrain after 100 days at 75°C. The adhesive bond is sensitive to moisture changes, and the gage itself interferes with local sample moisture equilibration.

Precision end-length measurements, such as gage block comparators employing linear variable differential transducers (LVDTs), are reviewed in Ref. 38. Such methods are not contactless and may be accompanied by problems when used with polymeric matrix composites. Capacitance methods are contactless and have excellent resolution. However, the method is a differential one and depends on the stability of the sample support system materials. There is also an inherent nonlinearity, as well as temperature sensitivity and adiabatic heating and cooling effects.

Laser interferometry, principally Michelson, Fizeau, and Fabry-Perot, offers real-time strain monitoring to a very high resolution [107]. Frequency-stabilized lasers are needed, and only the Michelson approach offers contactless measurement for arbitrarily shaped samples [5,103,105]. Fabry-Perot techniques are particularly suited for temporal stability measurements [84,107]. Holographic interferometry has also been used to measure microcreep behavior via surface deformation due to thermal cycling [108].

Prediction of Creep Strains

Accurate predictive models are needed, especially when extrapolation on the order of years is anticipated. For viscoelastic matrices, there are theories based on superposition principles and on multiple integral representations, but a generally preferred approach is that of Schapery [55], who developed a constitutive equation based on thermodynamics which has the simplicity of a single integral and also retains the capability to model complex dependence on nonlinear stress, temperature, etc. This theoretical approach, together with the concept of accelerated (e.g., high temperature) testing, is the preferred approach to creep predictions (see, e.g., Refs. 72, 74, and 109). The emphasis has been on short-term observation of unidirectional materials at various temperatures. Additional approaches include incremental methods coupled with finite element methods [24], use of shear compliance data, and use of data from several layups. Computer-assisted accelerated creep data, moisture, and stress relaxation effects are presented by Crossman [57,75]. The design approach to long-term data acquisition is discussed in Ref. 110. A method of predicting the microyield strength of viscoelastic matrix composites as well as incremental and cumulative recovery strains after creep loading is given in Ref. 99.

Ernest G. Wolff

[*See also* CREEP RUPTURE; VISCOELASTICITY.]

References

1. K. H. Boller, *Plast. Techn.*, 1956, p. 808.
2. D. Peretz and Y. Weitsman, *J. Rheol.*, *27*(2), 97 (1983).
3. N. P. Freund, in *Composite Reliability* ASTM STP 580, 1975, p. 133.
4. H. Lilholt and M. Taya, in *Proceedings of the ICCM-VI*, 1987.
5. E. G. Wolff and R. C. Savedra, Report SD-TR-84-26, The Aerospace Corporation, El Segundo, CA, August 1984.
6. J. L. Rutherford and E. J. Hughes, *Proceedings of the 24th Natl. SAMPE Symposium*, 1979, pp. 728–737.
7. "Creep, Stress Rupture and Stress Relaxation Testing," in *ASM Metals Handbook*, 9th ed., Vol. 8, *Mechanical Testing*, American Society for Metals, Metals Park, OH, 1985, p. 299.

8. M. McLean, in H. Lilholt and R. Talreja, Eds., *Fatigue and Creep of Composite Materials,* Risø National Laboratory, Roskilde, Denmark, 1982, p. 77.

9. T. G. Nieh, *Metall. Trans., 15A,* 139 (1984).

10. W. R. Mohn and D. Vukobratovich, *SAMPE J., 24*(1), 26 (1988).

11. S. W. Beckwith *SAMPE Q., 11*(2), 8 (1980).

12. R. F. Foral and W. D. Humphrey, paper presented at the 23rd AIAA Structures Conference, New Orleans, 231, May 1982.

13. G. J. Svob and K. W. Bills, Air Force Rocket Propulsion Laboratory Technical Report AFRPL-TR-84-643, September 1984.

14. J. S. Cartner, W. I. Griffith, and H. F. Brinson, "The Viscoelastic Behavior of Composite Materials for Automotive Applications," VPI-E-78-15, July 1978.

15. K. E. Hofer "An Investigation of the Fatigue and Creep Properties of Glass Reinforced Plastics for Primary Aircraft Structures," IITRI, AD652415, April 1967; Final Report Contract 65-0425-5.

16. T. Ho, W. J. Renton, and R. A. Schapery, in E. Lenhoe, Ed., *Fibrous Composites,* Plenum Press, New York, 1980, p. 781.

17. Z. Jeffries and R. S. Archer, *Chem. Metall. Eng., 27*(17), 833 (1922).

18. H. and F. Le Chatelier *Compt. Rend., 171*(16), 695 (1920).

19. J. B. Speakman *J. Text. Inst., 17,* T472 (1926).

20. R. Meredith and J. W. S. Hearle *Physical Methods of Investigating Textiles,* Wiley-Interscience, New York, 1959.

21. V. F. Yatsenko "Strength and Creep of Laminated Plastics" (NASA translation) TT67-59014, NASA TT F461 (1970), from *Prochn. Polzuchest Sloistykh Plast.* (1966).

22. K. N. Street, in *The Properties of Fibre Composites,* IPC Science & Technical Press, 1971, p. 36.

23. V. K. Tewary *Mechanics of Fibre Composites,* Wiley, New York, 1978.

24. R. L. Foye, in *Composite Reliability,* ASTM STP 580, 1975, p. 381.

25. A. M. Skudra and F. Ya. Bulavs, in I. F. Obraztsov and V. V. Vasilev, Eds., *Mechanics of Composites,* 1982, p. 77.

26. R. H. Erickson *Fibre Sci. Technol., 7*(3) (1974); also *Metall. Trans. 4* (1973).

27. L. A. Feldman, Second Annual Report, Aerospace Report TOR-0084A (5728-02)-1, July 15, 1985.

28. Y. C. T. Yeung and B. E. Parker, in I. H. Marshall, Ed., *Composite Structures,* Vol. 4, Elsevier, New York, p. 1.309.

29. L. A. Feldman, Aerospace Corporation Report SD-TR-83-71, Dec. 1, 1983.

30. T. T. Chiao and R. L. Moore, *J. Compos. Mater., 5,* 124 (1971).

31. F. K. Ko, in A. R. Bunsell, Ed., *Advances in Composite Materials—ICCM-III,* Pergamon Press, Oxford, 1980, p. 719.

32. M. M. Schopee et al., Appendix II in AFML-TR-77-72, May 1977.

33. J. Cook, A. Howard, N. J. Parratt, and K. D. Potter, in H. Lilholt and R. Talreja, Eds., *Fatigue and Creep of Composite Materials,* Risø National Laboratory, Roskilde, Denmark, 1982, p. 193.

34. H. D. Wagner, S. L. Phoenix, and P. Schwartz, *Proceedings of the ICCM-V,* 1985, p. 245.

35. A. Howard and N. J. Parratt, *Proceedings ICCM-V,* 1985, p. 227.

36. H. Lilholt, *Compos. Sci. Technol., 22,* 277 (1985).

37. R. Warren and C.-H. Andersson, in H. Lilholt and R. Talreja, Eds., *Fatigue and Creep of Composite Materials,* Risø National Laboratory, Roskilde, Denmark, 1982, p. 335.

38. C. W. Marschall and R. E. Maringer, *Dimensional Instability,* Pergamon Press, Oxford, 1977, p. 263.

39. B. K. Min and F. W. Crossman, *J. Compos. Mater., 16,* 188 (1982).

40. R. M. Hill and L. A. Dissado, *J. Mater. Sci., 19,* 1576 (1983).

41. A. Kelly and W. R. Tyson, *J. Mech. Phys. Solids,* 1966, p. 177.

42. H. Lilholt, in H. Lilholt and R. Talreja, Eds., *Fatigue and Creep of Composite Materials,* Risø National Laboratory, Roskilde, Denmark, 1982, p. 63.

43. N. L. Hancox and D. C. C. Minty, *J. Mater. Sci., 13,* 797 (1978).

44. R. H. Erickson *Composites, 7,* 189 (1976).

45. H. Vangerko and A. J. Barker, *Composites, 16*(1), 19 (1985).

46. E. M. Heuse and G. Partridge, *J. Mater. Sci., 9,* 1255 (1974).

47. E. G. Wolff, *Introduction to the Dimensional Stability of Composite Materials,* Pergamon Press, Oxford, in preparation.

48. I. J. Toth, in ASTM STP 460, 1969, p. 236.

49. F. Y. Soliman, in ASTM STP 460, 1969, p. 254.

50. H. Lilholt, in G. Piatti, Ed., *Advances in Composite Materials,* Applied Science Publishers, London, 1978, p. 209.

51. R. J. White and M. G. Phillips, *Proceedings of the ICCM-V,* 1985, p. 1089.

52. X. R. Xico, C. C. Hiel, and A. H. Cardon, *Proceedings of the Sixth Int. Congress on Experimental Mechanics,* 1988, p. 695.

53. J. L. Sullivan, *Proceedings of the Sixth Int. Congress on Experimental Mechanics,* 1988, p. 683.

54. K. L. Jerina, R. A. Schapery, R. W. Tung, and B. A. Sanders, ASTM STP 772, 1982, p. 225.

55. Y. C. Lou and R. A. Schapery, *J. Compos. Mater., 5,* 208 (1971).

56. M. P. E. Desvaux and J. H. Greenwood, in H. Lilholt and R. J. Talreja, Eds., *Fatigue and Creep in Composite Materials,* Risø National Laboratory, Roskilde, Denmark, 1982, p. 199.

57. D. L. Flaggs and F. W. Crossman, *J. Compos. Mater., 15,* 21 (1981).

58. J. A. Kerr and J. F. Haskins, *Proceedings of the 23rd AIAA Structures Conference,* 1982, p. 101, paper 82-0657.

59. J. Mijovic, *J. Compos. Mater., 19,* 178 (1985).

60. A. Kelly, in H. Lilholt and R. Talreja, Eds., *Fatigue and Creep of Composite Materials,* Risø National Laboratory, Roskilde, Denmark, 1982, p. 51.

61. M. F. Ashby *Acta Metall., 20,* 887 (1972).

62. G. E. Dieter *Mechanical Metallurgy,* McGraw-Hill, New York, 1986, p. 446.

63. S. F. Dirnfeld and Y. Zuta, in H. Lilholt and R. Talreja, Eds., *Fatigue and Creep in Composites,* Risø National Laboratory, Roskilde, Denmark, 1982, p. 205.

64. G. Garmong, *Metall. Trans., 5A,* 1257 (1974).

65. P. M. Naghdi and S. A. Murch, *J. Appl. Mech.*, 321 (1963).

66. C. S. Wang and A. S. D. Wang, in A. R. Bunsell, Ed., *Advances in Composite Materials—ICCM-III*, Pergamon Press, Oxford, 1980, p. 569.

67. D. J. Plazek, in R. A. Fava, Ed., *Polymers, Part C, Physical Properties*, Academic Press, New York, 1980.

68. W. N. Findley, *SPE J.*, *16*, 57 (1960).

69. D. A. Dillard and H. F. Brinson, in ASTM STP 813, 1983, p. 23.

70. R. Mohan and D. F. Adams, University of Wyoming Report UWME-DR-301-107-1, December 1983.

71. S. Amijima and T. Adachi, in *Proceedings of the ICCM-IV*, 1982, p. 811.

72. R. D. Maksimov and E. Z. Plume, *Mech. Compos.*, *20*(2), 149 (1984).

73. H. F. Brinson and D. A. Dillard, in *Proceedings of the ICCM-IV*, 1982.

74. H. F. Brinson, D. H. Morris, and Y. T. Yeow, VDI-Berichte *313*, 395 (1978).

75. F. W. Crossman, *Compos. Tech. Rev.*, *1*(3), 8 (1979).

76. Y. Weitsman *J. Appl. Mech.*, *47*, 35 (1980).

77. Z. Hashin, E. A. Humphreys, and J. Goering, (MSC) AFOSR-TR-84-1124, AD-A149275 Final Report, Contract F49620-83-C-0103, August 1984.

78. D. E. Walrath, *Proceedings of the Sixth Int. Congress on Experimental Mechanics*, 1988, p. 698.

79. L. H. Miner, R. A. Wolffe, and C. H. Zweben, in ASTM STP 580, 1975, p. 549.

80. L. M. Brown "Creep of Metallic Composites Related to Deformation Mechanics," in H. Lilholt and R. Talreja, Eds., *Fatigue and Creep in Composite Materials*, Risø National Laboratory, Roskilde, Denmark, 1982, p. 1.

81. A. Kelly and K. N. Street, *Proc. R. Soc. London*, *328A*, 283 (1972).

82. A. R. T. de Silva, *J. Mech. Phys. Solids*, *16*(3), 169 (1968).

83. J. K. Keer, *J. Mater. Sci. Lett.*, *4*, 363 (1985).

84. J. W. Berthold, S. F. Jacobs, and M. A. Norton, *Metrologia*, *13*, 9 (1977).

85. P. S. Mangat and M. M. Azari *J. Mater. Sci.*, *20*, 1119 (1985).

86. J. J. Brennan and K. M. Prewo, *J. Mater. Sci.*, *17*, 2371 (1982).

87. E. Fitzer and R. Gadow, *Bull. Am. Ceram. Soc.*, *65*(2), 326 (1986).

88. R. K. Sinha and J. R. Blachere, *J. Mater. Sci.*, *15*, 1772 (1980).

89. Y. Weitsman and B. D. Harper, *J. Appl. Mech.*, *49*, 735 (1982).

90. F. X. Charentenay and M. A. Zaidi, in T. Hagashi, K. Kawata, and S. Umekawa, Eds., *Progress in Science and Engineering of Composites*, JSCM, Tokyo, 1982, p. 787.

91. J. F. Jansson, in H. Lilholt and R. Talreja, Eds., *Fatigue and Creep of Composite Materials*, Risø National Laboratory, Roskilde, Denmark, 1982, p. 229.

92. A. W. H. Morris and A. Burwood-Smith, *Fiber Sci. Technol.*, *3*(1), 53 (1970).

93. W. C. Harrigan and W. W. French, Aerospace Report ATR-75(9450)-2, March 10, 1984.

94. K. Friedrich, in H. Lilholt and R. Talreja, Eds., *Fatigue and Creep in Composite Materials*, Risø National Laboratory, Roskilde, Denmark, 1982, p. 211.

95. F. P. Gerstle, Jr., and S. C. Kunz, in ASTM STP 813, 1983, p. 263.

96. C. M. Howard and L. Hollaway, in *Proceedings of the ICCM-VI*, 1987, p. 4.232.

97. E. G. Wolff, B. K. Min, and M. H. Kural, *J. Mater. Sci.*, *20*, 1141 (1985).

98. M. Partl, C. Tinic, and A. Roli, *J. Mater. Sci.*, *17*, 970 (1982).

99. E. G. Wolff and S. T. Crane, *J. Comp. Tech. & Res.*, *10*, 165 (1988).

100. H. C. Wu and T. P. Wang, *J. Eng. Technol. (Trans. ASME)*, *104*, 280 (1982).

101. A. L. Cunningham, "Creep," ACDAP Report, Lockheed, GA, June 1971, Ch. 3.1.8.

102. M. N. Irion and D. F. Adams, *Composites*, *12*, 117 (1981).

103. E. G. Wolff, R. C. Savedra, and D. J. Chang, Report SD-TR-85-02, The Aerospace Corporation, El Segundo, CA, March 1984.

104. R. C. Allen, *SAMPE Q.*, *13*(3), 1 (1982).

105. E. G. Wolff, *Proceedings of the 9th Natl. SAMPE Technical Conference*, 1977, pp. 57–72.

106. H. S. Freynik and G. R. Dittbenner, *Exp. Mech.*, *16*, 155 (1976).

107. J. W. Berthold, S. F. Jacobs, and M. A. Norton, *Appl. Opt.*, *15*(8), 1898 (1976).

108. W. R. Goggin, *Appl. Opt.*, *13*(2), 444 (1974).

109. W. I. Griffith, D. H. Morris, and H. F. Brinson, in A. R. Bunsell, Ed., *Advances in Composite Materials—ICCM-III*, Pergamon Press, Oxford, 1980, p. 461.

110. M. P. E. Desvaux and G. Smith, *Fiber Sci. Technol.*, *18*, 53 (1983).

Creep-Rupture

The demand for high performance composite materials is continually increasing, especially for applications in which strength and weight of conventional structural materials are limiting factors. Fibrous composites often offer the advantages of high specific properties—properties measured while taking into account the weights of the materials—and the ability to align these properties in a specific direction or directions. In a composite material, the reinforcing fibers and the matrix form a synergistic group that gives the composite system strength, stiffness, and thermal stability often equal to or better than those of the more conventional materials.

Creep-rupture (variously called stress-rupture, static fatigue, or creep-fatigue) is the time-dependent failure of a material subjected to continuous loading at a level below its ultimate strength. Rocket motor casings, filament-wound pressure vessels, composite flywheels, pultruded beams, and gas centrifuge cylinders are examples of lightweight, filamentary composite structures that, over the past 25 years, have been developed to reliably sustain high tensile loads for long periods of time. Many of these are used in critical applications, and failures have disastrous consequences, including the loss of life. It is important therefore to establish appropriate design parameters that achieve high reliability.

In addition, advanced composites generally represent great potential for reducing energy costs in transportation systems. Application to commercial aircraft is happening at present, and application to automobiles and trucks can be expected when energy costs rise dramatically, composite fabrication costs decrease, or, as is likely, both occur. To realize this potential, however, material stresses must be as high as possible, while the structures remain highly reliable.

Because of natural variability in materials and manufacturing methods, the difficult problem is to develop a structure to carry a given load L such that over a period of time t^* the probability of failure is some small number P. Human safety and liability factors may require $P < 10^{-6}$ and $t^* > 10$ years, for example. It is not at all unusual for a newly introduced fiber or matrix material to be specified for a structural component that must have high reliability 10–20 years into the future. This poses major problems for composite materials. Because composites have been used for a relatively short period of time, we have neither the history nor the numbers from which to draw performance data. Indeed, many applications of composites, by their very nature, have been limited to a few or, at most, a few thousand pieces. Safety factors are regularly used with conventional materials, but it should be remembered that these have been developed largely from experience rather than from analytical models.

It is a fact that large statistical variation is *inherent* to both composite strength and lifetime. This variation begins with the manufacturing processes and materials used to produce the reinforcement fibers, and is carried right through to the manufacture of the finished part. Spinning processes used to produce the major reinforcement fibers—graphite, aramid, and glass—are well-known sources of material variability, with mechanical properties that vary as much as 7–10% being common. The impregnation step in which the fibers are embedded within the matrix also is the source of much variability, in the chemistry of the matrix as well as in the uniformity of its distribution around the filaments. Finally the manufacture (layup) of many composite parts is often done by hand, further inducing variability into the properties of the structure. Added to this, lifetime data with variability of 1000% or greater are not uncommon.

Because this variability and the lack of historical data, standard statistical methods are generally inadequate to accurately predict the lifetime for the required level of statistical confidence. In the absence of these parametric models, a "brute force" approach at the probability level P requires the testing of at least $10/P$ specimens, or in the case above more than 10^7 specimens. Even when the number of specimens is manageable, there may not be sufficient real time to do the testing, especially when t^* must be on the order of decades.

Creep-rupture in composite materials is an important phenomenon but, due to an inherently large degree of variation in the materials that, in turn, produces wide scatter in lifetime data, complex modeling and scaling techniques are necessary to analyze structures and predict failure times. Many of the above cited works have urged caution when attempting to predict lifetimes under constant stress using short-term techniques. Two of the many techniques currently used are proof testing and the testing of "matched" specimens with a go-no go decision made upon the failure strength of the partner specimen.

In the case of proof testing, because of the variability in fiber strength, almost surely there is the random breaking of individual fibers and the likely formation of clusters of broken fibers caused by the test itself. These lead to an actual **reduction** in lifetime, no matter what model is chosen to predict lifetime behavior. Thus we are presented with a Hobson's choice whether to not induce flaws but to pass an untested vessel, or to proof test the vessel with full knowledge that the testing itself is causing a reduction in the lifetime.

Oftentimes, the testing of matched specimens is seen as a way to avoid this choice. Unfortunately, because they depend, in large measure on different mechanisms of failure, static strength and lifetime are largely uncorrelated, and therefore a strong 'partner' specimen in no way ensures that its companion will enjoy a long service life.

Faced with the problems described above in the development of filament-wound pressure vessels, Lawrence Livermore National Laboratory (LLNL) began, in the late 1960s, a life testing program for S-glass–epoxy composites [1]. In view of the high cost and danger of performing tests on full-scale pressure vessels, it was decided to test large numbers of epoxy-impregnated strands, hoping that scaling laws could be used to accurately predict vessel life. In the early 1970s the testing was expanded to include graphite fibers [2] and du Pont's aramid polyparaphenylene terephthalamide PRD-49 [3] (preproduction Kevlar 49). In the mid-1970s a facility was built to life test aramid–epoxy pressure vessels. Experimental data [4,5] based on 250 aramid–epoxy vessels lined with aluminum (1100-O) bladders pressurized at four different levels resulted from those latter tests. Statistical methods [6,7] employing a two-parameter Weibull distribution were initially used to investigate pressure–failure time relationships, while more recent analyses [8] focused on spool-to-spool variations in the aramid filaments.

Working with full-scale models is very often expensive and dangerous. For example, in the case of a pressure vessel with an inner radius of 57.15 mm and an outer radius of 58.25 mm, made from Kevlar 49 impregnated with a DER-332/T-403 epoxy–hardener system at 0.65 fiber volume fraction, static failure occurs at a sustained internal pressure near 30 MPa, and the vessel will sustain a pressure of 24 MPa for just over 315 hours [9]. At these and often higher working pressures, failure is catastrophic and the entire test must be enclosed for safety. Indeed, composite materials generally are known for failing catastrophically, making post-mortem analyses difficult and often inconclusive. Thus it is important to develop models for composite failure processes in creep-rupture.

The major theoretical efforts that have evolved to predict lifetimes of composites (or composite structures) under sustained loading have followed three general paths: (a) fracture mechanics models based on viscoelastic, kinetic crack growth, (b) micromechanical models that incorporate kinetic approaches, and (c) elastic–viscoelastic models. The first two approaches have largely been statistical, while the third is a more classical, deterministic approach. There are a number of theories relating to the creep-life prediction (see, e.g., Refs 10–14), but in general the analyses fall into one of the three groups described above. (Note that in all the analyses below, the fibers are considered to be *continuous*.)

Examples of Theoretical Models

A Viscoelastic Crack Growth Model

One description of the essentials of the idealized crack-growth model for the creep-life of a composite material under constant load is due to Christensen [15] and follows the formalism of Schapery [16]. As shown in Figure 1, an initial flaw is assumed to exist in a homogeneous specimen, and a one-dimensional, constant state of stress is suddenly (but smoothly) applied. An associated problem, that of crack growth in a semi-infinite, viscoelastic medium, illustrated in Figure 2, was considered by Christensen and Wu [17]. Using global energy conservation methods, the criterion for crack growth under an imposed stress σ is of the form

$$\sigma^2 = \frac{2\Gamma/h}{J(h/\beta c)} \tag{1}$$

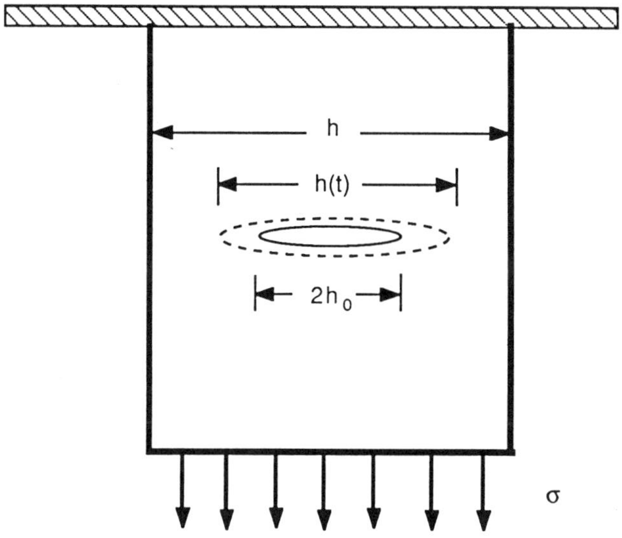

FIGURE 1 **Idealized creep-rupture problem.** (Reprinted from R. M. Christensen, *J. Rheol.*, *25*, 517 © 1981 John Wiley & Sons.)

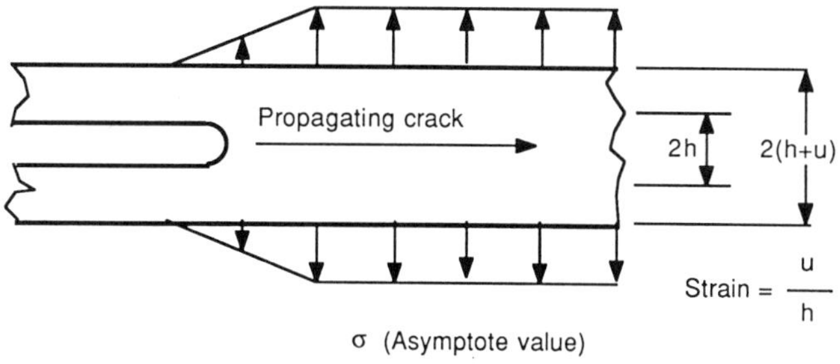

FIGURE 2 **Semi-infinite crack growth problem.** (Reprinted from R. M. Christensen and E. M. Wu, *Eng. Fract. Mech.*, *14*, 215 © 1981 Pergamon Press.)

where h is the strip half-width, Γ is the irrecoverable work generated in creating new crack surface, $J(\cdot)$ is the uniaxial creep function for the viscoelastic material, c is the crack velocity, and β is a parameter governed by the stress distribution gradient at the crack tip. The results in Eq. (1) are used to solve for the rate of growth for the crack shown in Figure 1, appealing to the usual stress intensity factorization of classical fracture mechanics.

To apply Eq. (1), two additional assumptions are made. The first is that any far-field stress will be taken in such a way that the interaction between the crack and the free edge can be ignored, allowing for the stress σ to be a constant, uncorrected value that does not depend on crack length. This presumes that most of the lifetime of the structure occurs when the flaw dimension is very small. The second assumption is that the creep function $J(\cdot)$ is of the power law form

$$J(t) = \boldsymbol{J}t^n + J(0) \tag{2}$$

where $\boldsymbol{J}$ and the exponent n are taken to fit the experimental data, and $J(0)$ is the initial value of the creep function. Finally, two additional restrictions should be considered [18]: (a) the stresses described in the model must be finite and bounded everywhere, including at the crack tip, and (b) the existence of initial damage or a craze zone at the crack tip must be assumed a priori.

The dimension of the crack at any time t, $h(t)$ can be written as

$$h(t) = h_0 + \int_0^t c(t)\,dt \tag{3}$$

Using $c = dh/dt$, the time t required for the crack to grow to the overall dimension h is given by

$$t = \beta\left[\frac{J(0)}{\boldsymbol{J}}\right]^{1/n} \int_{h_0}^{\Sigma} \left[\frac{2\Gamma}{J(0)\sigma^2 y} - 1\right]^{1/n} \frac{dy}{y} \tag{4}$$

The static strength of the material σ_s is now given by

$$\sigma_s = \frac{2\Gamma}{h_0 J(0)} = \frac{\gamma}{h_0} \tag{5}$$

where $\gamma = 2\Gamma/h_0$. Consistent with the condition that the crack reaches a critical length before it reaches the edge of the specimen, the dimension h is replaced by

$$h \rightarrow \frac{\gamma}{\sigma^2} = \frac{h_0 \sigma_s^2}{\sigma^2} \tag{6}$$

Equation (4) may now be written in a nondimensionalized form as

$$\tau = \int_1^{1/\tilde{\sigma}^2} \left[\frac{1}{\tilde{\sigma}^2 \zeta} - 1 \right]^{1/n} \frac{d\zeta}{\zeta} \tag{7}$$

where

$$\zeta = \frac{y}{h_0}$$

$$\tilde{\sigma} = \frac{\sigma}{\sigma_s}$$

$$\tau = t\beta^{-1/n} \left[\frac{J(0)}{\boldsymbol{J}} \right]^{-1/n}$$

Because of the inherent scatter in any strength data (and even larger scatter in lifetime data), Eq. (7) may be cast in statistical terms. It has been shown (see, e.g., Ref. 19) that the strength of individual filaments is well modeled by the two-parameter Weibull distribution, which may be written in the form

$$W(\sigma) = 1 - \exp\left\{ - \left[\frac{\sigma}{\sigma_0} \right]^b \right\}, \qquad \sigma \geq 0 \tag{8}$$

where $\sigma_0 > 0$ is called the scale parameter and $b > 0$ is called the shape parameter. Thus the probability of filament survival Λ at any stress σ is

$$\Lambda = 1 - W(\sigma) = \exp\left\{ - \left[\frac{\sigma}{\sigma_0} \right]^b \right\} \tag{9}$$

Combining Eqs. (7) and (9) we arrive at the desired statistical formulation

$$\tau = \int^{\left[\sigma_0^2 \frac{(-\ln \Lambda)^{2/b}}{\sigma^2} \right]} \left[\frac{\sigma_0^2 (-\ln \Lambda)^{2/b}}{\sigma^2 \zeta} - 1 \right]^{1/n} \frac{d\zeta}{\zeta} \tag{10}$$

Equation (10) may be used to predict the lifetime of a composite at a stress level σ when the strength distribution and uniaxial creep function are known (or assumed)—that is, when the Weibull parameters σ_0 and b and the creep exponent n are given.

A Micromechanical Model

Micromechanical models consider the eventual failure event in a composite to be the end result of a series of "microfailures," usually at the fiber level. Consider a large composite structure. Upon loading, fibers within the structure will break at random places. These fiber breaks, because of stress transfer through the matrix, cause adjacent fibers to become overloaded, perhaps causing some of them to break, and these in turn may cause others to break. If the load is sufficiently large, one of these clusters will grow (or, perhaps, through a longitudinal crack in the matrix, join up with a nearby cluster) to a critical size, and failure will occur. If the applied load is below that necessary for the formation of a critical cluster, and if the load is maintained, creep in the matrix will cause the load transfer length on the adjacent fibers to increase. The increase in length increases the probability of failure of the adjacent fiber. Eventually the entire structure fails when, as before, a critical size cluster is achieved. (This, of course assumes that there is zero creep in the fibers.) If the fibers themselves are subject to creep, the effects of both creep rates must be taken into account. For composites formed with epoxy matrices, matrix creep usually may be neglected because the fiber failure strain is usually much less than that of the matrix. An important consideration is the inherent randomness of the fiber breaks both with time and position along the fiber; geometric and morphological irregularities further compound the problem.

Models incorporating many of the parameters above have recently been developed by Phoenix and coworkers [20,21], Smith et al. [22], Batdorf and Ghaffarian [23], Manders and coworkers [24,25], Kuo and Phoenix [26], and Lifshitz and Rotem [27]. Necessary for the application of these models is a forehand knowledge of the behavior of the fibers and the composite in static tests. Following Phoenix et al. [28], the fiber strength is assumed to follow the classical two-parameter Weibull distribution given in Eq. (8). The scale parameter σ_0 is a function of the gage length tested, say l_0. To predict the strength at any gage length l, classical weakest link rules (see, e.g., Ref. 29) yield the same Weibull distribution except that the scale parameter σ_l is modified to

$$\sigma_l = \sigma_0 \left(\frac{l}{l_0} \right)^{-1/b} \tag{11}$$

For some types of fiber this scaling is somewhat severe; filaments with length $l > l_0$ are stronger than predicted, and shorter filaments are weaker. Independently, Watson and Smith [19] and Gutans and Tamuzs [30] have suggested that the exponent $-1/b$ be modified to $-\alpha/b$, where $0 \leq \alpha \leq 1$. Recent experimental data [29] for aramid fibers suggest $\alpha = 0.6$, and Watson and Smith [19] obtained $\alpha = 0.90$ for carbon fibers. In contrast, Schwartz et al. [31] found that for ultra-high strength polyethylene fibers $\alpha = 0$. Thus far, no physical interpretation has been assigned to the value α, although Watson and Smith [19] have speculated that random variation among diameters within a yarn cross section may play a role.

To describe short-term failure in a composite, one technique is to model the composite as a regular, hexagonally packed array. A model array illustrating, for clarity, only seven fibers (although in an actual composite,

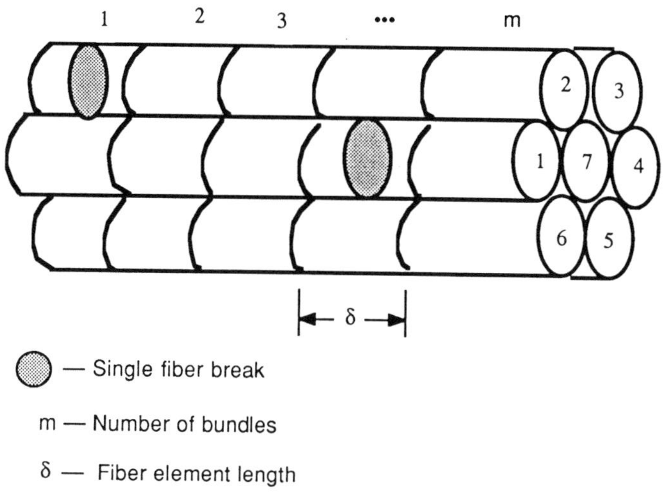

FIGURE 3 **Idealized model for seven-fiber composite (matrix omitted).** (Reprinted from S. L. Phoenix, P. Schwartz, and H. H. Robinson IV, *Compos. Sci. Technol., 32,* 81, © 1988 Elsevier Applied Science Publishers Ltd.)

the number of fibers n may be of the order of 10^6) is presented in Figure 3 [28]. As described above, under a moderate stress σ, weak portions in the fibers will lead to fiber failure at those locations, and, in the vicinity of the break, the stress in the broken fiber will fall to zero, but the load in an adjacent fiber increases locally to some value $K_1\sigma$ over some effective distance δ. These overloads will cause some of the adjacent fibers to break, further overloading neighboring fibers to another value $K_2\sigma$, $K_2 > K_1 > 1$. As the load is increased, eventually one cluster of adjacent breaks will grow to critical size k^*, and the overload $K_{k^*}\sigma$ is such that the probability of failure of surviving fibers is 1. The values K_i are called load intensity factors. Hedgepeth [32] has shown, for the elastic case that

$$K_i = \prod_{j=1}^{i} \frac{2j+2}{2j+1} \qquad (i = 1, 2, 3, \ldots, k^*) \tag{12}$$

Each subbundle of length δ is assumed independent; thus the fiber strength at the length δ (usually on the order of 5–30 fiber diameters) is of importance. If the fiber strength is known to follow the Weibull distribution given in Eq. (8), then the strength distribution at length δ, $W_\delta(\sigma)$, is given by

$$W_\delta(\sigma) = 1 - \exp\left\{-\left[\frac{\sigma}{\sigma_\delta}\right]^b\right\} \tag{13}$$

where, from Eq. (11), we have

$$\sigma_\delta = \sigma_0\left(\frac{\delta}{l_0}\right)^{-1/b} \tag{14}$$

where σ_0 is the Weibull scale parameter at test length l_0.

The maximum load in a fiber next to a cluster of i broken fibers is $K_i\sigma$, and it decays from this value to the nominal value σ outside the region δ; this stress increase is due to shear transfer through the matrix. Hedgepeth [32] and Hedgepeth and Van Dyke [33] were the first of many researchers to use a shear–lag analysis to study the stress profiles around fiber breaks. For more recent examples see Goree and Gross [34], Reedy [35], Batdorf and Ko [36], and Fukuda and Chou [37]. For the elastic case, Hedgepeth [32] showed that δ varies as $\sqrt{E_f/G_m}$, where E_f is the fiber tensile modulus and G_m is the matrix shear modulus. In reality, δ is also dependent on factors such as fiber packing, fiber spacing, matrix plasticity, and fiber–matrix debonding.

At this point, the composite of length L may be viewed as a chain of m independent bundles where $m = L/\delta$; for details see Phoenix and Smith [20] and Smith et al. [22]. Now let $F_n(\sigma)$ be the strength distribution for a single bundle and let $H_{m,n}(\sigma)$ be the corresponding distribution function for the whole composite. (Note: $V = mn$ is a measure of the total volume of fiber and, for a large composite may be of the order 10^{10}.) Assuming statistical independence between bundles, the standard weakest link relation

$$H_{m,n}(\sigma) = 1 - [1 - F_n(\sigma)]^m \qquad (\sigma \geq 0) \tag{15}$$

results.

In the calculation of $H_{m,n}(\sigma)$, not only is the number of adjacent fiber breaks important, but also the number of ways that they may occur must be taken into account. It can be shown [28] that

$$H_{m,n}(\sigma) = 1 - \exp\left\{-Vd_k\left[\frac{\sigma}{\sigma_\delta}\right]^{k^*b}\right\} \tag{16}$$

where d_k is a function of the load intensity factors and the number of possible arrangements of clusters of size k^*. If the strength distributions for both the individual filaments and the composite structure are known, then the critical cluster size k^* can be easily found as b_c/b, where b_c is the Weibull shape parameter for composite strength. Noting that $V = nL/\delta$, the Weibull scale parameter for composite strength σ_c may be written as

$$\sigma_c = \sigma_L(nd_k)\exp\left[\frac{-1}{(k^*b)}\right]\left(\frac{\delta}{L}\right)\exp\left[\frac{-(k^*-1)}{(k^*b)}\right] \tag{17}$$

where σ_L is the Weibull scale parameter for strength at length L. Solving Eq. (17) for δ yields

$$\delta = L(nd_k)\exp\left[\frac{-1}{(k^*-1)}\right]\left(\frac{\sigma_L}{\sigma_c}\right)\exp\left[\frac{k^*b}{(k^*-1)}\right] \tag{18}$$

If, as in the fracture mechanics case, we model the creep in power law form, the change in δ with time may be written as

$$\delta = \delta_0\left(\frac{t}{t_0}\right)^{r/2} \tag{19}$$

where δ_0 is the effective load transfer length at time $t = t_0$, and the exponent is divided by 2 because $\delta \propto \sqrt{1/G_m}$. Substitution of Eq. (19) into Eq. (17) gives

$$t_c = t_0\left(\frac{\sigma}{\sigma_c}\right)\exp\left[\frac{-2k^*b}{r(k^*-1)}\right] \quad (20)$$

the Weibull scale parameter for lifetime under applied stress σ.

An Elastic–Viscoelastic Model

In an attempt to predict the creep-rupture of a filament-wound, spherical pressure vessels, Hackett and Dozier [9] used an elastic–viscoelastic correspondence principle based on the creep of the epoxy matrix component. The entire composite structure is modeled as a transversely isotropic system, and the epoxy matrix as a linear viscoelastic material. This system is said to have failed when the circumferential strain ε_θ becomes equal to, or exceeds, the composite failure strain ε_f.

For a layered, anisotropic spherical vessel, the equations (in standard spherical coordinates r, θ, and ϕ) governing the system subjected to an internal pressure p fall into three categories as follows:

Equilibrium:

$$\frac{d\sigma_r}{dr} + \frac{2}{r}(\sigma_r - \sigma_\theta) = 0 \quad (21)$$

Kinematic:

$$\varepsilon_r = \frac{du}{dr} \qquad \varepsilon_\theta = \frac{u}{r} \quad (22)$$

Constitutive:

$$\sigma_r = C_{rr}\varepsilon_r + 2C_{r\theta}\varepsilon_\theta \qquad \sigma_\theta = C_{r\theta}\varepsilon_r + (C_{\theta\theta} + C_{\theta\phi})\varepsilon_\theta \quad (23)$$

where the σ_i are the stress components, the ε_i are the strain components, the C_{ij}, are the stiffness coefficients, u is the radial displacement, and r and θ define the radial and circumferential directions, respectively.

The boundary conditions for this case are:

$$\sigma_r = -p \text{ at the inner surface} \quad (r = r_i) \quad (24a)$$

$$\sigma_r = 0 \text{ at the outer surface} \quad (r = r_0) \quad (24b)$$

Combining Eqs. (21)–(23) gives

$$C_{rr}\frac{d^2u}{dr^2} + \frac{2C_{rr}}{r}\frac{du}{dr} + 2(C_{r\theta} - C_{\theta\theta} - C_{\theta\phi})\frac{u}{r^2} = 0 \quad (25)$$

which when solved using the boundary conditions gives

$$u = ar^{(n-1/2)} + br^{-(n+1/2)} \quad (26)$$

where

$$a = \frac{g}{f}r_0^{-2n}\,b \quad (27a)$$

$$b = \frac{p}{gr^{(n-3/2)}(r_i^{-2n} - r_0^{-2n})} \quad (27b)$$

$$n = \frac{1}{2}\left[1 - 8\left(\frac{C_{r\theta} - C_{\theta\theta} - C_{\theta\phi}}{C_{rr}}\right)\right]^{1/2} \quad (27c)$$

$$f = (n - \tfrac{1}{2})C_{rr} + 2C_{r\theta} \quad (27d)$$

$$g = (n + \tfrac{1}{2})C_{rr} - 2C_{r\phi} \quad (27e)$$

Combining, the maximum circumferential strain can be found to be

$$\varepsilon_{\theta;\max} = \frac{p}{g(\beta^{2n} - 1)}\left(\frac{g}{f} + \beta^{2n}\right) \quad (28)$$

where $\beta = r_0/r_i$.

The stiffness coefficients C_{ij} are calculated from the expressions

$$C_{r\theta} = \frac{1}{\pi}\int_0^\pi C_{1r}\,d\alpha \quad (29a)$$

$$C_{\theta\theta} = \frac{1}{\pi}\int_0^\pi C_{11}\,d\alpha \quad (29b)$$

$$C_{\theta\phi} = \frac{1}{\pi}\int_0^\pi C_{12}\,d\alpha \quad (29c)$$

where the 1, 2, r coordinate system and the angle α are defined as shown in Figure 4. The values of C_{ij} for each band can be found by standard transformation methods (see, e.g., Ref. 38) if the mechanical properties—moduli and Poisson's ratios—for each band are known.

The time-dependent strain is proportional to the applied stress σ and may be written as

$$\varepsilon(t) = \sigma J(t) \quad (30)$$

Using the standard rule of mixtures, a lower bound expression for the transverse modulus E_T of a band is found to be

$$E_T = \frac{1}{\dfrac{v_f}{E_f} + \dfrac{1 - v_f}{E_m}} \quad (31)$$

where v_f is the fiber volume fraction and E_f and E_m are the fiber and matrix stiffnesses, respectively.

The creep compliance of a linear viscoelastic material can be expressed in a series form

$$J(t) = D + \sum_{i=1}^{(} F_i \exp(-R_i t) + Ht \quad (32)$$

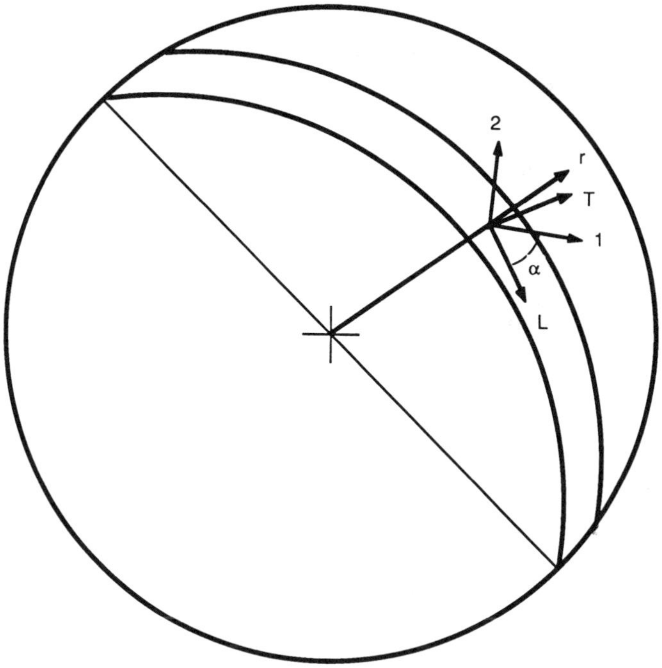

FIGURE 4 **Transformation angle between *L-T* and 1-2 coordinate systems for spherical pressure vessel.** (Reprinted from R. M. Hackett and J. D. Dozier, Paper 84-Mat-5 © 1984 American Society of Mechanical Engineers.)

where *D, F_i, R_i, and H* may be determined from experimental creep data for the matrix material.

Transforming into the Laplace domain (see Ref. 9 or 39 for details) and then performing an inversion on the associated elastic solution at maximum strain, the transverse modulus is found to be

$$E_T = E_T^1\left[1 - \frac{\varepsilon_{\theta;\max}(t) - \varepsilon^1_{\theta;\max}}{\varepsilon_f - \varepsilon^1_{\theta;\max}}\right] \tag{33}$$

where E_T^1 is the initial (undamaged) transverse modulus of a band and $\varepsilon^1_{\theta;\max}$ is the initial maximum circumferential strain. The time required for $\varepsilon^1_{\theta;\max}$ to reach ε_f is given by

$$\varepsilon_{\theta;\max}(t) = \varepsilon_{\theta;\max}(t_0) - \sum_{v=1}^{6} \eta\left\{1 - \exp\left[-\frac{t}{\alpha_v}\right]\right\} \tag{34}$$

where η is a constant and $\alpha_v = e^{(7-2v)}$. Note that the solution to this problem is found by iteration, as $\varepsilon_{\theta;\max}$ is a function of E_T which, in turn, is a function of $\varepsilon_{\theta;\max}$.

Experimental Results

Test Methods and Specimens

For a variety of reasons, few experimental data have been generated to study the creep-rupture of composite materials. Among the reasons for this are:

1. The generally long time required to run a single experiment to failure and the need to conduct experiments at different stress levels and/or ambient conditions.
2. The (usually) large number of samples required for statistical significance because of the large variability in the lifetime distributions.
3. The delicate nature of the experiments, especially when conducted at high levels of stress.
4. The (often) prohibitive cost associated with full-scale samples.

A number of techniques have been used to minimize the time and cost of creep-rupture experiments, among these are testing at accelerating conditions of high stress, high temperatures, or both; testing small, manageable specimens such as strands or small coupons, and using scaling laws to predict full-scale behavior; running standard tensile tests over a range of strain rates and extrapolating time-to-fail to longer time regimens; and running tests for shorter lengths of time, censoring surviving specimens using either type I (censoring after a fixed elapsed time) or type II (censoring after a fixed number of failures) rules, and using statistical techniques to account for the censored samples.

A simple, dead-weight loading method suitable for small-scale composites such as strands or small coupons is schematically illustrated in Figure 5 [39]. Timing is incorporated using a microprocessor to poll individual

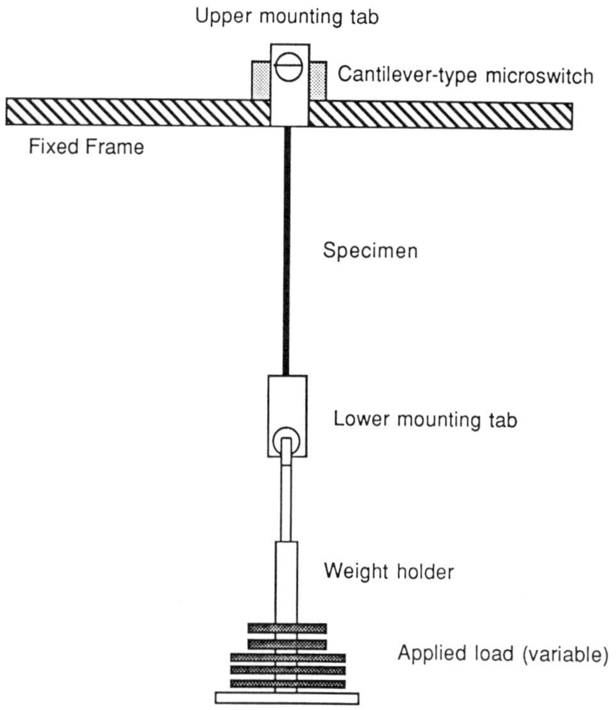

FIGURE 5 **Schematic diagram of one position on a creep-rupture setup for fibers and small strands.** (Reprinted from H. D. Wagner, P. Schwartz, and S. L. Phoenix, *J. Mater. Sci., 21,* 1868 © 1986 Chapman and Hall.)

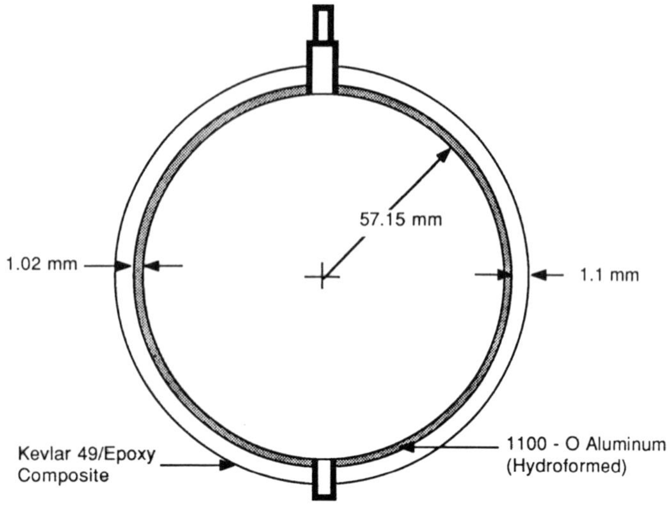

FIGURE 6 **LLNL creep-rupture spherical pressure vessel.** (Data from R. M. Hackett and J. Dozier, *J. Reinf. Plast. Compos., 6,* 126 © 1987 Technomic Publishing Company, Inc.)

TABLE 1
Properties of Kevlar 49–Epoxy Filament-Wound Spherical Pressure Vessels (v_f = 0.65)

Longitudinal modulus	89.40 GPa
Transverse modulus	6.65 GPa
Poisson's ratio	0.25
Shear modulus	6.65 GPa
Effective quasi-isotropic Young's modulus	35.80 GPa
Effective quasi-isotropic Poisson's ratio	0.27

microswitches designed to change polarity when the specimen fails. Other loading methods, including the use of springs or pneumatic or hydraulic cylinders, have been used, and are commercially available, and all have both advantages and disadvantages. One disadvantage of many such systems is time resolution in the early domain. Because of strength variability, immediate failures ($t = 0$) at a given stress level can occur, especially if that level is near the failure stress. In addition to these "instantaneous failures," there is the probability of early creep failures after a few seconds at load. Knowledge of the static strength distribution and the applied stress enables one to predict how many failures to expect upon loading, and this may be used as a measure of the smoothness of loading and help in the attribution of failure to static stress or to creep-rupture. Another problem with these systems is that, to avoid dynamic overloading, the load is usually applied over a finite length of time. Thus in the early time domain the specimen sees ramp loading rather than stepwise loading, as is usually assumed in many models.

Figure 6 [40] illustrates the configuration of filament-wound, aluminum-lined pressure vessels tested at LLNL. As tested, the vessels consisted of Kevlar 49 aramid fiber impregnated with a bisphenol A epoxy (DER-332) cured with an aliphatic primary triamine (Jeffamine T-403). The properties are characterized in Table 1. The vessels were pressurized to four different levels (29.7, 27.6, 25.5, and 23.4 MPa) and allowed to remain at that pressure until failure. In Table 2 [6] are presented the Weibull parameters for lifetime at each pressure level; and in Figure 7 [40] the data are presented using log time coordinates. The line labeled "Viscoelastic model" was obtained using Eq. (31), which does not take into account internal damage. The other is constructed from a series of iterative solutions using Eq. (34). Based on these assumptions, the

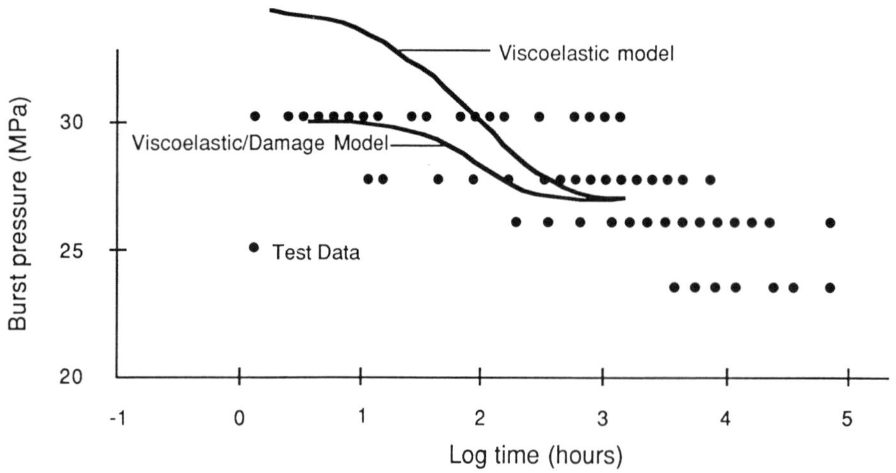

FIGURE 7 **Comparison of experimental and model-generated data for pressure vessel creep-rupture.** (Reprinted from R. M. Hackett and J. Dozier, *J. Reinf. Plast. Compos., 6,* 126 © 1987 Technomic Publishing Company, Inc.)

TABLE 2
Weibull Parameters for Lifetime of Kevlar 49/Epoxy Spherical Pressure Vessels at 23°C

Pressure (MPa)	Shape Parameter	Scale Parameter (hours)
29.7	0.52	1.64×10^2
27.6	0.87	8.39×10^2
25.5	1.01	8.28×10^3
23.4	0.88	7.09×10^4

model predicts a severe decrease in stiffness very early in the test.

In Figure 8 [6] the data for one complete lifetime run at 27.6 MPa are presented using Weibull probability coordinates, including the best fit to the data obtained using maximum likelihood estimation (MLE). The goodness of fit of these data to a straight line indicates that they are well approximated by the Weibull distribution. Figure 9 [40] plots the Weibull scale parameter at each stress level versus log time, allowing for extrapolation to higher levels of stress, or alternatively, longer lifetimes. Table 3 [6] presents lifetime data for Kevlar 49–epoxy (DER332-T403) strands produced using yarns taken from the same production lot as those used for the pressure vessels. These data indicate that the strands behave statistically much the same as the pressure vessels, even though the vessels have a much more complex geometry.

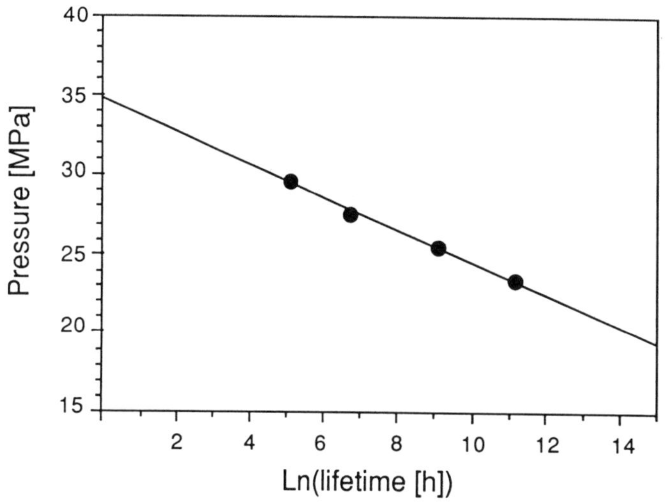

FIGURE 9 **Weibull scale parameters for pressure vessel lifetimes plotted using a log time scale.** (Data from R. M. Hackett and J. Dozier, *J. Reinf. Plast. Compos.*, *6*, 126 © 1987 Technomic Publishing Company, Inc.)

Bunsell and Valentin [42], using acoustic emission techniques, studied the fiber damage accumulation in pressurized carbon fiber, filament-wound tubes (Toray T6K fibers and epoxy resin "very similar" to Ciba Geigy 914C). They noted that the acoustic emission behavior of ±88° and ±75° wound tubes subjected to steady pres-

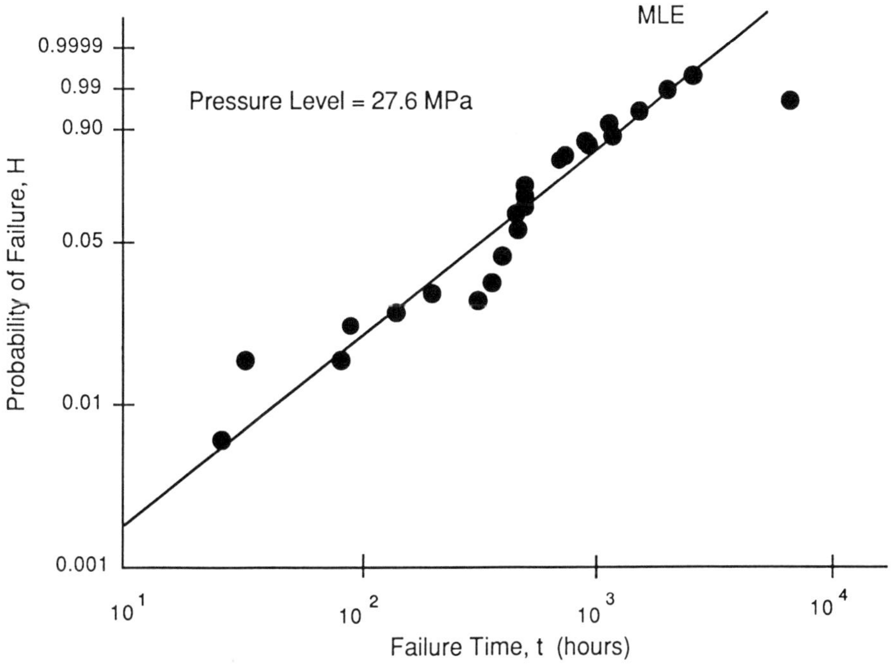

FIGURE 8 **Lifetime data plotted on Weibull paper for one complete pressure vessel experiment at 27.6 MPa internal pressure.** (Reprinted from R. M. Hackett and J. D. Dozier, Paper 84-Mat-5 © 1984 American Society of Mechanical Engineers.)

TABLE 3
Weibull Parameters for Lifetime of Kevlar 49/Epoxy Strands at 23°C

Fiber Stress (MPa)	Shape Parameter	Scale Parameter (hours)
	$v_f = 0.68$	
2780	0.77	4.73×10^1
2724	0.64	4.05×10^2
2537	1.23	2.64×10^3
2396	1.34	9.76×10^3
2165	0.94	2.94×10^5
	$v_f = 0.45$	
3129	1.85	1.20×10^1
2724	2.61	5.27×10^2

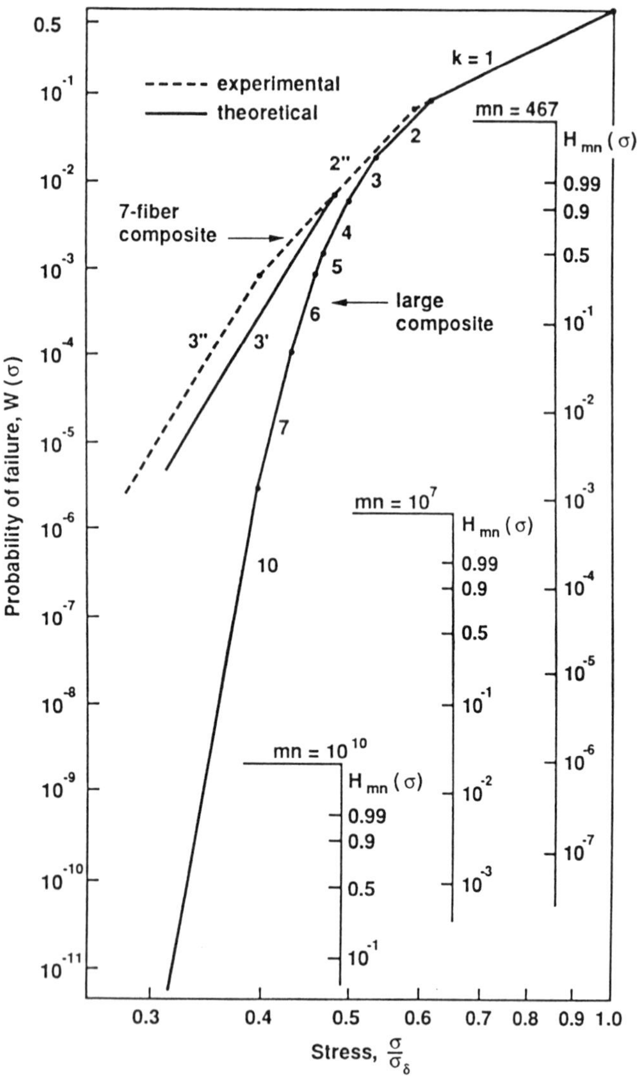

FIGURE 10 **Characteristic distribution function and distributions for strength for three structural volumes.** (Reprinted from S. L. Phoenix, P. Schwartz, and H. H. Robinson IV, *Comput. Sci. Technol.,* to appear, © Elsevier Applied Science Publishers Ltd.)

sure is similar to that of unidirectional and cross-plied plates subjected to steady loading, and thus suggest that lifetime data for plates may be used to predict the lifetime of tubes.

Phoenix et al. [28] report on creep-rupture experiments performed on hexagonally packed, seven-filament "microcomposites" (IM-6/DER 331-DEH 26) loaded to 95, 90, and 87.5% of their breaking strength. Their work indicated that failure occurred when a critical cluster size of two adjacent breaks was reached, and, following Smith et al. [22], they plotted (Fig. 10) the distribution functions for bundle strength $H_{m,n}(\sigma)$ for large bundle volumes mn, based on data for the seven-fiber composite $mn = 467$. They noted anomalies in the data that suggest that carbon fibers are susceptible to creep in real time, are not creep resistant, as is often assumed. Using matched specimens from the same short length of fiber, they also found creep-rupture lifetime to be independent of tensile strength; in other words, short-term strength measurements are not reliable predictors of long-time performance. If one assumes the statistical independence of the adjacent segments, as is commonly done in weakest link analyses, this result may easily be predicted.

Peter Schwartz

[*See also* CREEP: VISCOELASTICITY.]

References

1. T. T. Chiao, J. K. Lepper, N. W. Hetherington, and R. L. Moore, *J. Compos. Mater., 6,* 358 (1972).
2. R. L. Moore, M. A. Hamstad, and T. T. Chiao, *Fukugo Zairyo* (*Compos. Mater. Struc.*) *3,* 19 (1974).
3. C. C. Chiao, R. J. Sherry, and N. W. Hetherington, *J. Compos. Mater., 11,* 79 (1977).
4. R. H. Toland, R. J. Sanchez, and D. Freeman, "Stress-Rupture Life of Kevlar/Epoxy Spherical Pressure Vessels," Part 1, Lawrence Livermore National Laboratory, Report UCID-17755, March 1978.
5. R. H. Toland and T. T. Chiao, "Stress Rupture Life of Kevlar/Epoxy Spherical Pressure Vessels," Part 2, Lawrence Livermore National Laboratory, Report UCID 17755, June 1978.
6. S. L. Phoenix and E. M. Wu, "Statistics for the Time Dependent Failure of Kevlar 49/Epoxy Composites: Micromechanical Modeling and Data Interpretation," Lawrence Livermore National Laboratory, Report UCRL-53365, March 1983.
7. R. E. Barlow, R. H. Toland, and D. Freeman, "Stress Rupture Life of Kevlar/Epoxy Spherical Pressure Vessels, Part 3, Lawrence Livermore National Laboratory, Report UCID 17755, February 1979.
8. F. P. Gerstle, Jr., and S. C. Kunz, in *Long-Term Behavior of Composites,* ASTM Special Technical Publication 813, American Society for Testing and Materials, Philadelphia, 1983, p. 263.
9. R. M. Hackett and J. D. Dozier, ASME Paper 84-Mat-5, 1984.
10. S. N. Zhurkov, *Int. J. Fract. Mech., 1,* 311 (1965).
11. S. N. Zhurkov and V. E. Korsukov, *J. Polym. Sci., A-2, 112,* 385 (1975).
12. H. Eyring, *J. Chem. Phys., 4,* 283 (1936).

13. A. Tobolsky and H. Eyring, *J. Chem. Phys., 11,* 125 (1943).
14. B. D. Coleman, *J. Appl. Phys., 27,* 862 (1956).
15. R. M. Christensen, *J. Rheol. 25,* 517 (1981).
16. R. A. Schapery, *Int. J. Fract., 11,* 549 (1975).
17. R. M. Christensen and E. M. Wu, *Eng. Fract. Mech., 14,* 215 (1981).
18. R. M. Christensen and L. N. McCartney, *Int. J. Fract., 23,* R11 (1983).
19. A. S. Watson and R. L. Smith, *J. Mater. Sci., 20,* 3260 (1985).
20. S. L. Phoenix and R. L. Smith, *Int. J. Solids Struct., 19,* 479 (1983).
21. S. L. Phoenix and L.-J. Tierney, *Eng. Fract. Mech., 18,* 193 (1983).
22. R. L. Smith, S. L. Phoenix, M. R. Greenfield, R. B. Henstenburg, and R. E. Pitt, *Proc. R. Soc., London, A388,* 353, (1983).
23. S. B. Batdorf and R. Ghaffarian, *Int. J. Fract., 26,* 113 (1984).
24. P. W. Manders and T.-W. Chou, *J. Compos. Mater., 17,* 26 (1983).
25. P. W. Manders, M. G. Bader, and T.-W. Chou, *Fibre Sci. Technol., 17,* 183 (1982).
26. C. C. Kuo and S. L. Phoenix, *J. Appl. Prob., 24,* 137 (1987).
27. J. M. Lifshitz and A. Rotem, *J. Compos. Mater., 6,* 358 (1972).
28. S. L. Phoenix, P. Schwartz, and H. H. Robinson IV, *Compos. Sci. Technol., 32,* 81 (1988).
29. H. D. Wagner, S. L. Phoenix, and P. Schwartz, *J. Compos. Mater., 18,* 312 (1984).
30. J. Gutans and V. Tamuzs, *Mech. Compos. Mater.* (in Russian), *20,* 1107 (1984).
31. P. Schwartz, A. Netravali, and S. Sembach, *Text. Res. J., 56,* 502 (1986).
32. J. M. Hedgepeth, "Stress Concentrations in Filamentary Structures," NASA TN D-882, 1961.
33. J. M. Hedgepeth and P. Van Dyke, *J. Compos. Mater., 1,* 294 (1967).
34. J. Goree and R. S. Gross, *Eng. Fract. Mech., 13,* 395 (1980).
35. E. D. Reedy, *J. Compos. Mater., 18,* 595 (1984).
36. S. B. Batdorf and R. W. C. Ko, *AIAA J., 23,* 1749 (1985).
37. H. Fukuda and T. W. Chou, *J. Mater. Sci., 15,* 79 (1981).
38. J. J. Whitney, *J. Compos. Mater., 1,* 188 (1967).
39. H. D. Wagner, P. Schwartz, and S. L. Phoenix, *J. Mater. Sci., 21,* 1868 (1986).
40. R. M. Hackett and J. D. Dozier, *J. Reinf. Plast. Compos., 6,* 126 (1987).
41. H. T. Hahn and T. T. Chiao, in A. R. Bunsell, Ed., *Advances in Composite Materials,* Pergamon Press, London, 1980, p. 584.
42. A. R. Bunsell and D. Valentin, *Compos. Struct., 1,* 67 (1983).

Cure Monitoring

There are a large number of material variables and processing parameters that must be known and/or controlled during cure of a composite. In situ sensing techniques that can measure the actual processing properties such as viscosity and reaction advancement during cure in both a laboratory and a manufacturing tool are essential. In situ cure monitoring sensor techniques are useful for the following purposes: evaluating the cure processing properties of new resins; evaluating the variation of cure properties due to resin handling, age, and vendor batch; determining the cure cycle for new resins, complex parts, and differing tools; and verifying, refining, and testing analytical models. Ultimately, cure monitoring sensor technology combined with intelligent process model software are critical requirements for closed-loop intelligent production control.

Current efforts to develop and use cure monitoring sensor techniques can be grouped into three major areas: (a) frequency-dependent electromagnetic impedance measurements in the hertz–megahertz frequency region, (b) ultrasonic–acoustic measurements, and (c) optical techniques. Of these three, the first, frequency-dependent impedance measurements, has shown the most dramatic advances as well as the widest application and use in recent years. Various approaches to cure monitoring using ultrasonic–acoustic techniques have shown promising results but are currently limited by the absence of comprehensive theory and models for interpreting the data on a molecular level. Optical measurements, particularly Fourier transform–infrared spectroscopy (FTIR), are in the early stages of experimental development. Each of these types of cure monitoring measurement provides different information about the state of cure. Thus in general they are not replacements for each other, but rather will need to be made together to accurately monitor the complex and diverse variables that affect composite cure.

Frequency-Dependent Electromagnetic Impedance Measurements

Progress in the development and application of frequency-dependent electromagnetic impedance measurements to monitor cure processing properties has been dramatic over recent years. This progress is due to the emphasis placed on relating the chemistry and rheology of the cure process to the changing electrical properties of the polymer system through correlation of the time, the temperature, and the frequency dependence of the impedance with chemical and rheological characterization measurements.

Frequency-dependent electrical measurements made over a wide range of frequency are a sensitive and convenient means for monitoring the cure process in both thermosets and thermoplastics. Impedance measurements are particularly effective for quality control monitoring of the cure process in complex resin systems because impedance measurements are one of only a few instrumental techniques available for studying molecular properties in both the liquid and solid states. Furthermore, they are unique in their ability to monitor the cure process continuously—that is, to examine cure throughout the process of going from a monomeric liquid of varying viscosity to a cross-linked, insoluble, high temperature solid (Fig. 1).

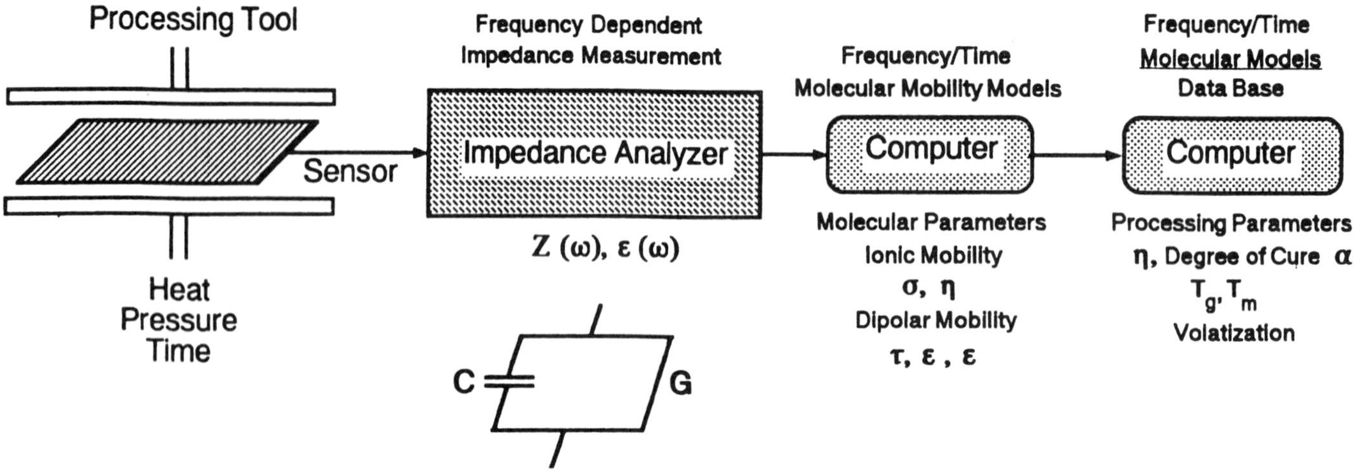

FIGURE 1 Block diagram of the frequency-dependent electrical apparatus.

Earliest Work

Interest in the use of impedance measurements to characterize materials has existed for more than 50 years, and there are many excellent books reviewing this literature. The extensive use of electrical measurements, primarily utilizing dielectric techniques, to characterize materials has led to the current interest in using electrical measurements to monitor cure in composites. A good reason for this interest is that electrical measurements have the ability to conveniently and continuously monitor the impedance over a very large range in magnitude, which reflects the resin's viscosity as it varies from less than 1 to $> 10^7$ Pa·s (10^2 to $> 10^9$ poise). On the other hand, equally important but neglected in the earliest work, is the potential to use the variation in the impedance with frequency to provide information on a molecular level about the changing chemical, physical, and rheological state of the composite resin.

The earliest work using electrical measurements to monitor composite cure fell short of expectations. This was due primarily to the failure to understand the need to make frequency-dependent measurements over a wide range of frequencies and to interpret the frequency dependence on a molecular basis in terms of dipolar and ionic diffusion processes. Advances in commercially available automated bridges and the recent development of geometry-independent microsensor measurement systems first by Senturia and then by Kranbuehl are significant recent advances in instrumentation, making in situ measurement over a wide range in frequencies much more convenient.

Experimental Theory

The impedance Z at a frequency f of the curing composite resin at any point in time can be represented by an equivalent parallel capacitance C and conductance G:

$$Z^{-1} = G + i\omega C \qquad (\omega = 2\pi f) \tag{1}$$

Measurements of the equivalent capacitance and conductance, both dependent on sample geometry, are used to calculate the intensive geometry independent complex permittivity $\varepsilon^* = \varepsilon' - i\varepsilon''$, from

$$\varepsilon'(\omega) = \frac{C(\omega)\text{ material}}{C_0}$$
$$\varepsilon''(\omega) = \frac{G(\omega)\text{ material}}{C_0 2\pi f} \tag{2}$$

where C_0 is the effective air-replaceable capacitance of the sensor. Both the real and the imaginary parts of ε^* have a dipolar and an ionic component:

$$\varepsilon'(\omega) = \varepsilon'(\omega)_d + \varepsilon'(\omega)_i$$
$$\varepsilon''(\omega) = \varepsilon''(\omega)_d + \varepsilon''(\omega)_i \tag{3}$$

The dipolar term is generally the major component of the impedance signal at high frequencies and in highly viscous media. The dipolar component arises from diffusion of bound charge or molecular dipole moments. The frequency dependence of the dipolar component may be analytically represented by a distribution of relaxation times, such as the Williams–Watts function, which is now widely accepted, or the Cole–Davidson function, which is similar and in many cases more convenient to use. Its form is:

$$\varepsilon^*(\omega)_d = \varepsilon_\infty + \frac{\varepsilon_0 - \varepsilon_\infty}{(1 + i\omega\tau)^\beta} \tag{4}$$

where ε_0 and ε_∞ are the limiting low and high frequency values of ε^*, τ is a characteristic relaxation time, and β is a parameter that measures the distribution in relaxation times. The relaxation time τ is an experimentally determined parameter that measures the rotational mobility of the dipoles.

The ionic component ε_i^* often dominates ε^* at low frequencies, low viscosities, and/or higher temperatures. The presence of mobile ions gives rise to localized layers of charge near the electrodes. Since these space charge layers are separated by very small molecular distances on the order of angstroms, the corresponding space charge capacitance can become extremely large, with ε' on the order of 10^6. Johnson and Cole derived empirical equations for the ionic contribution to ε^* that show ε_i' to be frequency dependent as a result of these space charge ionic effects and to have the form:

$$\varepsilon_i'(\omega) = C_0 Z_0 \sin\frac{(n\pi)}{2}\, \omega^{-(n+1)} \left(\frac{\sigma}{8.85 \times 10^{-14}}\right)^2 \tag{5}$$

where $Z^* = Z_0(i\omega)^{-n}$ is the electrode impedance induced by the ions and n is an empirical parameter between 0 and 1. We also have:

$$\varepsilon_i''(\omega) = \frac{\sigma}{8.85 \times 10^{-14}\omega} - C_0 Z_0 \cos\frac{(n\pi)}{2}\, \omega^{-(n+1)} \left(\frac{\sigma}{8.85 \times 10^{-14}}\right)^2 \tag{6}$$

where σ is the conductivity ($\Omega^{-1}\text{cm}^{-1}$), an intensive variable, in contrast to conductance G (Ω^{-1}), which is dependent on cell and sample size. The conductivity σ is an experimentally determined parameter that measures the translational mobility of the ions. The first term in Eq. (6) is due to the conductance of ions translating through the medium. The second term, which is due to electrode polarization effects, makes impedance measurements increasingly difficult to interpret and use as the frequency of measurement becomes lower.

Electrode polarization, represented by the second term in Eq. (6), in general is a significant and often difficult to account for factor at frequencies below 10 Hz and/or for high values of σ usually associated with a highly fluid resin state. The frequency dependence ε^* due to dipolar mobility is generally observed at frequencies in the kilohertz and megahertz regions. For this reason an analysis of the frequency dependence of ε^* (Eq. 3–6), in the hertz–megahertz range is, in general, optimum for measuring both the ionic mobility parameter σ and the dipolar mobility parameter τ.

The magnitude of the ionic mobility σ and the rotational mobility of the dipole τ depend on the extent of the reaction and the physical state of the material. As such, σ and τ, determined from the frequency dependence of $\varepsilon^*(\omega)$, provide two molecular probes for monitoring the reaction advancement and the viscosity during cure.

Instrumentation

The easiest and most direct method for measuring in a laboratory environment a material's complex permittivity is to place the material between two parallel plates of known geometry and to measure the impedance of the material-filled parallel-plate electrodes using a commercial impedance bridge. This technique continues to be the most useful and accurate measurement approach, particularly when guard electrodes and thin samples of known geometry can be used. Often, however, it is difficult to make parallel-plate measurements, particularly if the plate or material dimensions are changing with time and/or temperatures. Furthermore it is difficult to make a small parallel-plate sensor, since the sensitivity of such a device is determined by the ratio of the plate area to the separation. Thus in recent years, there have been various approaches to making and using small planar sensors consisting of a pair of interdigitated comb electrodes.

The first geometry-independent, disposable microsensor system for measuring the complex permittivity or impedance during cure was developed and patented by Senturia. The sensor consists of a set of small, planar, interdigitated electrodes fabricated with a pair of field-effect transistors into a silicon integrated microcircuit. The measurement involves integration of the current reaching the sensor electrode. This sensor instrumentation system is now commercially available. It can be used to measure cure of composite resins in the range of 5×10^{-3} to 10^4 Hz. Most measurements are made at 0.1 Hz or above, since the measurement time is quite long at the lowest frequencies. For many resins, when the resin is in a high flow, low viscosity state, the magnitude of the loss ε'' exceeds 10^3 and this sensor overloads. Thus a second type of sensor is required to monitor conductivity. The additional sensor is used to follow high flow, low viscosity time–temperature regions during cure of the composite resin.

A second approach to disposable, planar, geometry-independent microsensor measurements has been developed and patented by Kranbuehl. This sensor, which is now also commercially available, consists of only a fine array of two interdigitated comb electrodes. In this case the sensor is constructed from noble metals and high temperature ceramics; it does not contain solid state circuitry, which is vulnerable to the harsh processing environment. The sensor is designed to withstand the curing/tool temperature, pressure, and oxidative conditions during processing for temperatures exceeding 400°C and pressures of MPa. This single disposable sensor is capable of monitoring the entire range in magnitude (usually 10^{-2}–10^8) of both the real ε' and imaginary ε'' components of the permittivity continuously without interruption throughout the cure cycle. The sensor and the accompanying software are designed for use with the advanced impedance and conductivity bridges made by Hewlett Packard, Schlumberger-Solartron, Gen Rad, Keithley, and Tetrahedron over frequency–time scales of 10^{-5} Hz to 30 MHz.

Ultrasonic–Acoustic Techniques

Ultrasonic–acoustic cure monitoring techniques are based on the principle that the change during cure in a composite resin's high frequency acoustic properties directly reflects the resin's changing low frequency mechanical properties. In general, acoustic measurements are made at frequencies near 10^5–10^6 Hz compared with the much lower frequencies (10^{-2}–10^2 Hz) used in viscoelastic measurements. Thus acoustic measurements monitor the variation of the resin's modulus over shorter time scales and distances than conventional rheological measurements. The relationship between the two remains an open subject for experimental and theoretical investigations. Nevertheless recent experimental work and theoretical models have clearly demonstrated that these techniques are very promising and qualitatively sensitive to the buildup in modulus during cure.

The most widely used technique for monitoring cure using acoustic techniques involves sending a pulse through the material, then monitoring its echo and thereby the changing velocity of the acoustic wave in the curing resin. In another variation of this approach, an acoustic waveguide is embedded in the composite resin. Rather than monitoring the propagation properties of the wave in the material directly, the changing propagation properties of the acoustic wave in the resin–embedded waveguide system are monitored. This technique permits interrogation of larger volumes of the composite, since the attenuation of the wave is reduced in the waveguide. On the other hand, the problem of relating the changes in wave propagation to the viscoelastic modulus of the composite resin becomes more complex, since the wave is propagating in a waveguide–resin system rather than directly through the composite.

Both techniques have shown that the acoustic properties change dramatically as the resin cures, and simple models for interpreting the longitudinal velocity in the resin have shown promise for monitoring viscoelastic properties during cure.

Optical Methods

Two major approaches rely on optical cure monitoring techniques. The first involves in situ measurement of the fluorescent intensity, frequency, and relaxation time. The second focuses on in situ measurement during cure of the resin's FTIR spectra. While totally different in approach and in the information obtained, both techniques use optical fibers to detect in situ the changing optical properties of the composite resin during cure.

Fluorescence techniques monitor the variation in the fluorescent intensity during cure. The technique relies on the fact that the intensity varies with the fluorescent relaxation time, which in turn depends on the local viscosity and diffusion rate of the fluorescing molecules. Thus during cure, the fluorescence intensity increases as the viscosity increases. The relationship between the two is qualitative. The correlation is most direct in the high flow, low viscosity regions of cure, where the rotational relaxation time is often directly dependent on the resin's viscosity. The technique requires either that one of the molecules in the resin fluoresce or that a dye molecule be added to the resin near the optical fiber. Internal standards, dyes with a constant viscosity-independent intensity, have been shown to enhance the precision of the measurement. Measurement of the fluorescent frequency provides an additional parameter for monitoring the advancement of the cure reaction (Fig. 2).

In situ FTIR fiber optic measurements are just beginning to be developed for cure monitoring. Progress has been reported on developing a fiber with high transmission properties that can withstand 200–300°C temperatures. Preliminary measurements have shown that the technique can be used to monitor the FTIR absorbance spectrum during cure in an epoxy prepreg. This absorbance spectrum permits direct qualitative monitoring of the changing concentration of the reacting chemical moeties during cure. FTIR optical fiber techniques are unique in that they directly monitor the chemistry of resin during cure. On the other hand, they do not monitor the viscosity and rheological changes during cure.

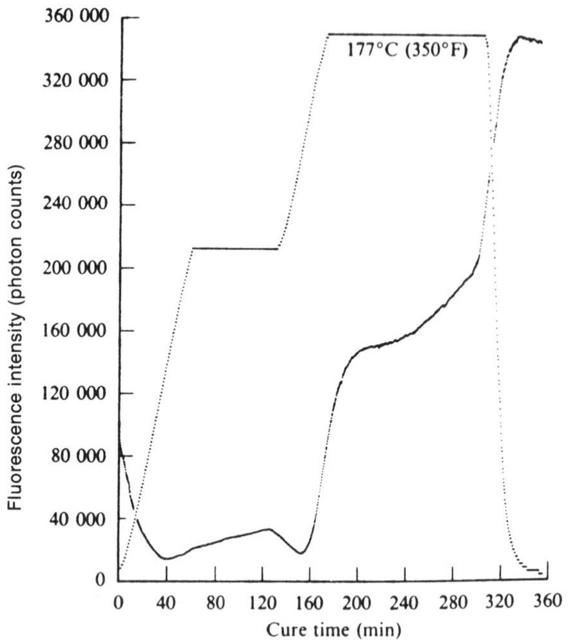

FIGURE 2 **Fluorescence intensity signal during curing of Hercules 3501-6 laminate.**

A combination of techniques such as frequency-dependent electromagnetic impedance fluorescence and acoustic measurements, which monitor the viscosity and changing physical state of the curing resin, along with FTIR spectroscopy, which directly monitors the concentration of reacting chemical groups, should be ideal for fully and continuously characterizing both the viscoelastic changes and chemistry of the composite resin during cure.

FDEMS Monitoring of Cure Processing Properties of Thermoplastics

Frequency-dependent electromagnetic spectroscopy (FDEMS) measurements can provide a sensitive and convenient sensing technique for monitoring the changes in the chemical and physical state of thermoplastics during processing. FDEMS sensors are able to monitor the viscosity, glass transition T_g, crystal melting, recrystallization, residual solvent–moisture content, solvent evolution, and degradation. The technique has been used to examine these processing properties in a number of thermoplastic resins including polysulfone, polyphenylene sulfides, polyarylene ethers polyether etherketone

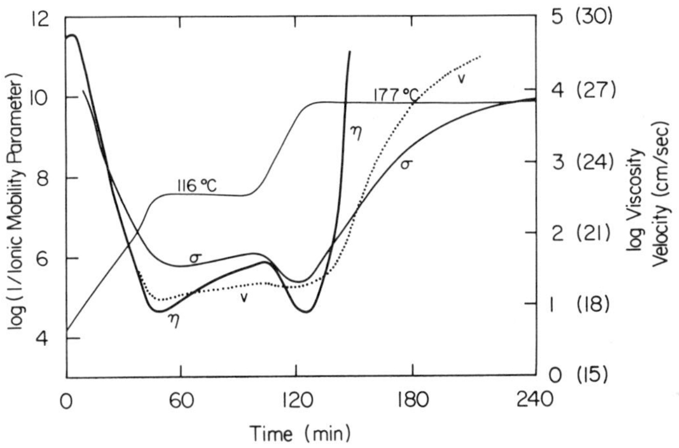

FIGURE 3 **Simultaneous measurement of viscosity η, wave velocity, v, and FDEMS conductivity σ during the standard cure cycle of a TGDDM amine-cured epoxy, Hercules 3501-6.**

(PEEK), and aromatic polyimides such as LaRC-TPI. One advantage of the technique is that the measurement can be made on a slow time–temperature ramp, whereas with differential scanning calorimetry (DSC), fast rates are often needed to detect T_g. The measurements can be made continuously over the entire processing cycle and can be made both in the laboratory and in situ at various positions in the part during fabrication in the mold. This cure monitoring capability is particularly important for semicrystalline thermoplastics, since variations in the extent of supercooling can build in stresses within the part and variations of the crystal morphology (Fig. 3).

Glass Transition

A plot of ε' versus time for a polyarylene ether (Fig. 4) shows a sharp transition at the glass transition both on heating and cooling. The ability of the FDEMS to measure T_g is shown by the agreement of the DSC-determined T_g of 150°C and the rise of ε' at the same temperature. An advantage of this sensing technique in addition to its in situ production tool capability is that the magnitude of the inflection is not decreased for slow heating and cooling rates.

Viscosity and Free Volume as Measured by Ionic and Dipolar Rate

A plot of ε'' times 2π frequency versus time (Fig. 5) shows the dipolar and ionic contributions to the loss term. When the ε'' times 2π frequency curves differ at each frequency, the dominant molecular process contributing to ε'' is due to rotational diffusion of dipoles. On the other hand, overlapping curves indicate that translational diffusion of charged species is the dominant molecular process determining the loss, ε''.

Below the glass transition, the rate of translational diffusion of ions is extremely slow and the value of the loss, ε'', is small. A frequency-dependent dipolelike relaxation can be observed just above T_g. The maximum in ε'' at a given frequency corresponds to the relaxation rate $1/\tau$ of the dipolar rotational diffusion process. At the temperature T of the $\varepsilon''(f)$ maximum, the relaxation time τ is $1/(2\pi f)$. As the temperature increases, the resin becomes more fluid, permitting the dipoles to orient faster. Thus ε''_{max}, the peak in ε'' for the higher frequencies, occurs at higher temperatures.

As the temperature increases, ionic translation contributes to the loss at increasingly higher frequencies as seen by the overlap of the $\varepsilon''2\pi$ frequency lines. The results are identical for the heat-up/cool-down curve, confirming that no reaction has taken place and the material is a true thermoplastic.

The magnitude of the specific conductivity σ ($\Omega^{-1}cm^{-1}$), which reflects the rate of translational diffu-

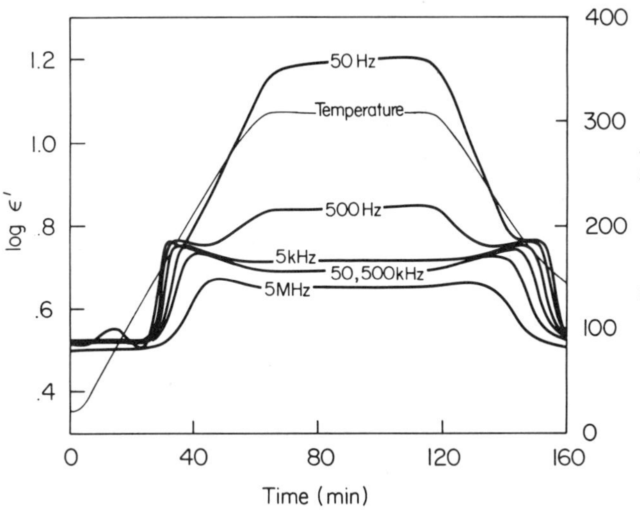

FIGURE 4 Plot of ε' versus time for a polyarylene ether thermoplastic showing the glass transition upon heat-up and cool-down.

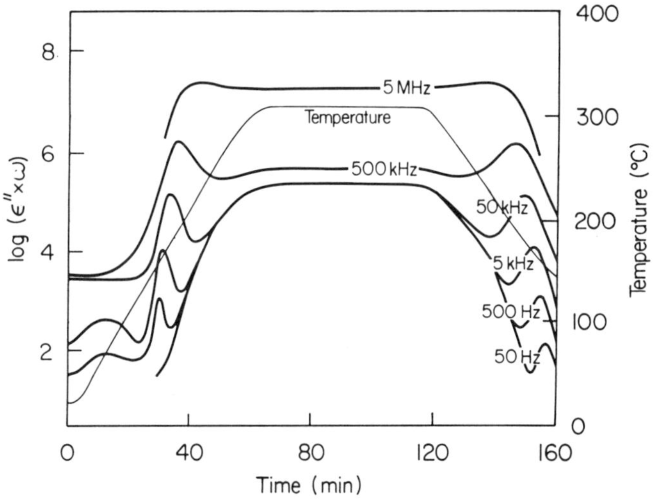

FIGURE 5 Plot of $\varepsilon'' \times \omega$ versus time for a polyarylene ether thermoplastic showing the ionic and dipolar contributions to the signal upon heat-up and cool-down.

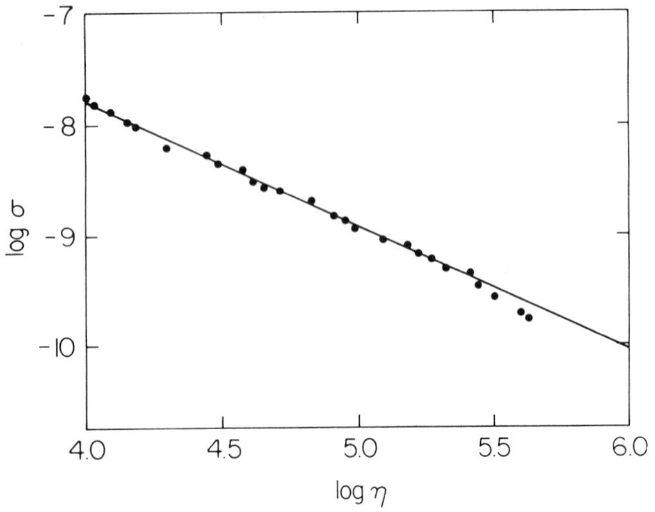

FIGURE 6 **Plot of the ionic mobility σ versus the viscosity η for a polyarlyene ether thermoplastic; data taken simultaneously.**

sion of charge, may be determined from the first term of equation (6), $\sigma = 8.85 \times 10^{-14}\ \omega\ \varepsilon''(\omega)$, since charge polarization is usually small when f exceeds 10 Hz. The usefulness of σ to measure viscosity is shown by the results of simultaneous measurement of σ and viscosity (Fig. 6).

The dipolarlike relaxation process observed just above T_g in Figure 5 is the α relaxation process. This process, associated with the cooperative polymer chain motions, dominates diffusion and viscosity in the region between T_g and $T_g + 50$°C. The relaxation time τ_α associated with the α relaxation process and its temperature dependence is determined as discussed above from $\varepsilon''_{max}(\omega)$.

The temperature dependence of σ and τ (Fig. 7) can be fit to both an Arrhenius and a Williams–Landel–Ferry (WLF) dependence. The results of such a fit for a polyarylene ether are shown below.

Arrhenius Fit

E_{ionic}(kcal/mol)
31.8

$E_{dipolar}$(kcal/mol)
95.4

WLF Free Volume Fit

C_1^I	C_2^I	C_1^D	C_2^D
16	26	11	50

Values of the activation energy are generally higher for the dipolar diffusion processes, reflecting the large size of the dipoles relative to most ions. The values of the WLF coefficients can be used to examine in greater detail the fractional free volume in the thermoplastic, which in turn may be related to toughness.

Crystalline and Melt Phase Transitions

Figure 8 is a plot of the ε''^* frequency data of the poly(aryl ether ether ketone) semicrystalline thermoplastic PEEK. The resin was heated to 360°C, held 30 minutes, and then cooled. The DSC data on a similarly treated sample are shown in Figure 9. The FDEMS sensor is able to monitor T_g at 145°C, melting at 342°C, recrystallization of the supercooled melt at 295°C, and the cooling glass transition at 145°C. It is important to note that the variation in the magnitude of ε'' is due to changes in the ionic mobility (e.g., Eq. 6). As such, the magnitude of ε'' is quite sensitive to these physical changes of state. However, additional experiments such as DSC are required to verify the molecular phenomena causing the sudden change. The FDEMS measurement technique has several advantages over DSC measurements insofar as FDEMS can be used to make in situ point location measure-

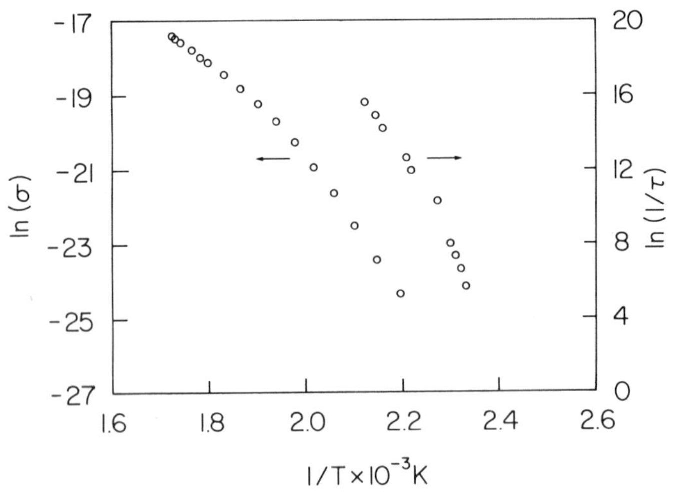

FIGURE 7 **The temperature dependence of both the ionic mobility σ and the dipolar mobility τ for the polyarylene ether thermoplastic.**

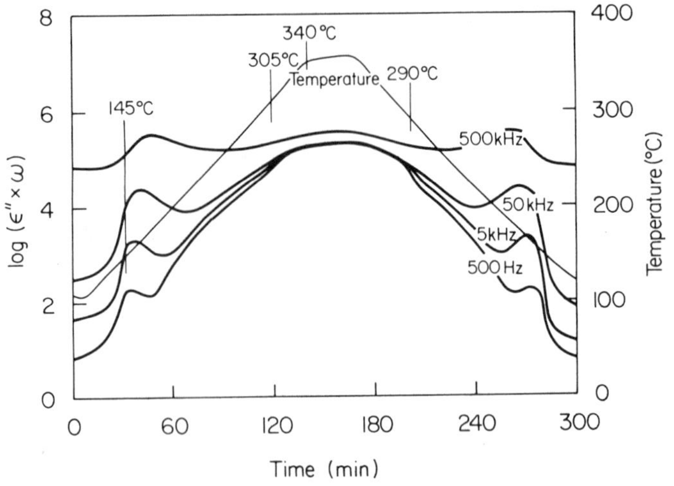

FIGURE 8 **Plot of $\varepsilon'' \times \omega$ versus time for the polyether ether ketone (PEEK), upon heat-up and cool-down, showing the glass transition, onset of melt, melting, and recrystallization.**

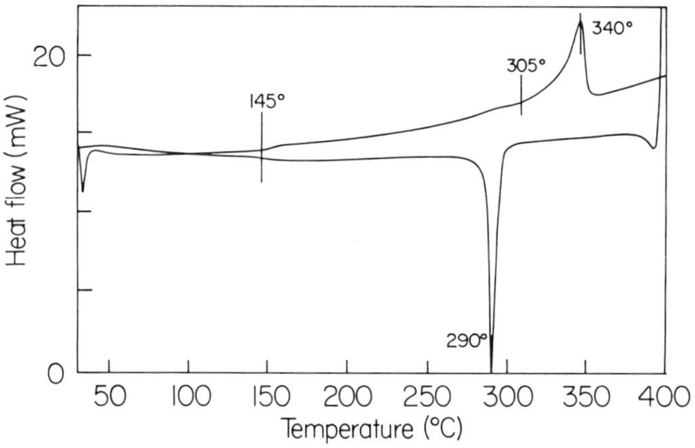

FIGURE 9 **A DSC trace of PEEK upon heat-up and cool-down showing glass transition, onset of melt, melting, and recrystallization.**

ments in the manufacturing mold regardless of the speed of the heat-up/cool-down process.

Solvent–Moisture Content

Many polymers are hydroscopic. Solvents are used to increase their processibility, and difficult-to-dissolve polymers are often mixed with solvents to form slurries. As an example, Figure 10 displays the magnitude of ε'' on heat-up and cool-down for an aromatic thermoplastic polyimide LaRC-TPI film cast from a diglyme slurry. Two peaks are observed on heat-up, reflecting the evolution of residual moisture (90–100°C) and residual diglyme (110–120°C). On cool-down the peaks are absent. A second heating cycle would not show the diglyme peak because the solvent has been driven off, but it could show a moisture peak resulting from readsorption of moisture between cycles. The magnitude of the moisture peak is dependent on the elapsed time before reheating and the ambient humidity.

FDEMS Monitoring of the Cure Processing Properties of Thermosets

Measurement of the Viscosity and Degree of Cure Using the Ionic and Dipolar Diffusion Rate

The variation in the magnitude of ε'' with frequency and with time for a tetraglycidal amine-cured epoxy held at 165°C is shown in Figure 11. The magnitude of ε'' changes by more than 10^5 during the course of the cure reaction. The FDEMS measurement sensitivity is 10^{-2}, as shown in the nonlog plot of the raw data (Fig. 12). The sensitivity and stability of FDEMS permit changes in ε'' to be used to monitor the long-time, final advancement in cure. A plot of $\varepsilon'' \times$ frequency (Fig. 13) is a particularly informative representation of the cure because as discussed, from Eqs. (1)–(6), overlap of $\varepsilon''(\omega) \times$ frequency for differing frequencies indicates that ionic translational diffusion is the dominant physical process affecting the loss (first term of Eq. 6). Similarly the peaks in $\varepsilon'' \times \omega$ for individual frequencies indicate when dipolar rotational diffusion processes are contributing to ε'' (Eq. 4).

The frequency dependence of the loss ε'' can be used to find σ by determining from a computer analysis or a plot of $\varepsilon'' \times$ frequency (Fig. 13), the frequency region where $\varepsilon'' \times$ frequency is a constant. In this frequency region the value of σ is determined from the first term in Eq. (6). The ionic contribution ε''_i is subtracted from ε'' measured to determine the dipolar component ε''_d. The time at which a peak occurs in the dipolar portion ε''_d for a particular frequency is used to determine the time of occurrence of the corresponding mean relaxation time $\tau = 1/(2\pi f)$.

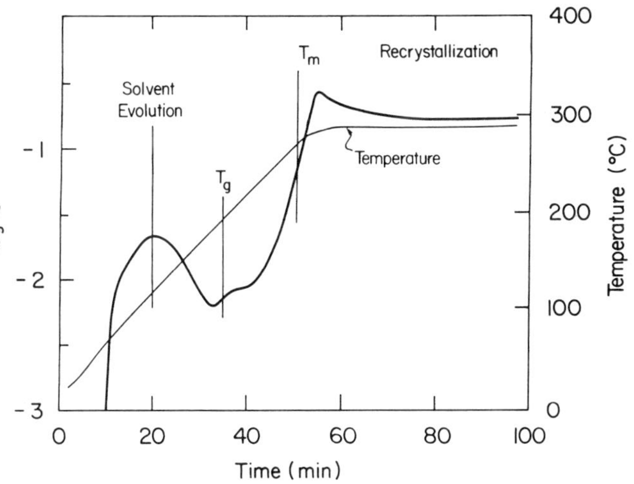

FIGURE 10 **Plot of ε'' versus time for the high temperature polyimide thermoplastic LaRC-TPI, showing solvent evolution, glass transition, melt, and recrystallation.**

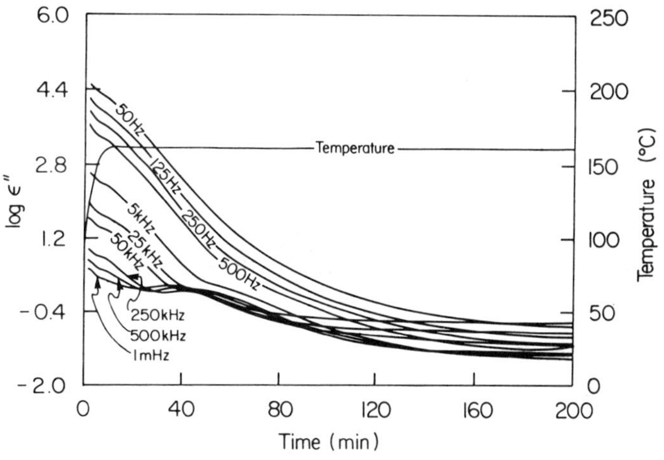

FIGURE 11 **Plot of ε'' versus time for a TGDDM epoxy Hercules 3502 during isotheral cure at 165°C.**

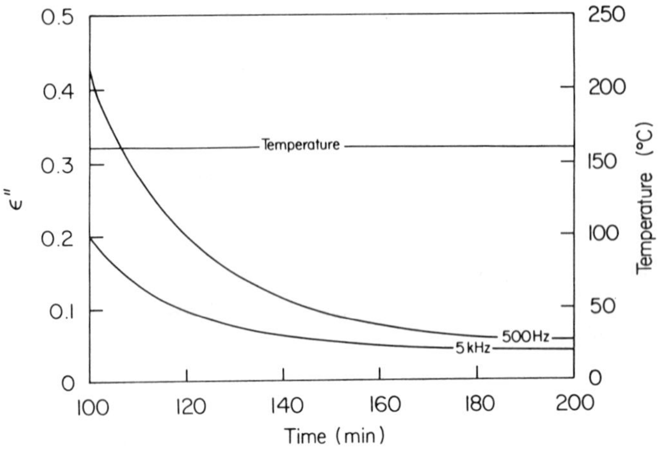

FIGURE 12 Plot of ε″ versus time for a Hercules TGDDM epoxy on a non log scale, showing the long-time cure and sensitivity of the FDEMS technique during a 165°C isothermal cure.

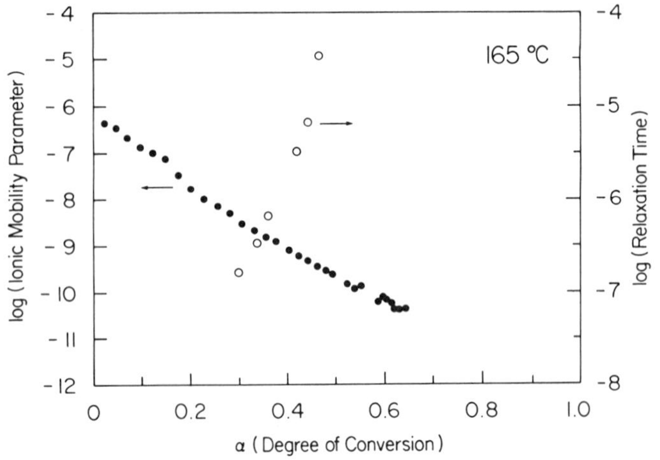

FIGURE 14 Plot of the ionic mobility and relaxation time versus the degree of conversion α for the epoxy of Figures 12 and 13 during a 165°C isothermal cure.

The relationship of ln σ and ln τ to degree of advancement α is shown in Figure 14. The degree of advancement at a particular time is measured by dividing the heat of reaction at that time by the total heat of reaction. For a similar degree of advancement α, the ionic mobility σ increases with temperature. Similarly, for the same value of α, the dipolar mobility increases. An increase in dipolar mobility corresponds to a shorter relaxation time. Thus τ decreases as temperature increases. There is no evidence of a break in the ln σ or ln τ dependence on α, which would indicate gel.

The rate of change of log σ with time approaches zero as the advancement of the reaction approaches completion. The point at which *dσ/dt* approaches zero indicates T_g. The values of α at this point are in general in agreement with α_{T_g} values determined by DSC scans.

The correlation during cure of −log σ with log η (viscosity) permits frequency-dependent ε*(ω) measurements to be used to measure the viscosity and to detect points of maximum flow during cure. The qualitative correlation of −log σ and log η is shown in Figure 15 for a conventional amine–epoxy cure cycle. A quantitative calibration between log σ and log η needs to be determined for each resin, as in Figure 16. In general, there is a break in the σ–η dependence as the resin approaches a viscosity near gel. The break in the relationship at gel is due in part to the fact that ionic and dipolar diffusion processes reflect a molecular viscosity, while the viscos-

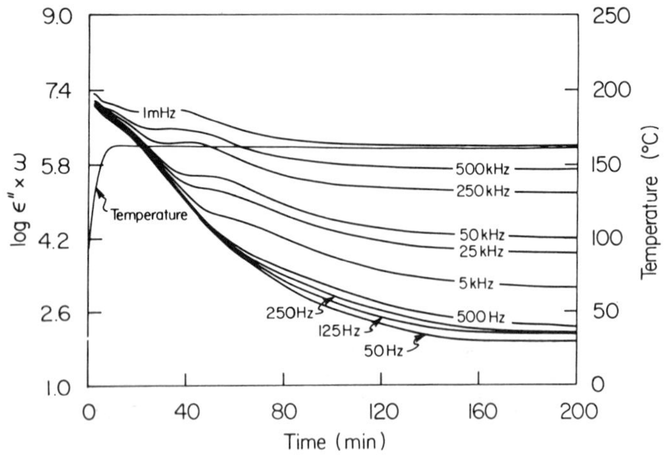

FIGURE 13 Plot of ε″ × ω versus time for the epoxy represented in Figure 12, showing both the ionic and dipolar contributions to the complex permittivity during a 165°C isothermal cure.

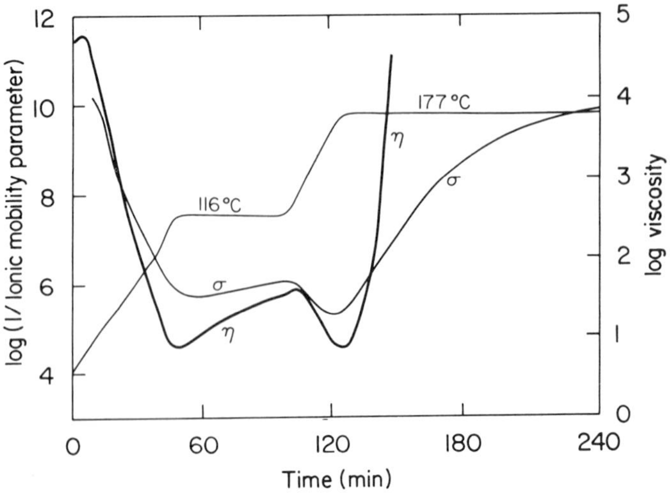

FIGURE 15 Simultaneous measurement of the ionic mobility and the viscosity of a catalyzed TGDDM epoxy (Hercules 3501-6) showing the qualitative correlation between the ionic mobility and the viscosity.

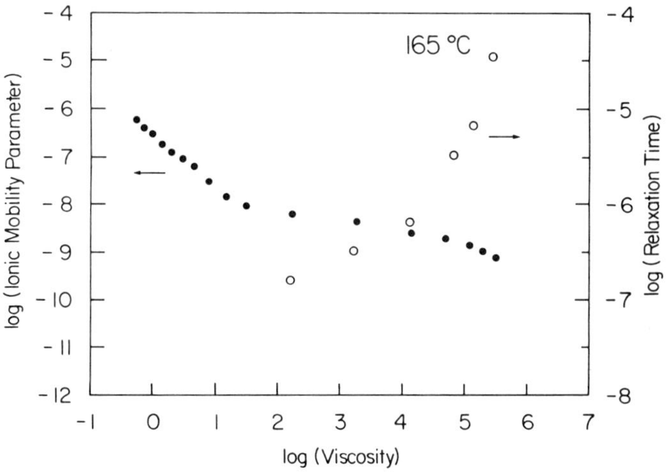

FIGURE 16 **The quantitative correlation between the ionic mobility and the relaxation time to the viscosity for a TGDDM epoxy (Hercules 3502) during a 165°C isothermal cure.**

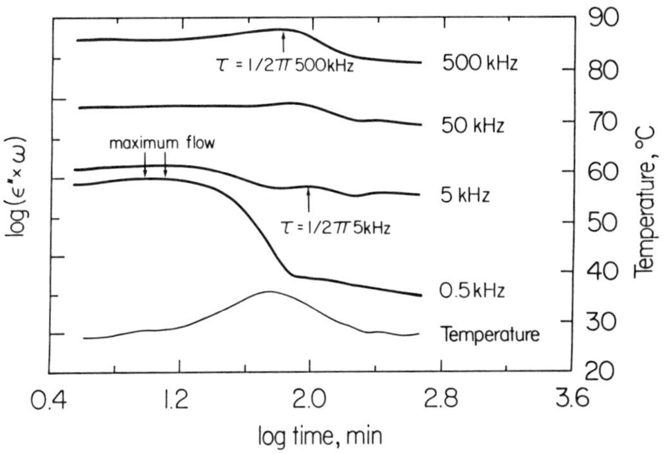

FIGURE 17 **Plot of $\varepsilon'' \times \omega$ versus time of an unsaturated polyester resin (1.9 mm thick, 0.50% initiator) showing the maximum flow region and relaxation times.**

ity measured by the rheometer reflects a macroscopic resistance to flow. At gel, the macroscopic viscosity rises rapidly with small changes in α. In contrast, the α dependence of the molecular viscosity or the α dependence of the ionic diffusion rate remains unchanged, as seen in Figure 14. A similar calibration relation between $\log \tau$ and $\log \eta$ can be determined for a resin during cure, as shown in Figure 16.

Measurement of Reaction Onset and Completion

The ionic diffusion rate as measured by the conductivity σ is in turn a sensitive measure of changes in viscosity. Thus low frequency measurements of both ε'' and ε' as well as σ, can be used to monitor the onset of the reaction and reaction completion. Reaction onset is marked by a sudden drop in $d\varepsilon''/dt$ reflecting the decreasing mobility of the ions with molecular weight buildup. Reaction completion is reflected in the approach of $d\varepsilon''/dt$ and $d\varepsilon'/dt$ to zero.

Monitoring Cure Under Exothermic Conditions

The ionic and dipolar mobility are able to monitor processing properties under many exothermic complex nonisothermal fabrication conditions. Figure 17 shows the trace of $\varepsilon'' \times \omega$ for an unsaturated styrene polyester resin, 75 mils thick and with 0.50% initiator. The effect of the exotherm is seen in the rise in $\varepsilon'' \times \omega$ at 0.5 kHz at the onset of the reaction. This rise reflects the initial increase in σ and the corresponding decrease in viscosity due to the exotherm. The time at which the low frequency values of $\varepsilon'' \times \omega$ are a maximum is the time at which ionic mobility and resin fluidity are a maximum. It occurs after about 10 minutes and near the onset of the exotherm. Shortly thereafter, the low frequency values of both ε'' and σ drop. Then dipolar peaks in the higher frequency values of ε'' start to occur at increasingly lower frequencies and larger elapsed times, reflecting the continuing buildup in viscosity as the resin approaches T_g.

Differing processing effects such as doubling the catalyzed concentration or doubling the thickness can be examined using FDEMS techniques. In Figure 18, run II differs from run I in that the initiator concentration was increased 50%. Run III differs from run II in that the thickness was doubled. Figure 18 shows differences in the onset of cure as indicated by the time at which $\varepsilon'' \times \omega$ rises. These times are in good agreement with the onset of the temperature rise and are an accurate measure of the onset of cure. The FDEMS data show that the cure reaction has stopped for resin III after 100 minutes

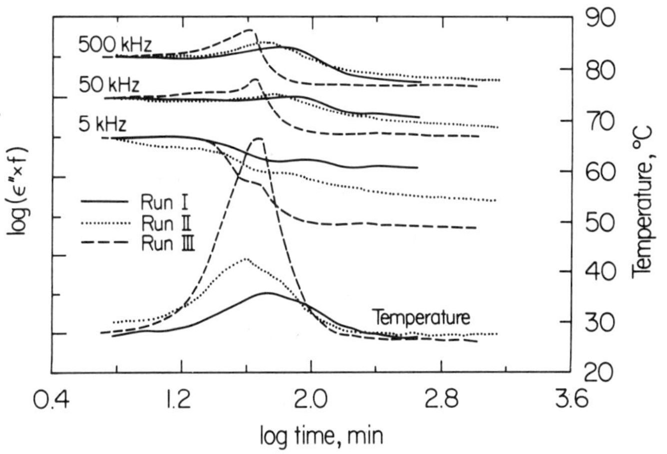

FIGURE 18 **A composite plot of three unsaturated polyester resin runs: run I, 1.9 mm mils thick, 0.50% initiator; run II, 1.9 mm thick, 1.0% initiator; run III, 3.8 mm mils thick, 1.0% initiator.**

TABLE 1
FDEMS Characterization of Unsaturated Polyester Resin Cure

	Time to Exotherm (min)	Time to Dipolar Relaxation (min)		Window (min)
		To τ_1 (η_1)	To τ_2 (η_2)	
Run I	58	62	89	27
Run II	55	58	77	19
Run III	56	41	44	3

as a result of the large exotherm. On the other hand, resins I and II are still reacting after 1000 minutes, as shown by their continuing drop in ε^*.

Table 1 summarizes the ability of the dipolar relaxation time, as determined from the ε''_{max} peaks, to measure the buildup in viscosity as the reaction advances from gel to glass formation. The relaxation time data show that the resin achieves the shortest relaxation time (500 kHz), corresponding to a particular viscosity, at 62 minutes in cure I, 58 minutes in cure II, and 41 minutes in cure III. Additional time was needed for the viscosity and relaxation time to increase the value associated with 5 kHz: 27 minutes in cure I, 19 minutes in cure II, and 3 minutes in cure III. The FDEMS data suggest that increasing the initiator concentration 50% increases the cure rate 50%. However doubling the thickness increases the cure rate by 600%. Overall, the combination of ionic and dipolar mobility provides an insightful and precise in situ means of monitoring the point of maximum fluidity, the rate at which the viscosity increases, and the approach to reaction completion in a nonisothermal fabrication environment.

Monitoring and Controlling Autoclave Cure

A major advantage of the FDEMS technique is the capability of monitoring and quantitatively measuring processing properties in situ during cure in manufacturing tools such as autoclaves and presses.

For example, sensors can be placed in a thick 192 epoxy–graphite laminate at multiple positions (Fig. 19) such as at the tool surface, at 32 plies, at 64 plies, and in the center at the 96th ply. The 192-ply composite layup with the embedded sensors can then be cured in an autoclave. The output at each sensor can be measured automatically at intervals throughout the cure cycle by multiplexing the four sensors through a computer-controlled switch to the impedance bridge. This multiplexed sensor–bridge–computer system can make more than 40 permittivity measurements in less than 2 minutes.

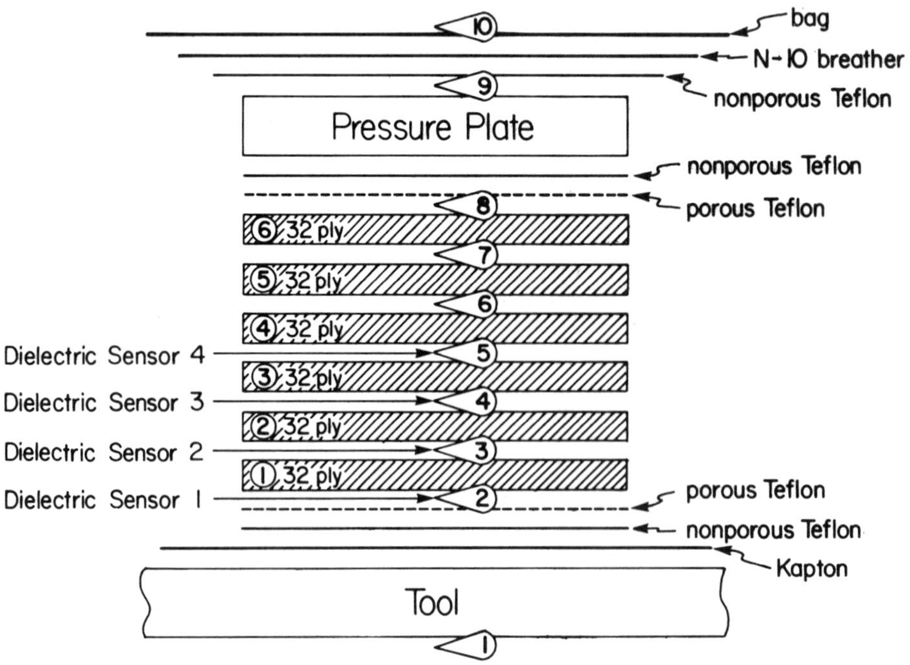

FIGURE 19 The layup of a thick TGDDM epoxy (Hercules 3501-6) graphite laminate autoclave run showing the positions of the dielectric sensors and thermocouples.

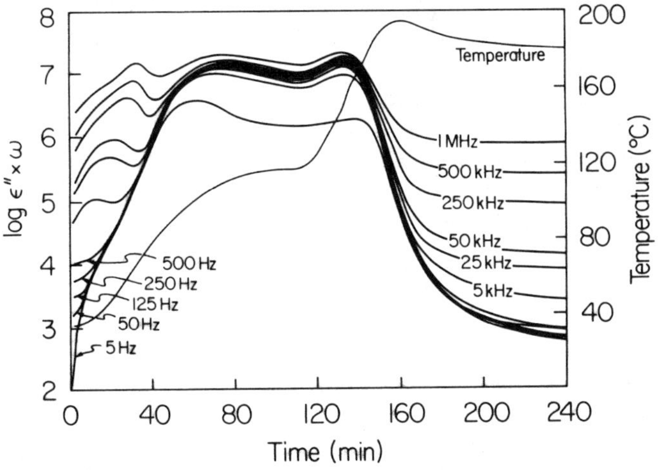

FIGURE 20 **Plot of ε″ × ω versus time of the sensor output at the 64th ply of the thick epoxy–graphite laminate during cure in the autoclave.**

Generally measurements are recorded continuously throughout the cure cycle without interruption over the 10^6–10^{-2} Hz range in magnitude of ε' and ε'' (Fig. 20). The frequency dependence of ε'' is used to determine the conductivity parameter σ. Use of calibration curves relating σ and τ to viscosity for representative time–temperature cure cycles permits the values of $\varepsilon''(\omega)$ and in turn σ to be used to quantitatively monitor viscosity.

Figure 21 is a plot of the viscosity determined by the sensor at each of four positions in the thick laminate during cure in the autoclave. The FDEMS sensor data show that the middle ply achieves its first viscosity minimum 20 minutes after the plies on the surface of the tool plate. The middle plies continue to lag the surface ply until the second ramp to 177°C. At this point the exothermic epoxy reaction starts heating the laminate from the inside. As a result, the heat generated at the center ply causes the viscosity to catch up with the viscosity of the surface plies. In this way, FDEMS sensor output measurements of the viscosity at the center and surface can be used to evaluate the cure cycle. In this example the center and surface plies achieve their viscosity minimum in the high temperature ramp at roughly the same time. At all earlier times the center lags the reaction at the surface.

The sensor output can also be used to test the validity of processing models such as the Loos–Springer model. Sensor measured values of η can be compared with the Loos–Springer model predictions. Figure 22 is a comparison of the model's predictions and the measured values at the 64th ply. Agreement in the viscosity's time dependence and magnitude with the predictions of models is essential if the model is to be used with confidence.

The sensor output can also provide useful input for intelligently controlling the cure cycle. For example, one can chose to use as a guide the observation that for this epoxy the most widely recommended and successful cure cycle forces the second viscosity minima to occur at the same time throughout the laminate's thickness. Thus one can hypothesize that an efficient and effective cure cycle causes the viscosity minima at the surface and center to occur at the same point, and as rapidly as possible. Accordingly, one might propose raising rapidly the air temperature in the FDEMS sensor-controlled run until the exotherm at the center causes the viscosity at the center ply to start to catch up to the surface viscosity. At this point the air temperature would be rapidly lowered, to hold the surface viscosity in this high flow condition while the center viscosity catches up. At such point as the center viscosity goes through its viscosity minimum and advances beyond the surface plies, the air temperature would be set to the final 177°C hold. The 177°C hold would continue until the sensor output indicated

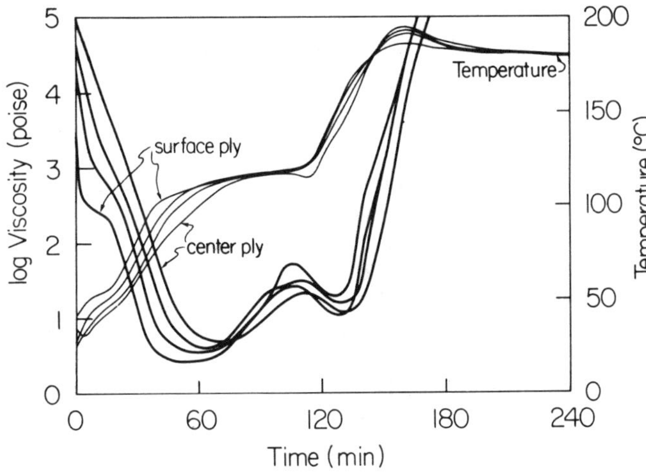

FIGURE 21 **The viscosity of the thick laminate as determined from the frequency dependence of the FDEMS sensors at the surface, 32nd, 64th, and center plies.**

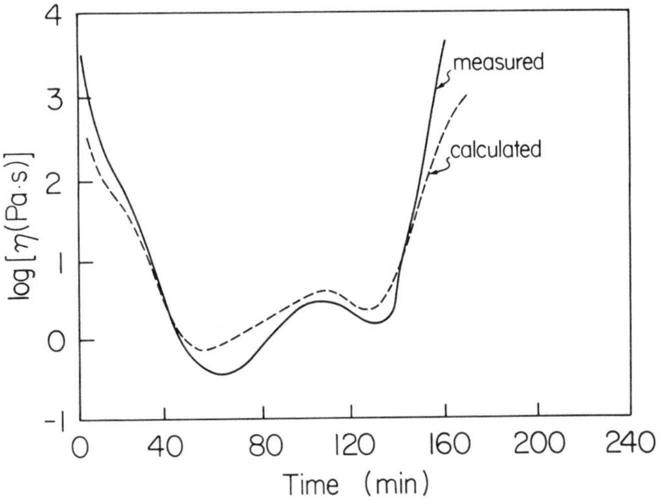

FIGURE 22 **Comparison of the viscosity at the 64th ply as predicted by the FDEMS sensor and the Loos–Springer model.**

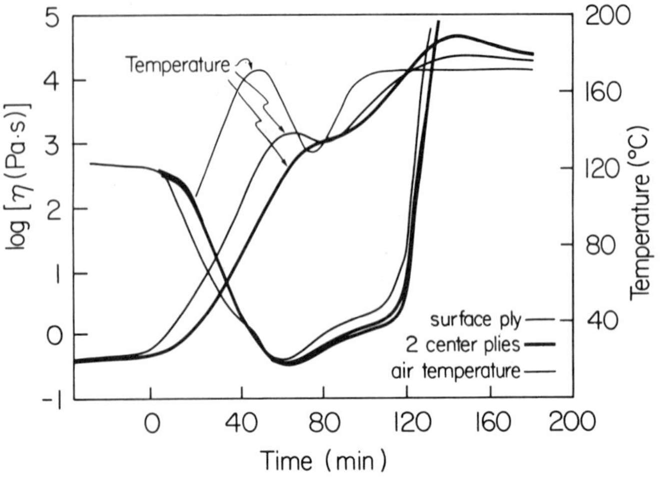

FIGURE 23 **The viscosity at each sensor position of a 192-ply graphite–epoxy composite during an FDEMS sensor-controlled autoclave cure.**

through $d\varepsilon''/dt$ approaching zero that the reaction was complete.

FDEMS sensor-measured viscosities from two sensors at the center ply and one sensor at the surface ply, obtained using the foregoing rules as an example, are shown in Figure 23. Air autoclave temperatures and the temperatures at the surface and center ply are also shown. The starting time for the FDEMS sensor-controlled autoclave run (Fig. 23) and the manufacturer's cure cycle run (Fig. 19) are defined as the time at which the tool surface temperature starts to increase.

Figure 22 shows that the FDEMS sensor-controlled run can significantly reduce the time lag and viscosity difference between the center ply and the tool surface ply. The amount of flow as measured by the magnitude of the viscosity minimum is significantly greater in the FDEMS sensor-controlled run. The approach of $d\varepsilon''/dt$ to zero is used to determine cure completion. The total cure time of 200 minutes in this FDEMS controlled run is 40 minutes less than the conventional cure cycle.

Summary

Three techniques—frequency-dependent electromagnetic measurements, ultrasonic–acoustic measurements, and fluorescent measurements using optical fibers—have been successfully used to monitor the viscosity in amine-cured epoxy systems. The latter two techniques are presently more limited in the extent to which theory is available for interpreting the output on a molecular level and the range of resins—problems that have been successfully addressed. FTIR optical fiber measurements have only very recently been developed to the point of where monitoring experiments have been conducted and reported.

Major advances have been made during the past several years in cure monitoring, particularly using FDEMS microsensors. Overall, a combination of FDEMS and acoustic and fluorescent sensors that monitor the reaction advancement, viscosity, and changes in other physical properties of the composite resin before and throughout cure, along with FTIR sensors that monitor the concentration of specific molecular groups, will undoubtedly be sought to fully and completely characterize the cure process.

David E. Kranbuehl

Bibliography

Frequency-Dependent Electromagnetic Measurements

Hedvig, P., *Dielectric Spectroscopy of Polymers,* Wiley, New York, 1977.

Hill, N., Vaughan, W., Price, A., and Davis, M., *Dielectric Properties and Molecular Behavior,* Van Nostrand, London, 1969.

McCrum, N., Read, B., and Williams, G., *Anelastic and Dielectric Effects in Polymeric Solids,* Wiley, London, 1967.

Cure Monitoring Overview

Kranbuehl, D., *Developments in Reinforced Plastics,* Vol. 5, Elsevier Applied Science Publishers, New York, 1986, pp. 181–204.

May, C., "Chemorheology of Thermosetting Resins," *Polymer Materials Science and Engineering,* ACS Symposium Series 227, American Chemical Society, Washington, DC, 1983.

Senturia, S., and Sheppard, N., *Adv. Polym. Sci., 80,* 1 (1986).

Instrumentation

Kranbuehl, D. E., Delos, S. E., and Jue, P. K., *Polymer, 27,* 11 (1986).

Sheppard, N., Gavericle, S., Day, D., and Senturia, S., *Proceedings of the 26th Natl. SAMPE Symposium,* 1981, pp. 65–76.

Thermoplastics

Kranbuehl, D., Delos, S., Hoff, M., Weller, L., Haverty, P., Seeley, J., and Whitham, B., *Proceedings of the 32nd Int. SAMPE Symposium,* 1987, pp. 338–348.

Thermosets

Bidstrup, S., and Prime, B., Cure monitoring and cure modelling issue of *Polymer Engineering and Science,* September 1988.

Dickie, R., Labana, S., and Bauer, R., *Crosslinked Polymers: Chemistry, Properties and Applications,* ACS Symposium Series 367, American Chemical Society, Washington, DC, 1988, p. 100.

Loos, A., Kranbuehl, D., and Freeman, W., in E. Wadley, Ed., *Intelligent Processing of Materials and Advanced Sensors,* Metallurgical Society, New York, 1987, pp. 191–211.

Cure Control

Day, D., *Proceedings of the 33rd Int. SAMPE Symposium,* 1988, pp. 594–602.

Kranbuehl, D., Hoff, M., Haverty, P., Loos, A., and Freeman, T., *Proceedings of the 33rd Int. SAMPE Symposium Ser.*, 1988, pp. 1276–1284.

Ultrasonic–Acoustic Monitoring

Winfree, W., and Parker, F., *Rev. Prog. NDE, 5*, 1055 and 1063 (1986).

Harrold, R., and Sanjana, Z., *Rev. Prog. NDE, 6*, 1277 (1987).

Optical Monitoring

Fluorescence

Wang, F., Lowry, R., and Grant, W., *Polymer, 25*, 690 (1984).

Wang, F., Lowry, R., and Franconi, B., in *Polymer Materials Science and Engineering*, ACS Symposium Series 53, American Chemical Society, Washington, DC, 1985, p. 180.

Levy, R. L., and Schwab, S. D., in *Crosslinked Polymers: Chemistry, Properties and Applications*, ACS Symposium Series 367, American Chemical Society, Washington, DC, 1988, p. 113.

FTIR

Druy, M. A., Elanjan, L., and Stevenson, W. A., "Composite Cure Monitoring with Infrared Transmitting Optical Fibers," *Fiberoptic Smart Structures and Skins*, SPIE, Vol. 989, 1988, p. 130.

Curing Agents

When considering the final cured characteristics of a composite material and its processing method, attention must be given not only to the type and amount of reinforcement used but also to the types of resin and curing agent employed. Curing agents are substances or mixtures of substances that serve to initiate and control the chemical reaction. This is accomplished by producing cross-linking of the initially linear or branched polymer or oligomer. The amount of cross-linking and the manner in which it occurs directly influence the strength of the resin matrix and its final properties. In very basic terms, curing agents are generally divided into two types, depending on their involvement in the reaction. If the curing agent is not permanently affected by the reaction, it is known as a "catalyst" or "catalytic hardener." On the other hand, if the agent does enter into the reaction (also referred to as an "addition reaction") it is called a "reactive curing agent," "initiator," or "hardener." The resulting polymerization obtained by a hardener is termed "heteropolymerization." In this case, the hardener acts as a bridge for cross-linking several resin molecules. Agents of these types are generally used in greater amounts than catalysts. The catalyst type of curing agent encourages self-polymerization or "homopolymerization" of the resin [1].

There are hundreds of types of hardener and catalyst systems available. Because composites are chemically engineered to meet a broad range of performance and processing requirements, curing agents are often special preparations. They can be formulated by the supplier as custom designs, by a prepreg compounder, or by the composite part molder using a generic base curing agent. Among the parameters affected by the curing agent are the shelf life or storage life, the gel time, the rate of reaction, the curing time and temperature, the processing method, the final properties, and the ultimate applications. There are distinct benefits and limitations associated with each curing agent. One way to describe the different types is by investigating the relationship between the curing agent and the resin system of interest.

Polyesters

Polyesters are addition-type resins. The curing agents generally used with these materials are organic peroxides. Most are stable when not in the presence of the resin. However, they break down into free radicals when combined with the resin and particularly when any source of energy is applied. These curing agents initiate polymerization, by activating the monomer, and set off a molecular chain reaction resulting in a polymer. A complete cure is attained when no unreacted monomer exists. The various peroxides available allow for a range of processing temperatures. Accelerators and promoters also can be added to influence the breakdown of the catalyst into its free-radical form. A combination of hydroperoxide and cobalt naphthenate is an example. Violent reactions, however, can occur if the reagents are mixed incorrectly. Organic-soluble cobalt salts and peroxides should never be mixed. Since these organic peroxide curing agents are directly related to the resin processing parameters and influence them, it becomes evident that no one material will meet all the requirements. Therefore, combinations usually are needed. However, when considering processing temperatures, these curing agents generally can be separated into three convenient divisions (see Table 1). For processing up to 49°C, acetyl benzoyl peroxide, cyclohexanone peroxide, cyclohexyl hydroperoxide, cumene hydroperoxide, and methyl ethyl ketone peroxide can be considered. When an intermediate processing temperature [$\leq$ 93°C] is required, acetyl peroxide, benzoyl peroxide, *t*-butyl hydroperoxide, lauroyl peroxide, methyl amyl ketone peroxide, and methyl cyclohexyl hydroperoxide are the agents of consideration. For higher processing temperatures [$>$ 93°C], the possibilities include *p*-chlorobenzoyl peroxide, di-*t*-butyl peroxide, and dibenzal diperoxide [2].

The most commonly used catalyst for unsaturated polyester resins with the addition of heat is benzoyl peroxide. Some of its advantages include stability at room temperatures, long storage life, monomeric styrene solubility, efficiency, and ease of use. It also aids in the full cure of the resin by causing high peak exothermic temperatures. Depending on the resin and monomer used, a range of 0.5–2.0% usually is required [3]. The reaction obtained by methyl ethyl ketone is quite different from that of the benzoyl peroxide. MEK will not leak to a full

TABLE 1
Peroxide Curing Agents for Three Process Temperature Ranges

Low (≤ 49°C)	Medium (≤ 93°C)	High (> 93°C)
Acetyl benzoyl peroxide	Acetyl peroxide	*p*-Chlorobenzoyl peroxide
Cyclohexanone peroxide	Benzoyl peroxide	Di-*t*-butyl peroxide
Cyclohexyl hydroperoxide	*t*-Butyl hydroperoxide	Dibenzal diperoxide
Cumene hydroperoxide	Lauroyl peroxide	
Methyl ethyl ketone peroxide	Methyl amyl ketone peroxide	
	Methyl cyclohexyl hydroperoxide	

From Ref. 2.

cure at room temperature without the addition of an accelerator.

Phenolics

When considering thermosetting B-staged novolac resins, hexamethylene tetramine may be the curing agent of choice. It decomposes when in the presence of heat and moisture (produced by the condensation process) and results in a C-staged final product.

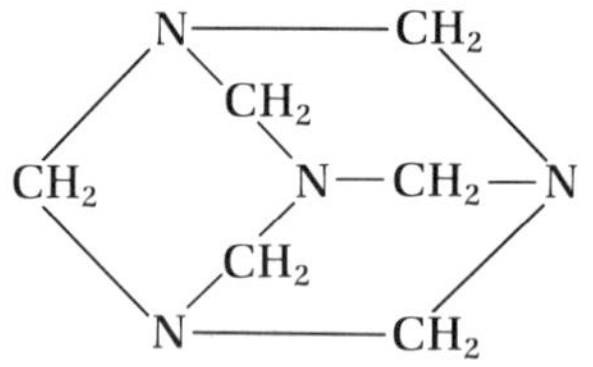

Hexamethylene tetramine (cystamine)

Epoxies

Epoxies are addition-type resins. The curing agents used for epoxies can, generally, be divided into six major categories. These categories are aliphatic amines, aromatic amines, acid anhydrides, latent curing agents, catalysts, and flexibilizing curing agents [3,4]. The most common aliphatic amines used to cure epoxy resins include diethylene triamine (DETA), triethylene tetramine (TETA), ethylene diamine (EDA), tetraethylene pentamine (TEPA), diethylaminopropylamine (DEAPA), and dimethylaminopropylamine (DMAPA) (see Table 2). These agents have the advantage of curing at room temperatures. They may, however, develop very high temperatures during the cure process. In these cases, mass becomes critical. Another consideration is that aliphatic amines tend to reduce the physical and electrical properties of epoxies at high temperatures.

The types of aromatic amines in use include metaphenylene diamine (MPDA), *p*, *p*′-methylene dianiline (MDA), 4,4′-diaminodiphenylmethane (DADM), and 4,4′-diaminodiphenyl sulfone (DDS or DADS). In special applications, benzidine, toluene diamine, 2,6-diaminopyridine (DAP), and 1-aminobenzylamine can be used (see Table 3). One advantage of aromatic amines is that they tend to retain a longer working life than the aliphatic amine cured epoxies. Also, epoxies cured with these agents can be used at higher temperatures (≤ 135°C) than those of aliphatic amine cured materials. They retain good overall electrical and chemical properties. The disadvantages are that they require moderate temperatures to cure, and many of them must be melted (from their natural solid state) into the epoxy. Formulations containing these curing agents are often cured for 2 hours at 80°C followed by 2 additional hours at 150°C [3].

Acid anhydrides, the third class of curing agents, are relatively easy to handle, have lower toxicity than amines, and result in optimum high temperature properties of the final resin. Because of the low incidence for exotherms, very large castings can be produced. Some do require premelting, however, and the high temperature cure cycles are very long. (Often, tertiary amines are

TABLE 2
Aliphatic Amine Curing Agents for Epoxies

Curing Agent	Formula	Molecular Weight	Boiling Point (°C)	Form
Diethylene triamine (DETA)	$NH_2(CH_2)_2NH(CH_2)_2NH_2$	103	206	Liquid
Triethylene tetramine (TETA)	$NH_2[(CH_2)_2NH]_2(CH_2)_2NH_2$	146	277.5	Liquid
Ethylene diamine (EDA)	$NH_2(CH_2)_2NH_2$	60	116	Liquid
Tetraethylene pentamine (TEPA)	$NH_2[(CH_2)_2NH]_3(CH_2)_2NH_2$	189	134	Liquid
Diethylaminopropylamine (DEAPA)	$(CH_3CH_2)_2N(CH_2)_3NH_2$	130	169	Liquid
Dimethylaminopropylamine (DMAPA)	$(CH_3)_2N(CH_2)_3NH_2$	102	134	Liquid

From Refs. 4, 8, and 12.

TABLE 3
Aromatic Amine Curing Agents for Epoxies

Curing Agent	Formula	Molecular Weight	Melting Point (°C)	Boiling Point (°C)	Pot Life
Metaphenylene diamine (MPDA)	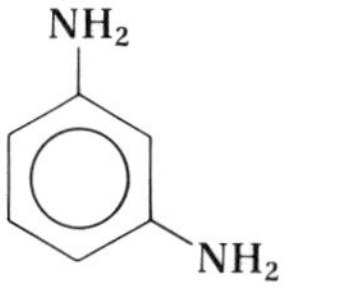	108.14	63.5	283	4–6 hours at RT
p,p′-Methylene dianiline (MDA)	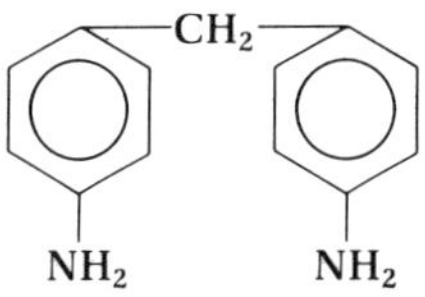	198.3	85–90	265[a]	4–6 hours at RT
4,4′-Diaminodiphenyl methane (DADM)					
4,4′-Diaminodiphenyl sulfone (DDS or DADS)	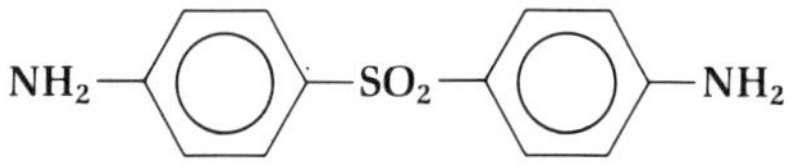	248	60	170	—
Benzidine	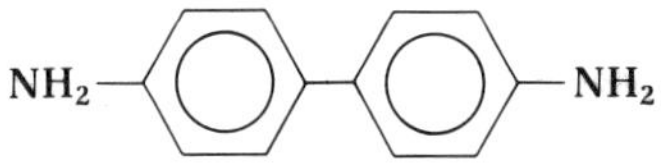	184.24	93–94	205	—
Toluene diamine	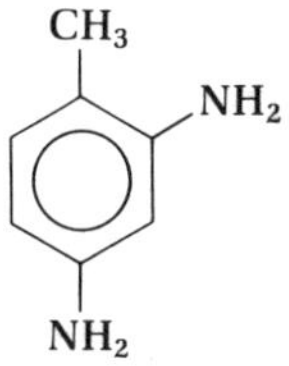	122.17	—	99	—
2,6-diaminopyridine (DAP)	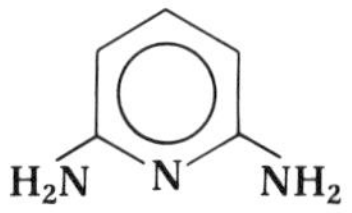	—	—	—	—
1-Aminobenzylamine	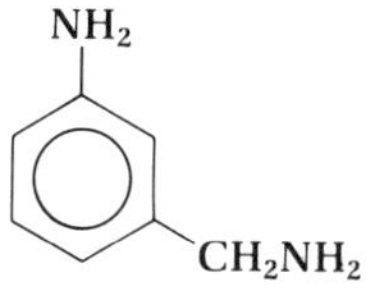	—	—	—	—

[a] At 25 mm Hg.
From Refs. 4, 8, and 12.

added to shorten the cycle.) Among the many types of acid anhydride that can be used, the most common include phthalic anhydride (PA), tetrahydrophthalic anhydride (HET), methyl nadic anhydride (MNA), pyromellitic dianhydride, and trimellitic anhydride. Other types are benzyldimethylamine, dimethyl aminomethylphenol, tris(dimethylaminomethyl)phenol, 2-ethyl-4-methylimidazole, and triethanolamine (see Table 4).

Latent curing agents have the advantages of long room temperature shelf lives and rapid cure at elevated temperatures. Some materials are insoluble in epoxy but will go into solution at moderate temperatures. An example of a latent curing agent is dicyandiamide. Boron trifluoride complexes can also be considered latent curing agents.

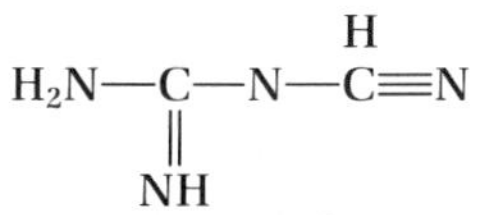

Dicyandiamide (cyanoguanidine)

The most popular class, catalytic curing agents, includes the Lewis acids and Lewis bases. Boron trifluoride, dimethylaminomethyl phenol, tris(dimethylaminomethyl)phenol, and benzyldimethylamine all fall into the category of catalytic agents. Piperidine produces epoxy castings with superior high temperature stability. Unlike the aliphatic amine curing agents, these materials have a longer working life. They do, however, require a cure temperature of 95°C or above. The mass of the resin can

TABLE 4
Acid Anhydride Curing Agents for Epoxies

Curing Agent	Formula	Molecular Weight	Melting Point (°C)	Boiling Point (°C)
Phthalic anhydride (PA)	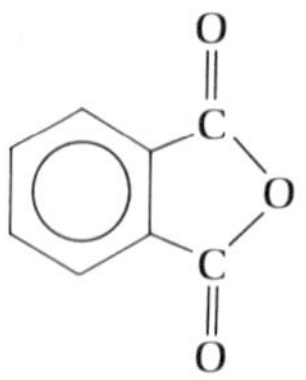	148.11	132	—
Tetrahydrophthalic anhydride (HET)	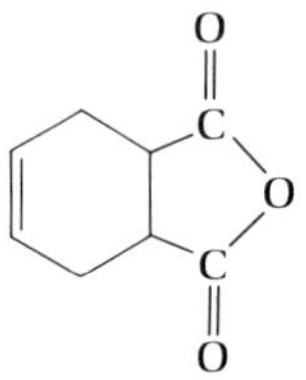	154.2	66.5	—
Hexahydrophthalic anhydride	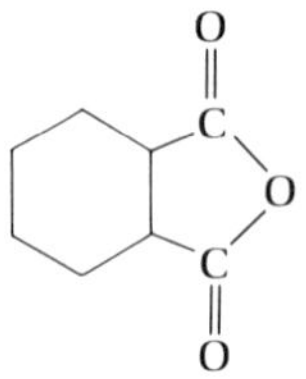	—	35.5	—
Chlorendic anhydride	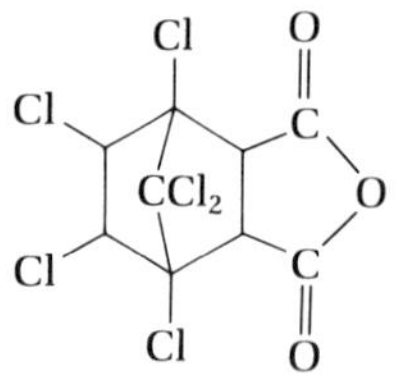	371	240	—
Methyl nadic anhydride (MNA)	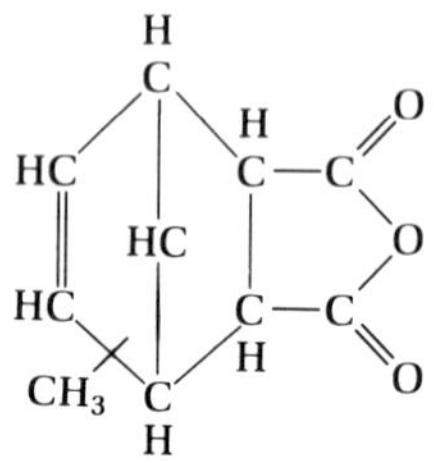	178.2	12	—
Pyromellitic dianhydride	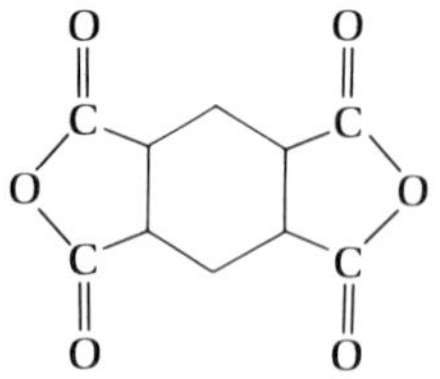	218	285	—
Trimellitic anhydride	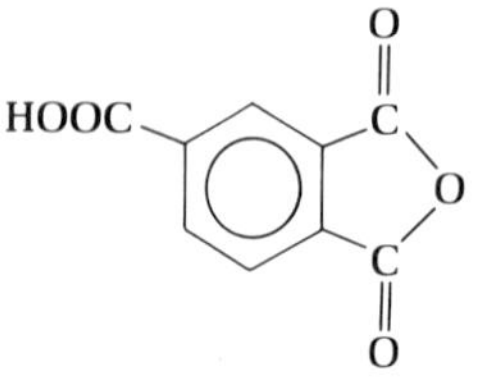	—	168	—
Benzyldimethylamine	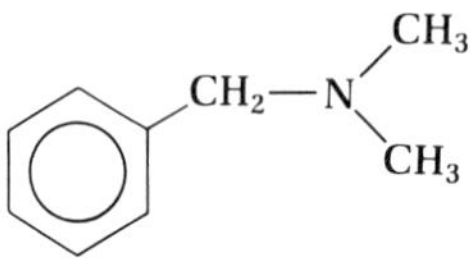	135	—	181

TABLE 4 (*continued*)

Curing Agent	Formula	Molecular Weight	Melting Point (°C)	Boiling Point (°C)
Dimethyl aminomethylphenol	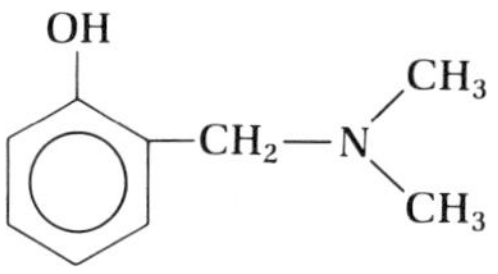	151	—	100[a]
Tris(dimethylaminomethyl)phenol	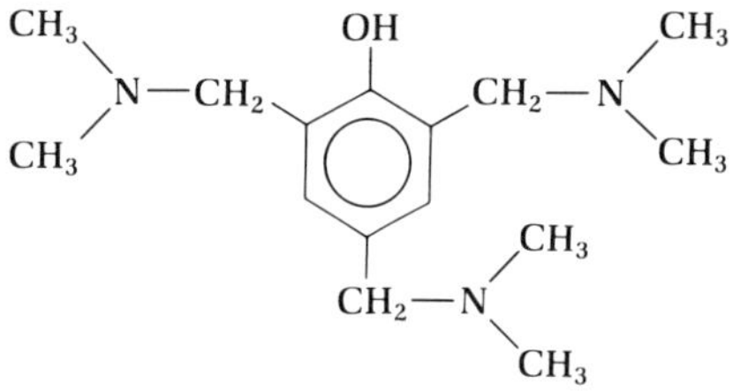	—	—	—
2-Ethyl-4-methylimidazole	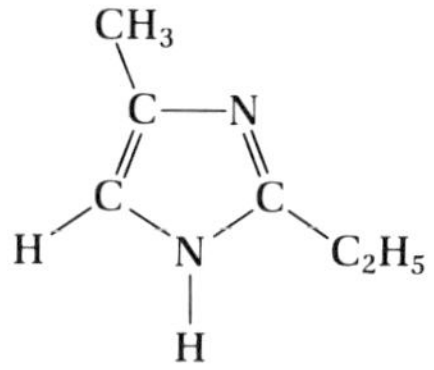	110	—	—
Triethanolamine	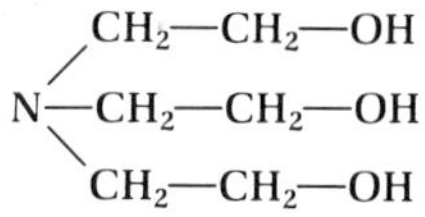	149	—	—

[a] At 2 mm Hg.
From Refs. 4, 8, and 12.

affect the exothermic reaction produced when using these catalytic agents. Often, the tertiary amines are used to accelerate the curing of anhydride–epoxy mixtures.

Flexibilizing curing agents generally are amino acids attached to long aliphatic chains. They cure at room temperatures but only very slowly. One advantage of these materials is that the ratio of curing agent to resin is not critical.

Melamines

Melamines are the product of condensation-type reactions, liberating heat and moisture. Substituted ammonium compounds are used to catalyze these materials.

Silicone Resins

Silicones are catalyzed with metal-organic salts such as naphthanates or octoates of zinc, iron, cobalt, and lead or with amines.

Availability

There are many suppliers or basic formulating manufacturers of curing agents. Listed below are a few suppliers provided through the listings in the International Plastics Selector, from *Additives for Plastics* and from correspondence with Composites Services Corporation, Demarest, New Jersey [5,6].

For unsaturated polyesters:

ABCO Industries, Inc., Roebuck, SC
Akzo Chemie America, Noury Chemical Division, Chicago, IL
Catalyst Resources, Inc., Houston, TX
Freeman Chemical Corporation, Port Washington, WI
Montedison USA, Inc., New York, NY
The Norac Company, Azusa, CA
Pennwalt Corporation, Lucidol Division, Buffalo, NY
Reichhold Chemicals, Inc., White Plains, NY
Witco Corporation, Argus Division, Brooklyn, NY

For epoxy resin systems:

Aerojet Strategic Propulsion Company, Van Nuys, CA
Air Products & Chemicals, Inc., Allentown, PA
Akzo Chemie America, Chicago, IL
Anhydrides & Chemicals, Inc. Newark, NJ
BASF Chemicals Division, Parsippany, NJ
Ciba-Geigy Corporation, Ardsley, NY

Diamond Shamrock Chemicals Company, Morristown, NJ
E. I. du Pont Company, Wilmington, DE
Ethyl Corporation, Baton Rouge, LA
Henkel Corporation, Ambler, PA
Hexcel Corporation, Chemical Products Division, Dublin, CA
Interez, Inc., Louisville, KY
Nuodex Inc., Piscataway, NJ
Pacific Anchor Chemical Corporation, Los Angeles, CA
Shell Chemical Company, Houston, TX
Sylvachem, subsidiary of Arizona Chemical Corporation, Panama City, FL
Texaco Chemical Company, Bellaire, TX
Uniroyal Chemical Company, Inc., Naugatuck, CT

Suppliers of accelerators and other additives:

American Cyanamid Company, Wayne, NJ
E. F. Whitmore & Company, San Gabriel, CA
Fairmount Chemical Company, Inc., Newark, NJ
Focus Chemical, Woburn, MA
Nor-Am Chemical Company, Wilmington, DE
NL Chemicals, Hightstown, NJ
Toyo Soda USA, Inc., Atlanta, GA
Union Carbide Corporation, Danbury, CT

Judy Morgon Thuen

References

1. L. R. Whittington, *Whittington's Dictionary of Plastics*, Technomic, Westport, CT, 1978, pp. 51, 77, 169.
2. N. E. Beach, *Plastic Laminate Materials: Their Properties and Usage*, Foster, Long Beach, CA, 1967, pp. 27–28.
3. C. A. Harper, *Handbook of Plastics and Elastomers*, McGraw-Hill, New York, 1975, pp. 8–11 to 8-17.
4. M. T. Gossey, *Plastics for Electronics*, Elsevier Applied Science Publishers, London/New York, 1985, pp. 108–117.
5. J. Thuen, *Additives for Plastics*, 1st ed., International Plastics Selector, San Diego, CA, 1987, pp. 351–359, and on-line data base access.
6. Joseph S. McDermott, president, Composites Services Corporation, Demarest, NJ, Dec. 11, 1987 (personal communication).
7. T. Richardson, *Composites: A Design Guide*, Industrial Press, New York, 1987, pp. 88–89.
8. H. R. Simonds and James M. Church, *The Encyclopedia of Basic Materials for Plastics*, Van Nostrand Reinhold, New York, 1967, pp. 171–176.
9. J. A. Brydson, *Plastics Materials*, 4th ed., Butterworths, London, 1982, pp. 682ff.
10. G. Lubin, *Handbook of Fiberglass and Advanced Plastic Composites*, Van Nostrand Reinhold, New York, 1969, pp. 23–113.
11. R. C. Weast, Ed., *CRC Handbook of Chemistry and Physics*, 68th ed., CRC Press, Boca Raton, FL, 1987, pp. C-42 to C-553.
12. J. W. Weeton, D. M. Peters, and K. L. Thomas, *Engineer's Guide to Composite Materials*, American Society for Metals, Carnes Publication Services, Metals Park, OH, 1987, pp. 6–9.

Cyanate Ester Resins

Cyanate ester resins, also known as cyanate resins, cyanic esters, or triazine resins, are a family of monomers and oligomers with the reactive cyanate end group—OCN on an aromatic ring. The most useful resins, like epoxies, are derived from bisphenols or polyphenols. Commercial monomers can all be represented by the structural model illustrated below and in Table 1.

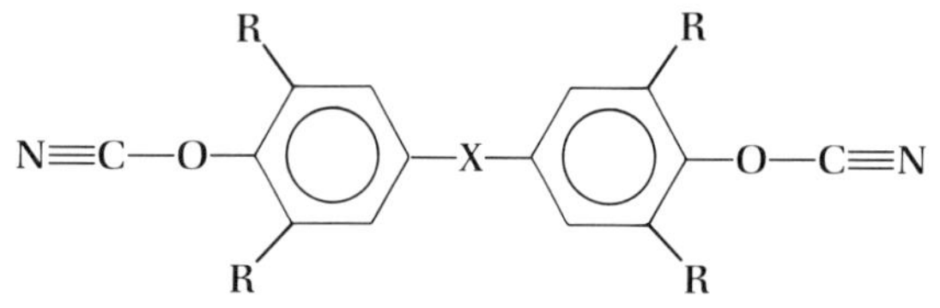

Monomer Synthesis

Grigat and Putter [1] reported the first synthesis of cyanate ester monomers in 1963 by reacting phenolic compounds with cyanogen chloride in the presence of an acid acceptor. They reported near-quantitative yields of stable cyanate esters (with no by-products), which could be distilled or recrystallized to highly pure materials. This basic method is still used to make commercial cyanate esters. The general reaction for monomer synthesis is:

$$\text{HO—Ar—R—Ar—OH} + \text{ClCN} \xrightarrow[-10^{\circ}\text{C}]{\text{base}} \text{NCO—Ar—R—Ar—OCN}$$

Polymerization

Oligomers, polymers, and cross-linked polymer networks are formed by heating the monomers. The polymer structure is a network of oxygen-linked triazine rings (cyanurates) and bisphenol esters/ethers. This structure results through the cyclotrimerization of the —R—OCN functionality. Fully cured cyanate resins are classified as polycyanurates or a cross-linked polyarylate. The cyclotrimerization process is shown in Figure 1. Stable oligomers and polymers can be recovered as liquids, semisolids, hard resins, or solutions. Continued heating of the oligomers yields thermoset plastics with cured-state properties equal to those obtained by uninterrupted heating of the monomer. Cure catalysts and other additives and modifiers are usually compounded with the oligomers and polymers before final curing to a cross-linked matrix.

Bauer et al. [2] studied the kinetics of cyanate ester polymerization and reported an autocatalytic process in which phenolic impurities in the aryl dicyanates acted as the polymerization initiator according to the following reaction scheme:

TABLE 1
Chemical Structures of Monomeric Dicyanates; Refer to Model Compound:

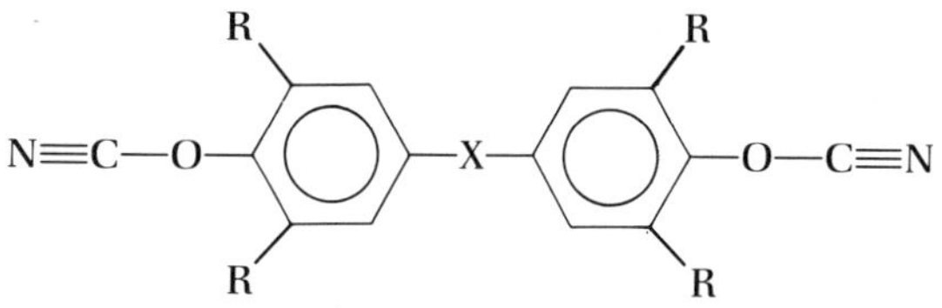

Product Designation	Acronym	Linkage (X)	Substituent (R)	Source
RDX 64826	BPADCy	$(CH_3)_2C$	H	Interez, Inc. Mitsubishi Gas Chemical
RDX 64833	METHYLCy	$—CH_2—$	CH_3	Interez, Inc.
RDX 76678	THIOCy	—S—	H	Interez, Inc.
XU-71787.00		Cycloaliphatic	H	Dow Chemical

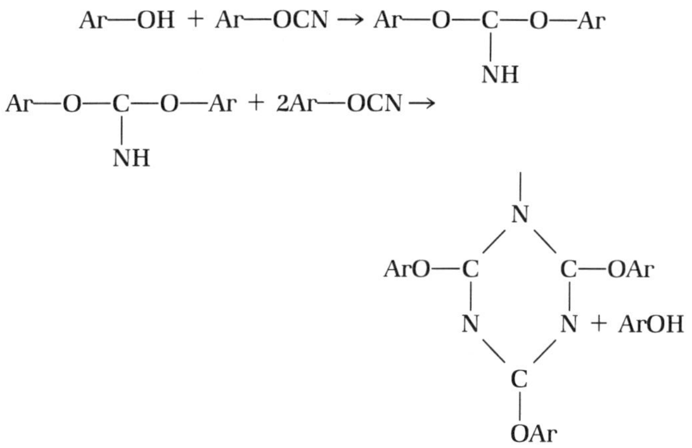

The investigators [2] explained the polymerization as a multi-step process.

1. Initiation by $ArOH/H_2O$ impurities catalyzes formation of a small quantity of triazine rings. The water vaporizes under the thermal conditions used and causes this reaction to become negligible.
2. ArOCN reacts with ArOH (with triazine as a catalyst?) to form carbonic ester imide structures.
3. Carbonic ester imide reacts with two other OCN groups to form a triazine and an ArOH.
4. The ArOH reacts with ArOCN as in step 2, thus propagating the cyclotrimerization reaction until all the ArOCN groups are reacted.

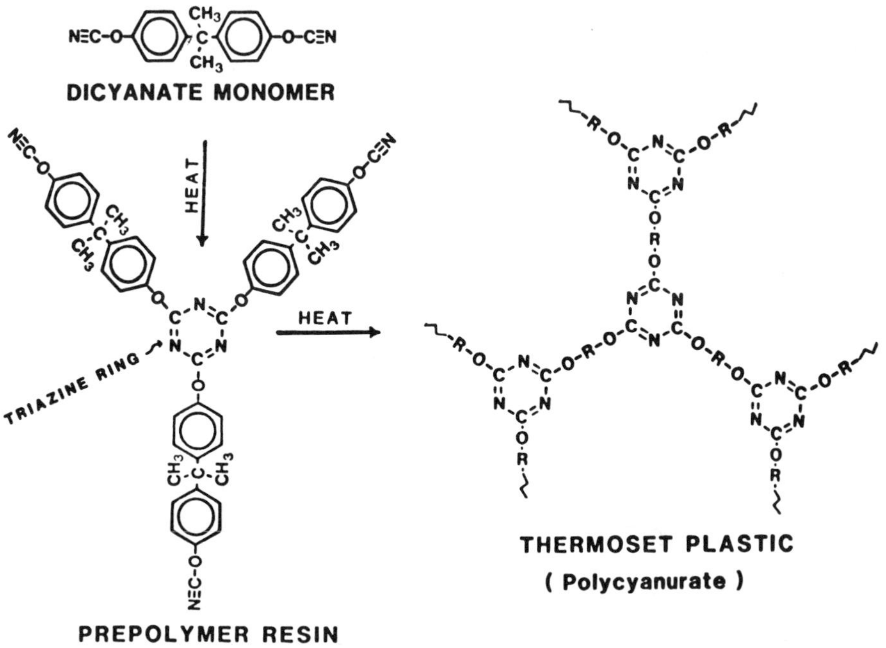

FIGURE 1 Curing via cyclotrimerization; R = bisphenol unit.

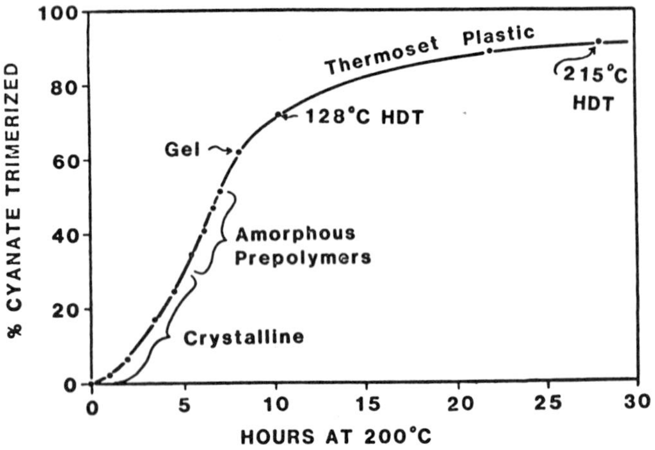

FIGURE 2 Isothermal polymerization of bisphenol A dicyanate (uncatalyzed).

Shimp [3] also noted the autocatalytic polymerization effect as long as the triazine polymer retained its mobility (before gelation). He noted that the apparent restricted mobility of molecules after gelation makes it difficult to react 100% of the OCN groups (Fig. 2). Note that gelation does not occur until about > 60% of cyanate groups are trimerized.

Shimp [3] also proposed a trimerization mechanism for transition metal carboxylates—active hydrogen-catalyzed systems. As illustrated in Figure 3, he suggests that the metal acts as a clustering agent with the active hydrogen source providing reactive carbonic imide functionality. The coordination metal catalyst is not very effective after gelation, when trimerization proceeds mainly by the step growth mechanism shown in Figure 4.

Commercial Products

Polymers and oligomers of bisphenol A dicyanate (BPADCy) were first offered by Bayer in Europe and Mobay Chemical in the United States in 1976 for use as laminating resins for high performance printed wiring boards. Blends of cyanate ester resins and bismaleimides were introduced in 1978 by Mitsubishi Gas Chemical (BT Resins). In 1986 Hi-Tek Polymers introduced dicyanate monomer and polymer variants using specialty bisphenols. Dow Chemical Company recently introduced a cycloaliphatic modified bisphenol dicyanate monomer.

Annual worldwide production of cyanate resins in 1986 has been estimated at about one million pounds. Prices in the late 1980s ranged from \$10 to \$25 per pound, reflecting developmental scale costs and high value-added properties.

Product Characteristics

Features

Cyanate ester monomers, resins, and ring-forming polycyanurates have these unique properties:

- Crystalline monomers are commercially available with > 99% assay [4] and ionic impurities < 10 ppm.
- Cyclotrimerization rates are essentially dependent on the catalyst used. A variety of transition metal carboxylates and chelates have been reported [3,5,6], which provide cure response ranging from 1 minute (reaction injection molding class) to shelf-stable prepregs curable at 177–250°C.
- Toughness/T_g properties, as indicated by impact strength G_{IC}, strain at break, and adhesive peel strength, are high for 250°C T_g resins [7]. This unusual combination of properties is attributed to the following structural features: ring spacing one atom apart, ether–oxygen linkages, low cross-link density, and apparent high free volume in cured state.
- After gelation, the volume increases as conversion increases above 65% (Fig. 5). This tends to eliminate stress-induced shrinkage at cure temperature.
- Dielectric constant (2.7–3.2) and dissipation factor

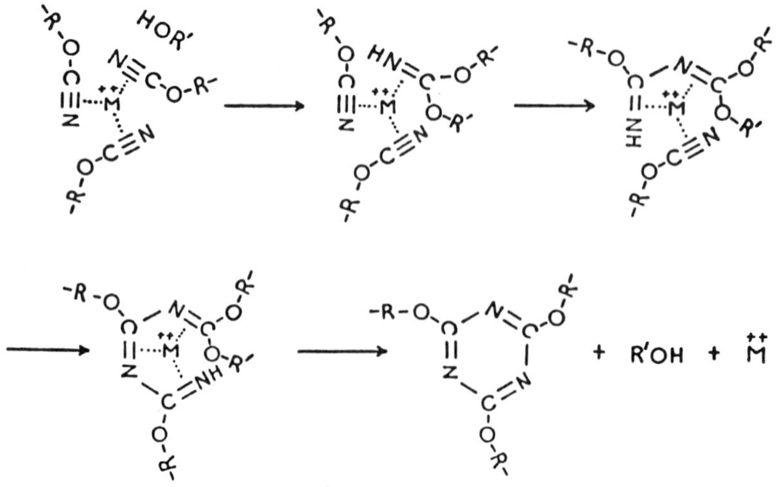

FIGURE 3 Trimerization catalyzed by transition metal carboxylates and active hydrogens.

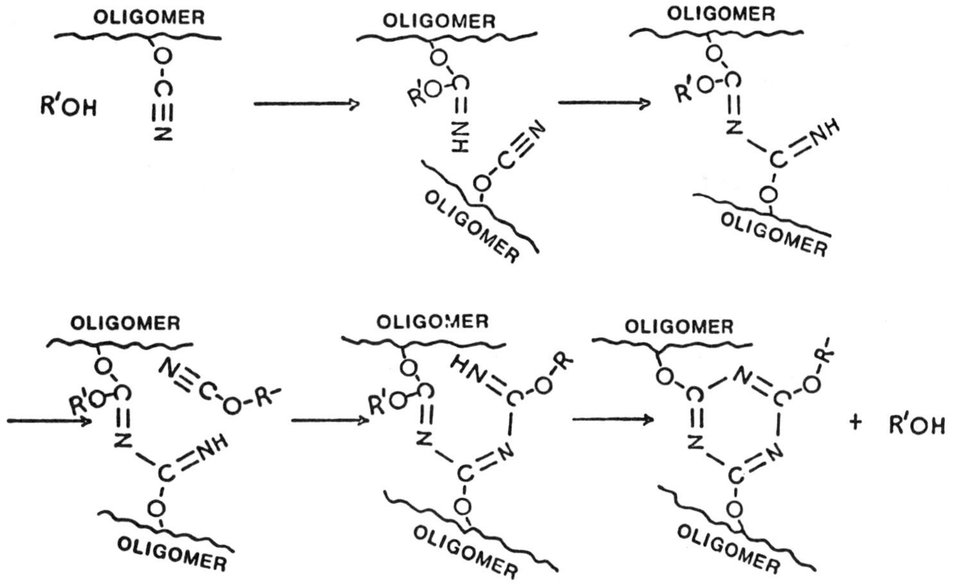

FIGURE 4 Trimerization by step growth mechanism.

($1–5 \times 10^{-3}$) are unusually low for high T_g resins [8,9]. High free volume and relatively weak dipoles may contribute to these low loss properties.

- Weight gain due to moisture pickup (1.3–2.4%) in boiling water is lower than epoxy and bismaleimide resins [10]. Long-term stability (> 500 hours) in 100°C water has been achieved with epoxy modification or orthomethylation of the cyanate monomer [11].
- Copolymers with epoxy resins provide hybrid formulations with lower cure temperatures, increased steam resistance, and lower cost [11,12]. The copolymers contain oxazoline rings (cyanate–epoxy resin copolymer), triazine rings (cyanate cyclotrimerization), and epoxy homopolymer (polyether).
- Several engineering thermoplastic resins have been used as modifiers for cyanate esters [13]. Low levels of thermoplastic resin modifiers can be solubilized in dicyanate; then efficient phase separation can be achieved during the cyclotrimerization step to provide a high degree of toughness [7].

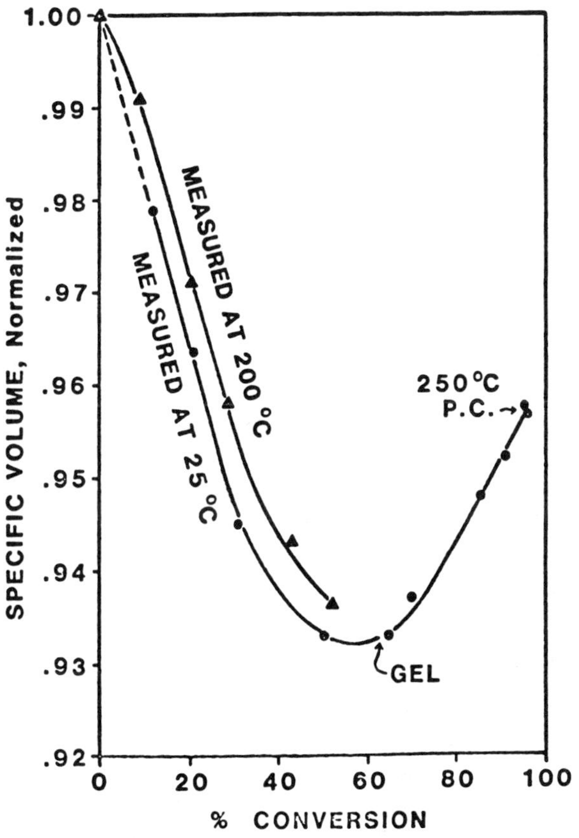

FIGURE 5 Volume changes during cyclotrimerization of BPADCy at 200°C.

Matrix Properties, Unreinforced

Typical cured-state properties associated with cyanate ester homopolymers, toughened bismaleimides, and epoxy resins cured with aromatic amines are listed in Table 2. Note that polycyanurates have the toughness and adhesion of the diglycidyl ether type of epoxies with thermal properties intermediate between brittle tetrafunctional epoxies and toughened bismaleimides (BMI). Polycyanurate advantages appear to include lower moisture absorption, less cure shrinkage, and lower dielectric loss properties. In addition, polycyanurates have superior adhesive strengths up to 250°C and solvent resistance intermediate between difunctional epoxies and tetrafunctional epoxies or BMI resins. This may be due to lower cross-link density of the polycyanurate network.

Structure–property relationships of four dicyanate resin homopolymers are given in Table 3. Modification of the bisphenol precursor by orthomethylation or cycloaliphatic modification, which appears to increase the hydrophobicity of the cured matrix, yields very low dielectric loss properties and improved moisture resistance.

TABLE 2
Matrix Properties Associated with Thermoset Resin Families

Property	Epoxy[a]	Cyanate	Toughened Bismaleimide
Residual chlorine, ppm	200–2000	<10	<10
Specific gravity	1.2–1.25	1.1–1.35	1.2–1.3
Tensile strength, MN/m^2	48–90 (7–13)	69–90 (10–13)	35–90 (5–13)
Tensile modulus GN/m^2	3.1–3.8	3.1–3.4	3.4–4.1
Tensile strain-at-Break, %	1.5–8	2–5	1.5–3
G_{IC},[b] J/m^2	70–210	105–210	70–105
Water absorption, %, saturated at 100°C	2–6	1.3–2.5	4.0–4.5
HDT, °C			
Dry	150–240	230–260	> 250
Water saturated at 100°C	100–150	150–200	200–250
Coefficient of thermal expansion, ppm/°C	60–70	60–70	60–65
TGA onset, °C	260–340	400–420	360–400
Flammability, UL-94			
Unmodified rating	Burns	V-O—Burns	V-O—Burns
% Br for V-O	18–20	0–12	0–12
Dielectric constant at 1 MH$_z$	3.8–4.5	2.7–3.2	3.4–3.7
Dissipation factor at 1 MH$_z$	$2–5 \times 10^{-2}$	$1–5 \times 10^{-3}$	$3–9 \times 10^{-3}$
Cure temperature, °C	150–220	177–250	220–300
Mold shrinkage,[c] mm/mm	0.0006	0.004	0.007

[a] Diepoxides and tetraepoxides cured with aromatic diamines.
[b] Double-torsion method.
[c] Transfer-molded parts filled 65 wt % with fused silica, cured at 177°C, and measured at 25°C.

Solvent resistance and toughness appear to be improved for homopolymers derived from dicyanates with single-atom linkages connecting the bisphenol unit. Self-extinguishing homopolymers can be obtained by replacing the aliphatic linkages with a sulfur group (THIOCy).

Modifying BPADCy resins with epoxy resins can lead to the properties shown in Table 4. The data indicate that cure temperatures to obtain greater than 95% conversion can be lowered 20–50°C. The modified systems have lower ultimate T_g values but superior moisture resistance, with no significant loss of physical strength properties. Self-extinguishing flammability ratings are achieved at 10–16% bromine content, depending on level of epoxy modification.

Epoxy resin modified cyanate ester resins not only absorb less water than polycyanurate homopolymers and tetrafunctional epoxies, but also retain a higher percentage of heat deflection temperature (HDT) and reach moisture content equilibrium sooner. Several materials are compared in Figure 6. The best performance for long-term boiling water resistance appears to be formulations with 1.0–1.8 epoxide equivalents per monomer cyanate group. Optimum cure for epoxy-modified cyanates is offered by copper acetylacetonate with 2 phr nonylphenol catalyst systems.

Resin Forms

Commercial quantities of cyanate ester monomers and resins are available in several forms, including crystalline monomers, amorphous semisolid prepolymers or oligomers, hard prepolymers in powder or lump form, solvent solutions, and blends. Toxicity screening test for all forms indicate low health hazards for ingestion or skin contact with bisphenol-type dicyanates. Table 5 gives typical toxicity screen data. In solution products, the principal hazards are associated with the solvents. This low toxicity is in contrast to that of highly irritating and sensitizing properties of isocyanates, acutely toxic cyanides, and nitriles. Material Safety Data Sheets should be consulted for specific handling recommendations.

Monomers are available that melt to low viscosity liquids (< 50 cP) at 79–106°C. Monomers are useful for formulating high solids systems and to control viscosity of impregnant systems for transfer molding or reaction injection molding (RIM) processing. Formulations con-

TABLE 3
Cyanate Homopolymer Matrix Properties

Representative Resins	BPADCy[a]	METHYLCy[a]	THIOCy[a]	Cycloaliphatic, Modified[b]
Monomer	RDX 64826	RDX 64833	RDX 76678	
Oligomer				XU 71787.00
Prepolymers	RDX 80352	RDX 80359	RDX 80371	
	RDX 76680	RDX 76681	RDX 76682	
	RDX 80351			
Cured state properties[c]				
Tensile strength, MN/m^2	83	73	79	77
Tensile modulus, GN/m^2	3.2	3.1	3.2	3.2
Tensile strain-at-break, %	3.5	2.5	3.6	2.3
G_{IC}, J/m^2	140	175	158	61
T_g (DMA), °C	289	252	273	250
CTE, ppm/°C	64	71	68	68
TGA onset, °C	411	403	400	405
Dielectric constant at 1 MHz	2.91	2.75	3.11	2.83
Dissipation factor at 1 MHz	5×10^{-3}	3×10^{-3}	3×10^{-3}	2×10^{-3}
Water absorption, % saturated at 100°C	2.4	1.3	2.2	1.4
$MeCl_2$ absorption,[d] %	6.6	9.4	0.6	21.8
Volume expansion,[e] %	2.5	4.1	1.6	Not available
Flammability, UL-94	Burns	Marginal V-1	V-0	Burns

[a] Interez, Inc.
[b] Dow Chemical Company.
[c] Catalyzed with zinc naphthenate and 2 phr nonylphenol or comparable accelerator; 0.125 in. thick castings postcured 2 hours at 250°C.
[d] Immersed in methylene chloride 3 hours at 25°C.
[e] Volume increase from gelation to more than 95% conversion.

TABLE 4
Matrix Properties of BPADCy Cyanate Resin Blends

	Blend			
Modification/Cure	25% Br Epoxy	57% Epoxy	57% Epoxy	10% BMI BT 2160
Epoxy resin	Epi Rez 5163	Epi Rez 509	Epi Rez 509	
Type	Brominated	Liquid	Liquid	
Catalyst, phr				
Copper naphthenate	0.24			
Copper acetylacetonate		0.05	0.05	
Zinc octoate				0.02
Nonylphenol	2	2	2	
Epi Rez 509				5
Maximum cure temperature, °C	215	177	200	232
Cured State Properties				
HDT, °C				
Dry	221	192	196	248
Wet[a]	181	178	173	170
Water Absorption,[a] %	1.65	1.31	1.31	2.02
Flexure strength, MN/m^2	173	117	114	131
Flexure modulus, GN/m^2	3.3	3.1	3.1	3.3
Flexural strain, %	6.6	4.4	4.3	4.4
Flammability, UL-94	V-0	Burns	Burns	Burns

[a] Moisture conditioned for 64 hours at 92°C and relative humidity exceeding 95%.

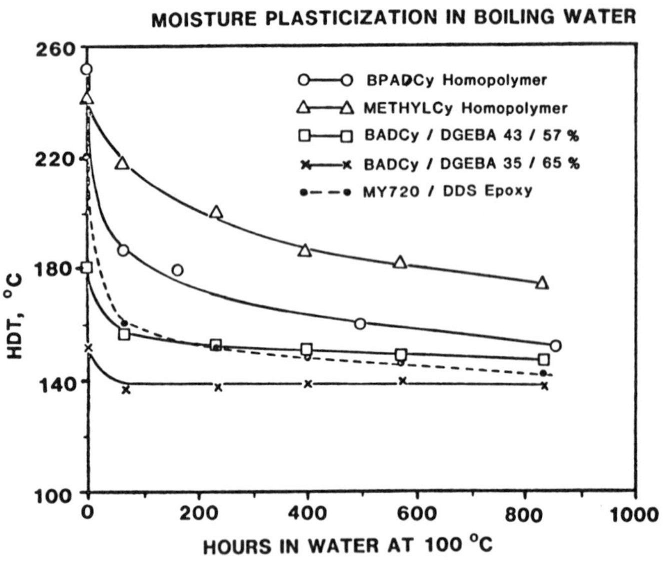

FIGURE 6 Boiling water resistance of polycyanurates and epoxy modifications.

taining more than 35% monomer should be avoided for prepreg usage because crystallization can occur when the material is stored below the monomer melt temperature.

Prepolymers are partially polymerized (cyclotrimerized) amorphous resins consisting of several oligomers and unreacted monomers. Conversion to trimerized oligomers can be stopped at the tacky semisolid stage (25–40% conversion; 1000–2000 MW range) or at the hard, grindable stage (40–55% conversion; 4000–9000 MW range). Semisolid resins are preferred for structural composites and film adhesives where tack and drape properties are required. Hard resins are preferred for compression molding and solution impregnation of woven fabrics such as prepregs for printed circuit boards. The melt viscosities of representative cyanate resins are shown in Figure 7.

Cyanate monomers and prepolymers are generally quite soluble in ketone solvent such as methyl ethyl ketone and acetone. However, monomer fractions in prepolymer solution can crystallize out of concentrated solutions (> 70% NV). Sulfur-linked (THIOCy) cyanate resins and blends of cyanates and bismaleimides usually

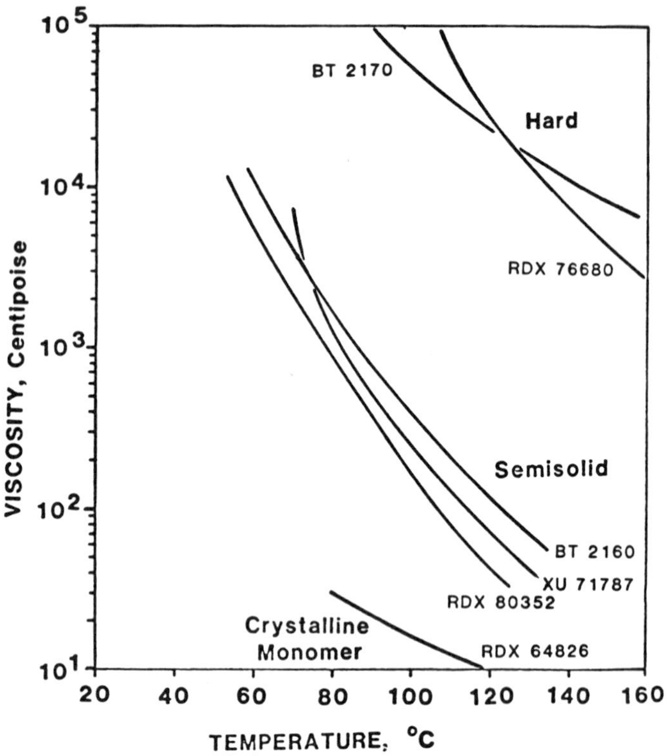

FIGURE 7 Melt viscosity of representative cyanate resins.

require stronger solvents such as dimethylforamide or cyclic ketones.

Processing

Optimum mechanical properties of cyanate resins are achieved with better than 85% conversion of cyanate groups to triazine structure [7]. Optimum chemical and moisture resistance properties are achieved at better than 95% conversion [3,6,7,10,11]. The references cited describe classes of catalysts that can be used to achieve efficient conversion of cyanate monomers and polymers:

- *Active hydrogen source:* this can be an alkylphenol, a bisphenol, an alcohol, an imidazole, an aromatic amine, or a hydroxyl-containing resin; many epoxy

TABLE 5
BPADCy Monomer Toxicity Data

Test	Result
Acute Toxicity	
LD_{50} rat, oral	> 2500 mg/kg
LD_{50} rat, subcutaneous	> 2500 mg/kg
Skin primary irritation	
Abraded skin, rabbit	Nonirritating
Intact skin, rabbit	Nonirritating
Skin sensitization: closed-patch repeated insult, guinea pig	Nonsensitizing
Mutagenicity: Ames screen	Negative

and phenolic resins can be used as the hydroxyl source.

- *Soluble metal carboxylate compounds:* these can be compounds such as copper naphthenate, zinc octoate, or cobalt acetylacetonate.

Most metal carboxylates have poor solubility in molten cyanate resins, but this problem can be overcome by dissolving the metal compound in a liquid component. Nonylphenol is an excellent solvent for most of the preferred metal catalysts.

Prepolymer solutions for laminating applications require very low levels of metal carboxylates. Typically the preferred level is 5–50 ppm of zinc, manganese, copper, or cobalt as octoate, naphthenate, or acetylacetonate. The level should be adjusted to provide a varnish gel time of 50–300 seconds at 177°C. If there is no epoxy resin in the varnish, 1–2 phr of bisphenol or nonylphenol should be added as a hydroxyl source. Prepreg drying tower conditions should be adjusted to volatilize the solvents with minimal advancement of the resin to higher molecular weight. Some slowly evaporating solvents such as dimethyl formamide (DMF) may be required to achieve desired level of fiber wetting and resin flow. Properly prepared boardy prepregs will have a storage life of several months at room temperatures. Final curing for laminates is usually done in a press at 177°C at a pressure that will provide full wetout of fabric and desired fiber volume.

Unidirectional compliant prepregs for advanced structural composite applications require metal catalyst levels of 20–200 ppm. The catalyst level should typically be adjusted to provide filming at temperatures that give a matrix viscosity of 0.5–50 Pa and gel times of 2–10 minutes at 177°C. These prepregs normally will require storage at 0°C or lower if not converted to composites within a few weeks. Most structural composites are molded and cured in an autoclave. The following conditions are typically used for cyanate resin based prepregs:

- 345–690 kPa clave pressure.
- Vent vacuum bag at 172 kPa.
- Heat to 177°C at 1–2°C/min.
- Hold 2–3 hours at 177°C then cool under full pressure to about 60°C.
- Demold and postcure at temperatures up to 250°C as required.

Fast-curing formulations (1–15 minutes) designed for RIM, RTM, or encapsulation of semiconductor devices may require higher catalyst levels. Typical catalysts might be 300 ppm Zn or 500 ppm Cu. This level of cata-

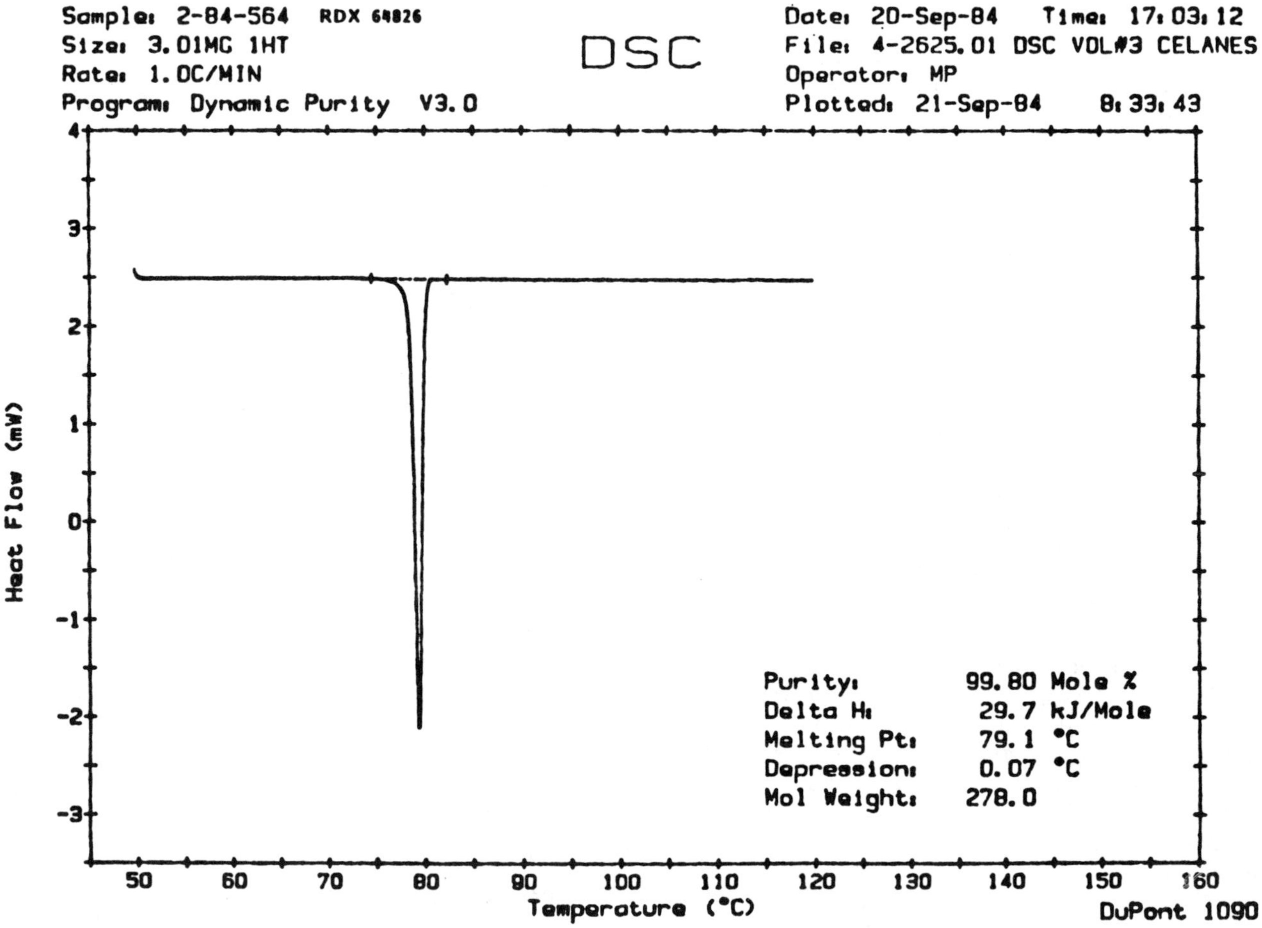

FIGURE 8 Monomer melting data.

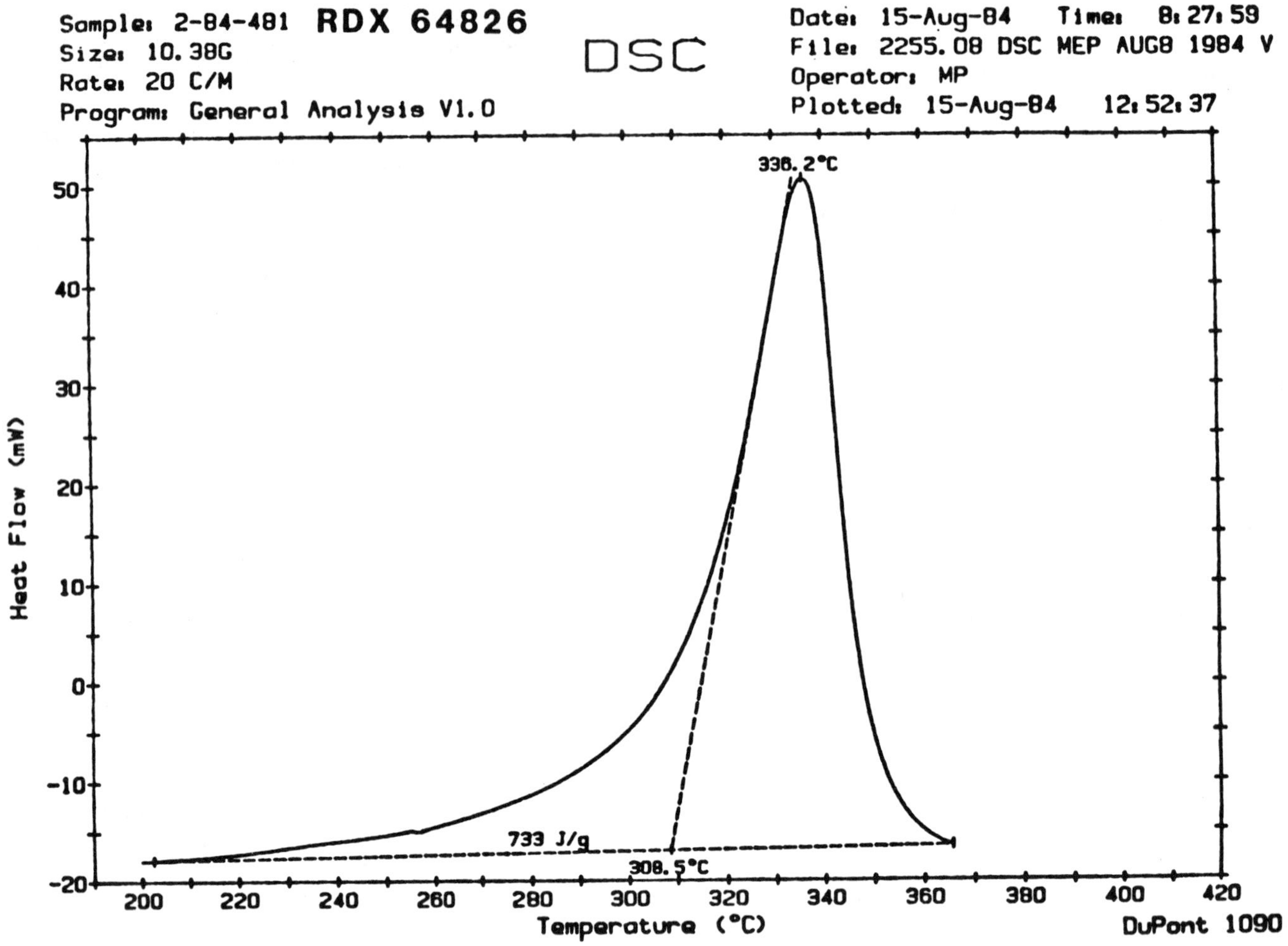

FIGURE 9 Heat of reaction.

lyst can be achieved by packaging with an epoxy component (RIM and RTM—resin transfer molding) or by extruder mixing at about 80°C (molding powders).

Analytical Techniques

Differential scanning calorimetry (DSC) can be used to determine monomer purity as illustrated in Figure 8. This technique is based on melting point depression and heat of melting. DSC can also be used to determine heat of polymerization as shown in Figure 9. The polymerization of cyanate esters is exothermic with about 102.6 ± 1/g-eq being liberated, about the same as for epoxy resins. DSC has also been used to study kinetics of the cyclotrimerization polymerization reaction [2].

Prepolymer molecular fraction analysis can be done using gel permeation chromatography (GPC) [4]. A typical chromatogram is shown in Figure 10. Note that separation of unreacted monomer ($n = 1$) from oligomers ($n = 3$ and $n = 5$) and higher molecular weight materials is fairly well defined.

The cyclotrimerization reaction can be followed with infrared (Fig. 11) and Fourier transform–infrared (FTIR) spectroscopy. The strong absorption bands for OCN at 2270 cm^{-1} and triazine at 1565 cm^{-1} can be used to study kinetics of the polycyclotrimerization reaction [11]. The results correlate well with the kinetics data obtained using DSC [2]. Copolymerization of cyanate esters with epoxy resins can also be followed using IR and FTIR [11] as shown in Figure 12.

Thermal stability of cured polycyanurates can be determined using thermogravimetric analysis (TGA). Typical TGA curves for homopolymers, bismaleimides, and

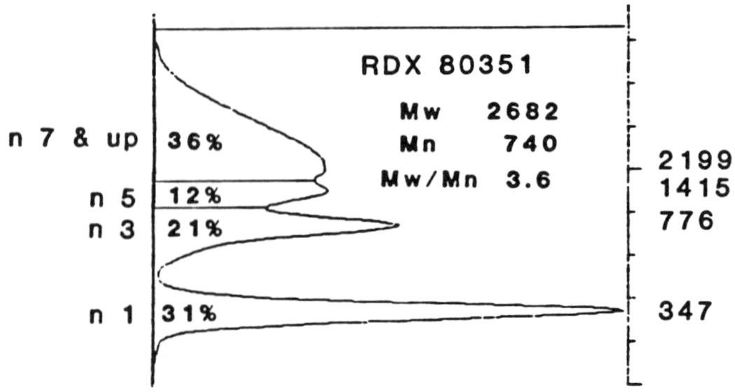

FIGURE 10 Prepolymer molecular fraction analysis by gel permeation chromatography.

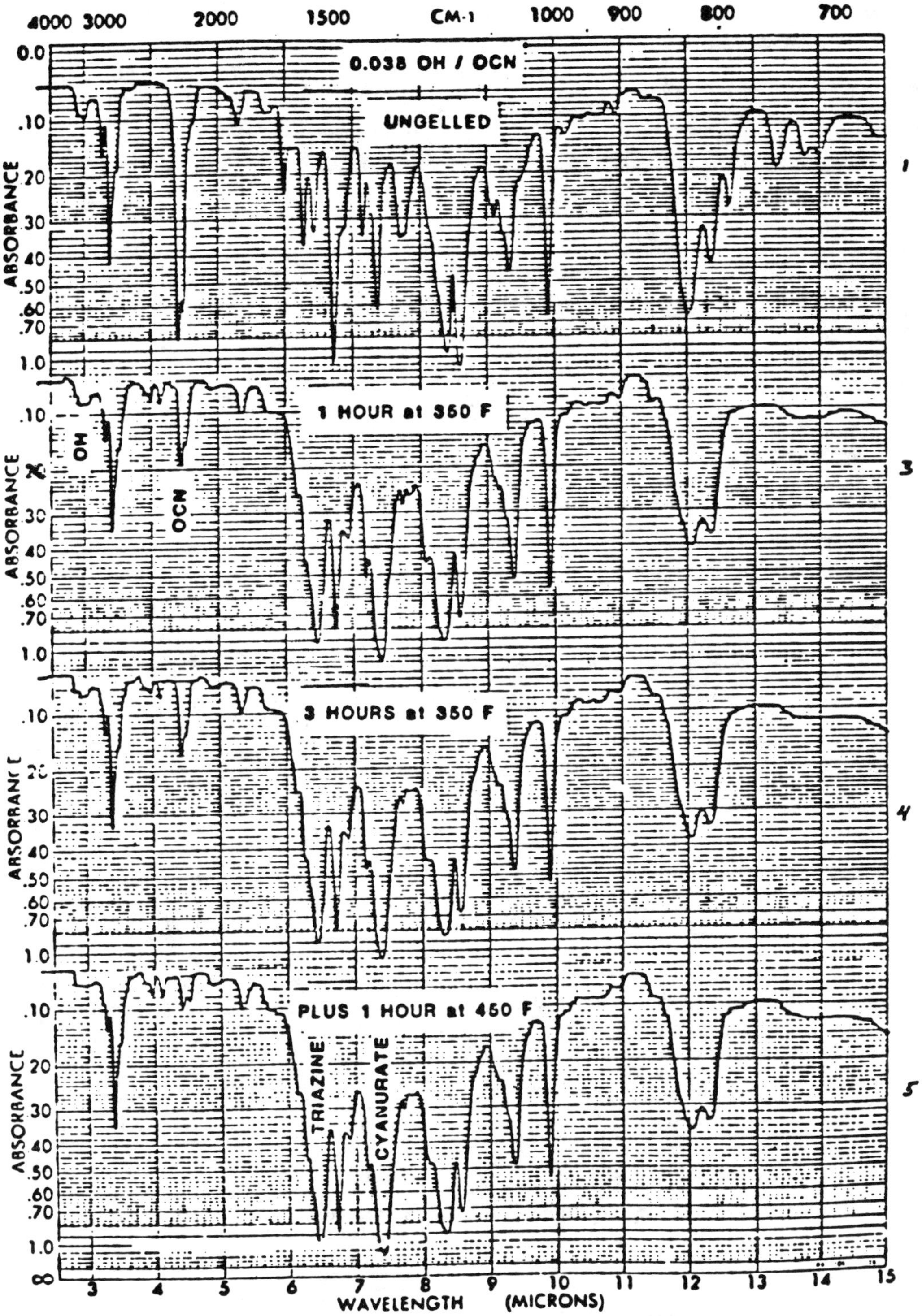

FIGURE 11 Tracking cyanate trimerization in BPADCy homopolymer by infrared spectroscopy.

tetrafunctional epoxies are compared in Figure 13 [10]. A cyanate ester brominated epoxy copolymer TGA curve is compared with a flame-retardant cyanate ester homopolymer (THIOCy) in Figure 14 [3].

Differential mechanical analysis (DMA) can be used to study mechanical properties such as loss modulus (E''), tan δ, and T_g. An interesting application of DMA is to determine degree of phase formation in cyanate ester/thermoplastic resin blends and alloys. This is illustrated in Figure 15 [7], where the addition of curing catalyst enhances the formation of a two-phase matrix at lower gel temperatures.

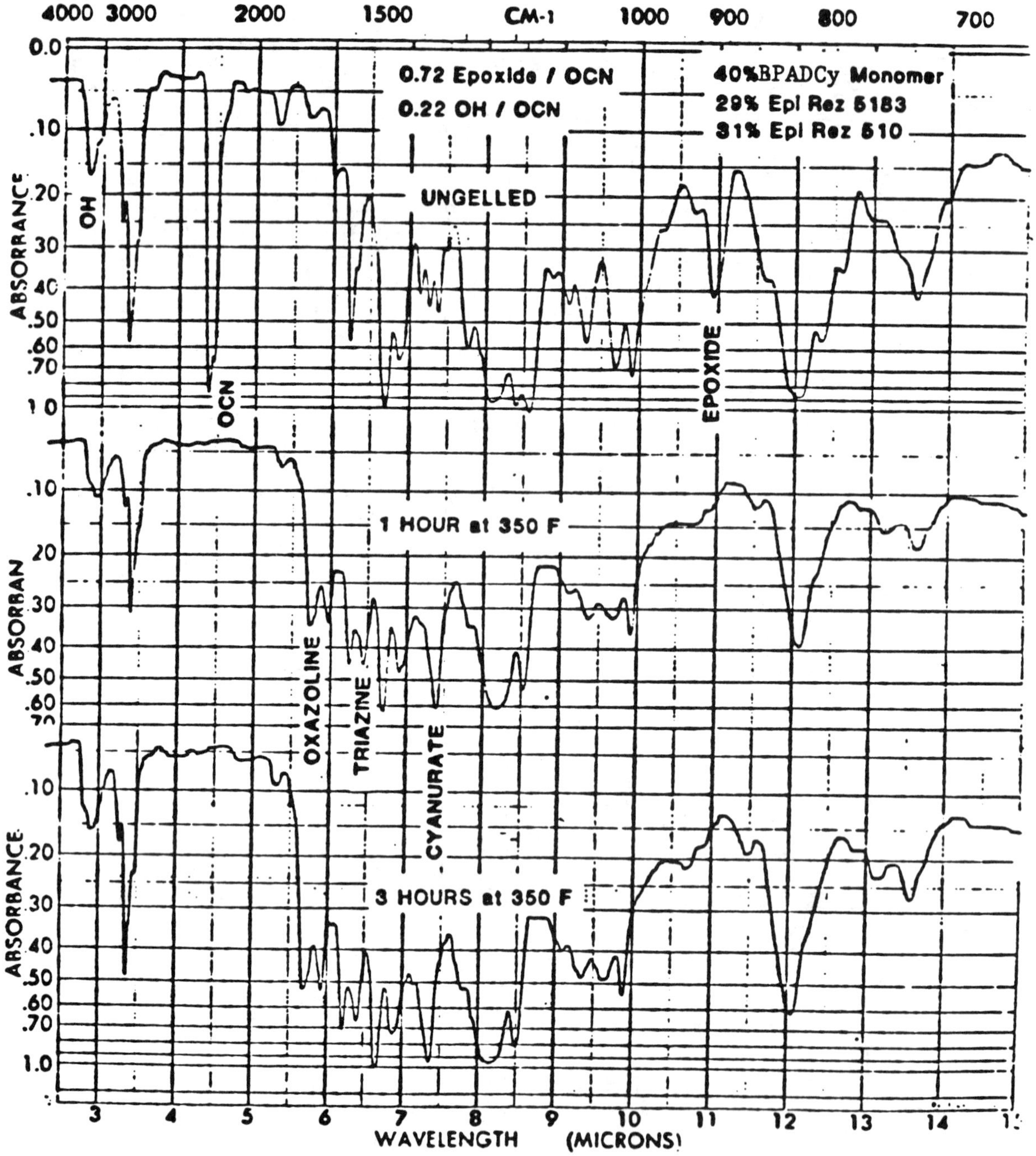

FIGURE 12 Tracking cyanate–epoxy curing reactions via infrared spectroscopy.

Applications

Printed Wiring Boards

The first commercial end use for cyanate resins, and currently the largest, is for printed circuit boards [8,9]. Cyanate resins are preferred over epoxy resins for use in high reliability multilayer boards for applications such as computer circuitry because of the following performance advantages:

- *Dimensional stability.* Multilayer circuit patterns interconnected between layers are prone to failure by differential movement caused by dissimilar coefficients of thermal expansion (CTE) and swelling by chemicals such as solvents, etchants, and strippers. Resins with high T_g, approaching molten solder temperature (250°C), have less thermally induced stress because CTE values below the T_g are lower by a factor of 2–5 times the CTE above the T_g.
- *Corrosion resistance.* Ionizable resin impurities, such as epoxy chlorohydrin residues, can attack metal conductors in hot–moist environments. Cyanate ester resins, which are largely free from ionizable impurities, are reported [14] not to be suscepti-

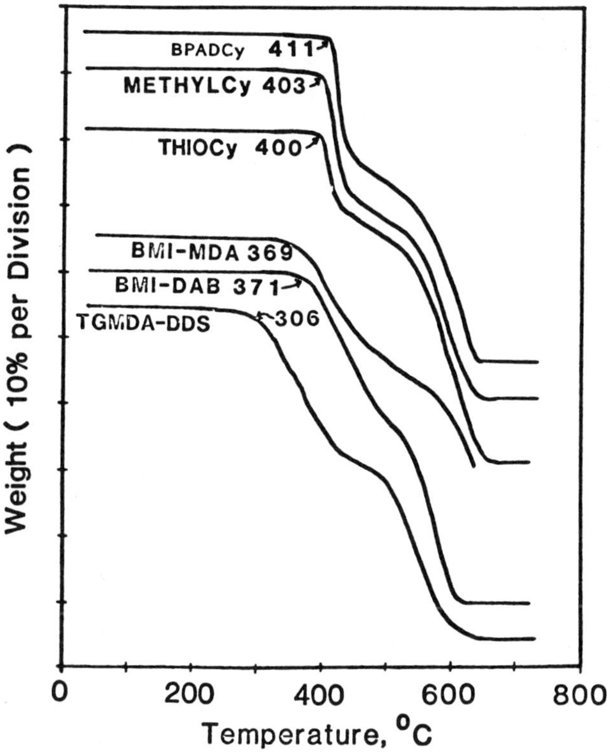

FIGURE 13 Thermogravimetric analysis in air.

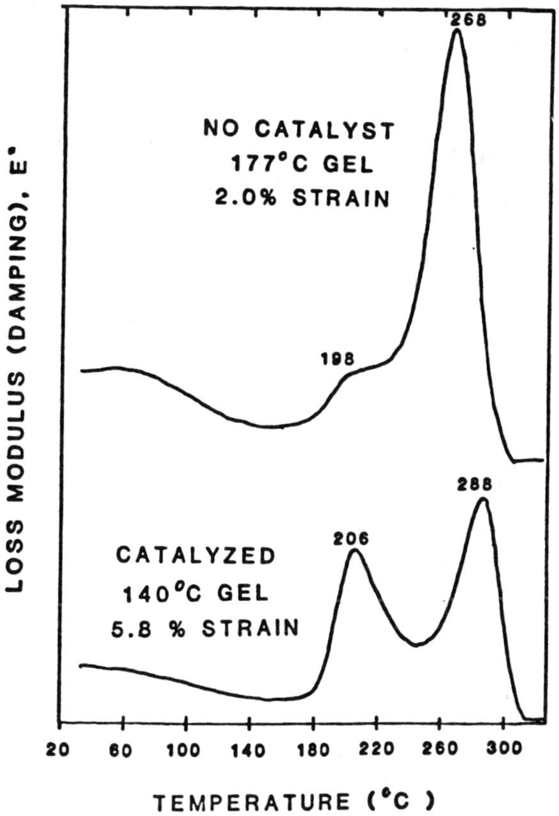

FIGURE 15 Thermosetting–thermoplastic phase development in 80/20 BPADCy-polyarylate alloys.

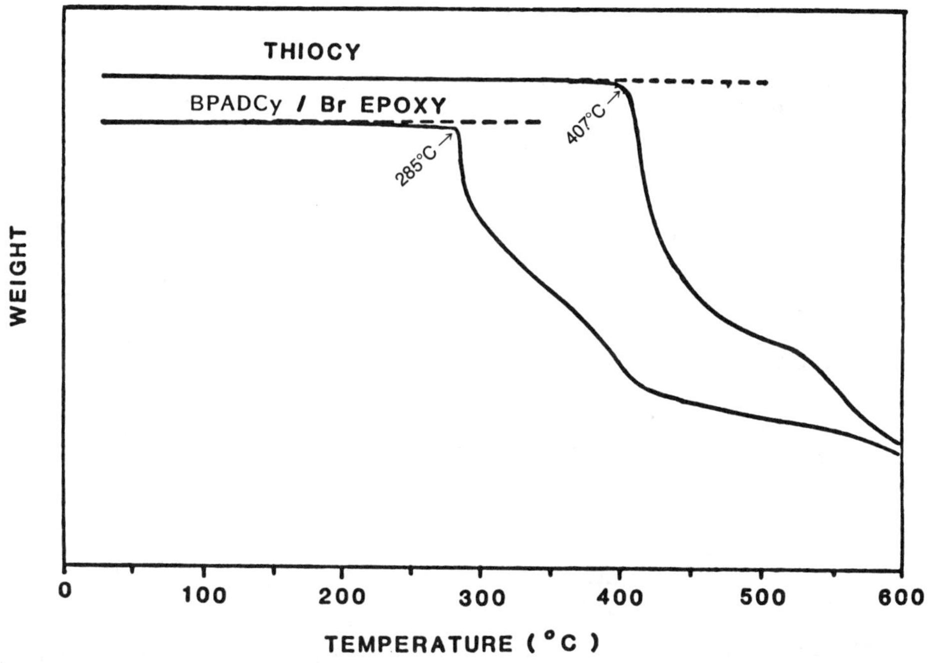

FIGURE 14 Thermogravimetric analysis of flame-retardant cyanate ester formulations.

TABLE 6
Properties of Fiberglass Printed Wiring Board Laminates

Properties	Unmodified				Modified	
	BPADCy	METHYLCy	THIOCy	XU 71787	BT 2170	BPADCy
Composition						
Modification					Epoxy	Epoxy
Modifier, wt %					45	60
Bromine, %	0	0	0	0	12	16
Glass reinforcement	7628[b]	7628[b]	7628[b]	NA	NA	7628
Resin content, wt %	40 ± 1	40 ± 1	40 ± 1	NA	NA	40 ± 1
Cure Temperatures						
Platen, °C	177	177	177	177	175	177
Postcure, °C	232	232	232	225	NA	None
Laminate Property						
T_g, °C	255	247	250	255	210	182
TGA onset, °C	405	427	400	425	305	300
Flexure strength, MN/mm²	448	414	443	NA	483	483
Peel strength, N/cm						
At 25°C	21.0	20.6	19.3	15.8	16.7	17.2
At 100°C	19.1	18.6	18.7	12.3	13.7	16.6
At 200°C	15.8	16.7	17.9	10.5	11.8	14.9
Steam/solder float, minutes[a]	90	90	90	NA	NA	120
Dielectric constant at 1 MHz	4.05	3.8	4.3	4.0	4.2	4.2
Dissipation factor at 1 MHz	0.004	0.003	0.003	0.003	0.008	0.011
Flammability, UL-94	Burns	V-1	V-0/1	Burns	V-0	V-0

[a] Minutes conditioning in 121°C steam passing 20 second float in 260°C solder.
[b] 7628 style glass cloth with Z-6040 finish.

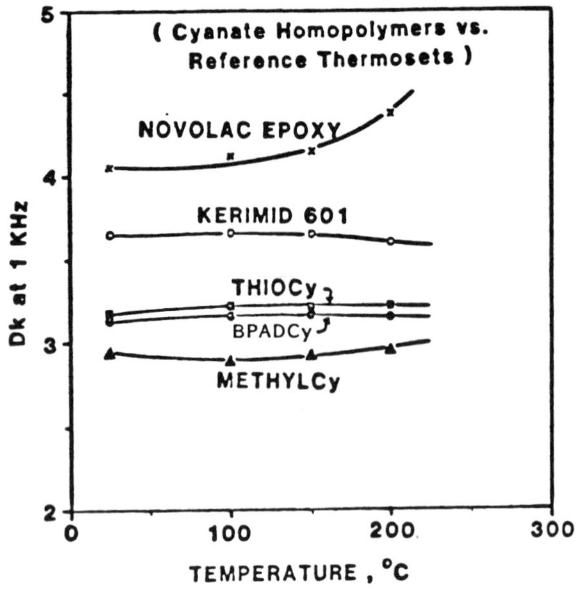

FIGURE 16 Short beam shear strength of carbon fiber composites.

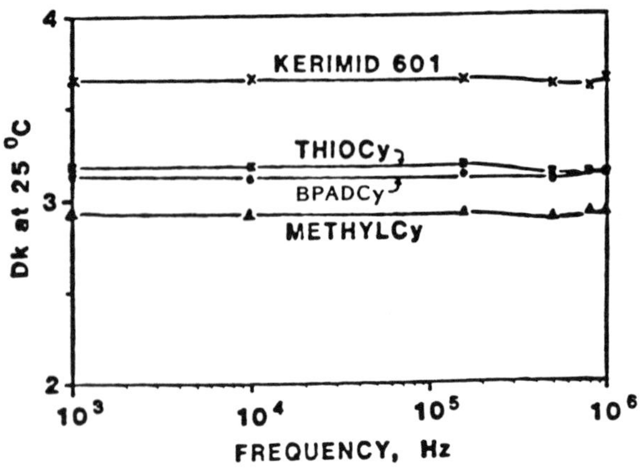

FIGURE 17 Fracture toughness response to thermoplastic resin concentrations in BPADCy alloys: PAr = polyarylate; PES = polyethersulfone; PEI = polyetherimide; PS = polysulfone; CPE = copolyester; TGMDA/DDS = tetrafunctional epoxy reference compound.

ble to filamentary growth of copper at anodic sites.

- *Low dielectric loss factors.* Low dielectric constant and dissipation factor is reported [15] to increase signal speed and signal strength, particularly in high frequency, elevated-temperature service. See Table 6 for data on selected fiberglass laminates used for printed wiring boards.
- *Field repair.* Manual soldering of mainframe computer boards often raises temperatures of laminates to 300–350°C for short periods. Decomposition of the resin matrix under these conditions can cause blistering and loss of adhesion. The high T_g and thermal stability of cured cyanate resins minimizes this type of failure during repair.

Moisture resistance of cured cyanate resins has been improved by development of high purity monomer (> 99%), development of superior cure catalysts [6], and development of more hydrophobic cyanate esters [9]. Modification of cyanate esters with bismaleimide resin is also reported to improve moisture resistance and thermal stability [16].

Structural Composites

Most of the data on use of cyanate resins in carbon fiber composites through 1987 has been published in research reports sponsored by the U.S. government [17,18]. With the introduction of high purity and specialty monomers and prepolymers [3,4,7], research and development activities have been increased to take advantage of the unique properties of these materials, which include the following.

- Prepolymers have hot-melt processibility combined with superior tack and drape retention properties. See Figure 7 for melt viscosity data of representative materials.
- Latent metal catalysts are reported that provide 3 weeks of prepreg stability at 25°C combined with ability to cure by vacuum bag/autoclave at 177°C followed by postcure at 210–250°C [6,7,10].
- Short beam shear strength of BPADCy-based composites appears to have an upper limit for hot–wet service of about 160°C. This property is plotted in Figure 16 as a function of temperature, conditioning, fiber form, and postcure.
- Epoxy resin modifications are reported that develop full cure at 177°C and have hot–wet properties superior to tetrafunctional epoxy resins cured with diaminodiphenylsulfone (DDS) [11] (Table 5).
- Dicyanate homopolymers when fully cured have 240–270°C T_g and higher fracture toughness and

TABLE 7
Dielectric Properties of Cyanate Esters

Dicyanate Homopolymer	Catalyst[a]	Metal (ppm)	Dielectric Constant at 1 MHz[b]	Dissipation Factor at 1 MHz	HDT (°C)
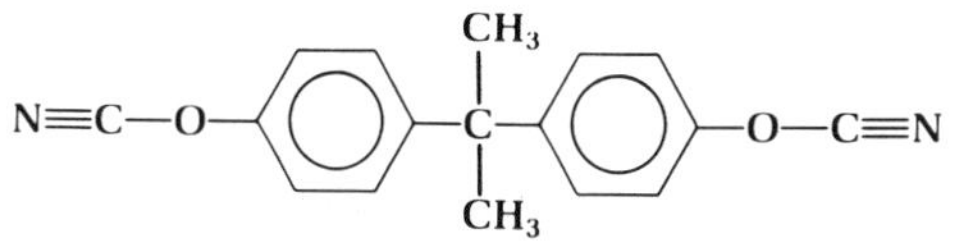 BPADCy	Zn naphthenate	100	2.91	5×10^{-3}	225
	Co naphthenate	90	2.91	5×10^{-3}	
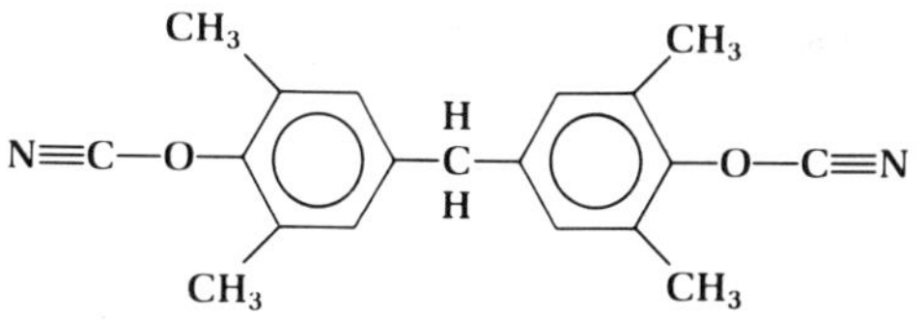 METHYLCy	Zn naphthenate	150	2.75	3×10^{-3}	243
	Cu naphthenate	400	2.75	3×10^{-3}	
	Co naphthenate	150	2.78	3×10^{-3}	
	Mn octoate	400	2.74	3×10^{-3}	
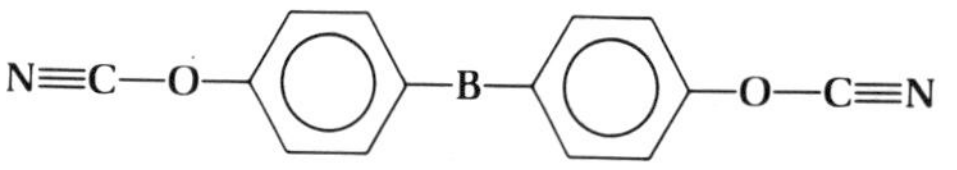 THIOCy	Zn naphthenate	60	3.11	3×10^{-3}	251
	Co naphthenate	120	3.12	3×10^{-3}	
References					
Kerimid 601			3.5	7×10^{-3}	> 270
Novolac epoxy			3.9	5×10^{-3}	180

[a] Nonylphenol (2 phr) used as cocatalyst; maximum temperature, 250°C.
[b] Electrical measurements made on a Genrad 1687B Digibridge equipped with LD-3 cell, two-fluid method.

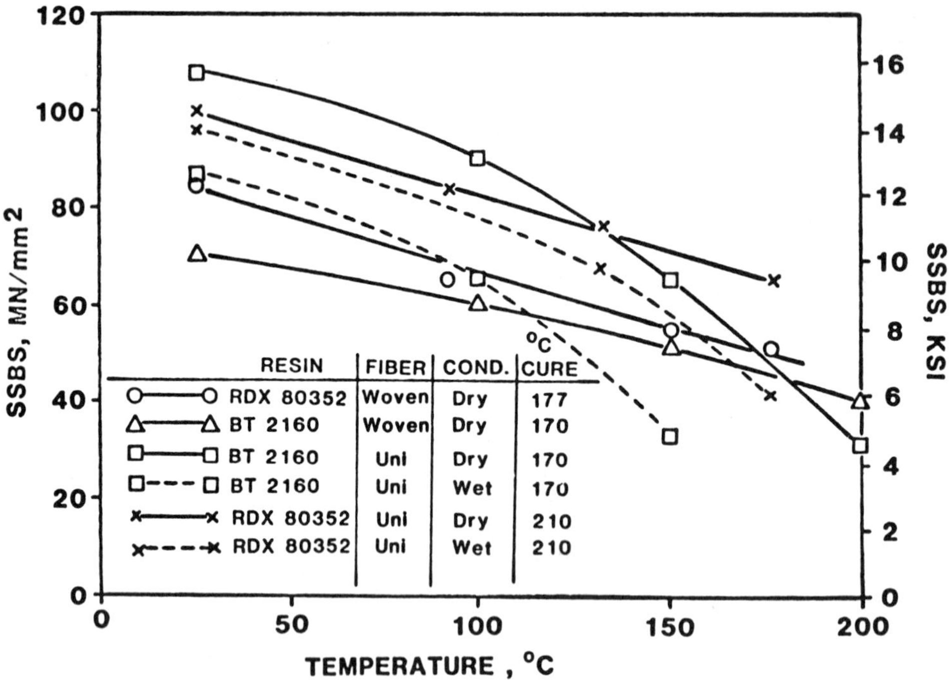

FIGURE 18 Temperature effect on dielectric constant. *Note:* D_k values in the 10^3–10^5 Hz frequency range were determined via ASTM D 150 using a General Radio 1615-A capacitance bridge. Above 10^5 Hz, a Q-Meter was used. These data provide valid comparisons; however, individual values range from 0.2 to 0.3 unit higher than measurements made with a Genrad 1687B Digibridge equipped with LD-3 cell (two-fluid method).

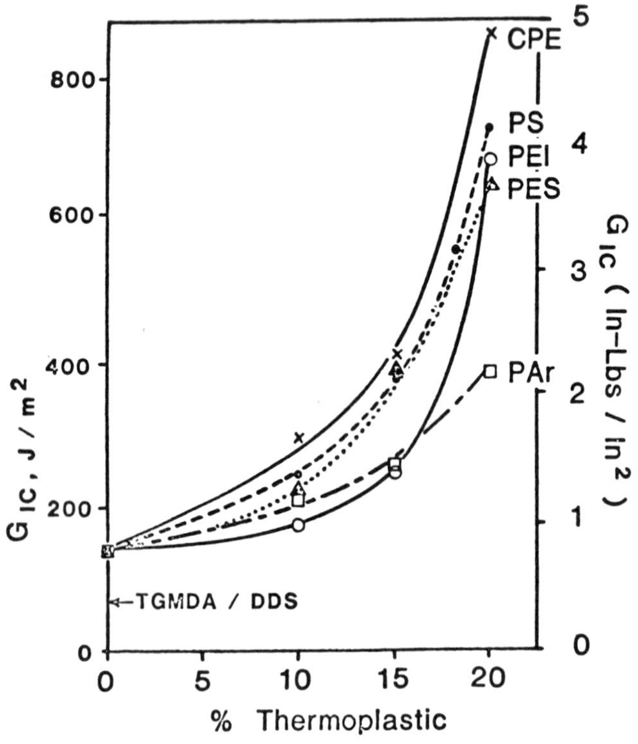

FIGURE 19 Frequency effect on dielectric constant. See note to Figure 18.

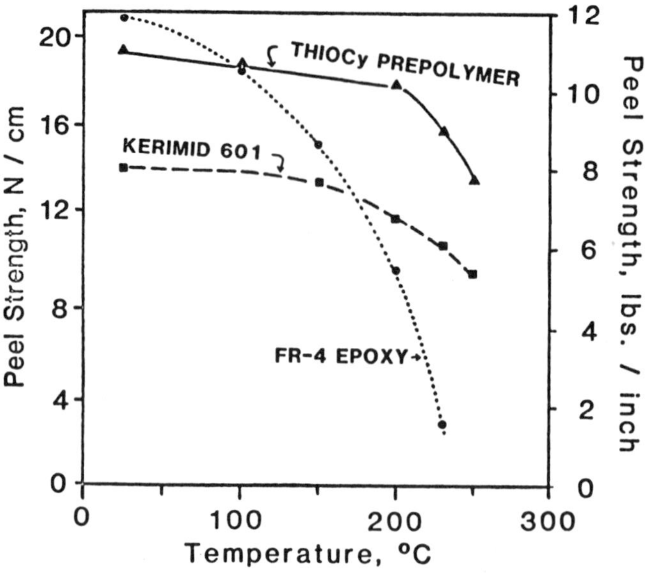

FIGURE 20 20 Peel strength (90°C) of copper-clad laminates.

strain-at-break than either epoxies or bismaleimides [10] (Table 2).

- High T_g thermoplastics may be used to modify cyanate esters systems to improve fracture toughness. It has been reported that 15–20% thermoplastic modification can increase fraction toughness by 5 times via an efficient phase separation during curing [10]. Figure 17 compares the effects of several thermoplastic resins on fracture toughness. Many thermoplastic resins are readily soluble in dicyanate monomers, making this modification relatively easy to achieve.

Future Trends

The low dielectric constants and low dissipation factors that can be achieved with polycyanurate matrices have accelerated commercial interest. The major applications are for obtaining faster signal speed and strength in computer circuitry and for microwave transparency in radar and communication housings. The use of low D_k reinforcements such as Kevlar, quartz, and high modulus polyethylene for cyanate resin composites is expected to grow rapidly. Table 7 lists capacitance properties of cyanate homopolymers and competitive polymers. Temperature and frequency effects on cyanate polymers and other polymers are compared in Figures 18 and 19.

Polycyanurates have excellent adhesion to copper-clad laminates as shown in Figure 20. However this property has not been fully exploited for other high value added commercial adhesive applications. Shear strengths of 27–48 MN/m^2 (4–7 ksi) have been reported [19] for aluminum, titanium, and ferrous metals up to 235°C. Potential adhesive applications of interest include structural adhesives for supersonic aircraft and die attach adhesives.

Ayano [16] discusses non-fiber-reinforced applications for cyanate ester resins, including class H motor windings and low friction bearings.

Suppliers

Hi-Tek Polymers, Inc., under license from Bayer AG, supplies three dicyanate monomers under the trademark AroCy: BPADCy based on bisphenol A, METHYLCy based on orthomethylated bisphenol F, and THIOCy based on sulfur-linked bisphenol together with semisolid and hard resin prepolymers in neat form and as solvent solutions.

Mitsubishi Gas Chemical, under license from Bayer AG, supplies BT resins, a series of bisphenol A dicyanate monomers and prepolymers modified with a bismaleimide resin. These resins are available as solutions and as neat resins.

Dow Chemical Company supplies XU 71787, a cycloaliphatic modified polycyanate resin, both as a semisolid and as a solution.

Suppliers of electrical-grade laminates and circuit boards using cyanate resins include Fortin Industries, Mica Corporation, Nelco Products, Inc., and Norplex/Oak.

Richard B. Graver

References

1. E. Grigat and R. Putter, *Angew. Chem. Int. Ed.*, *6*, 206 (1967).
2. M. Bauer, J. Bauer, and G. Kuhn, *Acta Polym. 37* (11/12), 715 (1986).
3. D. A. Shimp, *ACS PMSE Prepr. 54*, 107 (April 1986).
4. R. B. Graver, *ACS Polym. Prepr.*, *27*, 1, 491 (April 1986).
5. V. V. Korshak, S. V. Vinogradova, V. A. Pankratov, and A. G. Puchin, *VINITI*, 6457–6473 (1973).
6. D. A. Shimp, "Metal Carboxylate/Alkylphenol Curing Catalyst for Polycyanate Esters of Polyhydric Phenols," U.S. Patent 4,604,452 (Aug. 5, 1986), to Hi-Tek Polymers, Inc.
7. D. A. Shimp, F. A. Hudock, and W. S. Bobo, *Proceedings of the 18th Natl. SAMPE Technical Conference*, 1986, pp. 851–862.
8. K. K. Weirauch, P. G. Gemeinhardt, and A. L. Baron, *Soc. Plast. Eng. Tech. Pap. 22*, 317–321 (1976).
9. F. A. Hudock and S. J. Ising, presentation at IPC Fall 1986 meeting.
10. D. A. Shimp, *Proceedings of the 32nd Int. SAMPE Symposium*, 1987, pp. 1063–1072.
11. D. A. Shimp, F. A. Hudock, and S. J. Ising, *Proceedings of the 33rd Int. SAMPE Symposium*, 1988, pp. 754–766.
12. R. Kubens, H. Schulthier, R. W. Stammhein, and E. Grigat, U.S. Patent 3,562,214 (1971) to Farbenfabriken Bayer AG.
13. D. H. Wertz and D. Prevorsek, *Proceedings of the SPE ANTEC*, 1984, p. 483.
14. J. P. Mitchell and T. L. Welsher, *Printed Circuit World Convention* II, Munich, June 1981.
15. M. E. Lyssy, IPC 30th Annual Meeting, 1987.
16. S. Ayano, *Kunstoffe*, *75* (8), 475 (1985).
17. C. B. Delano, A. H. McLeod, and C. J. Kiskiras, NASA Contract Final Report FR-80-42, Contract NAS3-22025, 1980.
18. R. A. Brand and E. S. Harrison, NASA Contract Report 3615, Contract NAS1-15859, September 1982.
19. W. P. Fitzgerald, Jr., "XSR-10500 High Temperature Resin," MJ09167 Summarizing Report, 10/31/71, Whittaker Corporation, Oct. 31, 1971.